The Packaging Designer's Book of Patterns

Third Edition

George L. Wybenga
Lászlo Roth

John Wiley & Sons, Inc.

This book is printed on acid-free paper. ♾

Published by John Wiley & Sons, Inc., Hoboken, New Jersey

Published simultaneously in Canada

Library of Congress Cataloging-in-Publication Data:

Wybenga, George L.
The packaging designer's book of patterns / George L. Wybenga, Lászlo Roth. — 3rd ed.
p. cm.
Lászlo's name appears first on the earlier ed.
Includes bibliographical references and index.
ISBN-13: 978-0-471-73110-8 (paper)
ISBN-10: 0-471-73110-2 (paper)
ISBN-13: 978-0-471-77146-3 (paper/CD)
ISBN-10: 0-471-77146-5 (paper/CD)
1. Packaging—Design. 2. Paperboard. I. Roth, Lászlo. II. Roth, Lászlo. Packaging designer's book of patterns. III. Title.
TS195.4.R683 2006
688.8—dc22

2005030958

Printed in the United States of America

10 9 8 7 6 5 4 3 2 1

Contents

Preface

Fifteen years ago this was just an idea we had; draw paperboard packaging patterns, add a brief description, and submit it to a publisher. Never did we expect the world wide acceptance of The Packaging Designer's Book of Patterns nor that a third edition would be in its future.

I have added 64 more pages of patterns. Four pages cover zippers and tear strips, resealable top closures, and a number of corner lock variations. You will find their various applications throughout this book. The increase in warehouse style stores has resulted in an increase in the demand for stackable shipper/display containers which feature Stacking Tabs and Receptacles. I therefore have included new shipping containers in the display section since they serve both shipping and display functions.

It has been suggested that more detailed instructions for assembly of the various patterns be included. It is my belief that such explicity might stifle the creative process. Needless to say, every pattern has been scaled down to fit the page. Once the pattern rendering was completed I made a copy on oak tag and constructed the container. This process not only helped me check for accuracy of all the dimensions but also gave me the chance to discover new applications which are included in this book. I have attempted to only include 100% recyclable material in this volume. AGI/KLEARFOLD DUOFOLD® is a leading manufacturer of plastic and paperboard combination packaging. Patterns in this volume may be converted into such applications.

The reader is advised that patented features are protected by patent law and may not be used in practice without the express permission of the patent holder. For information about patents, contact the United States Patent Office.

I would like to thank my wife Betty for her support, King Kullen's manager Amos Glen for supplying me with many display ideas, and the reviewers who gave pointers for improvement of this work. My editors, Margaret Cummins and Rosanne Koneval at John Wiley & Sons were always there to guide me through the process of bringing this work to fruition.

George L. Wybenga

Preface from the First Edition

Here is the first definitive collection of the patterns and structural designs for paperboard packaging, point-of-purchase displays, and other three-dimensional graphic products. This workbook contains over 450 patterns and structural designs that can be adapted for packaging. The designer may choose among hundreds of alternative carton and box structures, which may be used as they are, modified, or adapted to create exciting new structures. The patterns are accompanied by a description of the historical and technical origins of the materials and methods used. Special attention is given to computer-aided design as an engineering tool that offers new insights into carton and box design and manufacturing.

Paperboard packaging has a profound impact on the U.S. economy. In 1988 about $16 billion was spent for paperboard packaging, accounting for 45 percent of the total expenditure for packaging. Today the United States leads the world in the use of paper and paperboard, with a consumption of over 600 pounds per person per year. There are more than 5000 plants in the United States that manufacture and convert paper and paperboard. In addition, there are about 530 carton manufacturers with 752 plants and a total employment of 80,000. The corrugated board industry includes 795 companies with 1427 plants and a total employment of 118,000.

The packaging materials industry serves every major retail environment—in many ways it has shaped the American lifestyle. If it is to continue to thrive, it must be creative and innovative. It must be market oriented and must continually supply new methodologies and new solutions to environmental problems.

Along with sensitivity to environmental problems, creativity should be the designer's most important consideration. This workbook may be used not only as a reference for structural design but also as a source of exercises in learning to stimulate creative skills. The cartons shown in this book are designed to depict generic styles whenever possible. Each illustration is accompanied by the name that is commonly used to describe the style. The grain direction of all folding cartons is horizontal unless otherwise indicated. The flute direction of all corrugated cartons is vertical unless otherwise indicated.

Some of the packages currently in use are patented or have patent-protected features for use on proprietary packaging machines. Where appropriate, patented features are indicated.

The reader is advised that patented features are protected by patent law and may not be used in practice without the express permission of the patent holder. For information about patents, contact the U.S. Patent Office.

ACKNOWLEDGMENTS

This book would not have been possible without the generous help, contributions, and advice of our colleagues in the packaging industries. We are especially indebted to Robin Ashton, editor-in-chief of *Packaging*, a Cahner's publication, for allowing us to adapt some of the diagrams and charts from the Packaging Encyclopedia and Technical Dictionary.

1

Introduction

17th Century Dutch Paper Mill

Paper is among the noblest of human inventions. It is worthwhile, therefore, to begin with a short history of papermaking.

Before books could be written and preserved, a writing surface had to be developed that was light, not too bulky, and was easily stored. The first great advance was the Egyptians' use of papyrus in the third millennium B.C. Sheets of beaten papyrus stems were fastened together into scrolls, some more than 120 feet long, that could be rolled up for storage. After papyrus came parchment, which was perfected in Asia Minor in the city of Pergamum (from which its name is derived) in the second century B.C. Animal skins had long been used as a writing surface in Greece and Rome, but it was in Pergamum that methods were evolved for the production of a durable, velvet-smooth parchment that could be written on on both sides.

Detail of an Egyptian papyrus scroll (ca. 2500 B.C.).

For hundreds of years all paper was made by hand from rag pulp. The use of wood fibers to make paper was discovered in the mid 1800s. In 1840 Friedrich G. Keller in Germany invented a way to grind logs into a fibrous pulp; this method produced a rather poor quality of paper, as all parts of the wood—not just the fibers—were used.

Paper as we know it today was first made in China in 105 A.D. Ts'ai Lun, a member of the court of Emperor Ho Ti, succeeded in turning husks of cotton fibers into paper pulp. This method spread throughout China, Korea, and Japan and as far west as Persia. In 751 A.D. Moslems captured a Chinese paper mill in Samarkand and learned the method of papermaking. They brought the method to Spain around 950 A.D., and by the thirteenth century paper mills had been established throughout western Europe, first in Italy and then in France, Germany, England, and Scandinavia.

The first paper mill in America was built in 1690 by William Rittenhouse near Philadelphia. Sheets of paper were produced one at a time until 1799, when Nicholas Louis Robert developed a continuous process. (This method was patented in England by the Fourdrinier brothers and is known by that

Egyptian scribe using paper made from papyrus.

Early Chinese print on paper (ca. 300 A.D.).

name.) In 1817 the first cylinder-type paper-making machine, which can produce a better quality of paper in a continuous process, was invented by John Dickenson.

MODERN PAPERMAKING

Today almost all paper is manufactured from wood. Cellulose fibers (which account for 50 percent of the content of wood) are the primary ingredient, followed by lignin (about 30 percent), which acts as a fiber binder or glue.

Water plays an important role in modern papermaking. The manufacture of 1 ton of paper requires about 55,000 gallons of water, most of which is recycled. The papermaking process also uses sulfur, magnesium, hydroxide, lime, salt, alkali, starch, alum, clay, and plastics (for coating). There are two basic types of paper: fine paper for writing and paper for printing and industrial use (packaging).

The first step in manufacturing paper from wood is to remove the bark. The cheapest way to separate the fibers is to grind up the wood by forcing the logs against grindstones submerged in water. The water carries off the wood fibers.

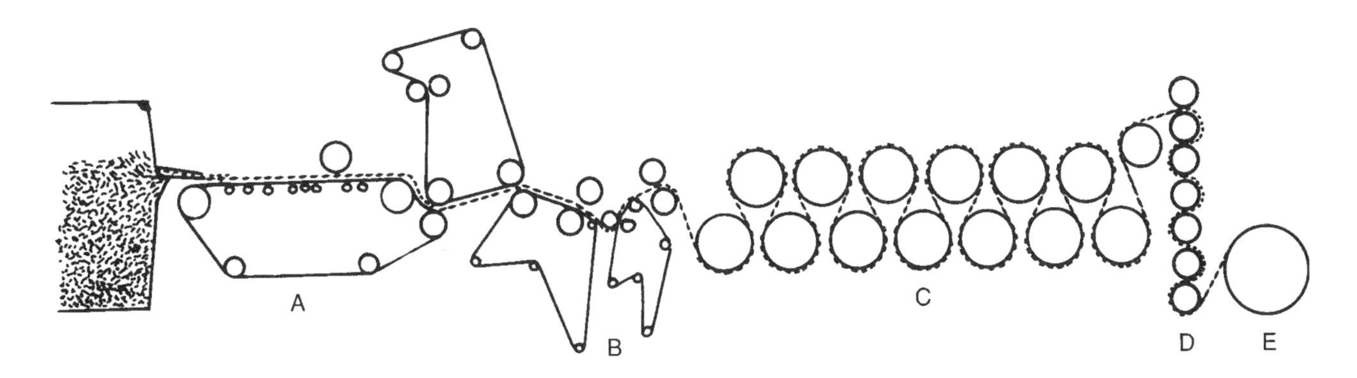

Diagram of a papermill (greatly simplified). The pulp is ejected in a thin layer onto the conveyer-sieve (A). The pulp is then pressed through a great number of cylinders (B) and dried by a series of heated cylinders (C). It is then calendered (D) and taken up by the web (E).

In this process everything is used, and the paper produced is of low quality. Another, more frequently used process is chemical pulping, in which the wood is chipped into small pieces, the fibers are extracted through a chemical process, and the unusable material is eliminated. Chemical pulping is more expensive, but it produces better-quality paper.

Chemical pulping creates a pulp, which is then refined by washing and separating the fibers. Refinement, a time-controlled process during which the manufacturer can add various chemicals to increase bonding, texture, and water resistance, increases the quality and strength of the paper. Pigments (for coloring) and coatings (plastics) can also be added at this stage.

Once the pulp is prepared, it goes to one of two types of machines: The Fourdrinier, or the cylinder machine. Modern papermaking machines are huge. They can be as long as a city block and several stories high. They produce paper up to 30 feet wide at a speed of 3,000 feet per minute, resulting in 800 miles of paper a day! The primary papermaking machine is the Fourdrinier. Most Fourdrinier machines make only one layer of material, although they can be equipped to make several layers.

Paper produced by a Fourdrinier machine is smoothed by a stack of highly polished steel rolls, a process known as calendering. The finished paper is then cut, coated, and laminated.

Another frequently used papermaking machine is the cylinder machine. This machine makes heavy grades of paperboard, generally using recycled paper pulp. The pulp is built up in layers. Since paperboard is much thicker than paper, the drying operation is far more extensive. Large steam-heated cylinders drive the excess moisture out of the paper. A coating is then added to create a smooth surface.

The great advantage of the cylinder machine is that it uses large amounts of recycled paper in thick layers to provide strength.

Paper is bought on the basis of the weight (or basis weight), in pounds, of a ream of paper. (A ream is equal to 3,000 square feet of surface.) The thickness of paperboard is expressed in caliper points, which are stated in thousandths of an inch (usually written in decimals). Since most papers are laminated or coated with other materials, caliper points are rarely used today to specify weight. The paperboard used in folding cartons is specified according to the size of the carton or, more often, the weight of the item that goes into it. A glass bottle for 3.5 fluid ounces of fragrance, for example, would require a folding carton with a thickness of approximately 18—24 points.

The thickness of paper can be controlled by means of calendering, pressing, and laminating. High-quality paper is up to 12 points thick; paperboard varies in thickness from 12 to 70 points.

About 20 million tons of fine papers are used for printing and writing annually. Five and a half million tons are used for packaging. Tables 1, 2, and 3 list the major boxboards and papers used in packaging. The uses, content, and characteristics of these packaging materials are described.

PRINTING, FINISHING, AND DIE-CUTTING

Folding cartons are manufactured using three main processes: printing, die-cutting, and finishing.

Printing Methods

Several methods of printing are available, they include letterpress, offset lithography, gravure, flexography, and silk screen. Each method is suitable for particular types of jobs.

The *letterpress* method transfers ink from a metal plate directly to the sheet paperboard. This is one of the oldest methods of quality

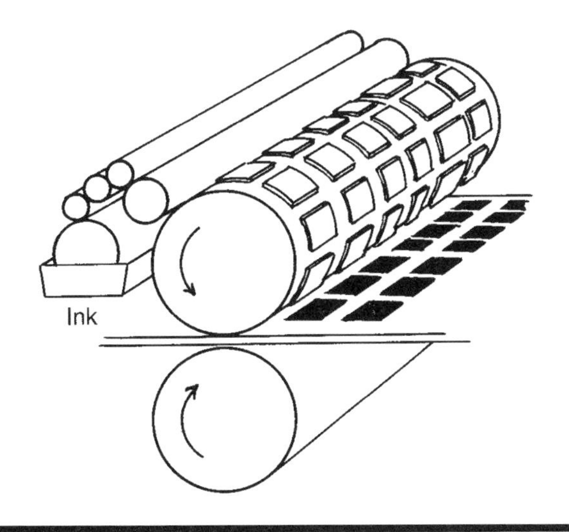

Letterpress: The ink is applied to the raised surfaces of the plate and transferred to the substrate.

printing. New technologies have rendered it almost obsolete. *Offset lithography* has replaced letterpress because of its production efficiency and high-quality color reproduction. New high-speed presses and computer-aided systems, along with technological advances in inks and coatings, have made "offset" the most popular process for printing on folding cartons. In this process specially sensitized metal plates are chemically treated to accept ink. The ink is transferred from the plate to a smooth blanket roller, which then transfers the image to the paperboard.

Gravure printing is used for high-quality reproduction in large-quantity runs (i.e., millions of copies). Specially etched printing cylinders have cells that accept and store inks. A "doctor blade" wipes off excess ink as the cylinder rotates to the impression cylinder, where the plate cylinder transfers the image to the paperboard. Gravure printing can be accomplished on an in-line web

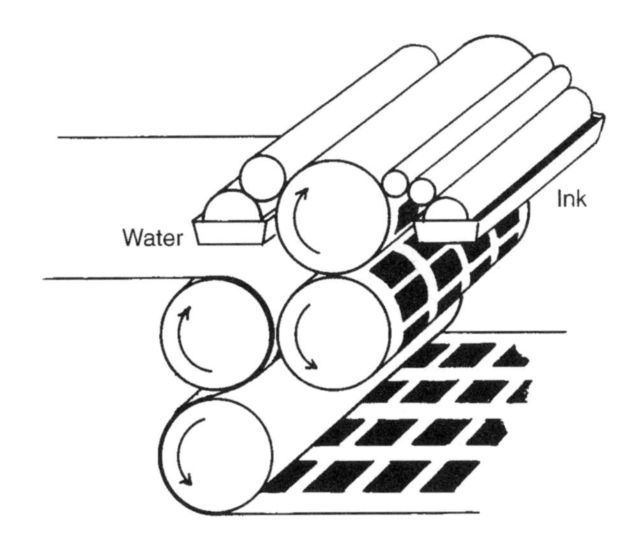

Offset lithography: The ink is picked up by the pre-wetted plate, which transfers the inked image to the offset blanket, which in turn transfers the image onto the substrate.

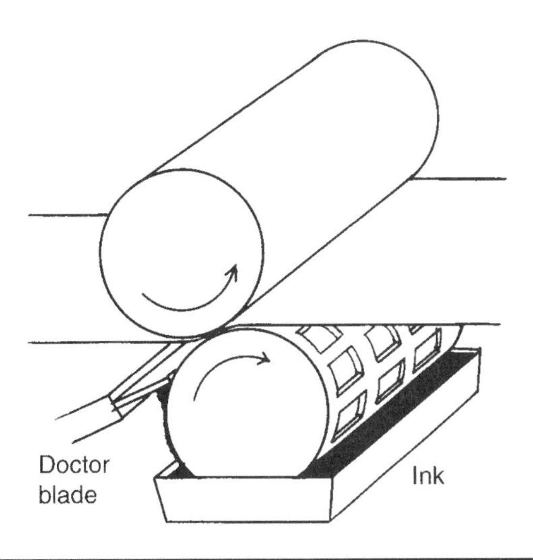

Gravure (Intaglio): The ink is applied over the entire plate, scraped off the surface by the doctor blade, and transferred from the ink wells onto the substrate.

press, which is known as *rotogravure,* or on a sheet-fed press, which is called *photogravure.*

Flexographic printing is similar to letterpress printing. It uses a raised positive composition plate made of rubber or plastic. High-speed in-line web presses are used. This process has been associated with low-quality simple line art printing, but recent technological breakthroughs with fast-drying inks have made flexographic printing a low-cost, high-quality method for medium production runs.

Silk screening is a simple method of color printing in which a fabric mesh stretched over a frame is used instead of a printing plate. A stencil-type design is adhered to the mesh and pigment is "squeegeed" through the stencil. A separate "stencil" is required for each color used. (See diagram.)

Printing technology has changed in recent years to better meet the needs of carton manufacturing. Special coatings, varnishes, lacquers, and inks are available to give a bright finish or provide a moisture-proof barrier. Environmental problems have been alleviated by the introduction of water-based coatings and inks.

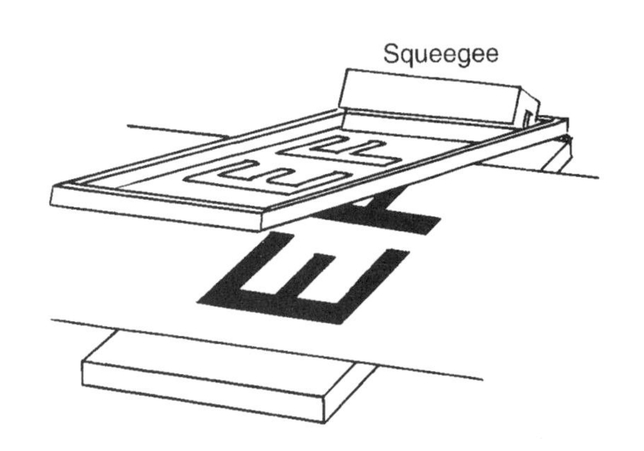

Silkscreen: The ink is squeegeed through a stencil adhered to the mesh of the silk onto the substrate.

Die-cutting

The process of *die-cutting* involves creating shapes using cutting and stamping dies. There are three methods of die-cutting. *Hollow* die-cutting is done with a hollow die, which looks like a cookie cutter. This method is used exclusively for labels and envelopes. *Steel rule* die-cutting is used when a close register is required. Steel rules are bent to the desired shape and wedged into a ¾" piece of plywood. The die is locked up in a chase on a platen of the die-cutting press. Several sheets can be cut at once. A flatbed cylinder press can also be used for die-cutting.

The third method of die-cutting uses *lasers,* which were invented by C. H. Townes and Arthur Schawlow in 1958. (The word *laser* is an acronym for "light amplification by simulated emission of radiation.") The laser beam, which can be concentrated on a small point and used for processes such as drilling, cutting, and welding, has become widely used in manufacturing, communications, and medicine. Since a laser beam is extremely sharp and precise, the resulting cut is very accurate and clean.

Laser cutout.

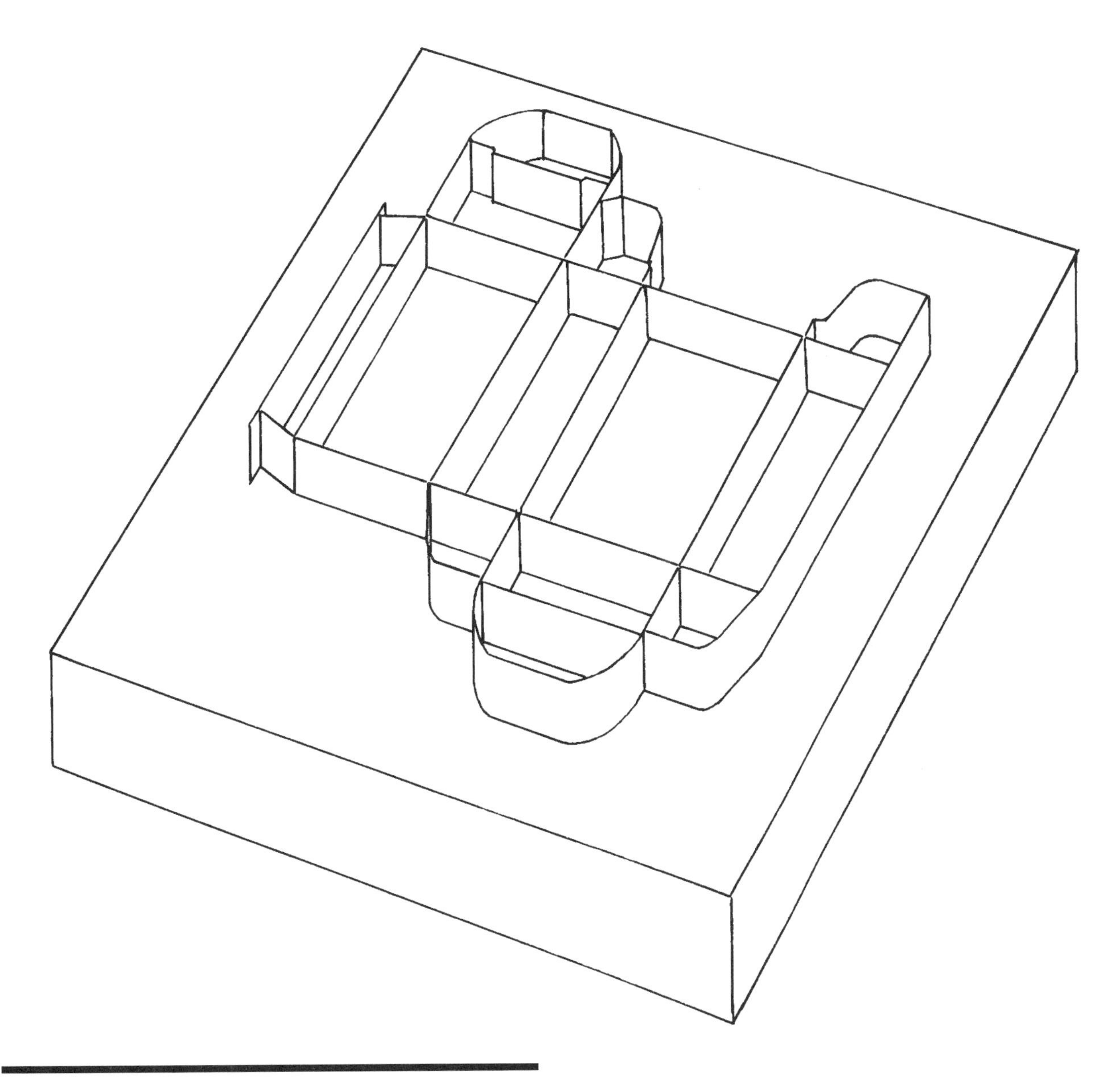

Steel rule die. This is a typical sample maker's die for a reverse-tuck carton. The die consists of scoring and cutting rules inserted into a sheet of plywood. A production die will have a great number of box layouts ganged so as to create as little waste as possible. Production dies are on either a flat sheet or a cylinder.

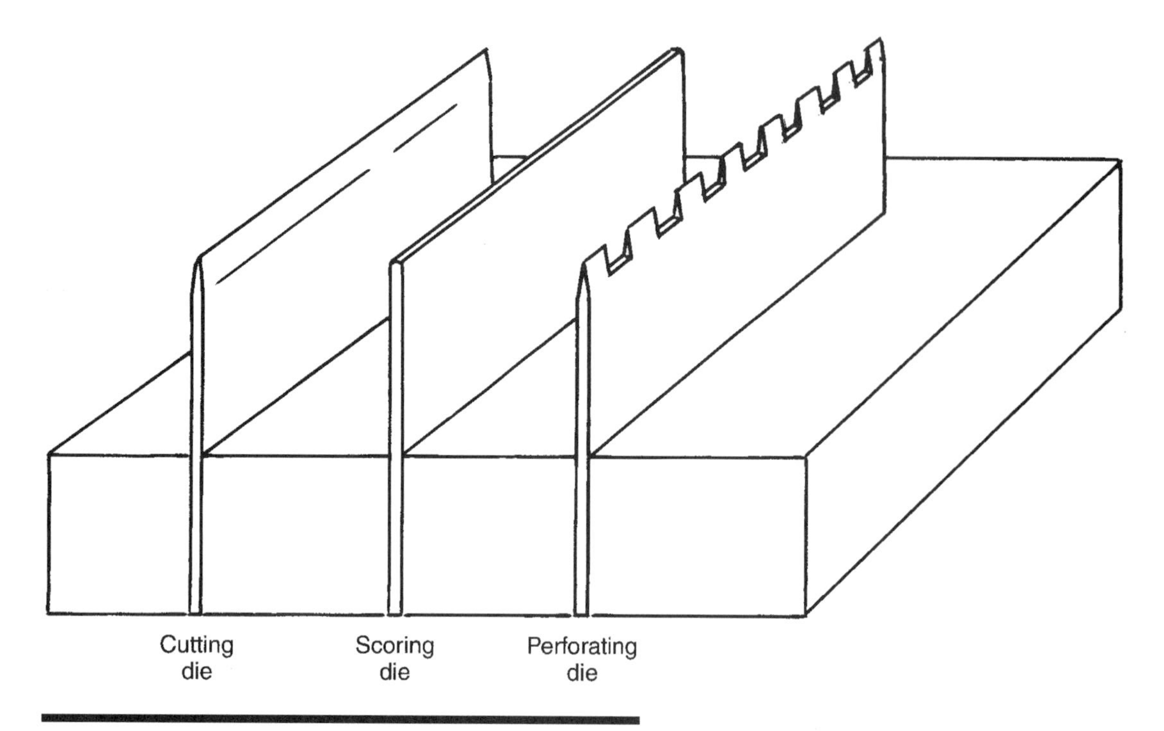

Dies for cutting, scoring, and perforating.

The steel die rule is completely embedded in the plywood, except where notches have been cut out of the bottom of the rule to prevent pieces of the plywood from dropping out.

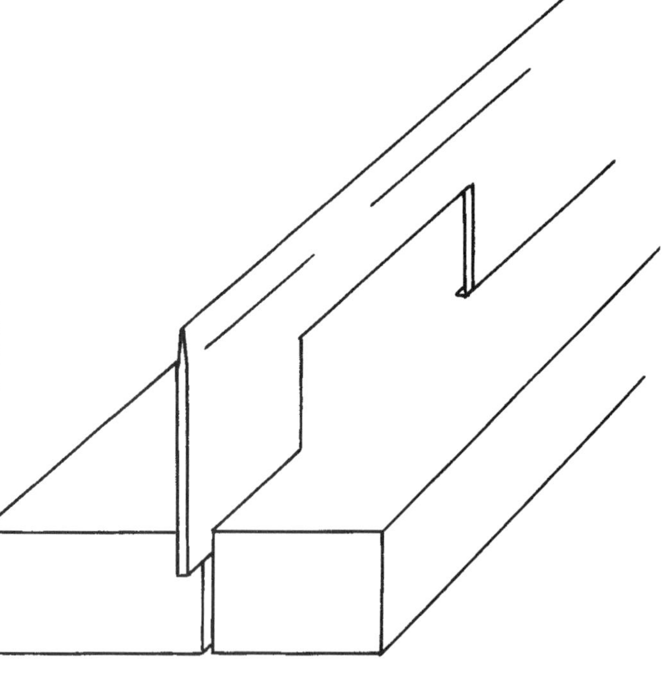

Windows may be die-cut out of the face panel or extended into the top or side panels. They may be made in any shape desired for the final graphic solution. When laying out the window shape, consider leaving several shapes that may be printed with seasonal messages and retail price suggestions when the package blank is printed and can be removed when not required.

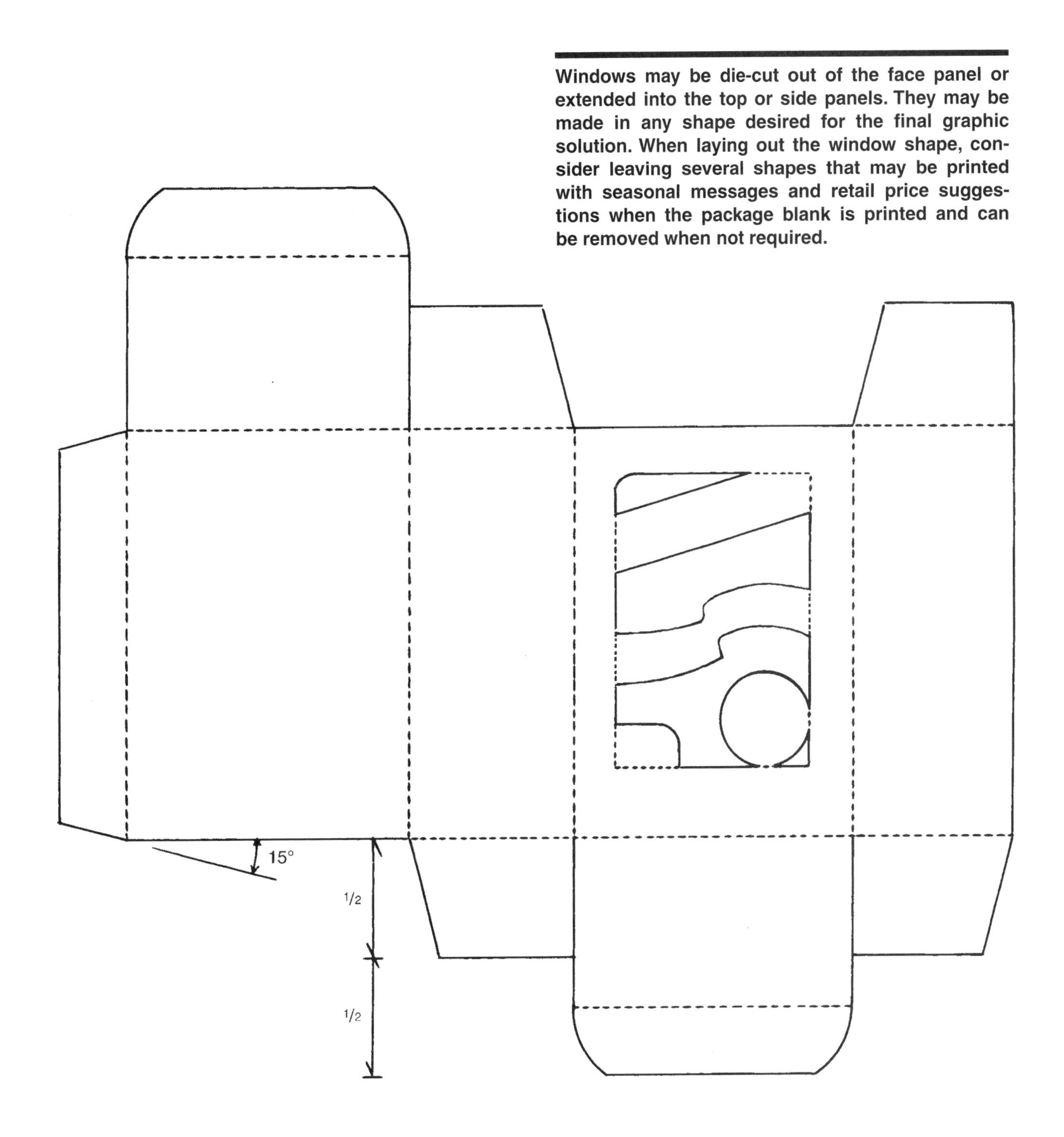

Embossing. Paper and board lend themselves to *embossing,* the process by which a design image is made to appear in relief. Embossing can be superimposed on printing or done on blank paper (blind embossing) for a sculptured three-dimensional effect. It is achieved by pressing a sheet of paper between a brass female die and a male bed, or counter, both of which are mounted in register on a press. Embossing is generally used on prestigious packages; on packaging for cosmetics, gifts, and stationery; and on promotional materials.

Finishing

Finishing operations include gluing (using adhesives), windowing (die-cutting), coating, and laminating.

There are several types of gluing equipment, each designed for specific purposes. Right-angle gluers are used for Beers-style and six-corner construction; straight-line gluers are used for side seams and automatic bottom-closured cartons. Automatic tray-forming devices are used in the manufacture of tapered trays, clamshells, and scoop-style cartons. Today computer-controlled multipurpose gluers are used in most large paperboard-manufacturing plants.

A wide variety of specially formulated adhesives are used for specialty cartons made of coated, laminated, and plastic materials. Environmental conditions such as high moisture levels, freezing temperatures, sterilization, and microwaving require specific types of adhesives. Many types of adhesives are available, including self-adhesive and pressure-sensitive labels, resin emulsion adhesive for coated boards, and cold- or hot-melt adhesives for plastics.

Scoring. Paper and board come in flat sheets or rolls. In all papers and boards the fibers are aligned in one direction, called the *grain.* If they are torn with the grain, the edges will be smooth. If they are torn against the grain, the edges will be ragged.

When scoring the fold lines on the comprehensive, only *compress* the fibers of the paper, do not crush or cut them. Use a dull knife or letter opener. With practice you should be able to score on one side of the paperboard and achieve clean folds. Use white glue sparingly on the glue flaps. A thin coat of rubber cement may be used for covering papers, foils, or fabrics. *Caution:* rubber cement is highly flammable and toxic; library paste is the preferred alternative.

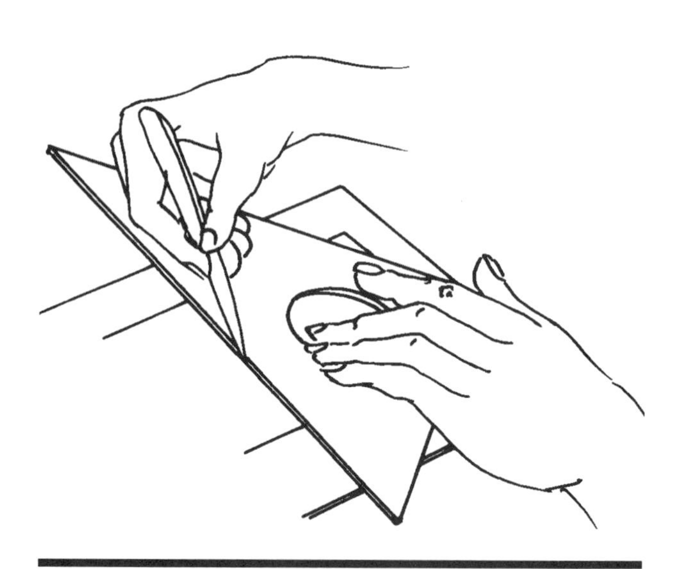

Scoring

Cutting. To facilitate the preparation of a packaging comprehensive, lay out the pattern carefully with triangle and T-square on the appropriate paper stock. Two-ply or three-ply are ideal for comps. Use a resilient backboard to score and cut; the chipboard on the back of drawing pads or a self-healing cutting board are great for this purpose. Use a steel straightedge for straight lines. To prevent sliding, glue a few strips of fine sandpaper to the back of the straightedge. An X-Acto® knife is easy to control for both straight and curved lines. Keep fingers well away from the cutting blade.

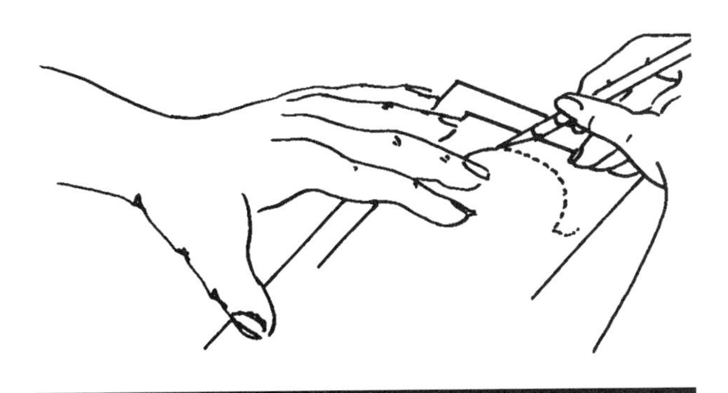

Cutting

Table 1. **BOXBOARDS USED IN PACKAGING: CYLINDER MACHINE GRADES**

Board	Uses	Chief Characteristics
Plain chipboard; solid newsboard; news vat-lined chip; filled news	Used in the manufacture of set-up boxes.	100% recycled. Lowest-cost board produced; not good for printing. color light gray to tan. Adaptable for special lining papers.
White vat-lined chipboard	Higher-grade set-up box with white liner.	Adaptable for color printing.
Bending chip	Folding cartons for light items.	100% recycled, excellent bending. Lowest-cost board for folding cartons.
Colored manila-lined bending chip	Same as bending chip.	Bright white liner, excellent for color printing.
White-line 70 newsback	Used for folding cartons, posters, displays.	100% recycled. Smooth white board.
Clay-coated boxboard	Used for quality cartons.	Very smooth white board with excellent printing surface.
Solid manila board	All carton uses, including food.	White liner and manila back.
Extra-strength plain kraft-type board	Used for hardware, automotive and other machine parts, toys.	Recycled. Available in various colors. Excellent bending ability.
Extra-strength white-lined or clay-coated kraft-type board	Used for heavy objects requiring durability and strength.	Recycled. Top liner is white, back is brown. Available in pastel shades.

Table 2. **BOXBOARDS USED IN PACKAGING: FOURDRINIER MACHINE GRADES**

Board	Uses	Chief Characteristics
Uncoated solid bleached sulfate	Waxed, polycoated, or plain frozen-food cartons.	Strong white board. Hard sized for water resistance and extended service under freezer conditions. Good bending. 100% sulfate pulp.
Clay-coated solid bleached sulfate	Cartons suitable for pharmaceuticals, cosmetics, hardware.	Excellent printability, scoring, folding, and die-cutting; ovenable. Outstanding merchandising appeal. Takes all types of coatings. 100% sulfate pulp.
Clay-coated solid unbleached sulfate	Heavy-duty packages; beverage carriers; folding cartons for food, cosmetics, textiles, housewares.	Strong, moisture-resistant board. Excellent printing surface.

Table 3. PAPERS FOR PRINTING, LABELING, AND DECORATIVE PACKAGING

Type	Uses	Characteristics
Flat or dull finish, coated and uncoated	Box wraps for gifts, cosmetics, jewelry.	Smooth, excellent printing surface. Good for embossing.
Glossy finish, supercalendered	Labels, displays, box coverings.	For quality printing. Embosses well.
Cast coated	Box wraps, gift wraps, labels.	Smooth, high gloss. Brilliant white and colors. Excellent printability. Scuff resistant.
Flint	Box and gift wraps, labels for cosmetics, gifts.	Extremely fine-quality surface, high gloss. Wide color range.
Friction glaze	Box covering. Good appearance at low cost.	High glaze. Scuff resistant. Economical.
Metallic-finish Argentine	Box wraps and overwraps, luxury items. Gift packages.	Has decorative effect of foil but is less costly. Lacquer lends gold or colored look.
Foil	Decorative packaging and labels.	Aluminum laminated to paper backing. Wide range of brilliant metallic colors. Expensive.
Gravure printed	Boxes, displays, labels, platforms.	Printed with metallic powders mixed with lacquer.
Half-fine, half-fine embossed	Specialty boxes for cosmetics.	Embossed, continuous metallic surface. Wide color range. Elegant, rich.
Pyroxylin	Box coverings, food wrappers for high-quality items.	Metallic tones. Bronze, aluminum, or copper ground into pyroxylin lacquer.
Vacuum metallized	Labels for canned goods, batteries, wrappers for food and confectionary products.	Produced by vacuum metal-vapor deposit method.
Flock	Platform coverings, linings for gift boxes.	Flock of cotton adhered to surface of paper to create soft, velvety look.
Foam paper	Protective cushioning for wraps, platforms, pads, box coverings.	Foamed polystyrene sheet laminated to paper. Lustrous finish; wide range of colors.
Glitter	Box coverings, platforms.	Specks of metal embedded in paper surface.
Iridescent or pearlescent coating	Luxury items, wraps, platforms.	Mother-of-pearl effect.
Tissue	Used for jewelry, flowers, fruit, hosiery. Specially treated and designed for visual appeal.	Treated for texture, water, resistance, printability.
Glassine	Laminated trays for candies and other food products.	Grease and oil resistant.
Parchment	Wrapper, liner for greasy or oily items.	Water resistant, high wet strength. Lacquered or waxed base.
Polyethylene and saran-coated kraft	Widely used for food products. Cereal and cracker cartons, bread wraps.	Excellent barrier to moisture.

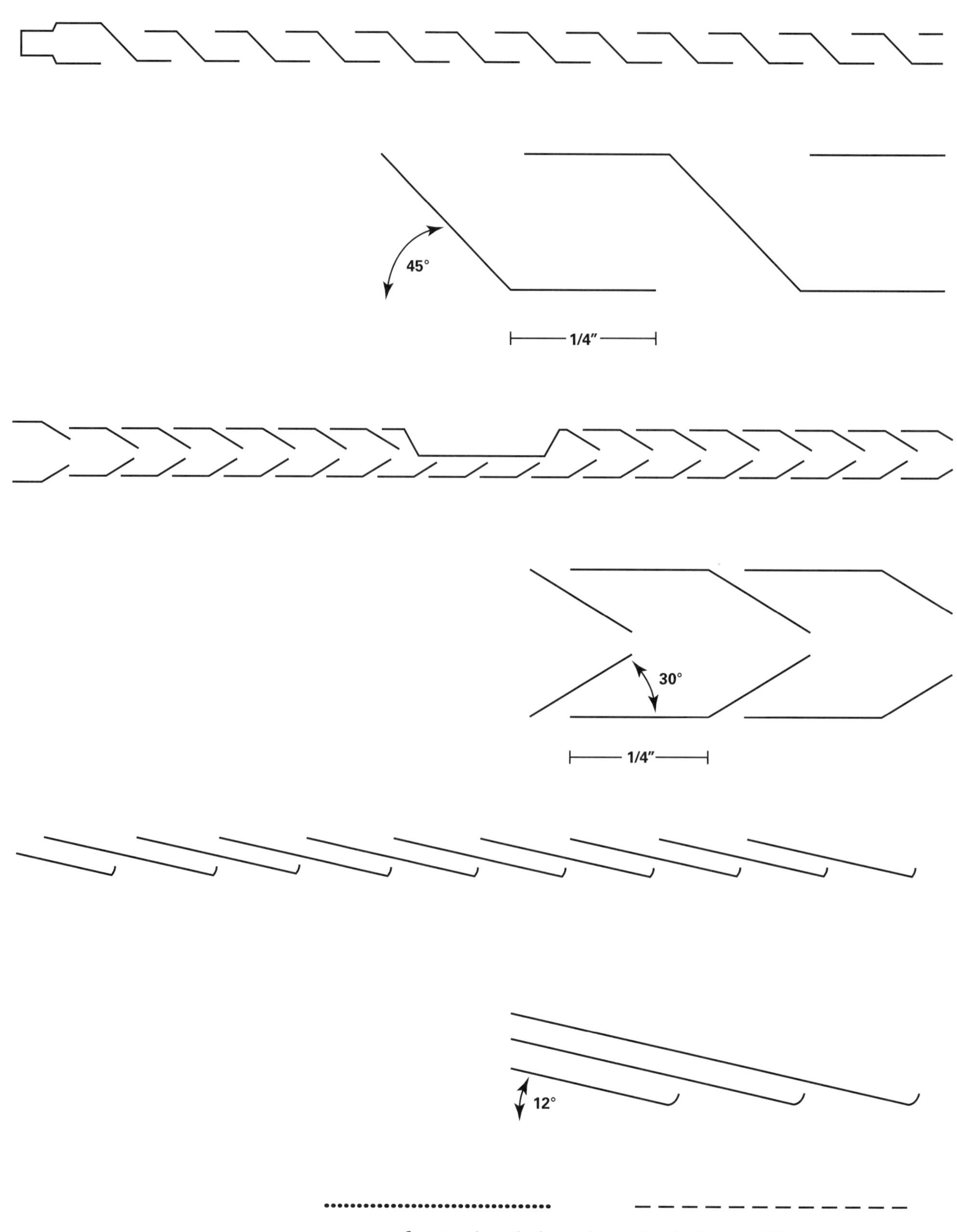

perforation length depends on the thickness of the card stock

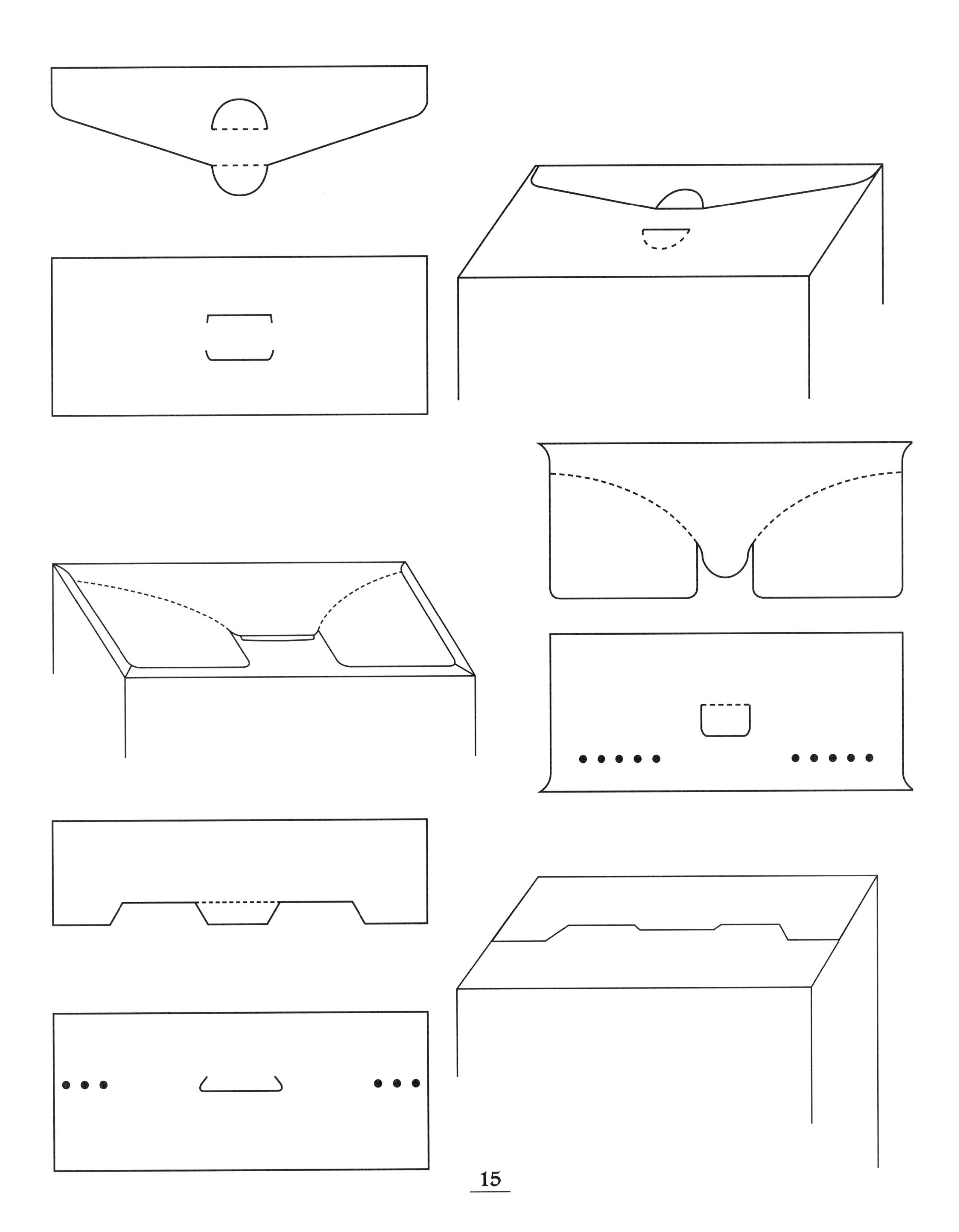

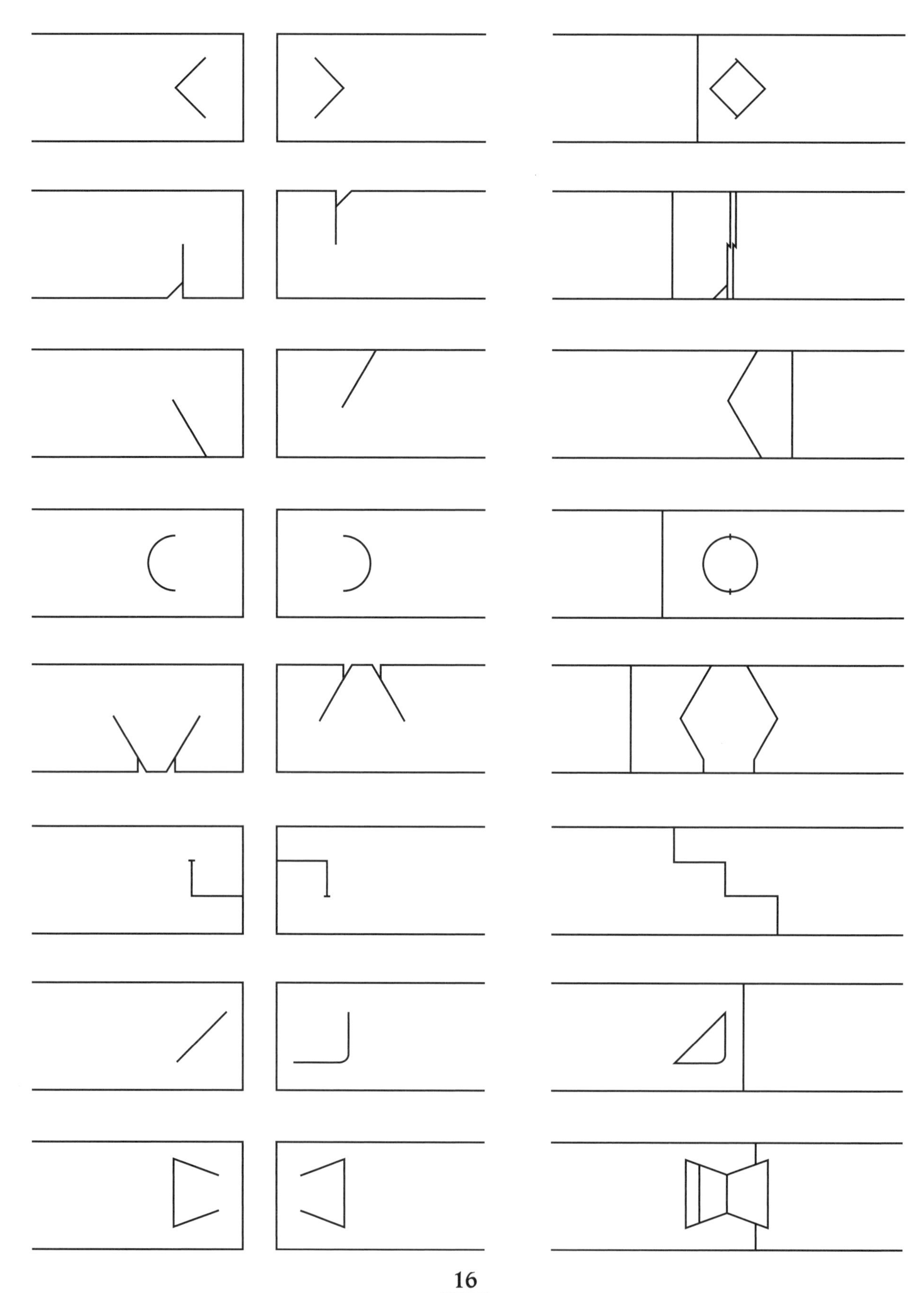

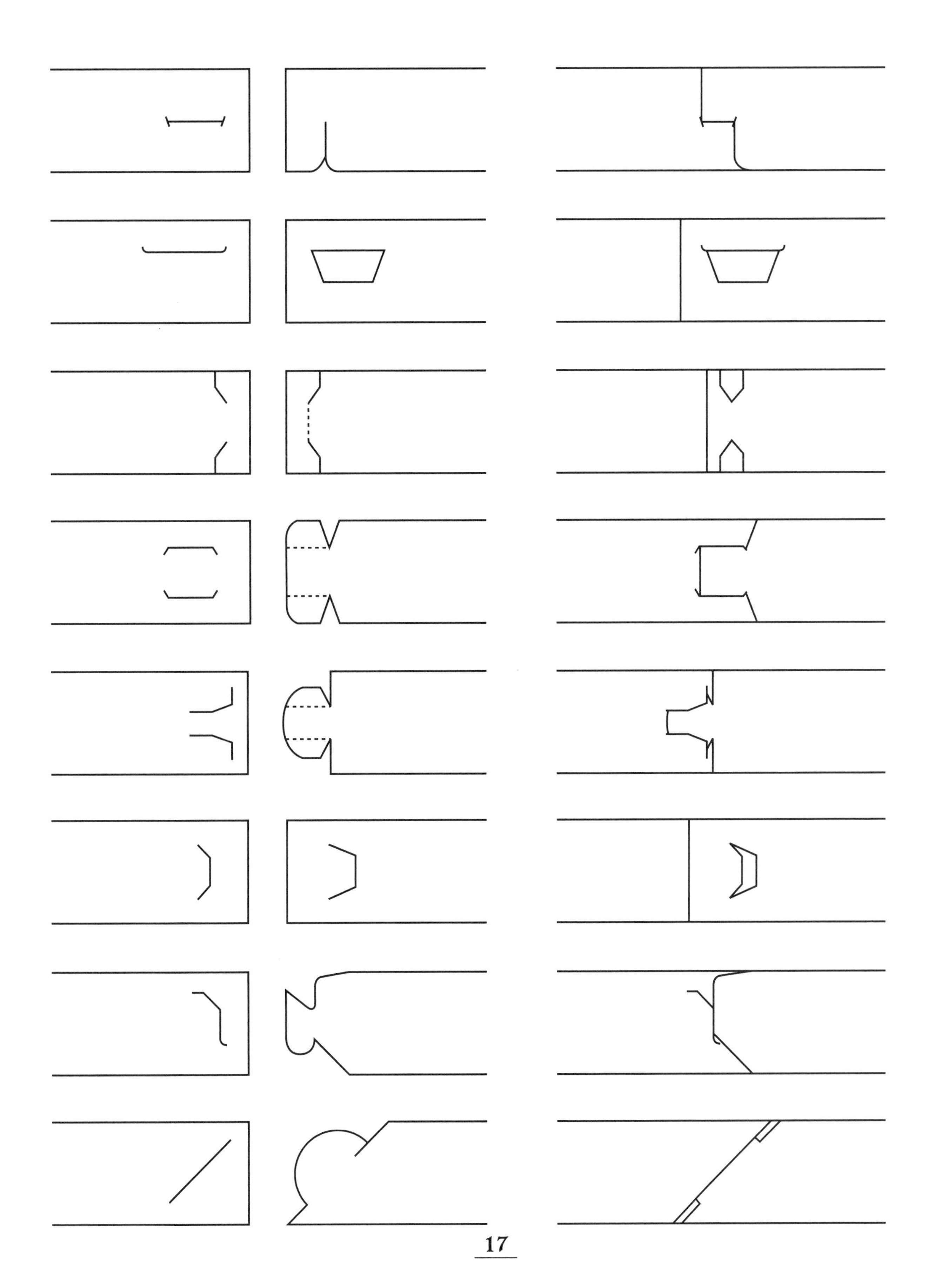

2

The Folding Carton

The cracker barrel and the flour sack were once the standard packages in the grocery trade. Today this method of "bulk packaging" has given way to a system in which small units are produced at manufacturing plants. The product is placed in paper containers made from a wide variety of boards.

Once in the hands of the consumer, a product may not be used all at once but only partially consumed. The package therefore must allow the consumer to remove a portion of the product without destroying the container. Dispensing of the contents over time is an important function of packages. Finding ways to protect and preserve products for future use is a primary concern of packagers.

With the advent of new methods of marketing and merchandising, packages now must be designed to be selected by the consumer without the help of a sales clerk. This development has given manufacturers and dealers an opportunity to enhance their products through the graphic and structural design of the package.

The shape, size, proportions, and material (board) of packages depend on the size and type of product they contain. To most people, the folding carton is the most familiar type of paper container. It is economical in terms of both material and production costs. Being collapsible, it takes up minimal space during shipment or storage. The versatility of the folding carton adds value and sales appeal to the product. The most advanced printing, embossing, and decorating techniques can be utilized to enhance the carton's appearance.

HISTORY OF THE FOLDING CARTON

The annals of invention are crowded with stories of fortunate accidents. Packaging is no exception. It happened in 1879 in the Brooklyn factory of Robert Gair, a successful printer and manufacturer of packaging materials. Gair was inspecting a printed seed package that had been accidentally slashed by a printing plate that had been improperly positioned on the press.

The revelation came to him: Why not create a die ruled for cutting and scoring a blank (paperboard) in a single impression? The idea worked, and the first machine-made folding carton was born.

In 1896 the first cartons of crackers appeared in stores, with a sealed inner bag to keep the product fresh. National Biscuit Co. was the first U.S. manufacturer to use this new packaging medium. The brand name was Uneeda, a name that is still in use.

By 1923 there were 200 manufacturers of folding cartons. By 1980 the number had grown to about 700 plants, which together use 3 ½ million tons of paperboard annually.

There are many variations in the construction of folding cartons. The modifications that can be made in the basic pattern are limited only by the designer's imagination.

Today folding cartons are precision-made. Low-cost packages are supplied in *knock-down* form and are known as *blanks.* When assembled, they become three-dimensional rigid packages. They can be filled by high-speed automatic, semiautomatic, or hand-operated equipment. Knock-down containers lend themselves to various types of marketing and retailing systems, notably those that handle food, gifts, pharmaceuticals, cosmetics, toys, hardware, and housewares.

THE TRAY-STYLE CARTON

One of the basic types of folding carton is the *tray.* In one type of tray-style carton, a solid bottom is hinged to wide side and end walls. The sides and ends are connected by a flap, hook, locking tab, or lock. These cartons have a variety of cover and flap parts extending from the walls and sides of the tray.

In another type of tray carton, two pieces, one slightly smaller than the other, form the base and cover of a two-piece telescoping box. Typical tray packages include cigarette cartons, bakery trays, ice cream cartons, pizza cartons, and garment carriers.

THE TUBE-STYLE CARTON

Another basic type of folding carton is the *tube.* The body of the tube-style carton is a sheet of board that is folded over and glued against its edges to form a rectangular sleeve. It has openings on the top and bottom that are closed with flaps, reverse or straight tucks, and locks.

Tube-style cartons give the product fully enclosed protection. They are used for bottled products, cosmetics, and pharmaceuticals. Often a window is added so that the consumer can see the product. Many unusual tube types and styles are available, including contoured, triangular, octagonal, and even rounded shapes. The following pages present a variety of folding-carton patterns. They can be constructed in any size from board or sheet plastic (acetate).

COSMETICS, PHARMACEUTICALS, AND PROMOTIONAL PACKAGING

The cosmetics industry was the first to effectively use packaging, and is still inspiration for the most creative types of packaging.

The history of cosmetics goes back thousands of years. The Bible describes the practice of anointing the head and body with oil. Primitive tribal symbols applied to the face and body were used to frighten enemies in various cultures in Africa, North and South America, and Oceania. Another traditional use of cosmetics is to attract members of the opposite sex. (The most famous figure associated with this practice is Cleopatra, the last queen of Egypt, who was renowned for her skill in making and applying cosmetics.) The ancient Greeks and Romans were great users of all types of cosmetics—hair dyes, creams, makeup, and the most important toiletry, soap, which was invented around 100 A.D. In the seventeenth century Queen Elizabeth I used dyes (red) for her hair as well as fragrances; some of the most well-known historical figures of this period never bathed, but used perfumes and rosewater to counteract body odors and dirt.

The wholesale manufacturing of cosmetics and toiletries began in France during the reign of the "Sun King," Louis XIV (1643-1713). During the Napoleonic era in the early 1800s, it became a major industry.

Cosmetics packages have the look of luxury and elegance. The standard cosmetics container is a folding or set-up carton embellished with rich colors, textures, and graphics. It attempts to portray the consumer as potentially beautiful, desirable, and elegant. From the marketing point of view, the cosmetics package is selling hope.

Pharmaceuticals are another important area for the packaging industry. There are about 1,000 drug-manufacturing companies in the United States. Pharmaceuticals packaging uses all types of packaging materials, including paper, board, and plastics.

Over-the-counter drugs must have sales appeal, and packaging plays an important role in providing that appeal. The folding carton encloses a glass or plastic container for the medication. The carton provides protection as well as serving as a promotional tool.

Pharmaceuticals packaging involves several factors, including moisture and oxygen control to preserve the potency of the medication; volatility and light protection to protect the drug against exposure to harmful ultraviolet rays; and protection against heat, which can cause deterioration of the product. Safety also plays a significant role. Many items (e.g., surgical dressings, injectables) must be kept sterile until they are used. The packaging must also be childproof and tamperproof.

The requirements for packaging of cosmetics and pharmaceuticals are very similar. In both cases manufacture and distribution is strictly regulated by the Food and Drug Administration.

Promotional packaging was originally used by the pharmaceuticals industry for the introduction and promotion of new products. The packages were sent to physicians. Gradually these packages became a total-communication effort. In addition to product samples, the package contained descriptive literature, reply cards, and videotapes. Manufacturers of other products and services soon followed the lead of the pharmaceuticals industry, and promotional packaging is now used by all major industries. Examples of mailers and other imaginative promotional packages are shown on the following pages.

UNUSUAL APPROACHES TO CARTON AND BOX CONSTRUCTION

Designers today are increasingly interested in technology. Computer-aided design, for example, enables them to create or alter concepts in minutes. Such technological advances have reduced the need for handwork and special skills. However, student and professional designers alike have a tendency to cling to familiar methods and skills, which give them a feeling of self-fulfillment.

Like new technologies, economic factors appear to limit the possibilities for design, yet they can actually stimulate innovation. An example is the revitalization of folding-carton design that occurred as a result of the prohibitive cost of set- up boxes and the increasing cost of hand labor.

Many of the following pages present examples of structural experiments. They reflect the exciting experimental work being done in schools of design. Innovations in design like those shown here can strongly influence the marketing of a product, often surpassing even the expectations of the client.

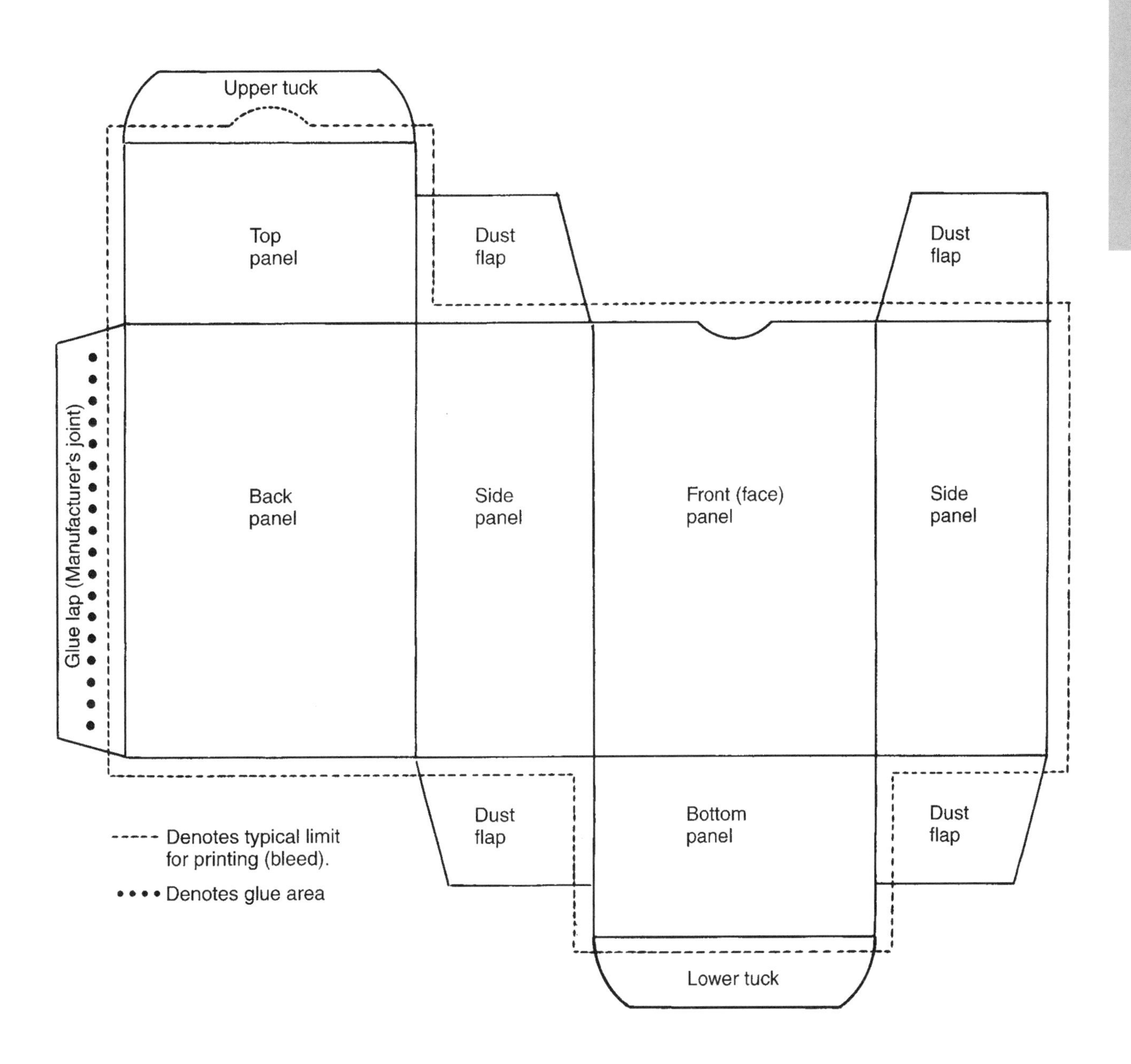
Upper tuck
Top panel
Dust flap
Dust flap
Glue lap (Manufacturer's joint)
Back panel
Side panel
Front (face) panel
Side panel
Dust flap
Bottom panel
Dust flap
----- Denotes typical limit for printing (bleed).
•••• Denotes glue area
Lower tuck

Trays

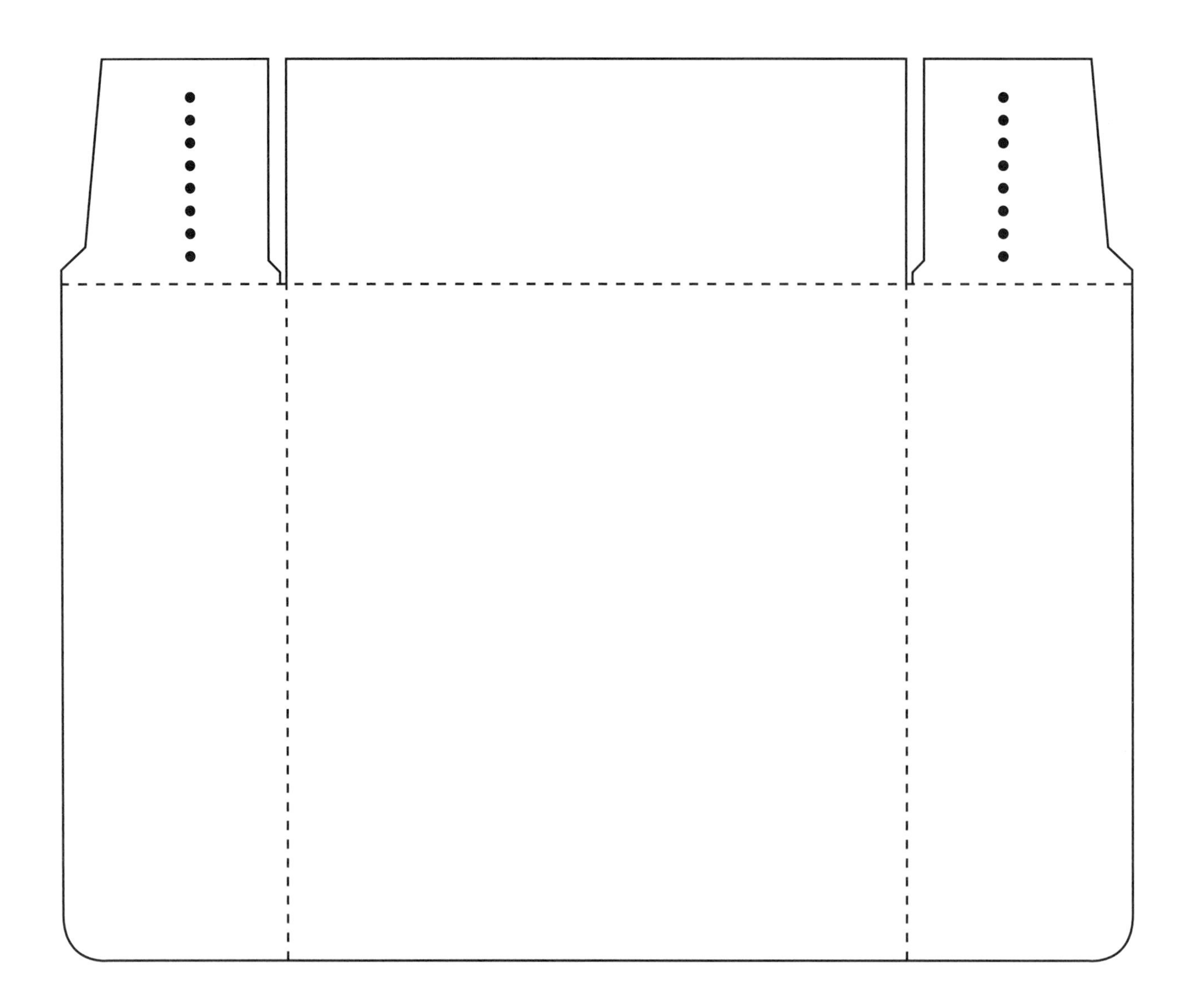

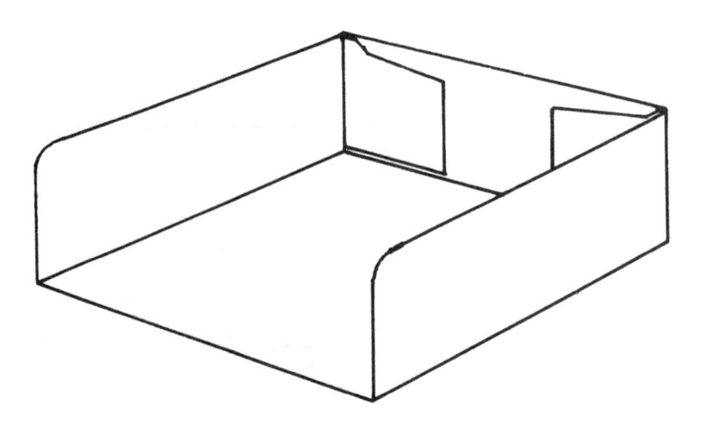

OPEN ENDED TRAY, TWO CORNER BRIGHTWOOD STYLE

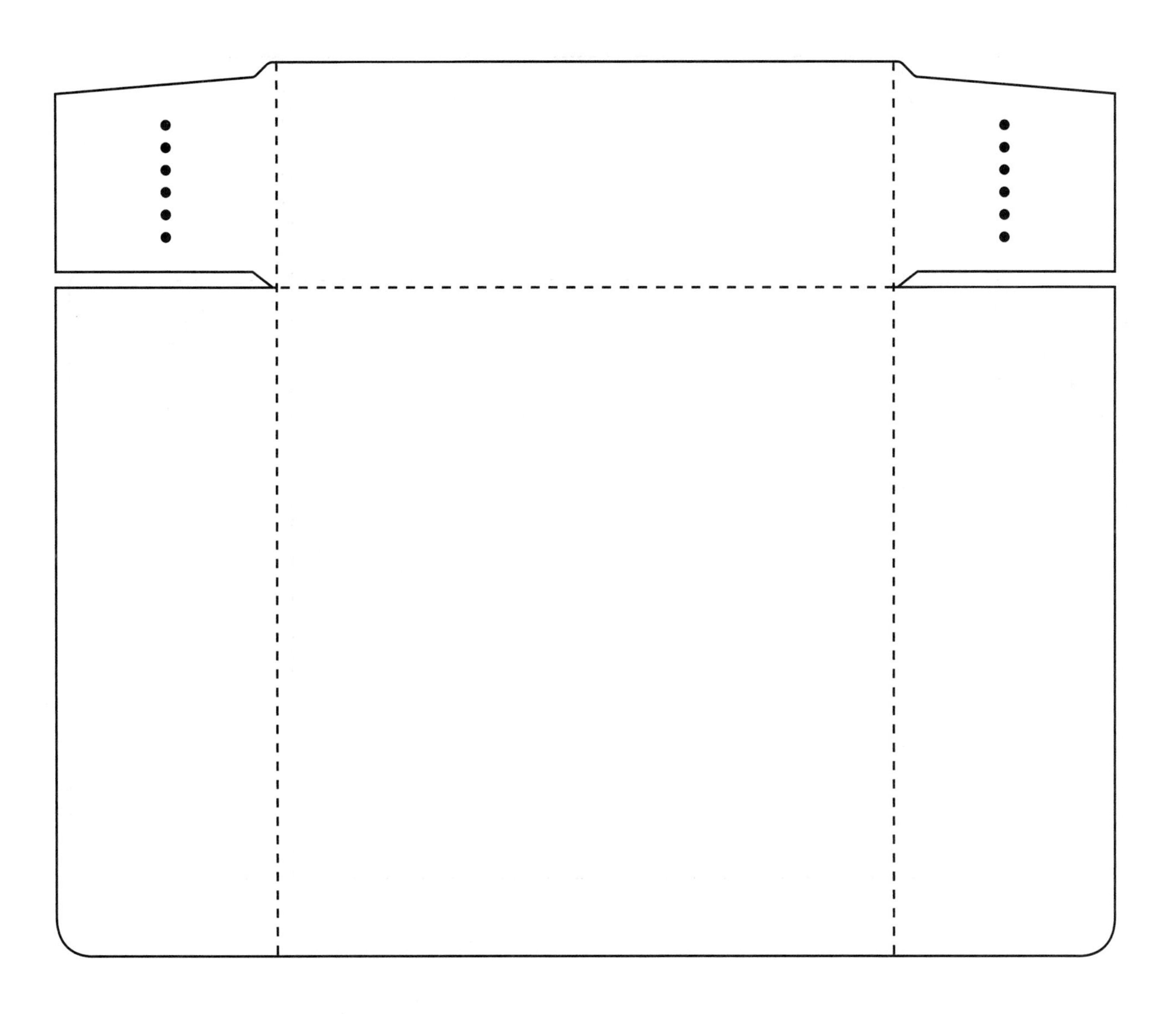

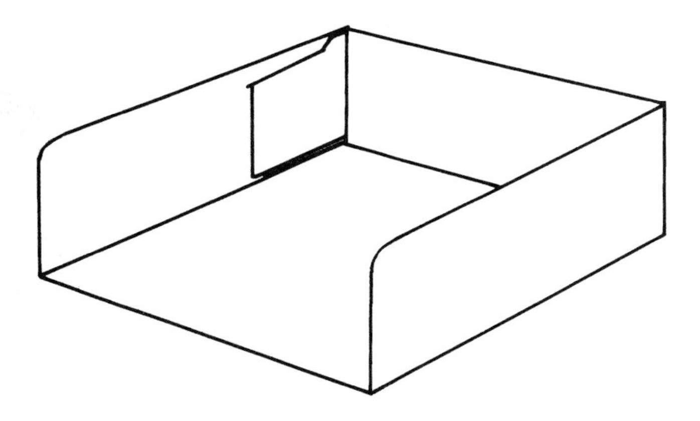

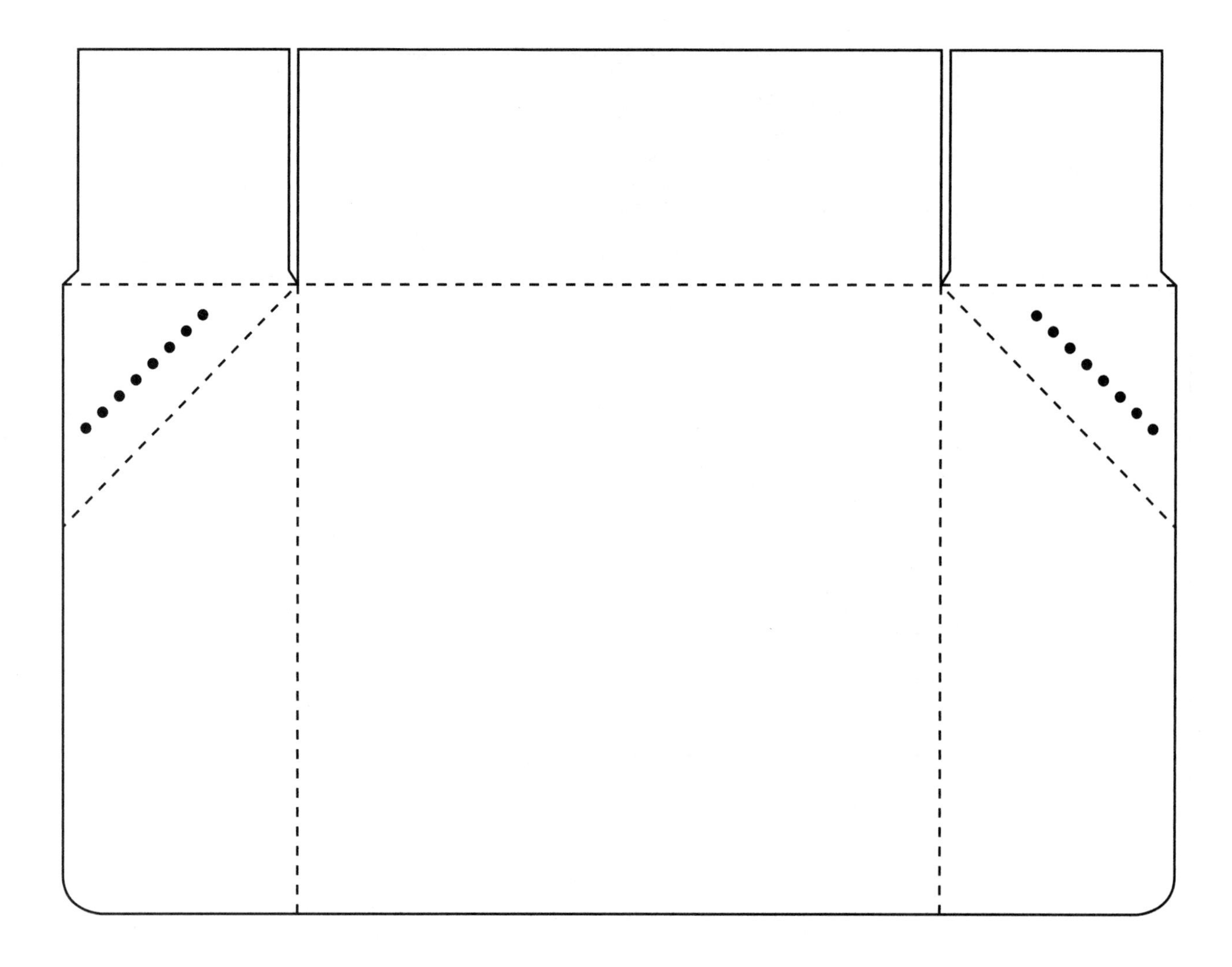

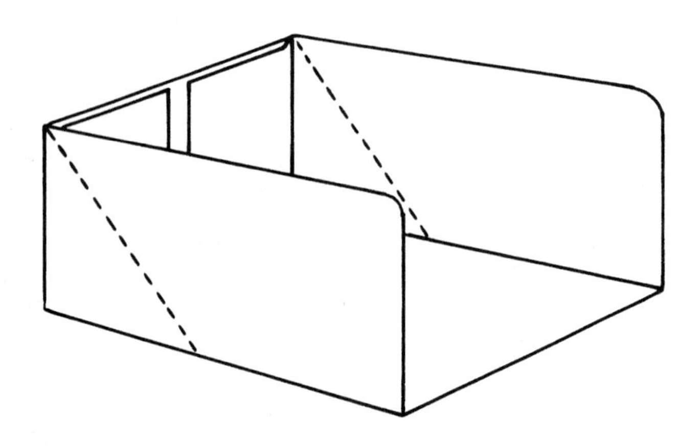

OPEN ENDED COLLAPSIBLE TRAY—BEERS STYLE

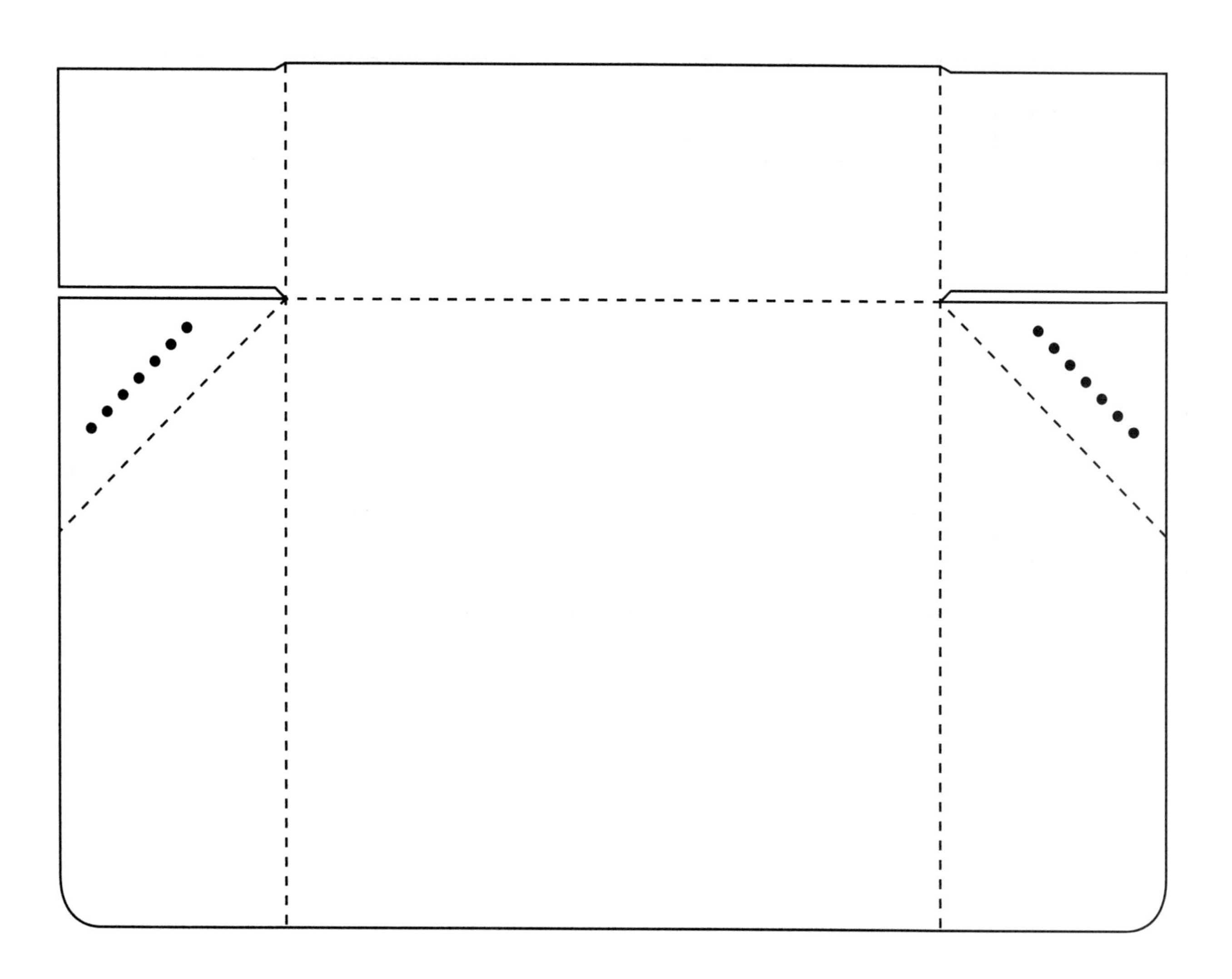

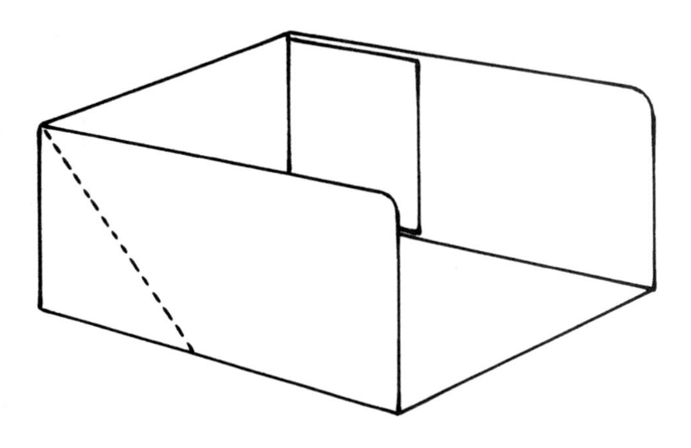

GLUED-UP TRAY WITH PROTECTIVE END PANELS

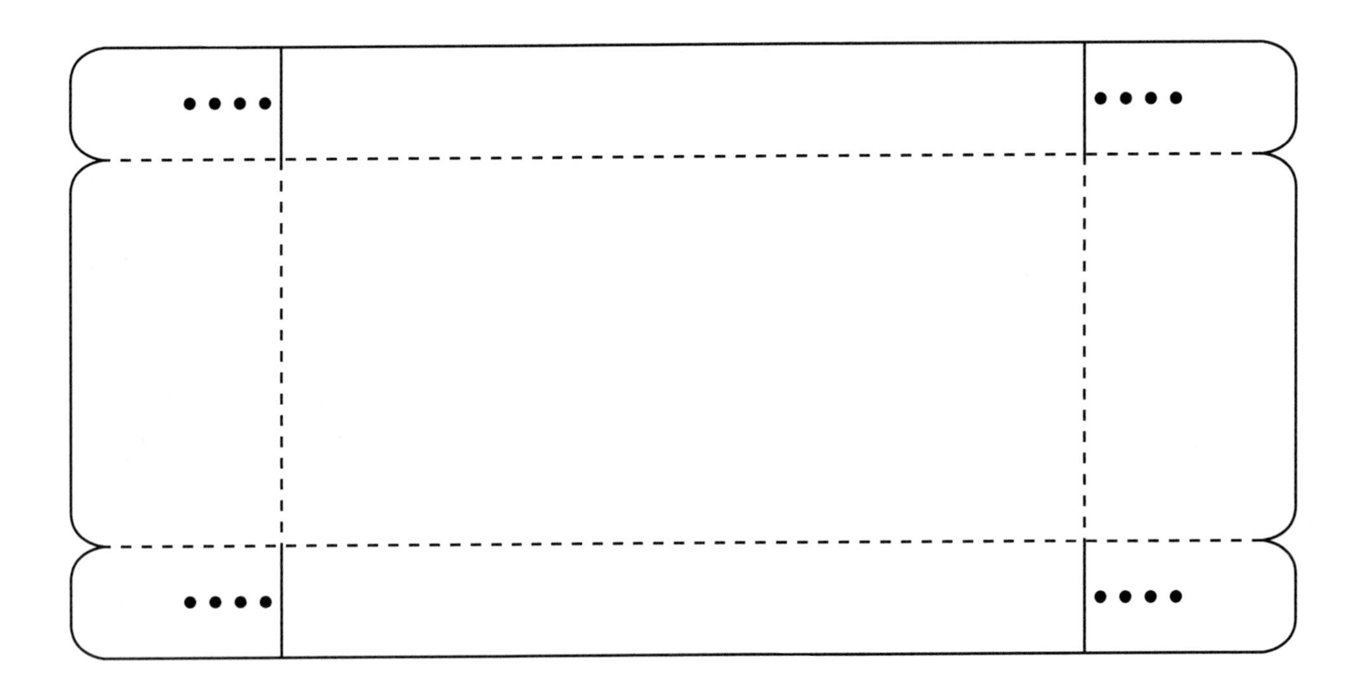

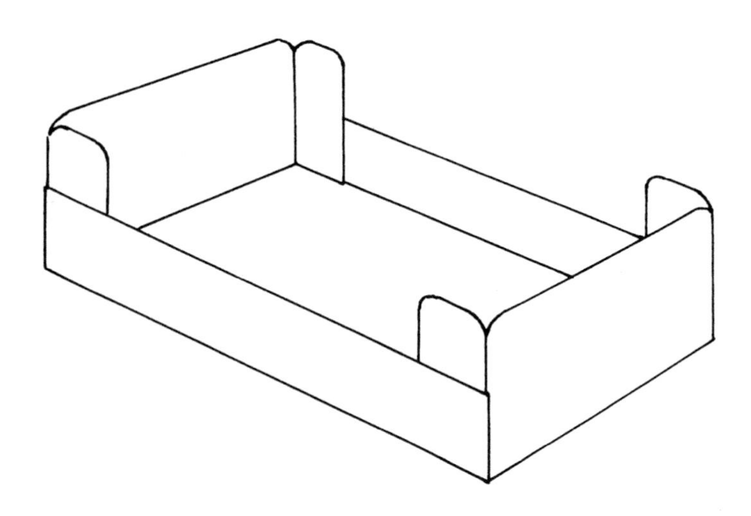

GLUED-UP TRAY WITH PROTECTIVE END PANELS AND TAPERED SIDE PANELS

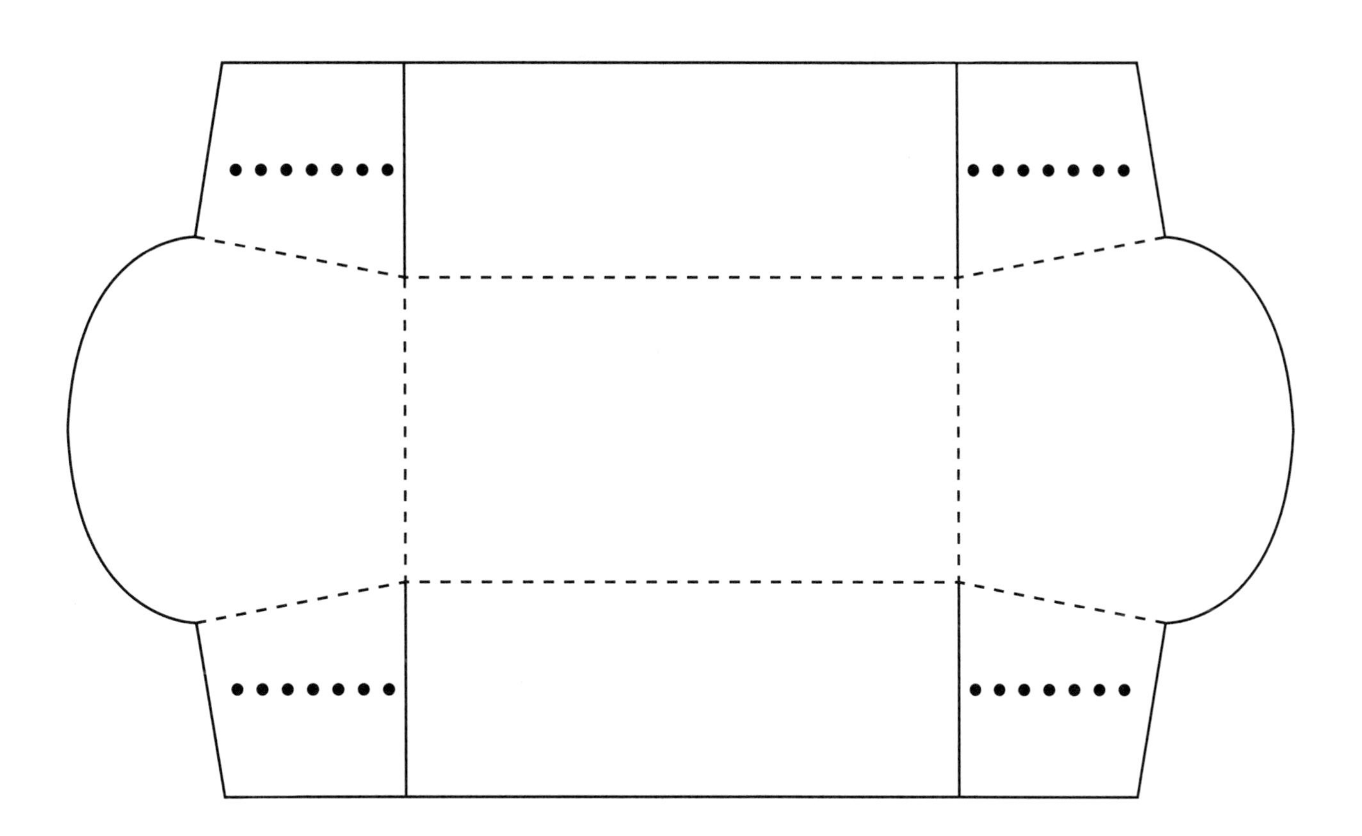

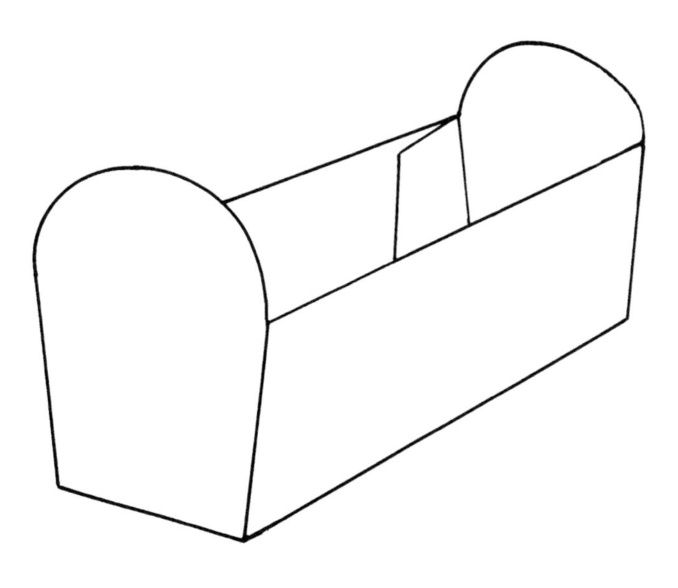

FOUR CORNER BEERS STYLE TRAY

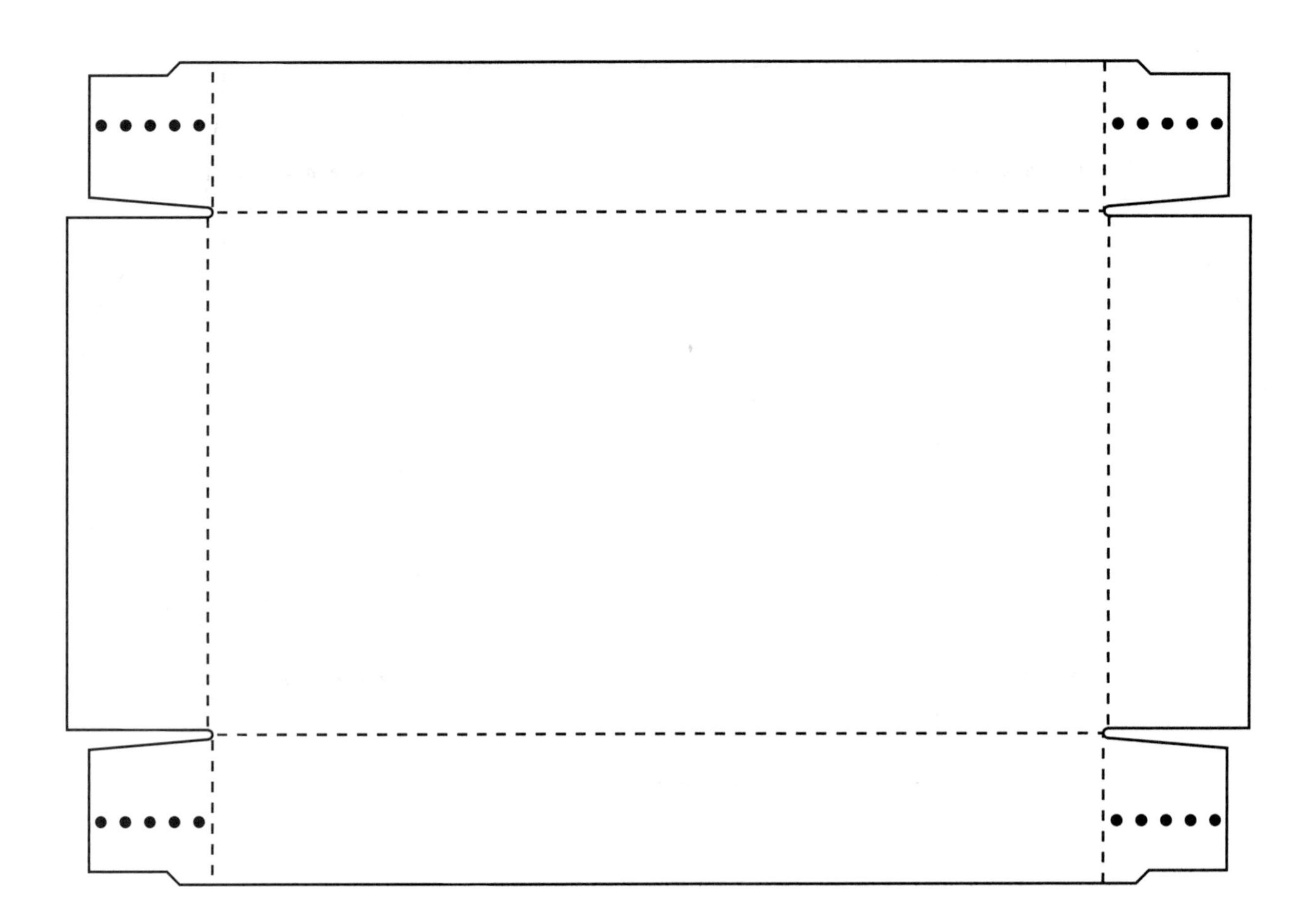

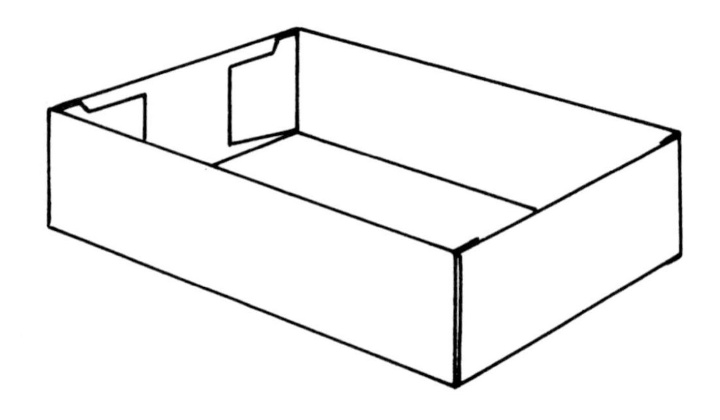

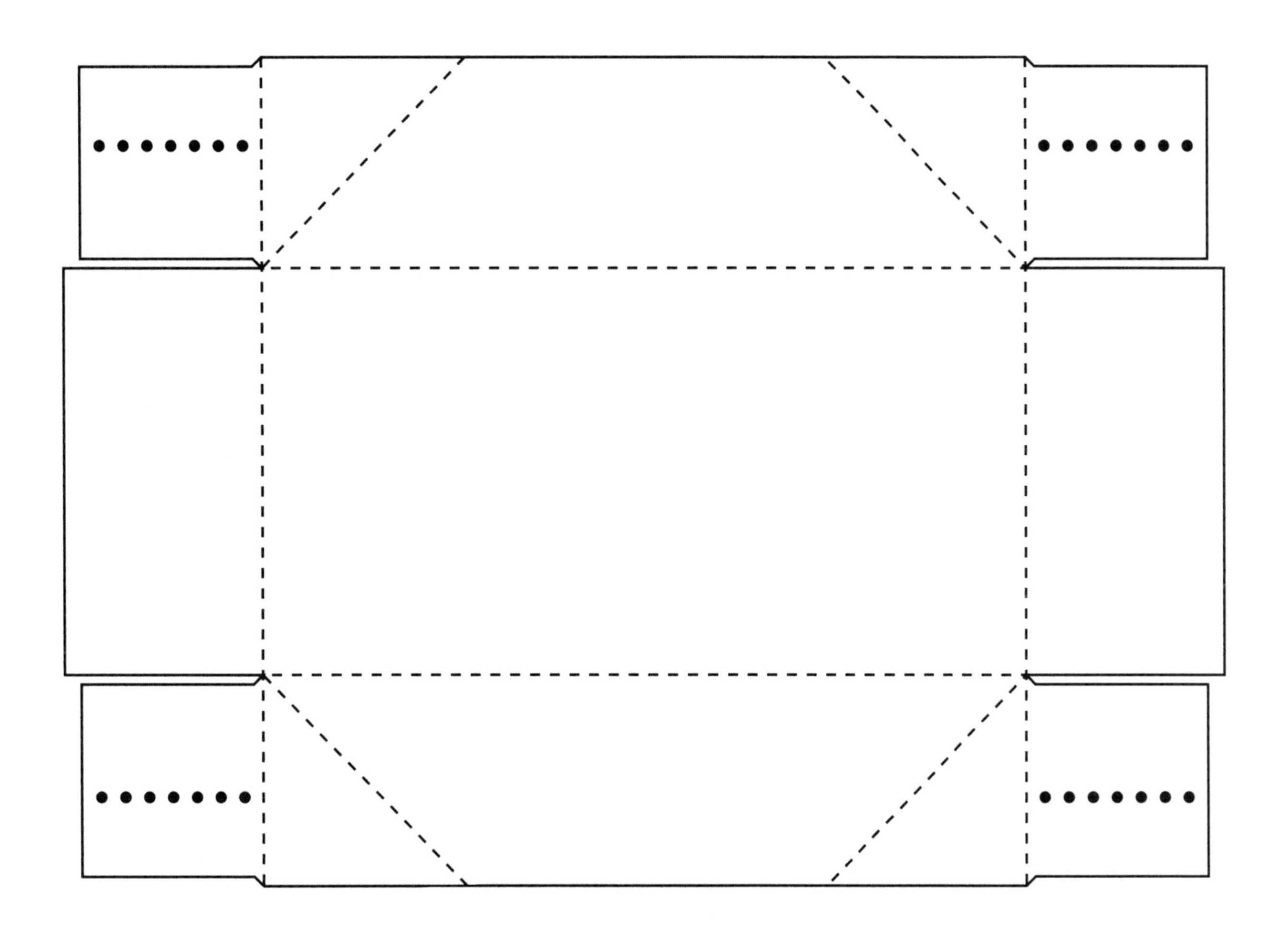

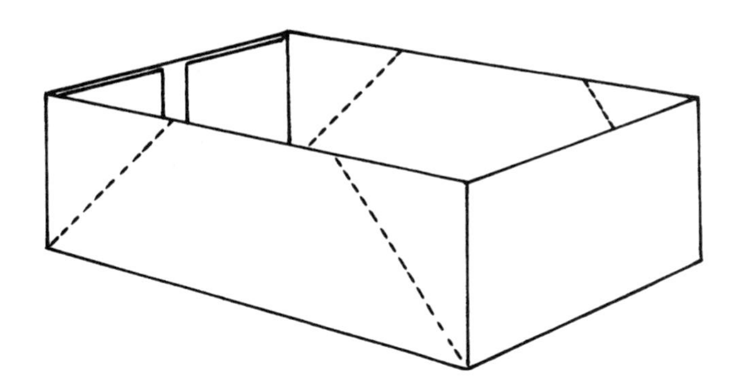

FOUR CORNER BRIGHTWOOD STYLE TRAY

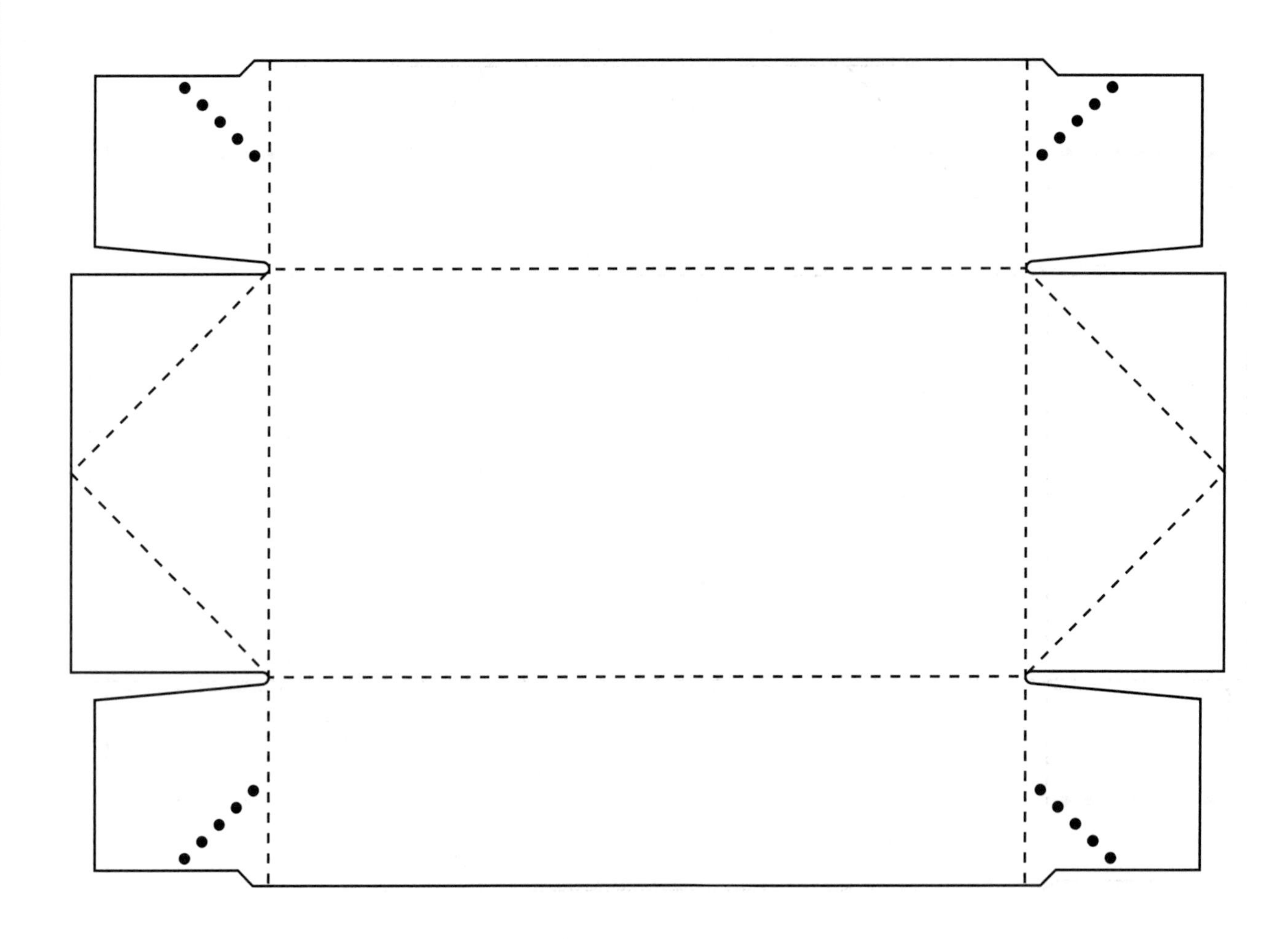

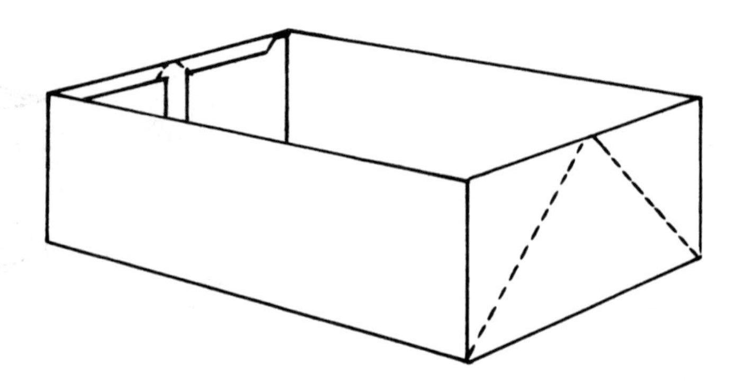

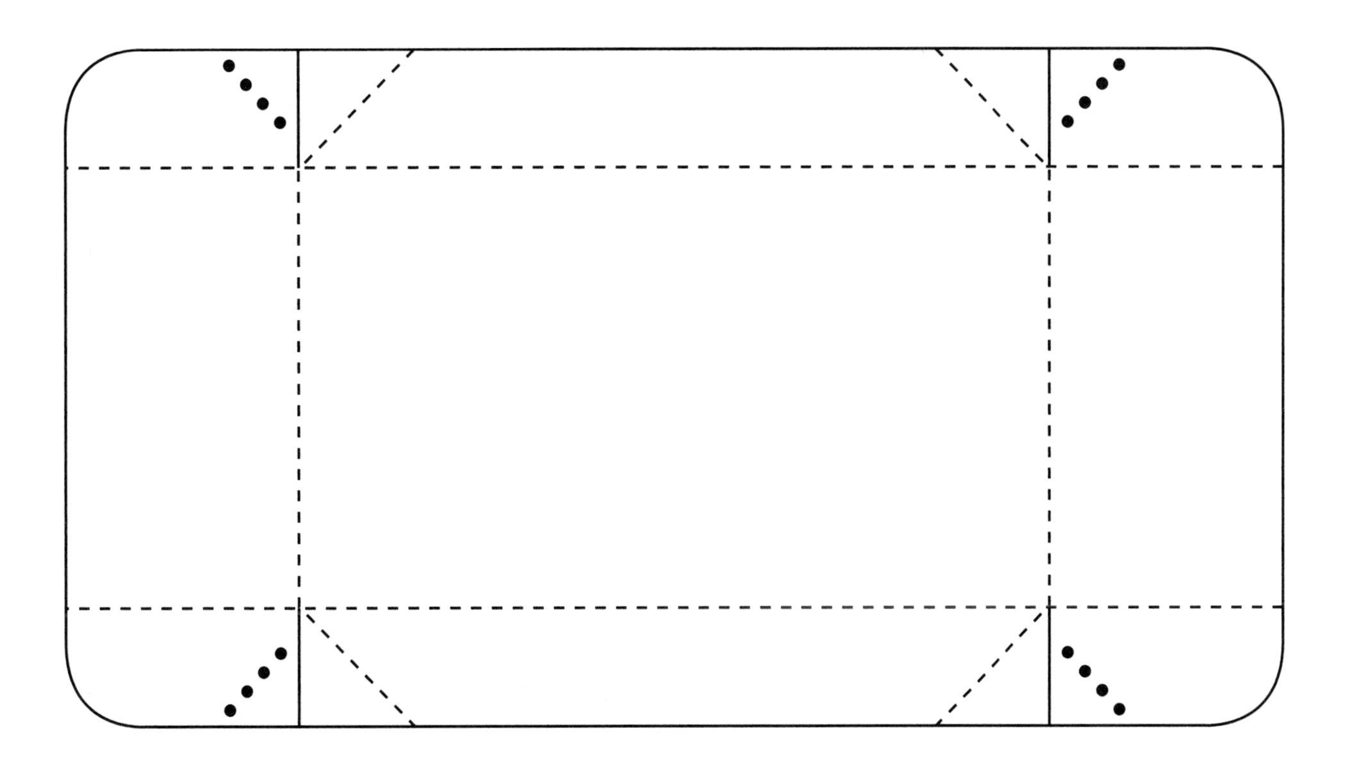

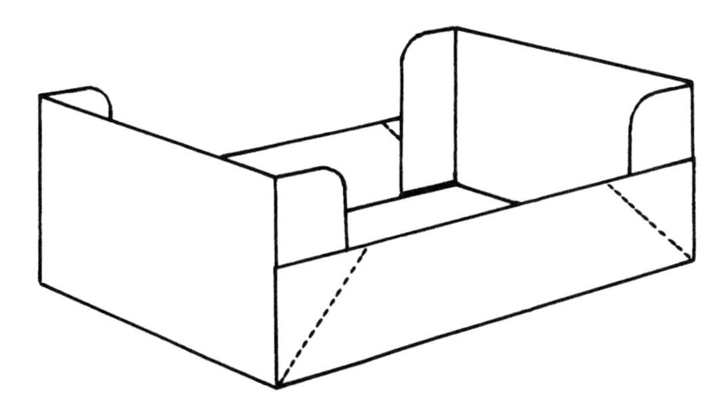

TRAY WITH CORNER LOCKS

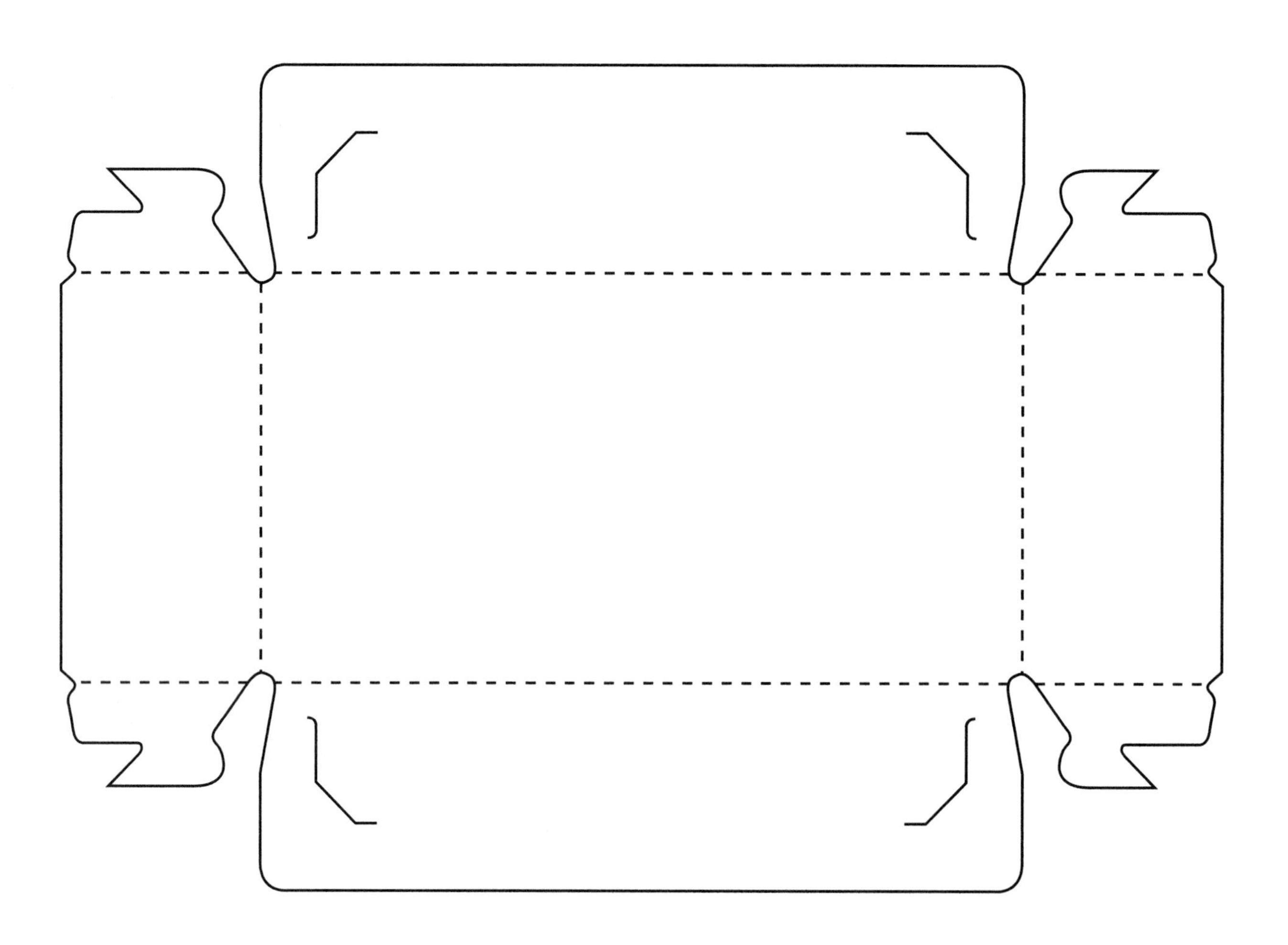

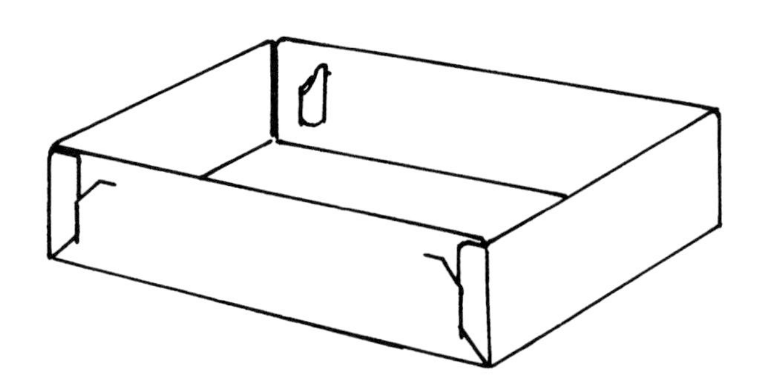

CORNER LOCK TRAY WITH TOP LABELLING FLAPS

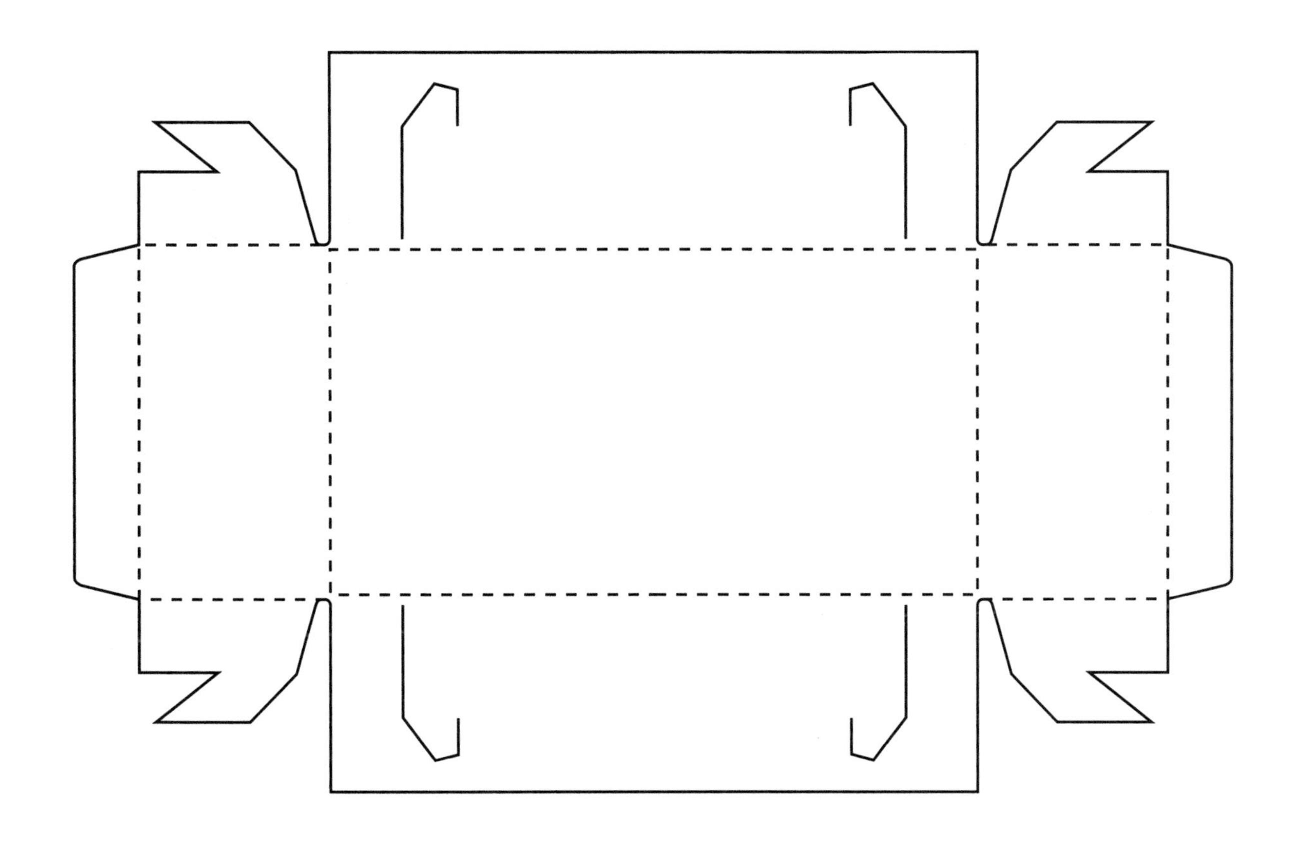

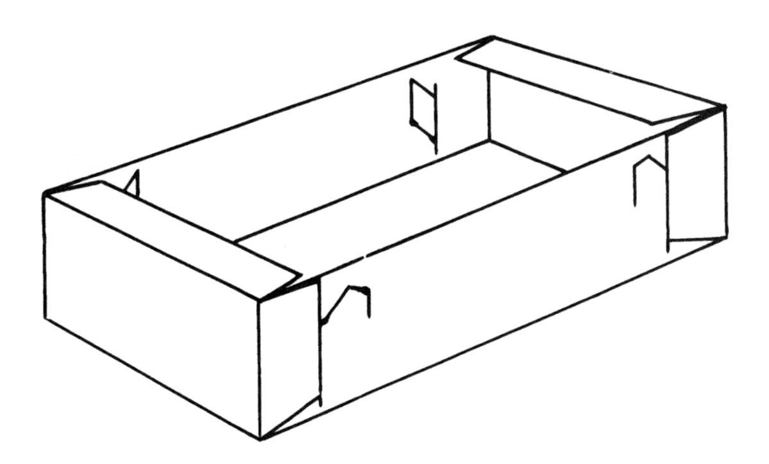

STRIPPER LOCK TRAY

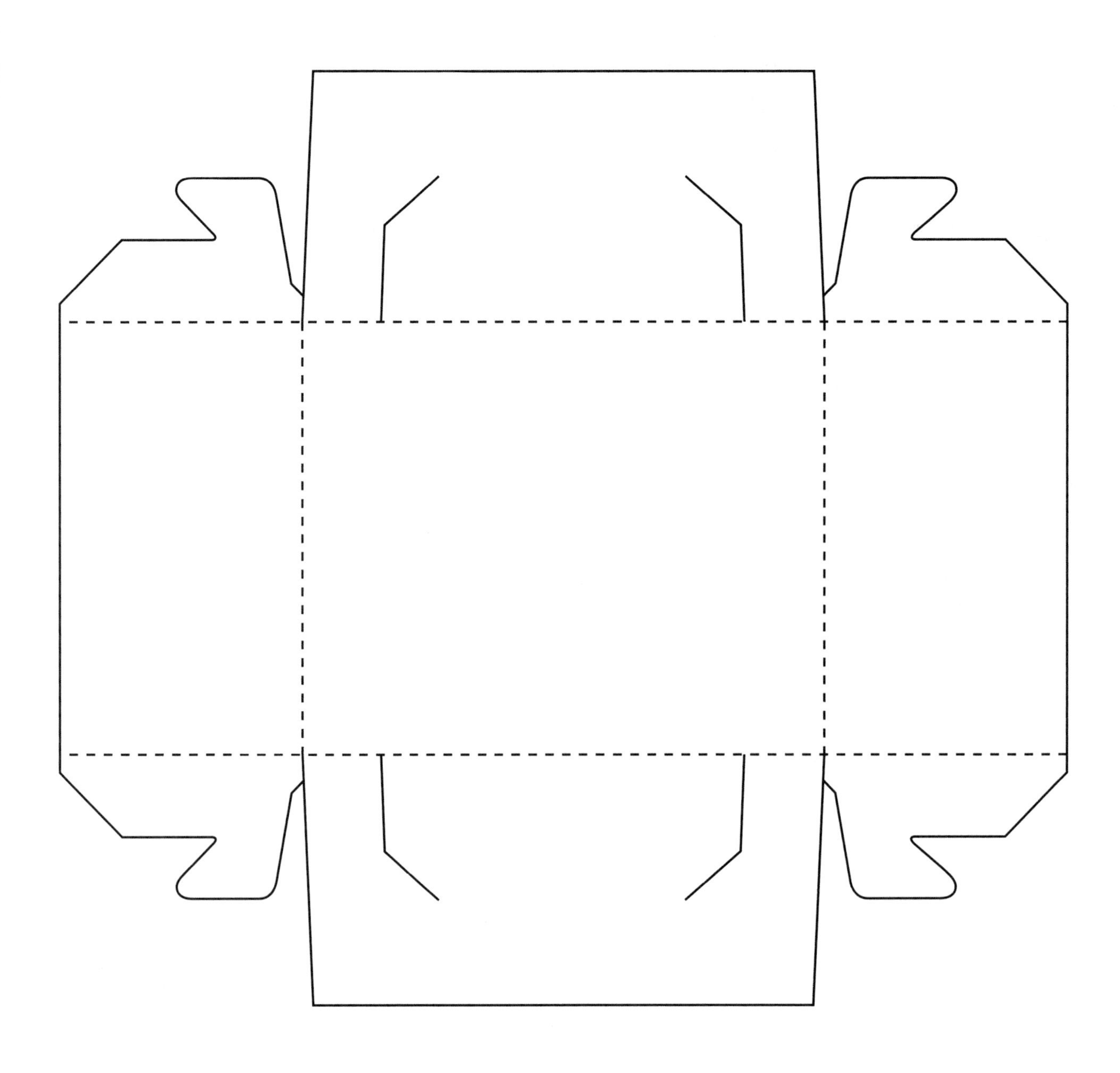

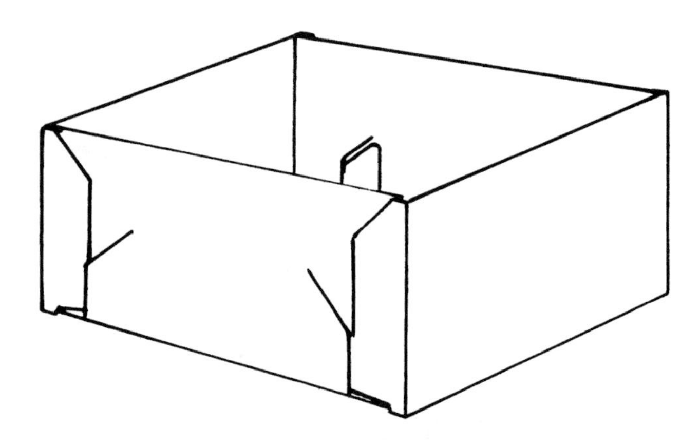

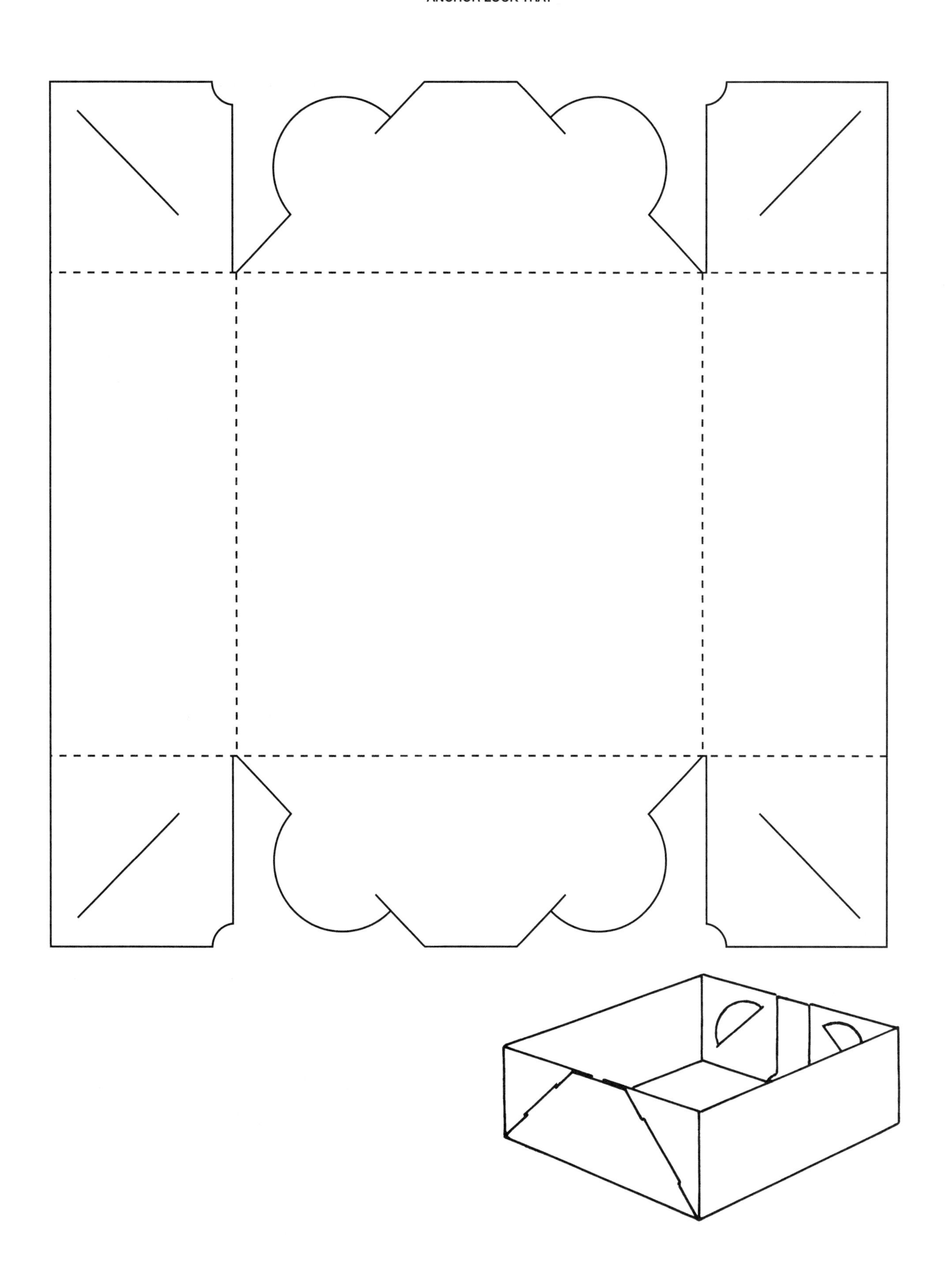

DOUBLE LOCK TRAY

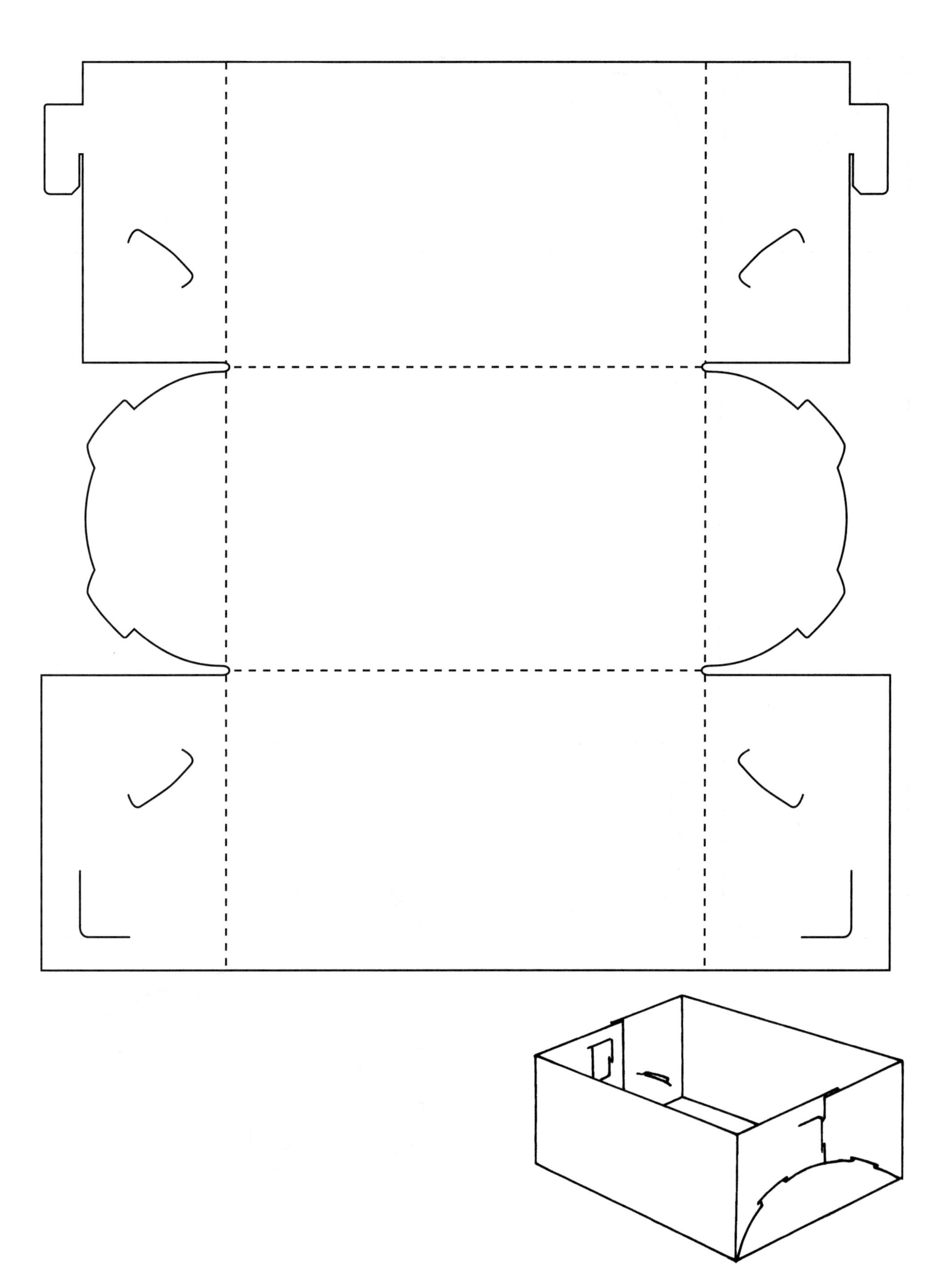

T-LOCK TRAY

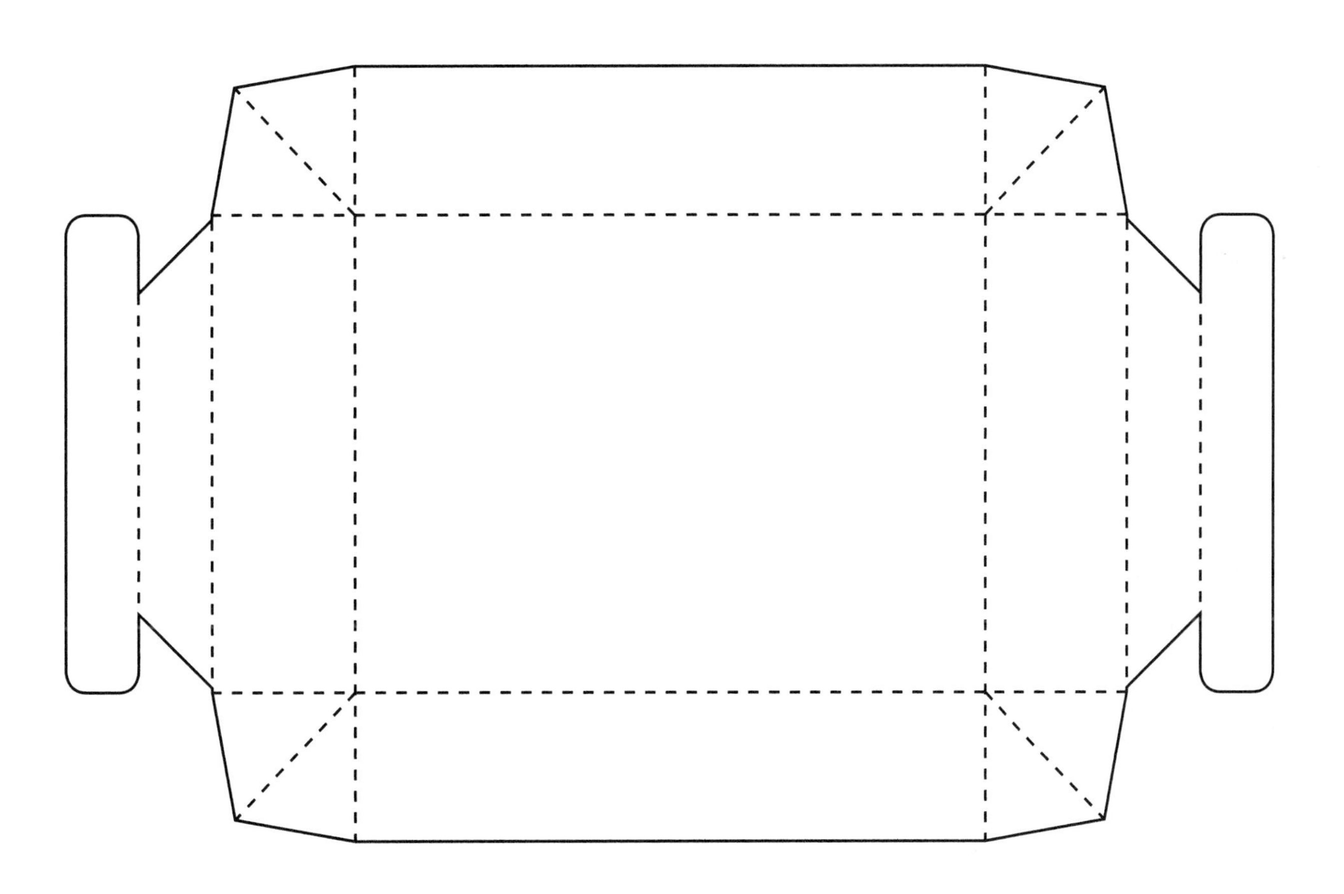

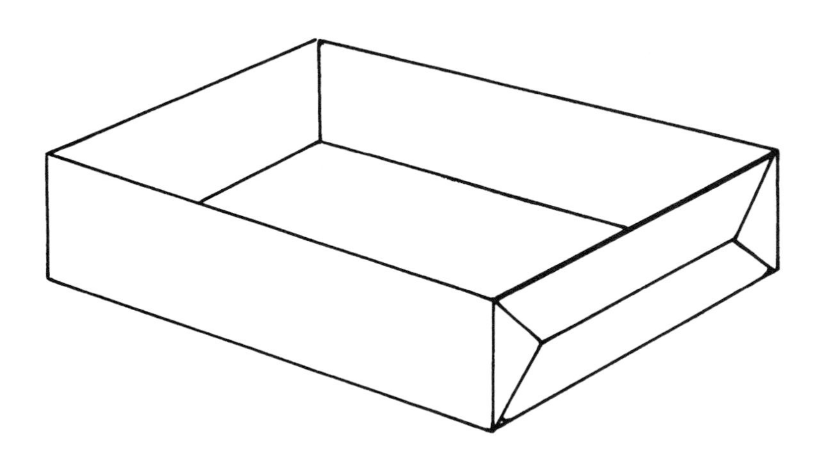

T-LOCK TRAY WITH BEVELED SIDE AND END PANELS

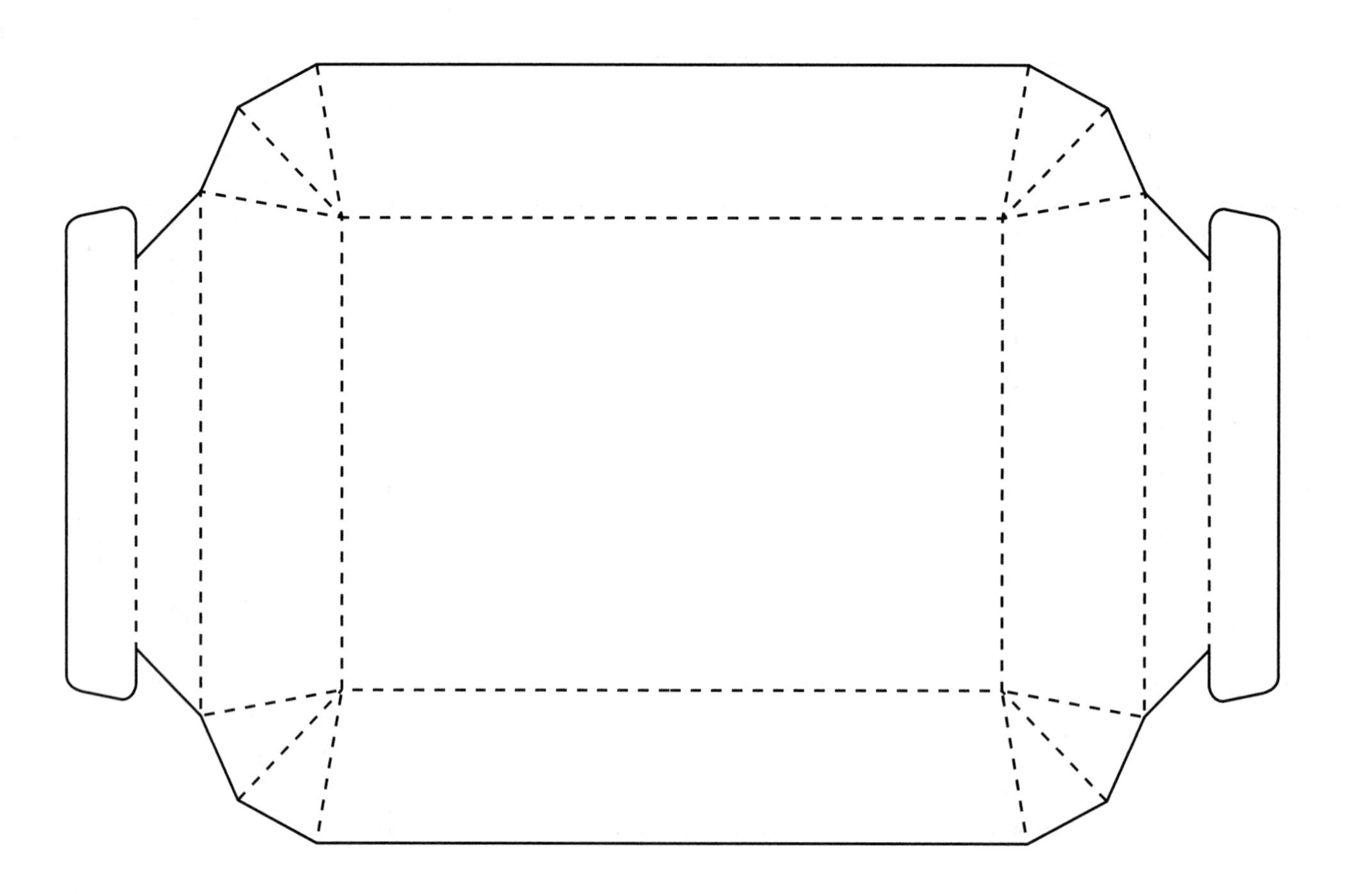

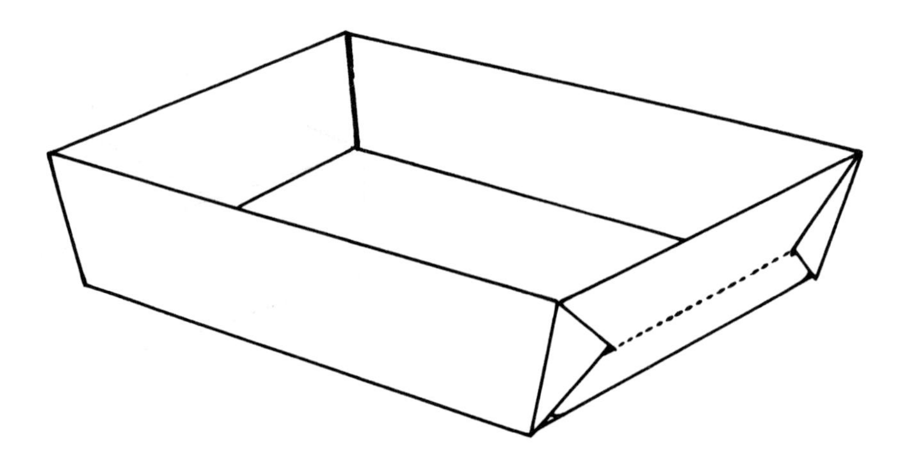

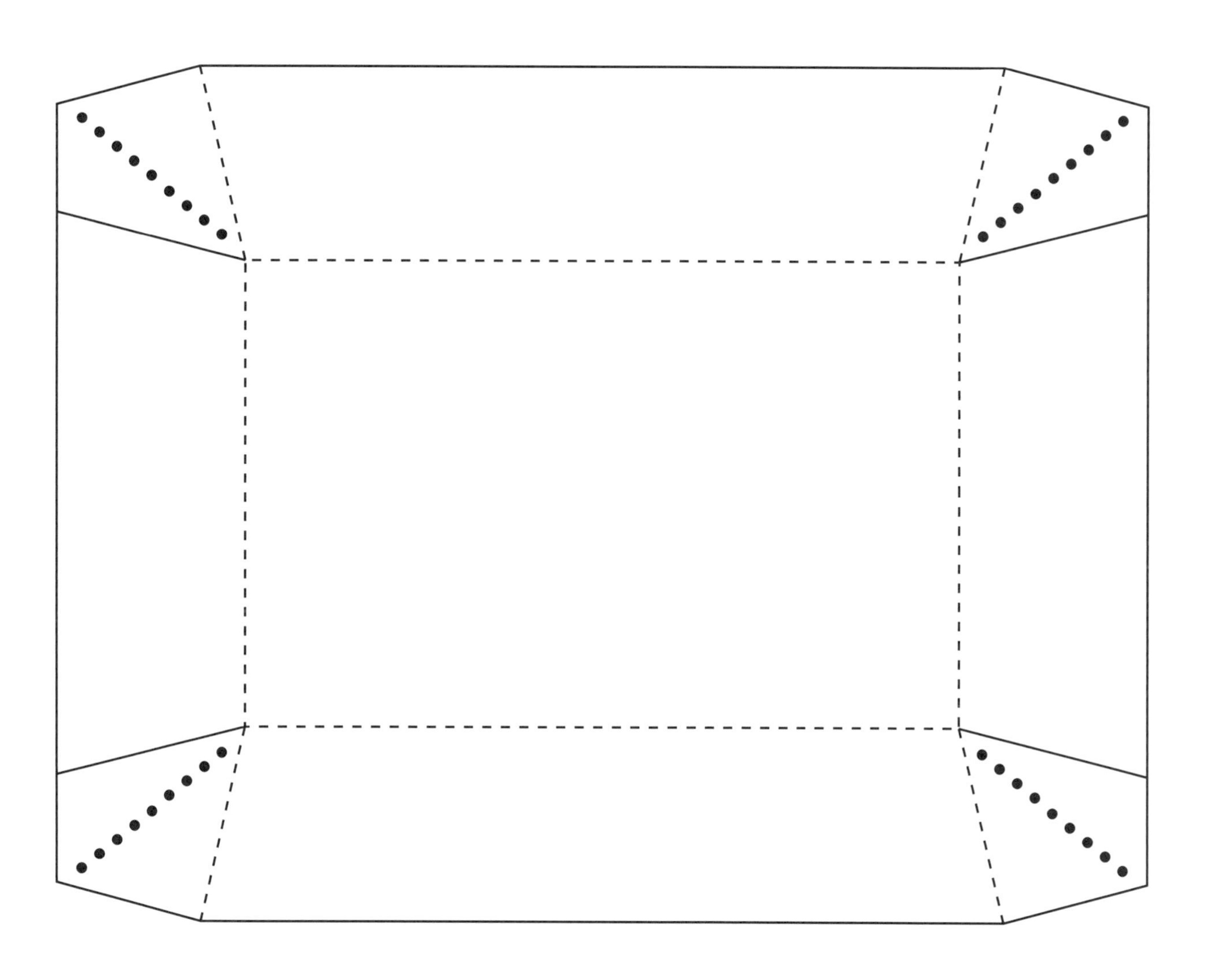

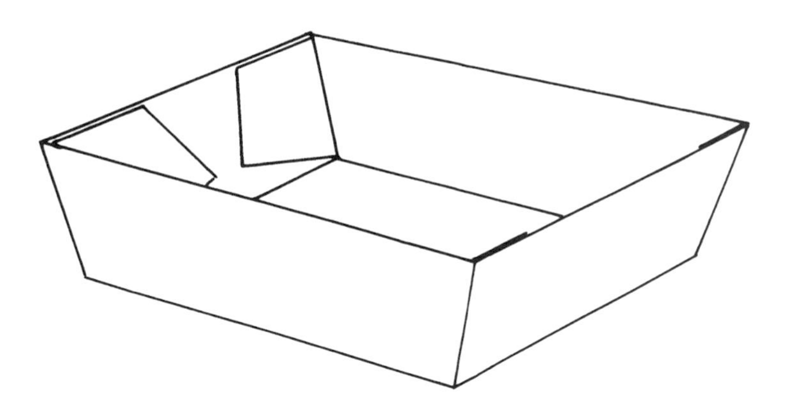

GLUE OR HEAT-SEALED TAPERED TRAY WITH FLANGE

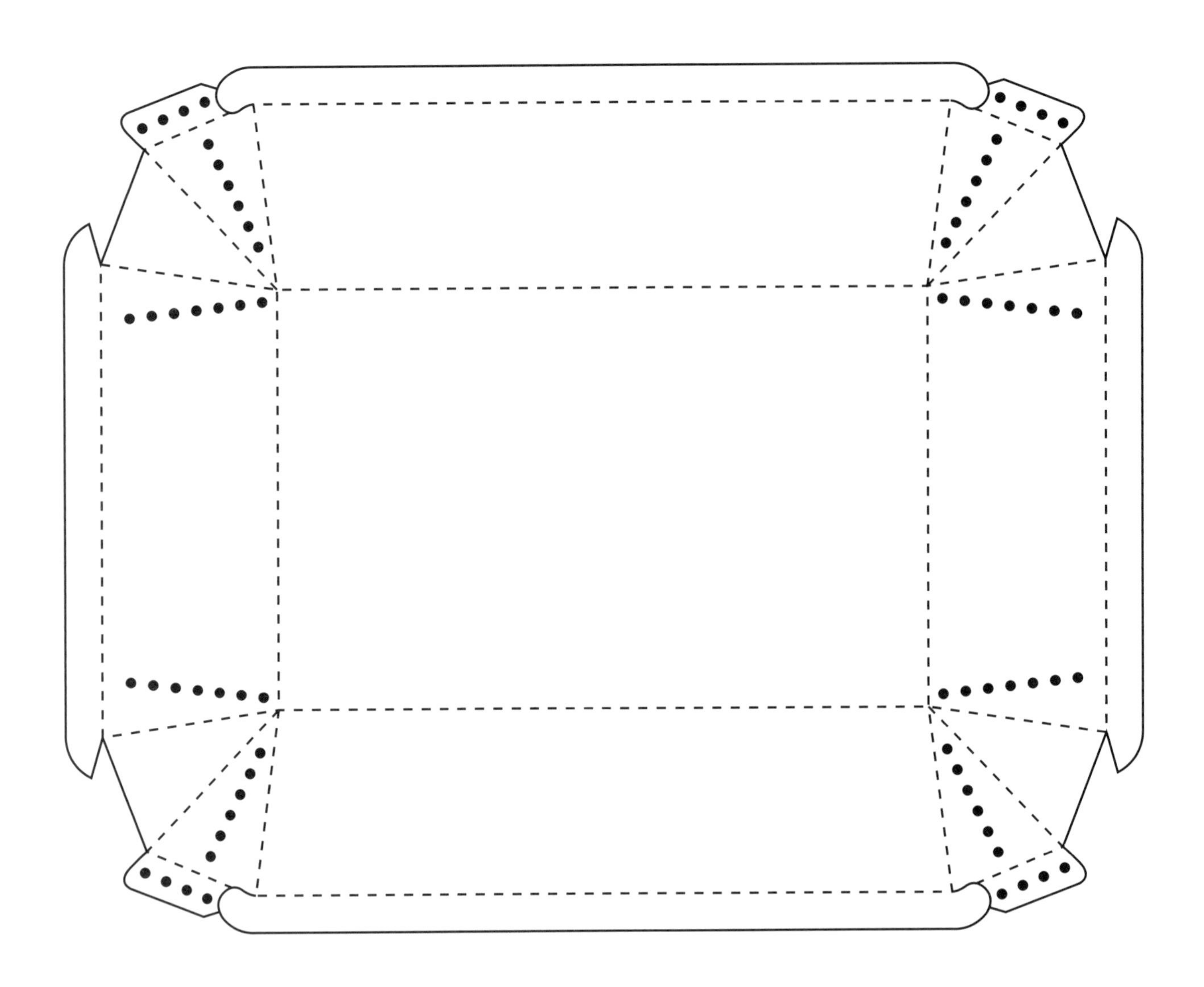

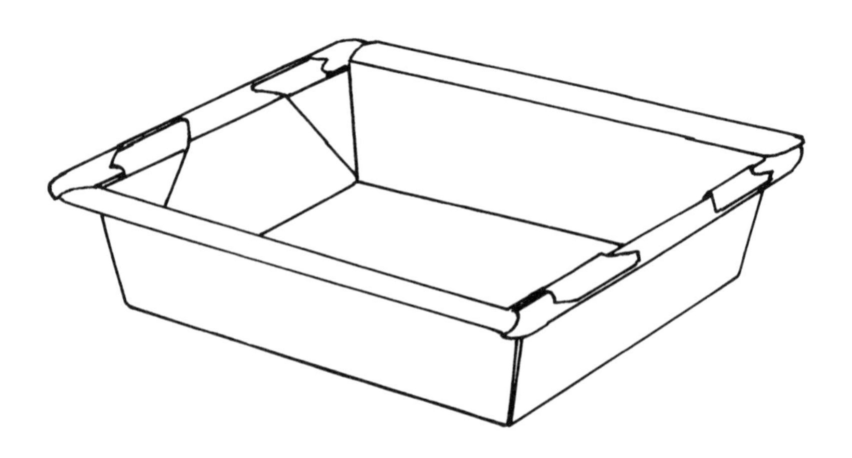

GLUED CORNER TAPERED WITH TOP-LOCK TAB

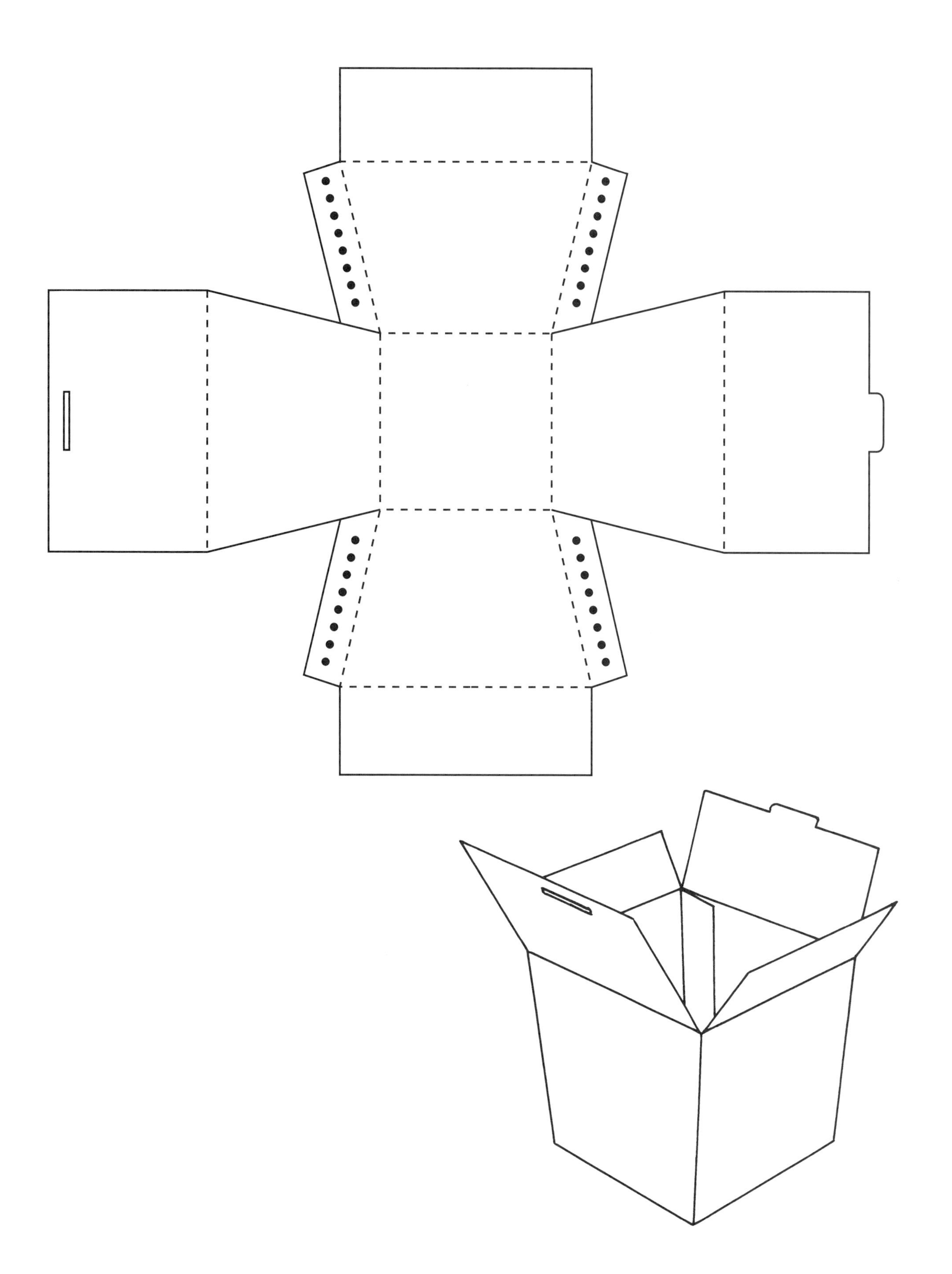

ONE PIECE BRIGHTWOOD TRAY WITH HINGED COVER AND DUSTFLAPS

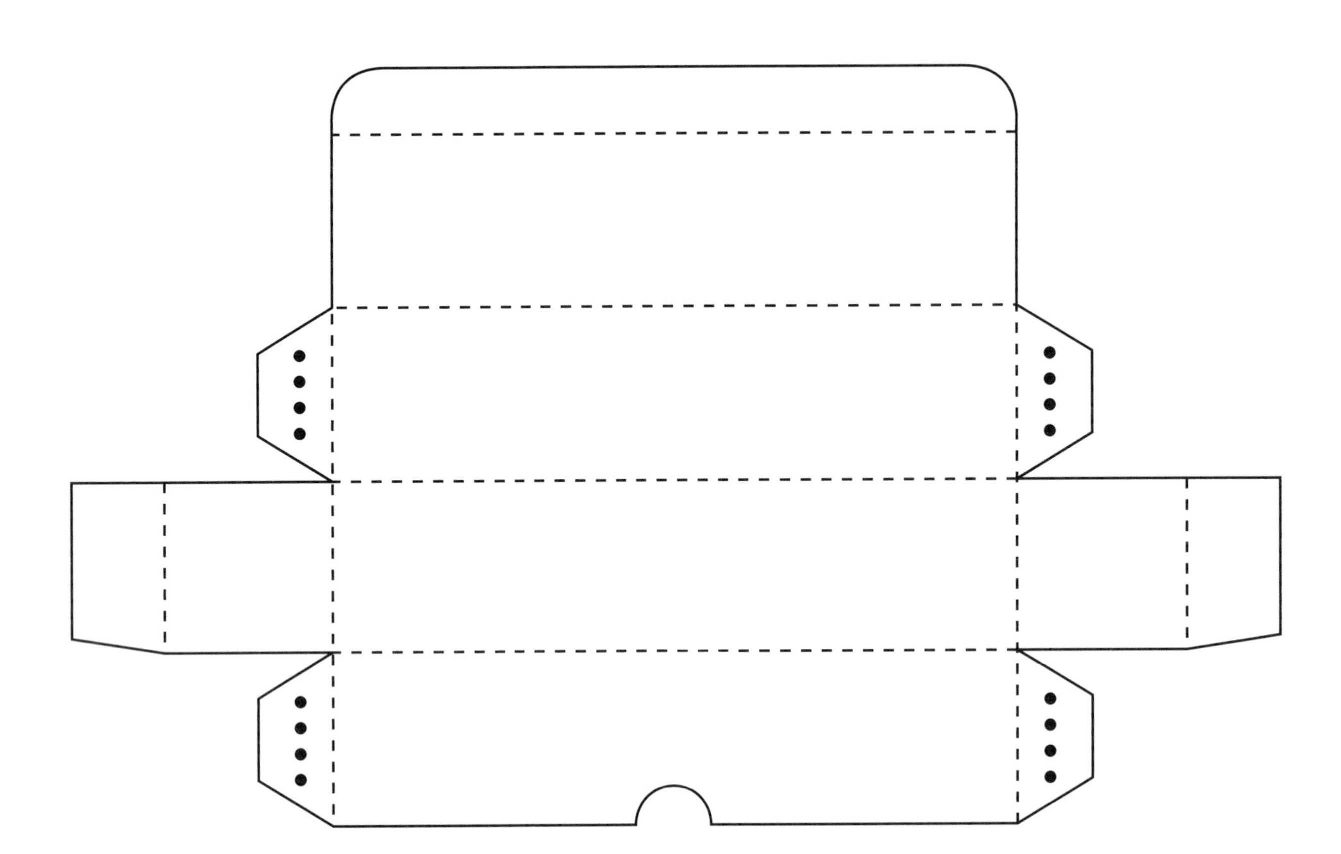

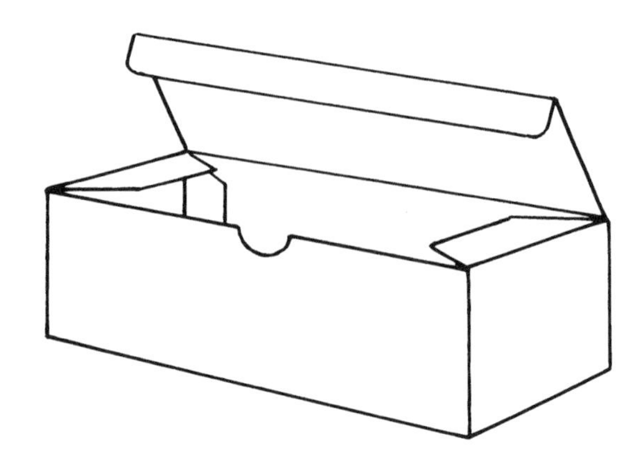

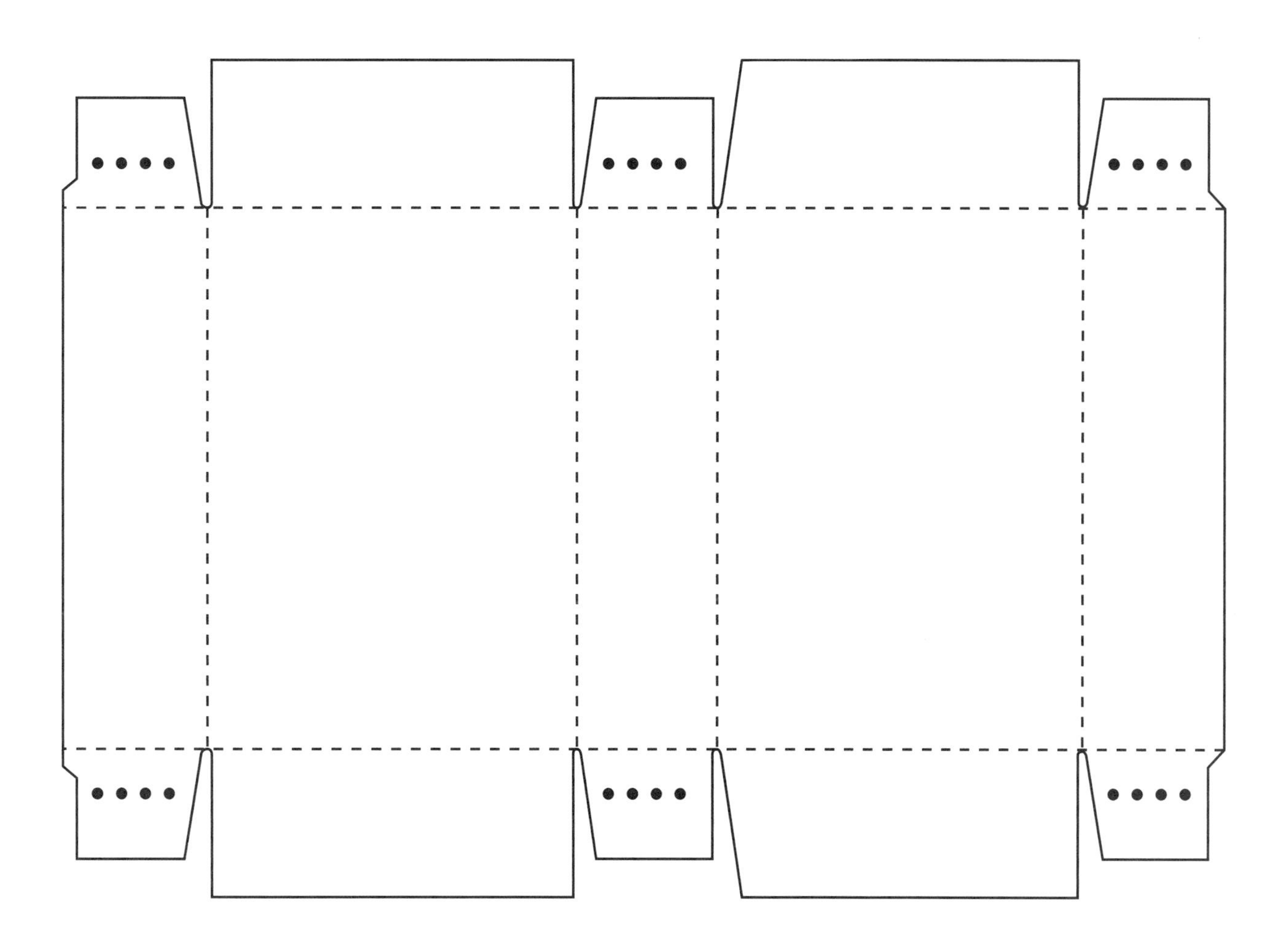

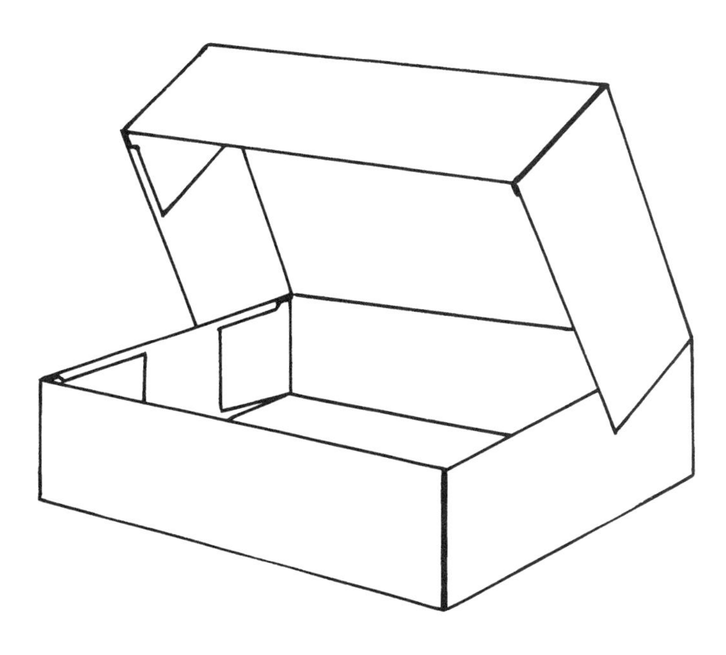

DOUBLE WALL TRAY WITH DOUBLE WALL LID

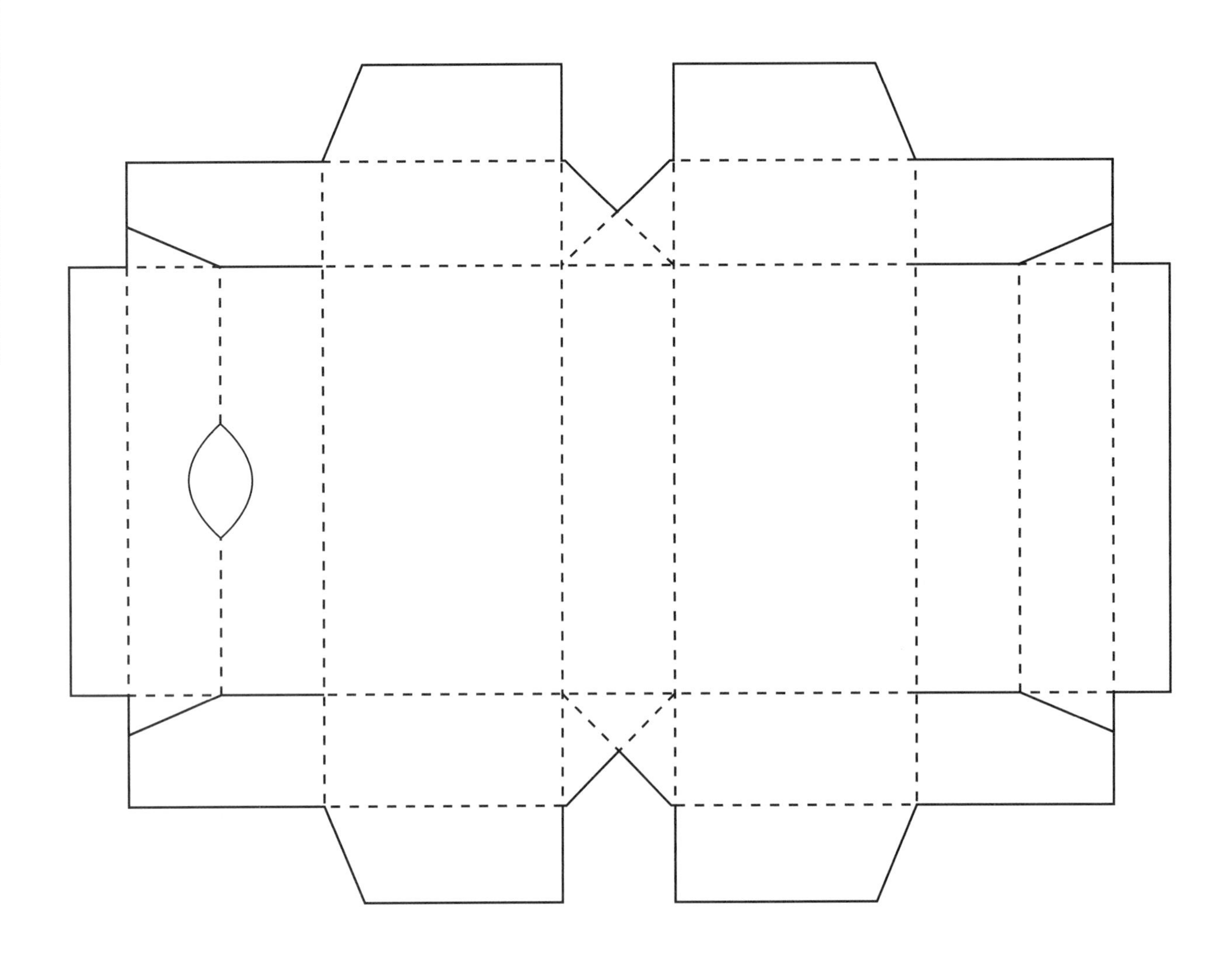

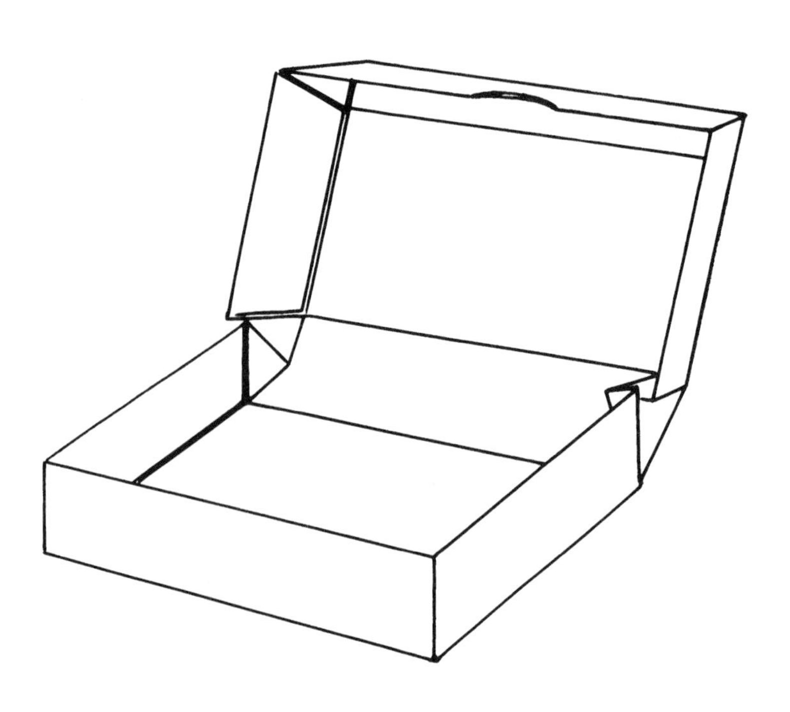

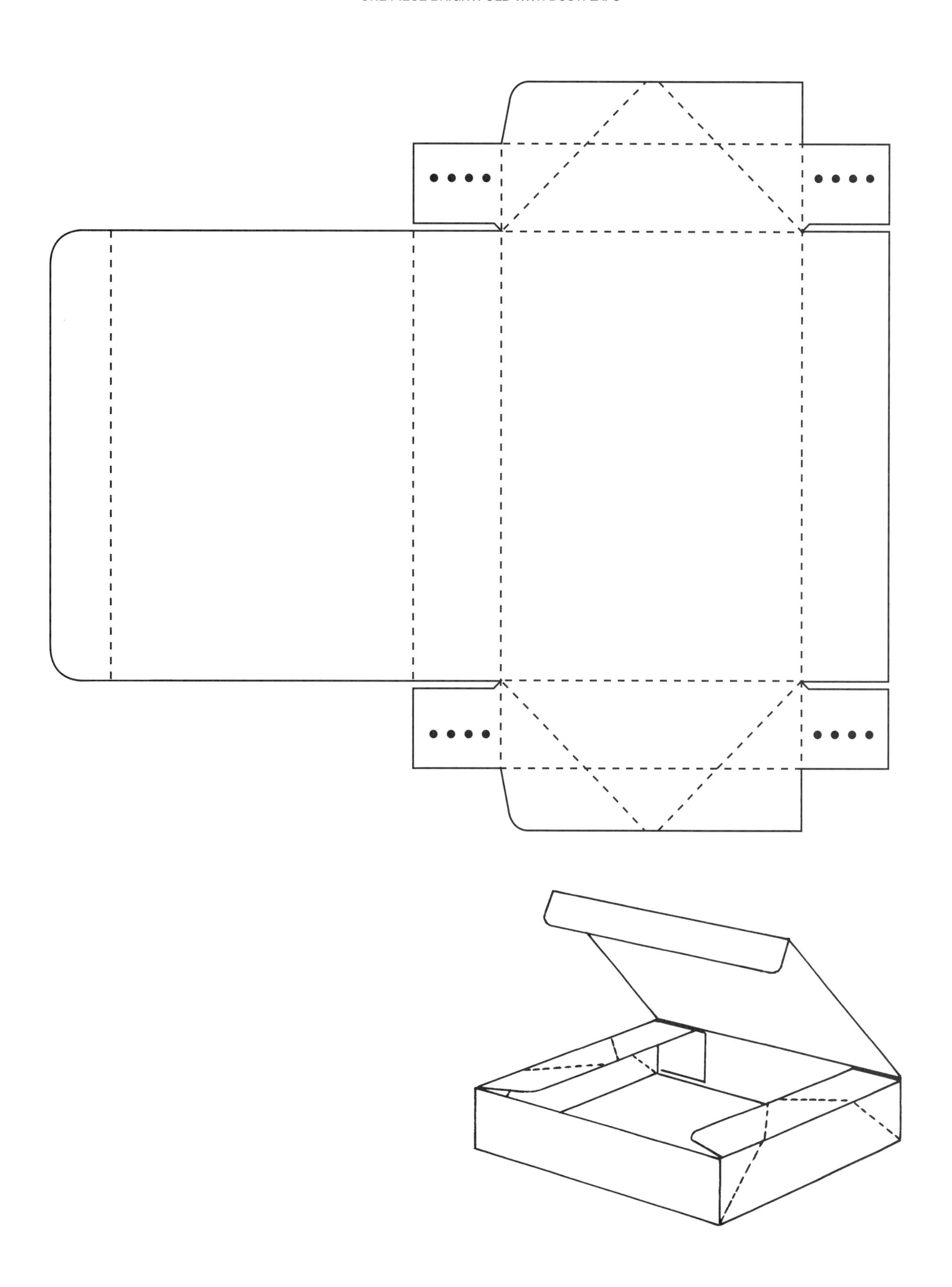

ONE PIECE BEERS INFOLD WITH DUSTFLAPS

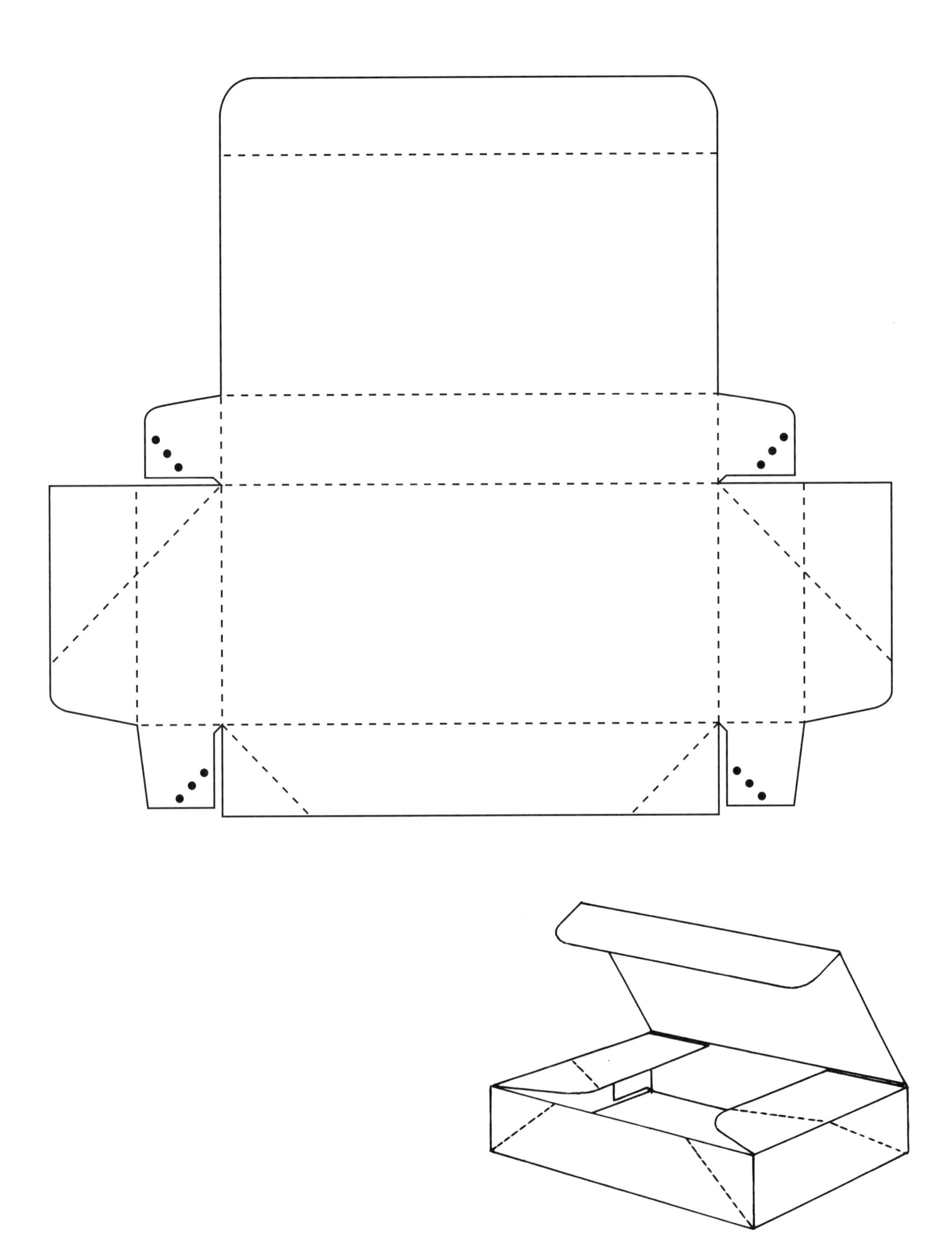

ONE PIECE BEERS INFOLD WITH DUSTFLAPS OFF COVER

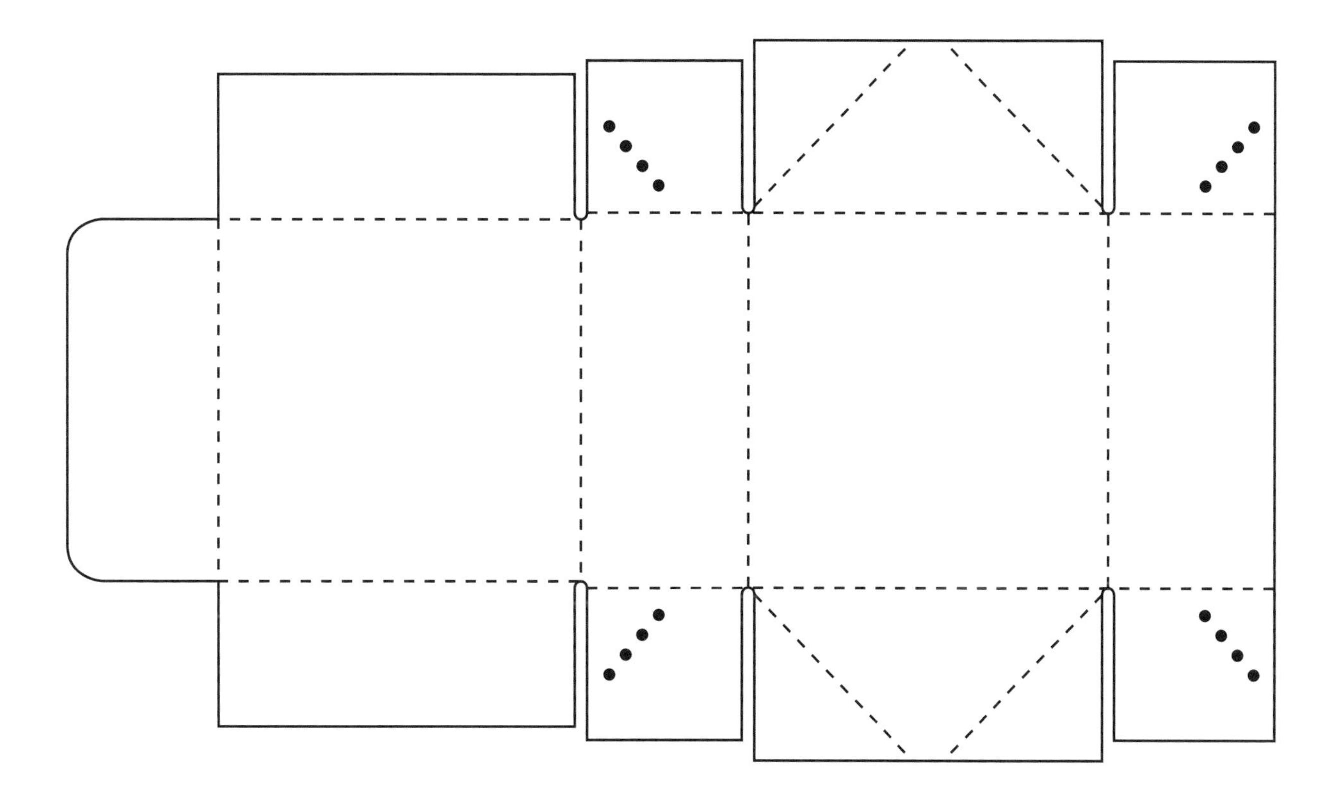

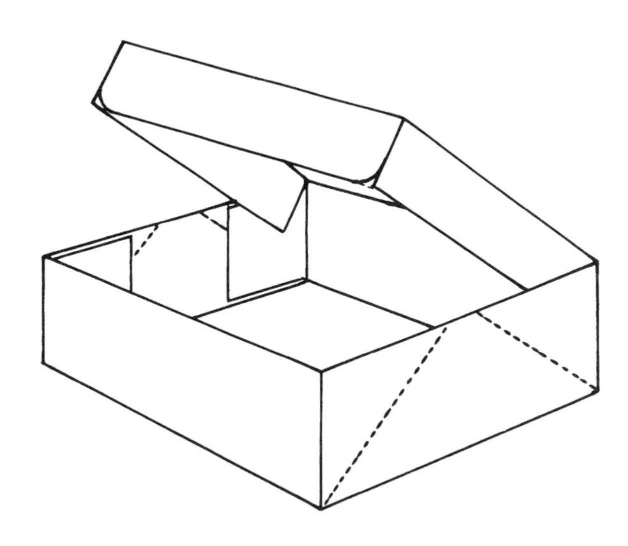

ONE PIECE BEERS FOUR CORNER INFOLD WITH COVER

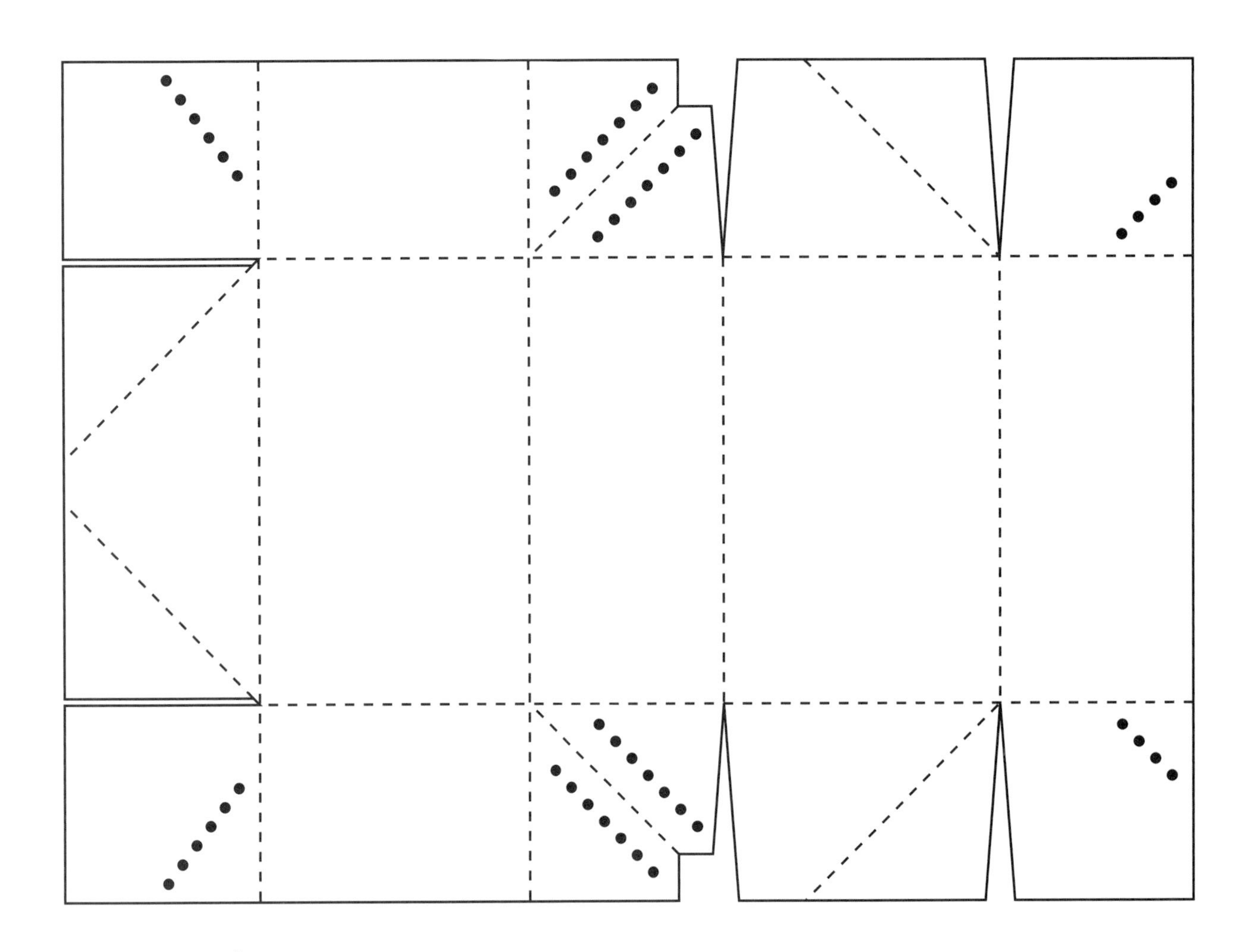

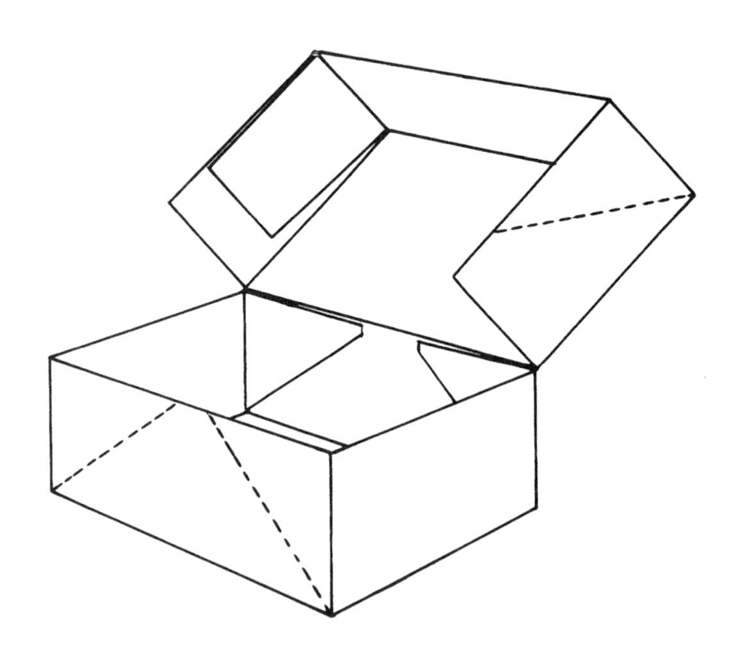

SIX CORNER GLUE STYLE TRAY

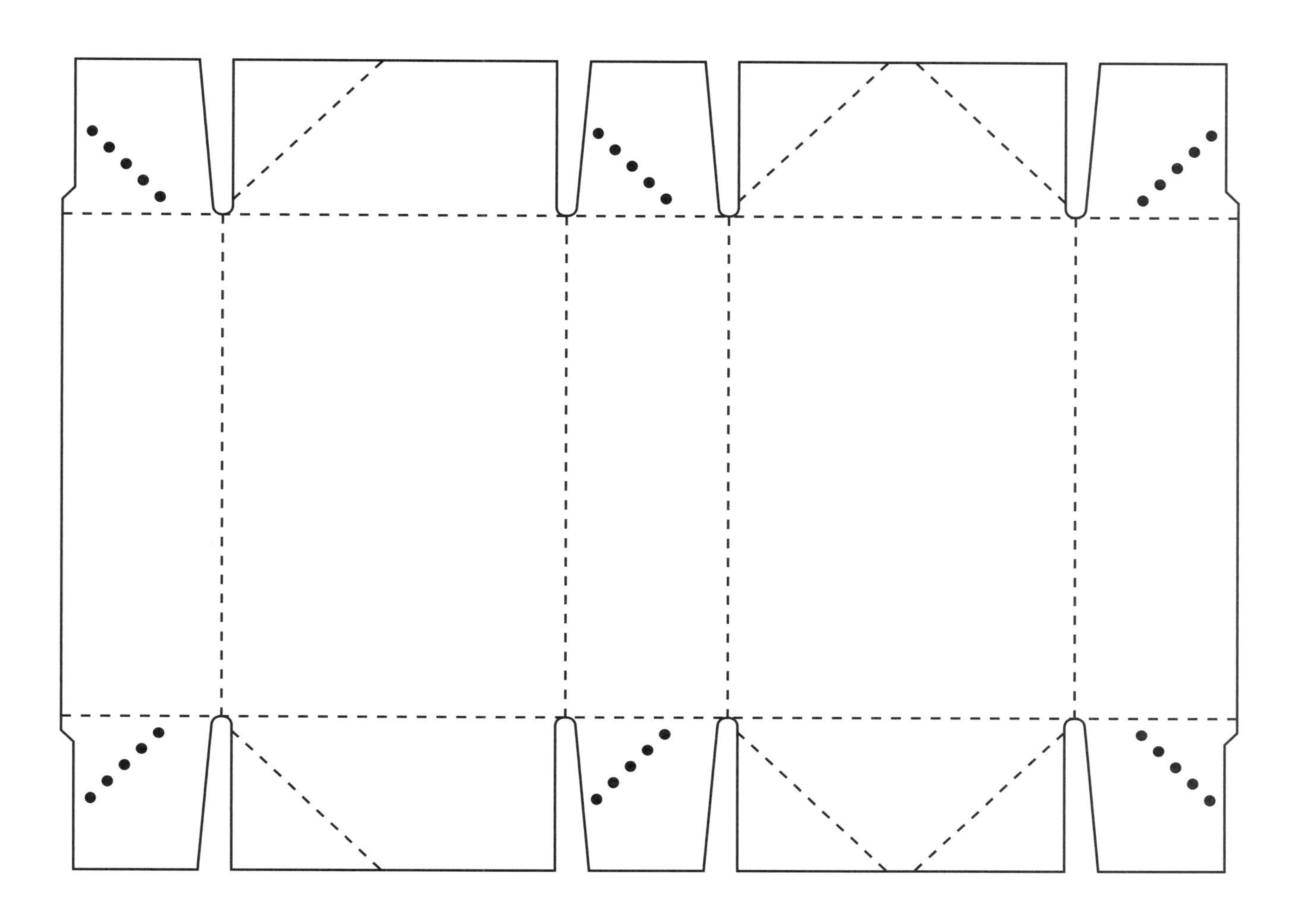

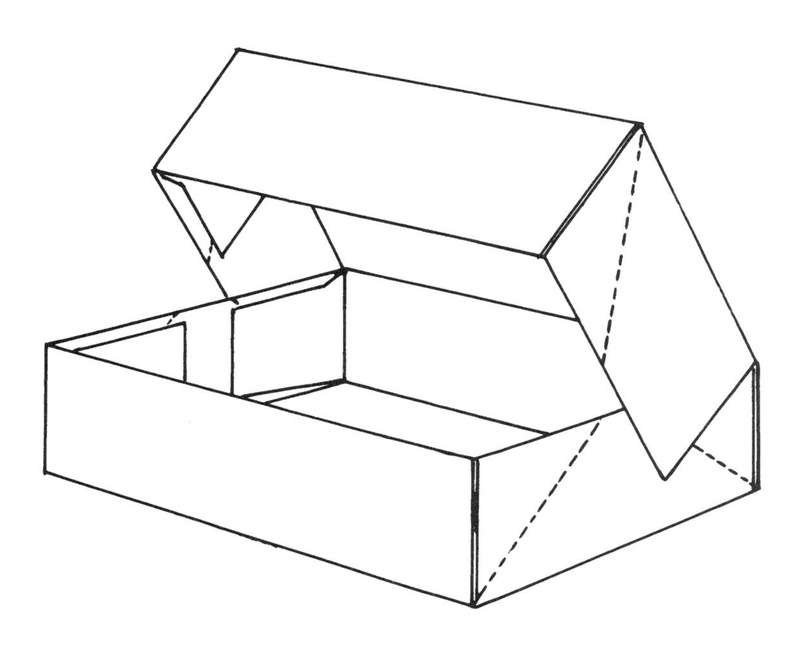

SIX CORNER GLUE STYLE TRAY WITH FLIP-TOP COVER

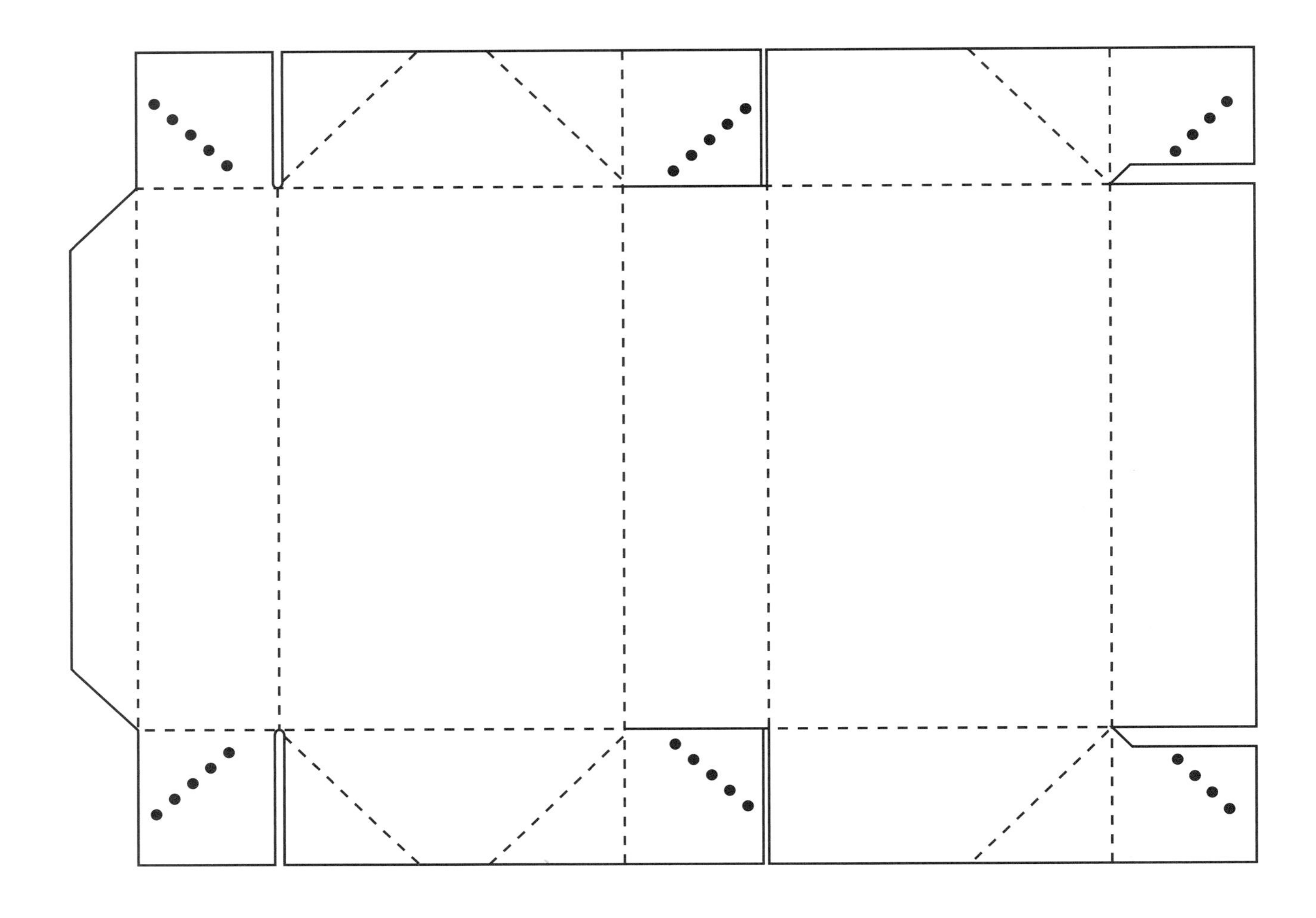

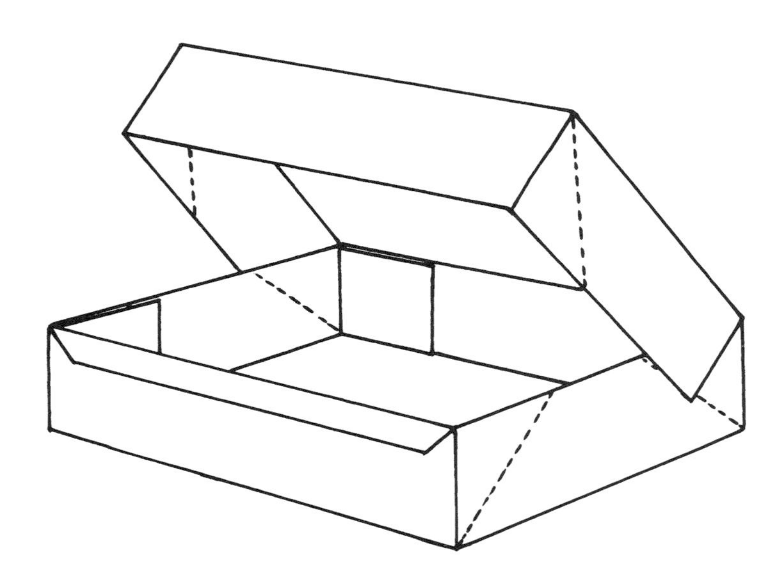

SIX CORNER BEERS STYLE TRAY WITH ANCHOR LOCK

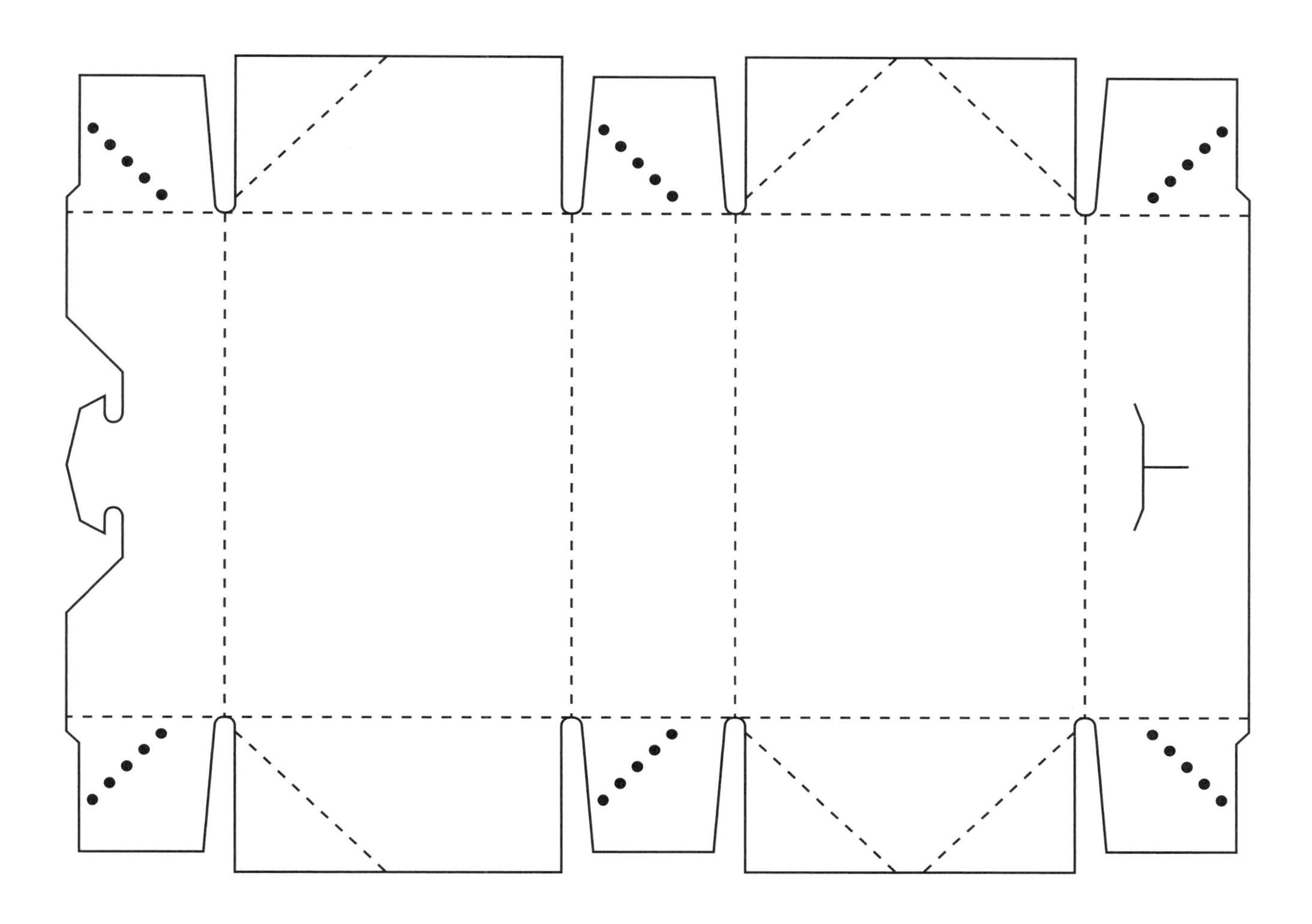

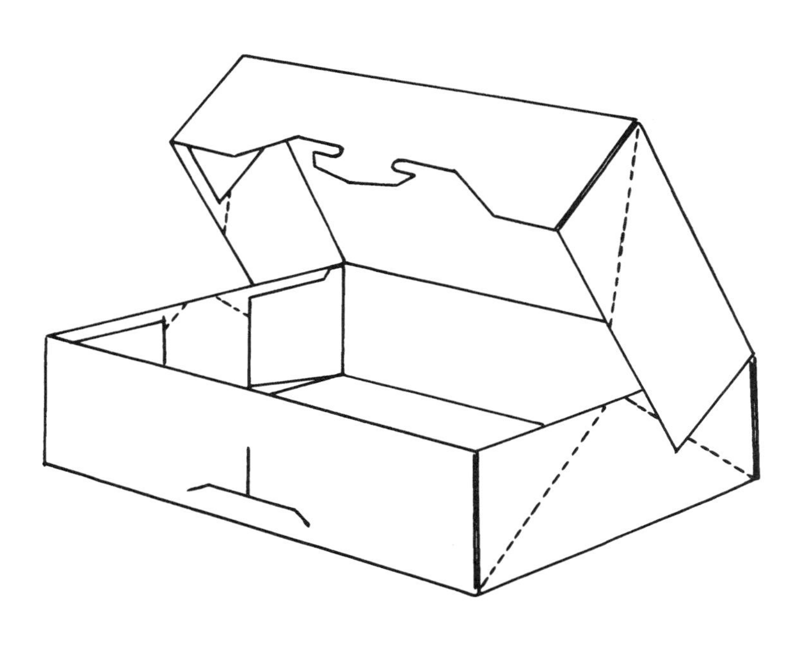

STRIPPER LOCK TRAY WITH HINGED COVER

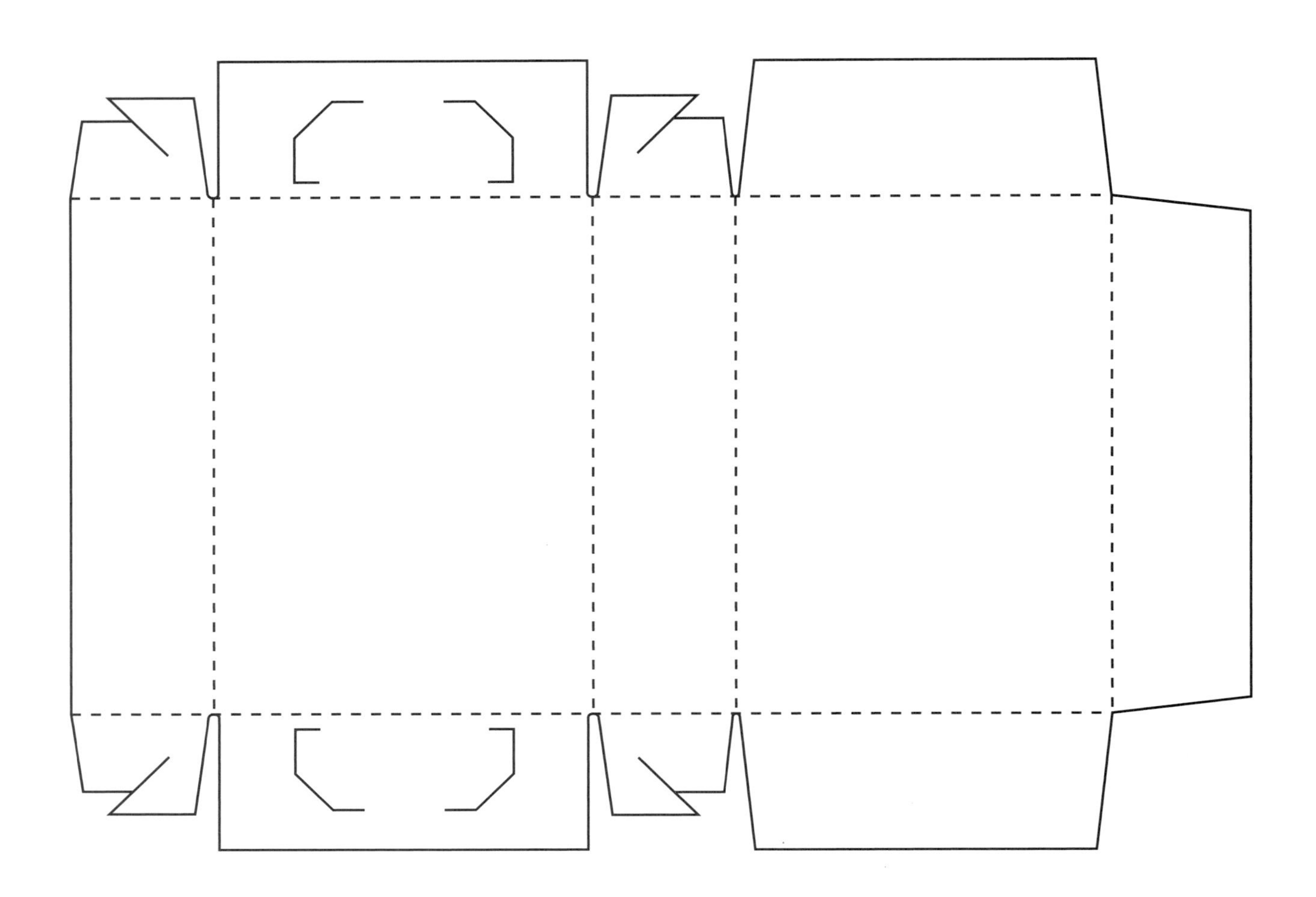

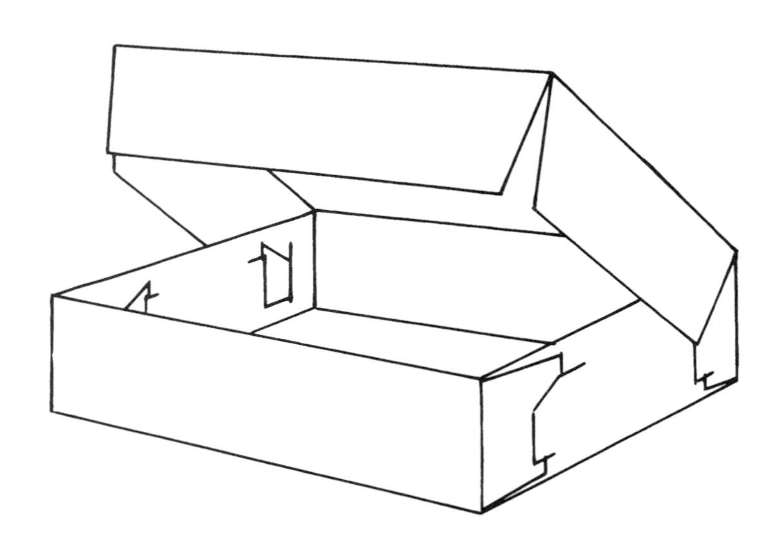

TWO PIECE FULL TELESCOPING BRIGHTWOOD STYLE TRAY

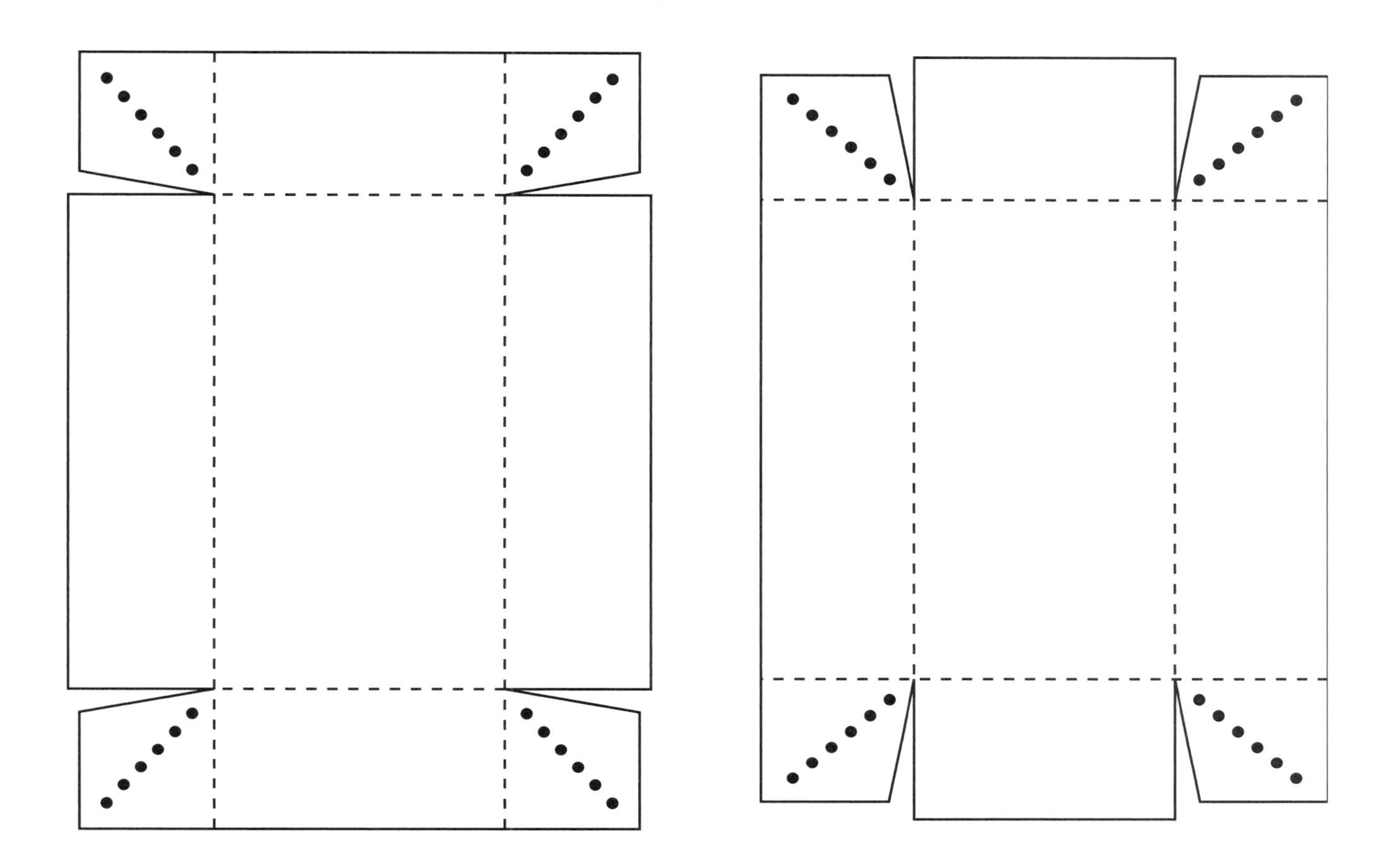

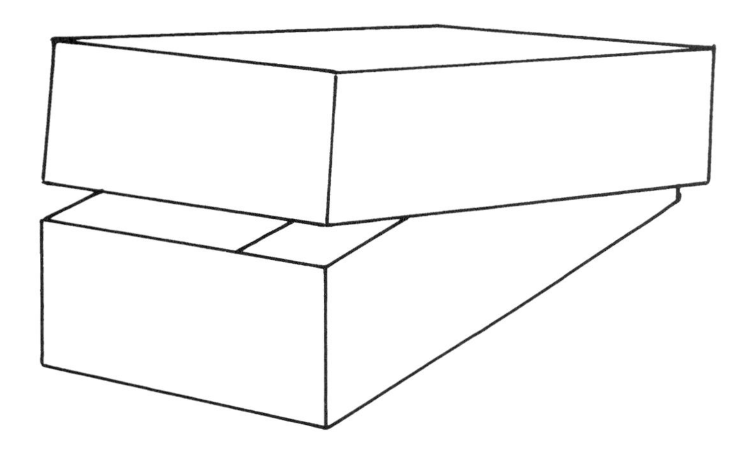

TWO PIECE FULL TELESCOPING FRICTION END TRAY

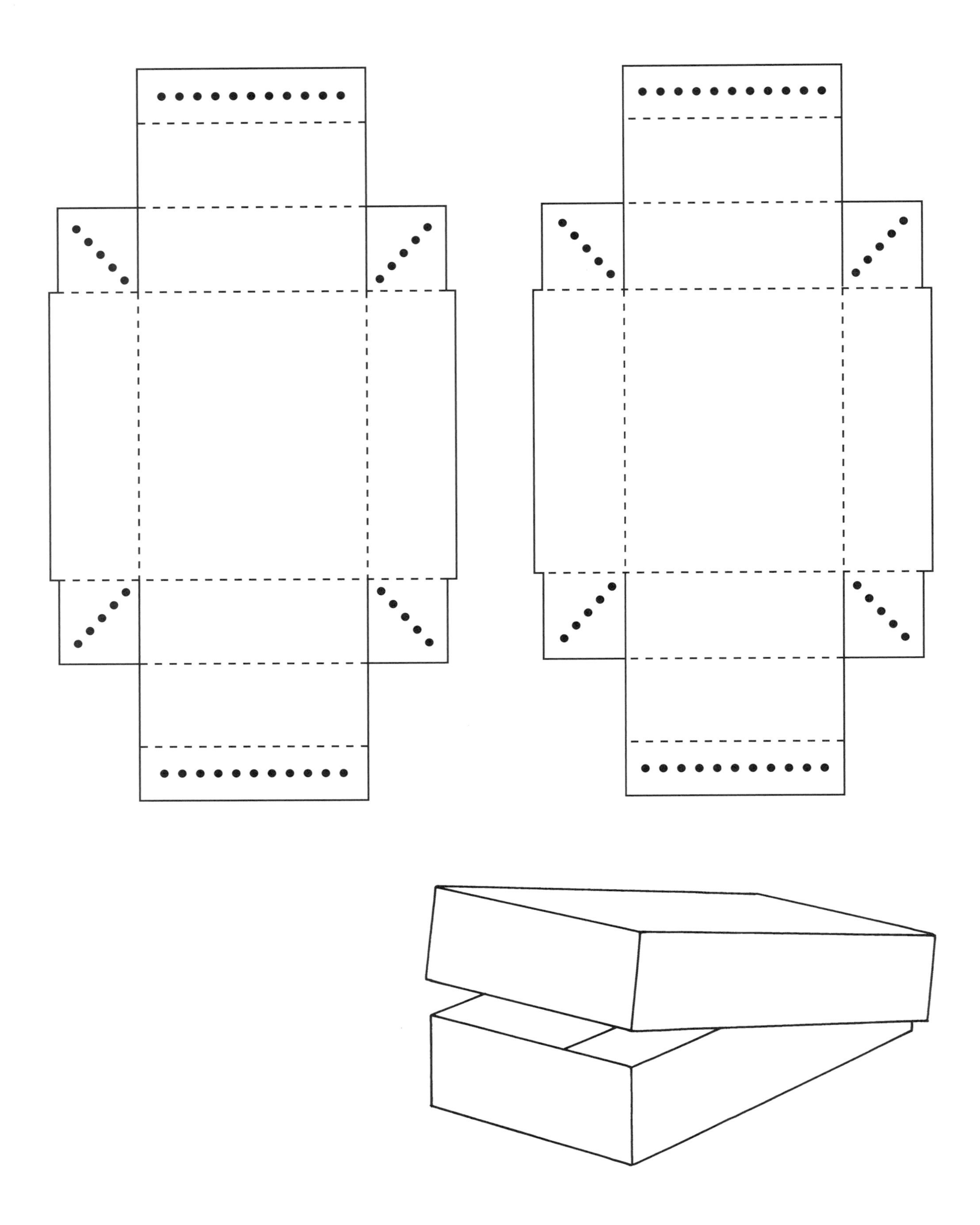

TWO PIECE FULL TELESCOPING HARDWARE STYLE TRAY WITH LOCK ENDS

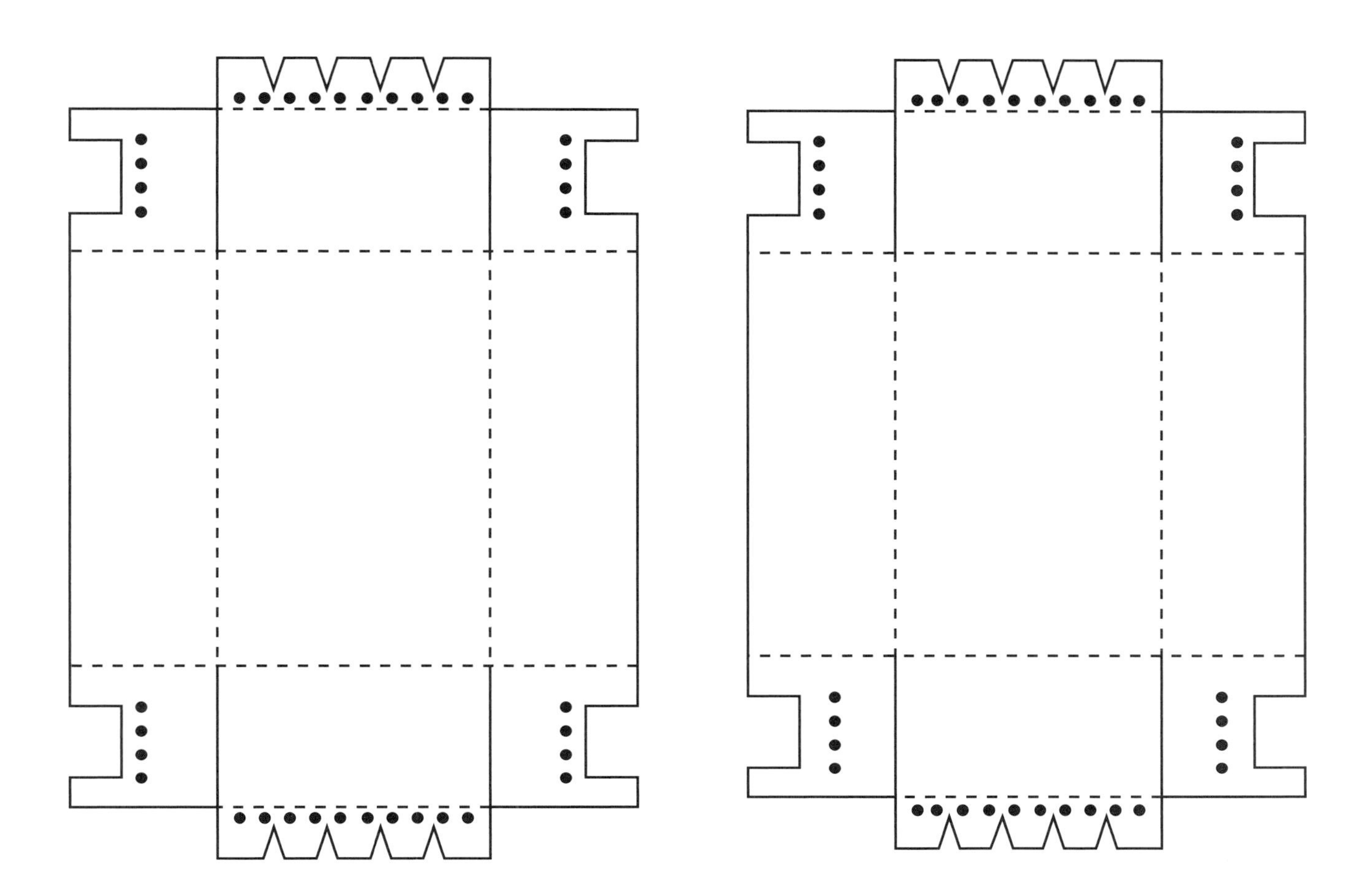

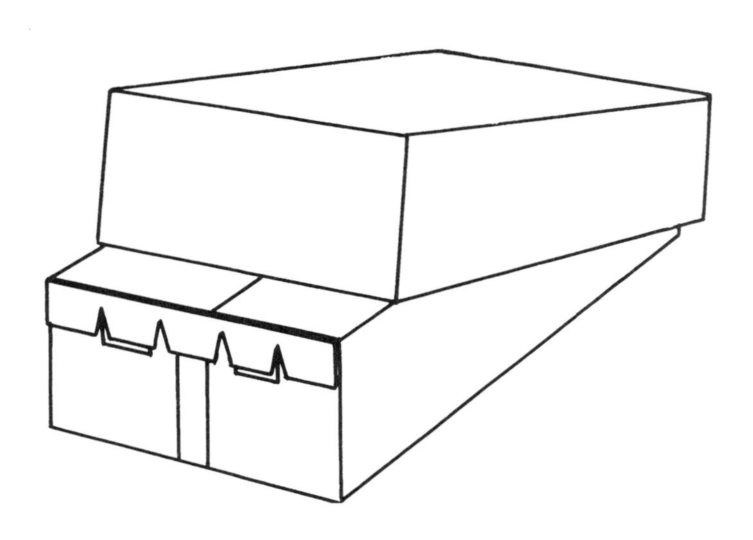

FULL TELESCOPING TRAY WITH LOCK CORNERS

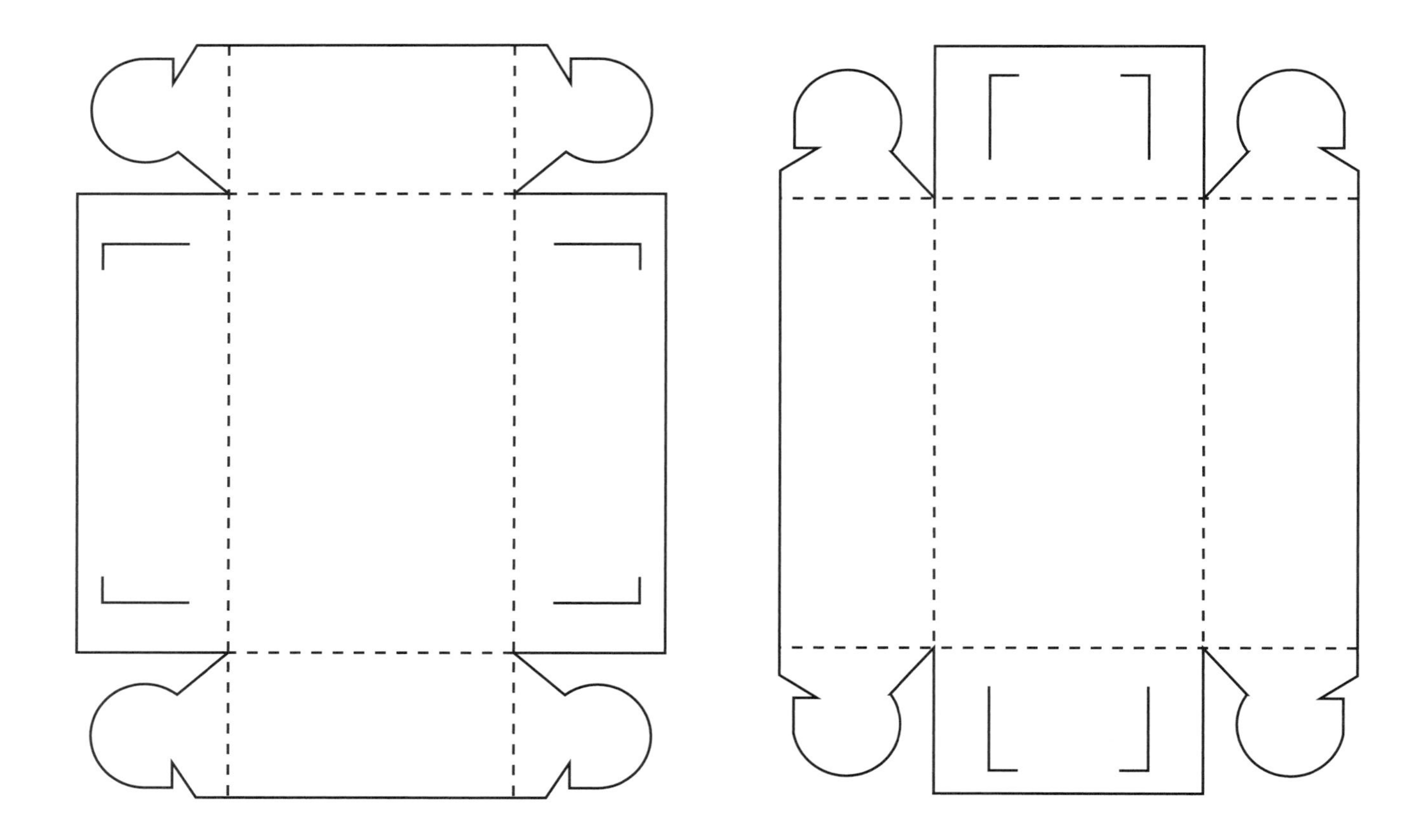

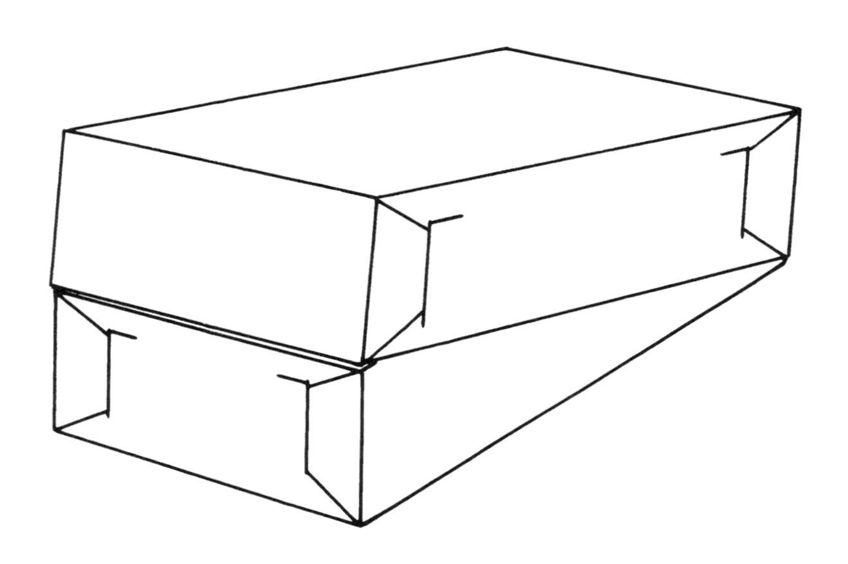

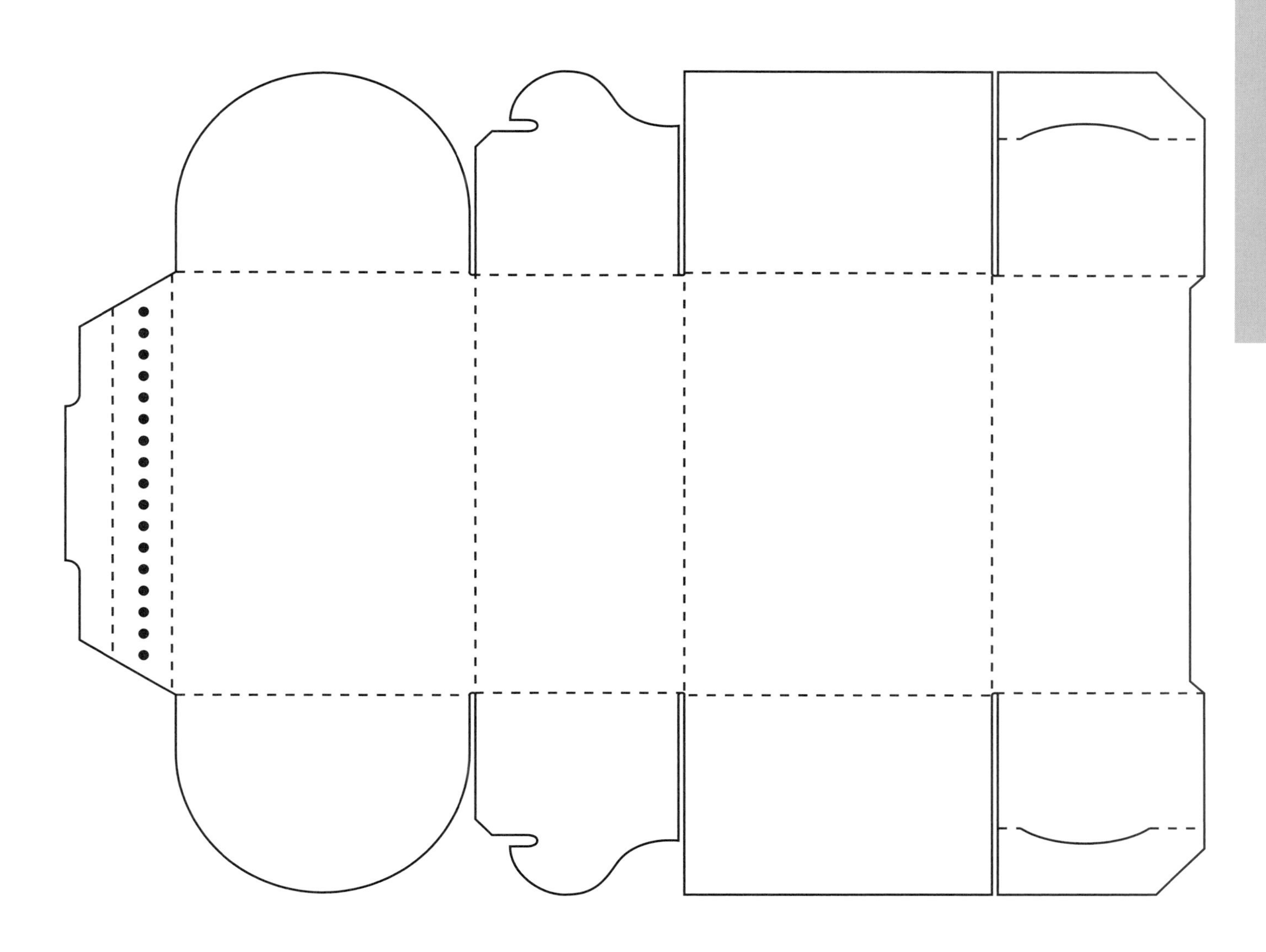

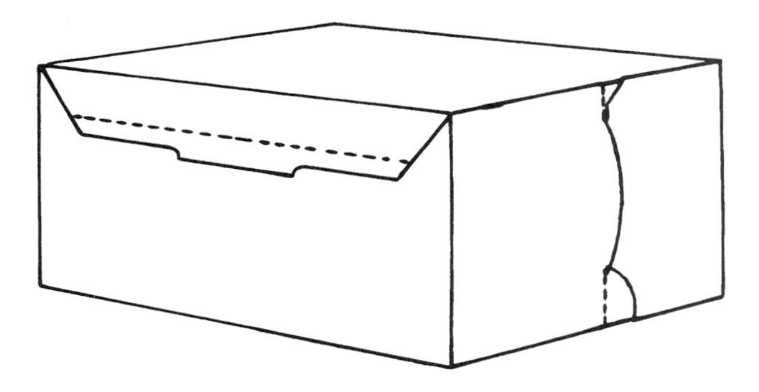

SEAL END STYLE (EASY OPEN) TRAY

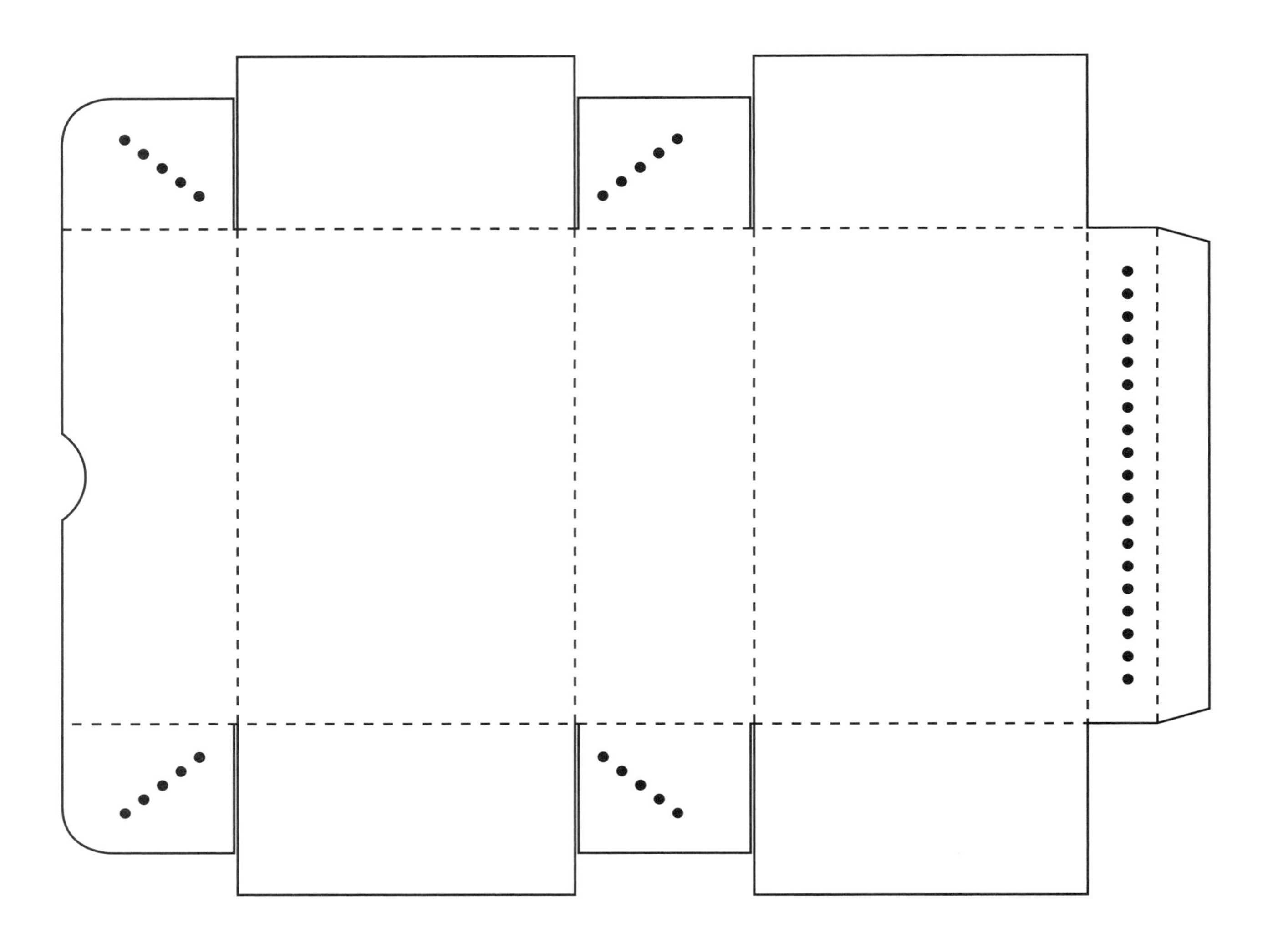

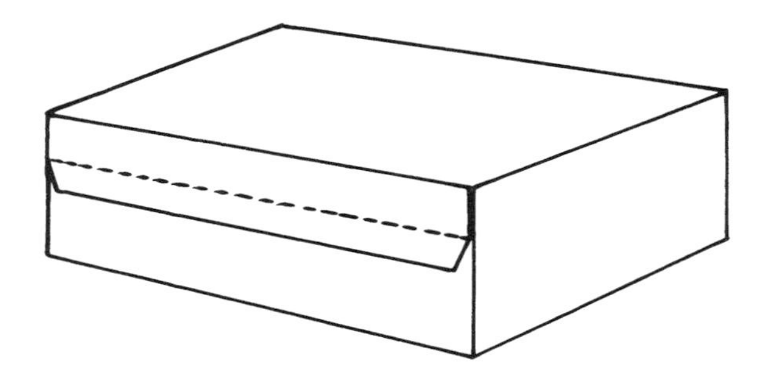

SEAL END STYLE TRAY WITH PERFORATED TEAR-AWAY TOP

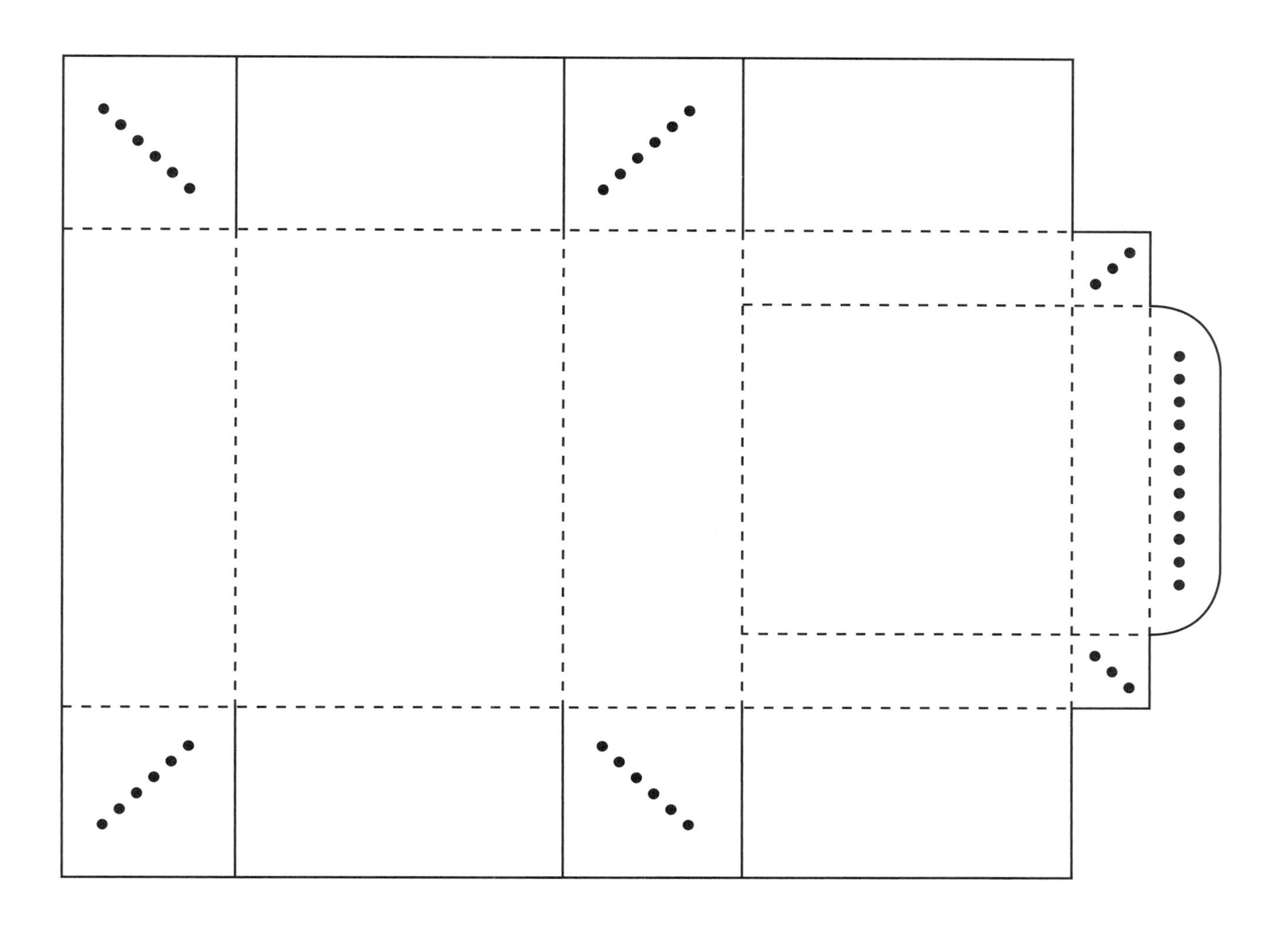

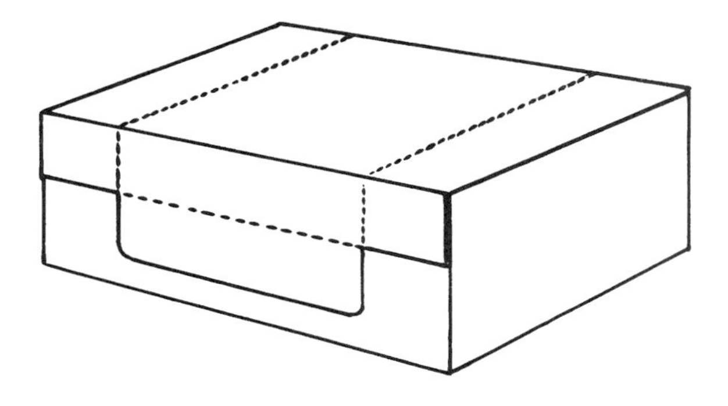

DISPENSER PACK

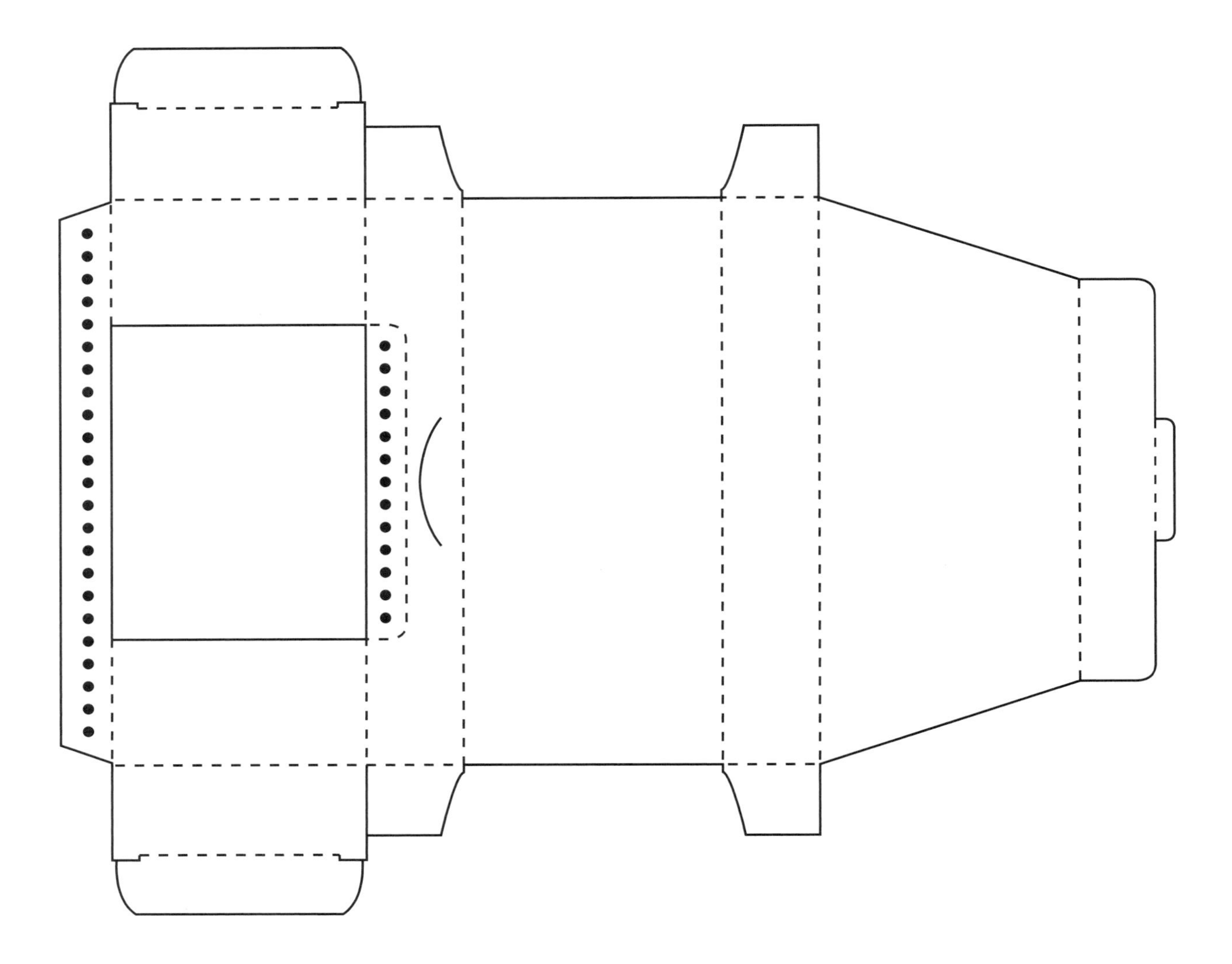

Ideal for a small tissue dispenser that fits in pocketbook or briefcase.

ONE PIECE ICE CREAM CONTAINER WITH DUSTFLAPS

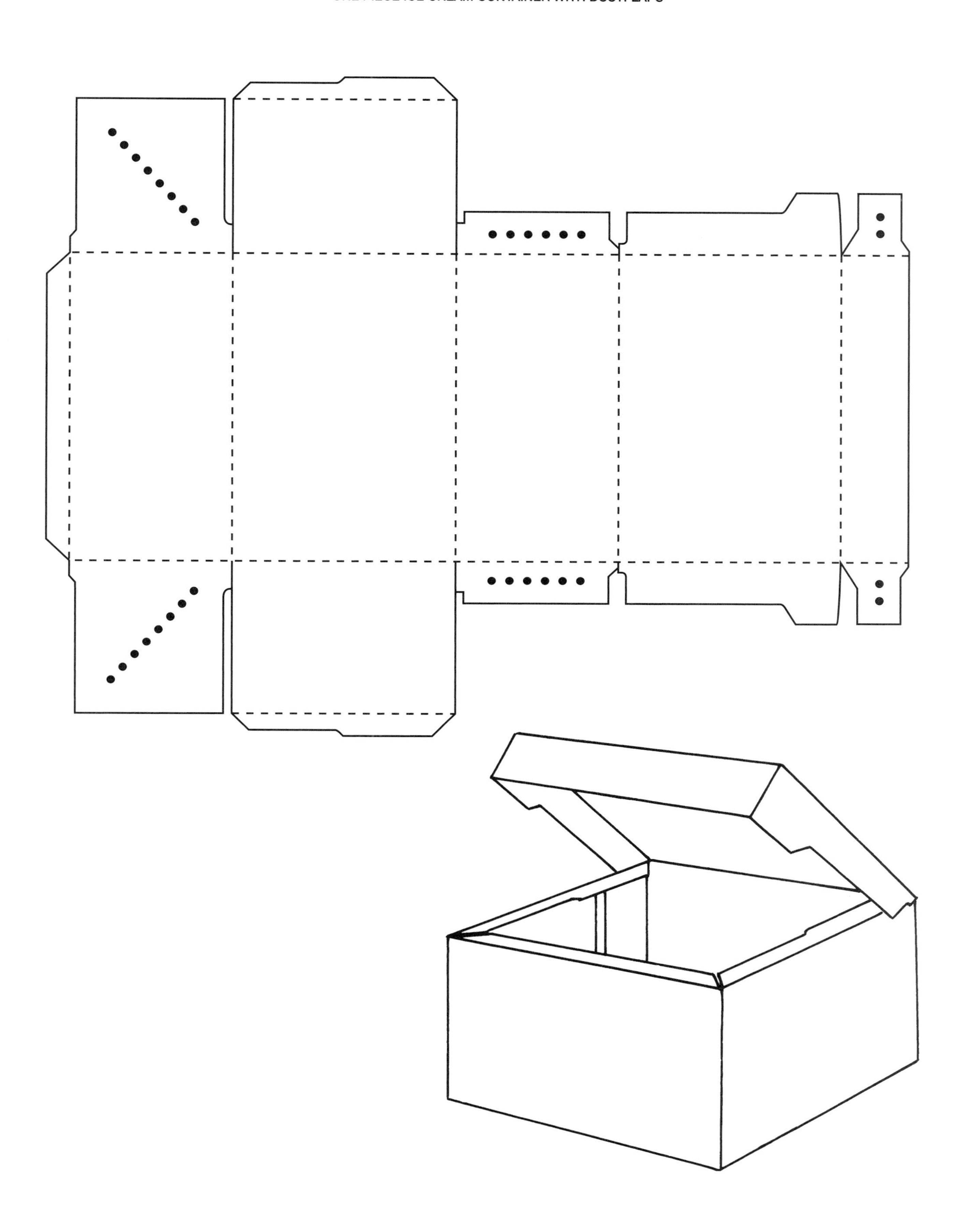

ICE CREAM CARTON

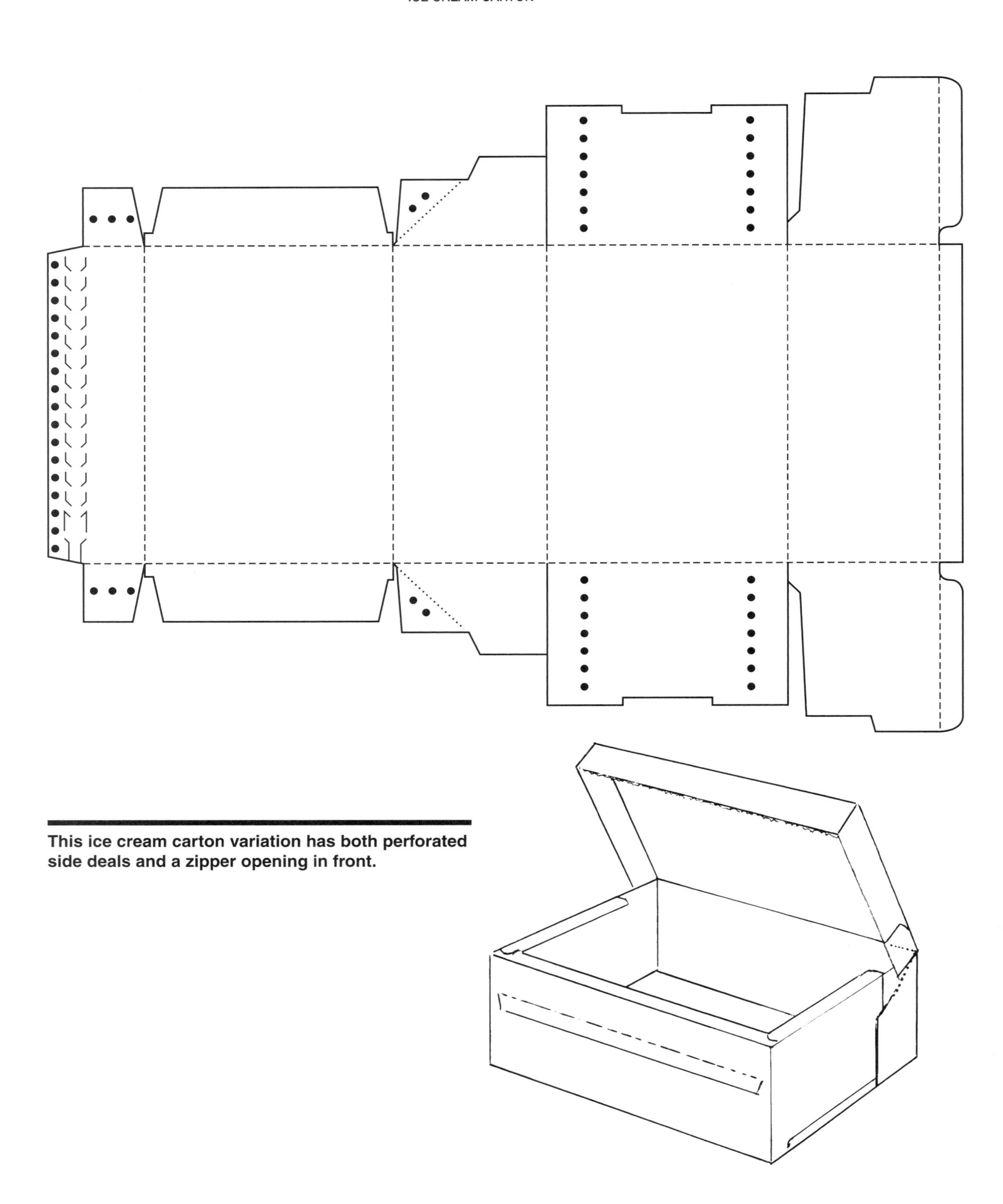

This ice cream carton variation has both perforated side deals and a zipper opening in front.

CORNER LOCK TRAY WITH TEAR-AWAY FEATURES ON LID SIDES AND FRONT

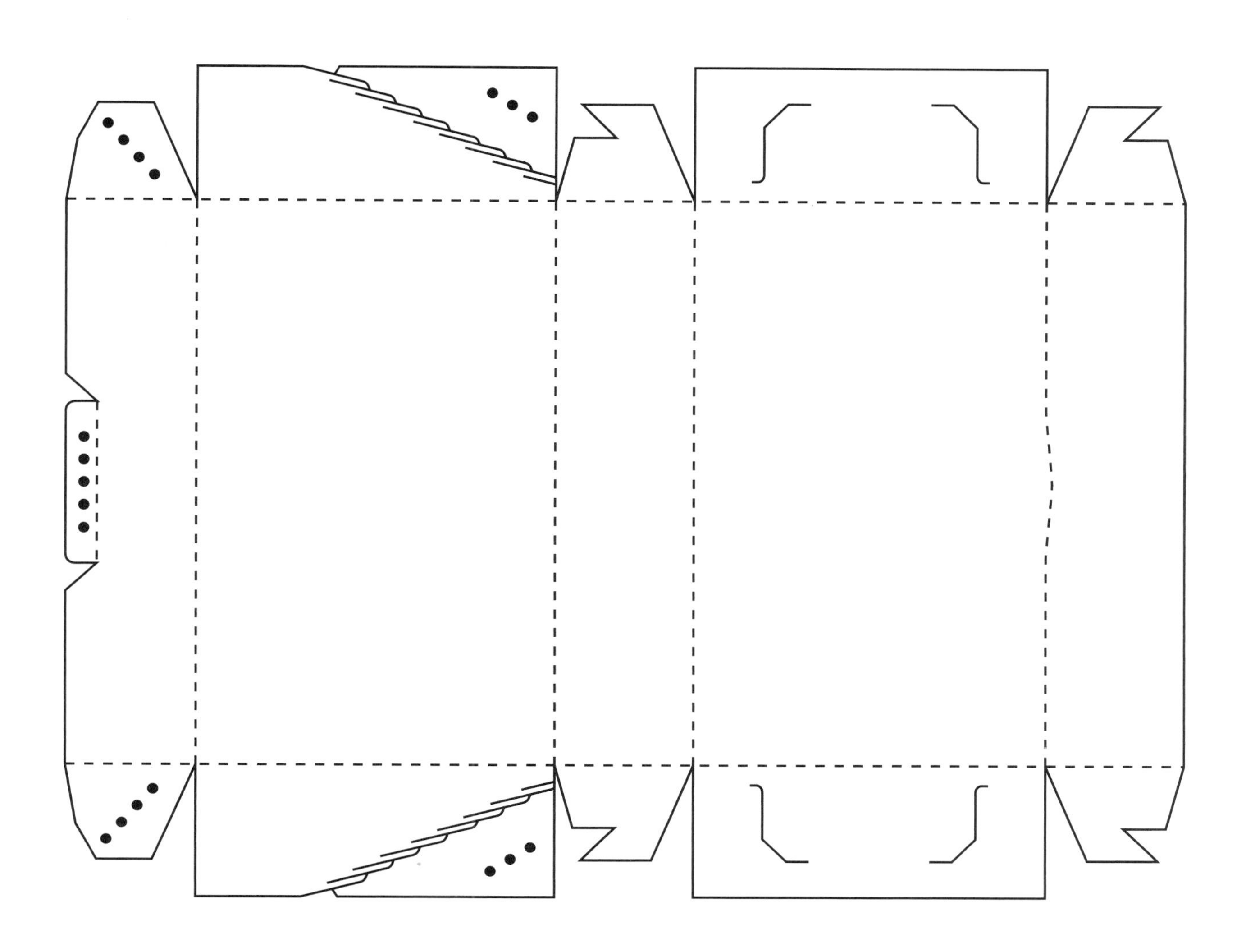

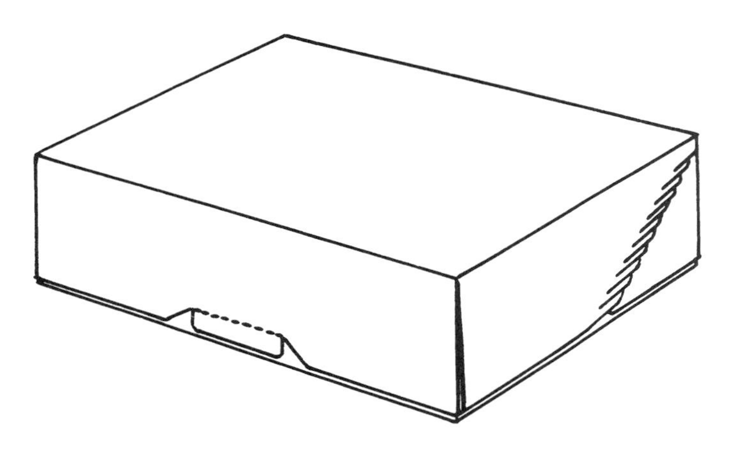

WALKER LOCK TRAY

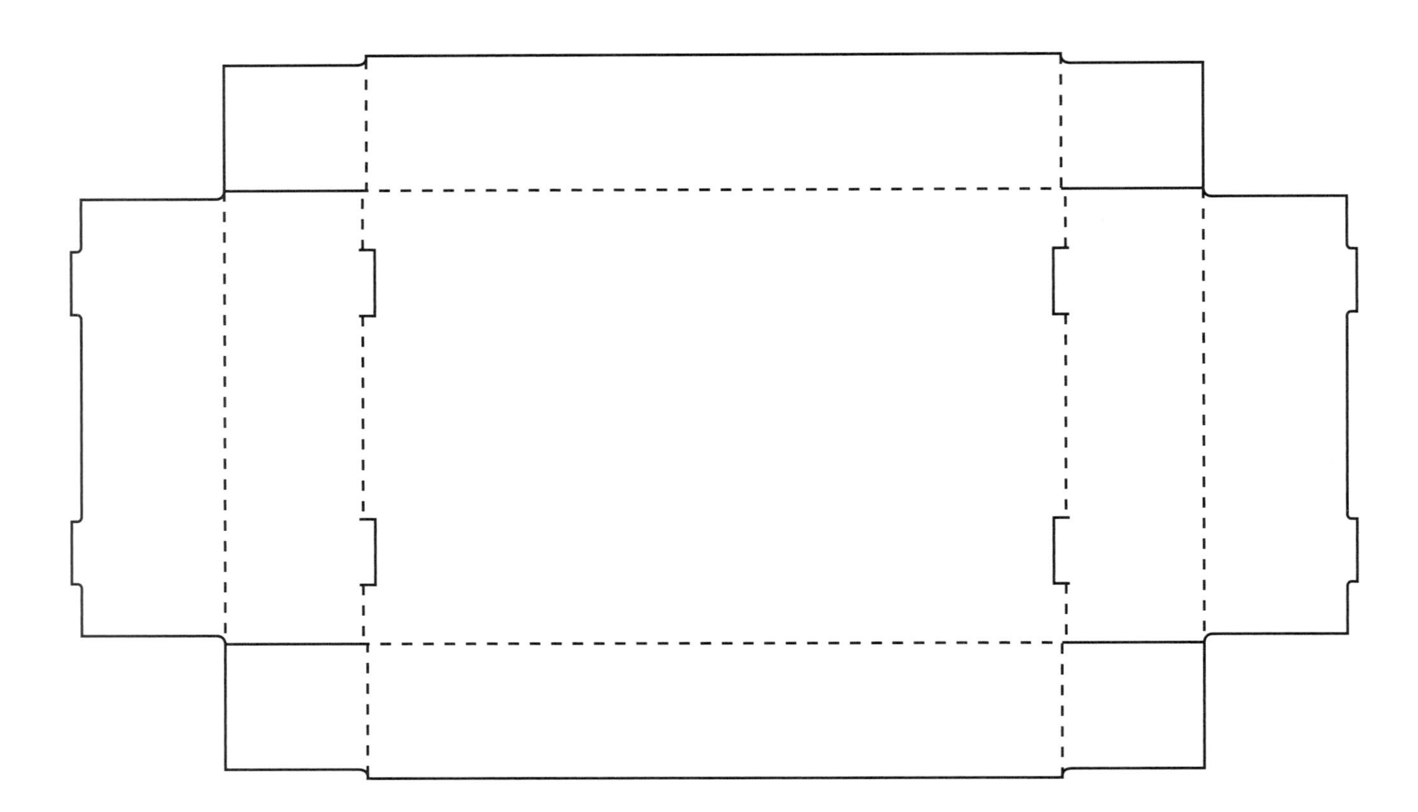

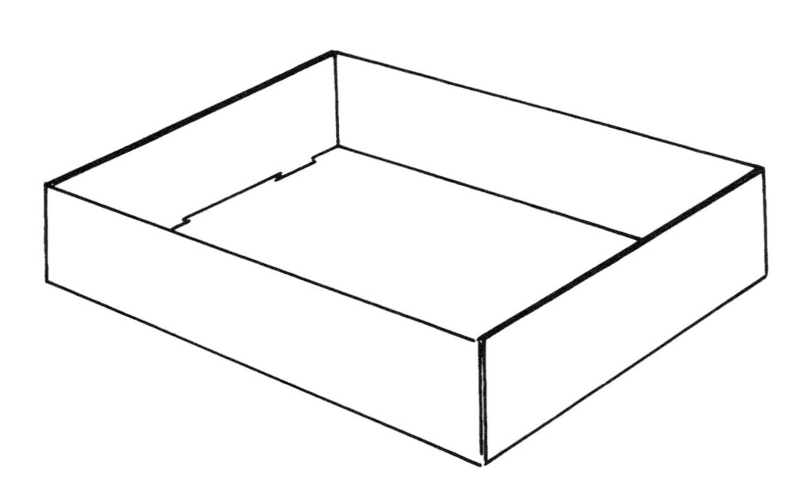

SINGLE SIDE WALL, DOUBLE END WALL PINCH LOCK TRAY

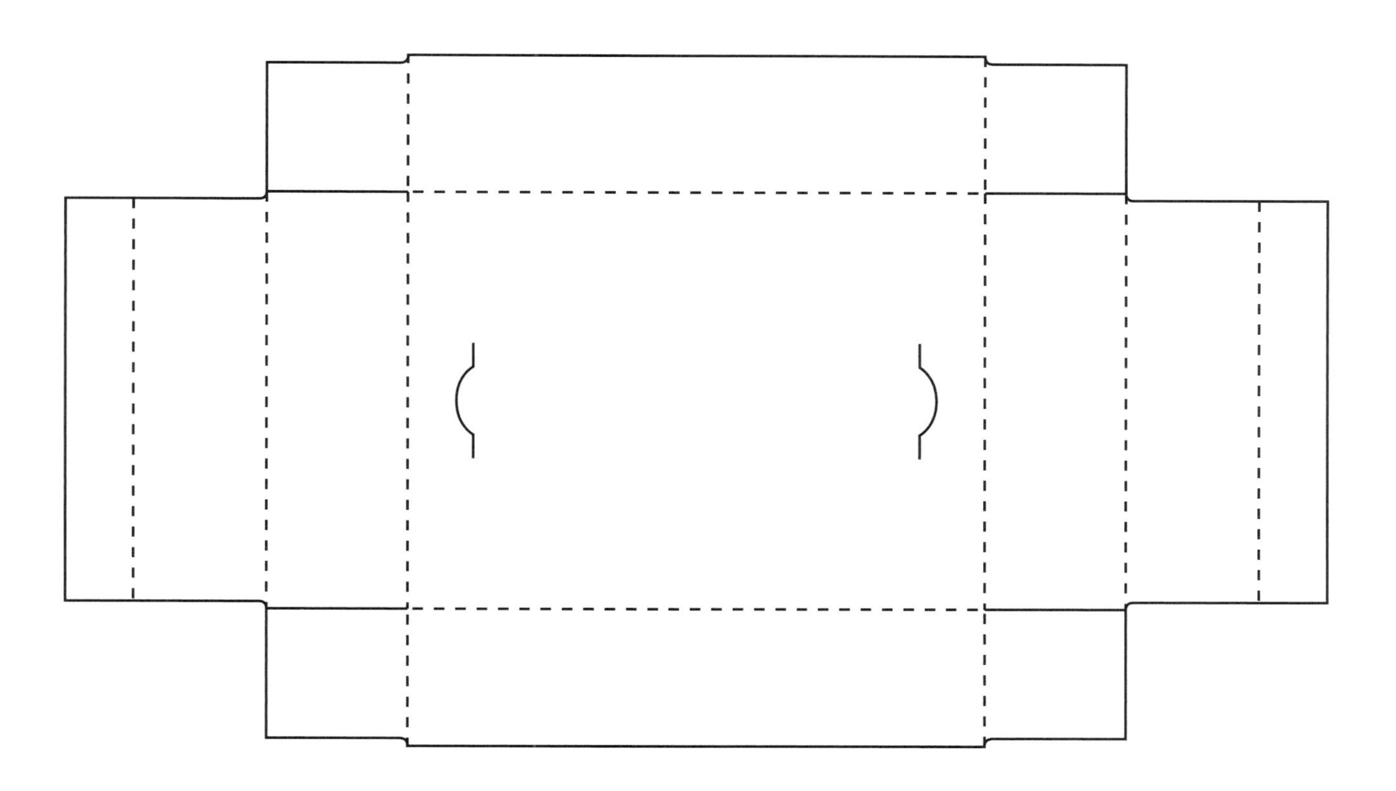

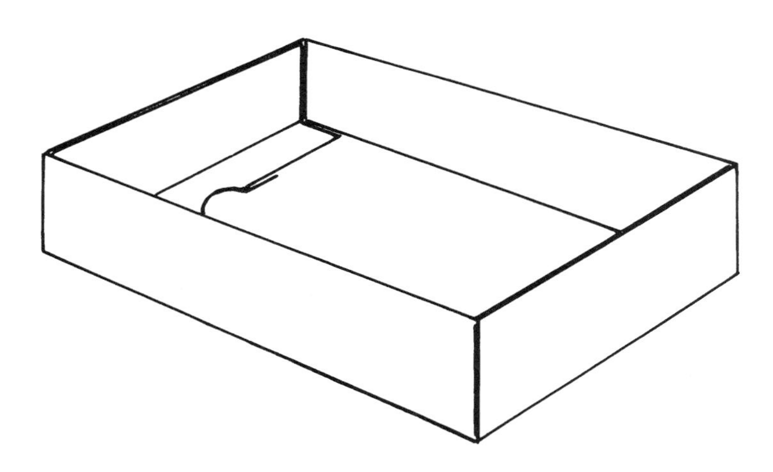

SINGLE END WALL, DOUBLE SIDE WALL TRAY

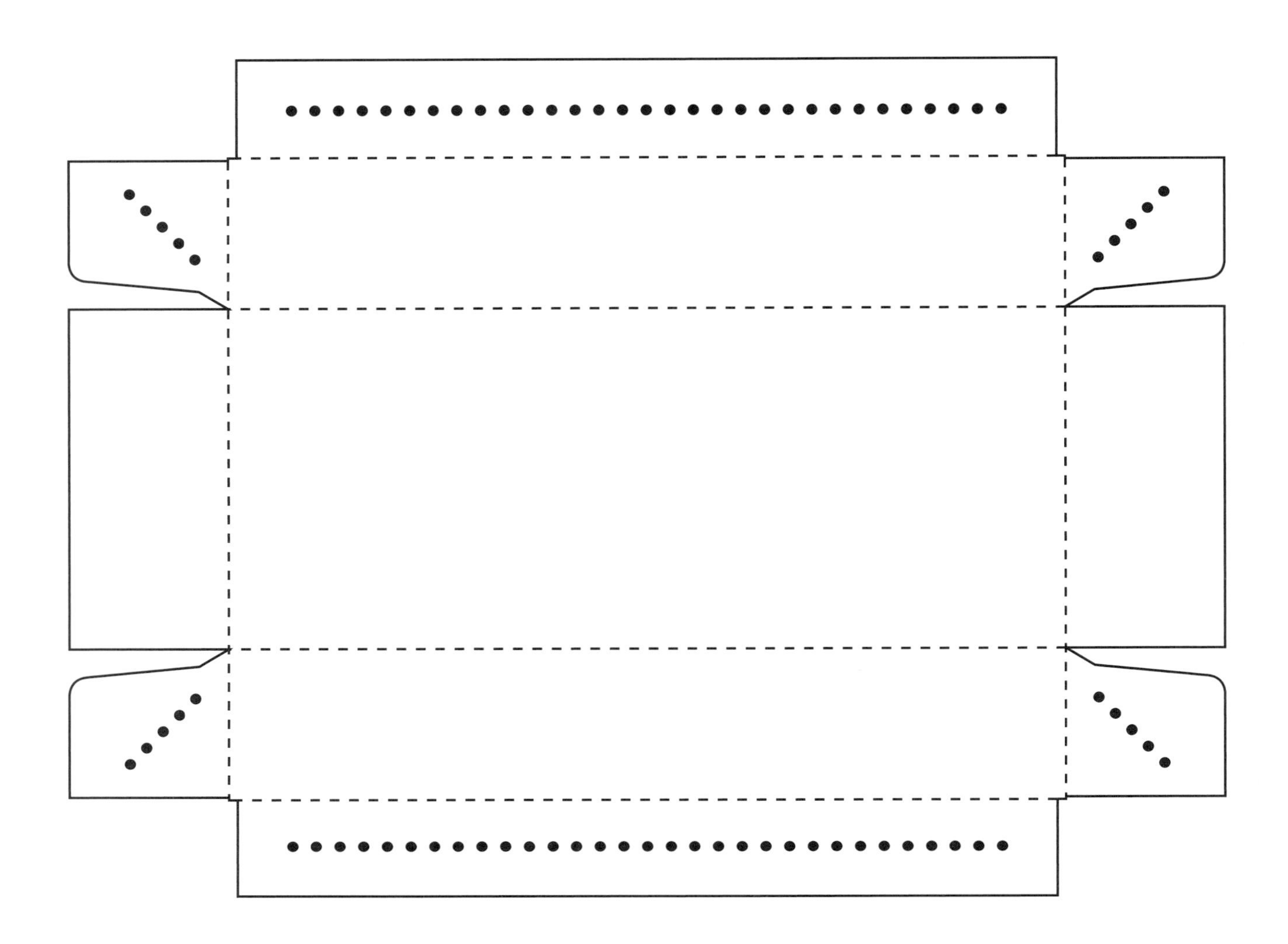

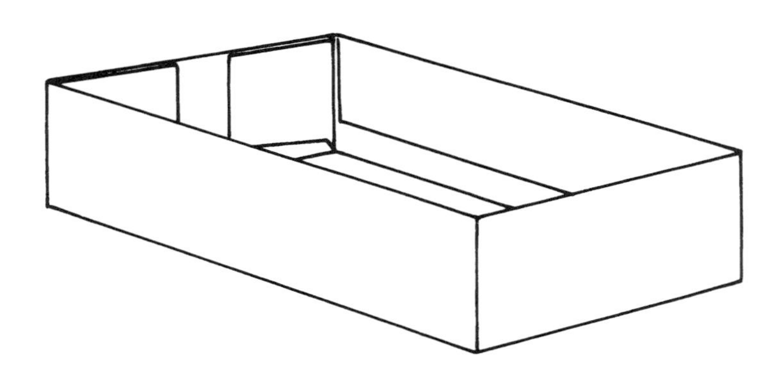

PARTIAL DOUBLE END WALL AND DOUBLE SIDE WALL TRAY

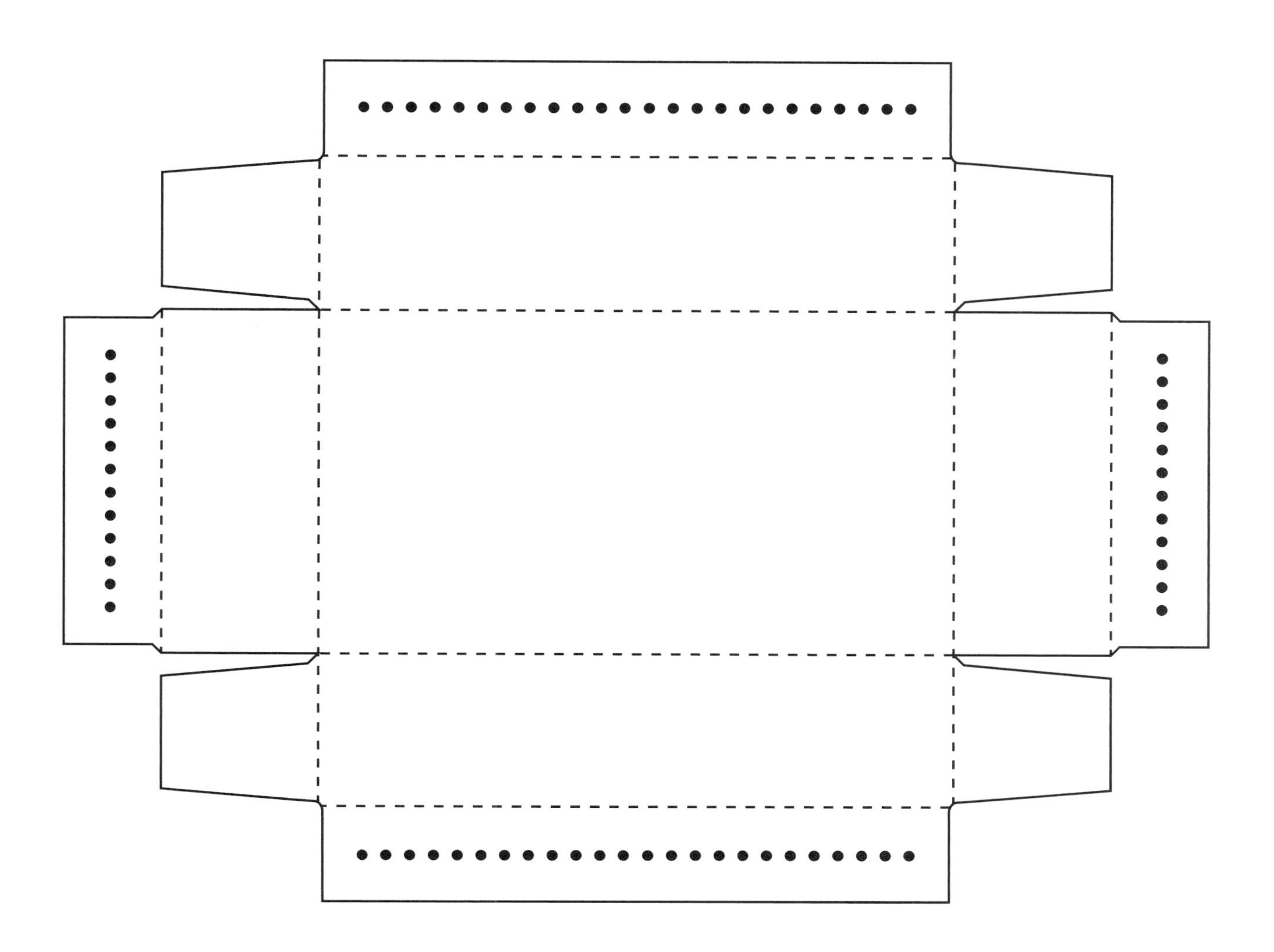

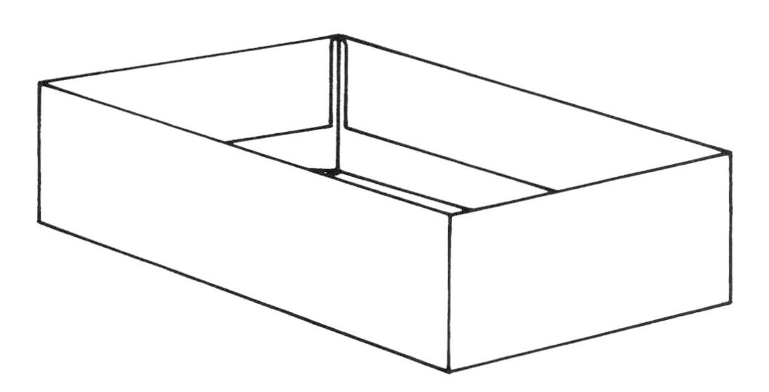

FOOT-LOCK DOUBLE WALL TRAY

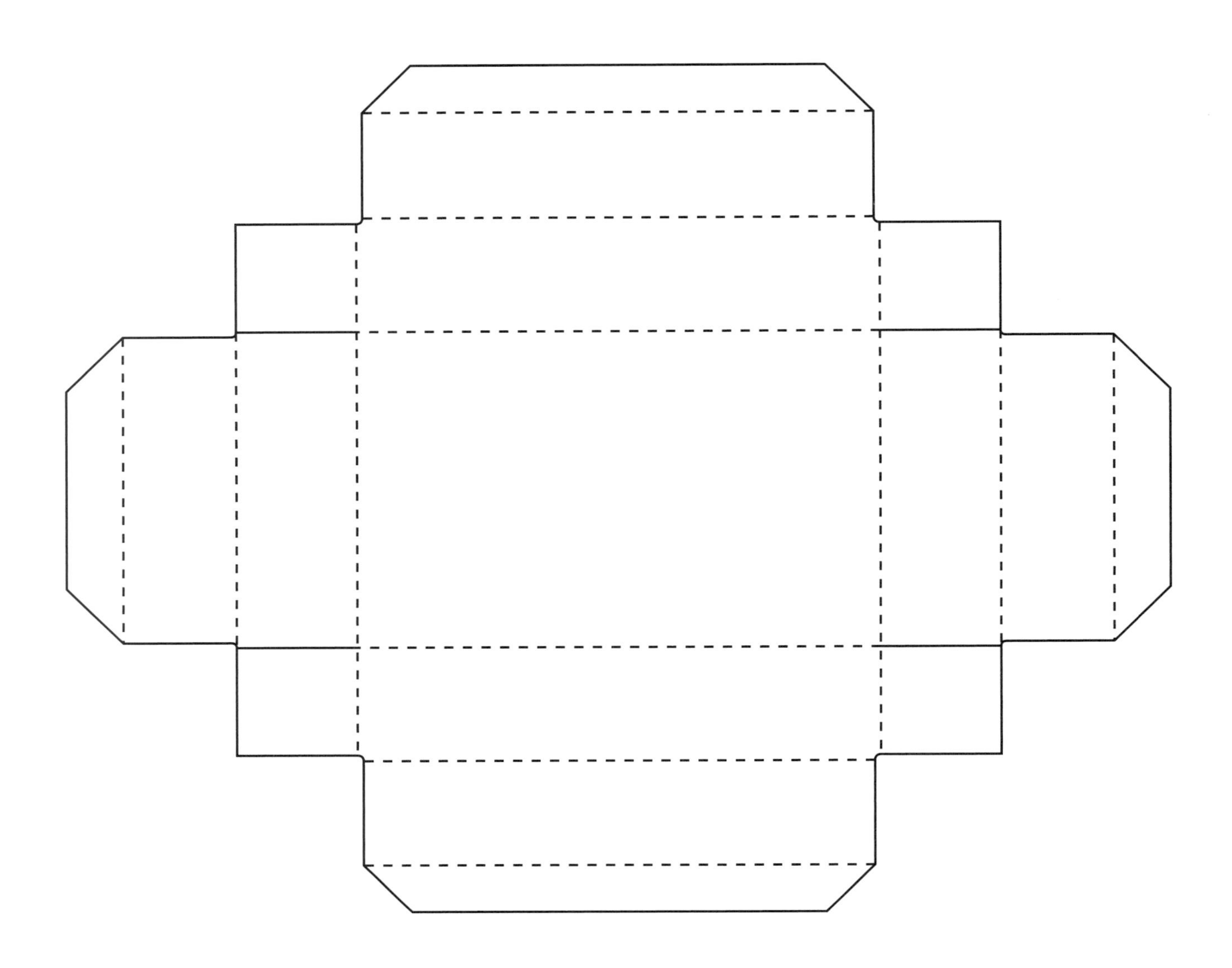

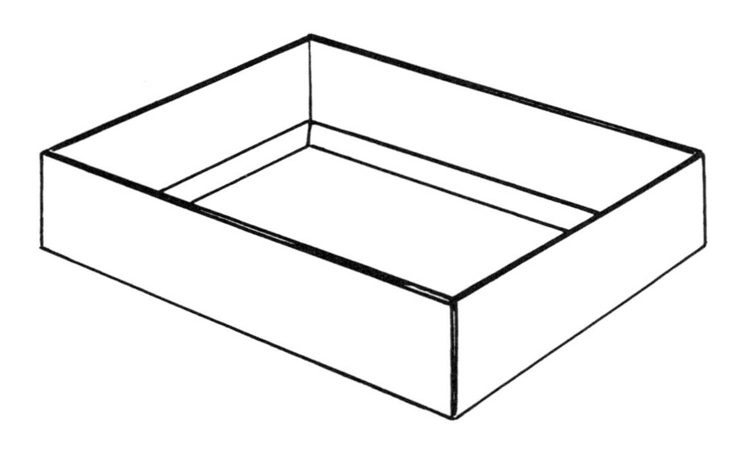

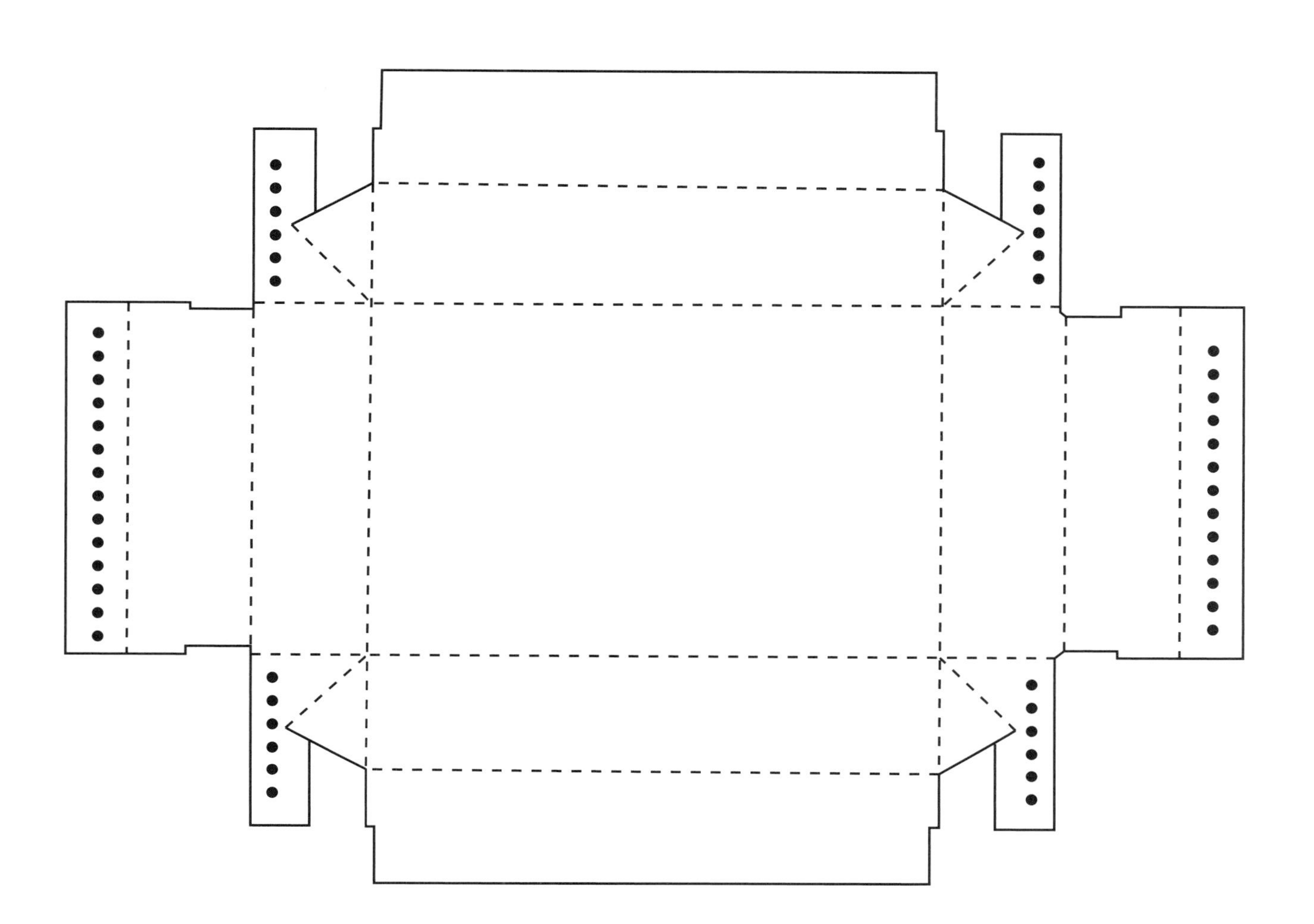

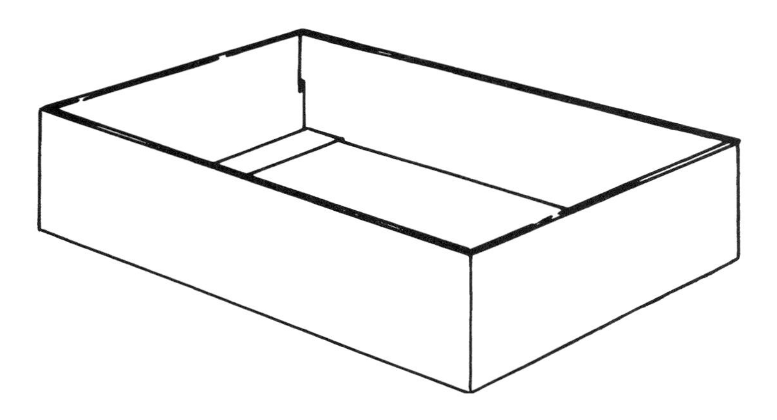

SIMPLEX STYLE TRAY (QUICKSET)

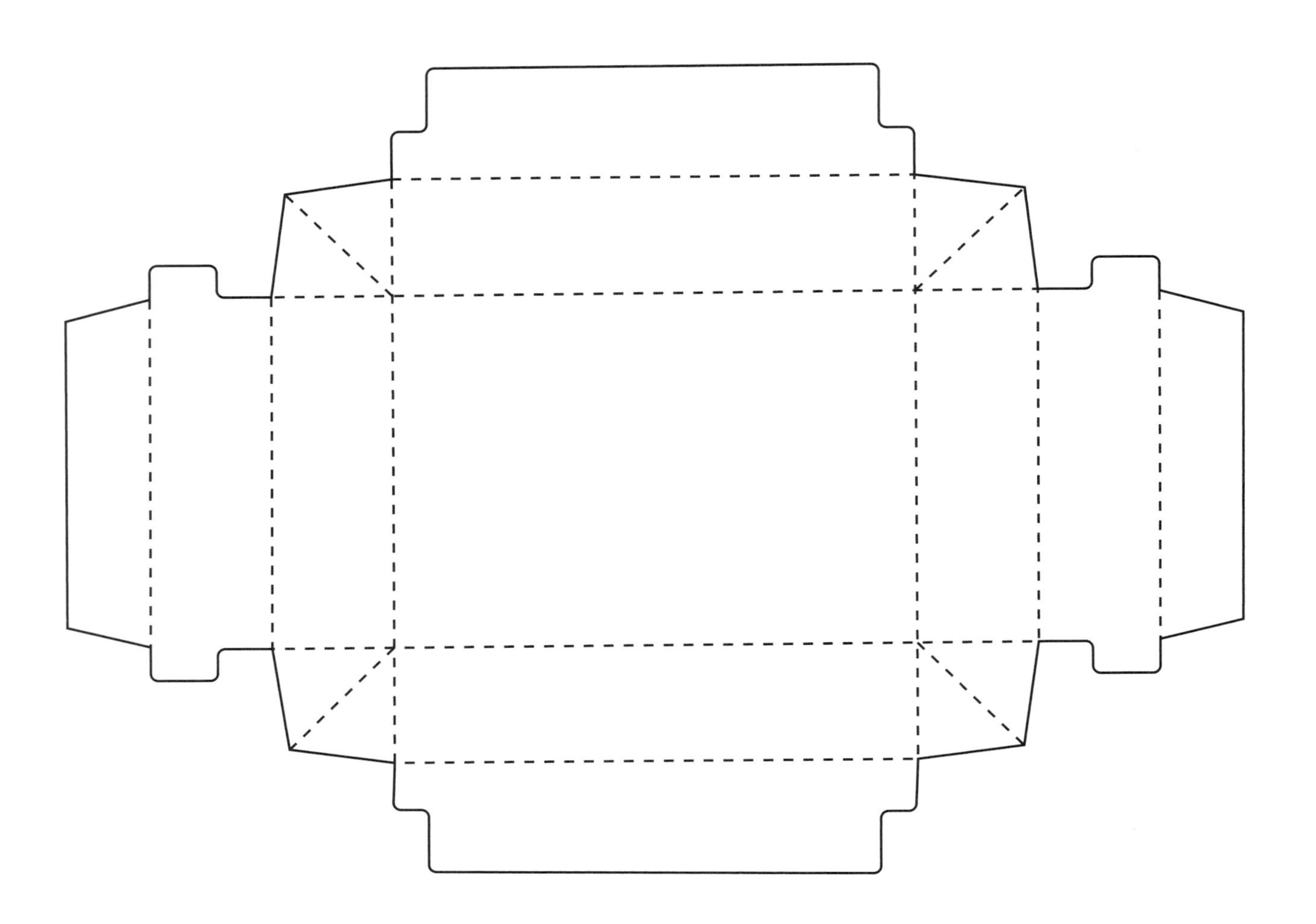

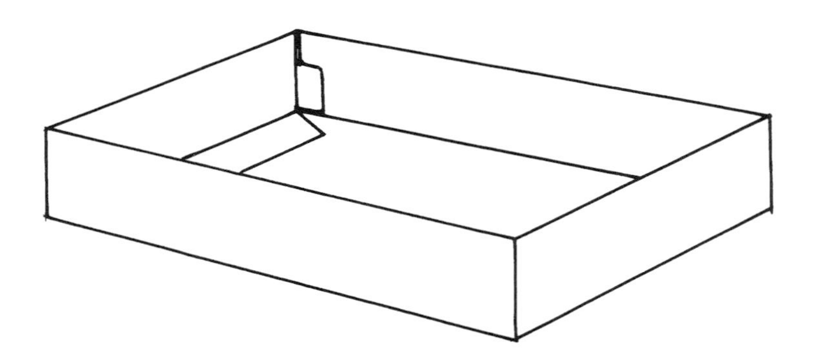

ARTHUR LOCK OR JONES LOCK TRAY

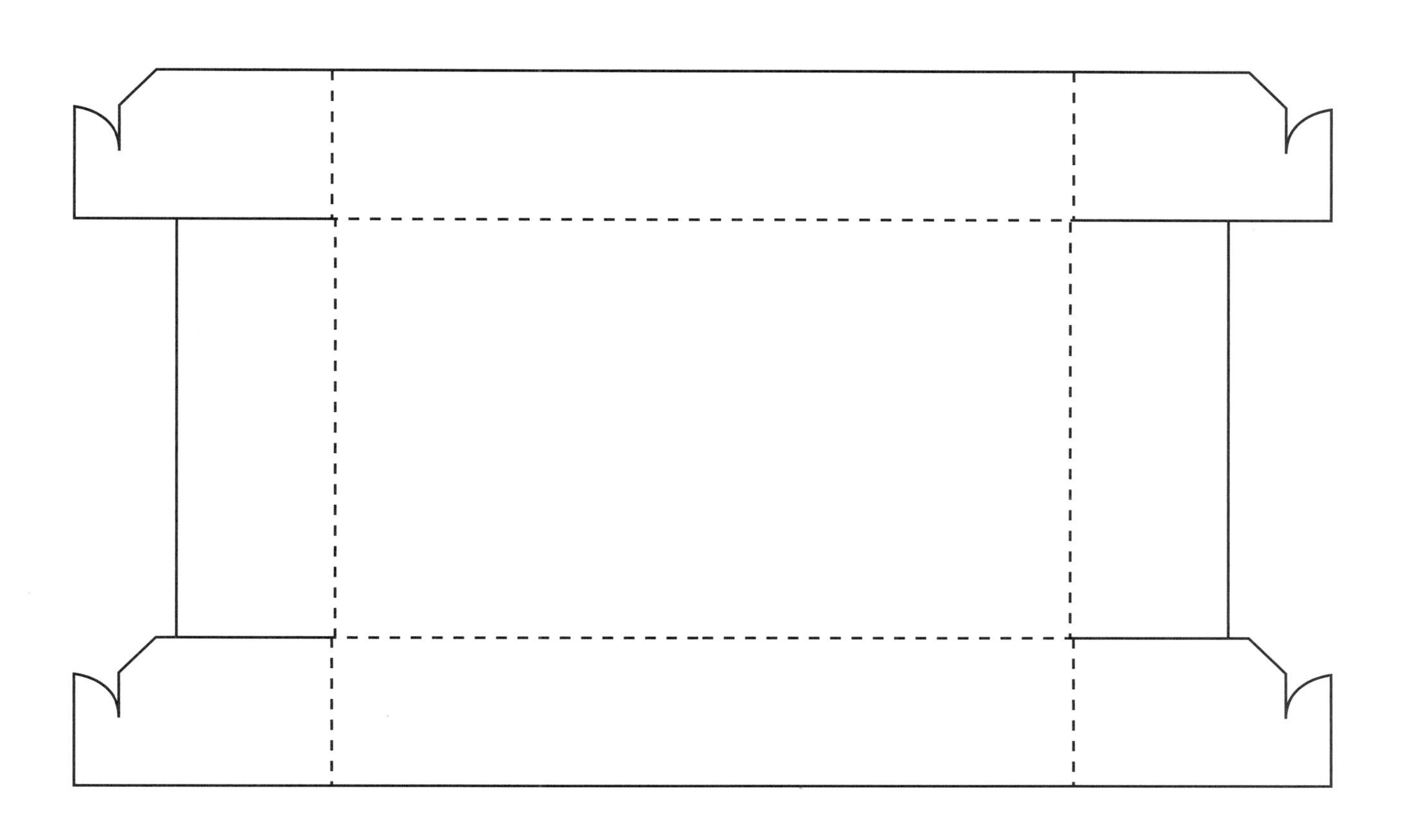

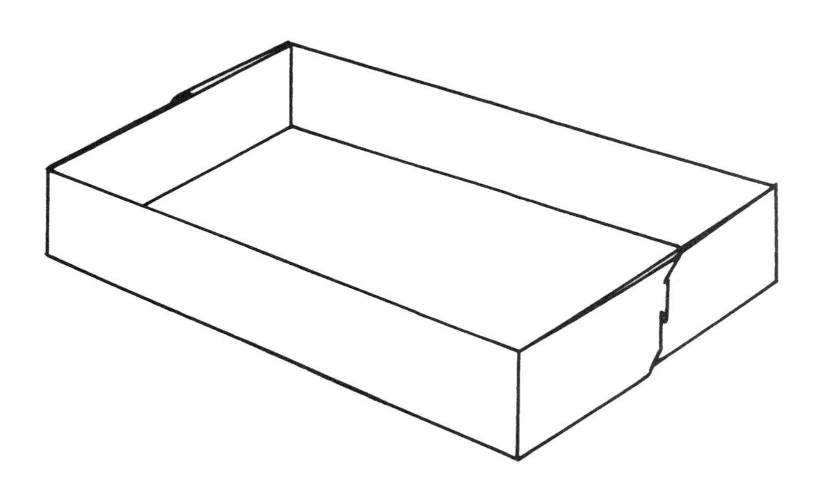

DOUBLE END WALL, SINGLE SIDE WALL TRAY

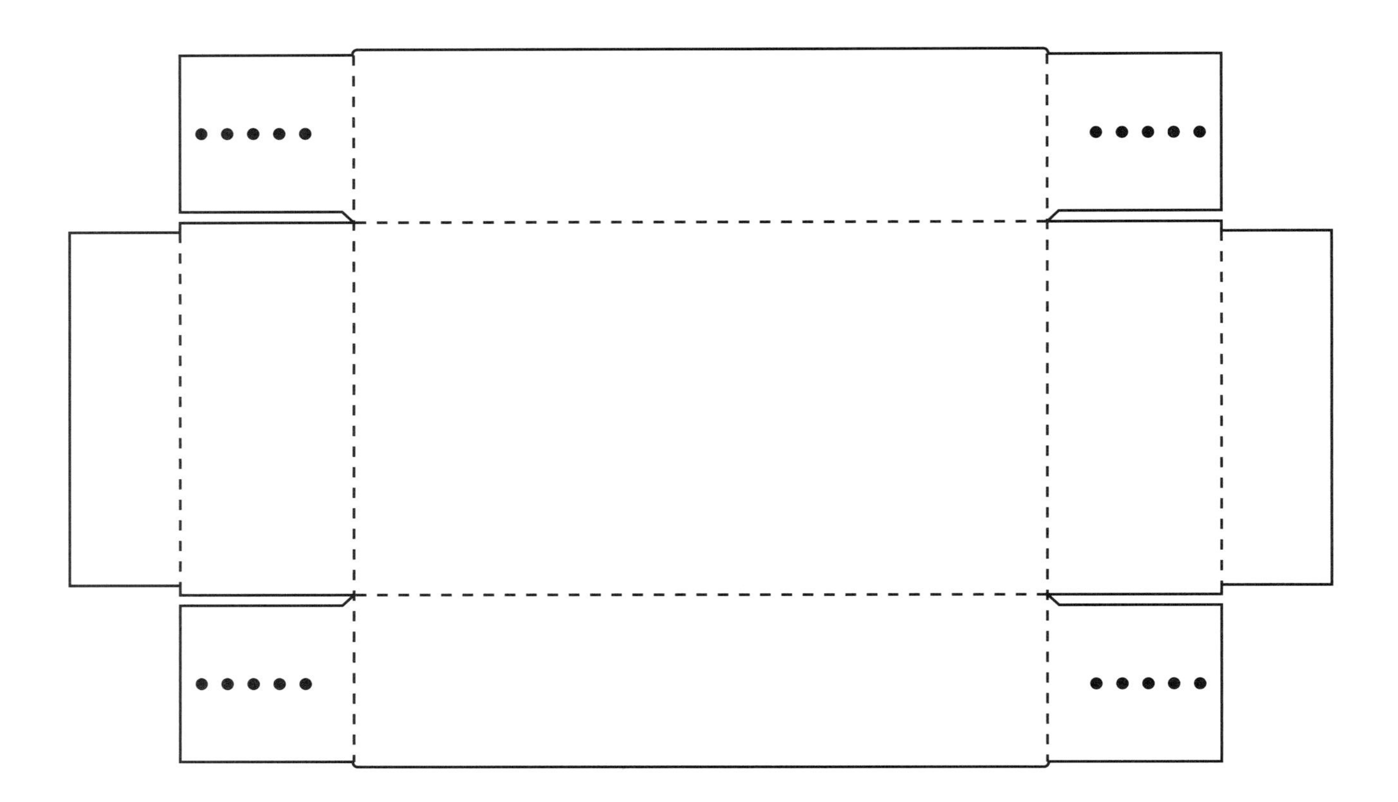

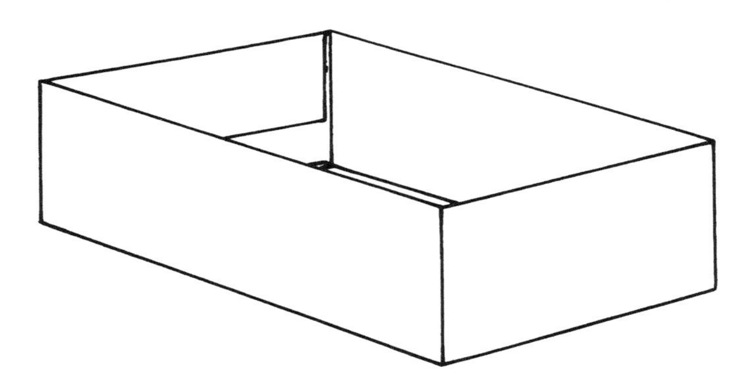

DOUBLE SIDE WALL AND DOUBLE BOTTOM TRAY

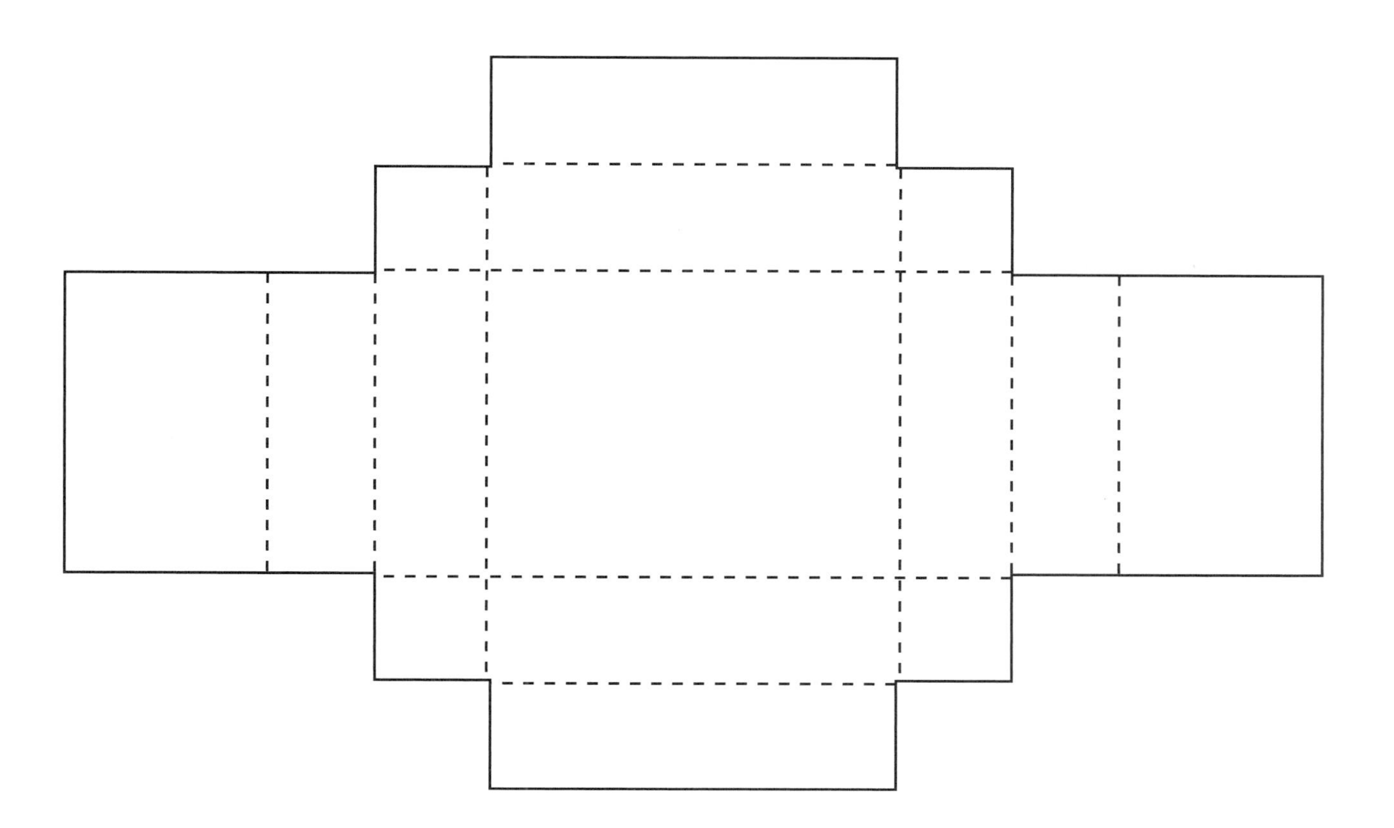

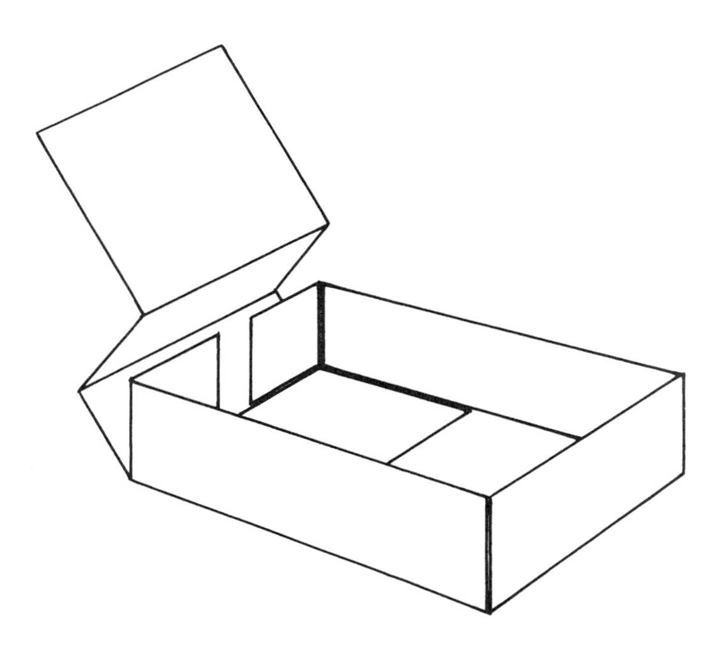

SIMPLEX STYLE TRAY, WEB CORNER SPOT GLUED

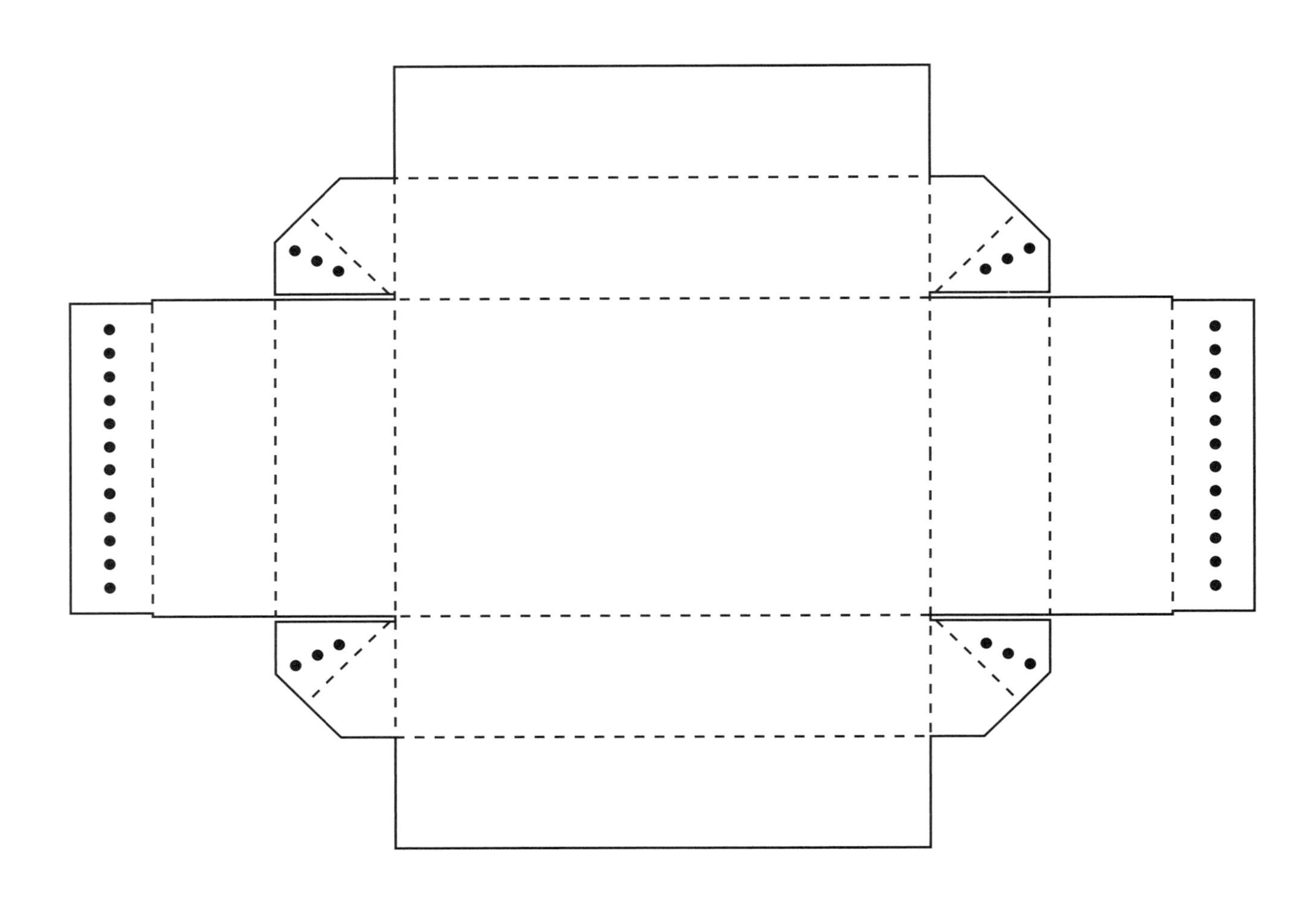

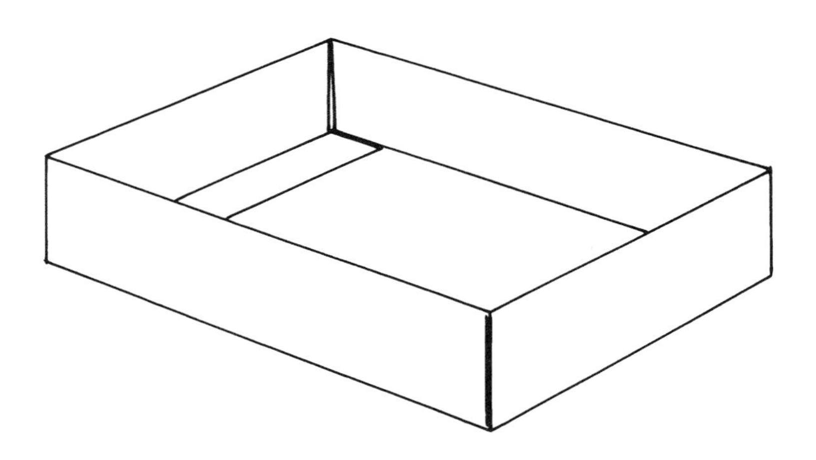

FRAME VIEW TRAY (VARIATION)

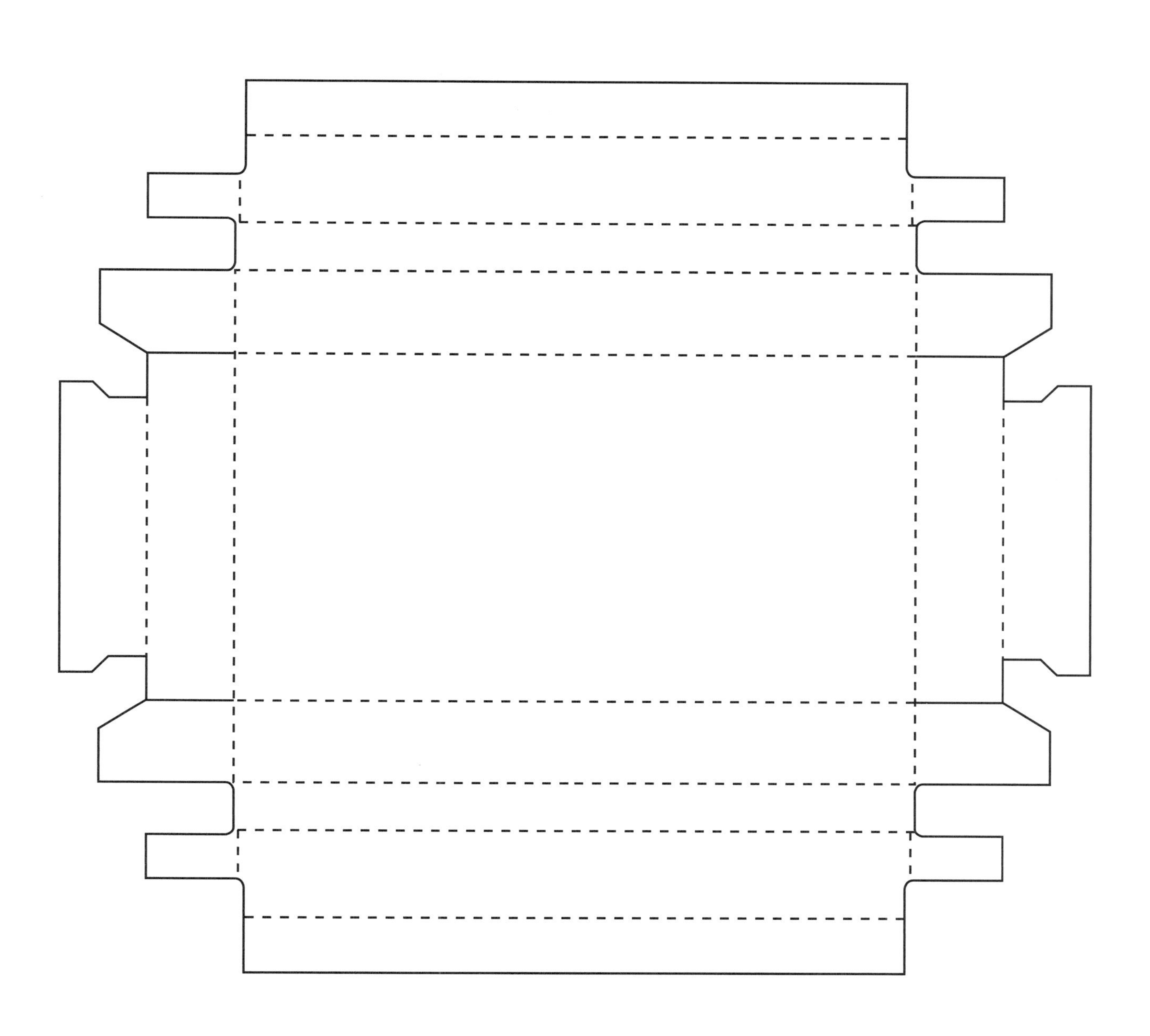

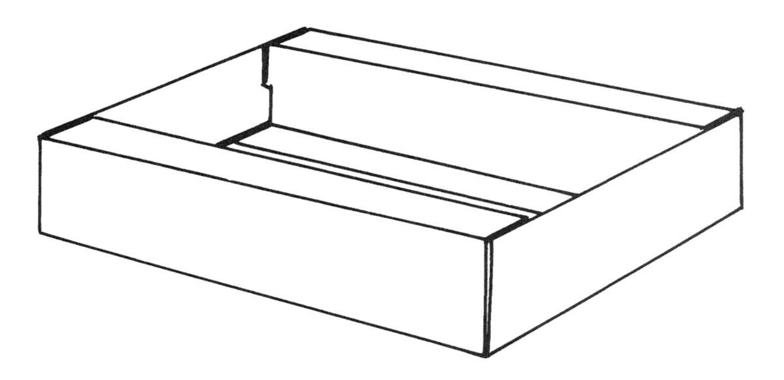

FRAME VIEW TRAY (VARIATION)

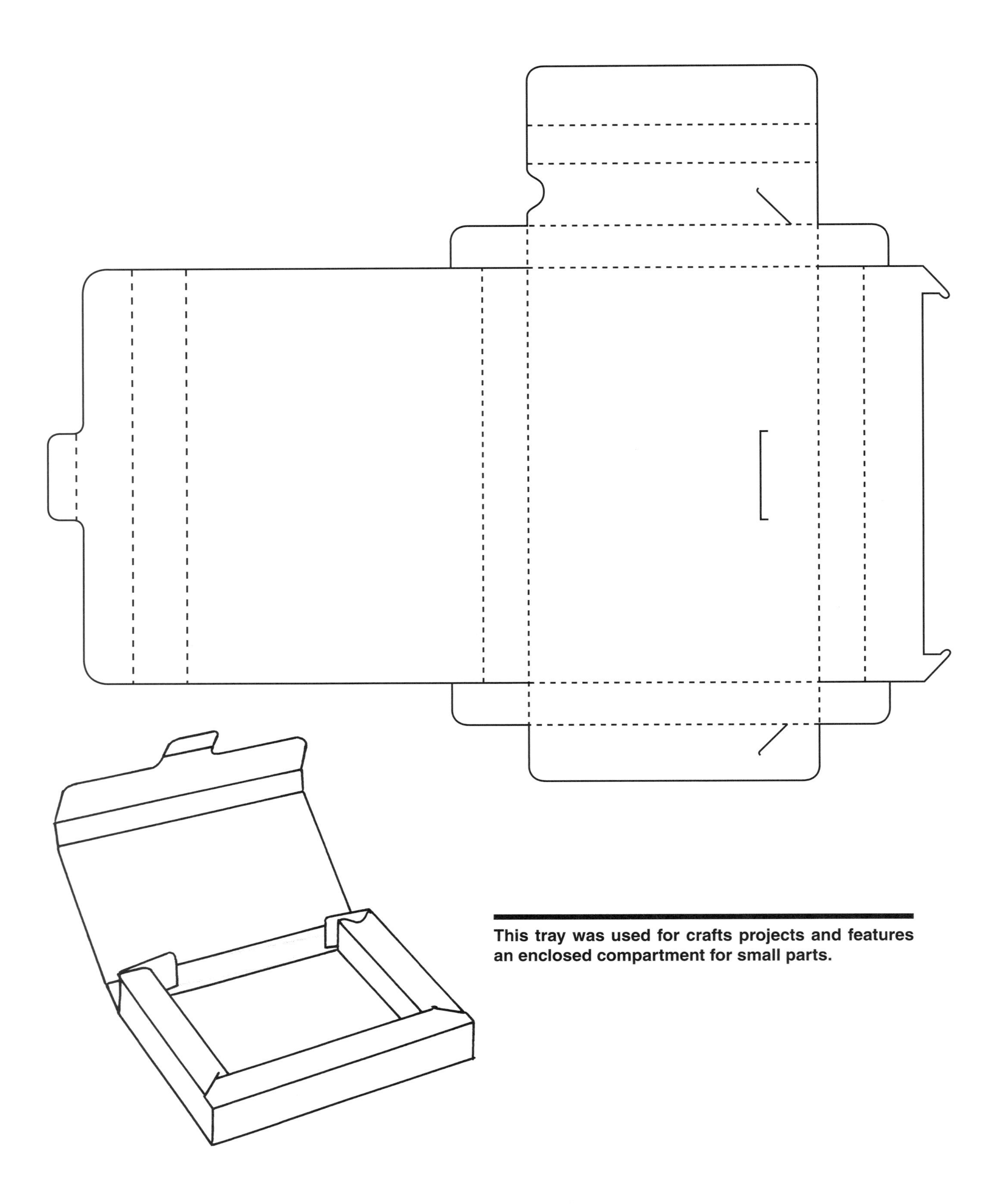

This tray was used for crafts projects and features an enclosed compartment for small parts.

FRAME VIEW TRAY (VARIATION)

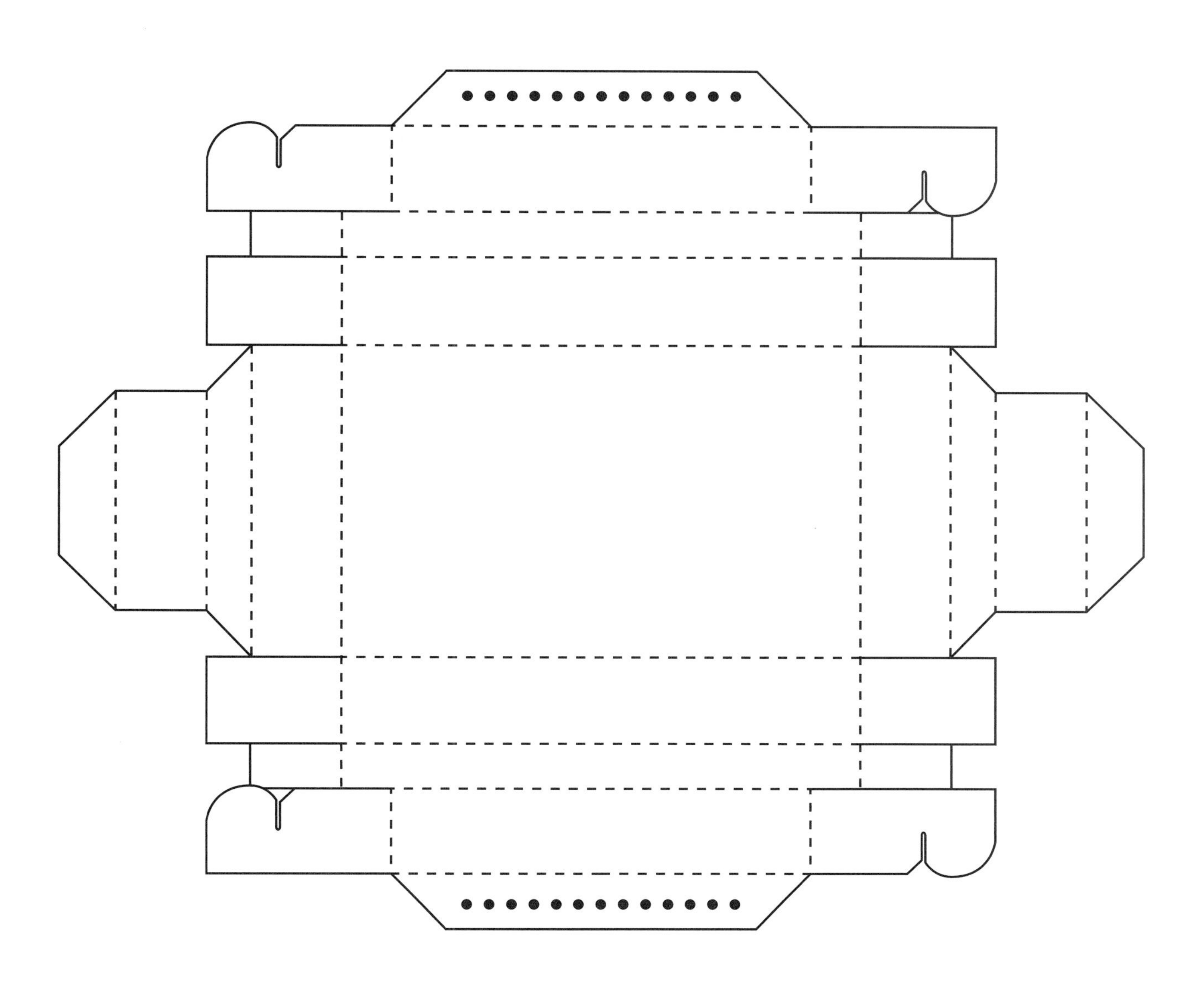

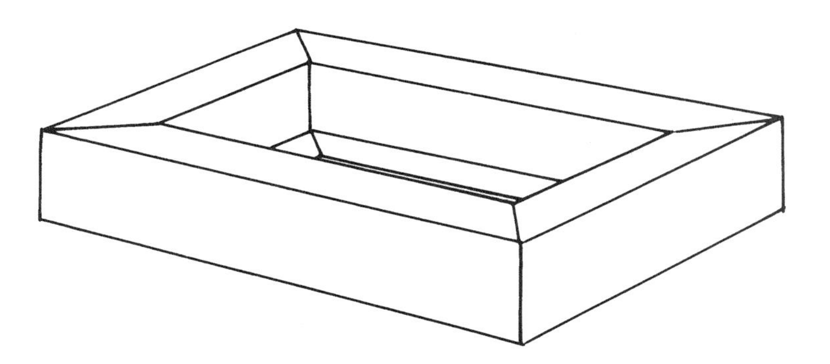

FRAME VIEW TRAY (VARIATION)

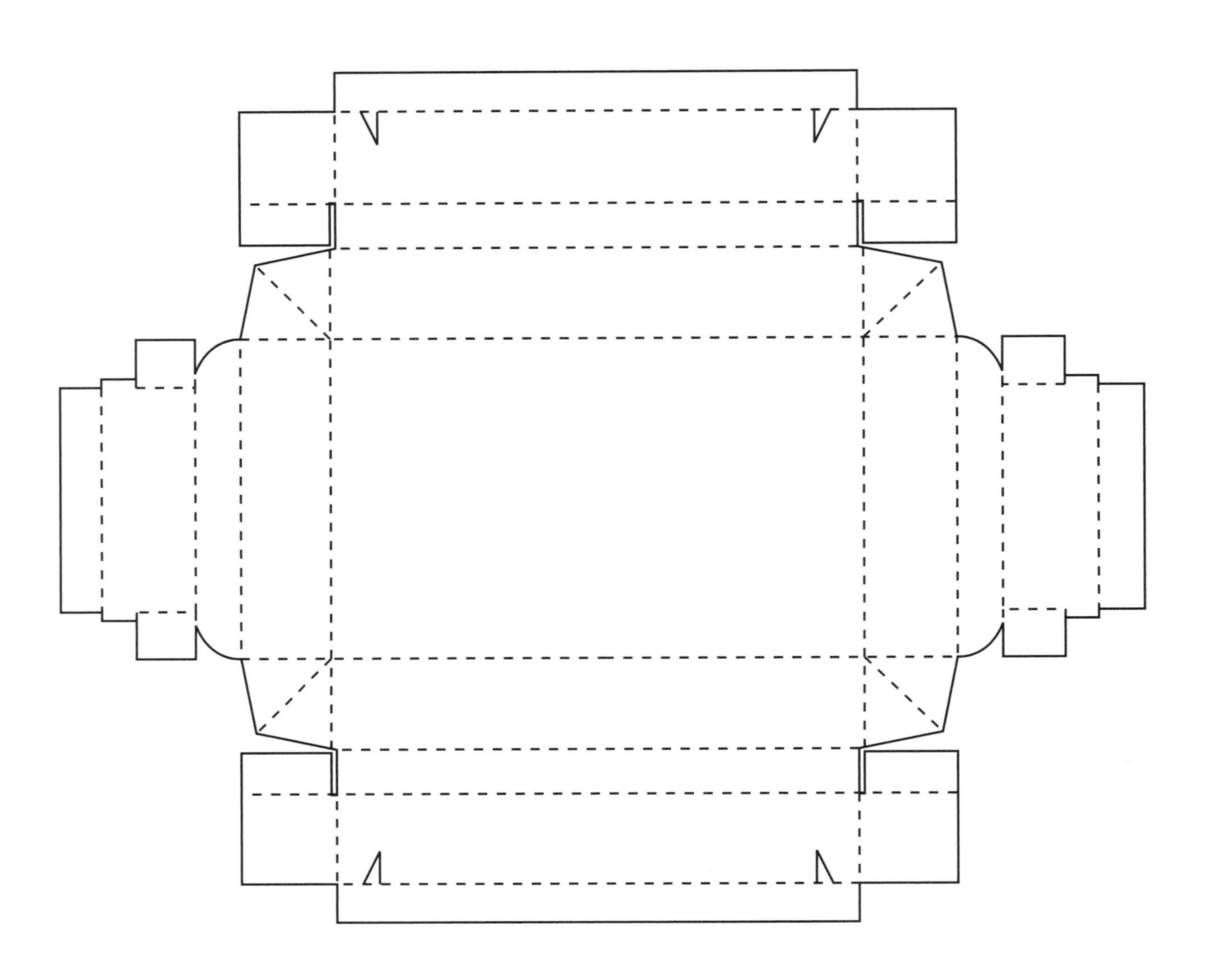

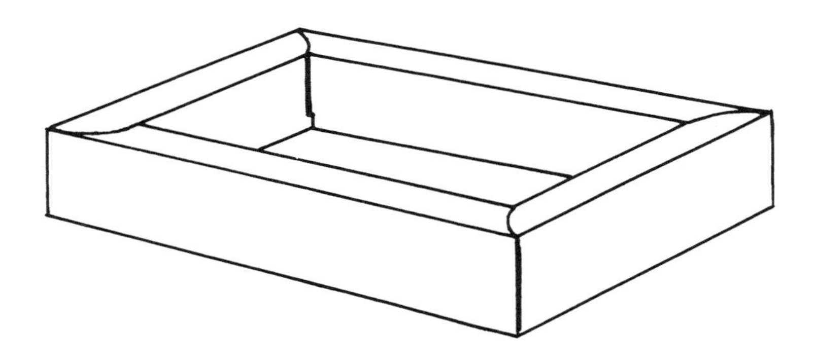

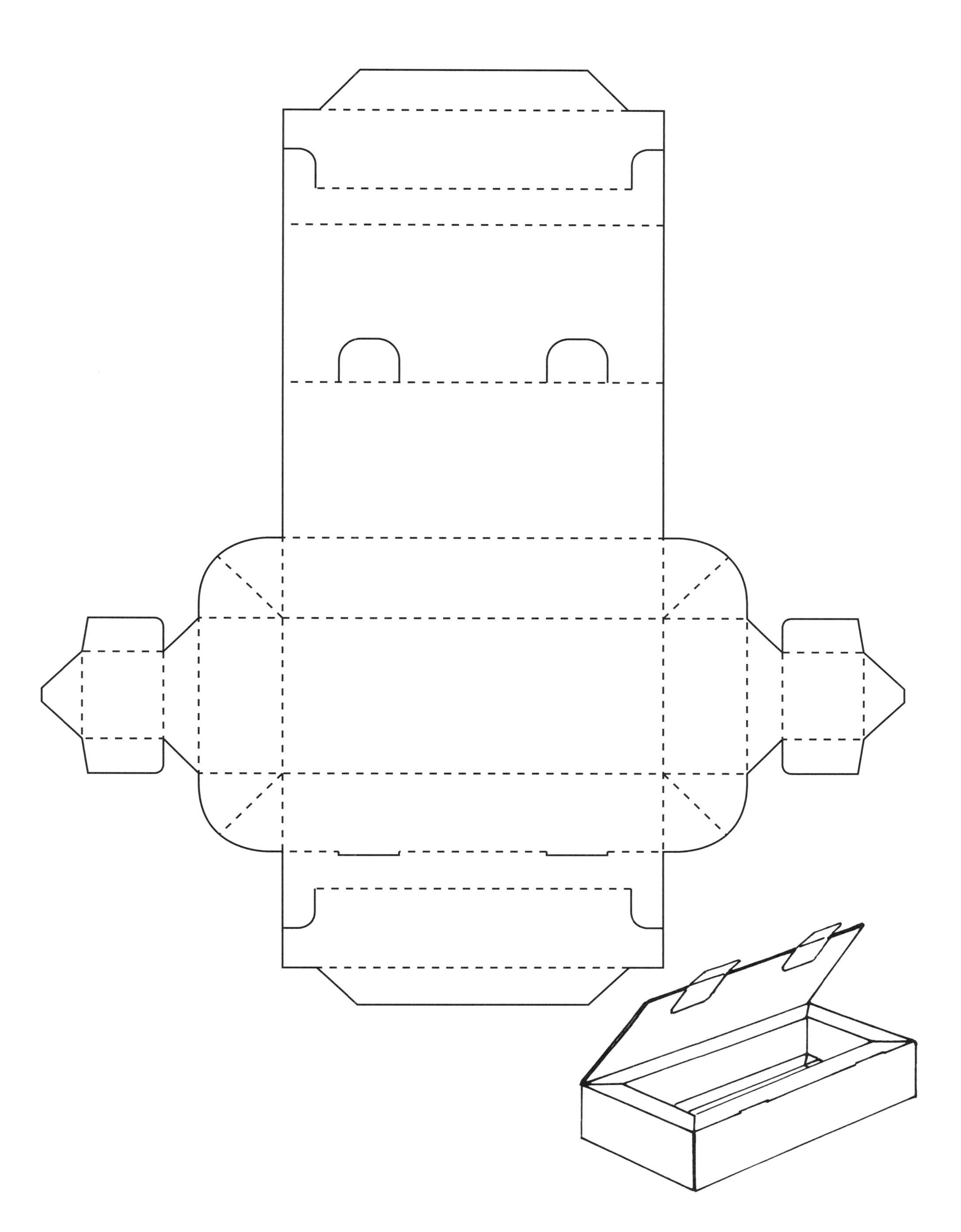

FRAME VIEW TRAY WITH DOUBLED COVER

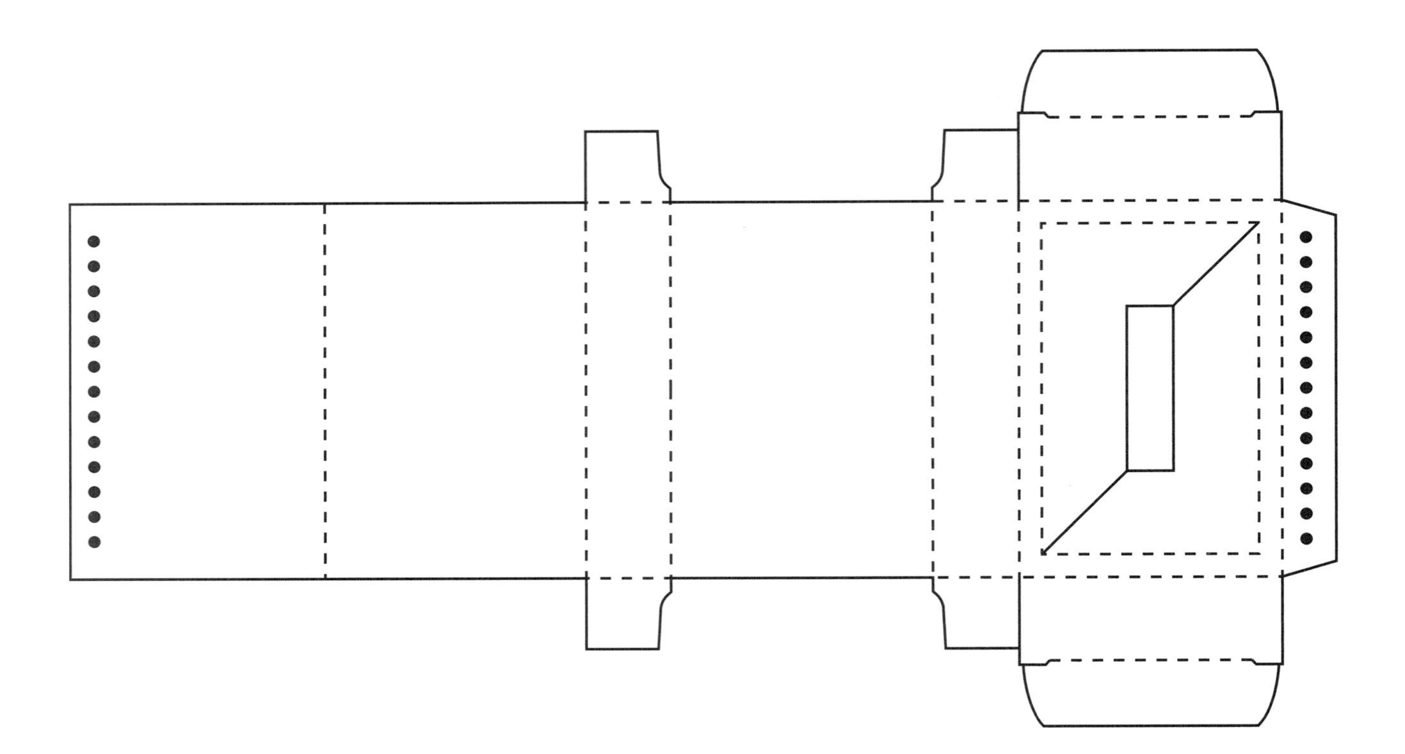

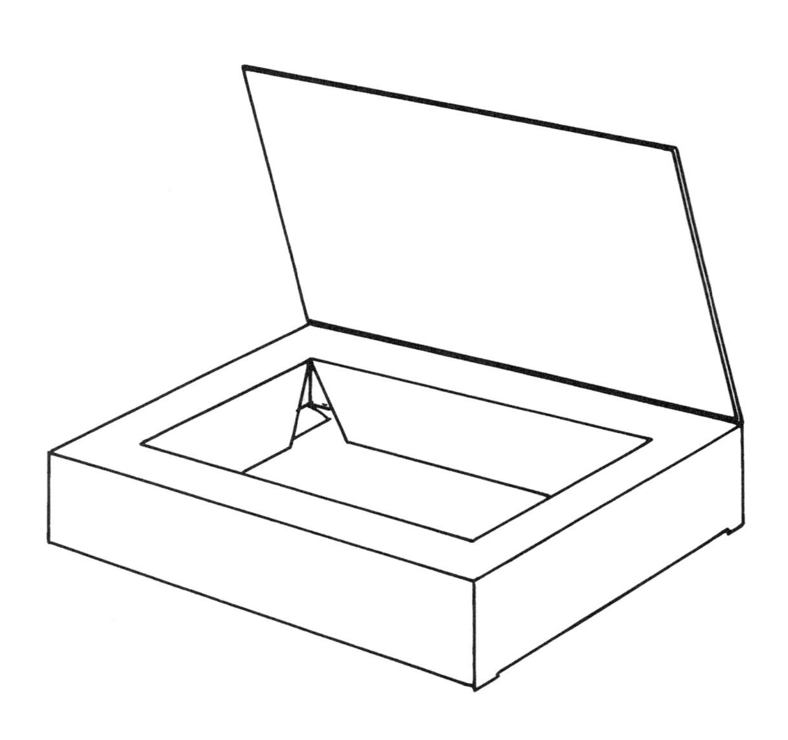

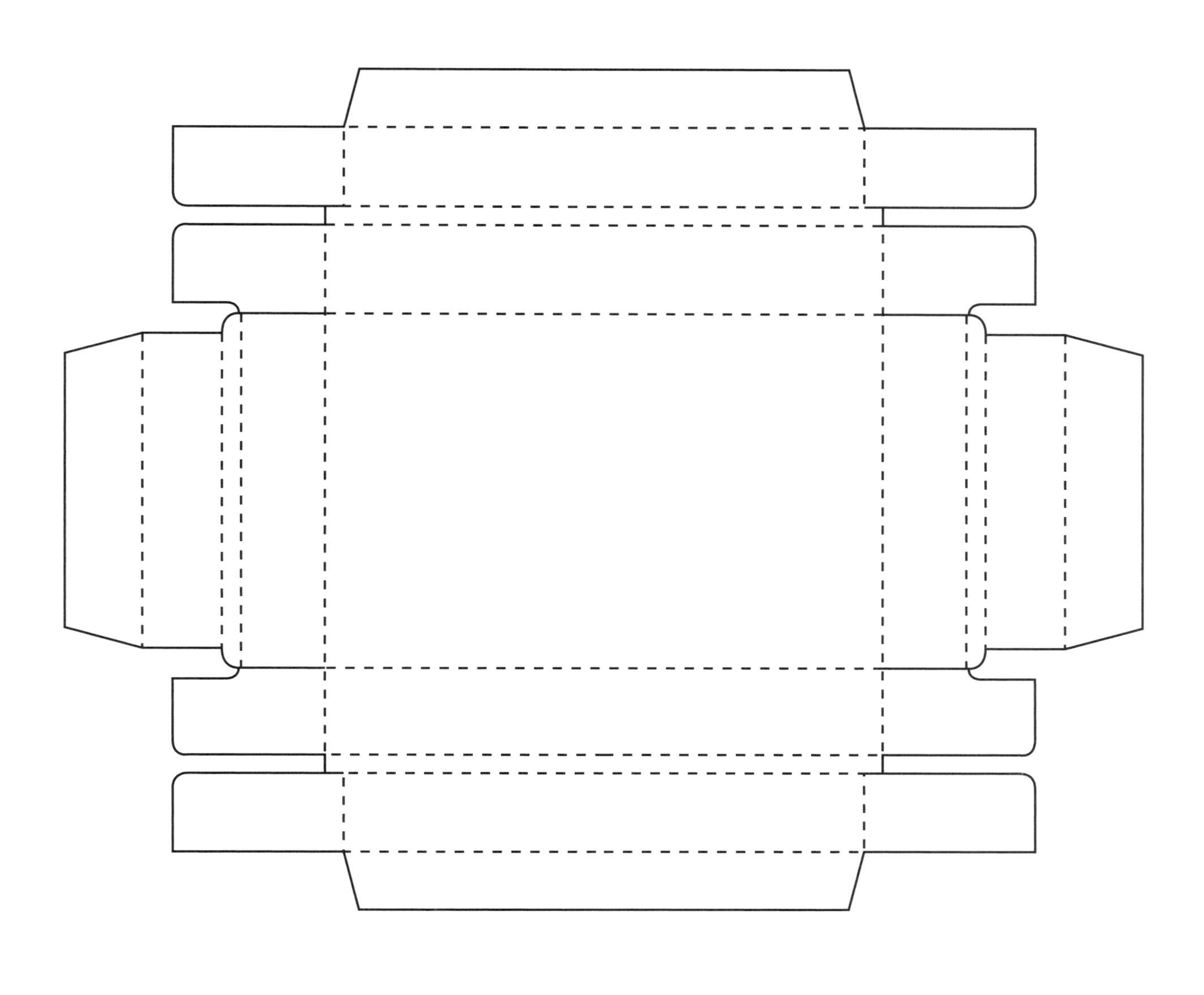

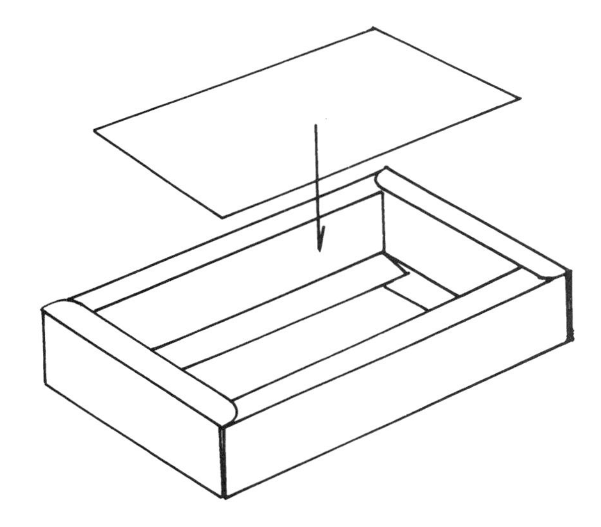

Combined with the overwrap pattern on the following page, this package lends itself to numerous graphic possibilities.

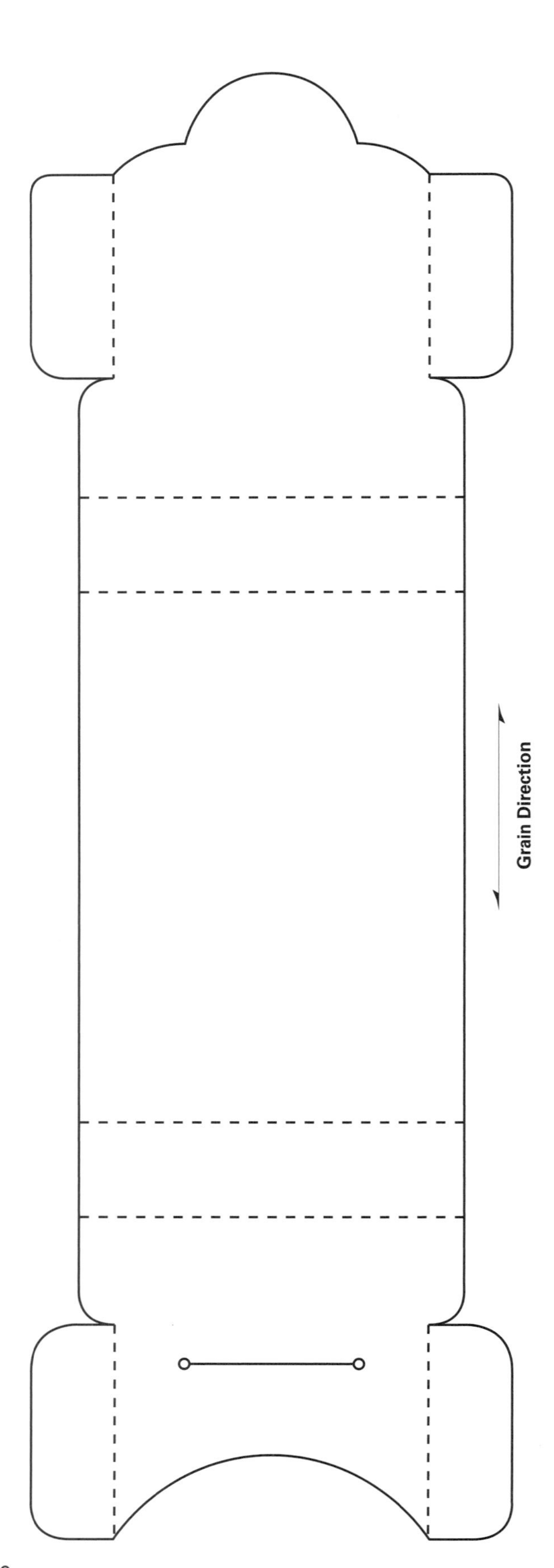

This wrap fits around the view tray of the preceding page; the four flaps fit neatly inside the tray.

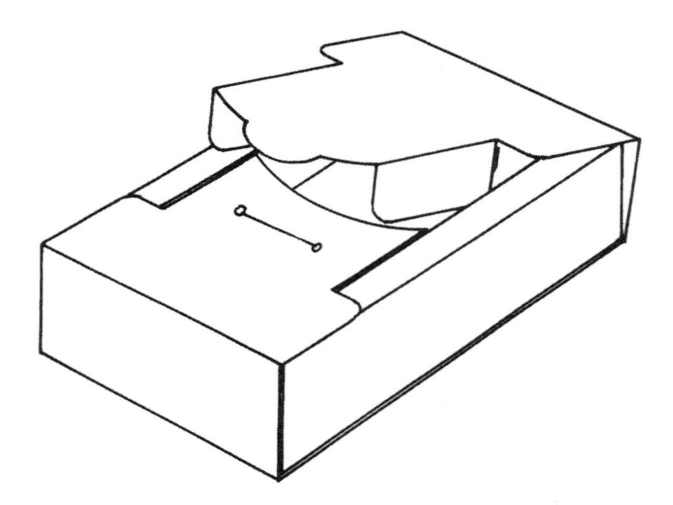

FRAME VIEW TRAY WITH COVER AND LOCK TAB

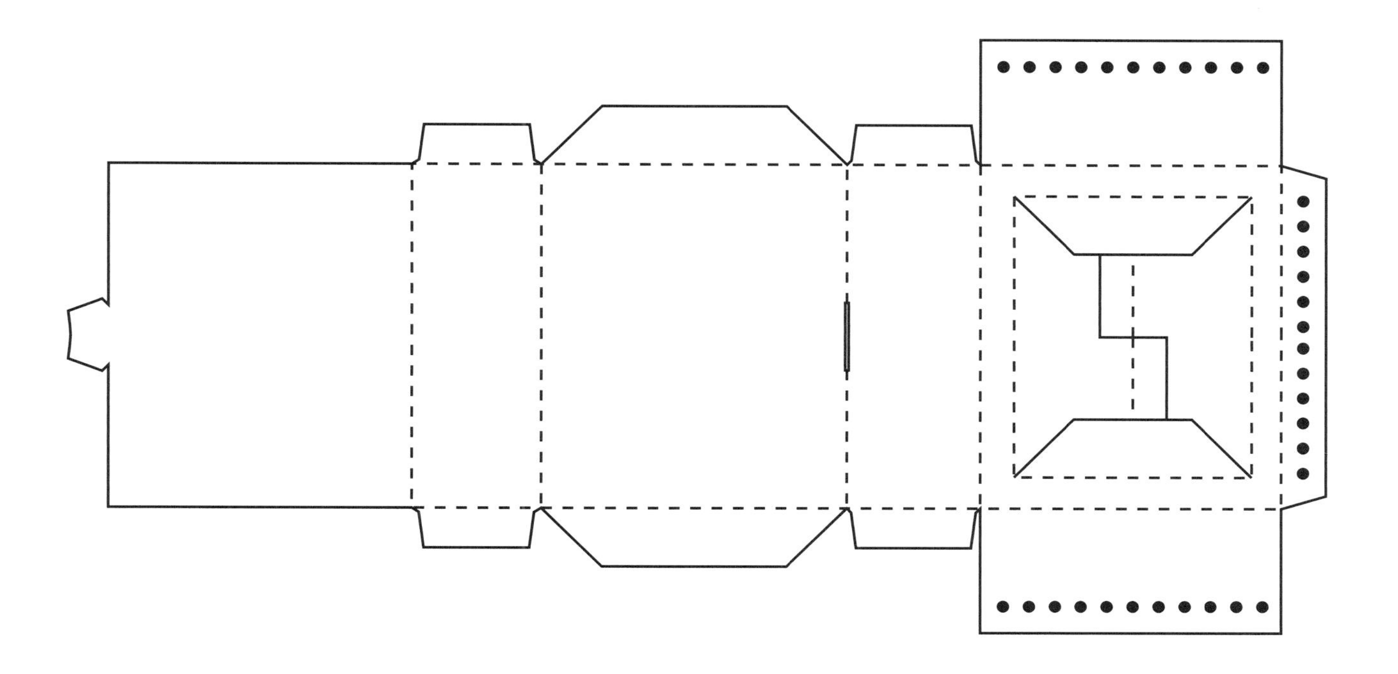

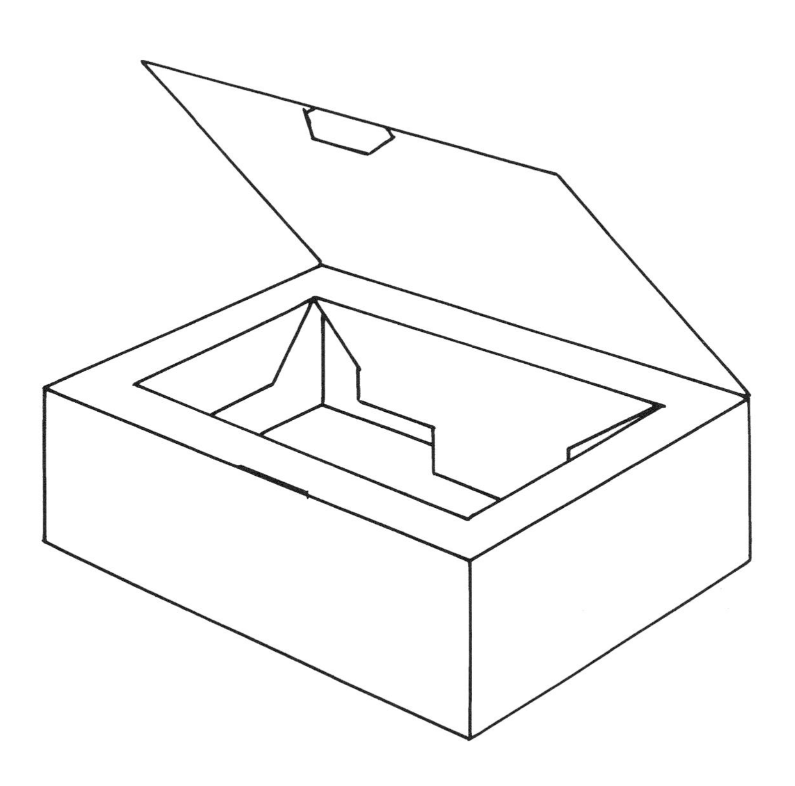

KLIKLOK® "CHARLOTTE" STRIPPER LOCK TRAY WITH HINGED COVER AND ANCHOR LOCK AND SIDELOCKS

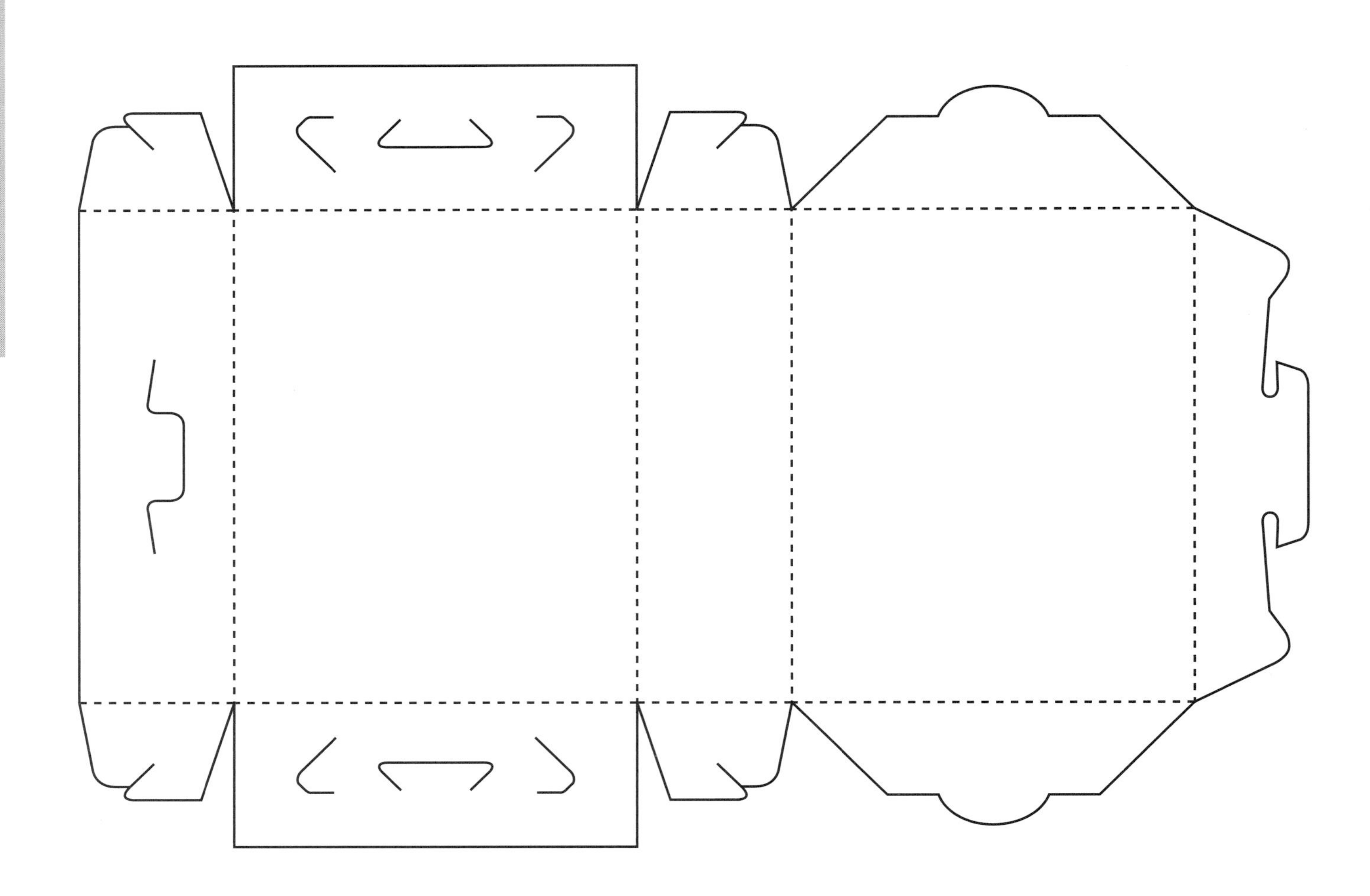

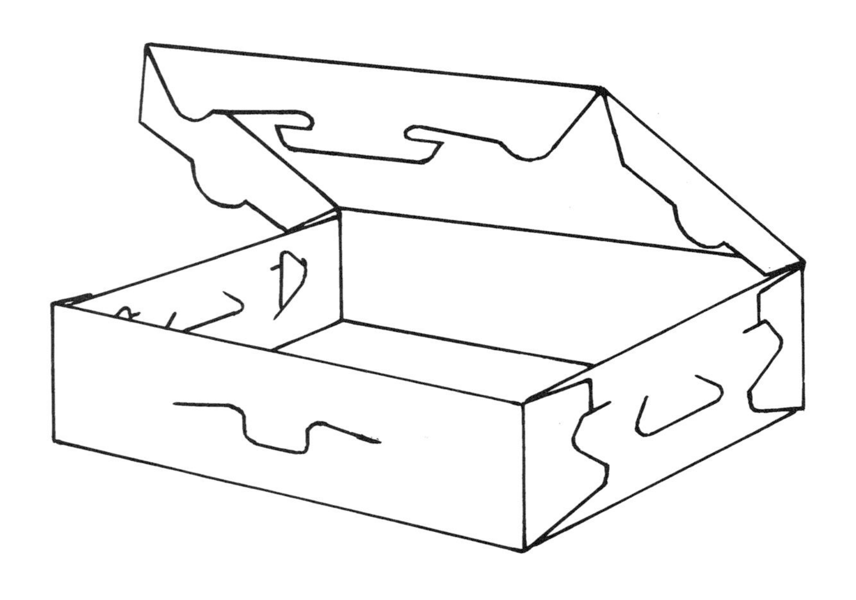

KLIKWEB® STYLE TRAY

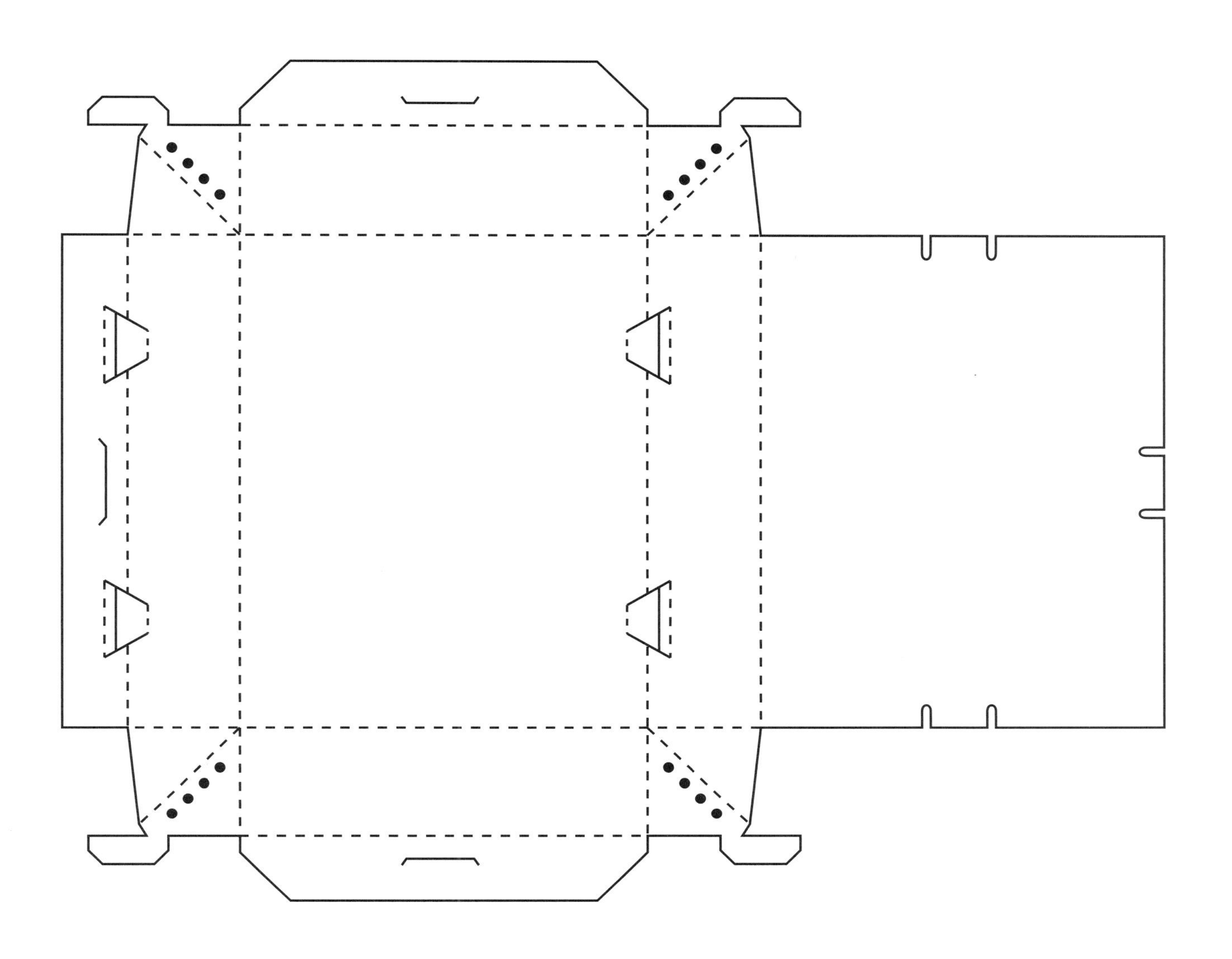

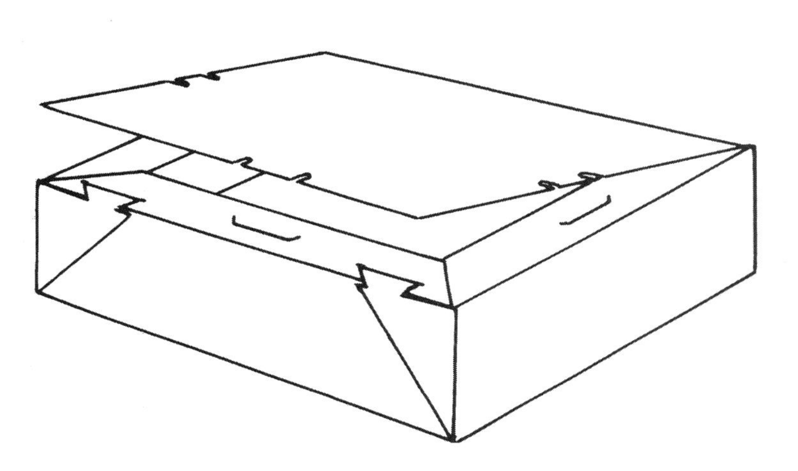

TAPERED TRAY WITH CONVEX FRONT COVER PANEL AND LOCK TAB

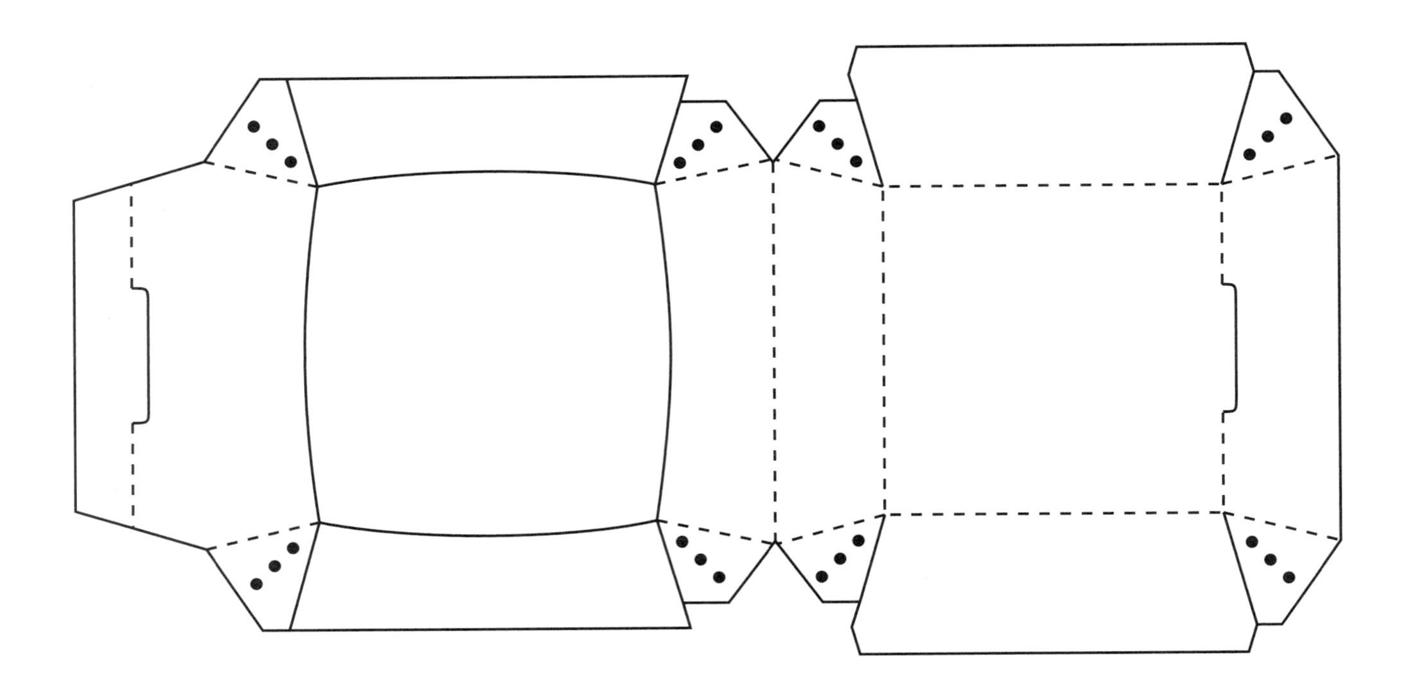

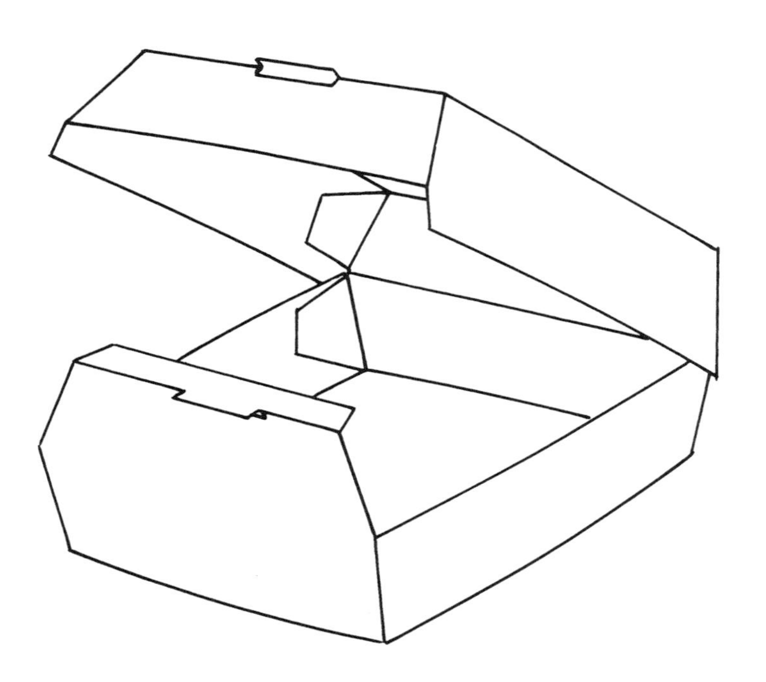

SIMPLEX STYLE VARIATION WITH TAPERED OUTSIDE END AND SIDE WALLS

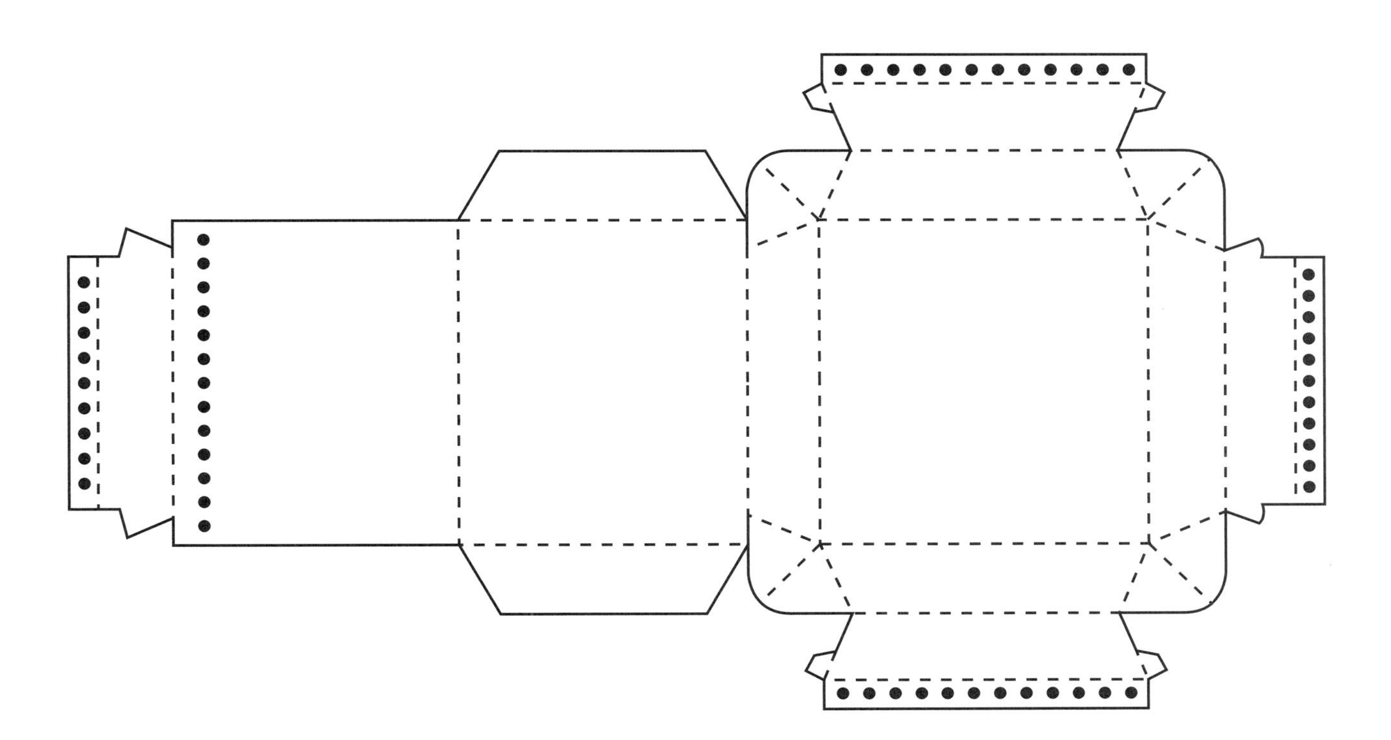

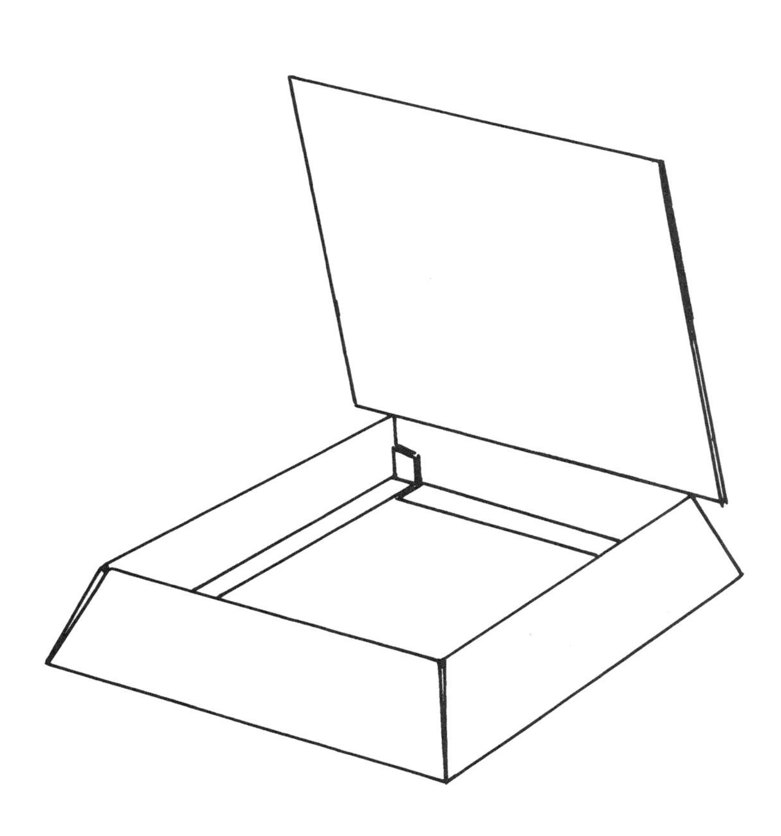

TAPERED SIDEWALL WITH ARTHUR LOCKS AND DOUBLE HOLLOW SIDE WALLS

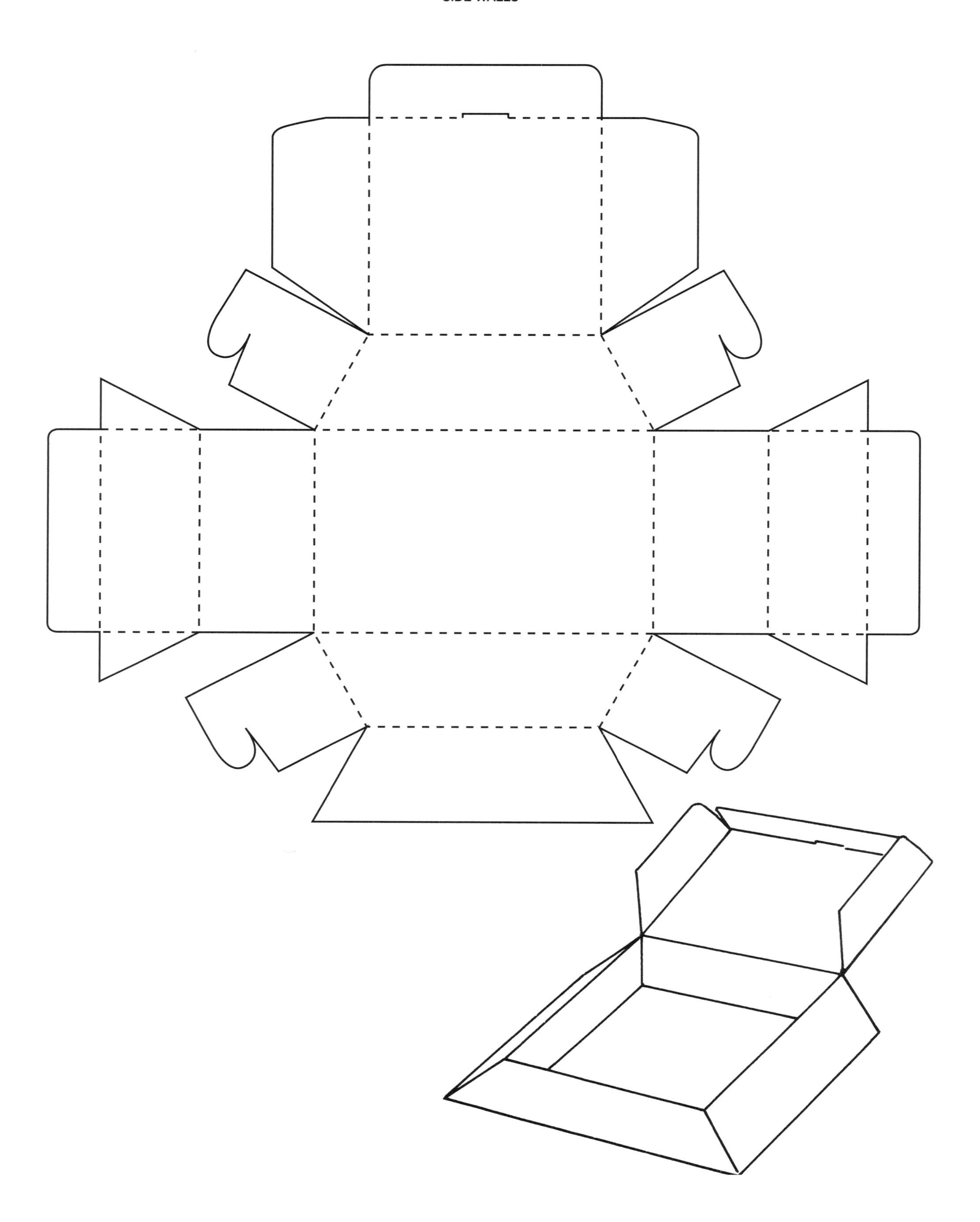

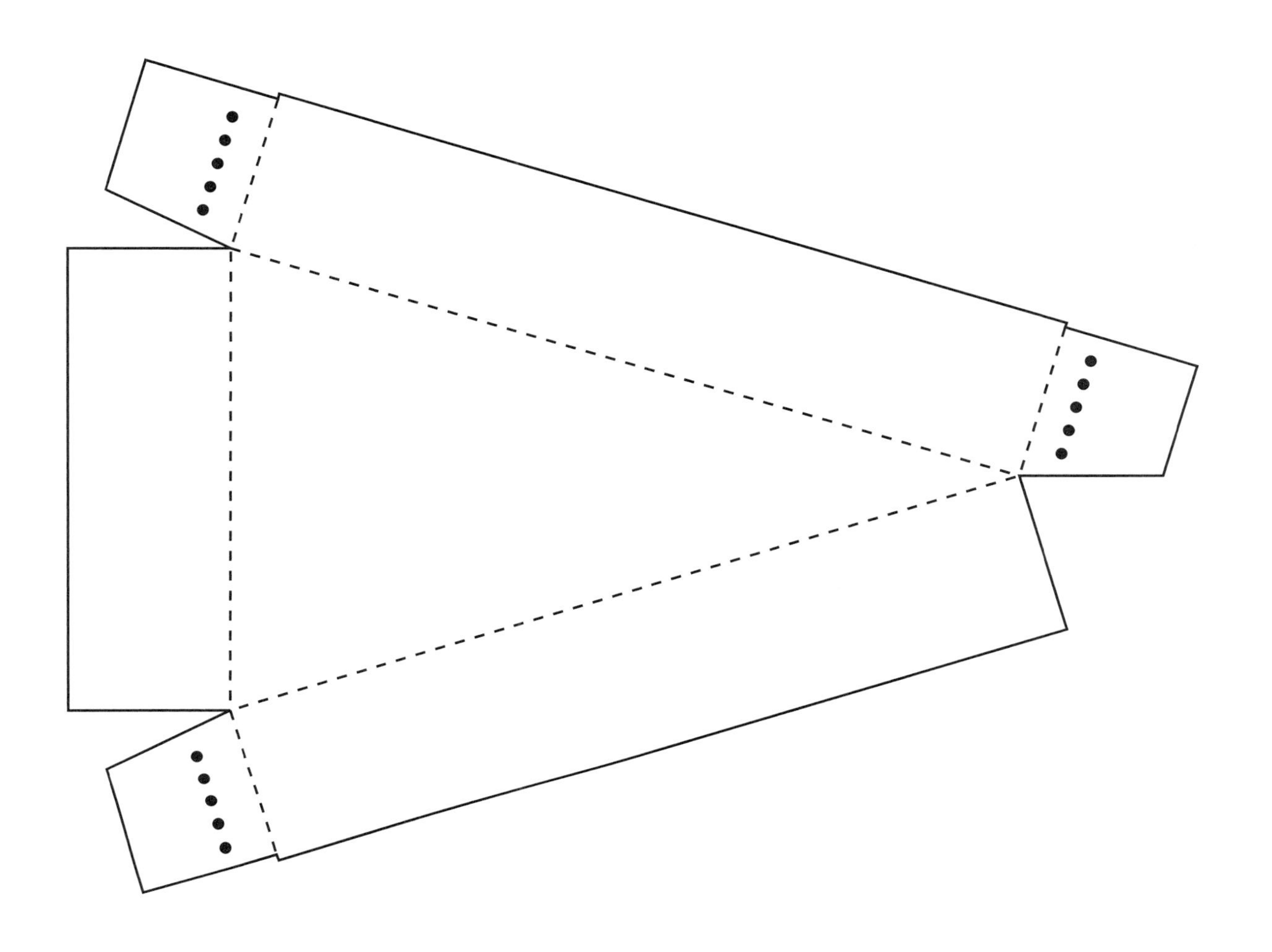

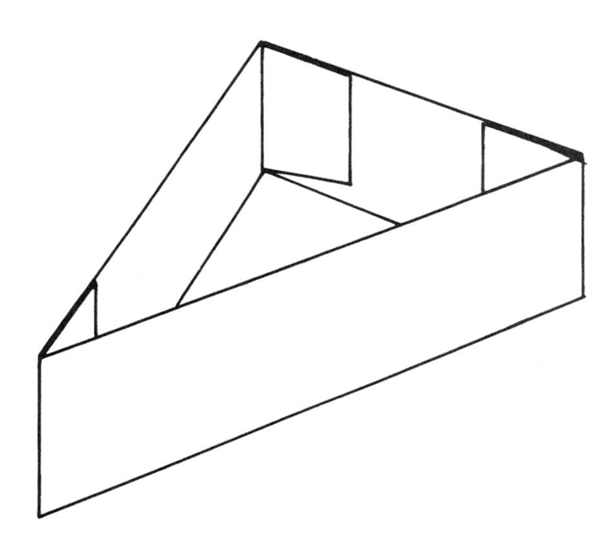

SELF-COVERED TRIANGLE WITH LOCK TABS OFF COVER

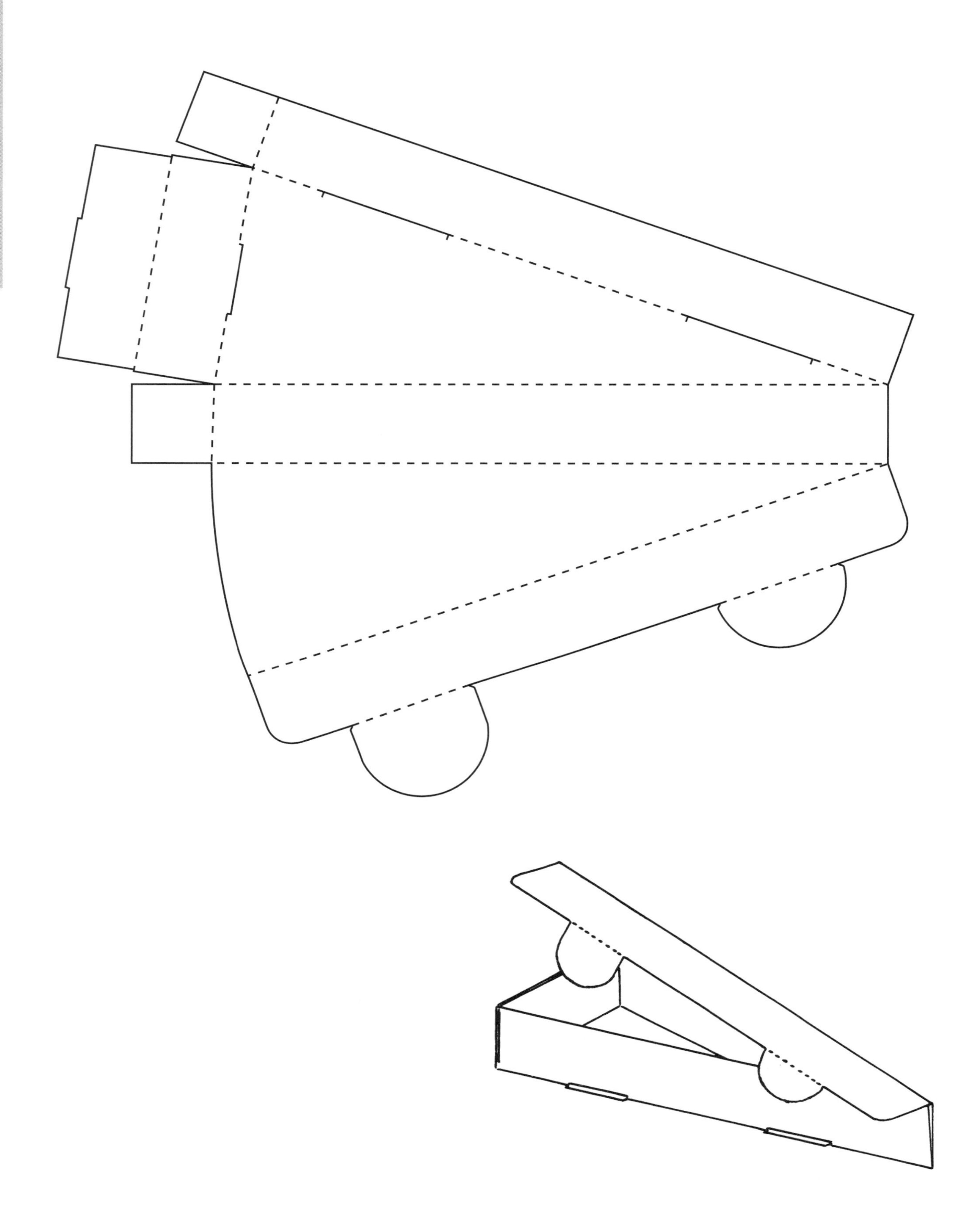

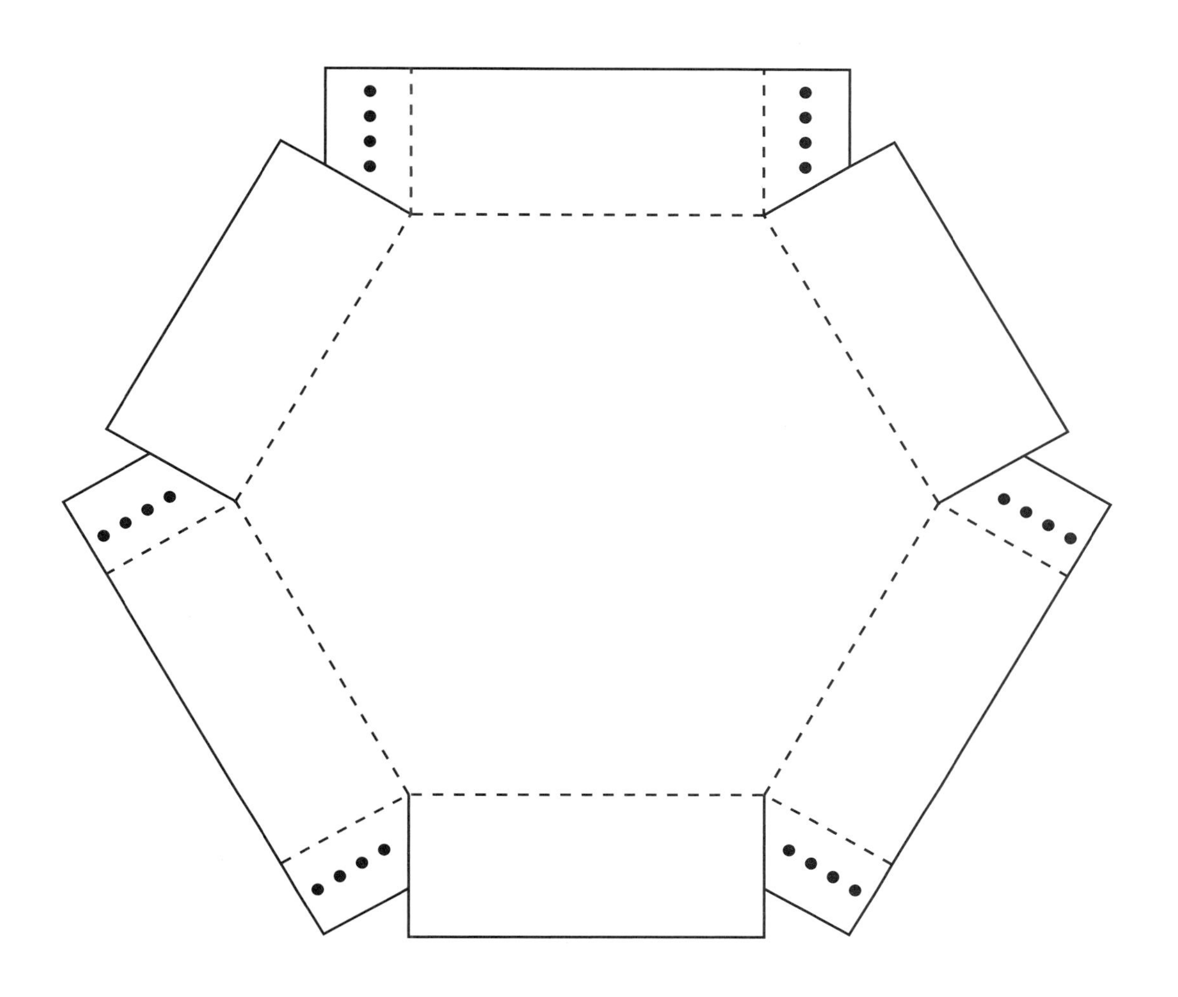

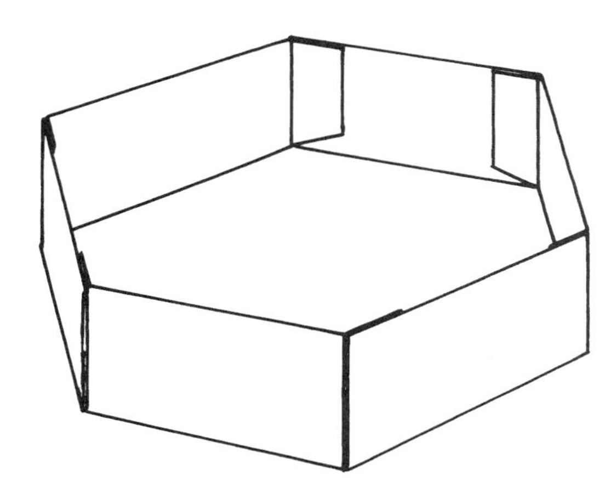

OCTAGON STYLE TRAY

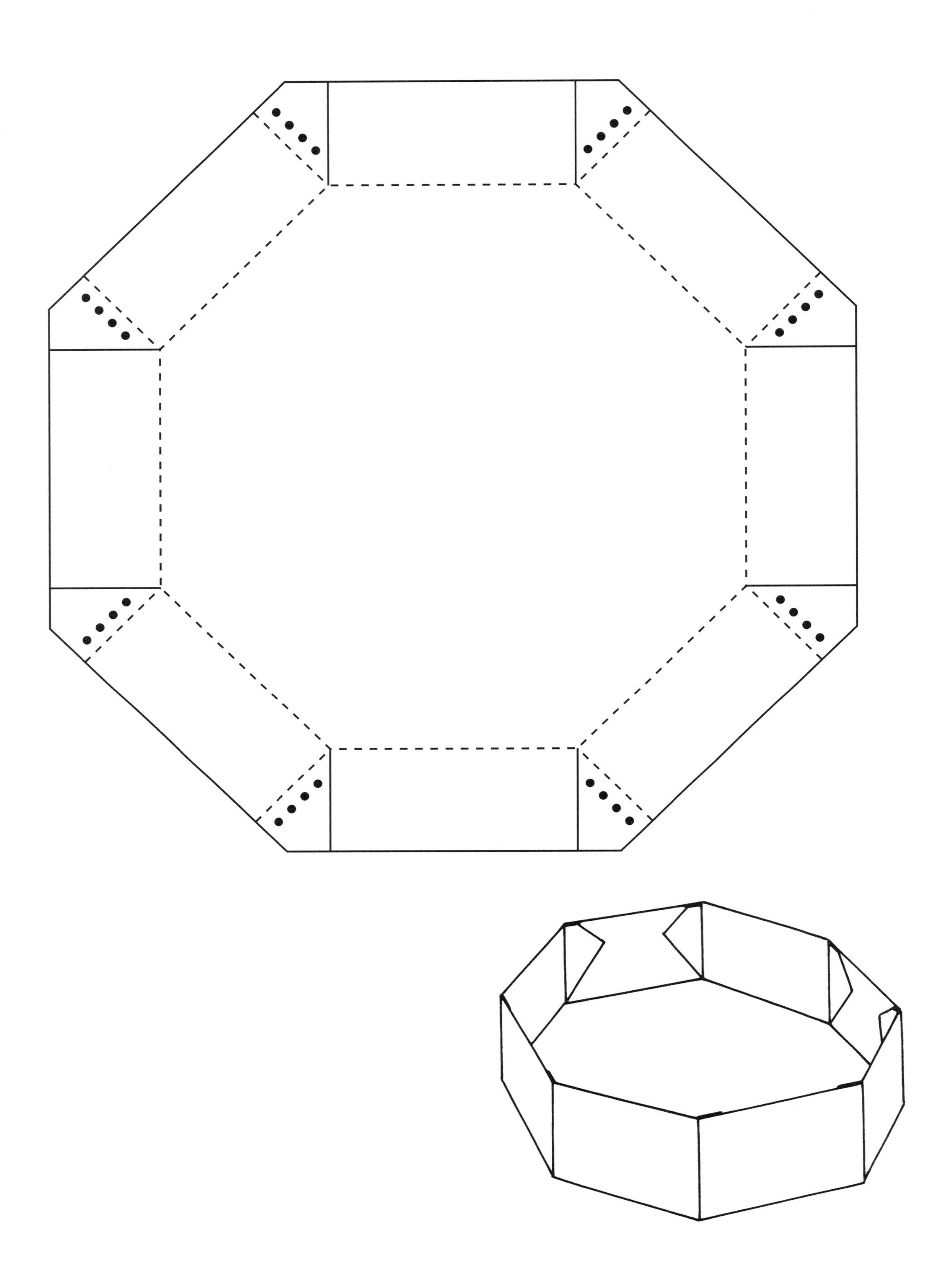

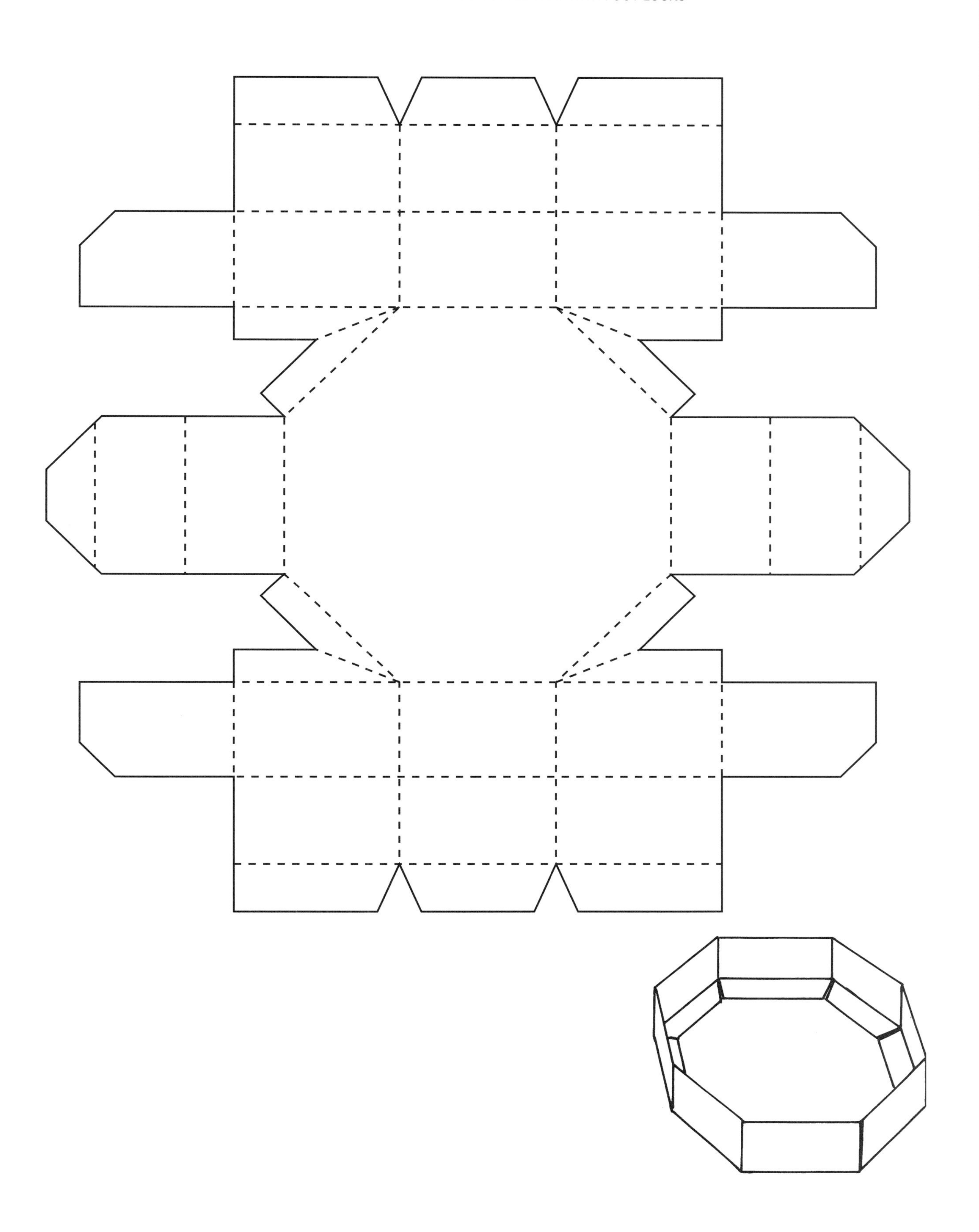

SIX-SIDED DOUBLE WALLED TRAY AND PARTIAL TELESCOPING COVER

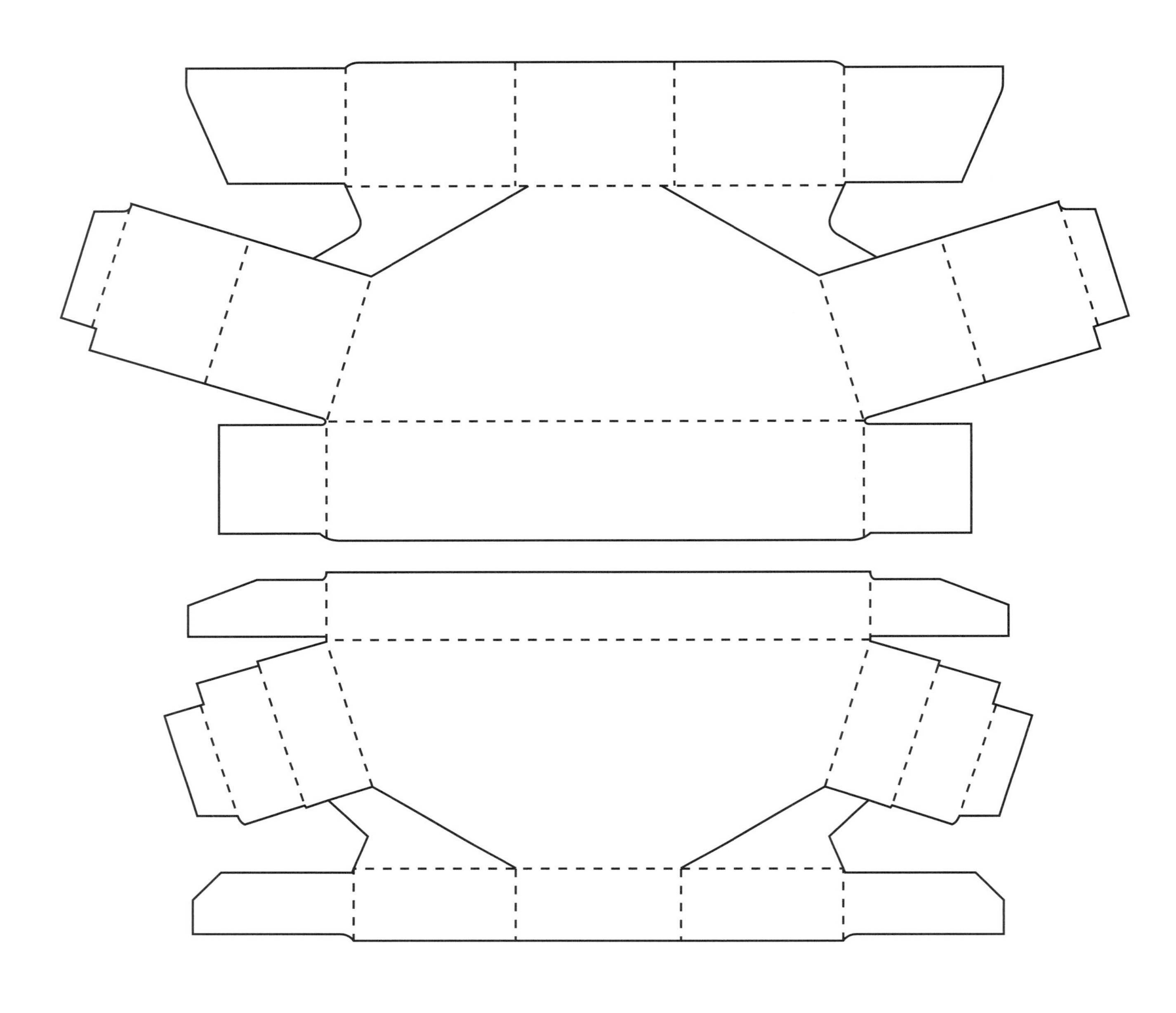

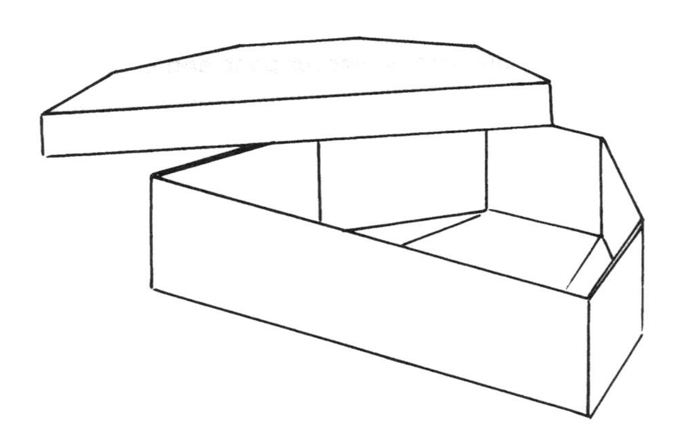

TRAY FOR WRITING IMPLEMENT

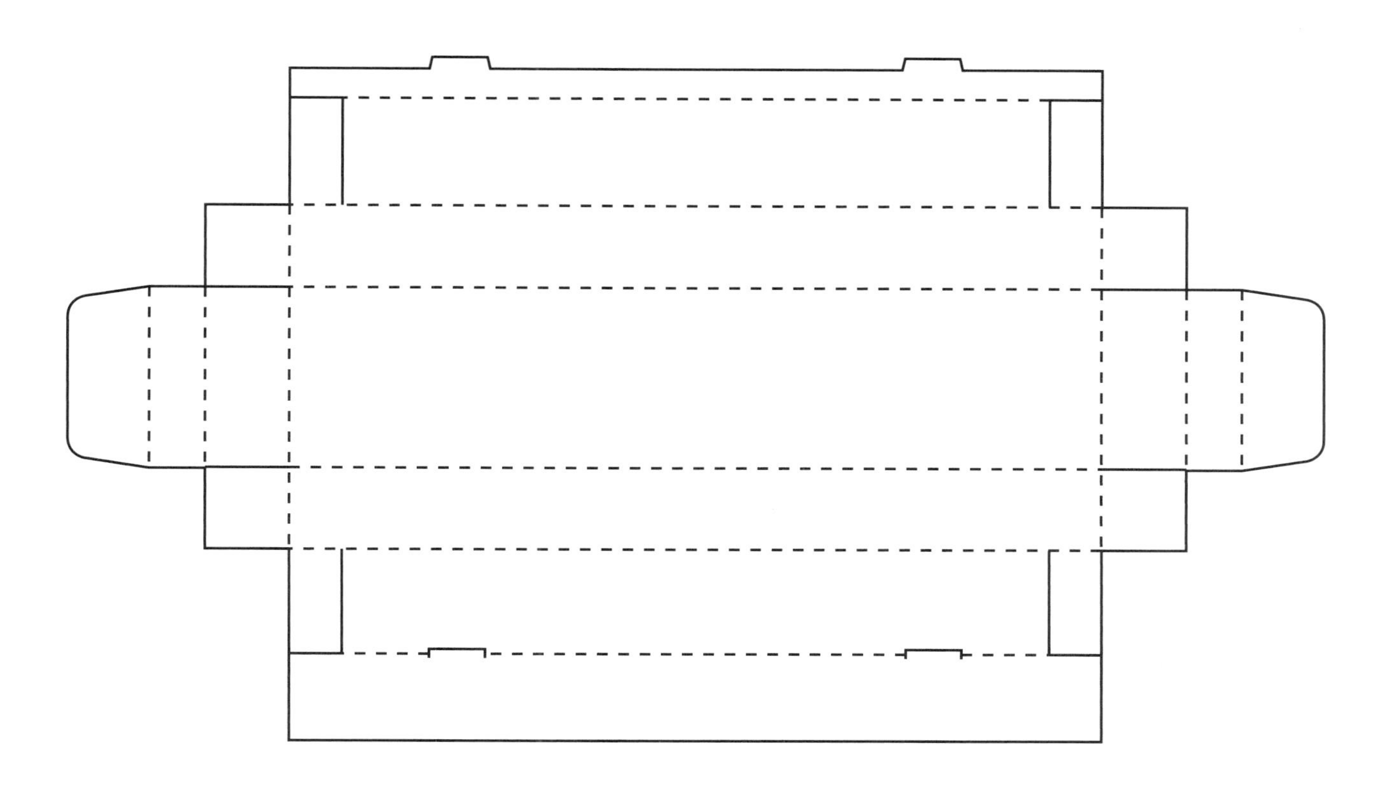

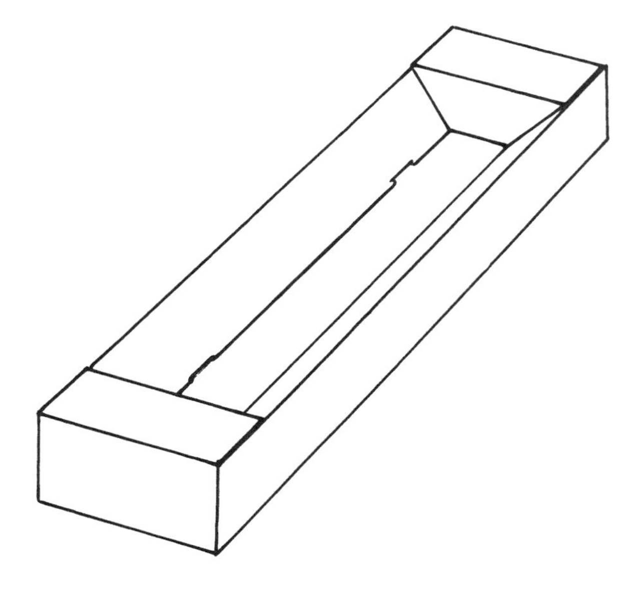

This specially constructed tray has a tapered interior that makes it an excellent container for pens and similar items.

PEN AND PENCIL TRAY

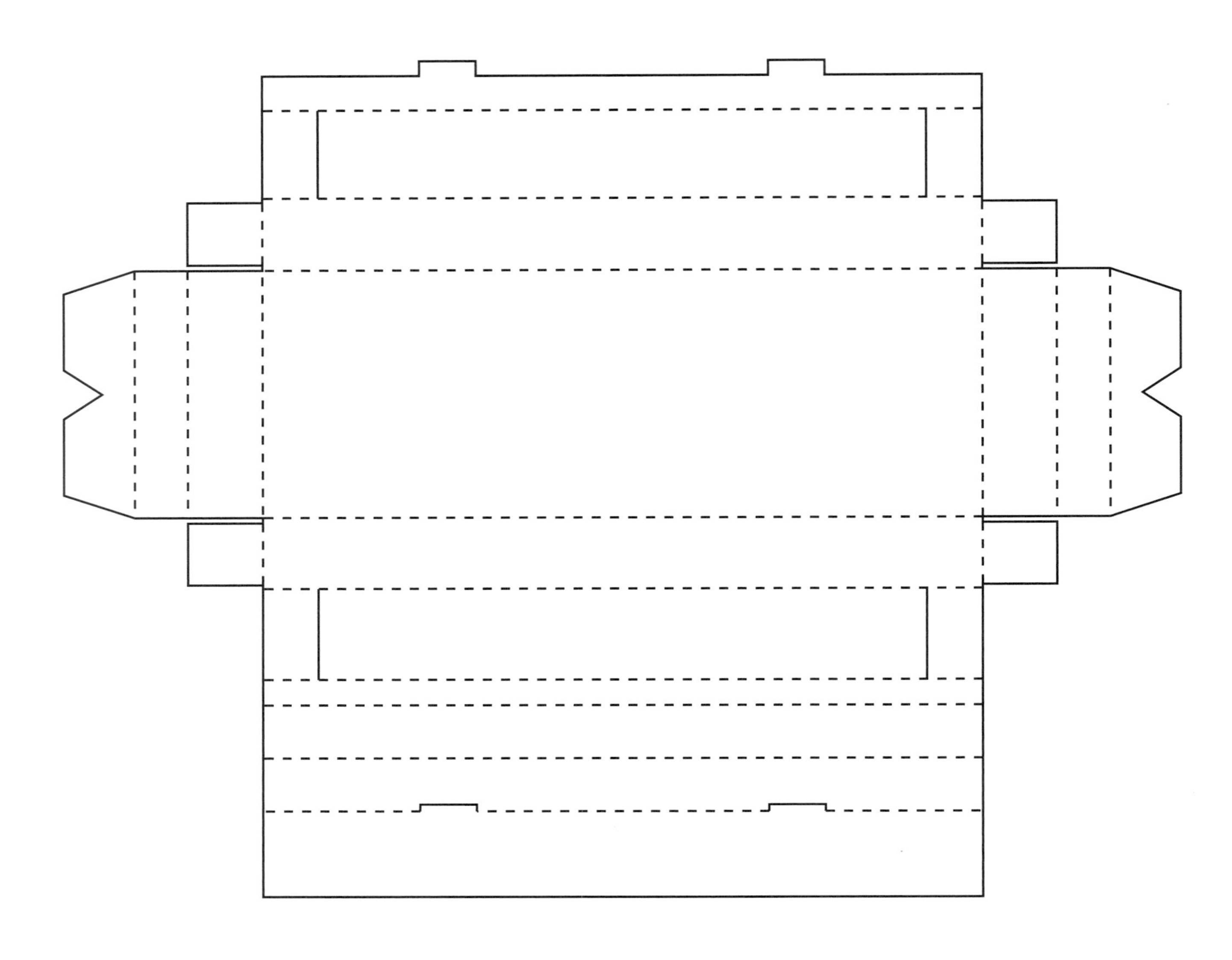

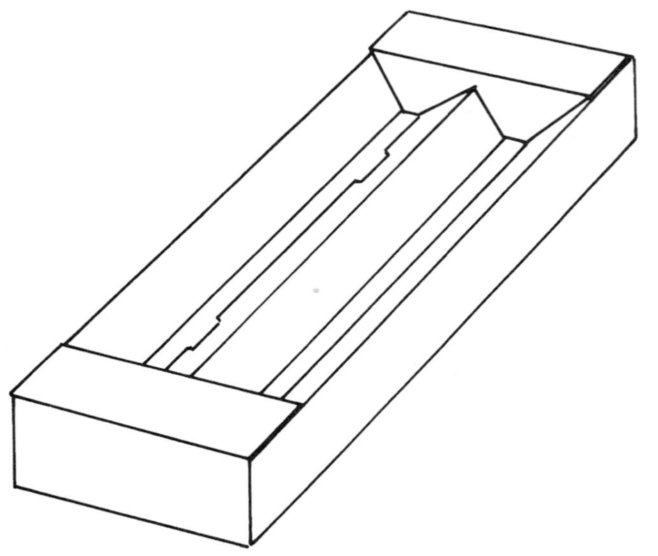

This variation of the preceding pattern is adaptable for multiunit packaging.

SIMPLEX STYLE TRAY

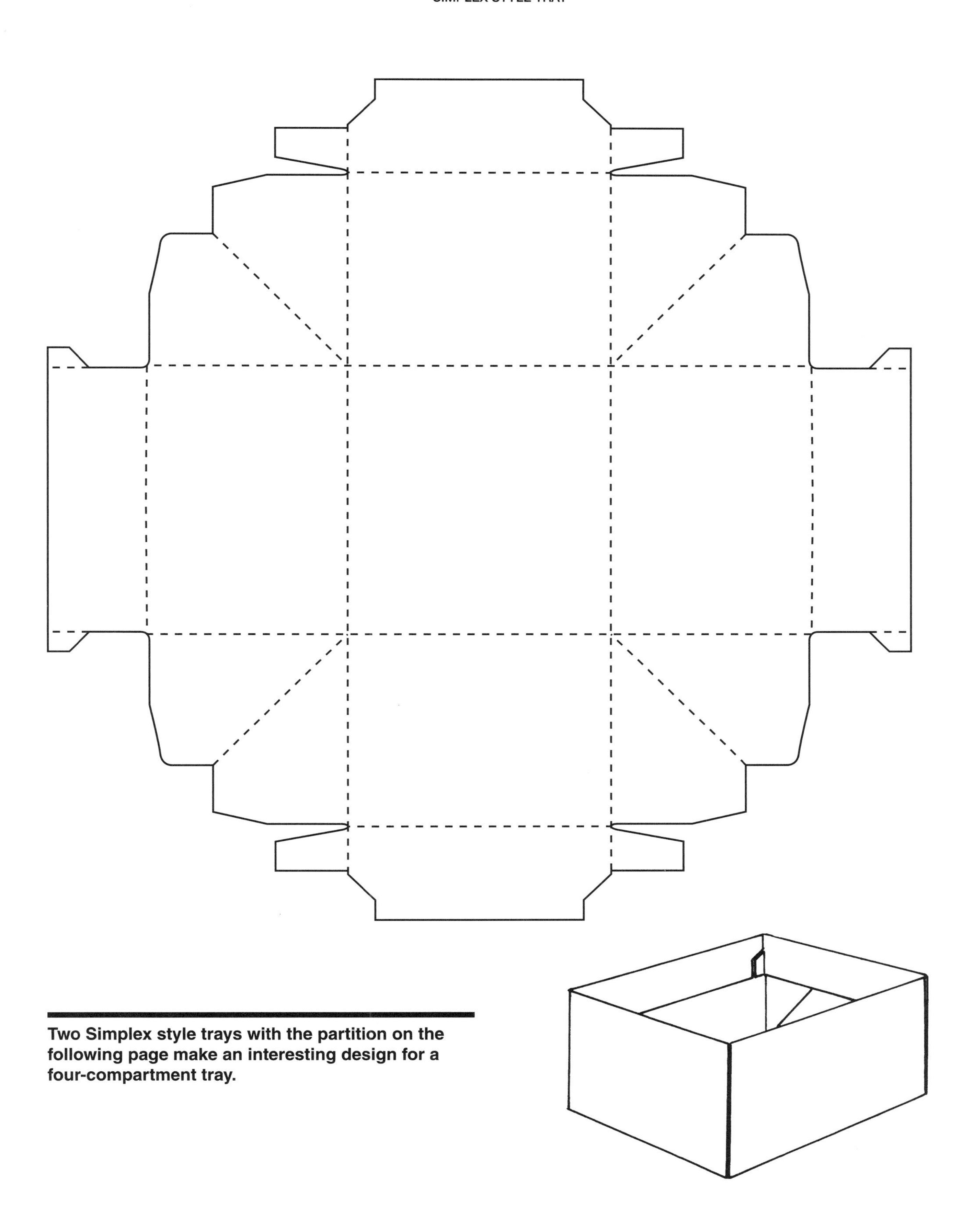

Two Simplex style trays with the partition on the following page make an interesting design for a four-compartment tray.

TRAY PARTITION

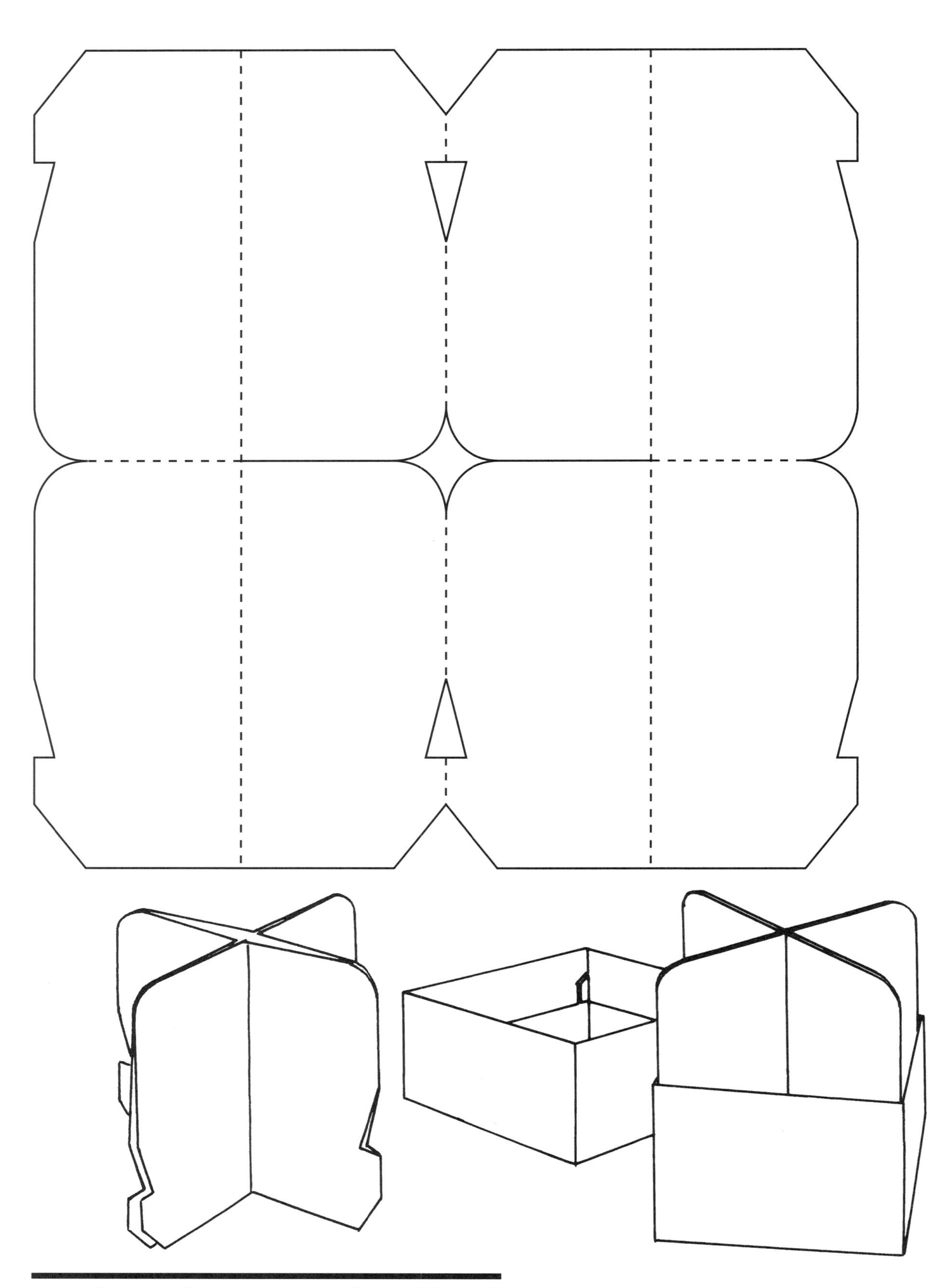

The notches in the lower part of the partition will lock into the partial double side walls of the tray.

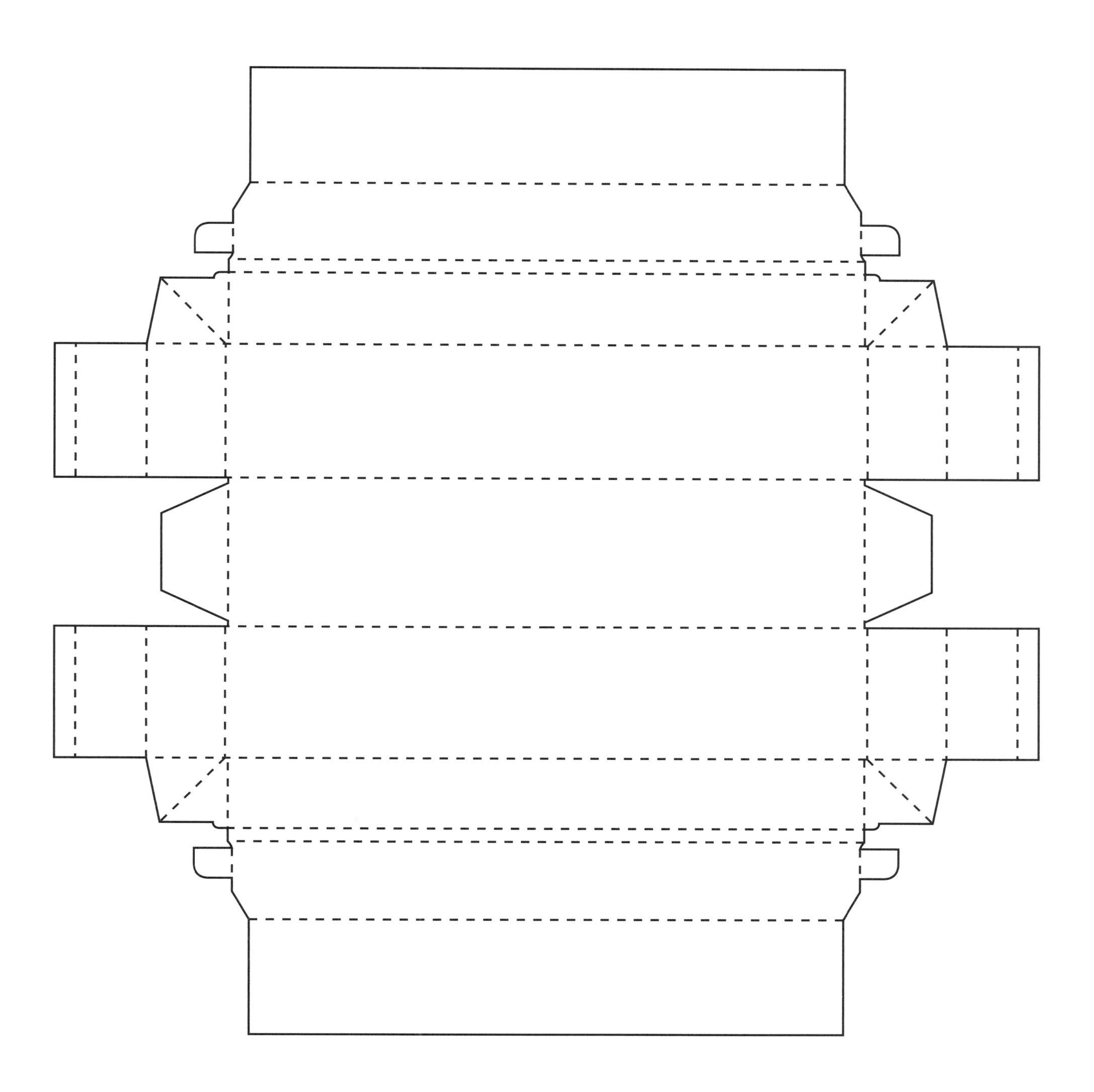

This variation of the Simplex style tray has two double walled cover sections that meet in the top center. The insert illustrated on the following page indicates the many possibilities inherent in these patterns. This package was sealed with a self-adhesive gold-foil sticker.

TWELVE-COUNT TRAY INSERT

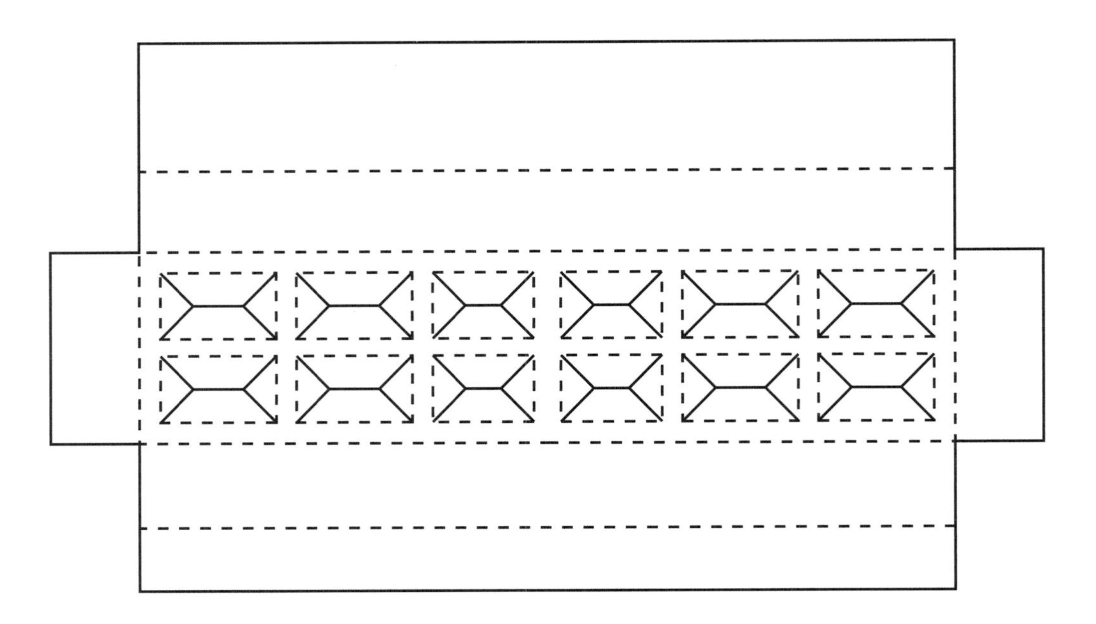

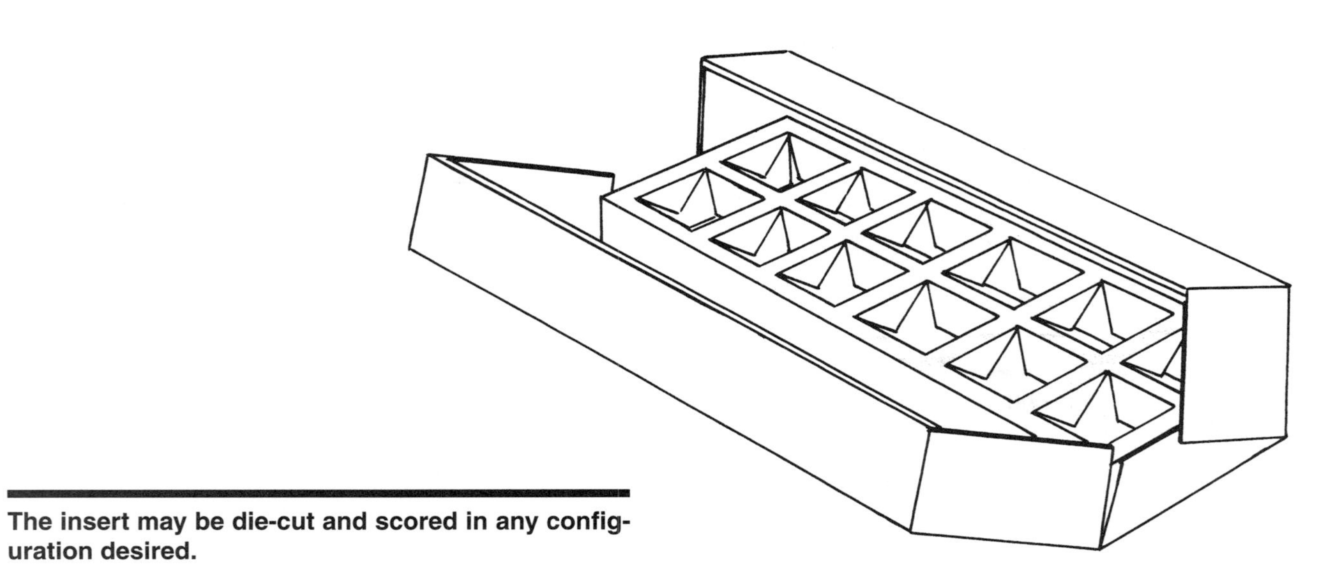

The insert may be die-cut and scored in any configuration desired.

PERFUME BOTTLE FRAME VIEW TRAY

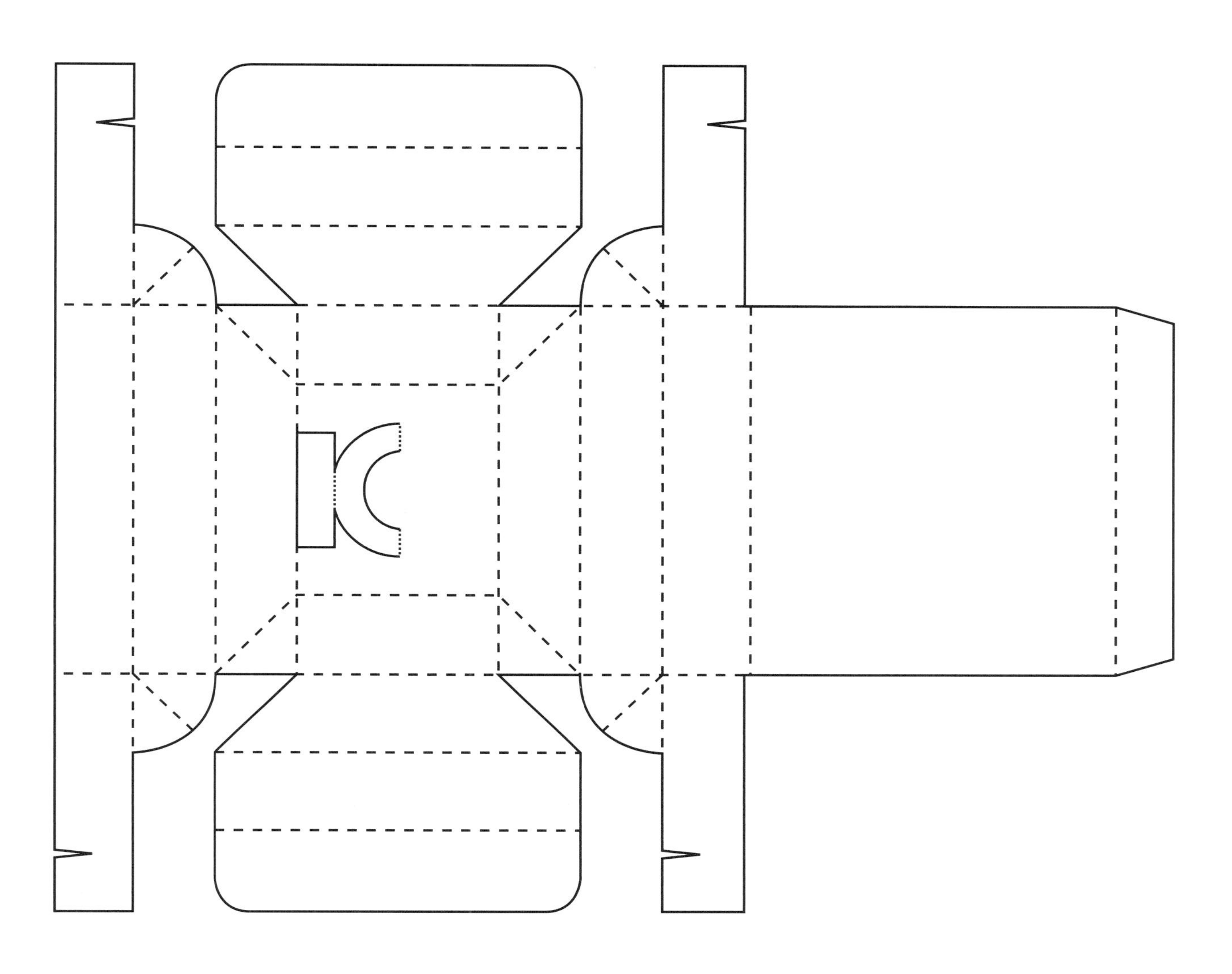

A more complex example of a frame view tray.

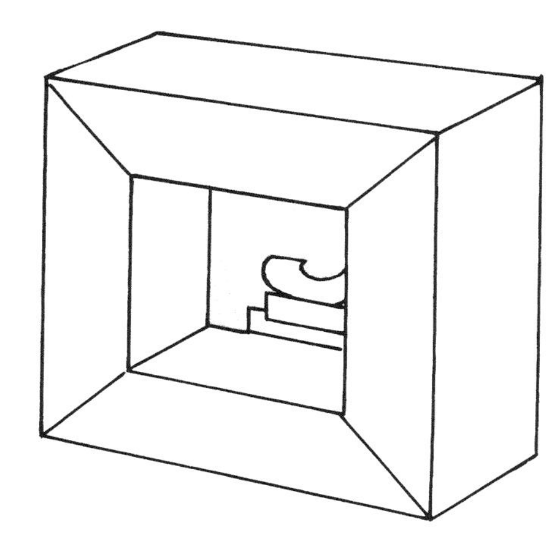

SIX-COUNT EGG OR FRUIT TRAY

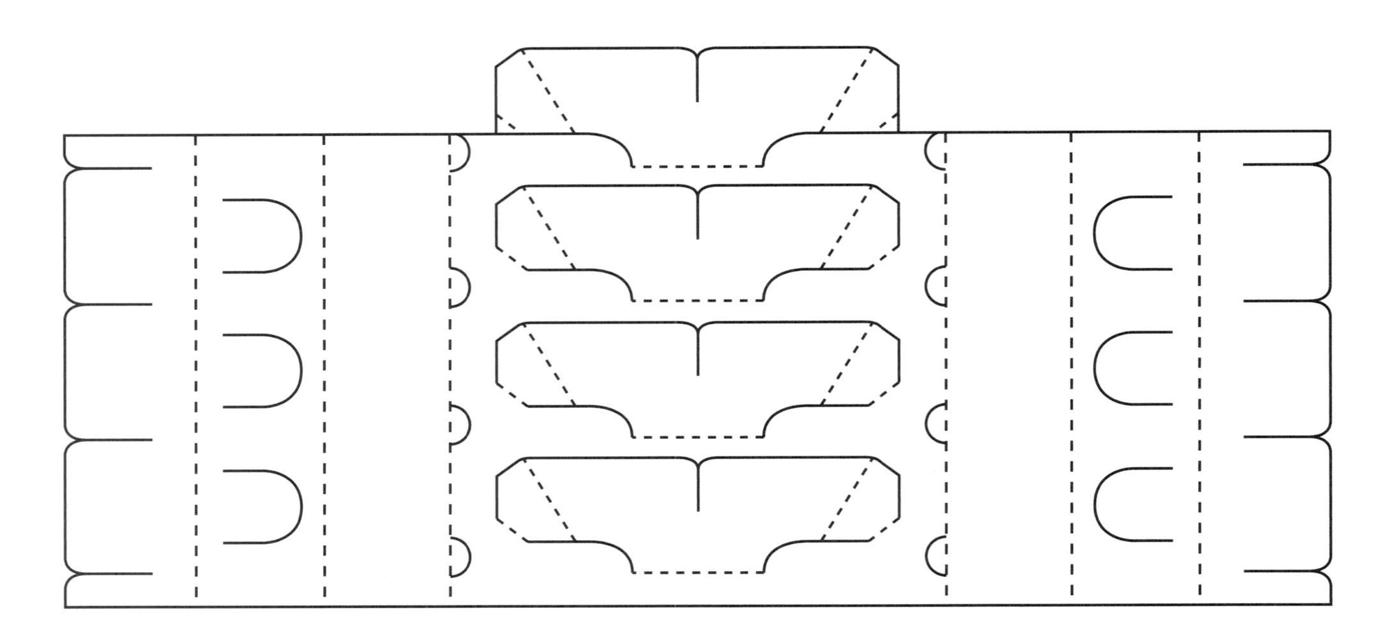

This one-piece scored and die-cut tray may be used for eggs, fruit, candy, and the like.

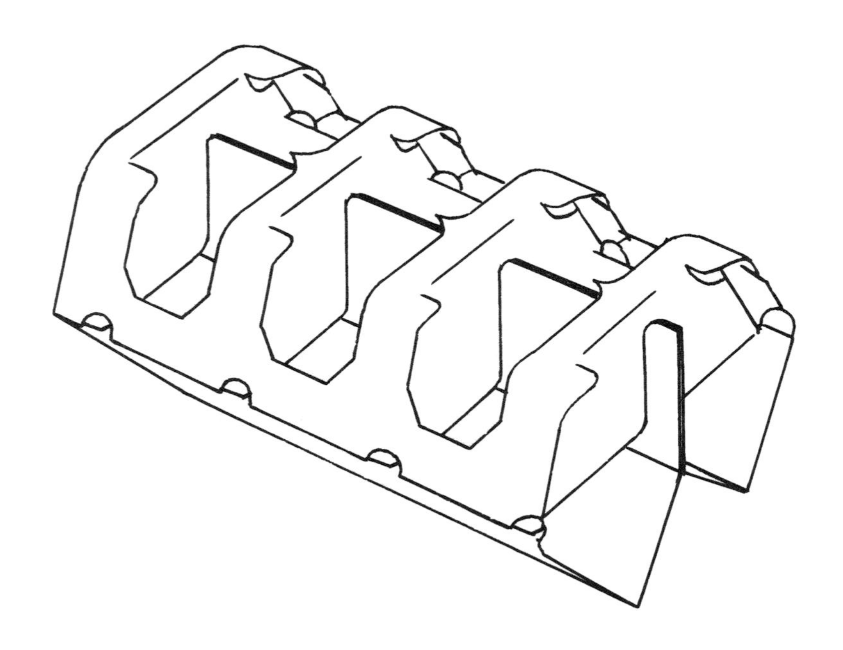

Tubes

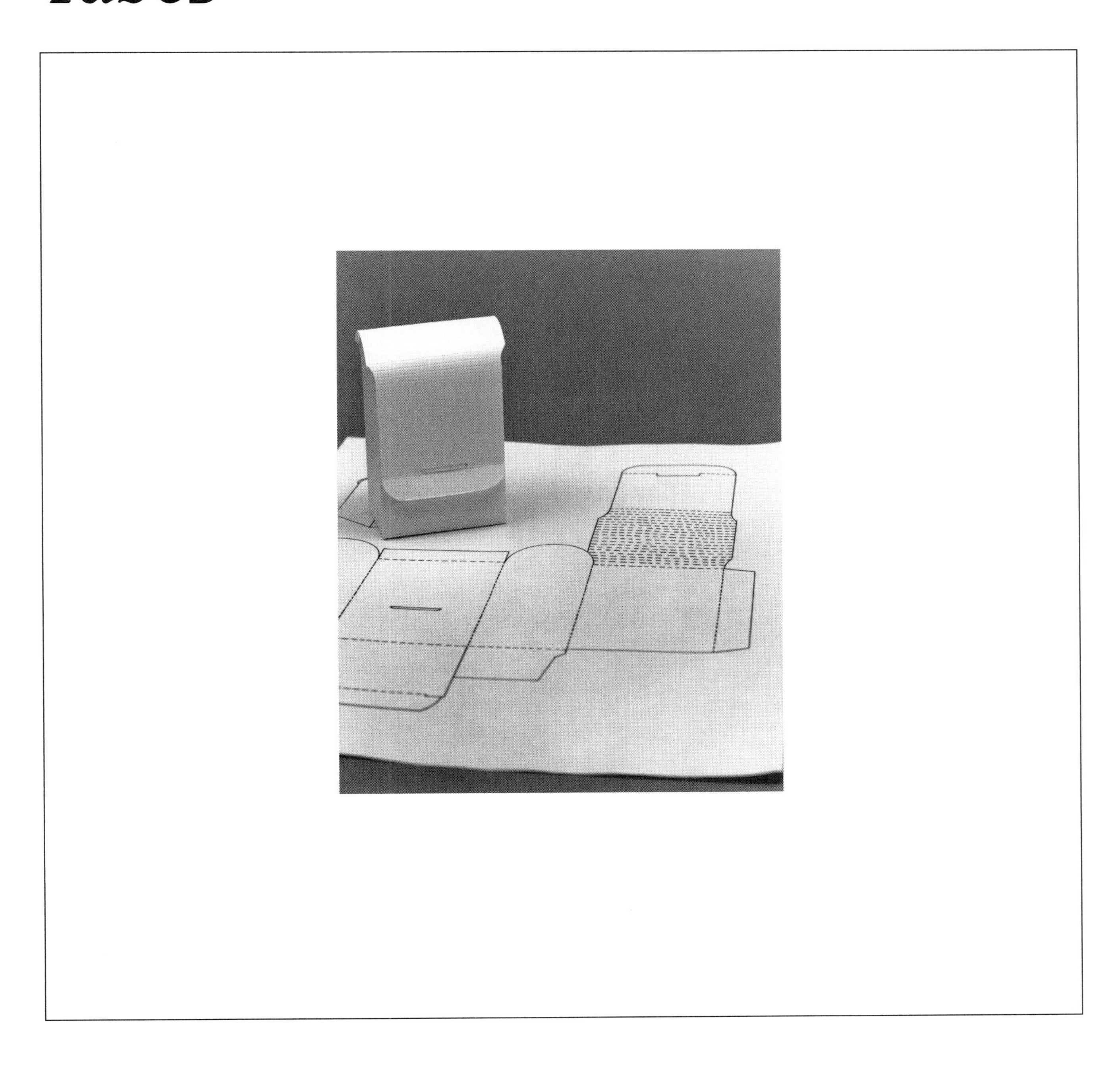

FULL OVERLAP SEAL END

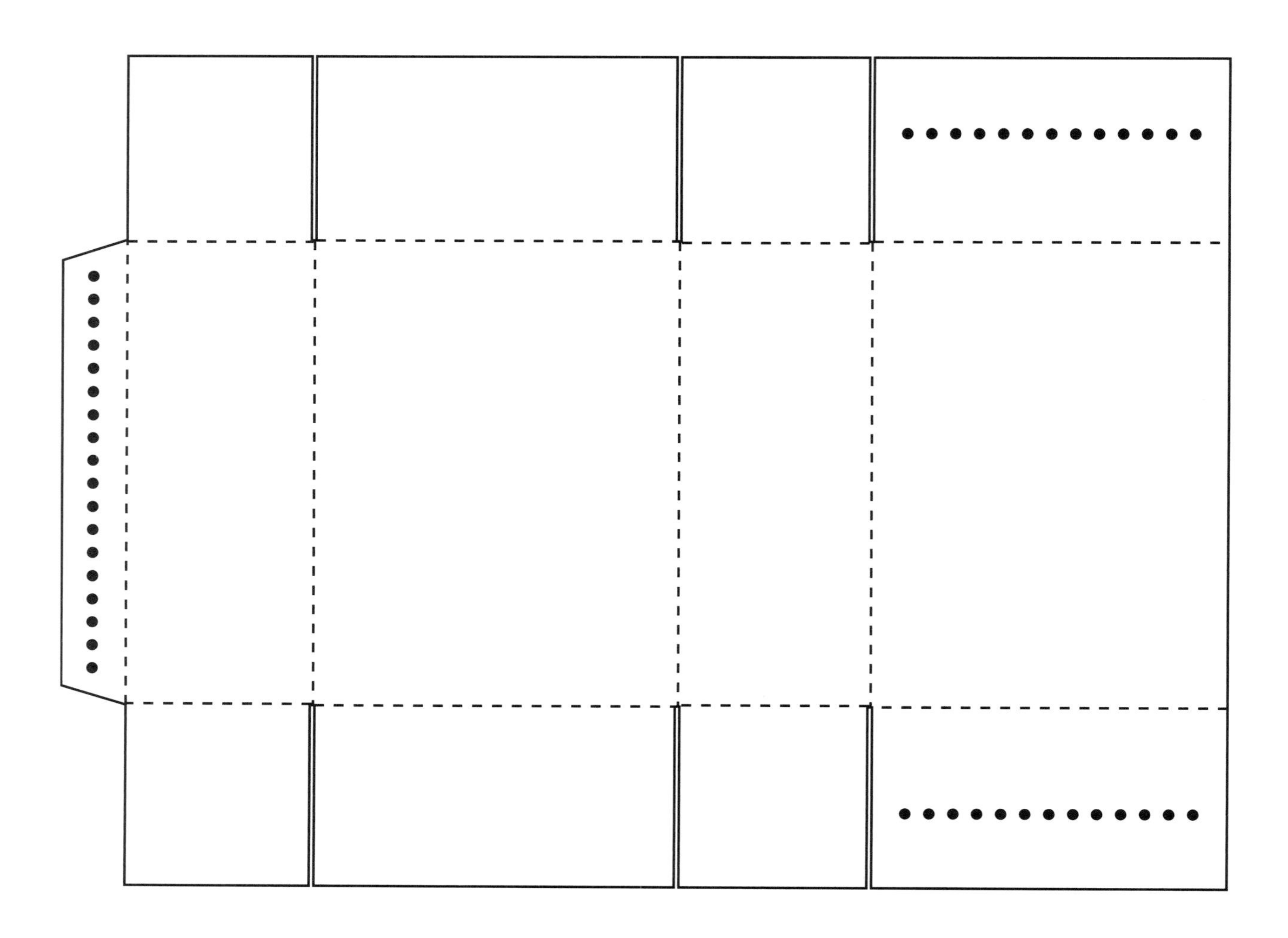

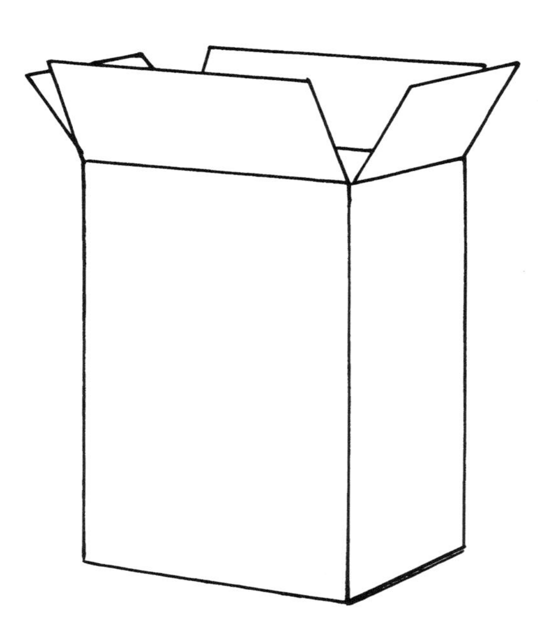

FULL OVERLAP SEAL END WITH VAN BUREN EARS AT BOTTOM

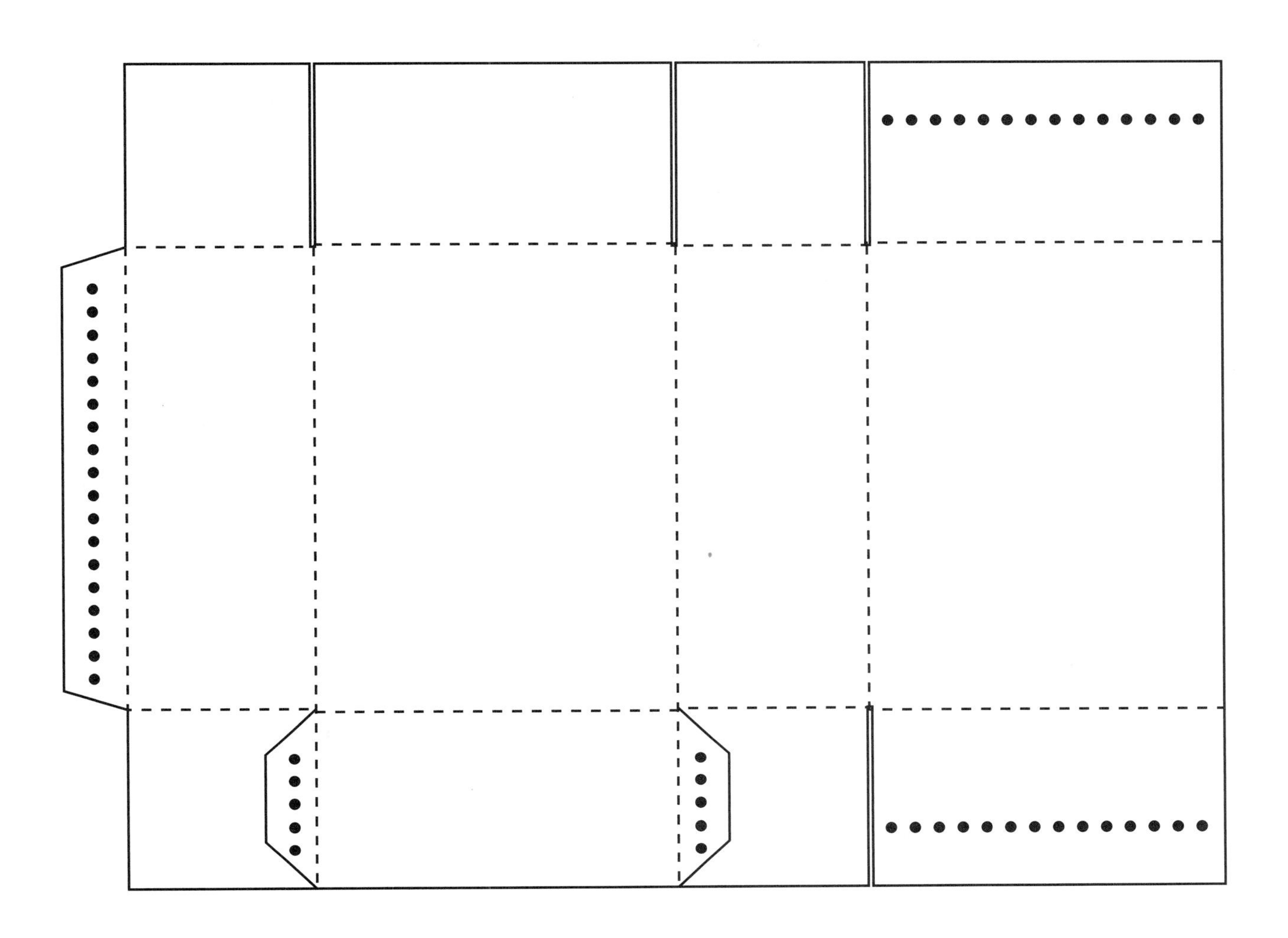

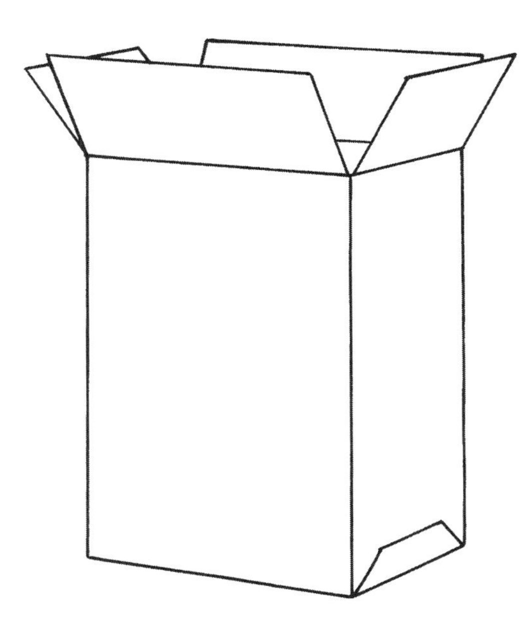

FULL OVERLAP SEAL END WITH PRE-PERFORATED POUR SPOUT

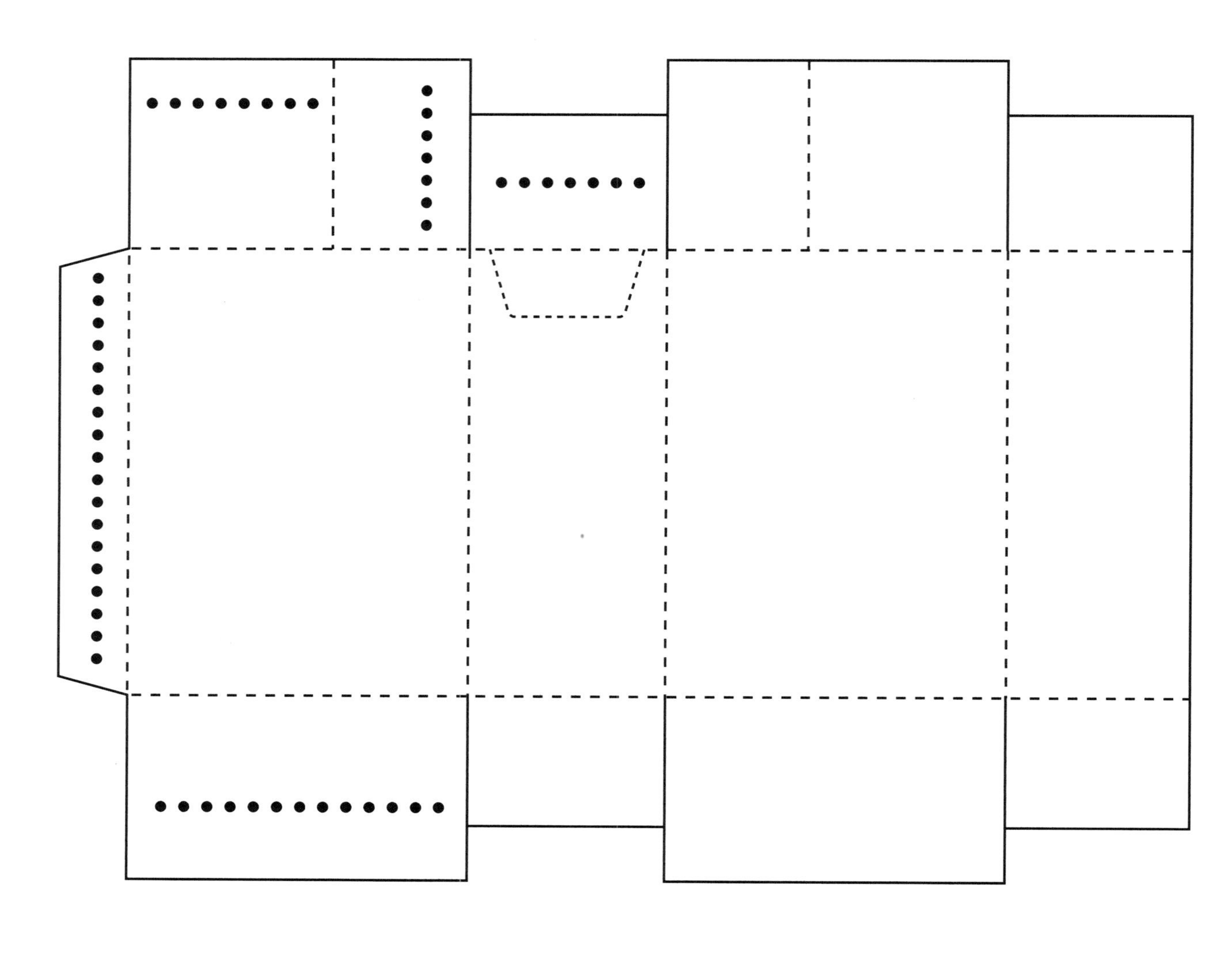

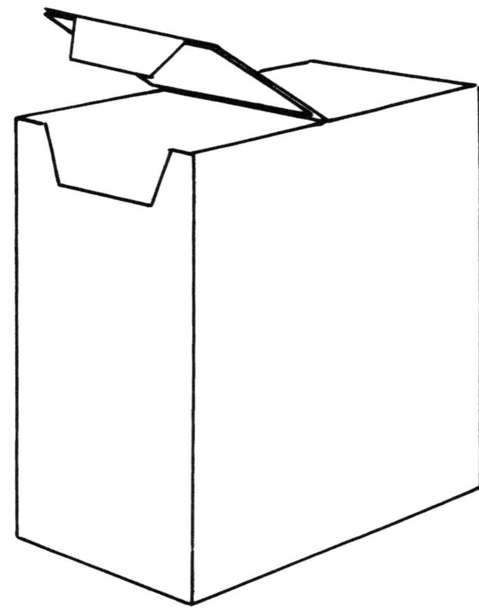

ECONOMY SEAL END

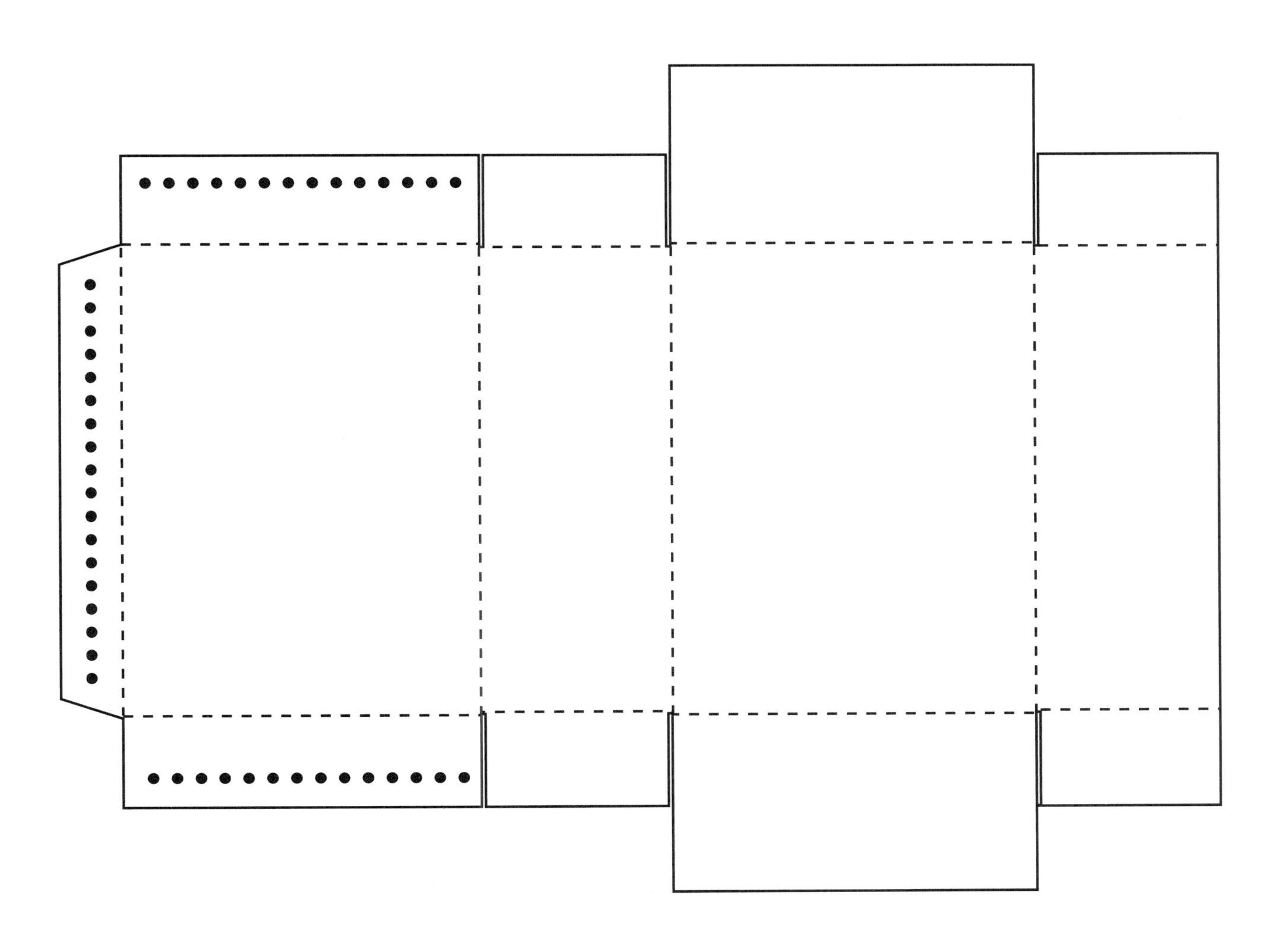

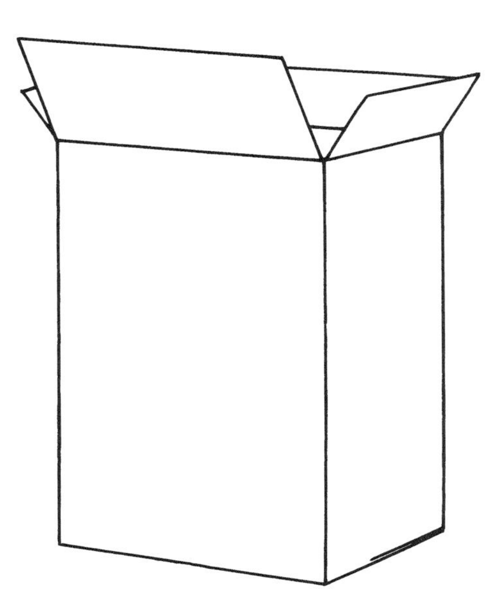

ECONOMY SEAL END (VARIATION)

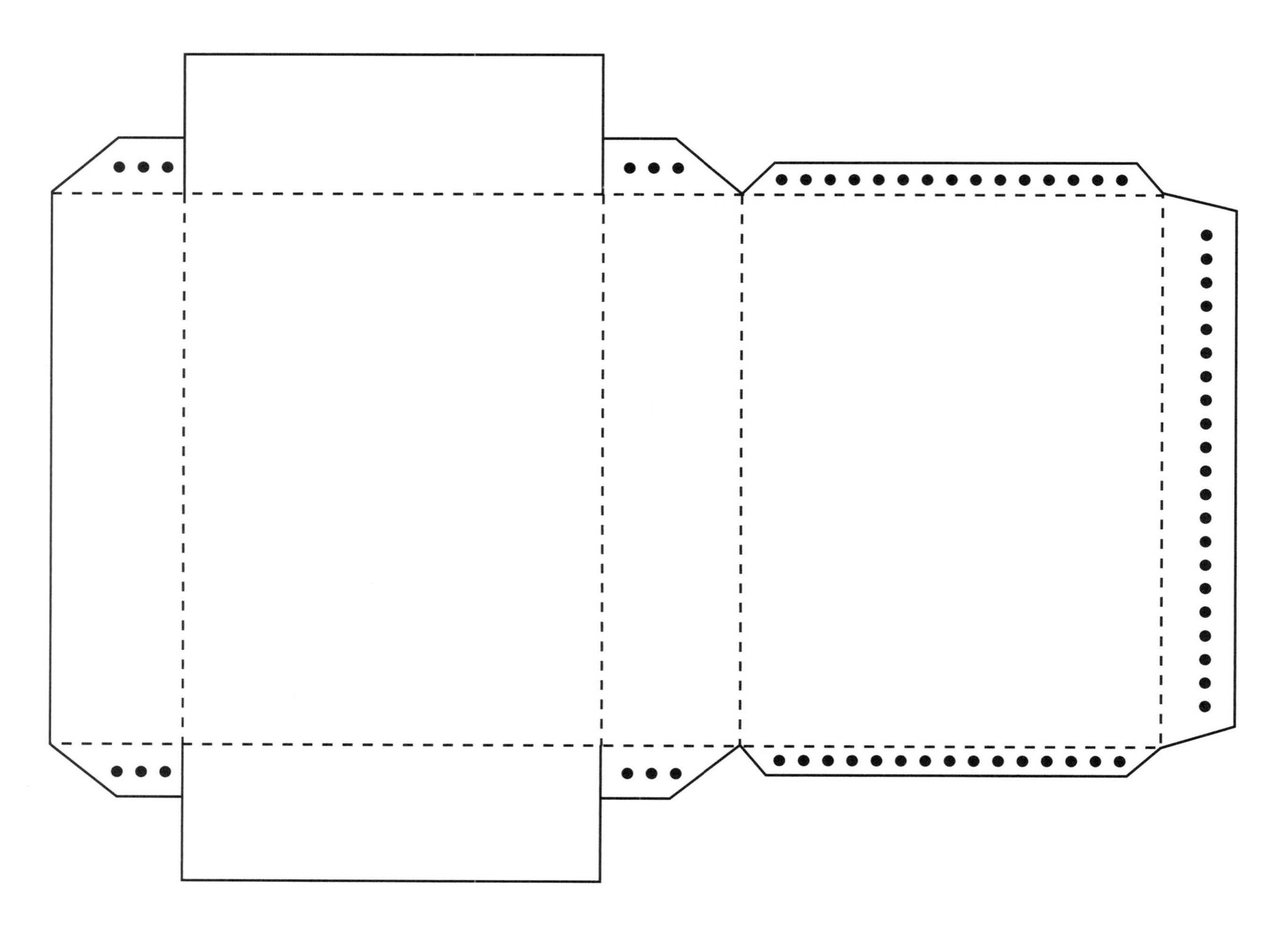

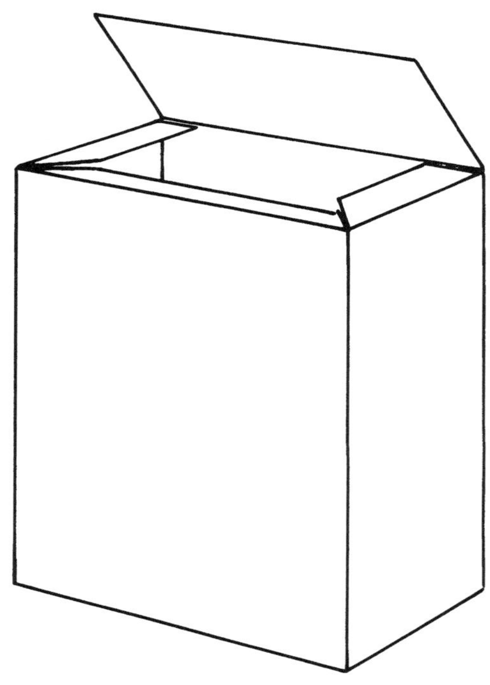

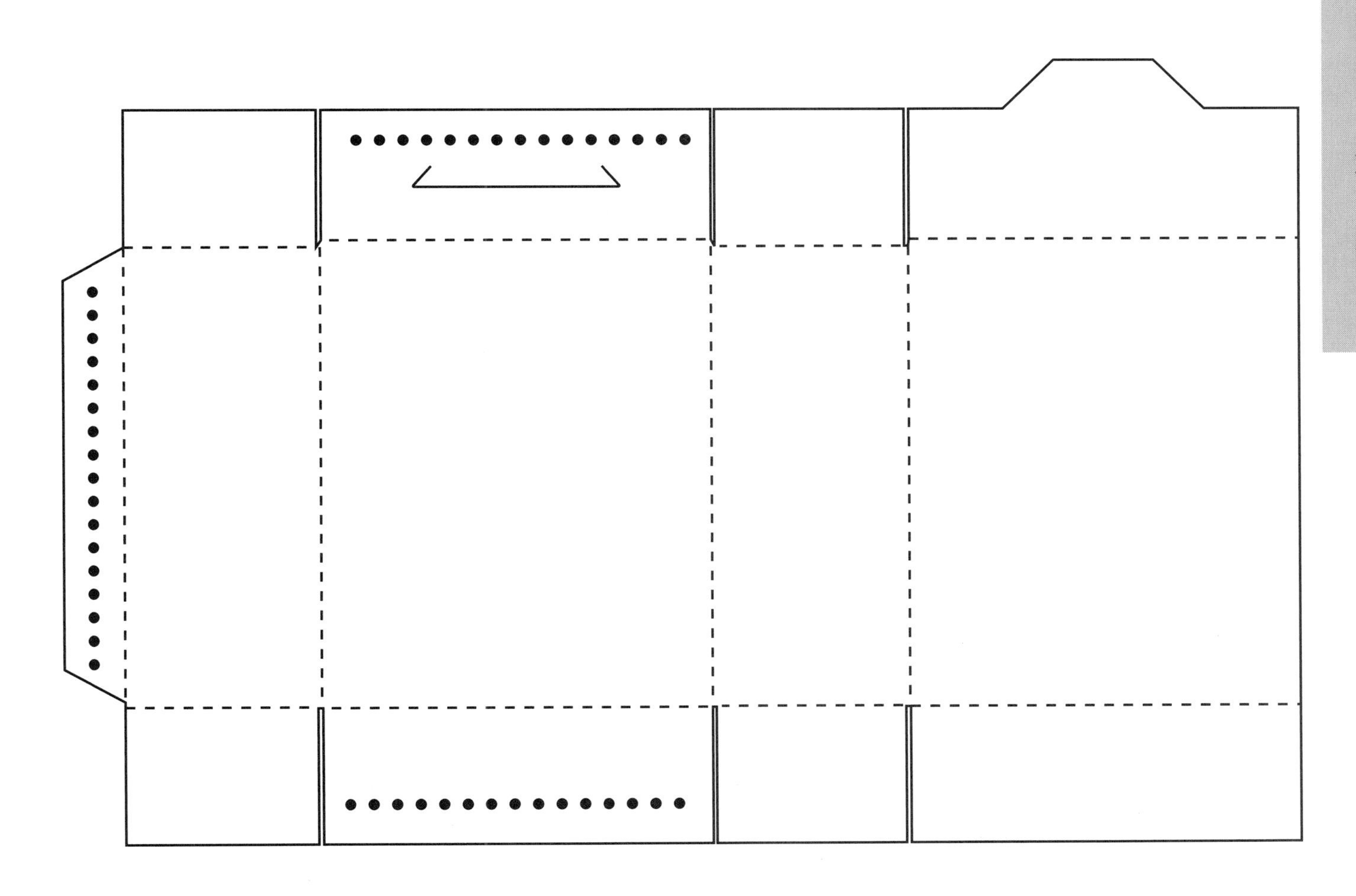

STRAIGHT TUCK WITH SLIT-LOCK TUCKS

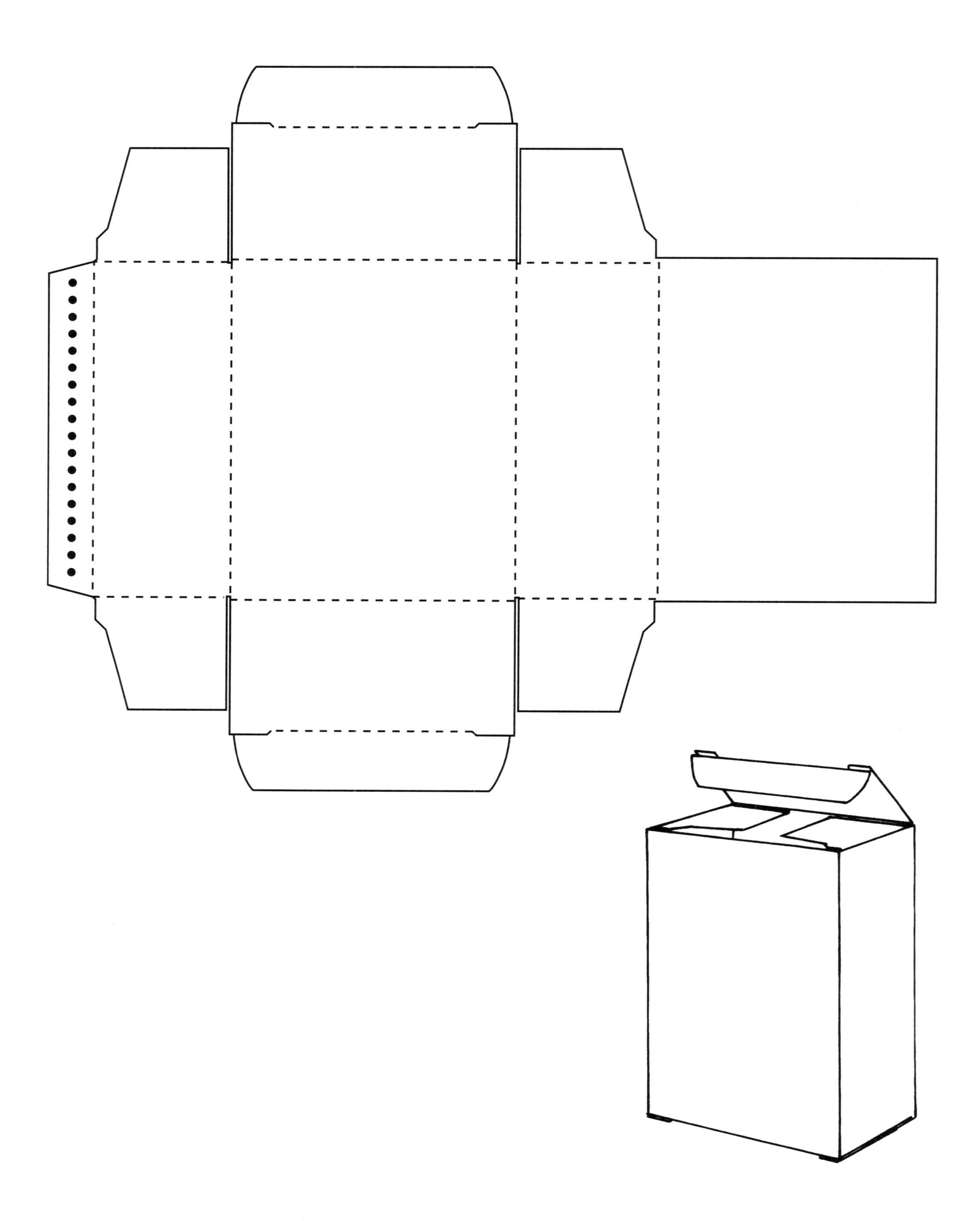

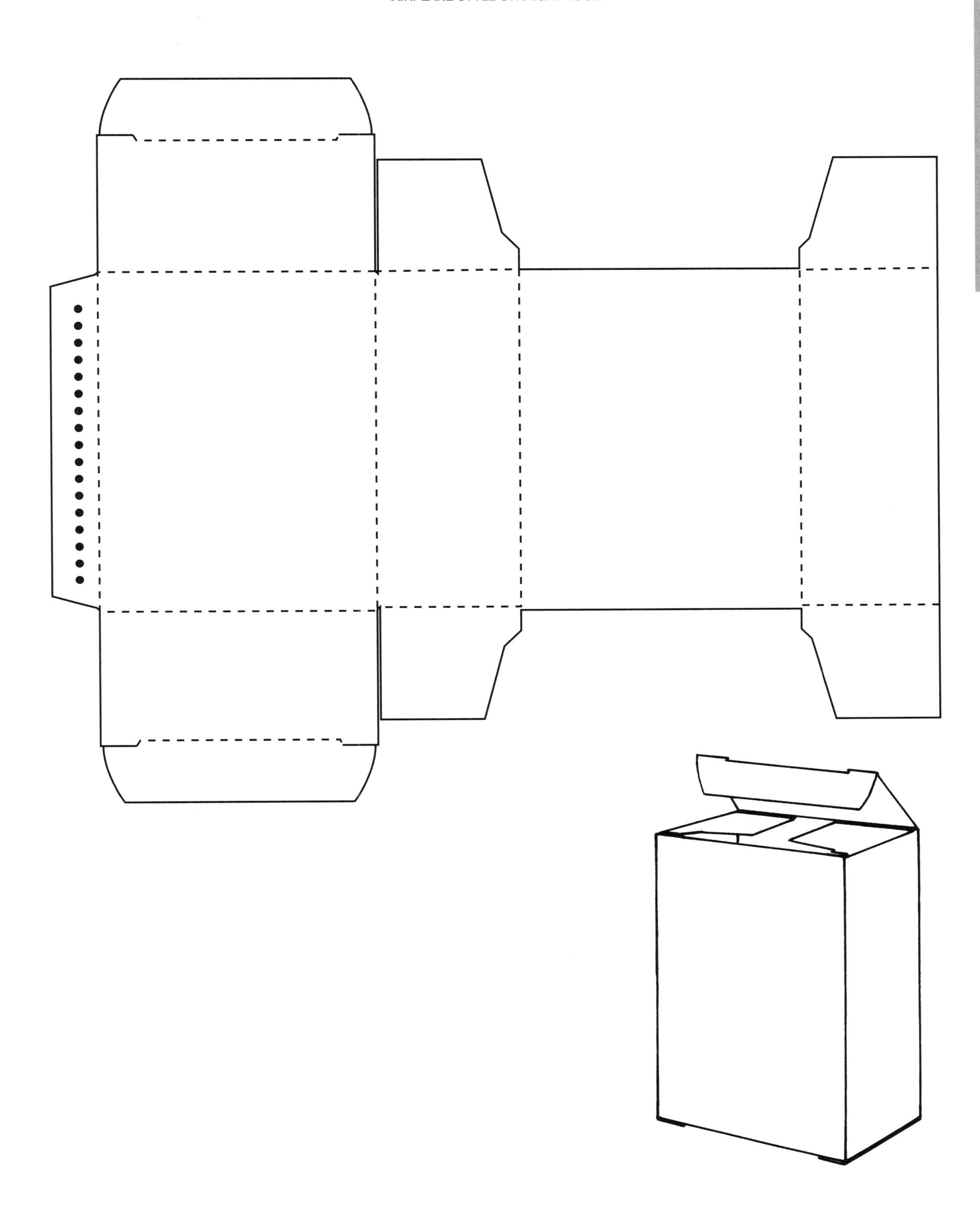

STANDARD REVERSE TUCK

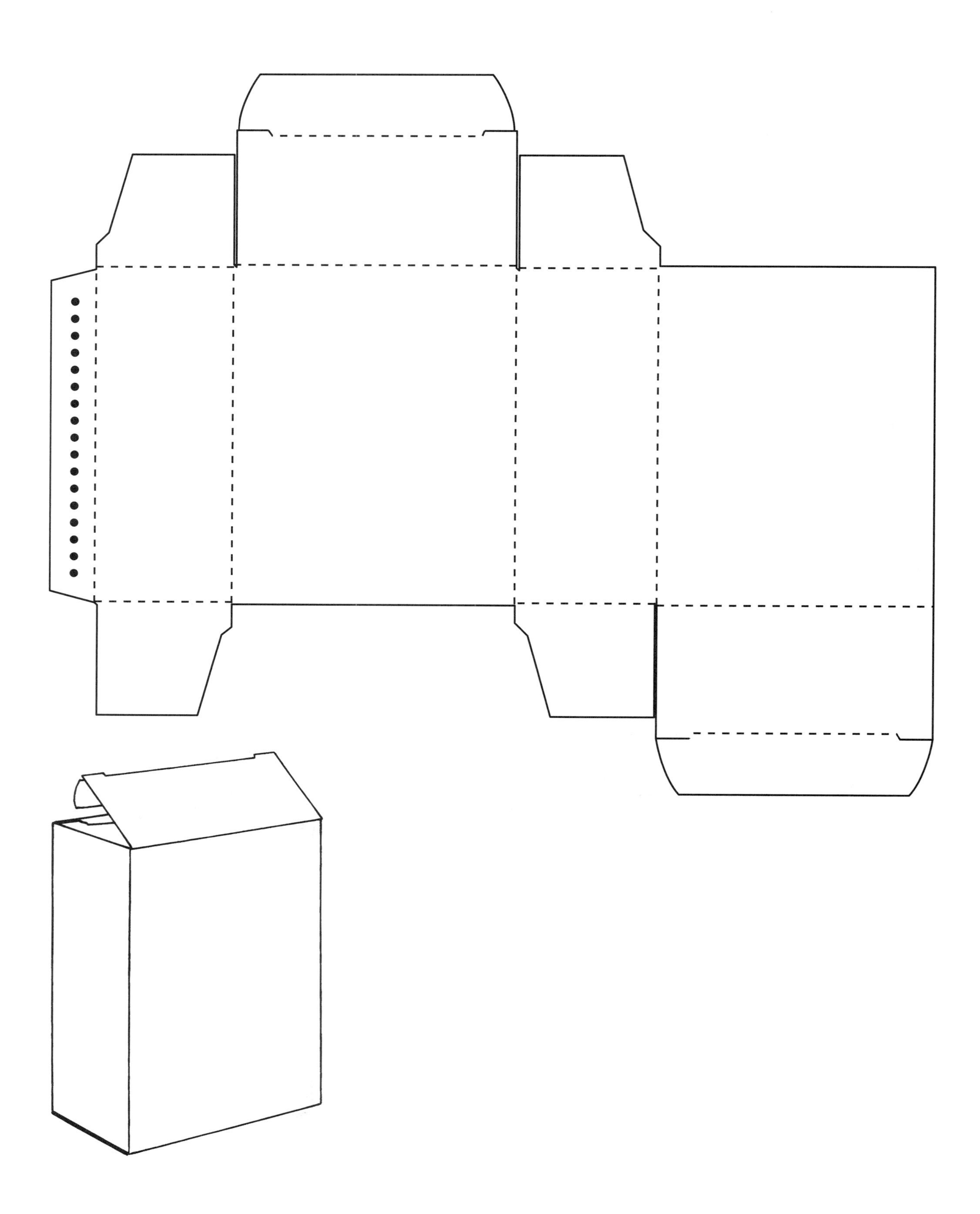

FRENCH REVERSE TUCK

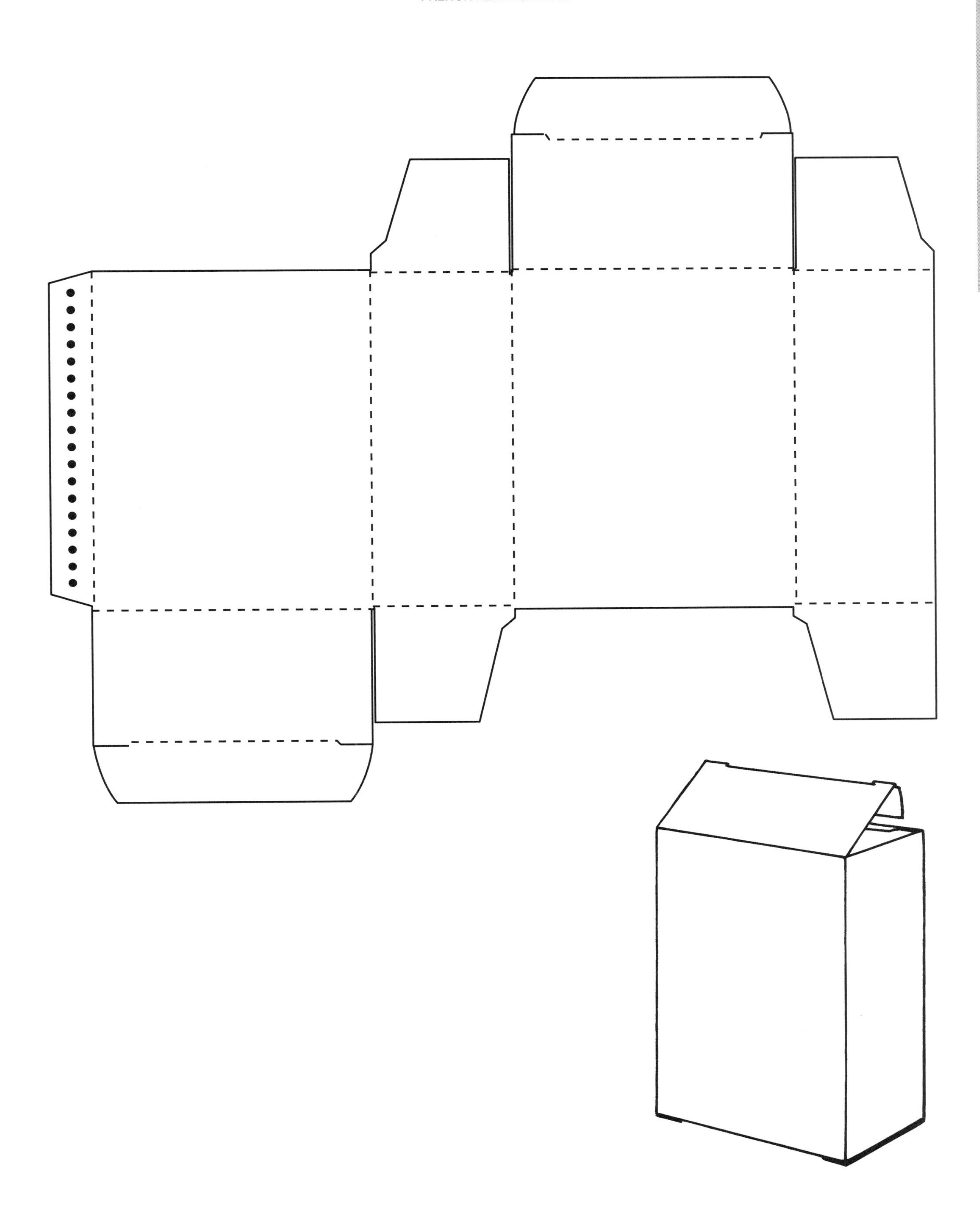

TUCK AND TONGUE

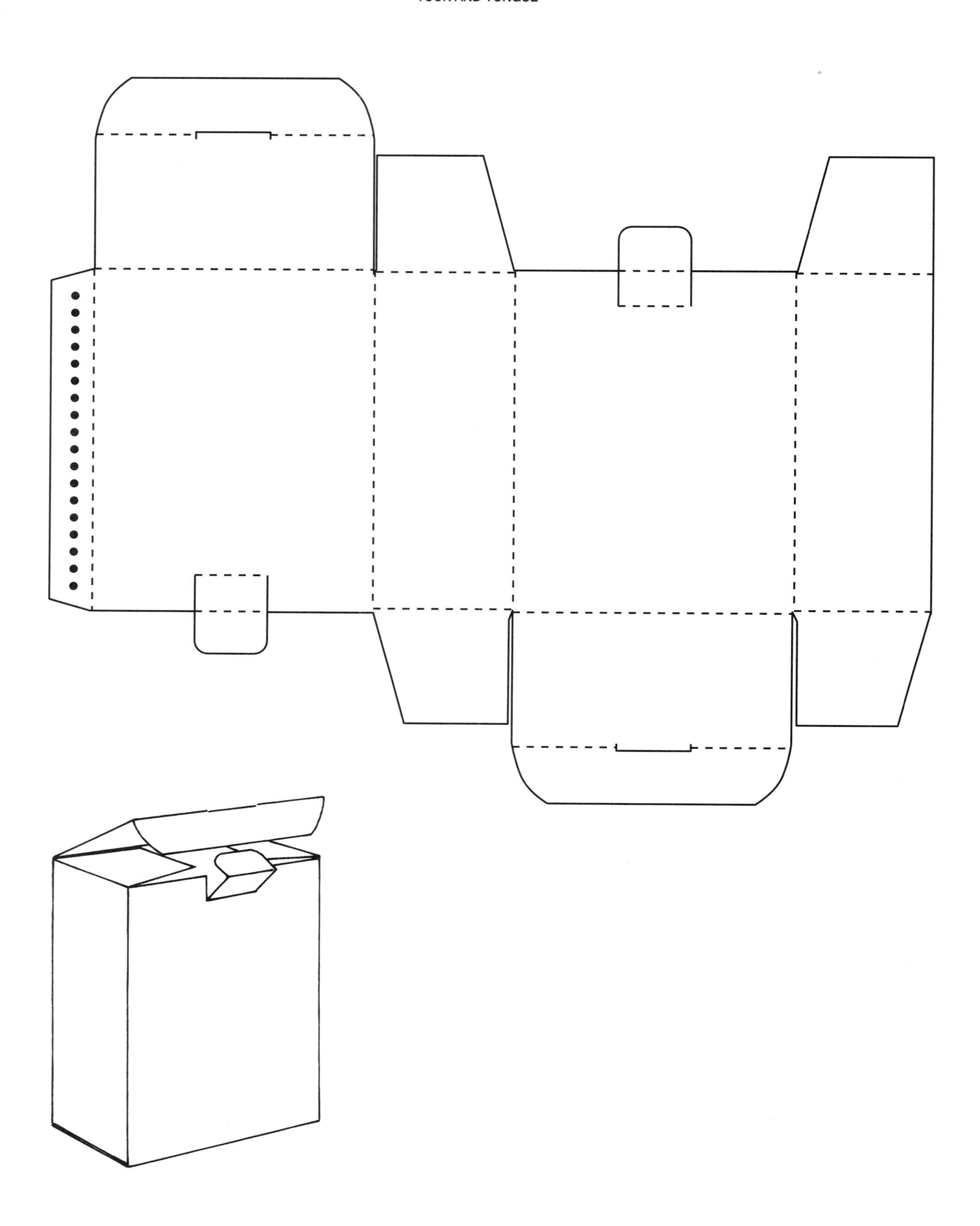

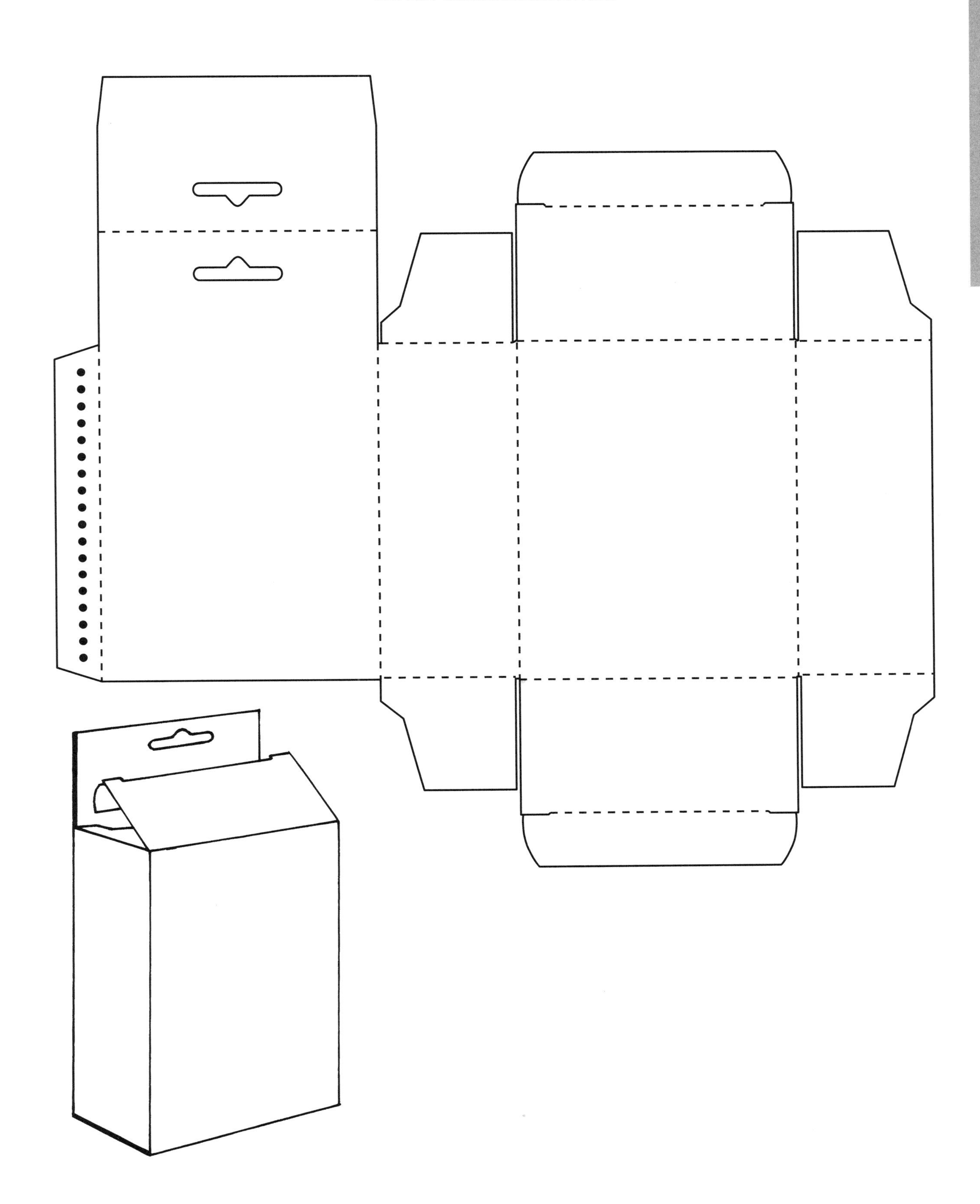

MAILER LOCK

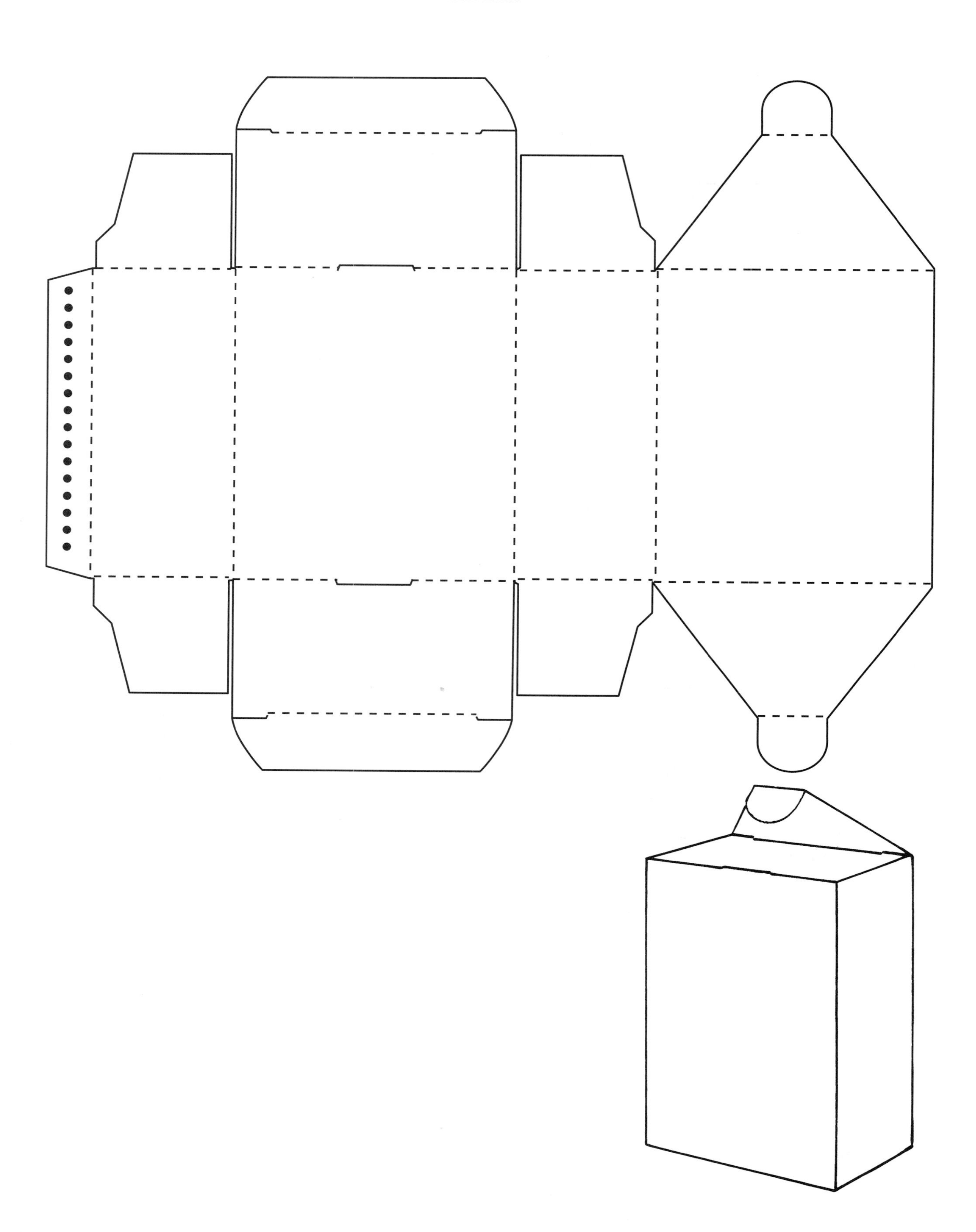

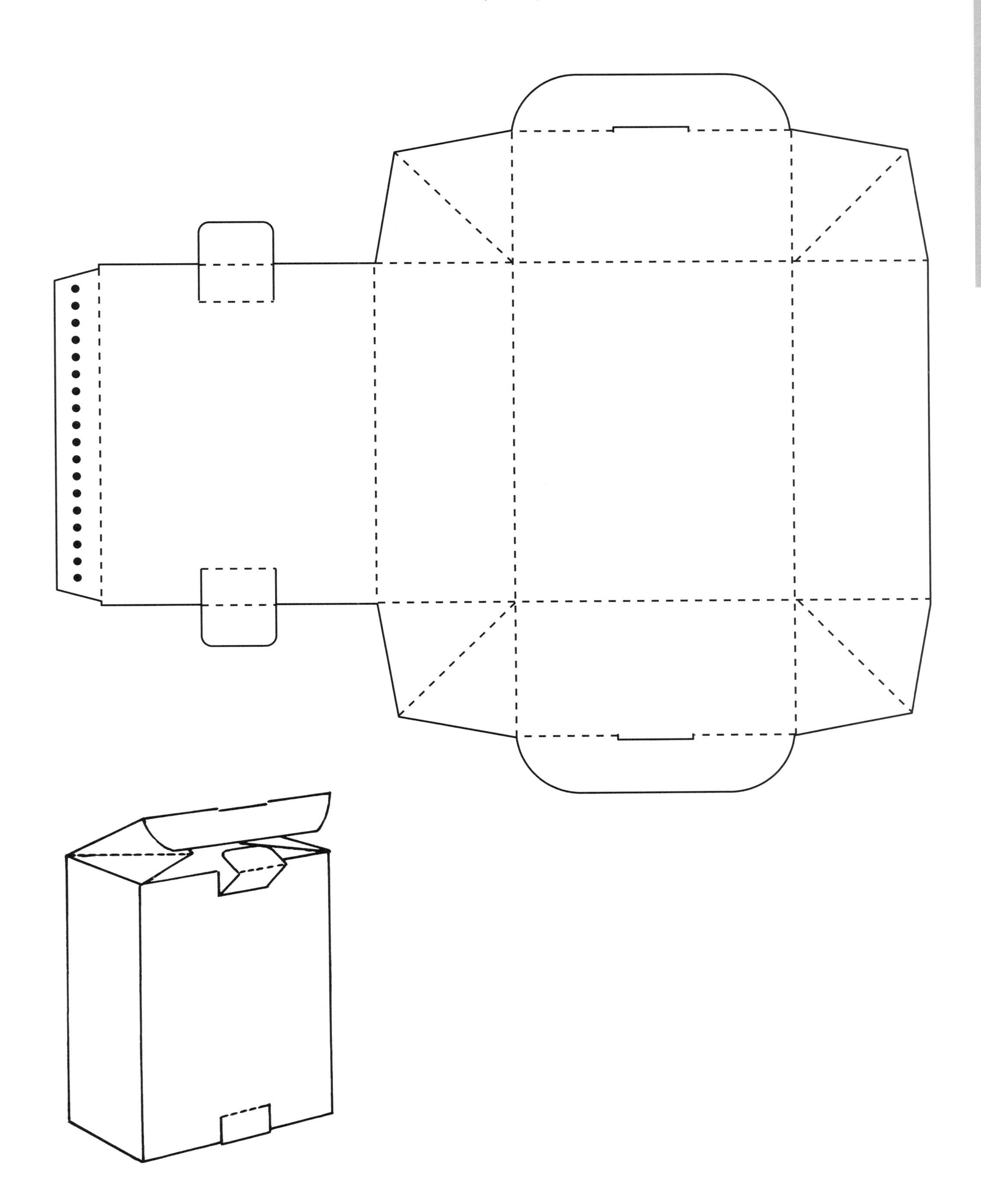

BELLOWS TUCK TOP, HIMES LOCK AUTOMATIC BOTTOM
(ALSO KNOWN AS CRASH LOCK)

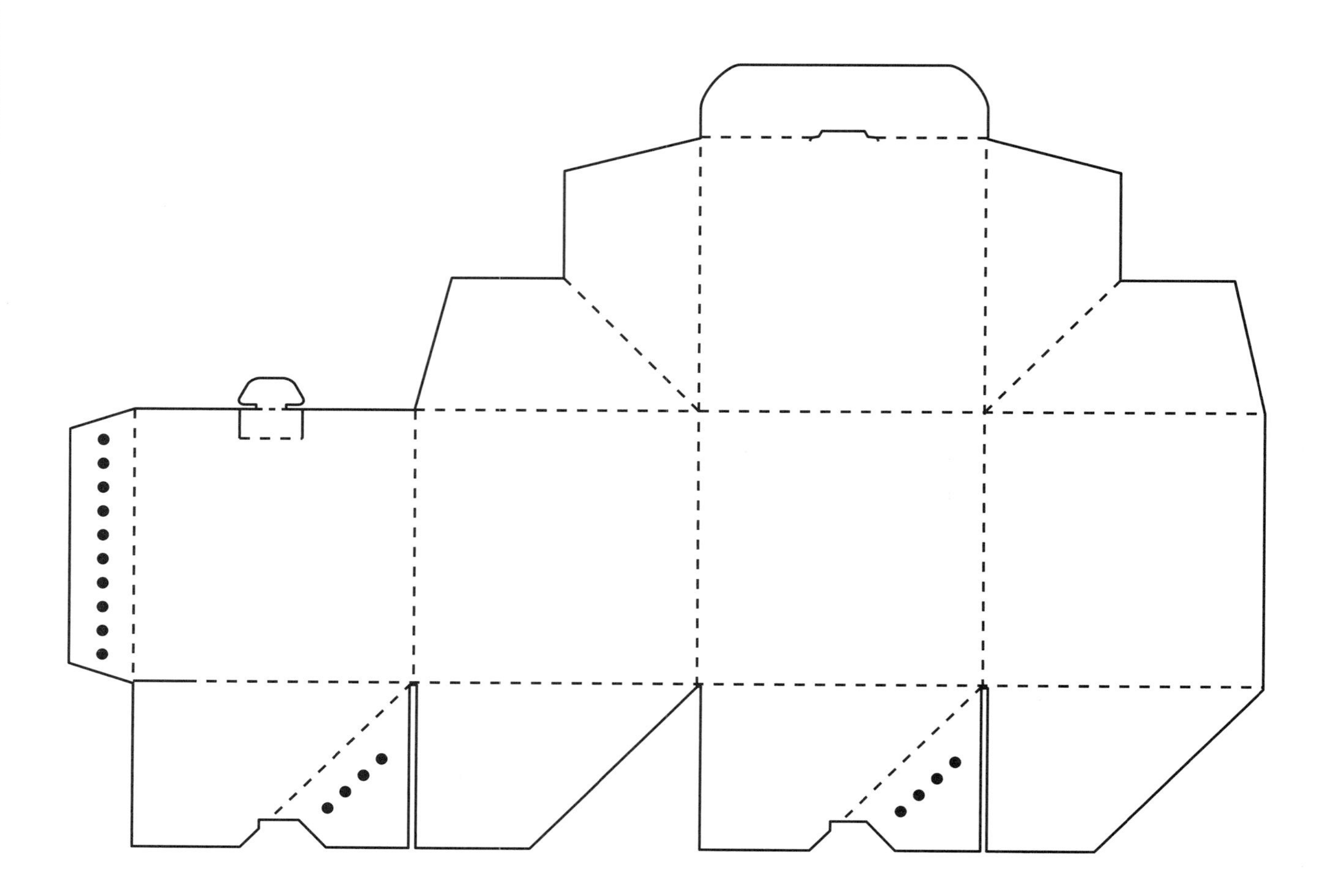

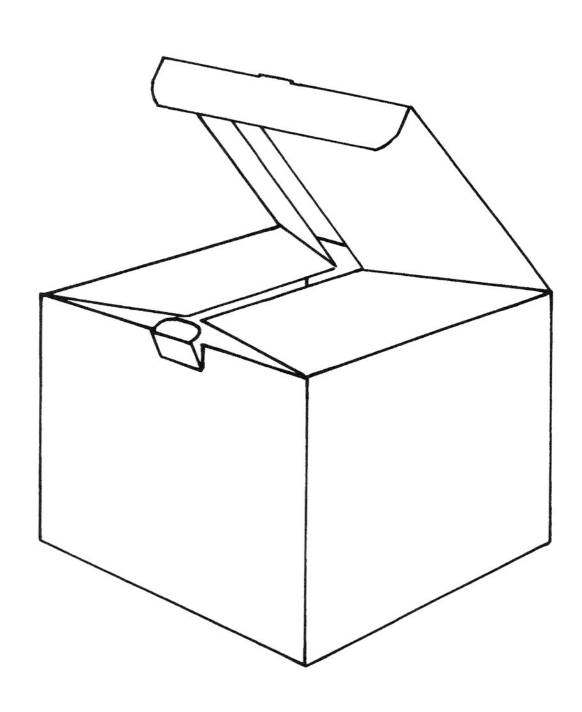

DOUBLE LOCK TAB TOP BELLOWS TUCK BOTTOM

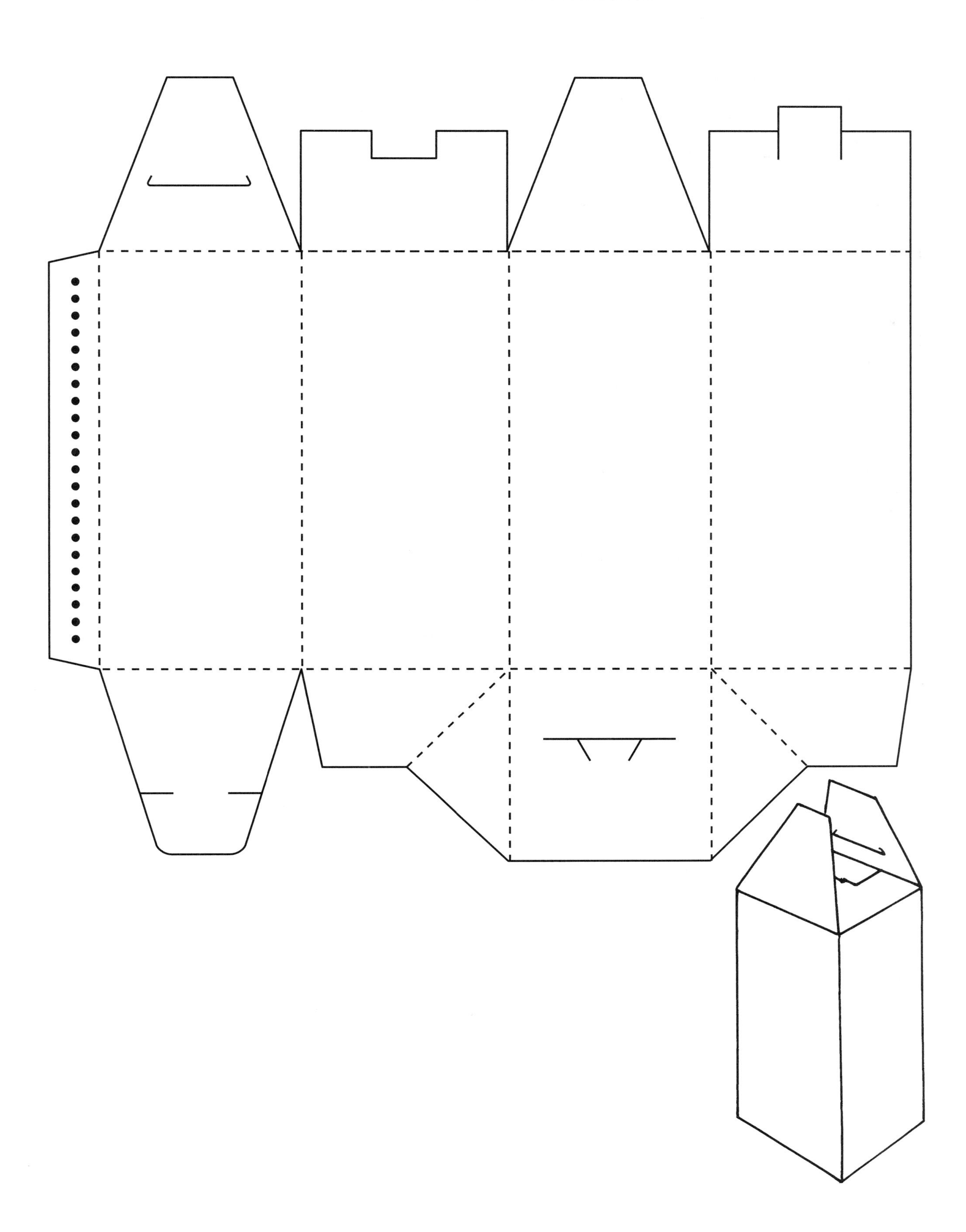

OPEN THROAT WITH TUCK LOCKS

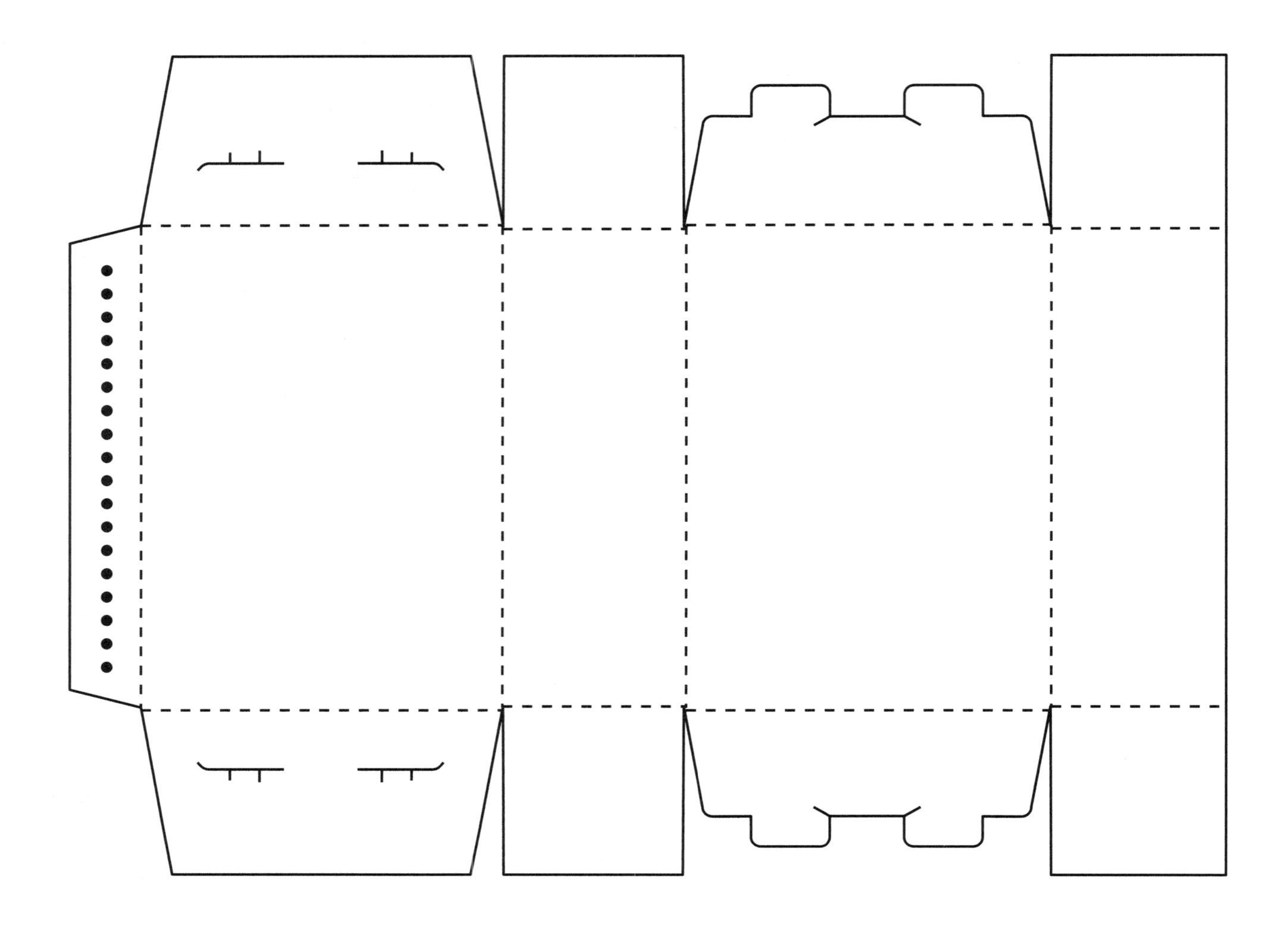

EDGE LOCK OR FROST LOCK

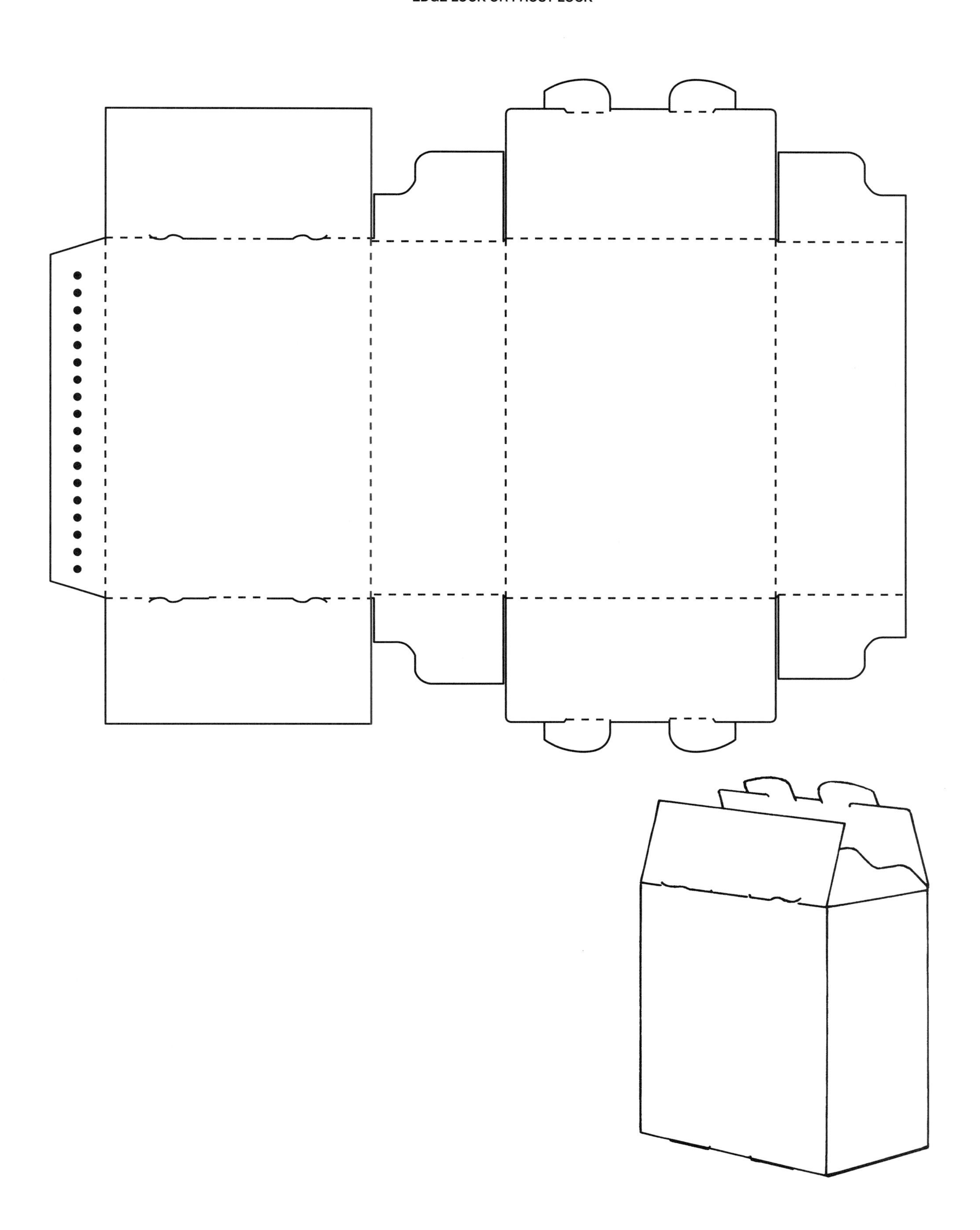

STRAIGHT TUCK WITH DRUGSTORE LOCK

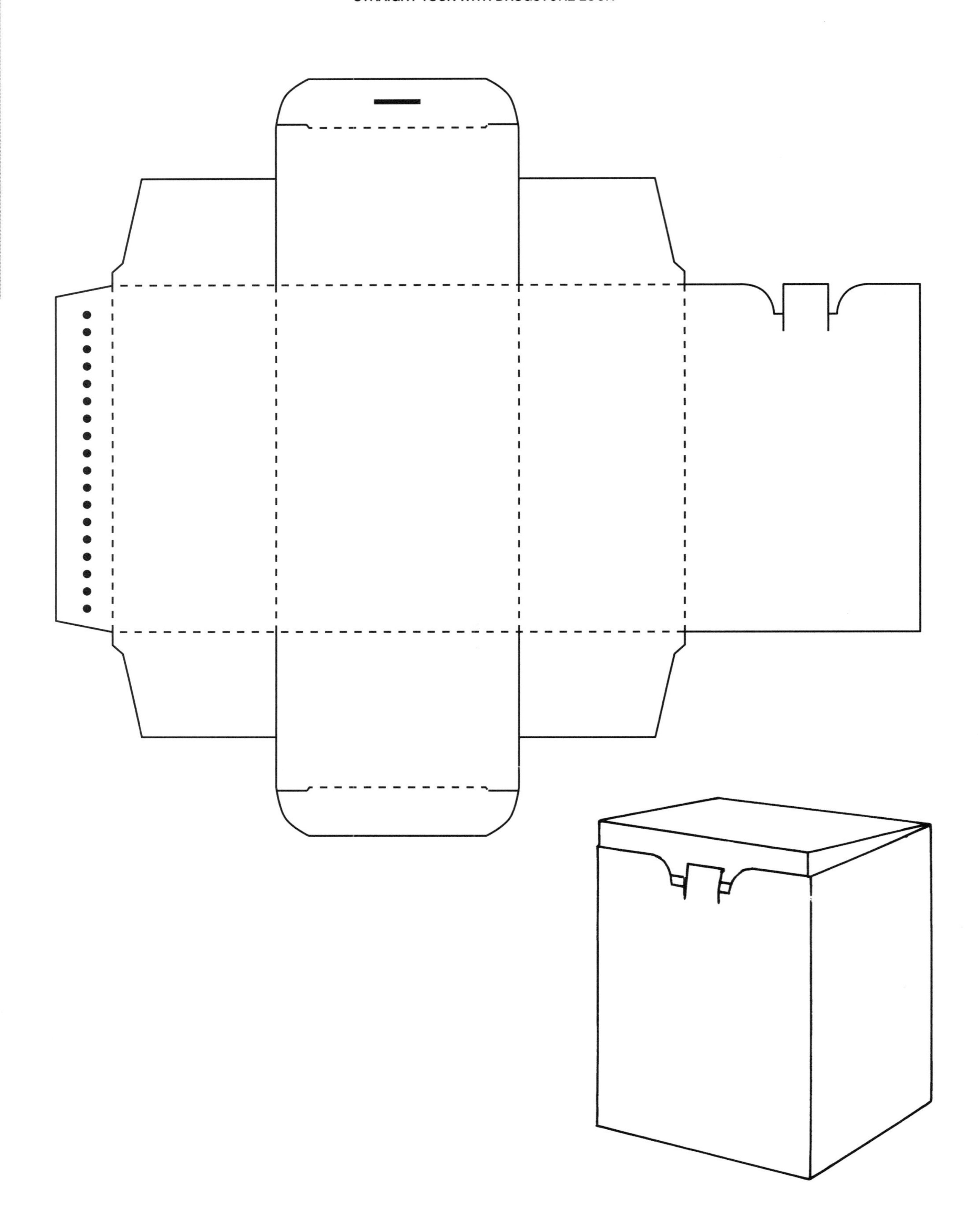

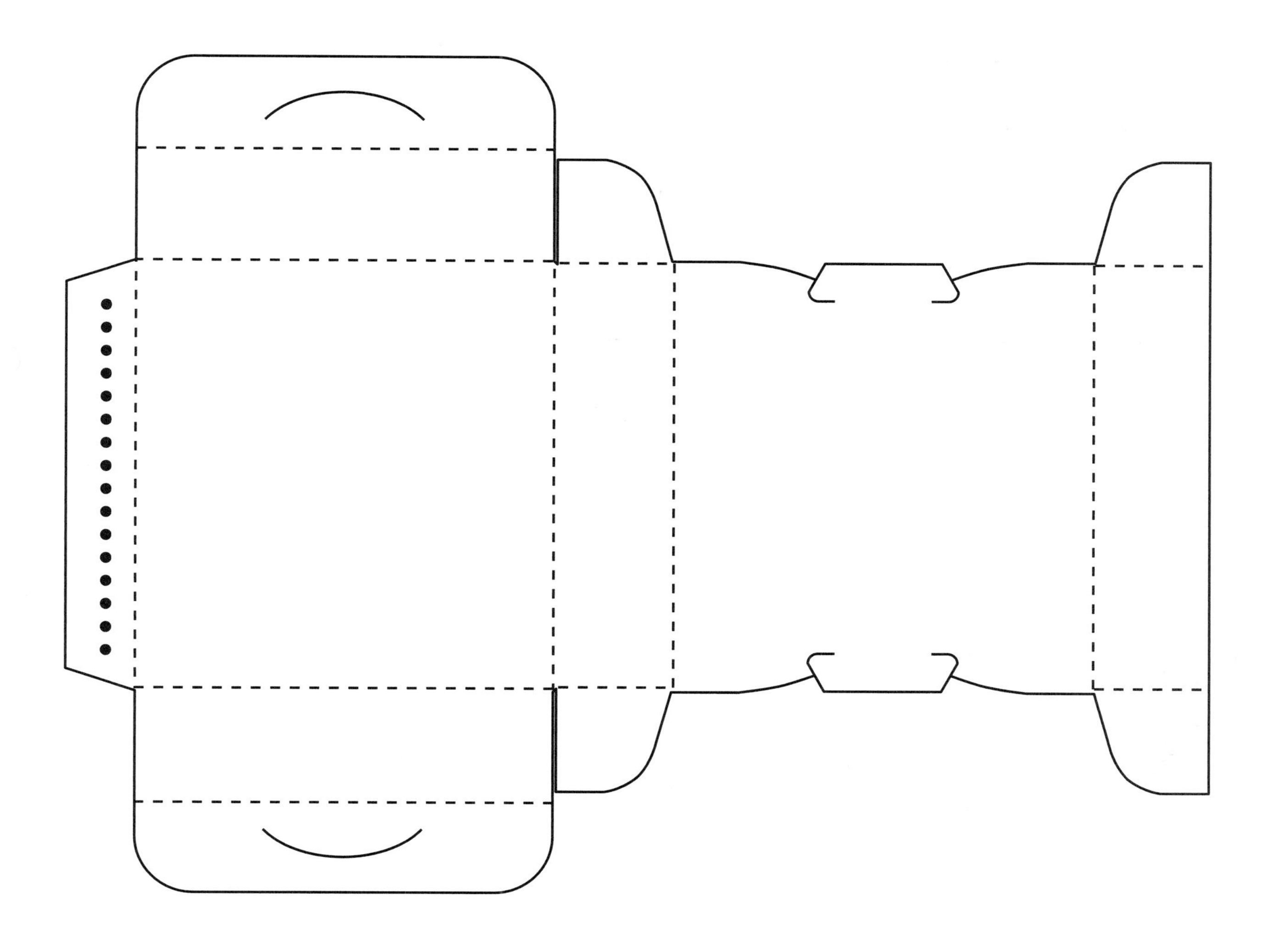

FRICTION TUCK TOP, SNAP-LOCK (OR HOUGHLAND) BOTTOM

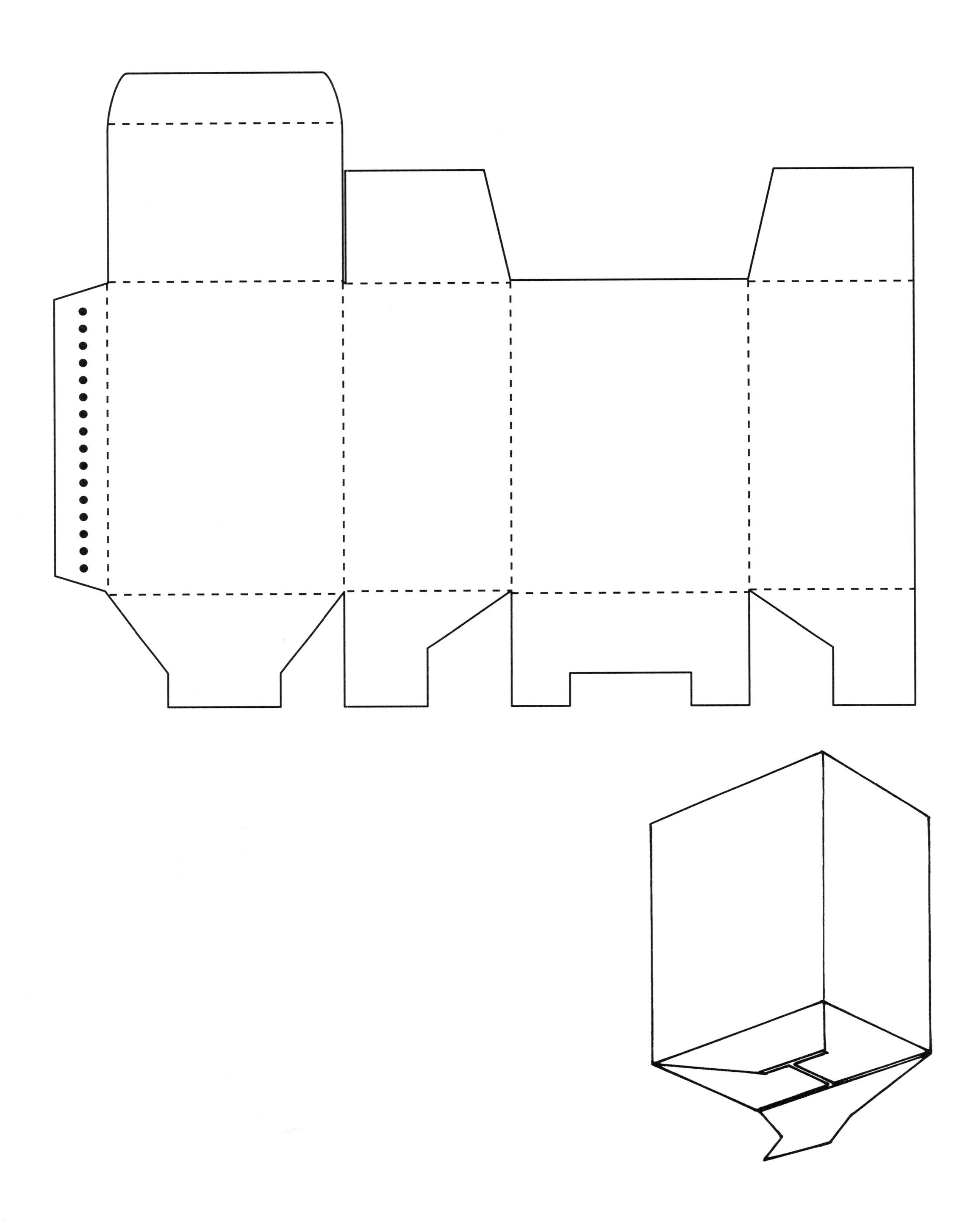

REINFORCED SNAP-LOCK (OR HOUGHLAND) BOTTOM

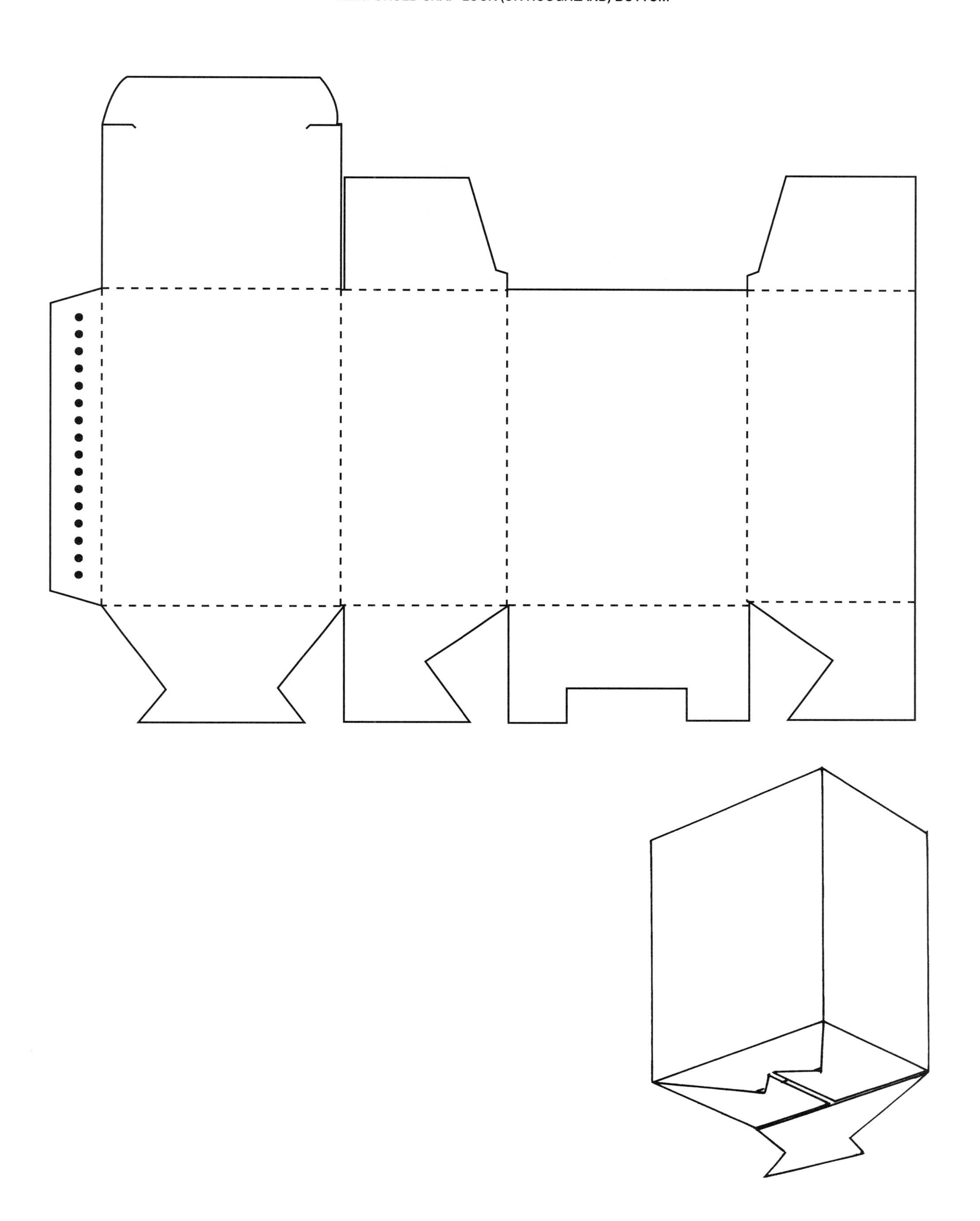

COMBINATION ARTHUR LOCK WITH REVERSE TUCK

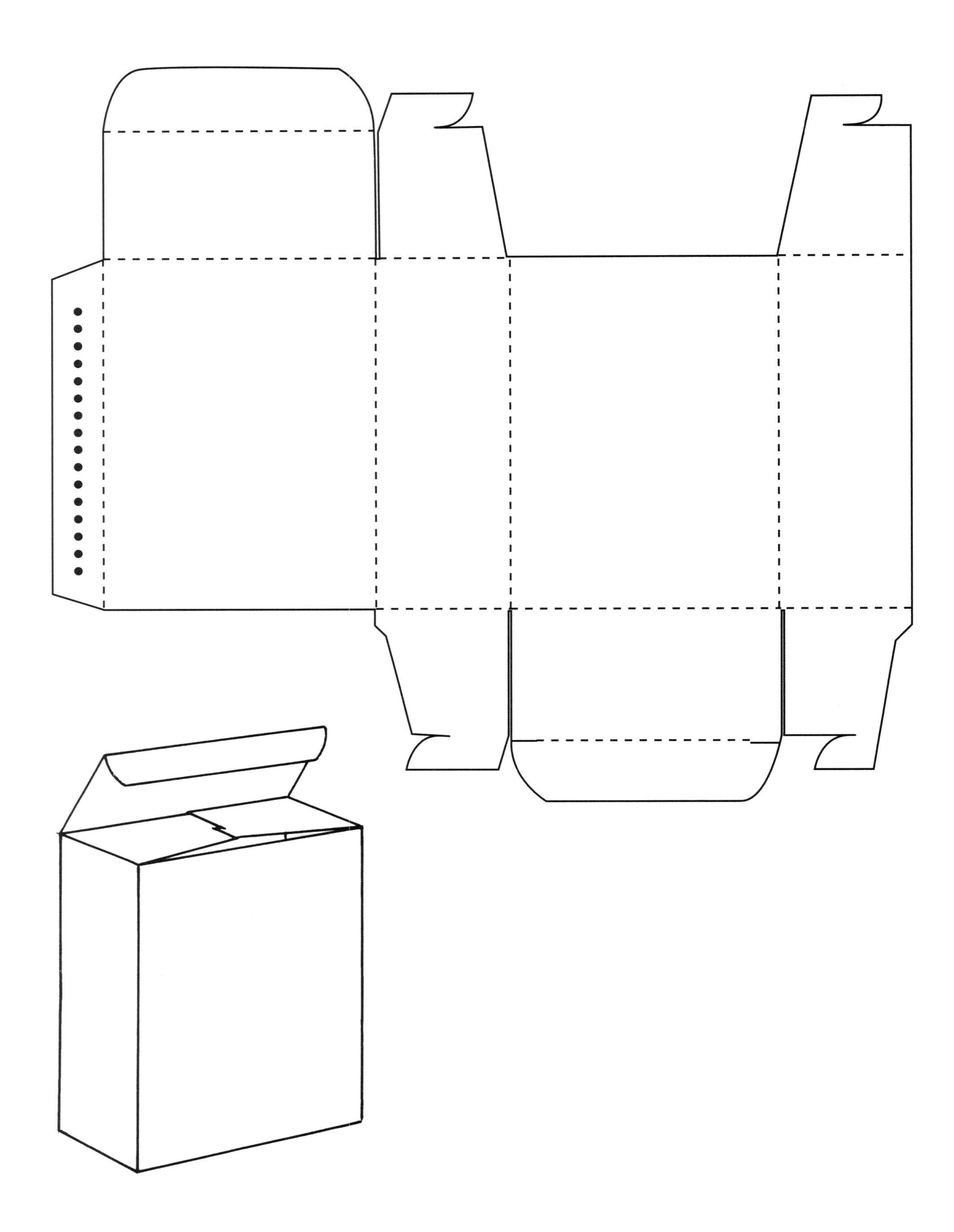

QUAD LOCK-EAR HOOK DOUBLE LOCK BOTTOM

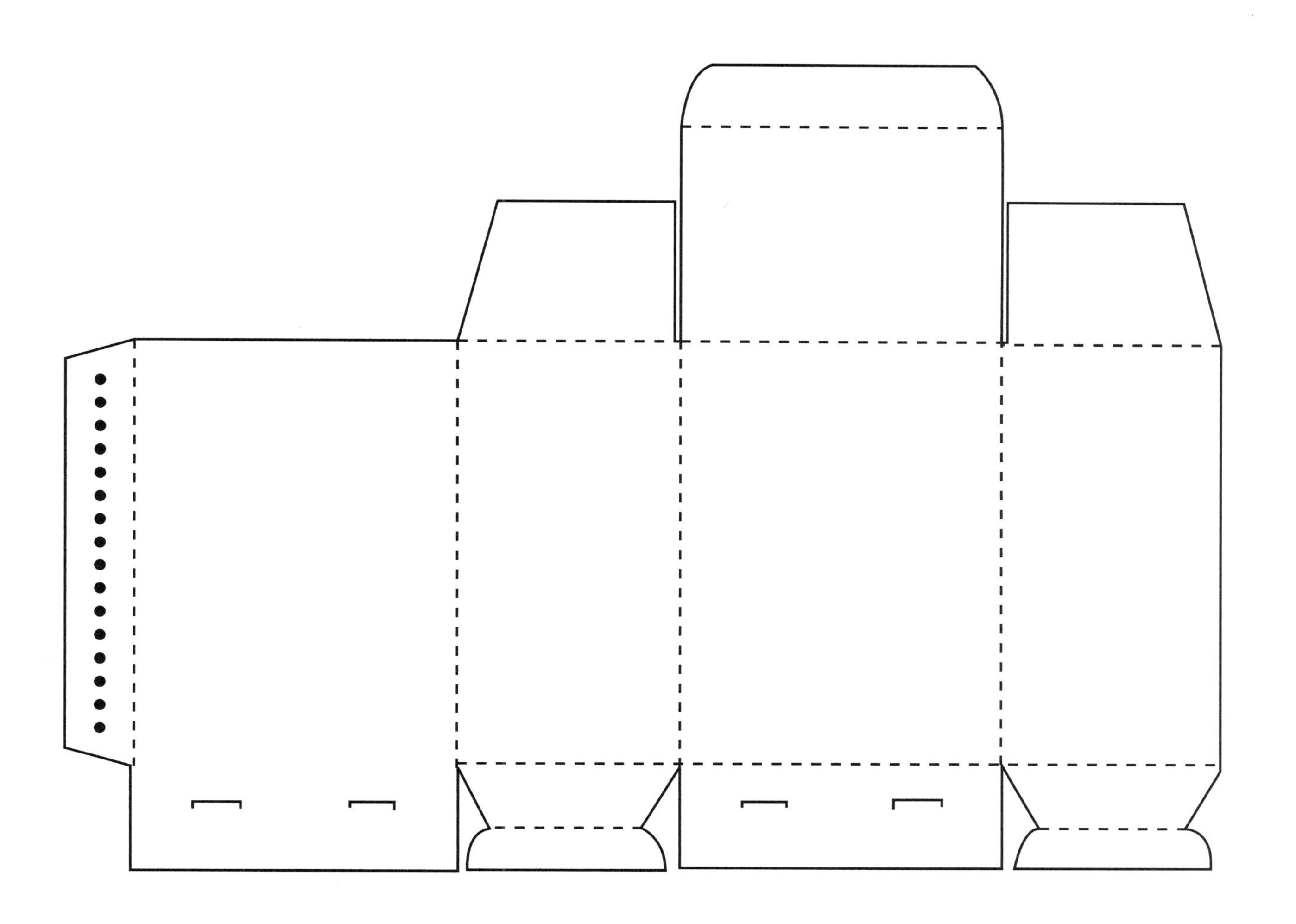

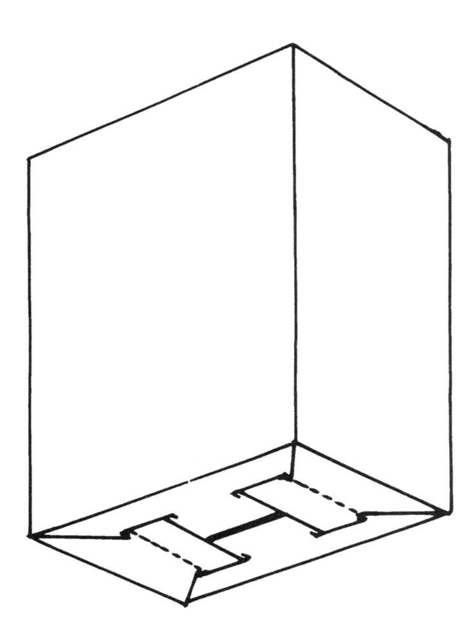

EAR HOOK BOTTOM VARIATION

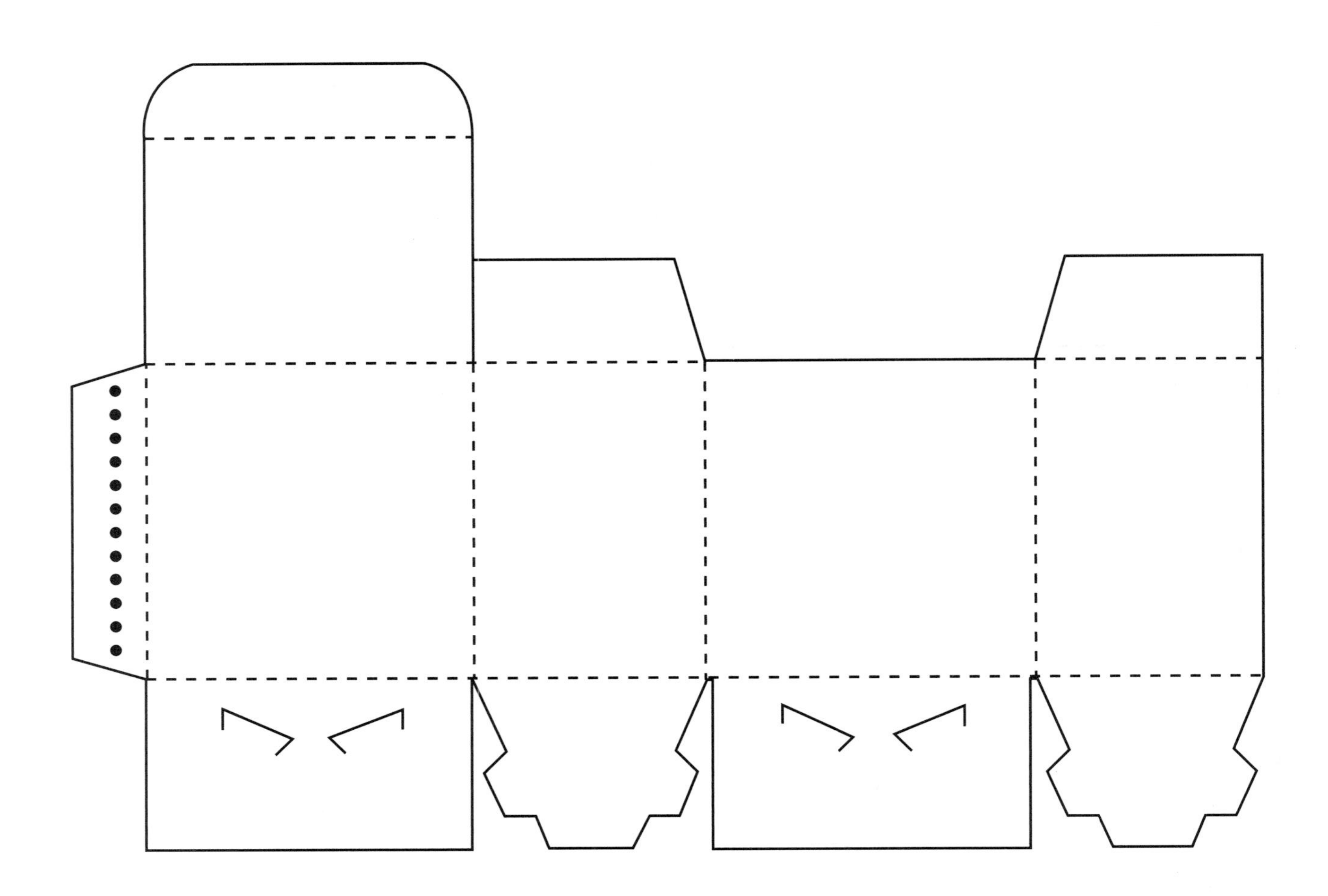

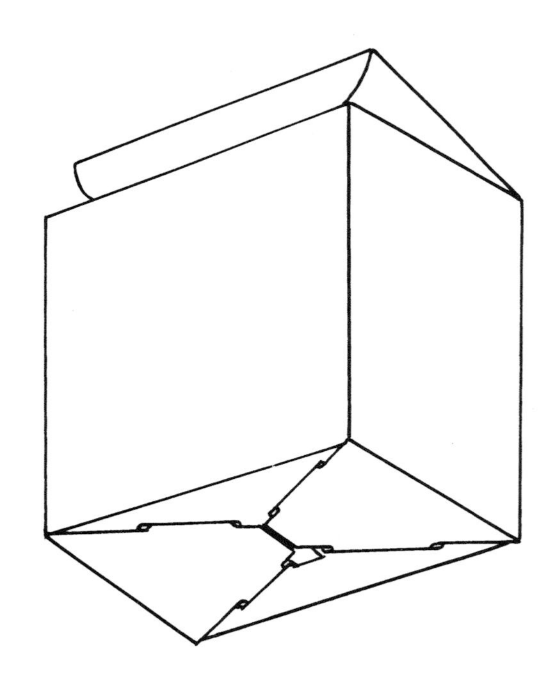

FULL FLAP AUTOMATIC BOTTOM

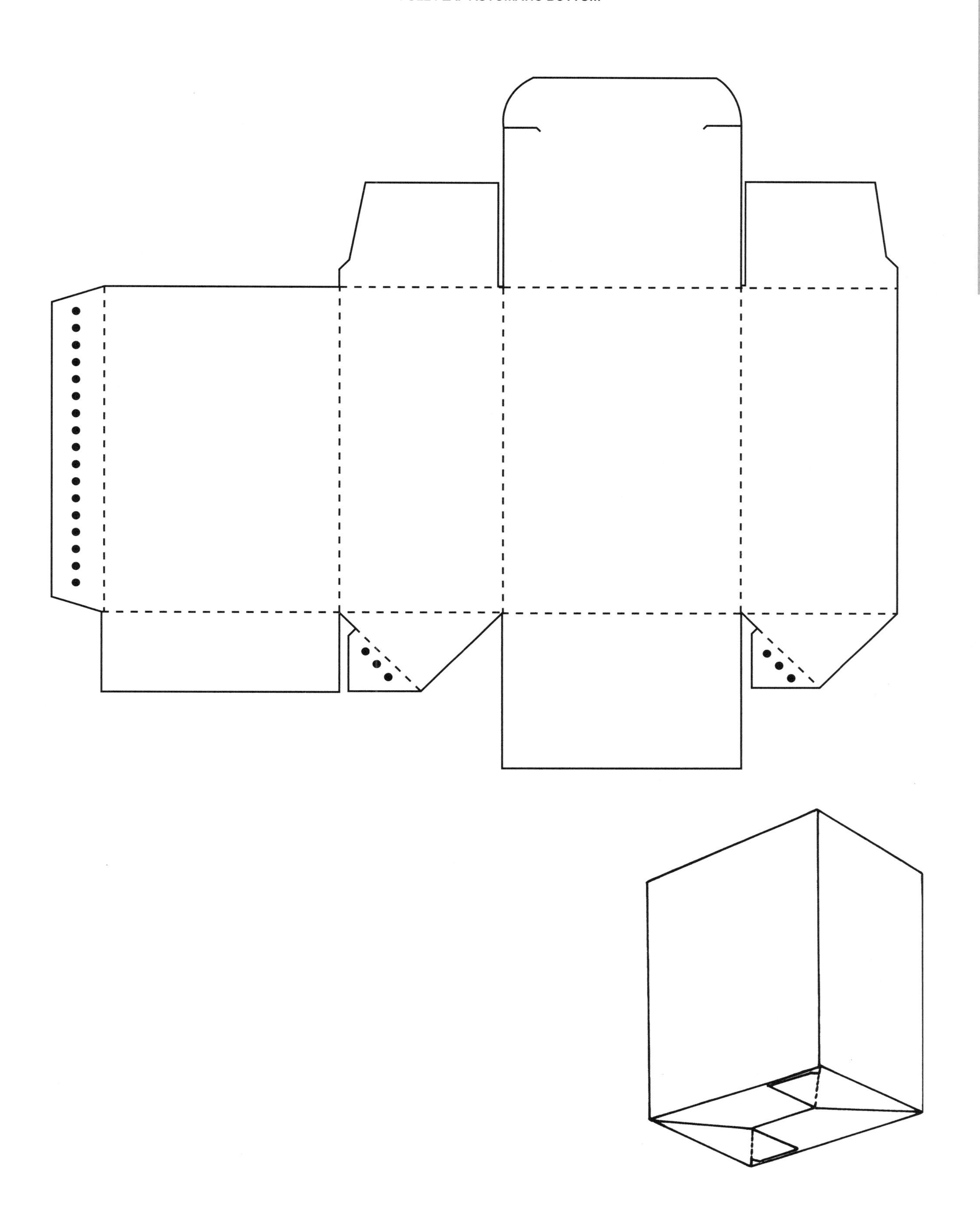

HIMES LOCK (POPCORN) BOTTOM

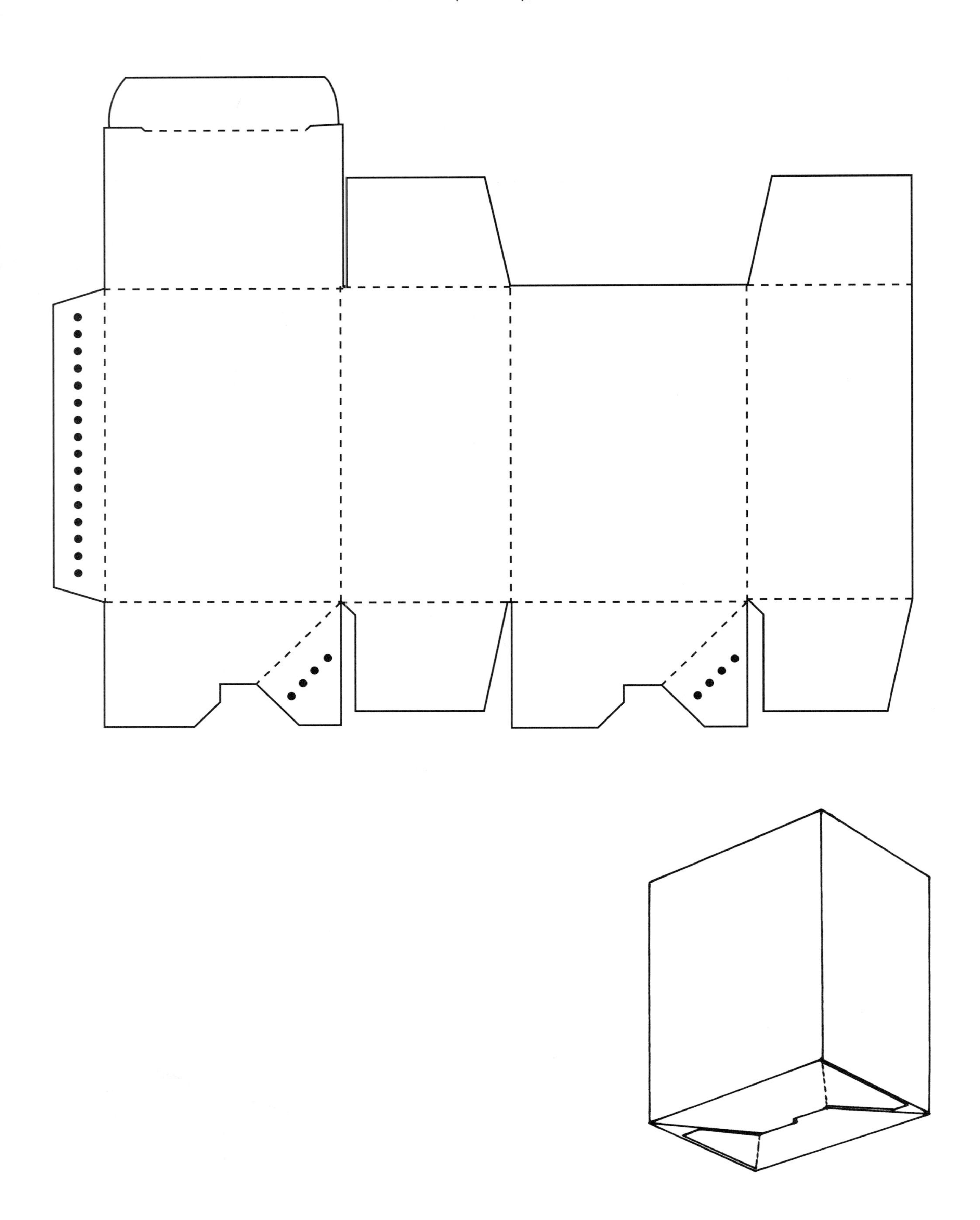

AUTOMATIC GLUED BOTTOM

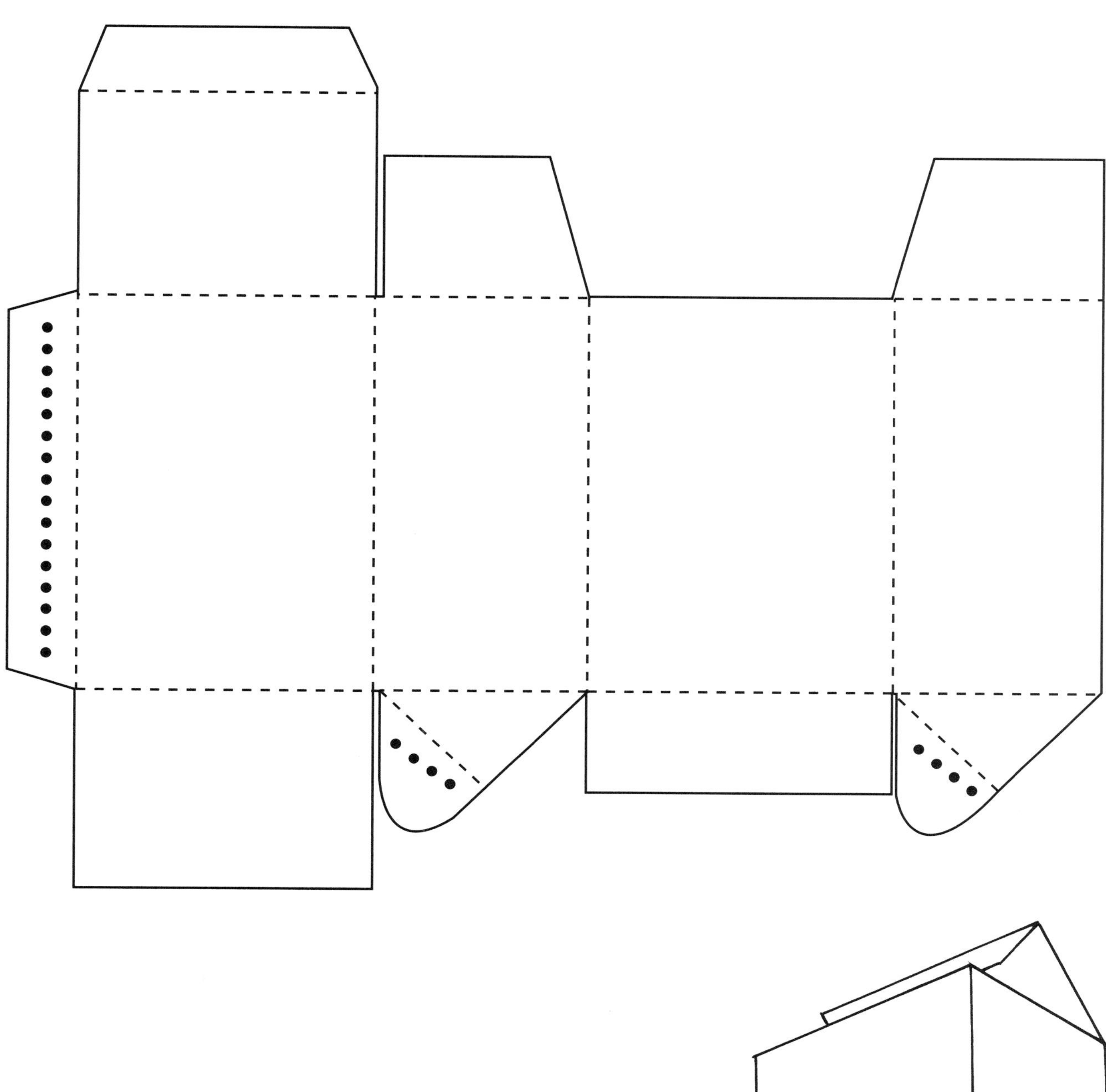

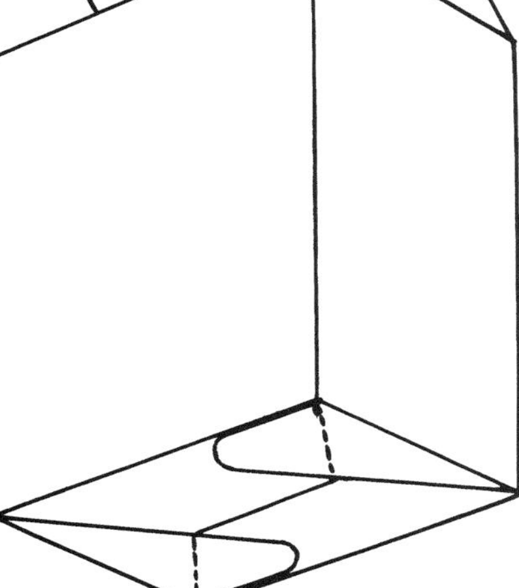

INFOLD AUTOMATIC BOTTOM

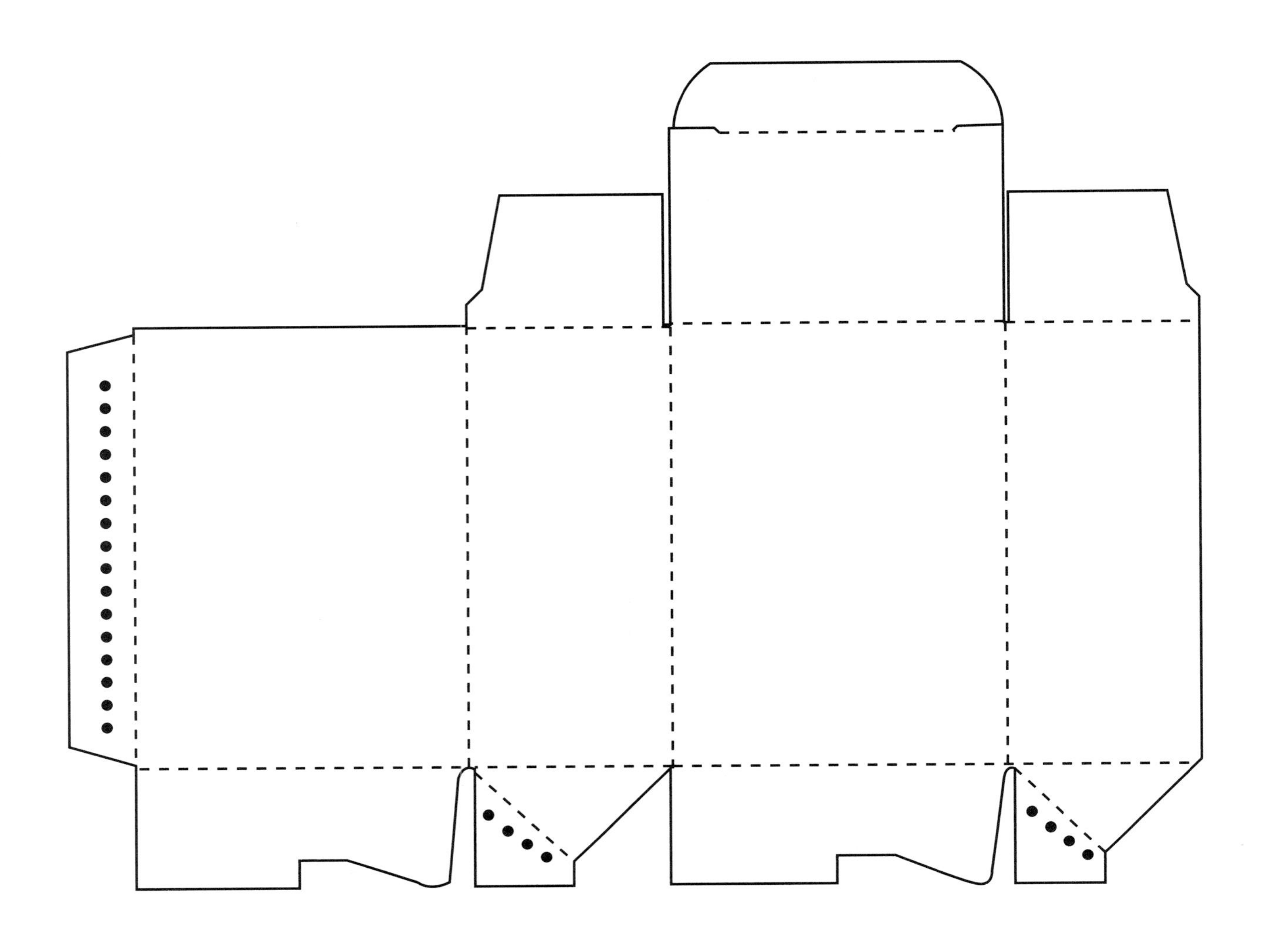

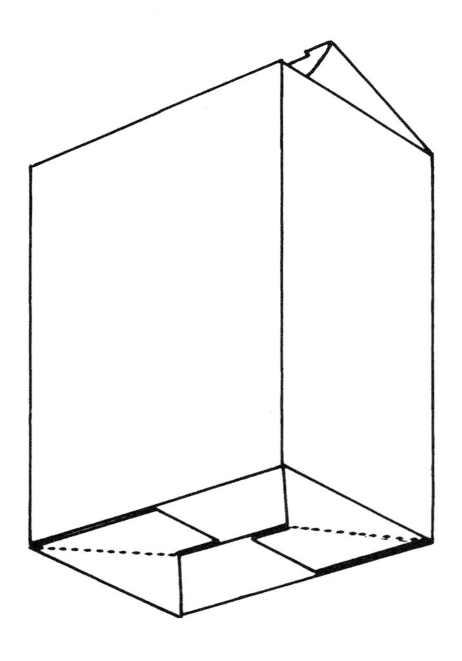

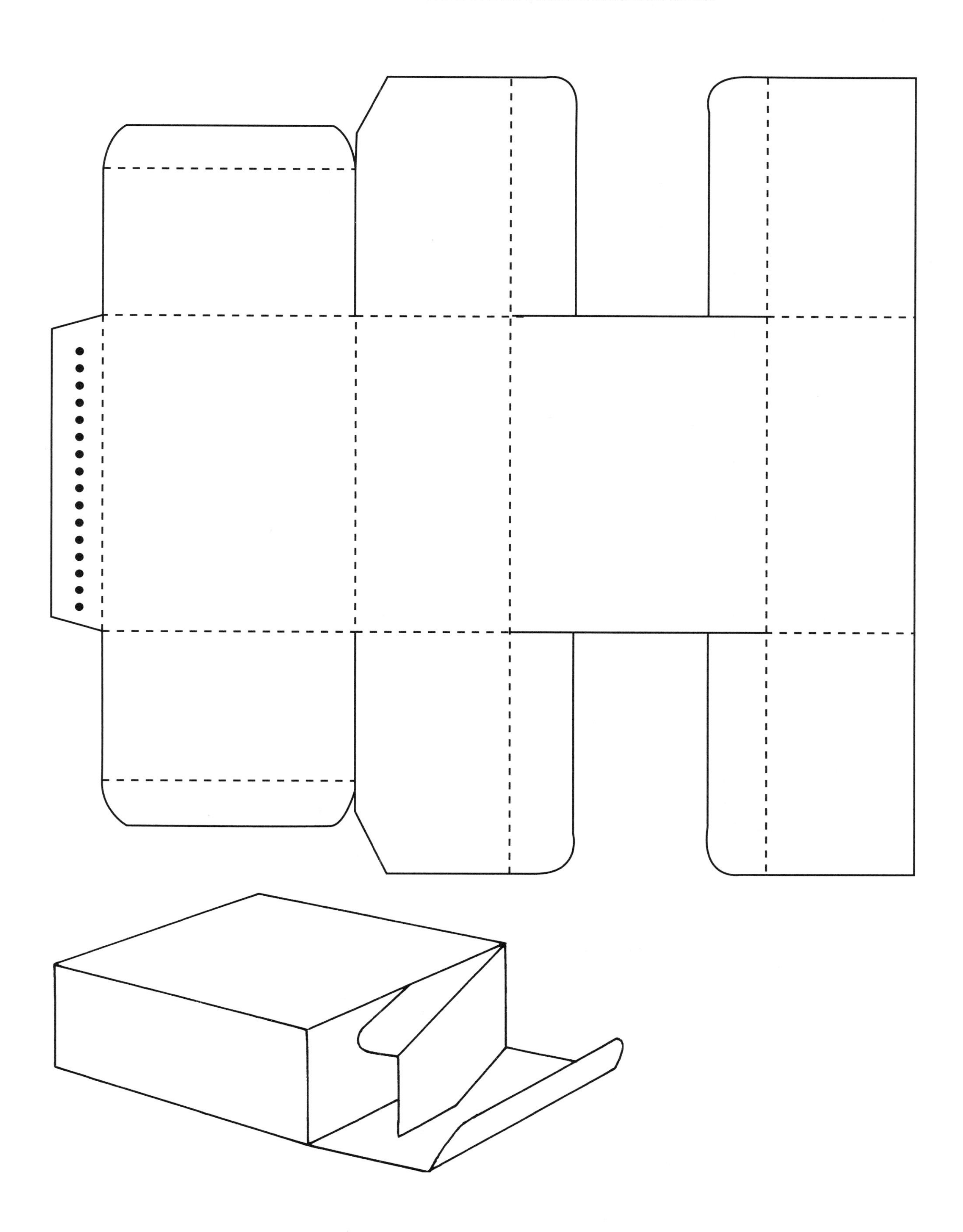

PETERS LOCK

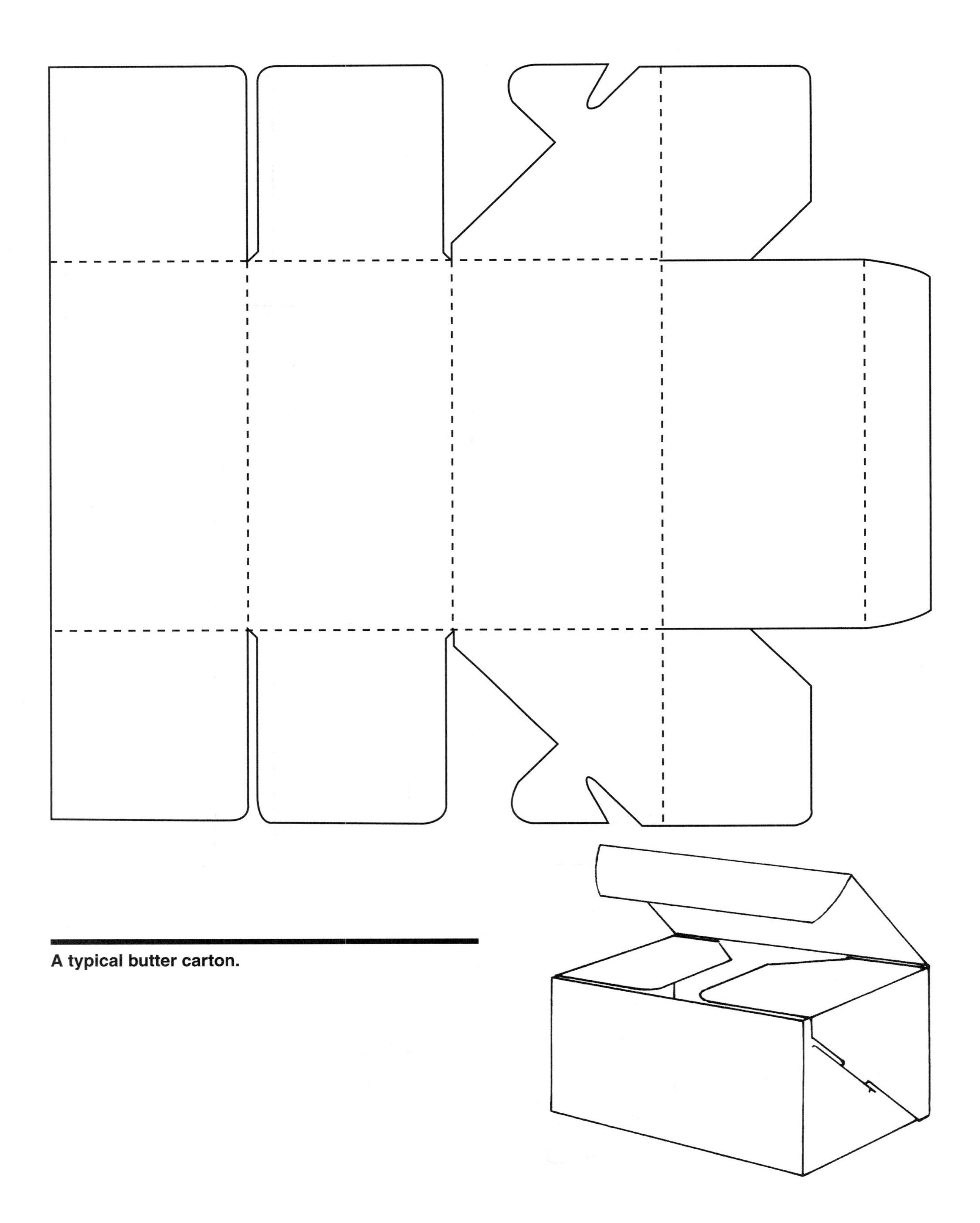

A typical butter carton.

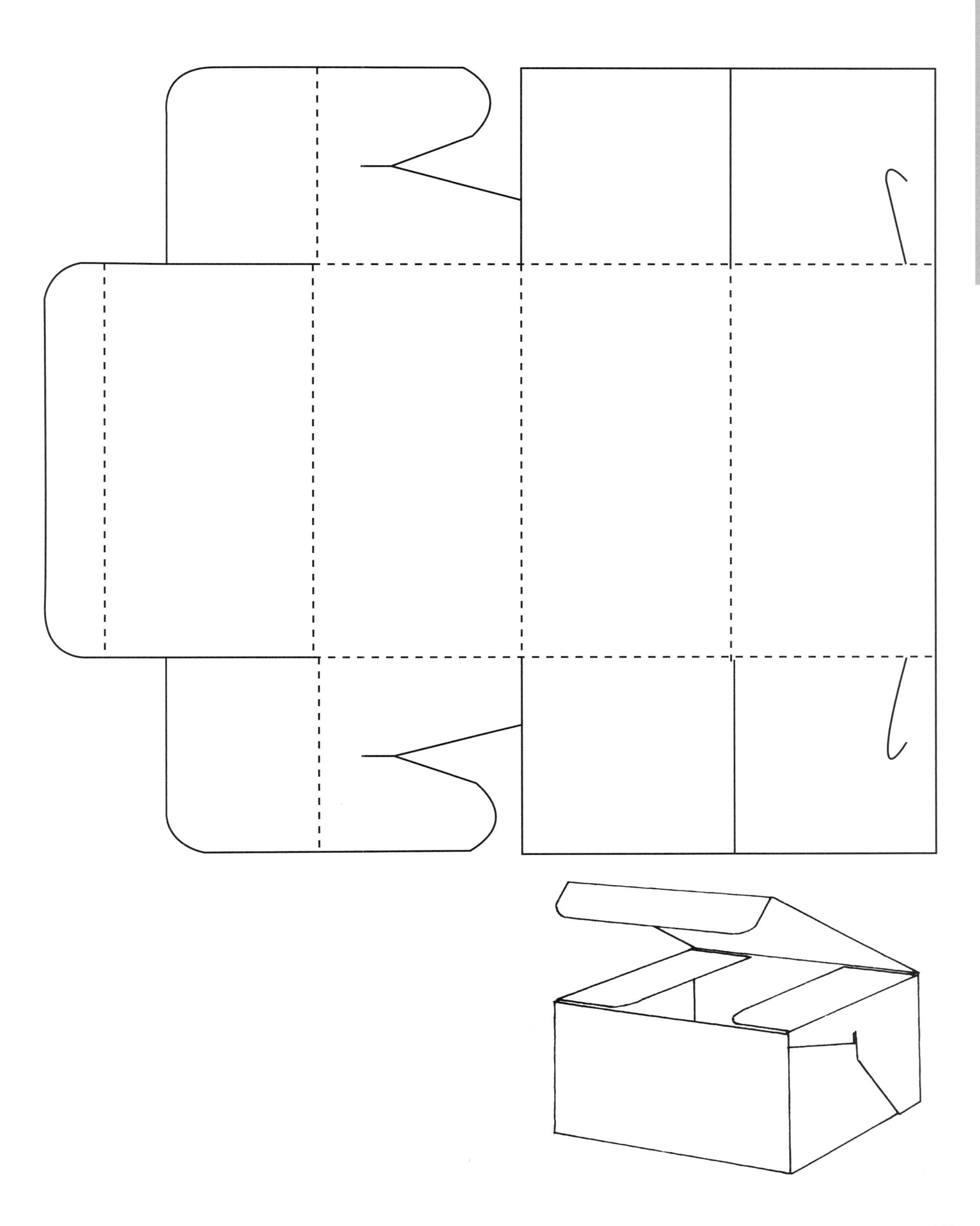

ELGIN STYLE BUTTER WRAP

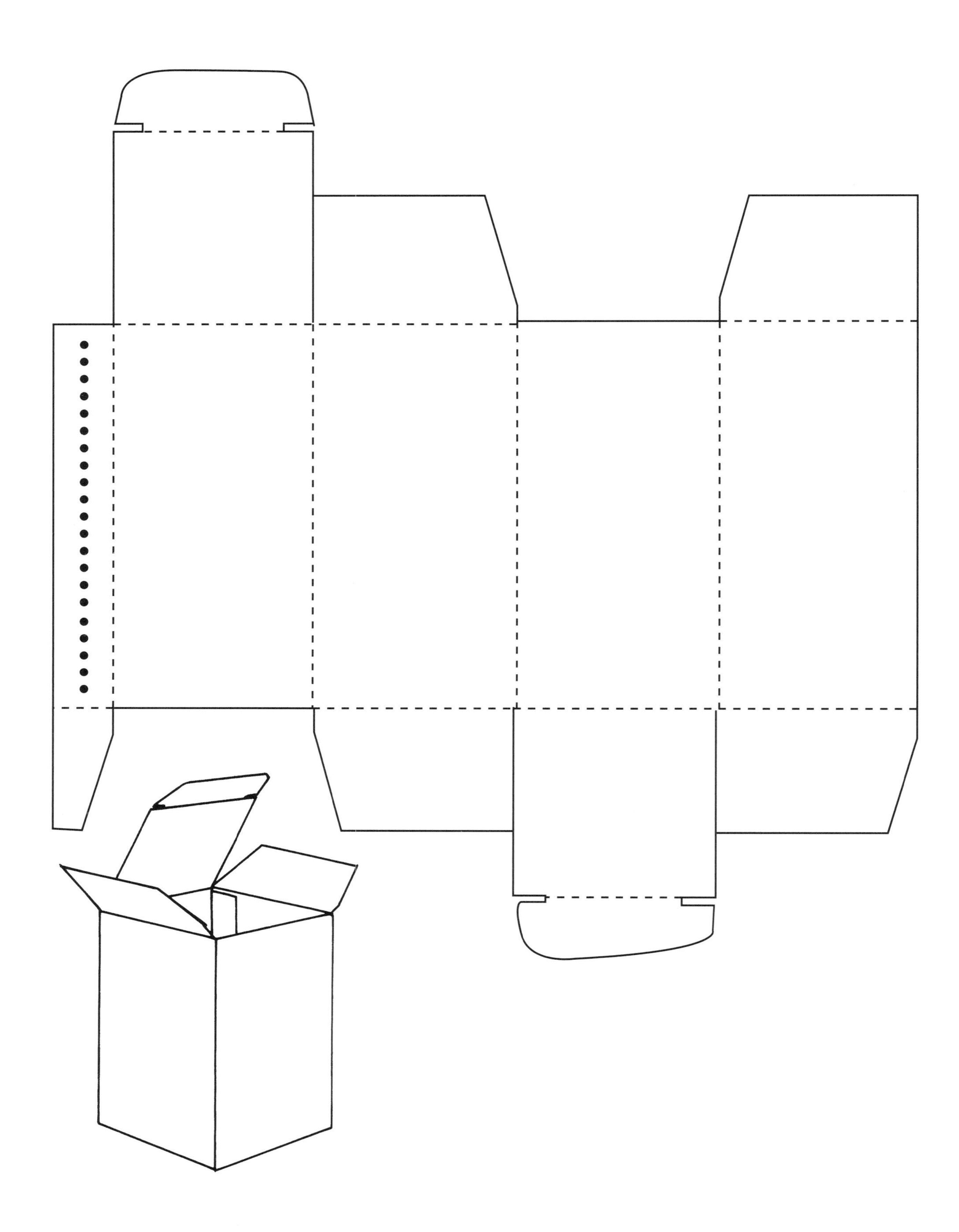

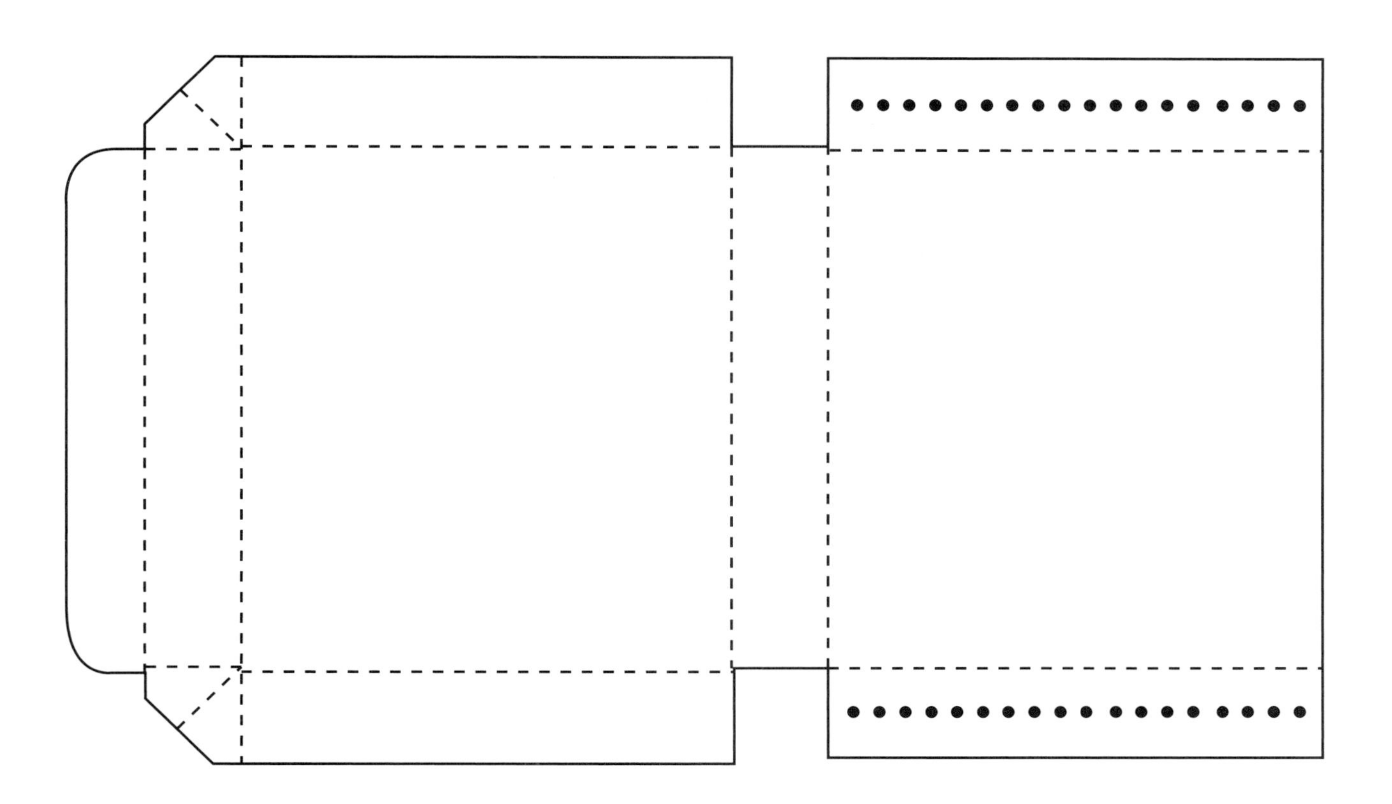

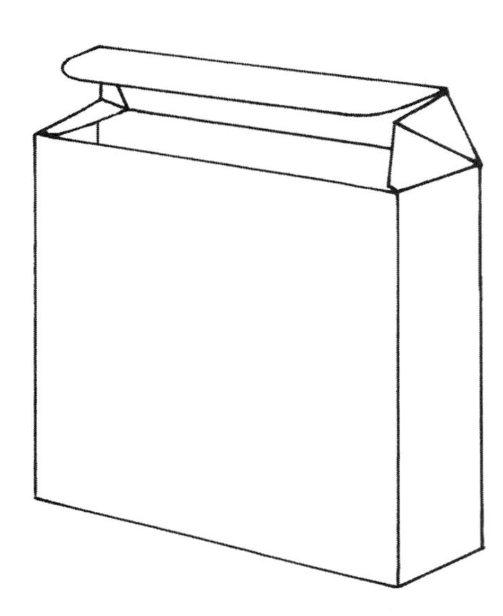

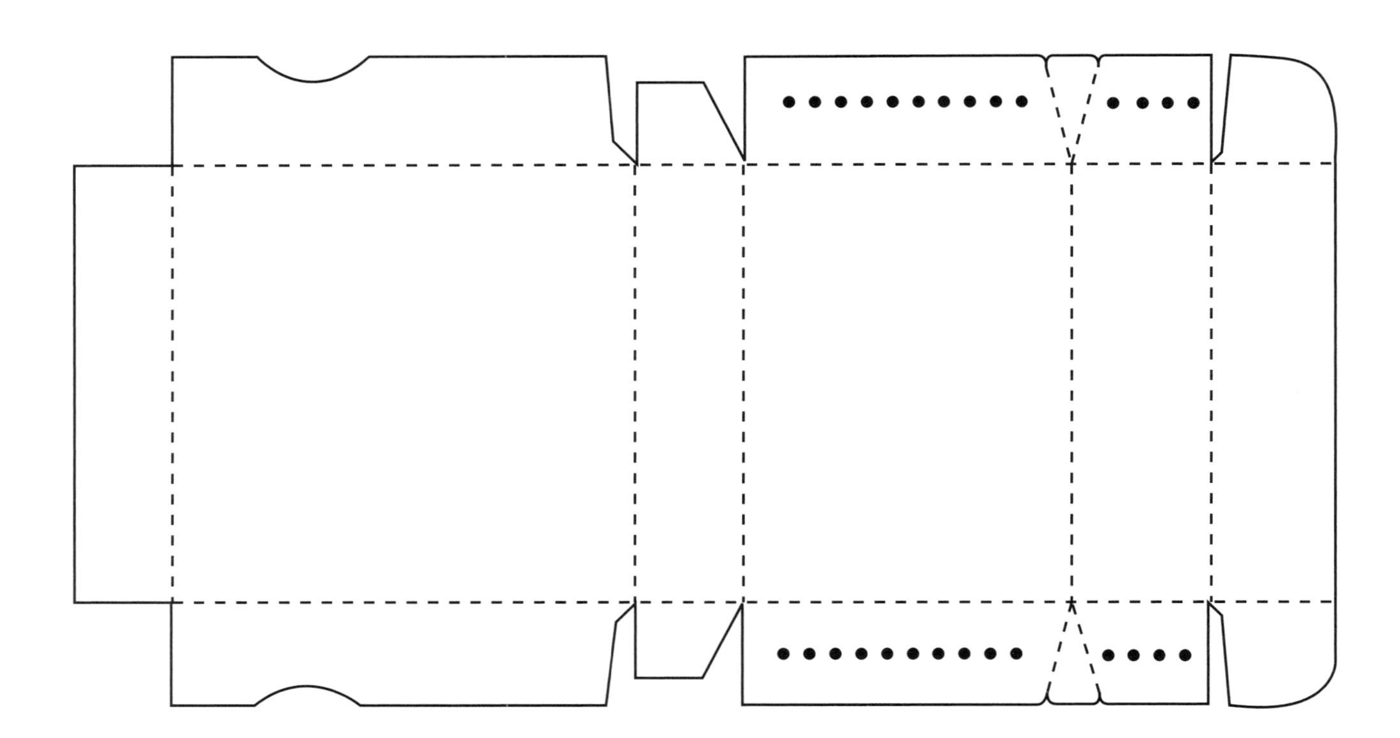

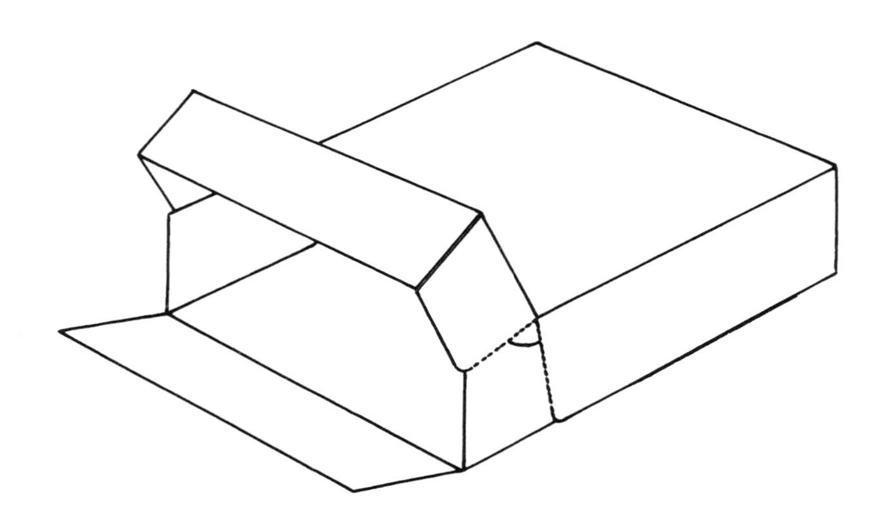

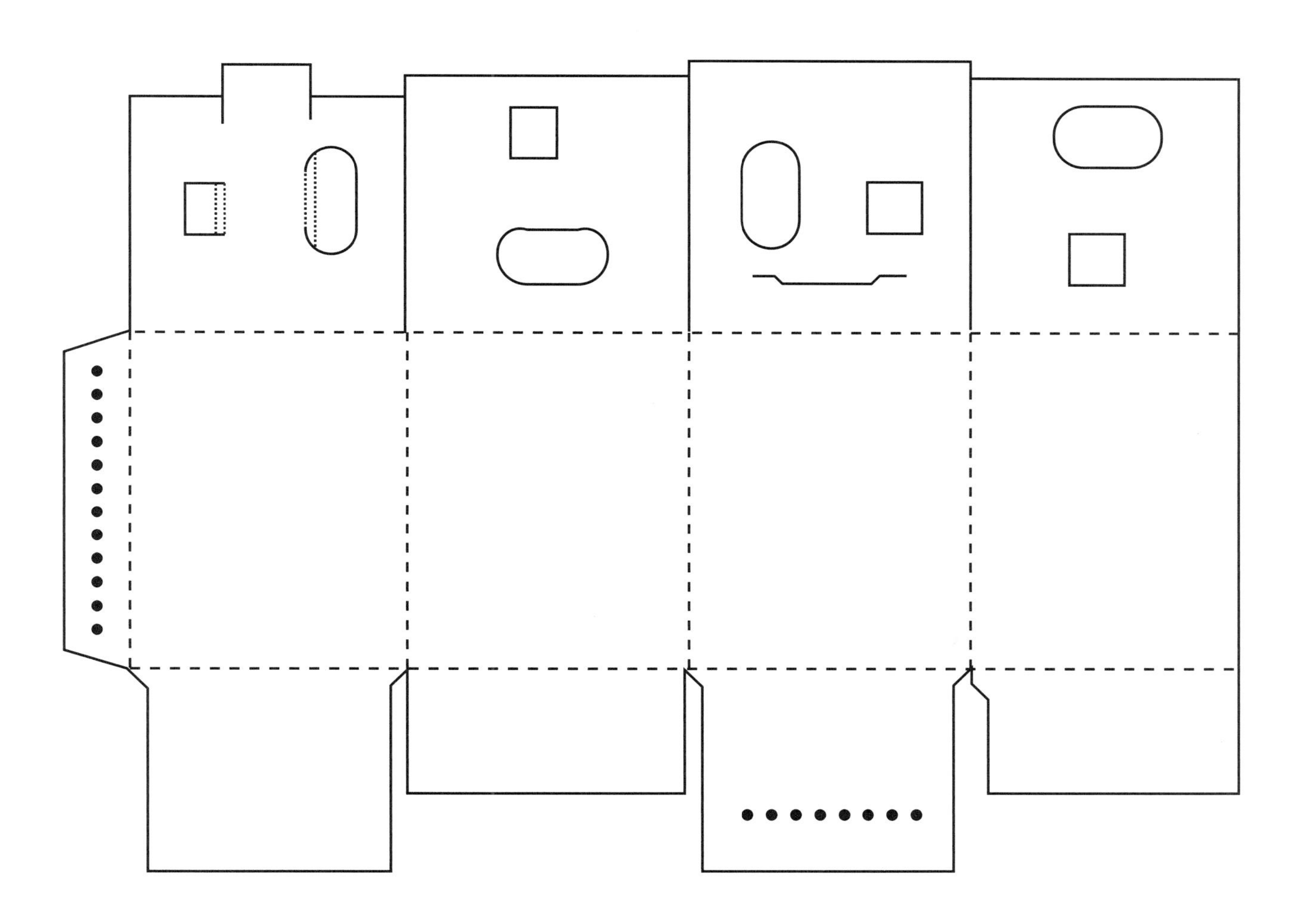

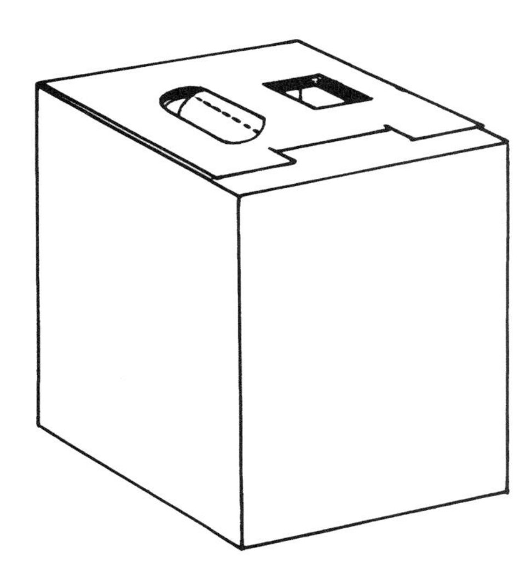

Especially useful for heavy loads.

STRAIGHT TUCK WITH EXTENDED TOP PANEL

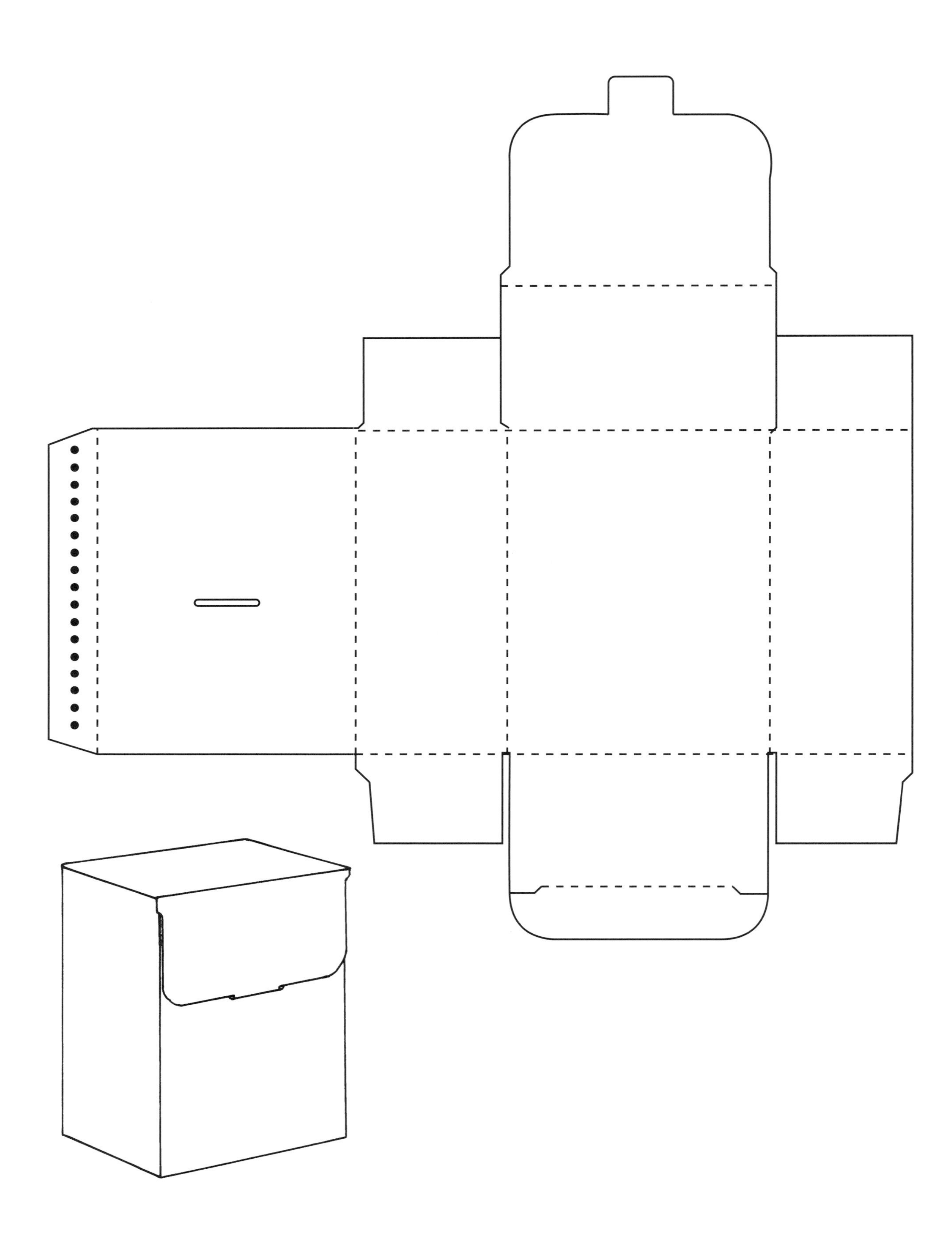

SEAL END WITH ZIPPER

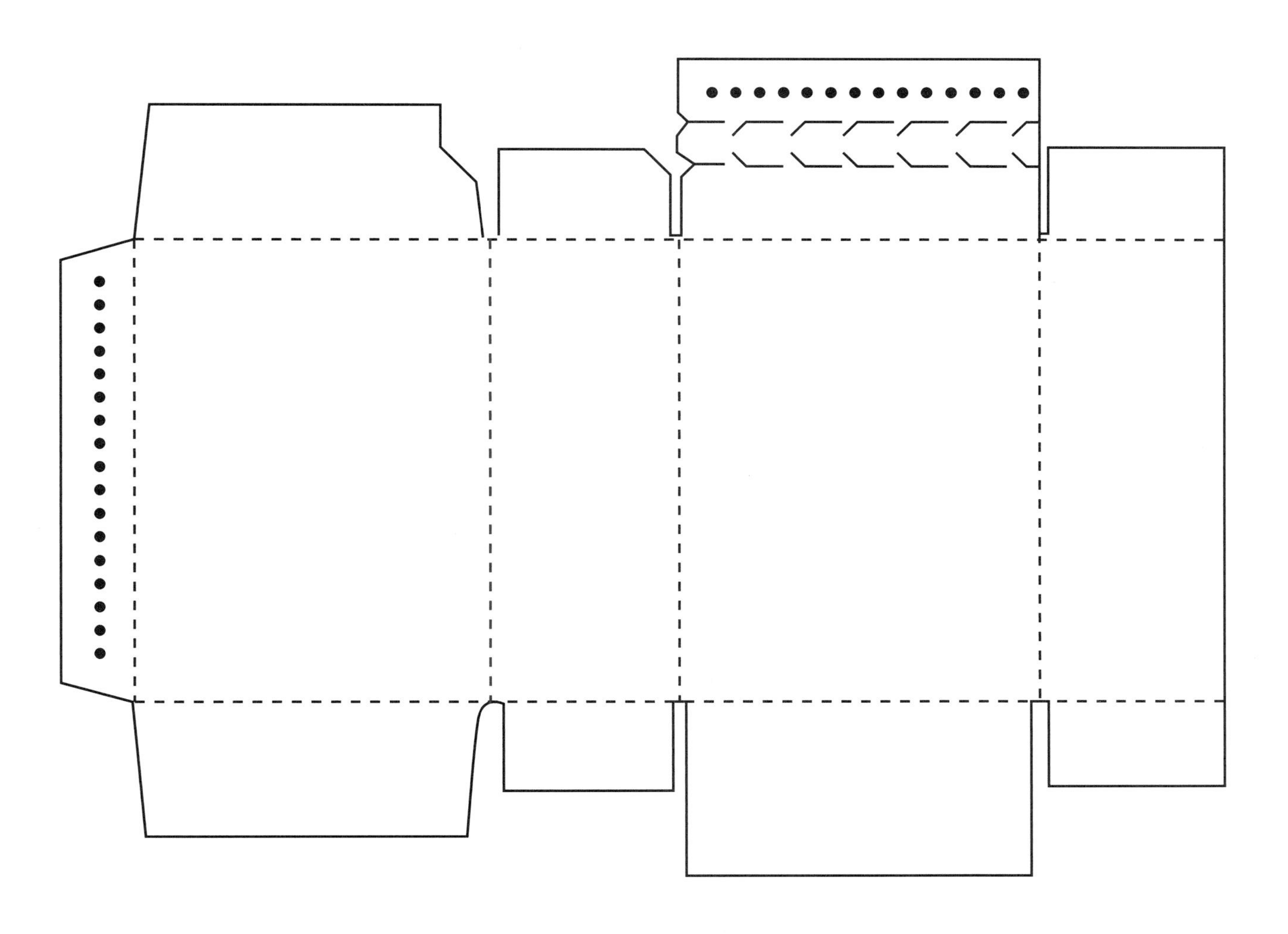

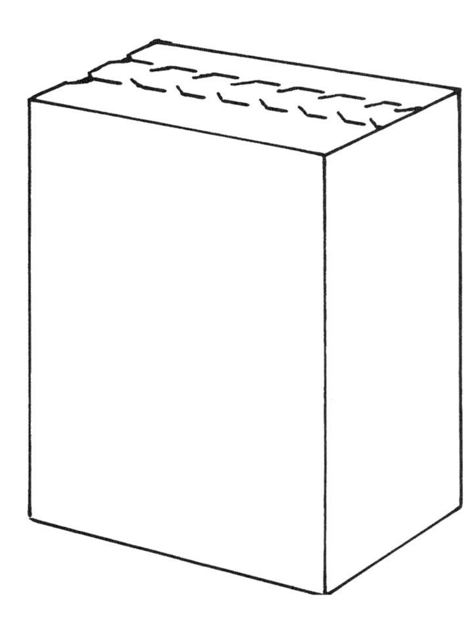

SEAL END WITH LOCK TAB AND ZIPPER

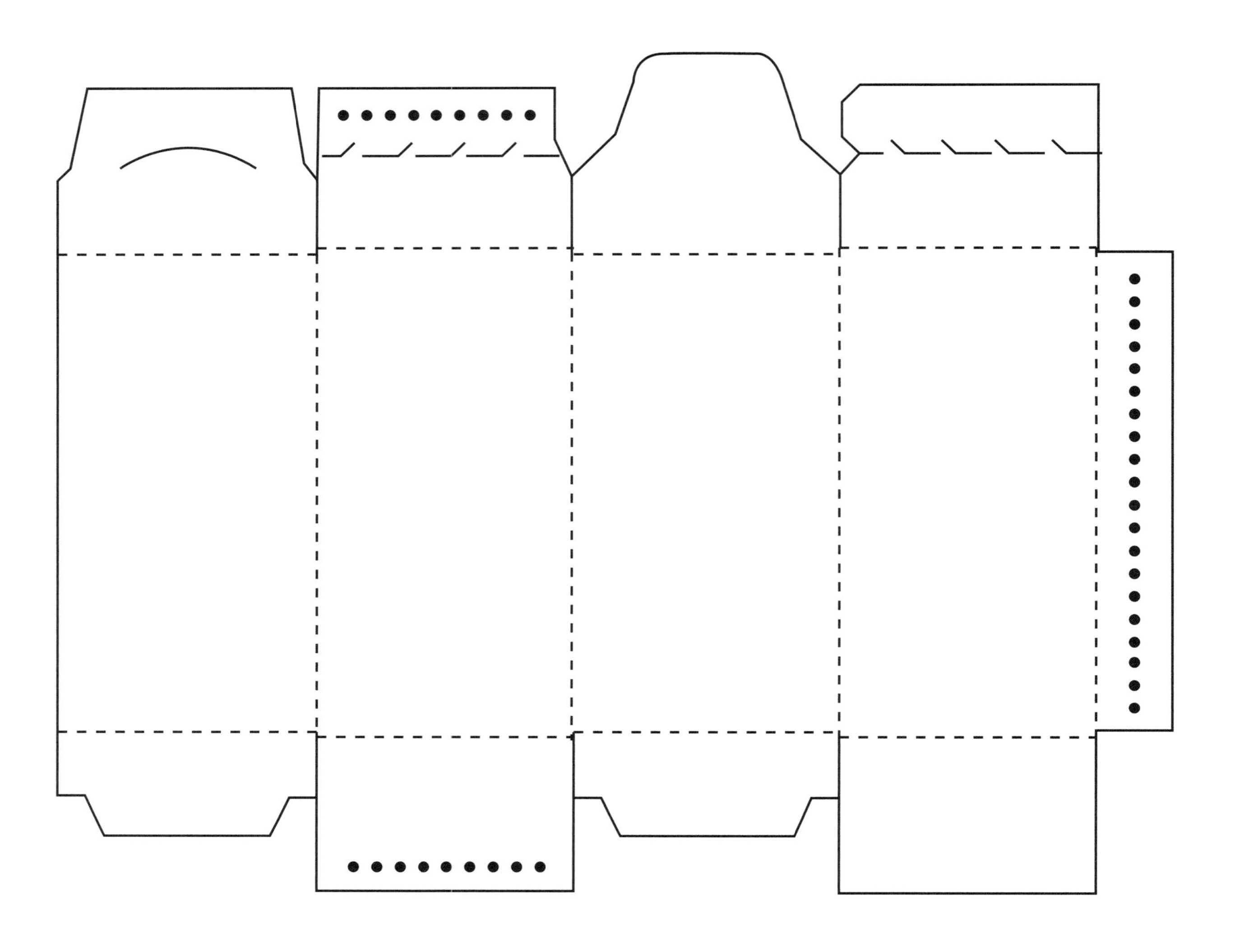

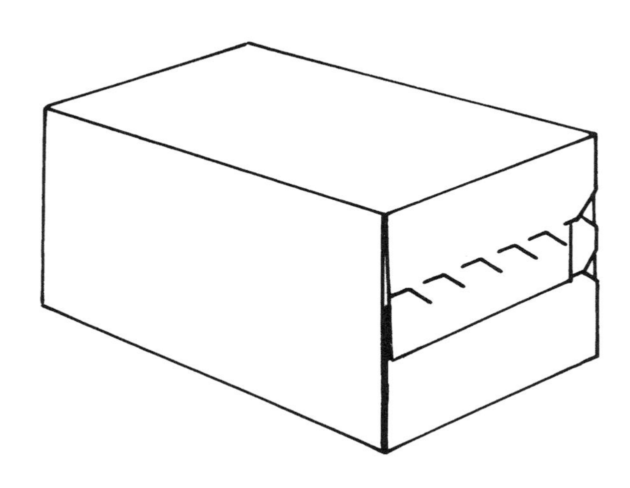

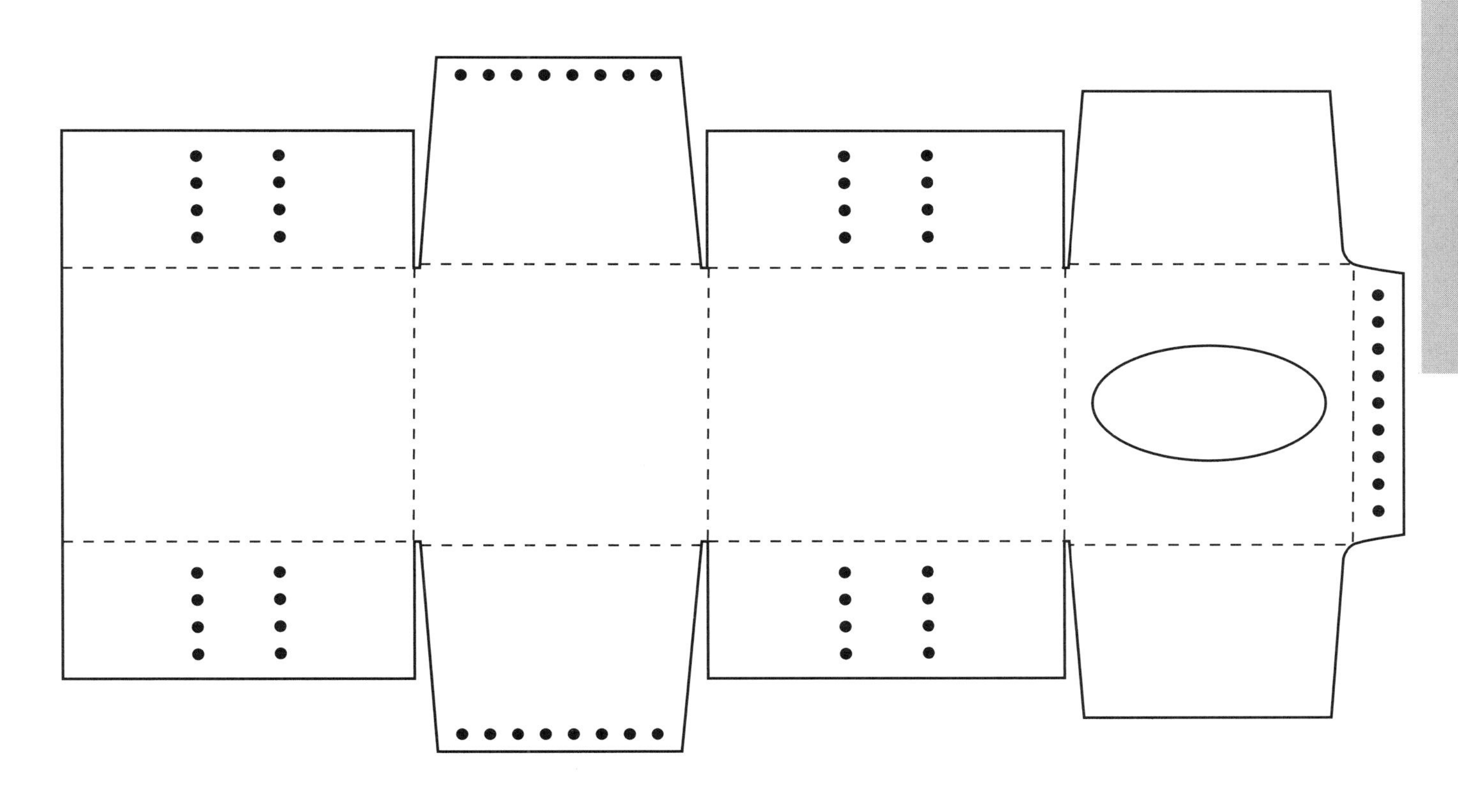

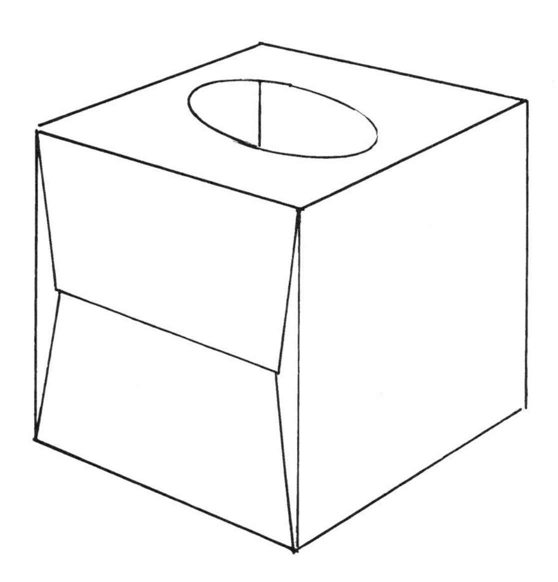

PARTIAL OVERLAP SEAL END WITH PERFORATED TOP

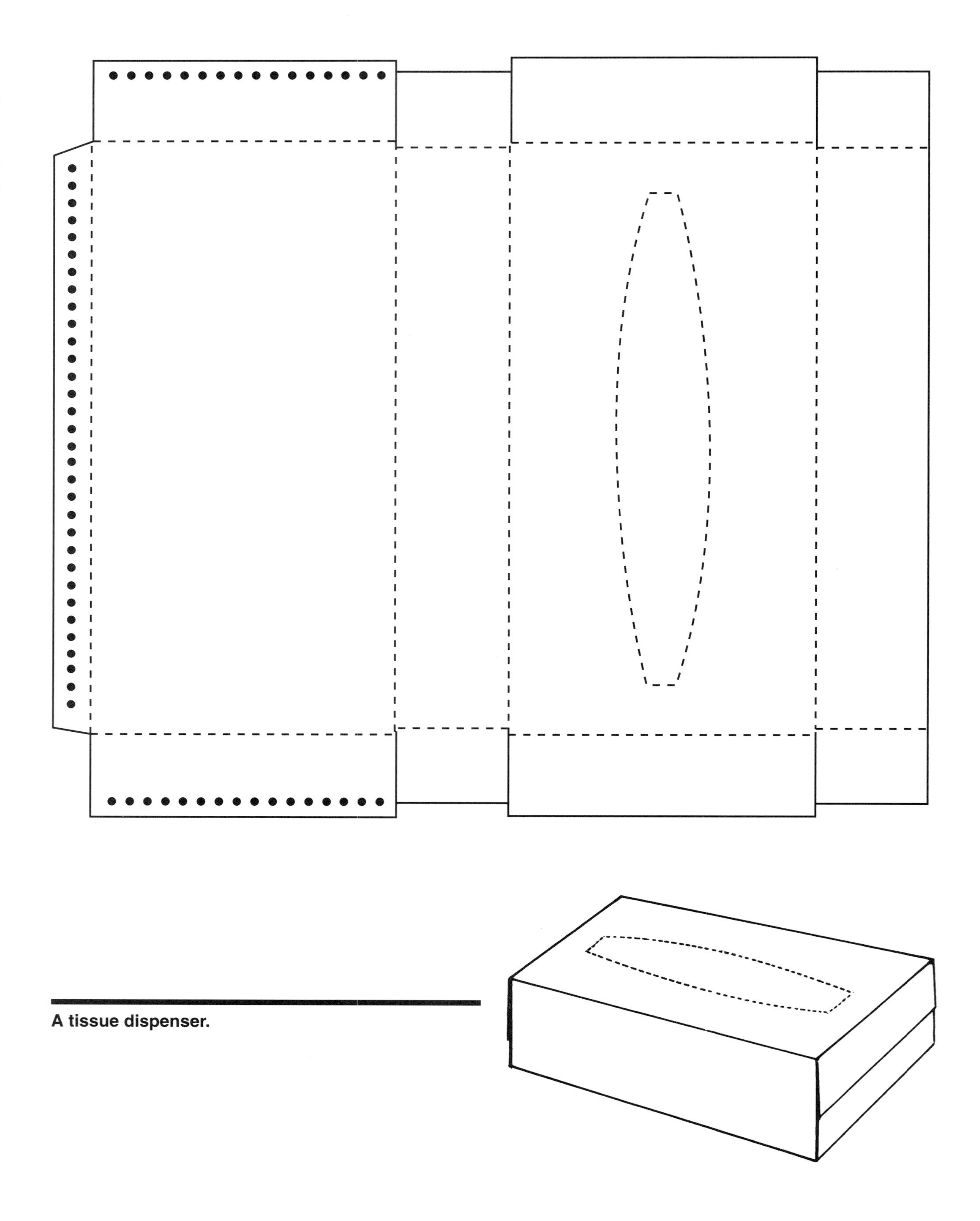

A tissue dispenser.

PARTIAL OVERLAP SEAL END WITH PERFORATED TOP AND FRONT PANEL

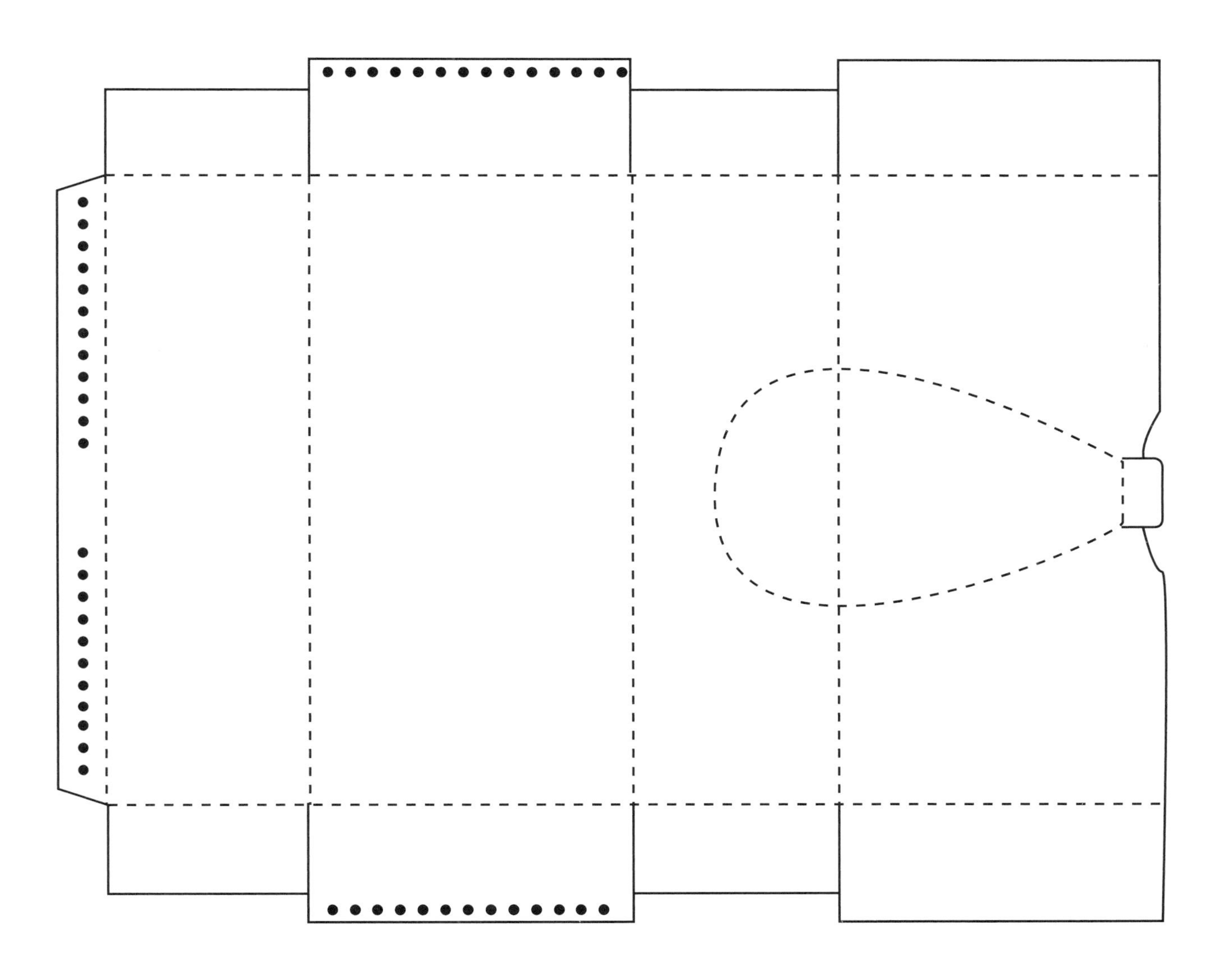

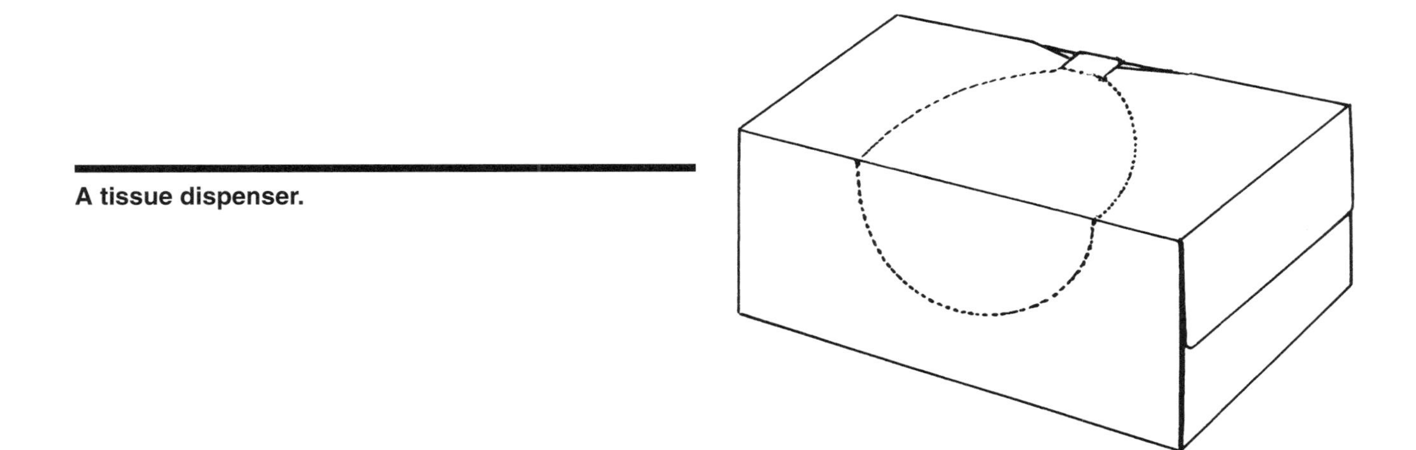

A tissue dispenser.

ELL-STRIP BACON PACK

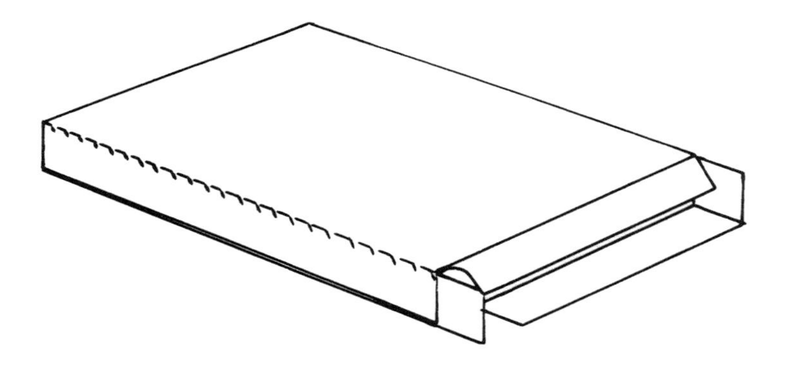

RIP-TAPE BACON PACK

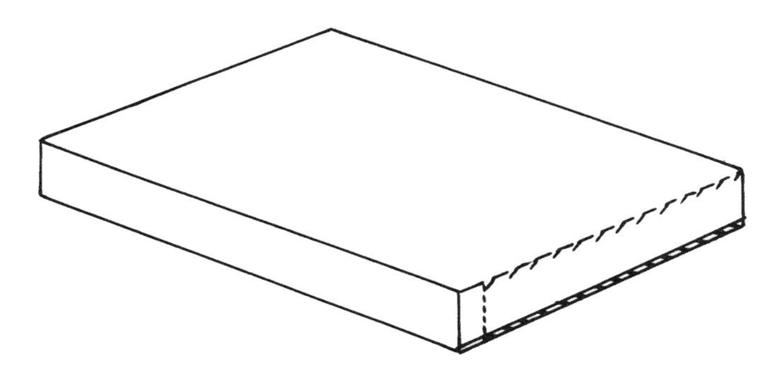

TOP SEAL BACON PACK

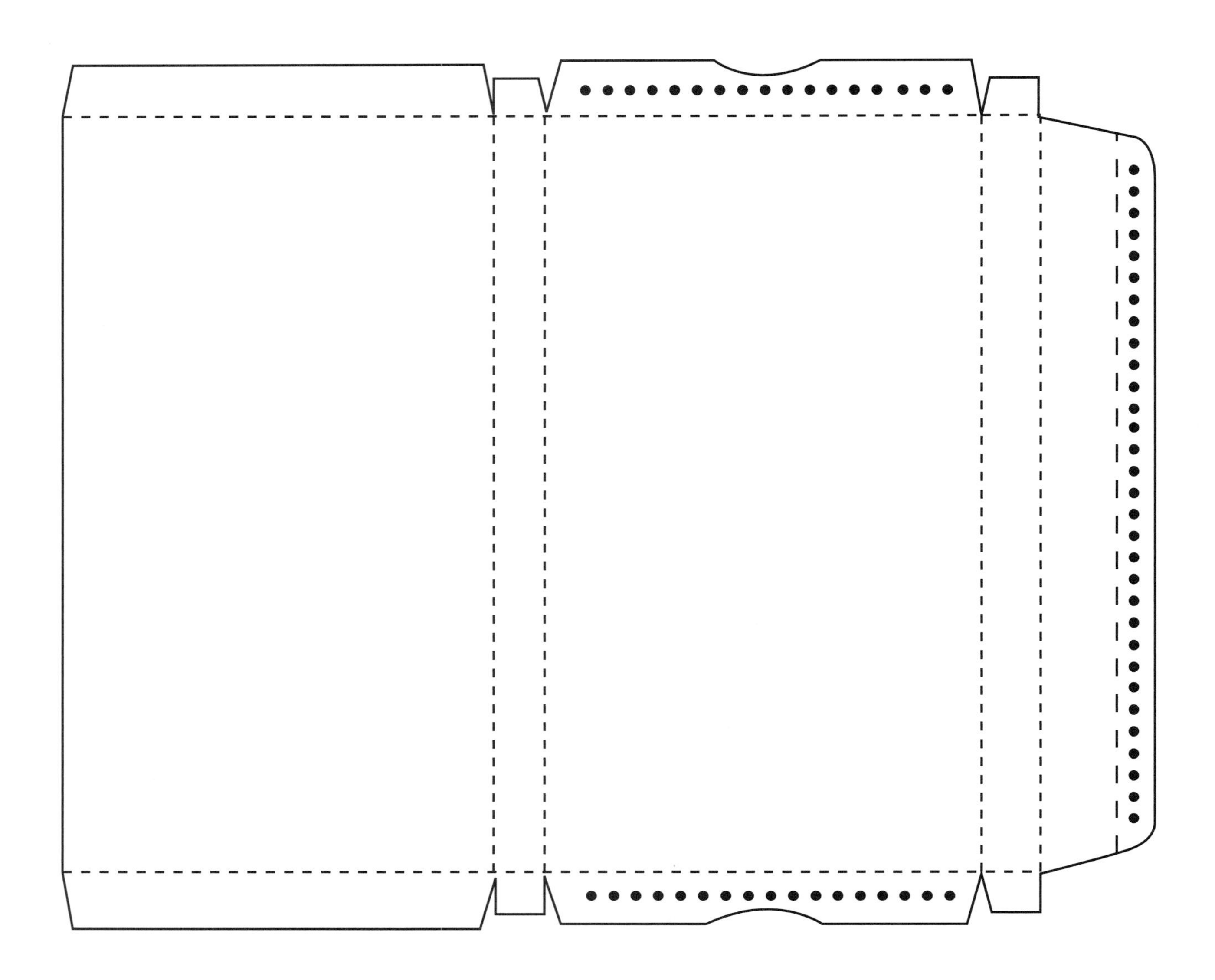

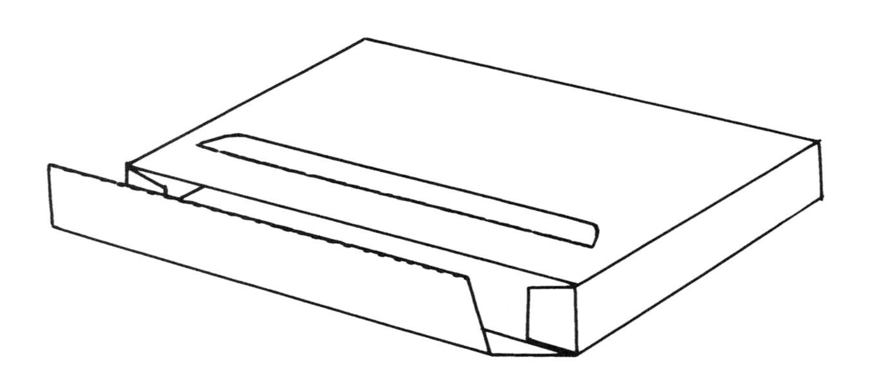

TUCK SEAL BACON PACK

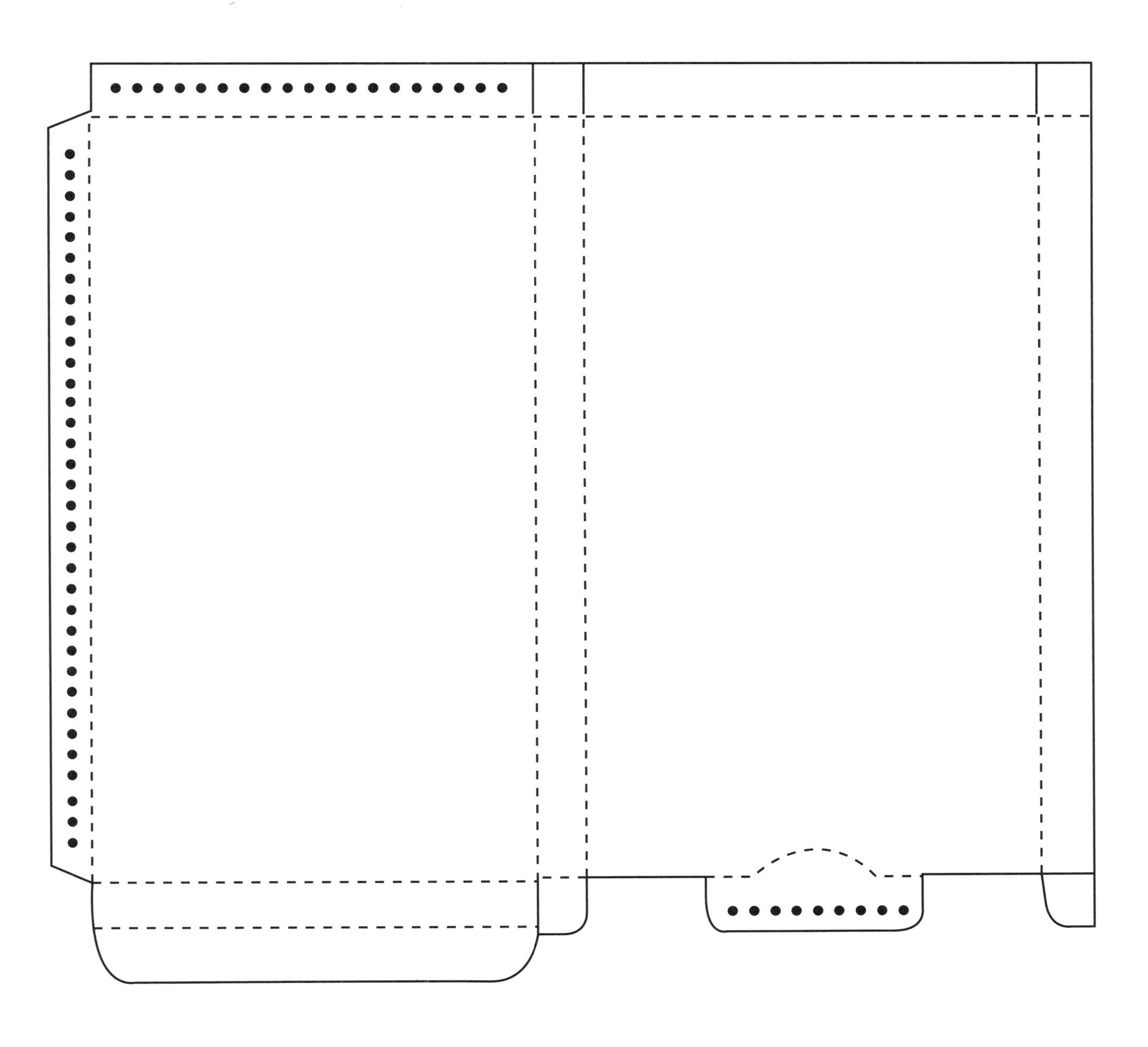

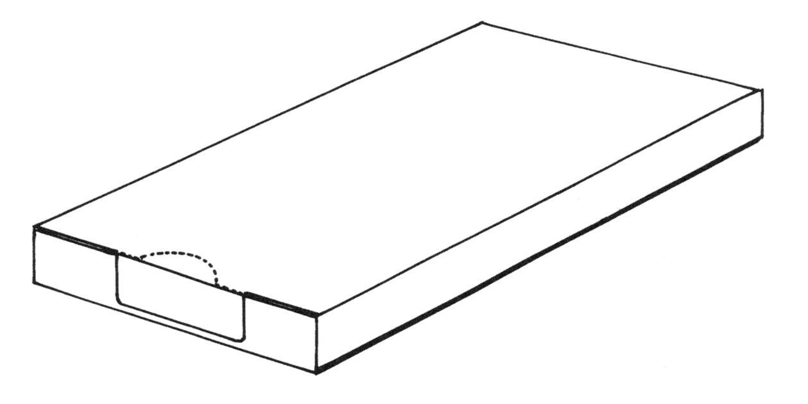

REVERSE TUCK WITH FACETED TOP

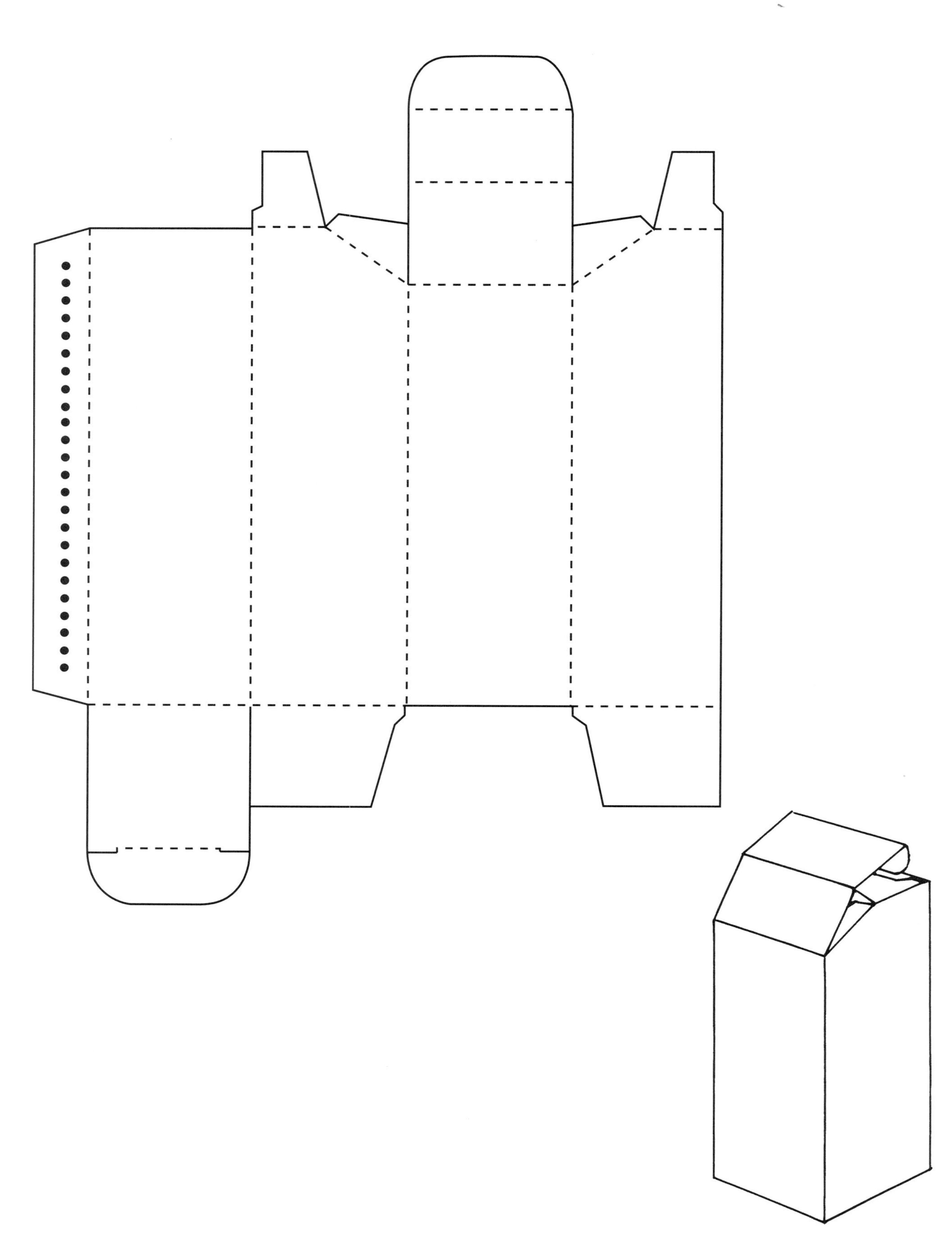

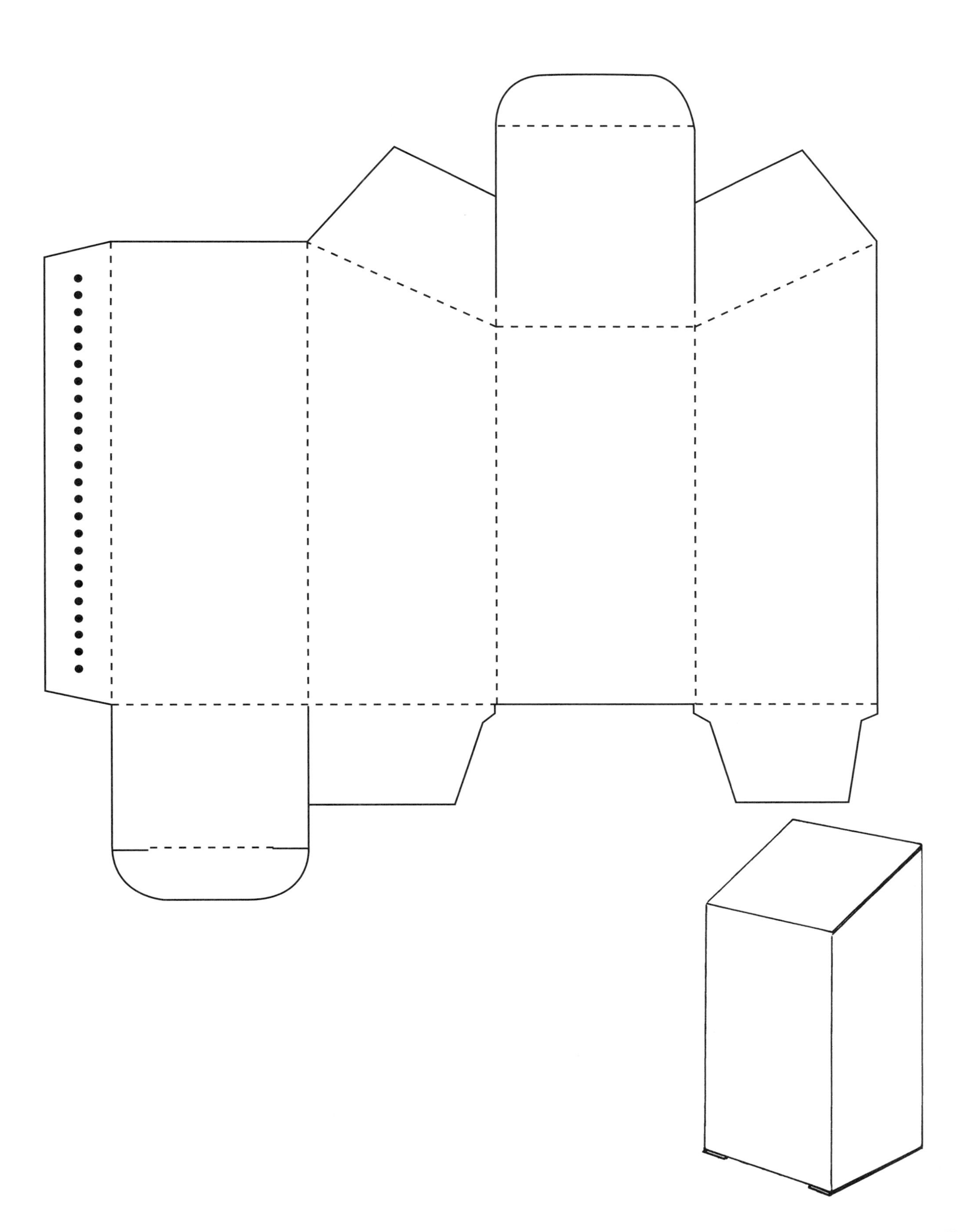

SLANTED TOP WITH HANG TAG

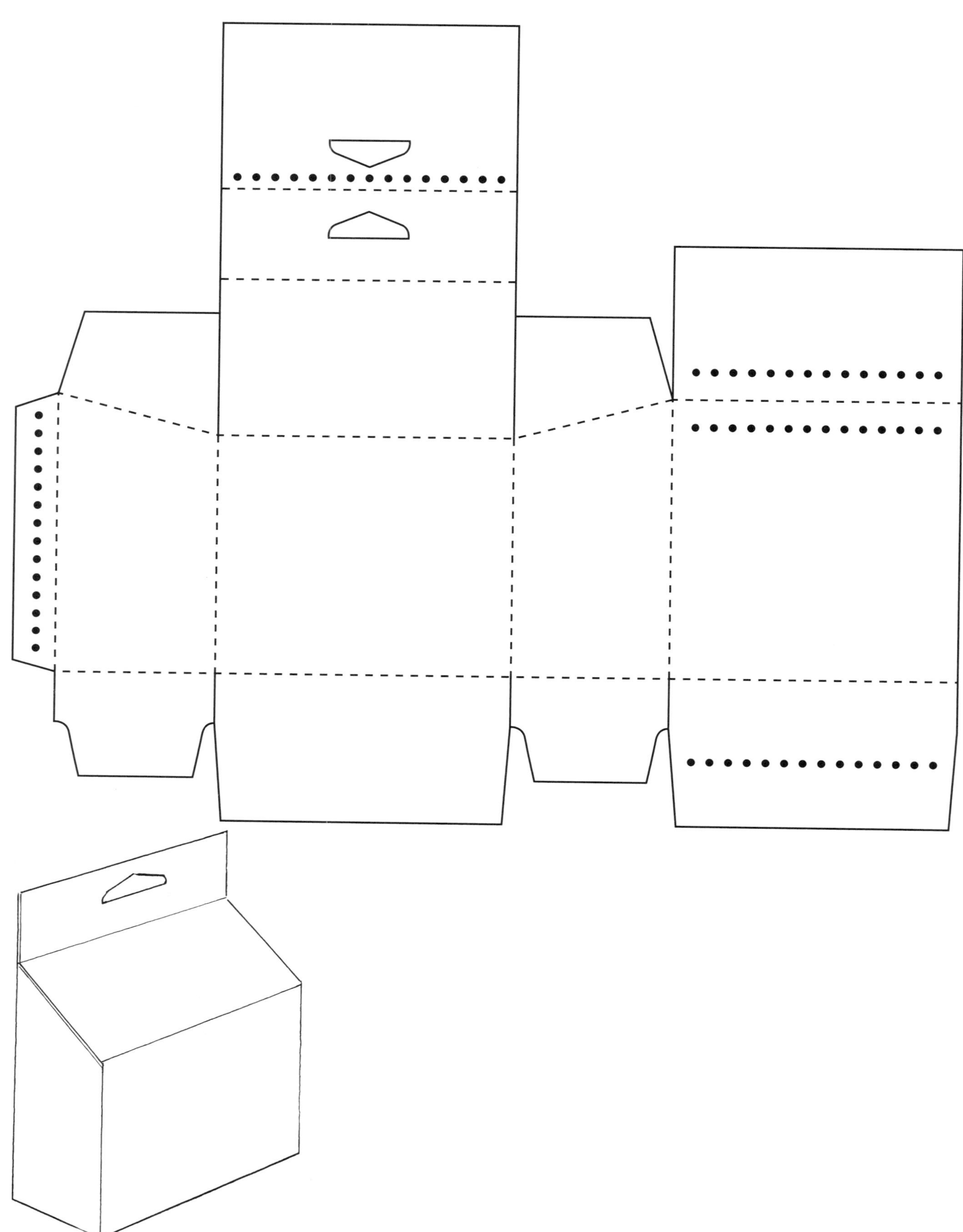

REVERSE TUCK WITH HEADER CARD OR FIFTH PANEL

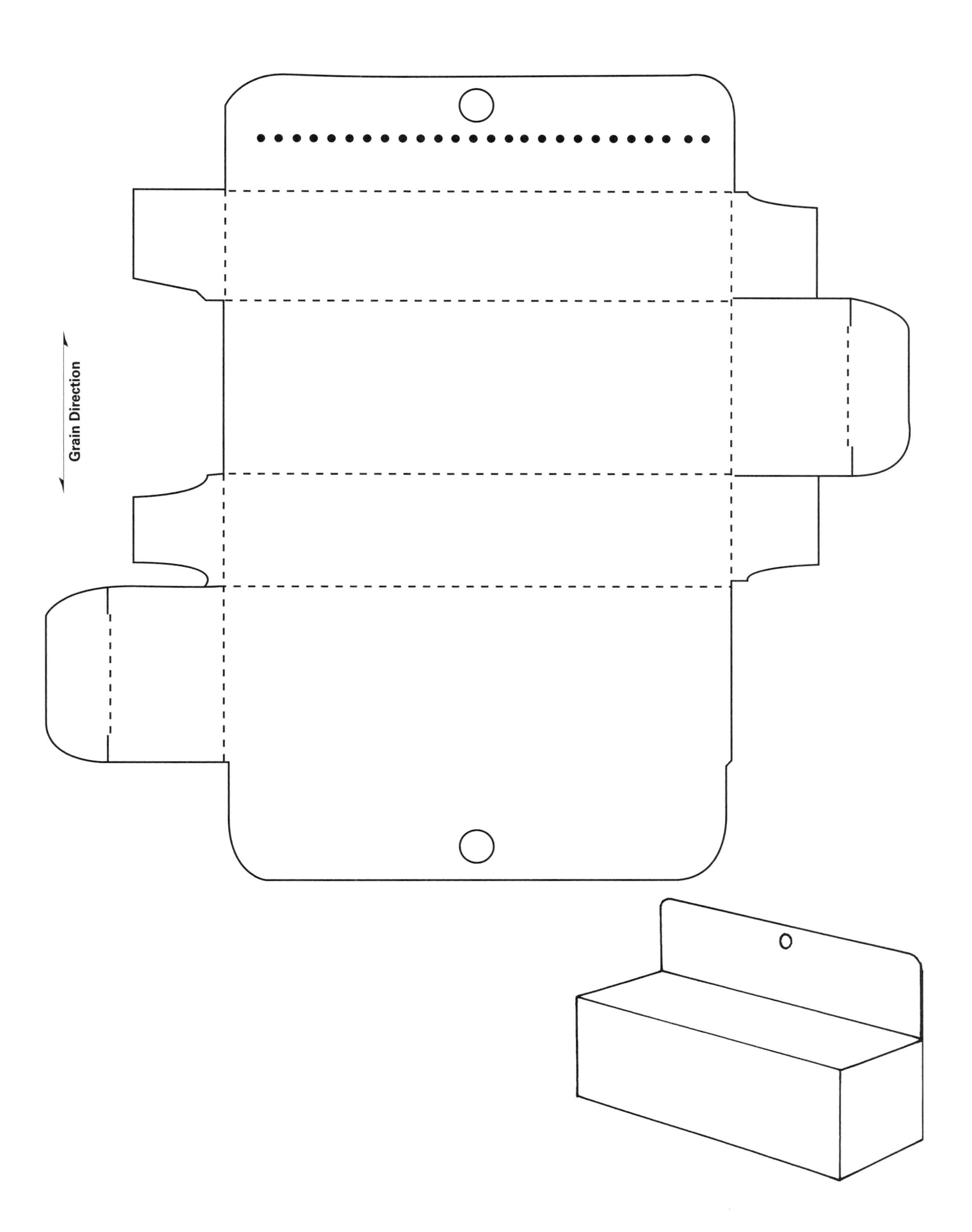

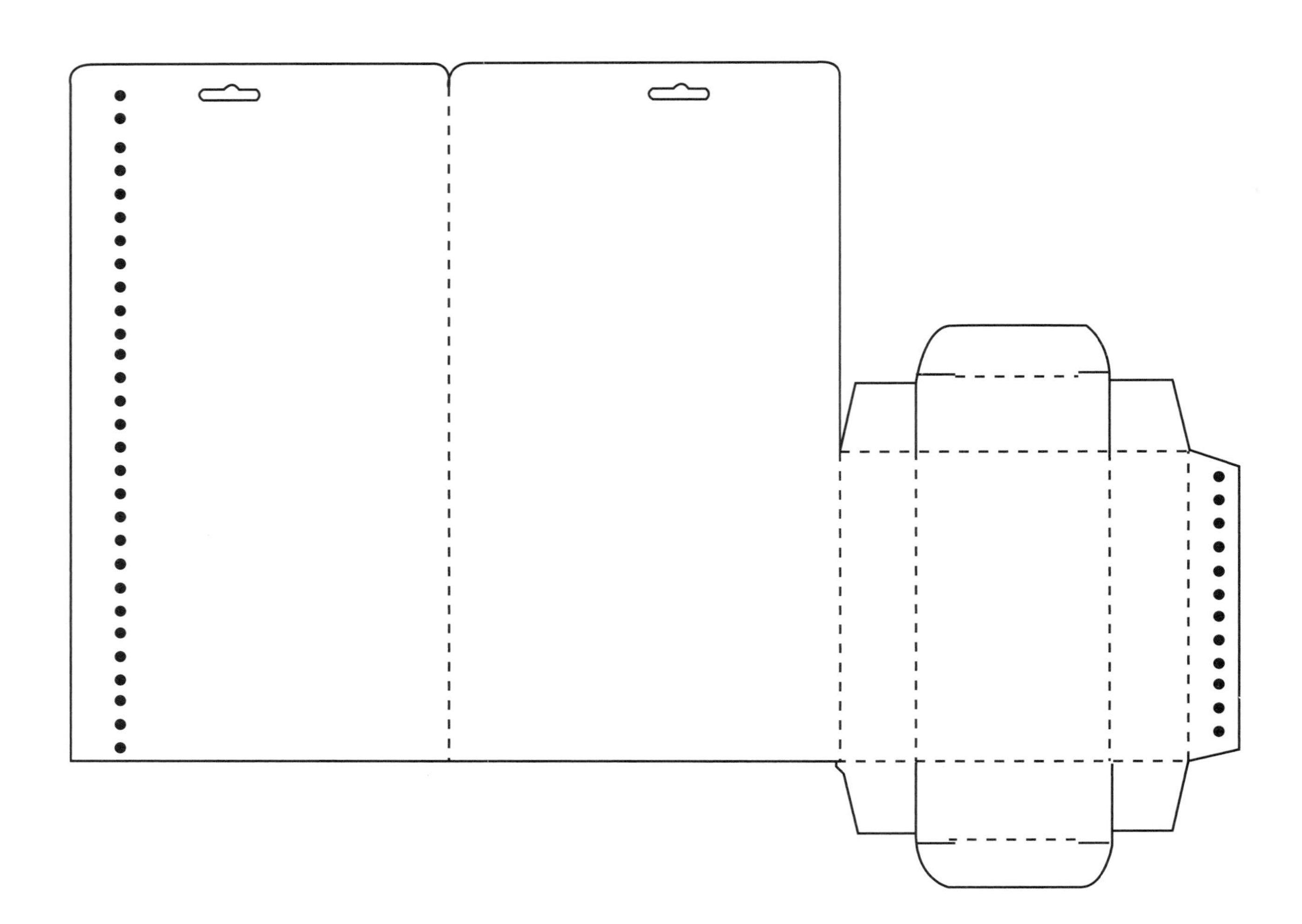

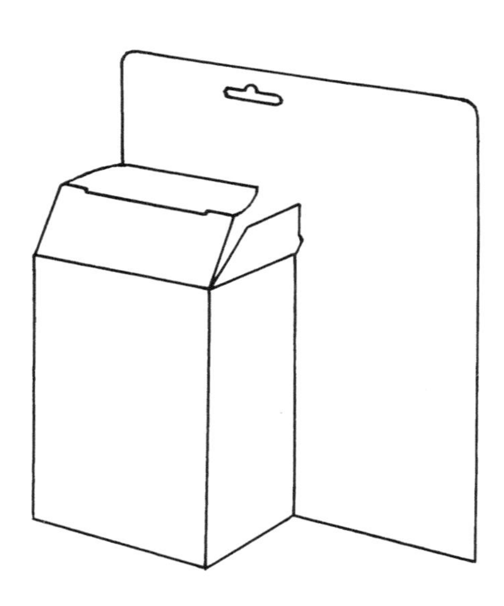

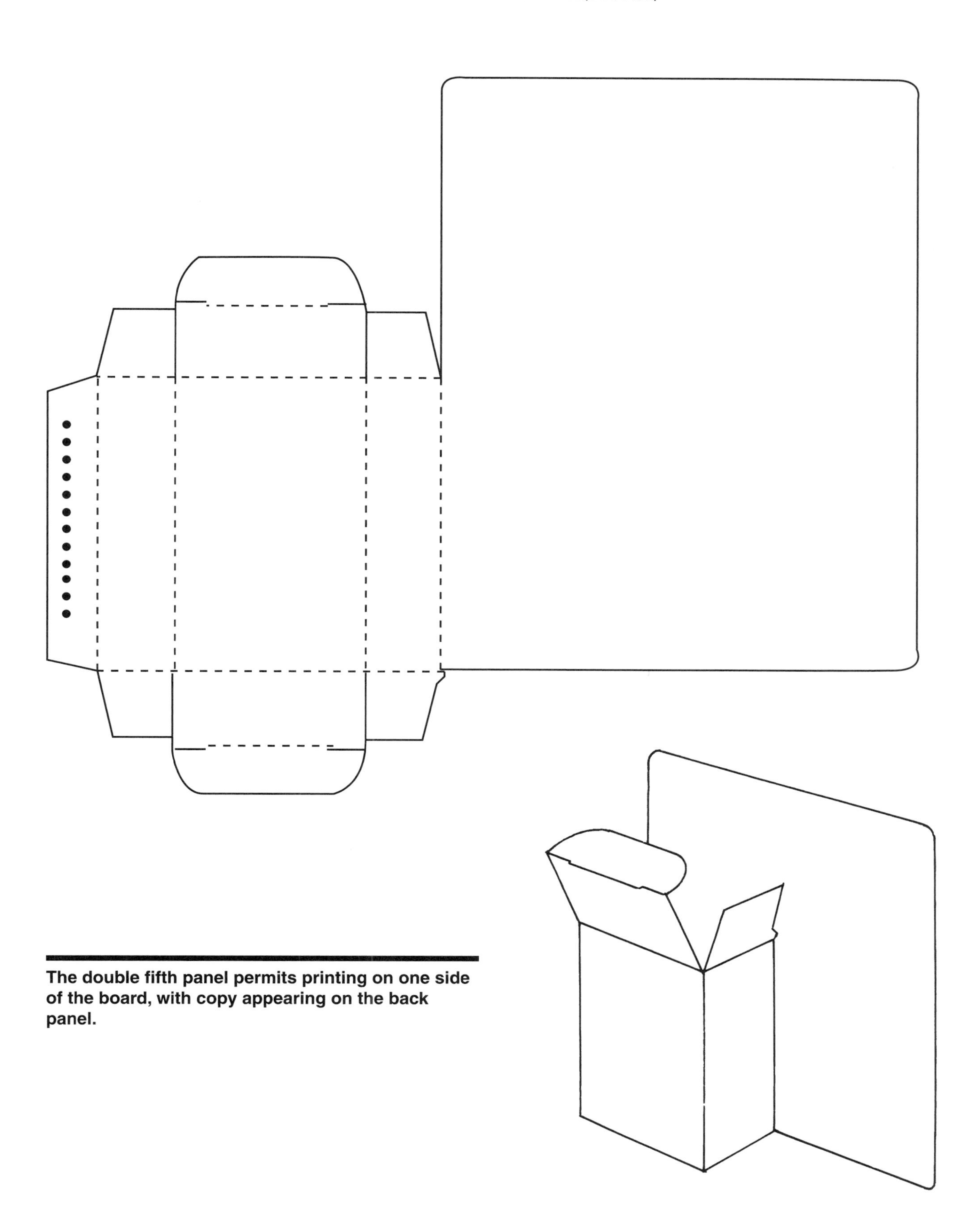

The double fifth panel permits printing on one side of the board, with copy appearing on the back panel.

FLIP-TOP DISPENSER BOX

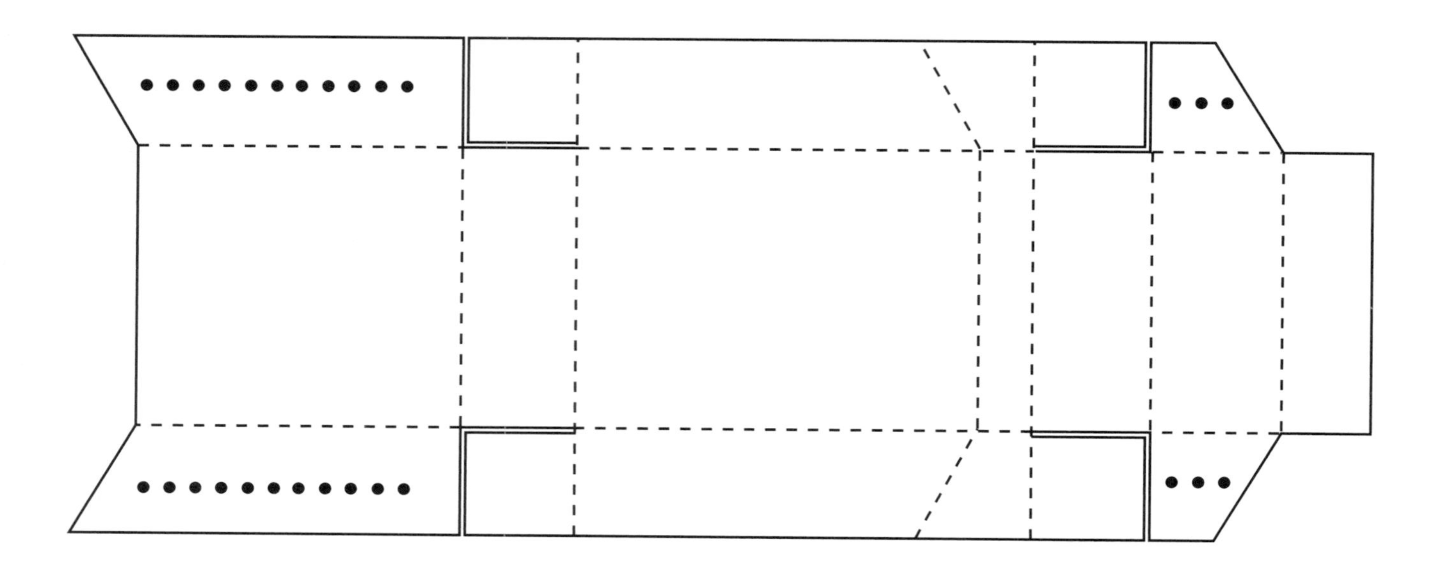

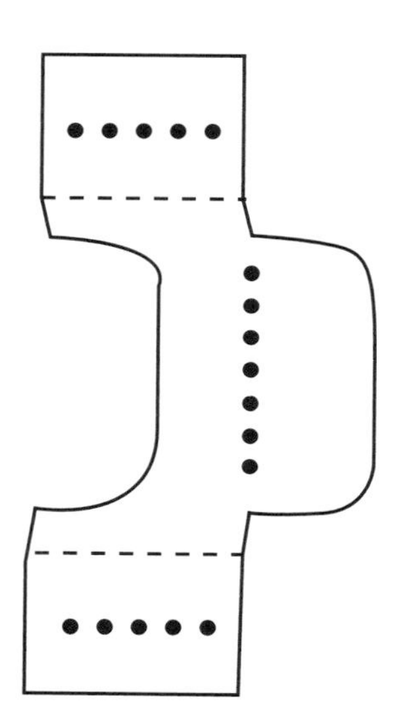

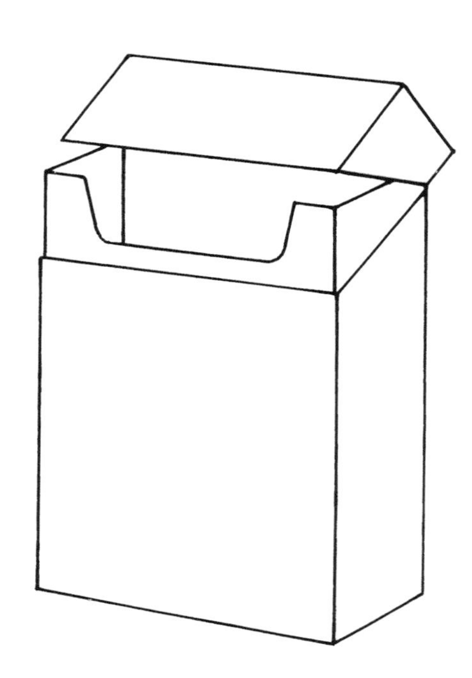

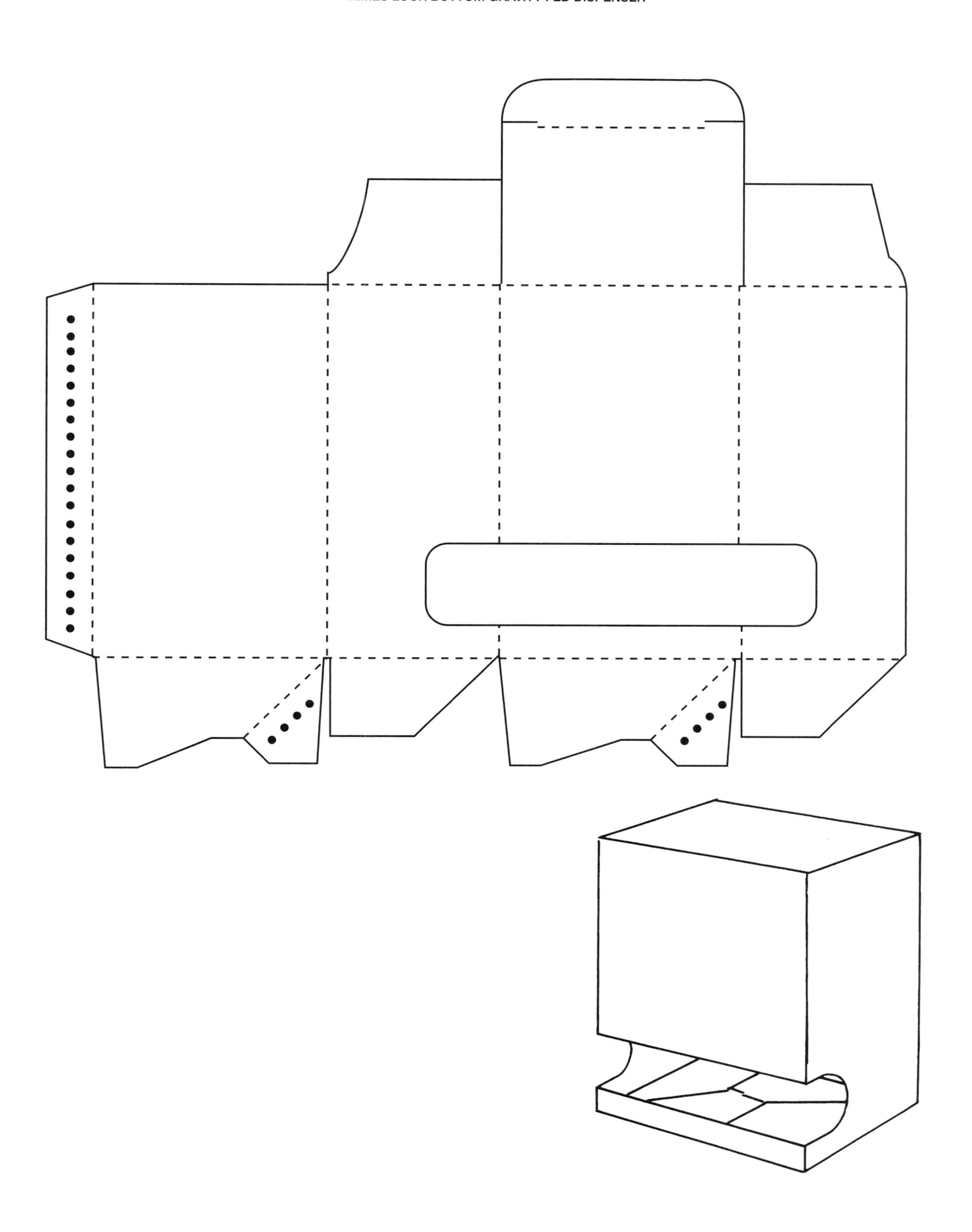

SEALABLE DISPENSER

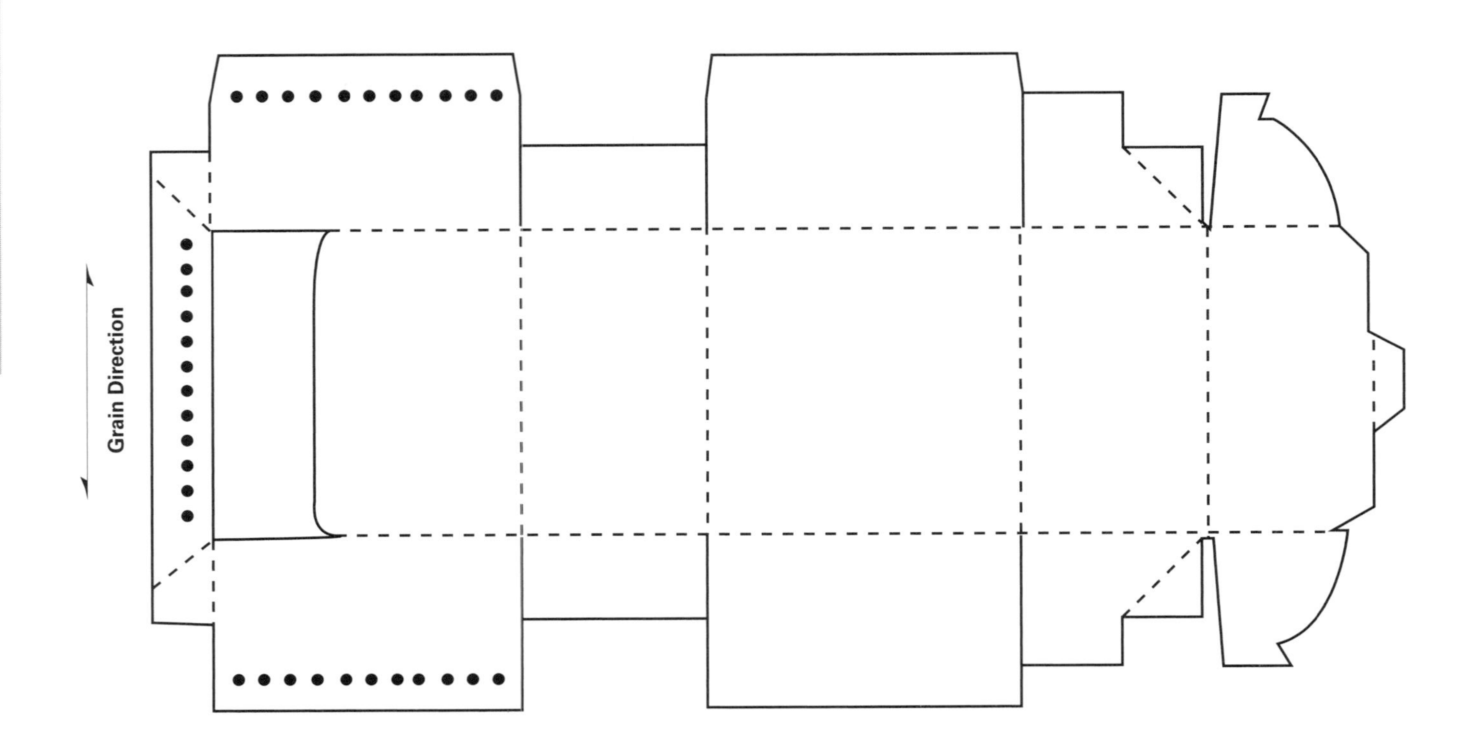

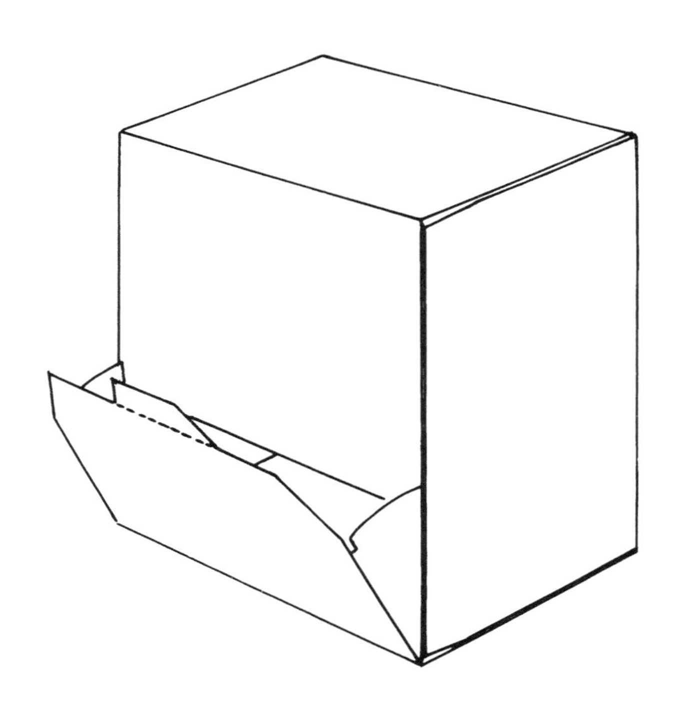

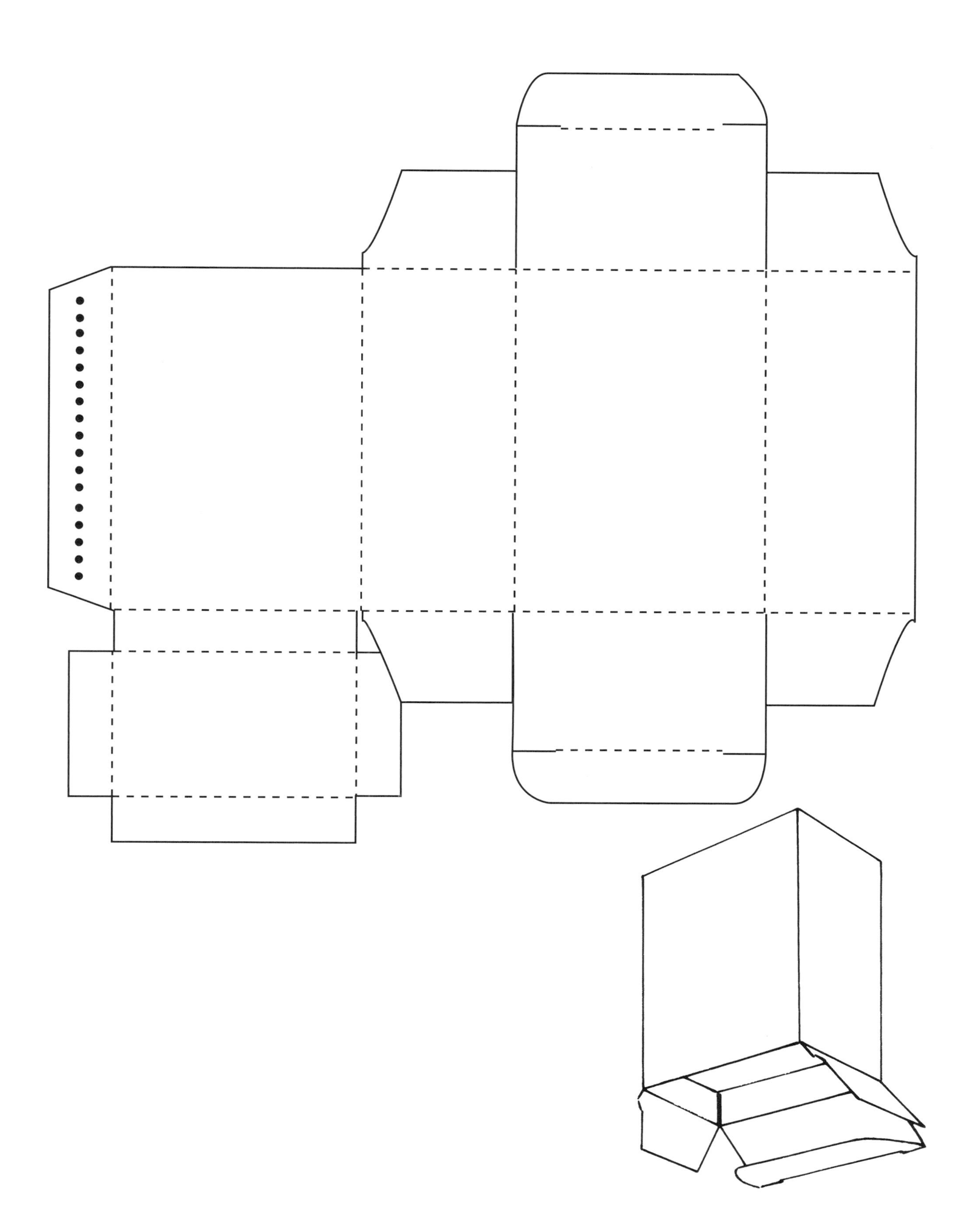

REVERSE TUCK WITH DIE-CUT BOTTOM AND HAND-TAG

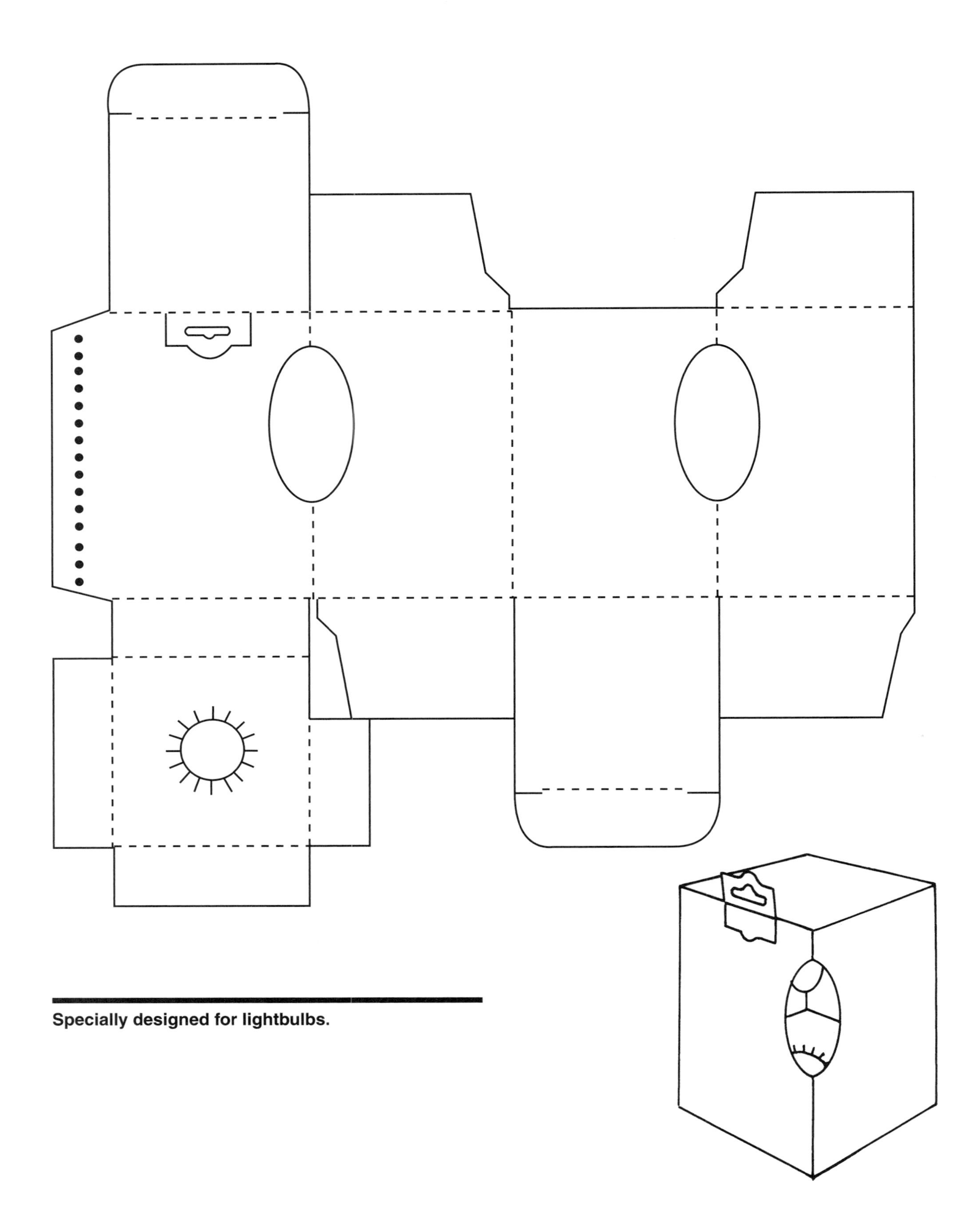

Specially designed for lightbulbs.

BOTTLE CONTAINER WITH RECESSED NECK RETAINER

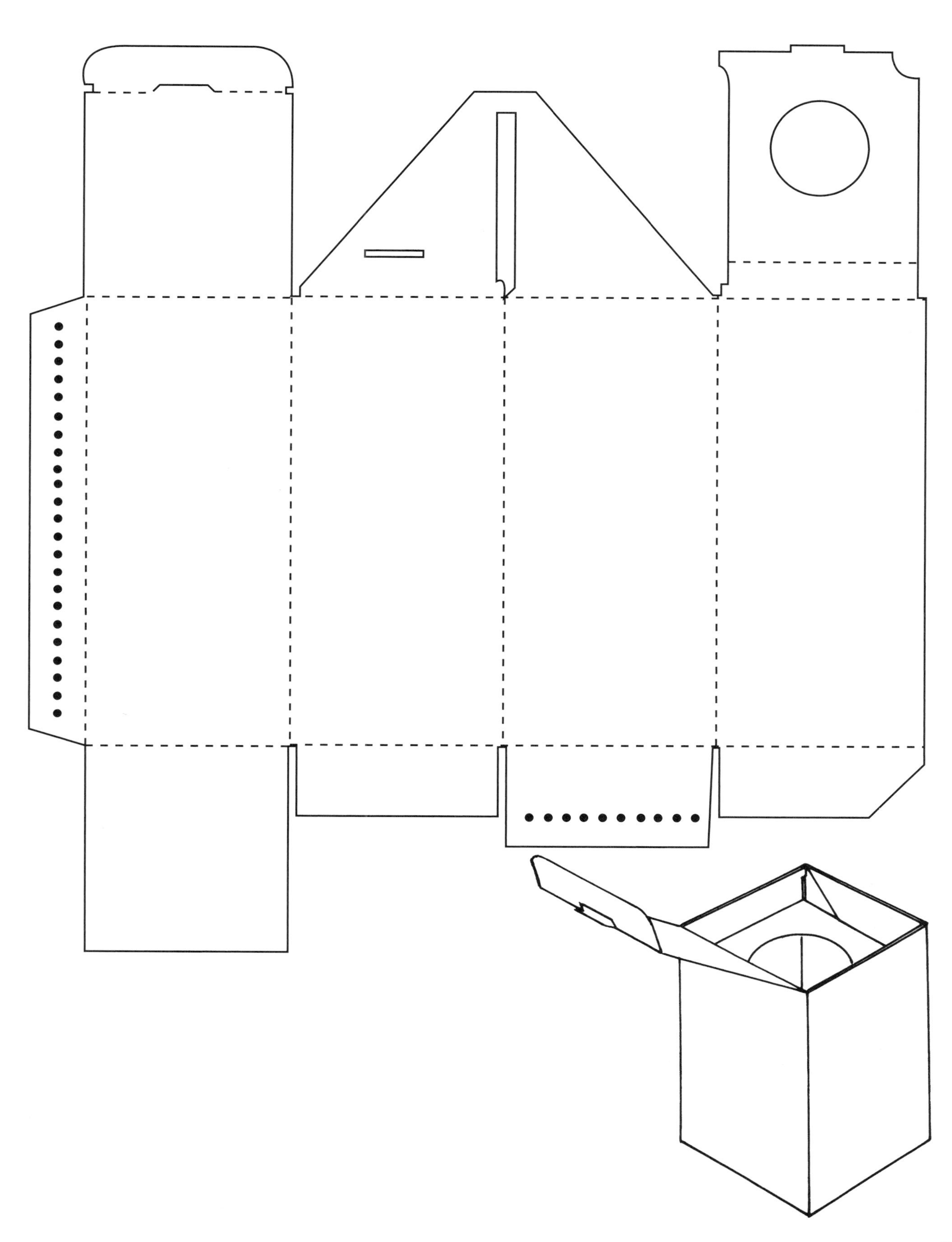

TWO-COUNT LIGHTBULB PACK

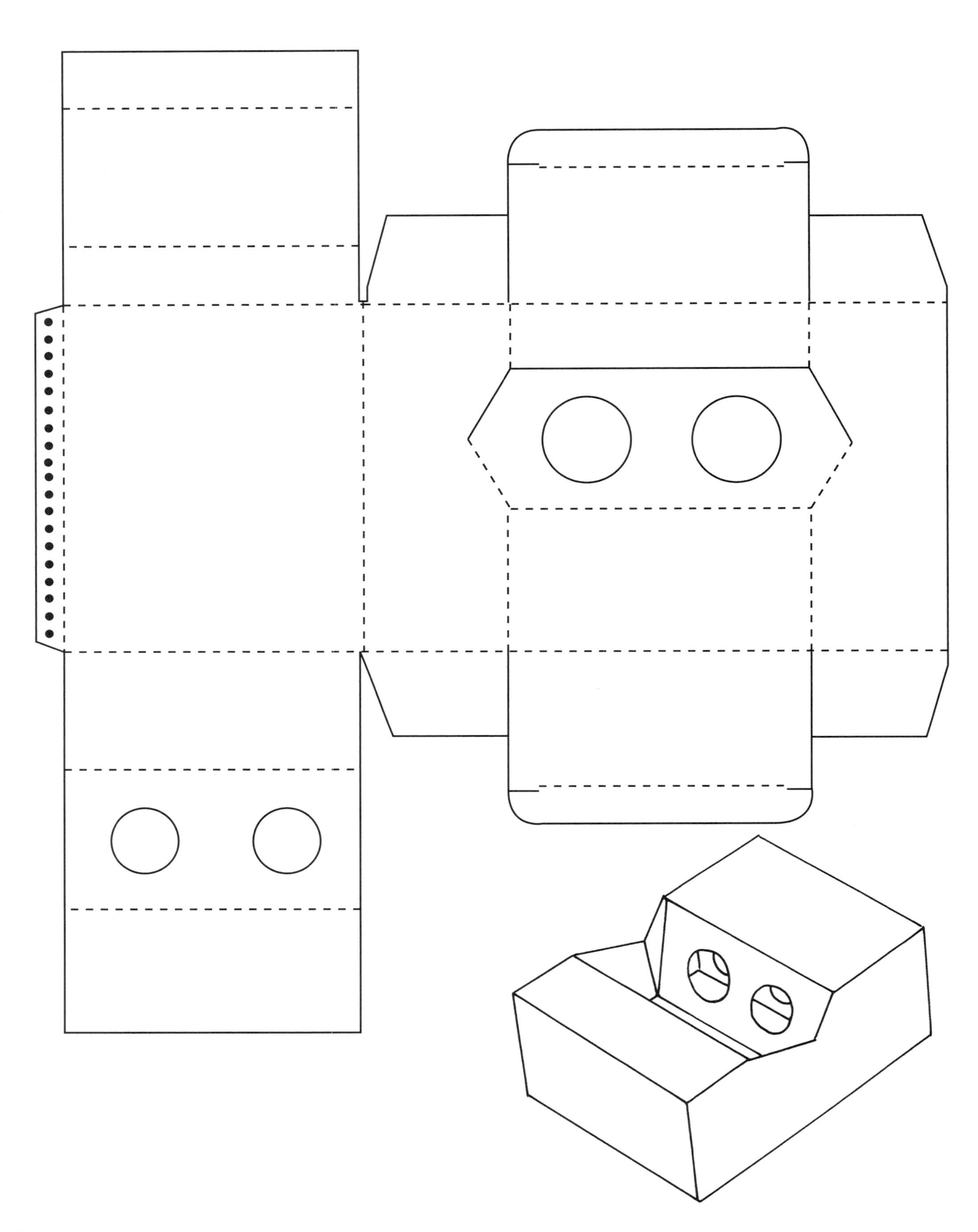

STRAIGHT TUCK WITH SPECIAL PANEL TO BE SEEN THROUGH WINDOW

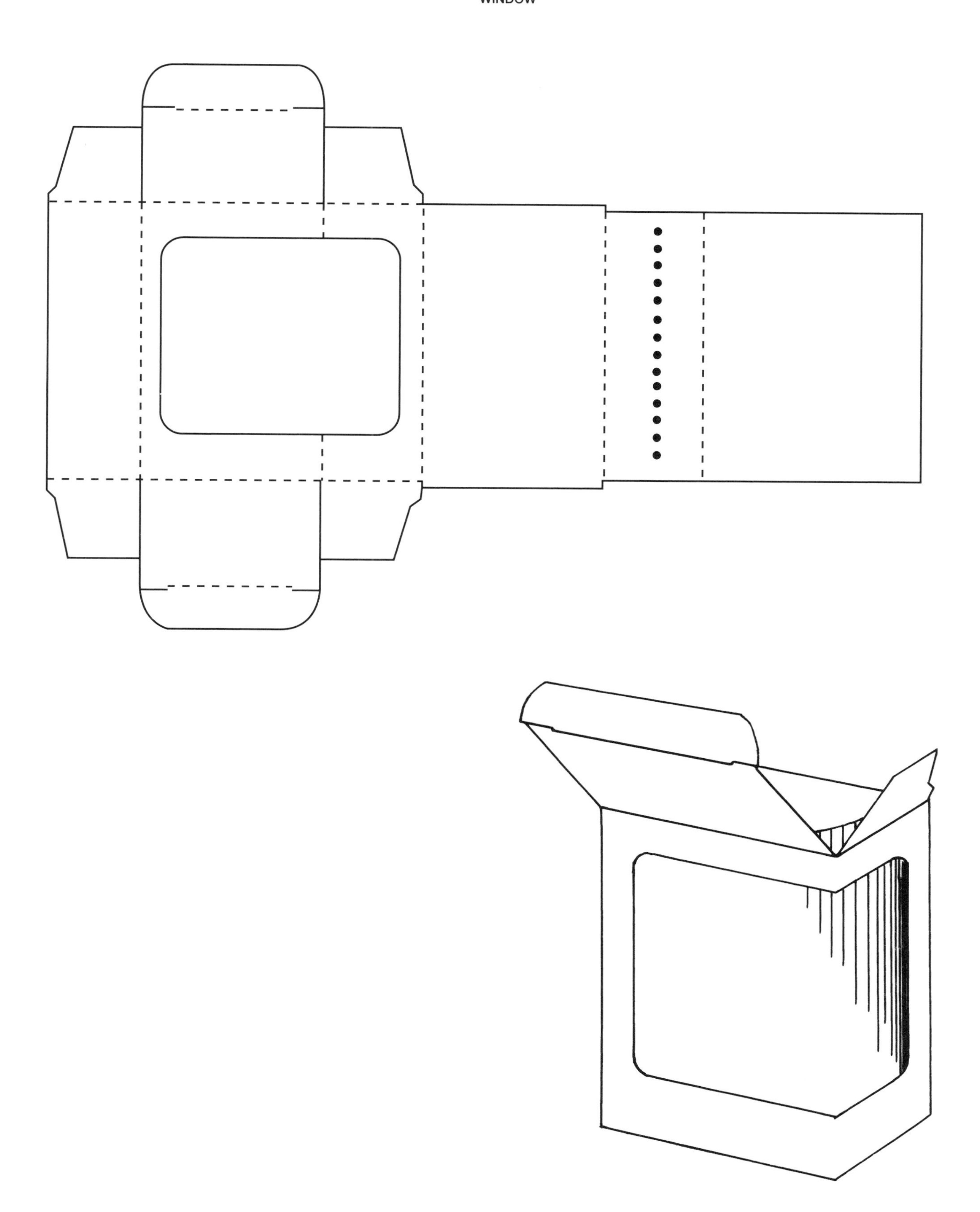

REVERSE TUCK WITH PARTITION FORMED BY THE DIE-CUT WINDOW

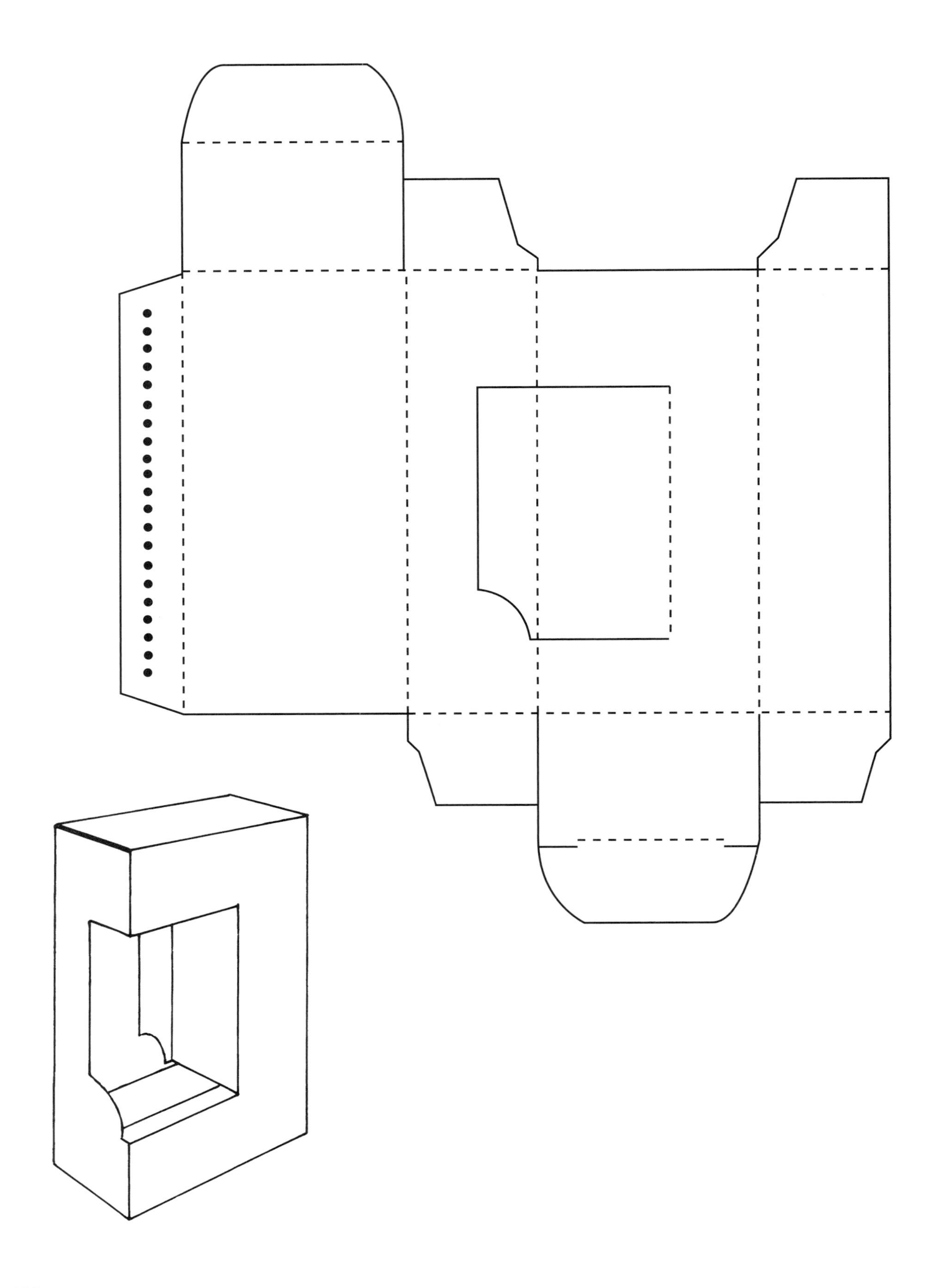

STRAIGHT TUCK WITH SPECIALLY FORMED PARTITION ATTACHED TO DUST FLAP

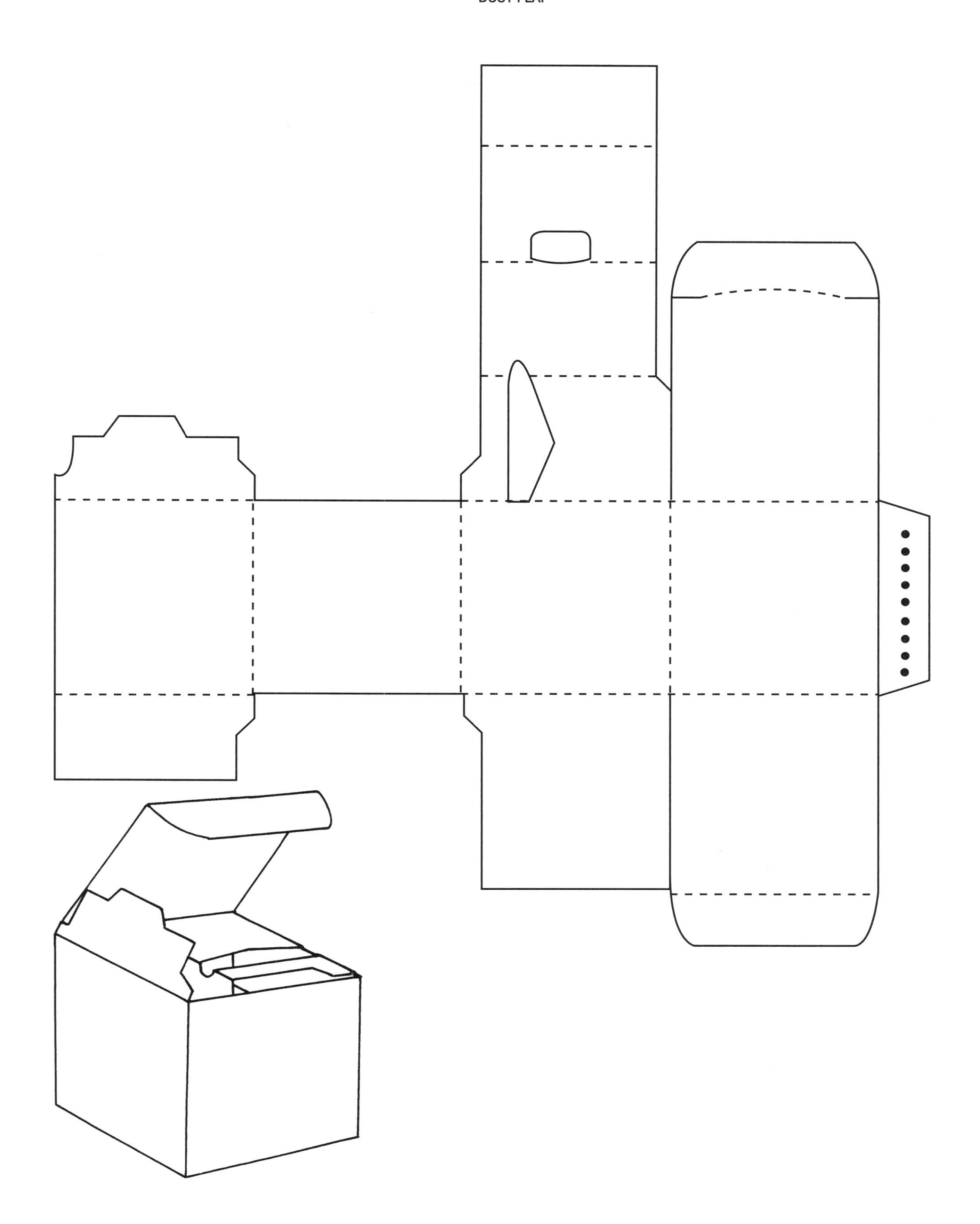

REVERSE TUCK WITH SPECIALLY CONSTRUCTED PARTITION

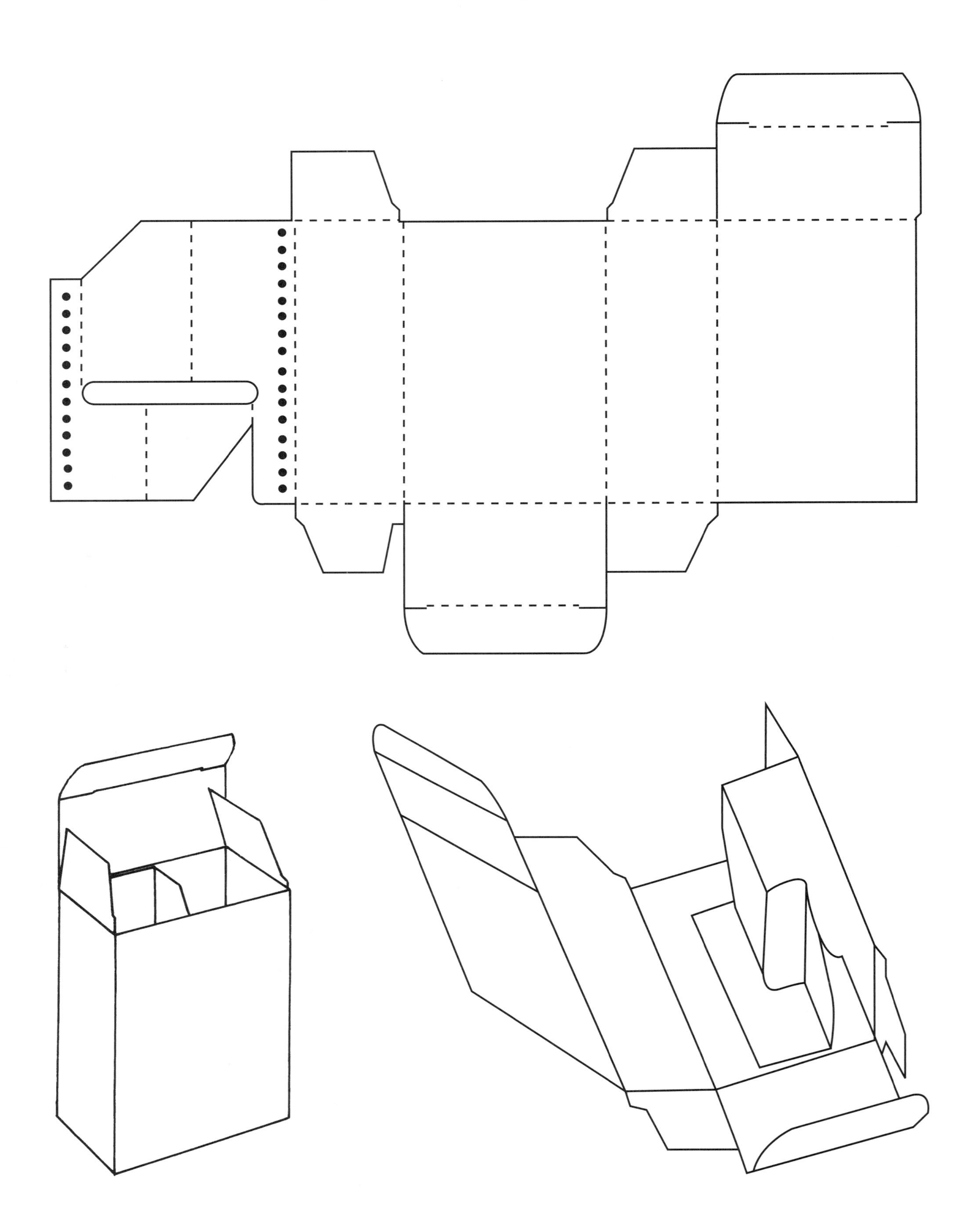

STRAIGHT TUCK WITH TWO COMPARTMENTS

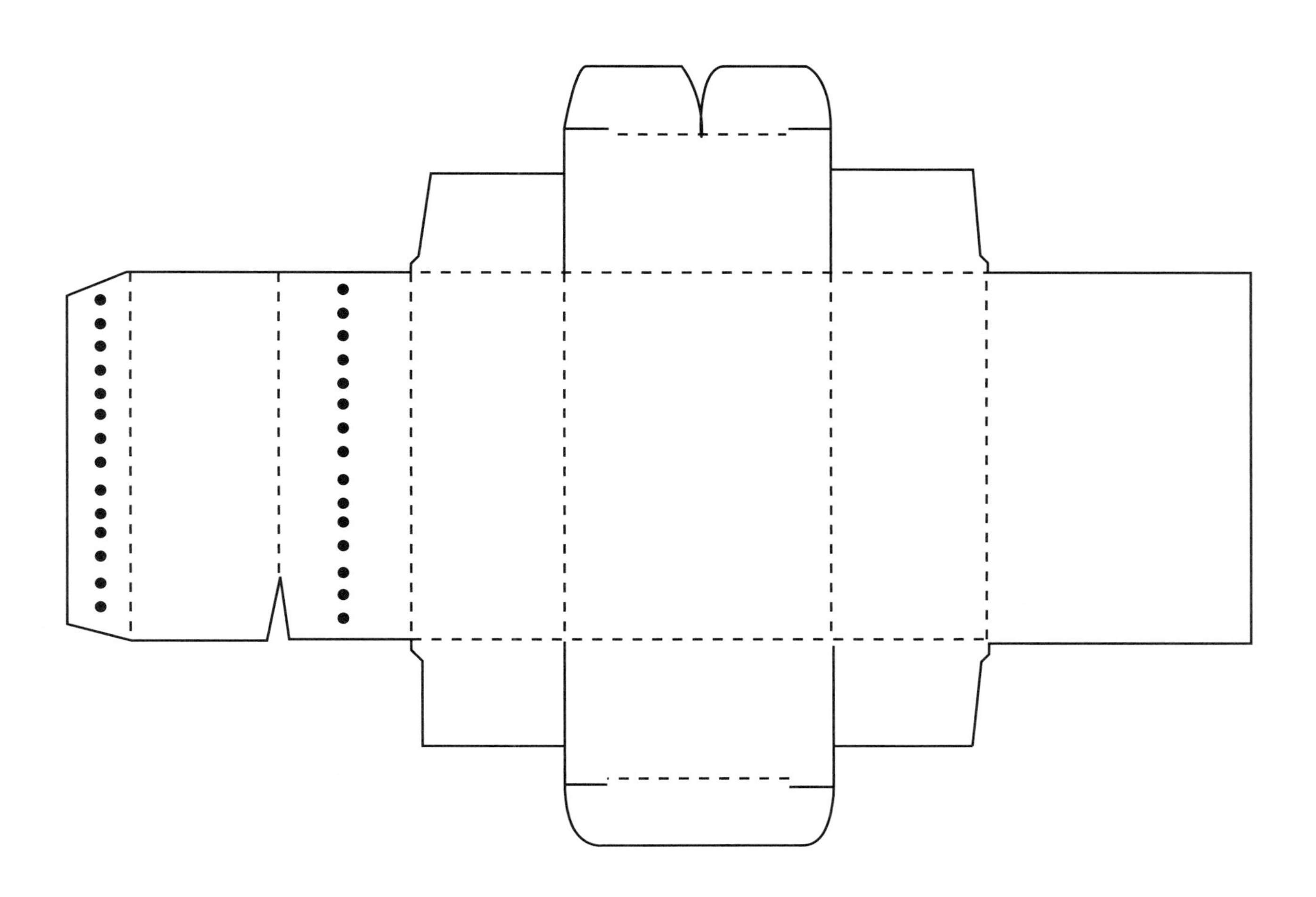

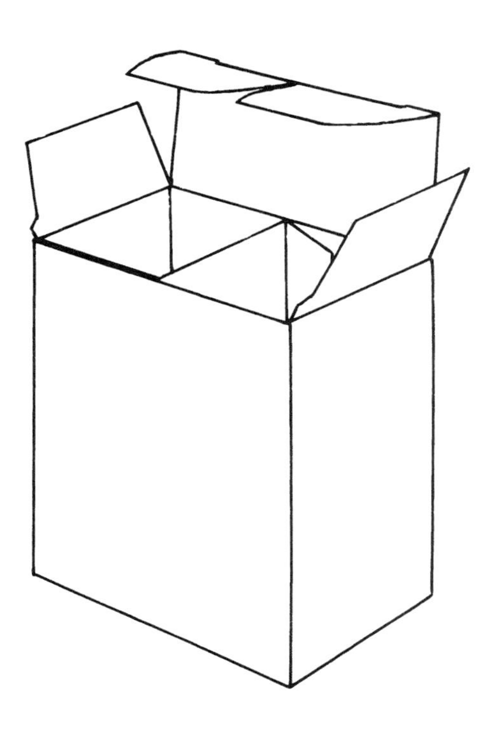

STRAIGHT TUCK WITH SPECIAL COMPARTMENTS

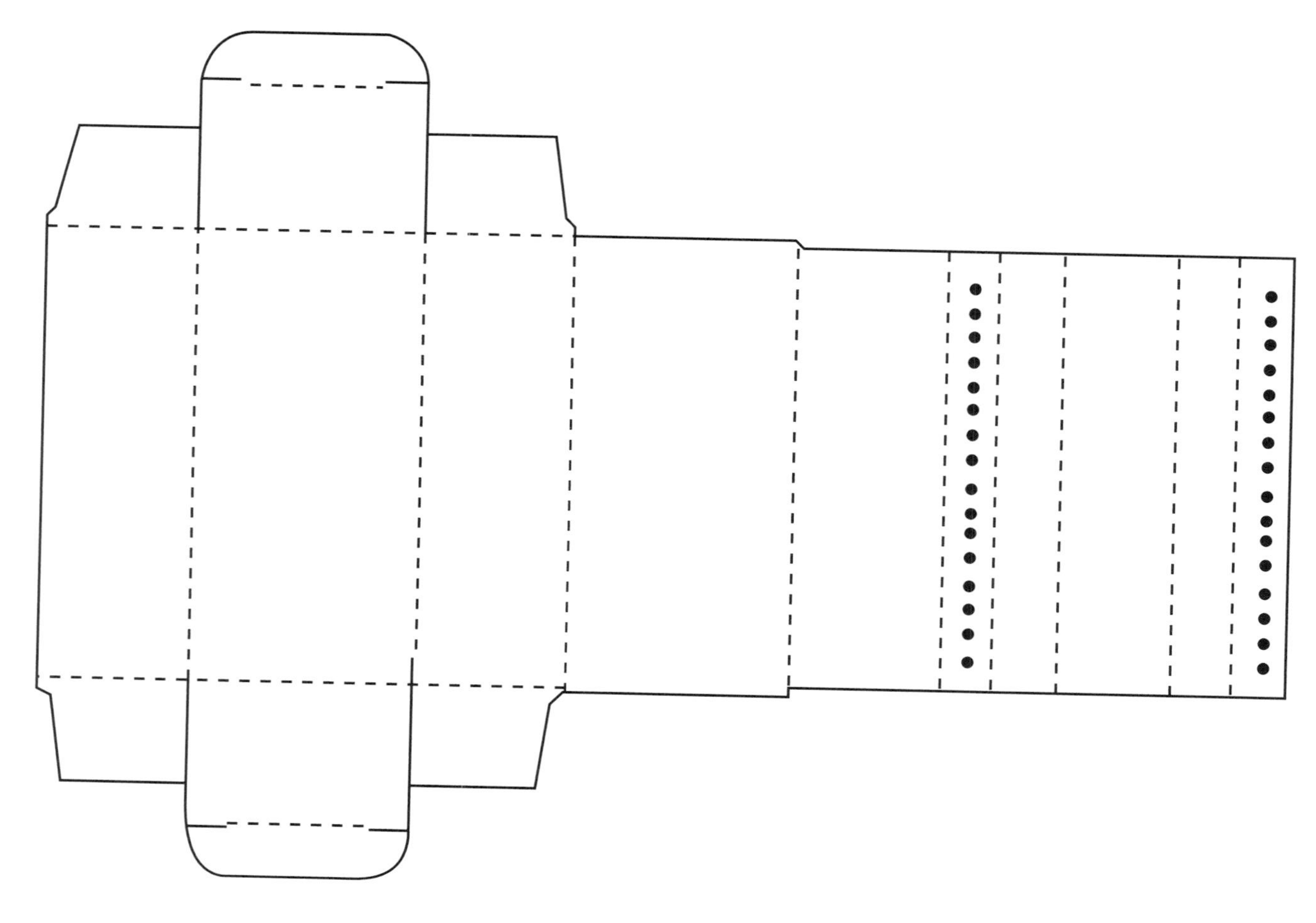

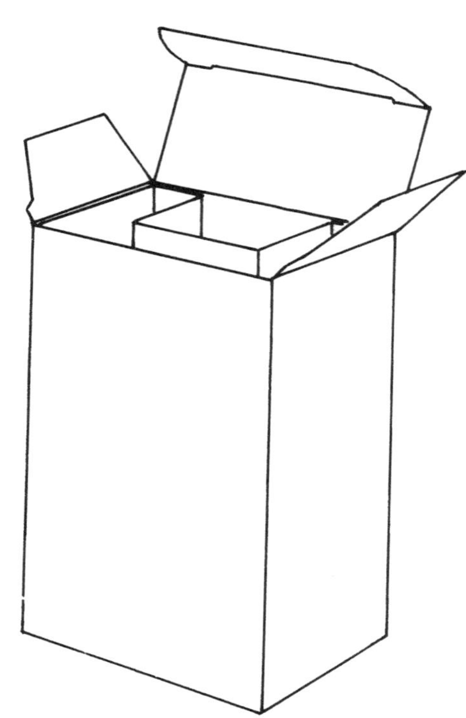

REVERSE TUCK WITH THREE COMPARTMENTS

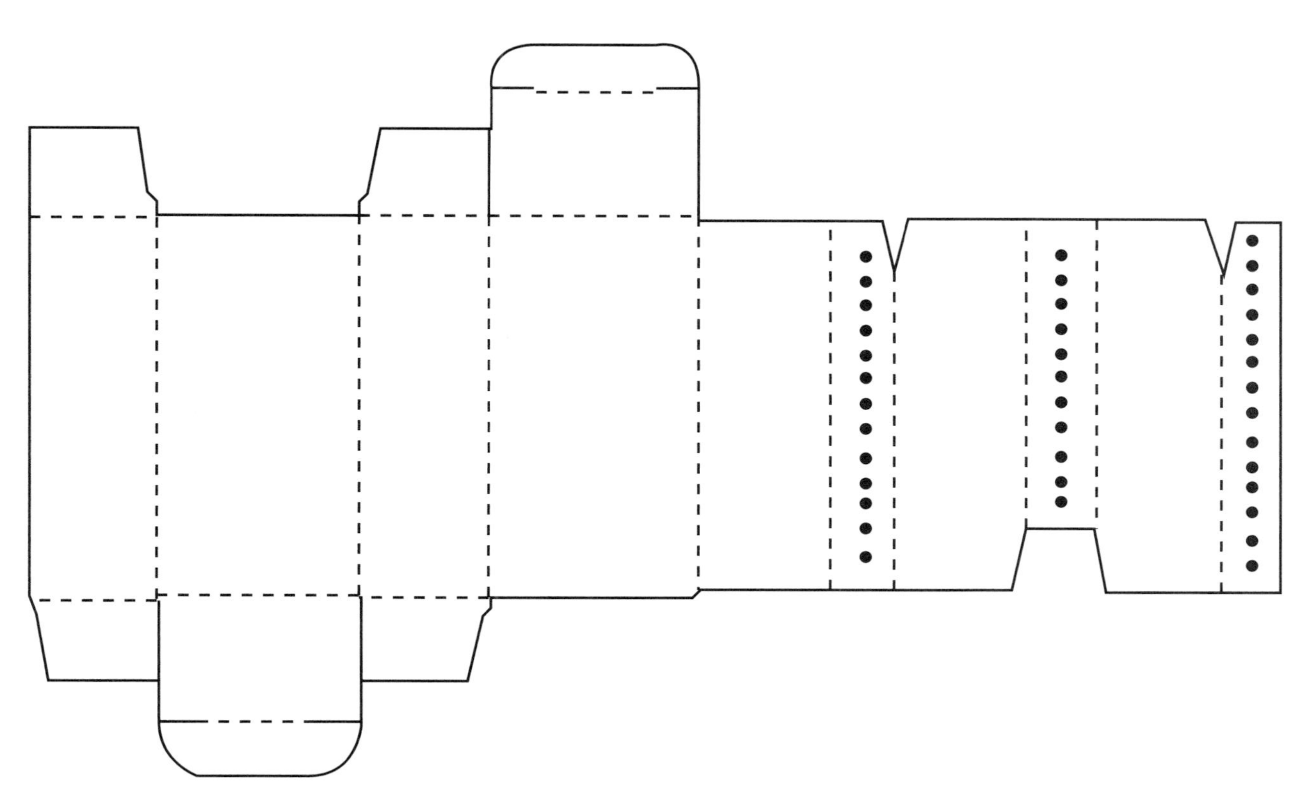

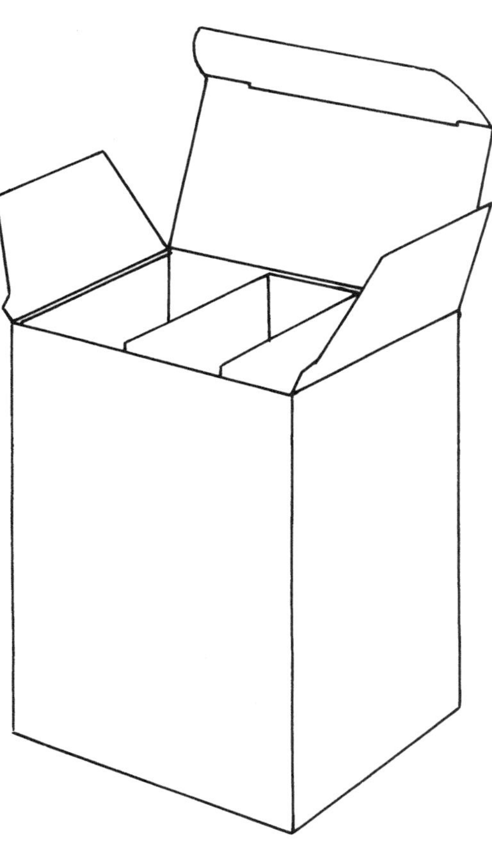

SIDE-BY-SIDE REVERSE TUCK/STRAIGHT TUCK CARTONS, VARYING IN HEIGHT

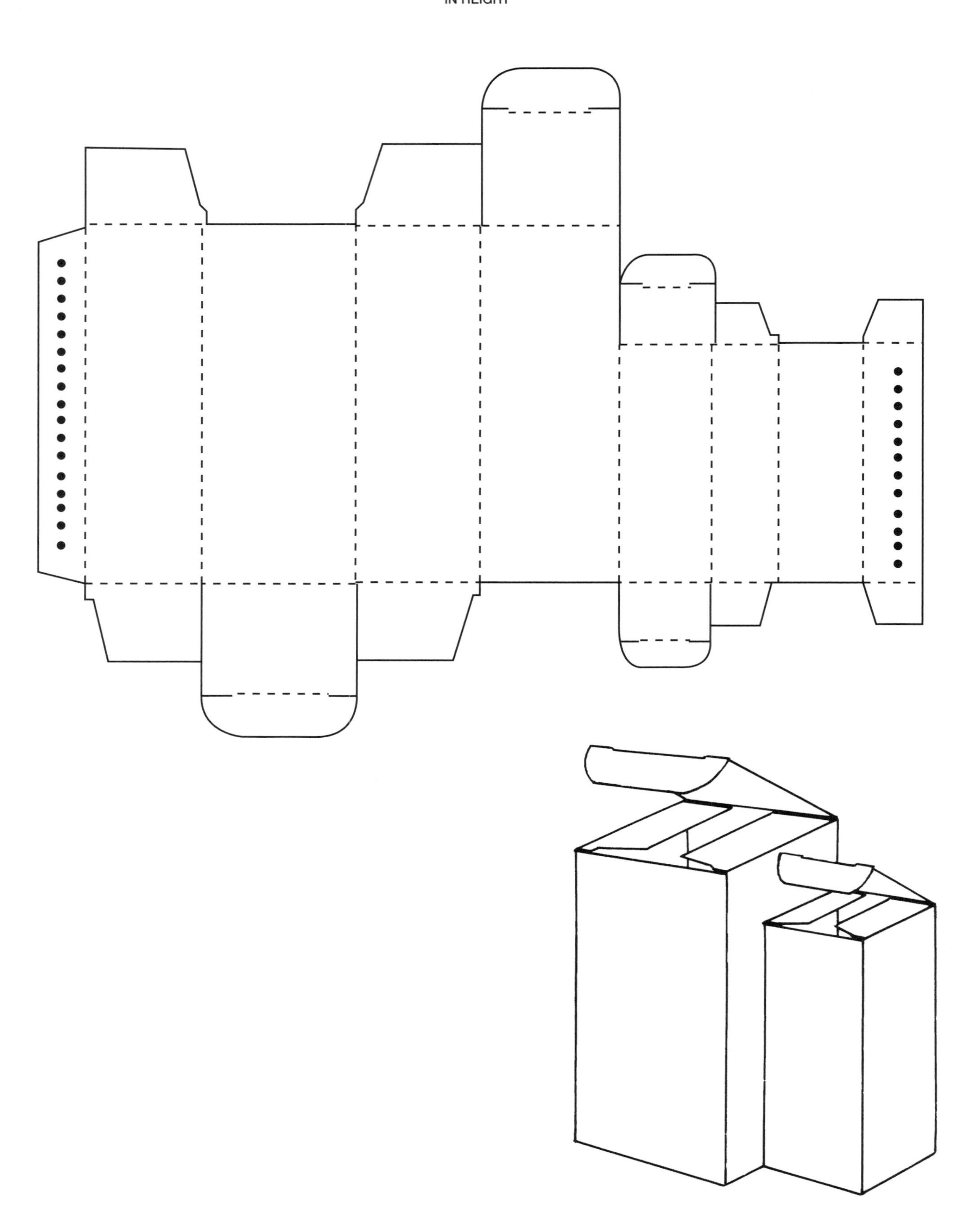

SIDE-BY-SIDE STRAIGHT TUCK CARTONS, IDENTICAL HEIGHT

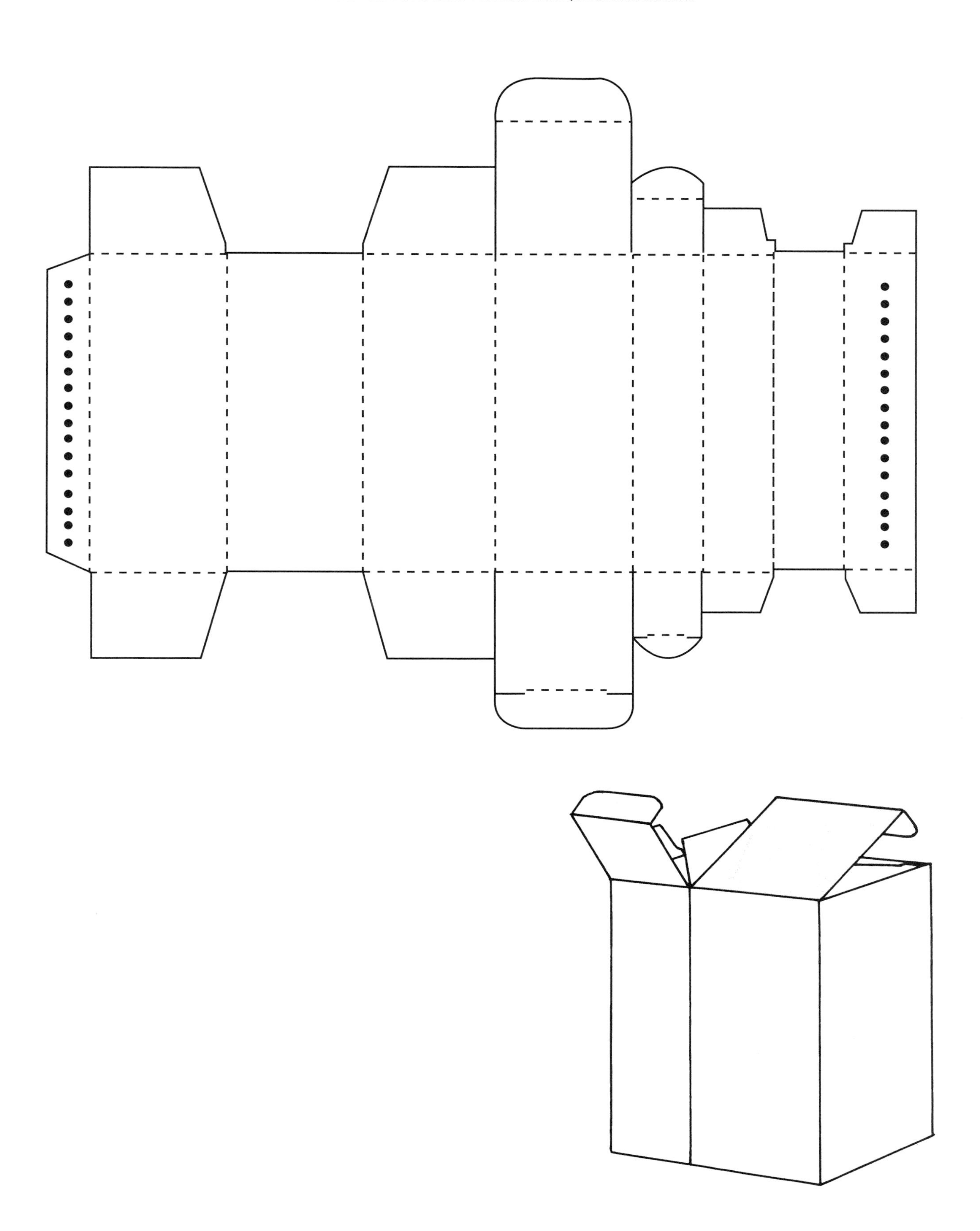

REVERSE TUCK SHADOW BOX

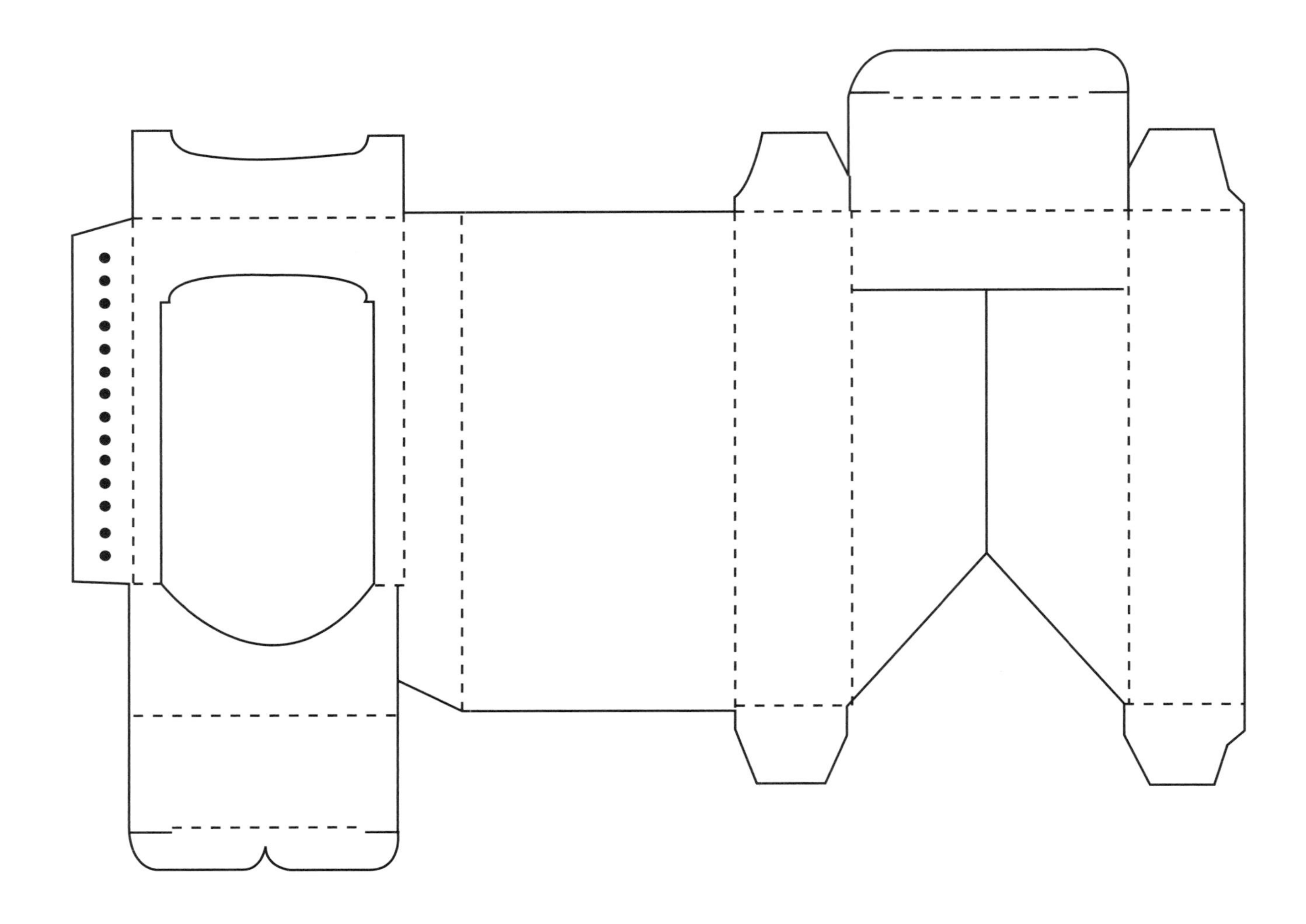

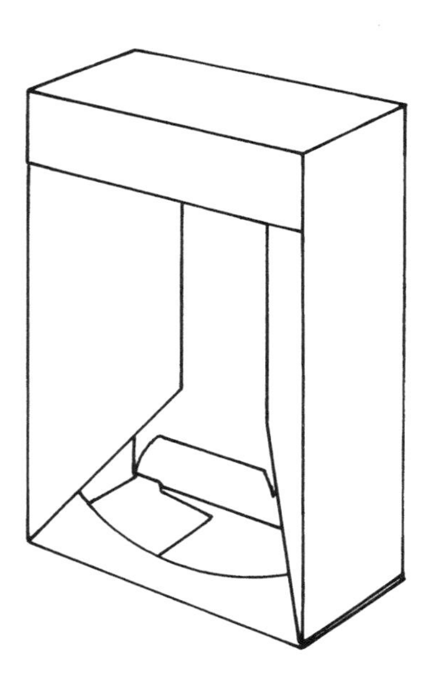

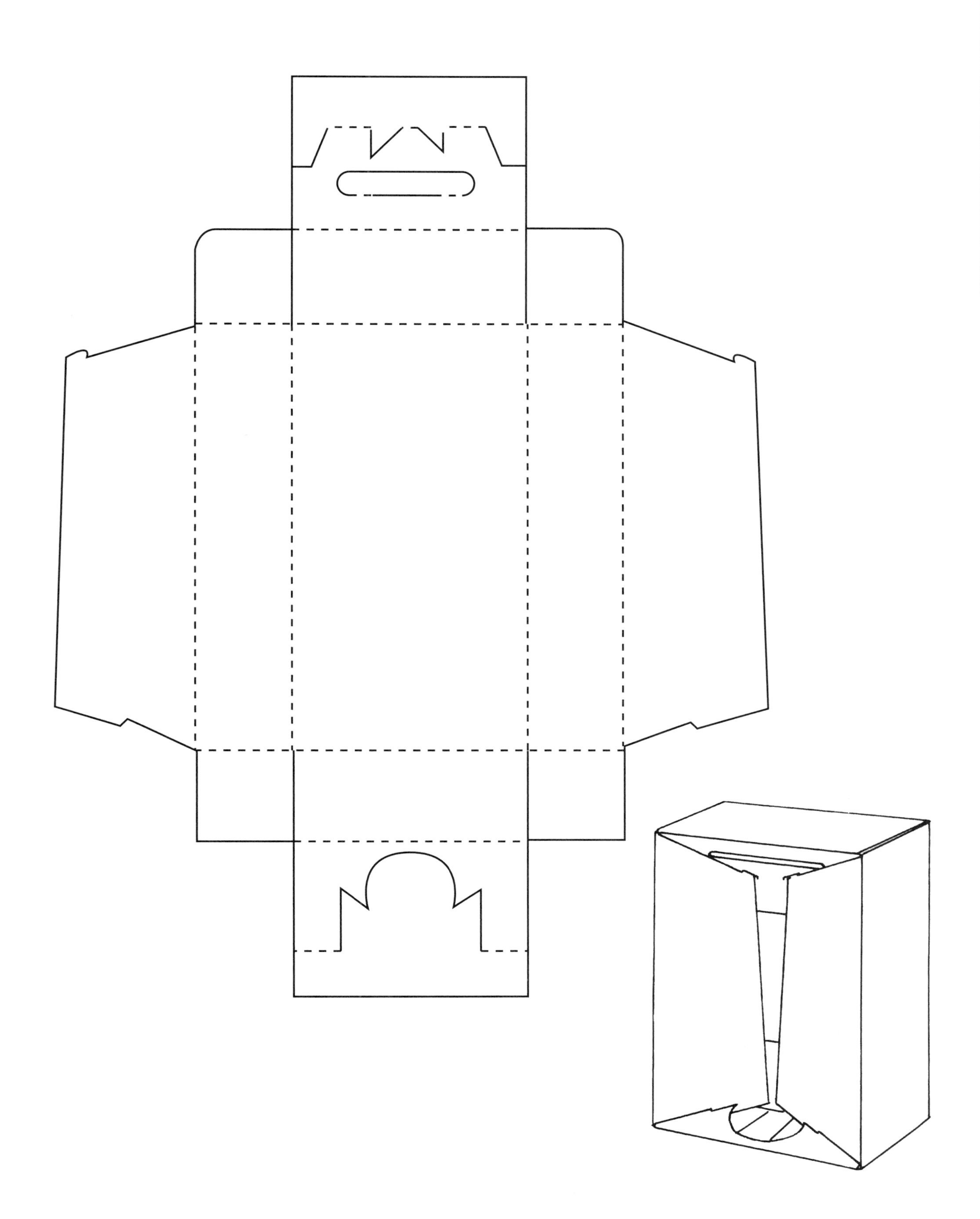

STRAIGHT TUCK SHADOW BOX

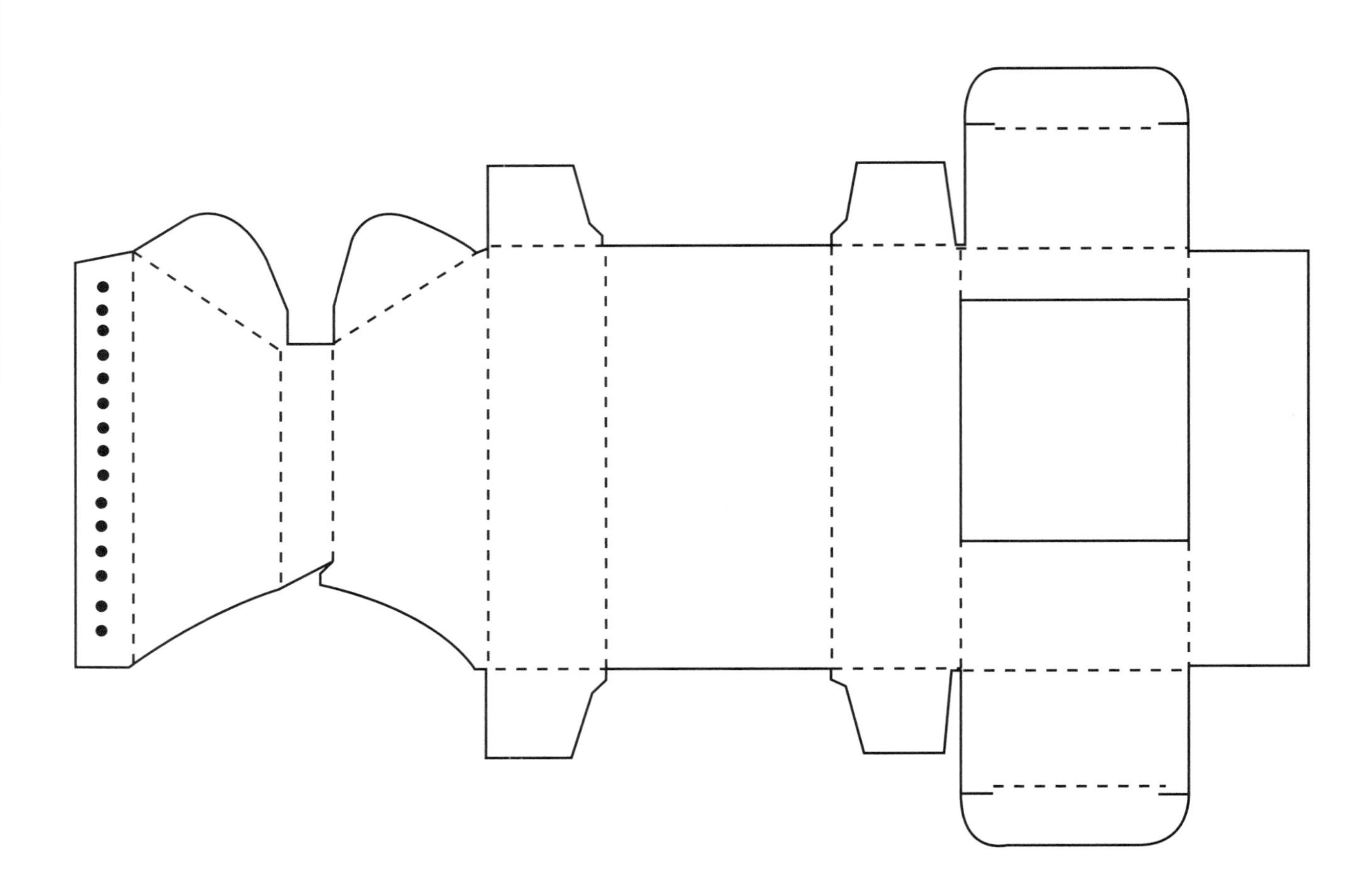

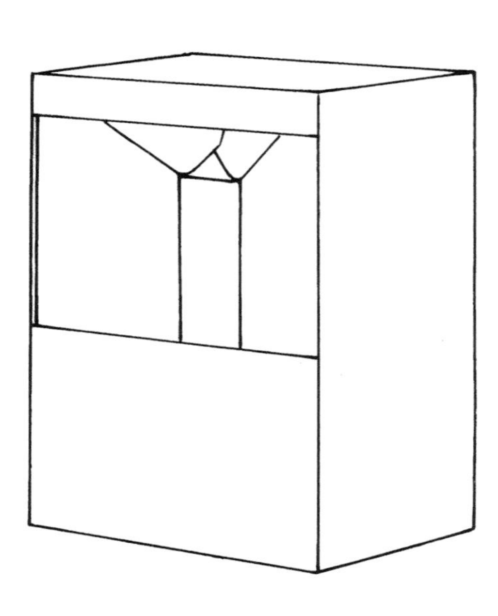

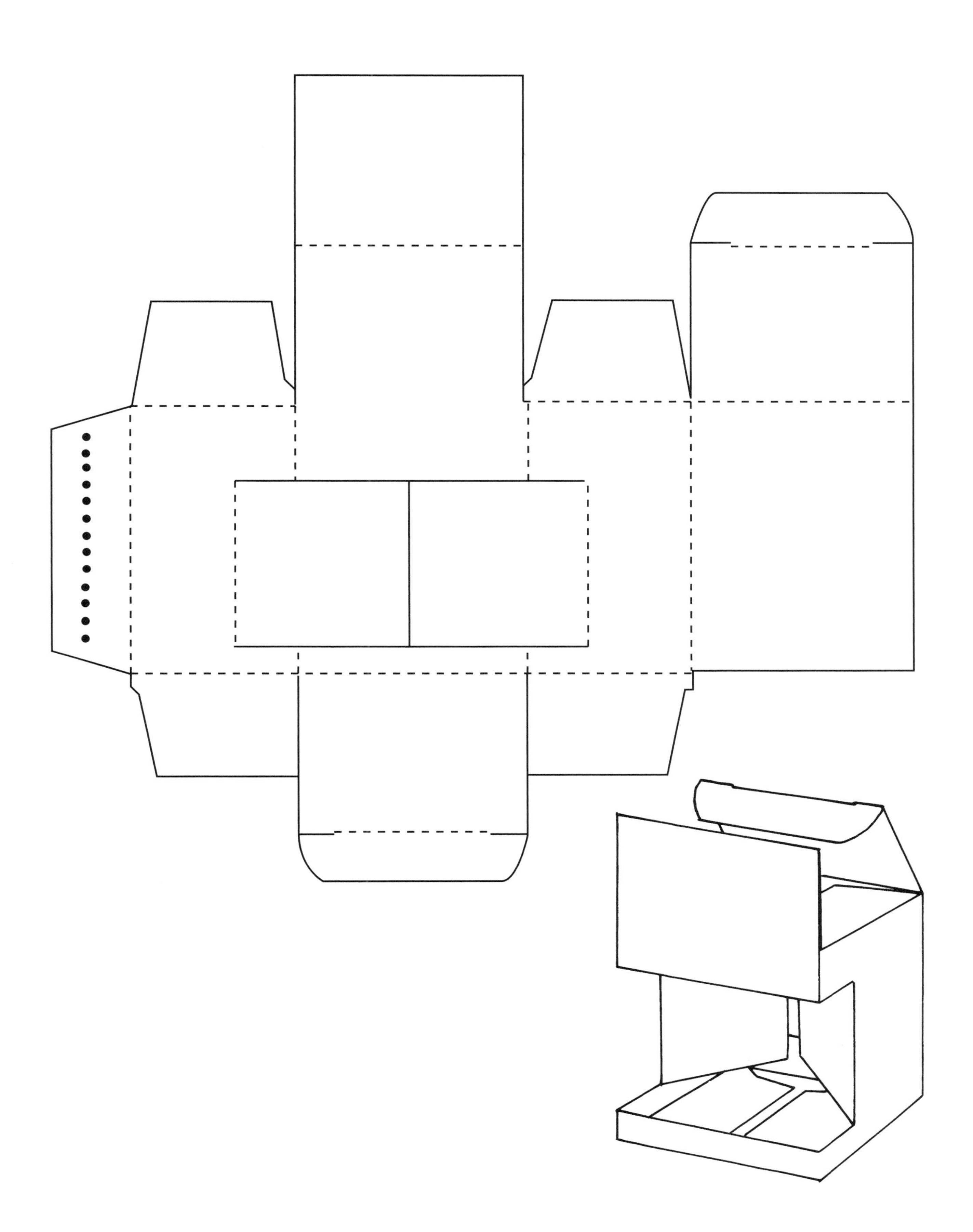

ONE PIECE OPEN VIEW DISPLAY CARTON

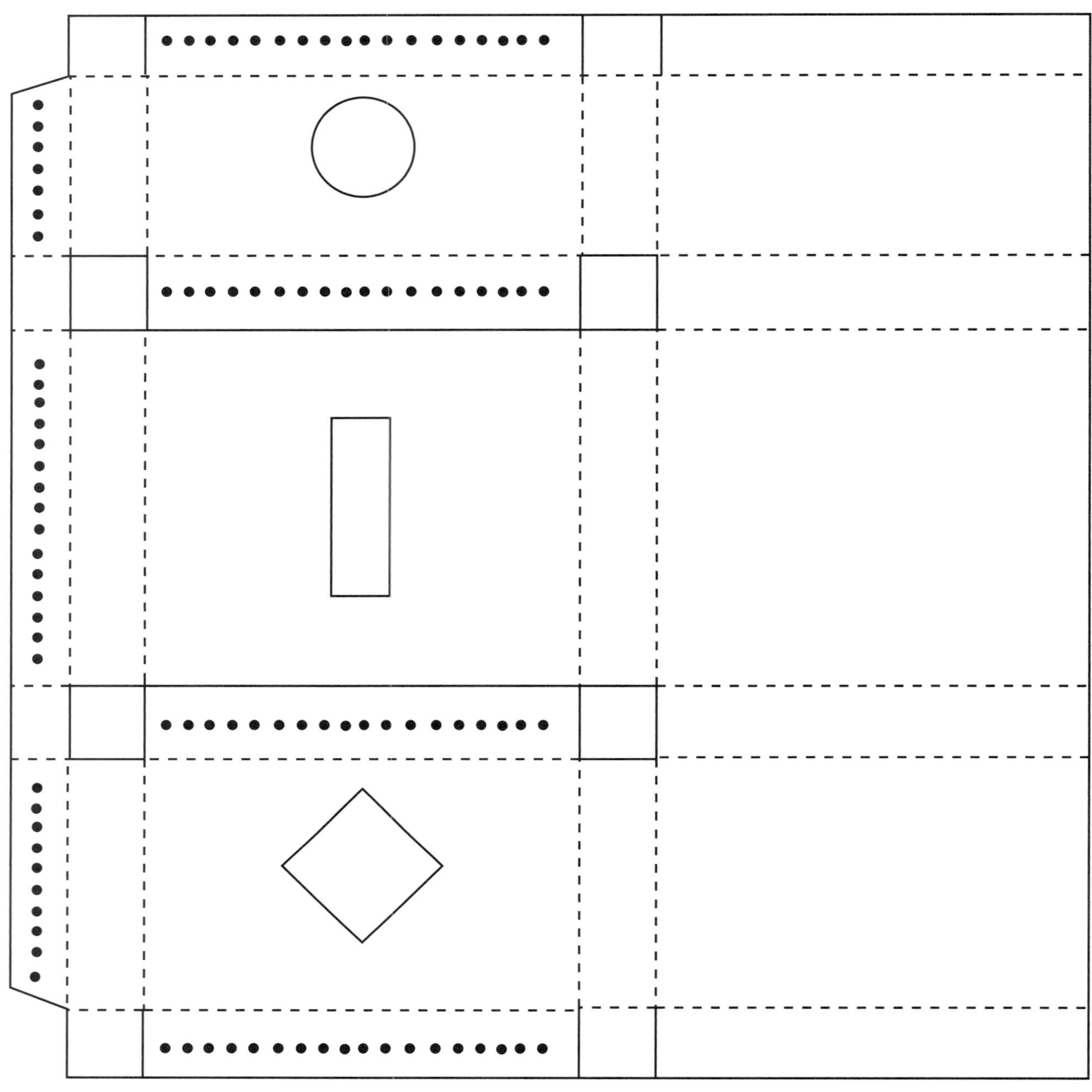

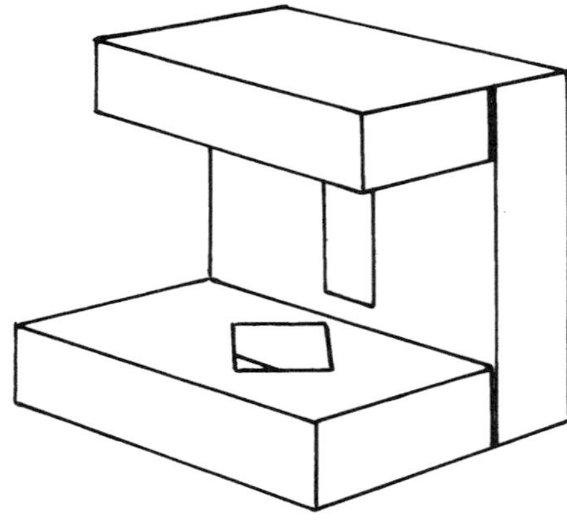

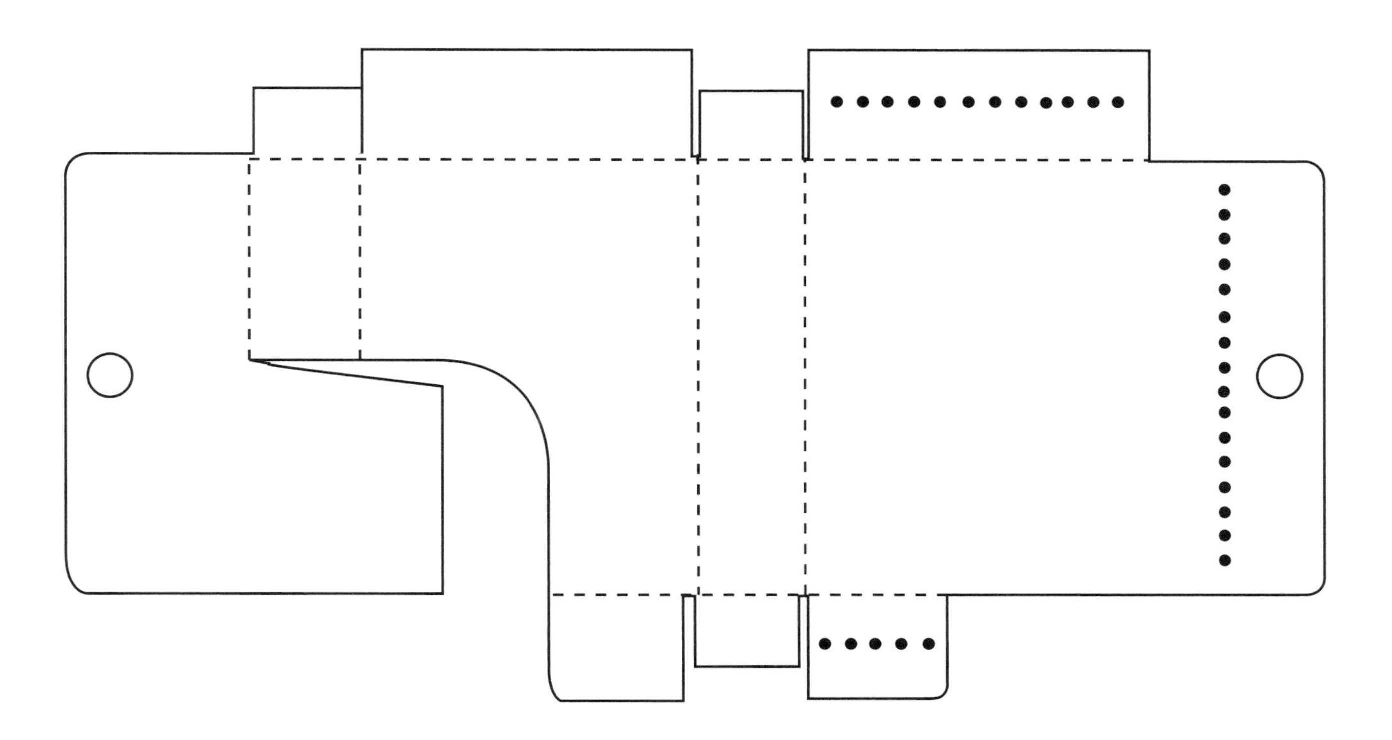

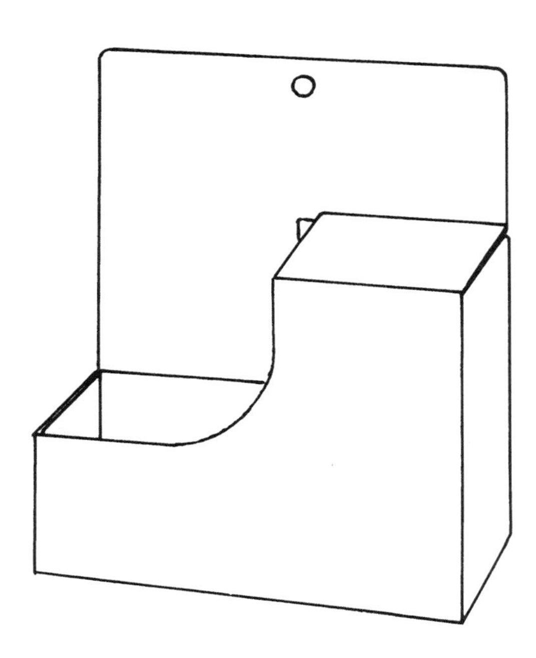

ONE PIECE SHADOW BOX

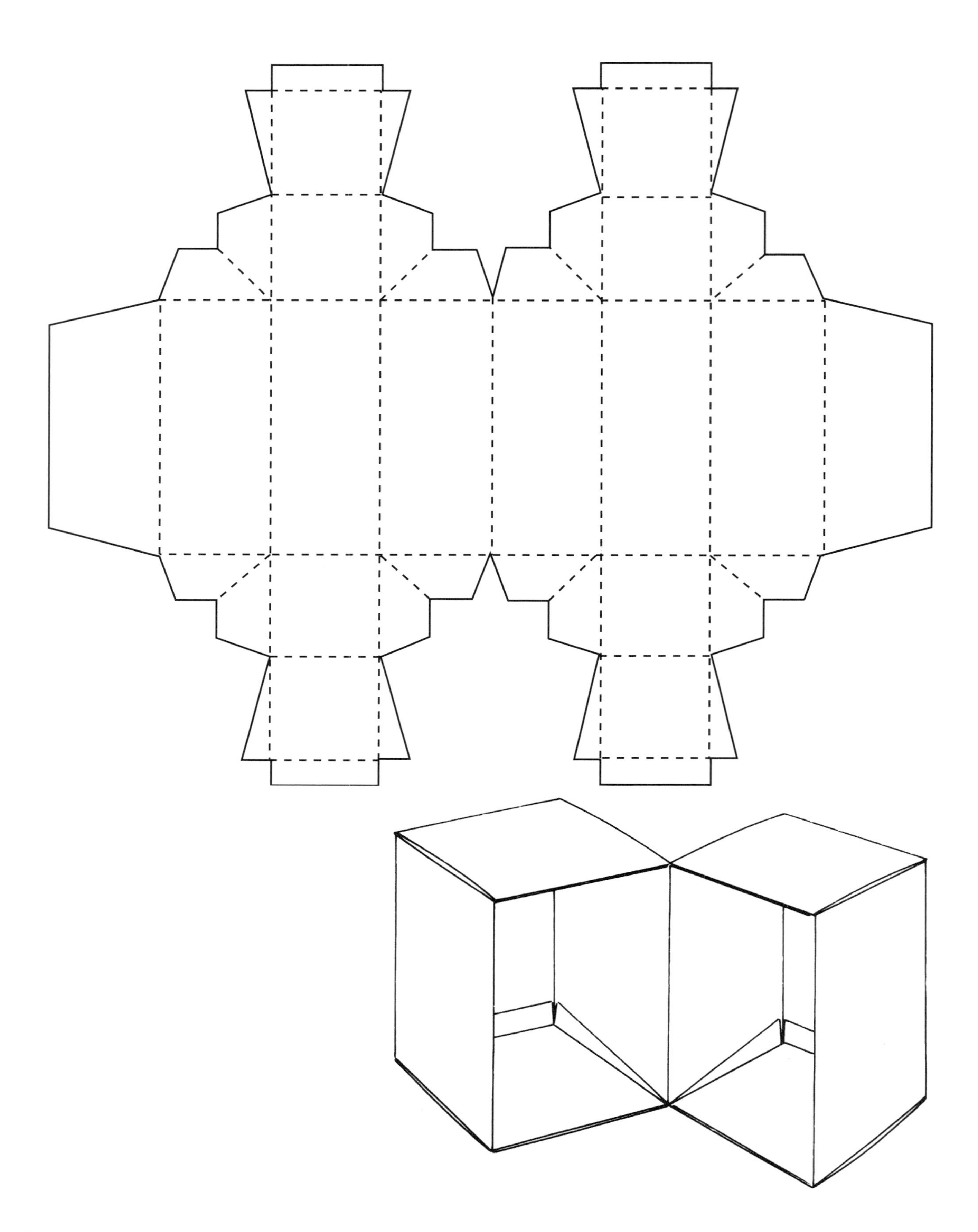

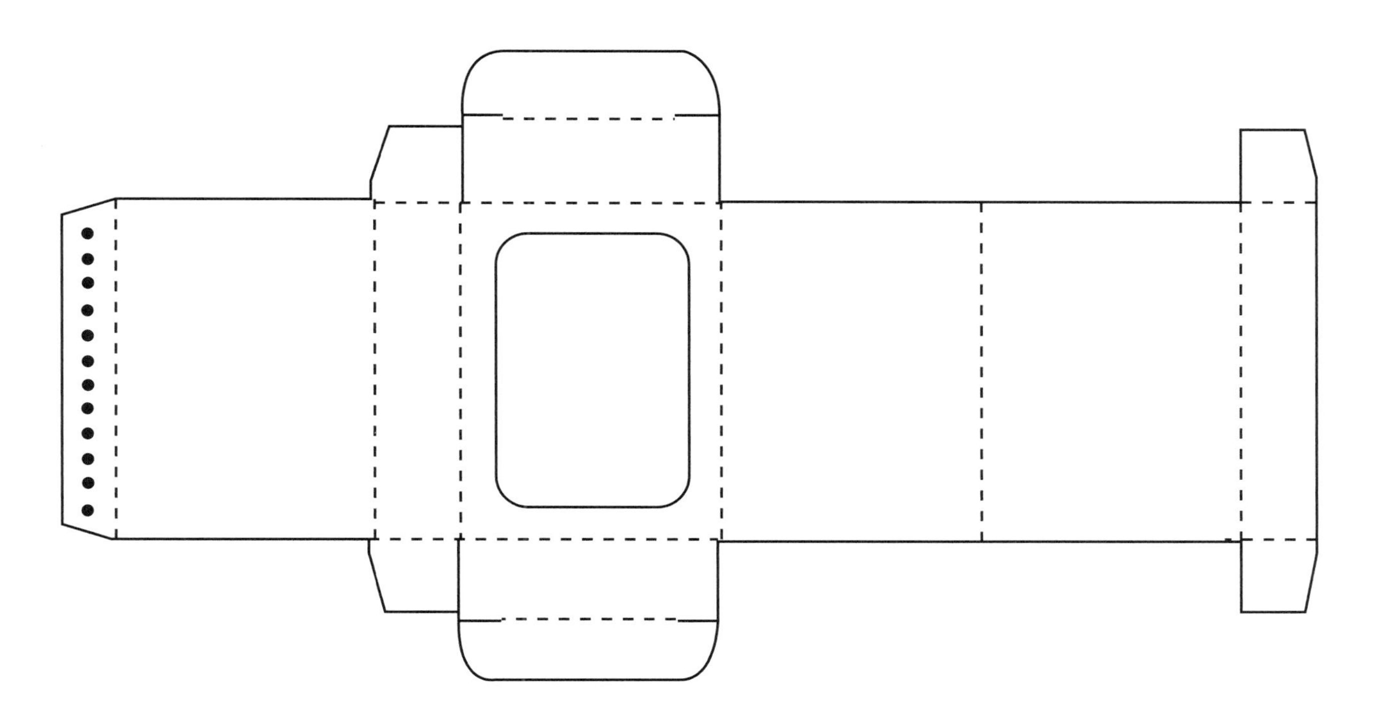

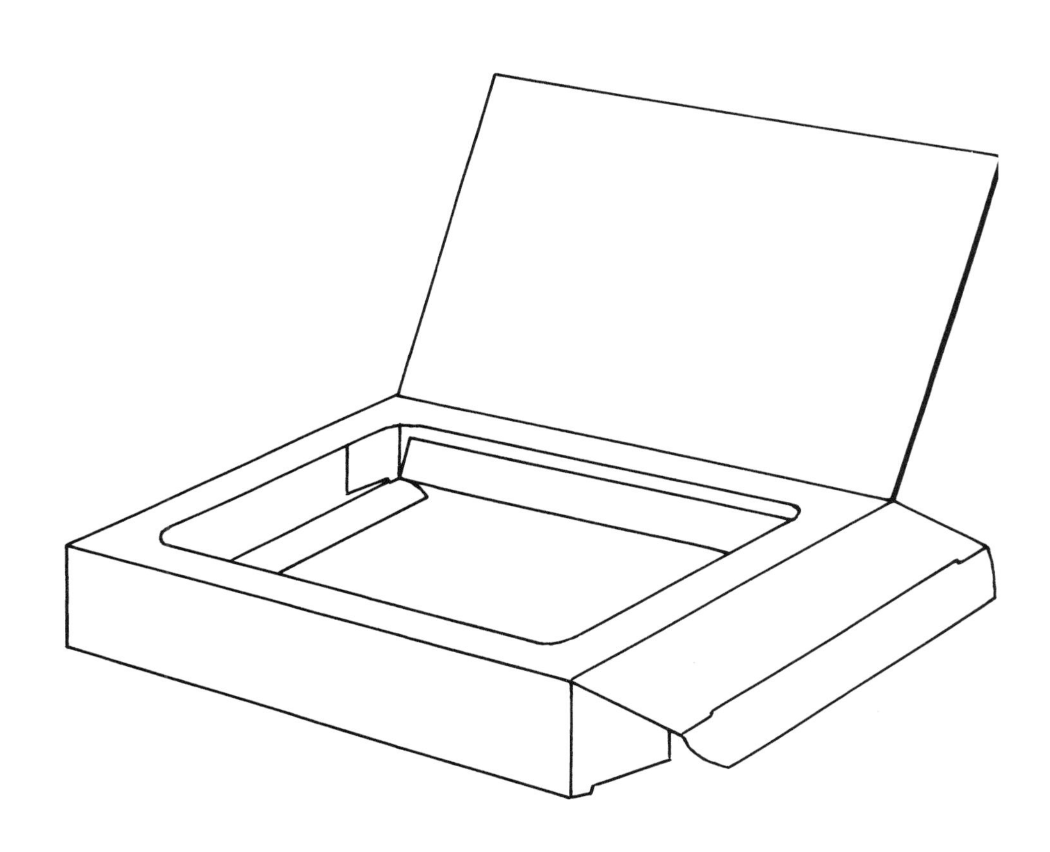

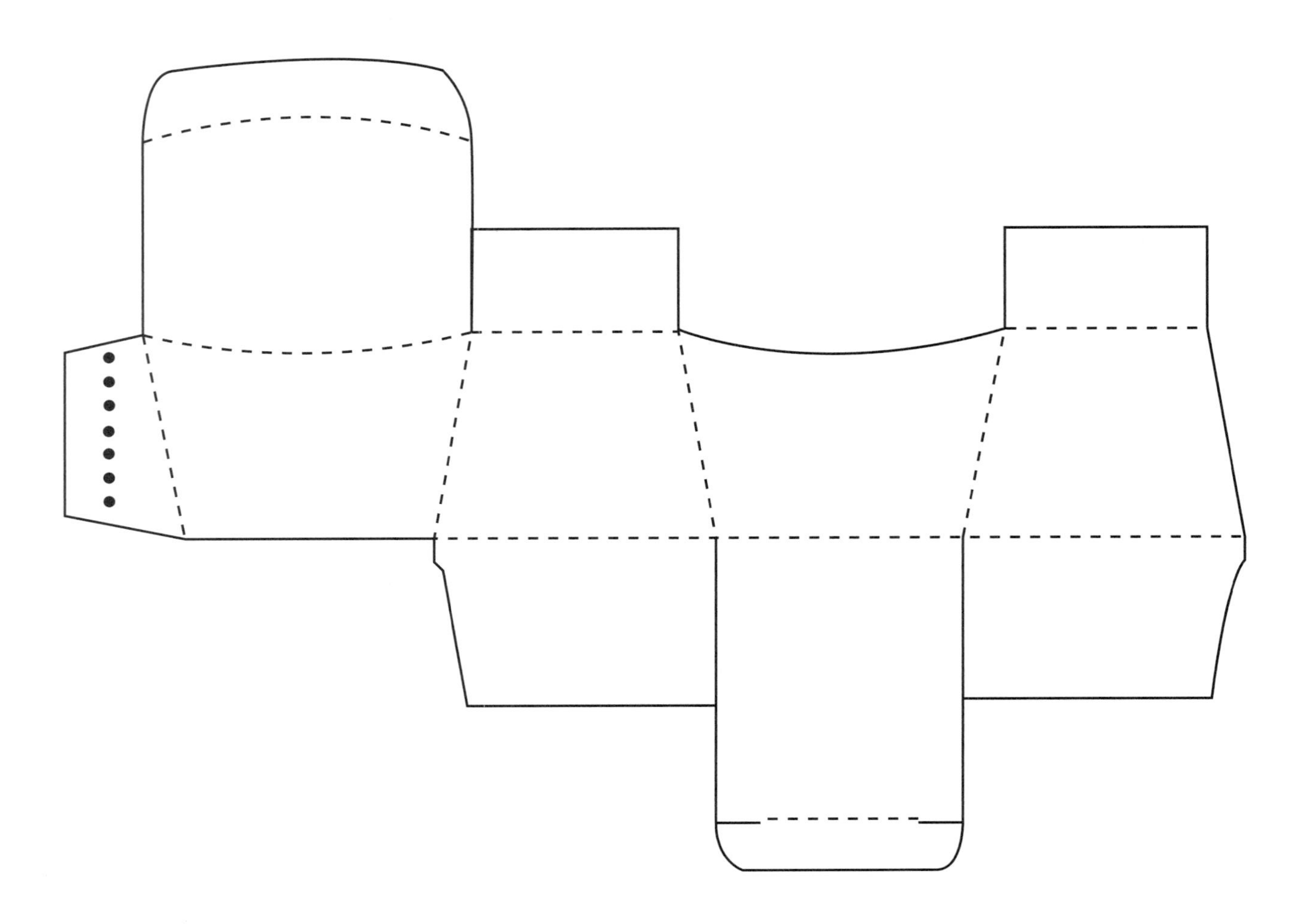

The special effect is created by the concave top and convex front or face panel.

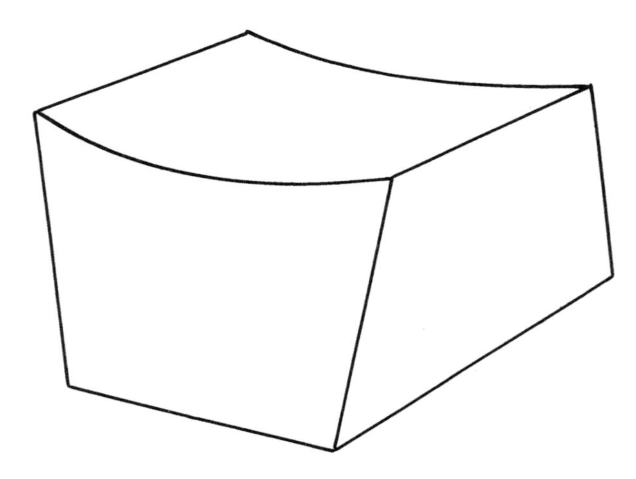

FRICTION SYTLE CLOSURE

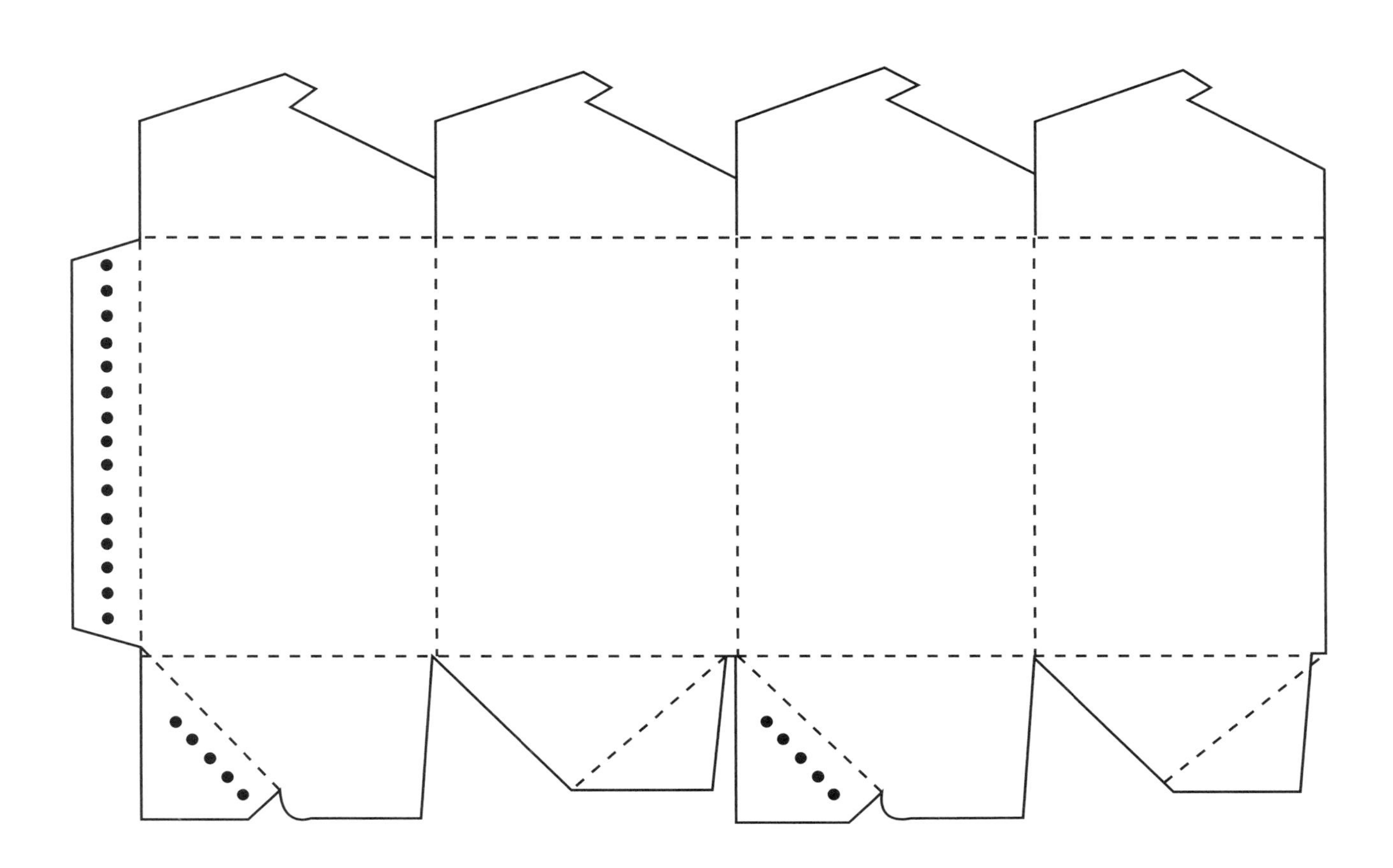

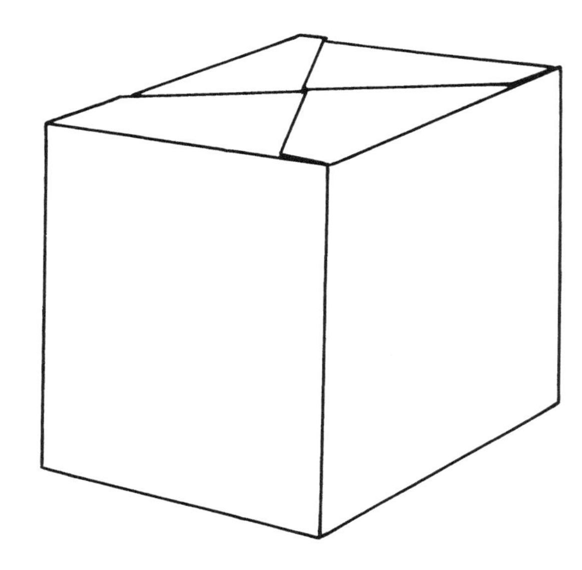

Alternating closure creates an interesting graphic effect that may be enhanced with color.

CONCAVE/CONVEX CARTON

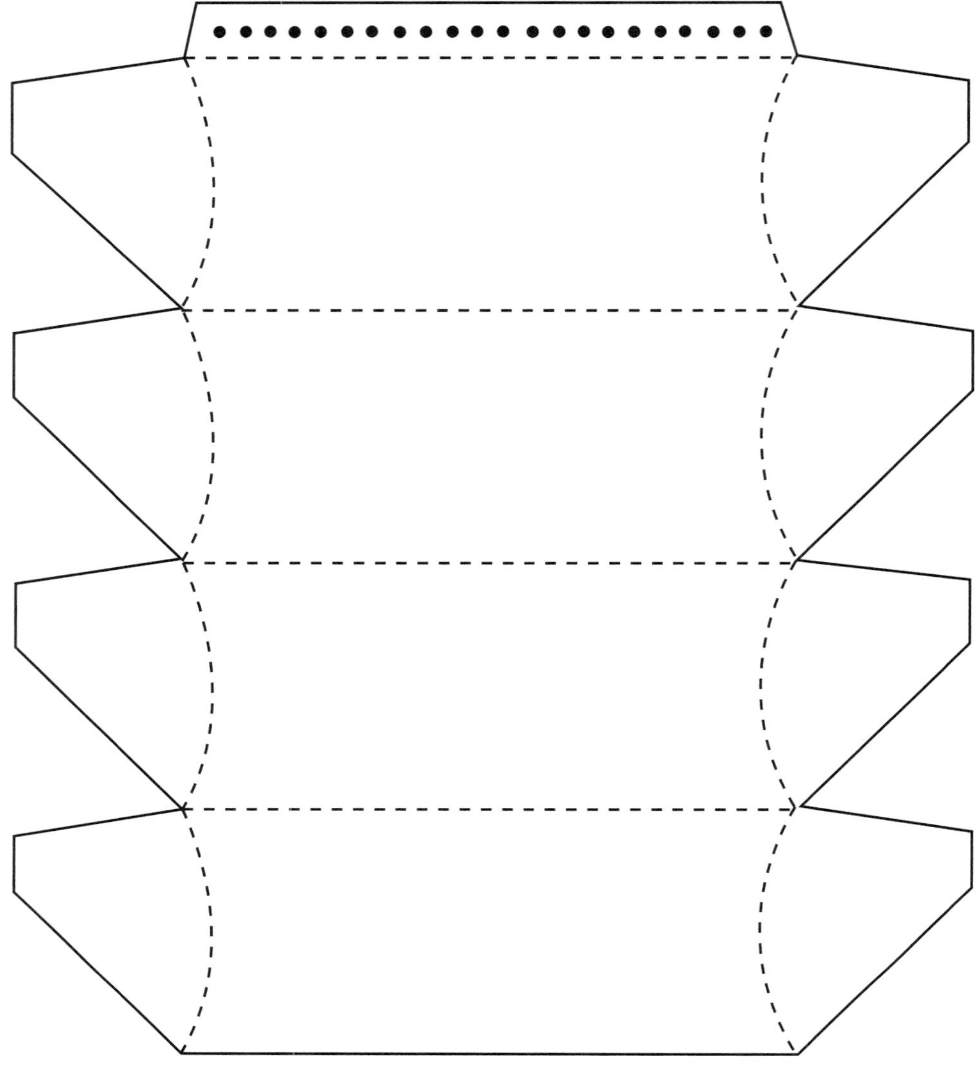

The curved scores on this pattern create a concave top and bottom and convex front, side, and back panels. The tension of the scores keeps both top and bottom close.

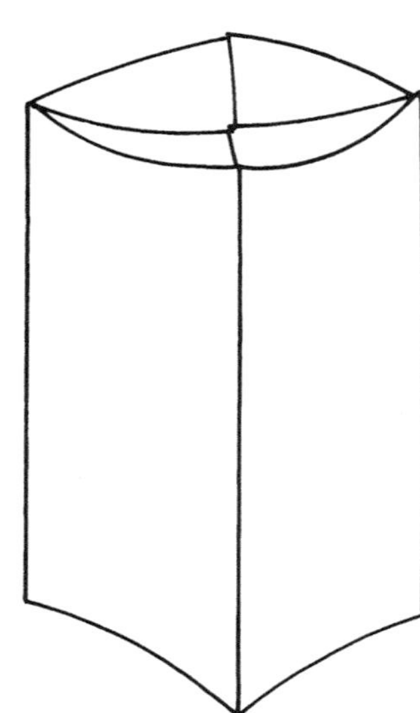

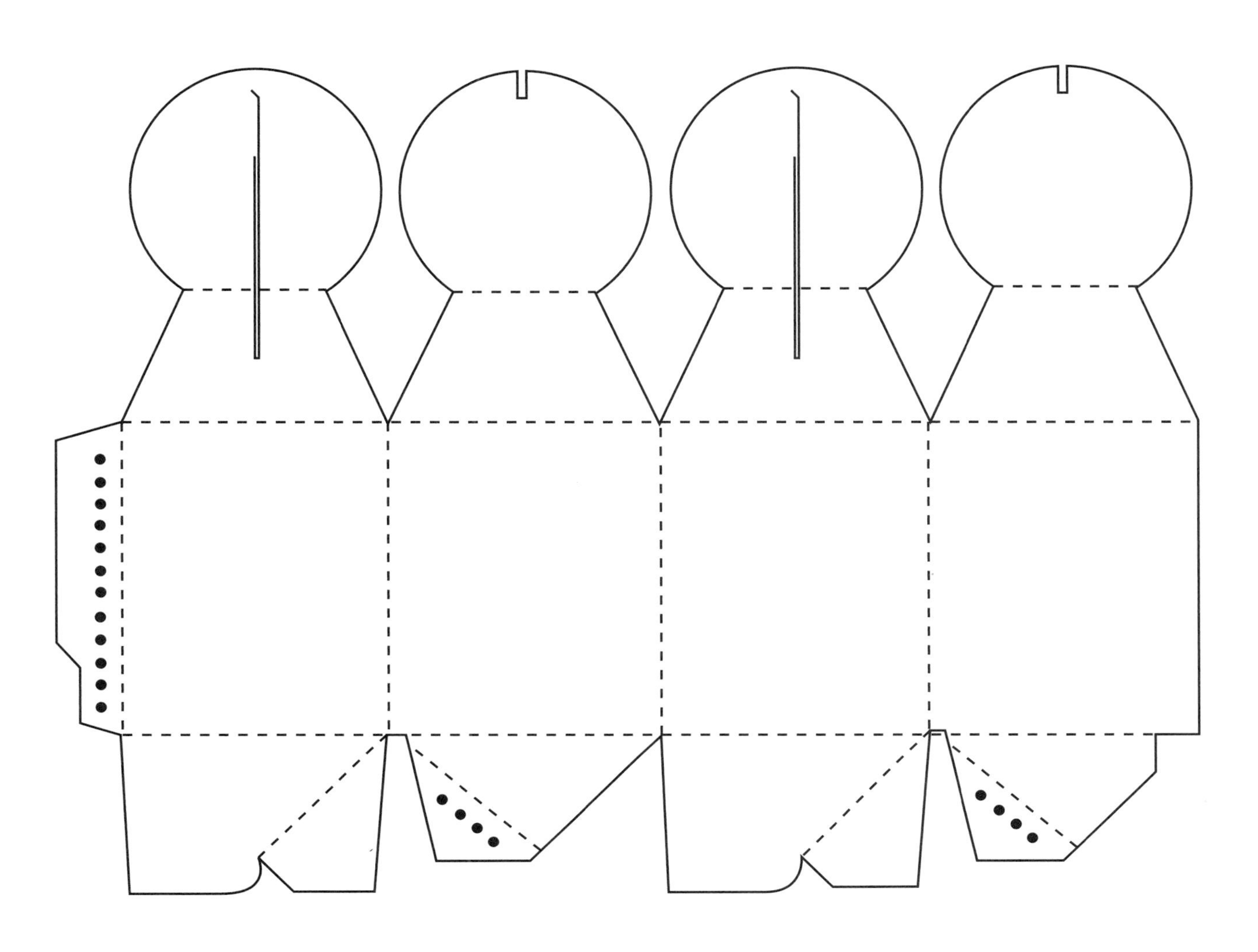

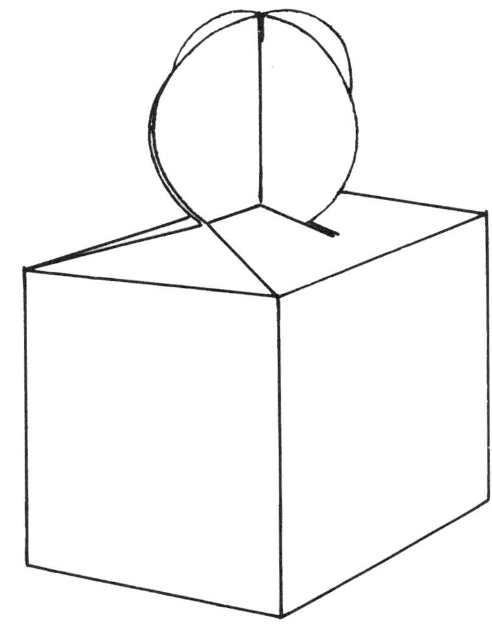

Specialty closure on top panels gives free rein to many graphic solutions.

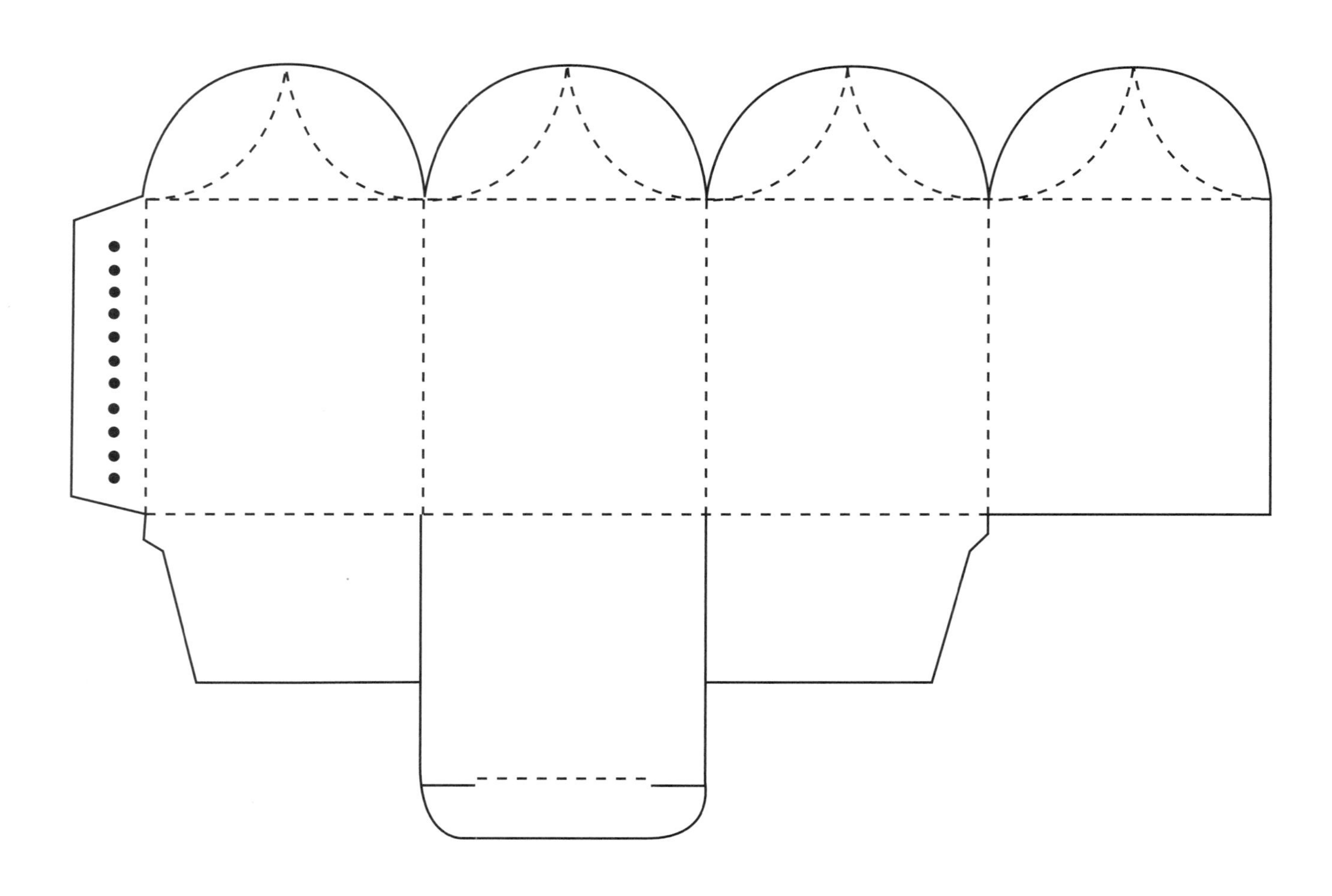

Specially shaped top closure creates an interesting concave/convex interplay.

OCTAGON WITH AUTO LOCK BOTTOM

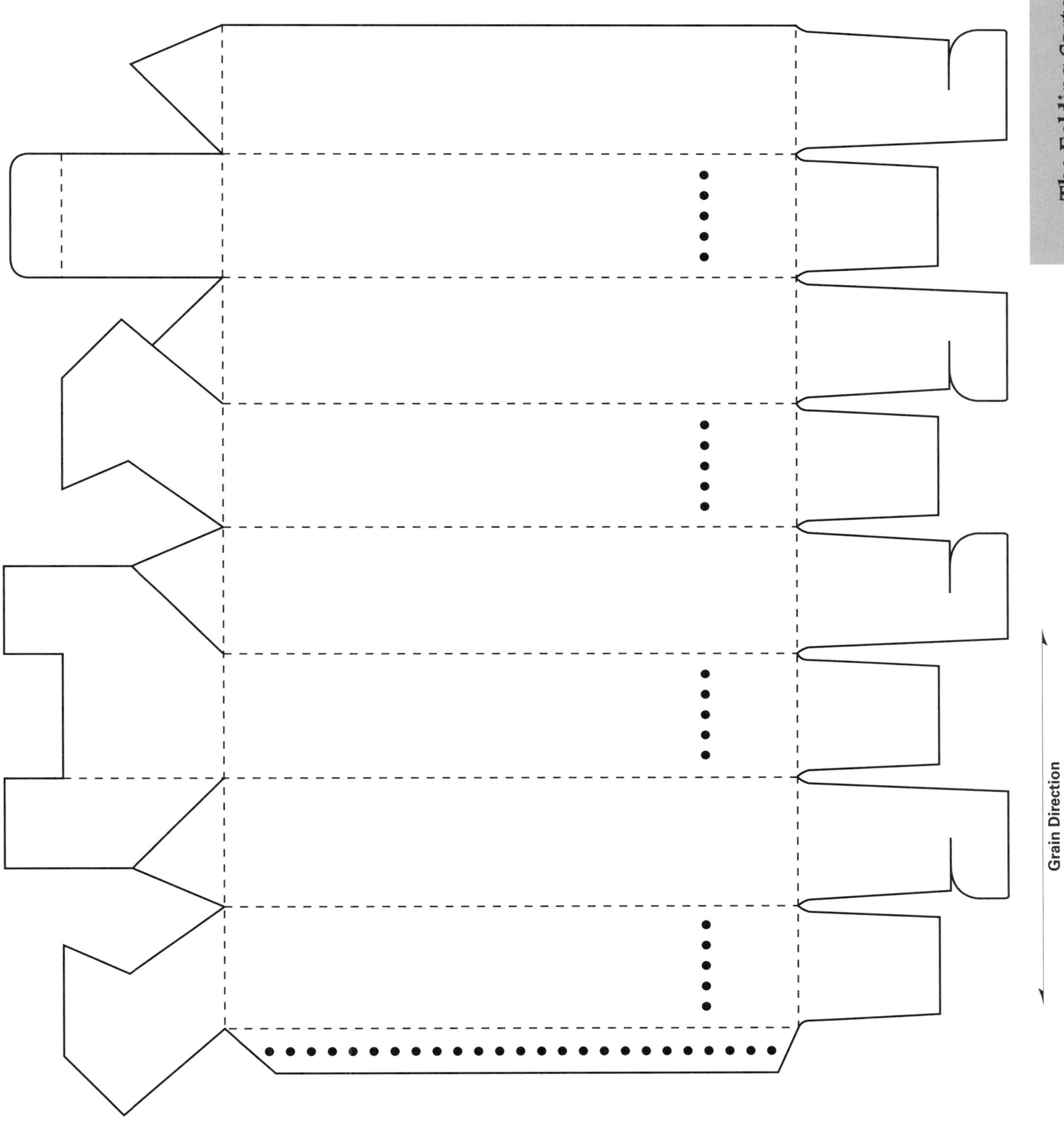

Combined with the decorative top on the following page, this makes a handsome package with many graphic design possibilities.

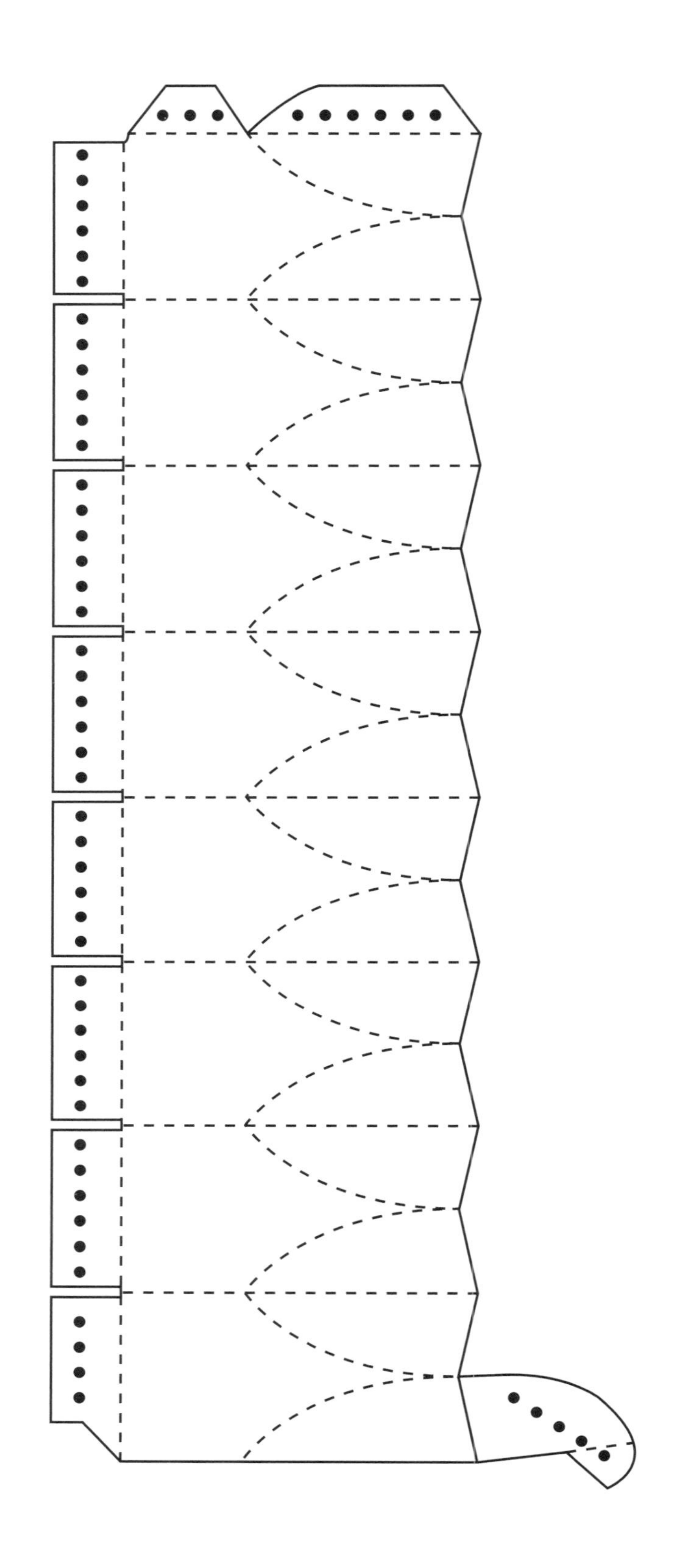

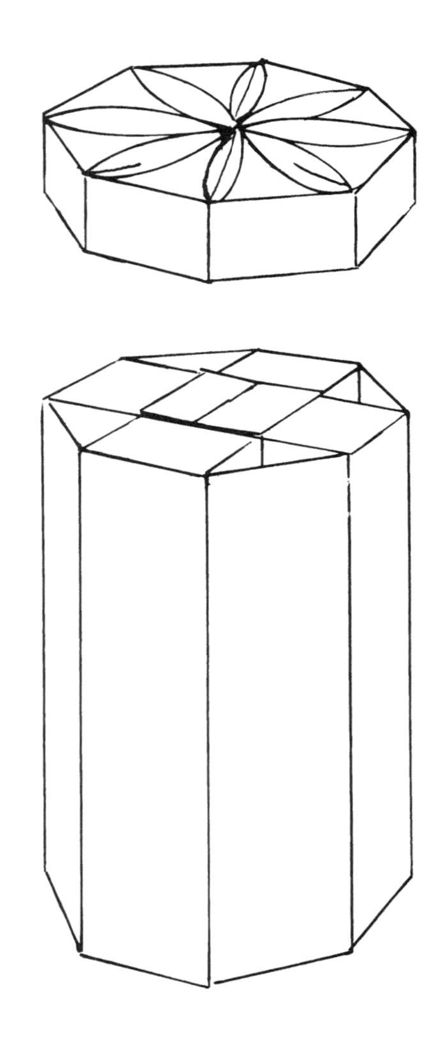

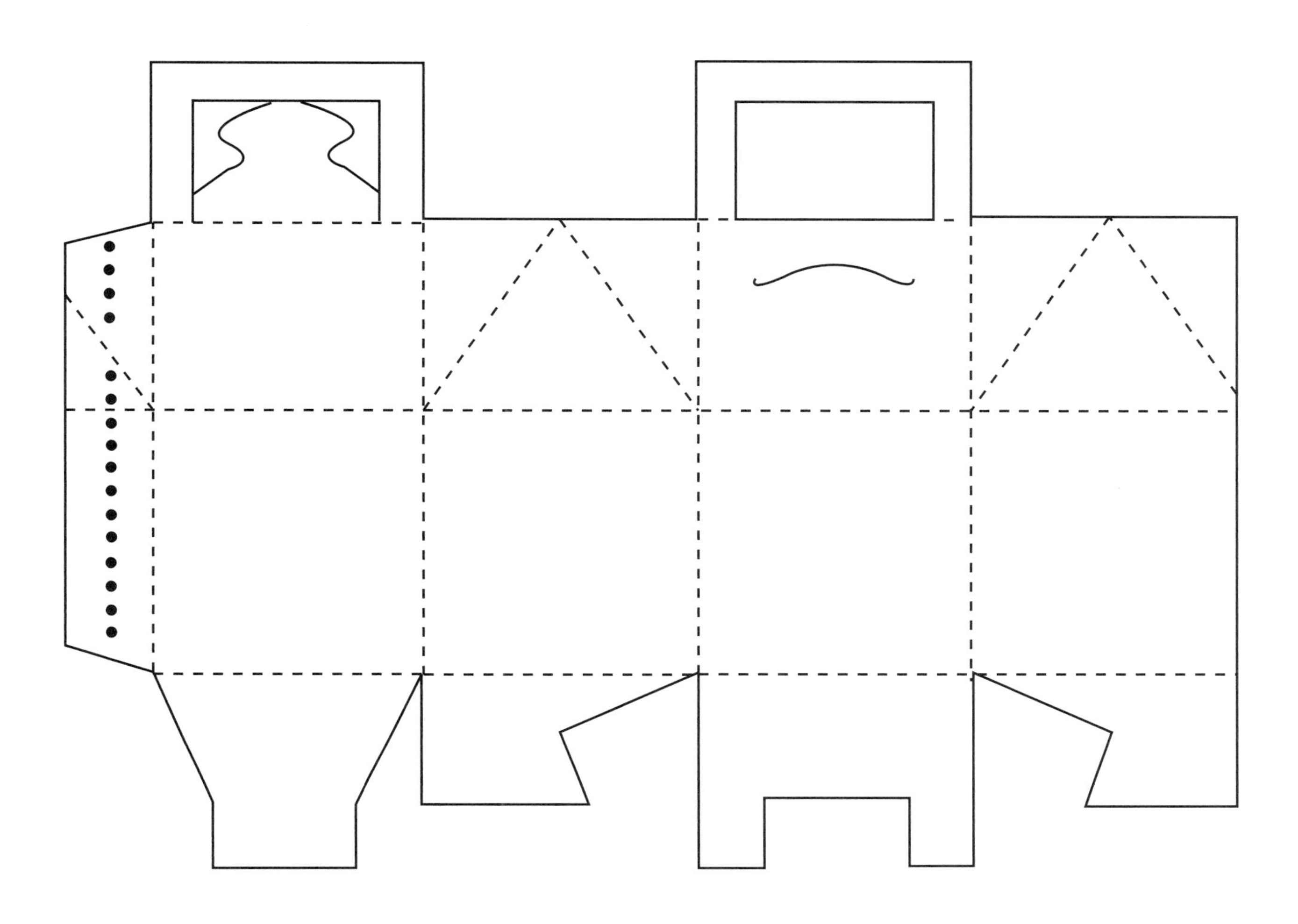

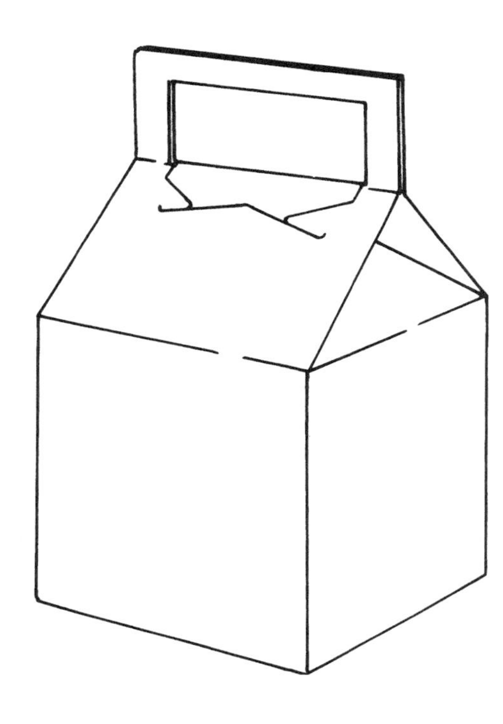

ACCORDION STYLE CLOSURE

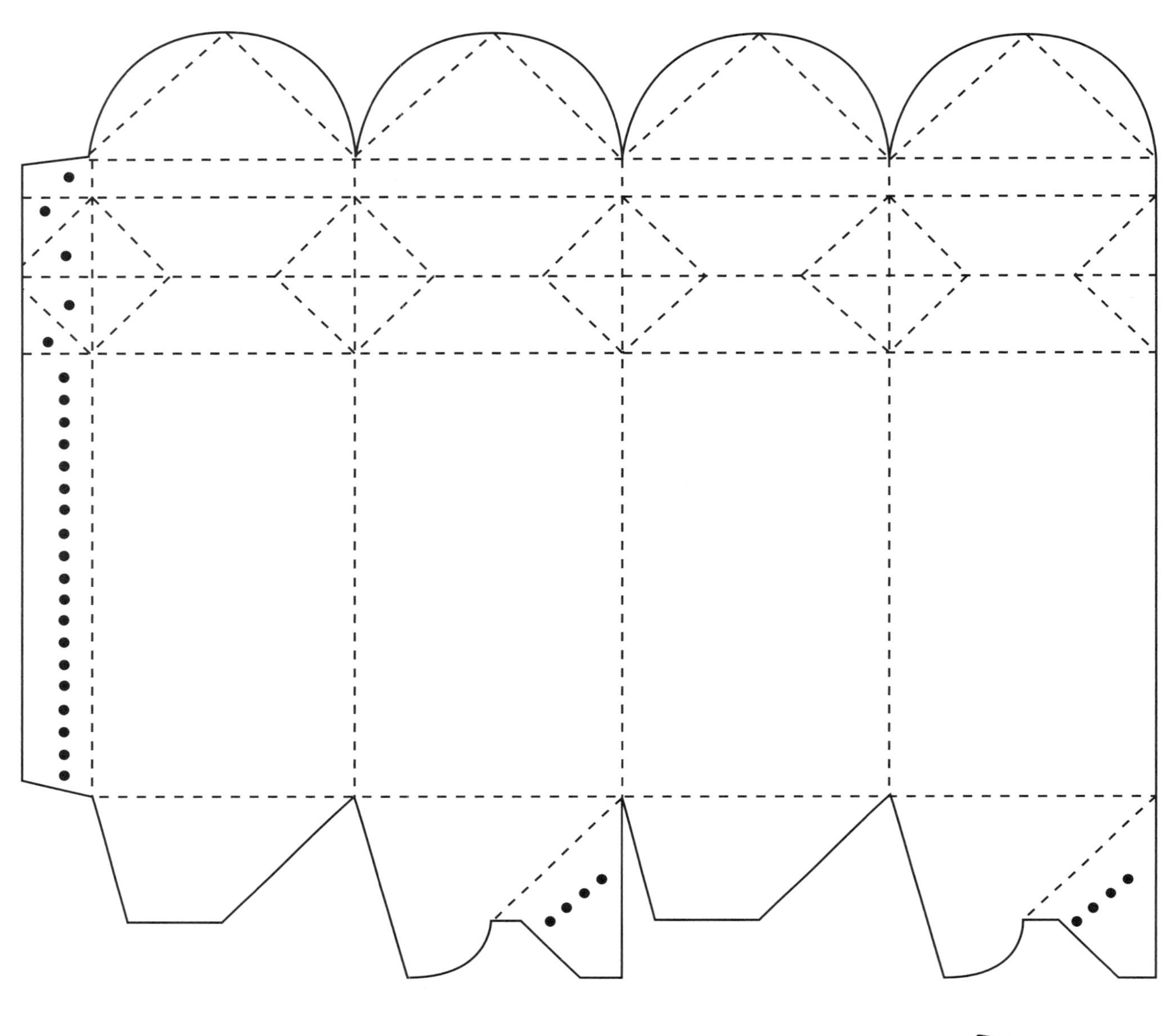

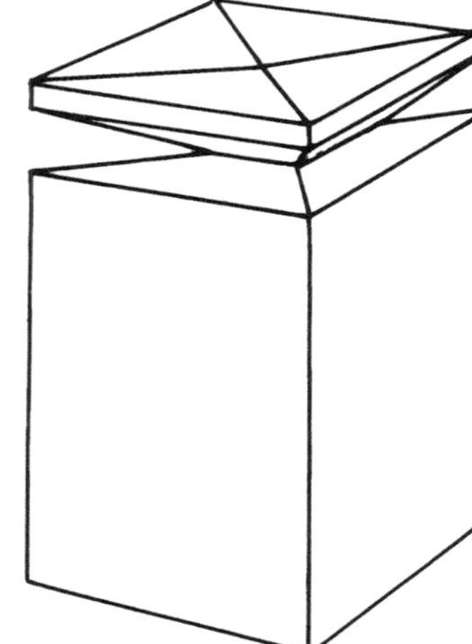

Scored front, back, and side panels lend themselves to cord or ribbon decoration.

FULL-FLAP AUTOMATIC BOTTOM, TAPERED TOP WITH LOCK TABS AND CARRYING HANDLE

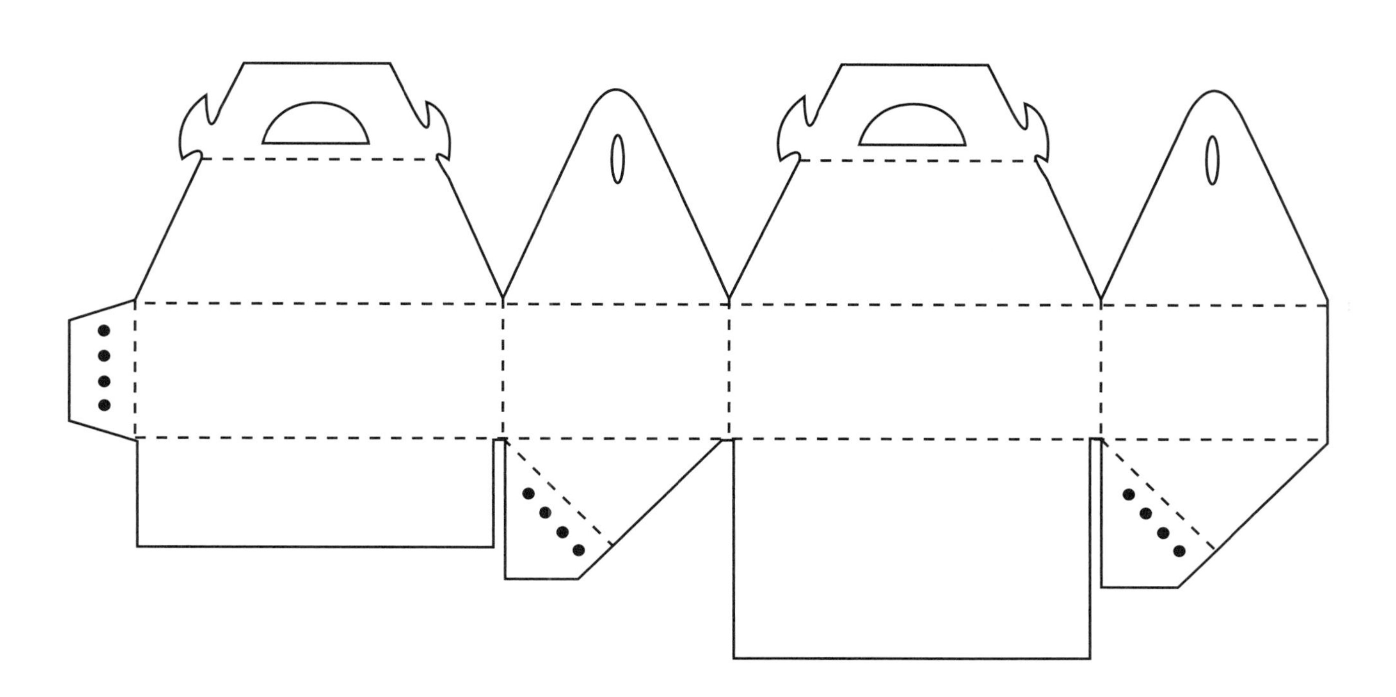

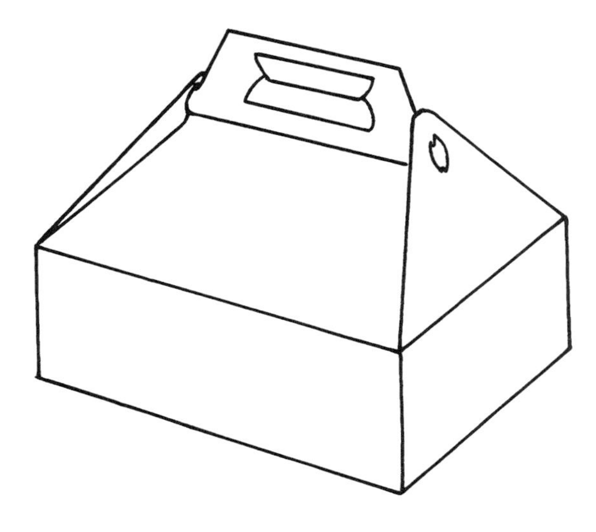

HOOK TAB CLOSURE WITH TAPERED SIDES

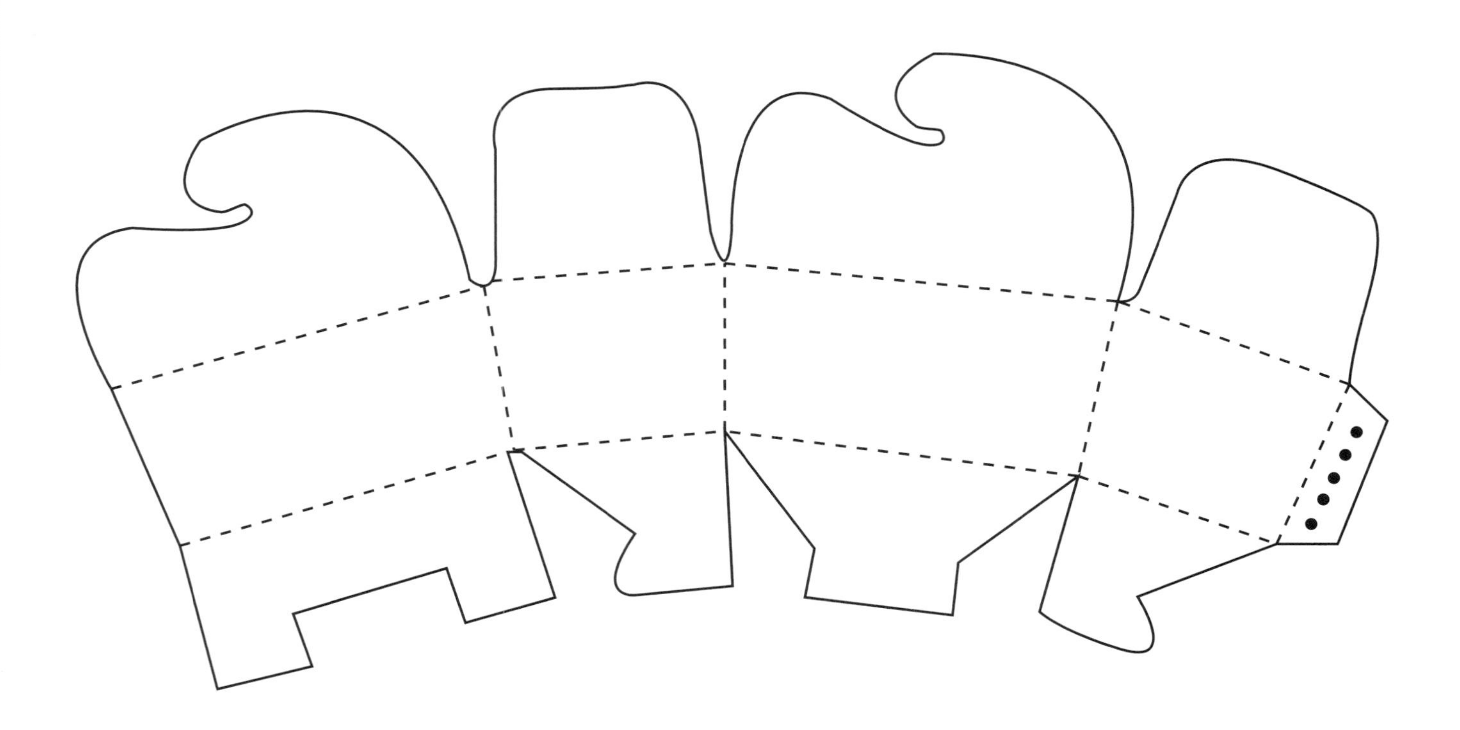

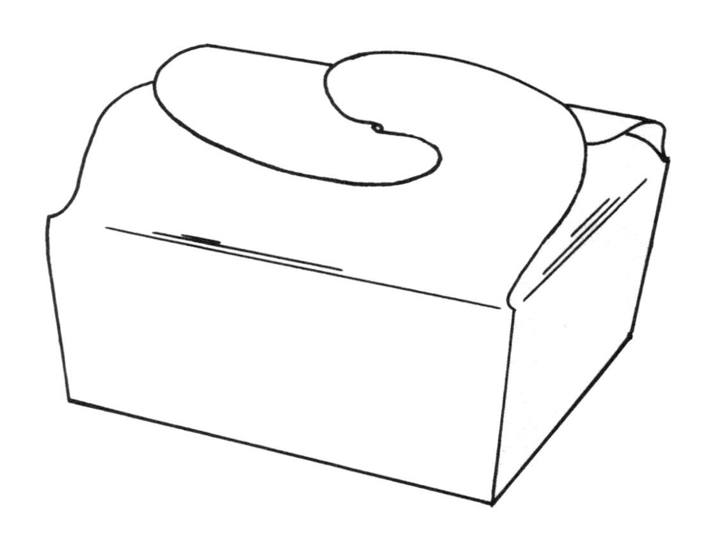

TAPERED SIDES, 1-2-3 BOTTOM TONGUE LOCK, TOP PANEL WITH CENTERED CARRYING HANDLE

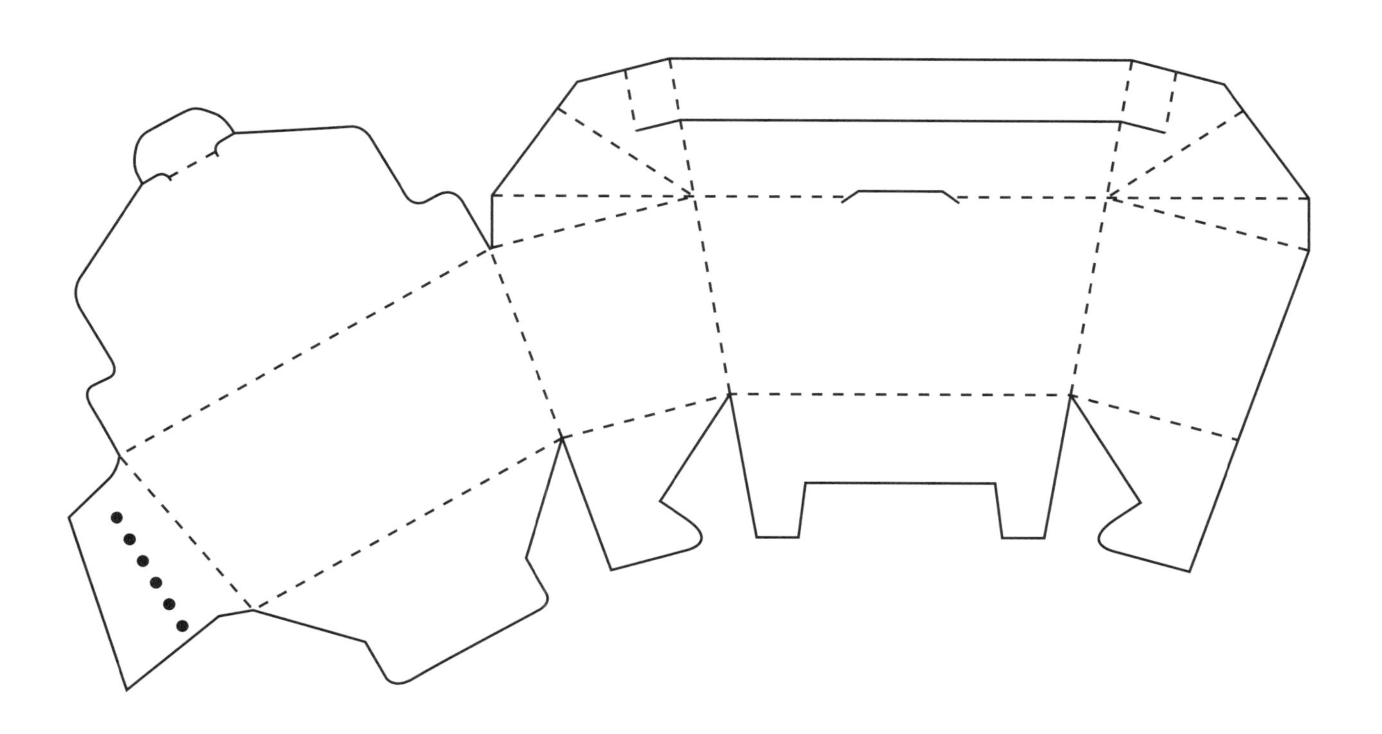

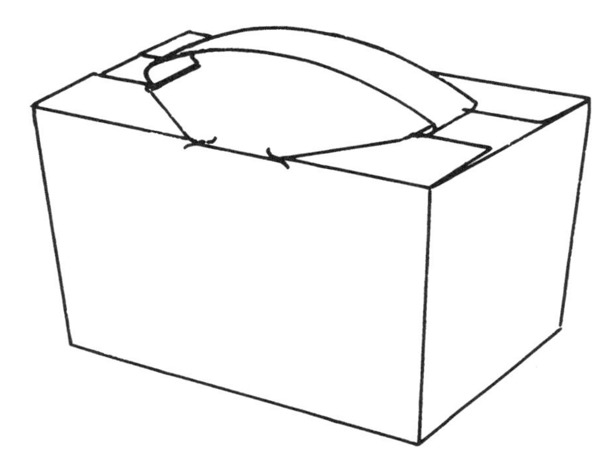

STRAIGHT TUCK WITH CENTERED HANDLE

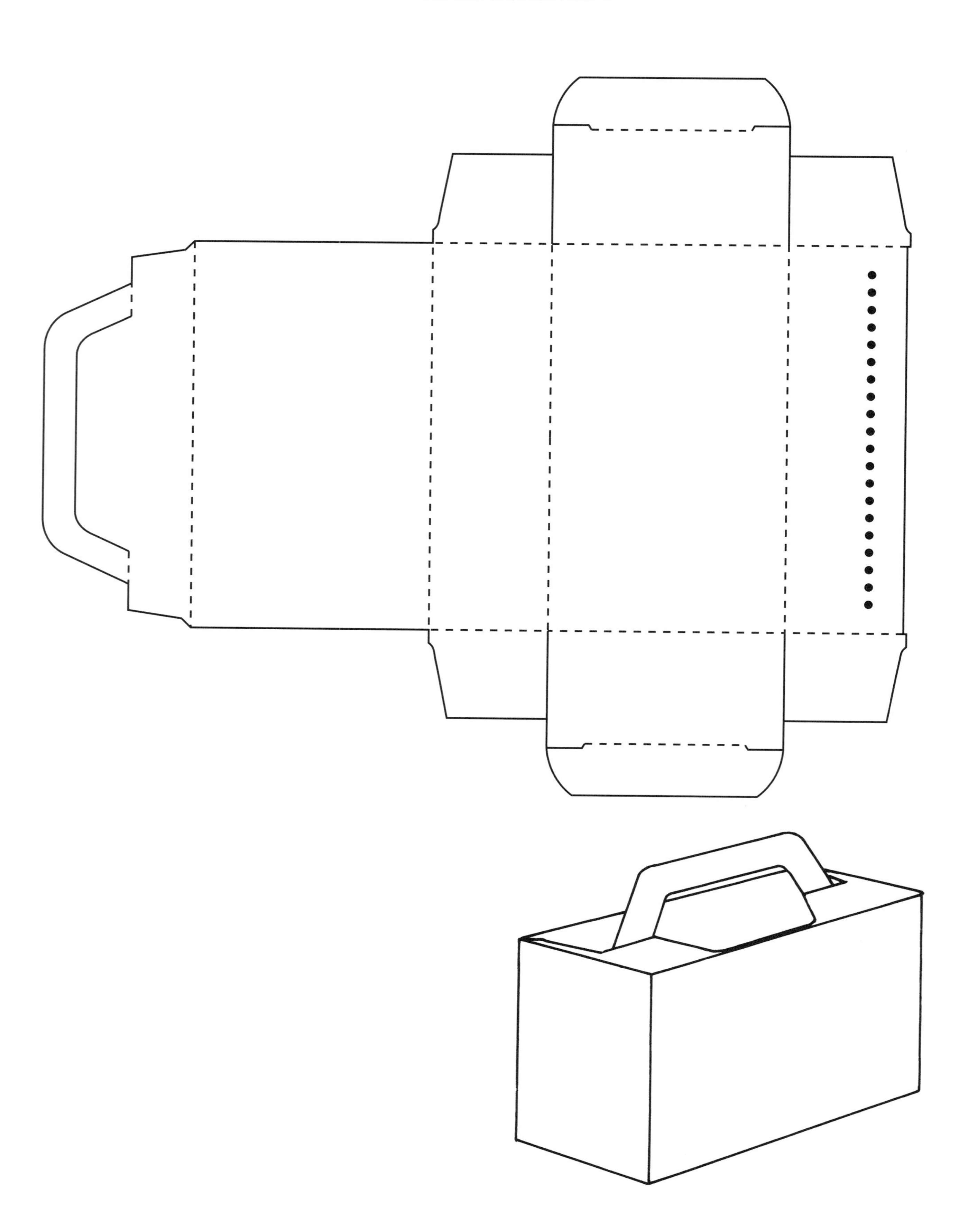

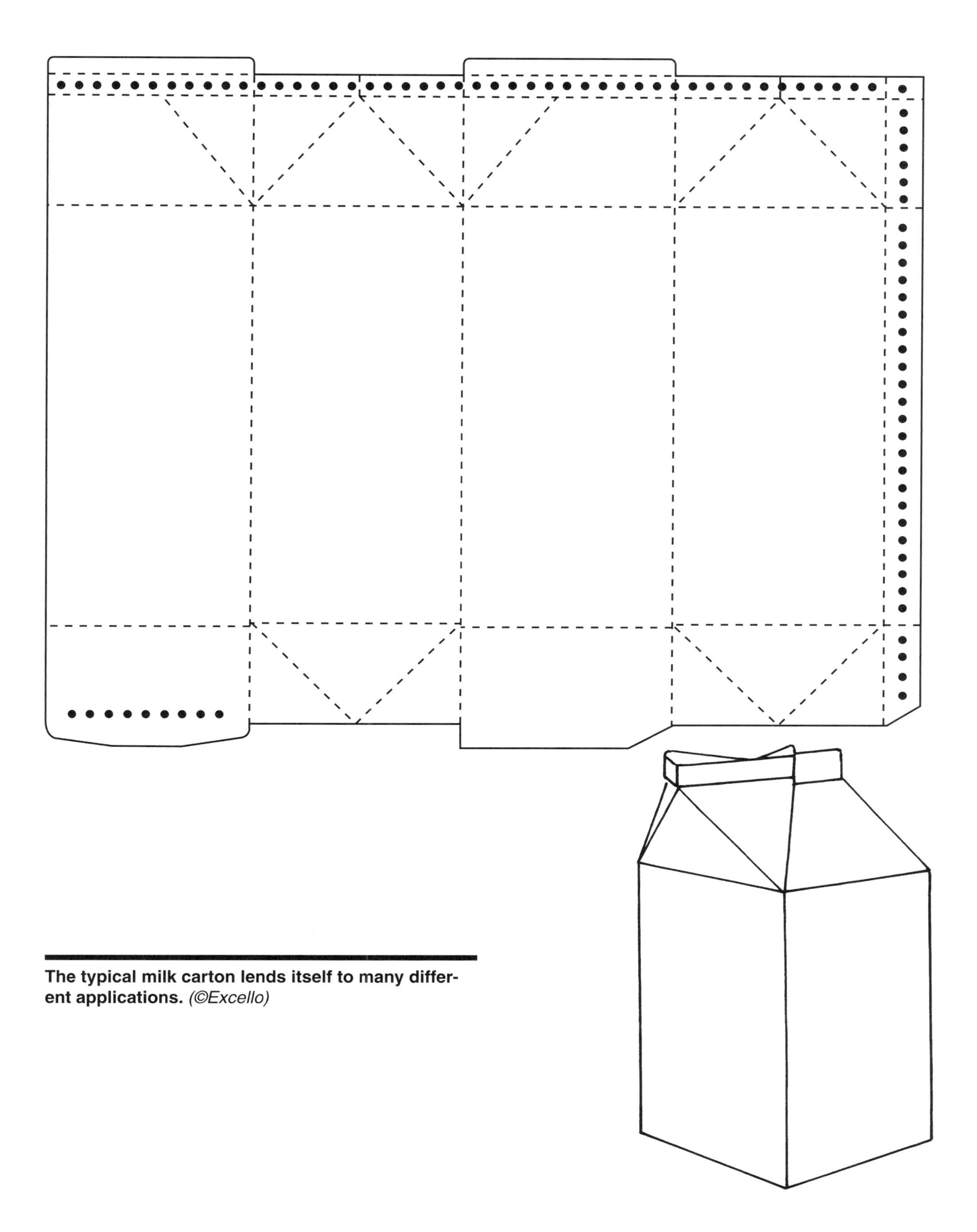

The typical milk carton lends itself to many different applications. *(©Excello)*

GABLE TOP CONTAINER

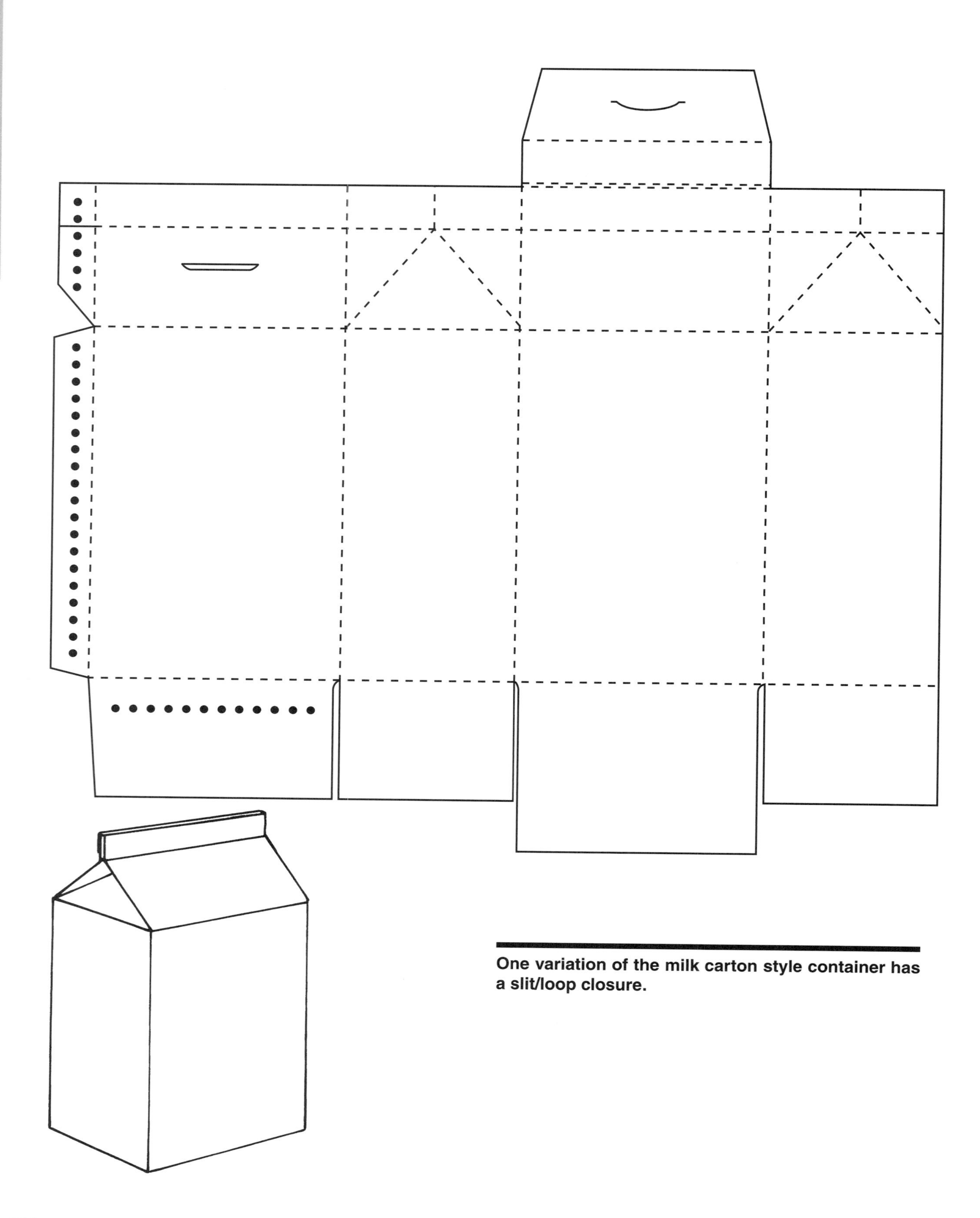

One variation of the milk carton style container has a slit/loop closure.

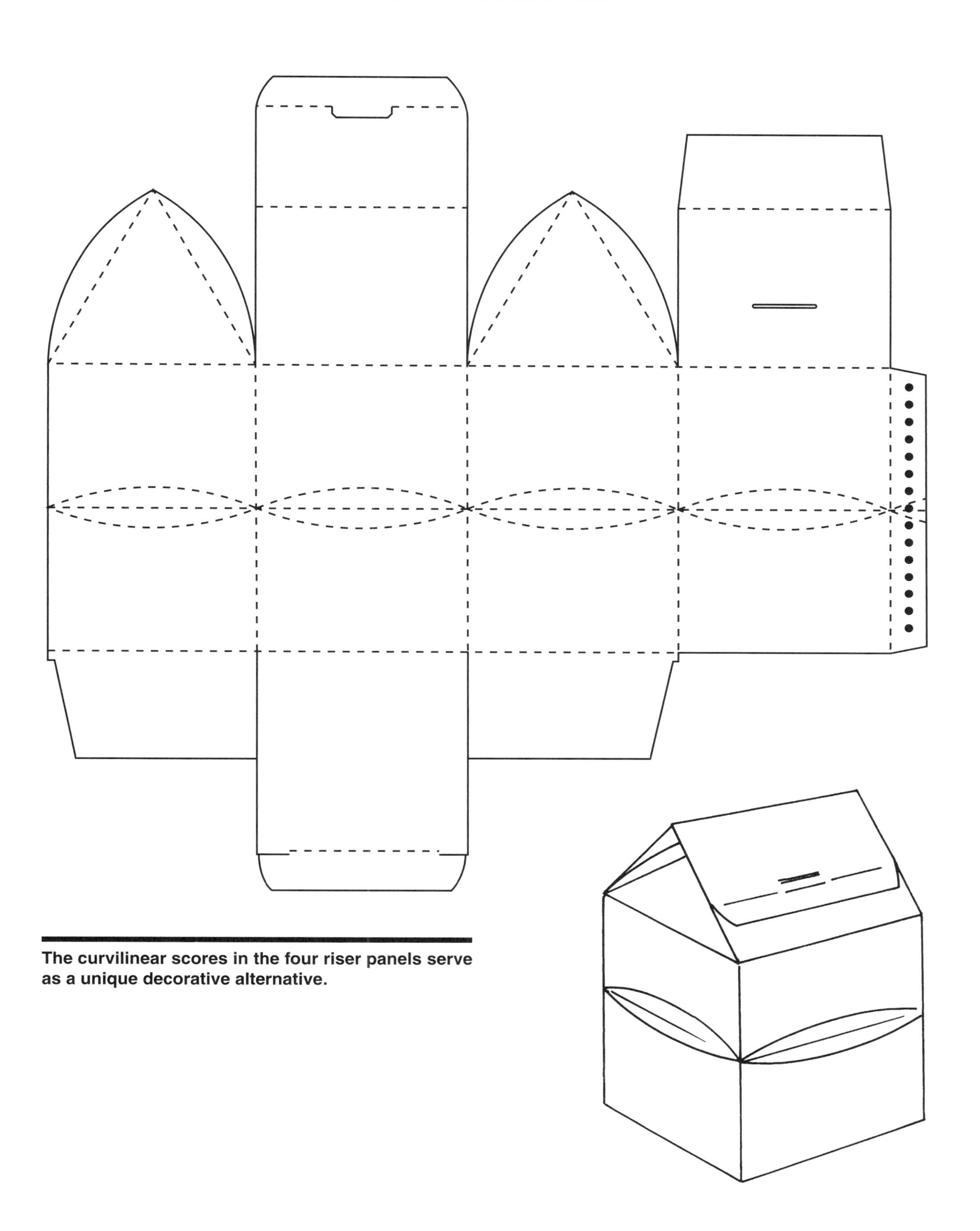

The curvilinear scores in the four riser panels serve as a unique decorative alternative.

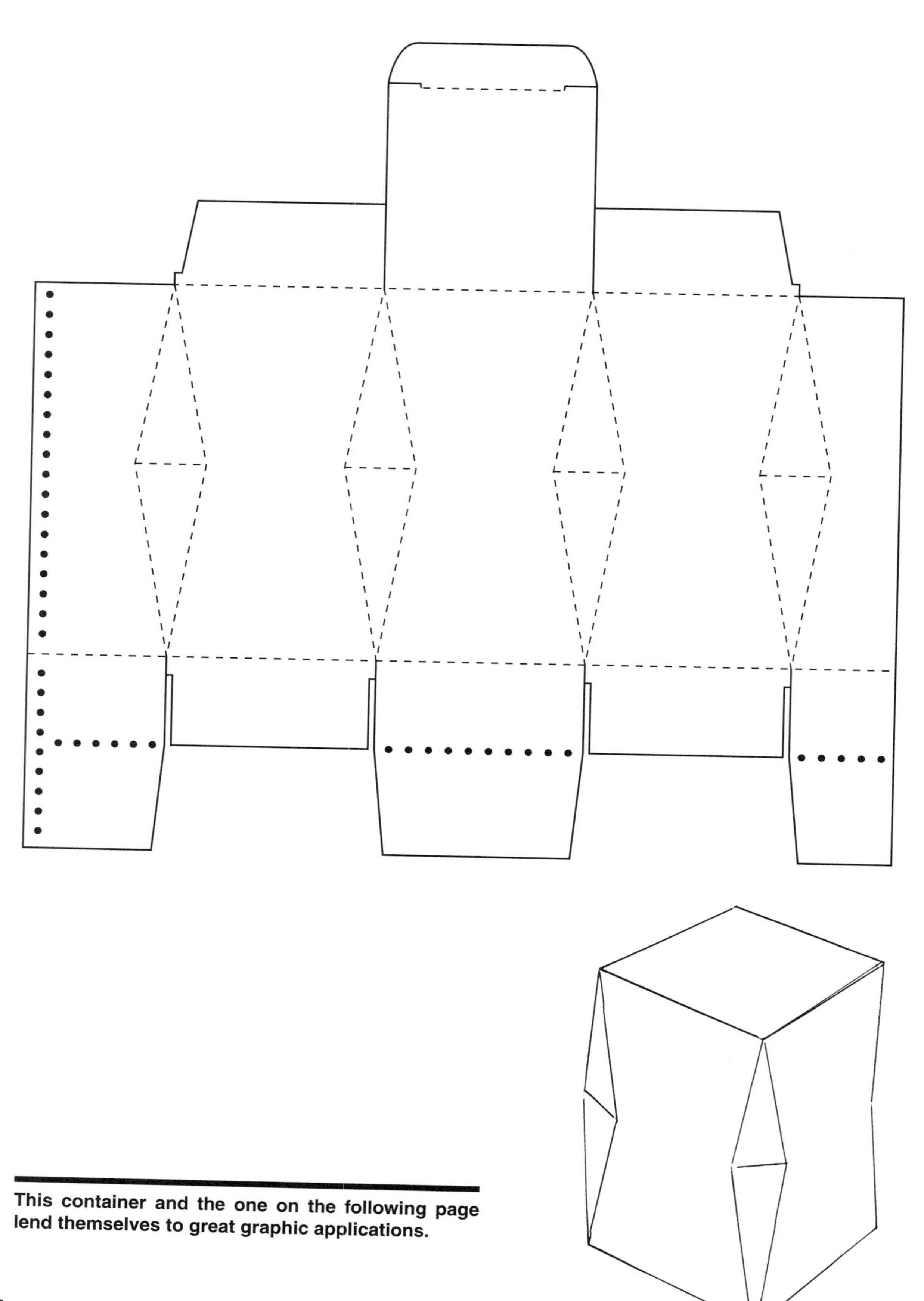

This container and the one on the following page lend themselves to great graphic applications.

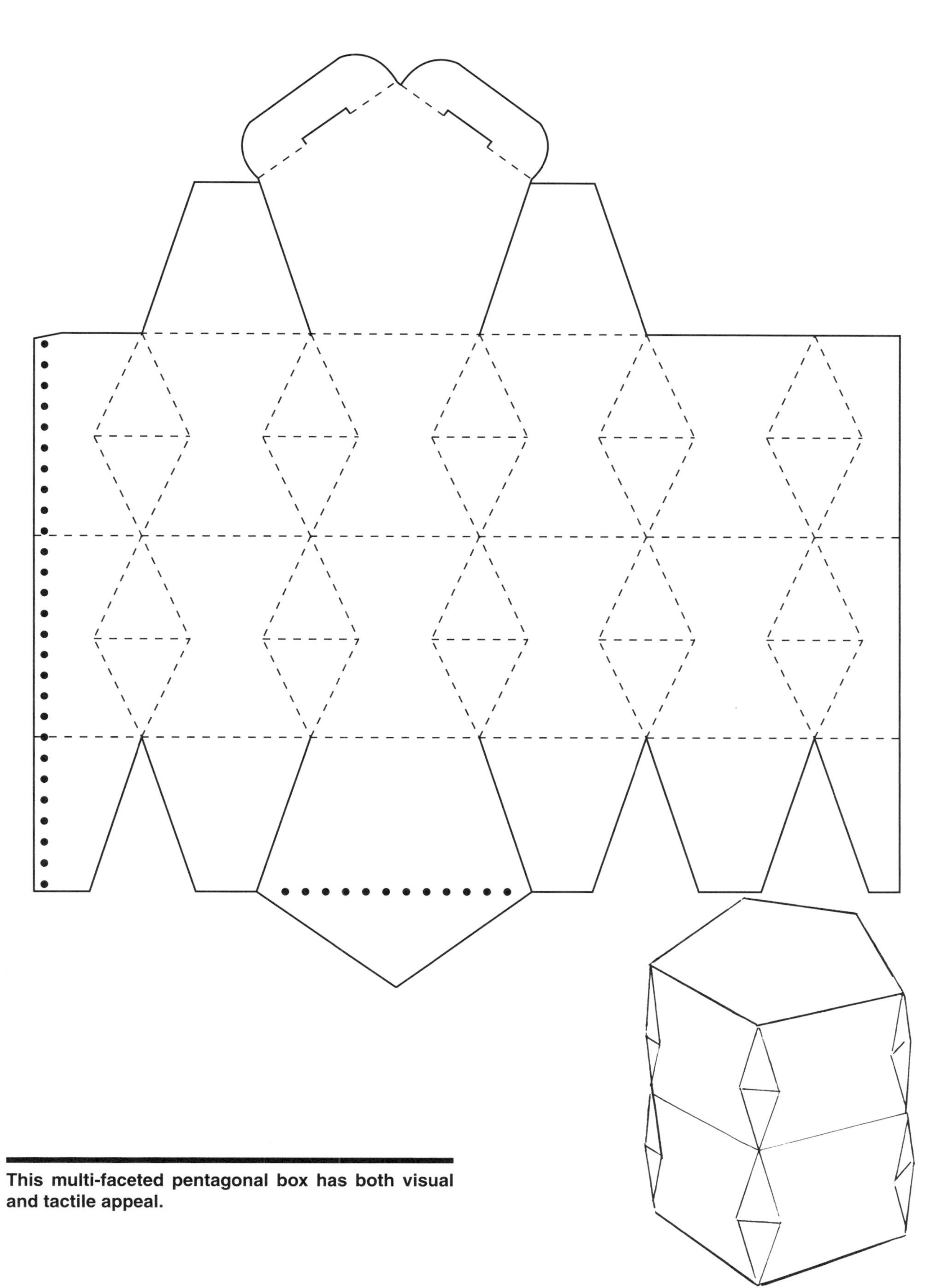

This multi-faceted pentagonal box has both visual and tactile appeal.

GABLE TOP CONTAINER VARIATION

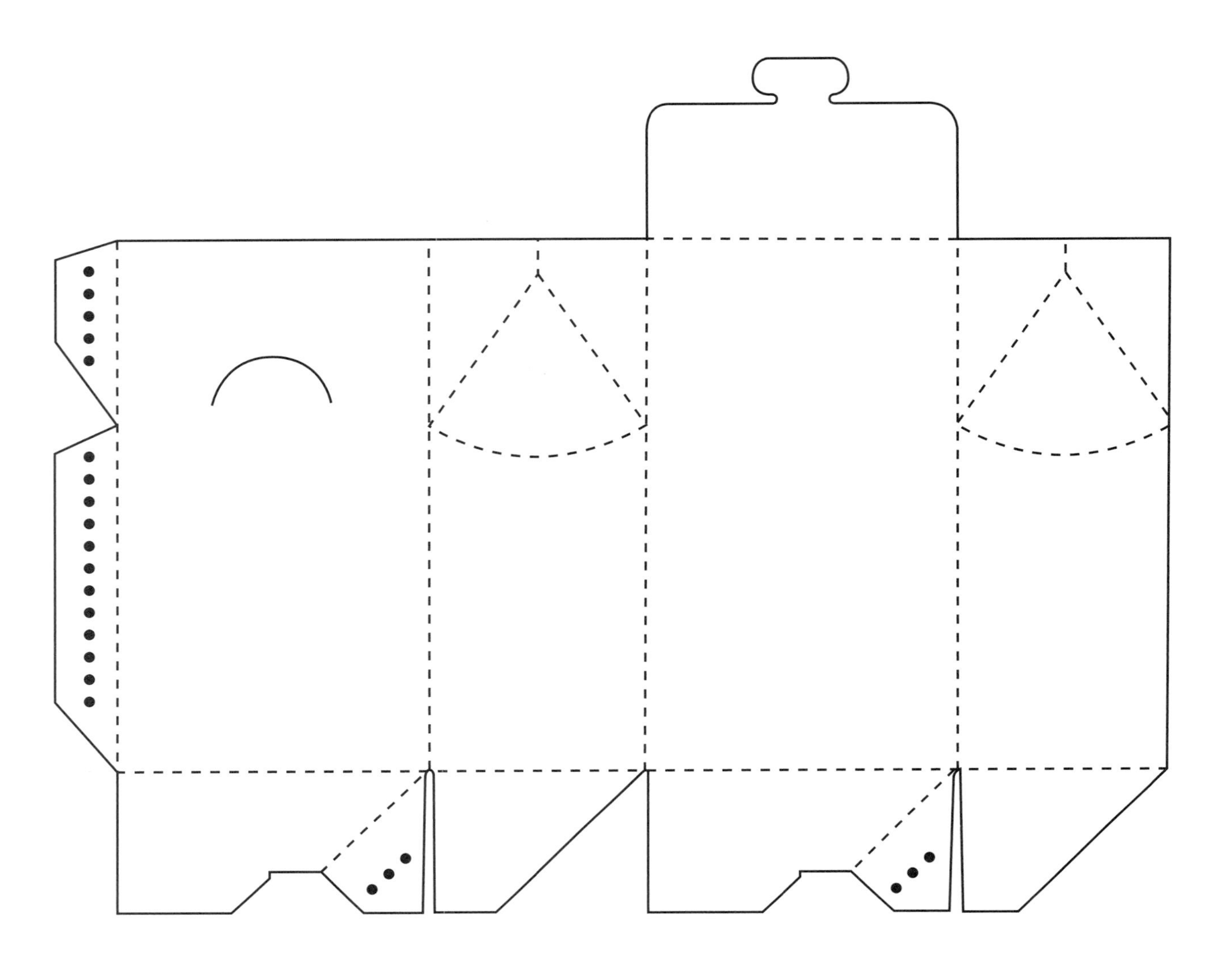

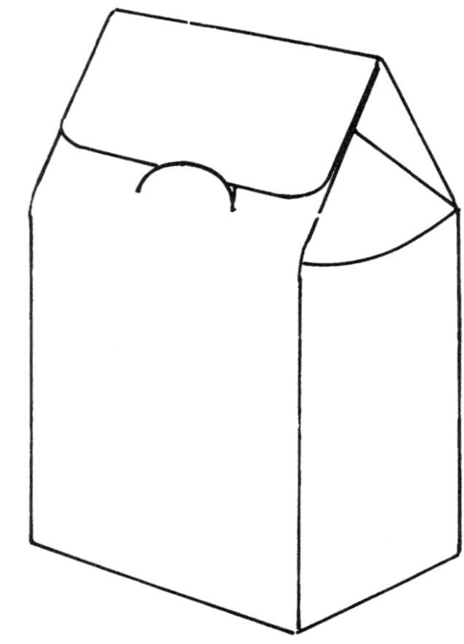

A different treatment of the top side-panel scores and closure tab. This carton features an auto-lock bottom.

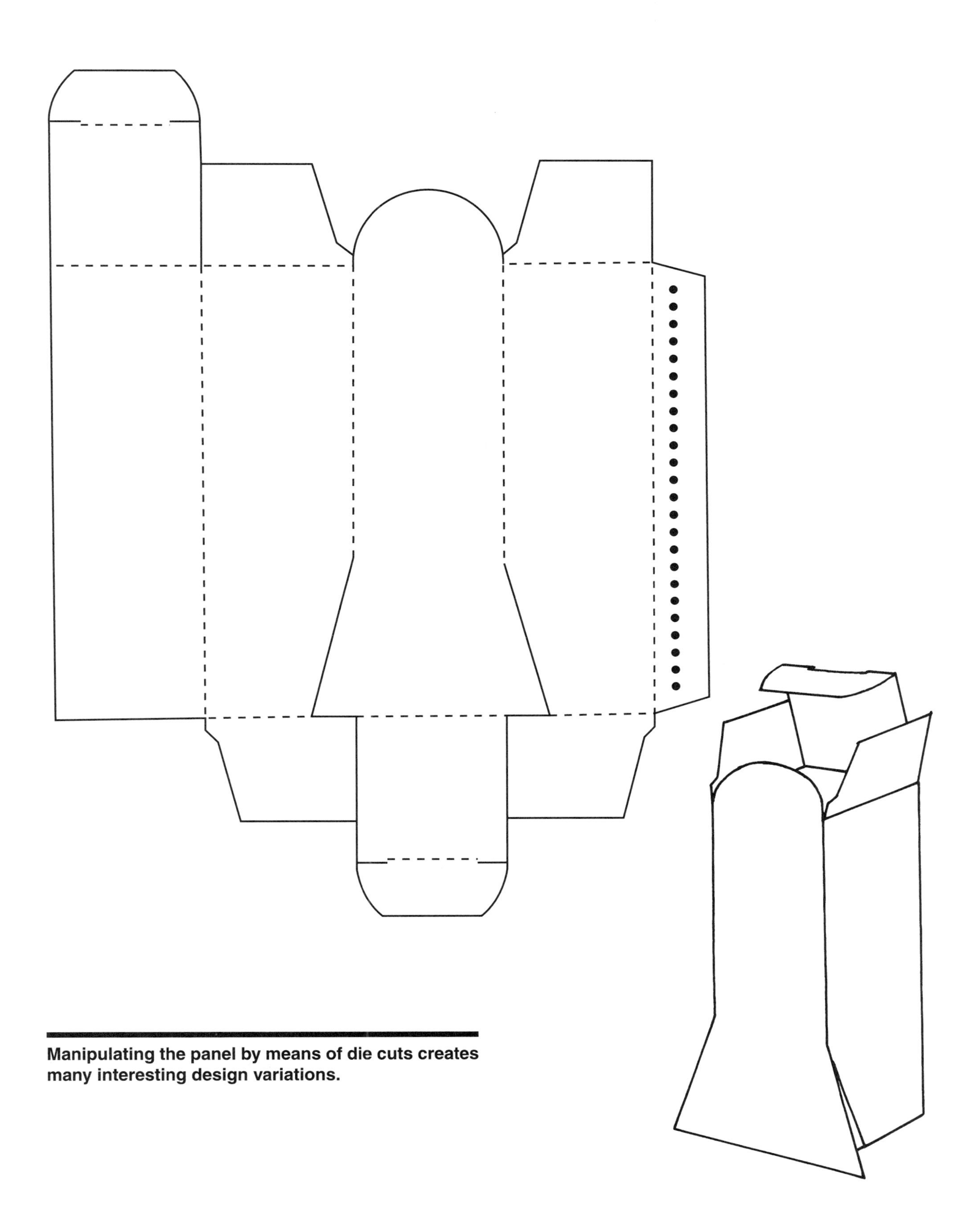

Manipulating the panel by means of die cuts creates many interesting design variations.

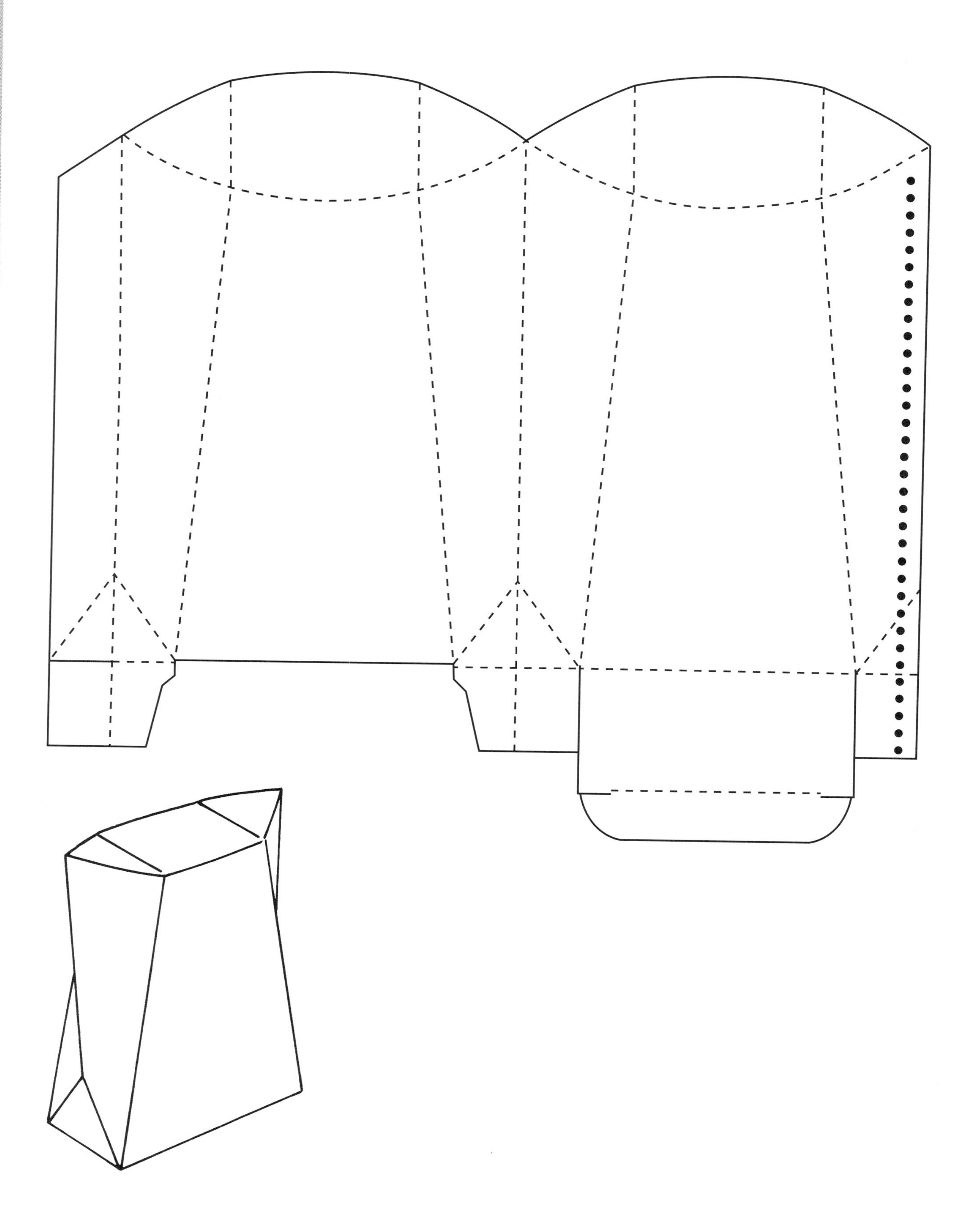

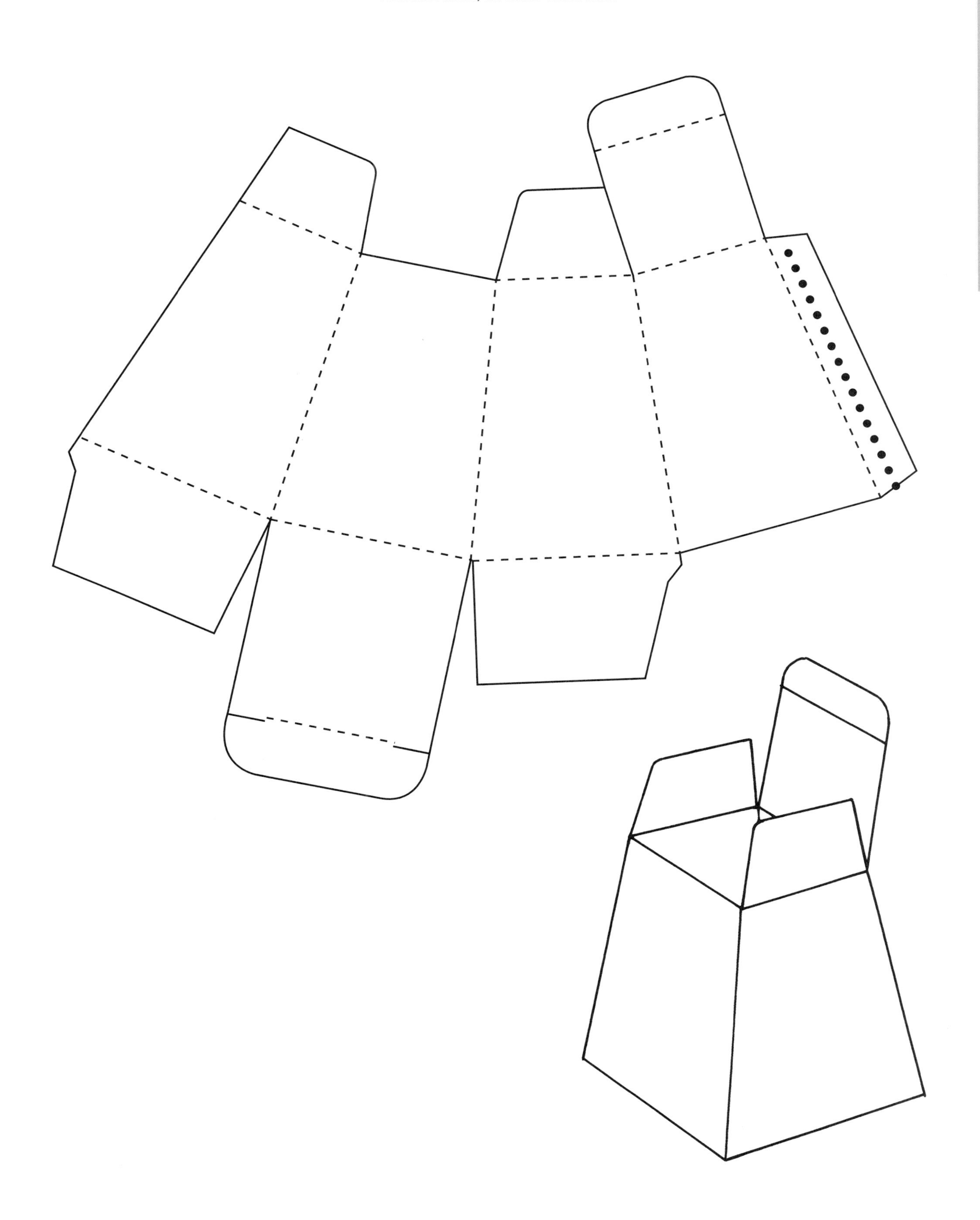

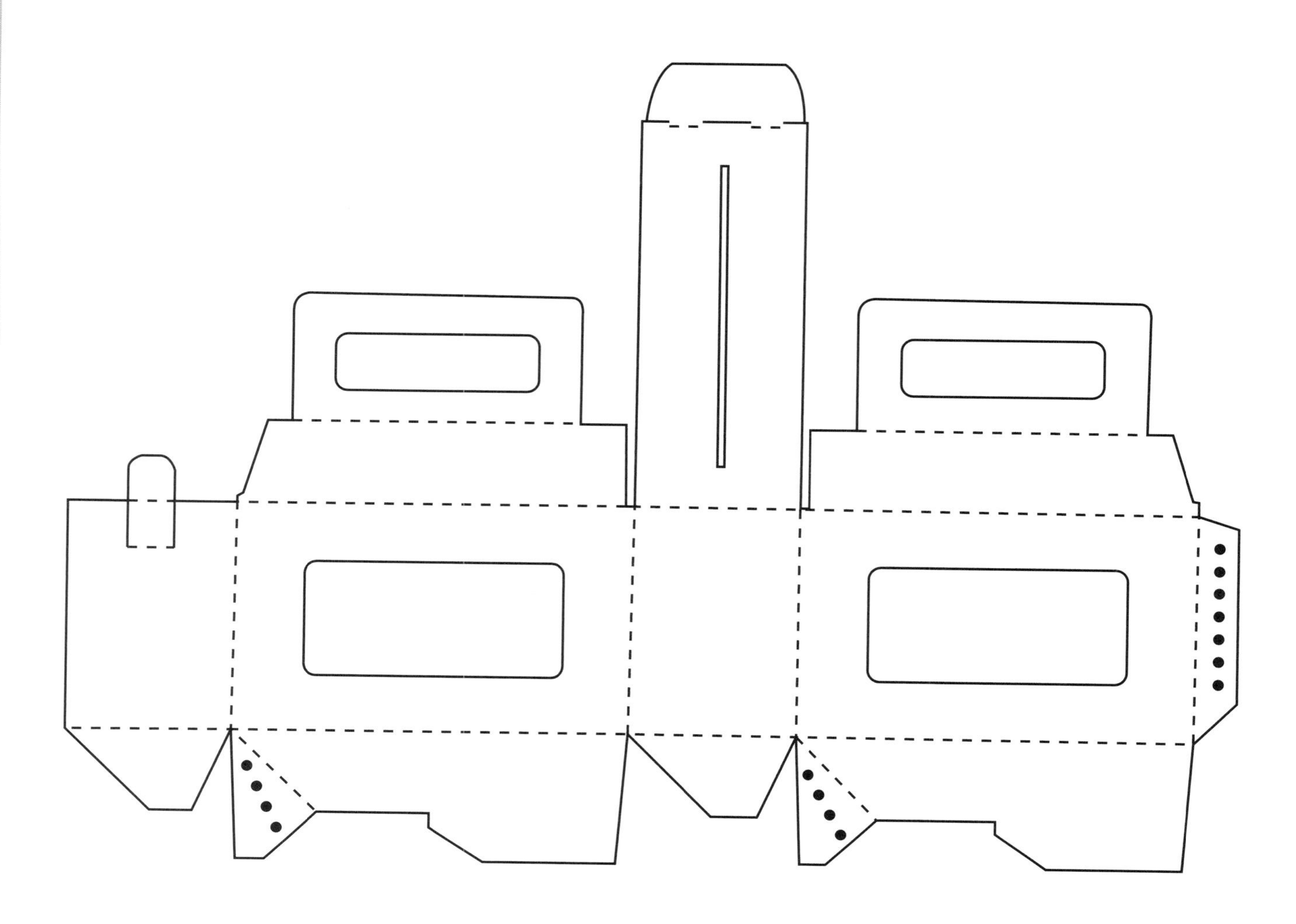

In combination with the patterns for the inserts on the facing page, this package with carrying handle was created for six small body lotion bottles.

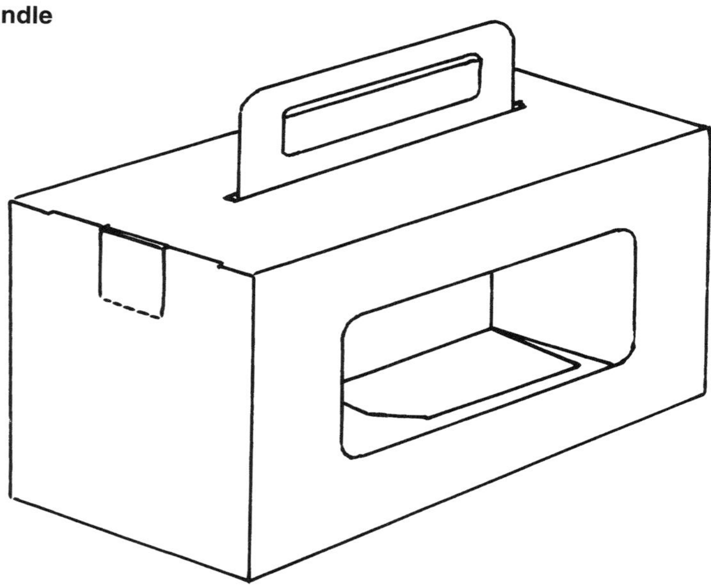

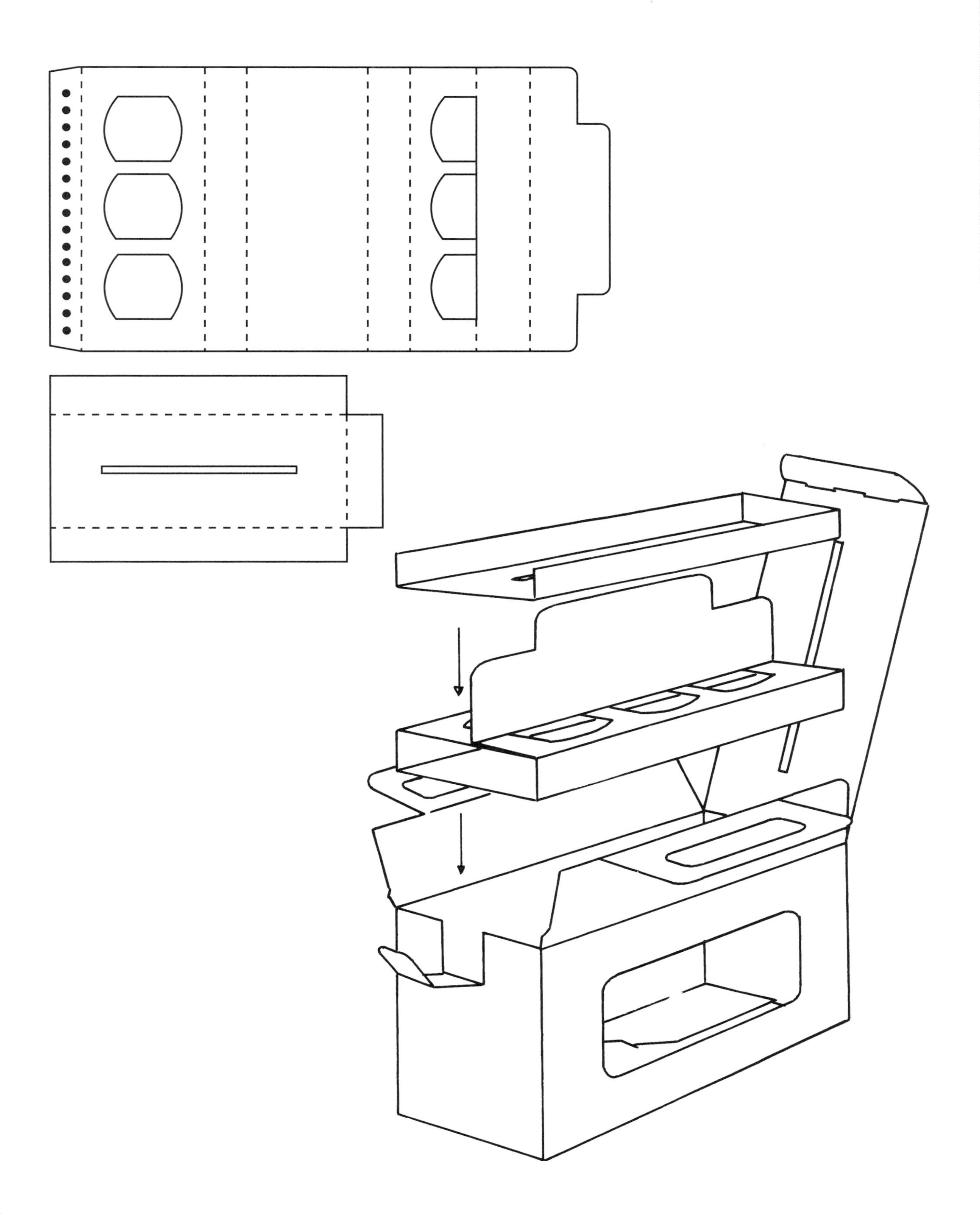

DOME TOP WITH SPECIAL TAB CLOSURE, GREENLEAF BOTTOM

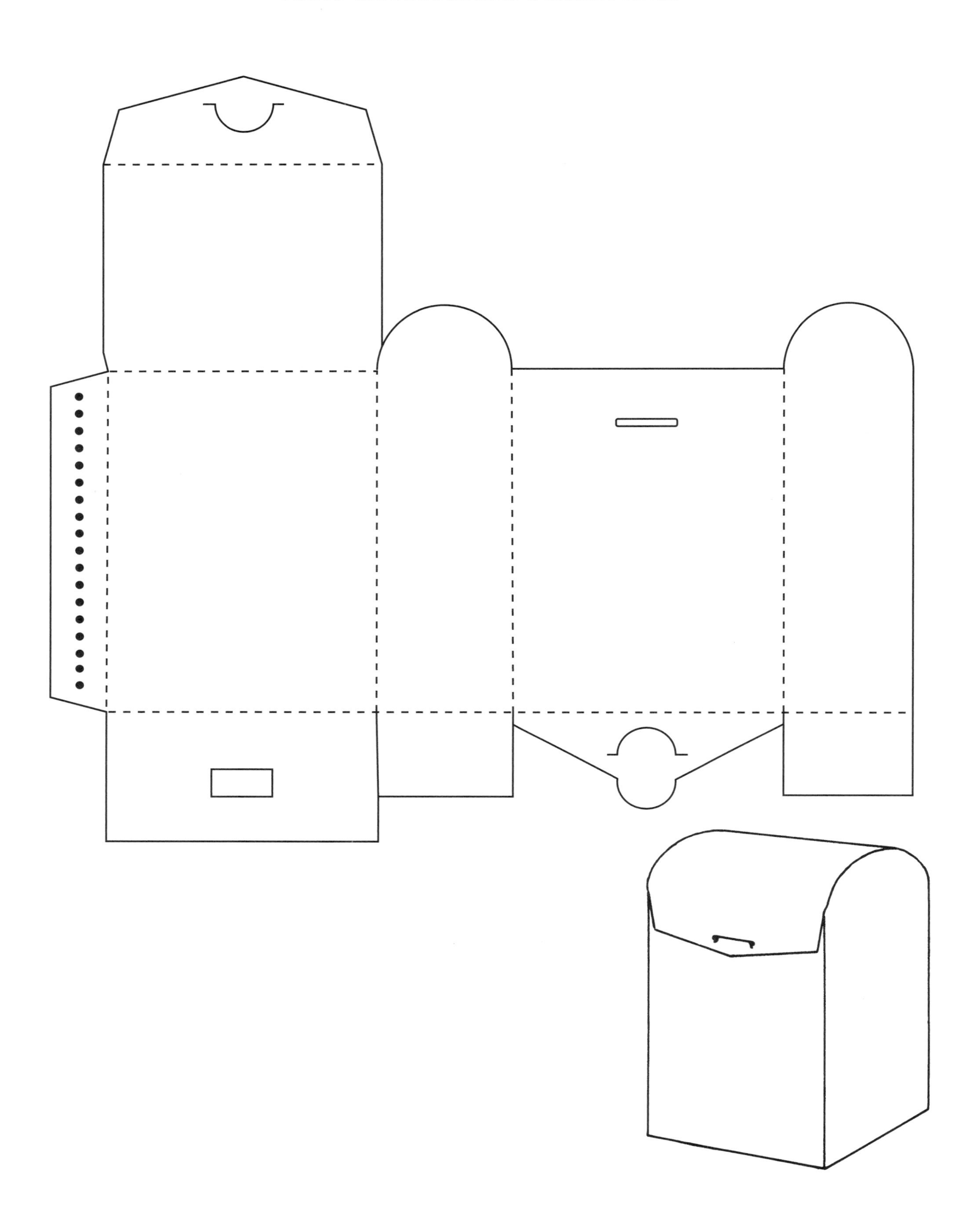

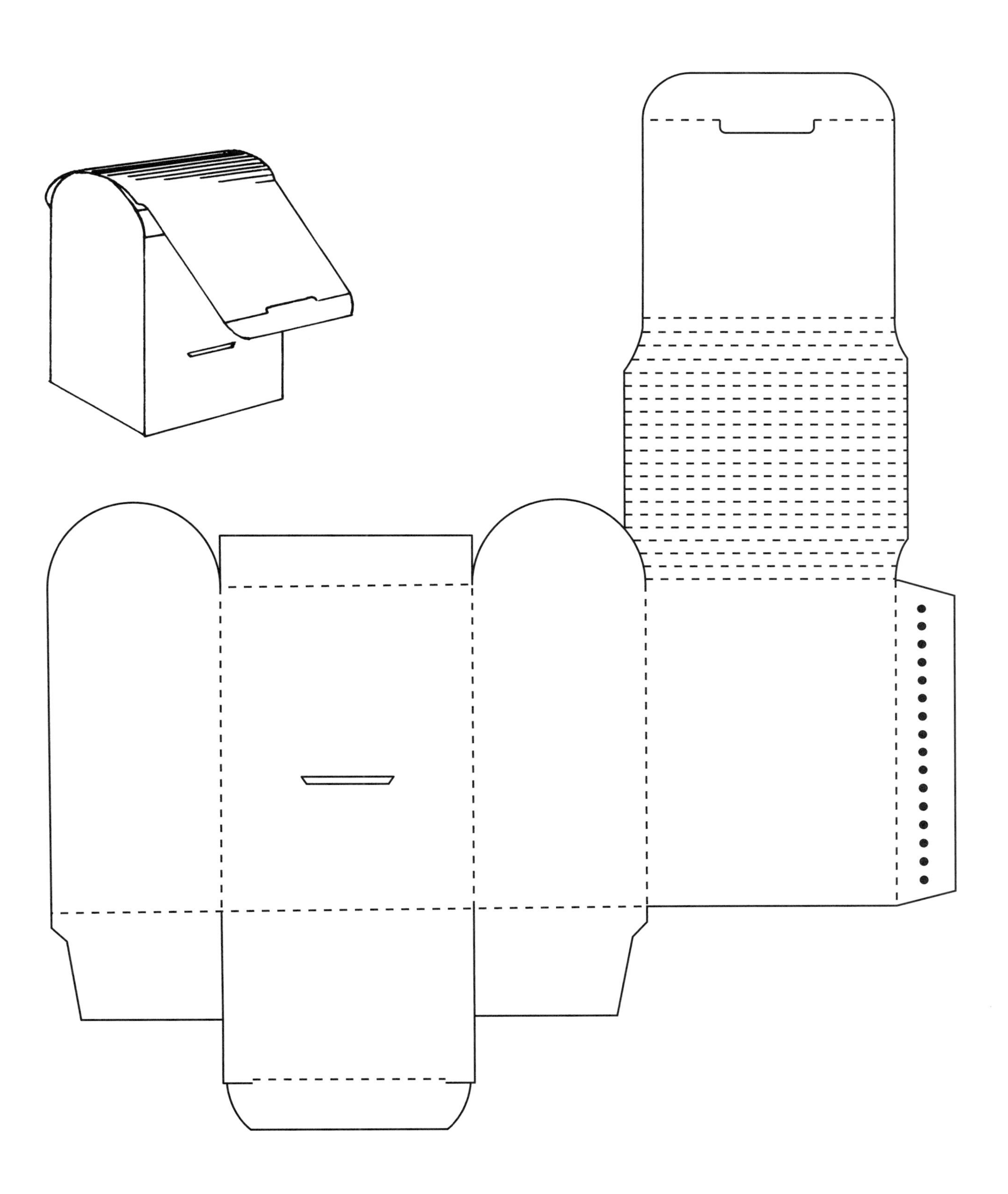

SPECIALLY SCORED TOP AND BOTTOM WITH FULL OVERLAP LOCK TAB AND SLIT

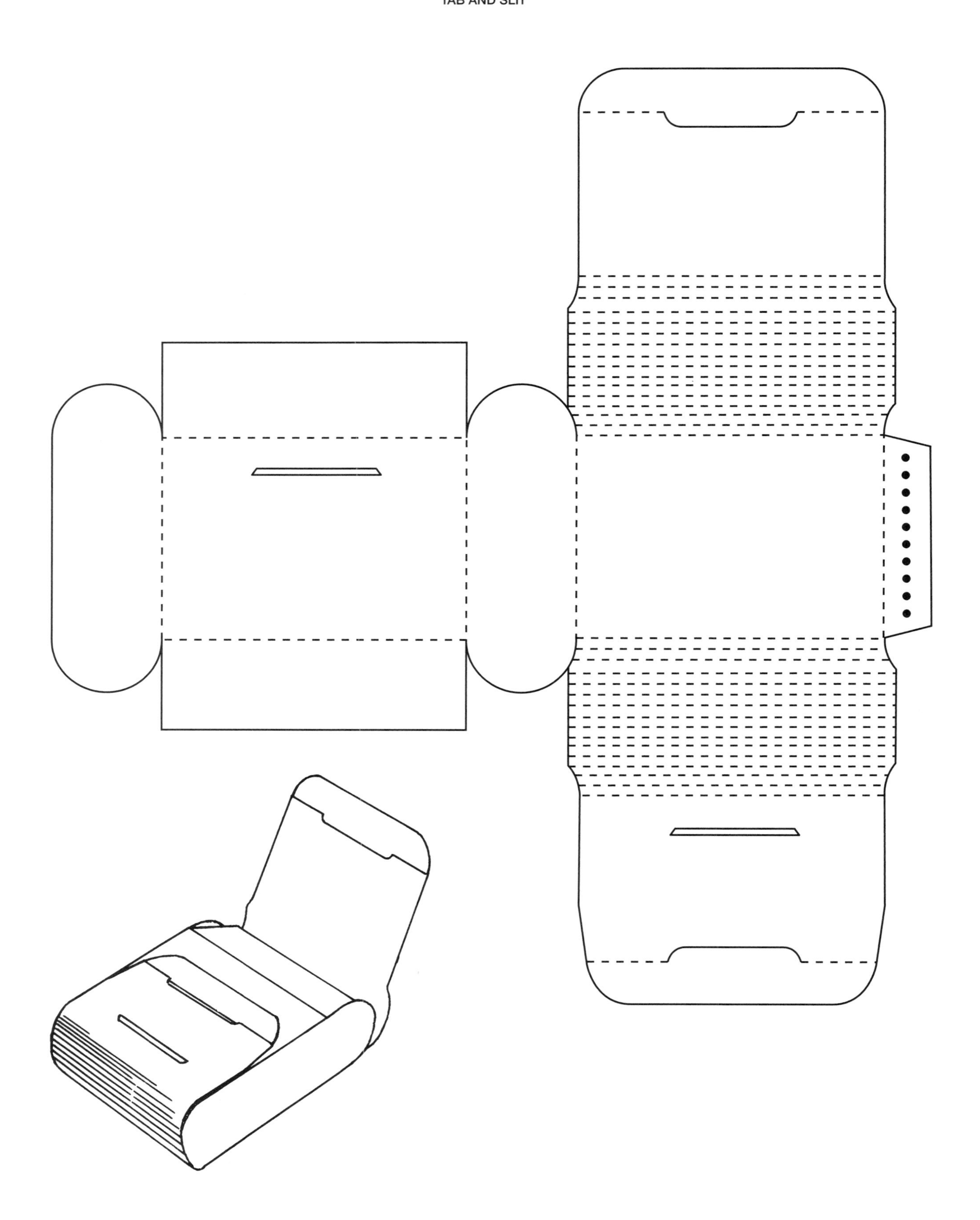

TRIPLE ARTHUR LOCK TOP CLOSURE

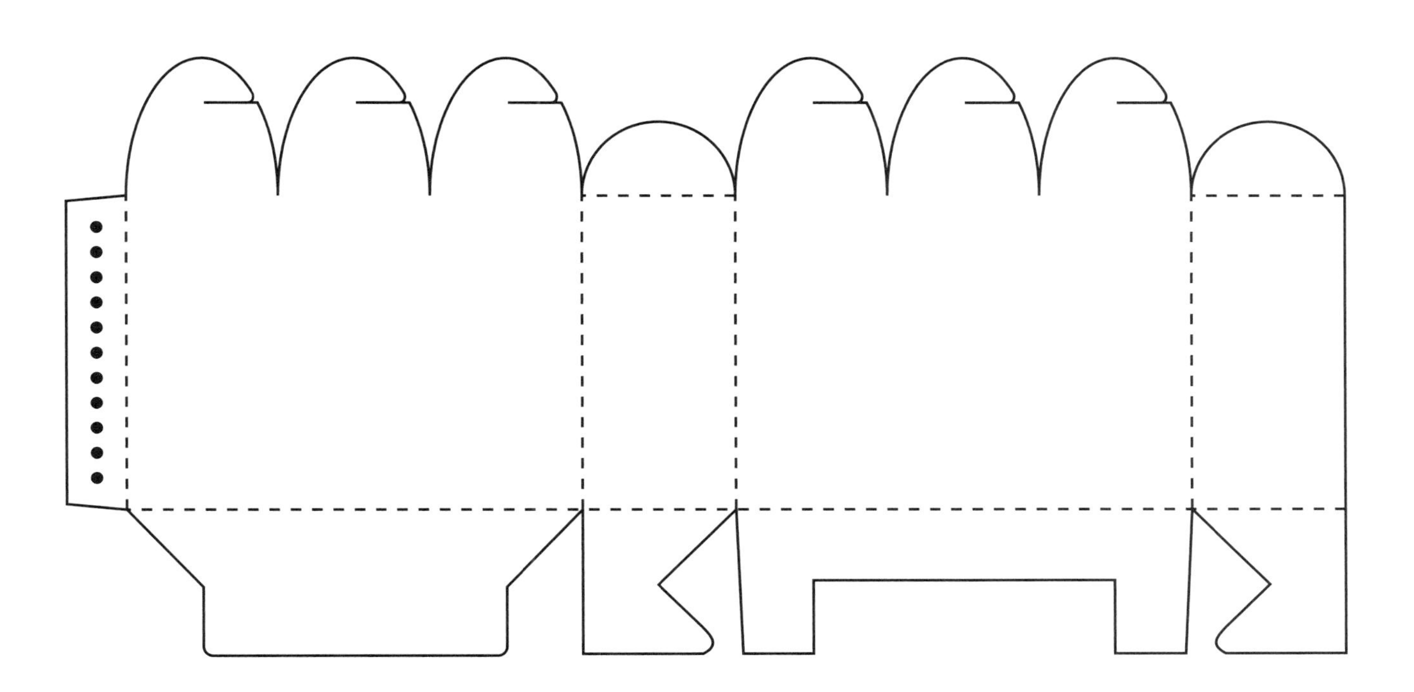

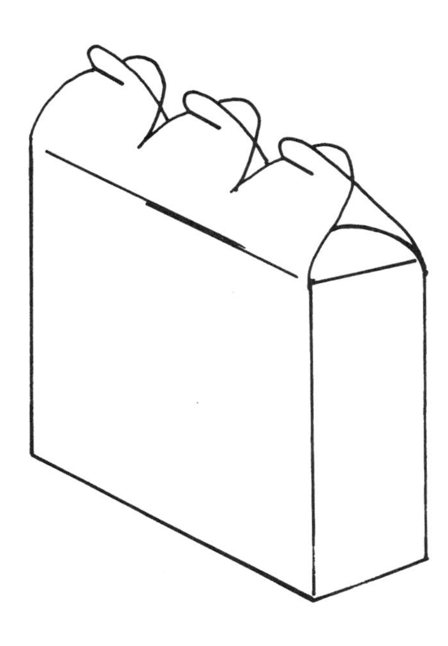

A carton with a variation of Arthur locks and 1-2-3 bottom develops into an attractive alternative packaging form.

PUSH-IN LOCK TOP WITH GREENLEAF BOTTOM

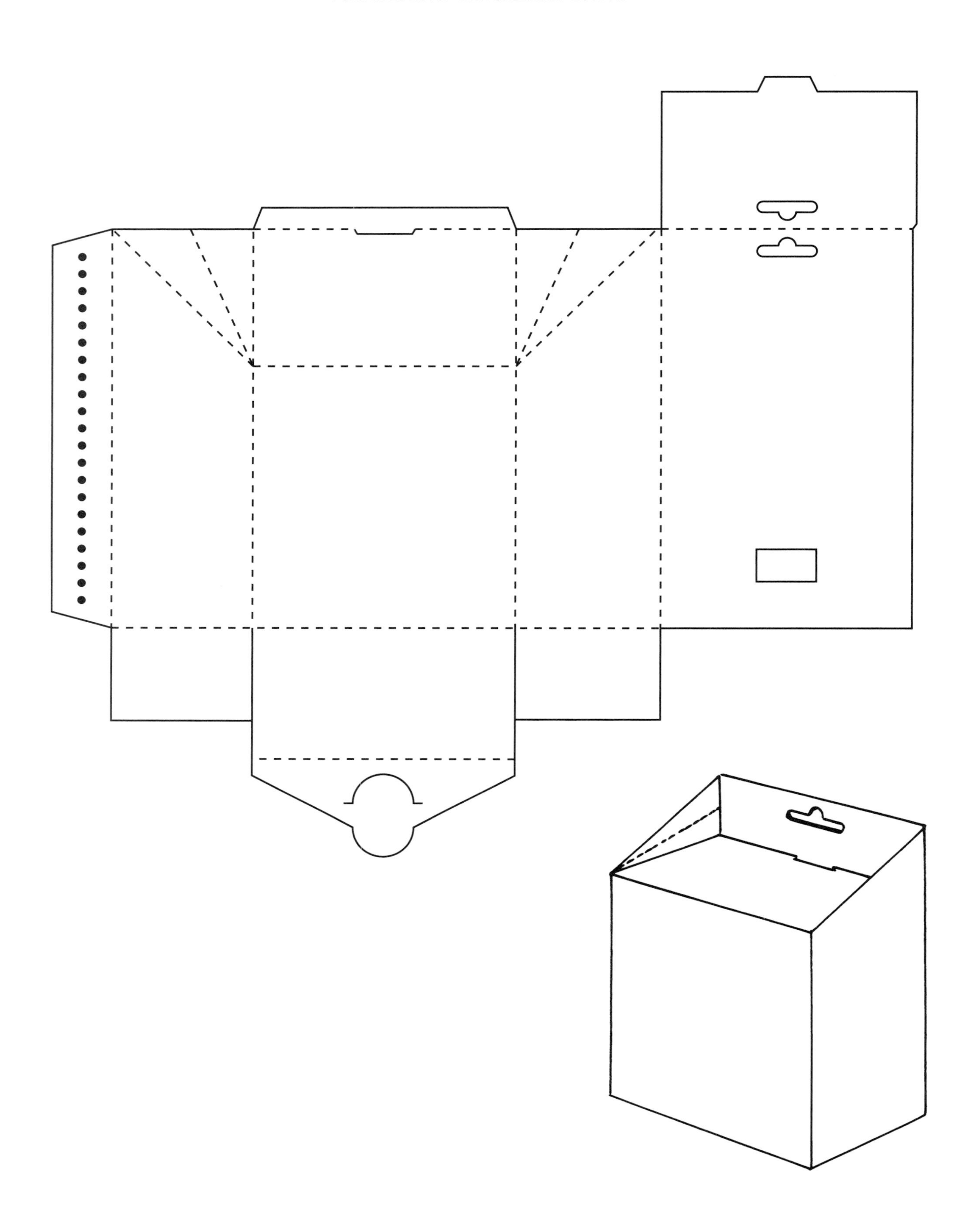

REVERSE TUCK WITH SPECIAL DISPENSER PANEL

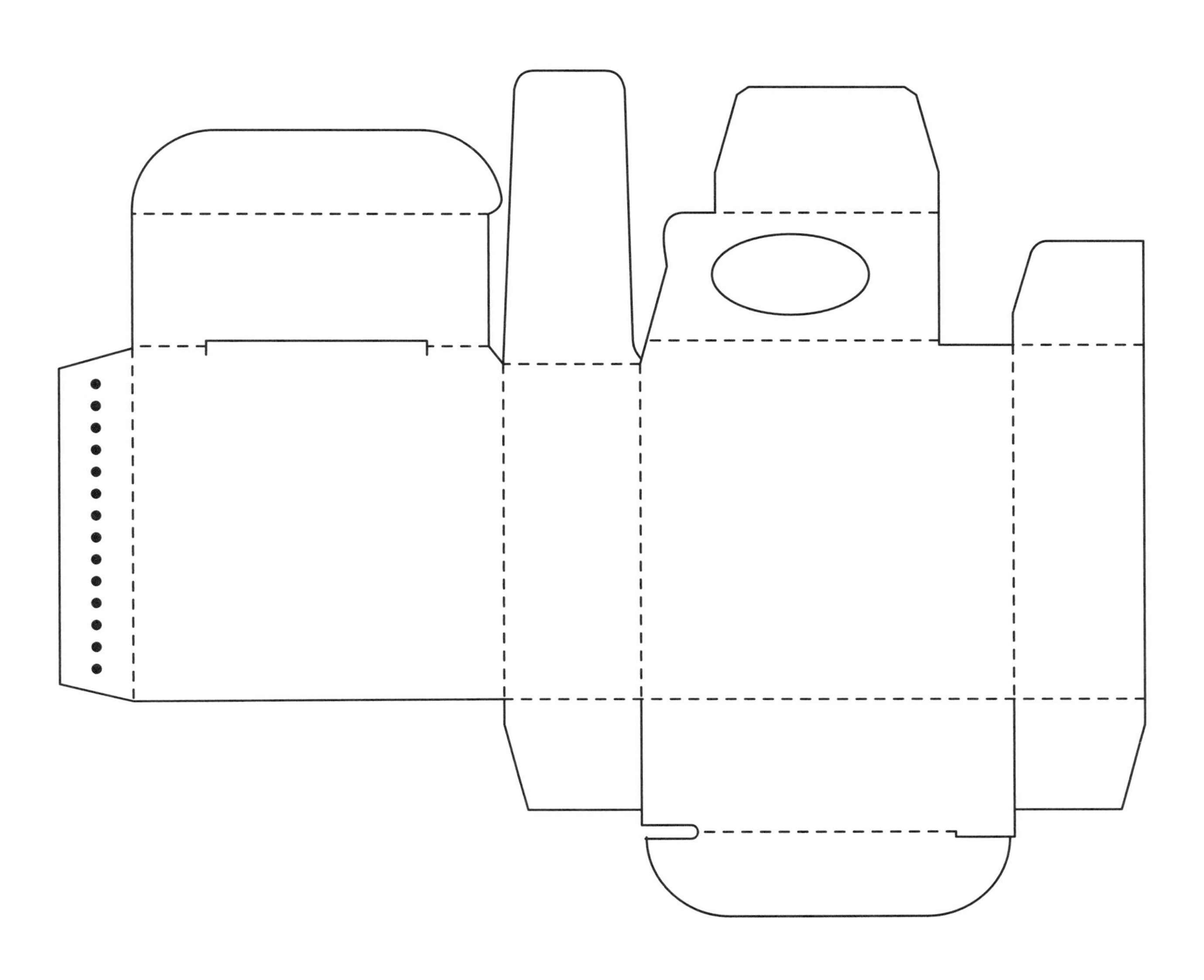

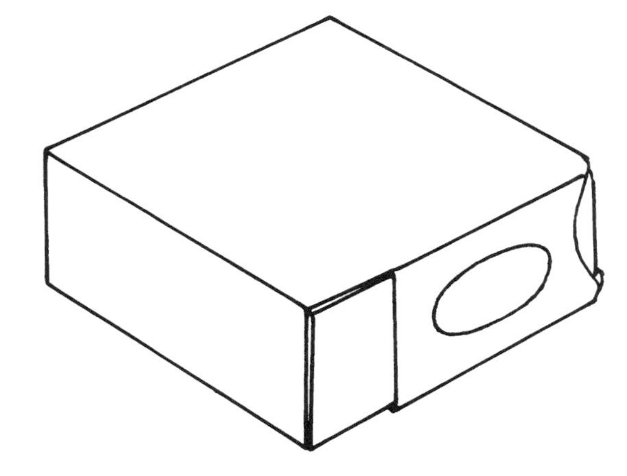

Ideal for ribbons, tape, etc.

TWO-COMPARTMENT DISPENSER WITH TEAR-AWAY CLOSURE

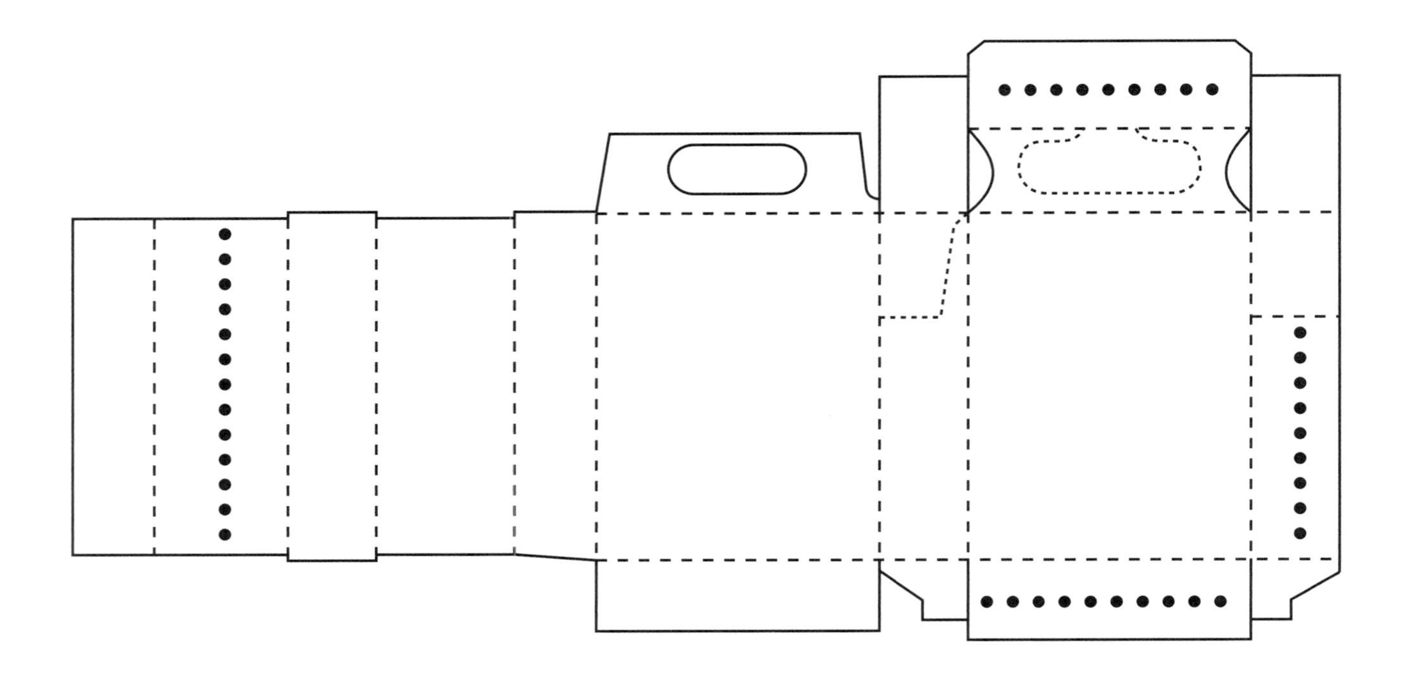

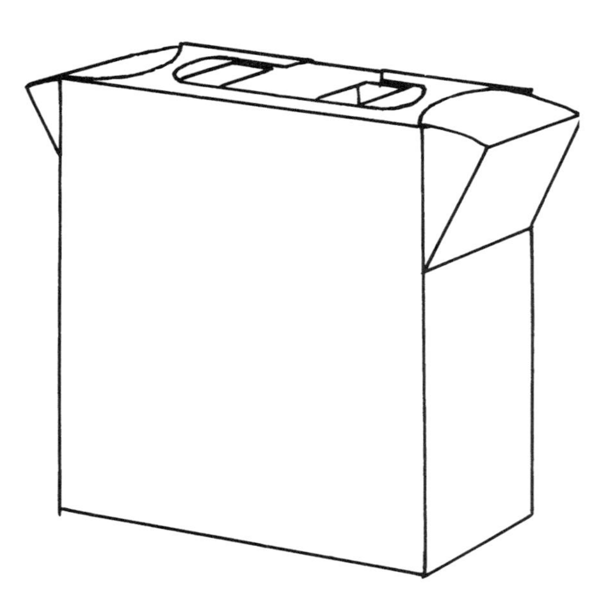

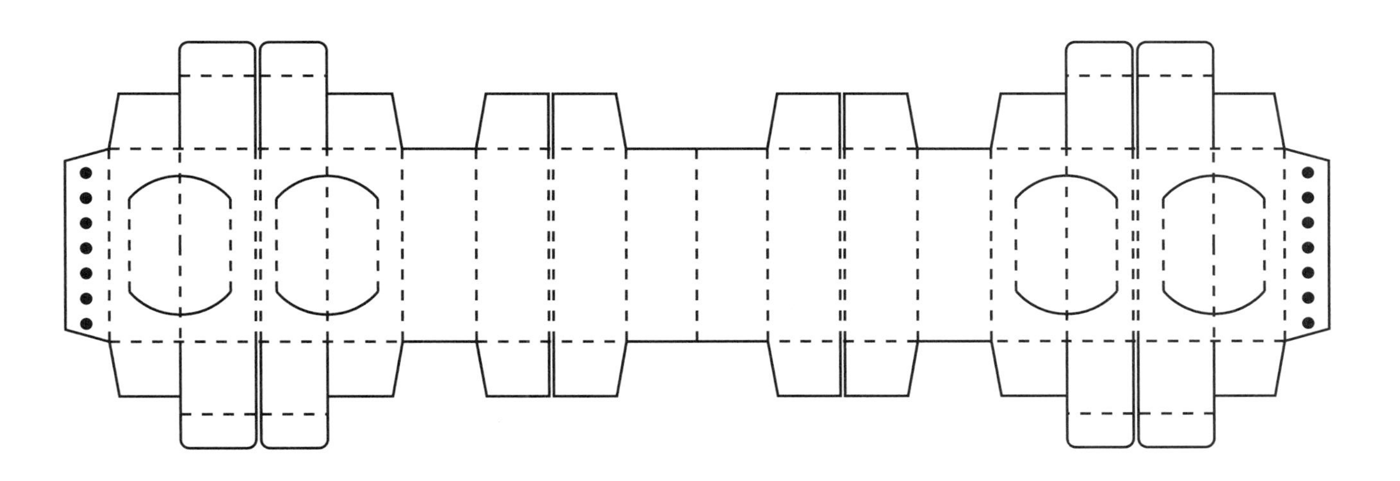

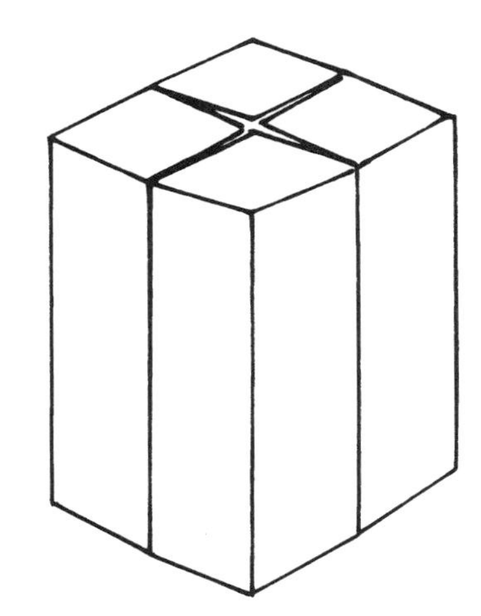

This quadruple box is ideal for sets of items such as cosmetics.

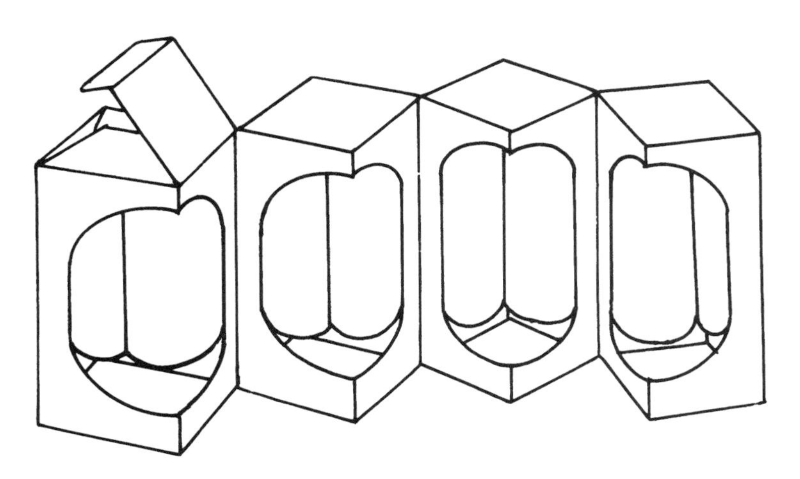

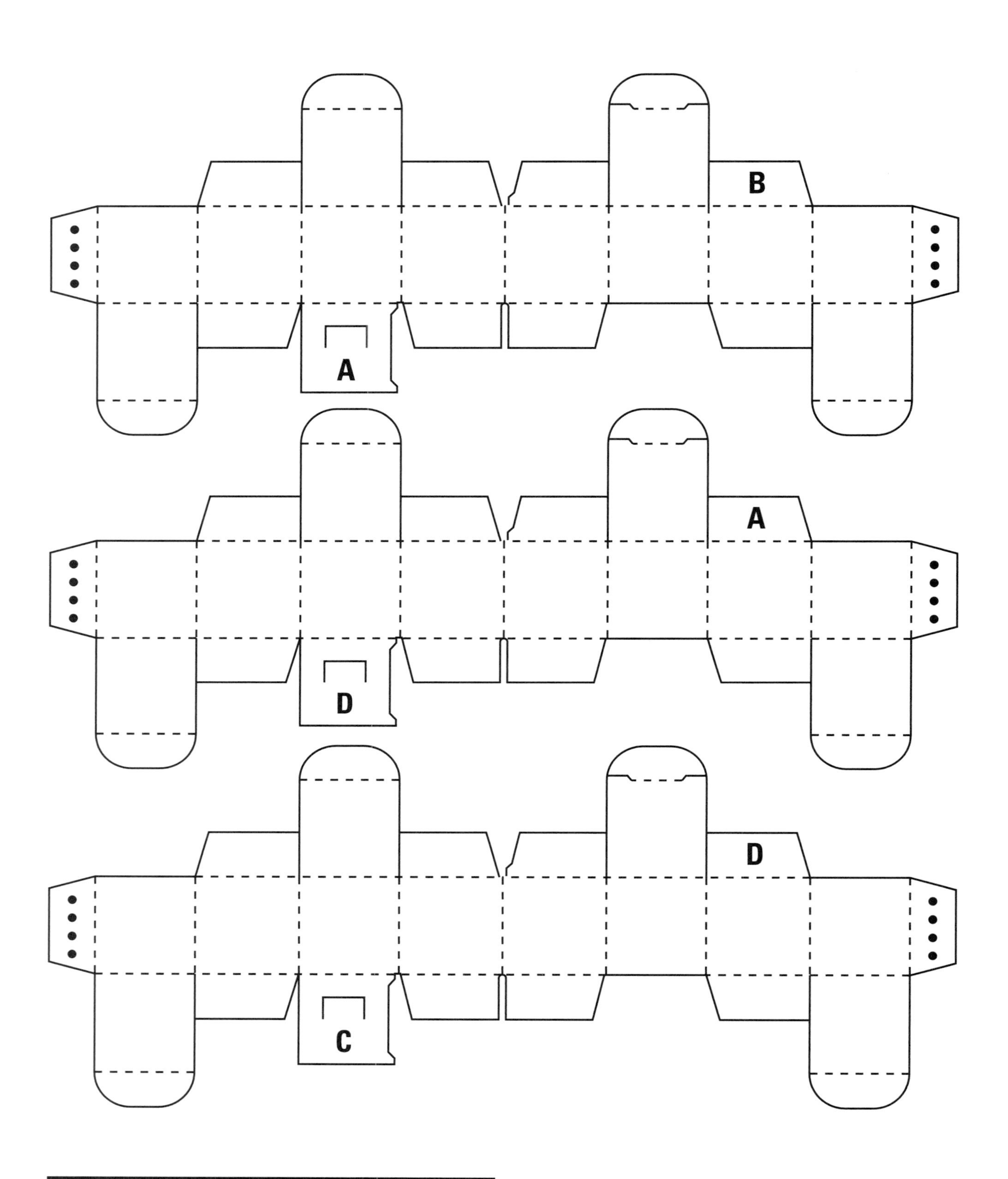

The eight cubes interlock at the dust flaps as indicated. This allows for many graphic innovations.

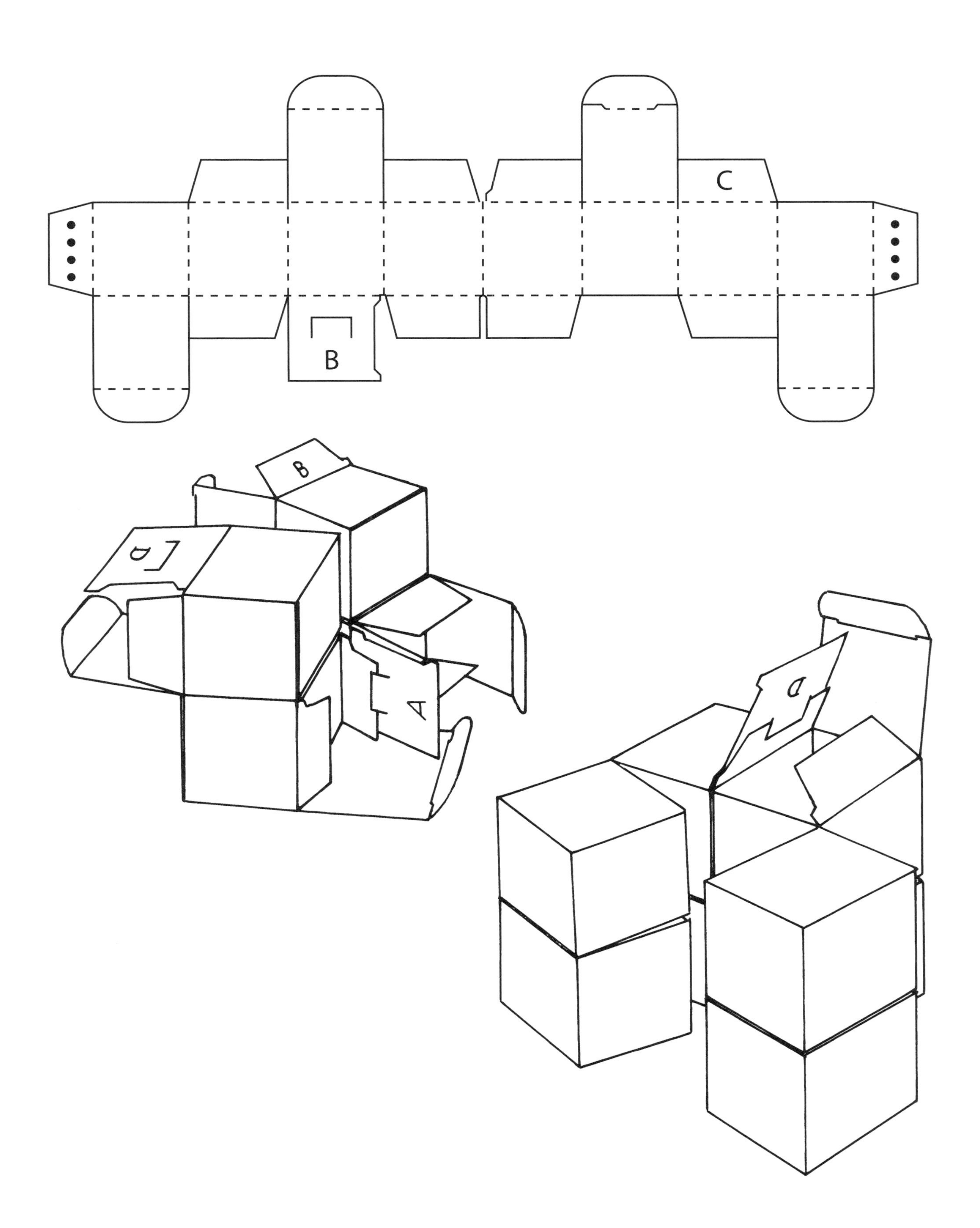

C
B
B
B
A
B

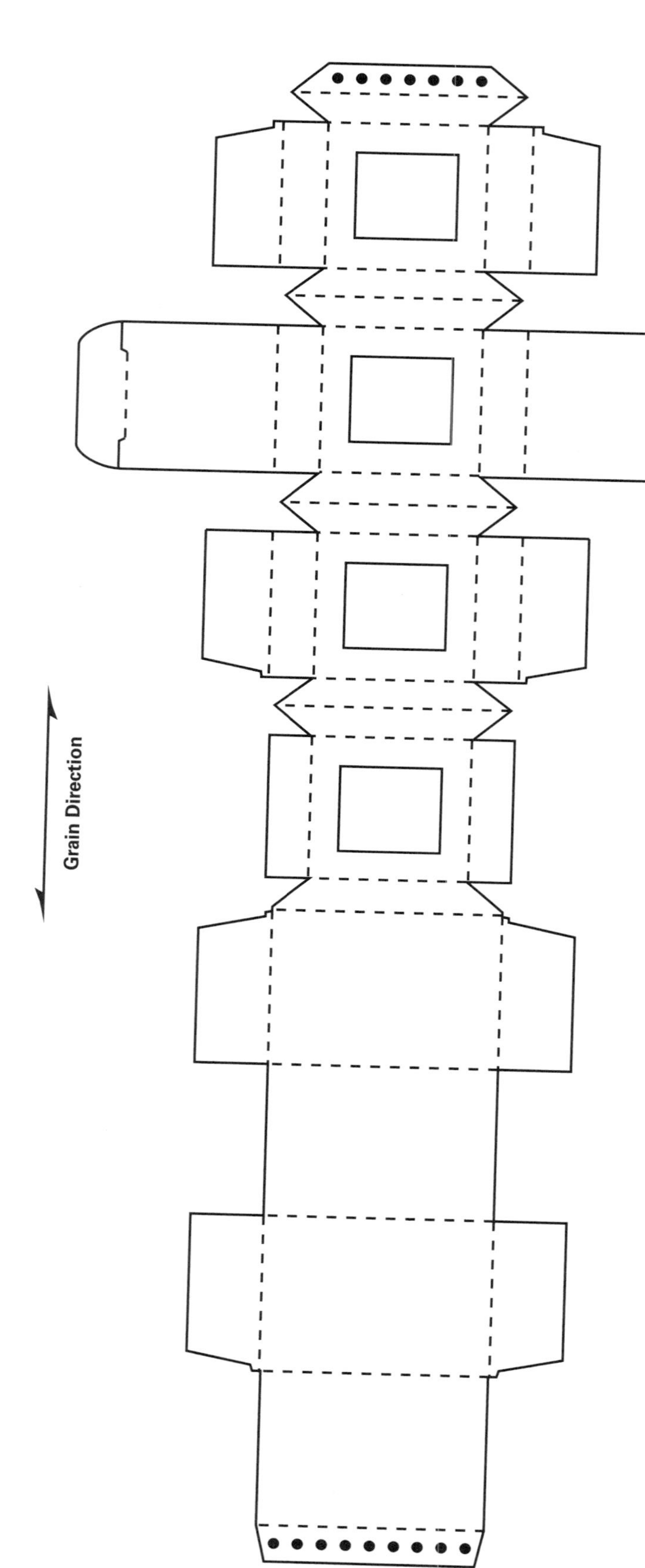

This one-piece shadow box, designed by Yick-Man Liu, lends itself to a great assortment of graphics and makes a beautiful gift package.

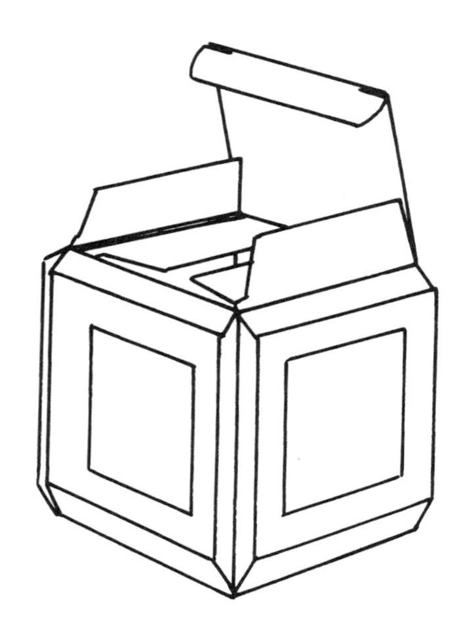

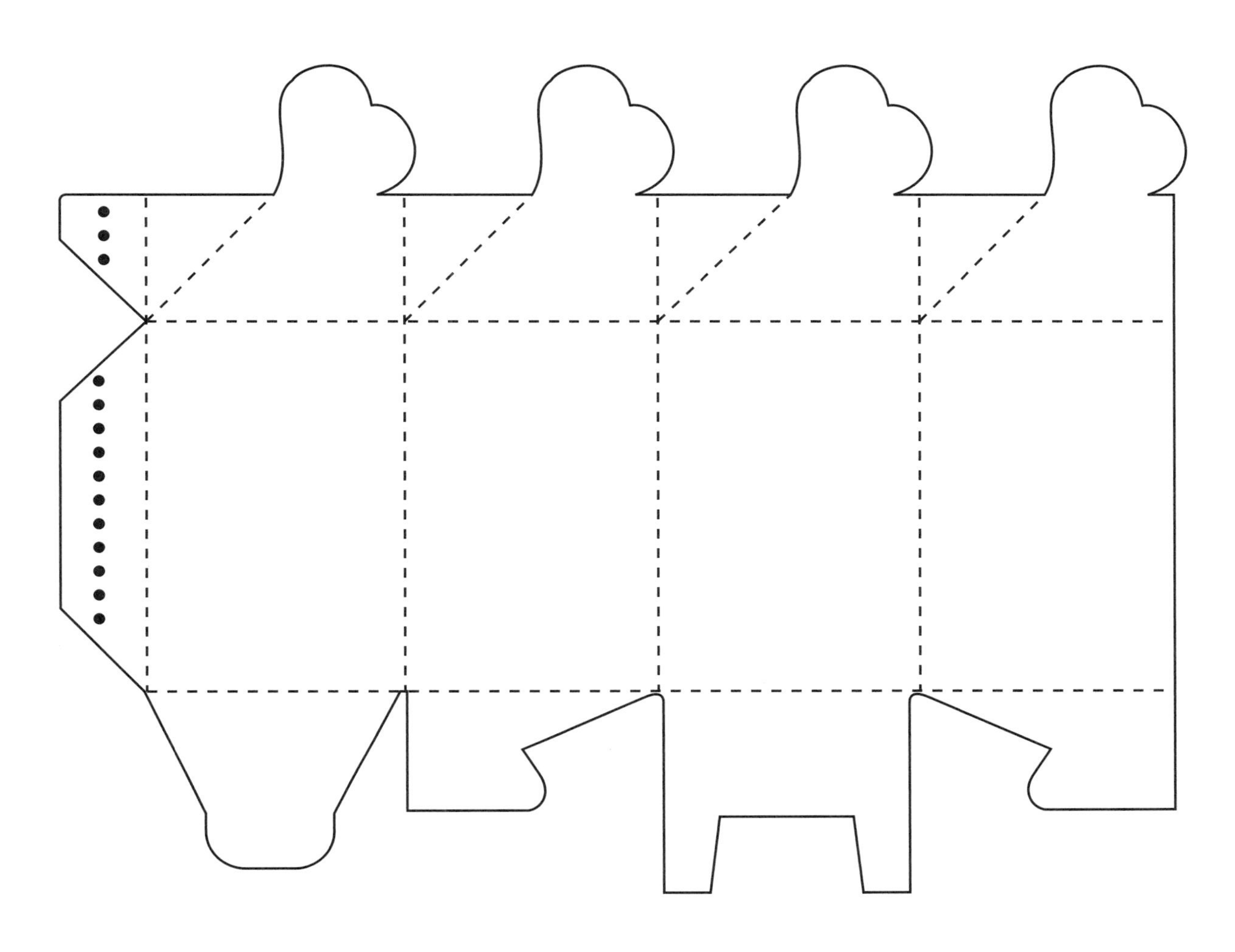

This is but one of the many design possibilities for top closure.

PUSH-DOWN OPENING TOP

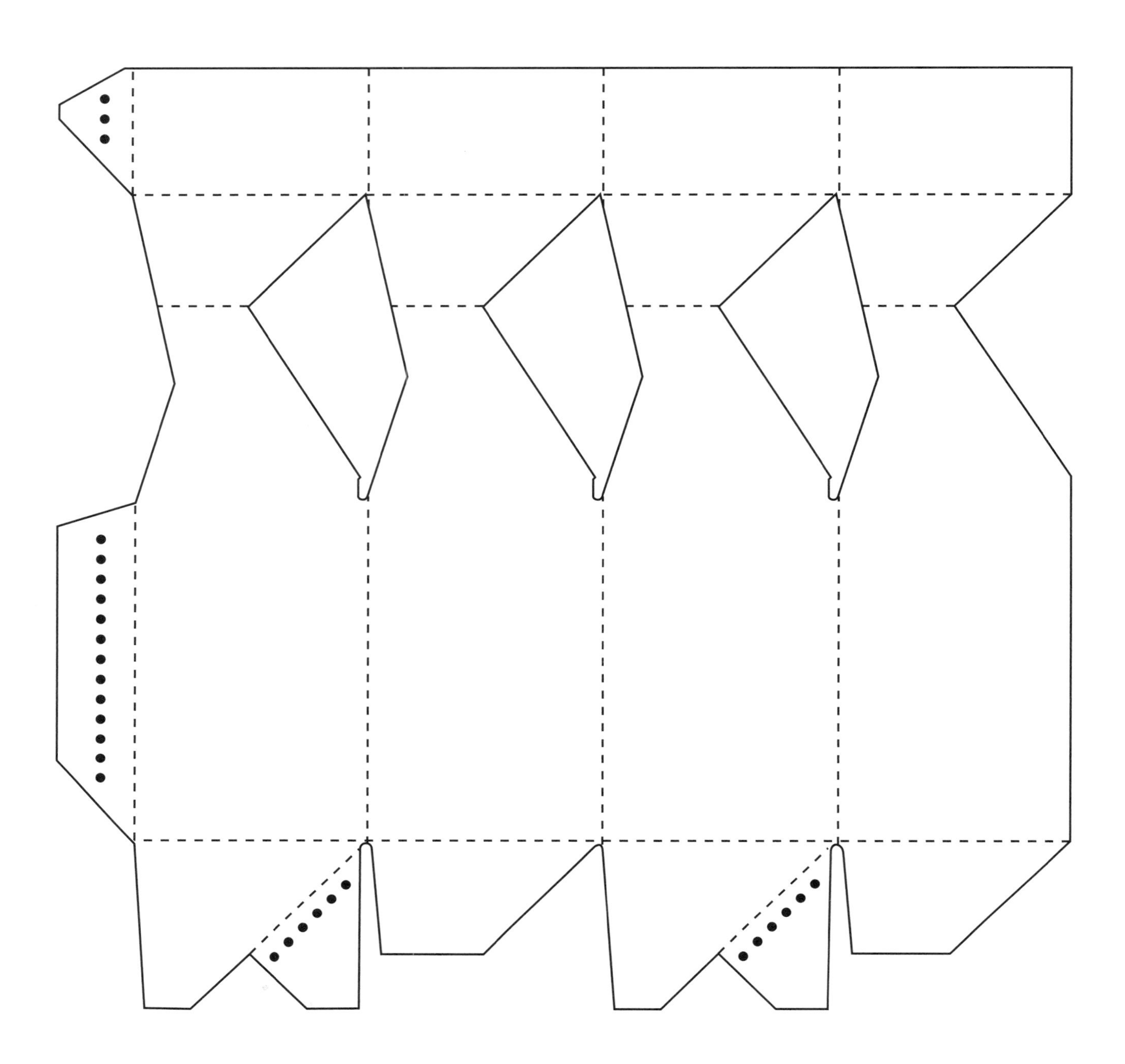

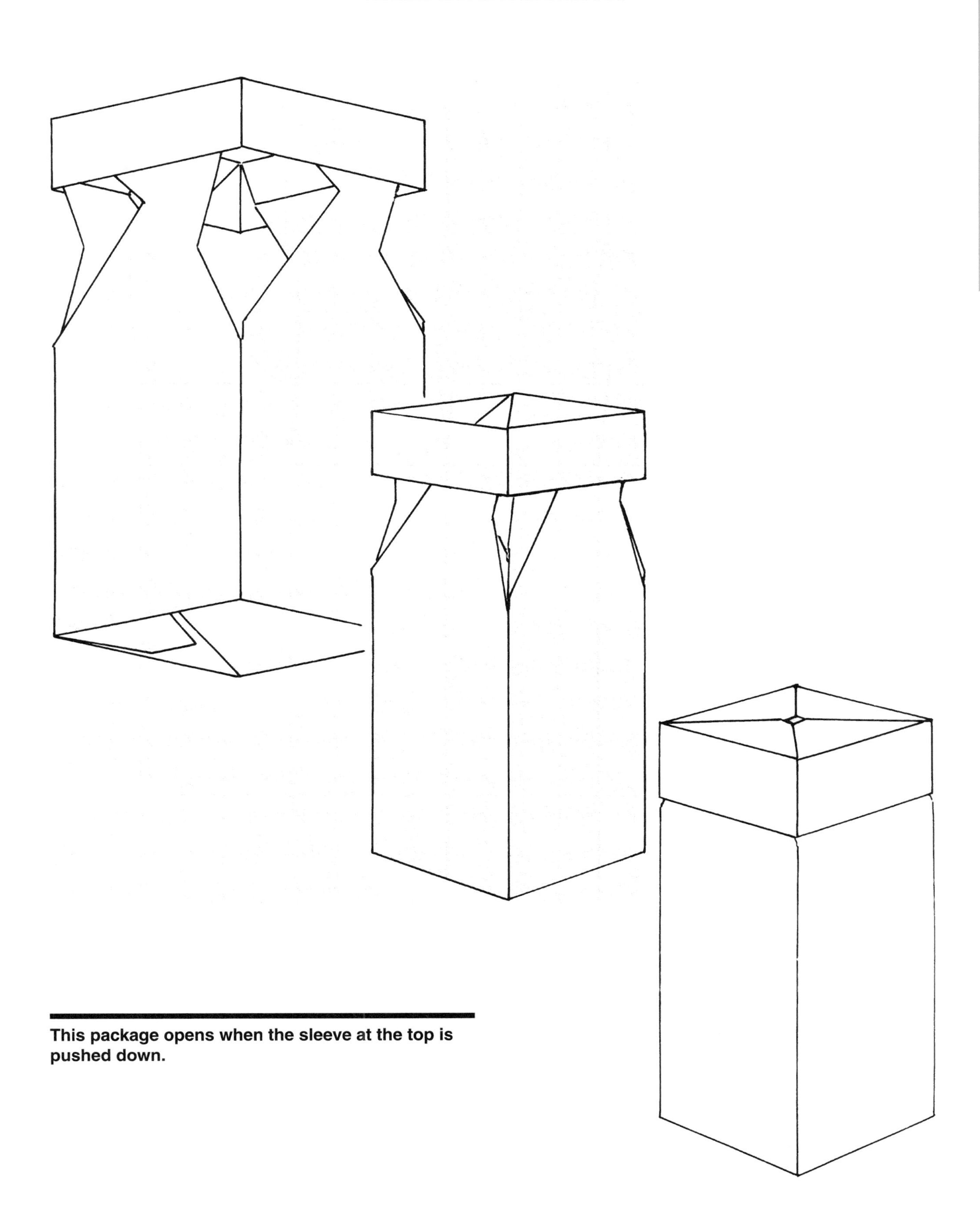

This package opens when the sleeve at the top is pushed down.

PUSH-DOWN OPENING TOP VARIATION

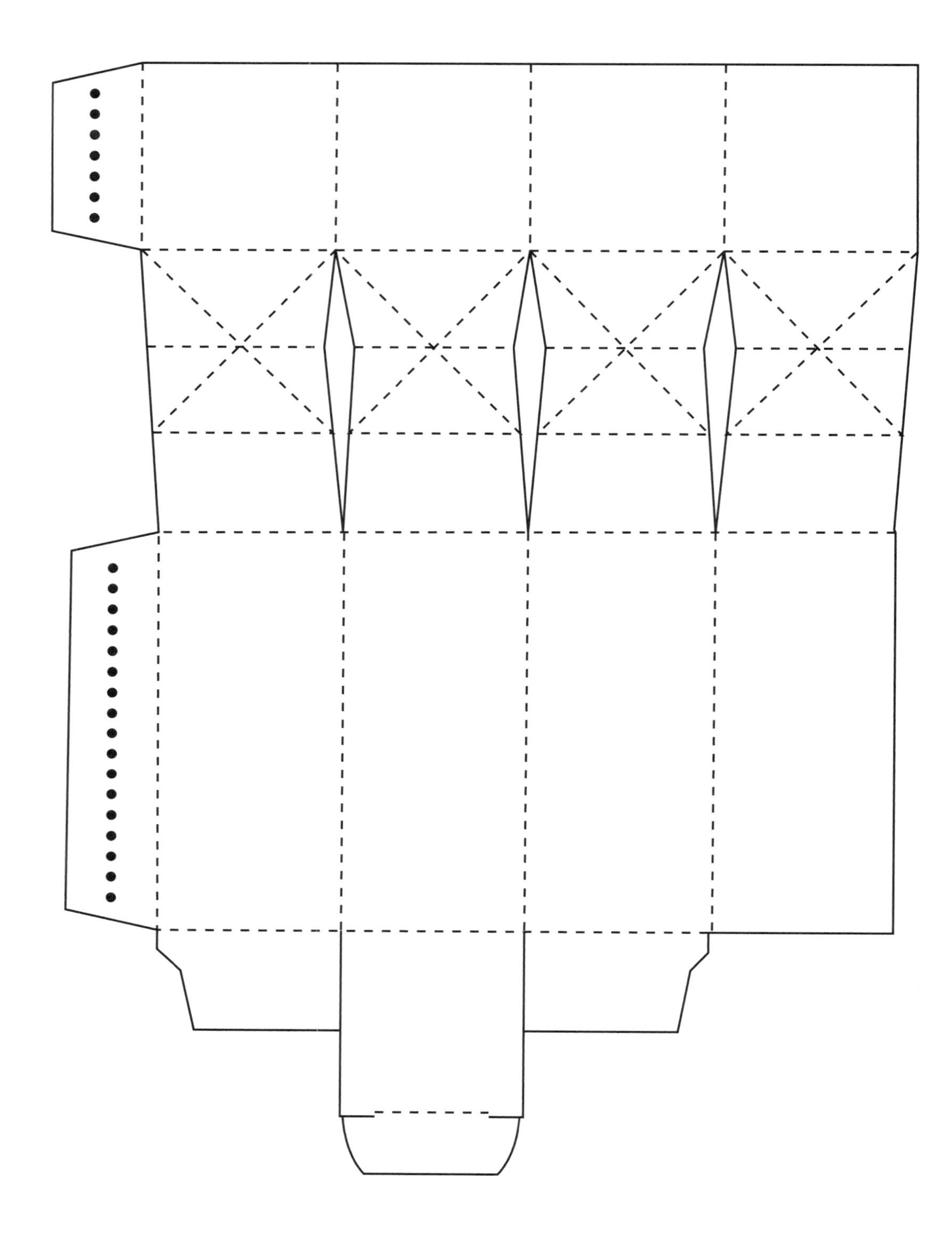

OPEN AND CLOSED VIEW OF PUSH-DOWN OPENING

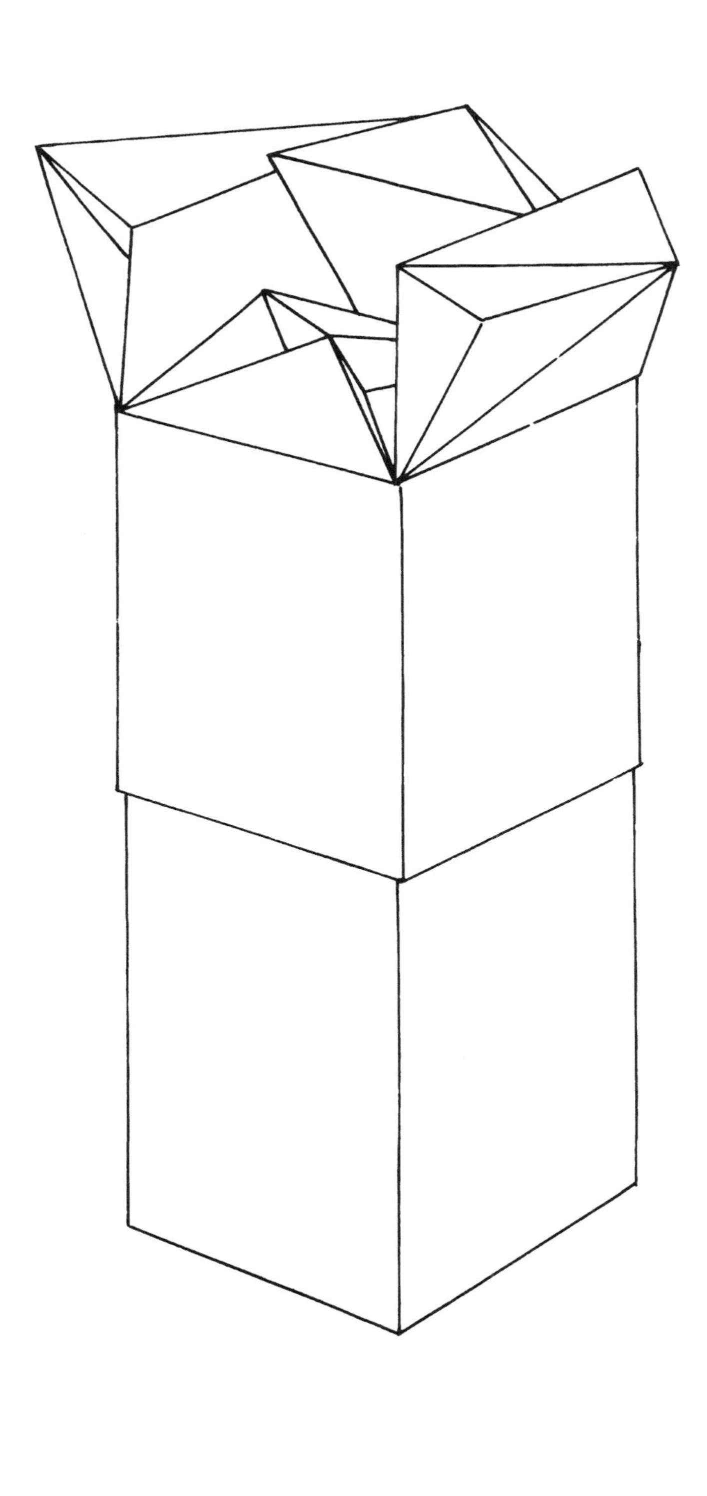

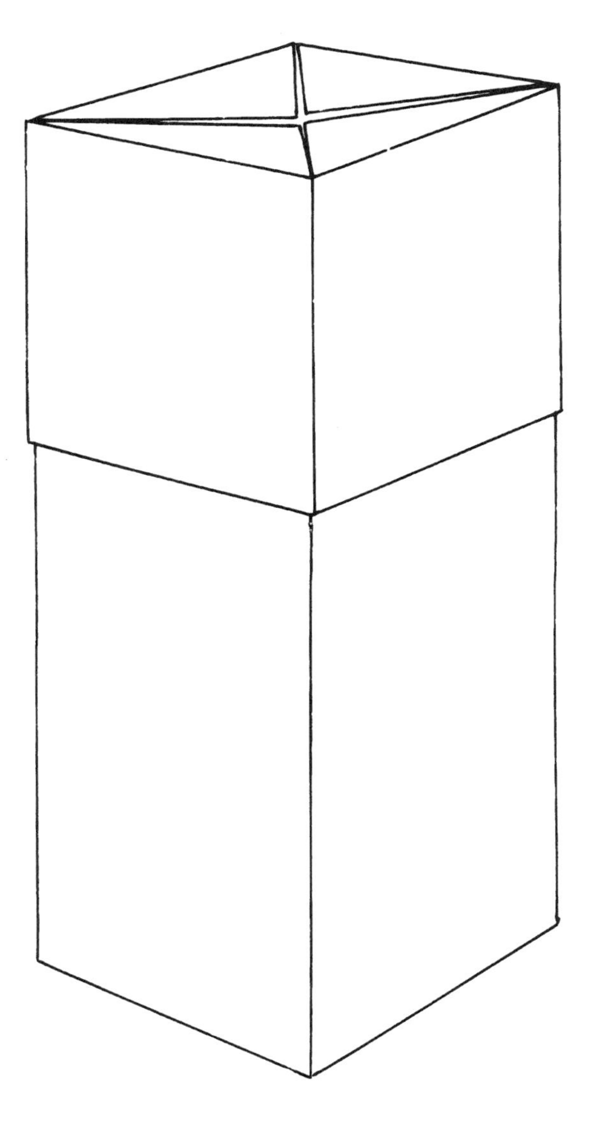

As the top overlapping sleeve is pushed down, the triangular closure shapes open. Sliding the sleeve up will close the top.

CONCAVE/CONVEX TOP AND SIDE PANELS

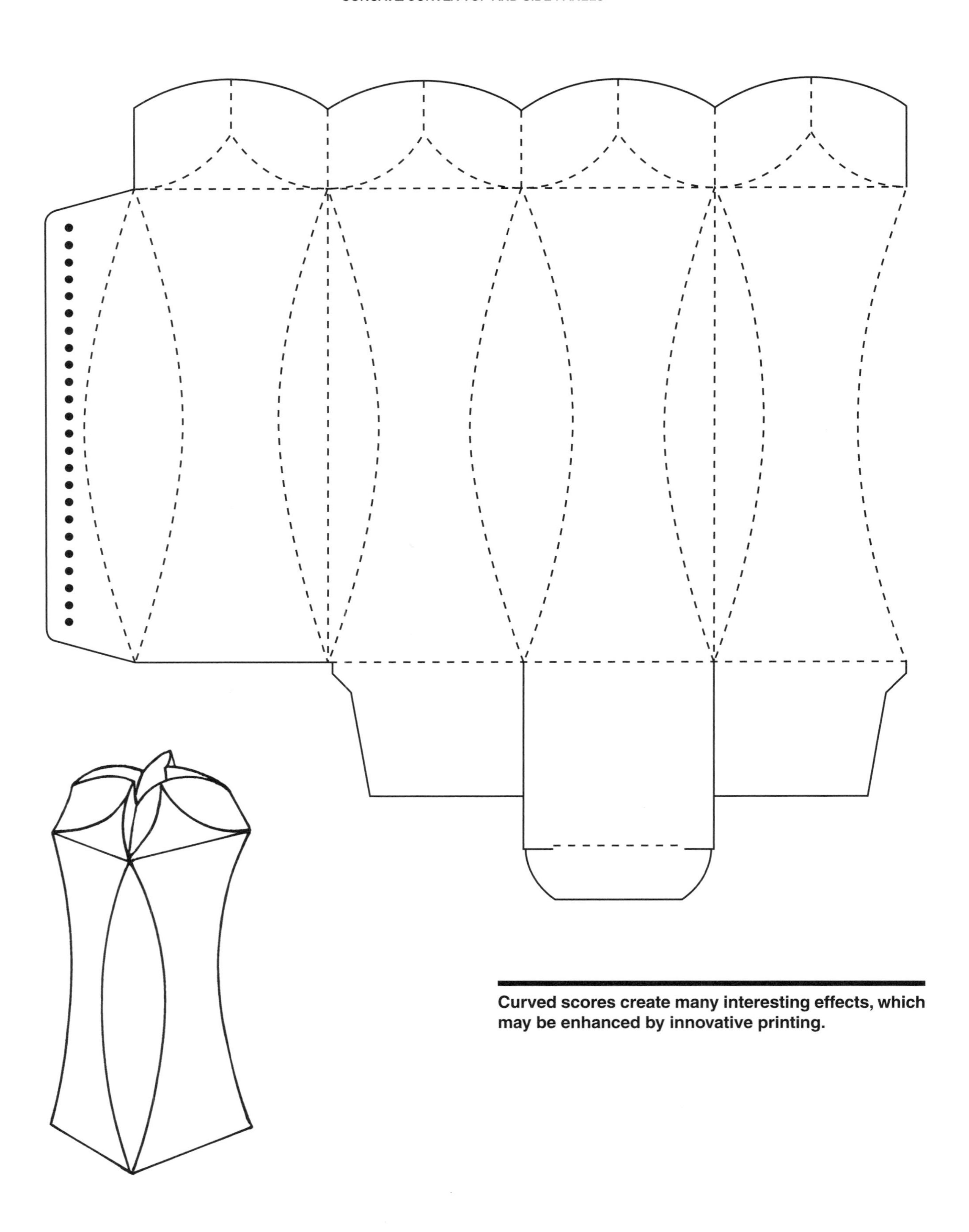

Curved scores create many interesting effects, which may be enhanced by innovative printing.

ROUND SNAP LOCK TOP

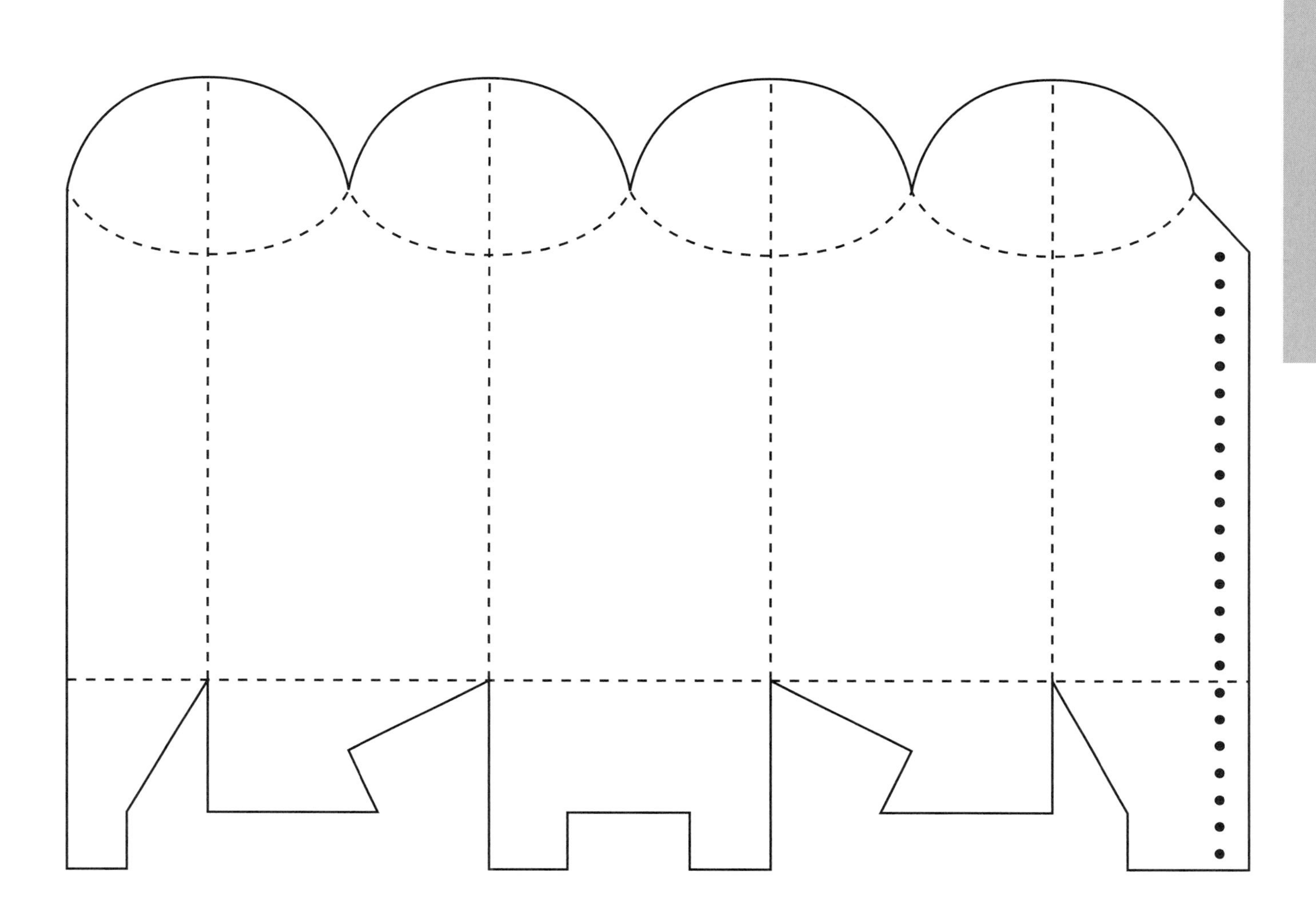

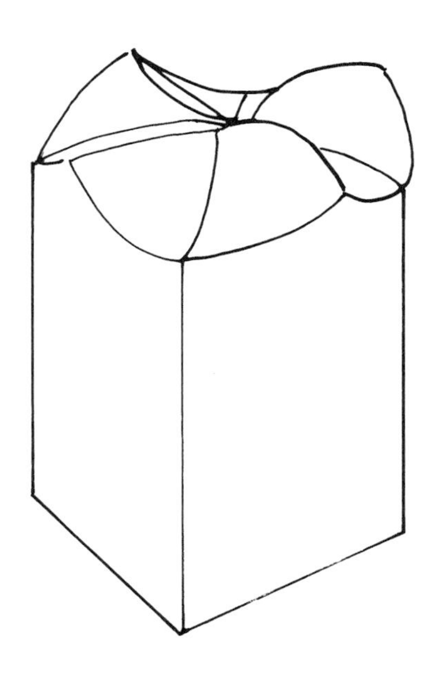

TRIANGULAR SNAP LOCK TOP

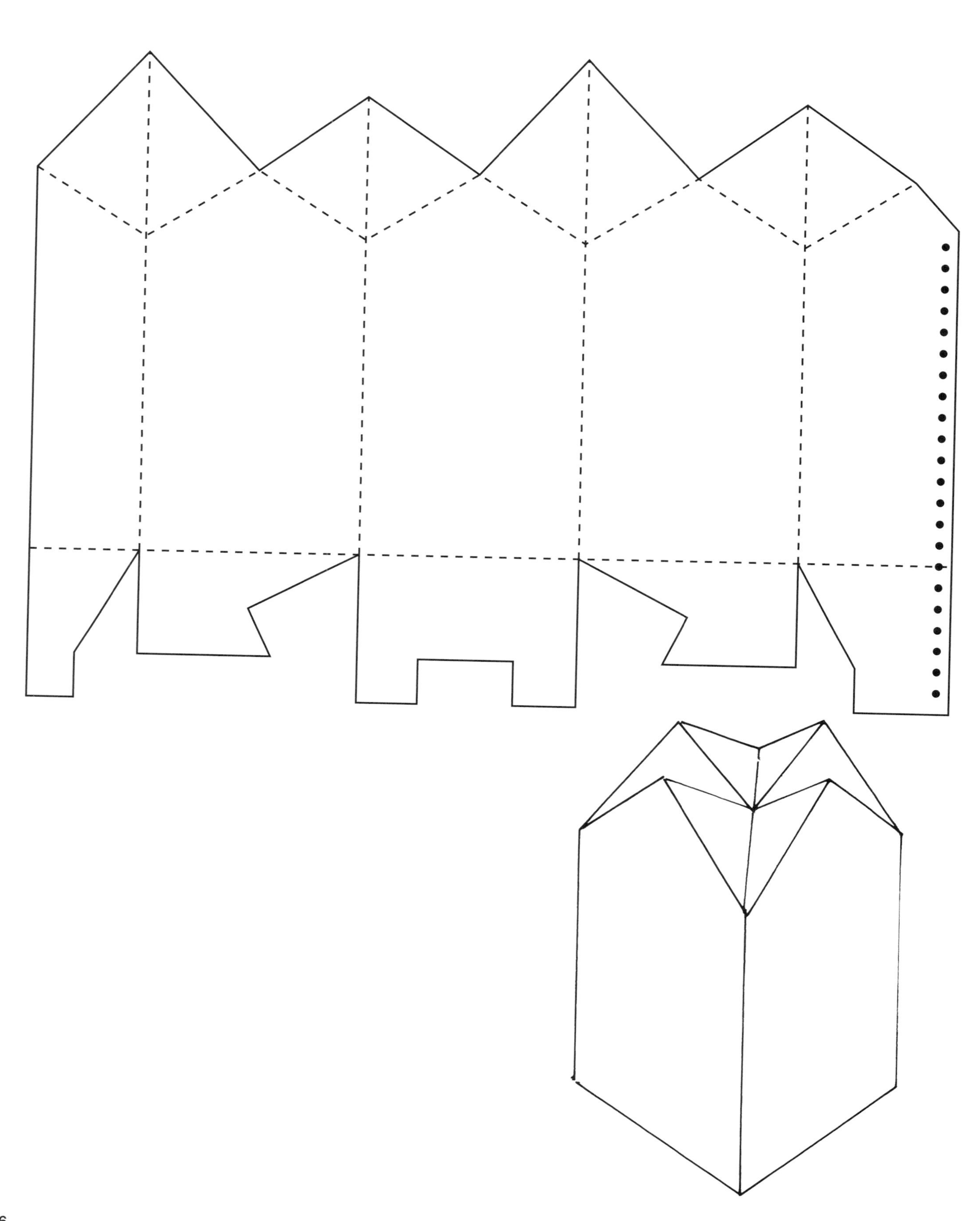

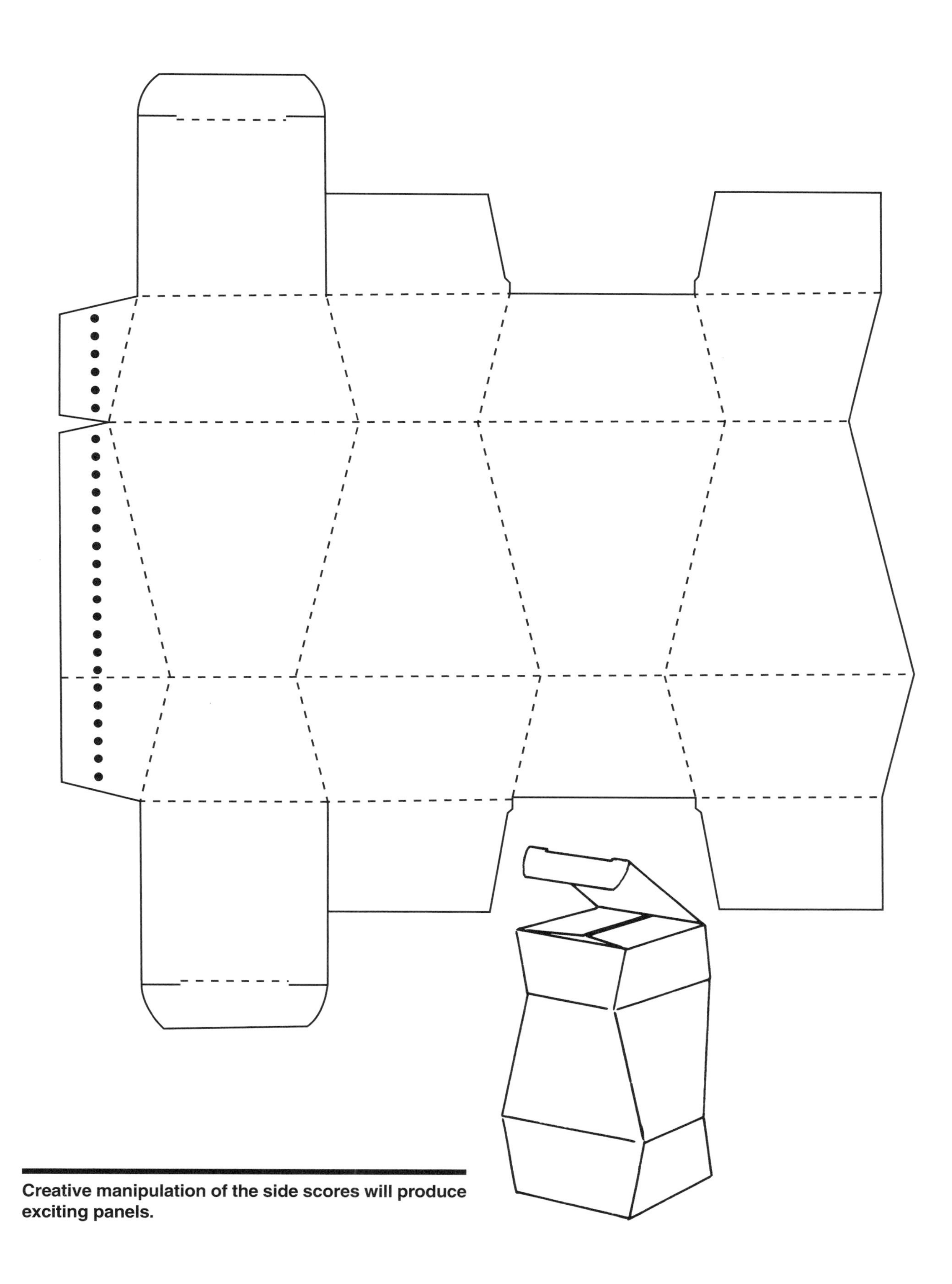

Creative manipulation of the side scores will produce exciting panels.

CONCAVE/CONVEX PANELS BOX

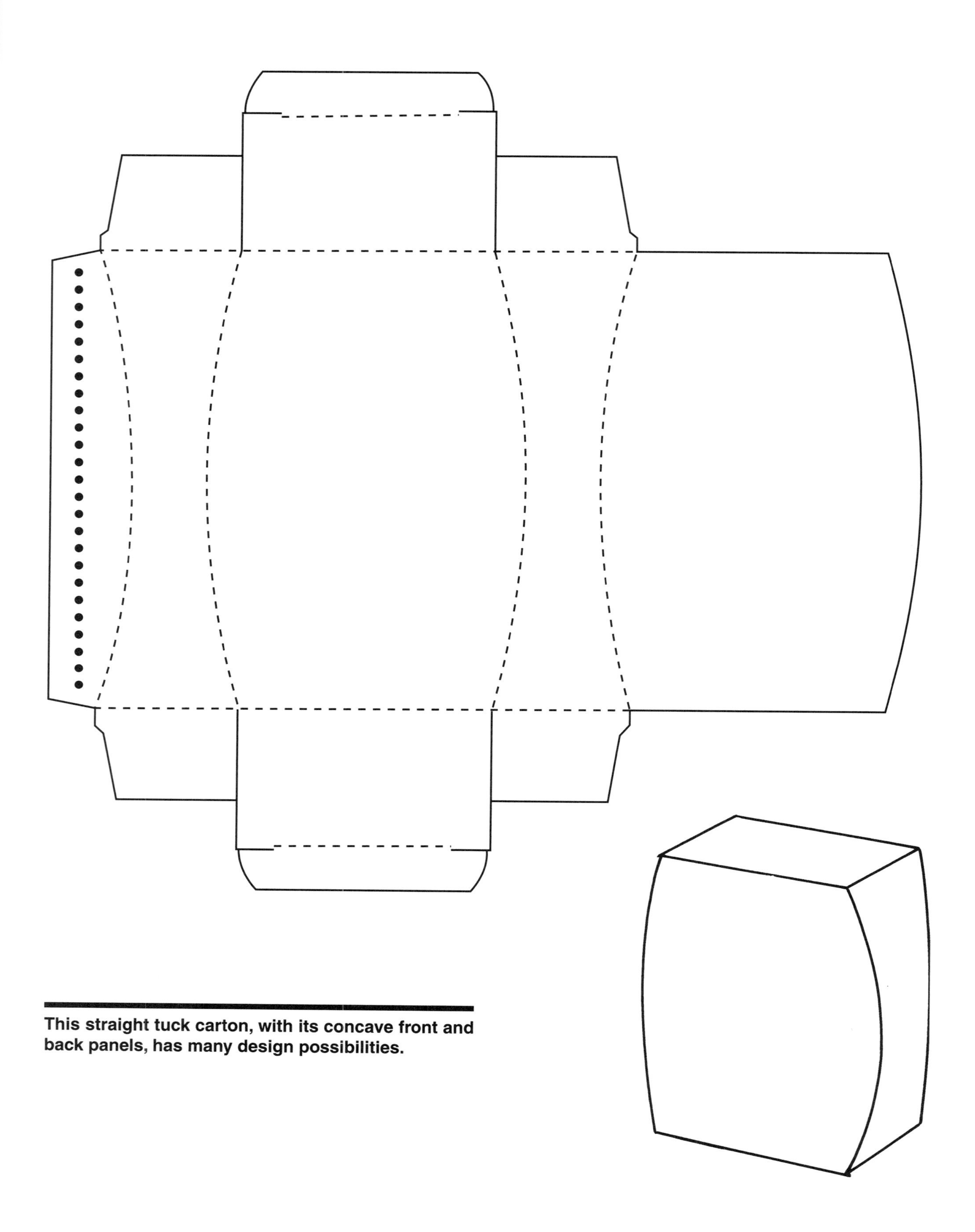

This straight tuck carton, with its concave front and back panels, has many design possibilities.

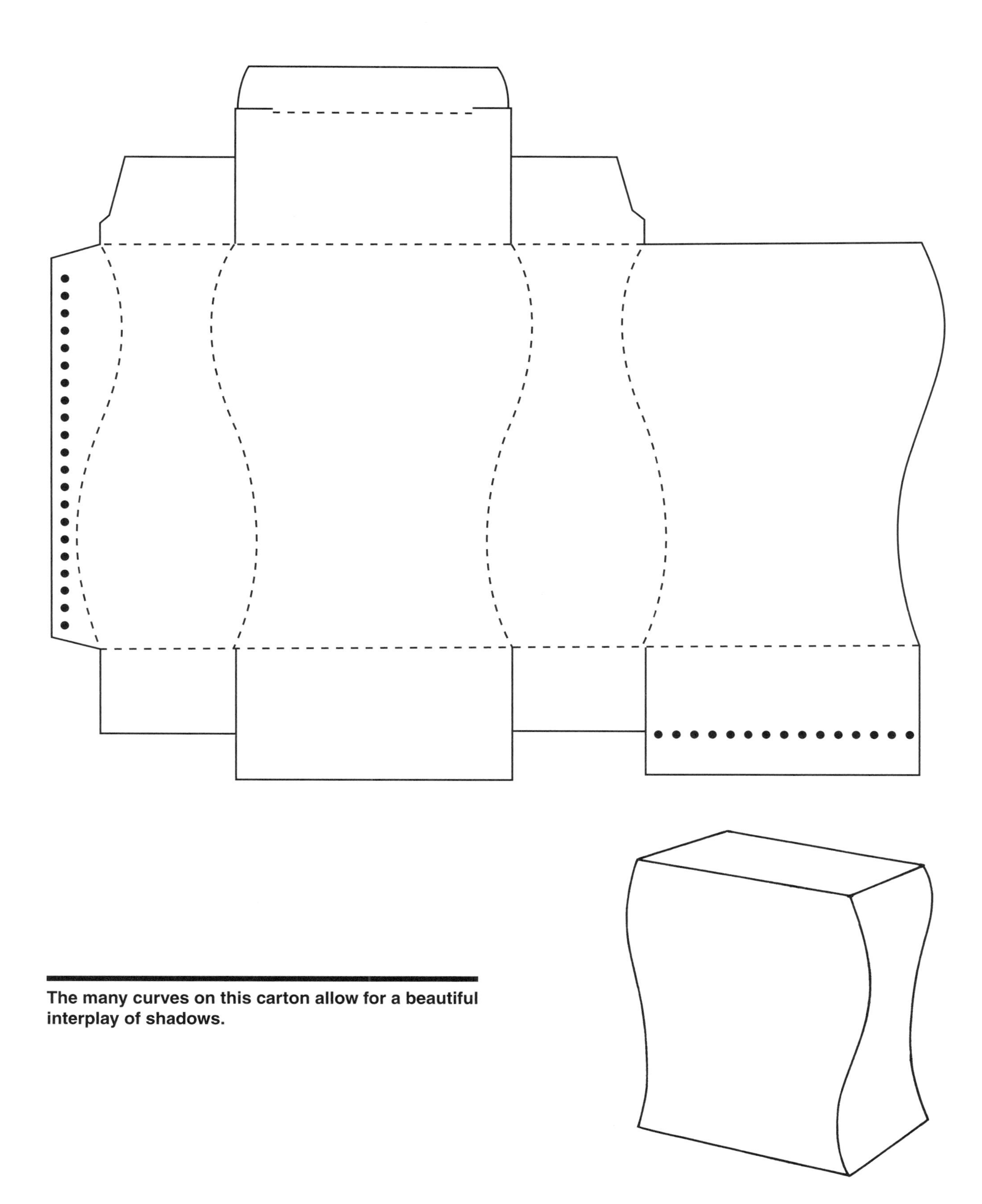

The many curves on this carton allow for a beautiful interplay of shadows.

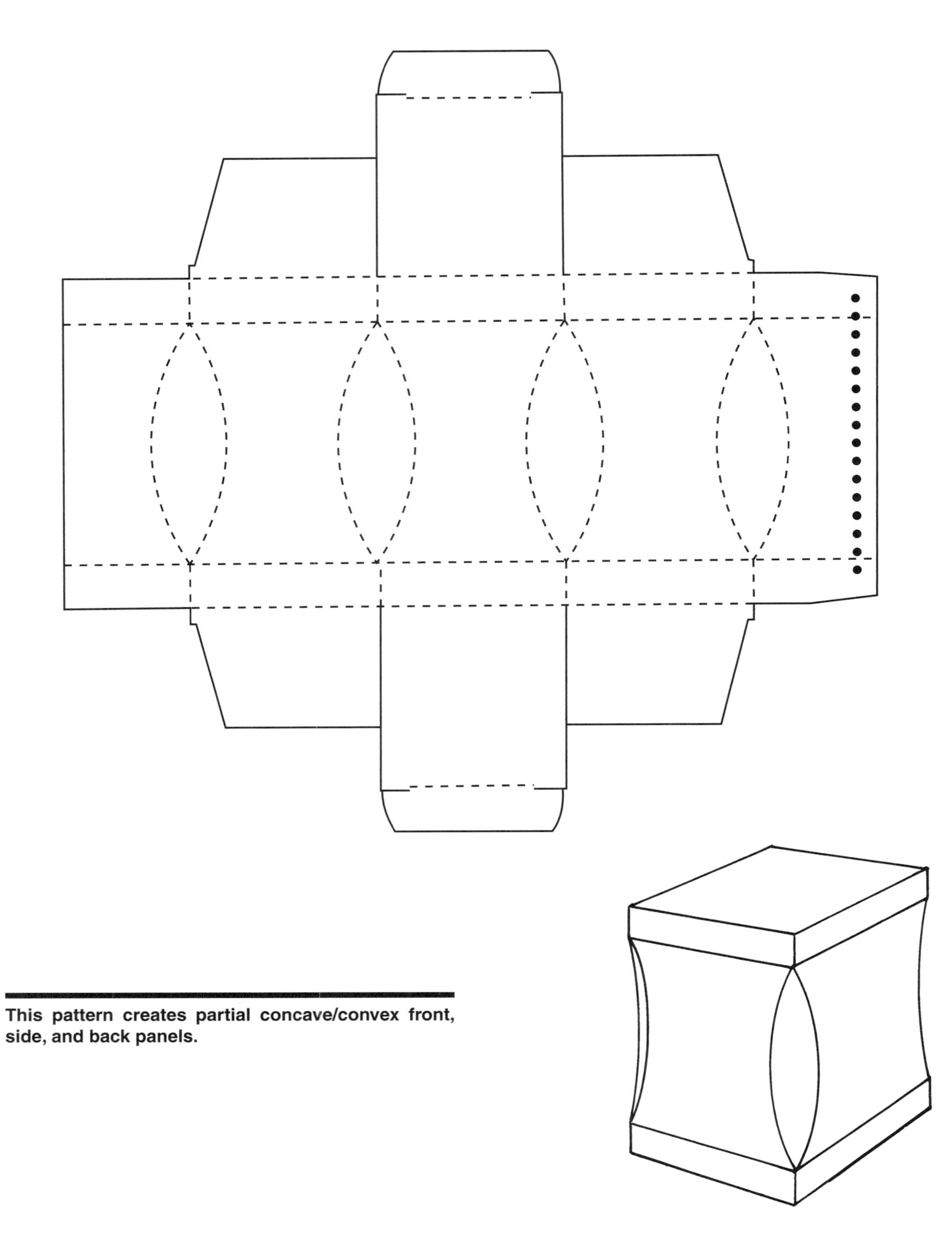

This pattern creates partial concave/convex front, side, and back panels.

TRIANGLE WITH TUCK CLOSURES AND TEAR-AWAY ZIPPER STRIP

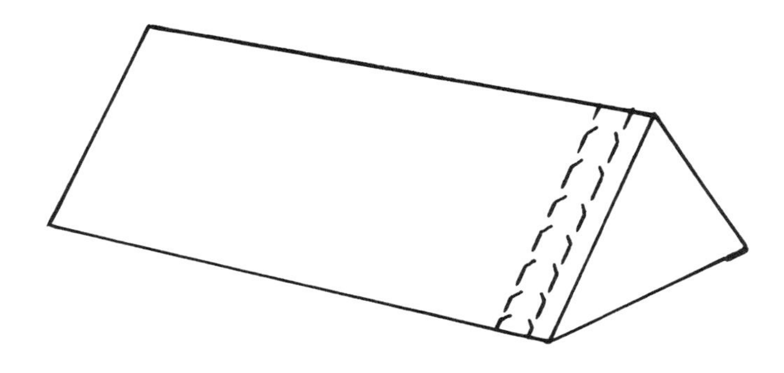

TRIANGLE WITH AUTO LOCK BOTTOM

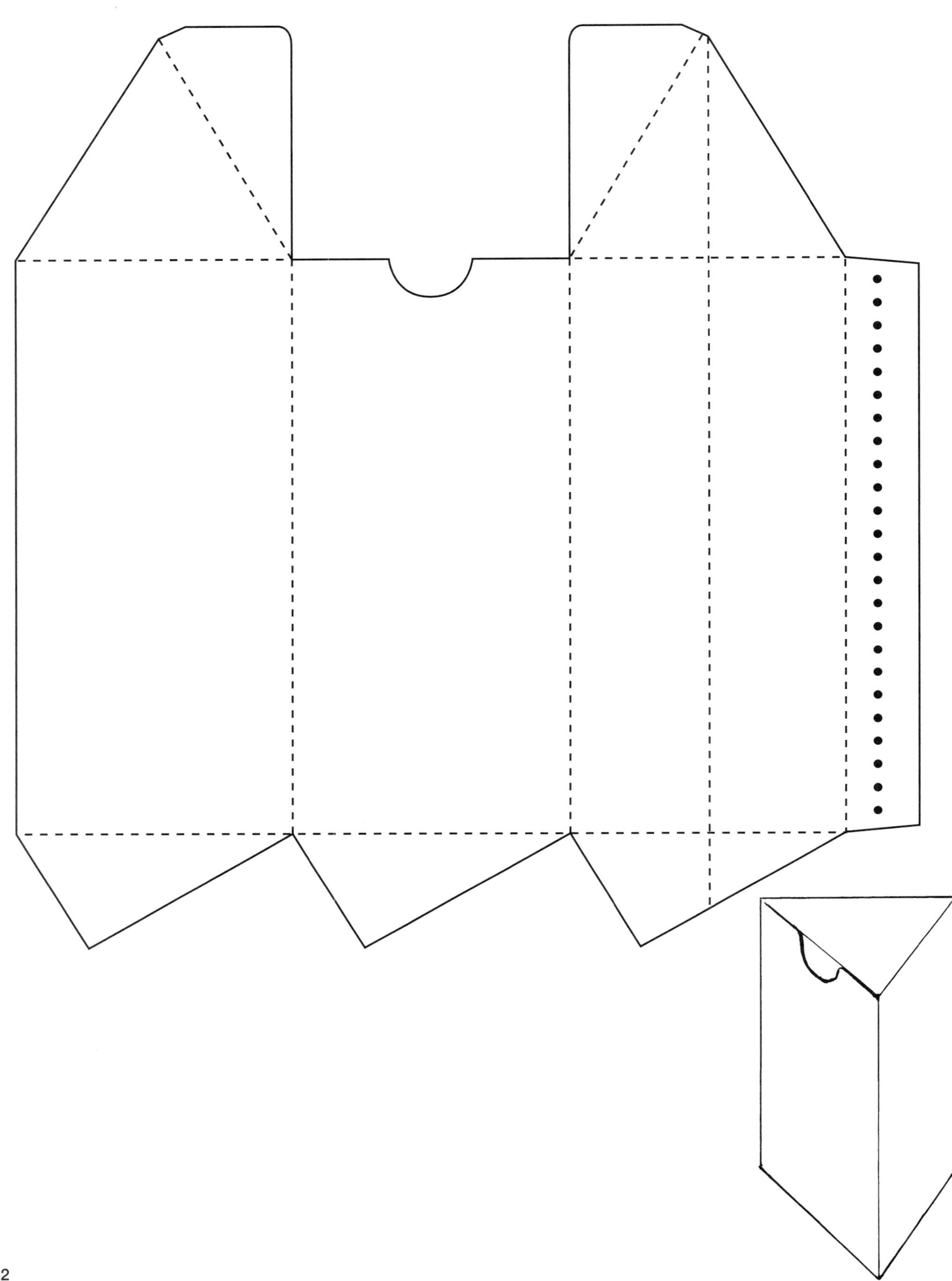

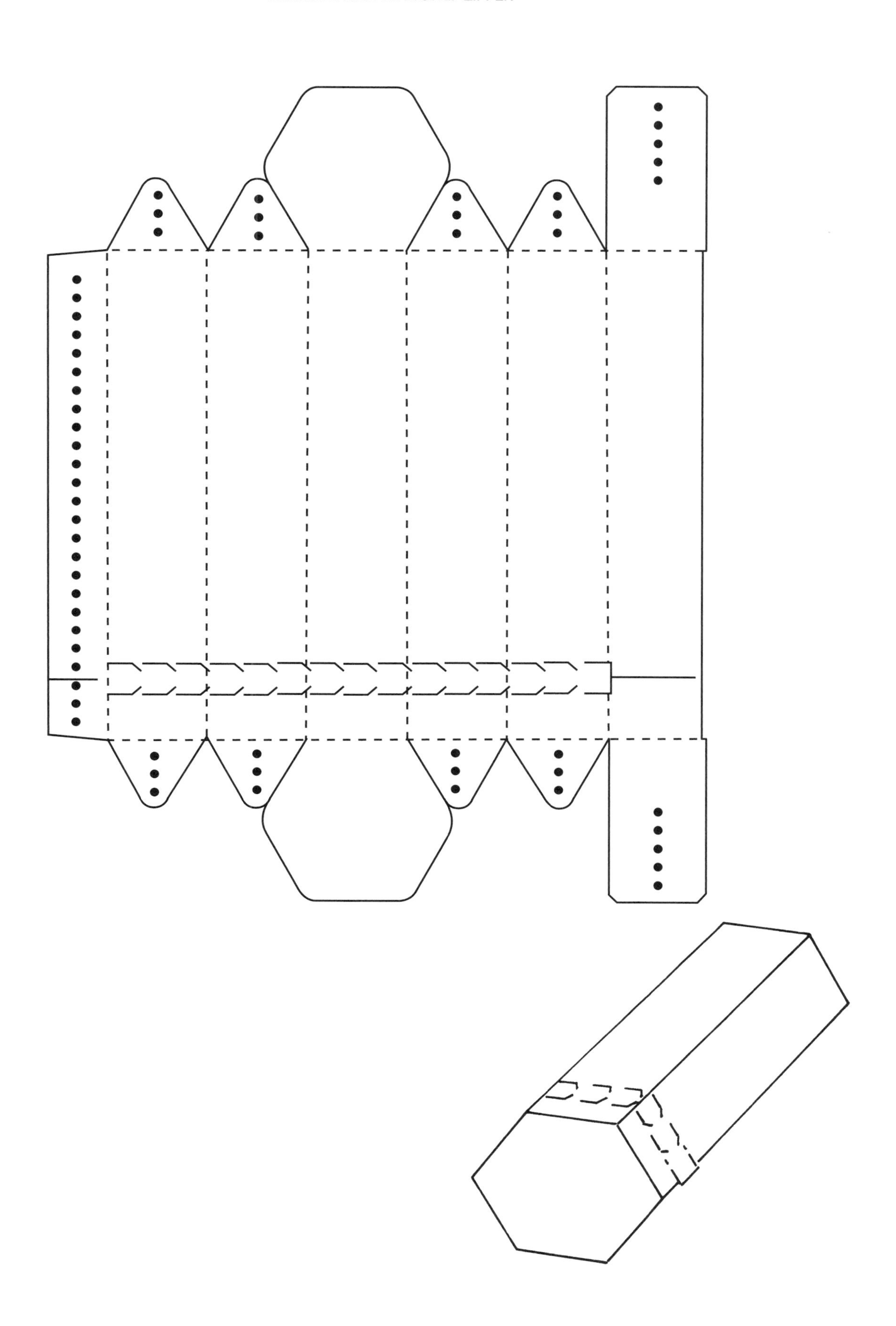

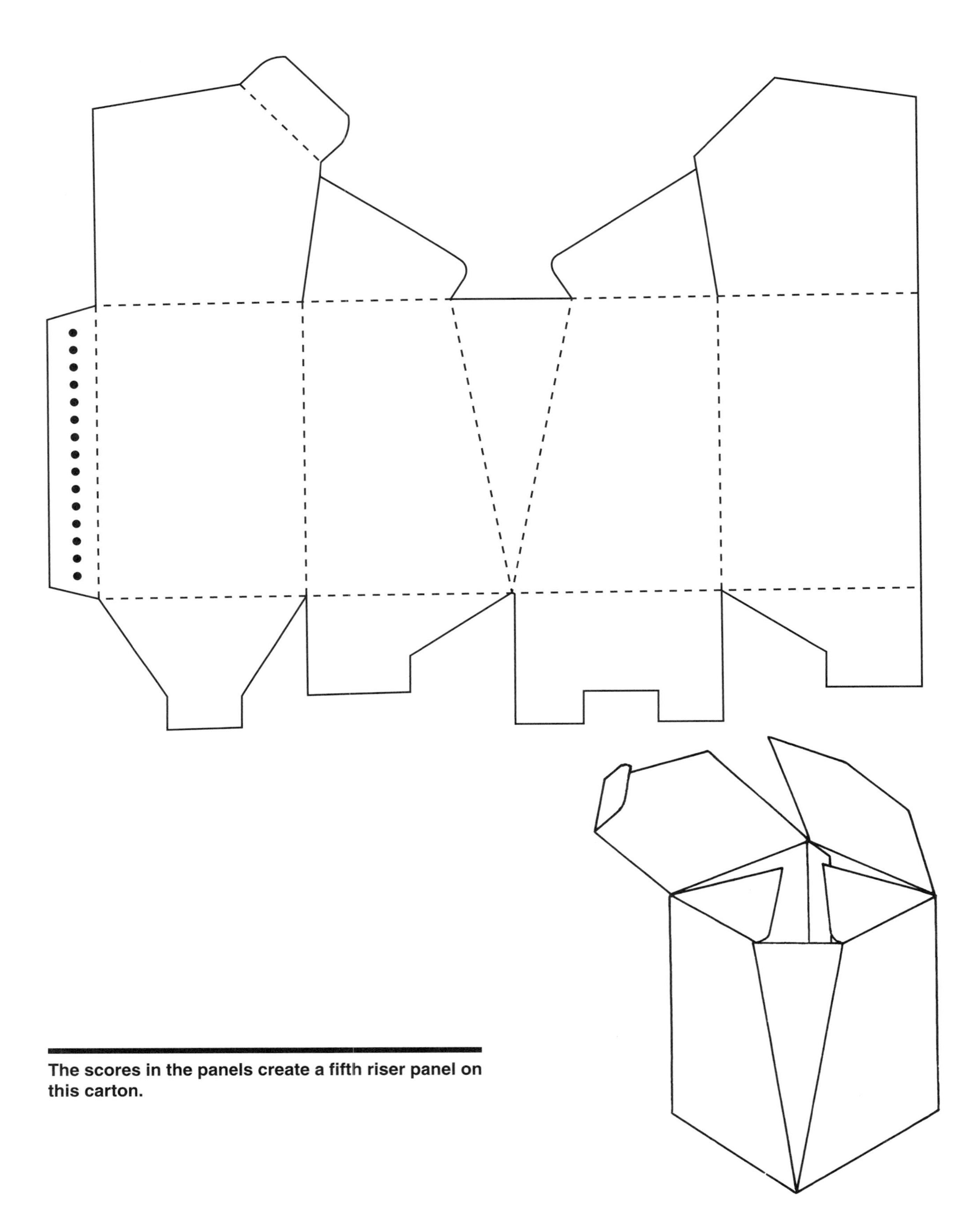

The scores in the panels create a fifth riser panel on this carton.

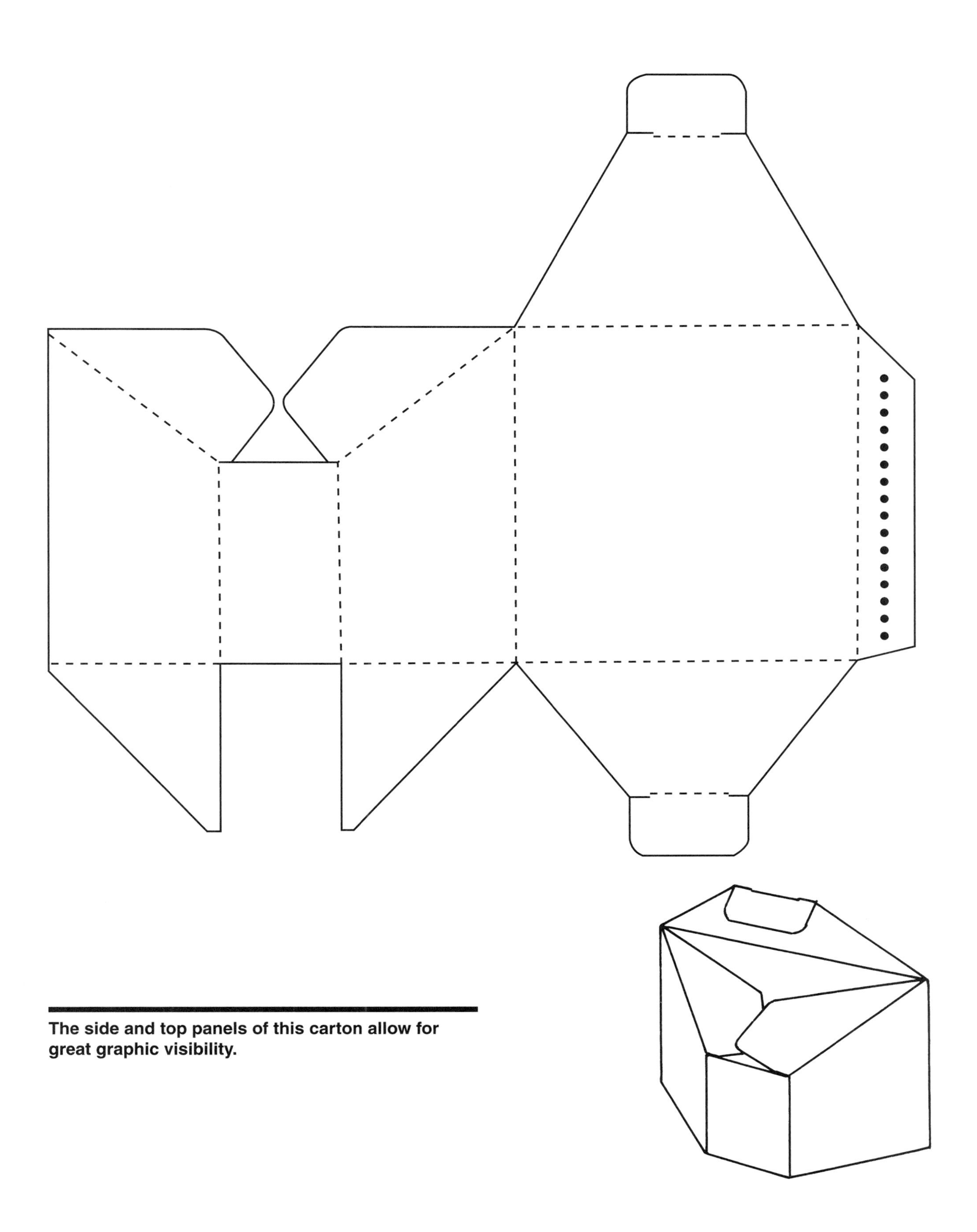

The side and top panels of this carton allow for great graphic visibility.

TRAPEZOID PACKAGE WITH REVERSE TUCKS

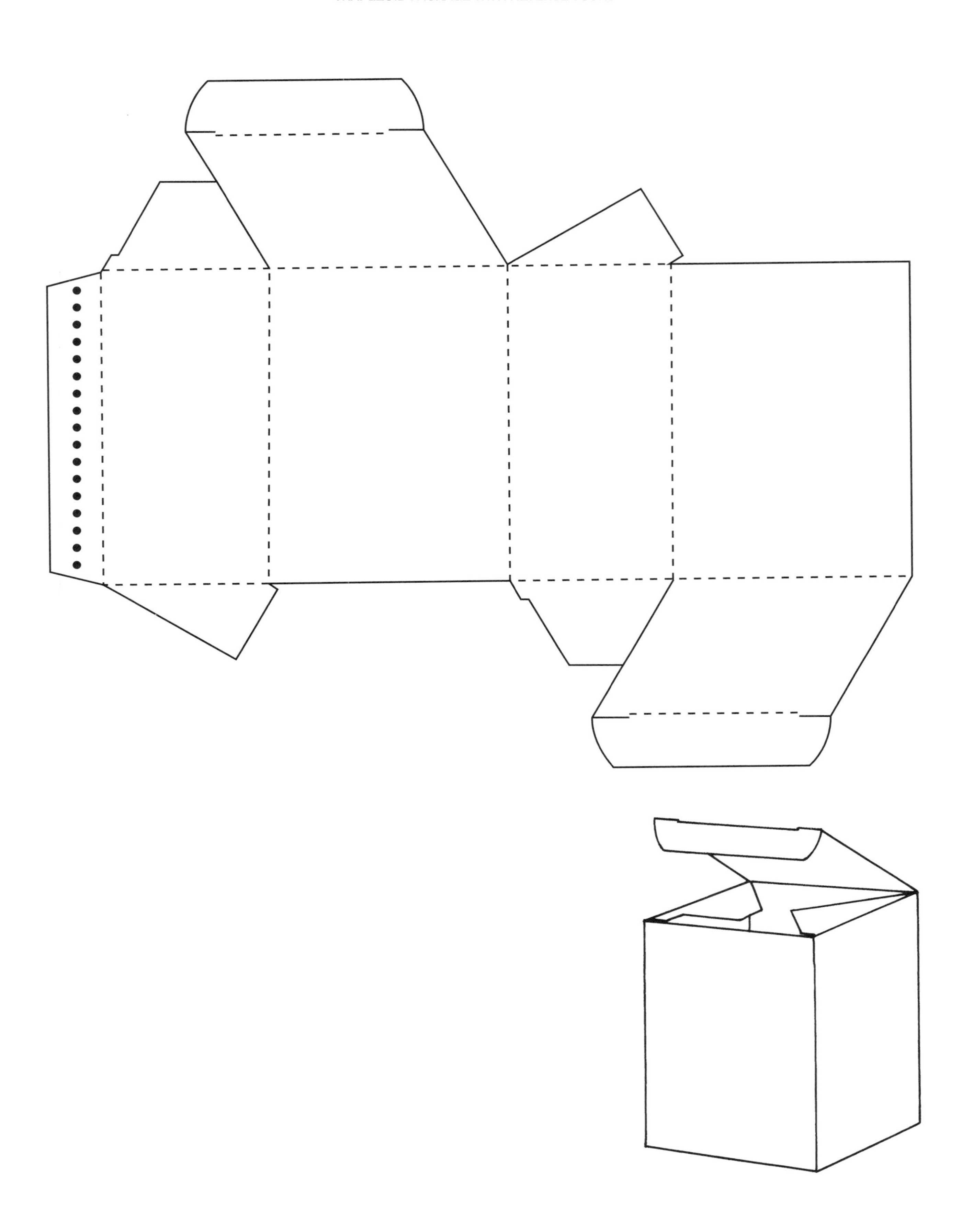

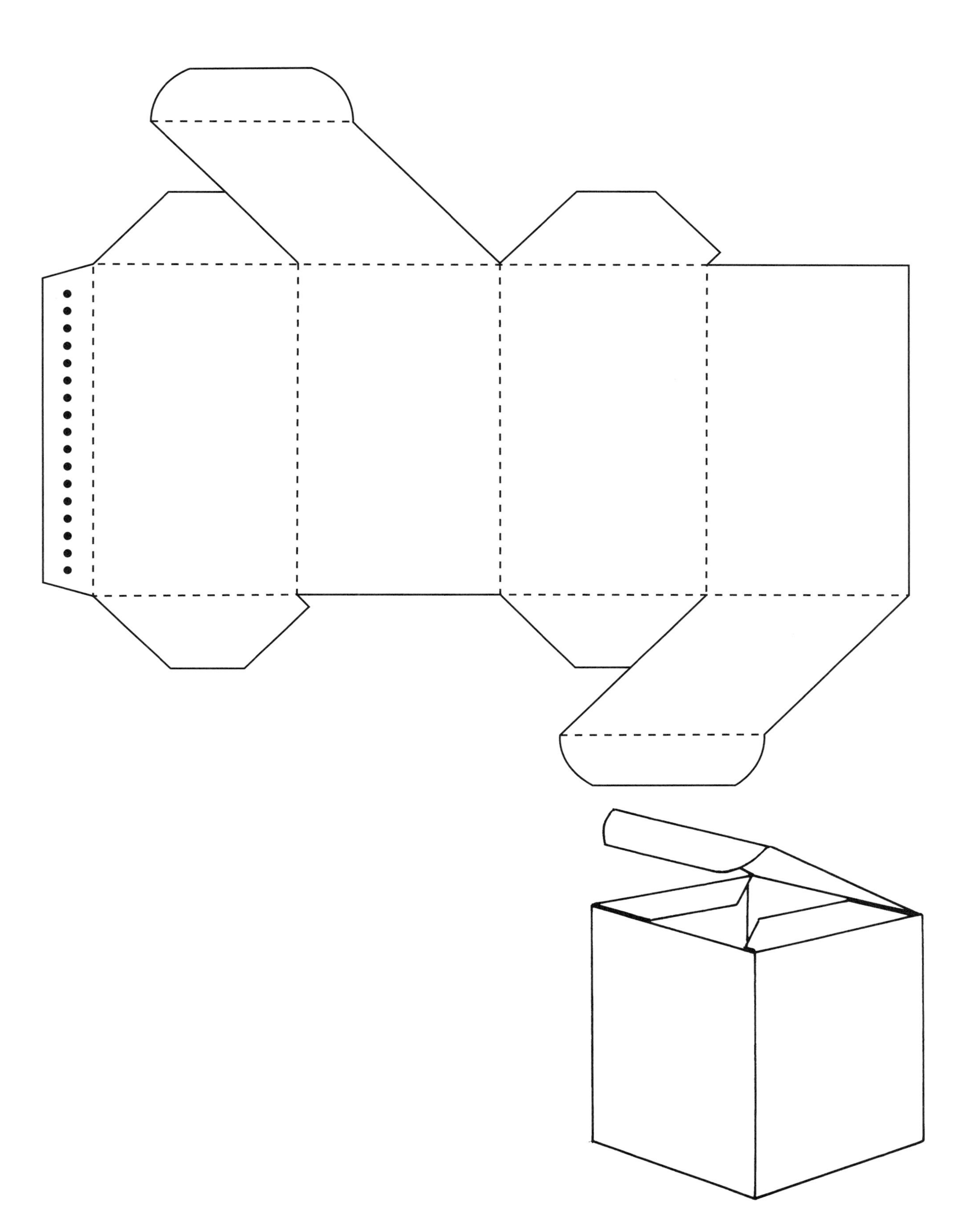

PENTAGON WITH TUCKS

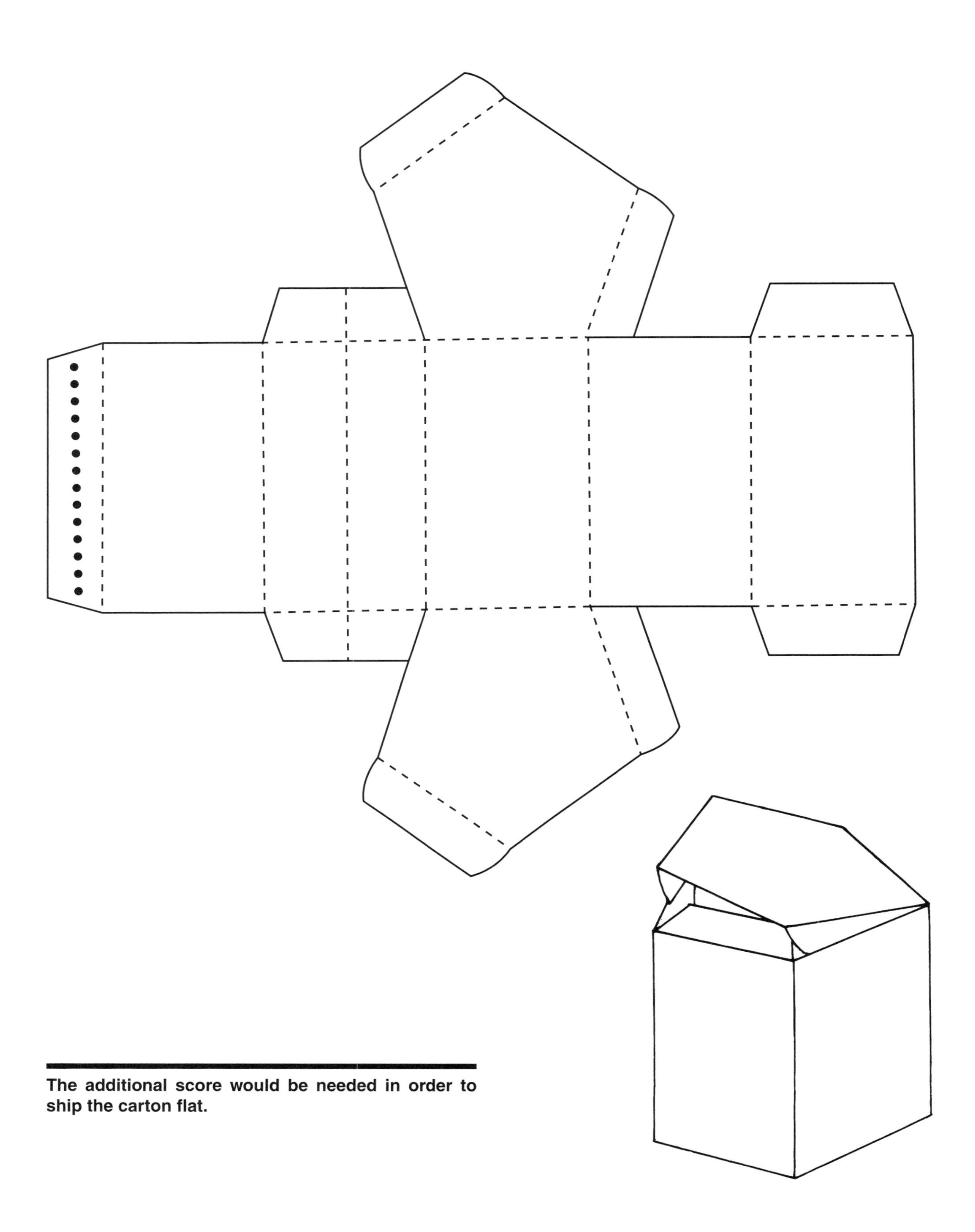

The additional score would be needed in order to ship the carton flat.

SIX-SIDED CARTON WITH STRAIGHT TUCKS
(PAT #3,148,443)

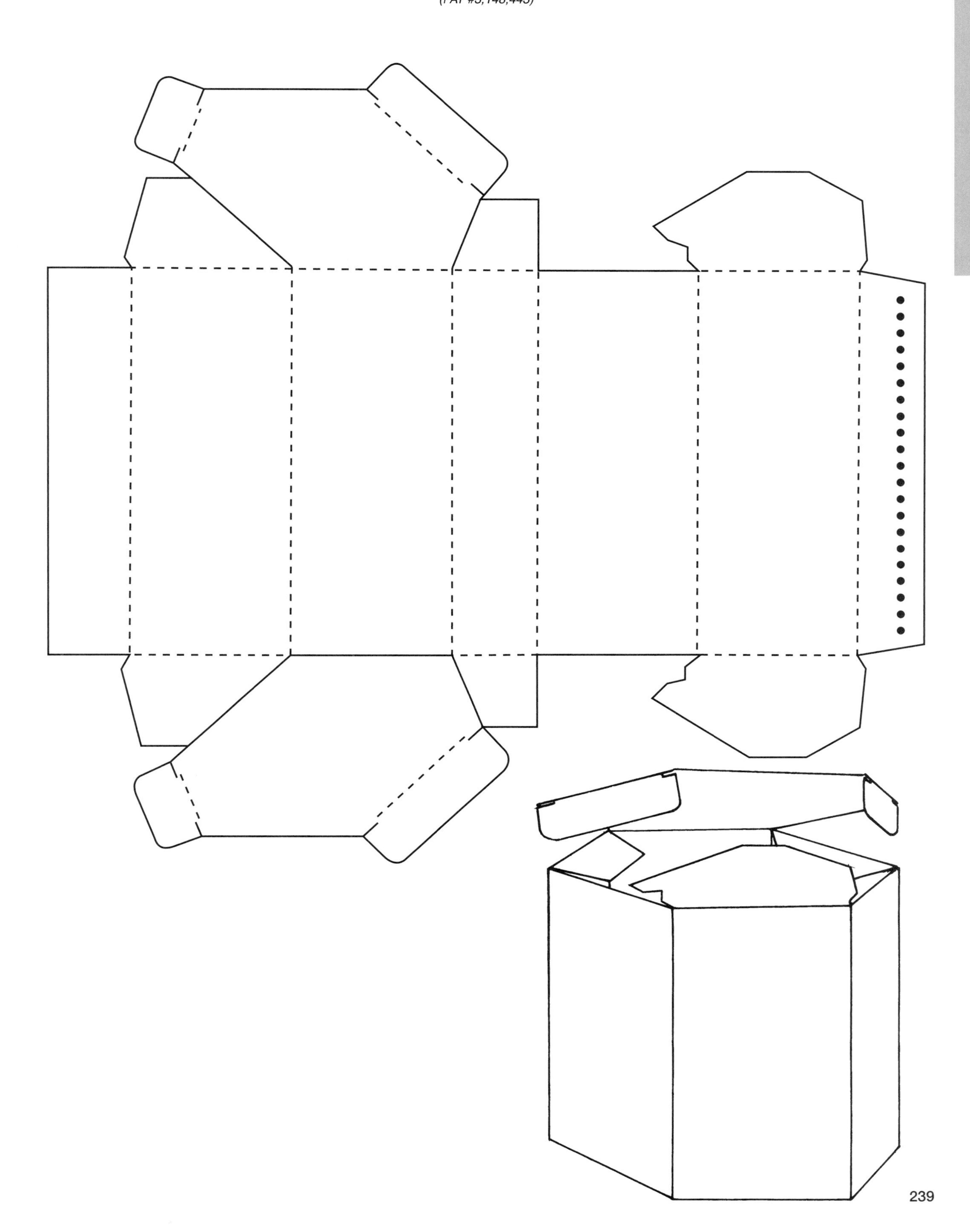

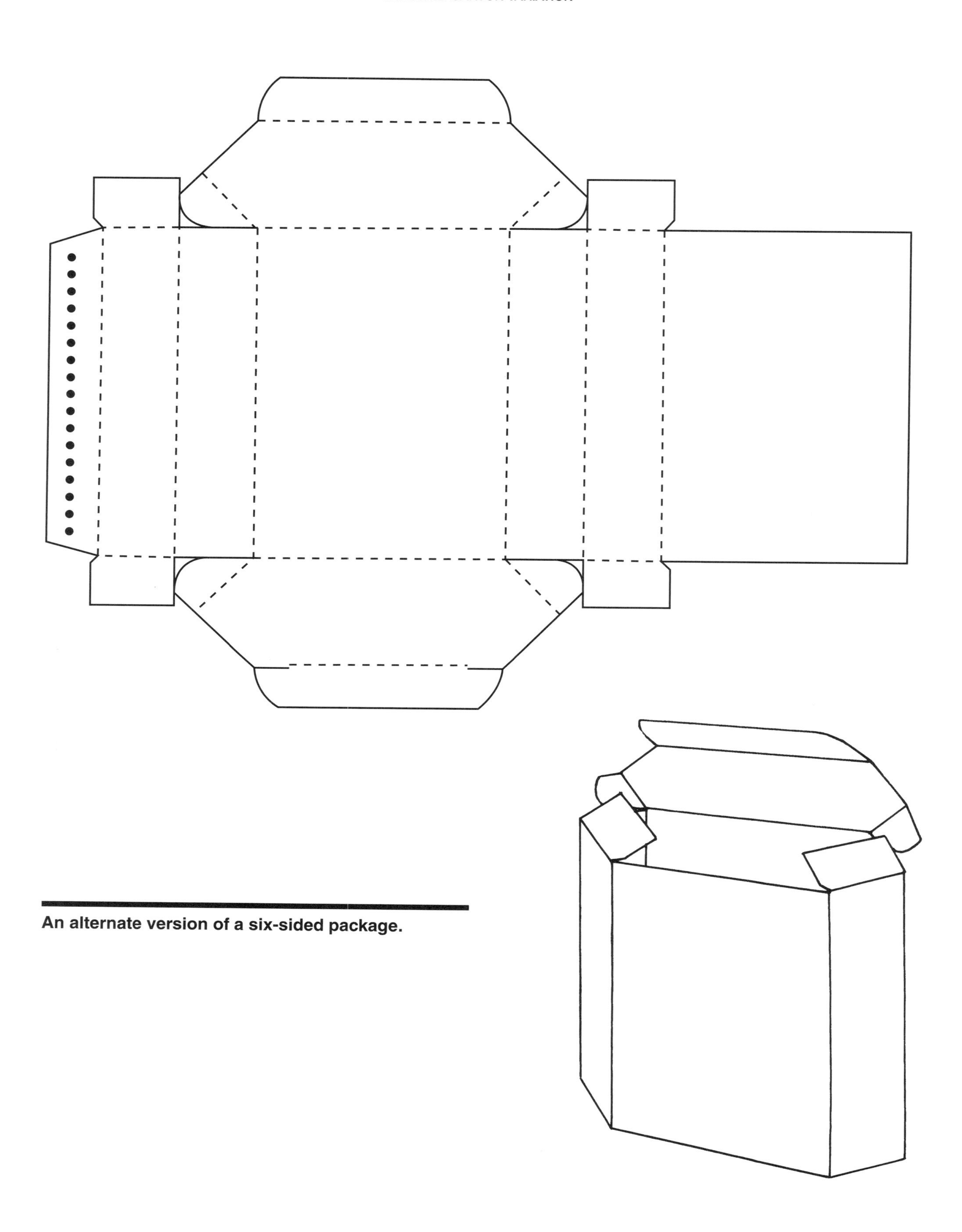

An alternate version of a six-sided package.

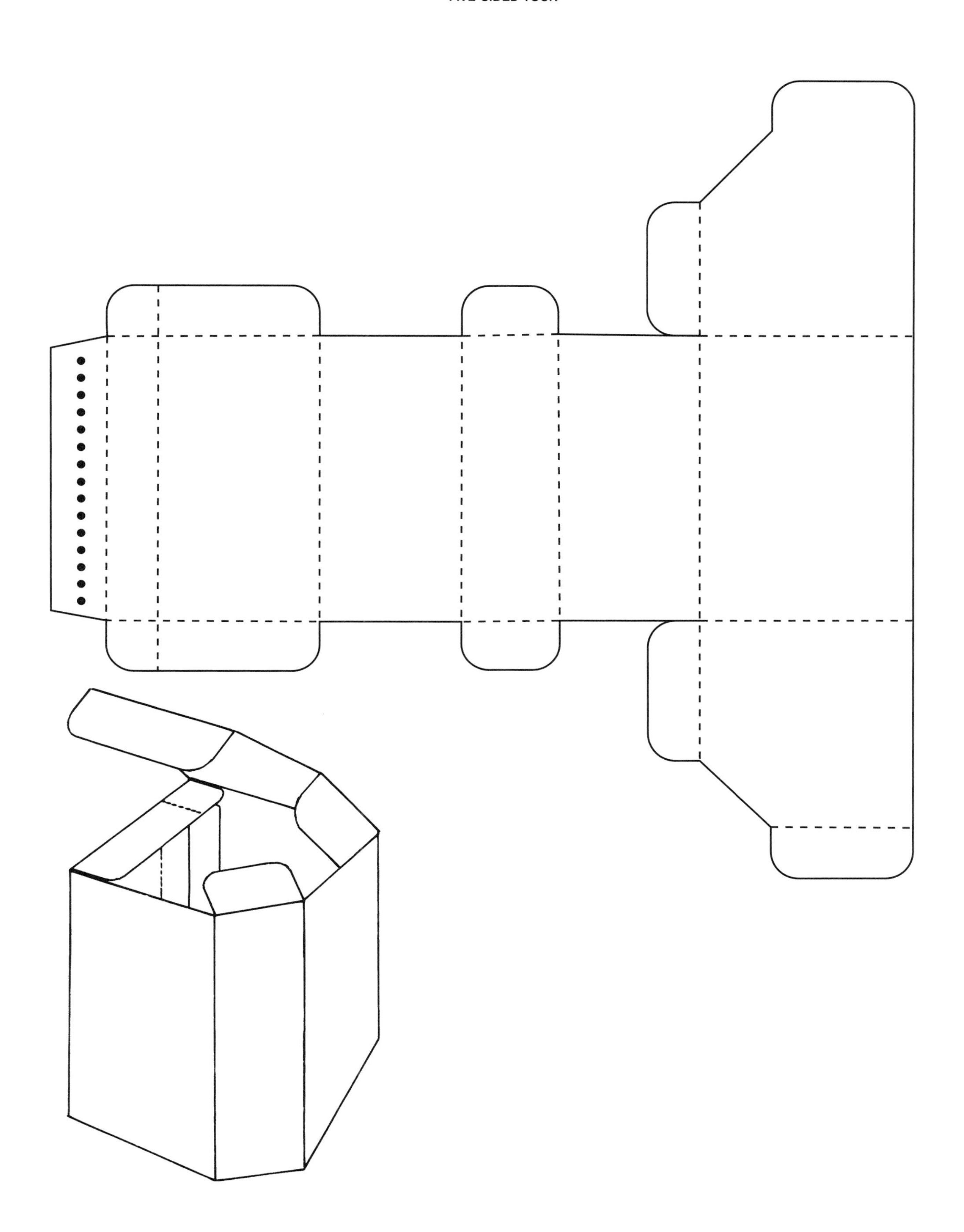

HEXAGON WITH GREENLEAF TOP AND BOTTOM

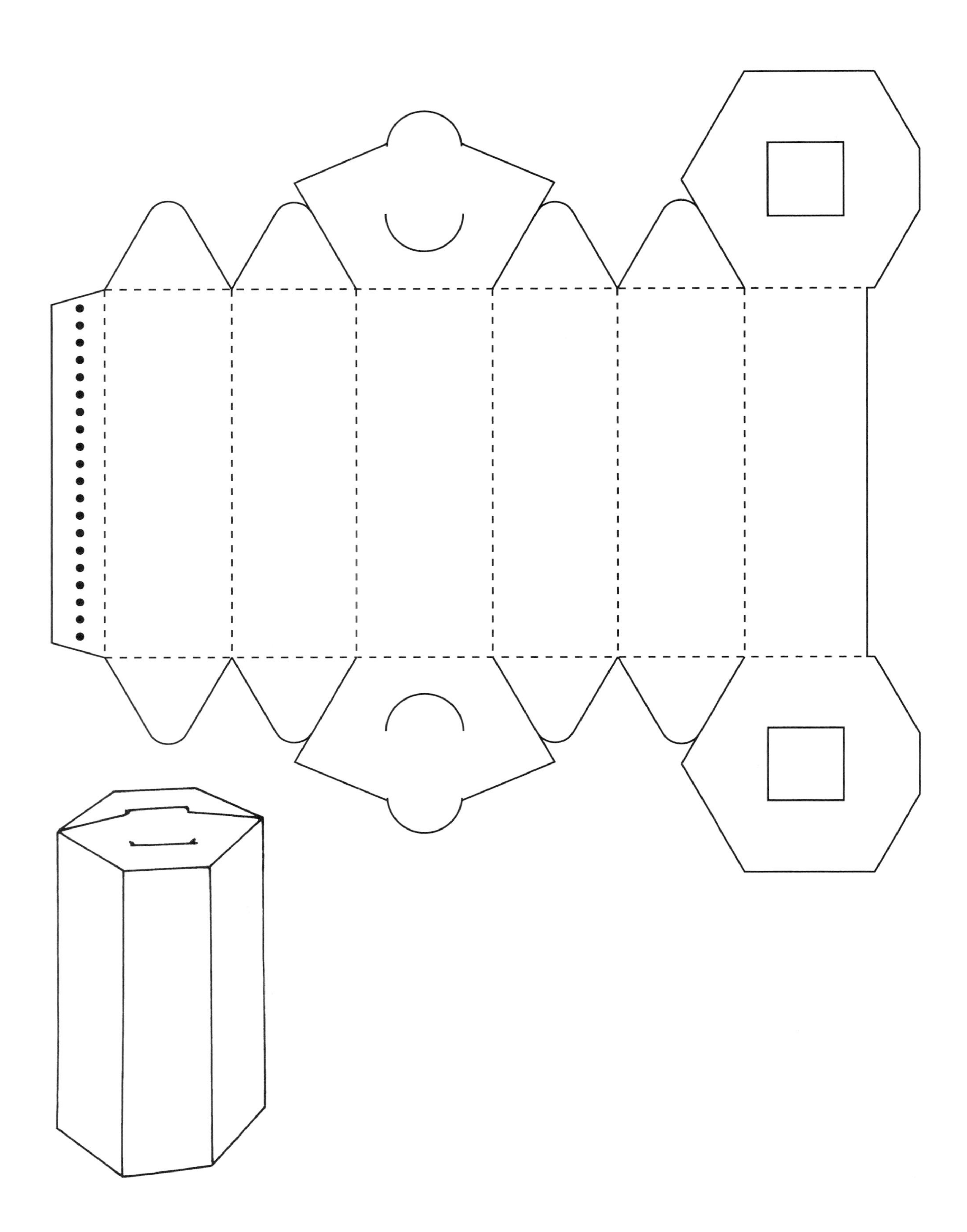

PAINT OR TOOTHPASTE TUBE PROTECTOR

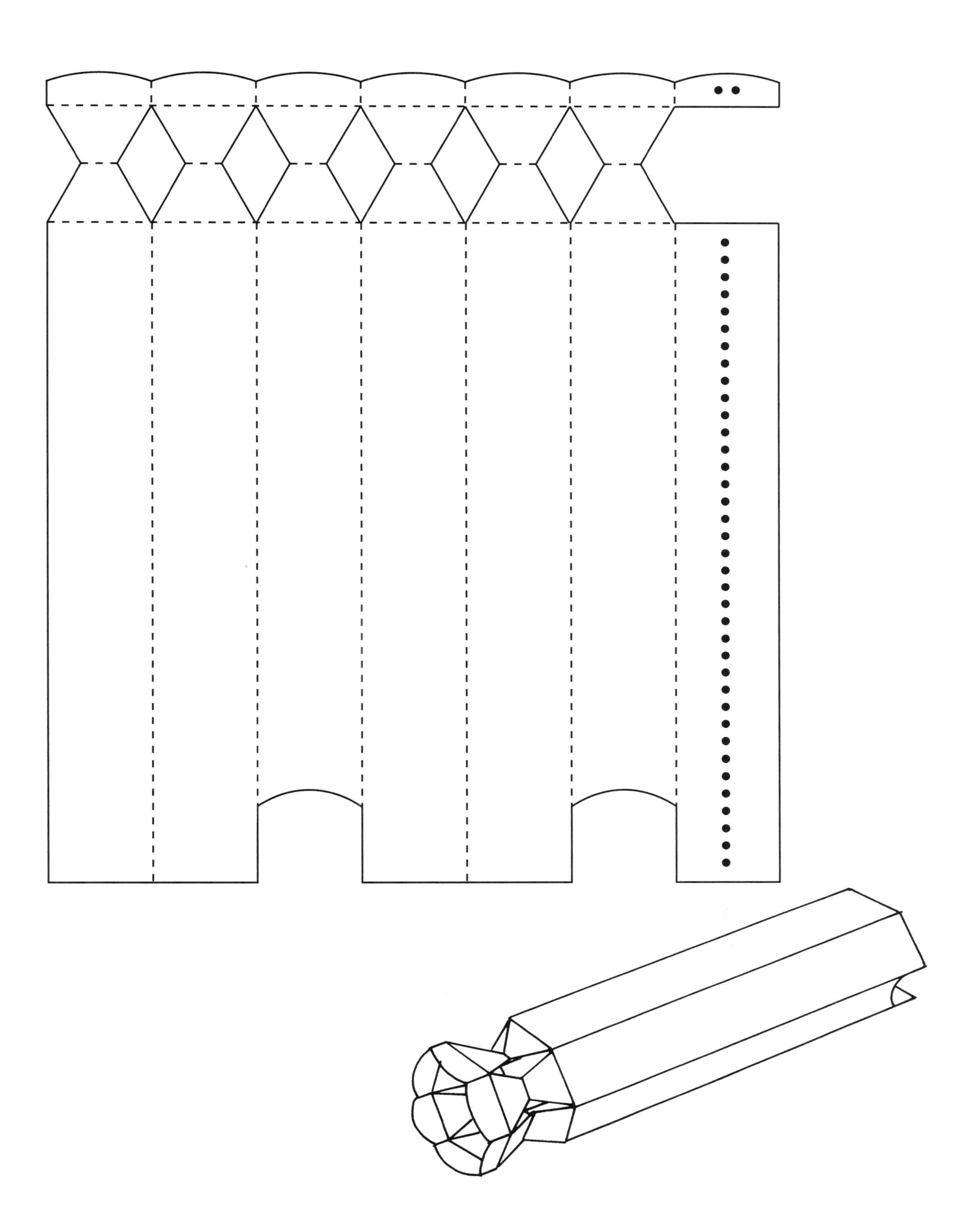

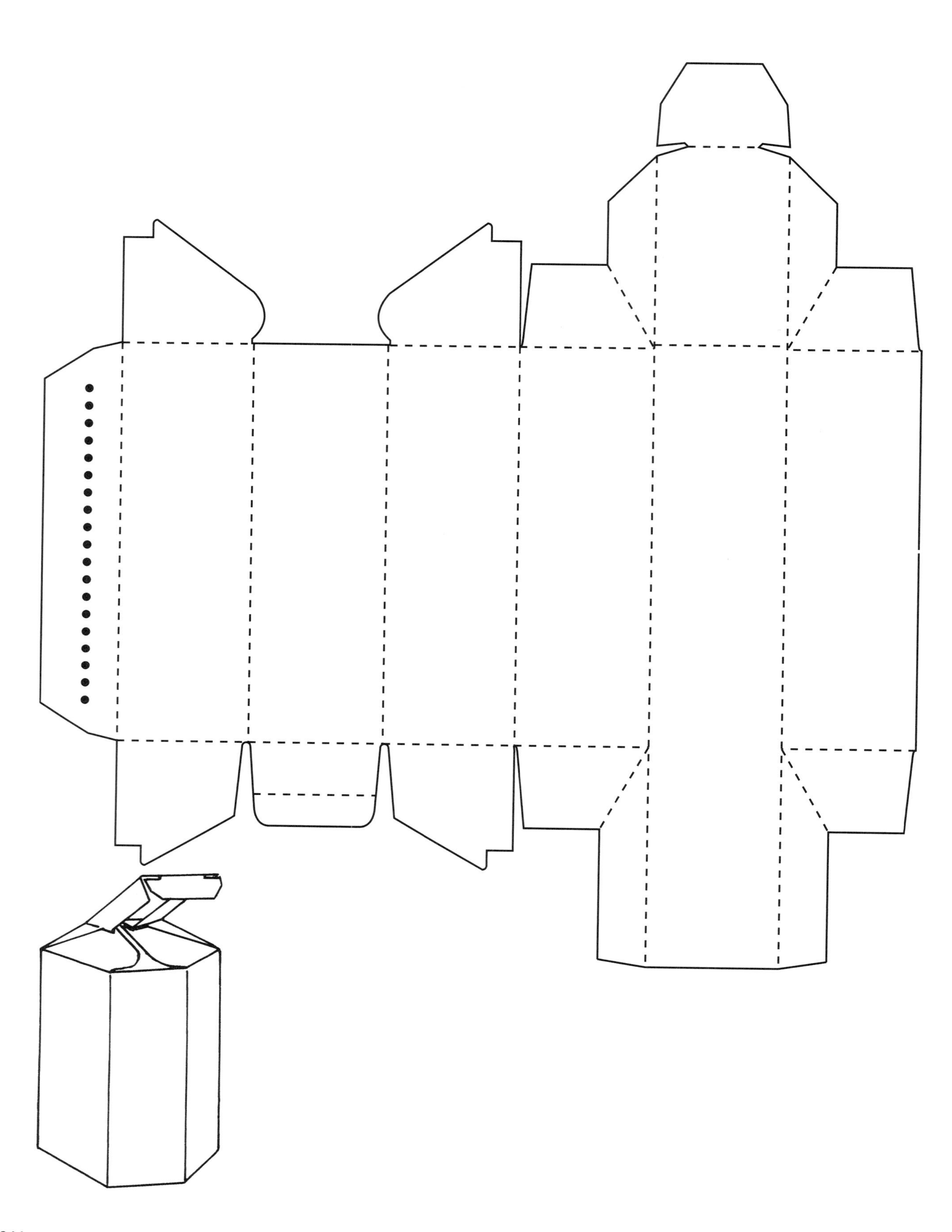

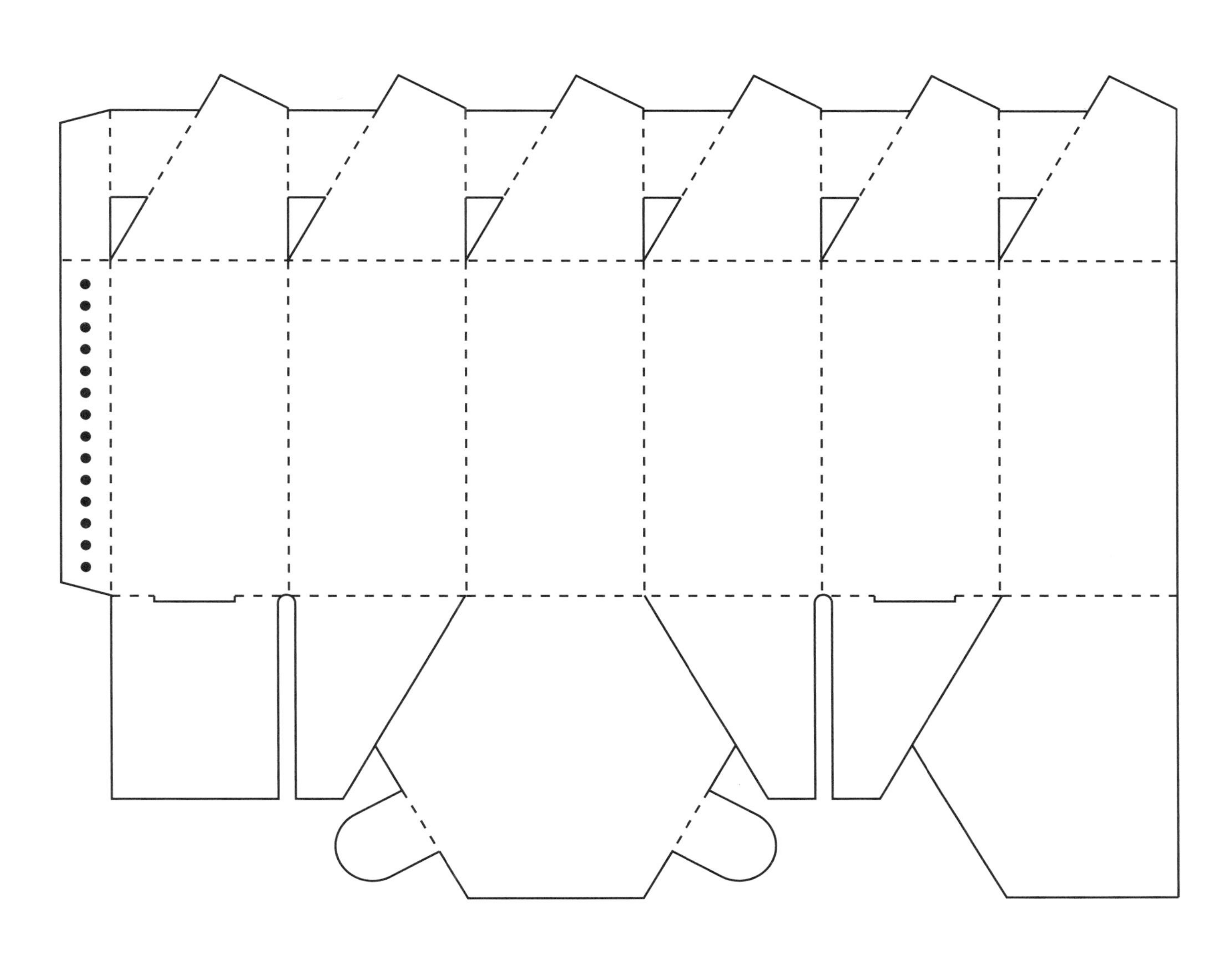

The small triangles have been die-cut out of the top closure to permit easier closing.

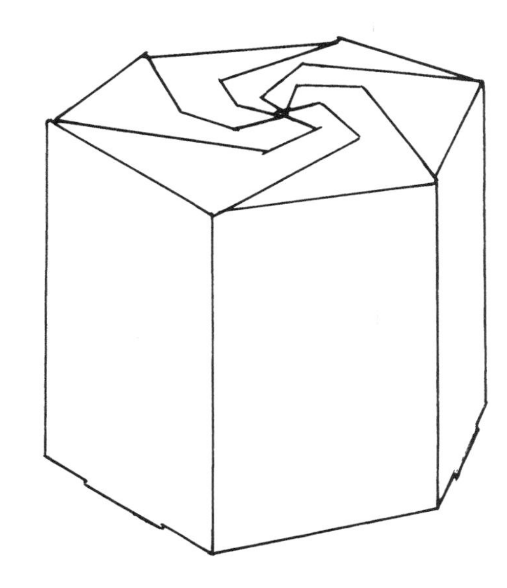

SIX-SIDED CARTON WITH PUSH-IN CLOSURE AND AUTOMATIC GLUED BOTTOM

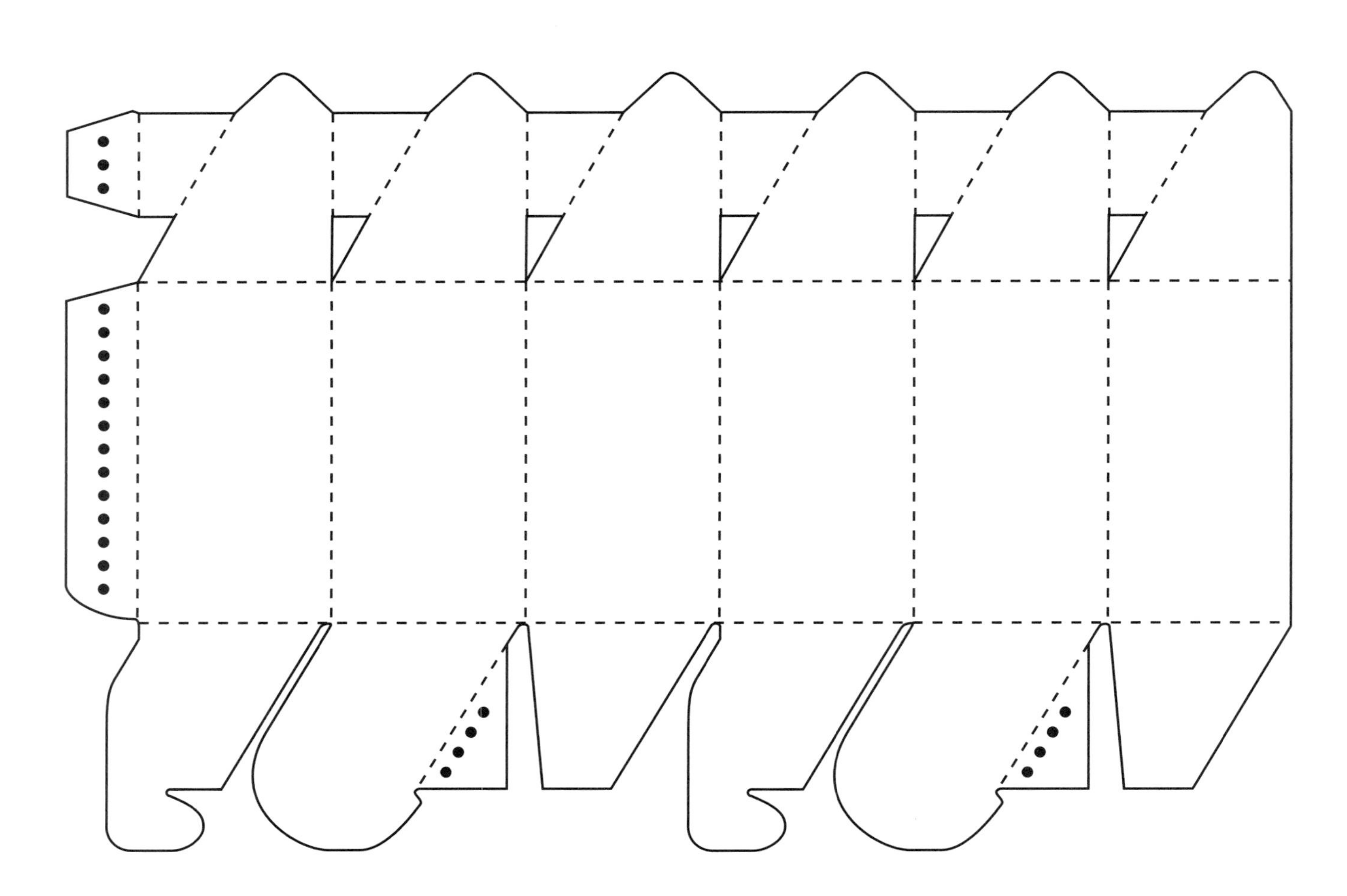

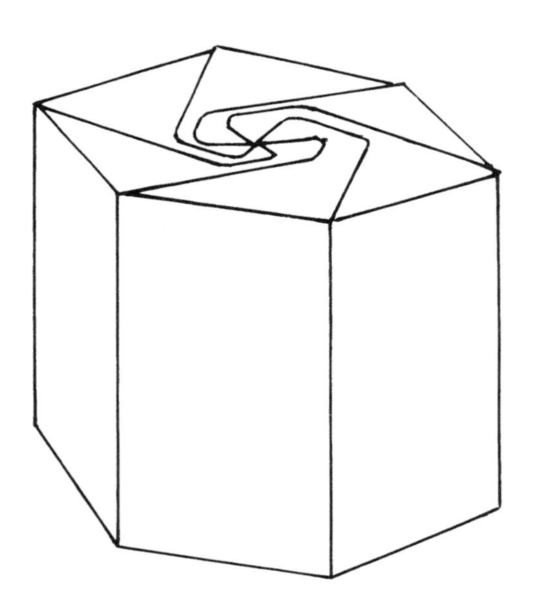

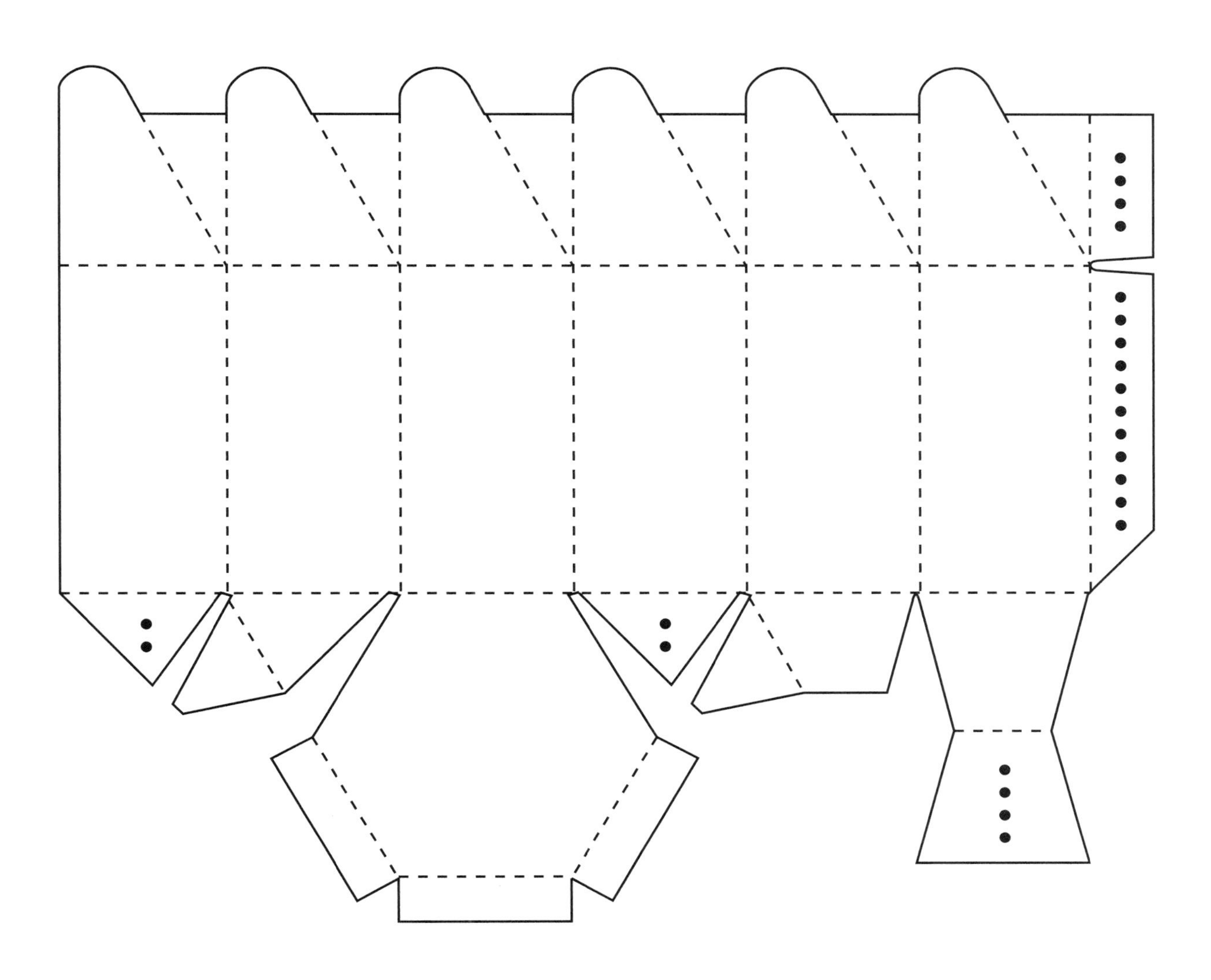

A variation of the pervious two patterns, this carton also has an auto-glued bottom.

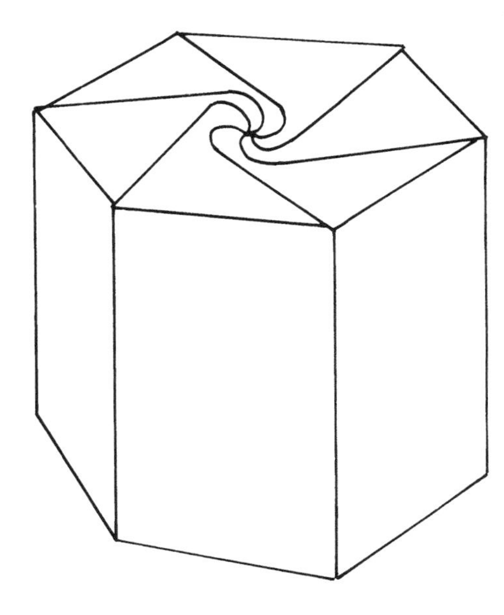

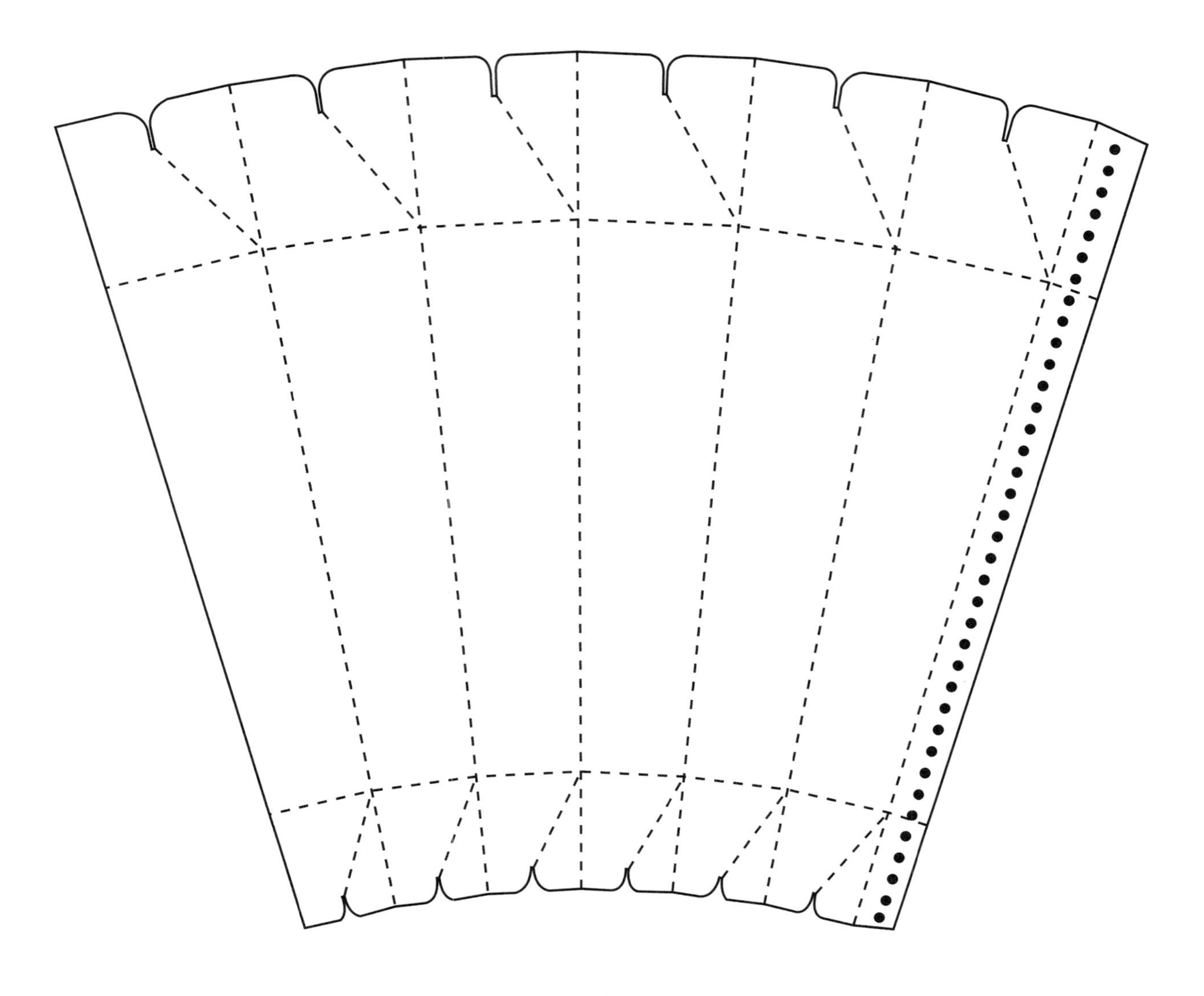

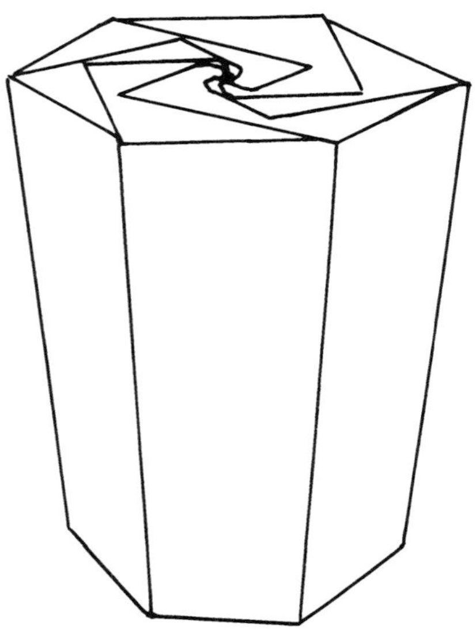

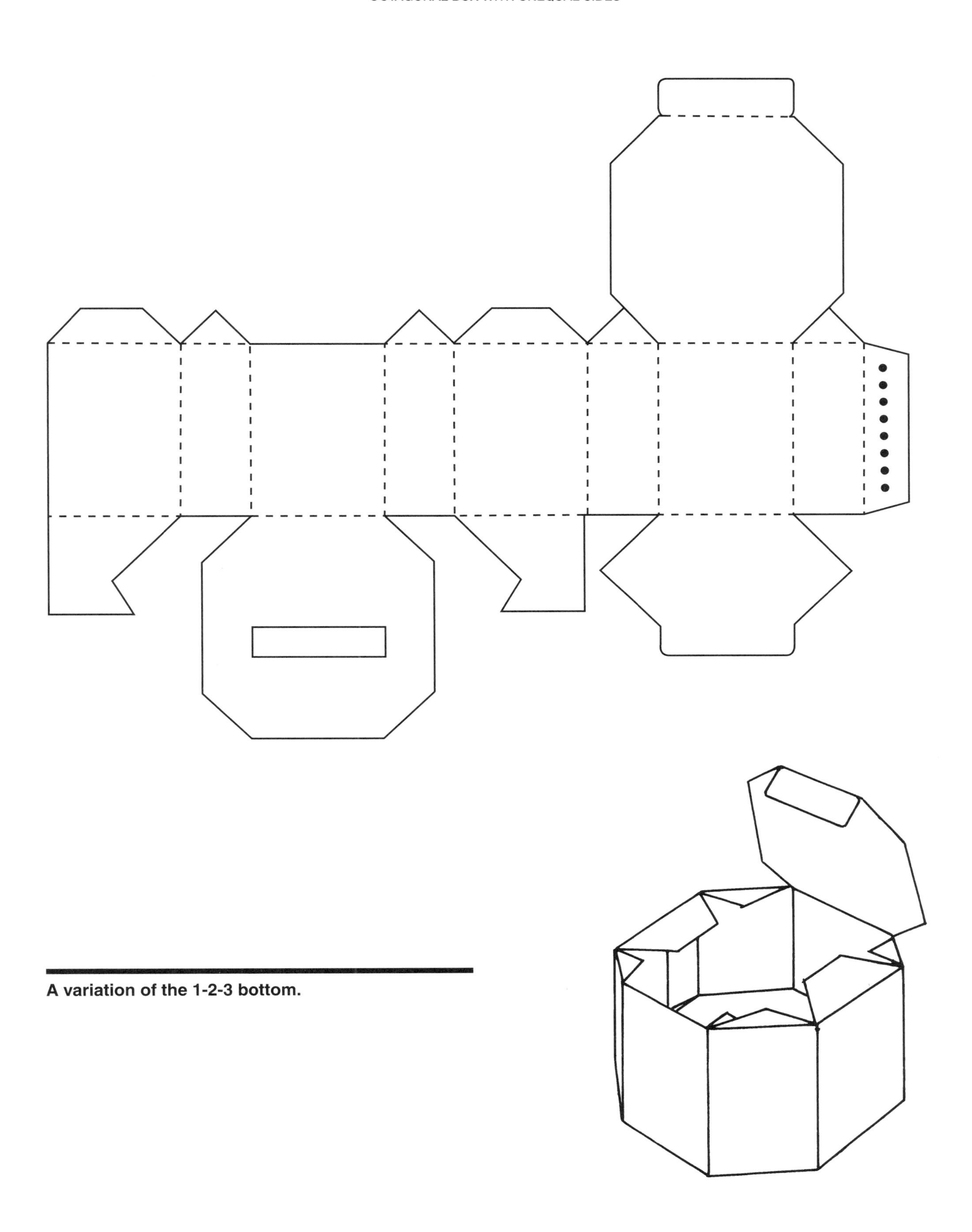

A variation of the 1-2-3 bottom.

SIX-SIDED DISPENSER PACKAGE WITH TEAR-AWAY OPENING
(Pat. Kliklok)

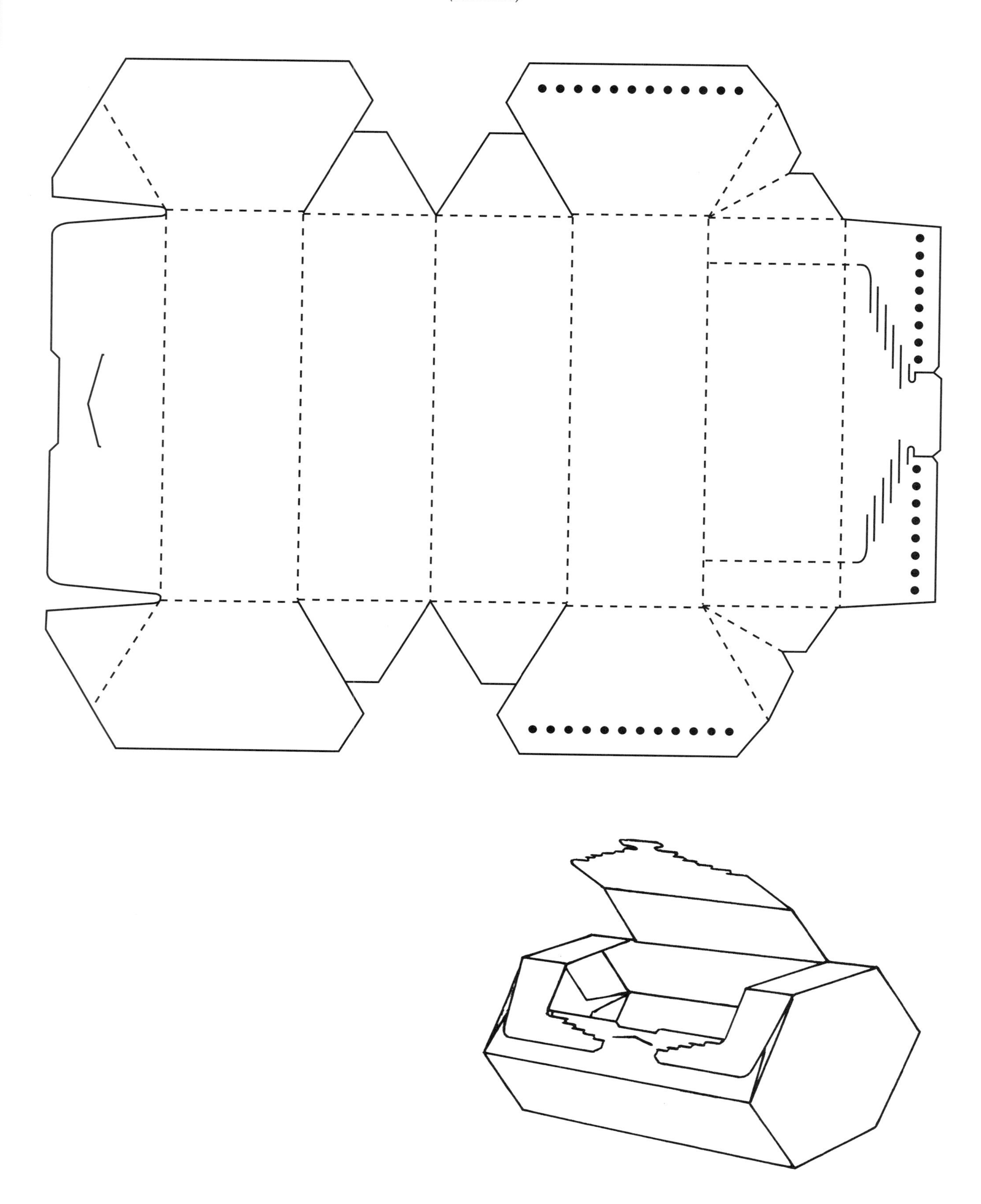

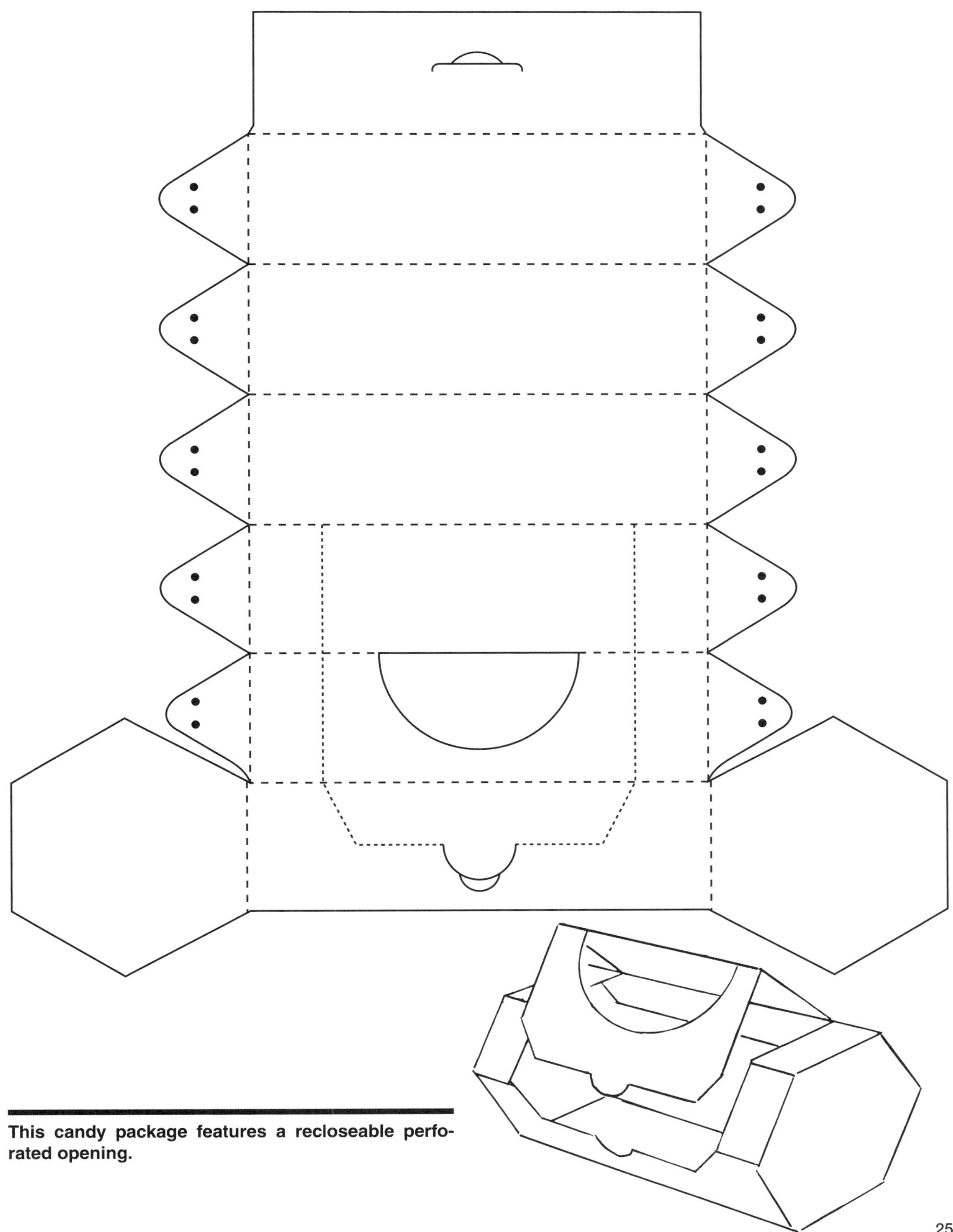

This candy package features a recloseable perforated opening.

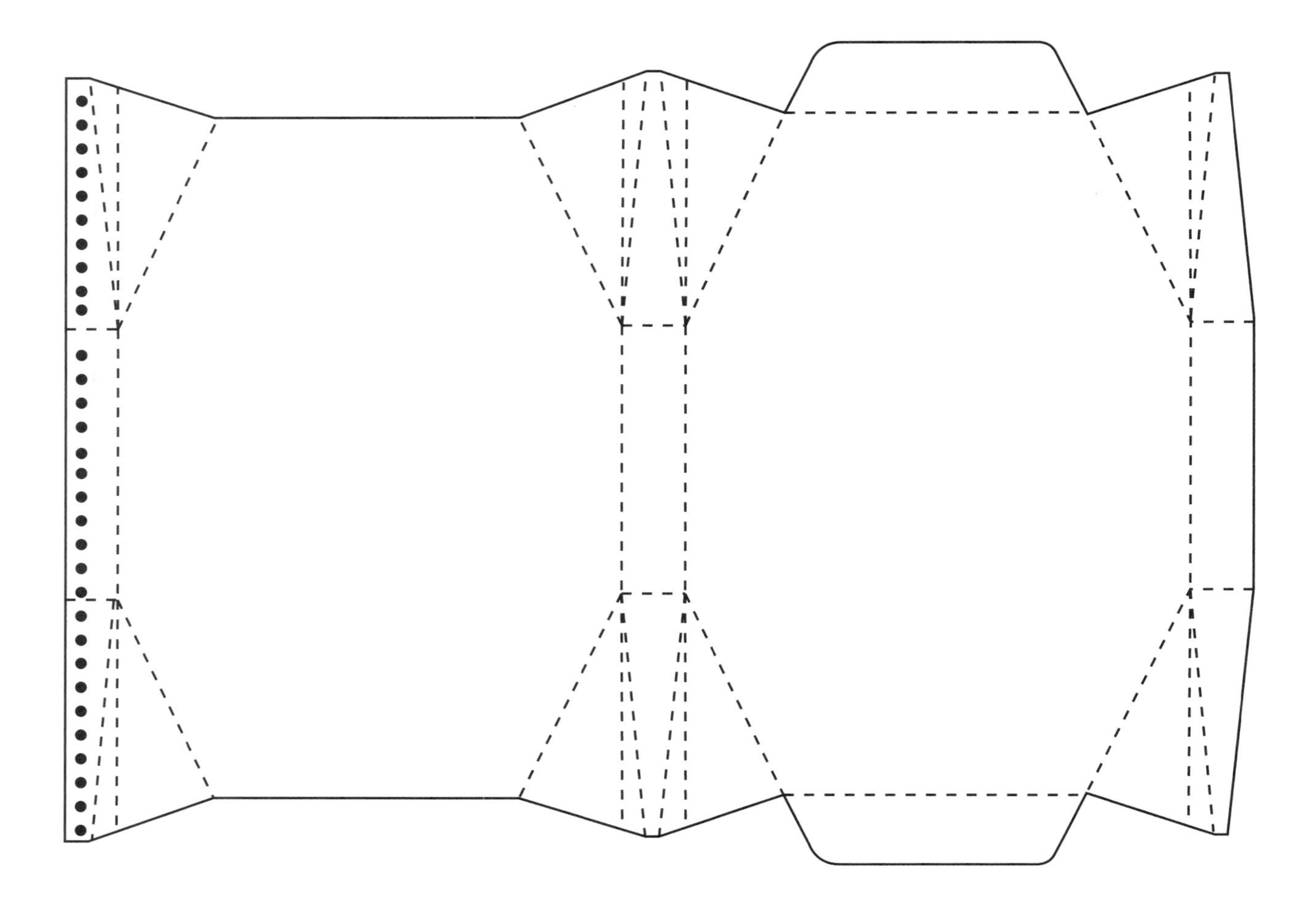

This carton has an intriguing shape created by the scoring of the riser panels.

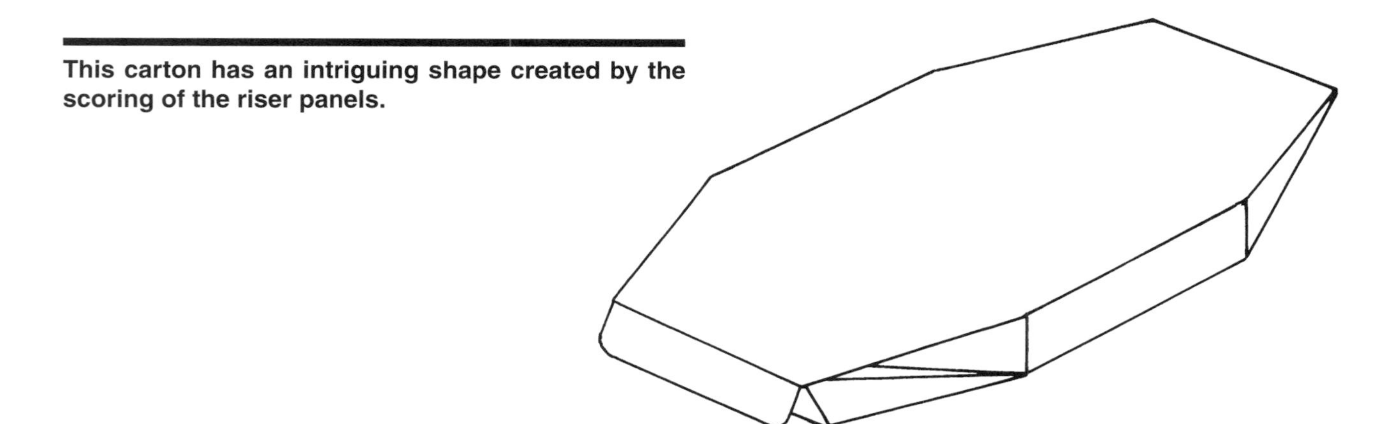

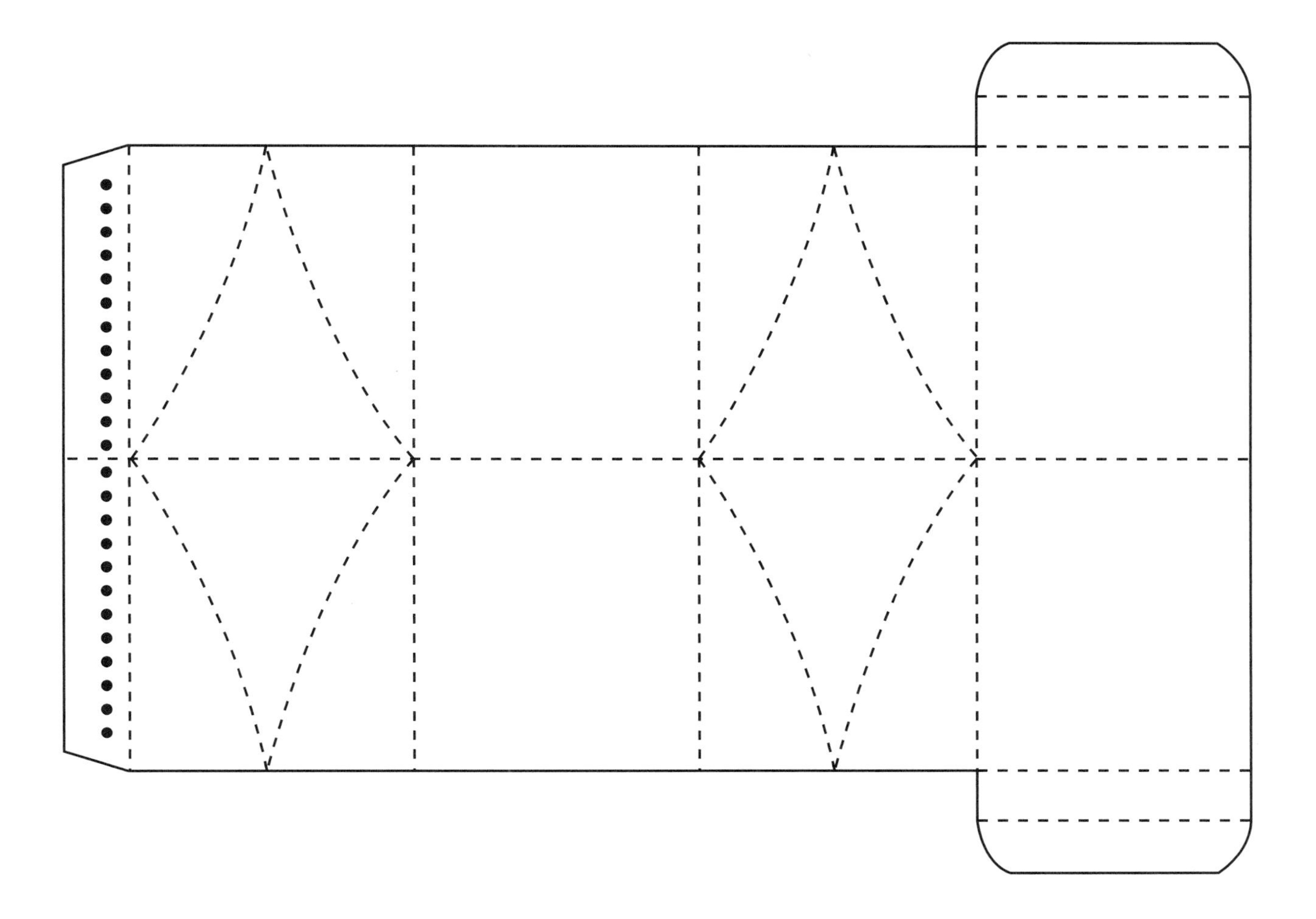

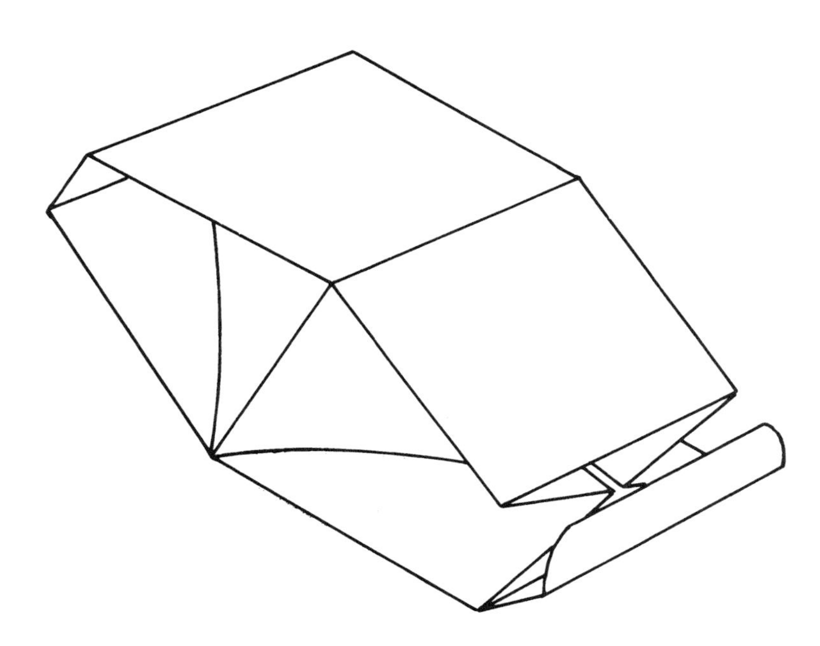

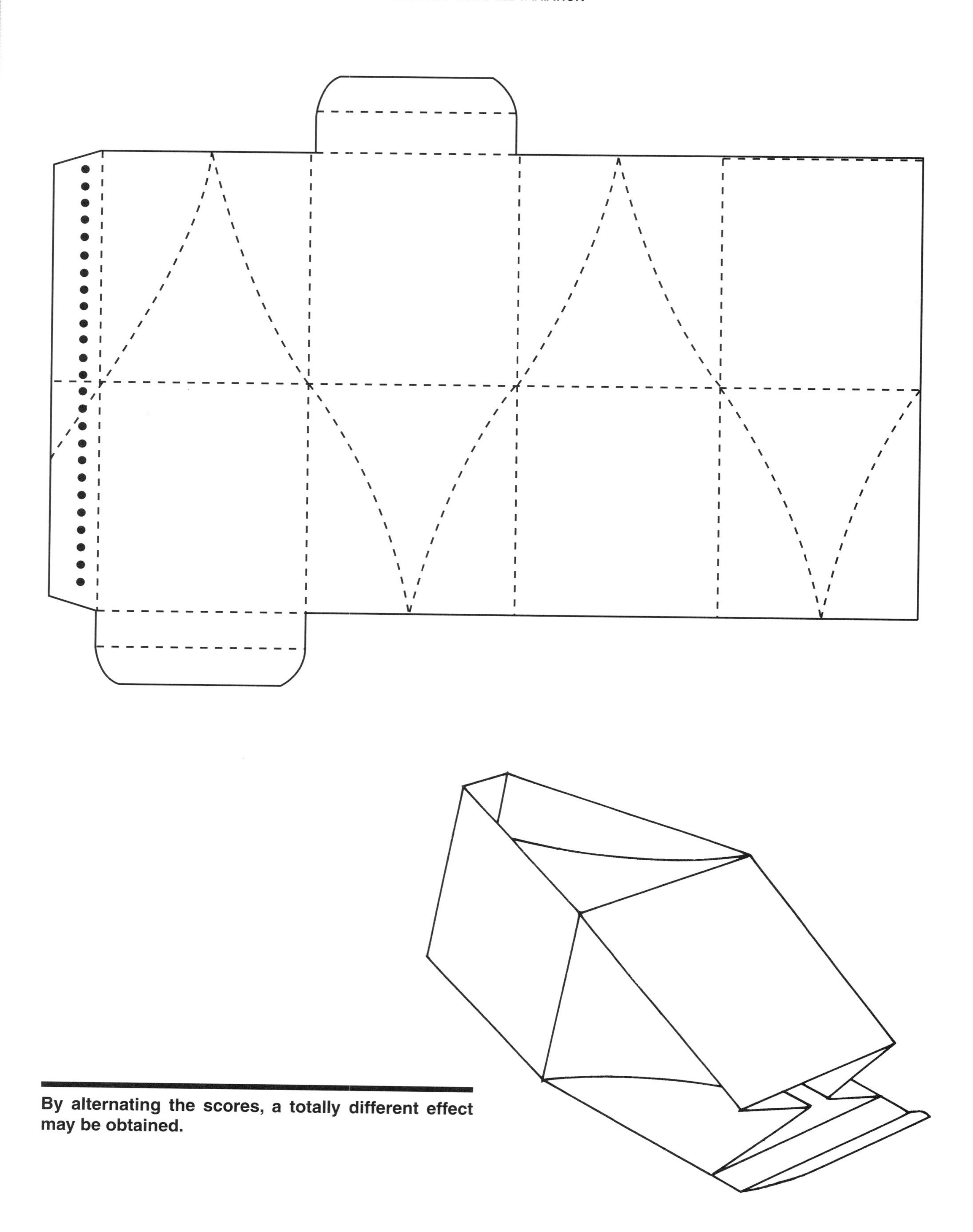

By alternating the scores, a totally different effect may be obtained.

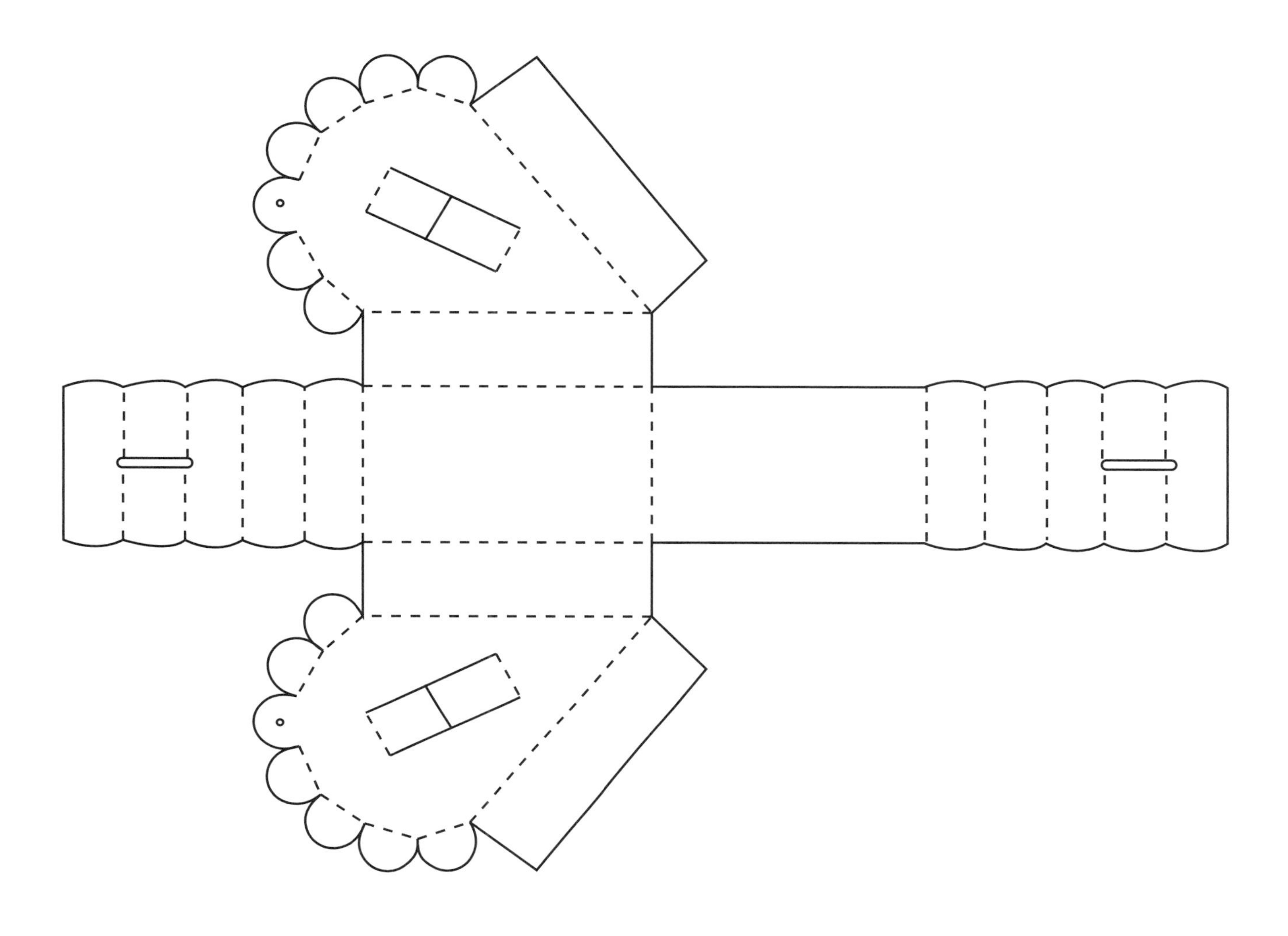

This and the following patterns may be exectued in many different stocks. With die-cut shapes to match their products, they make ideal holiday gifts/ ornaments.

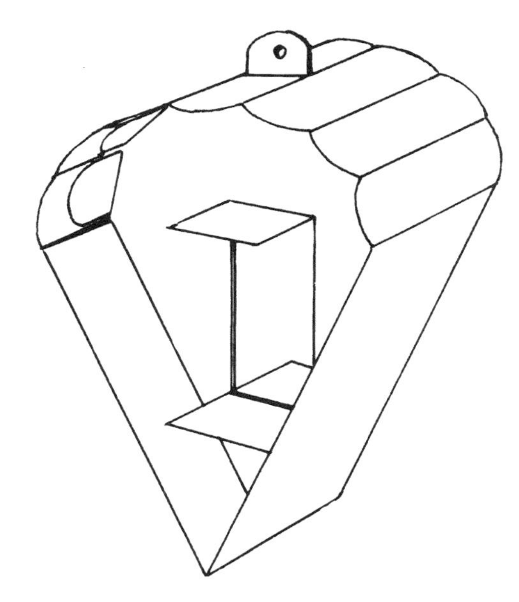

OCTAGONAL OPEN CONTAINER

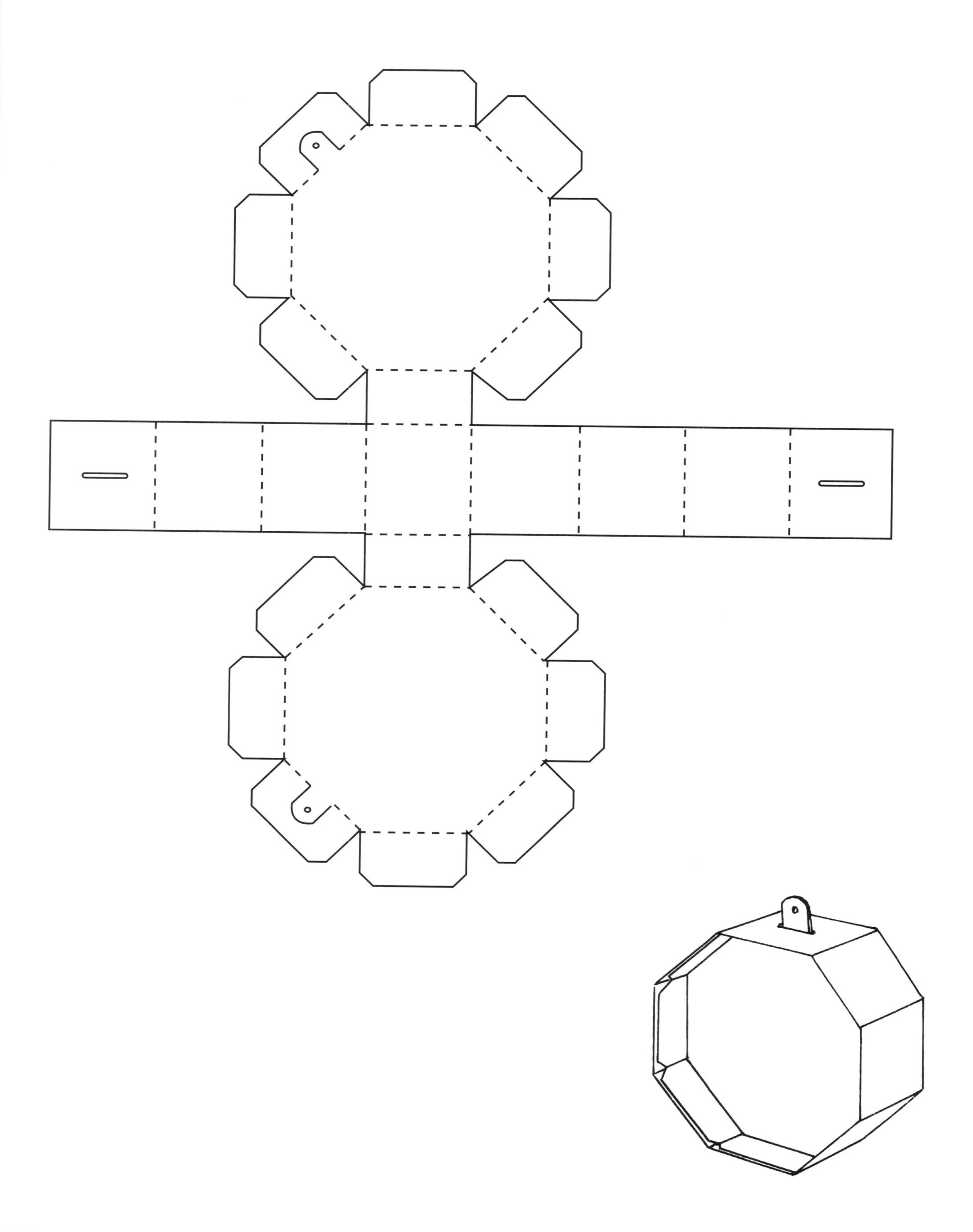

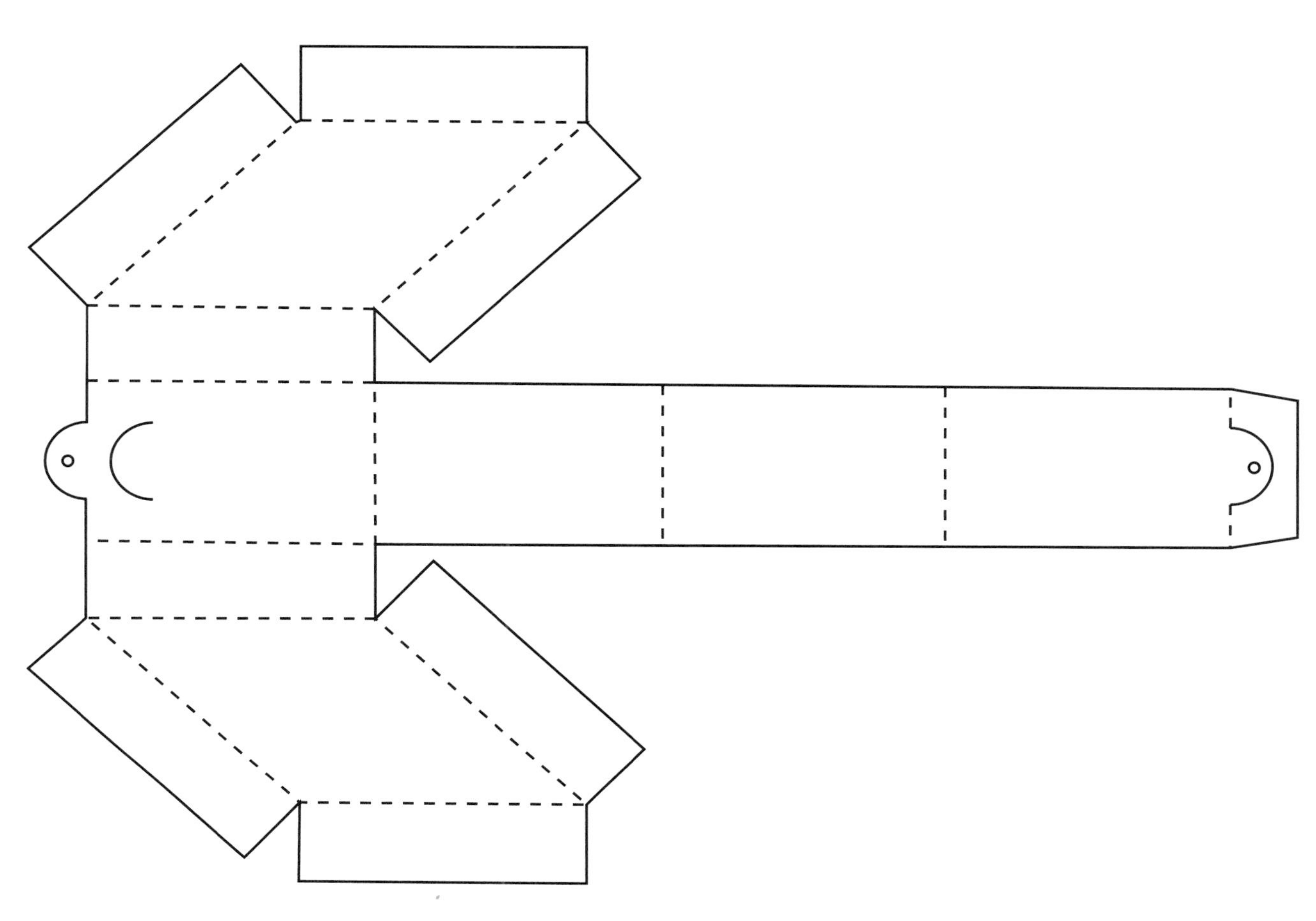

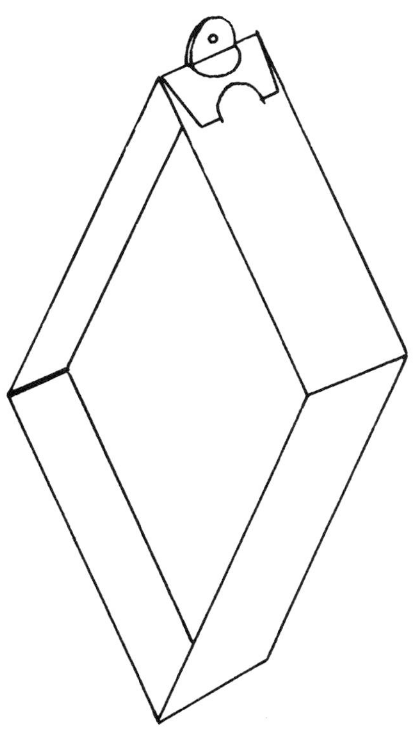

BIRDHOUSE SHAPED OPEN CONTAINER

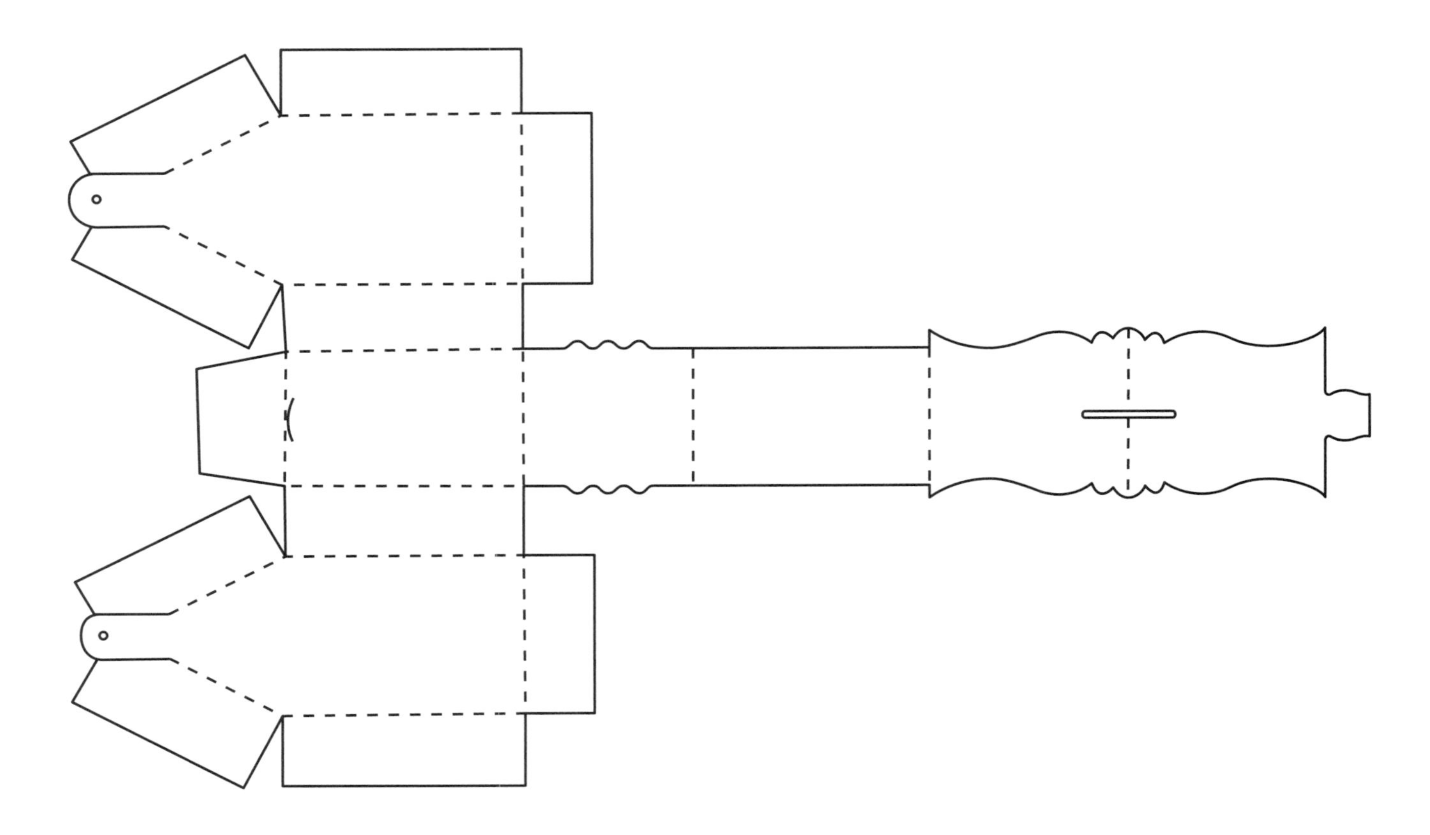

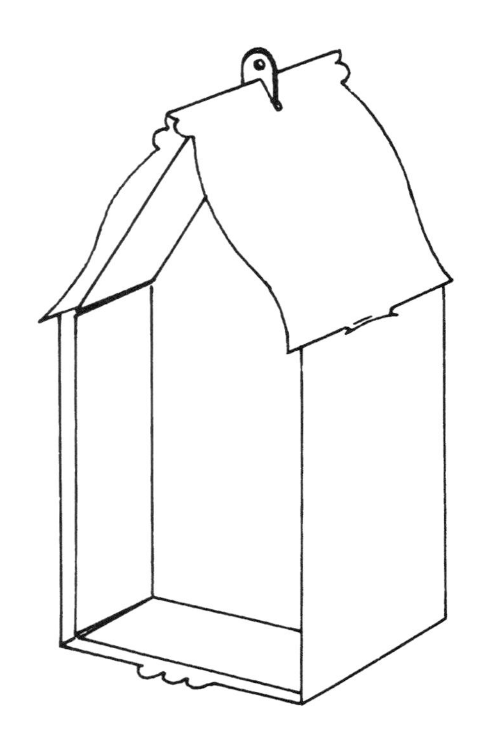

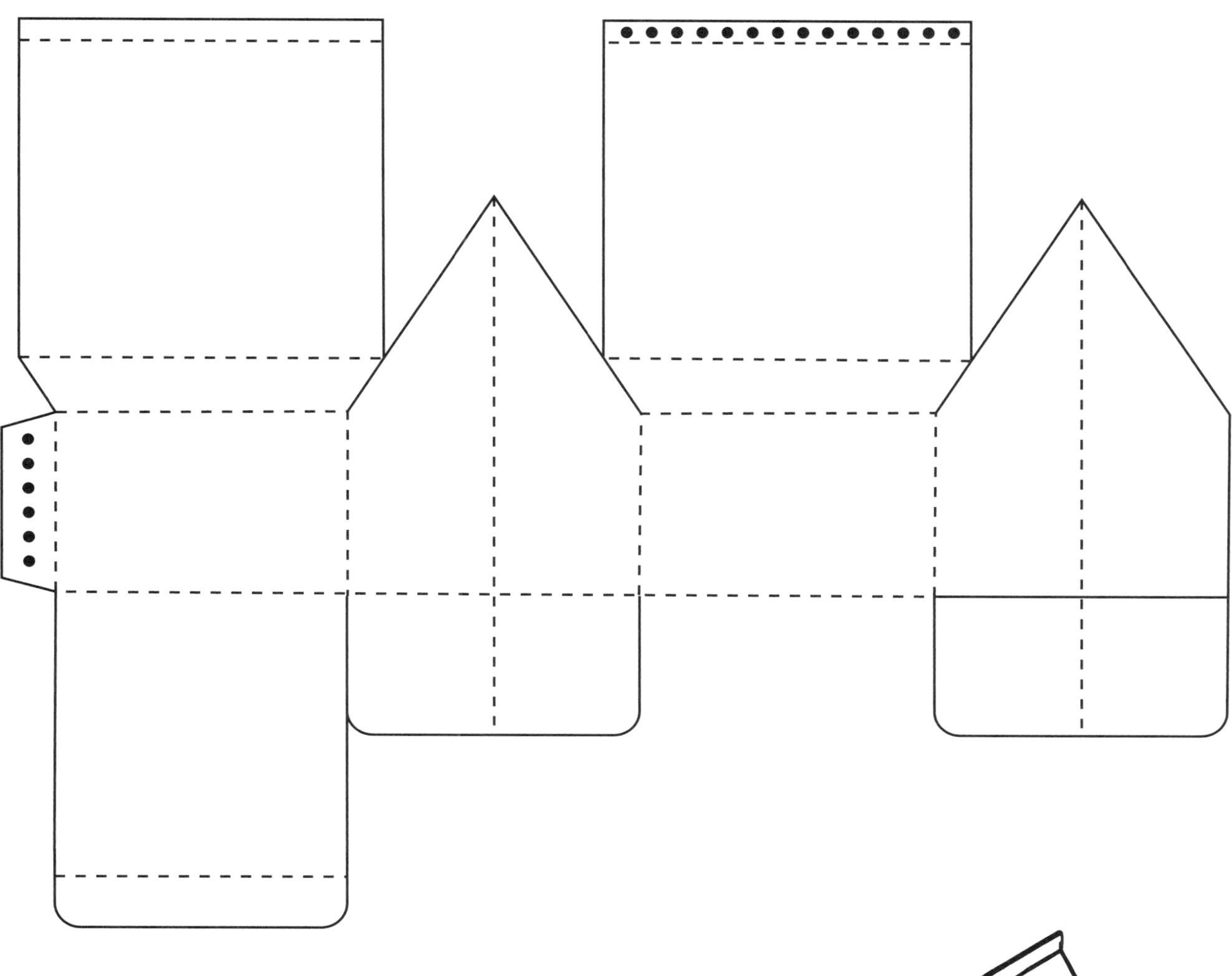

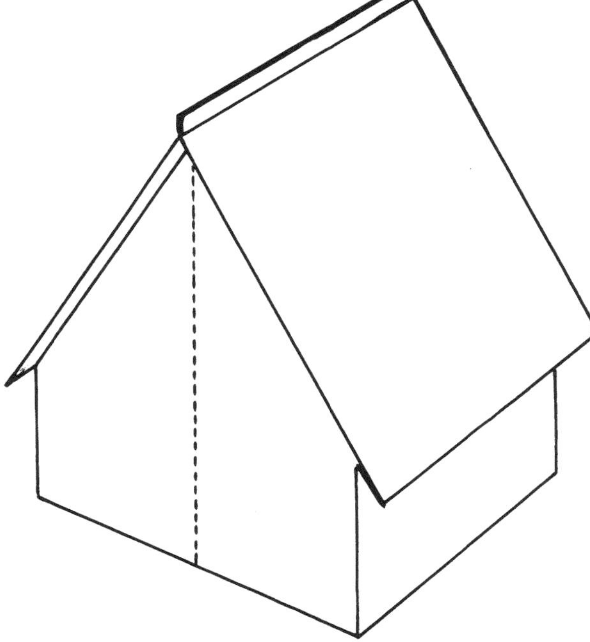

This house-shaped package only needs to be decorated to fit the specific needs of the product.

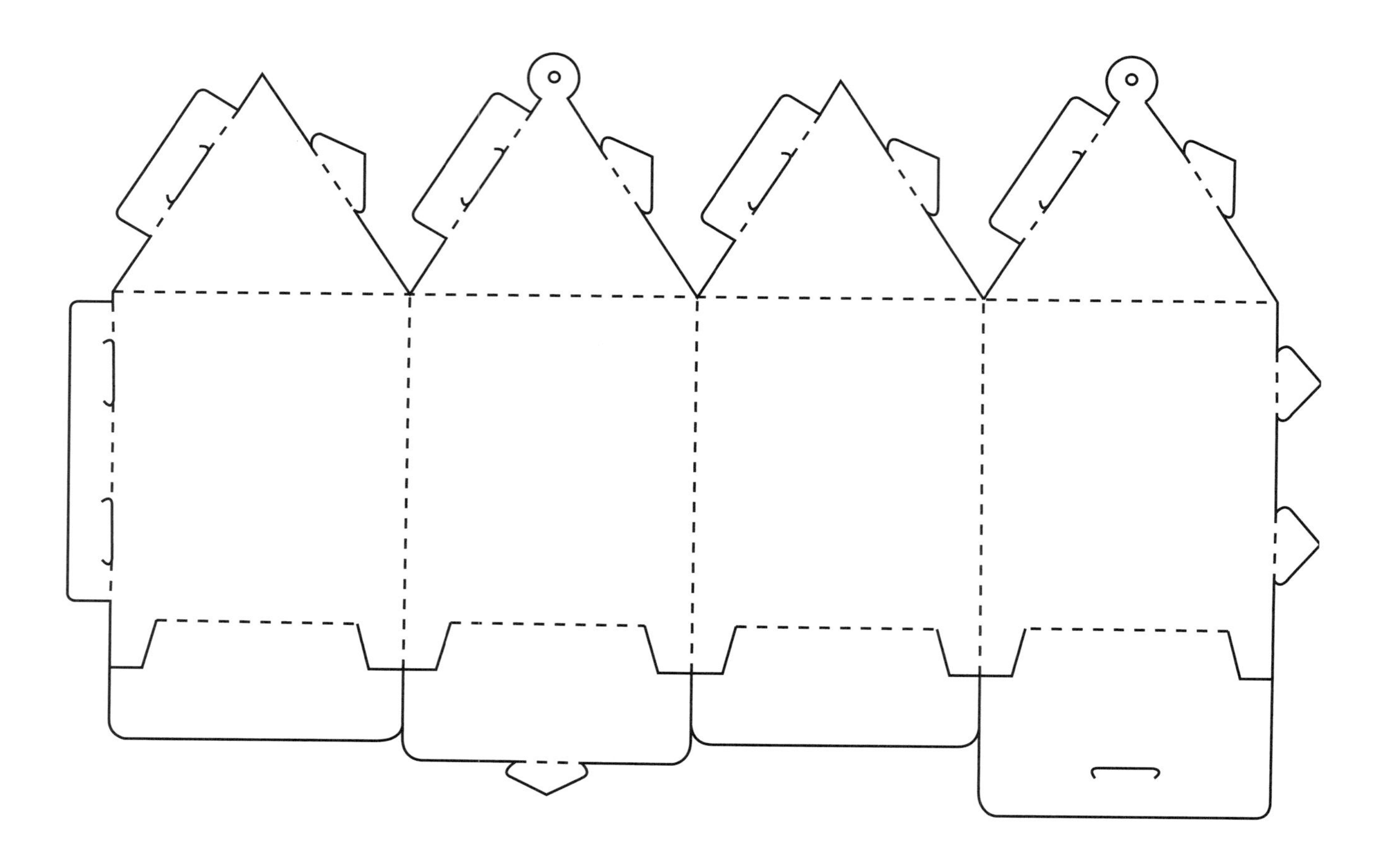

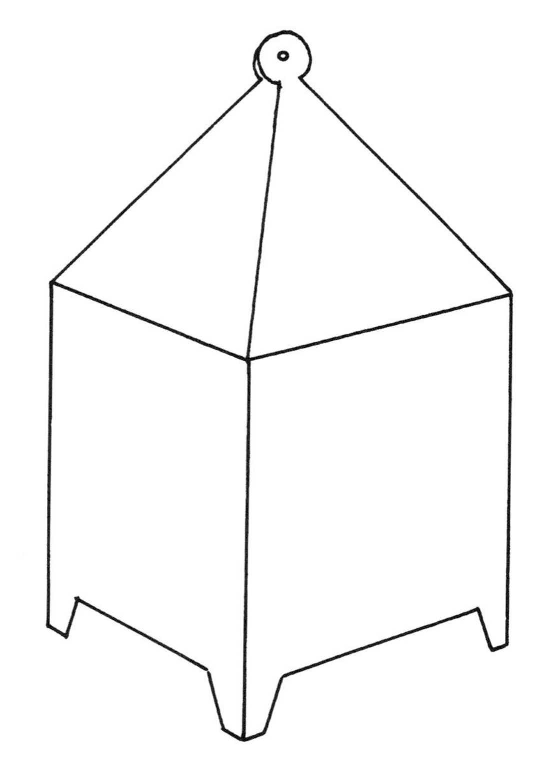

Lock tab and slit options make this package totally glue free.

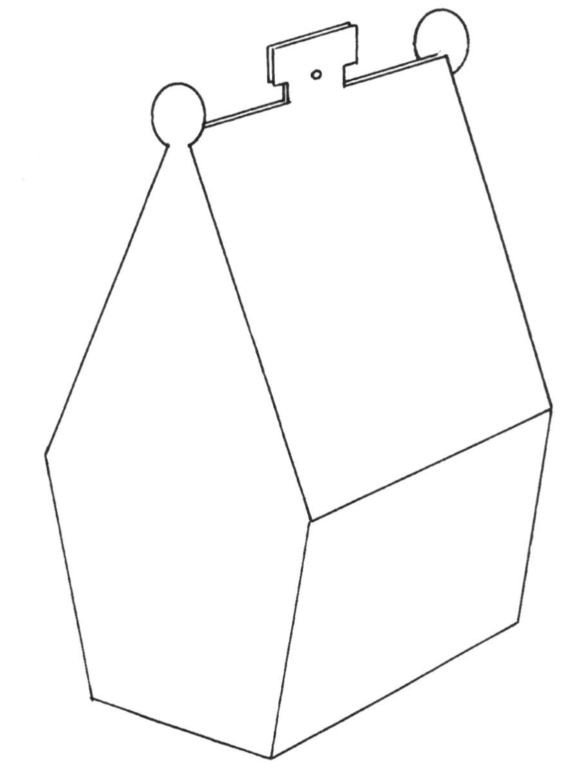

This package also requires no glue.

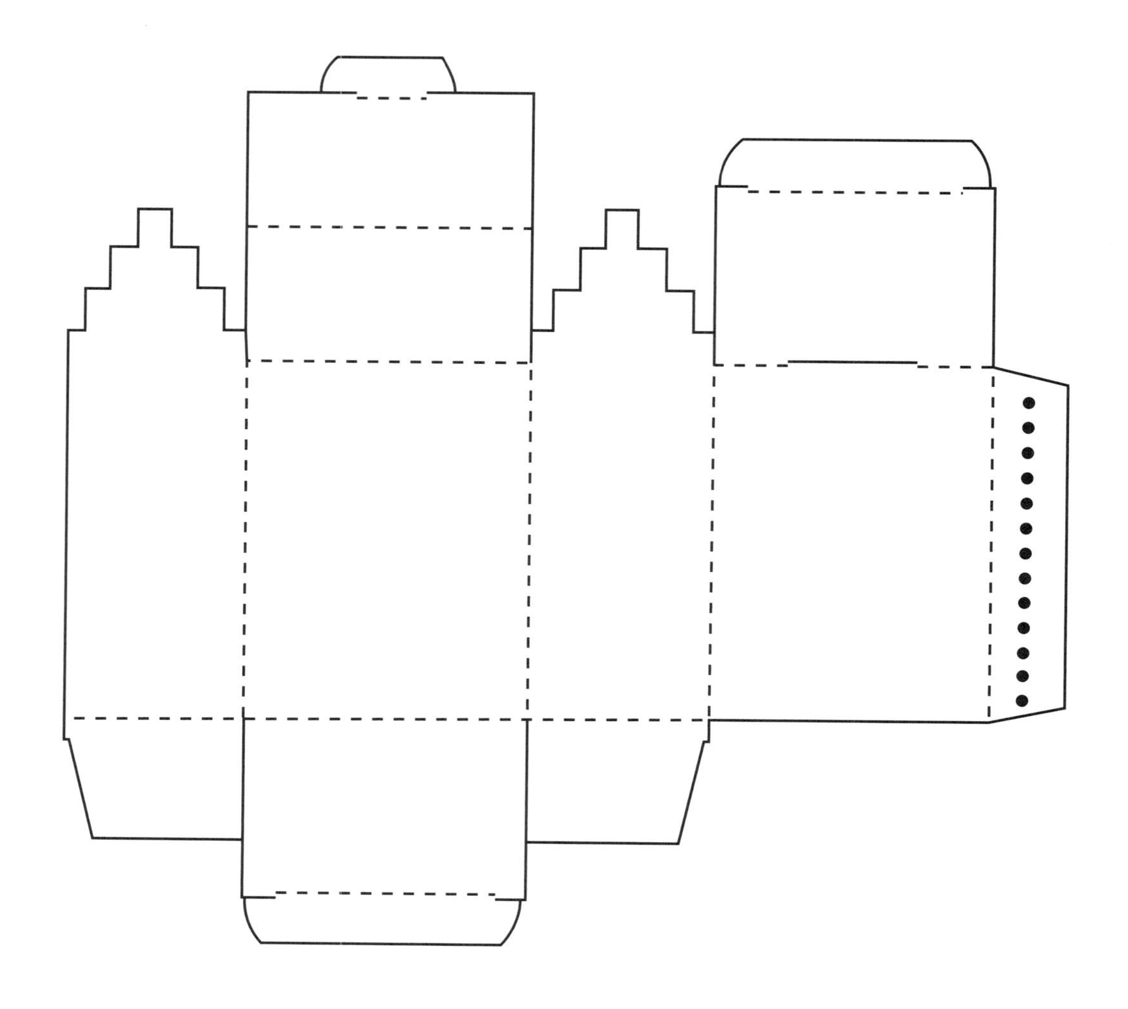

A container that depicts a traditional Dutch Burgher house makes an attractive keepsake.

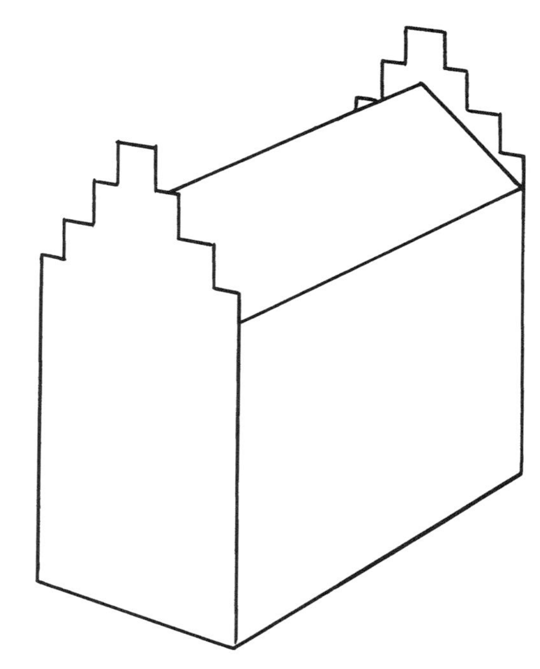

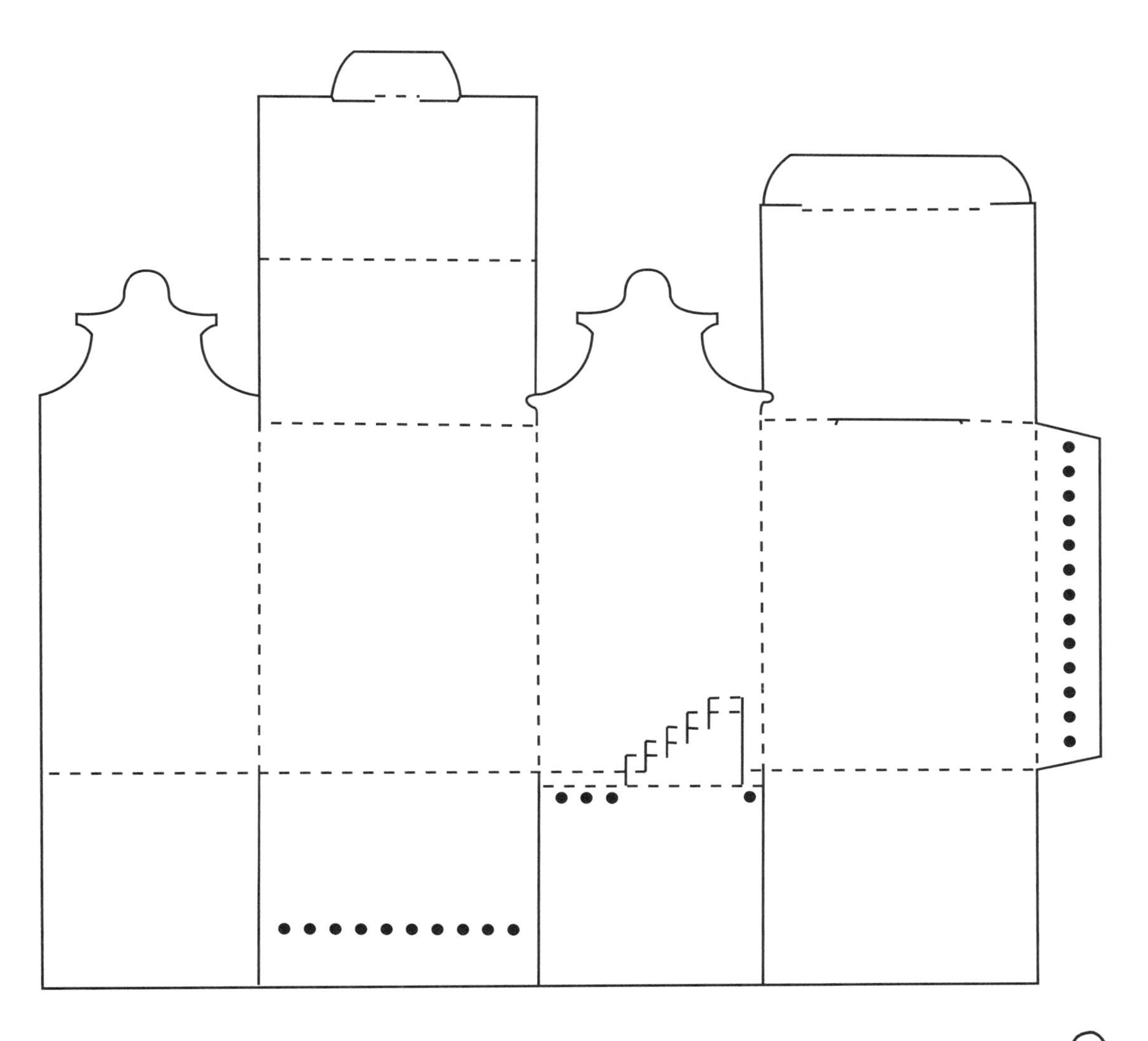

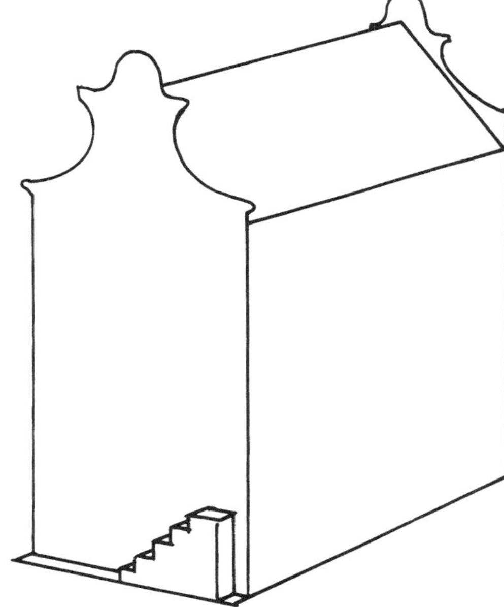

A somewhat more complex pattern could lead to a series of keepsakes.

HIP ROOFED BUILDING CARTON

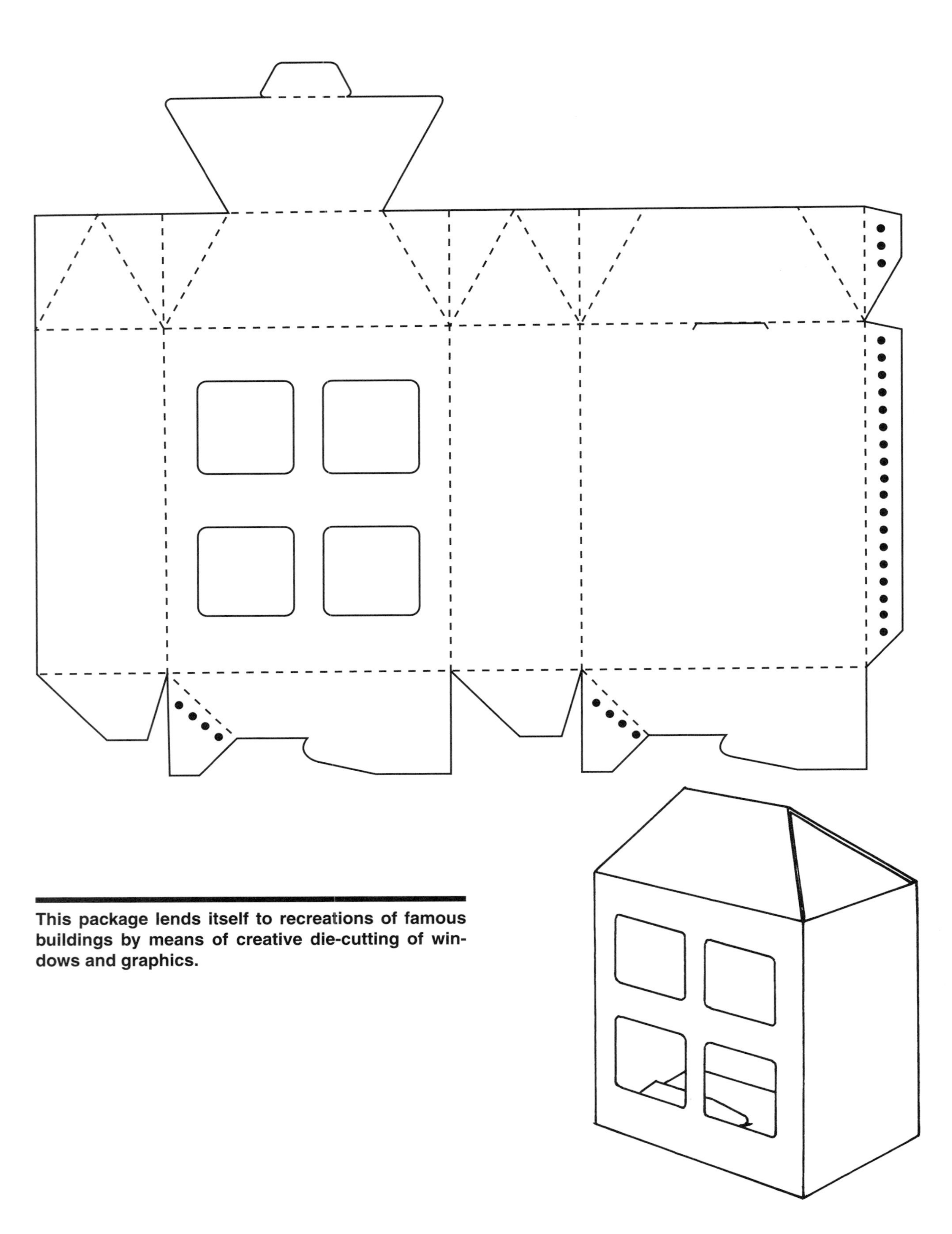

This package lends itself to recreations of famous buildings by means of creative die-cutting of windows and graphics.

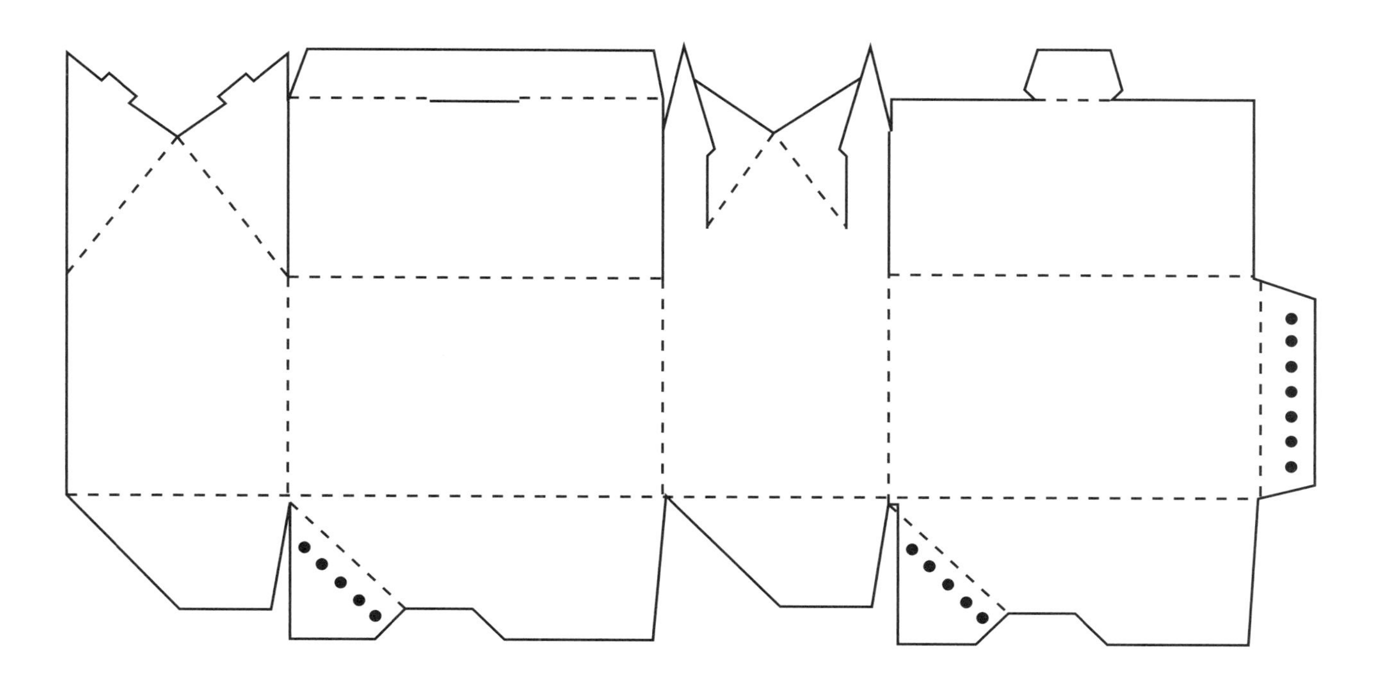

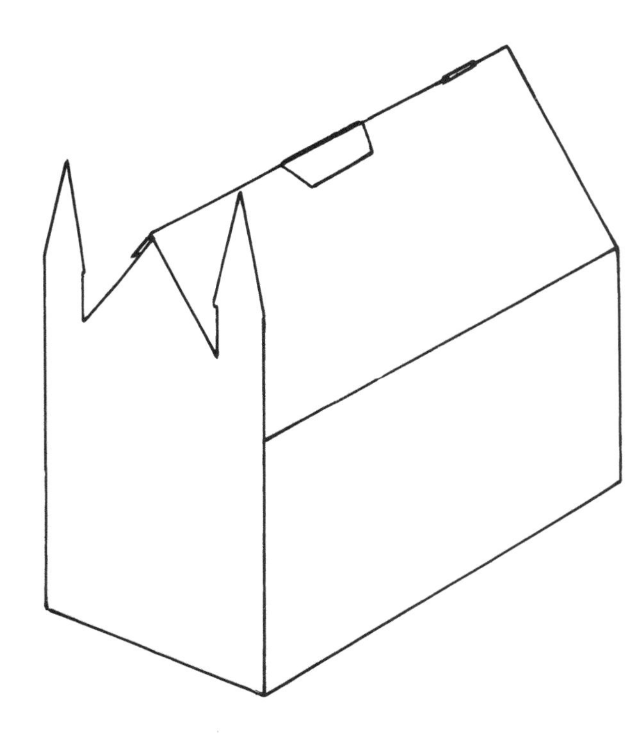

An alternative approach to the pattern on the preceding page.

FLOWER STALL CARTON

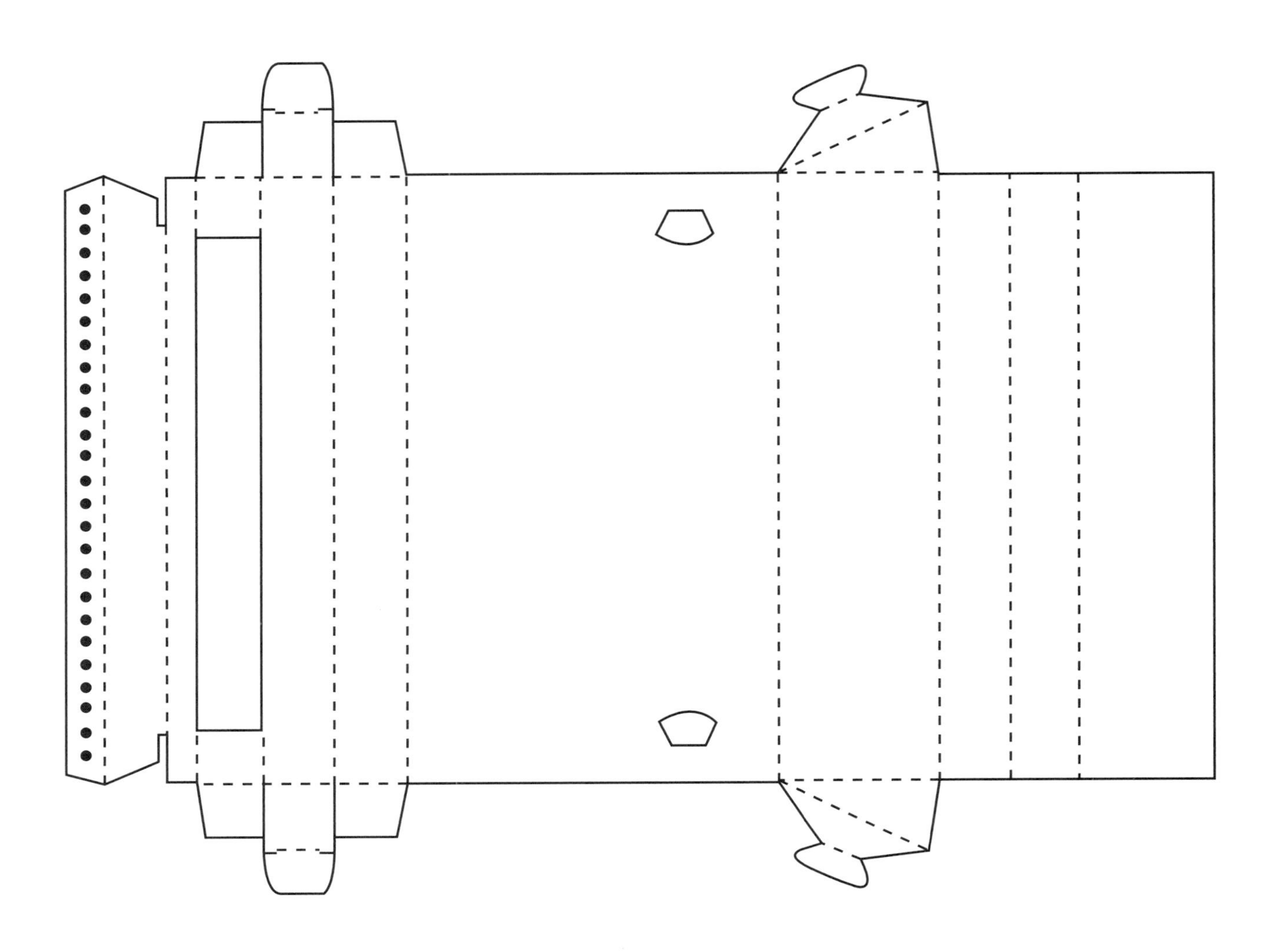

This little flower stall is an ideal package for cosmetics.

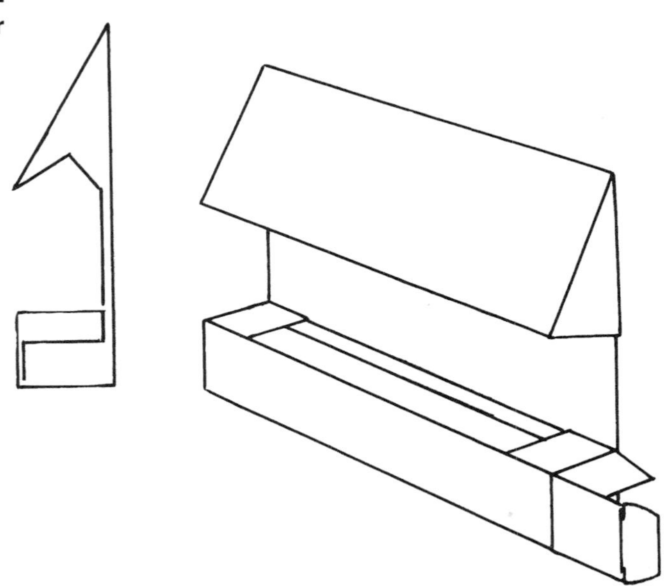

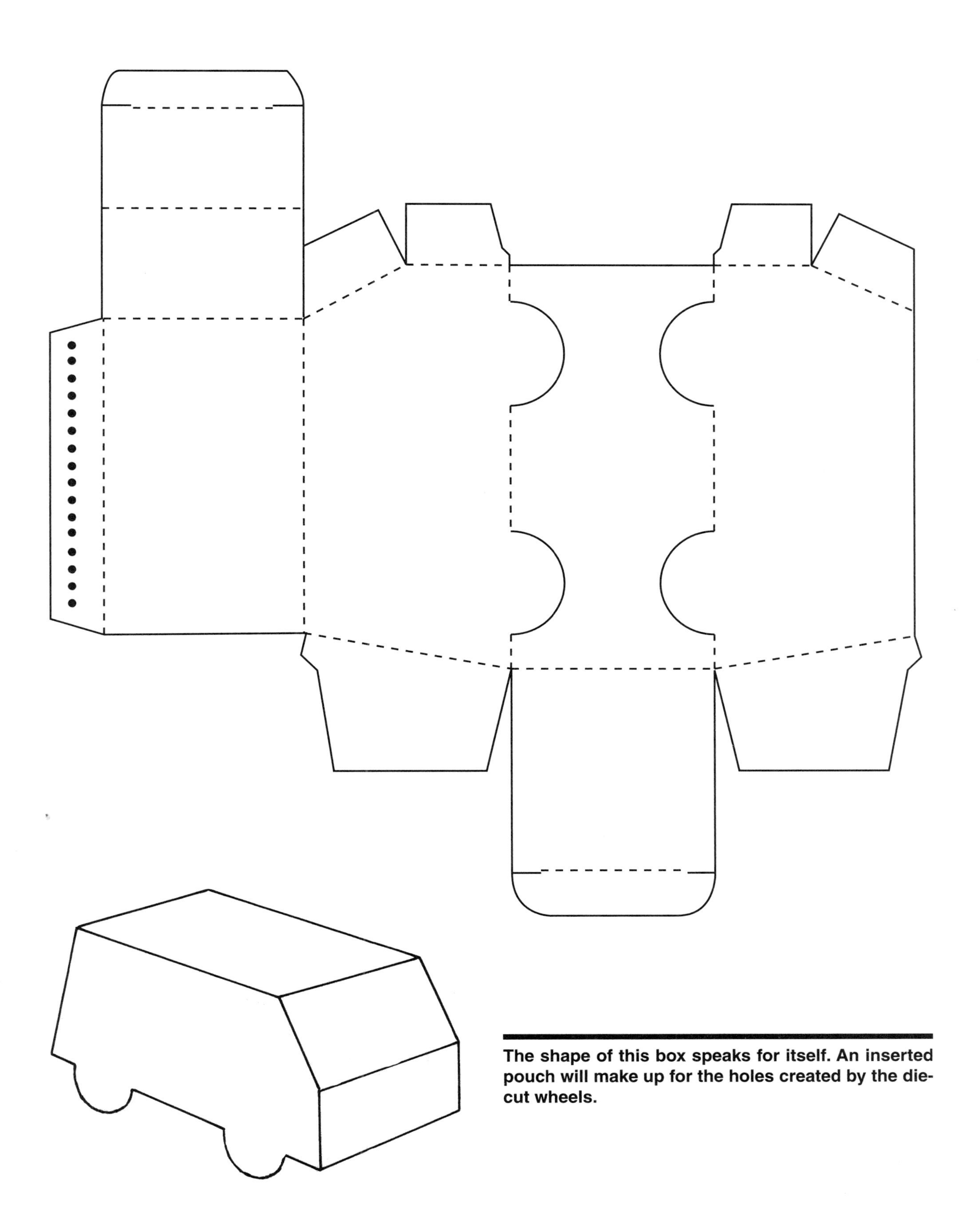

The shape of this box speaks for itself. An inserted pouch will make up for the holes created by the die-cut wheels.

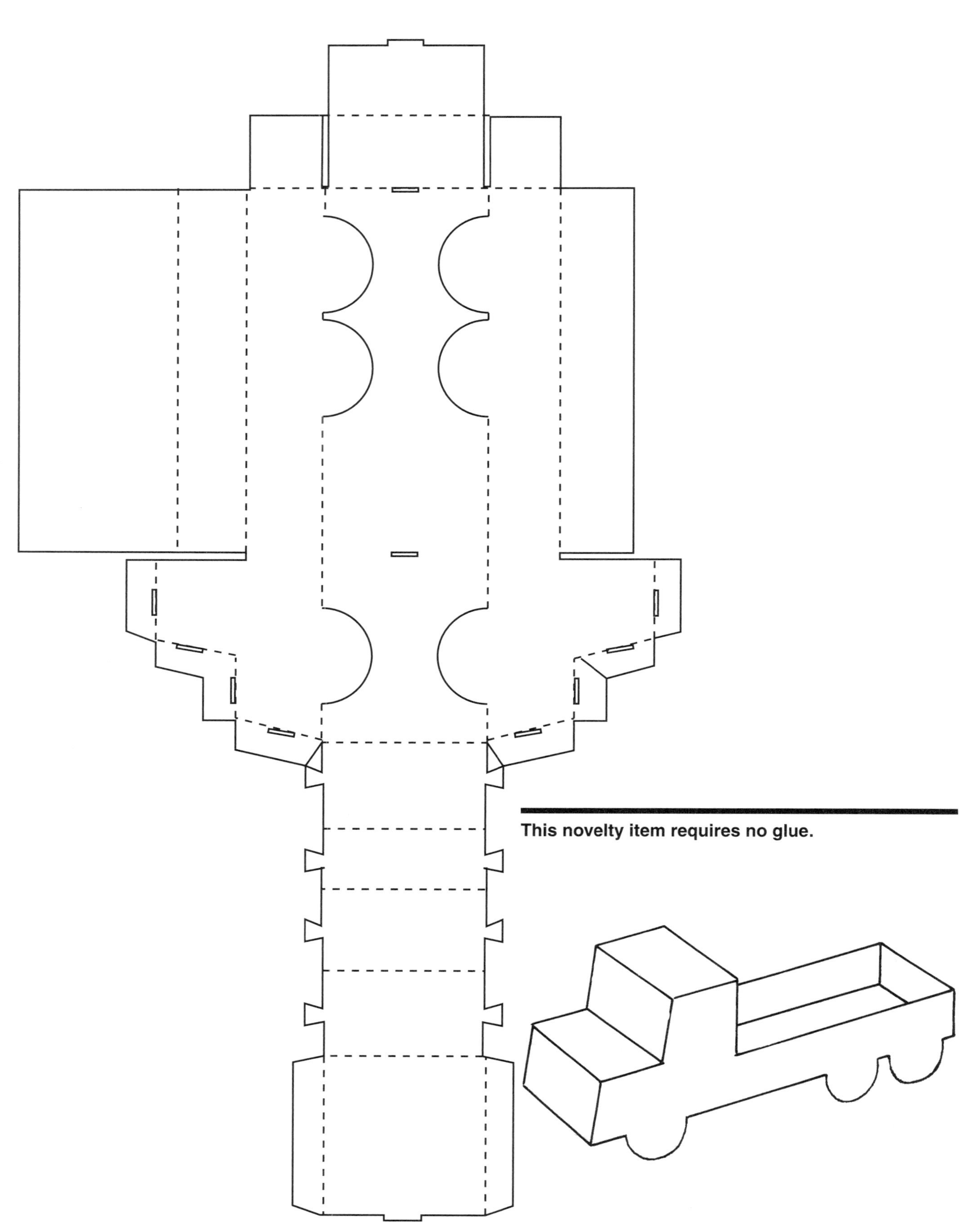

This novelty item requires no glue.

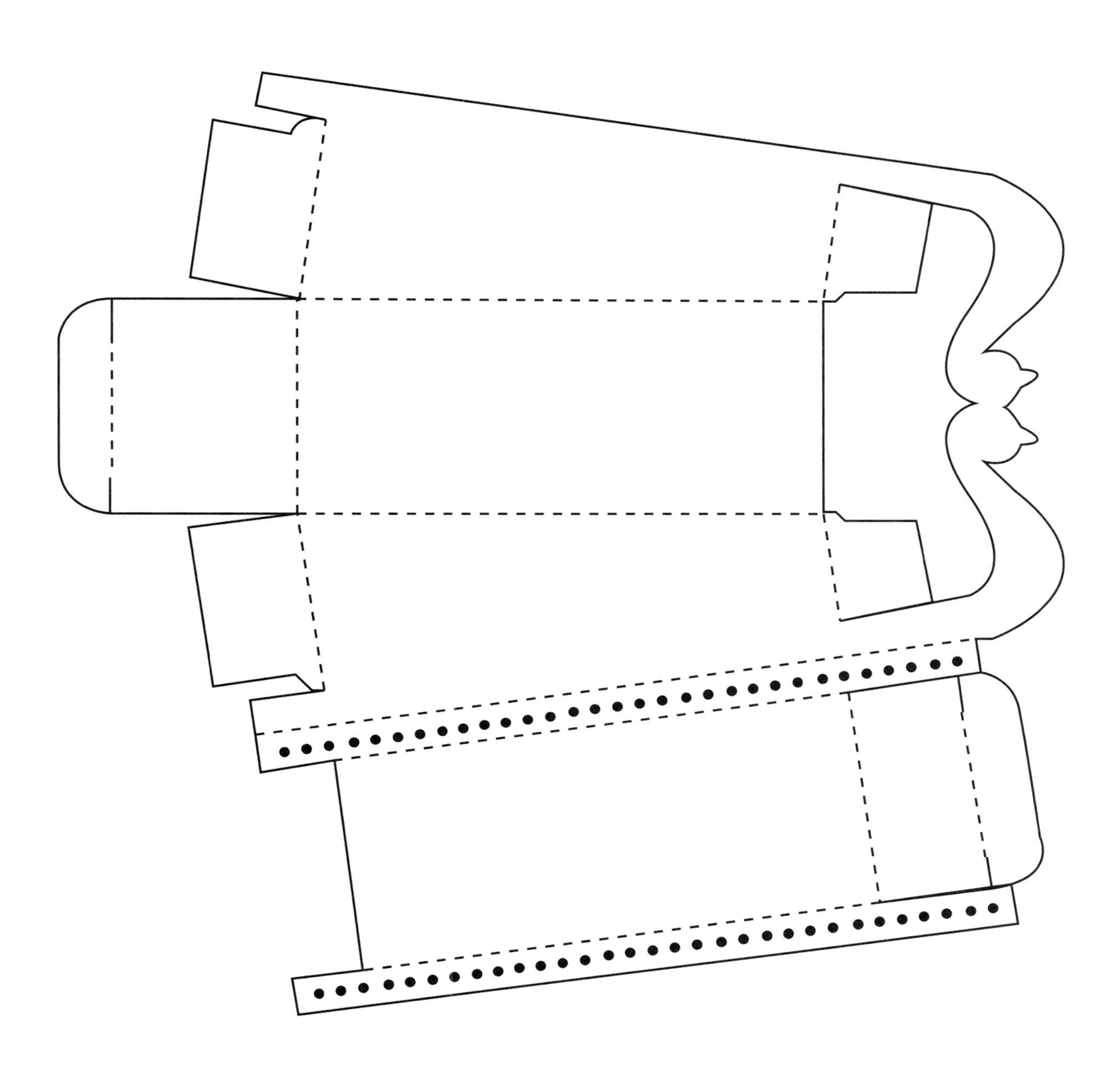

This old-fashioned sleigh has graphic application possibilities.

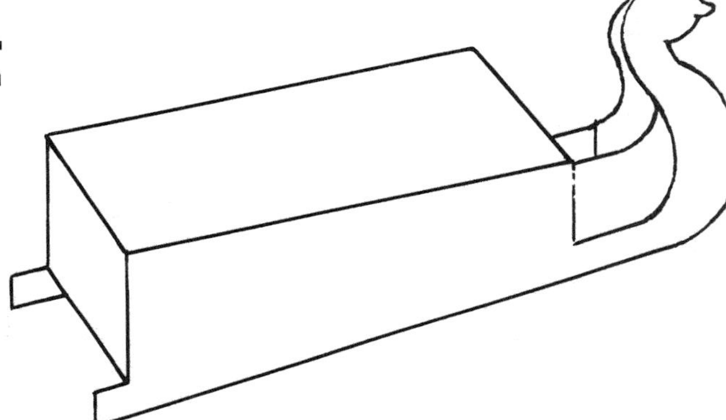

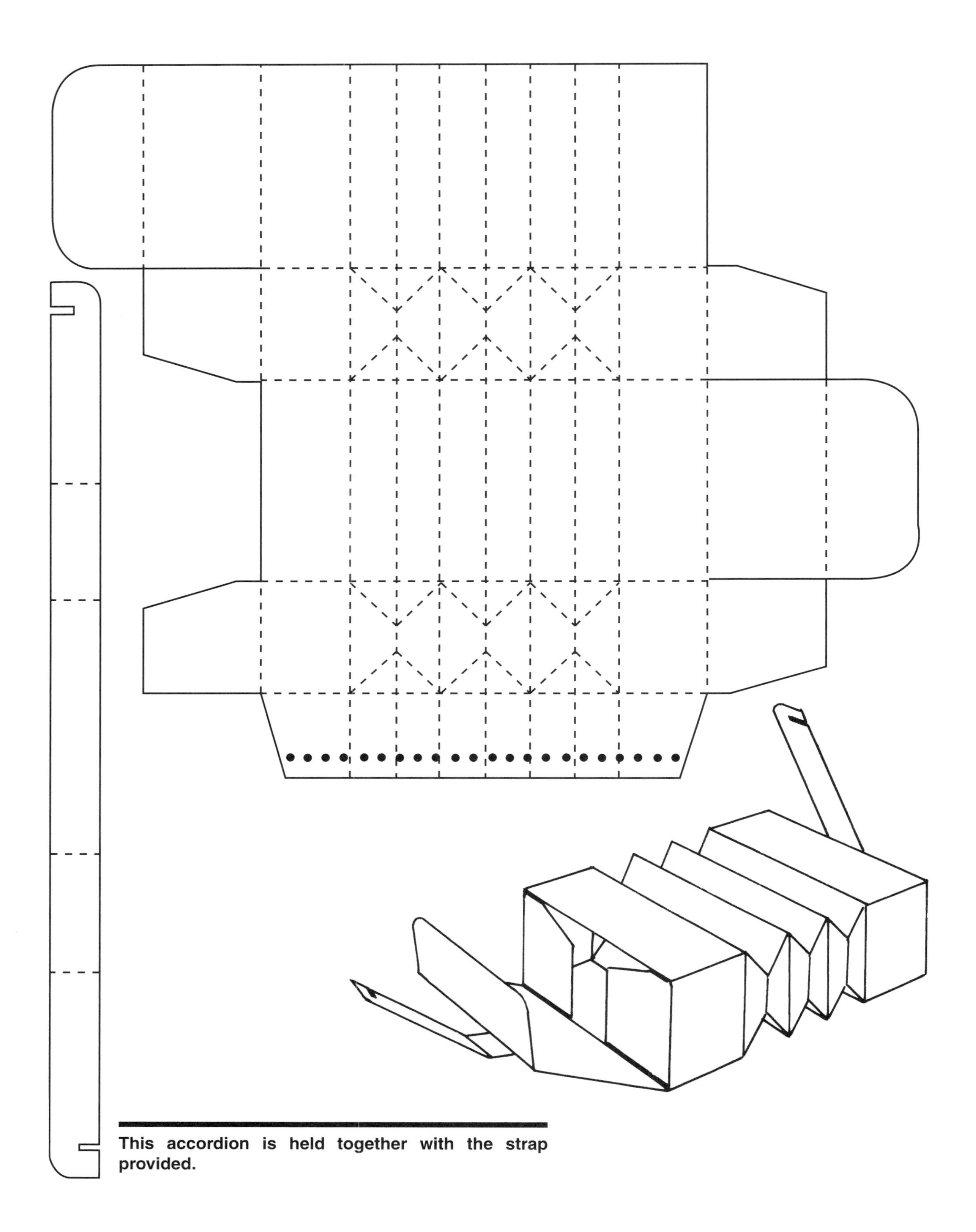

This accordion is held together with the strap provided.

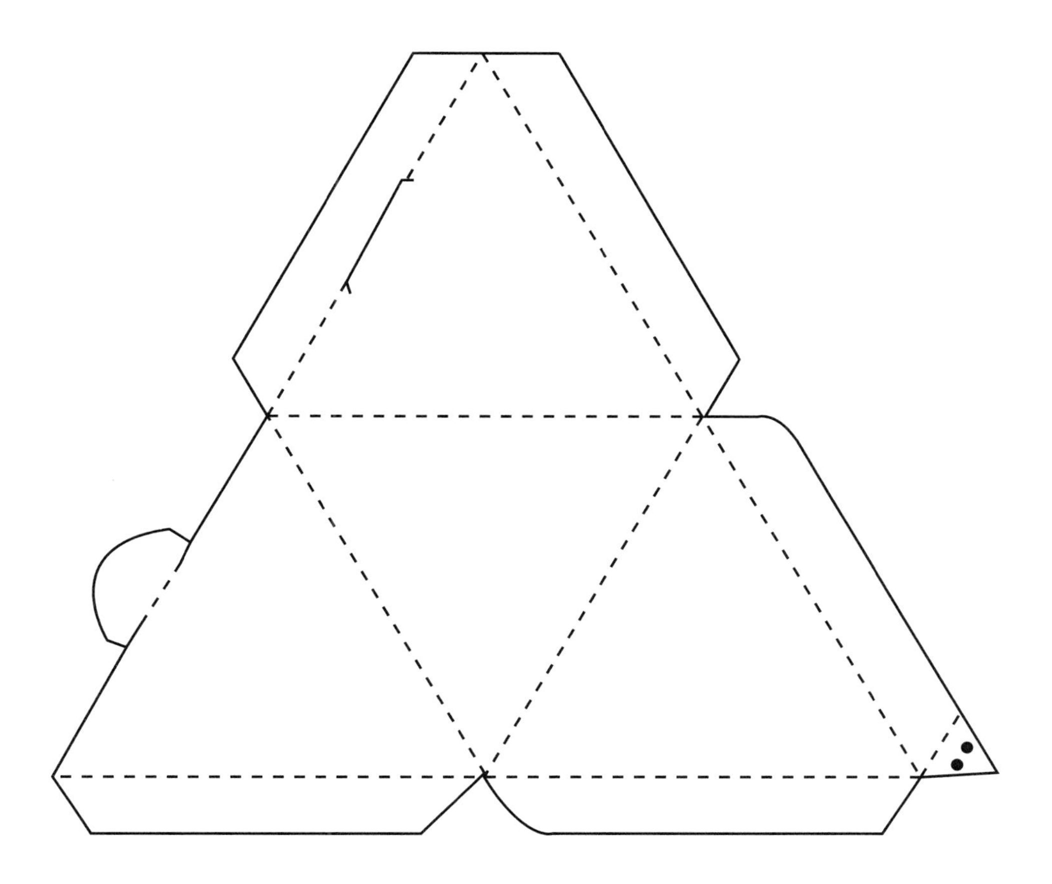

Spot gluing of the flap off the lid helps keep the package closed.

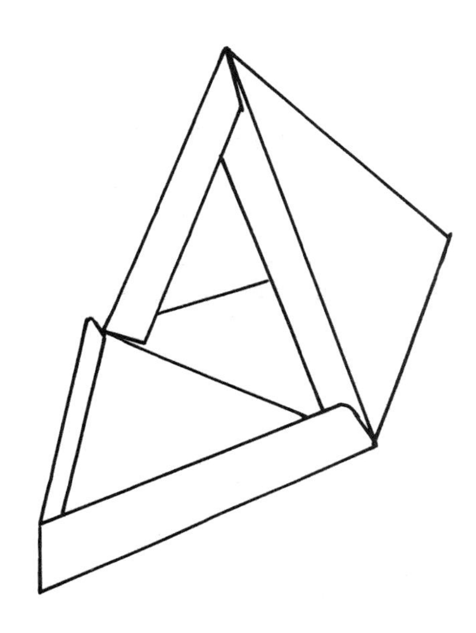

TRIANGULAR DIPYRAMID

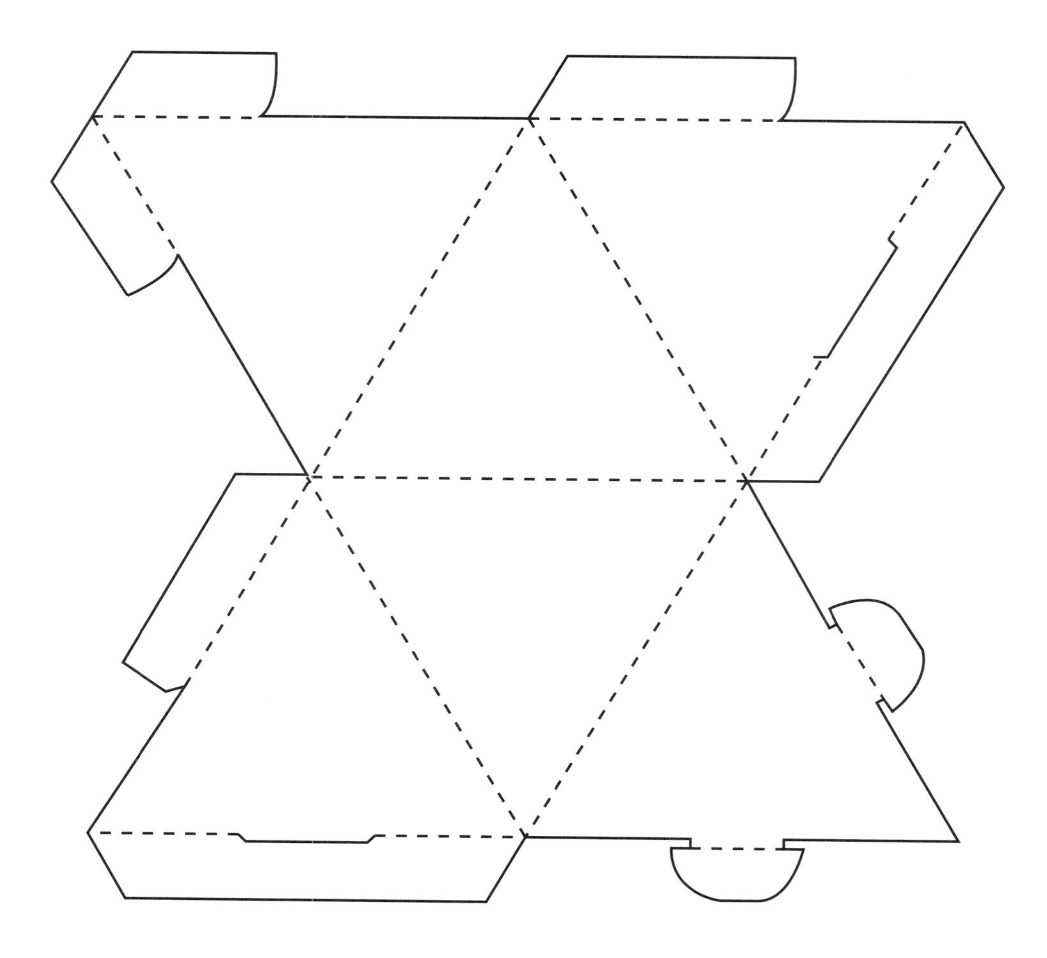

The flaps off the lid lock into the shortened dustflaps.

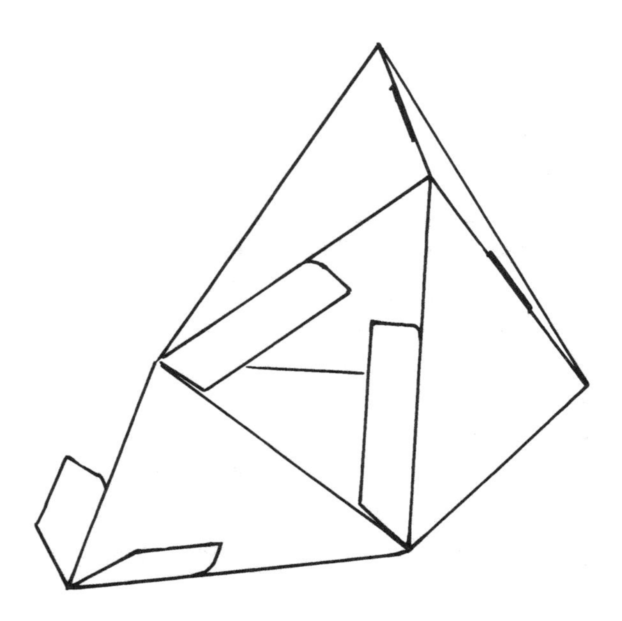

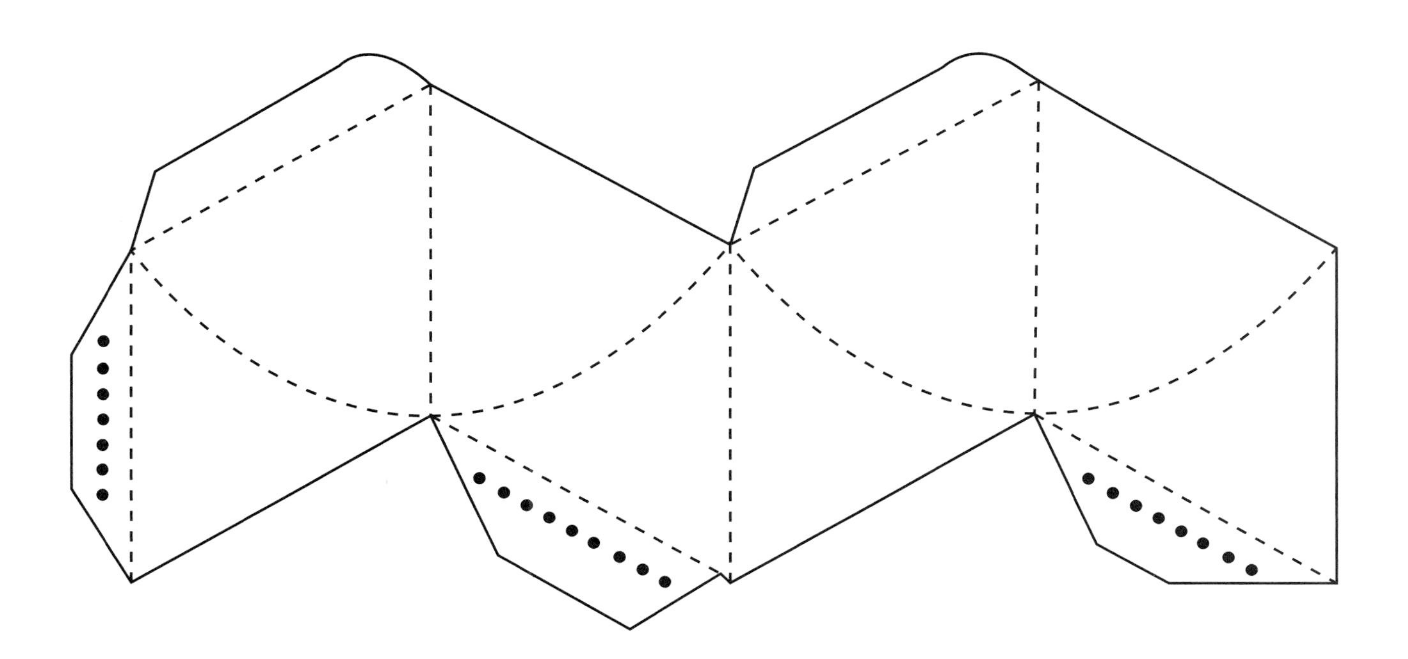

Straight and curved scores permit a wide range of variations.

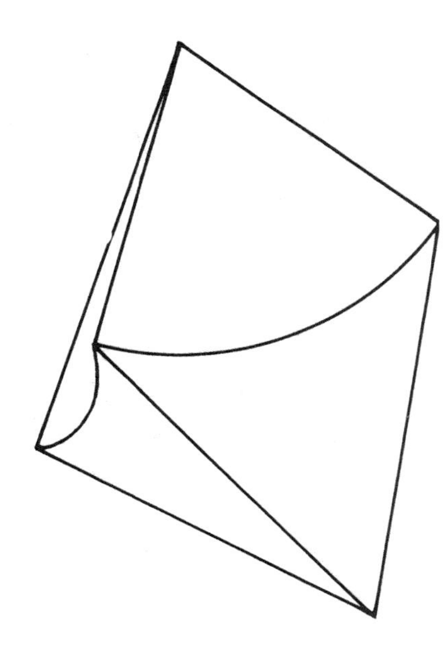

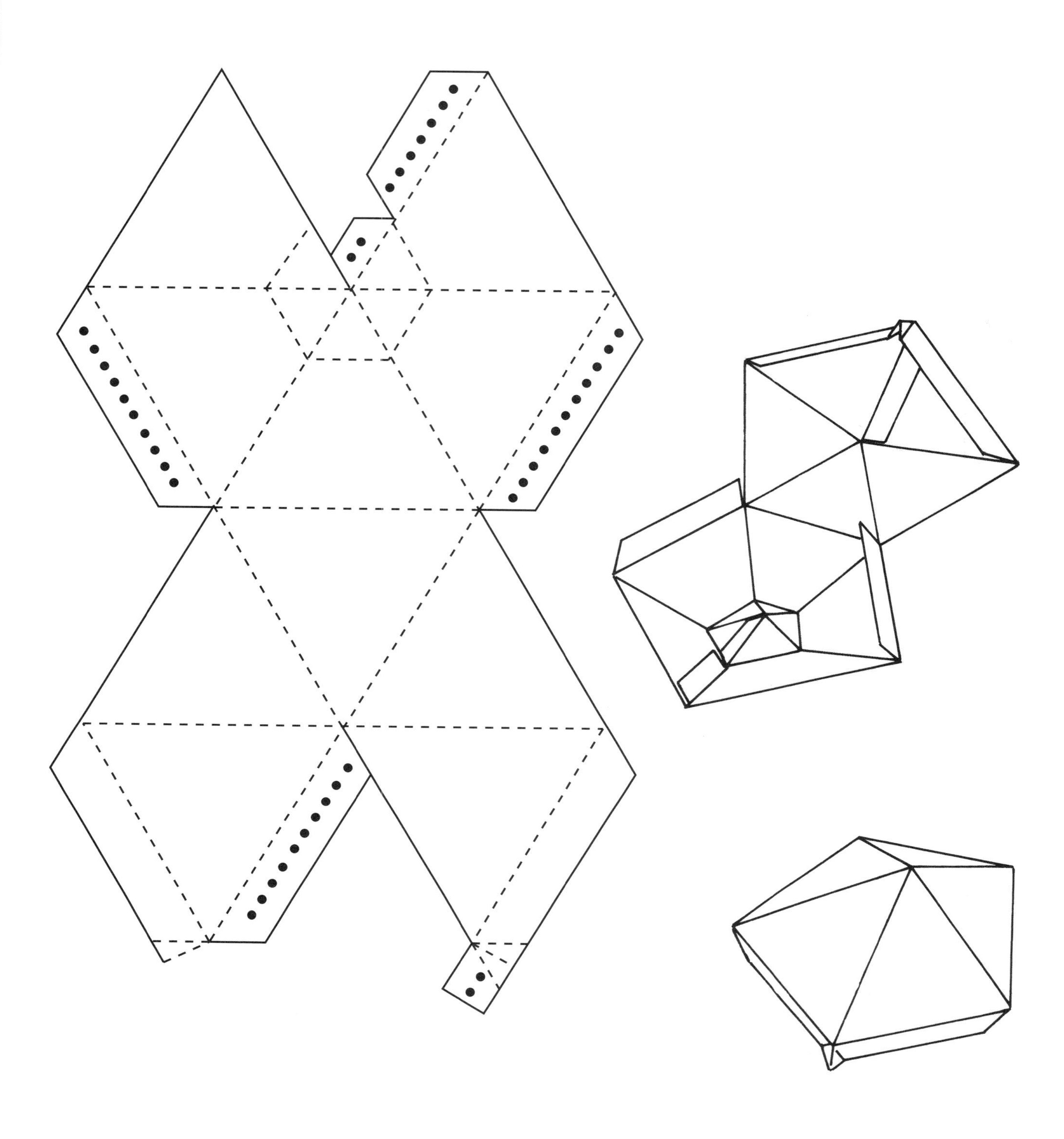

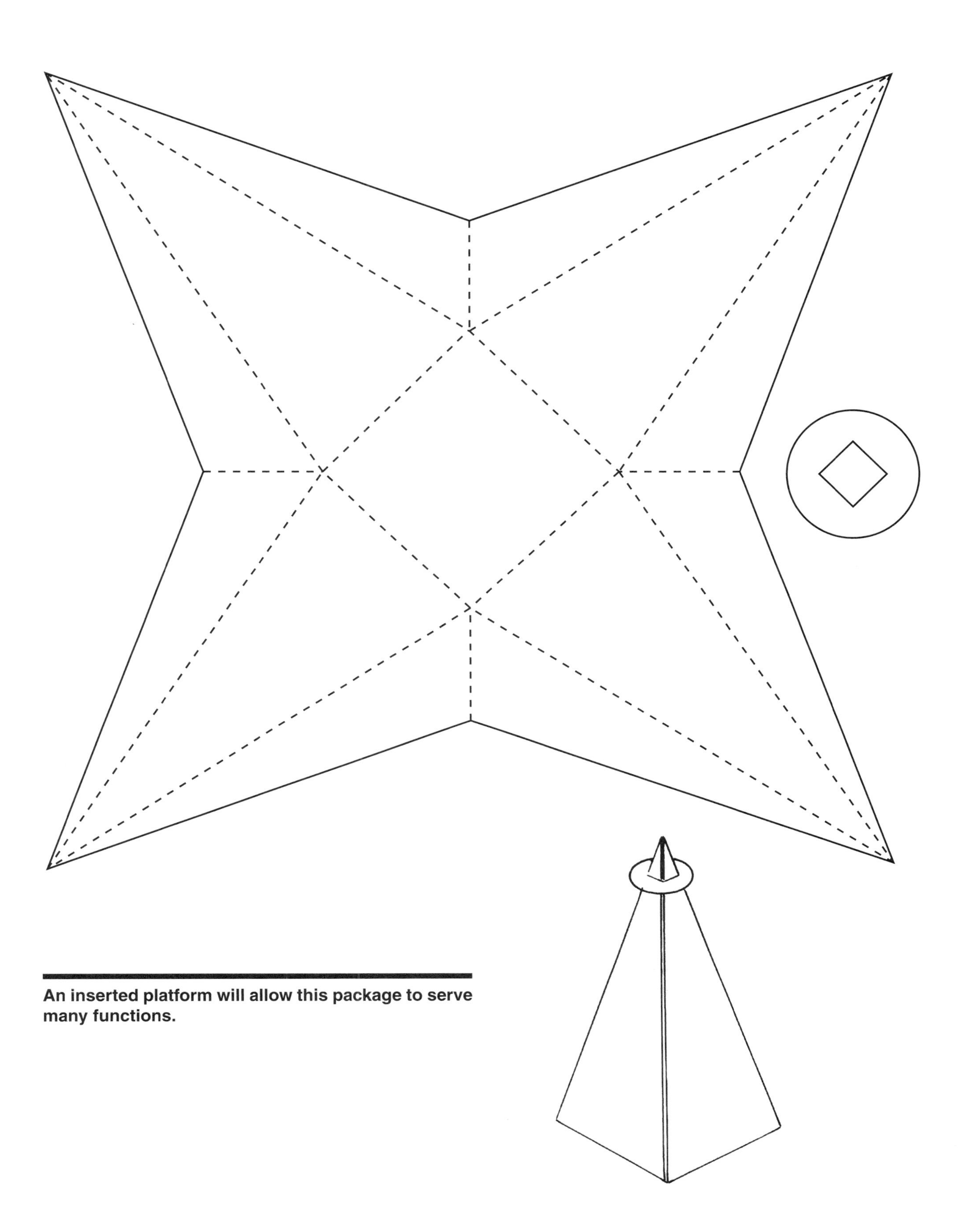

An inserted platform will allow this package to serve many functions.

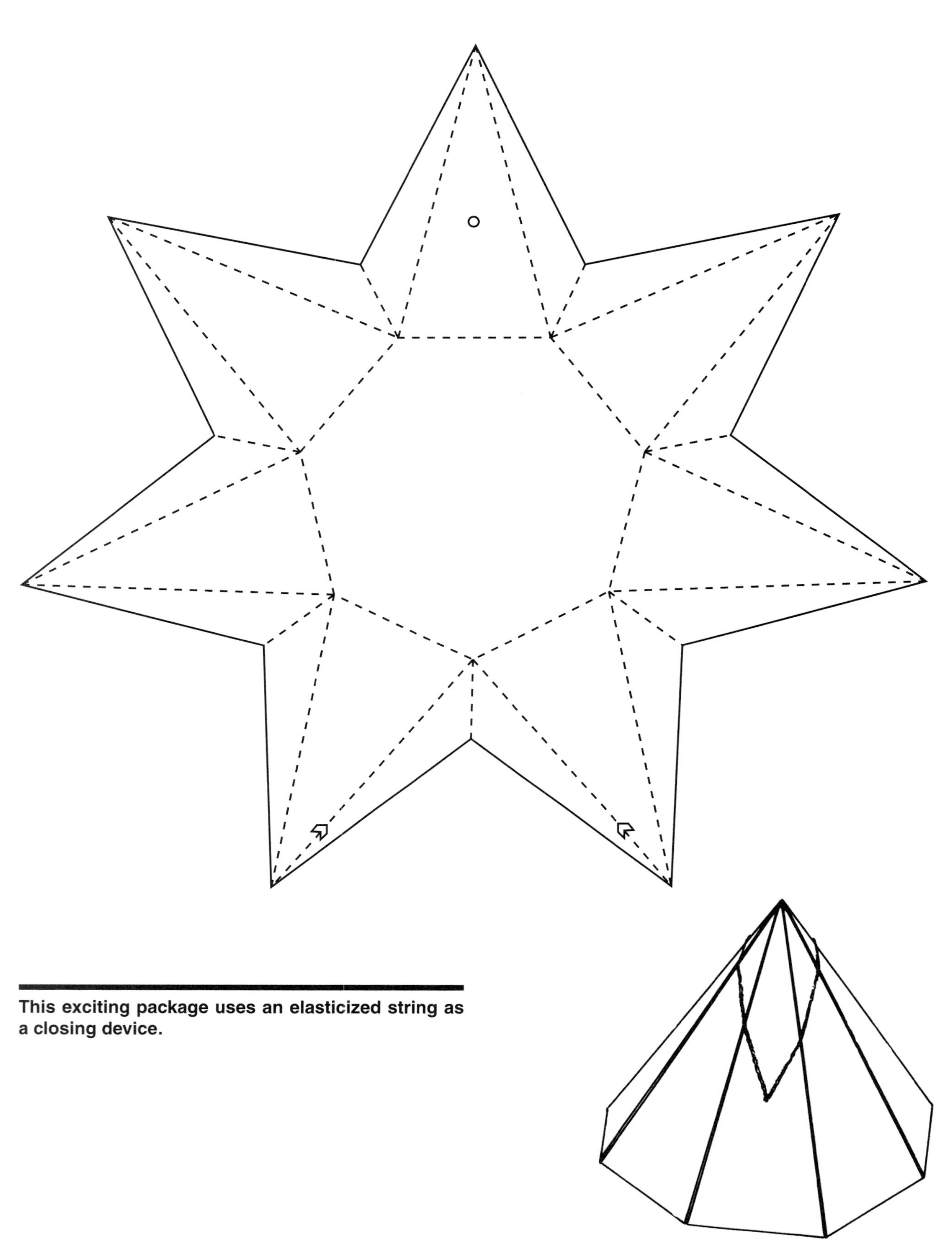

This exciting package uses an elasticized string as a closing device.

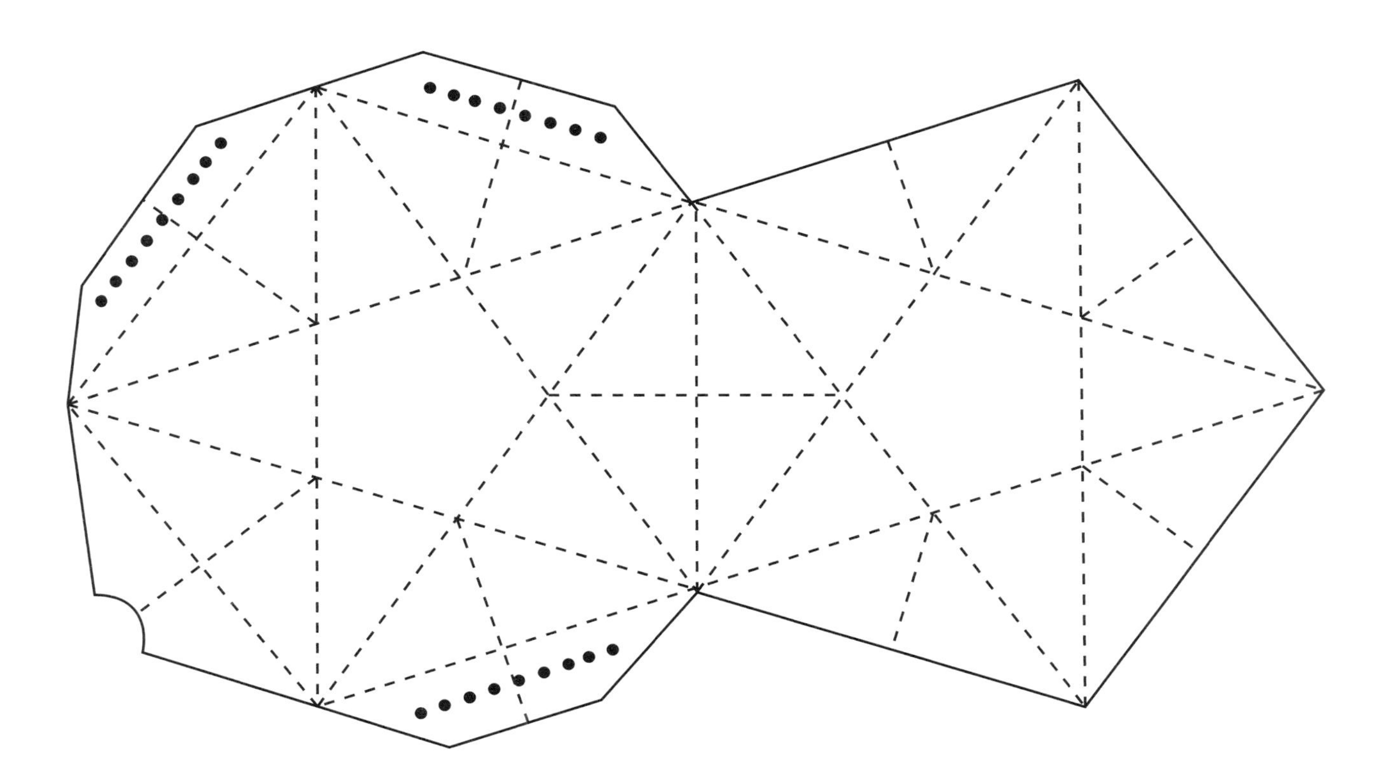

This beautiful candy package could be used as holiday ornament.

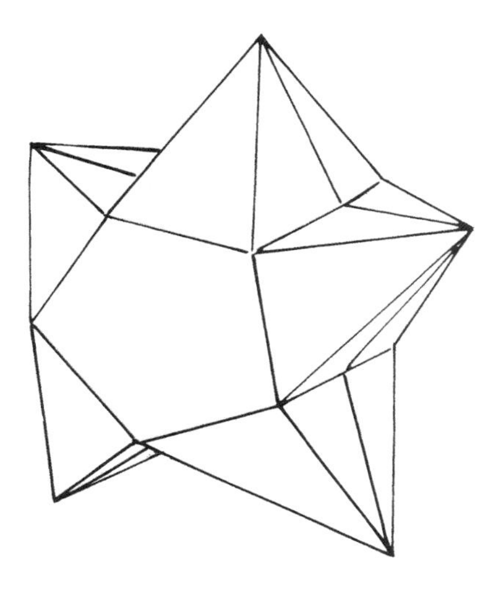

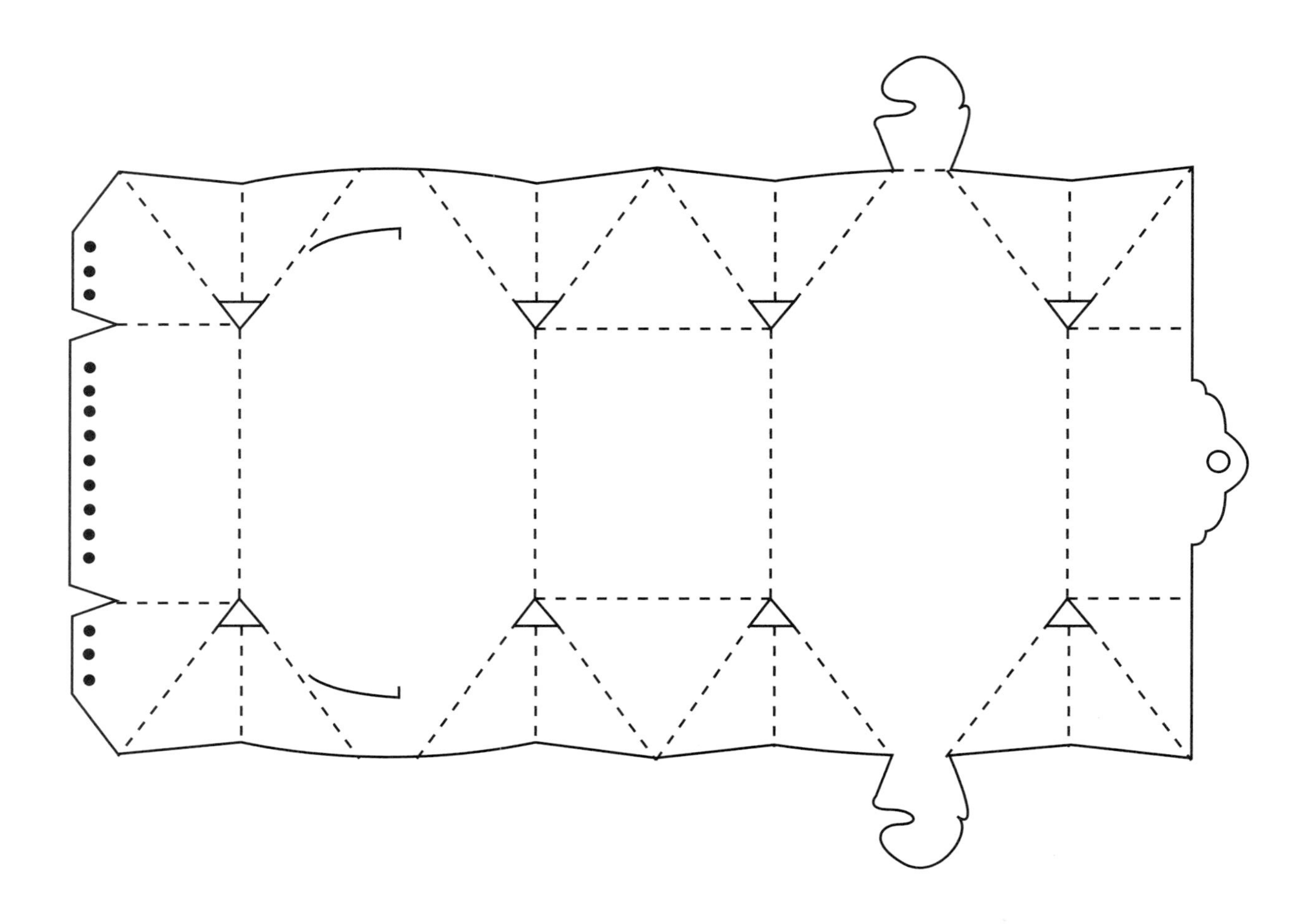

Designed by David James & Associates, this tea package lends itself to a variety of applications.

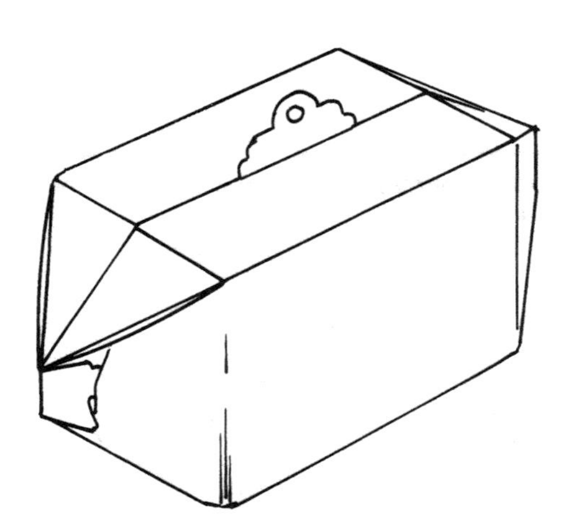

Sleeves, Wraps, and Folders

OPEN END SIMPLE GLUED SLEEVE

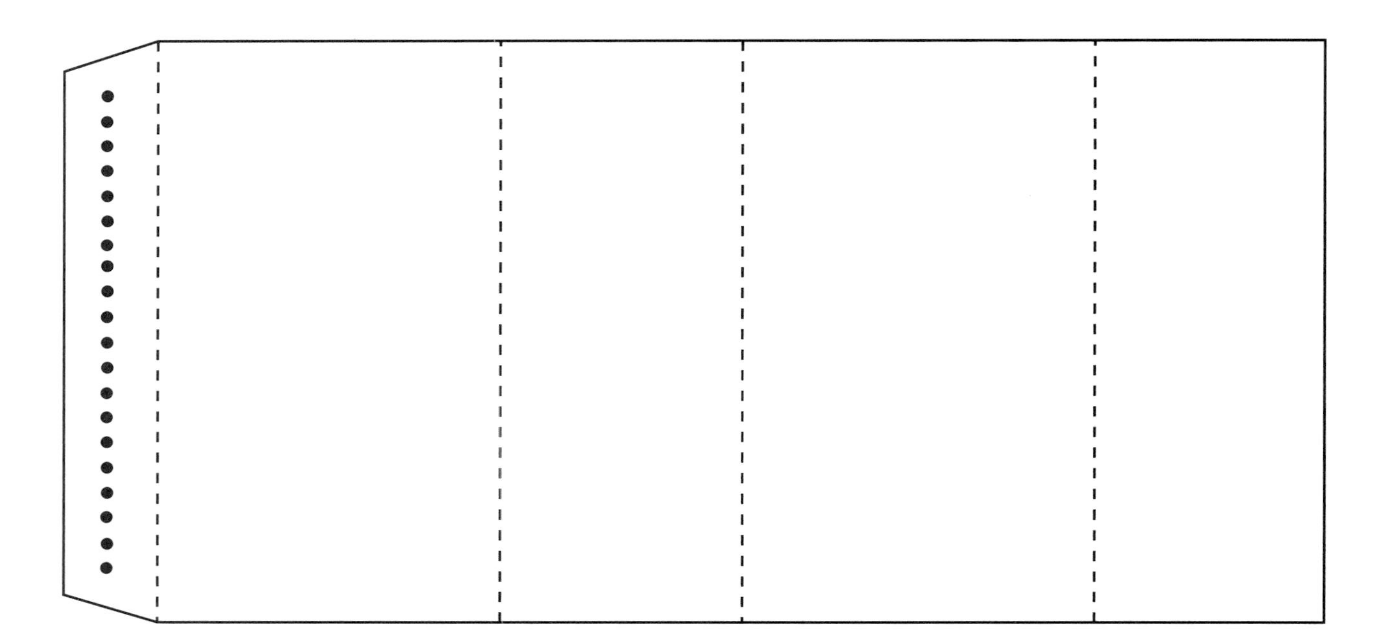

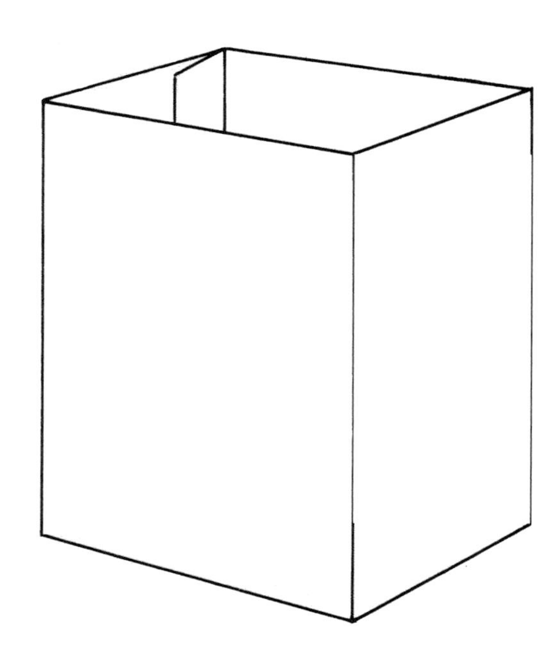

OPEN END SLEEVE WITH FINGER HOLES FOR EASY ACCESS

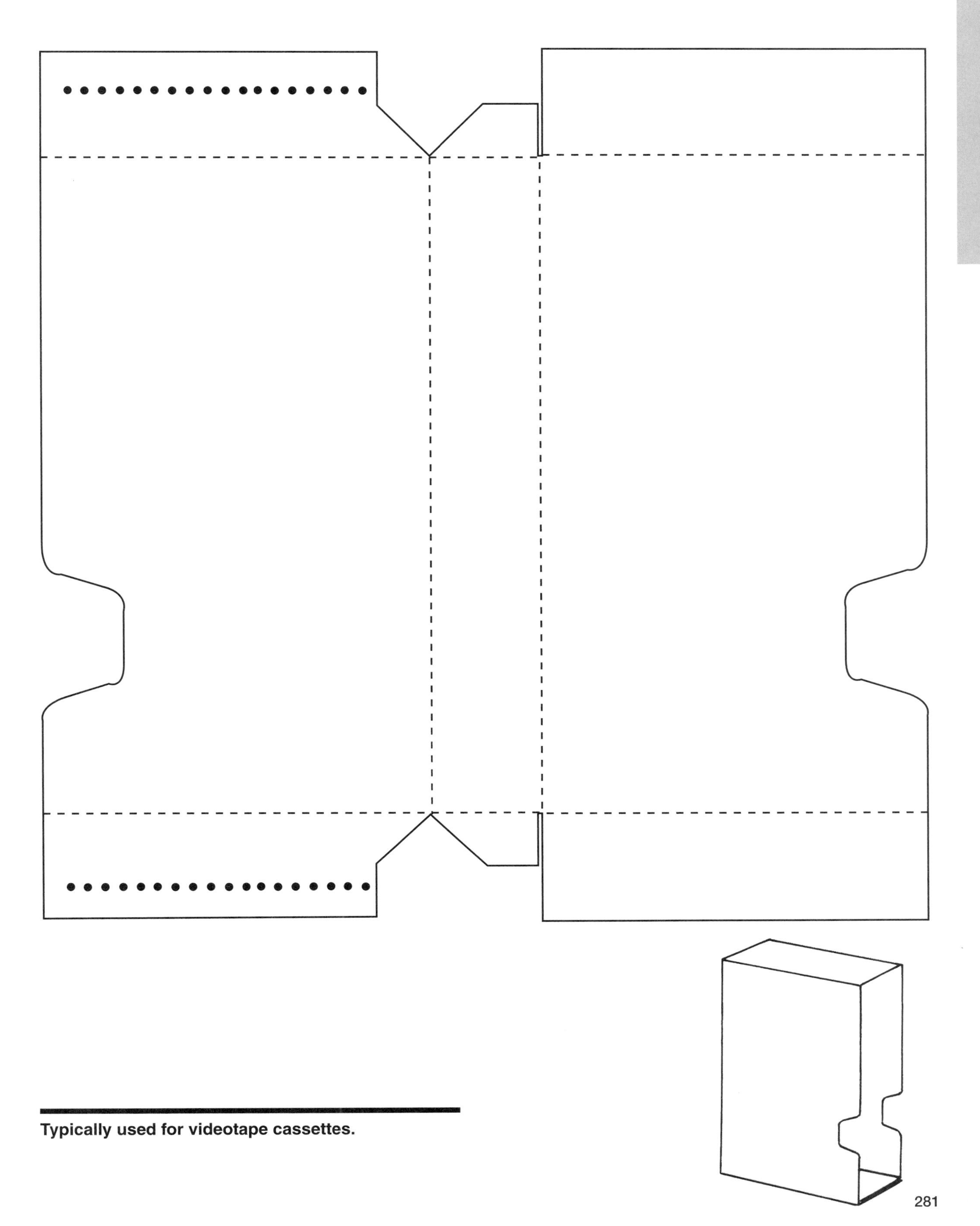

Typically used for videotape cassettes.

OPEN END SLEEVE WITH RECESSED PRODUCT RETAINERS

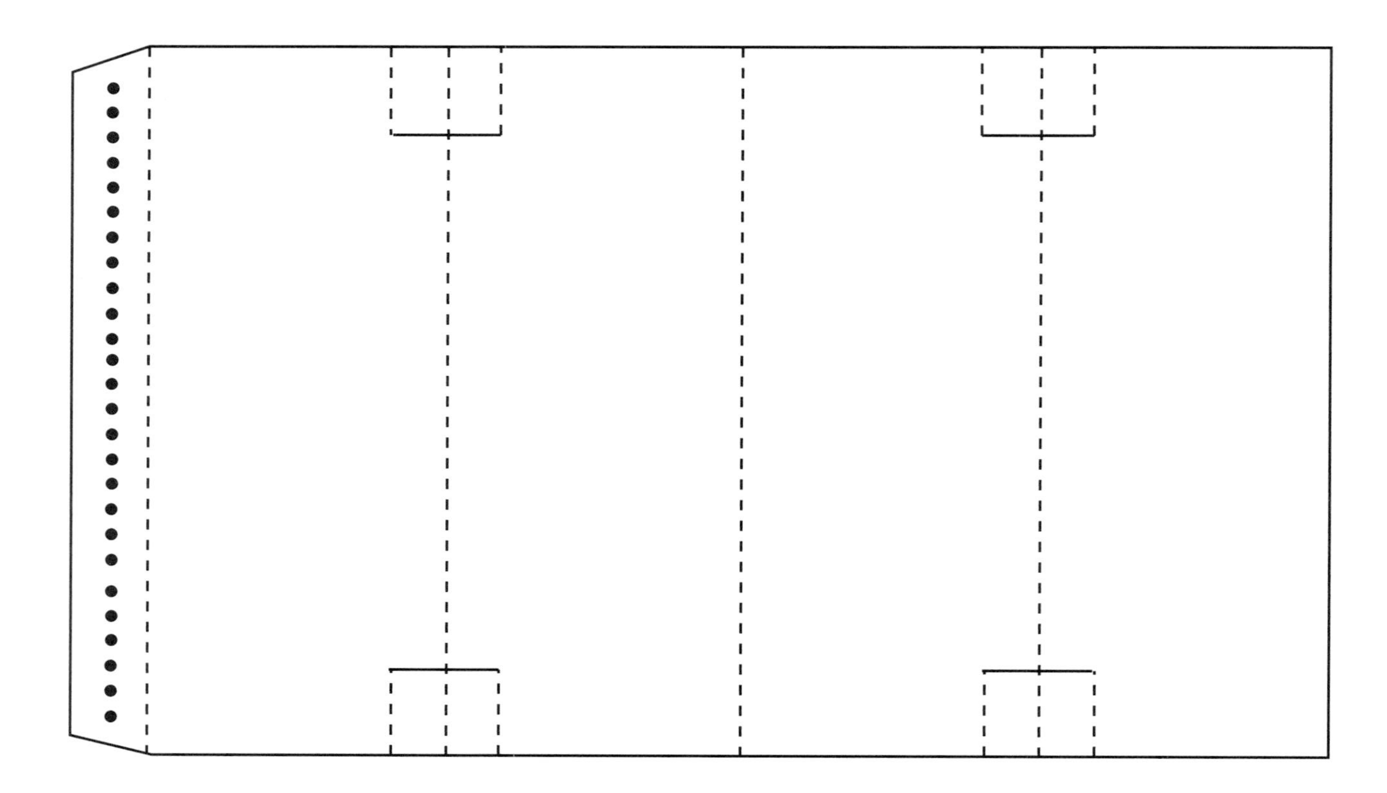

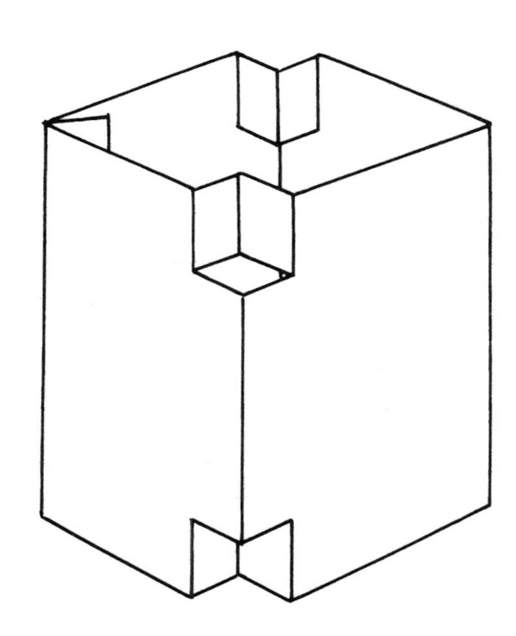

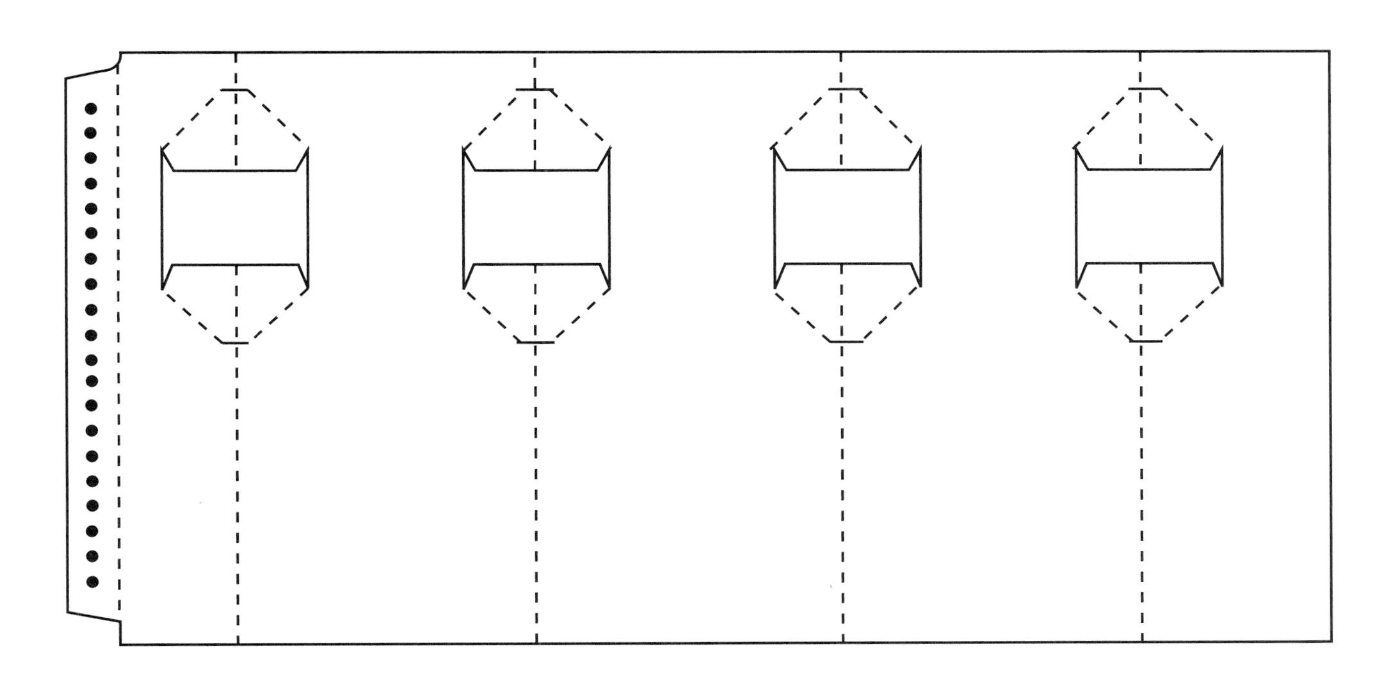

Typical contruction for a fragile product (e.g., light-bulbs). *(PAT. #3,082,931)*

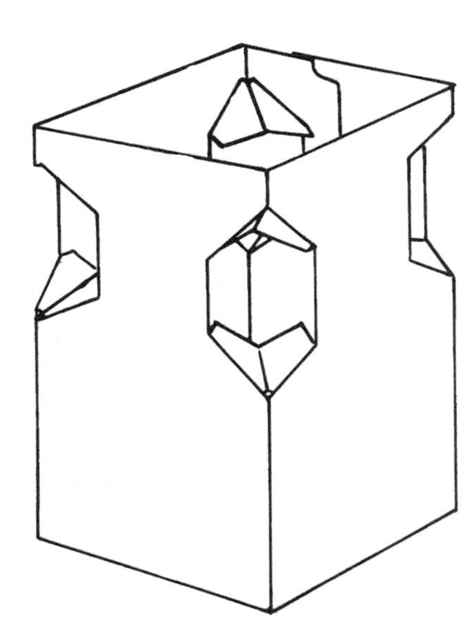

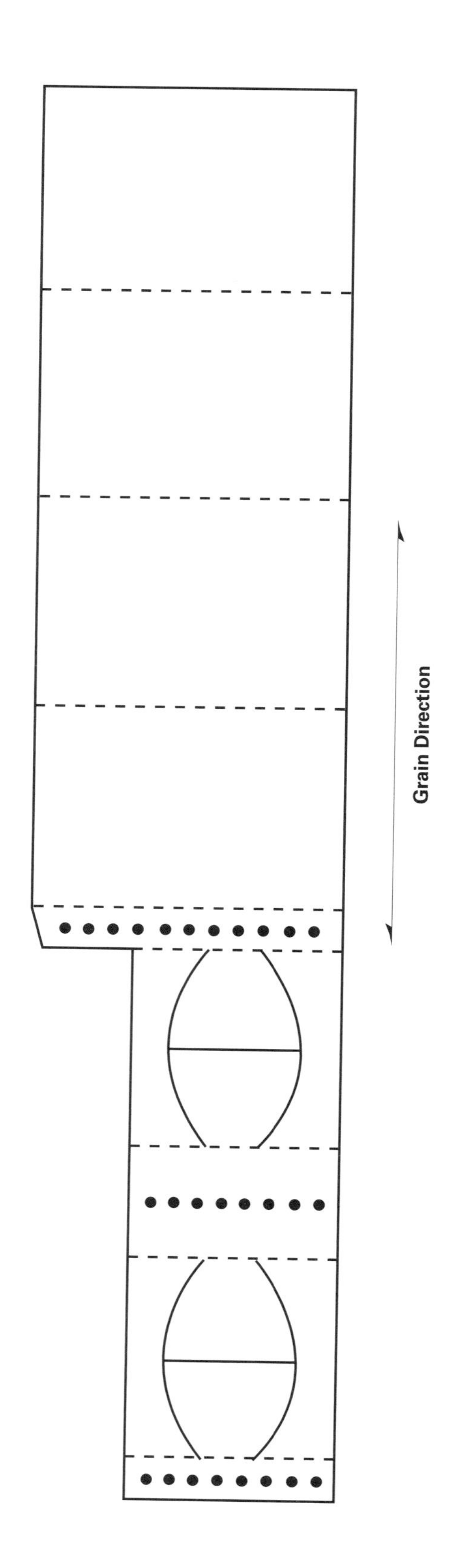

The tension of the inner panel die-cuts help keep the product in place.

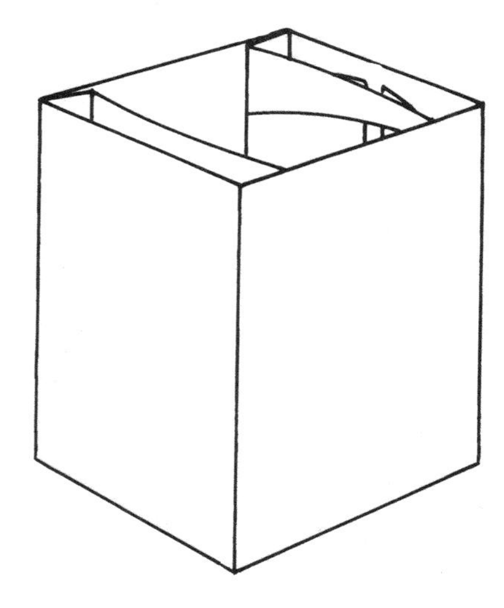

FIVE PANEL SINGLE LIGHTBULB SLEEVE

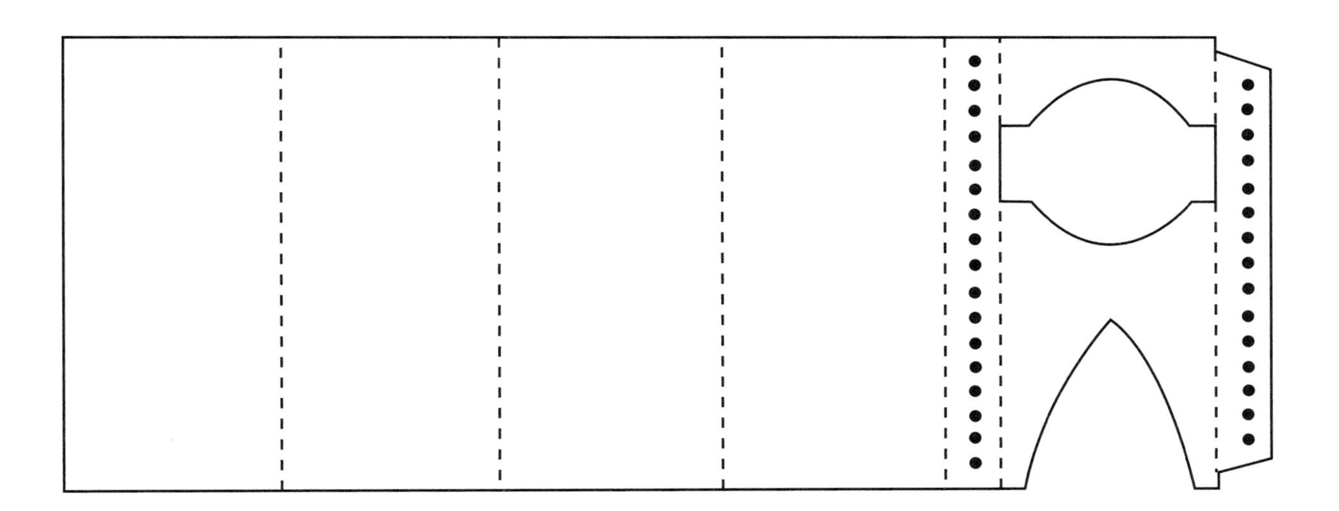

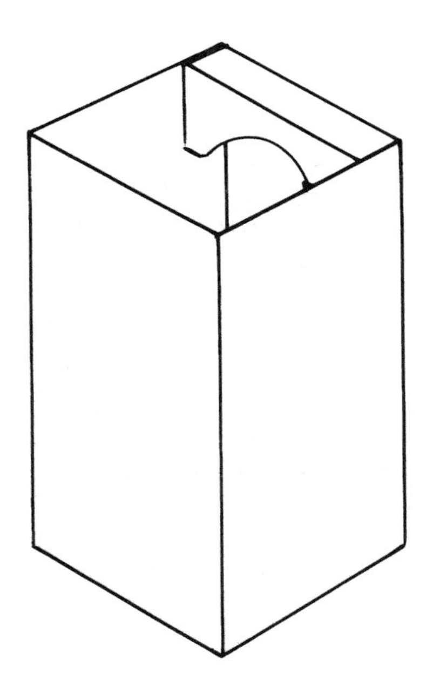

LIGHTBULB SLEEVE WITH SCORED INFOLDS AND PARTIAL COVER

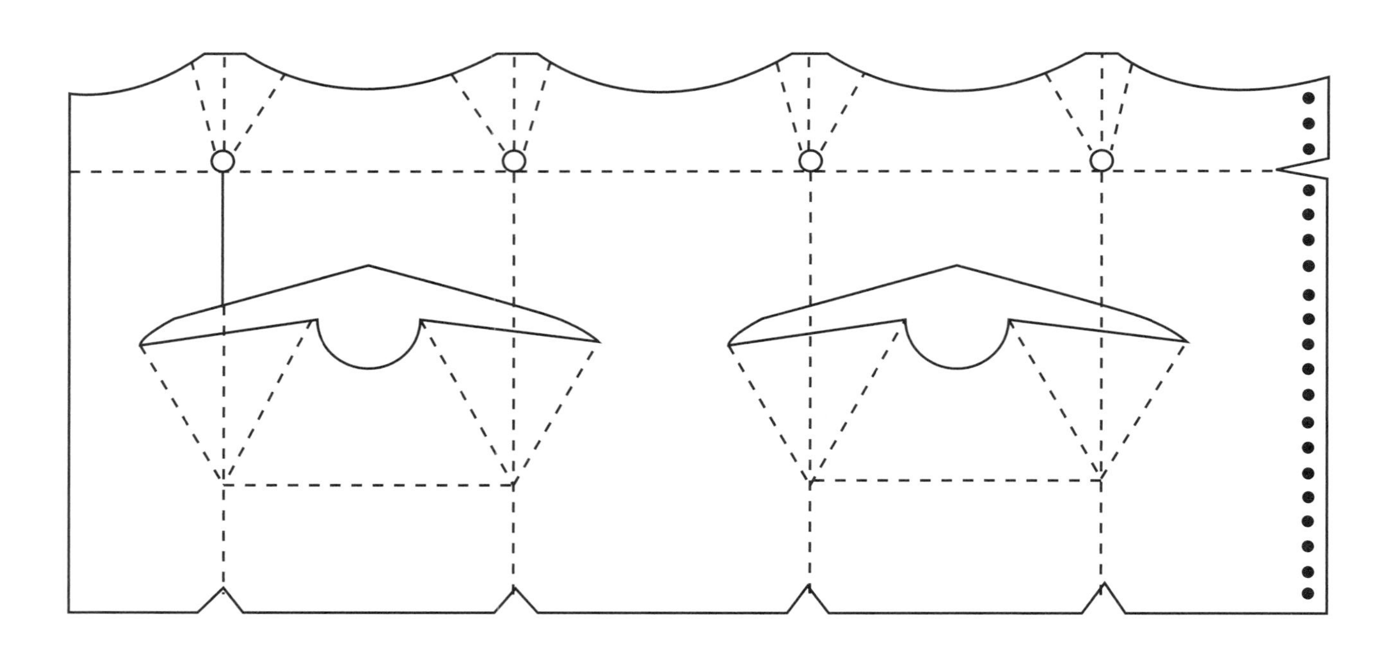

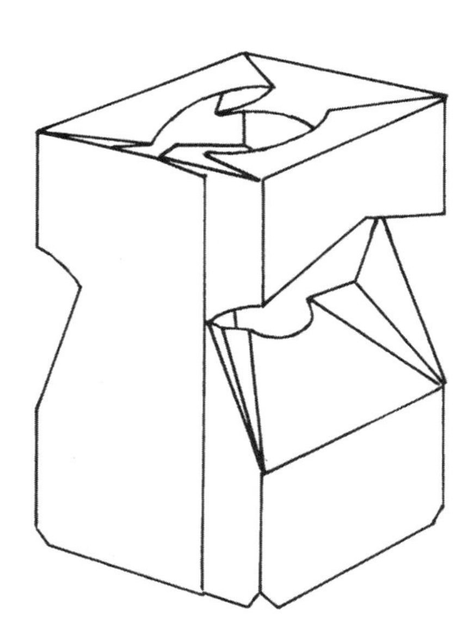

Both the cover and the infolds may be constructed to fit any product.

ALTERNATE CONSTRUCTION FOR LIGHTBULB SLEEVE

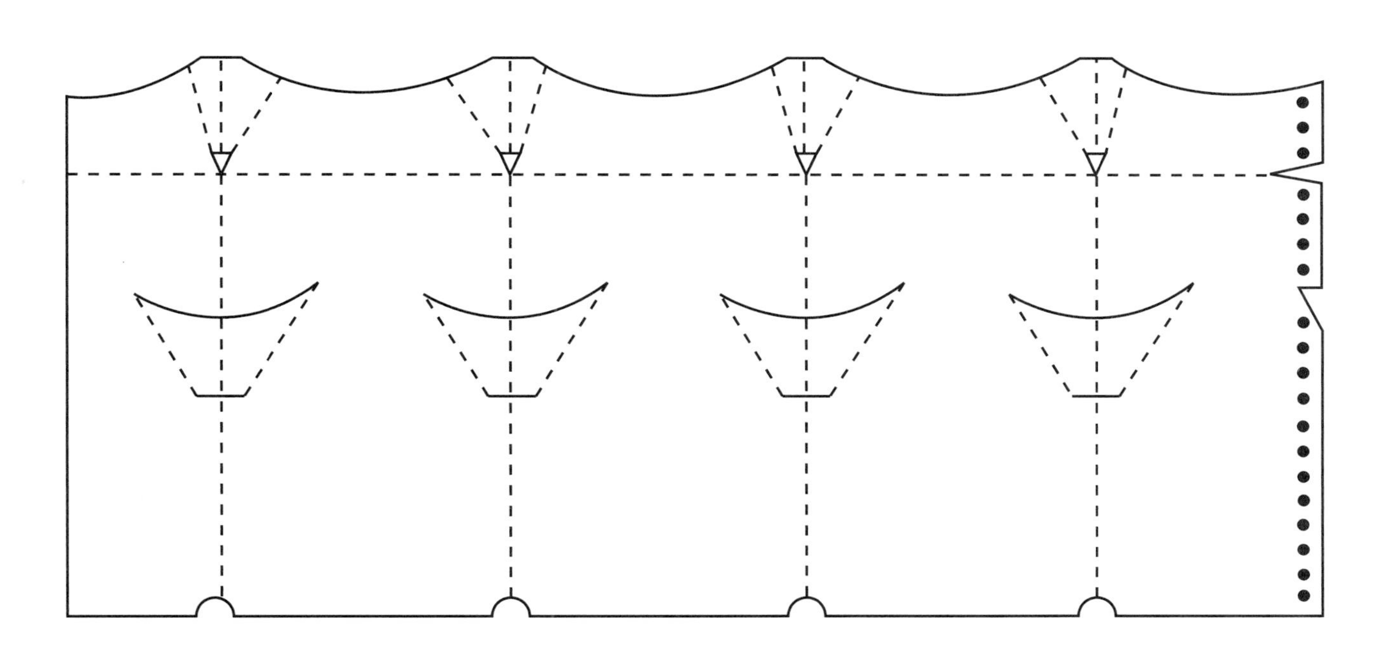

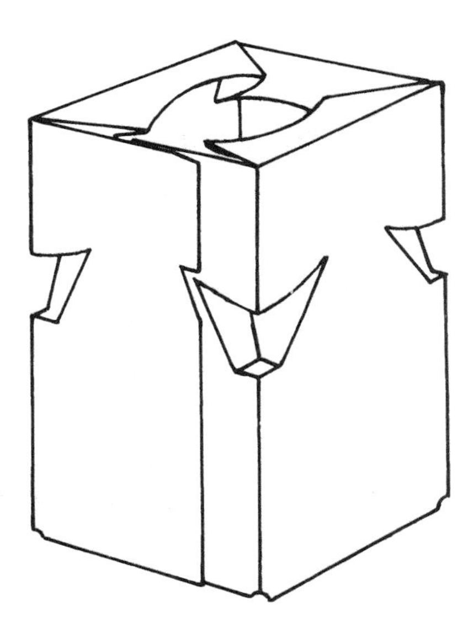

LIGHT BULB SLEEVE VARIATION

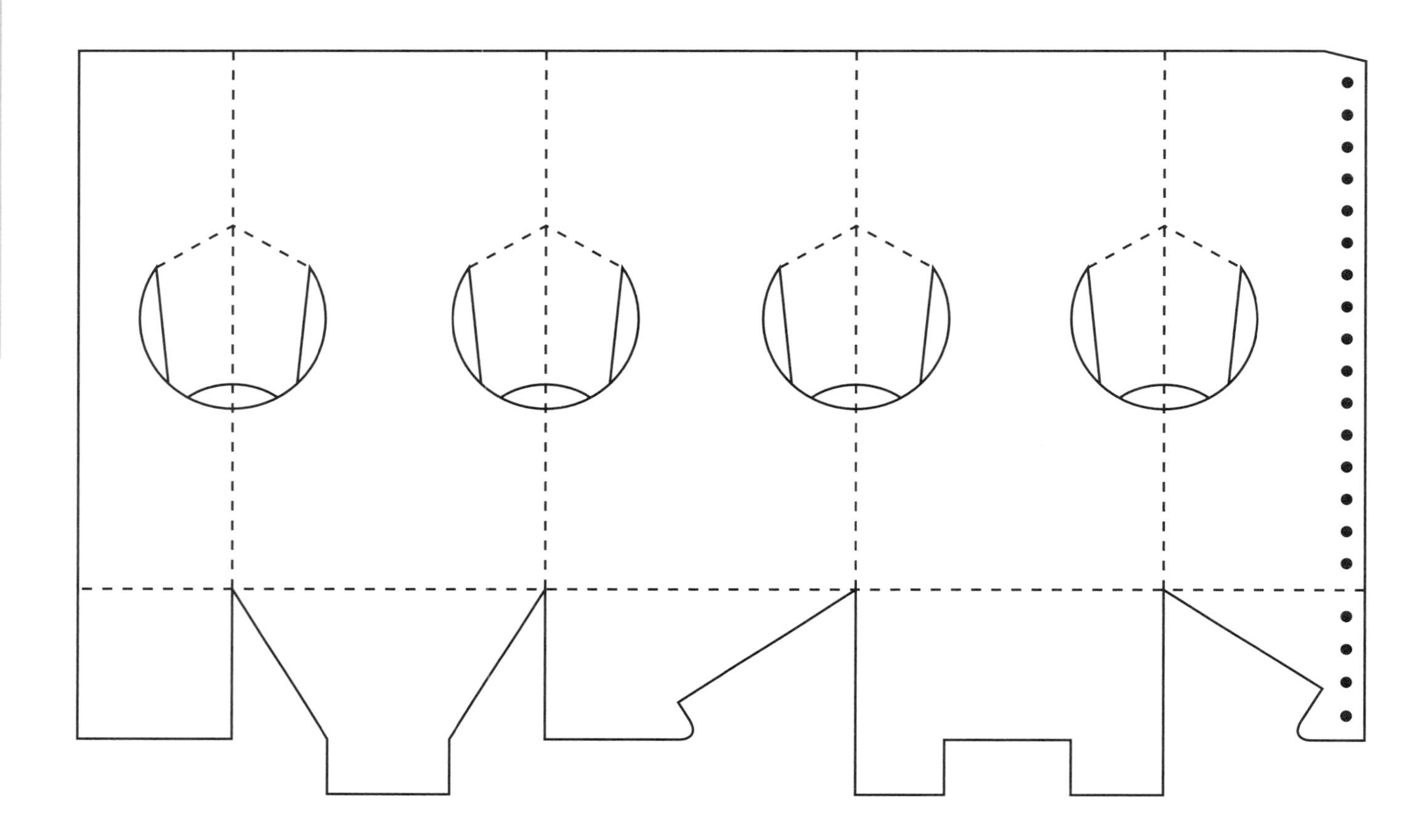

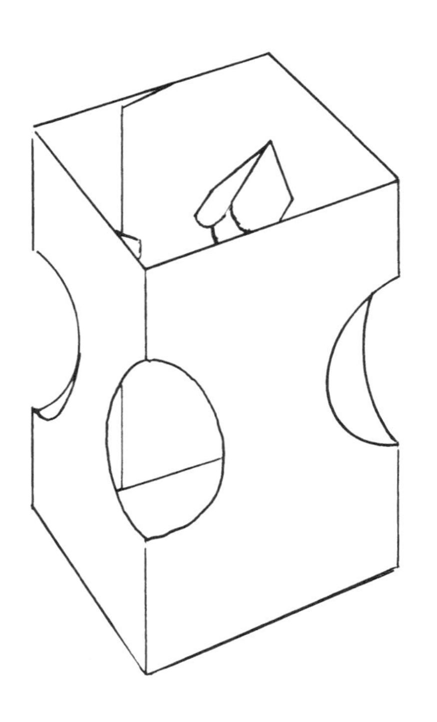

DECORATIVE LIGHT BULB SLEEVE

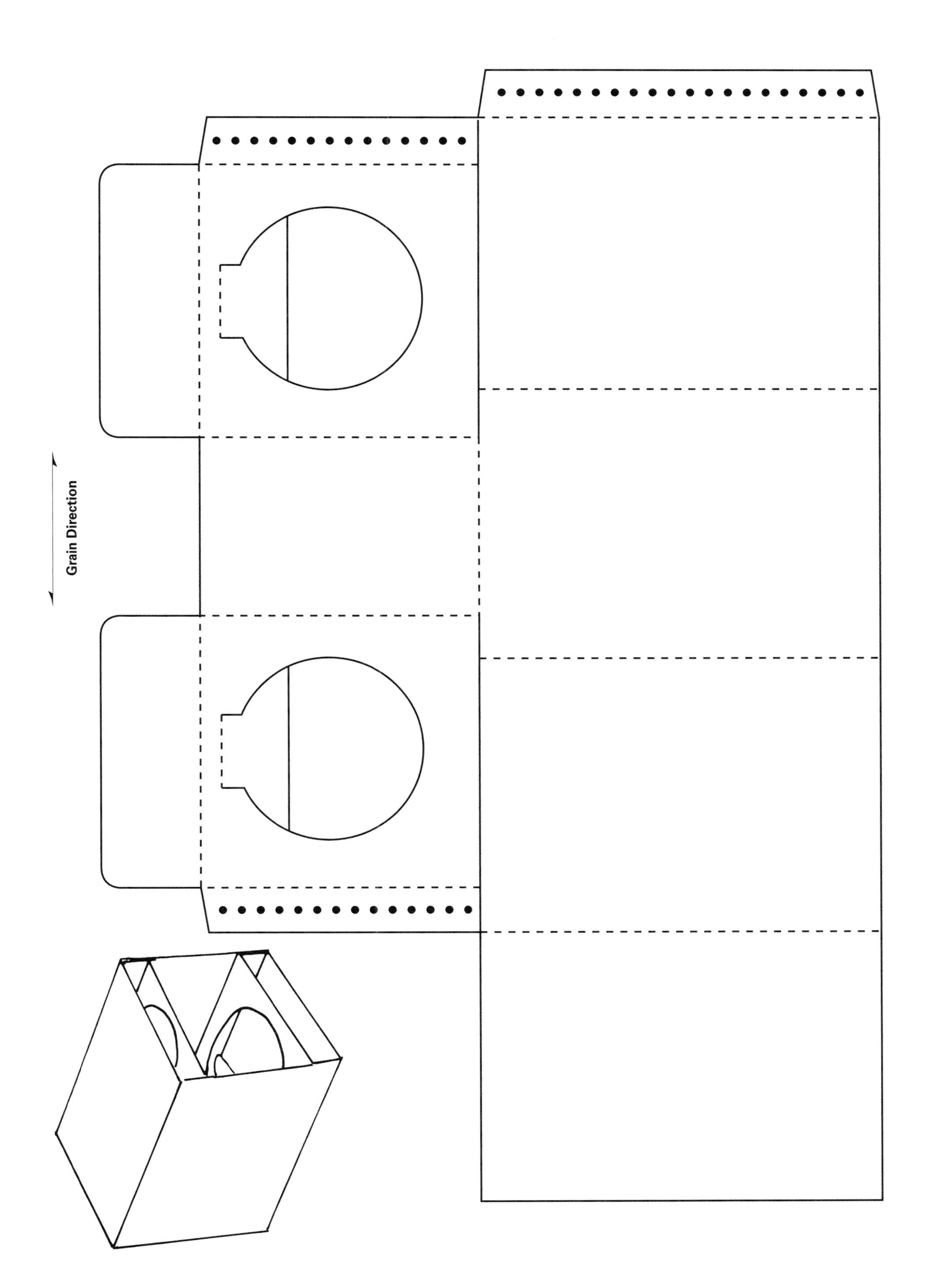

FOUR LIGHT BULB WRAP

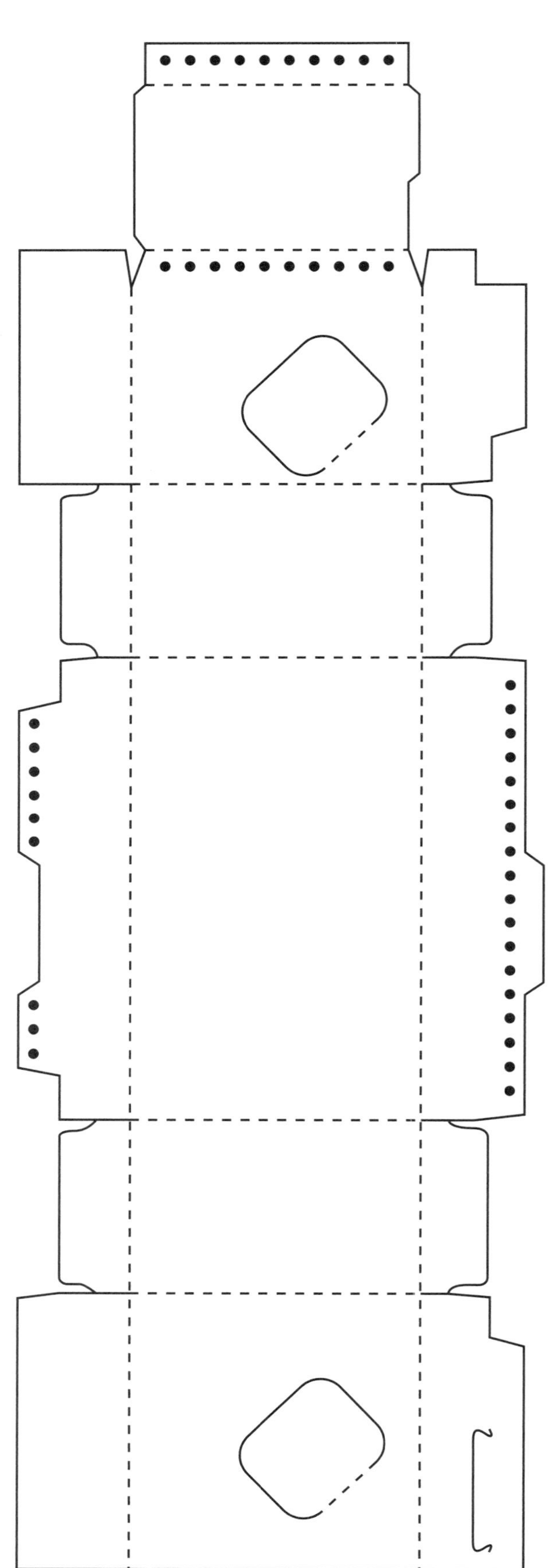

The die cut windows folded into the package act as separators.

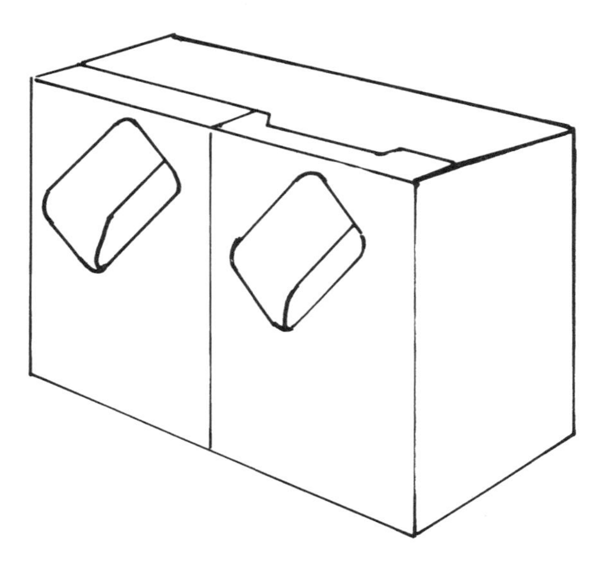

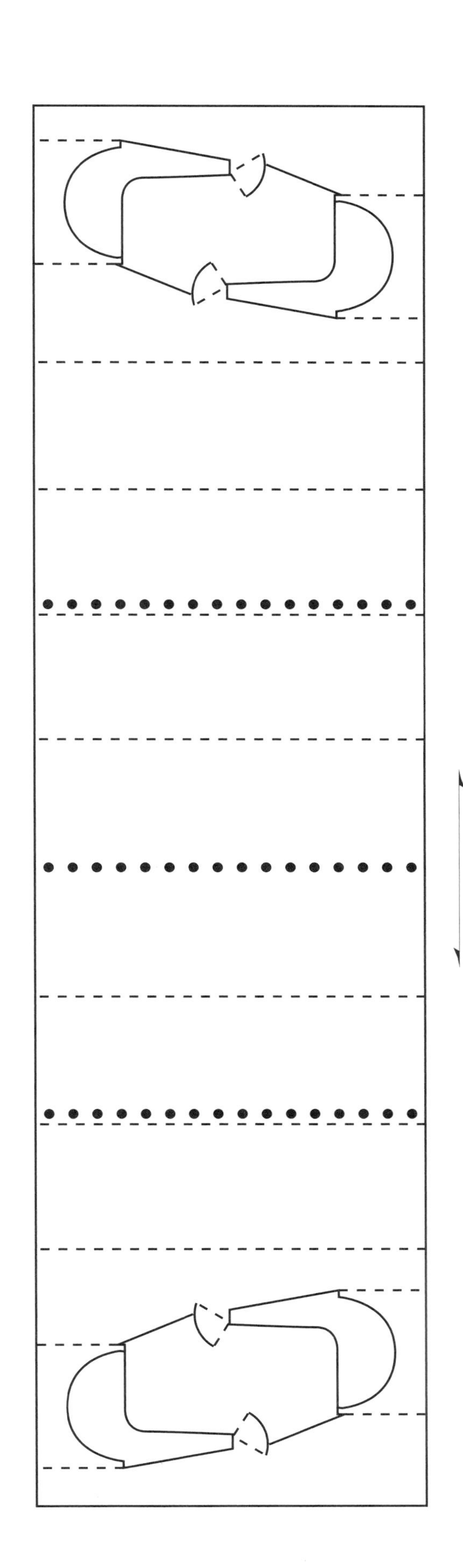

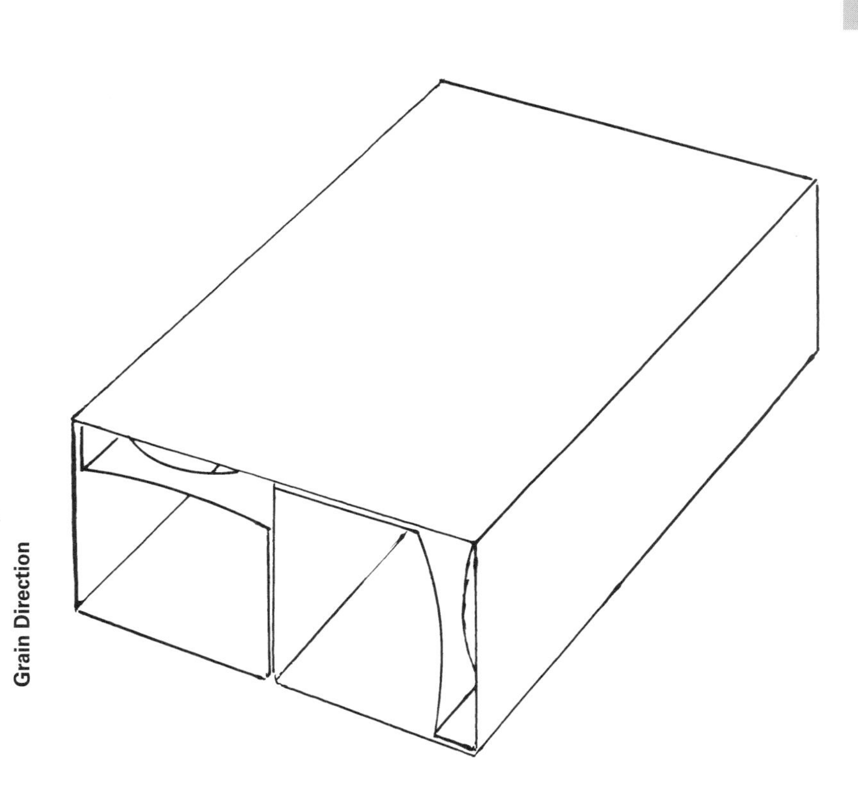

The swiveling panels at each end panel help to lock as well as protect the fragile glass bulbs.

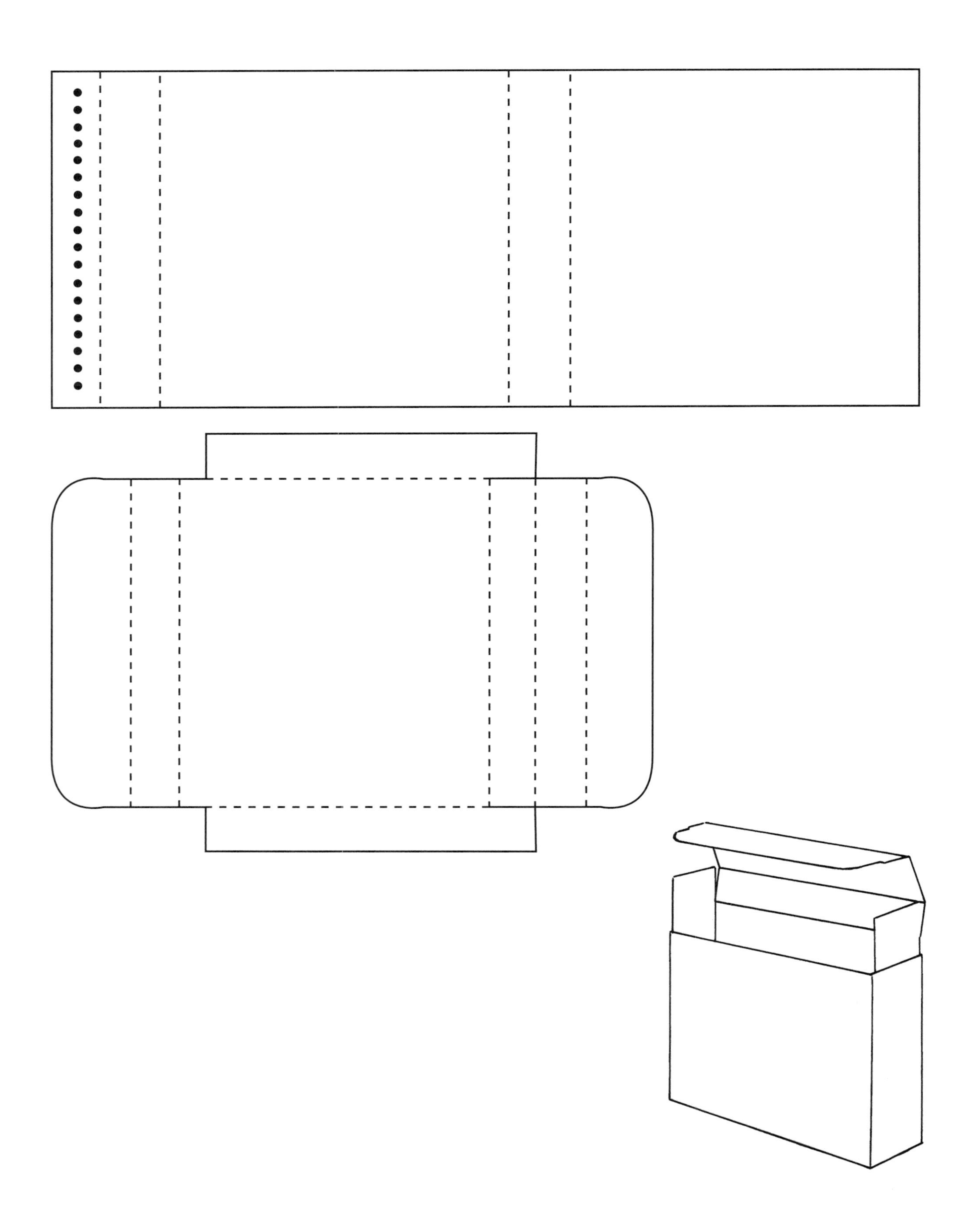

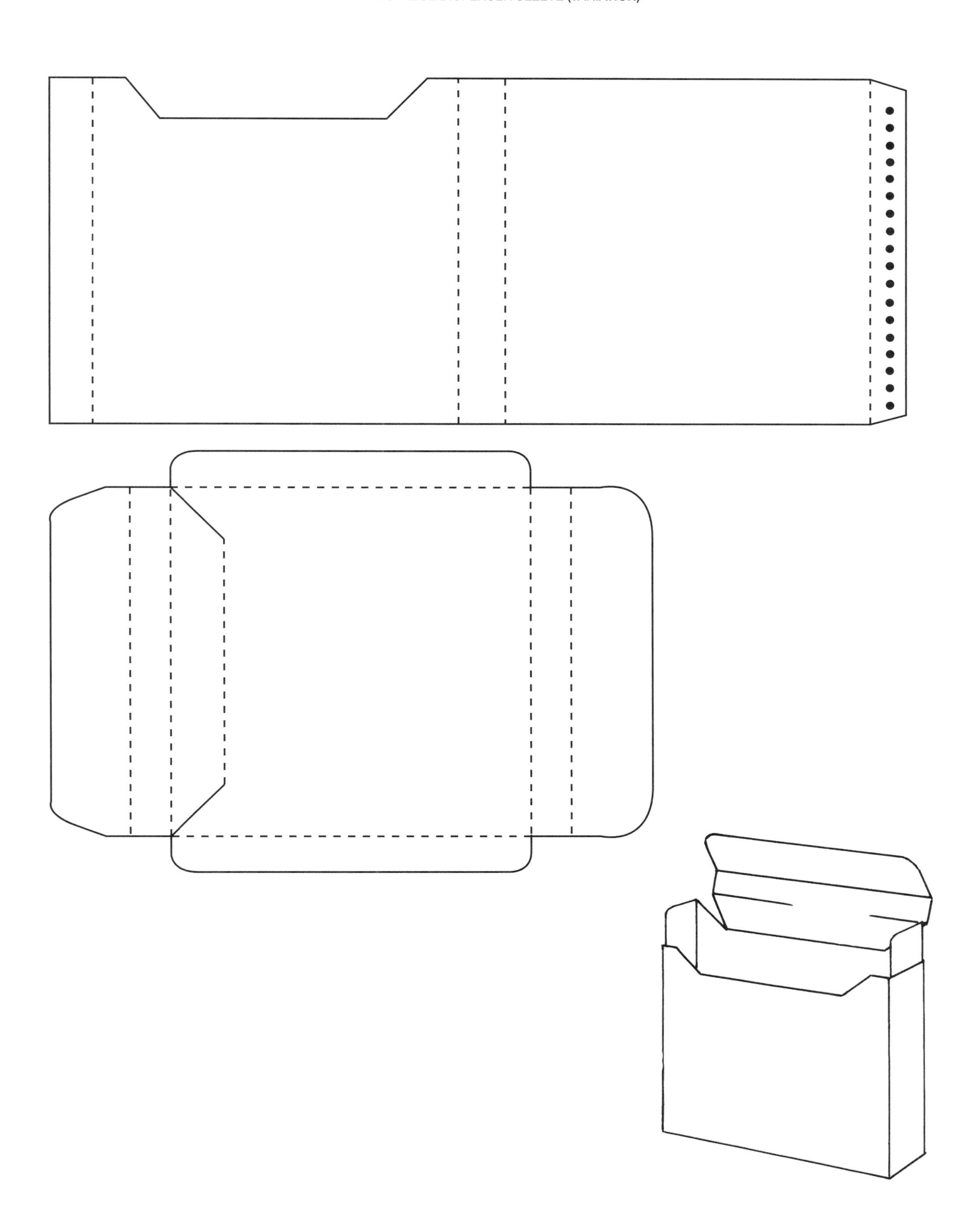

TYPICAL CIGARETTE SLEEVE

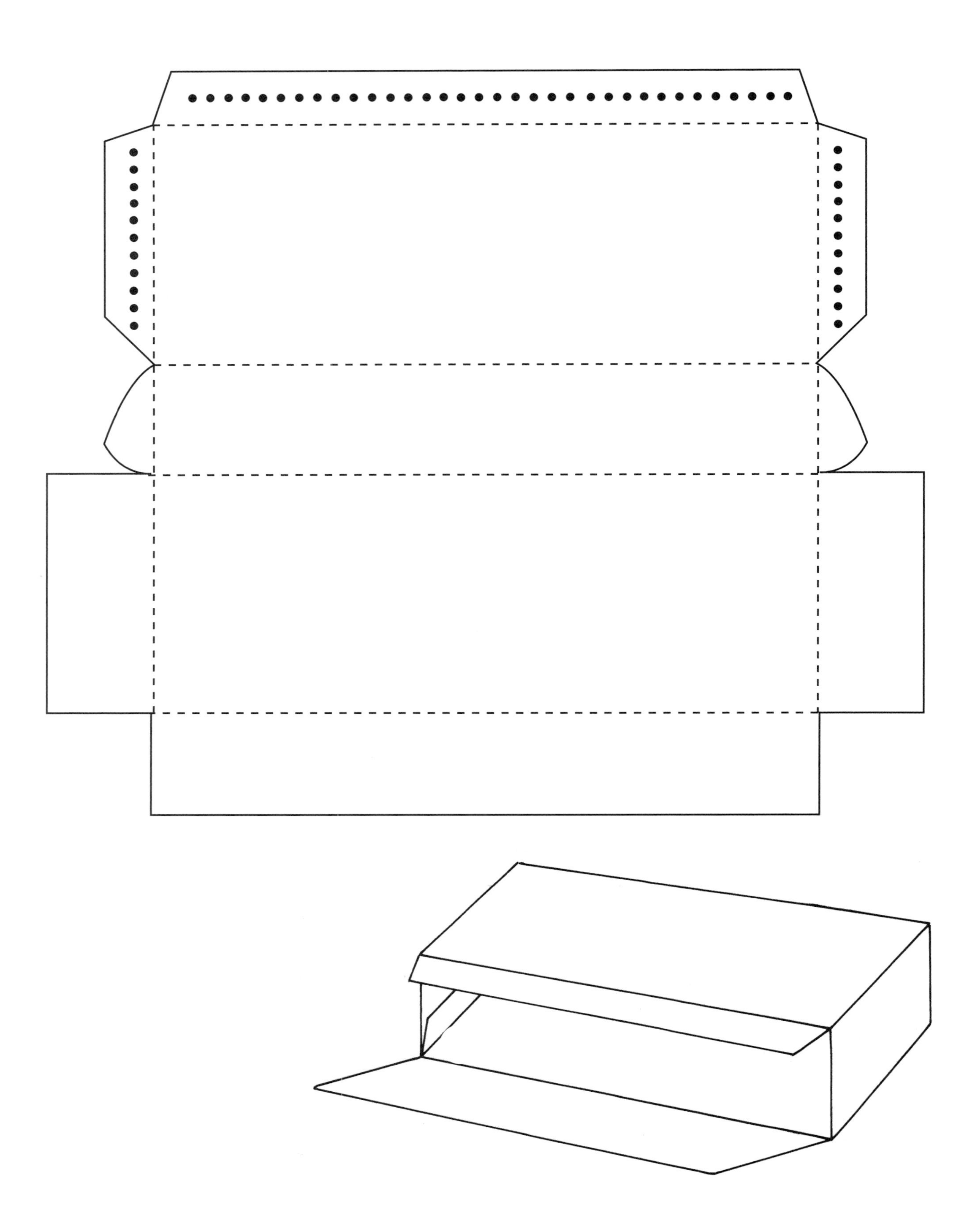

MANSARD ROOF SHAPED WTIH SLIDE-OUT TRAY

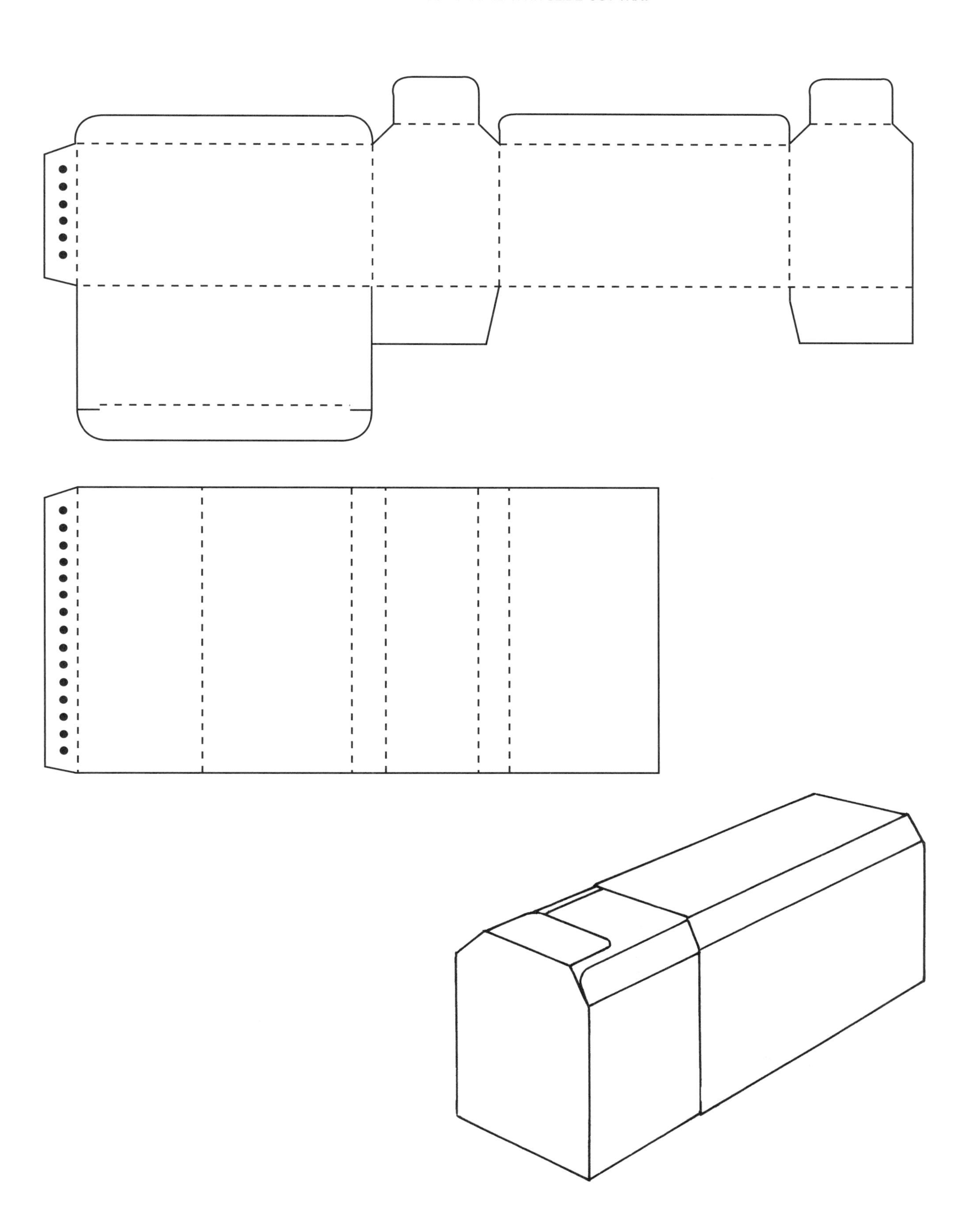

TWO PIECE SLIDE DISPENSER

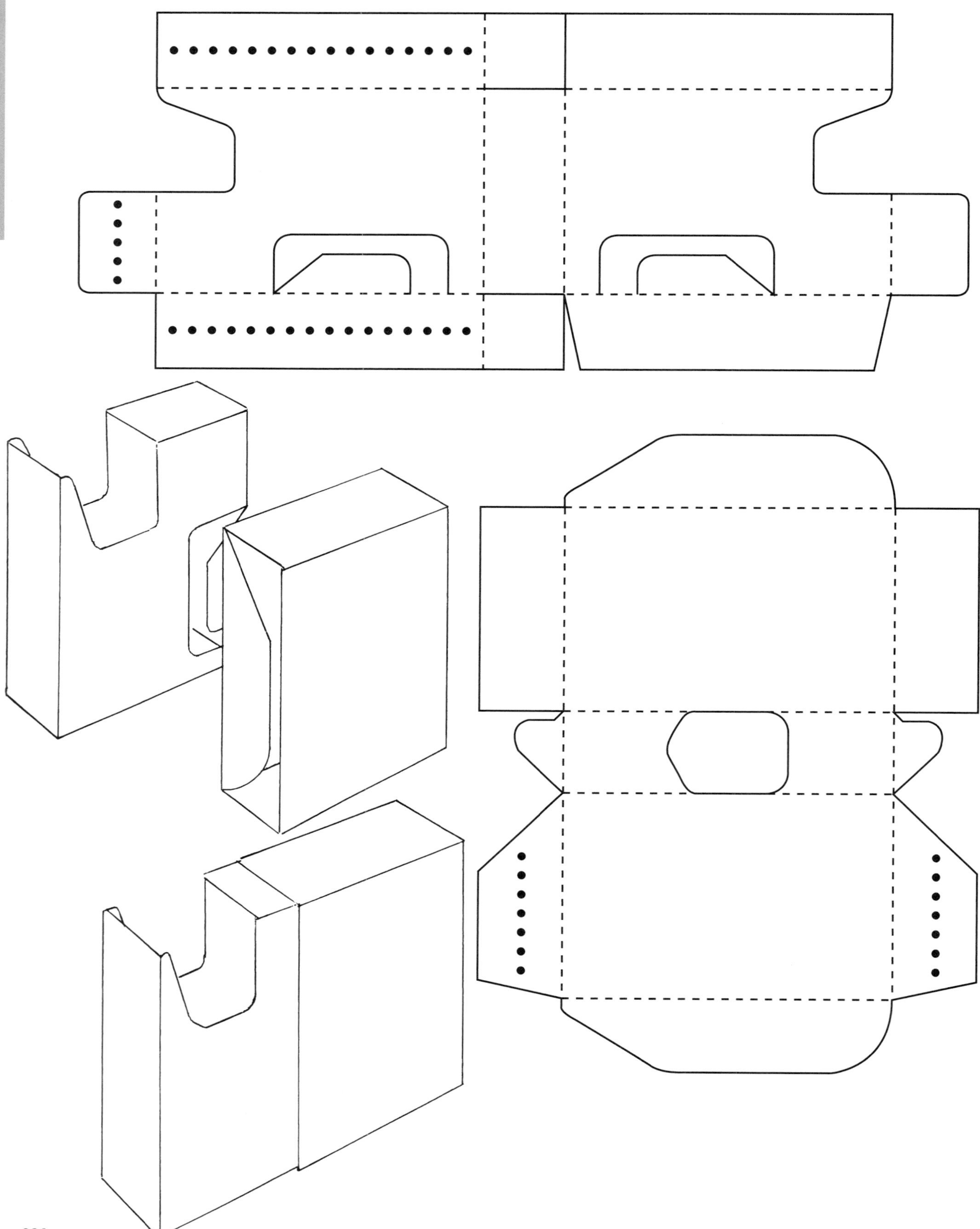

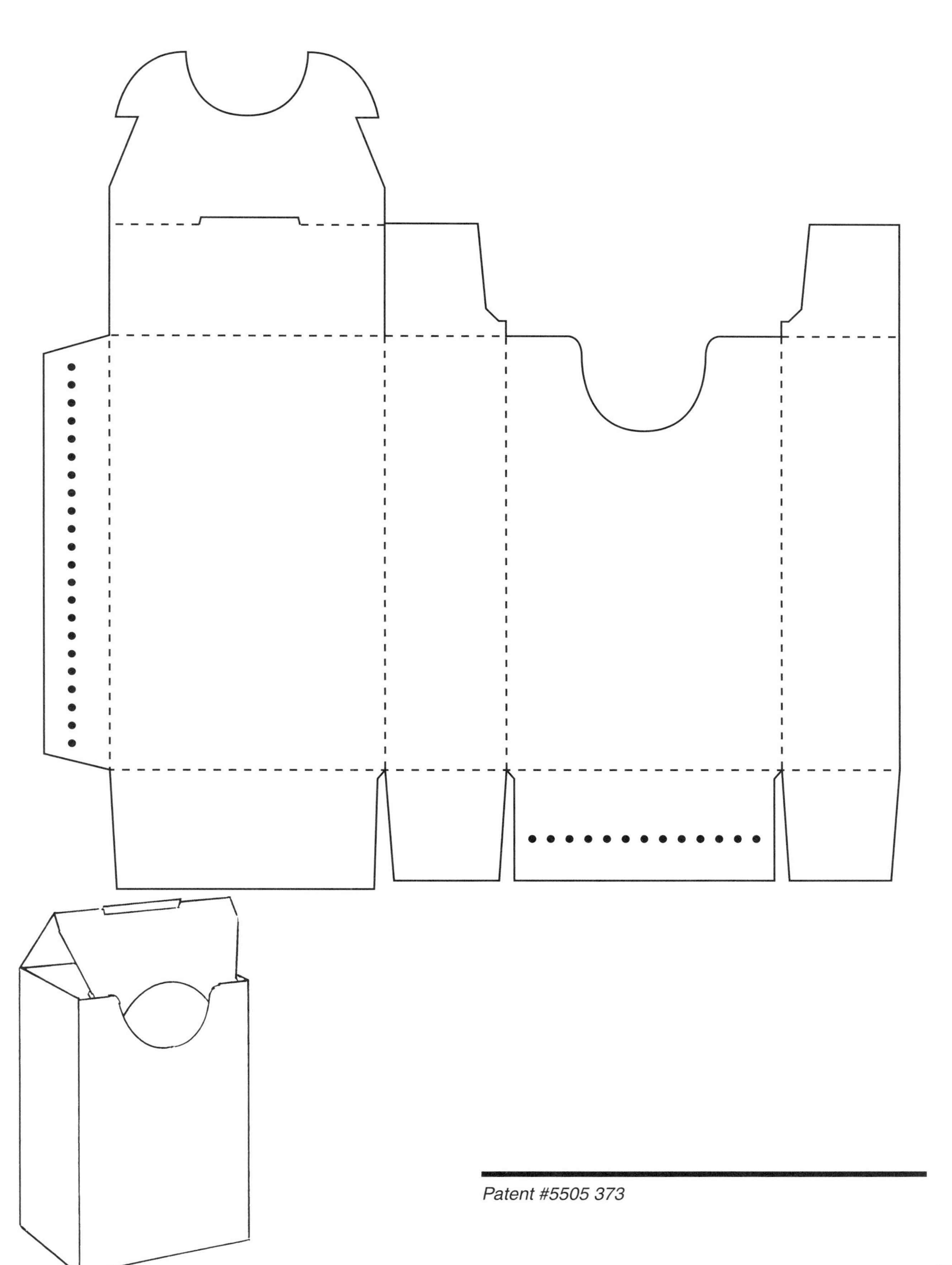

Patent #5505 373

CONTAINER SLEEVE FOR FRENCH-FRIED POTATOES

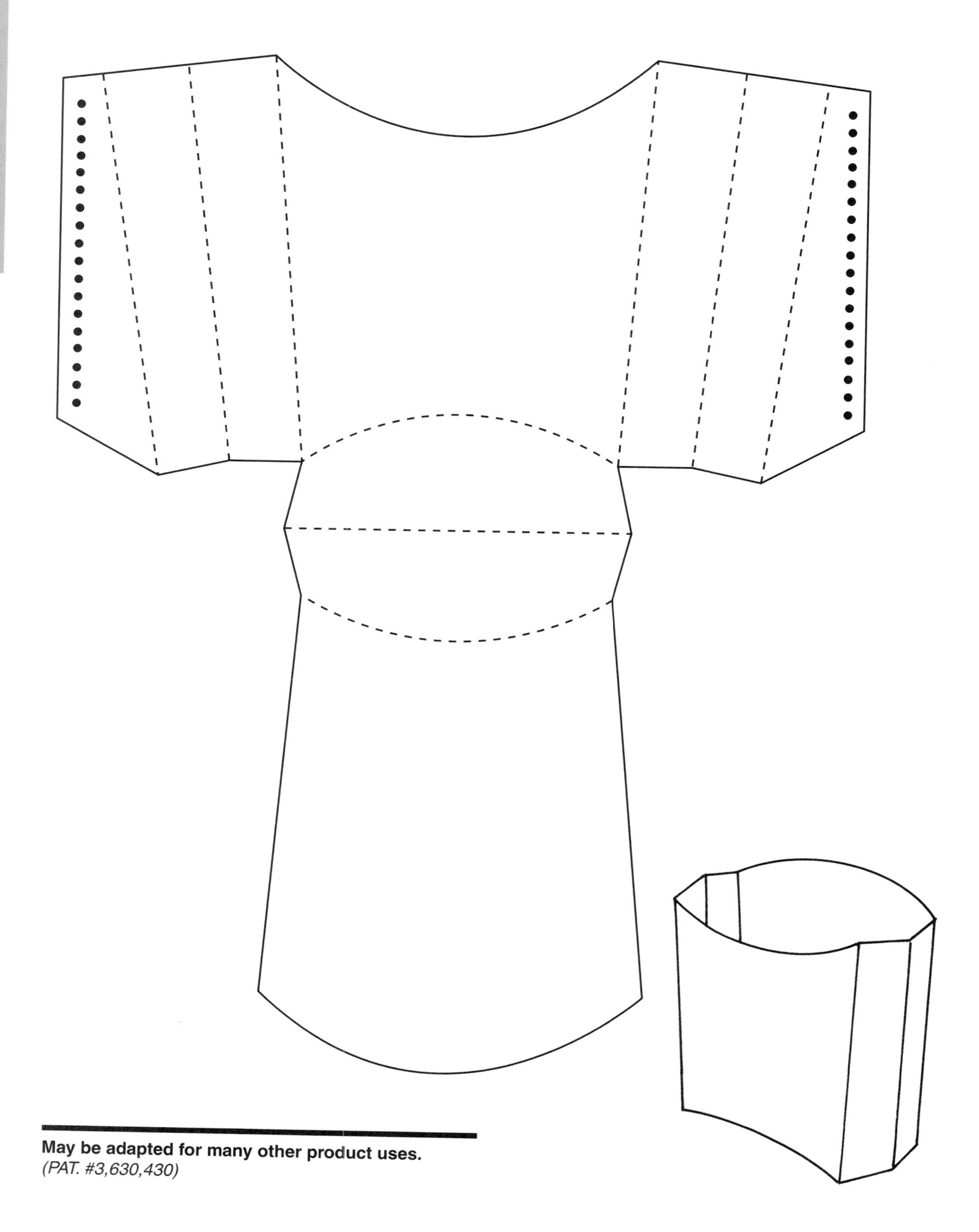

May be adapted for many other product uses.
(PAT. #3,630,430)

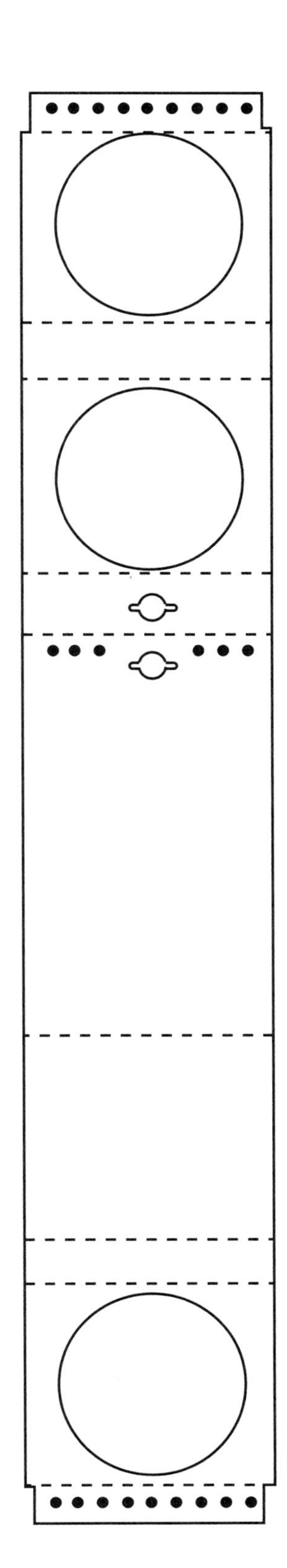

This one-piece construction lends itself to a large variety of lightweight product packaging with maximum product visibility.

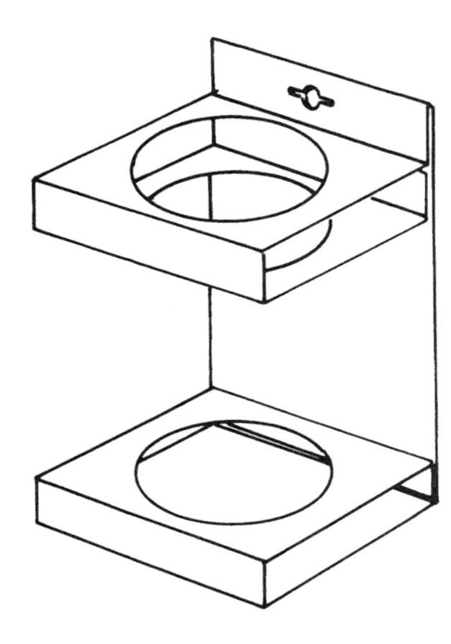

HARDWARE BOX

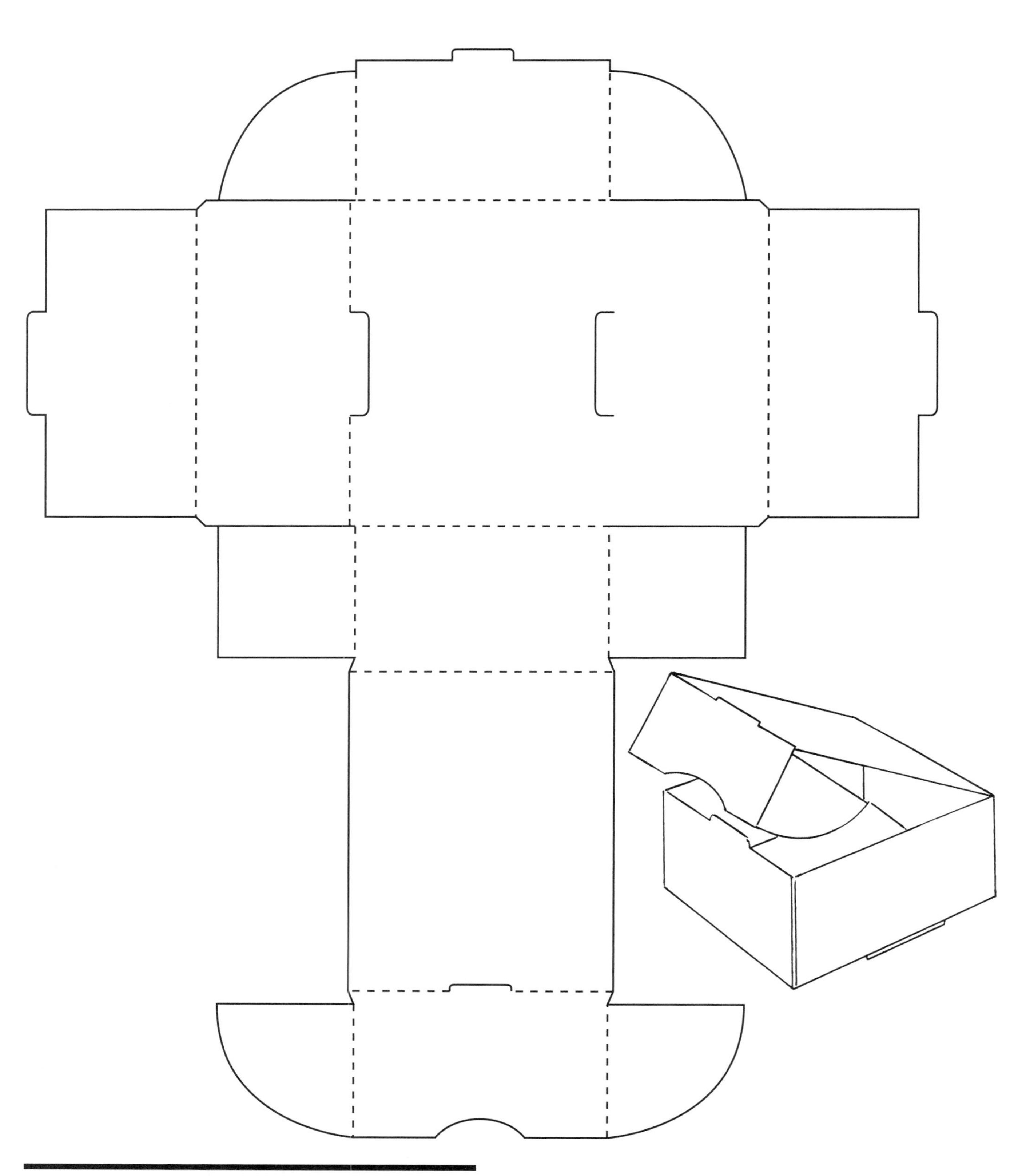

This double side wall/double end wall box is ideal for heavy hardware parts.

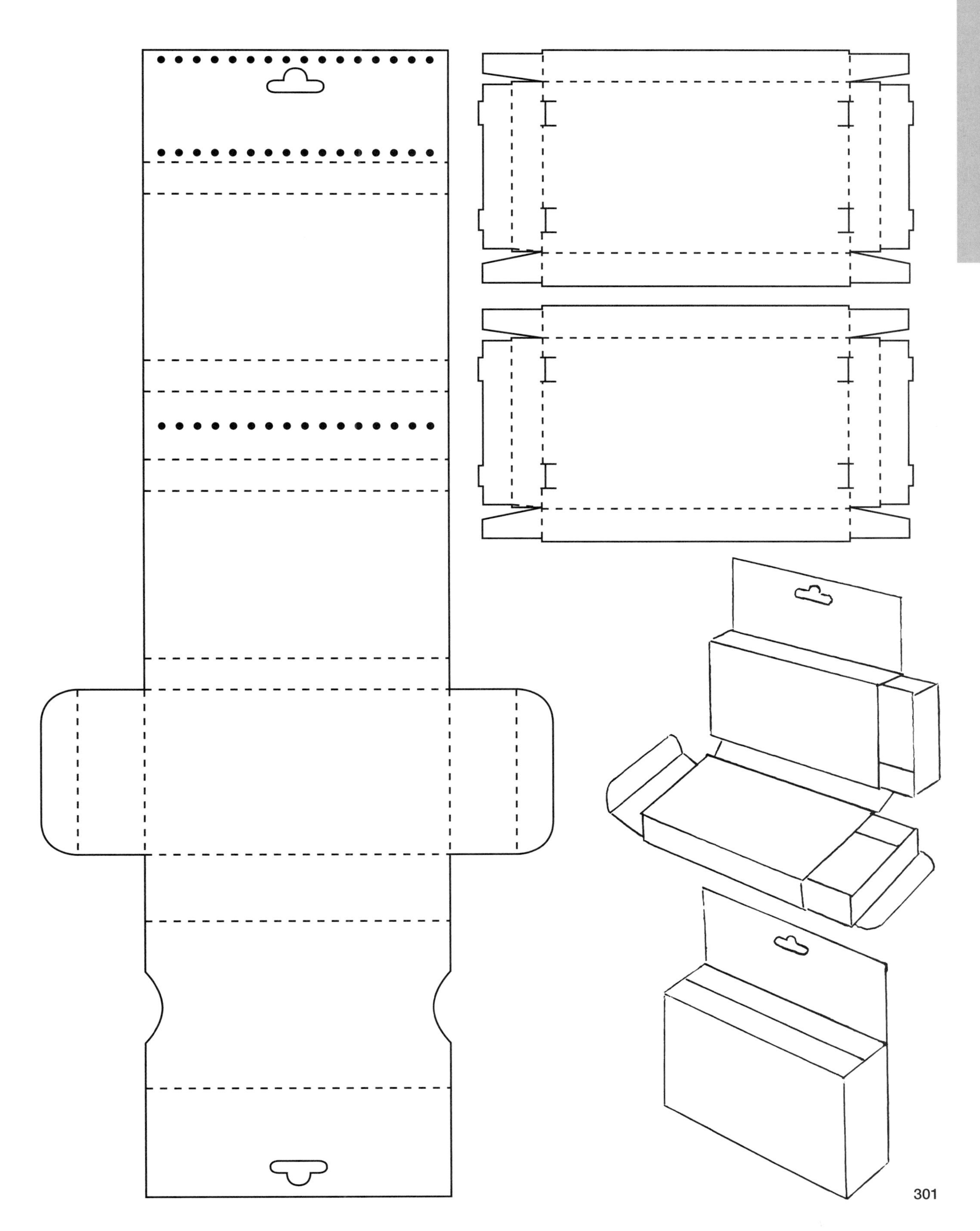

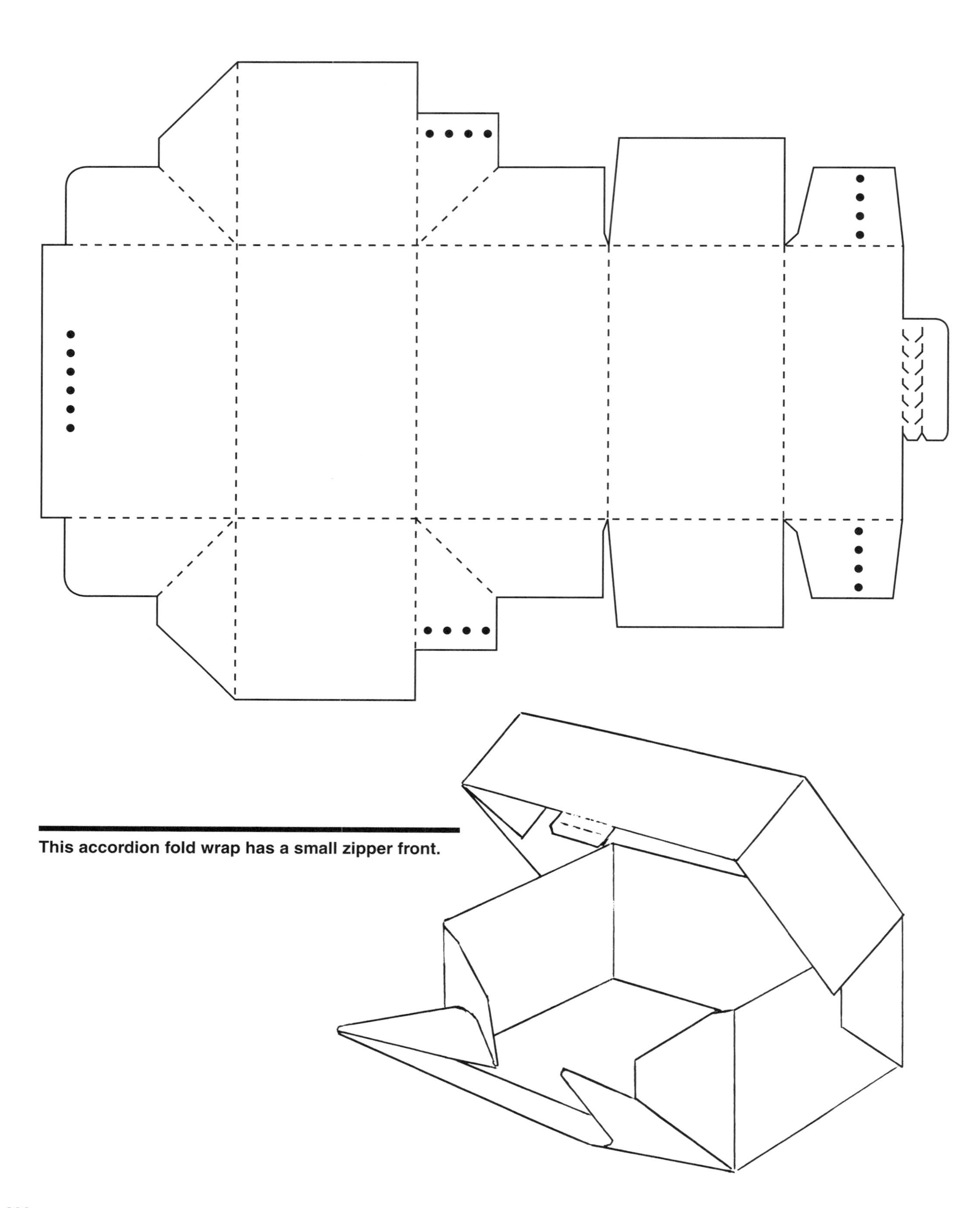

This accordion fold wrap has a small zipper front.

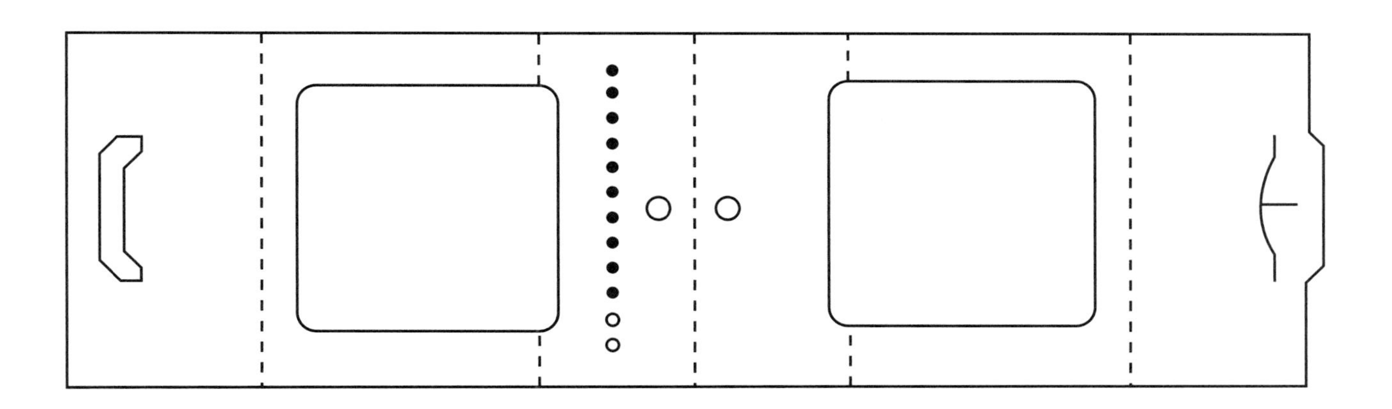

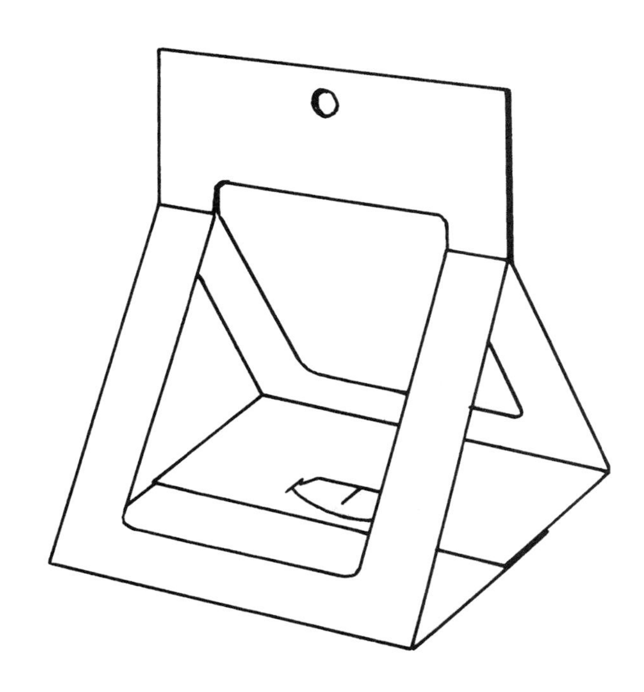

BOTTOM GLUED SINGLE TUB SLEEVE

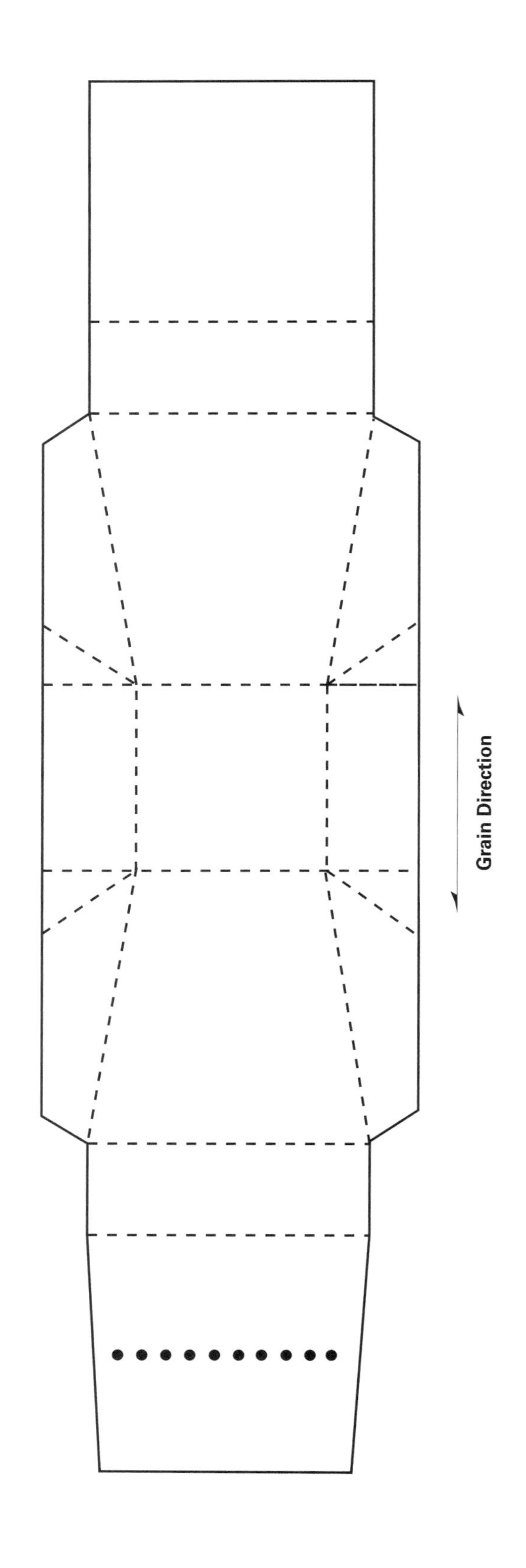

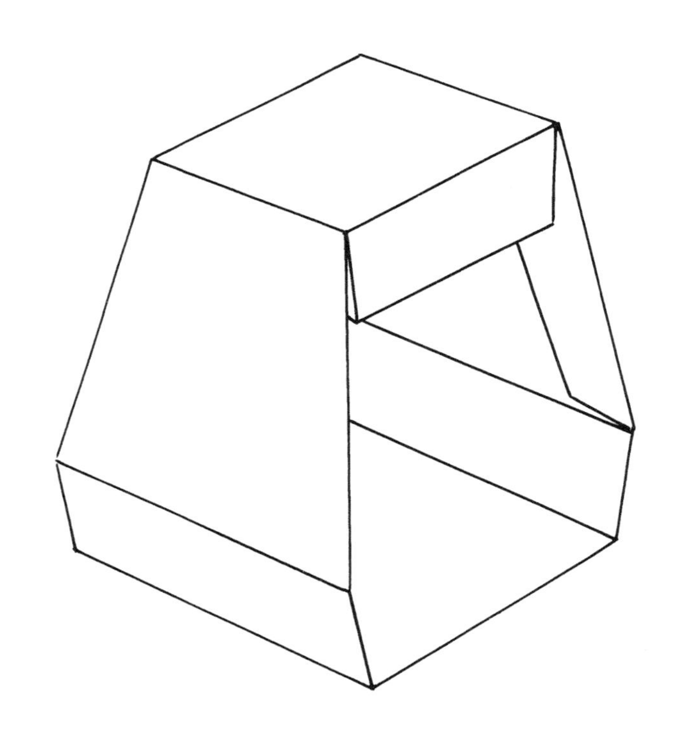

OPEN SINGLE TUB SLEEVE

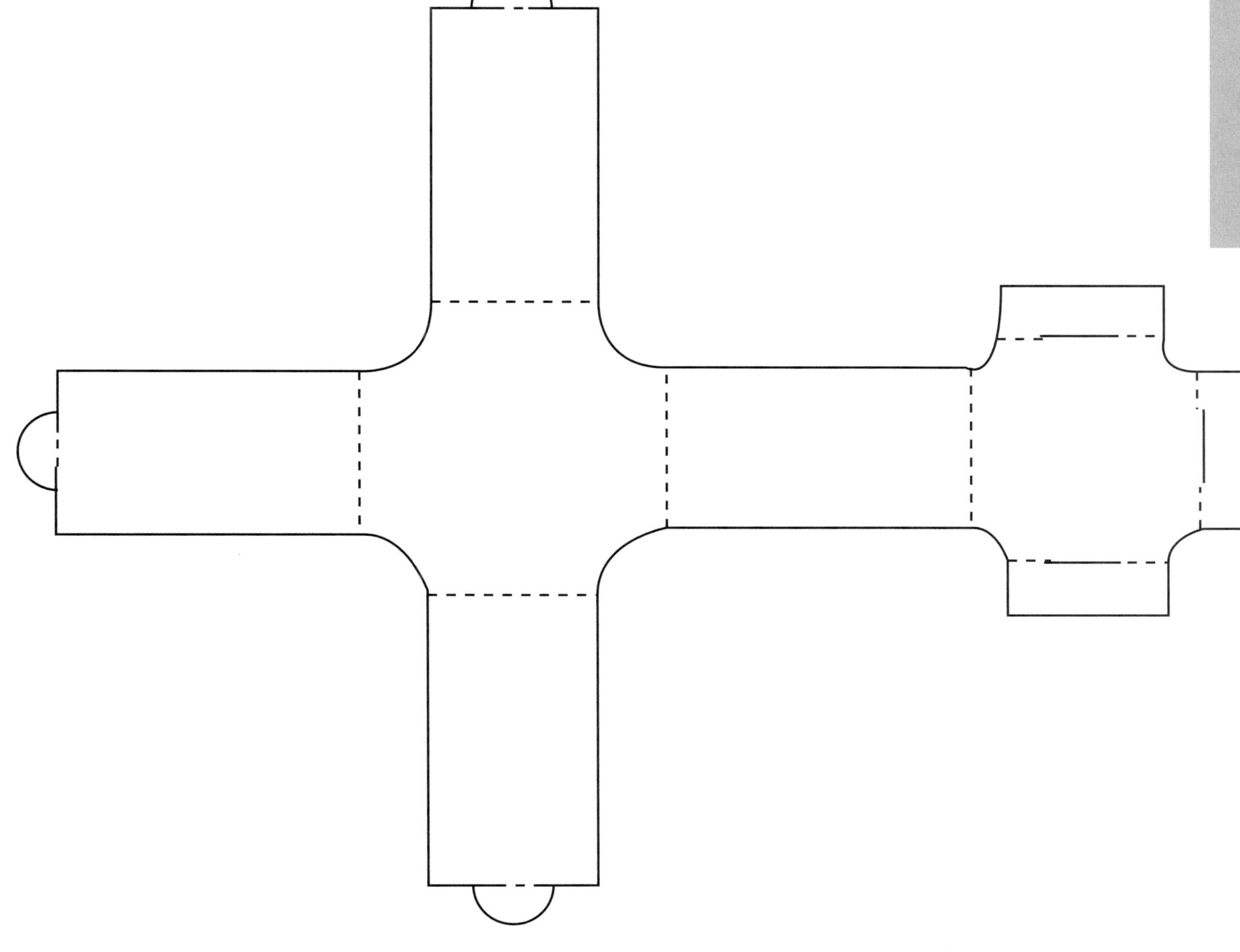

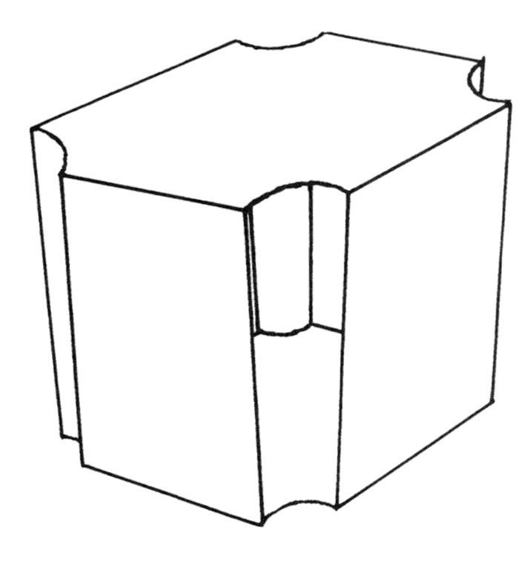

BOTTOM GLUED SLEEVE FOR SINGLE CUP OR TUB

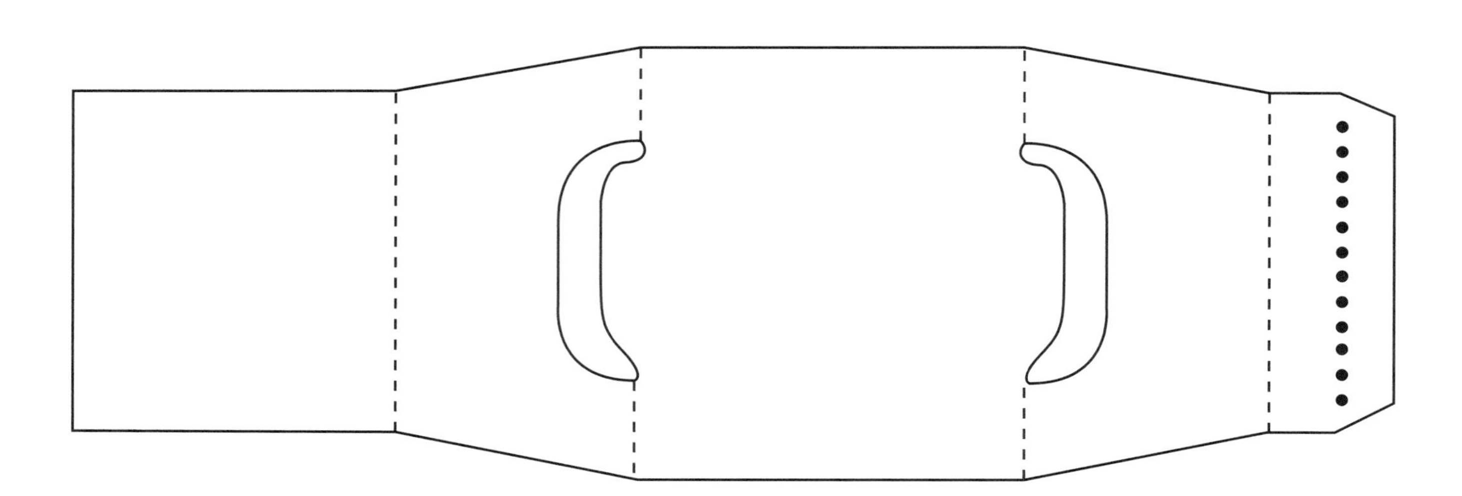

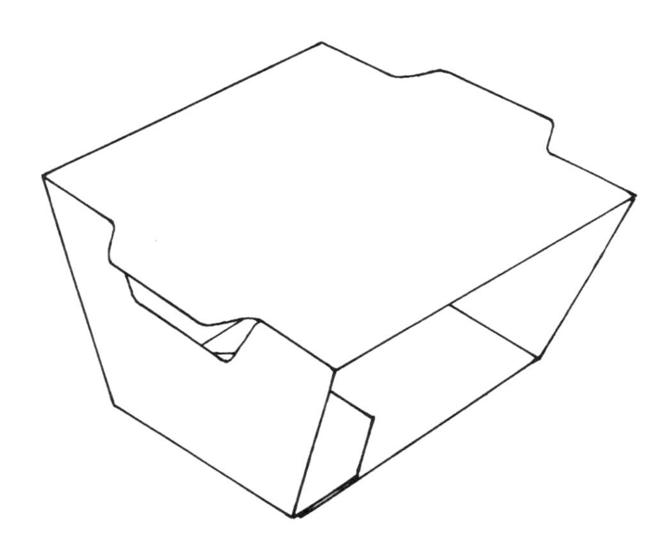

EDGE LOCK SLEEVE FOR SINGLE CUP OR TUB

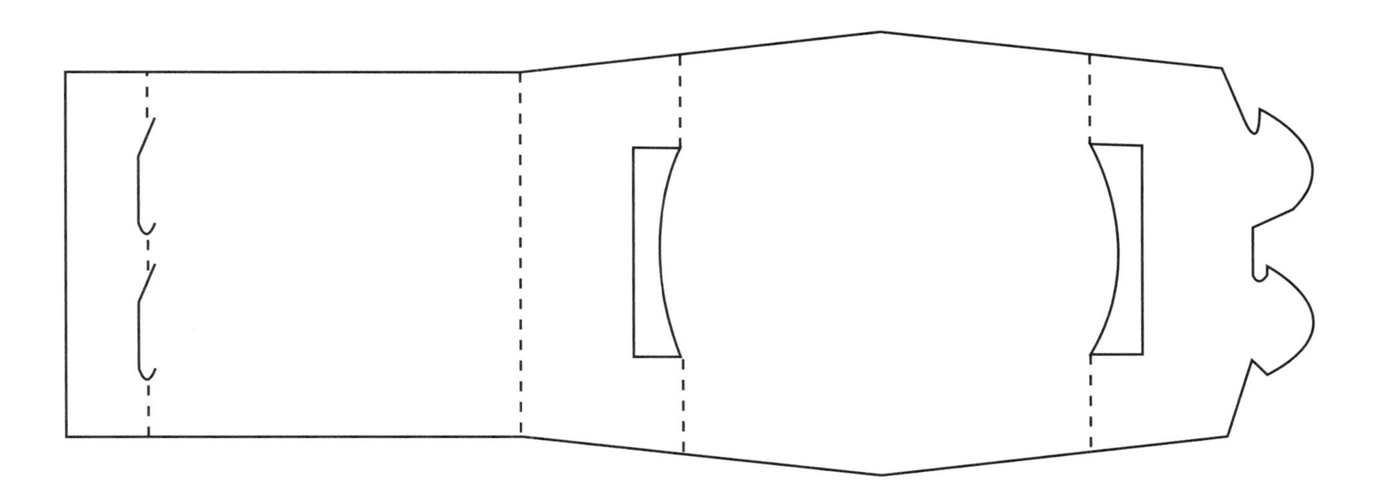

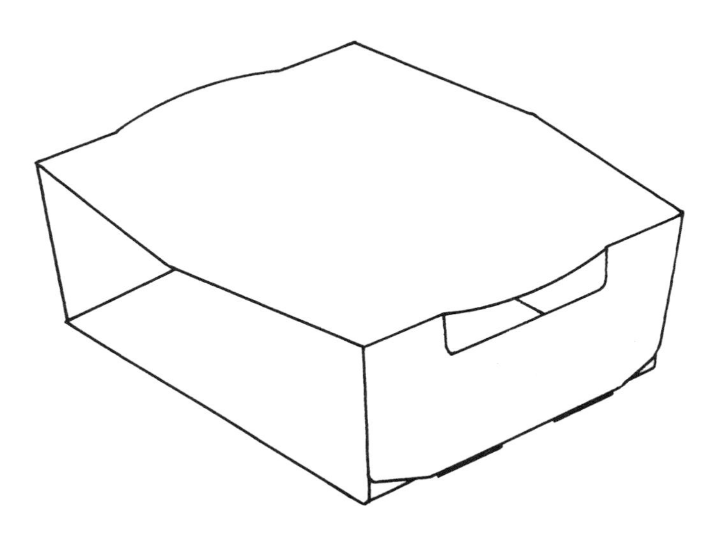

BOTTOM GLUED SLEEVE FOR TWO TUBS

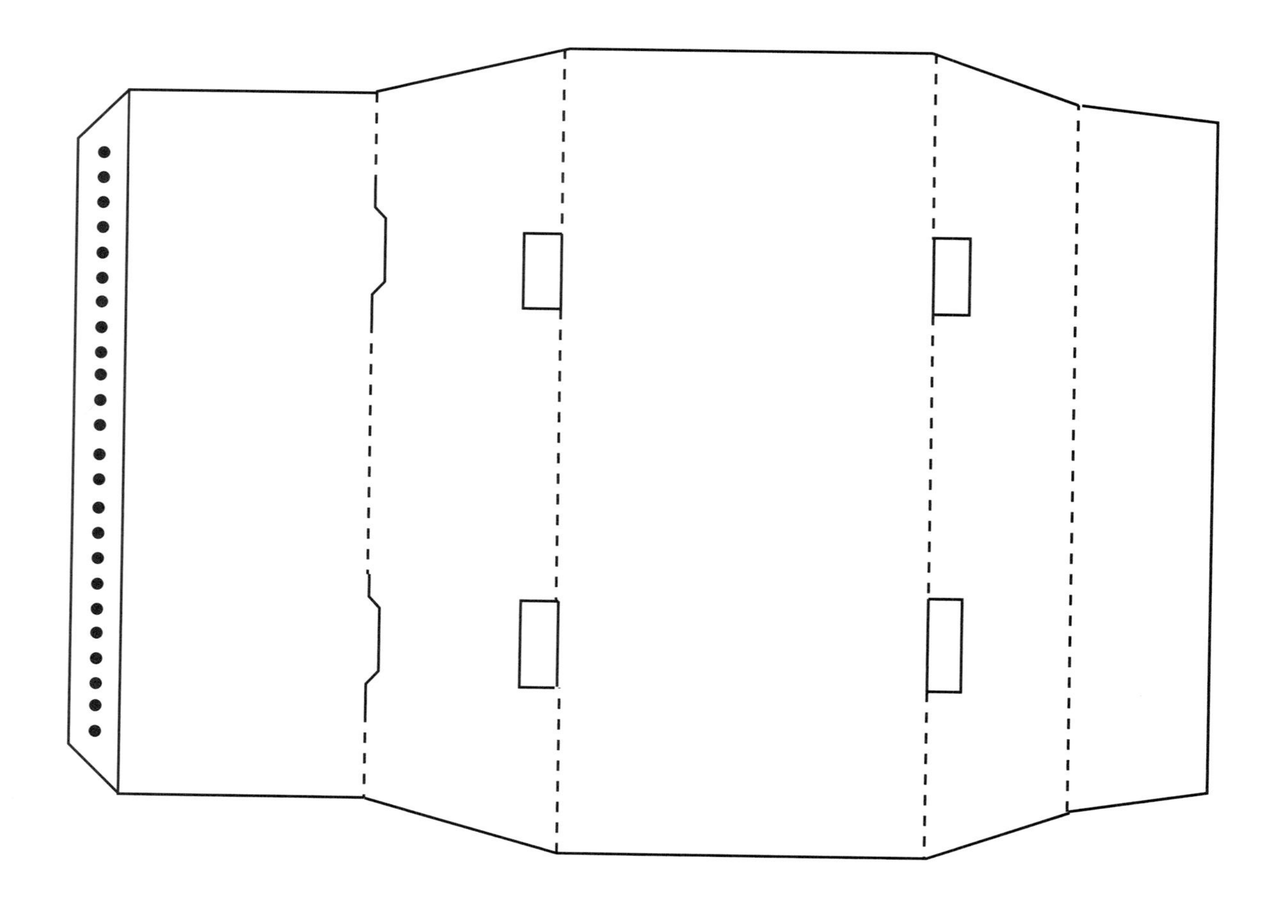

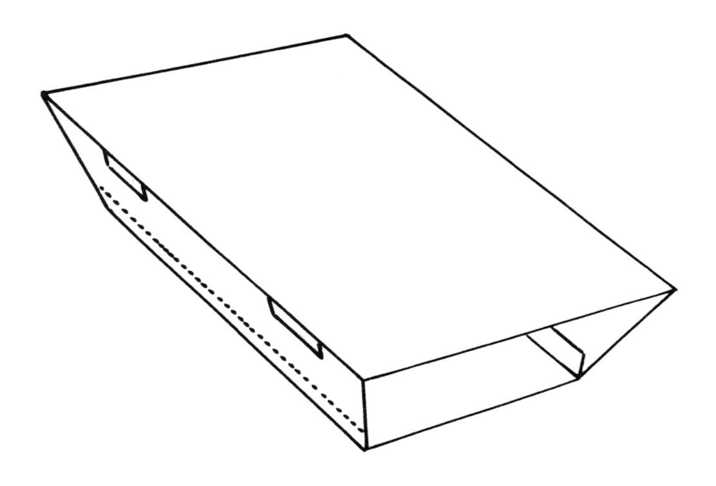

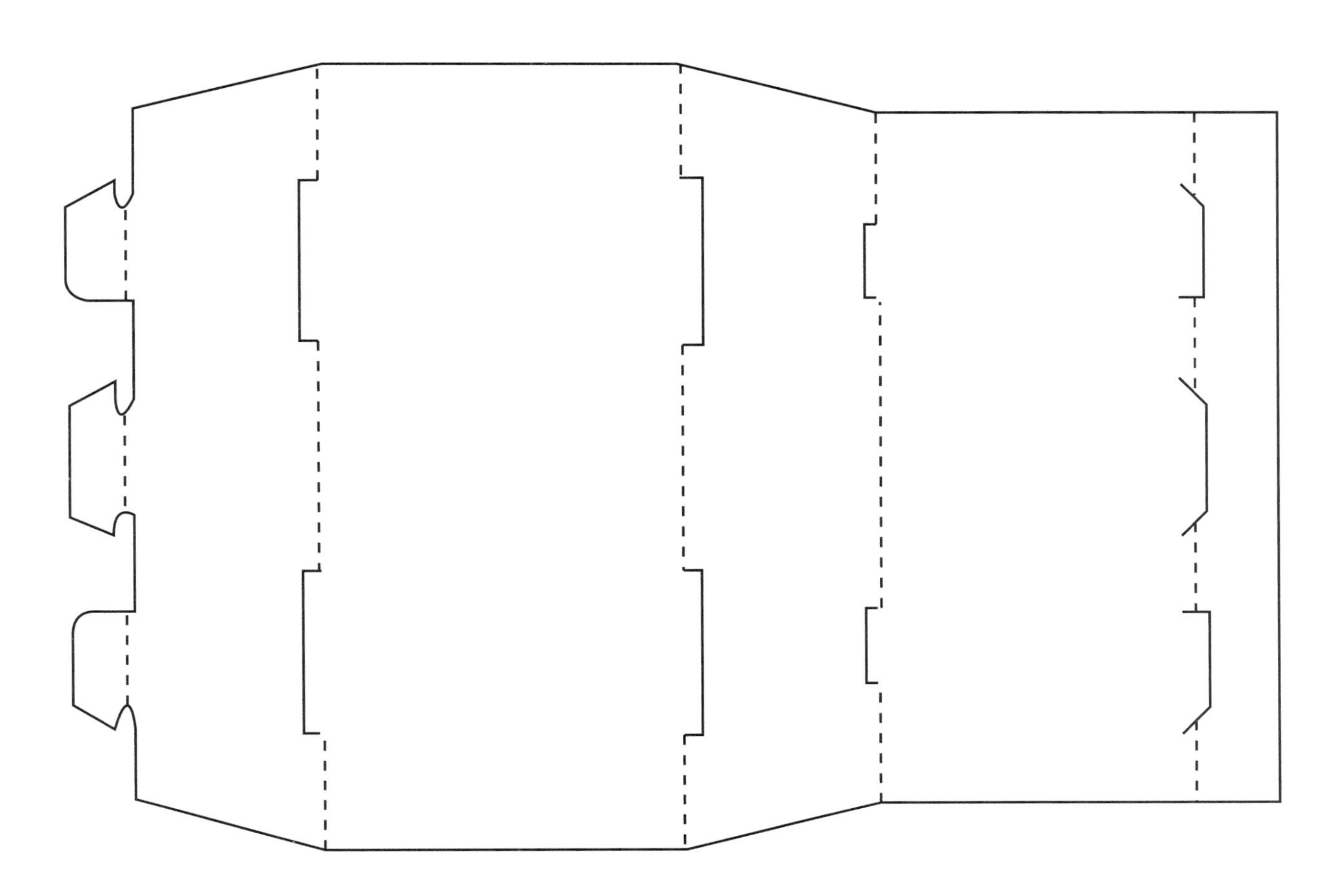

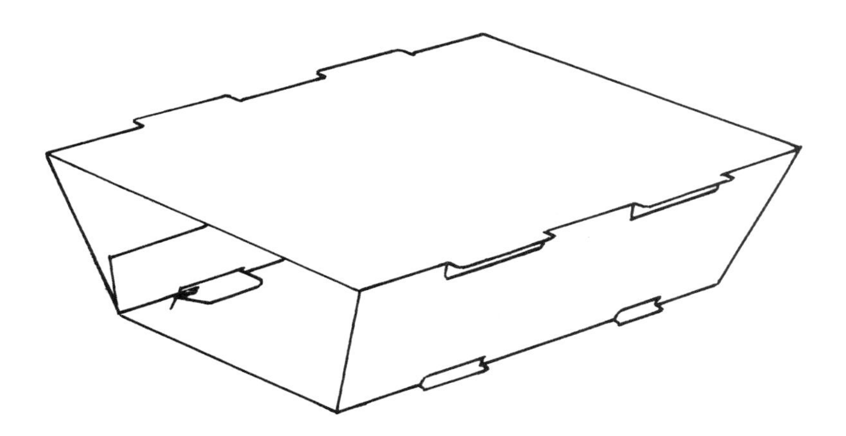

BOTTOM LOCK SLEEVE

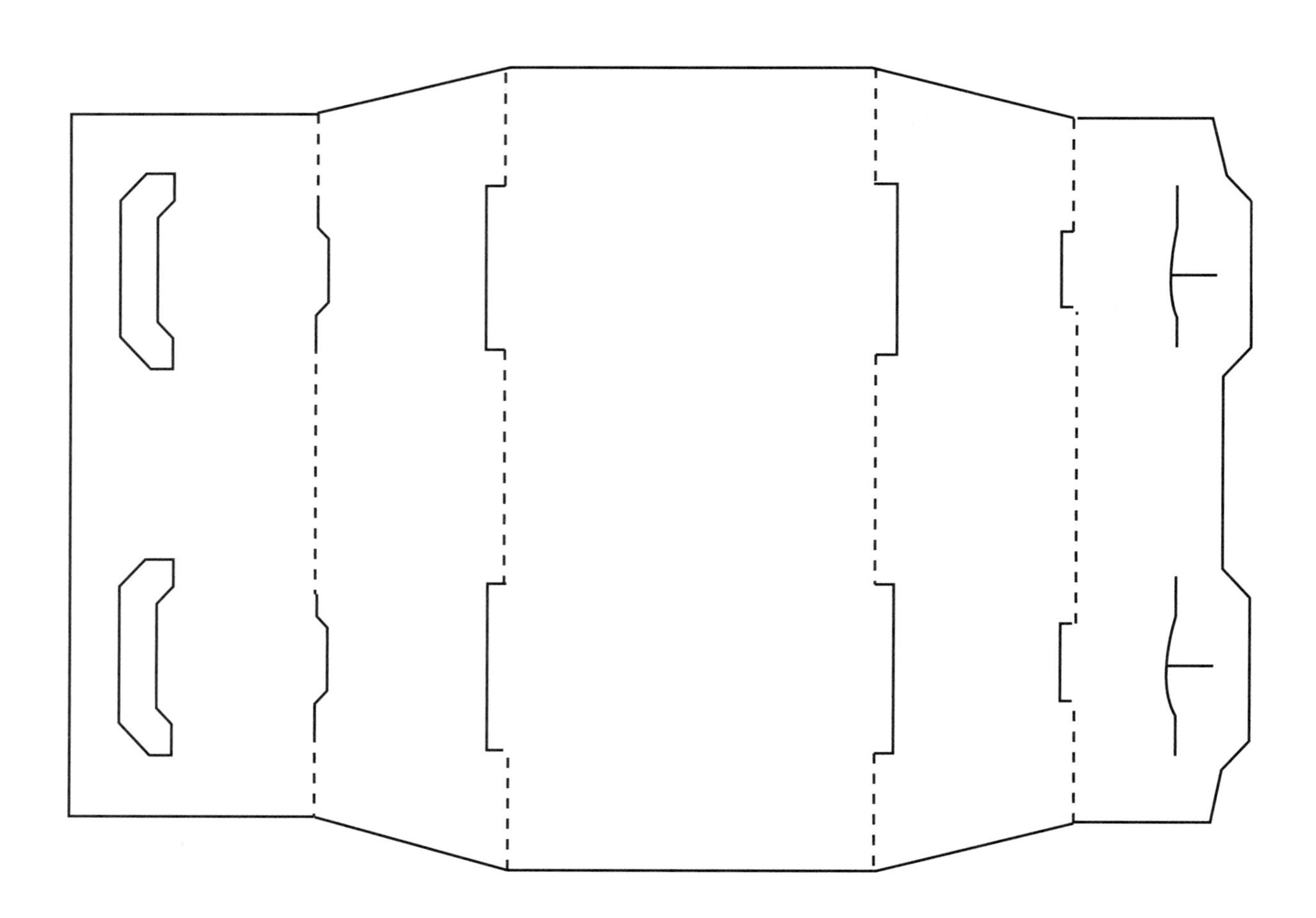

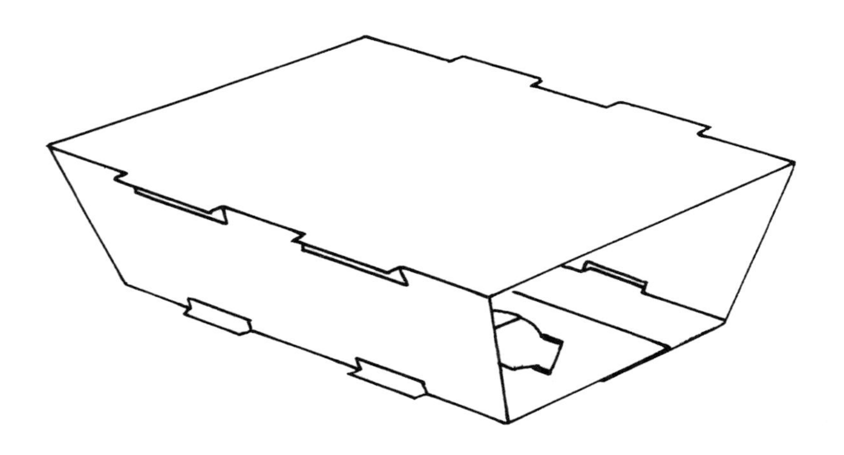

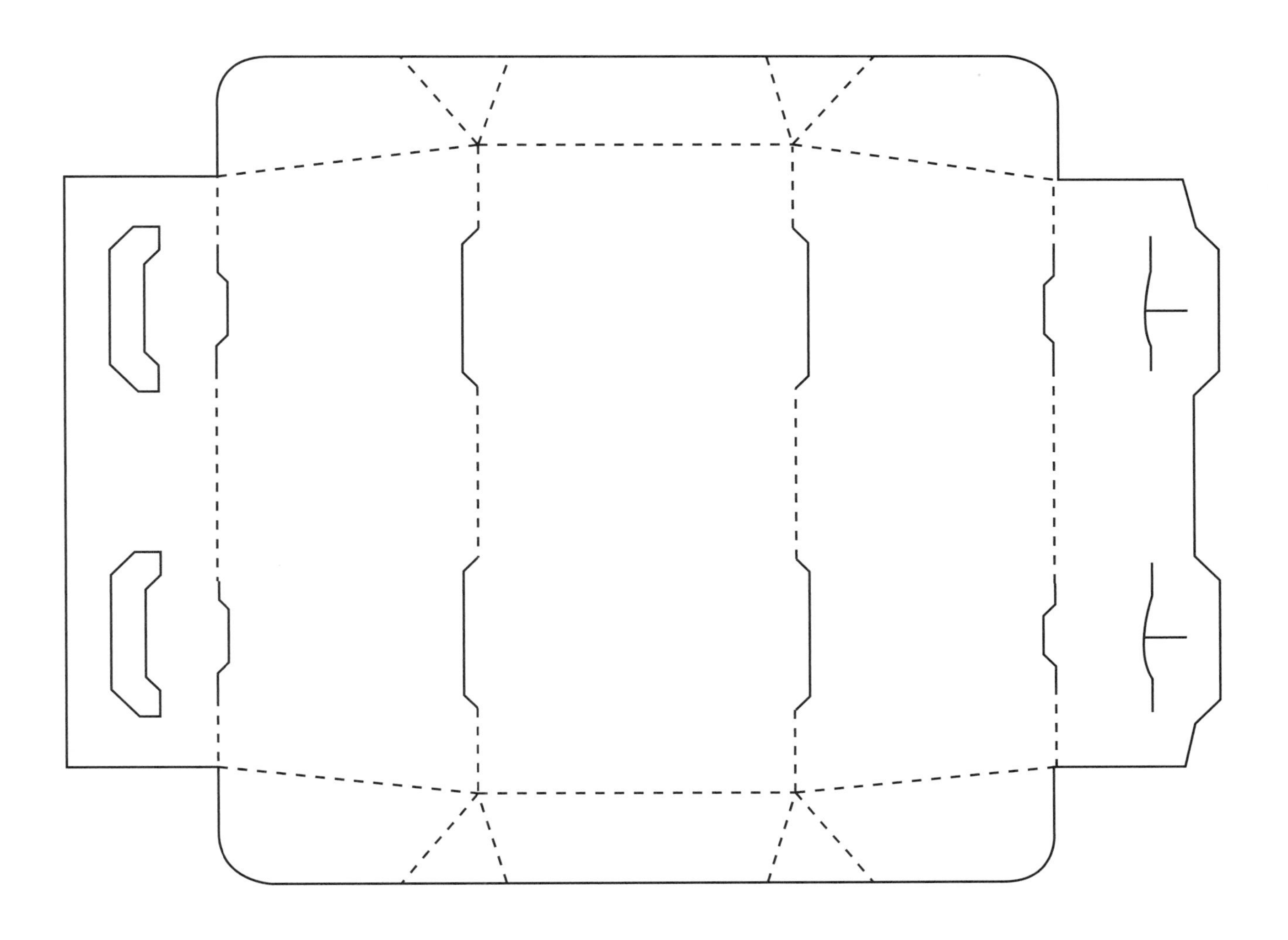

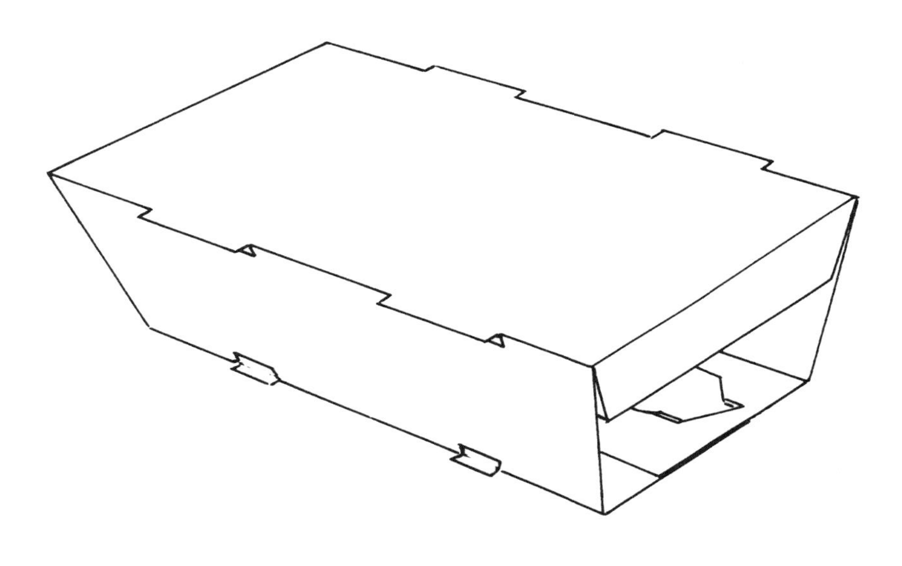

SIMPLE FOLDER FOR SHALLOW PRODUCTS

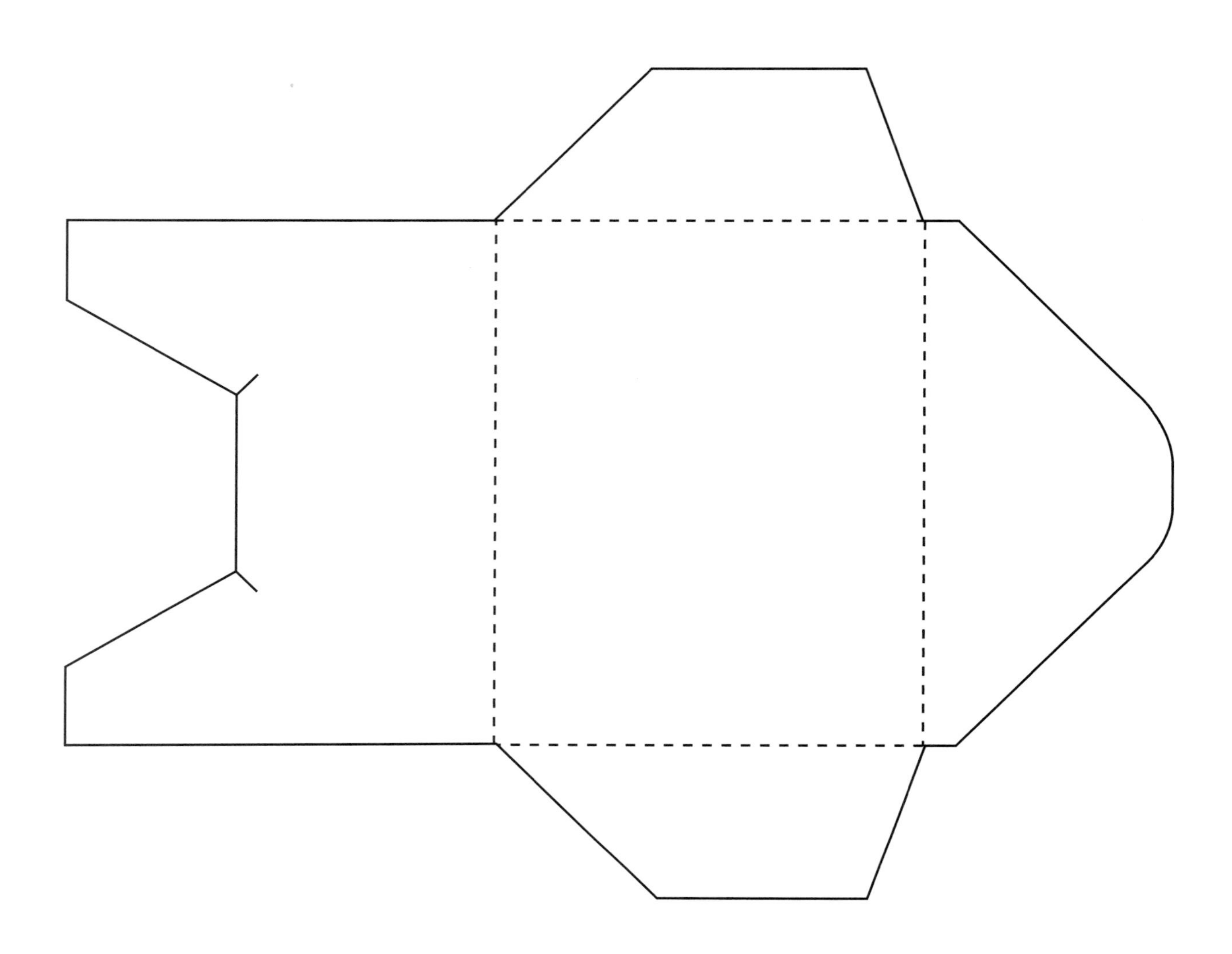

Used for scarves, hosiery, ect.

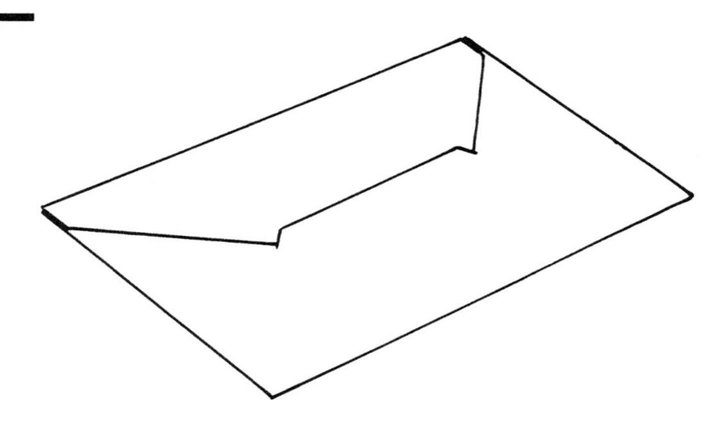

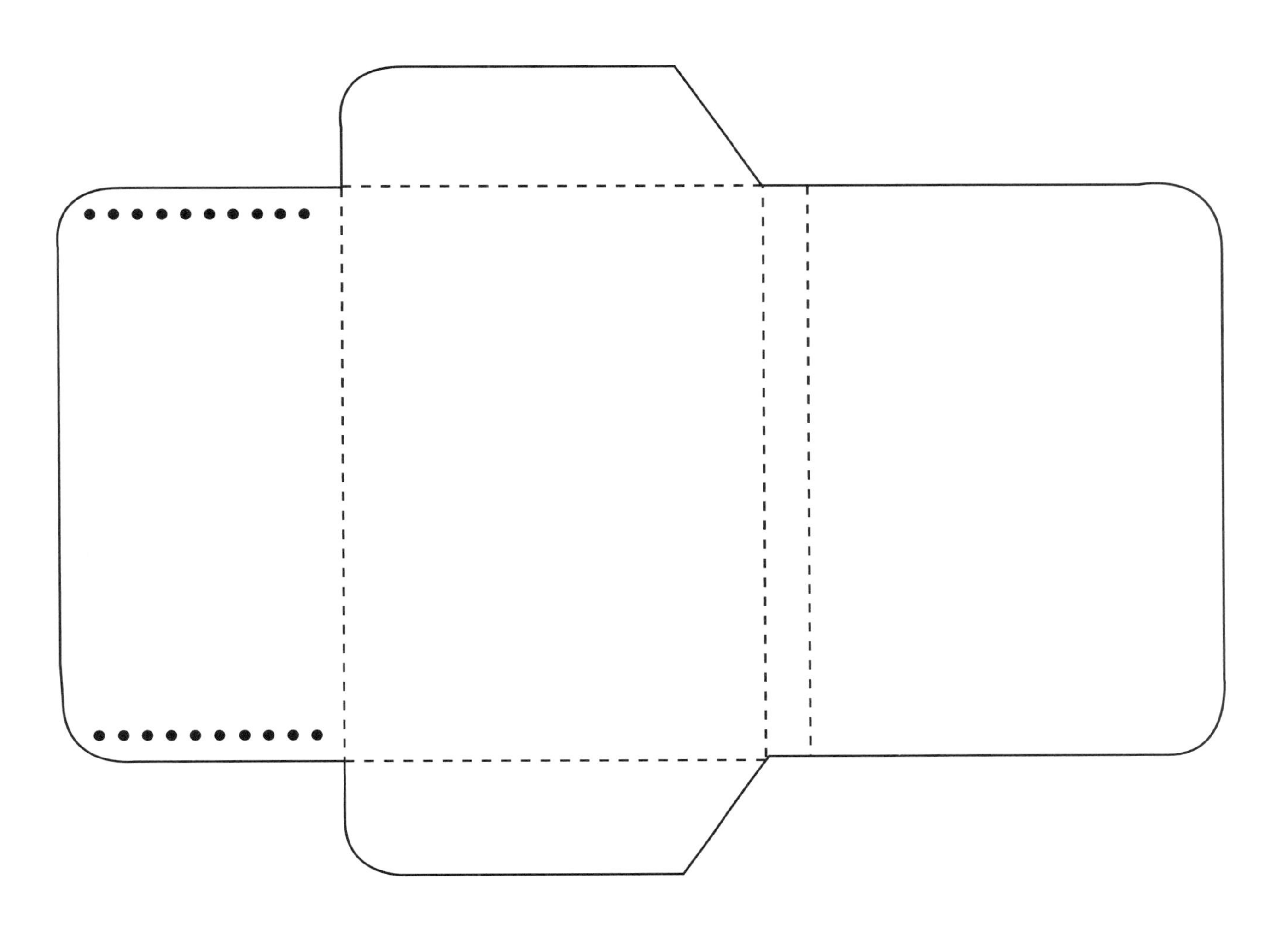

May be spot glued.

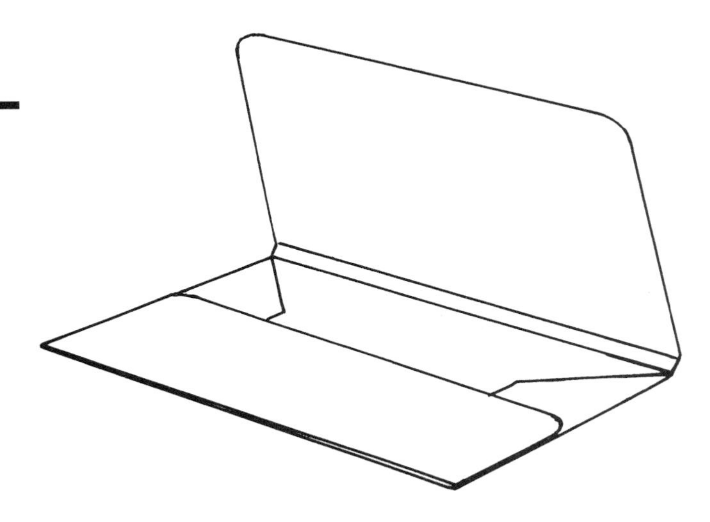

SHALLOW SIDE WALL FOLDER WITH WINDOW FOR PRODUCT INDENTIFICATION

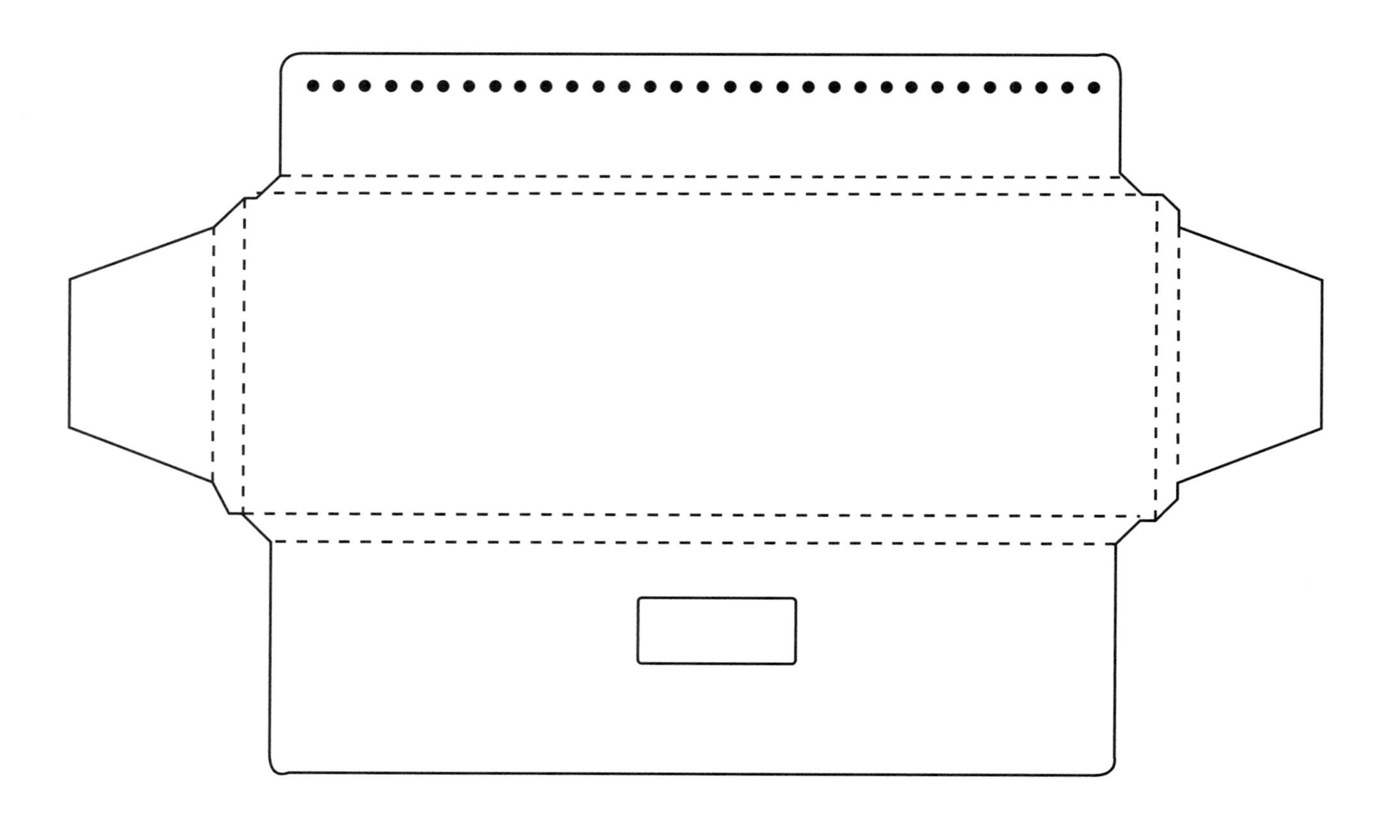

Perfect for hosiery.

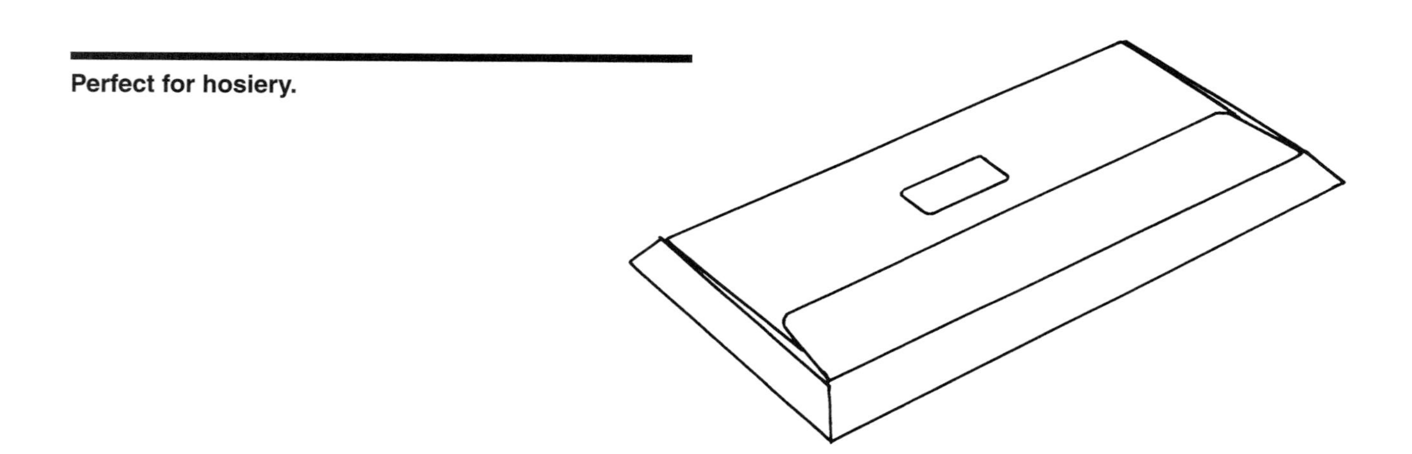

FOLDER TYPE SLEEVE WITH GLUED-UP CORNERS

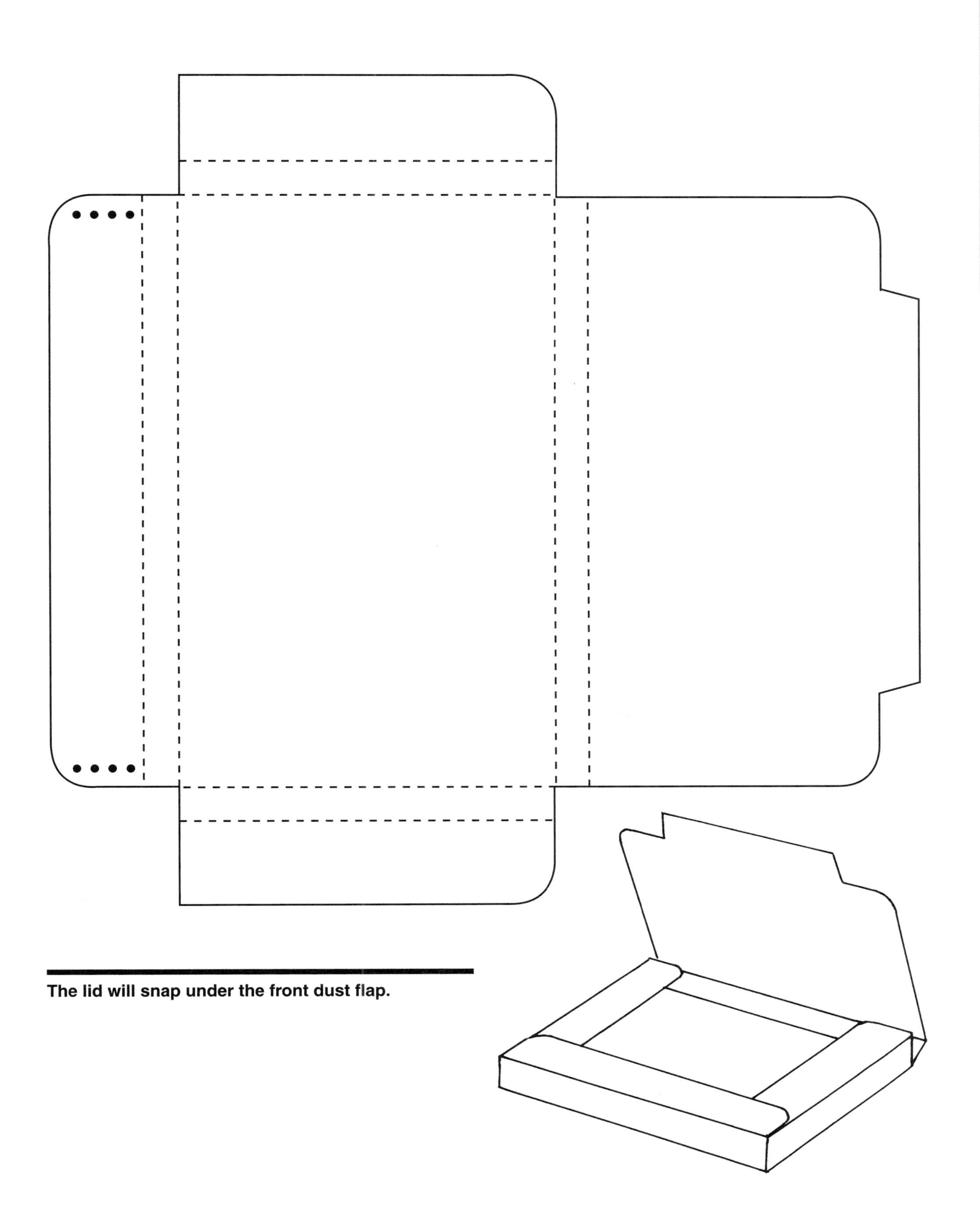

The lid will snap under the front dust flap.

TWO SECTION FOLDER

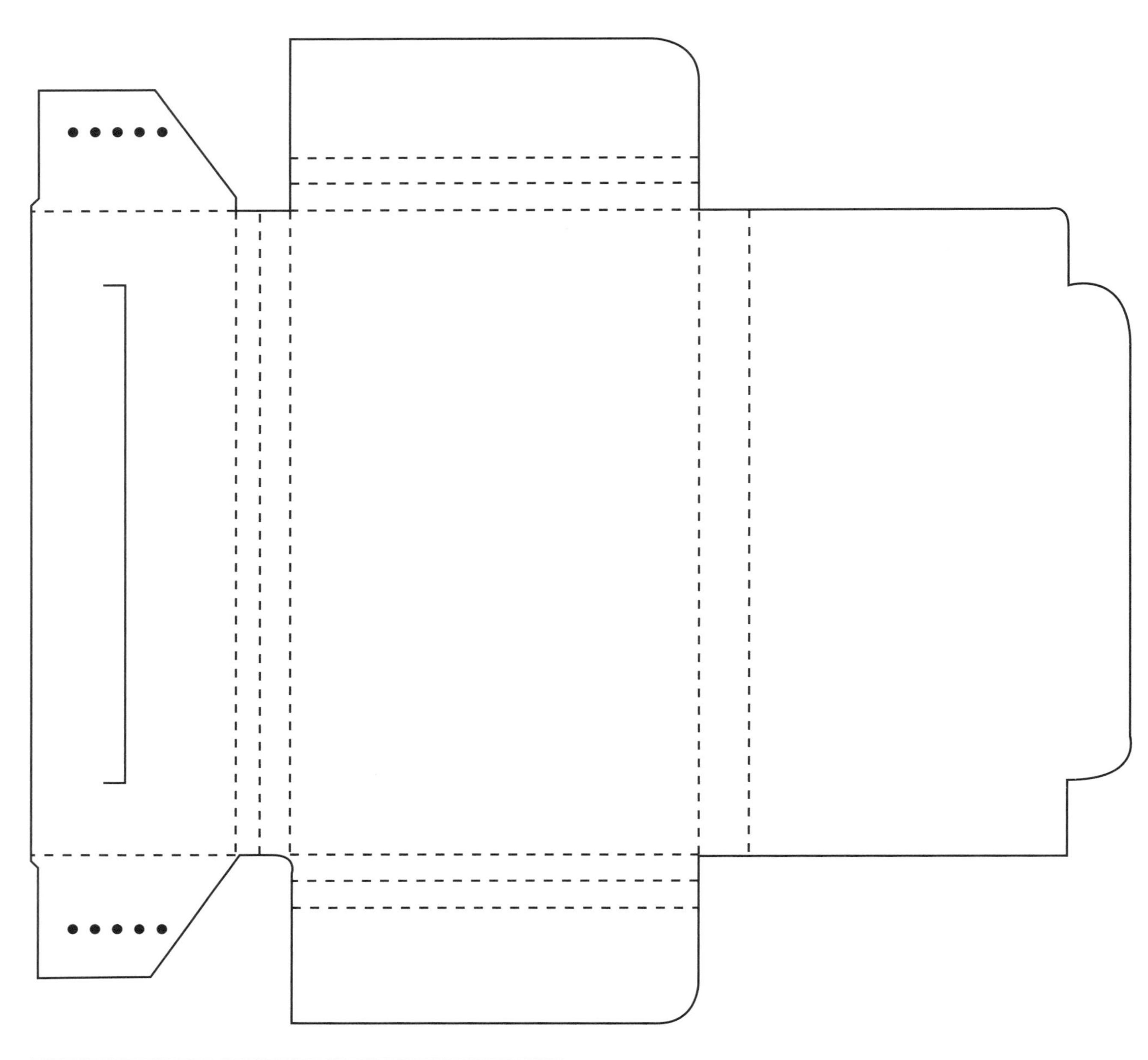

Typically used for photographic prints and negatives. The scored gusset allows for expansion.

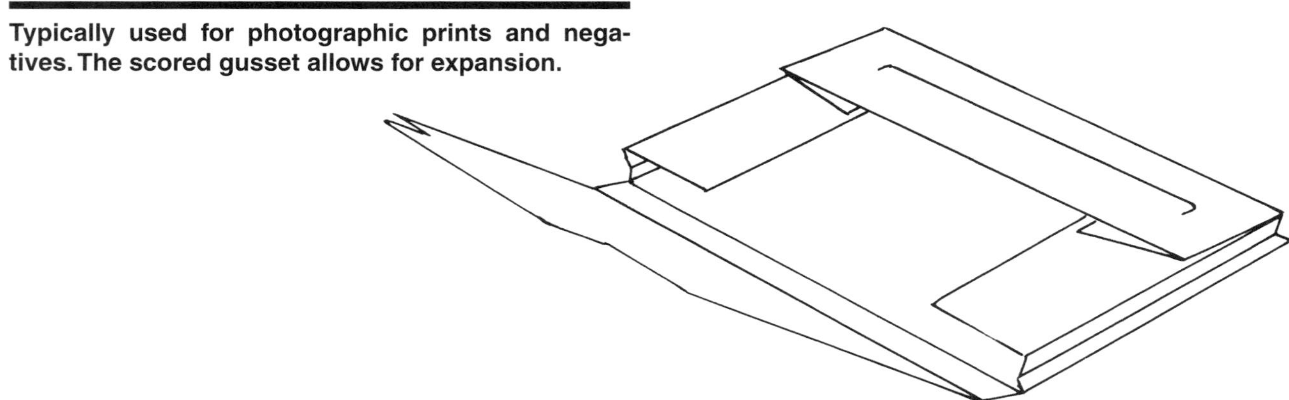

The Greenleaf style closure makes this sleeve easy to open for inspection of the product. A window may be die-cut in the package.

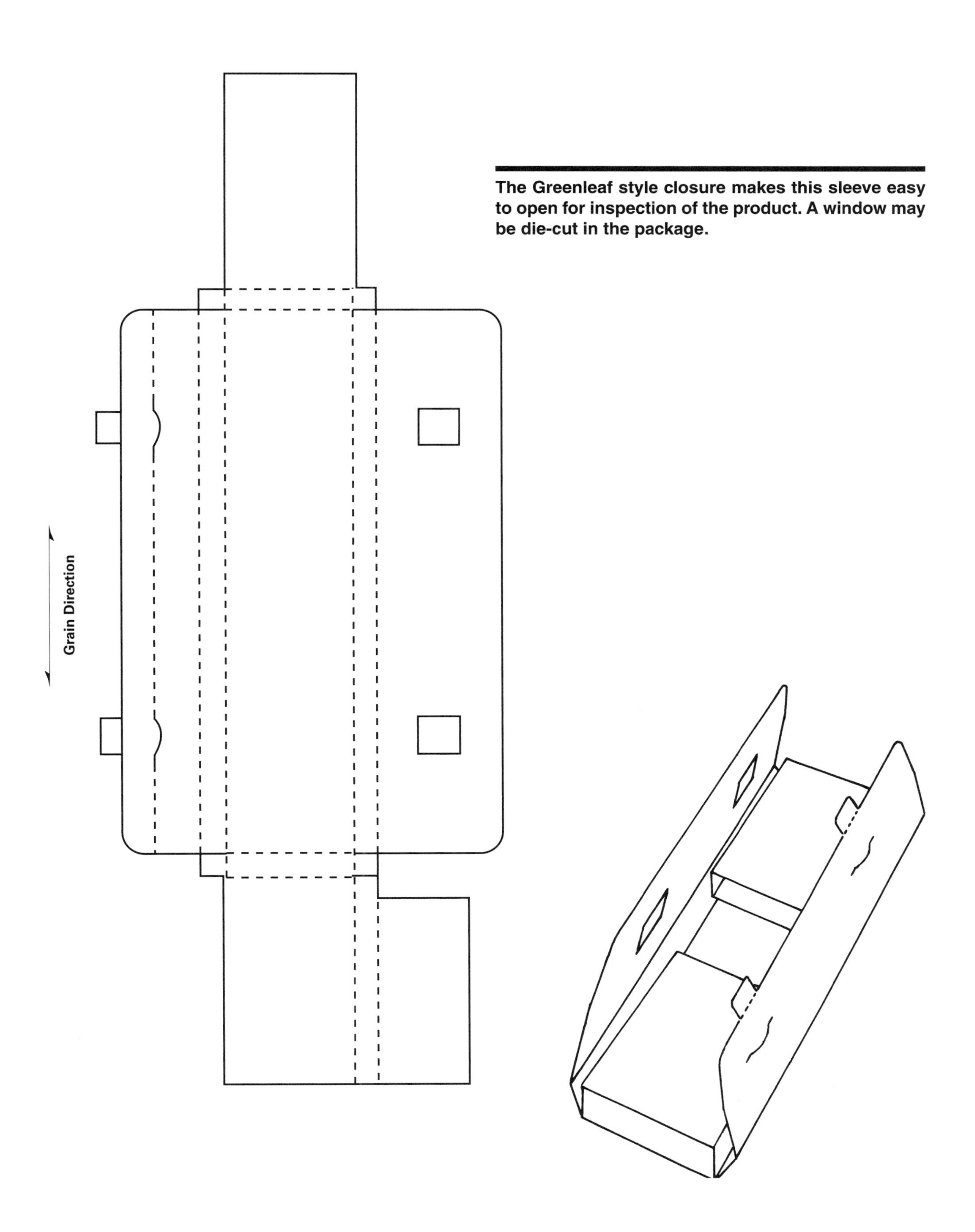

EAR-HOOK BACK PANEL WITH HANGING LOOP

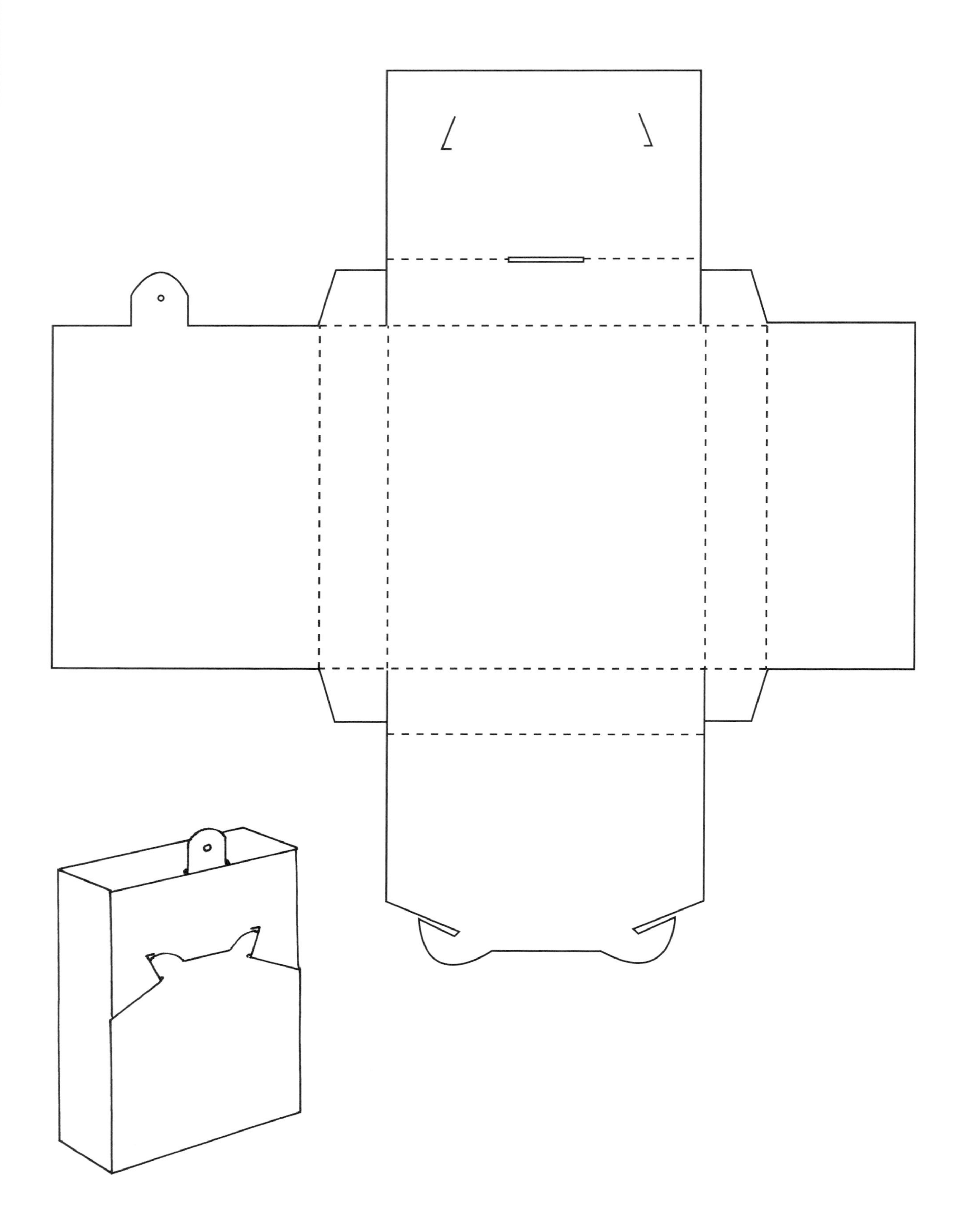

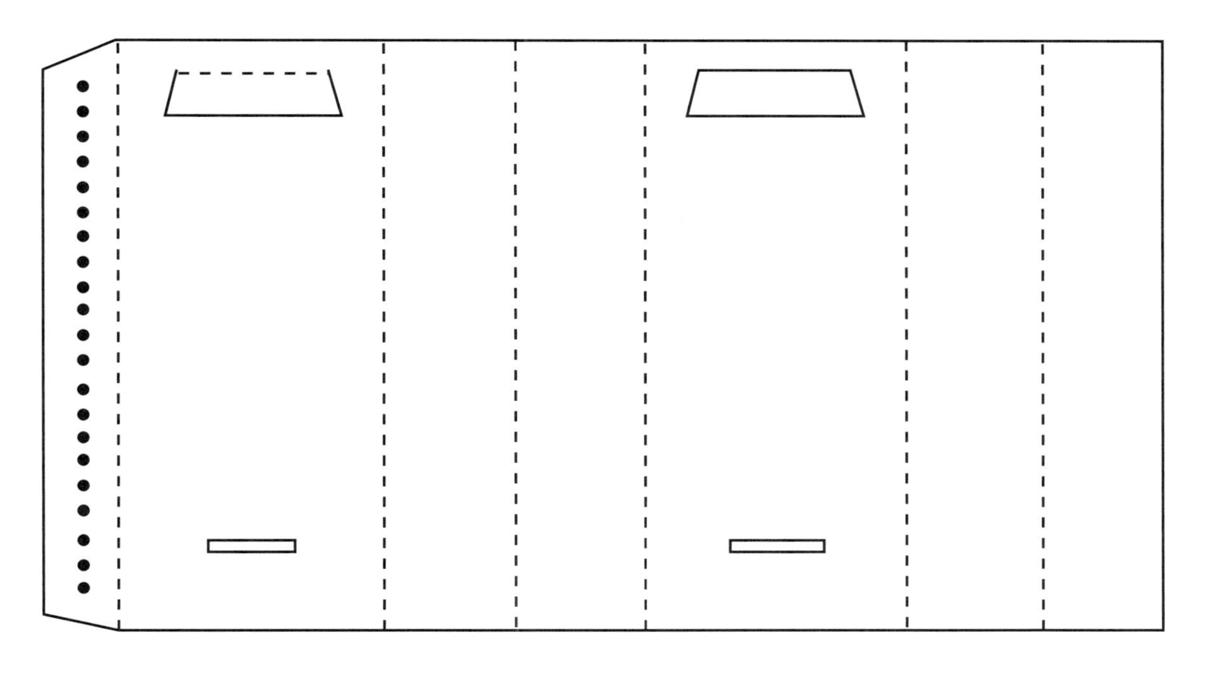

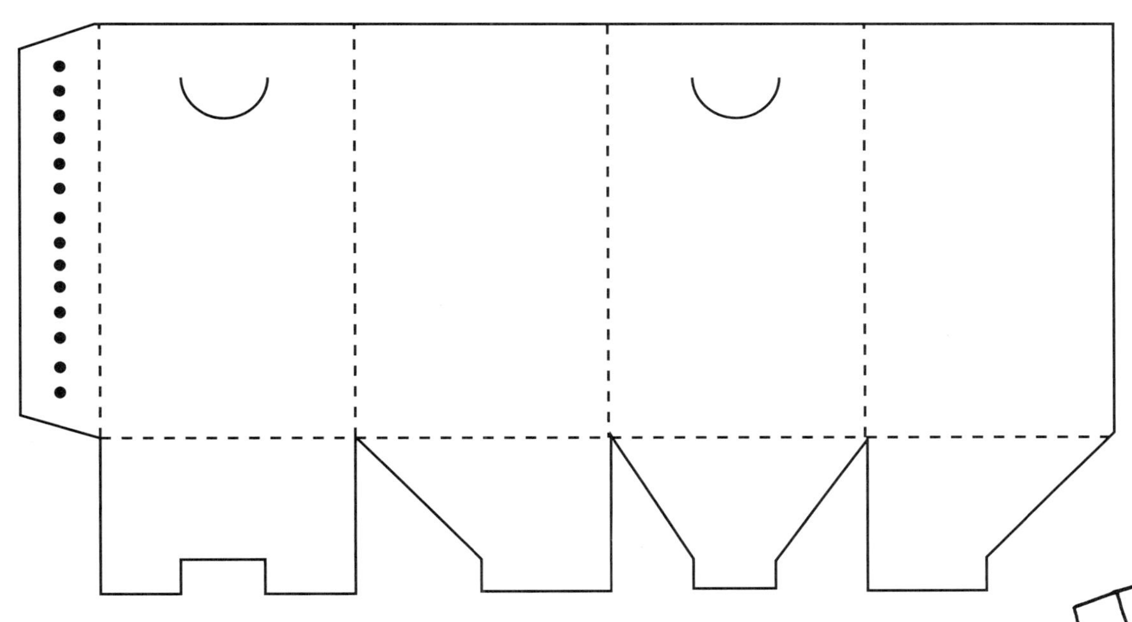

This package is developed from two interlocking sleeves with a carrying handle.

SIX-SIDED SLEEVE WITH AUTOMATIC LOCK BOTTOM

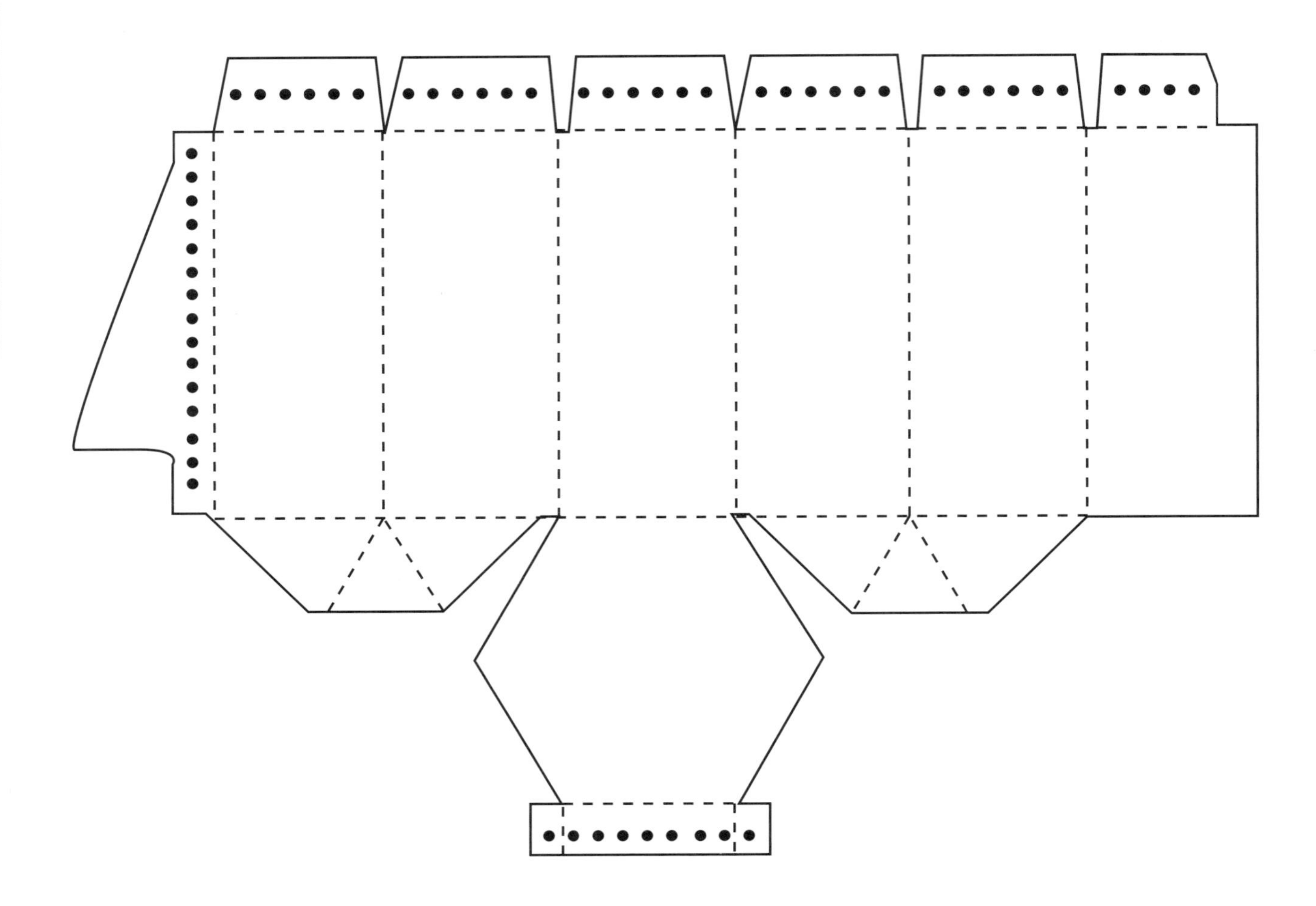

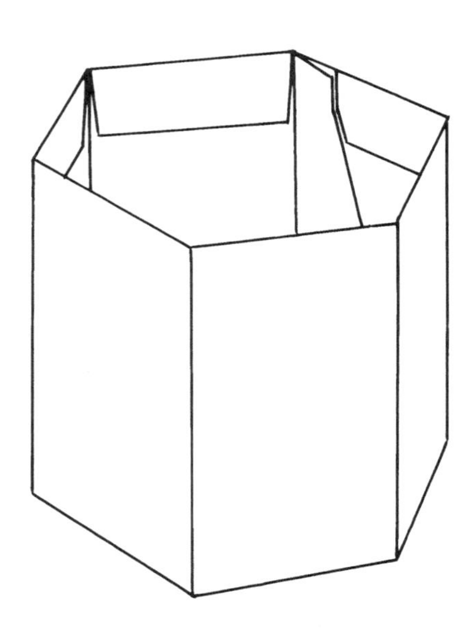

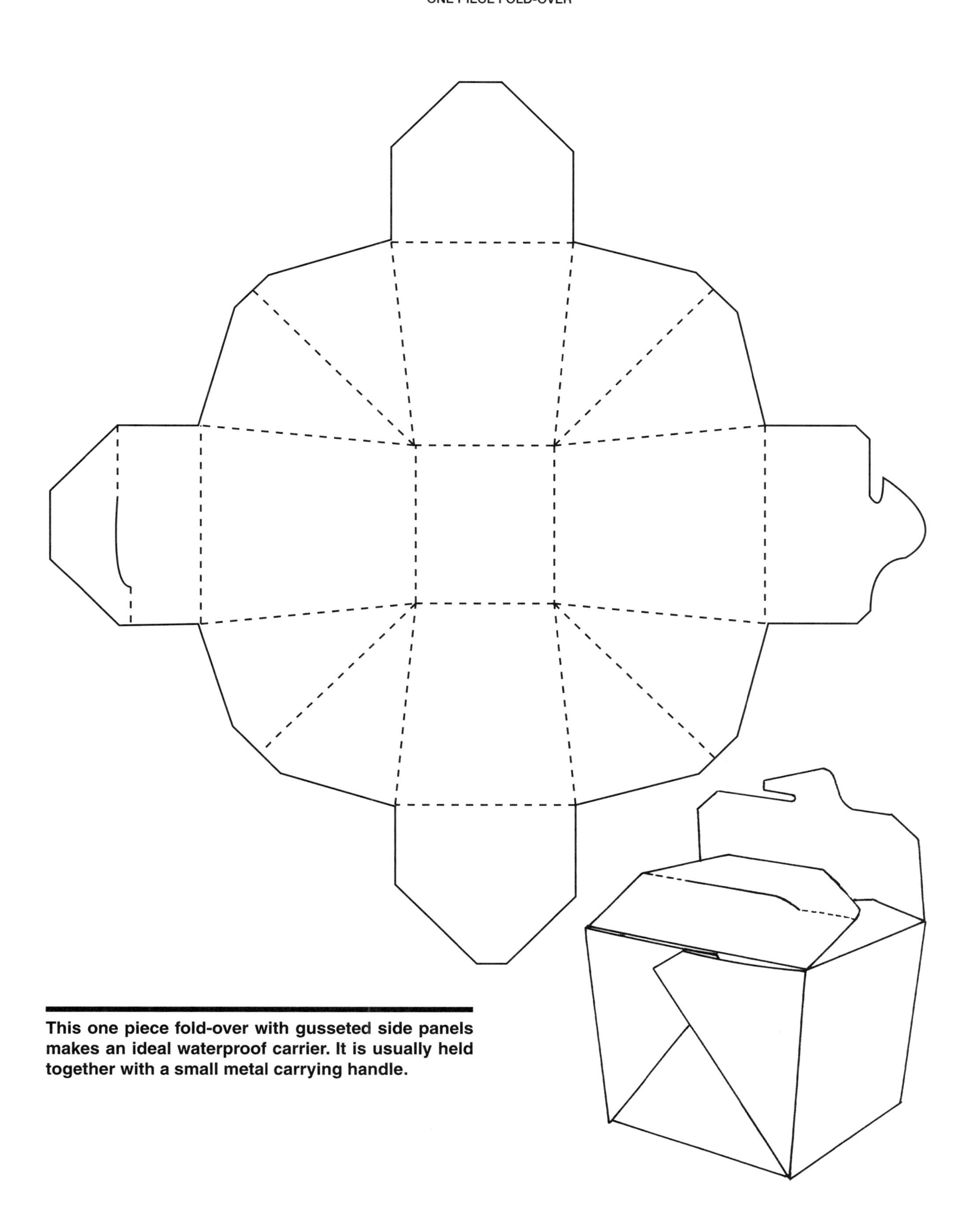

This one piece fold-over with gusseted side panels makes an ideal waterproof carrier. It is usually held together with a small metal carrying handle.

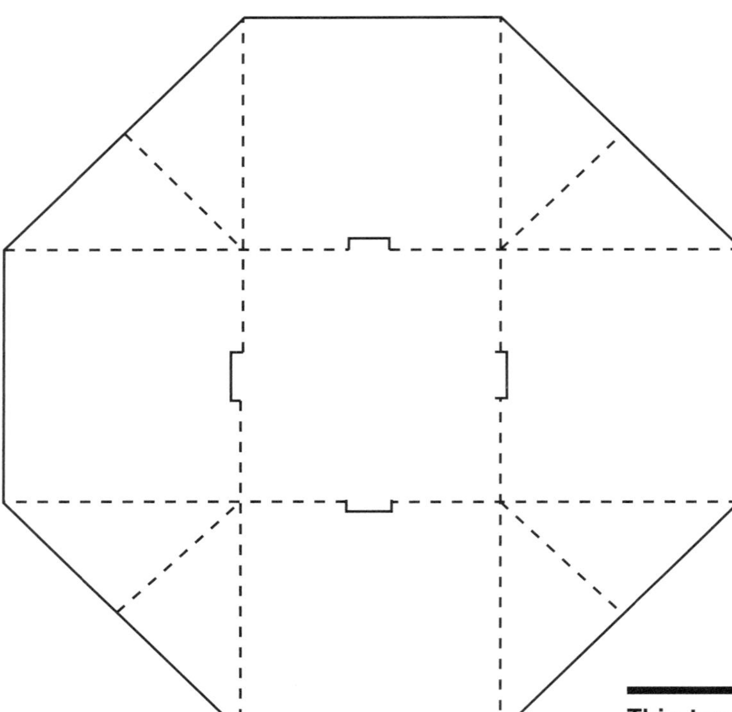

This two piece wrap needs no glue and makes an excellent gift box, expecially if made from different color board.

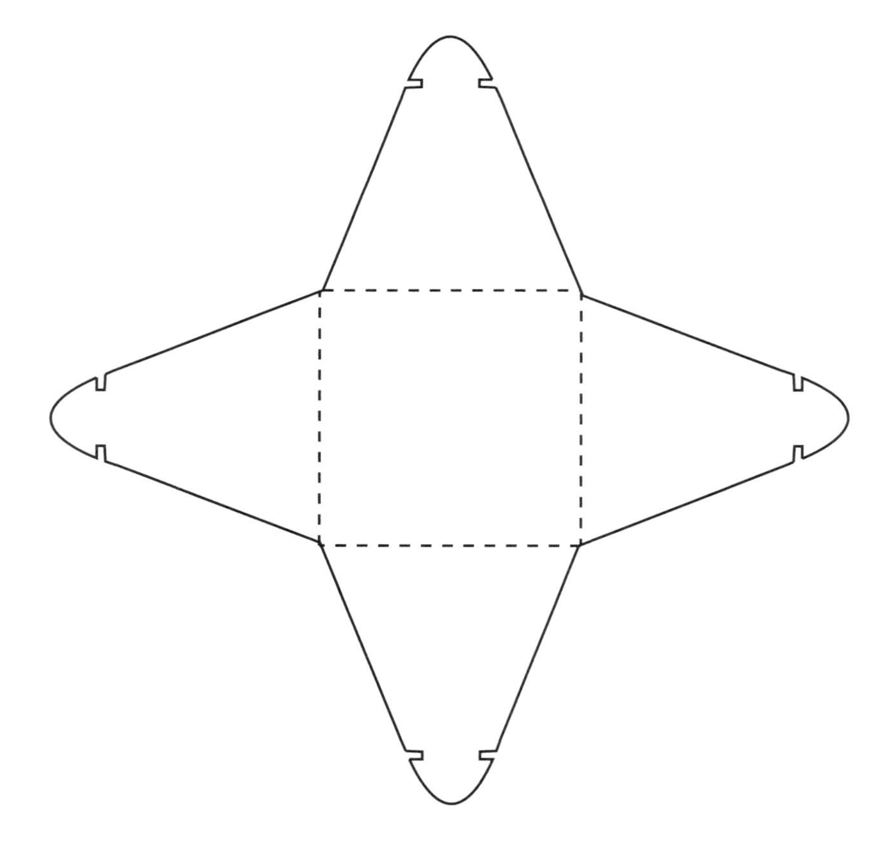

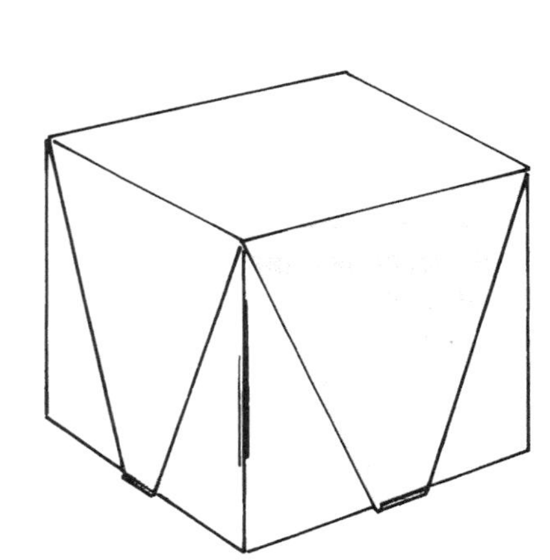

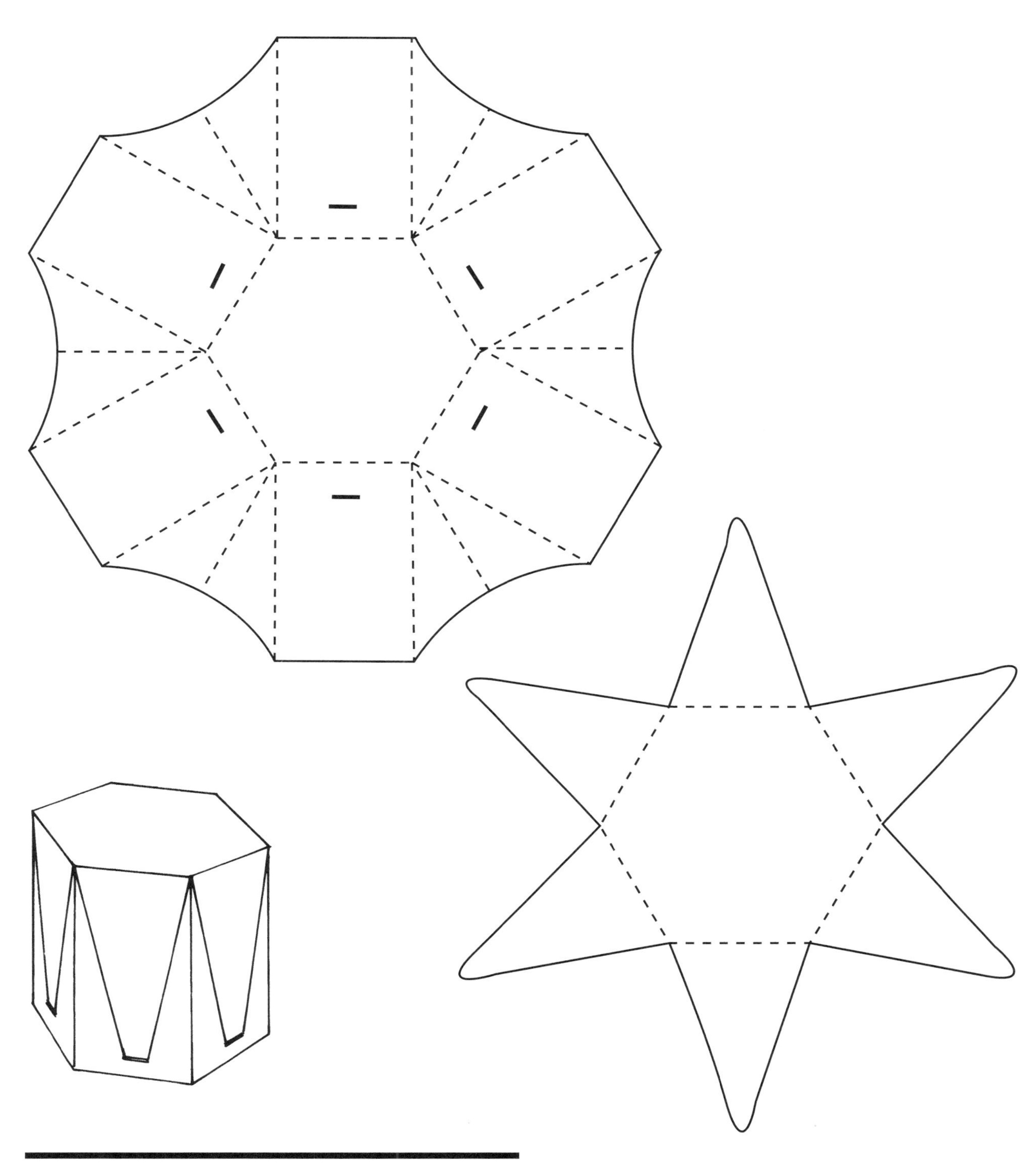

This six-sided two piece wrap is a variation of the previous pattern.

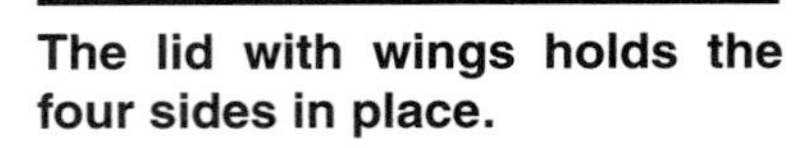

The lid with wings holds the four sides in place.

Adding four fold-over flaps on the riser panels as in the next pattern, will serve as a locking device for the lid.

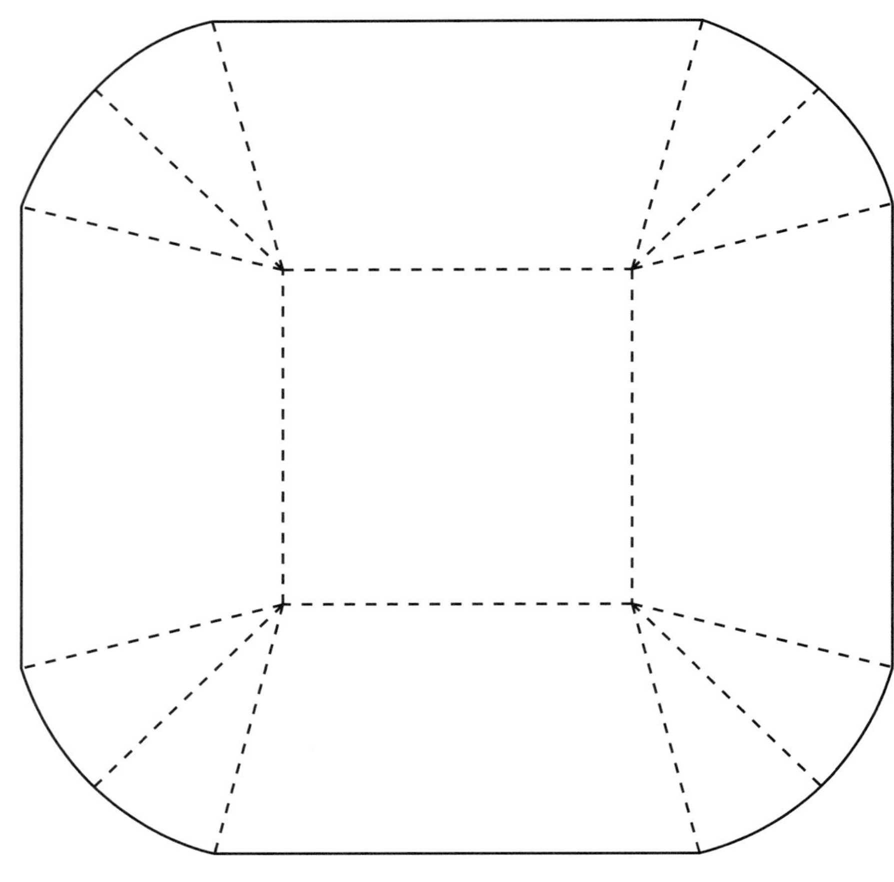

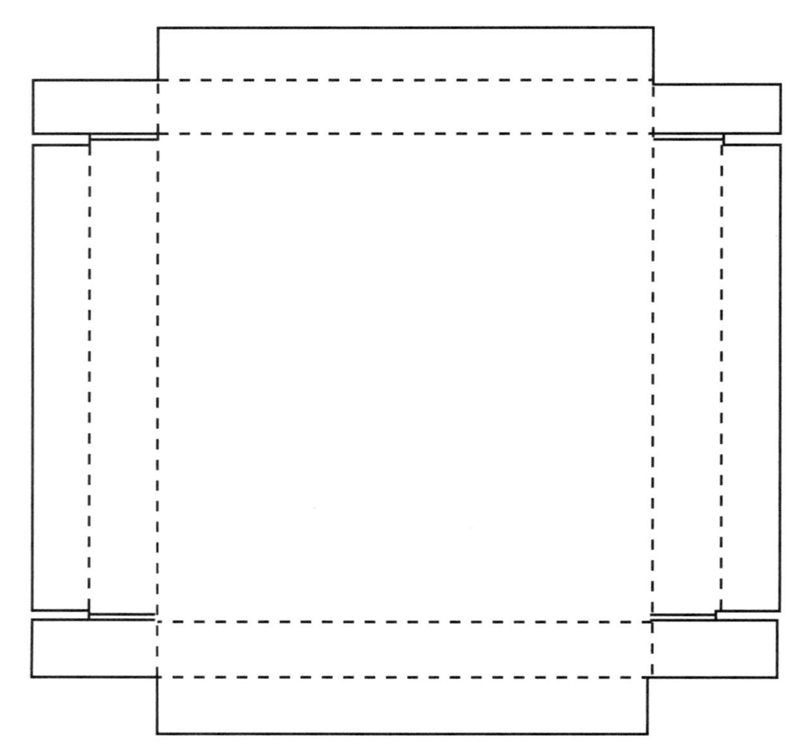

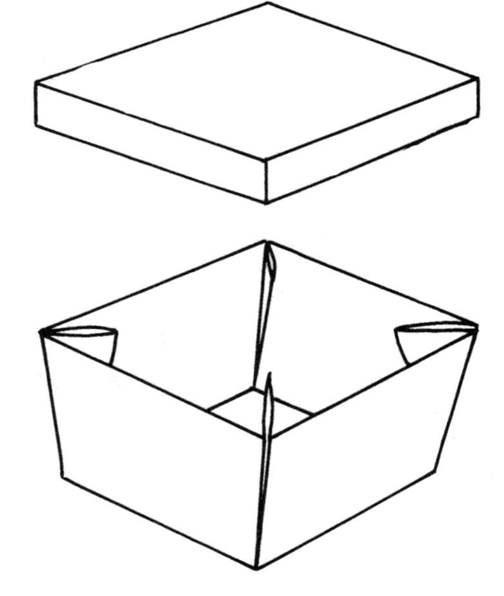

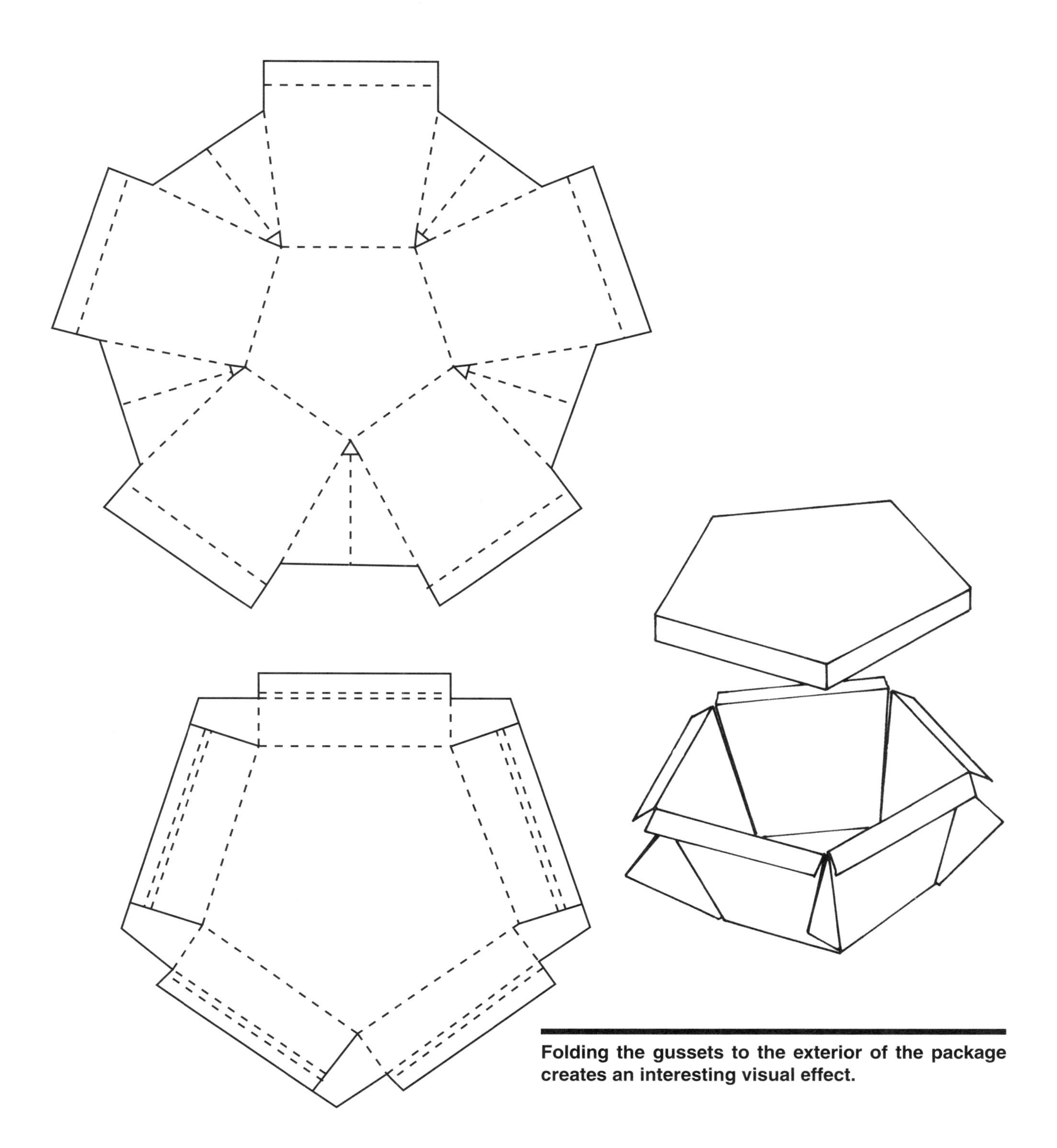

Folding the gussets to the exterior of the package creates an interesting visual effect.

TRIANGULAR FOLDING CARTON WITH TUCK LID

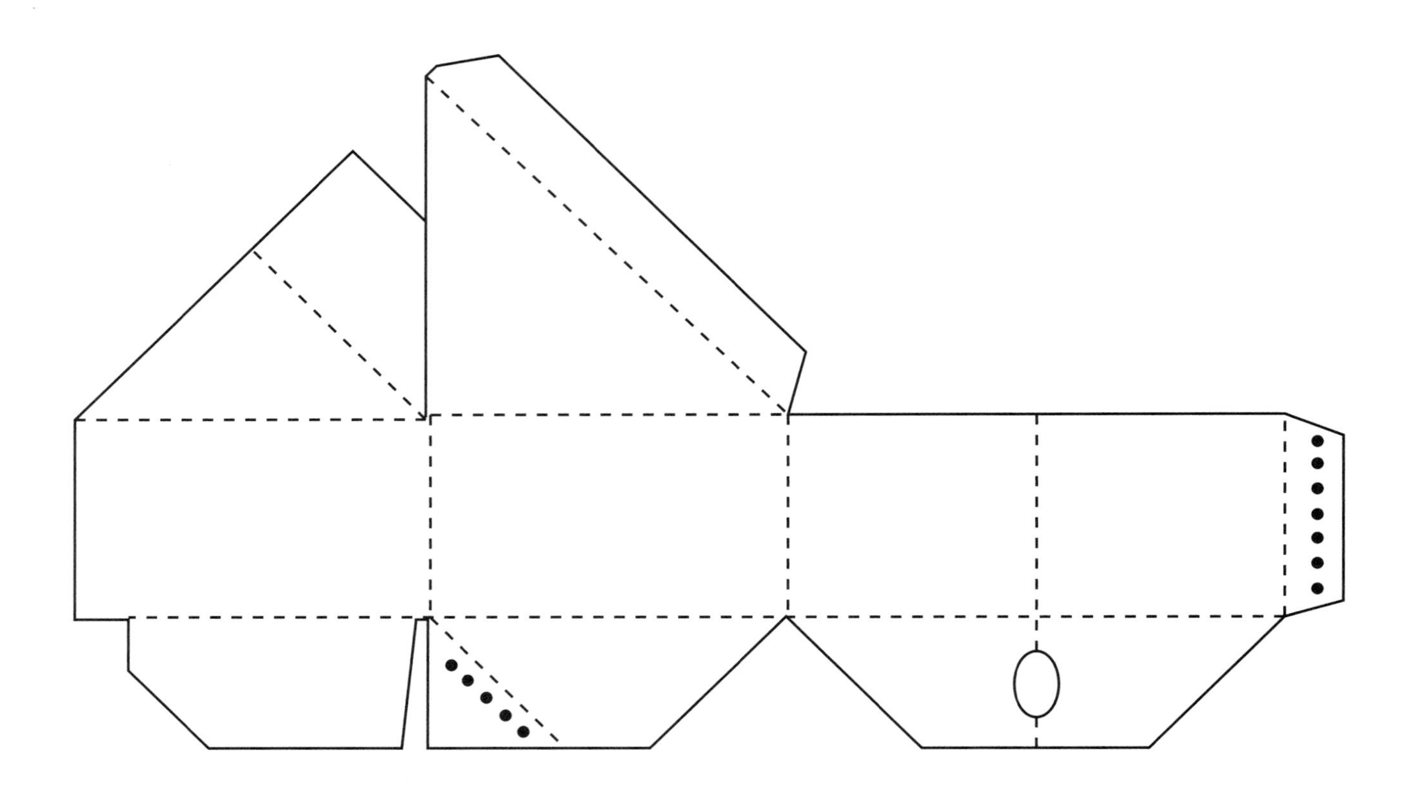

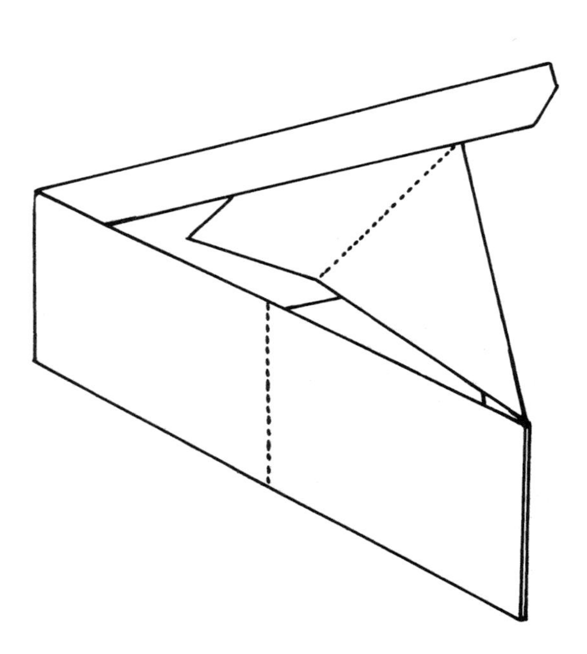

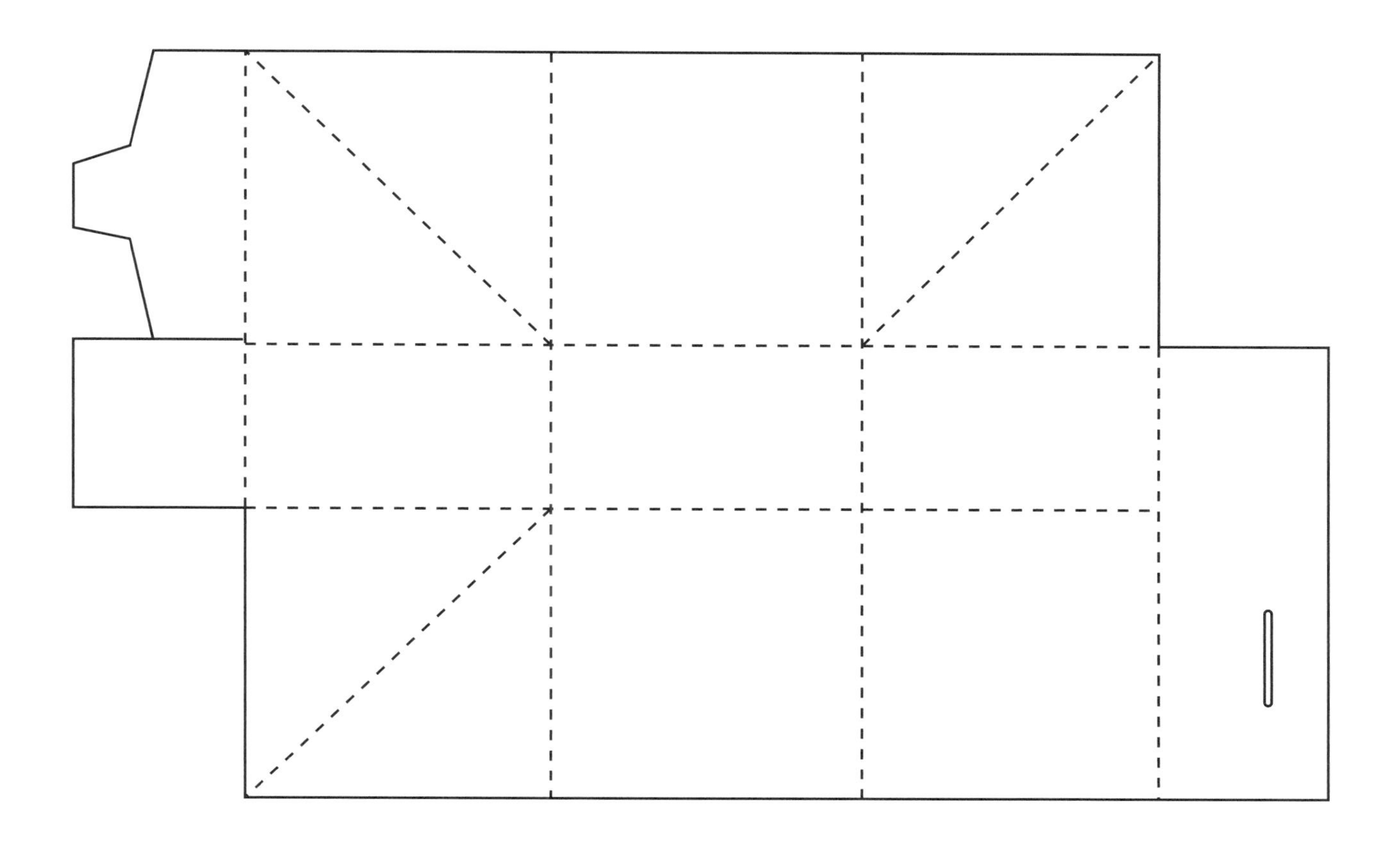

This wrap makes a handsome container, without the use of adhesives.

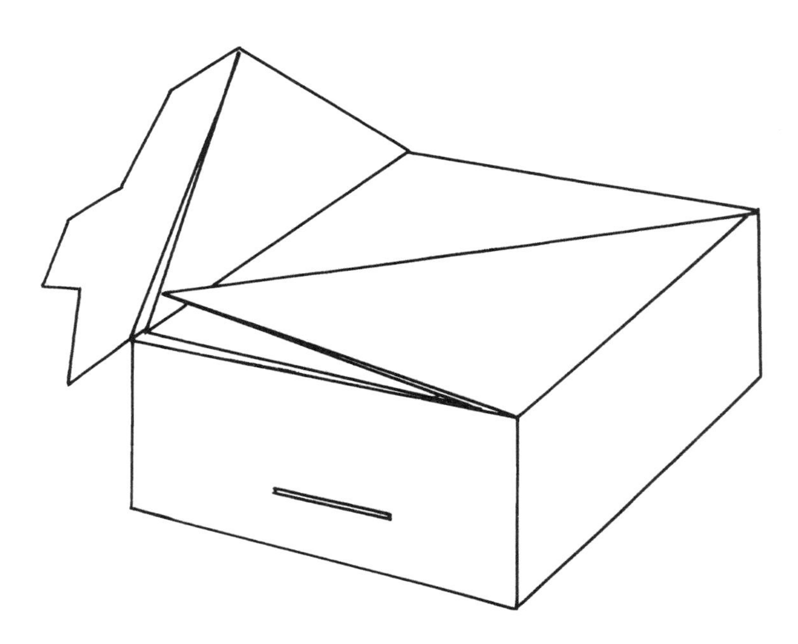

ONE PIECE WRAP

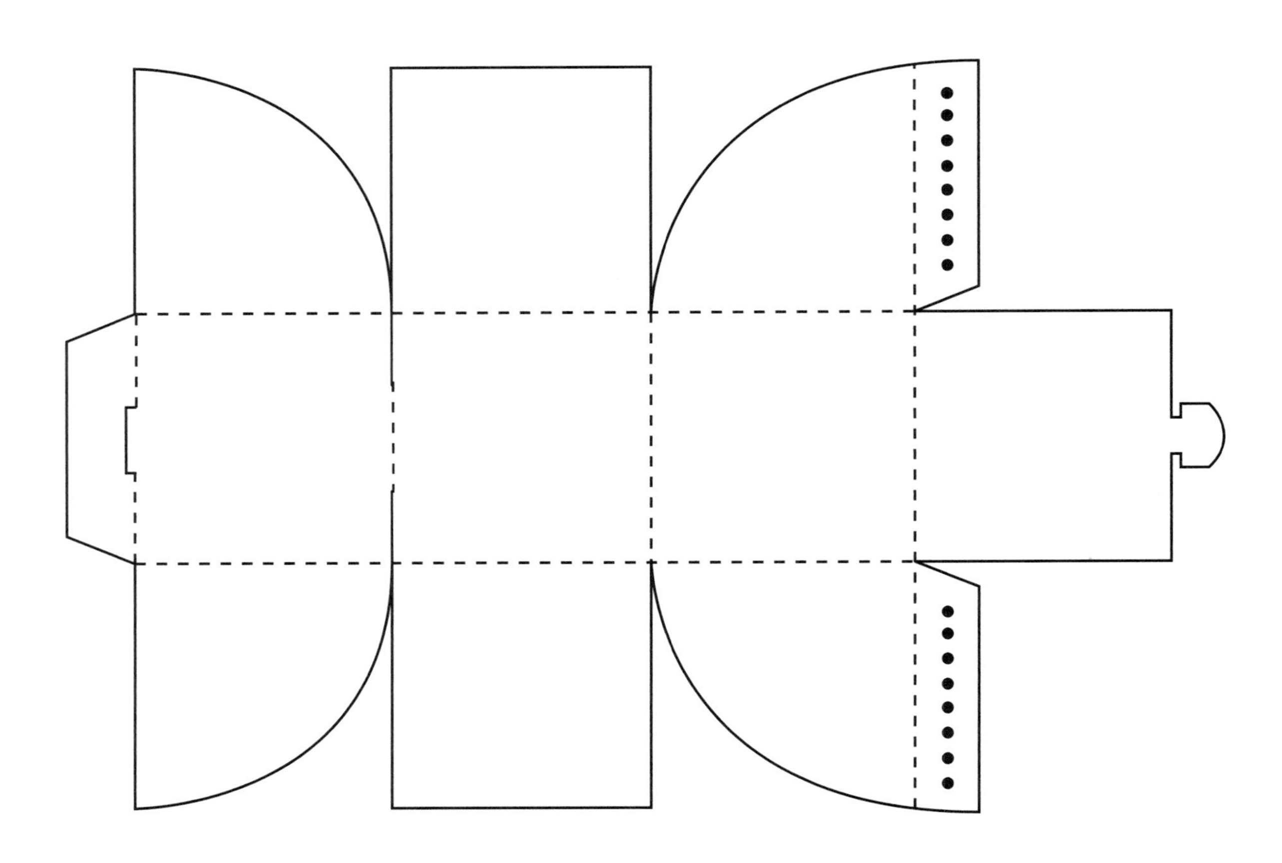

A one piece wrap that shows an interest-catching configuration on the sides.

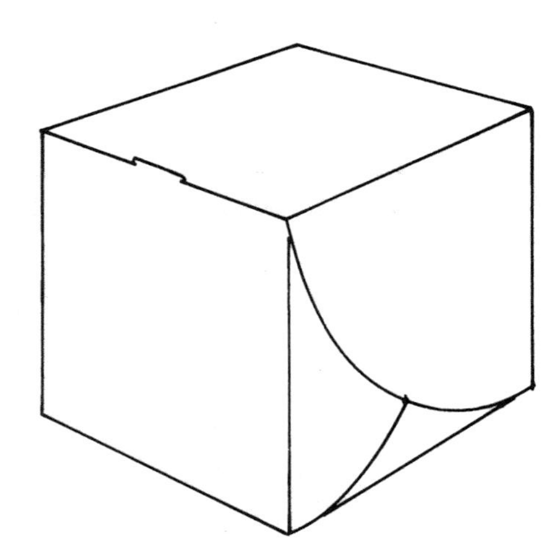

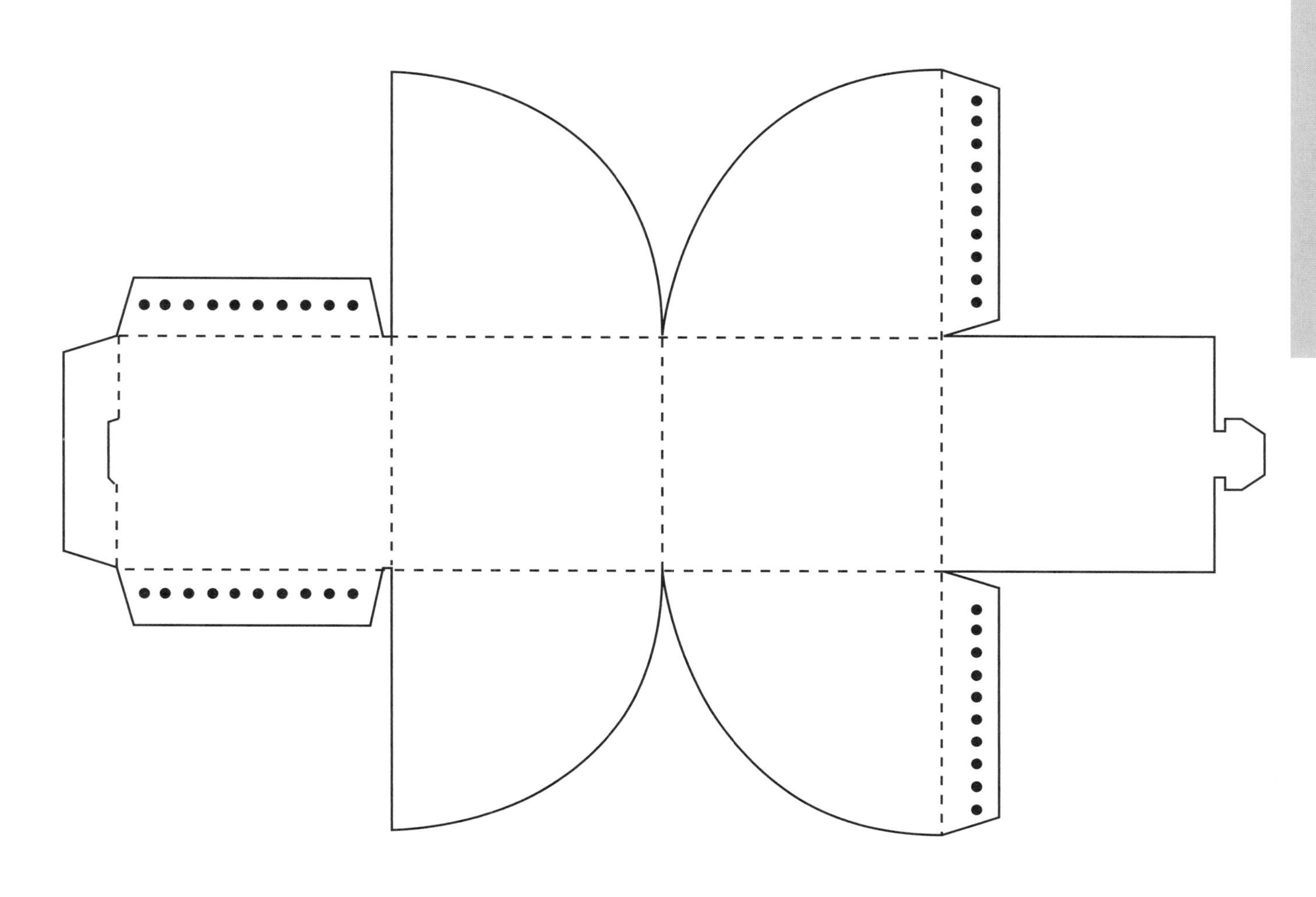

A rearrangement of the panels and glue flaps alters the appearance.

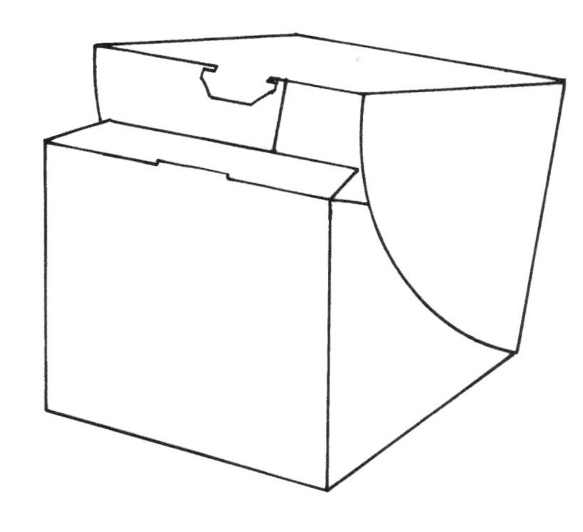

DECORATIVE ENDS WRAP

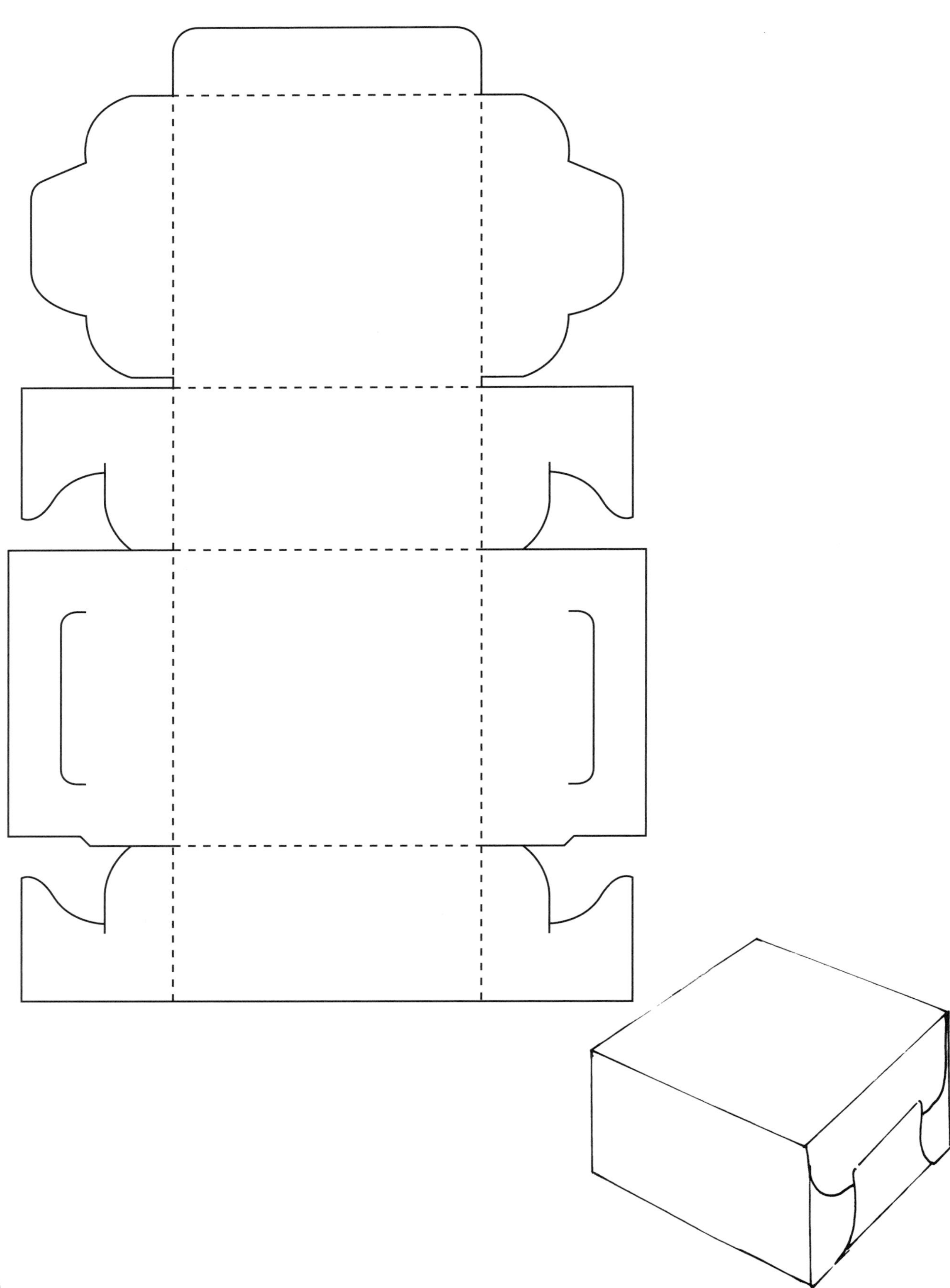

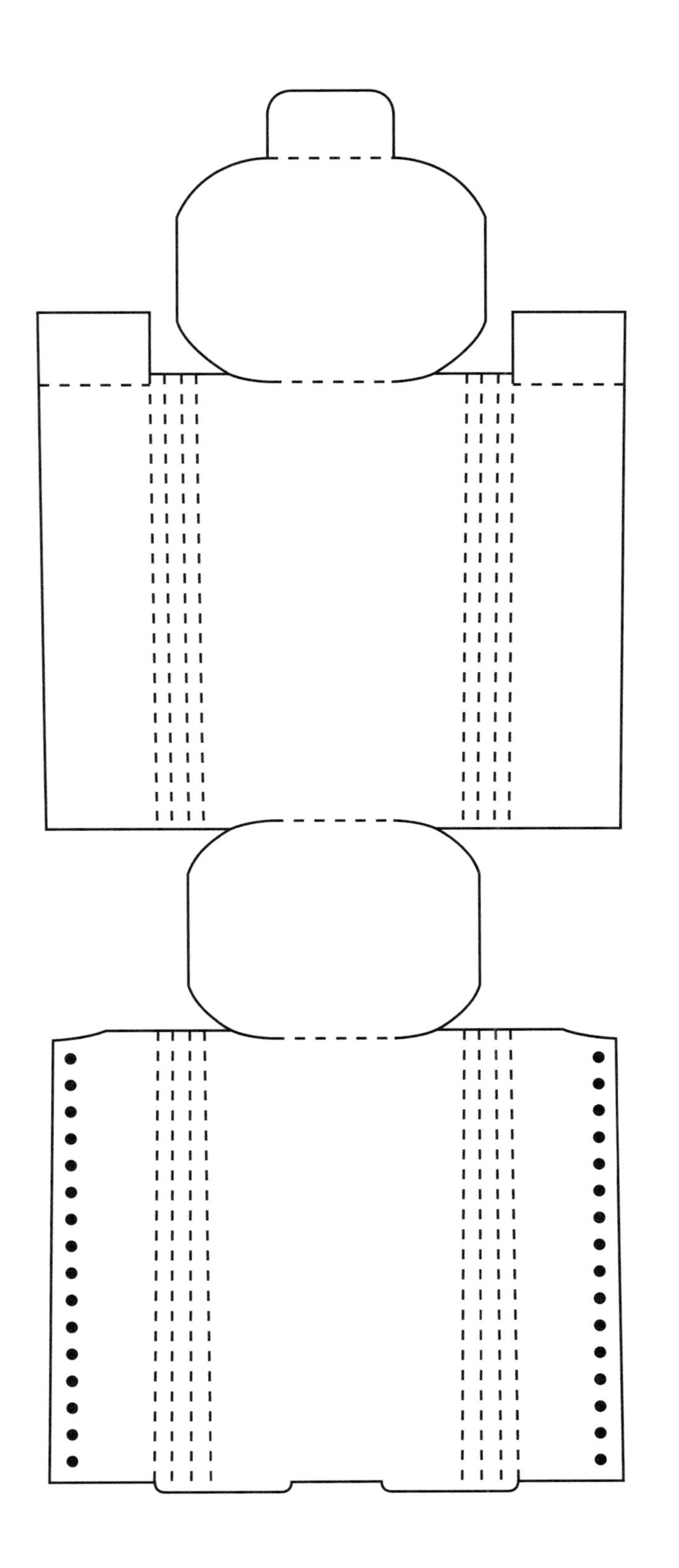

Grain Direction

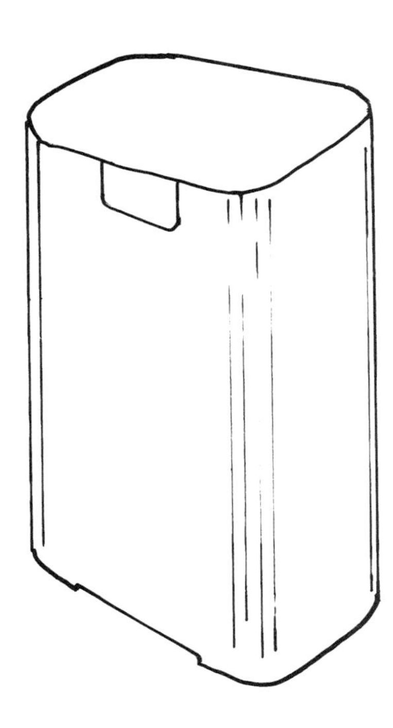

FOLD-OVER WITH TEAR STRIP

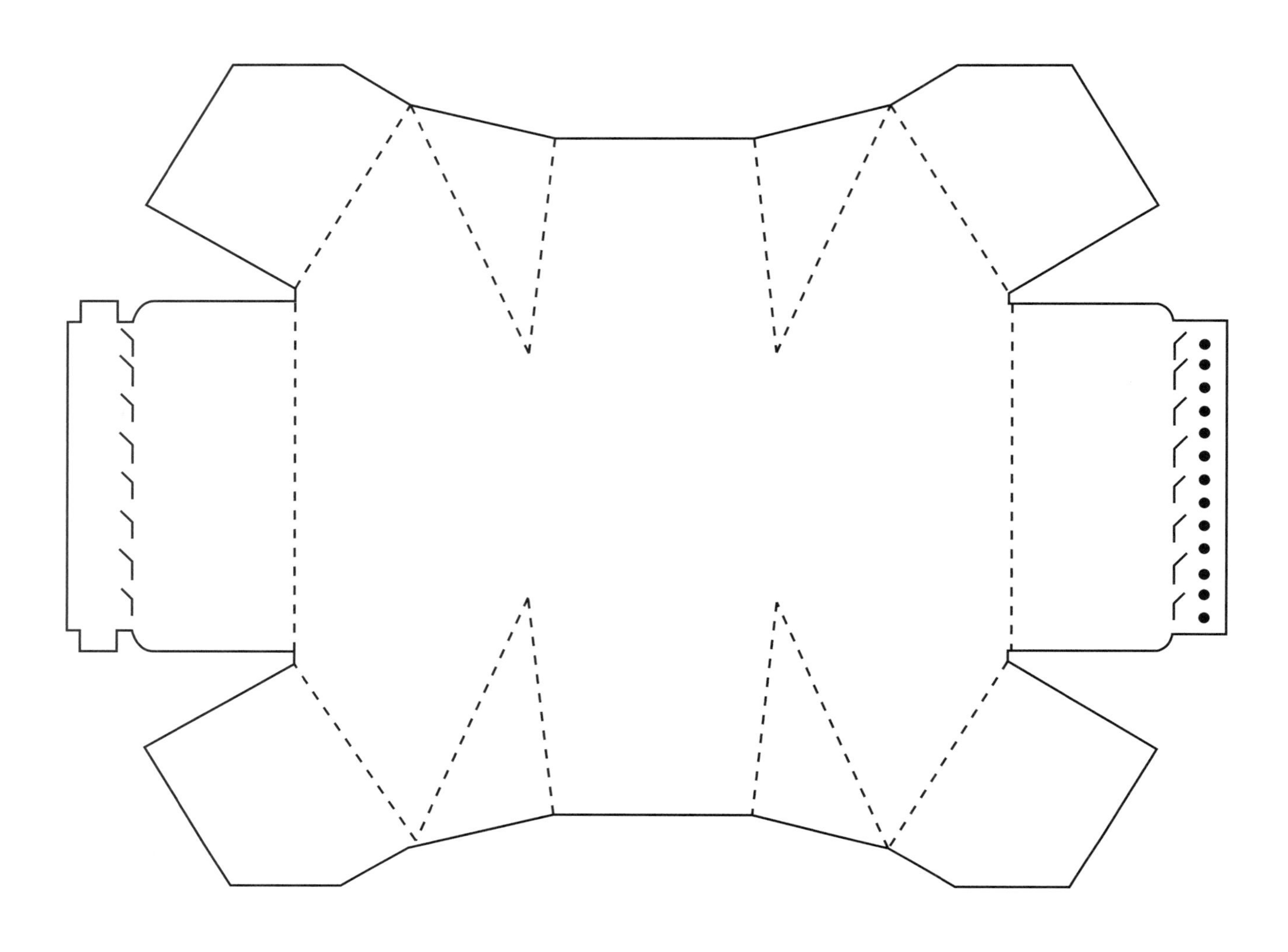

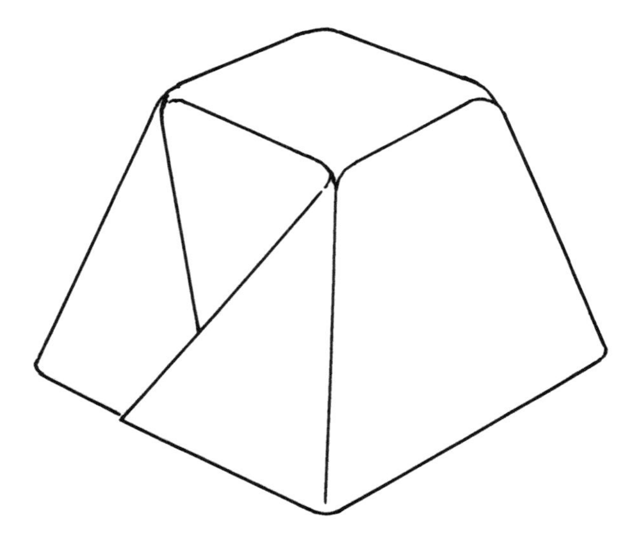

This fold-over wrap has been partially scored in order to create a soft, rounded top when assembled.

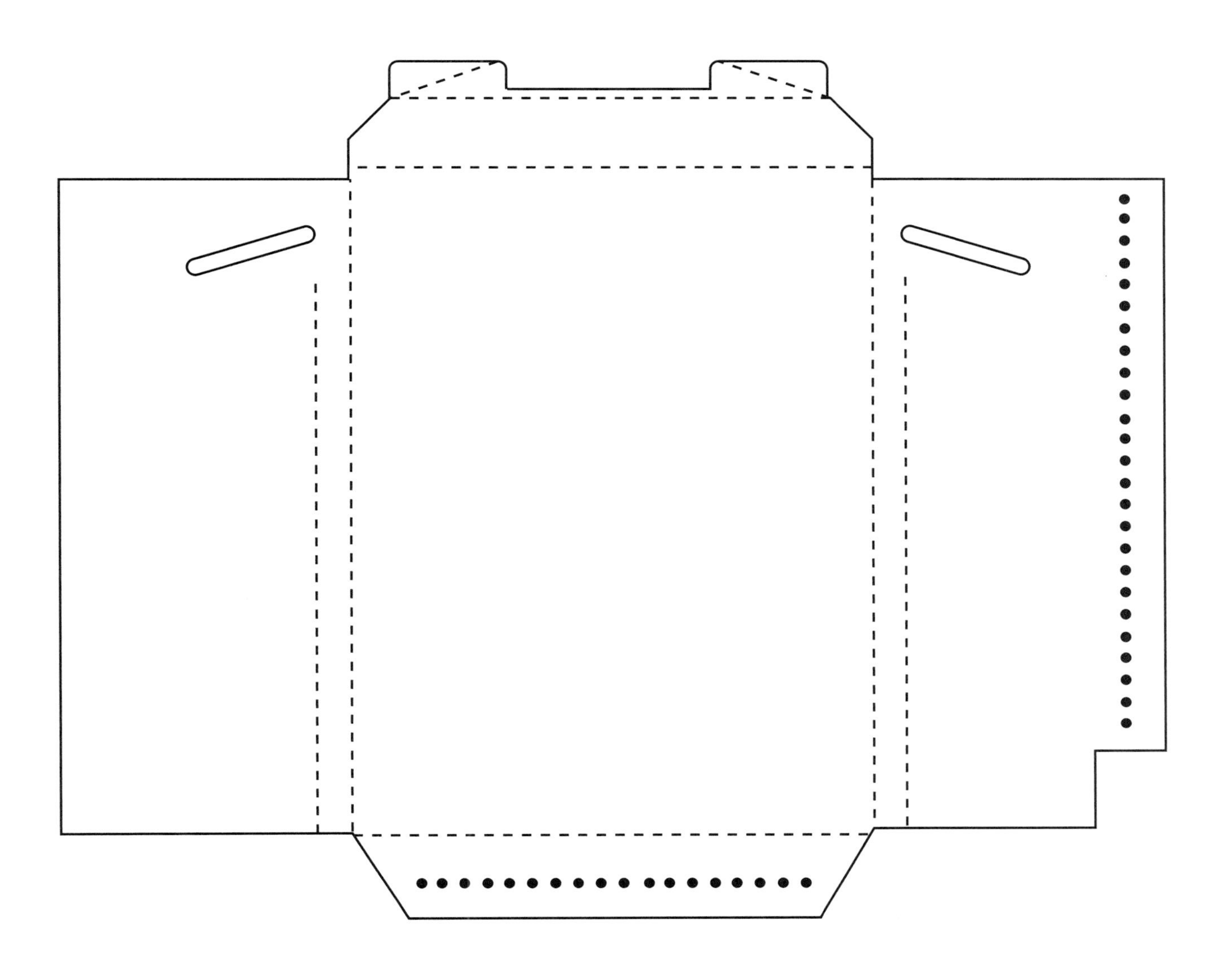

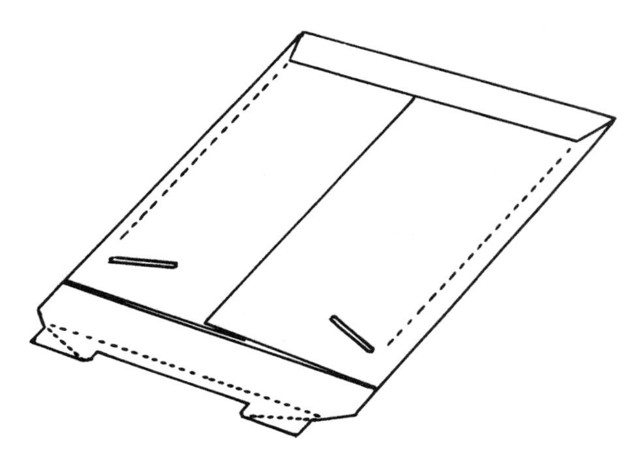

GIFT WRAP WITH INTERLOCKING COVER

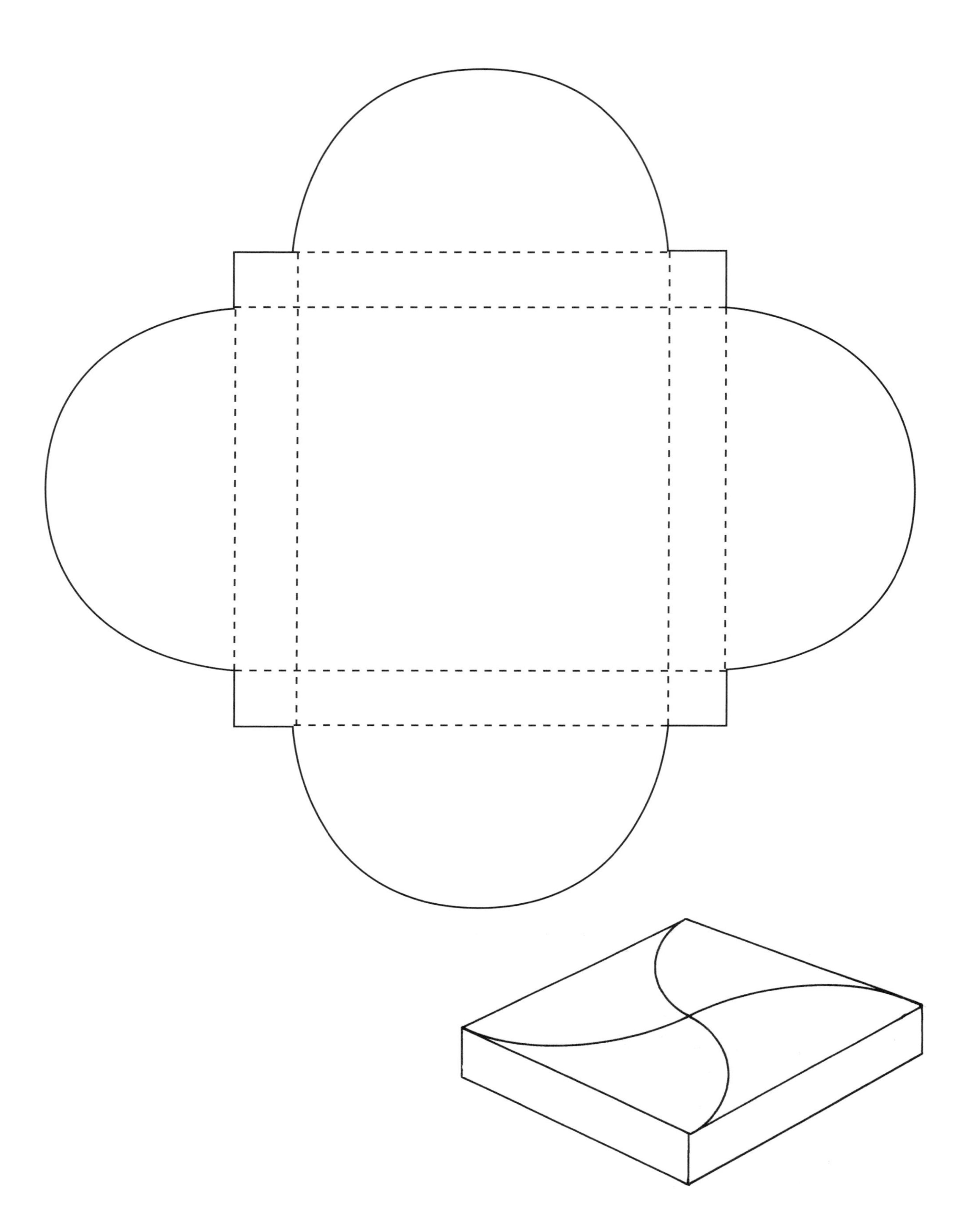

TYPICAL WRAP FOR CHOCOLATE BAR

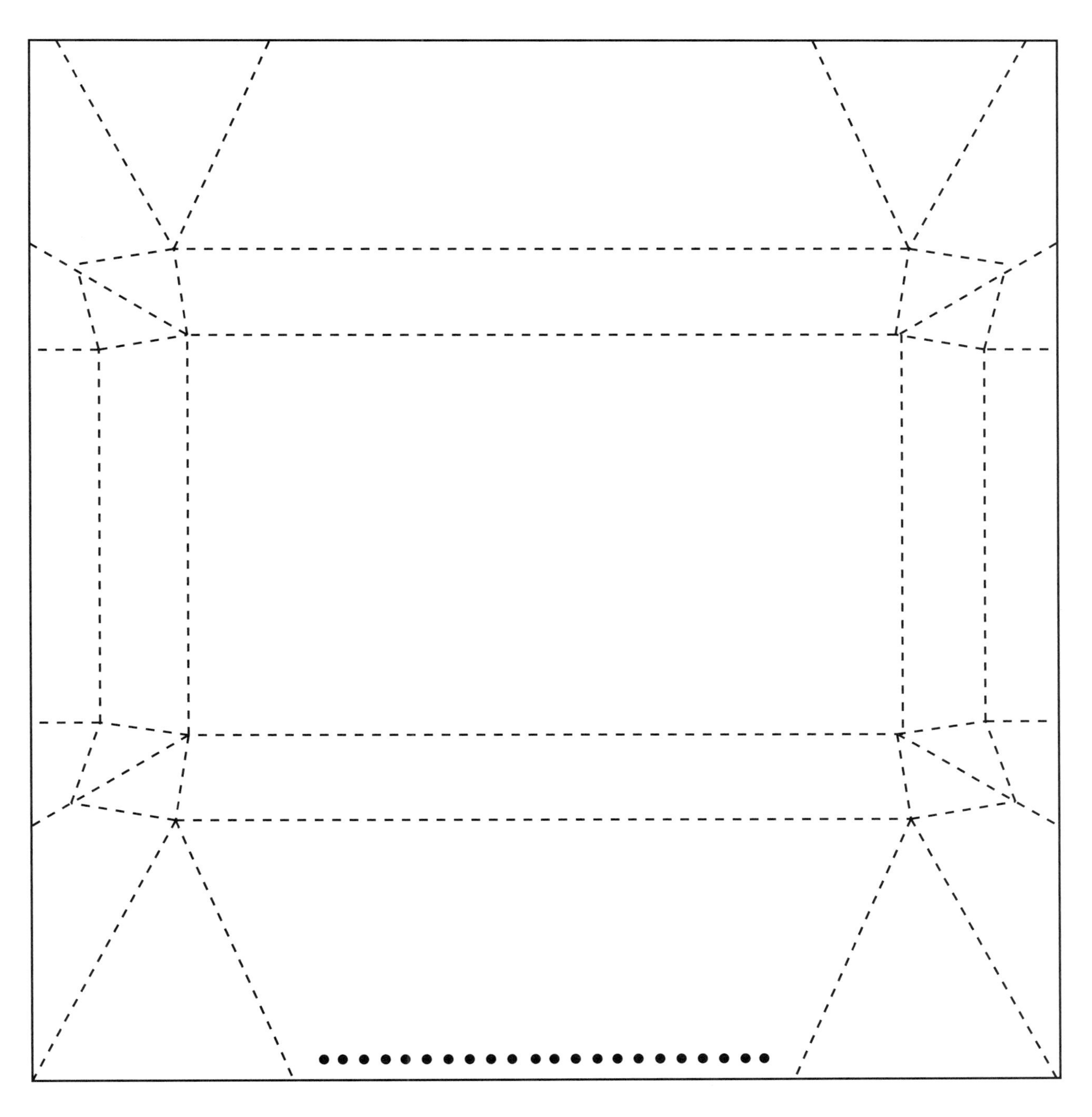

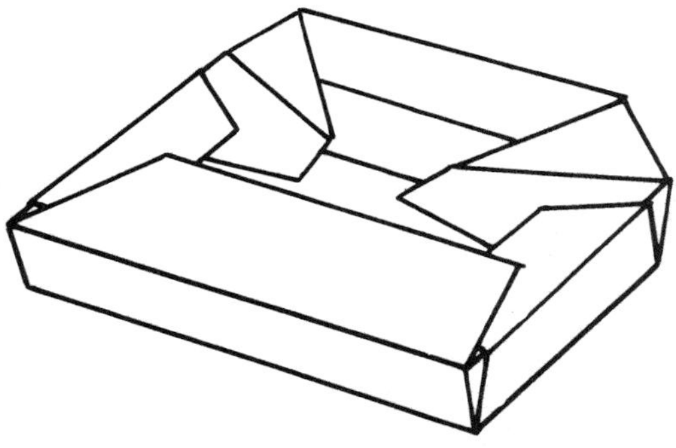

LAYOUT FOR SOAP BAR WRAP

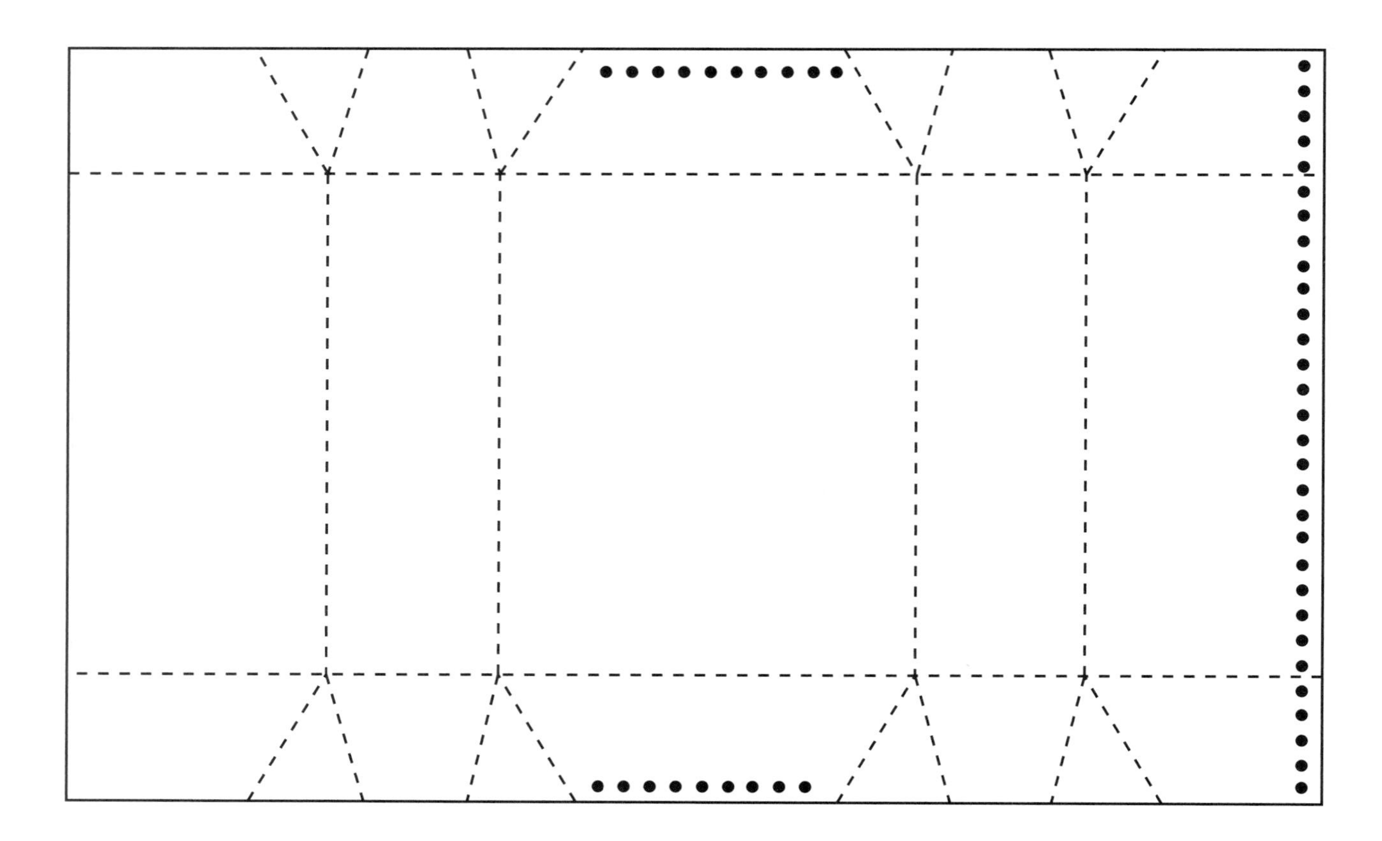

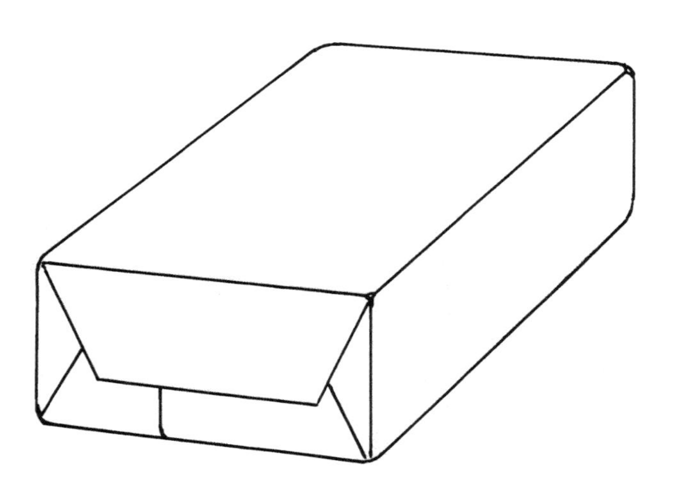

ONE PIECE FOLDER

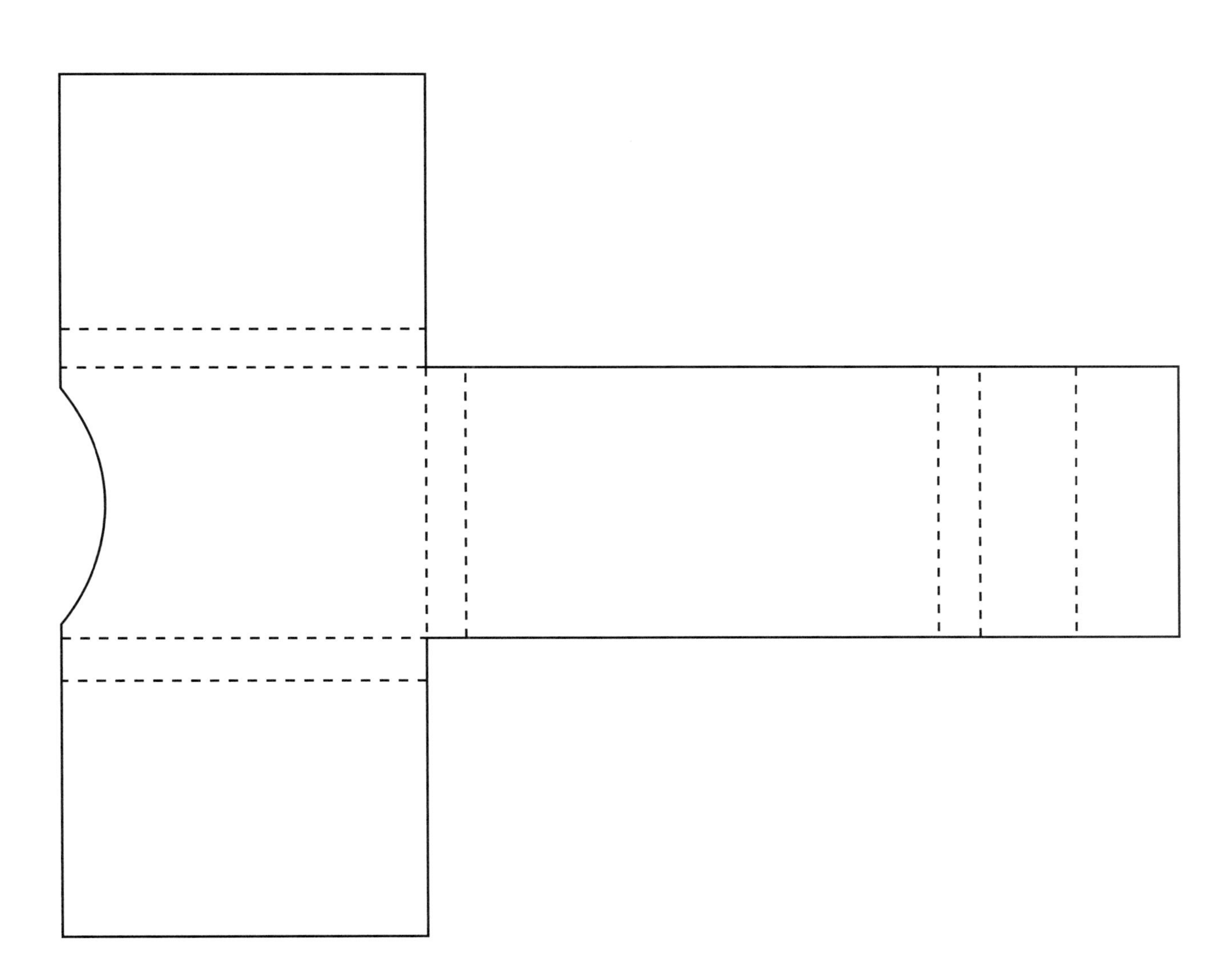

Suitable for cards as well as haberdashery items.

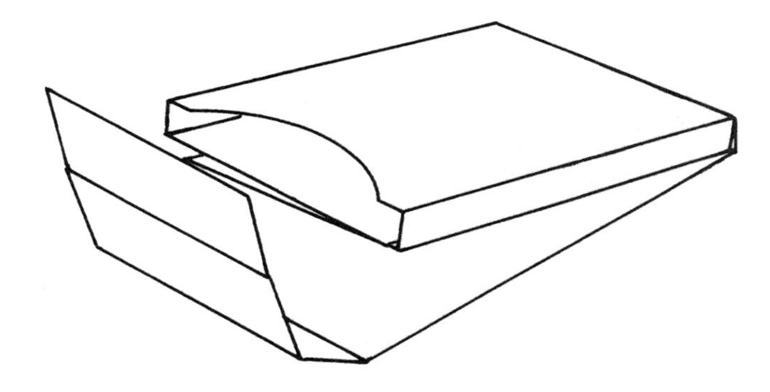

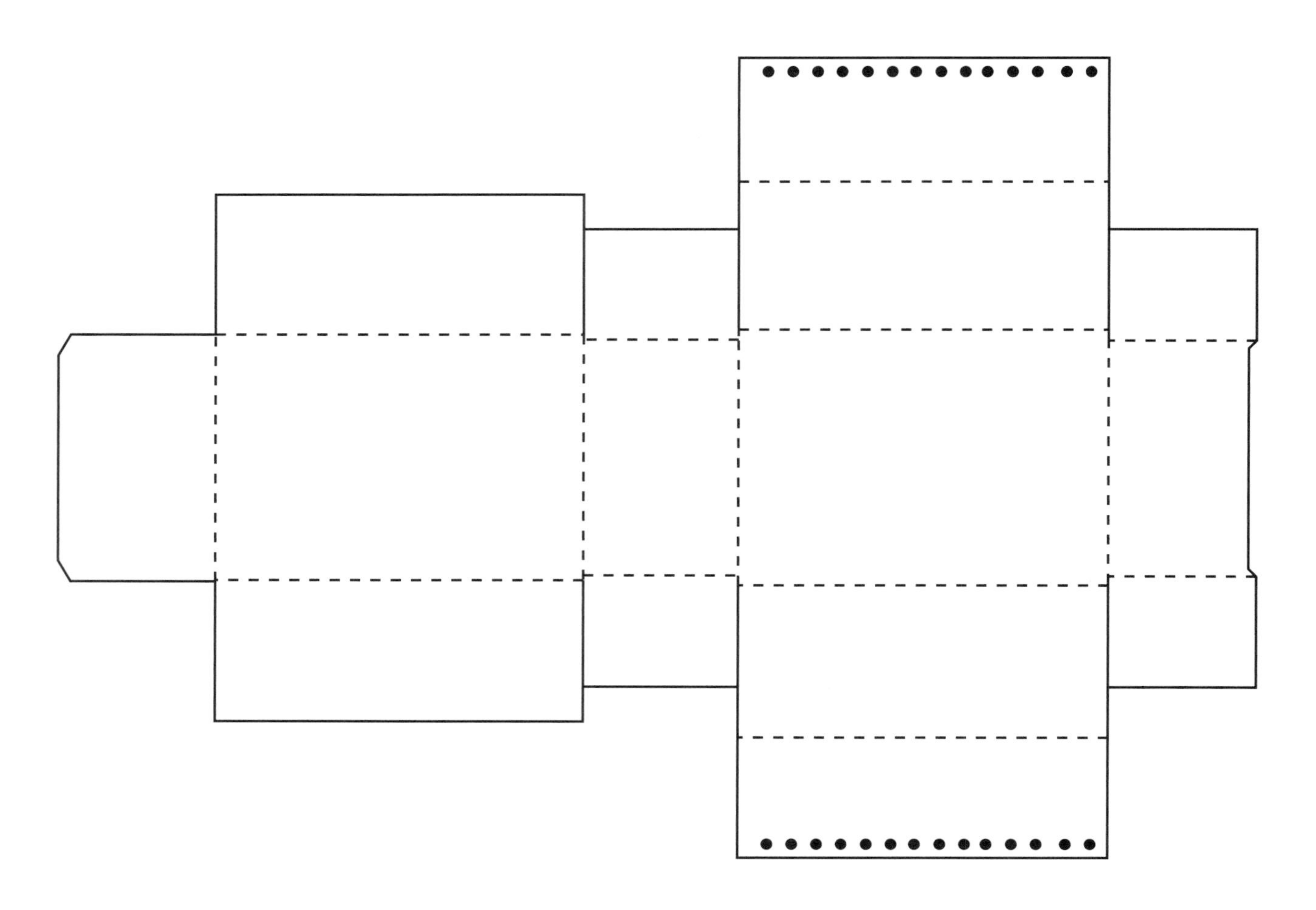

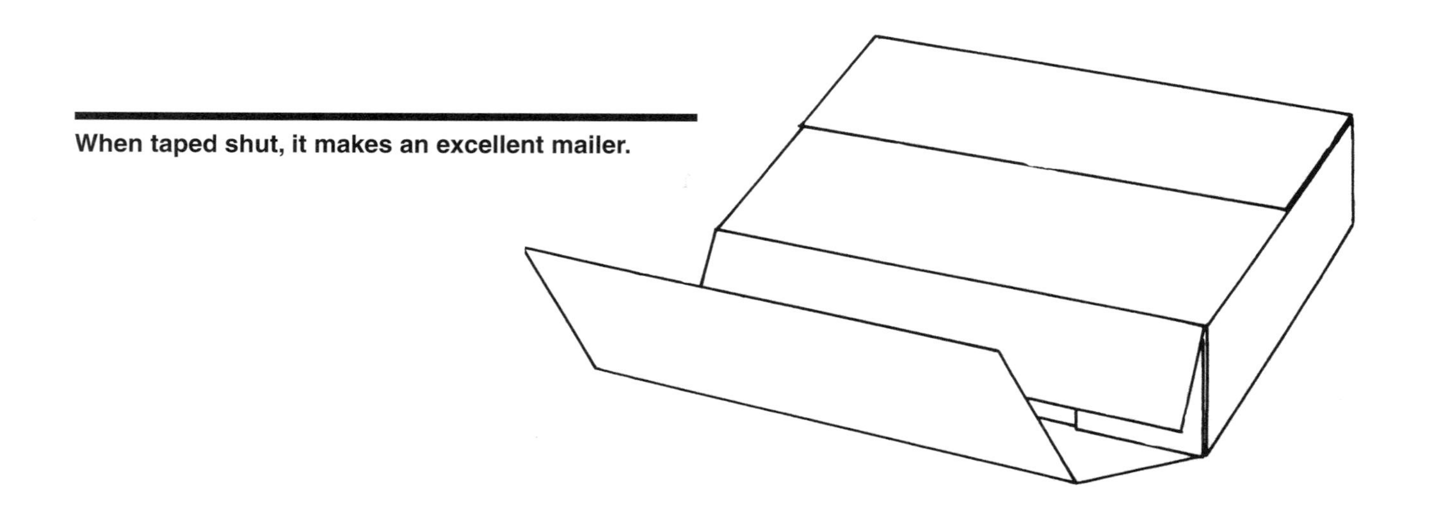

When taped shut, it makes an excellent mailer.

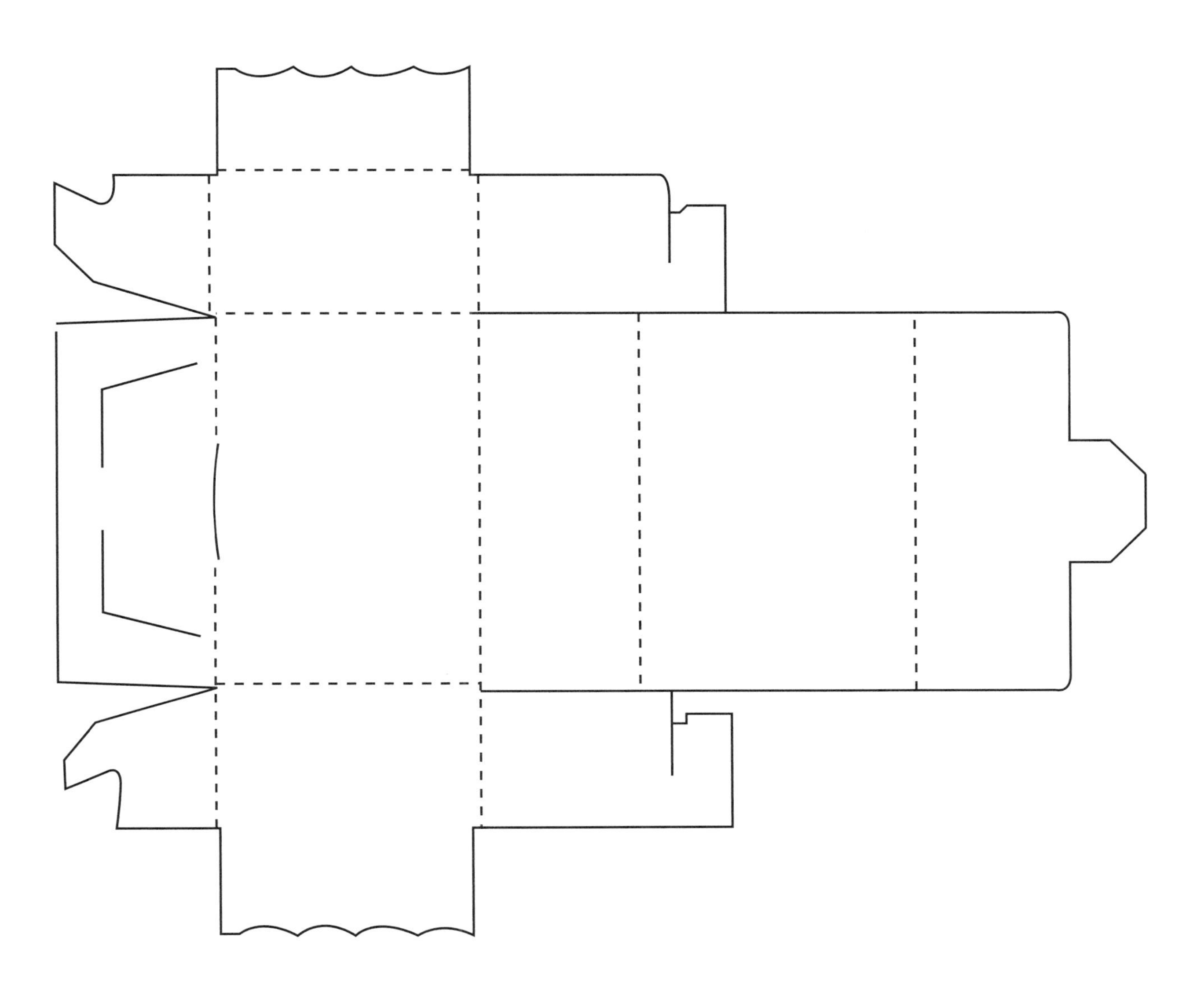

This box with Arthur locks and hook locks needs no glue. It makes an excellent container for jewelry items.

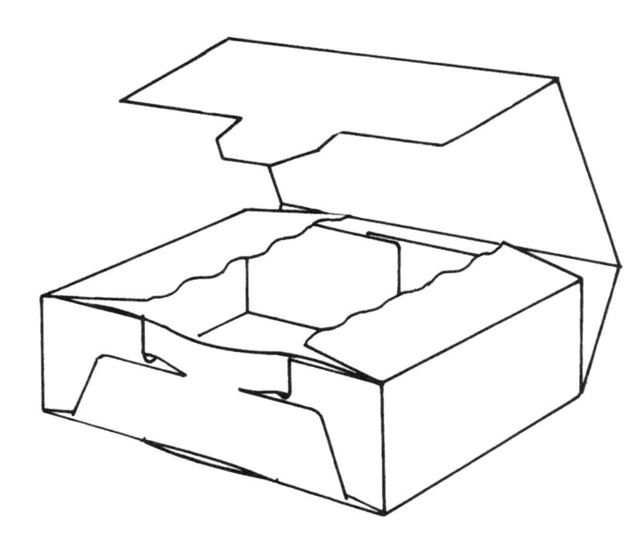

ONE PIECE WRAP WITH LOCK TAB

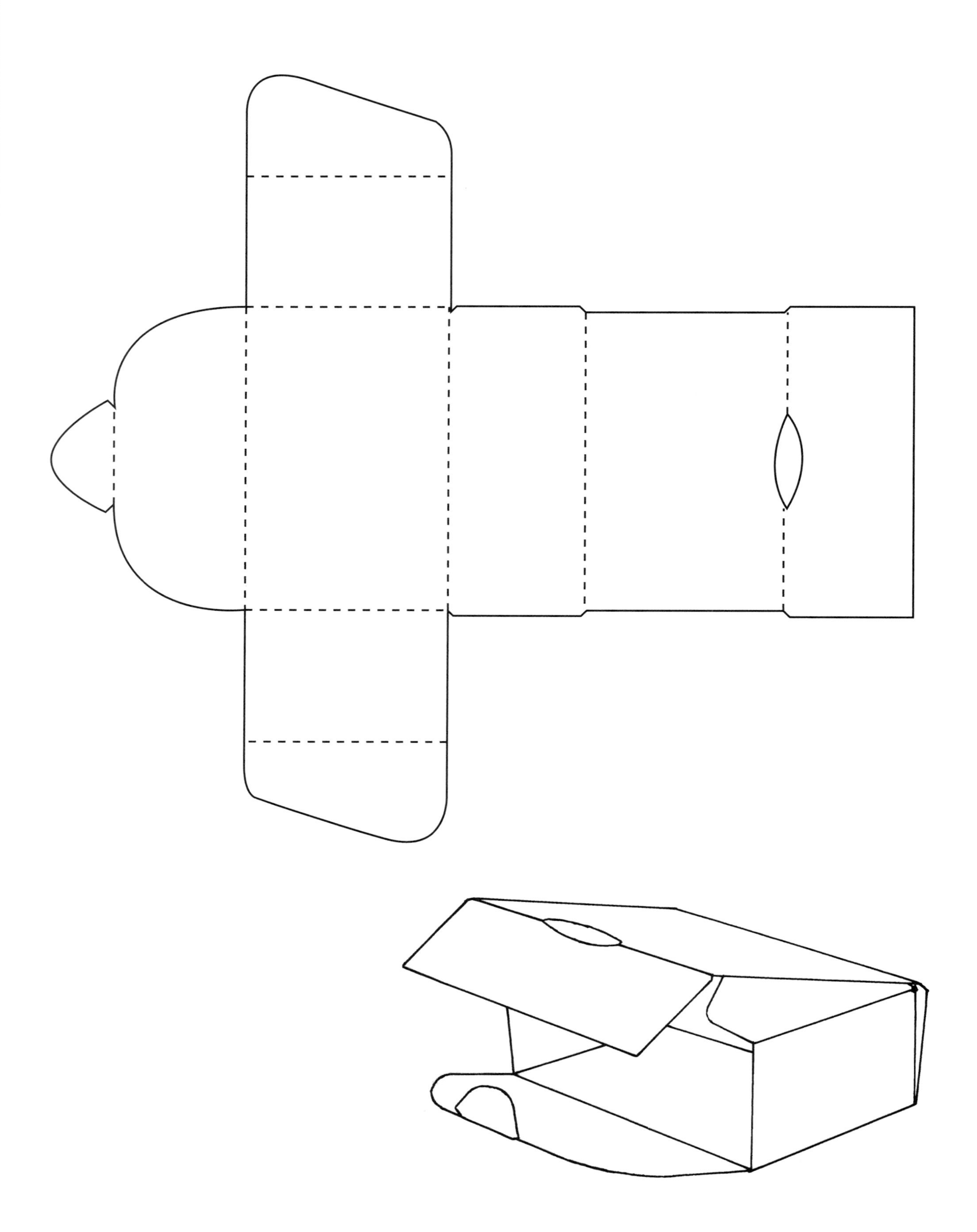

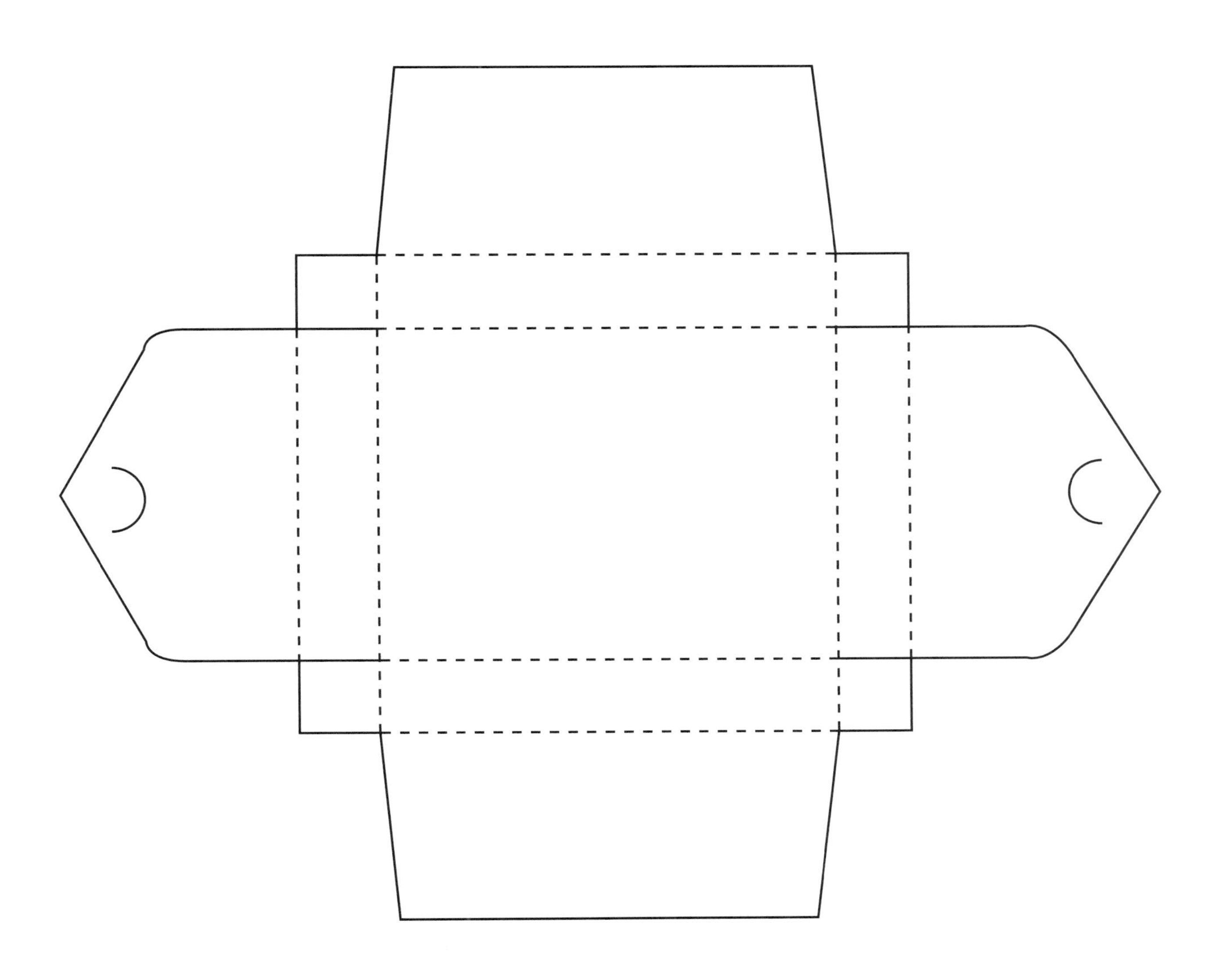

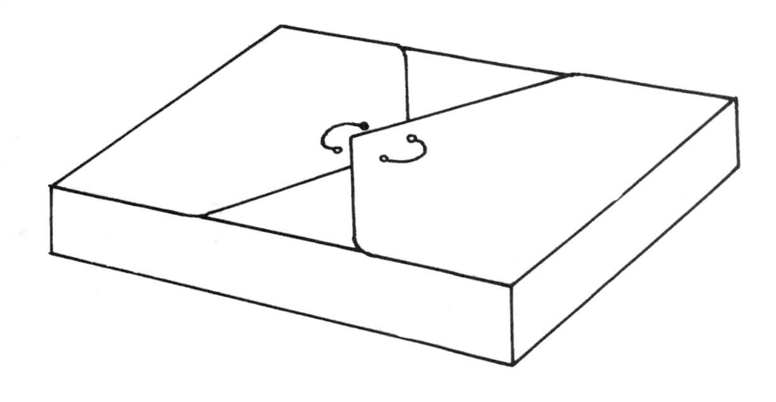

COMPACT DISK SLEEVE

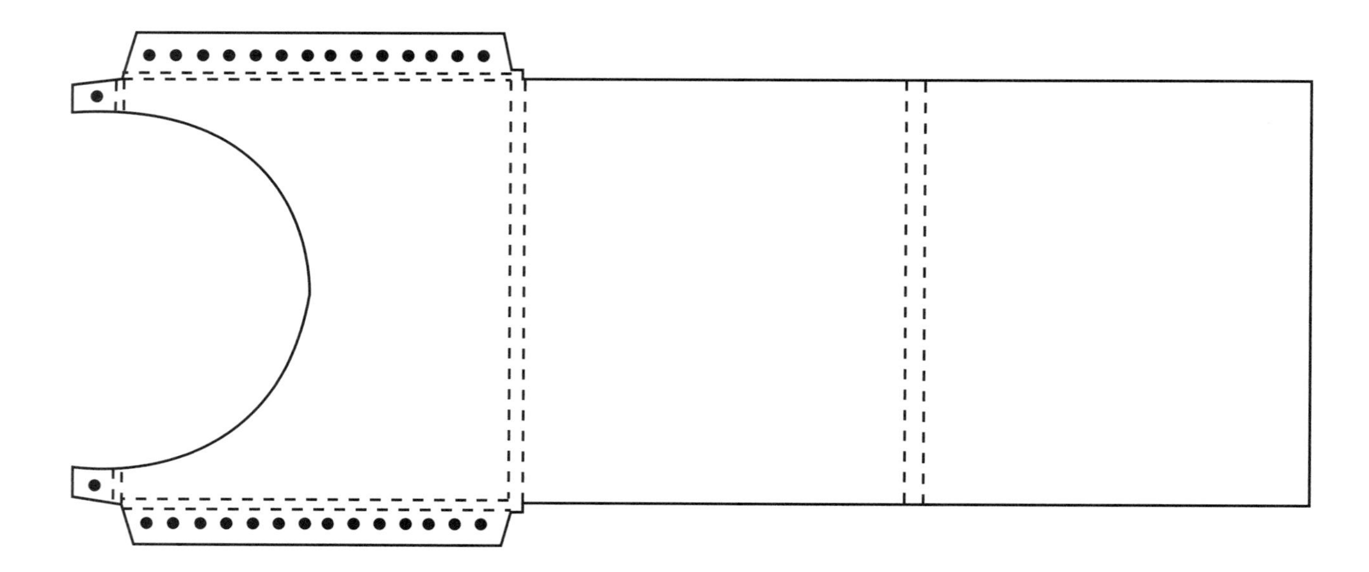

A simple, economical, one-piece compact disc folder with good product protection, easy product access, and sufficient printing surface for product information.

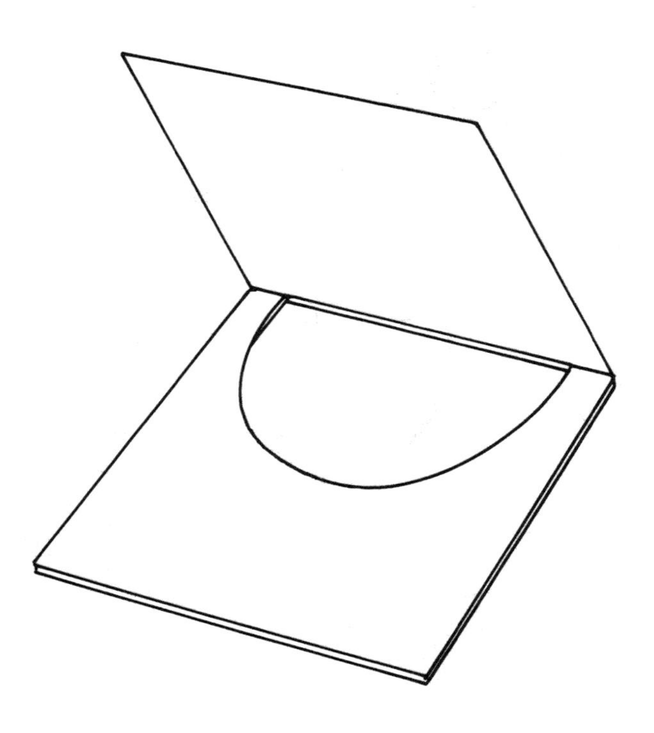

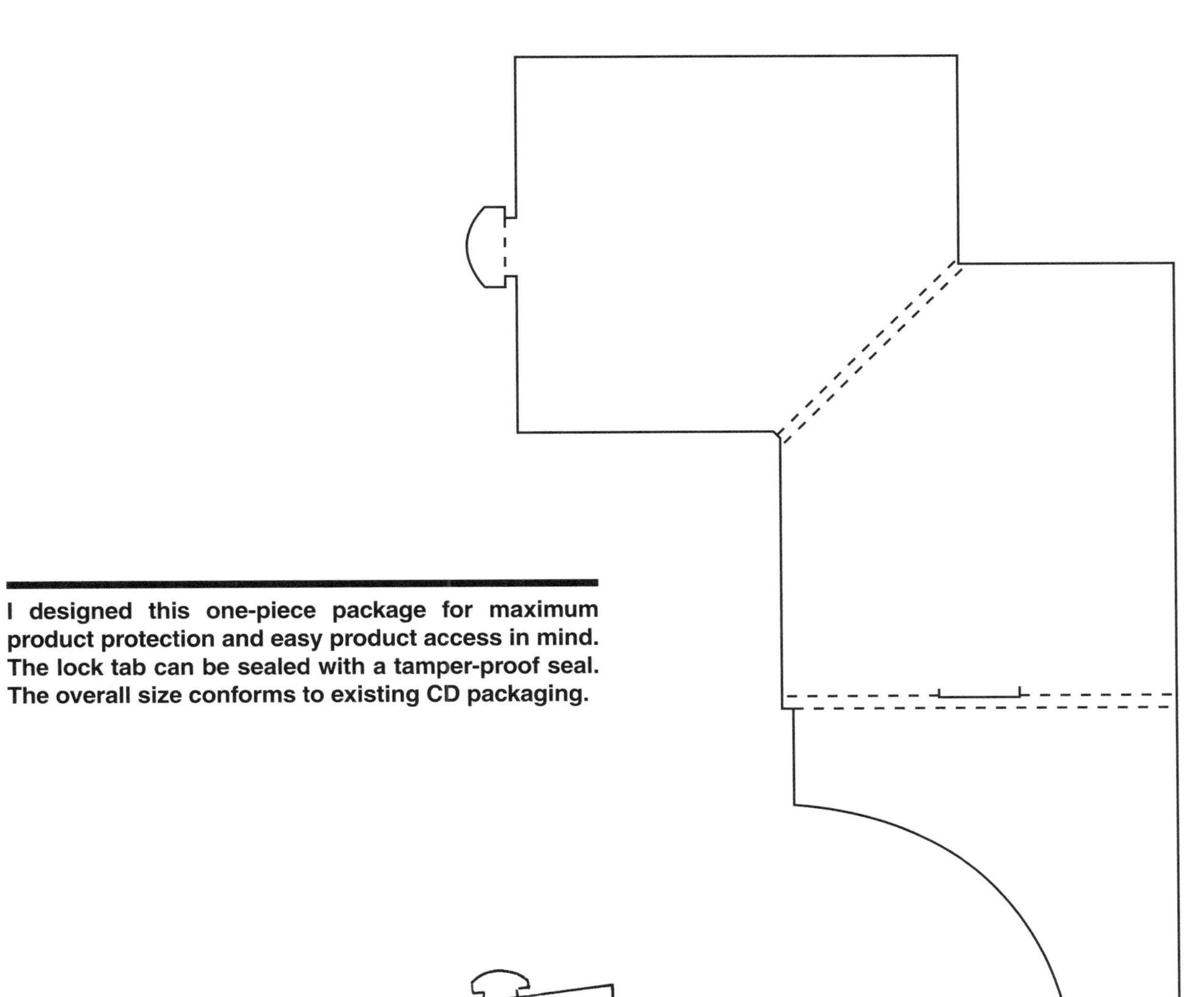

I designed this one-piece package for maximum product protection and easy product access in mind. The lock tab can be sealed with a tamper-proof seal. The overall size conforms to existing CD packaging.

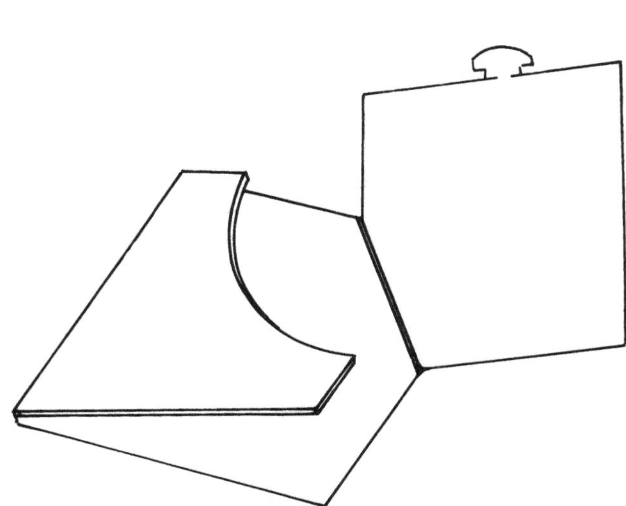

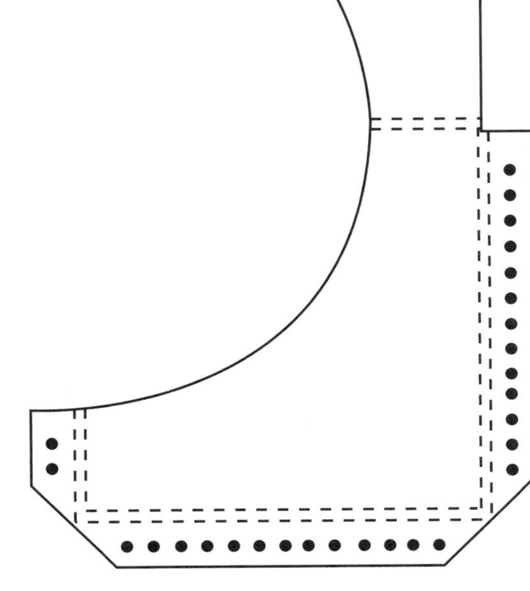

COMPACT DISC SLEEVE

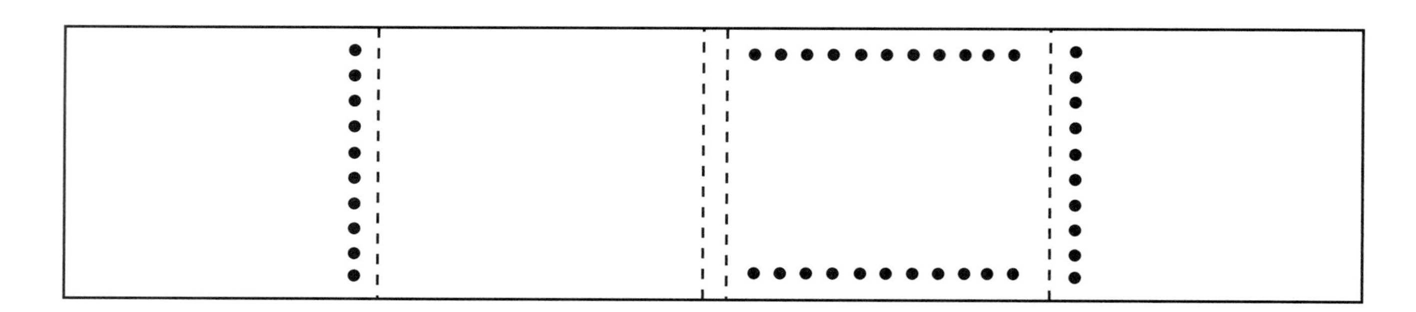

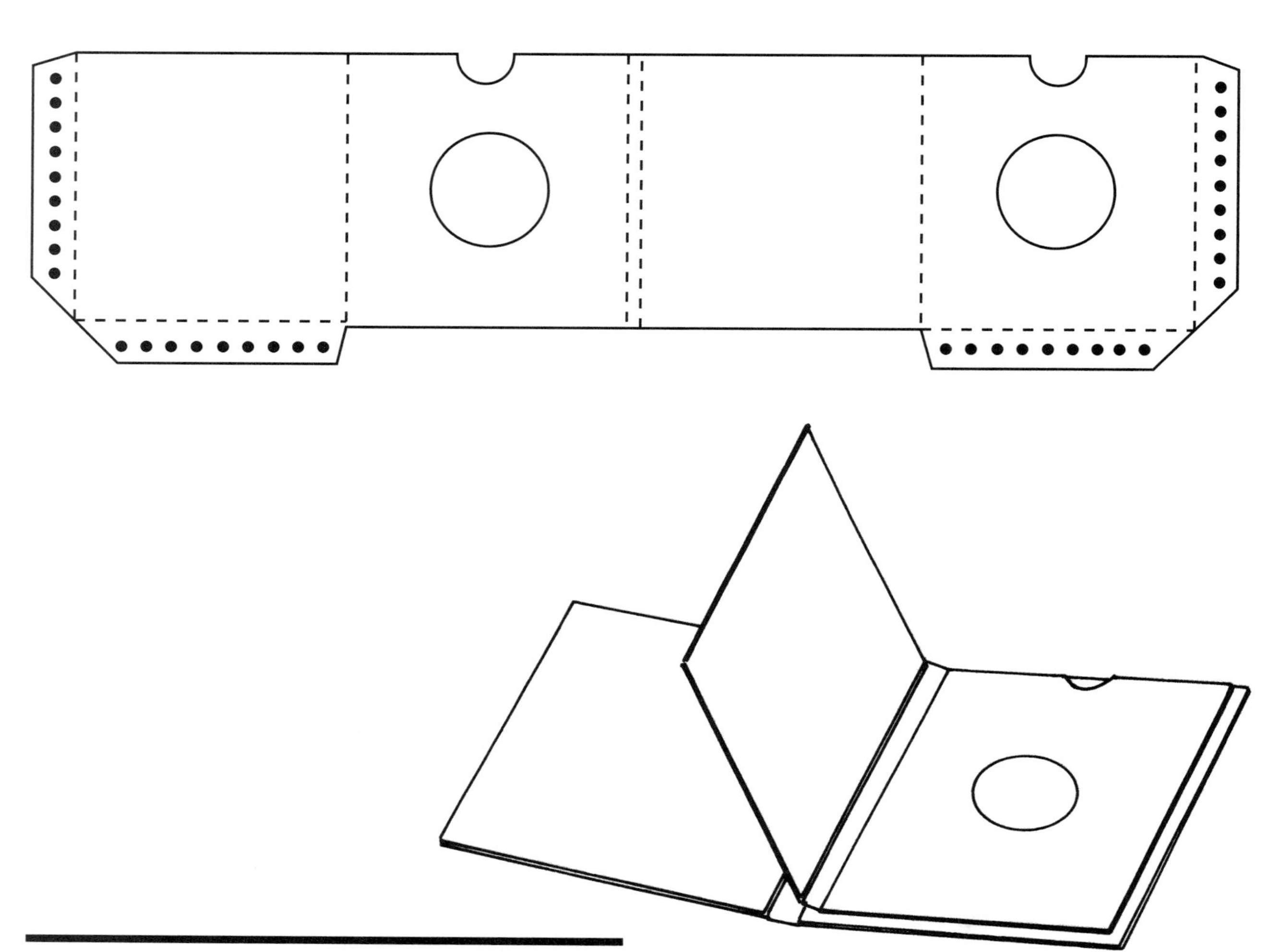

Two piece 2 CD sleeve based on the old vinyl gramophone record packaging.

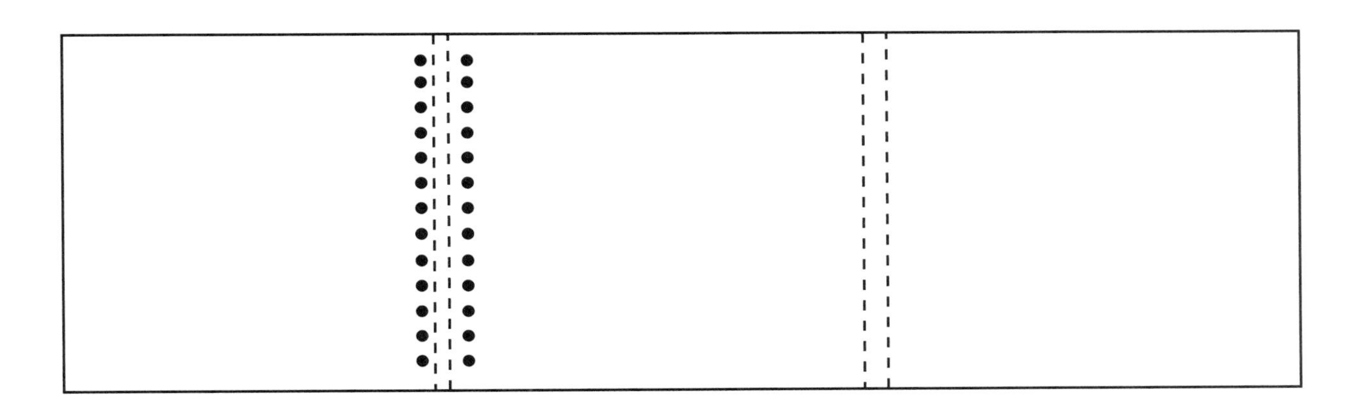

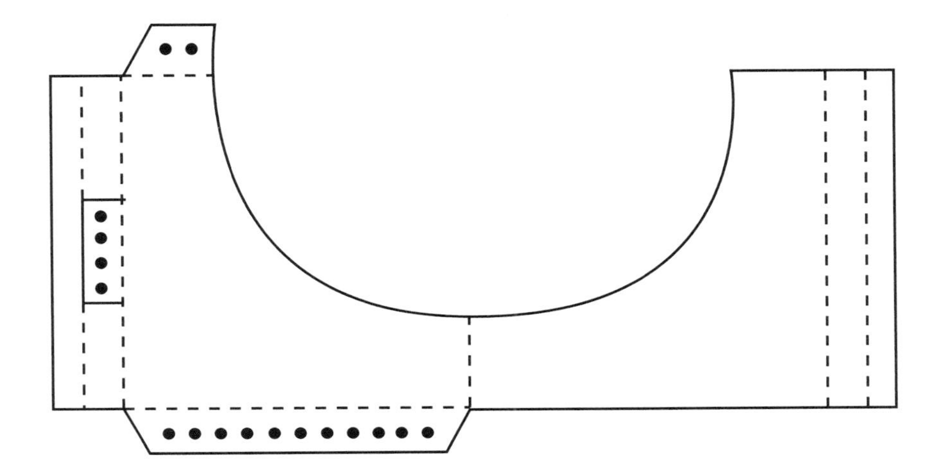

This GW compact disc folder pops up when the outer sleeve is opened. The CD may be accessed by its spindle hole, thus avoiding contact with grooves.

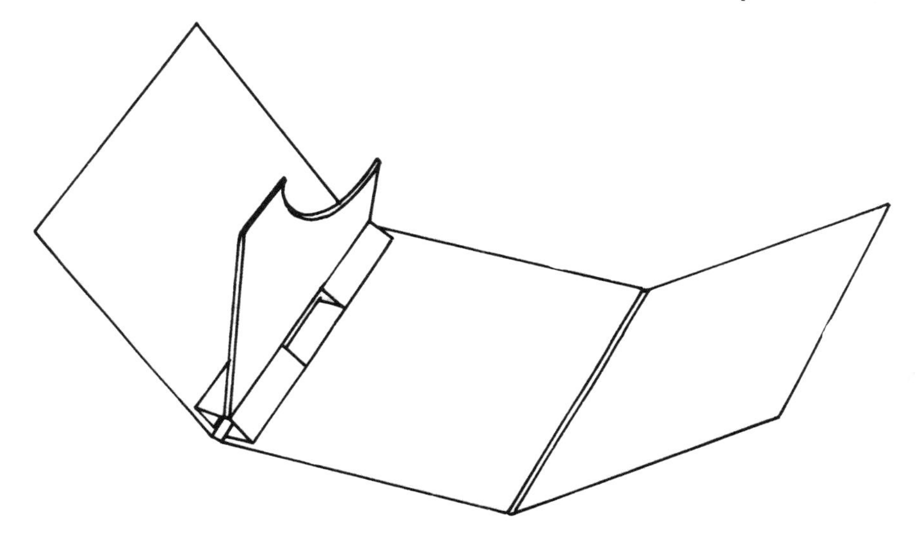

ASEPTIC PACKAGE

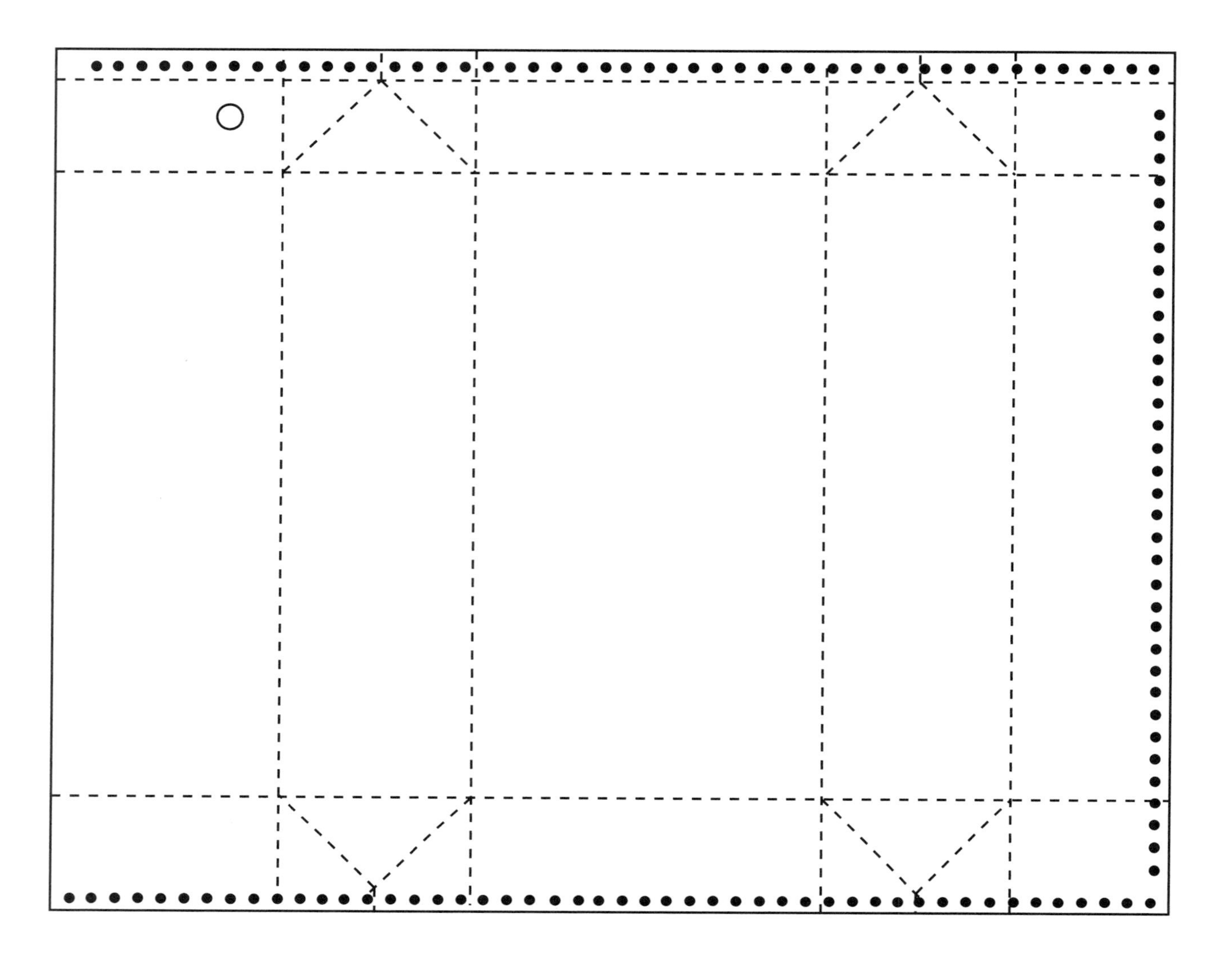

The package is formed and sterilized around the product.

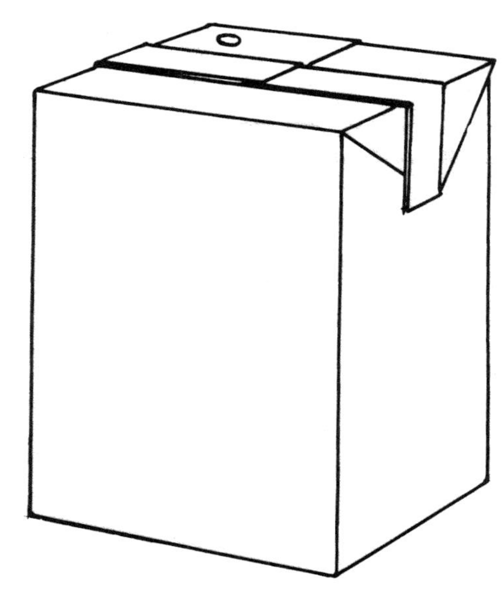

Also suitable for novelty items, hosiery, ect.

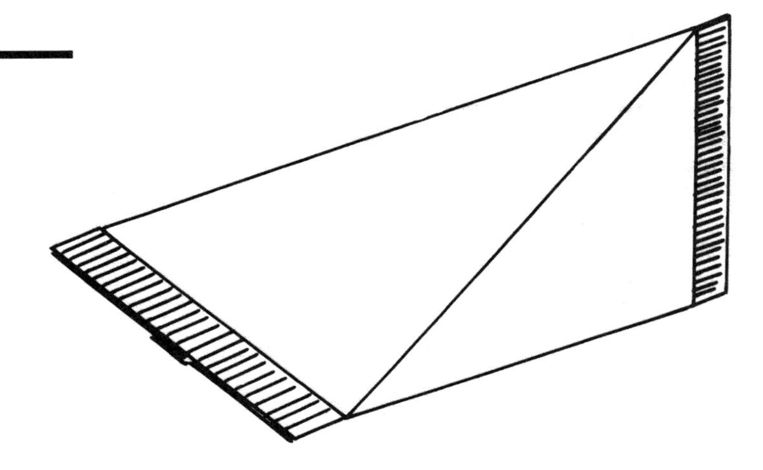

P.D.Q. OR ARCUATE FOLDER

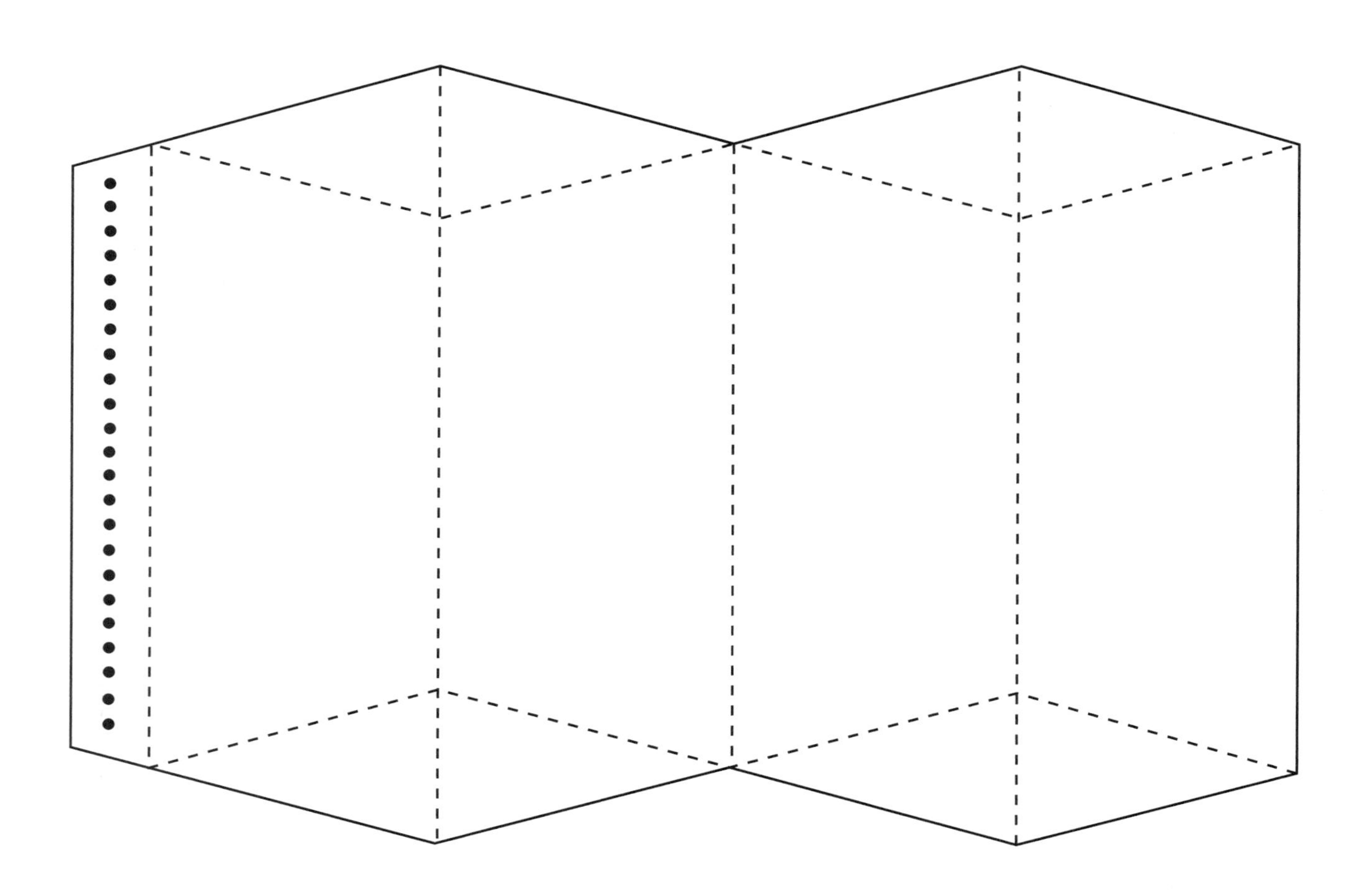

The end closures snap in.

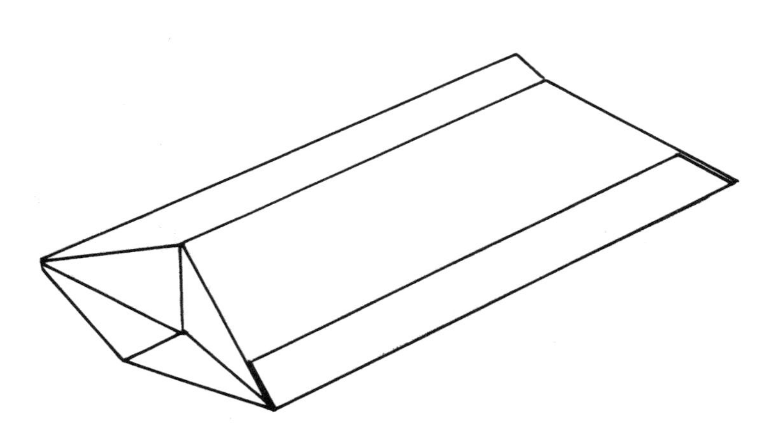

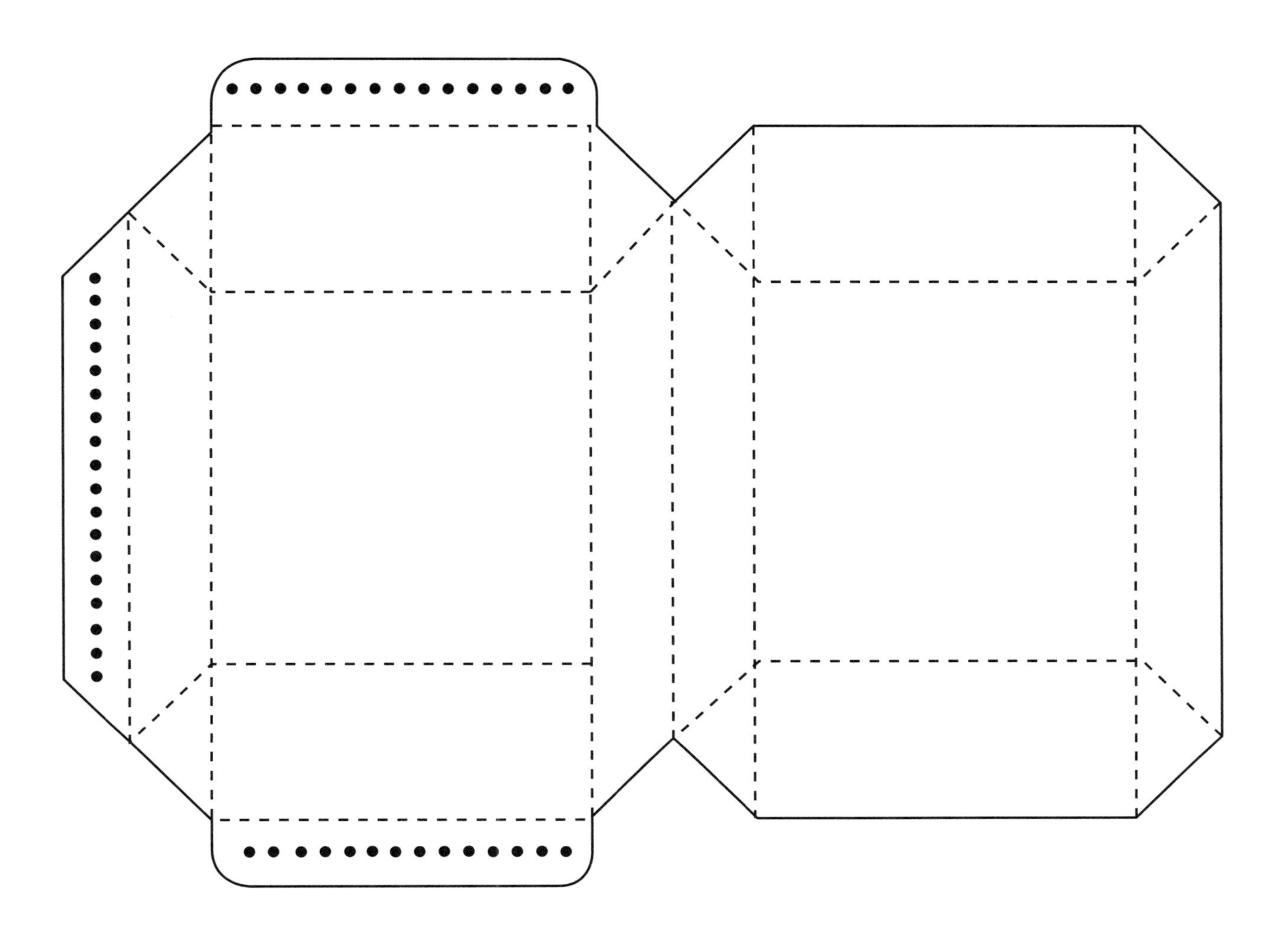

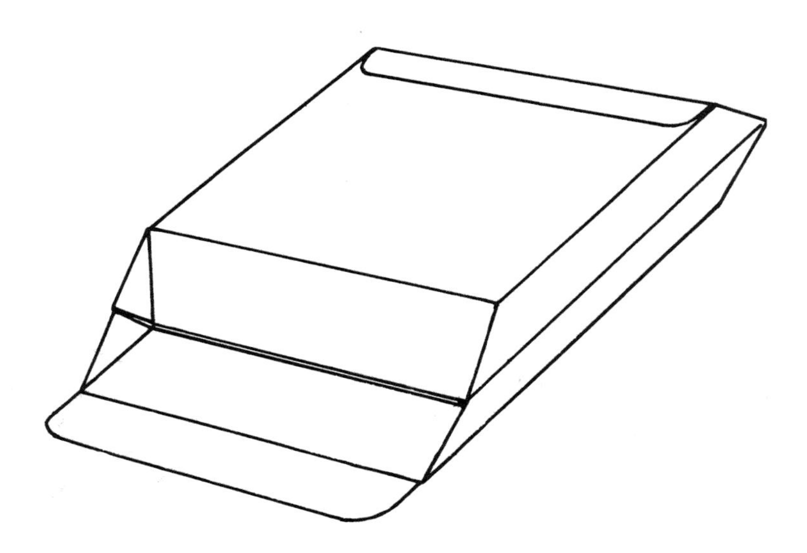

PILLOW PACK

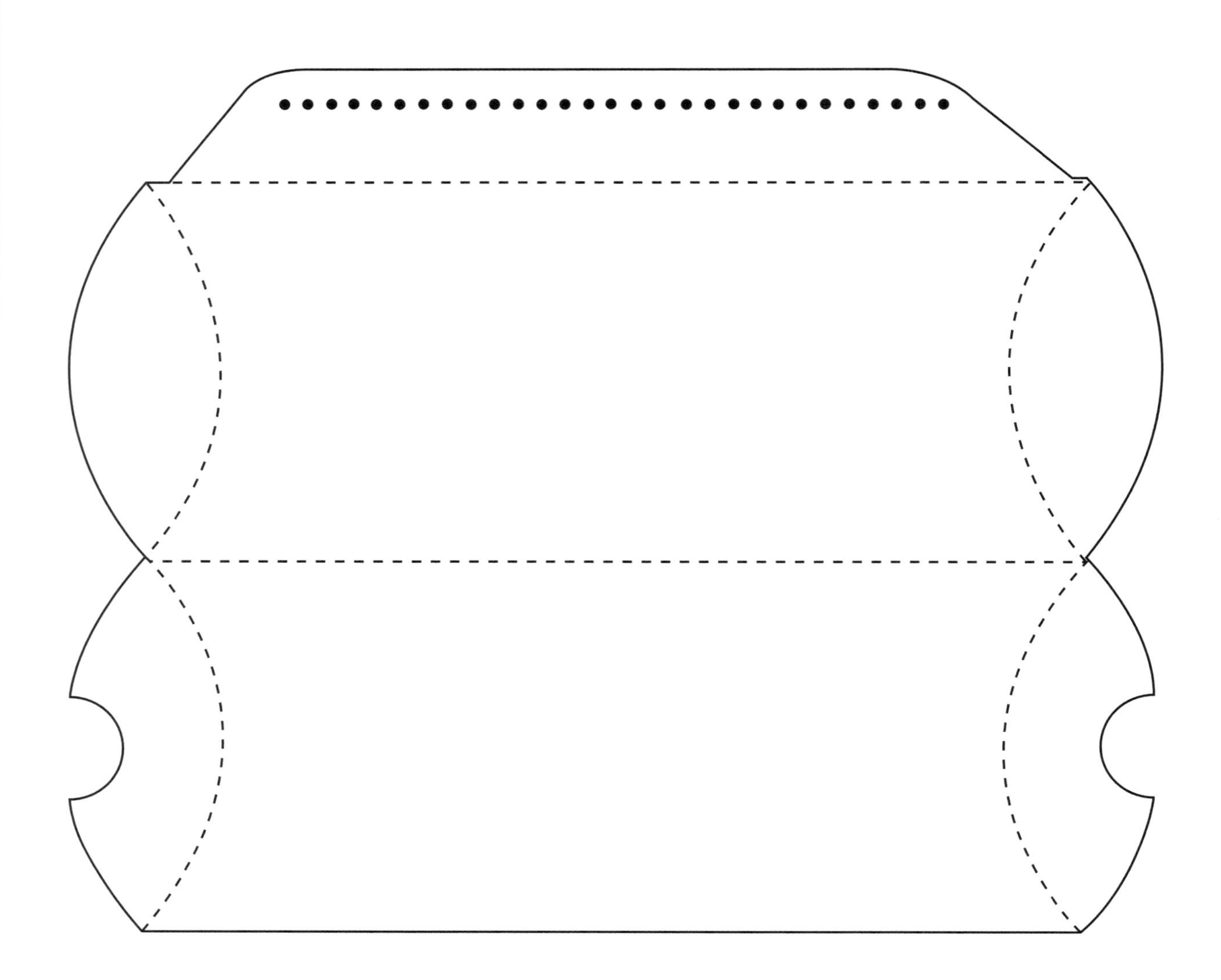

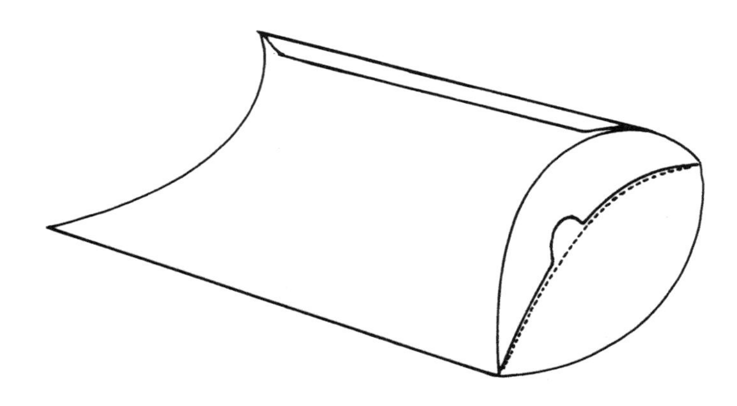

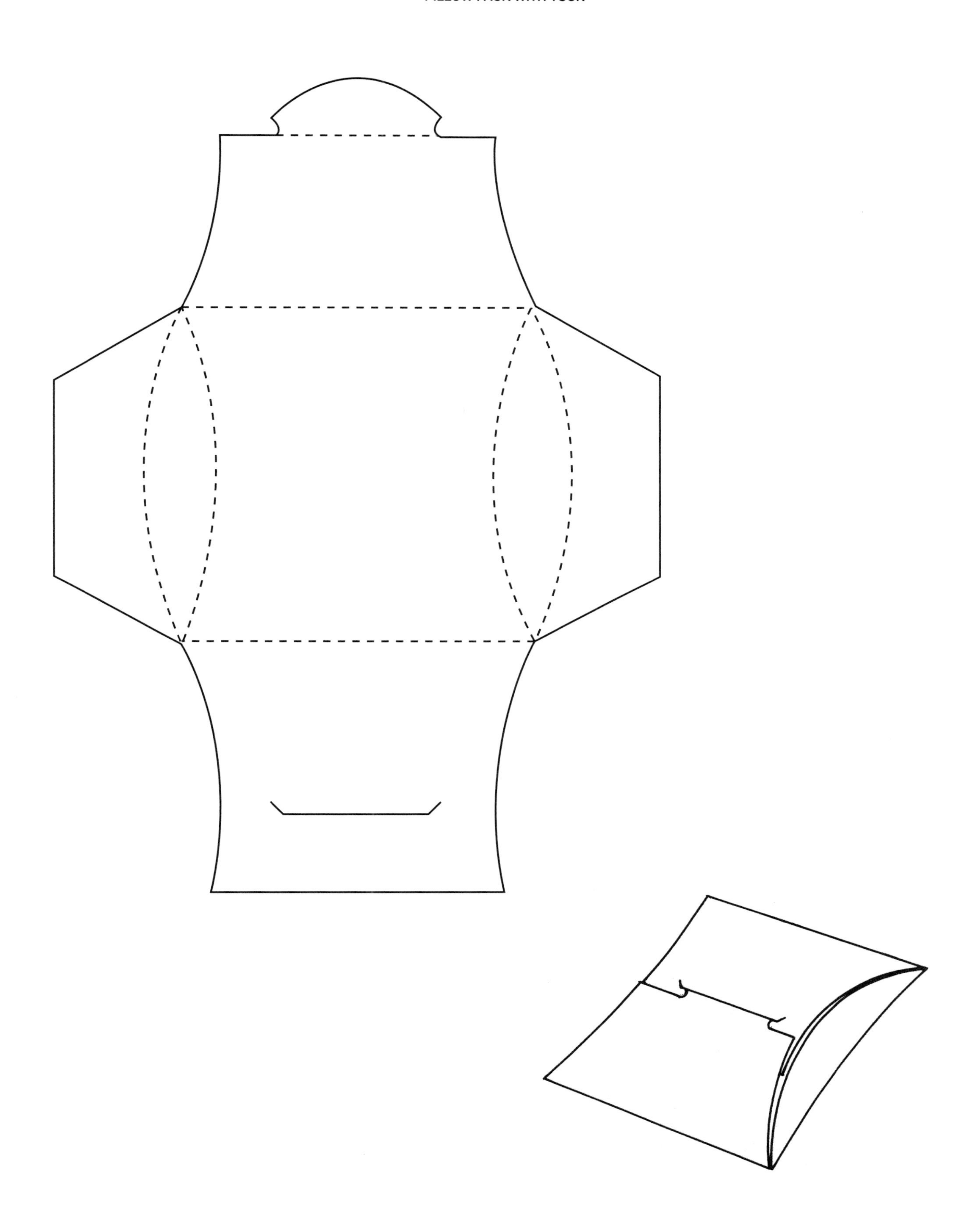

PILLOW PACK WITH HANDLE

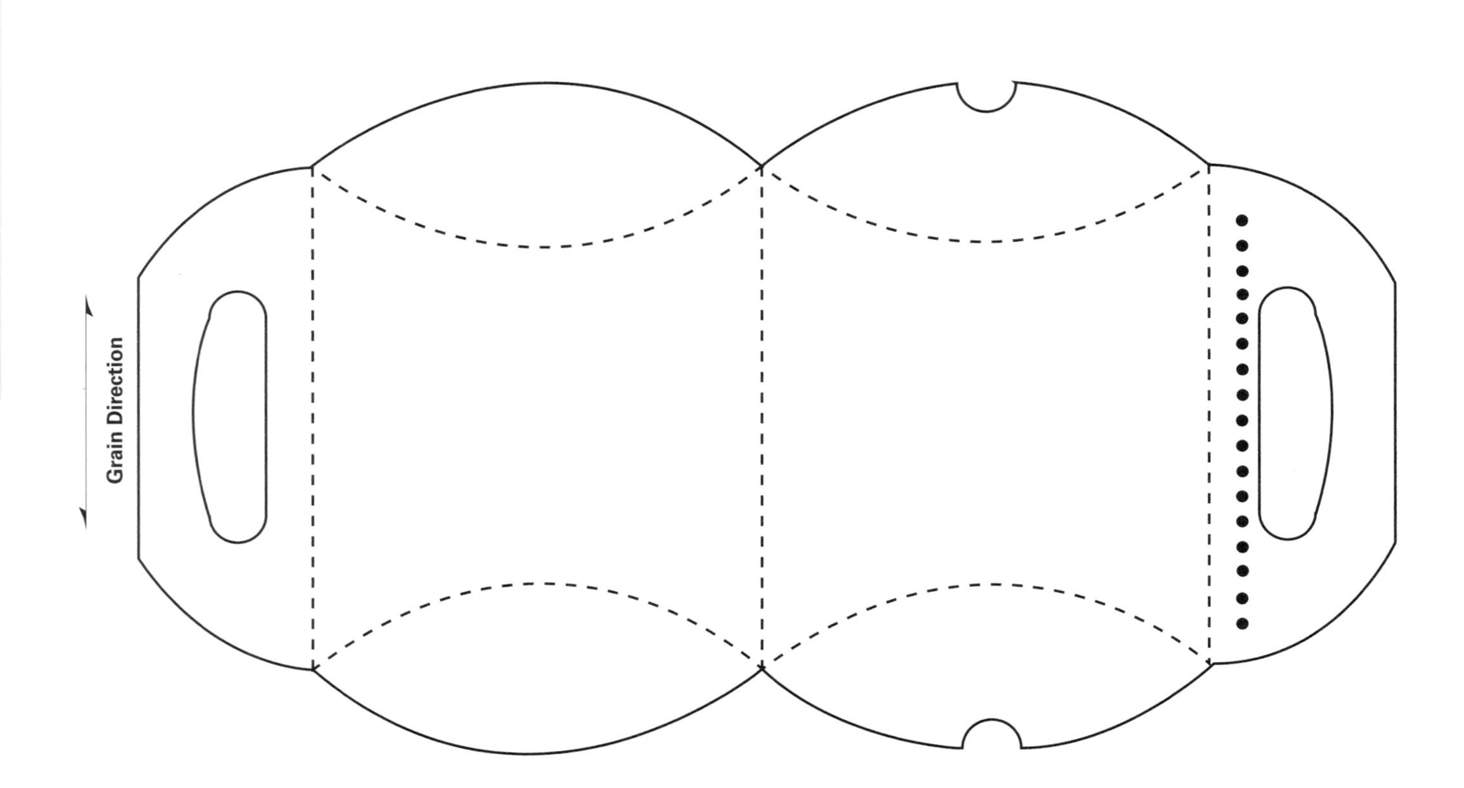

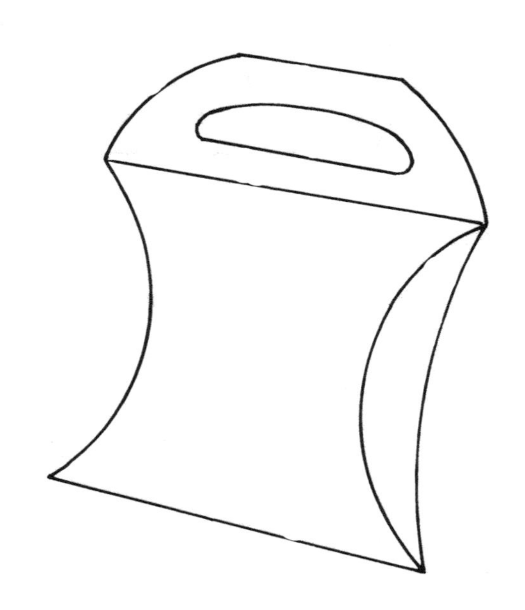

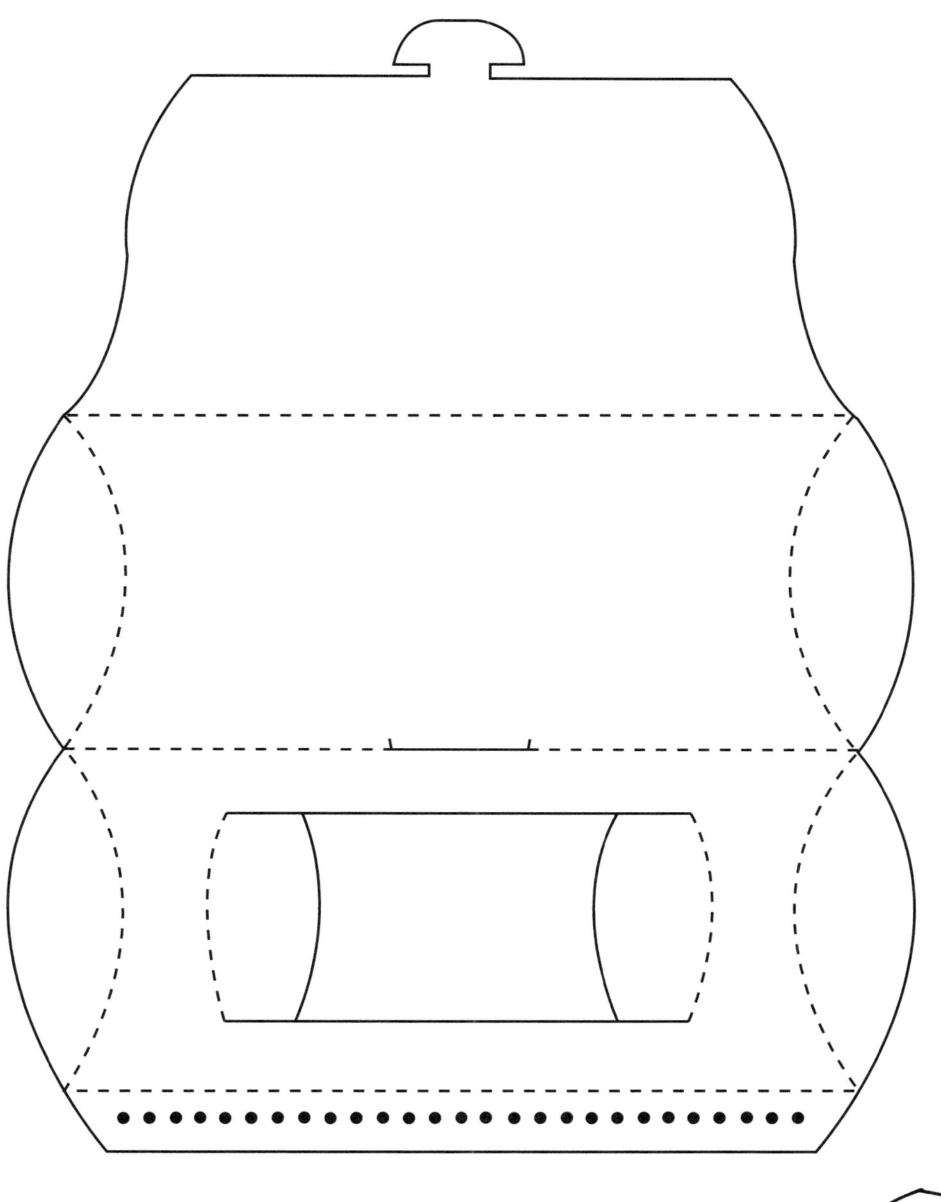

This Pillow Pack features a cushioned compartment.

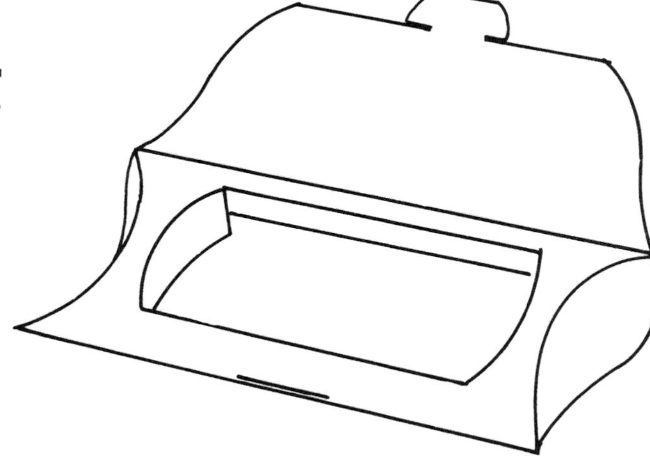

PILLOW PACK VARIATION

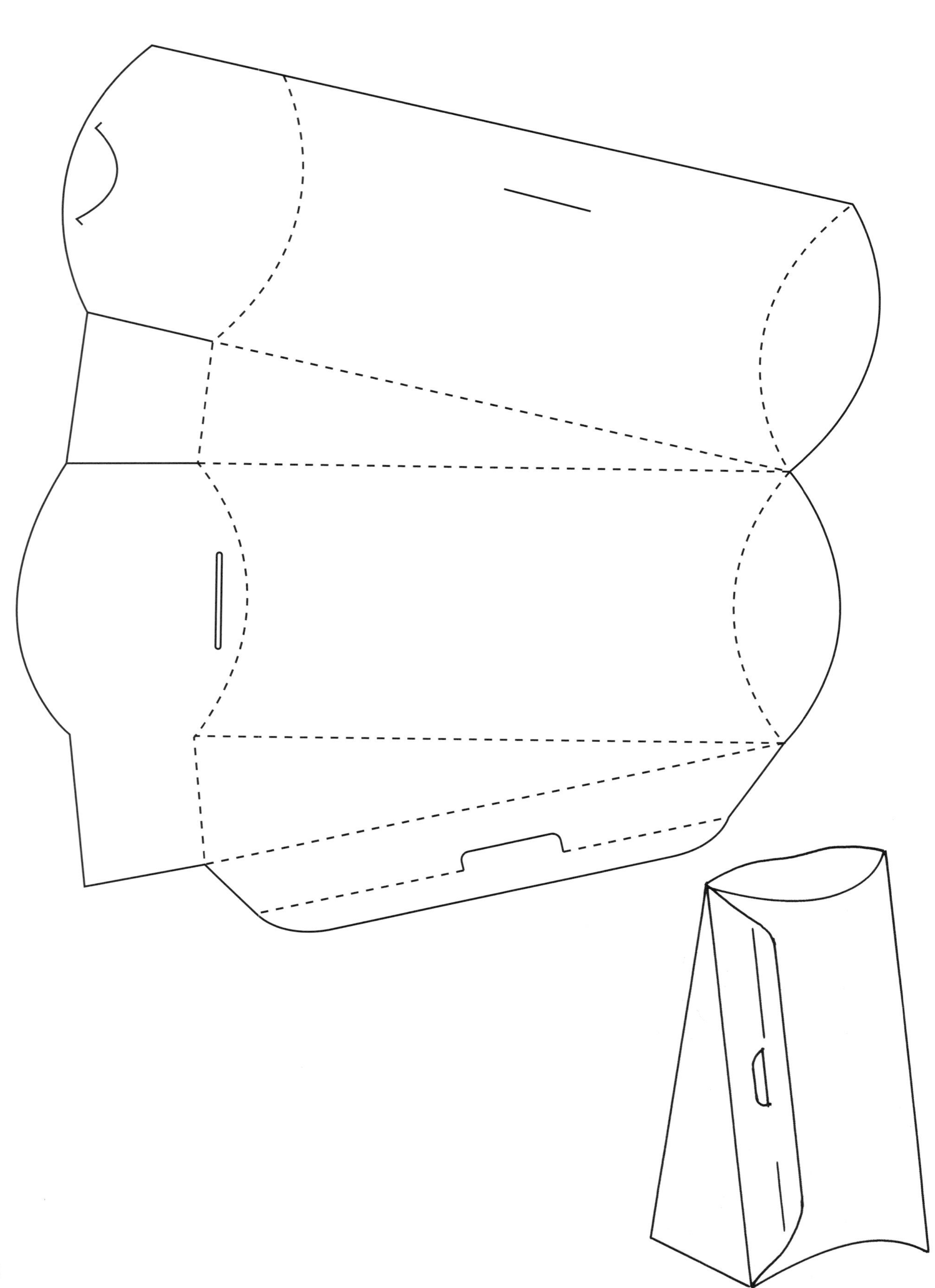

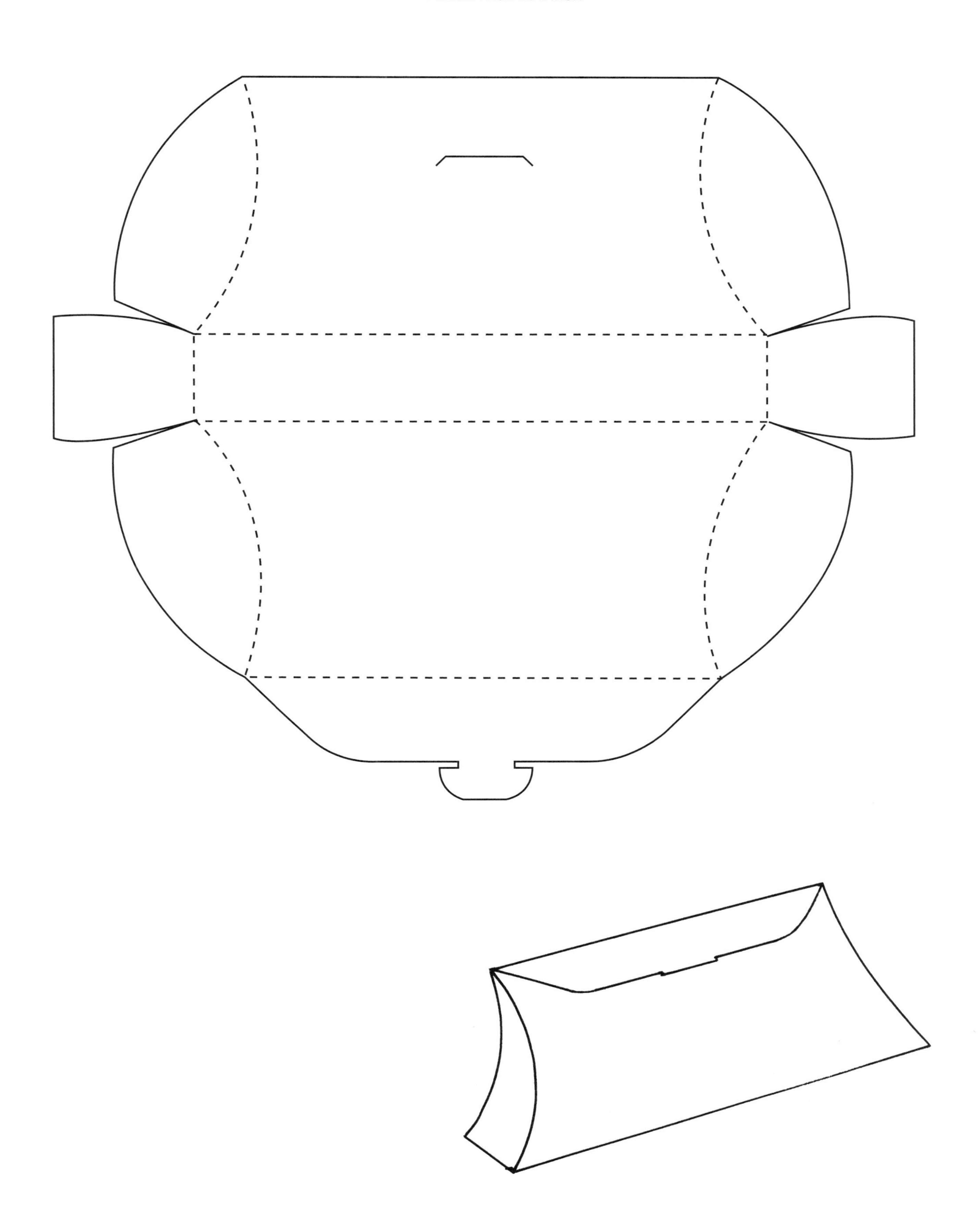

PILLOW PACK VARIATION

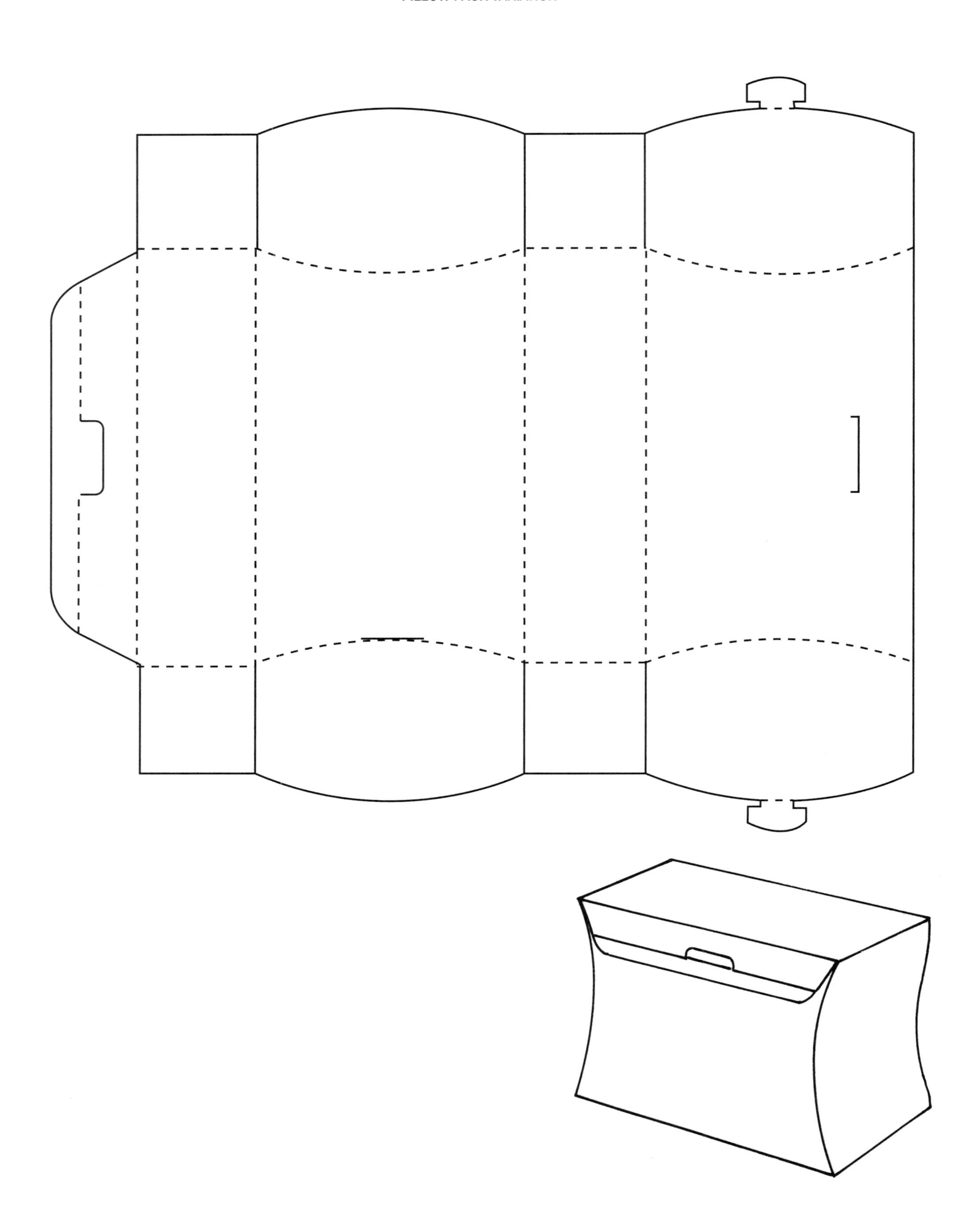

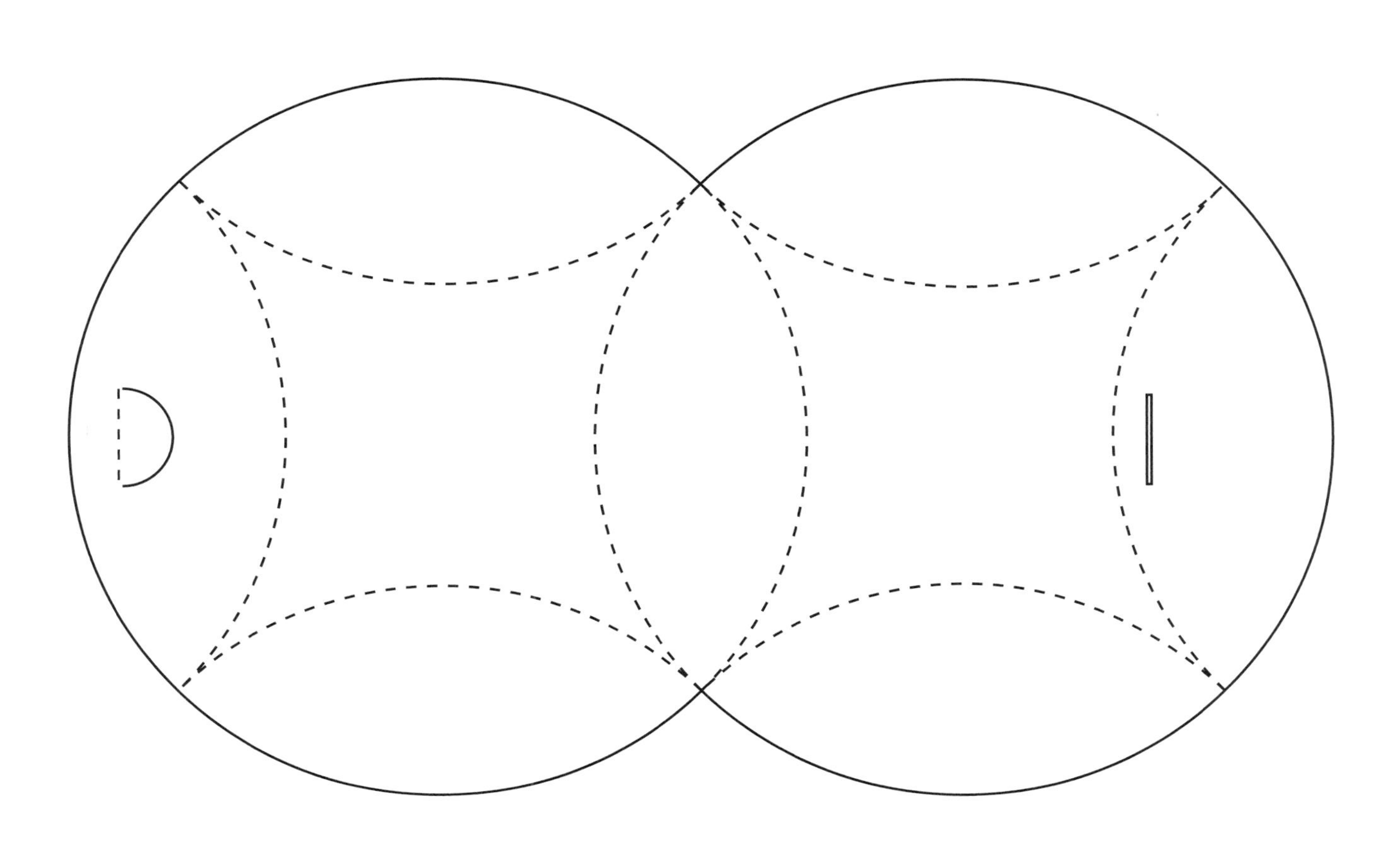

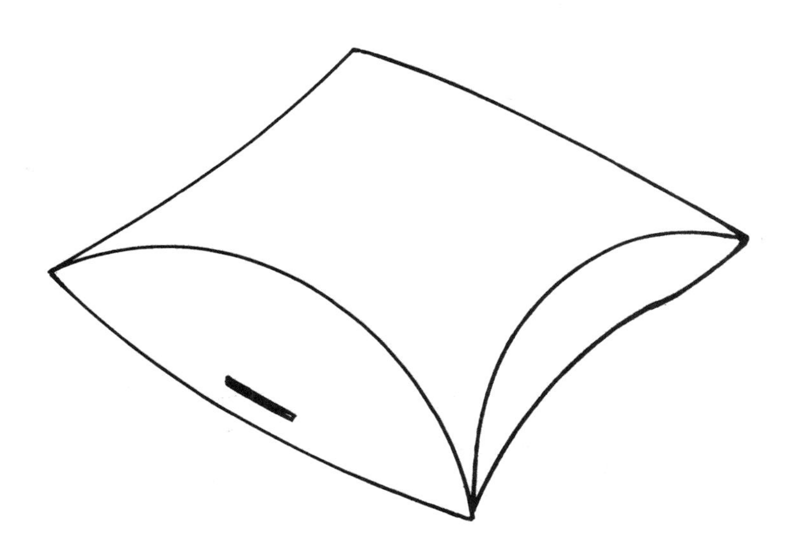

PILLOW PACK VARIATION

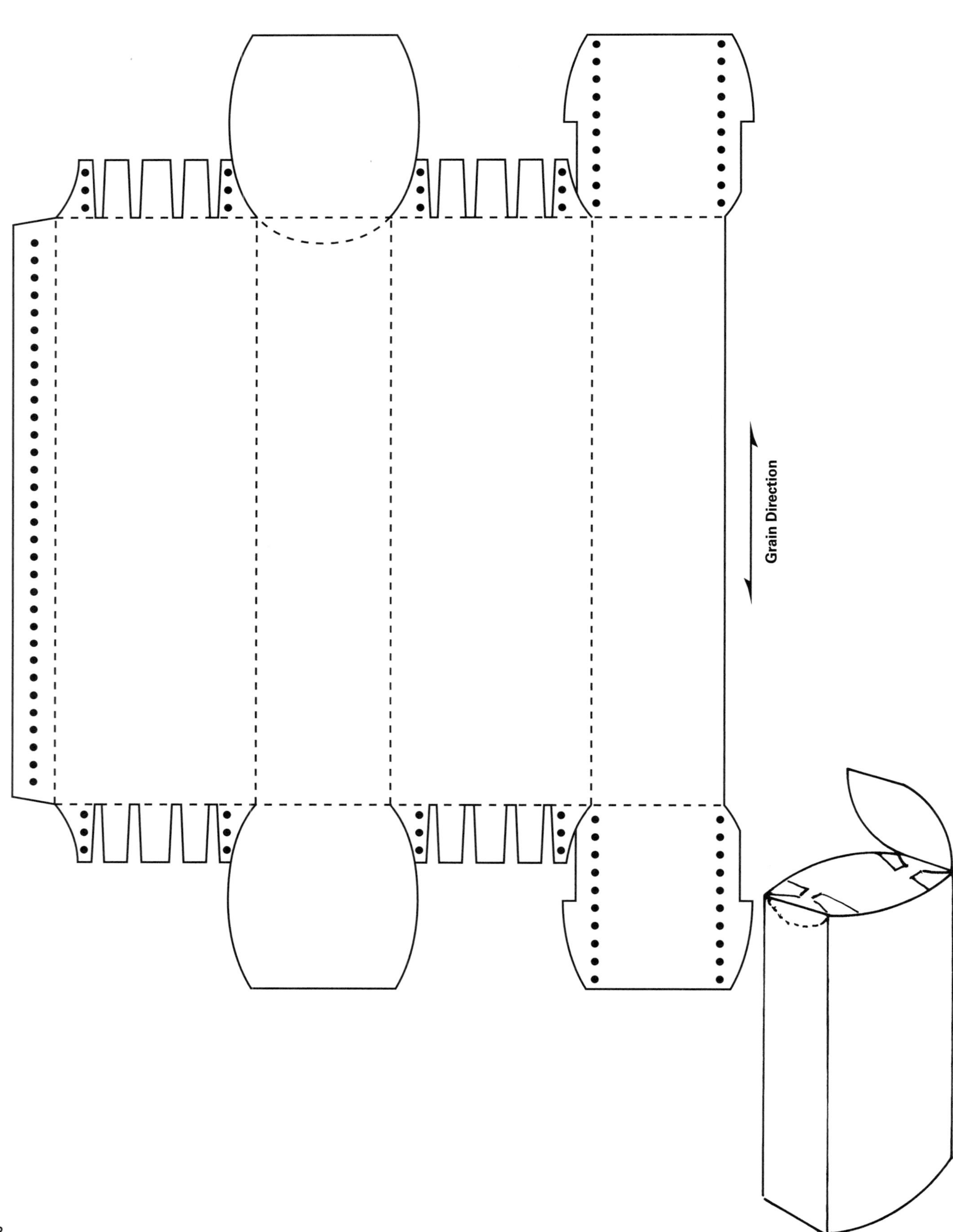

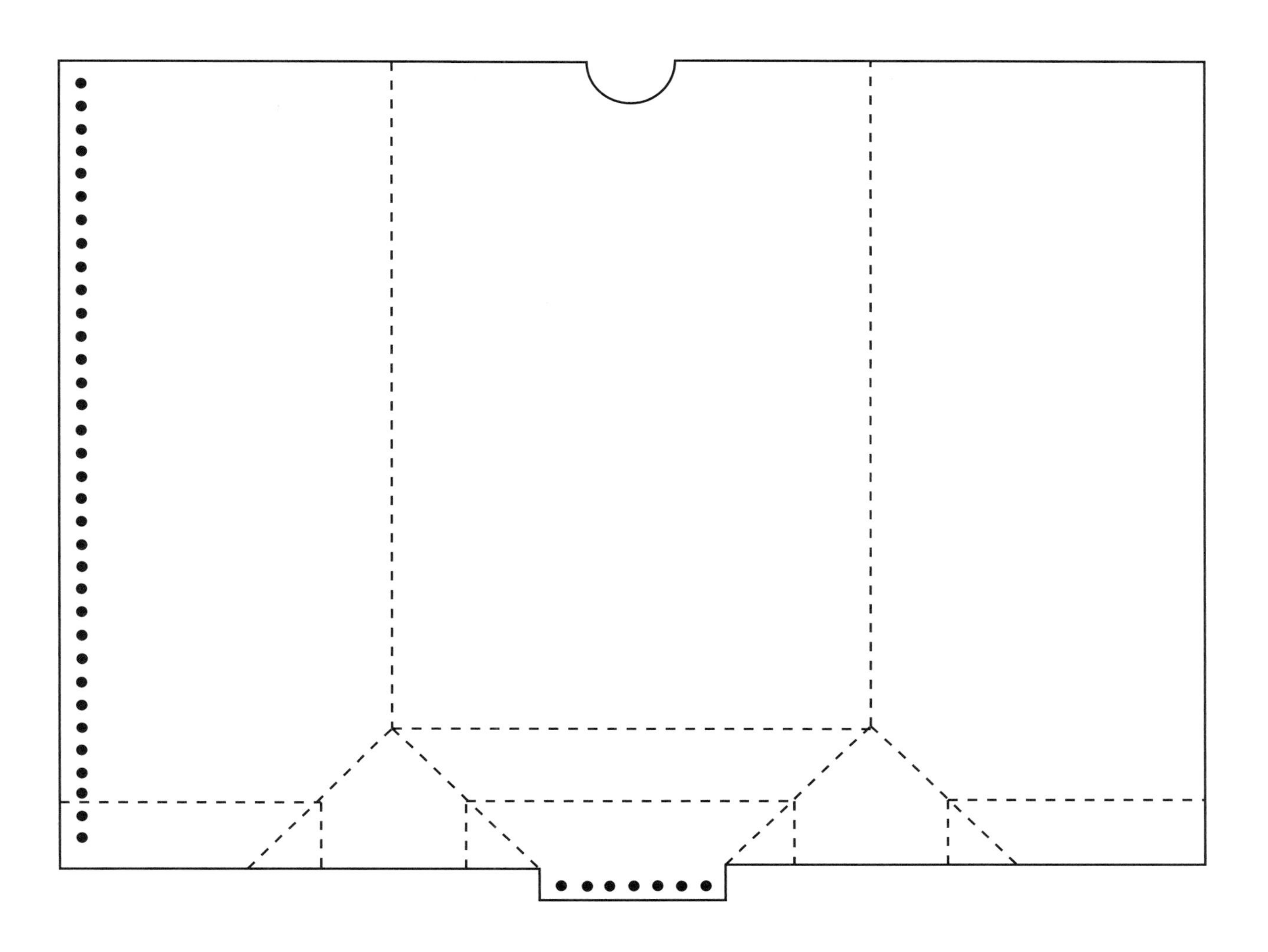

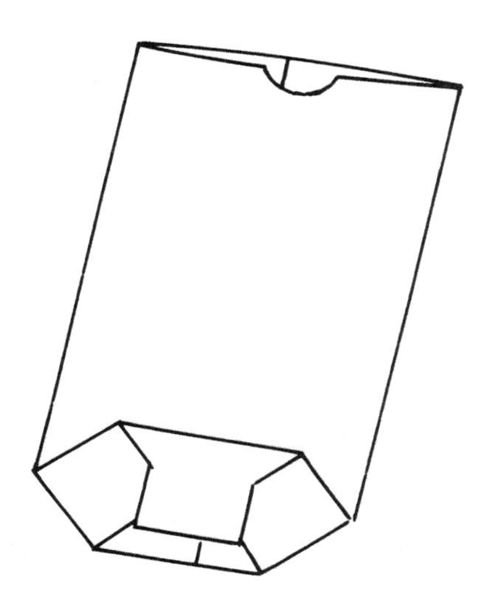

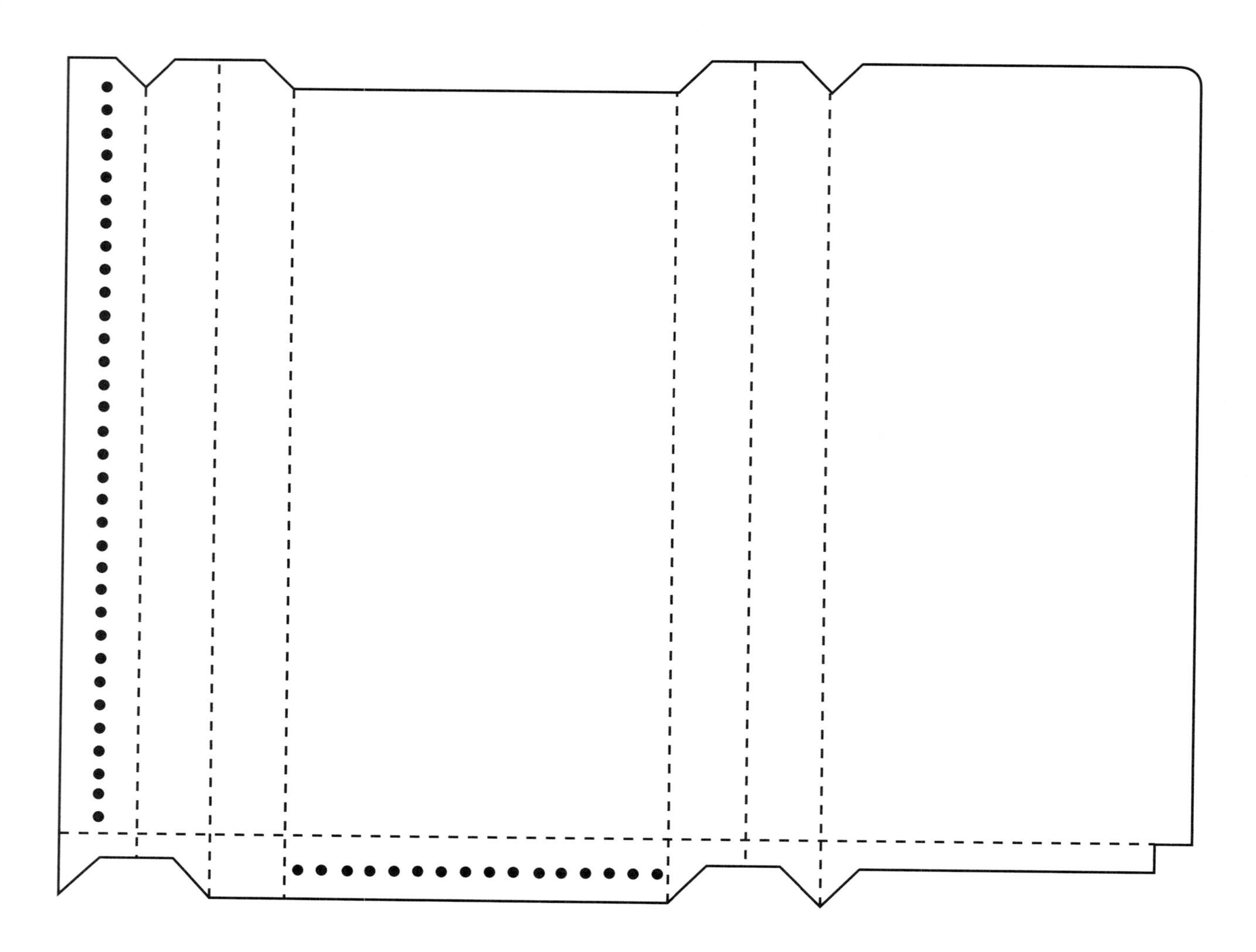

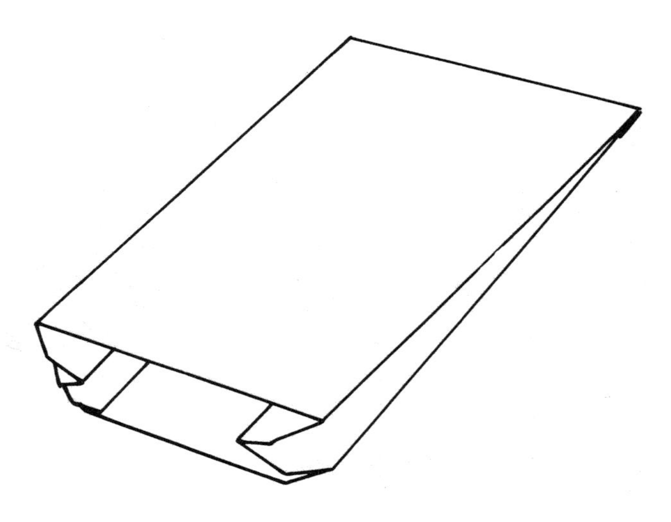

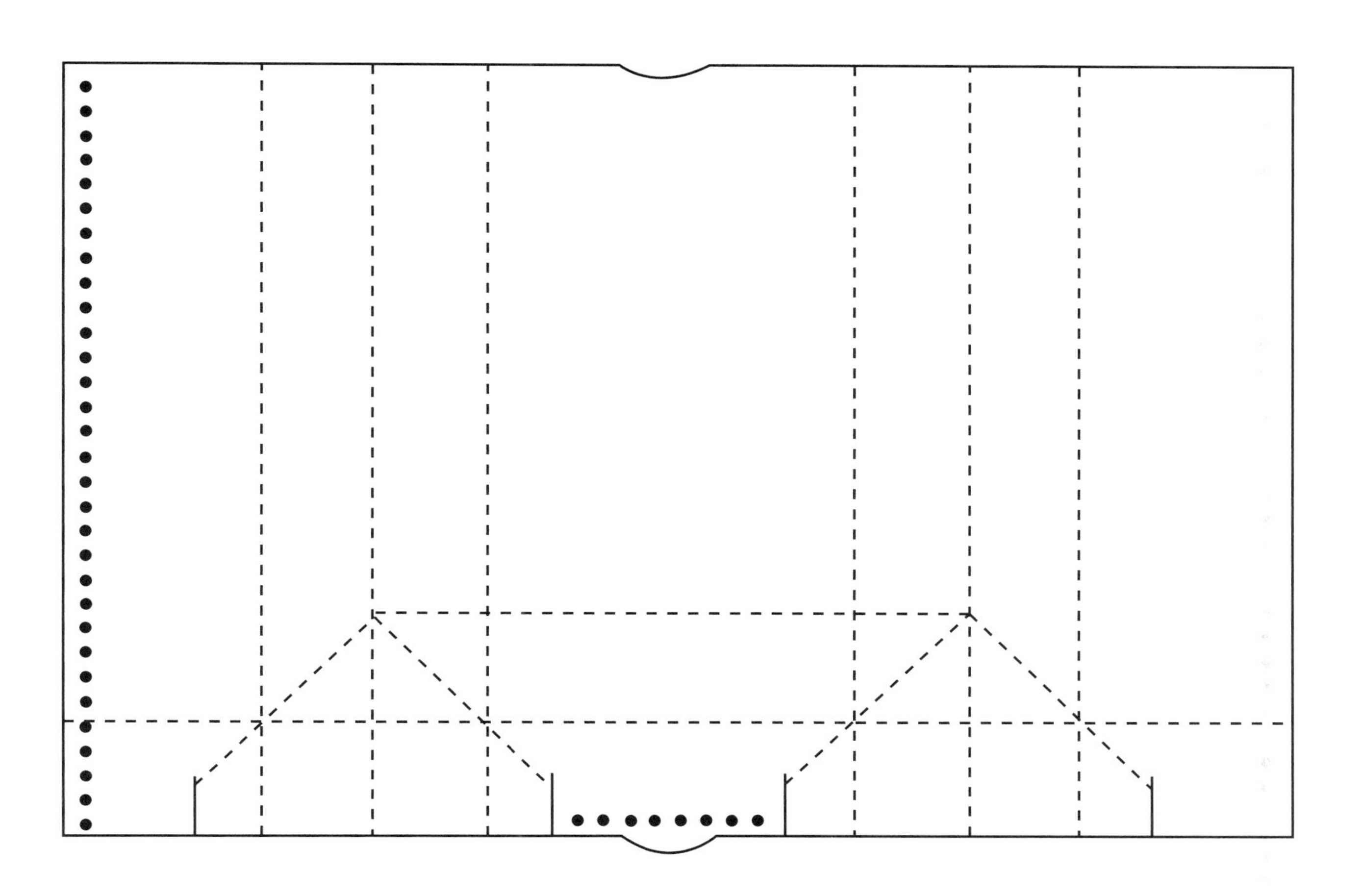

Typical prototype for grocery bag, shopping bag, etc.

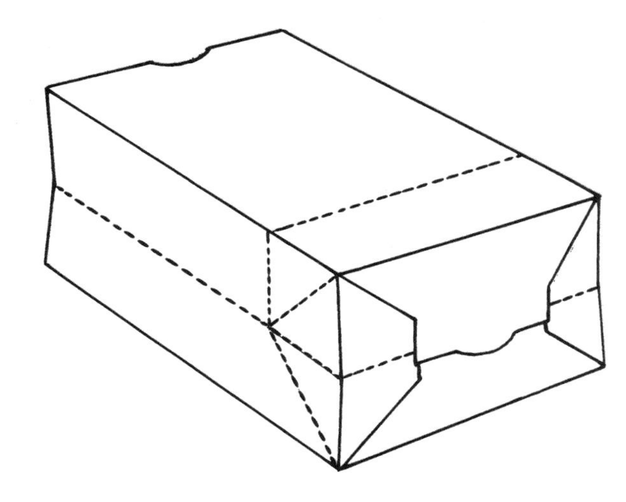

BAG WITH GLUED GUSSETED SIDES, CARRYING HANDLE, AND CLOSURE TAB

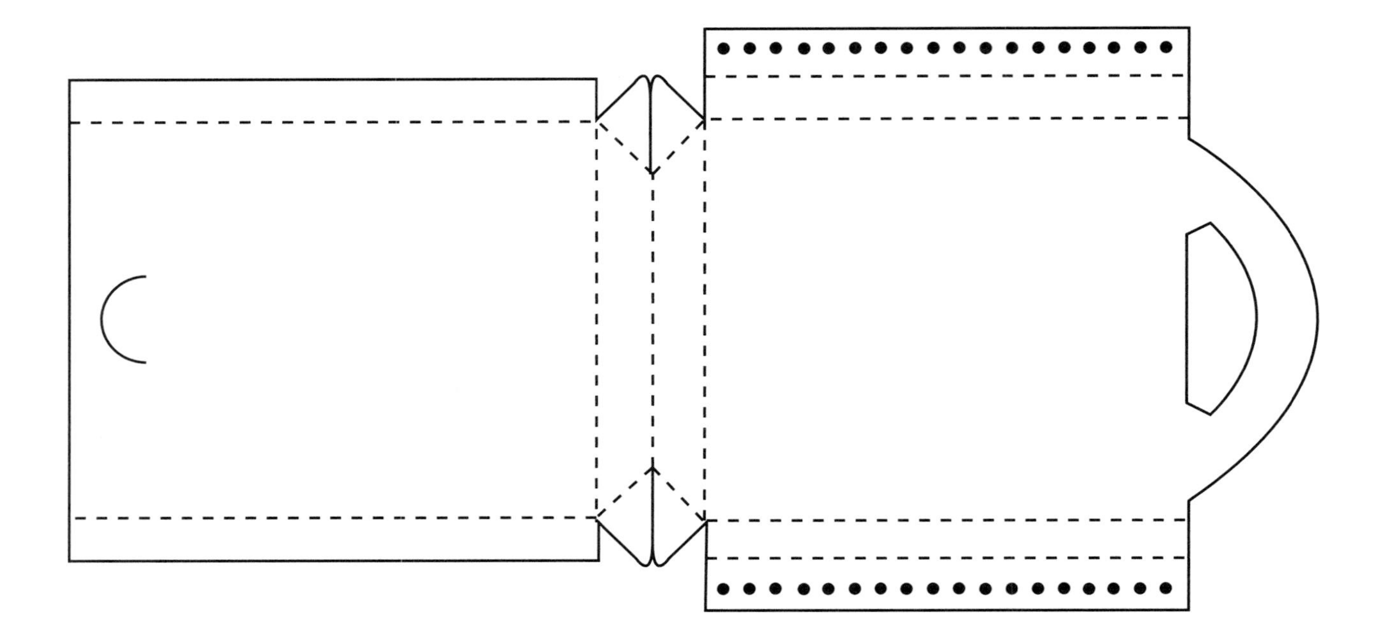

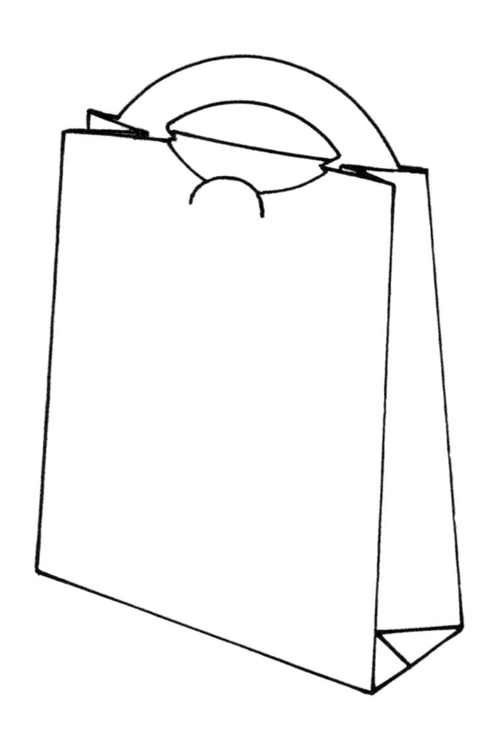

BAG WITH GREENLEAF BOTTOM, SIDE GUSSETS, AND HOOK LOCK CLOSURE

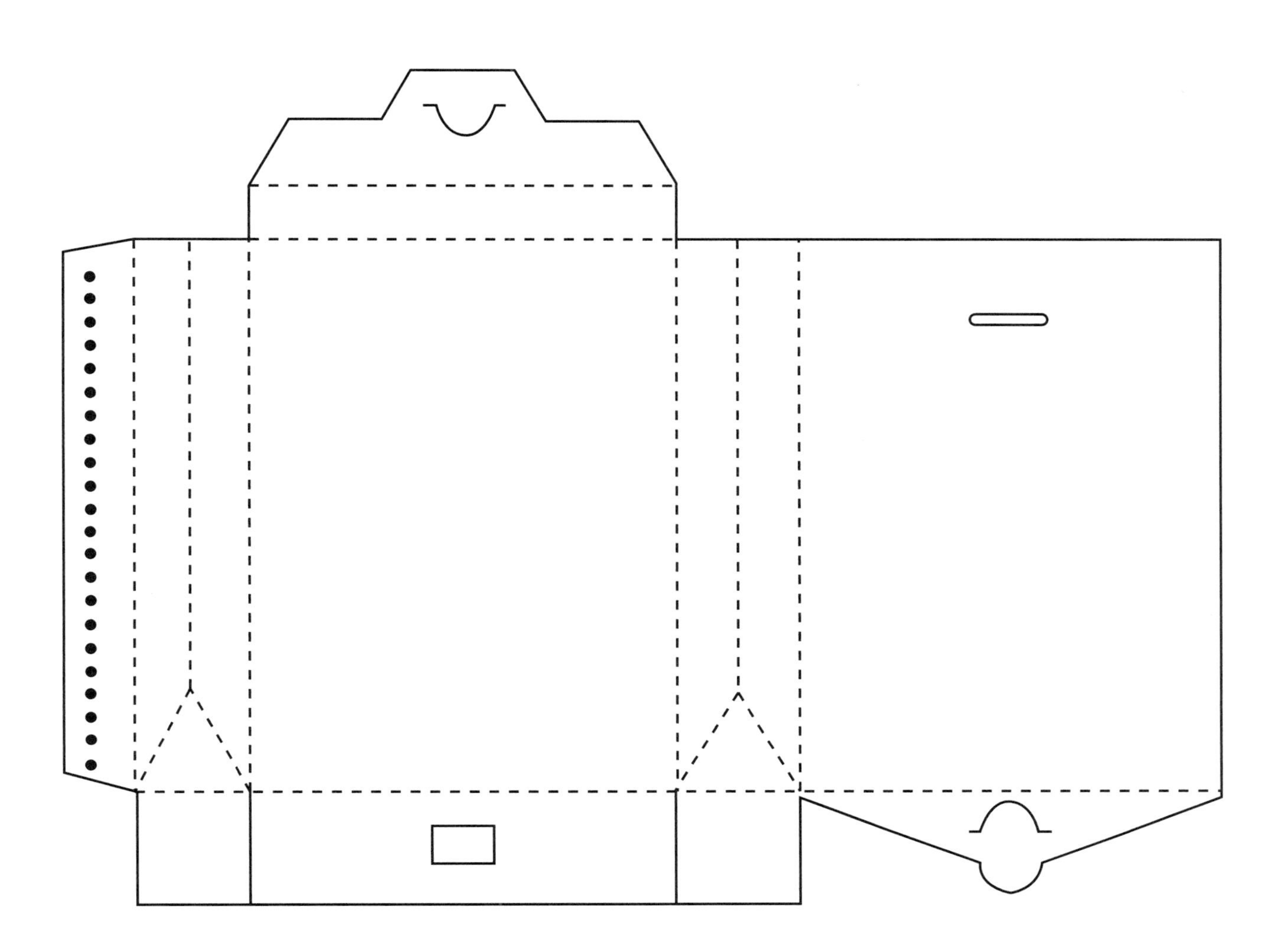

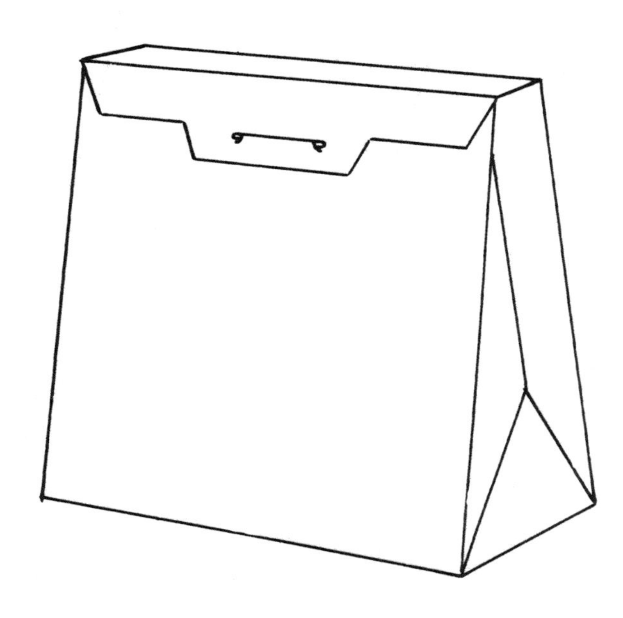

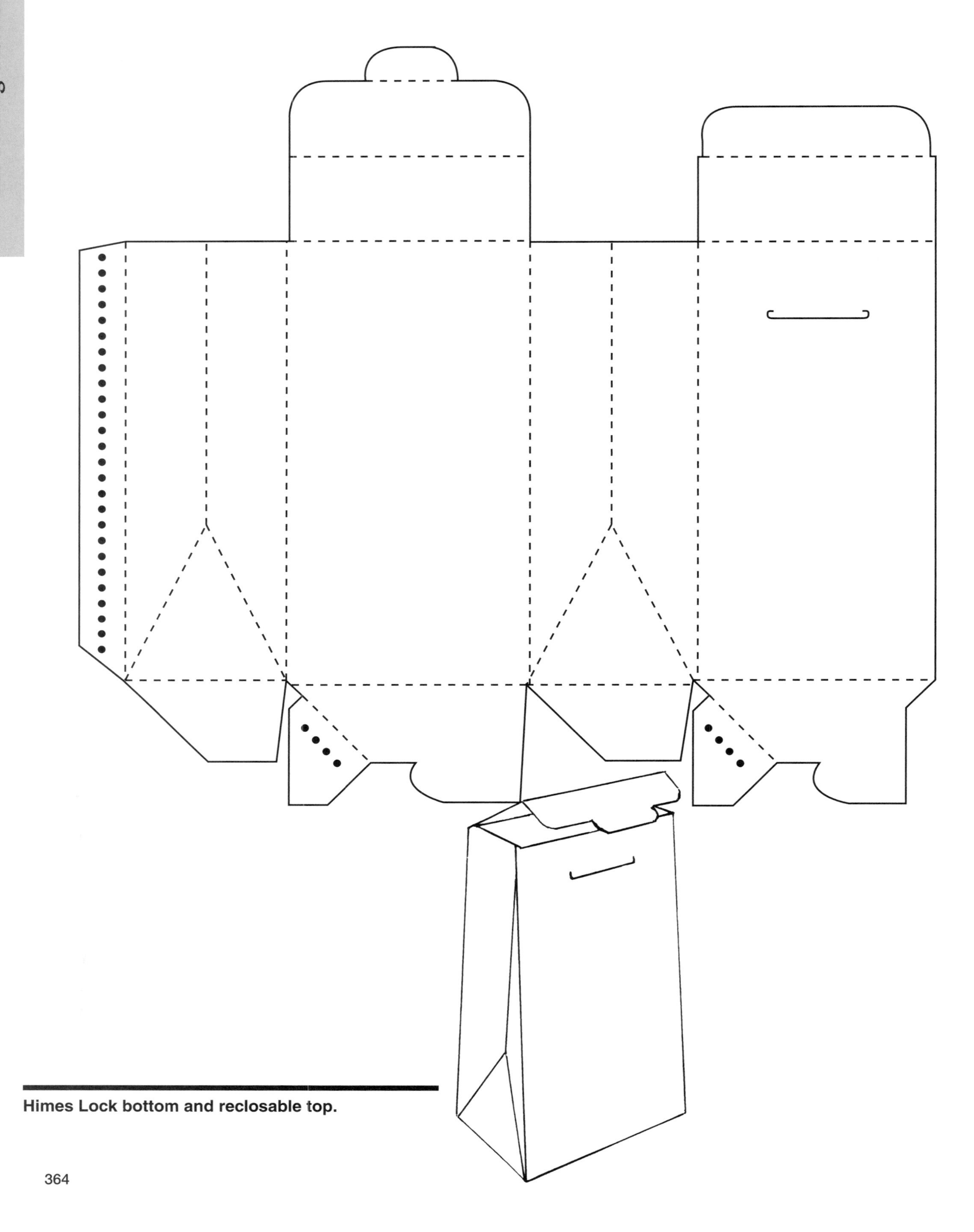

Himes Lock bottom and reclosable top.

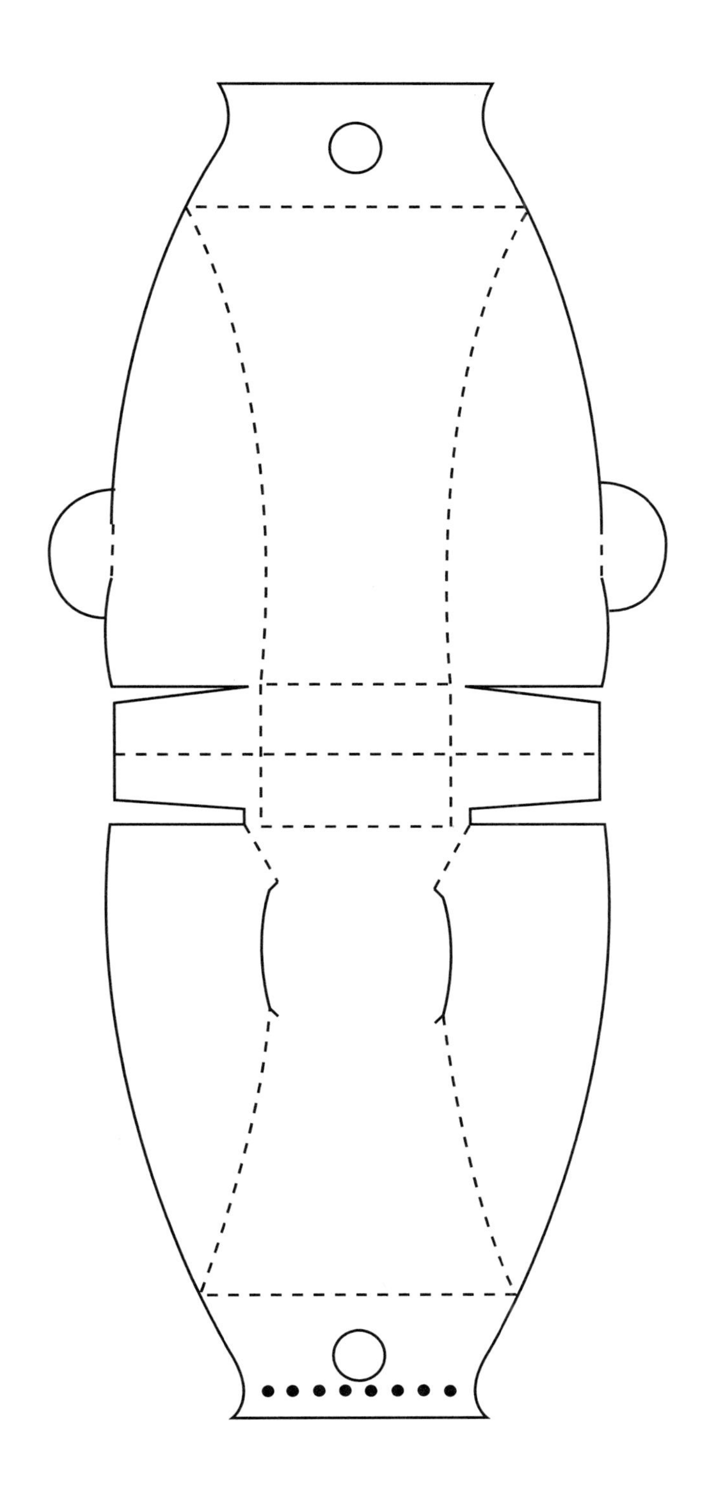

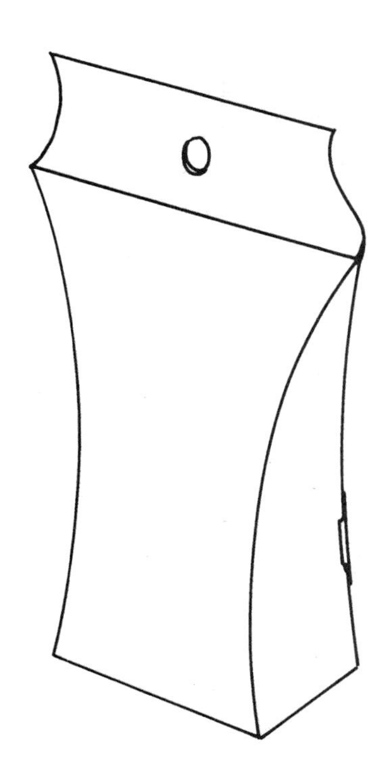

GUSSETED TAPERED BACK

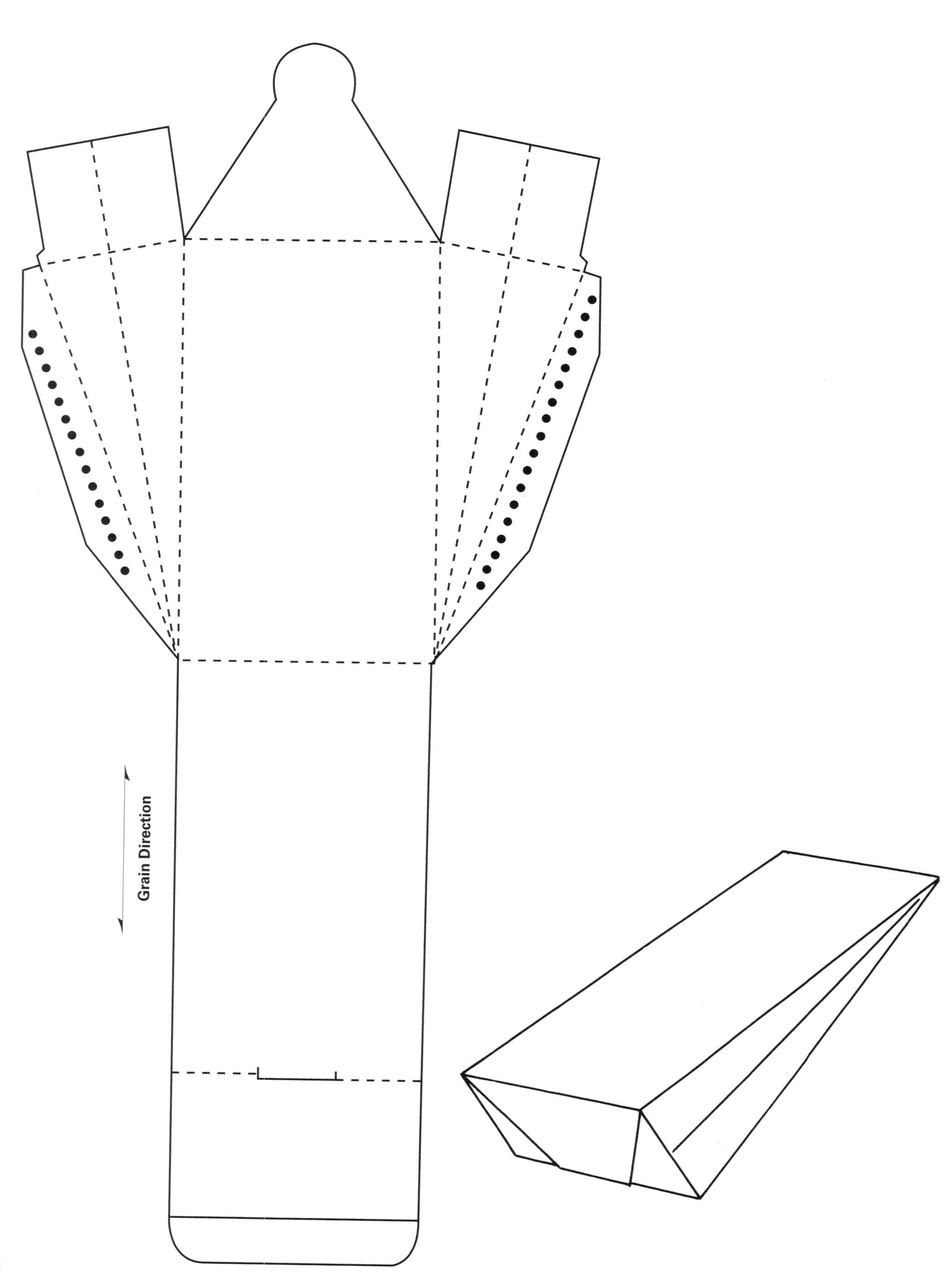

GUSSETED BAG WITH HIMES LOCK AUTOMATIC BOTTOM, GUSSETED SIDES, AND FLAPS ON COVER THAT LOCK INTO GUSSET

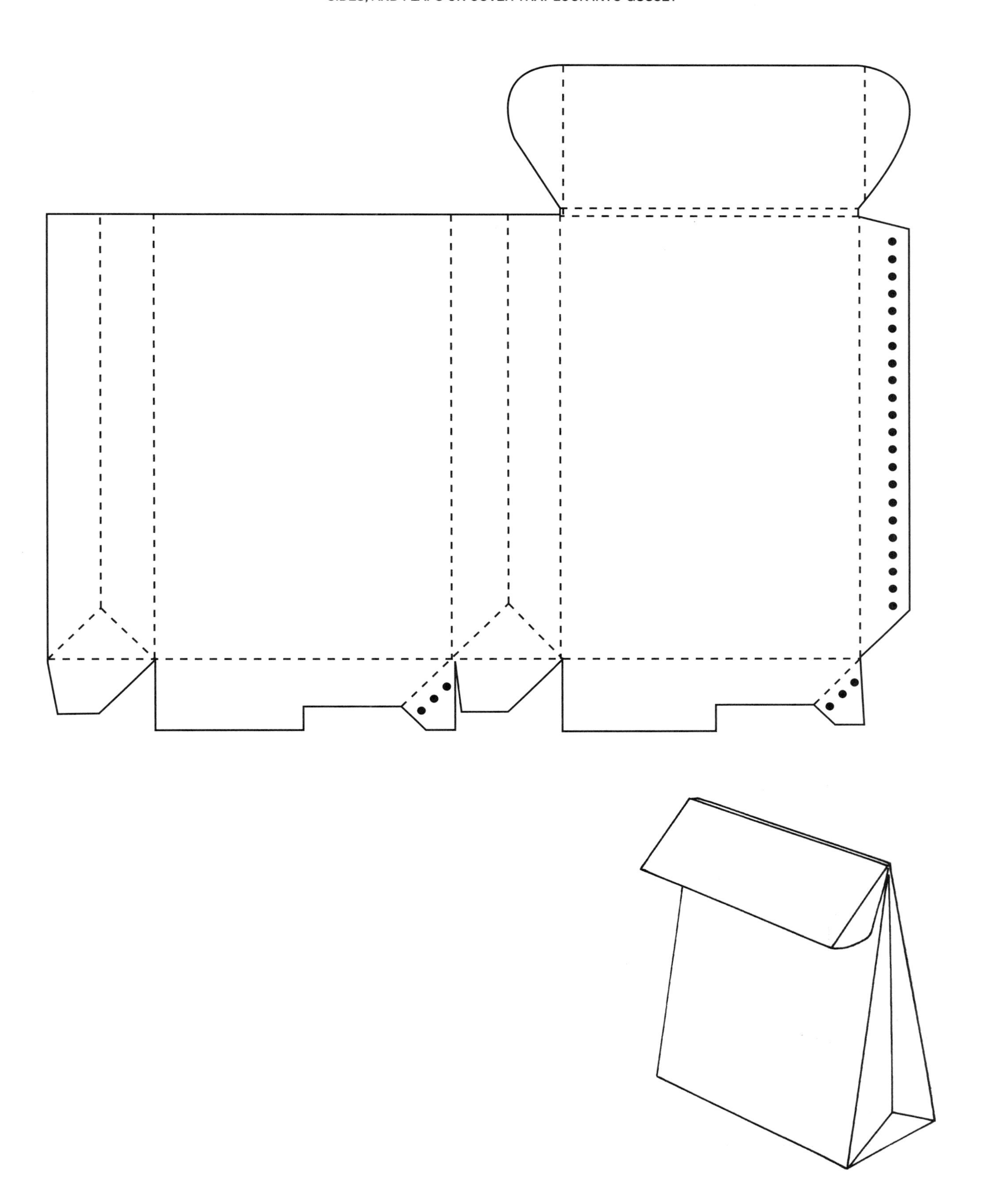

POUCH WITH HIMES LOCK AUTOMATIC BOTTOM, SIDE GUSSETS, AND REINFORCED SNAP-LOCK CLOSURE

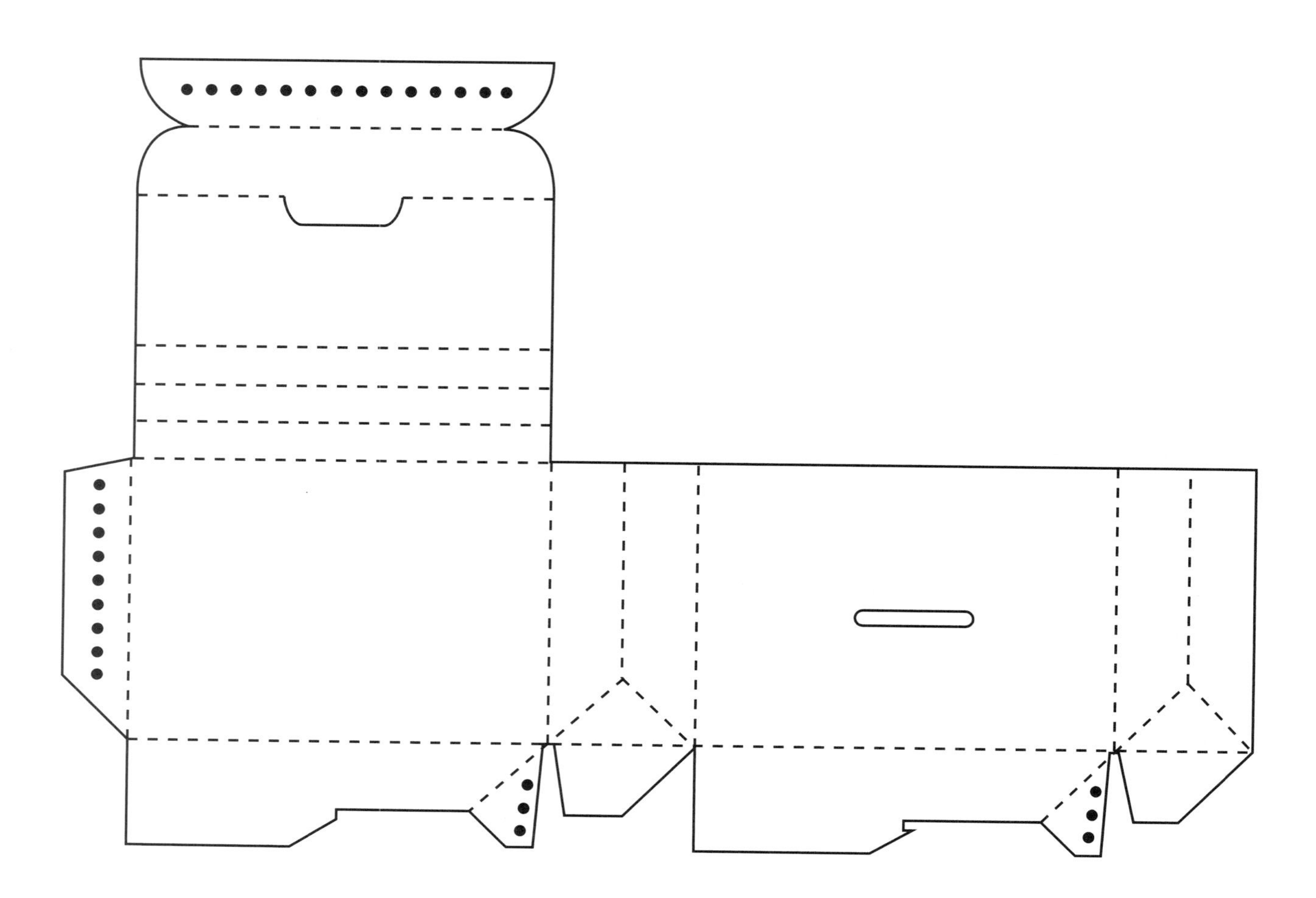

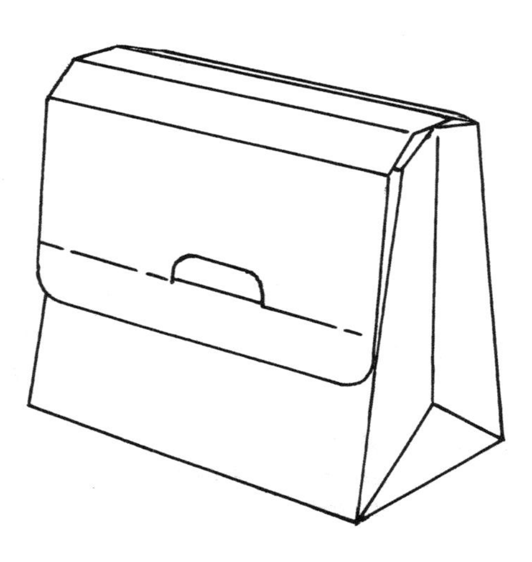

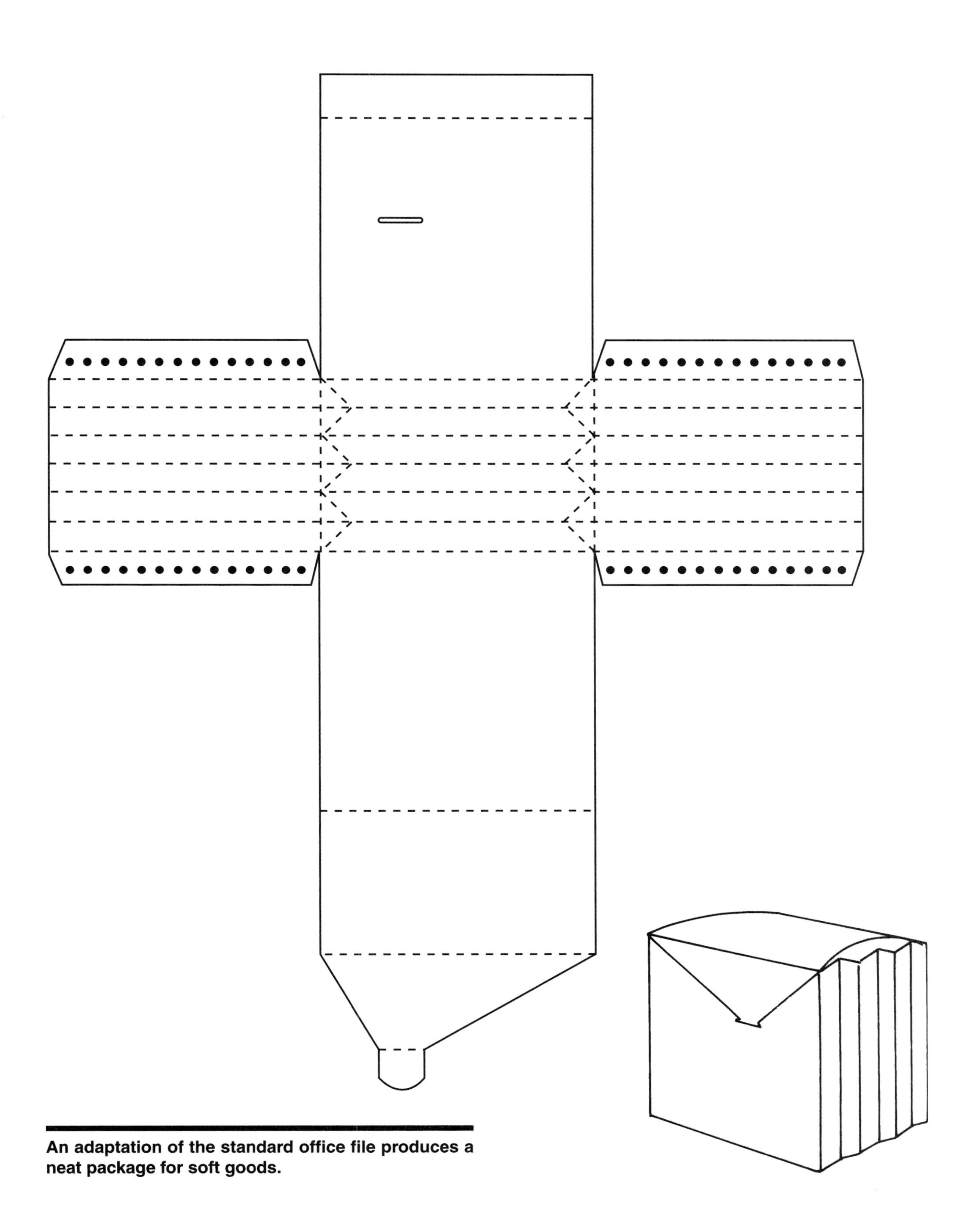

An adaptation of the standard office file produces a neat package for soft goods.

BAG WITH HIMES LOCK AUTOMATIC BOTTOM, GUSSETED SIDES, AND CARRYING HANDLE WITH SNAP-LOCK CLOSURE

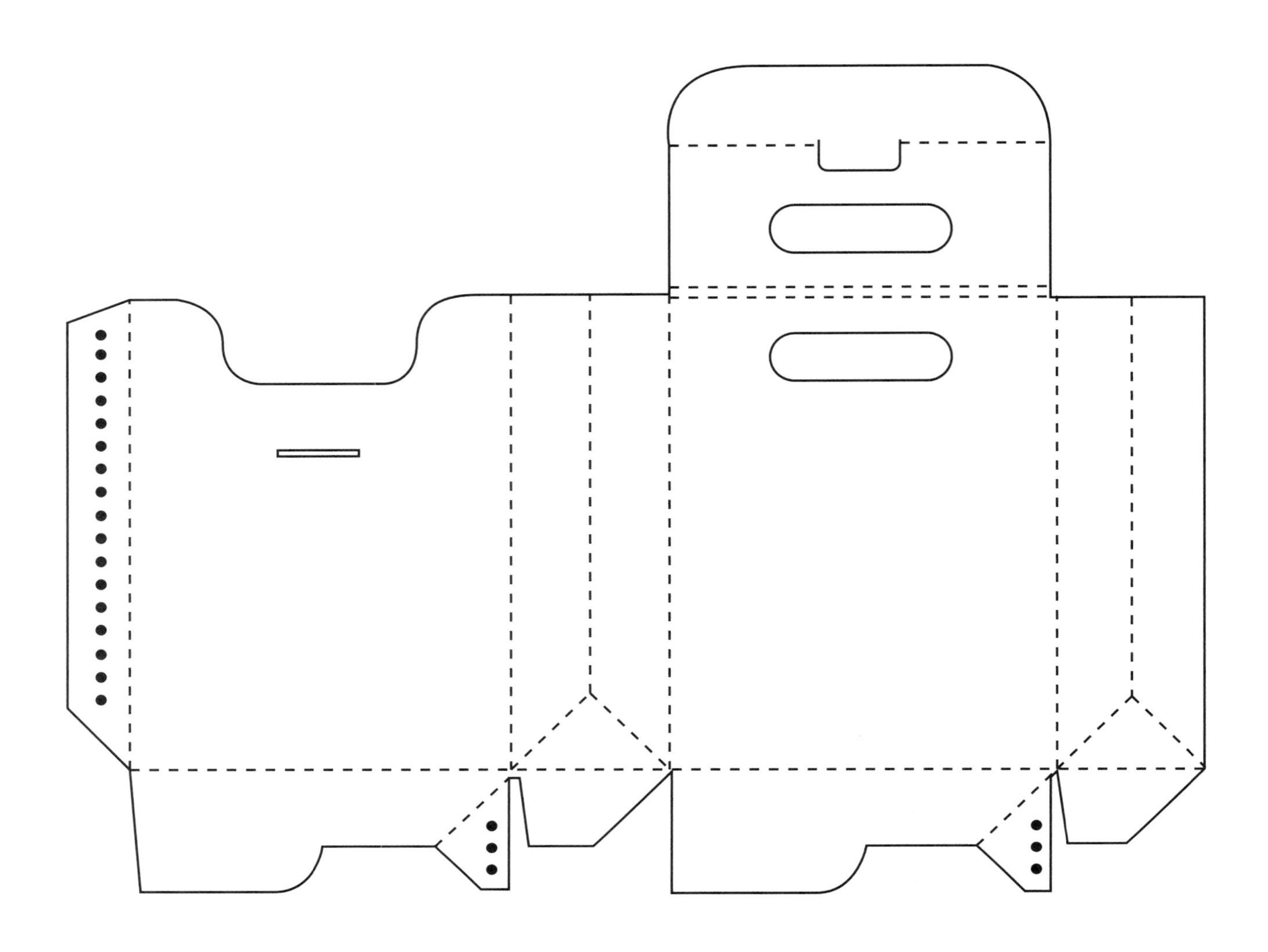

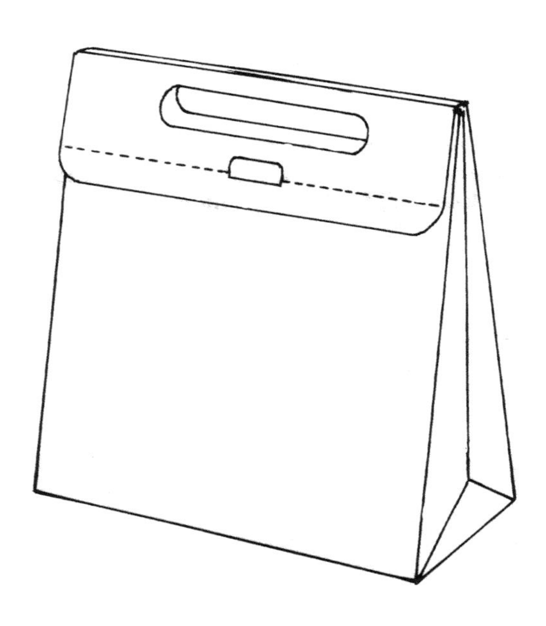

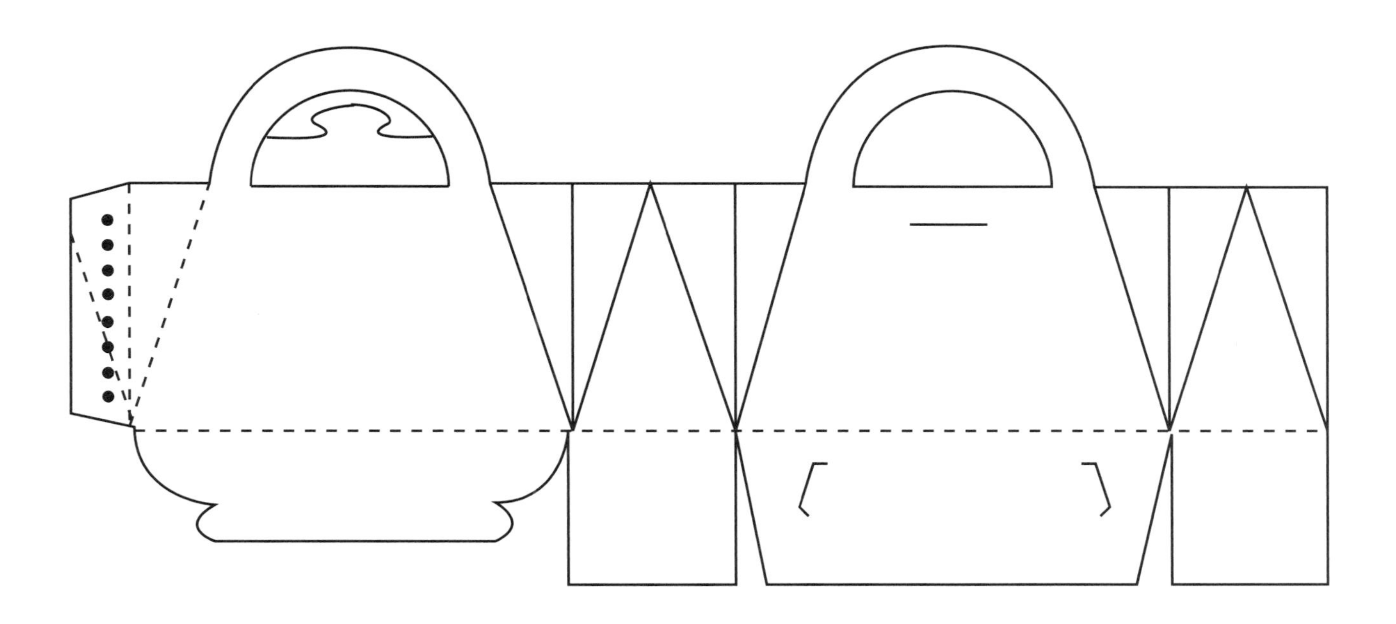

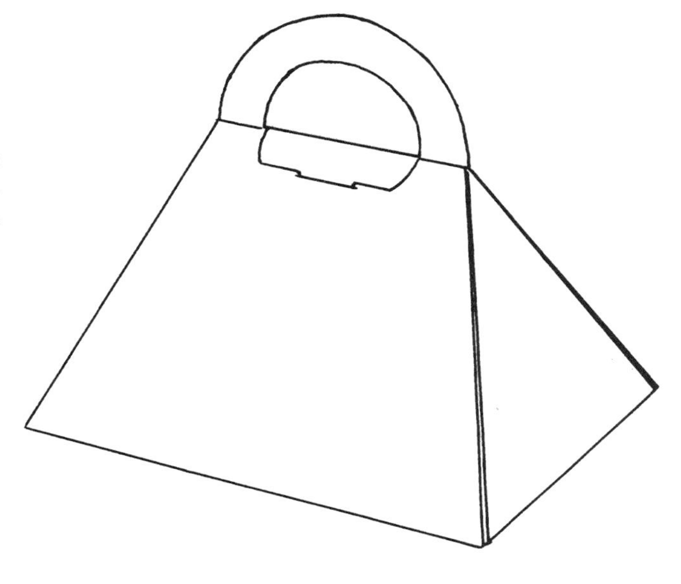

This carrying satchel with double Anchor lock bottom and lock-tab/slit top is one of many possible variations.

TRIANGULAR CARRYING CASE

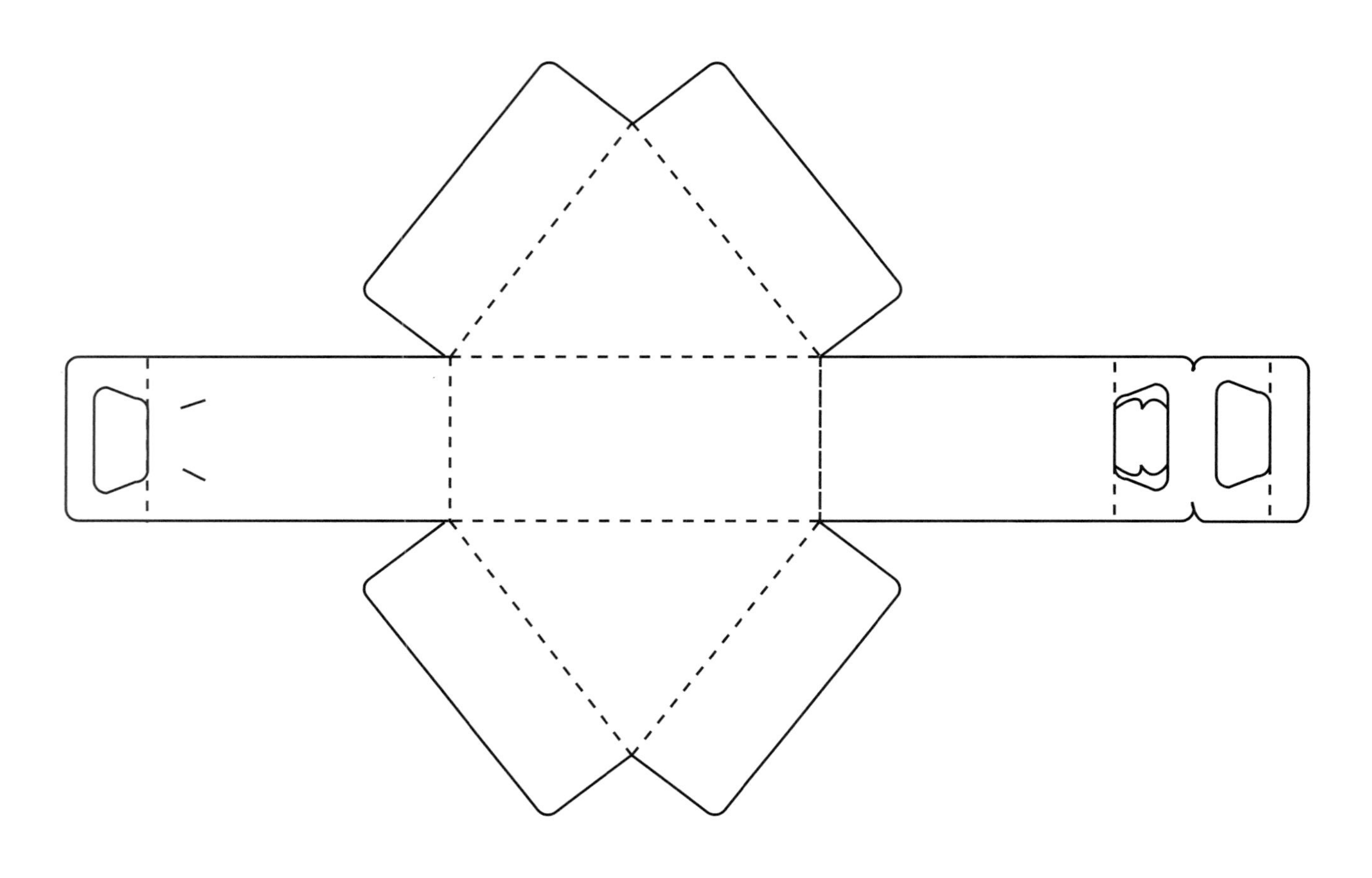

This triangular carry-all needs no glue and may be used for many different gift items.

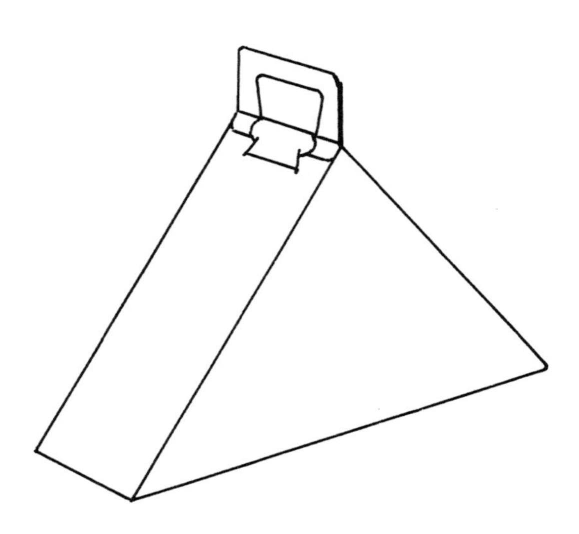

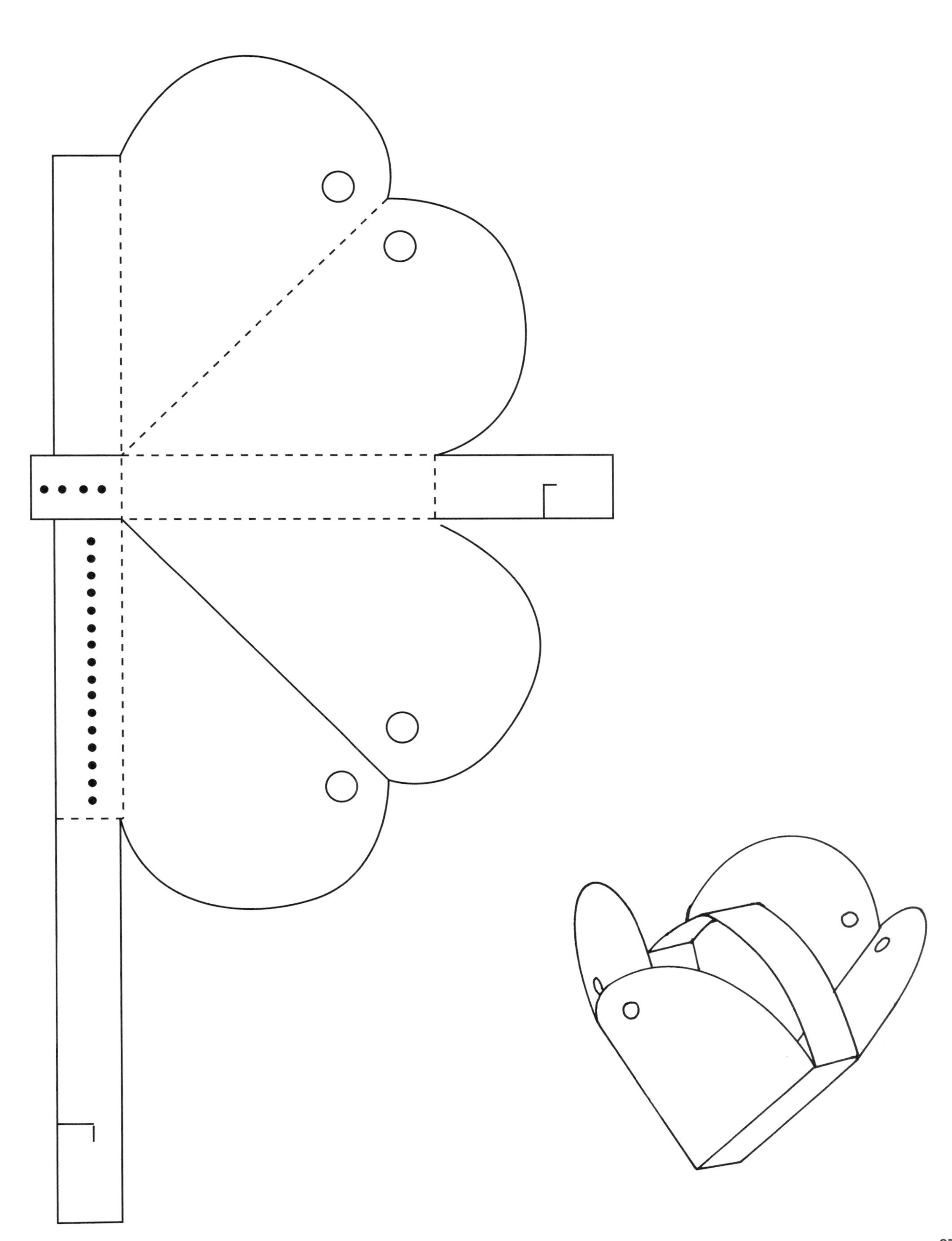

PEN/PENCIL SET WRAP

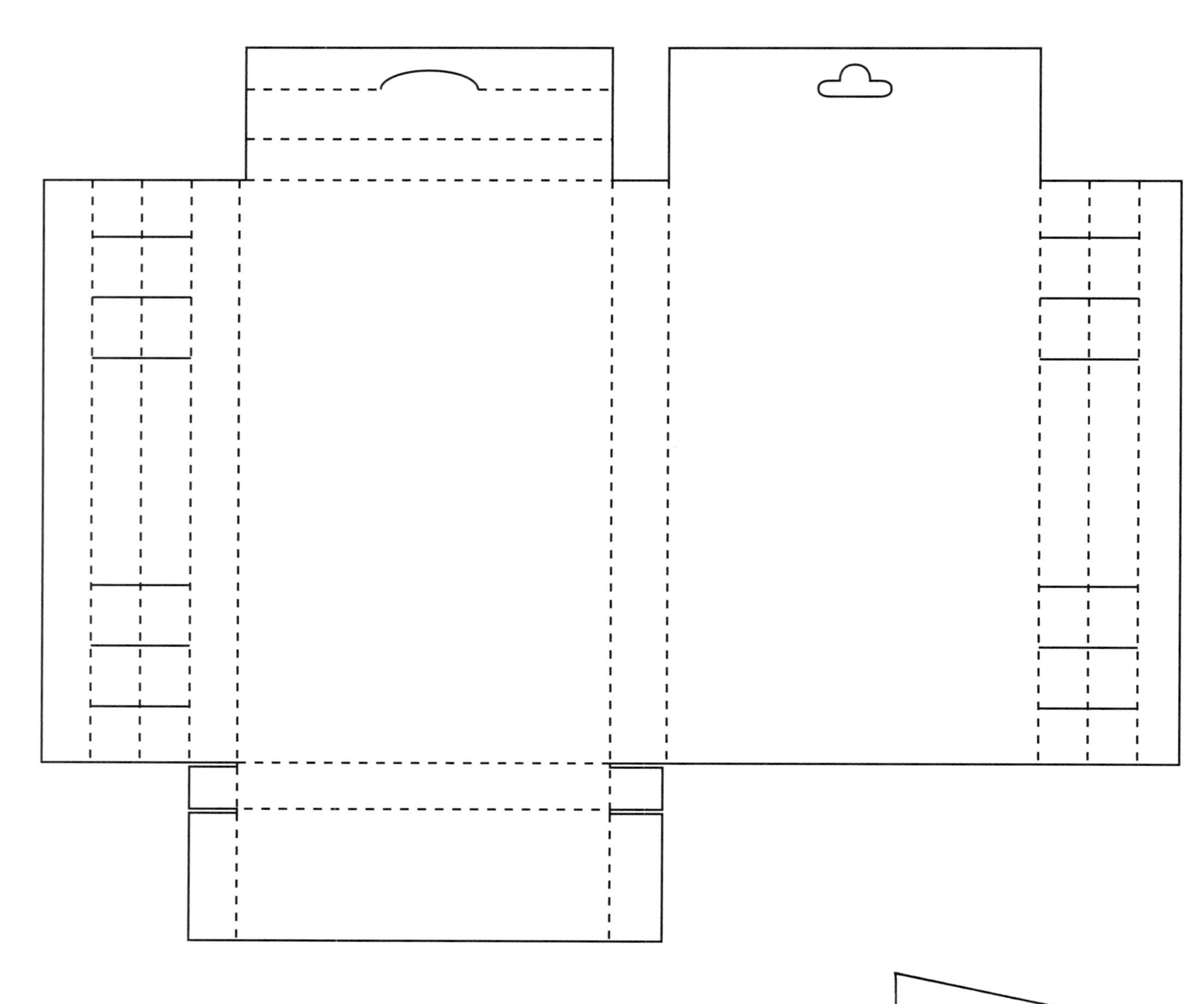

The tabs at the side hold a writing instrument which serves as a locking device.

TWO PIECE, SINGLE LAP CONTAINER WITH OUTSIDE ROLLED TOP AND INSIDE FOLDED BOTTOM

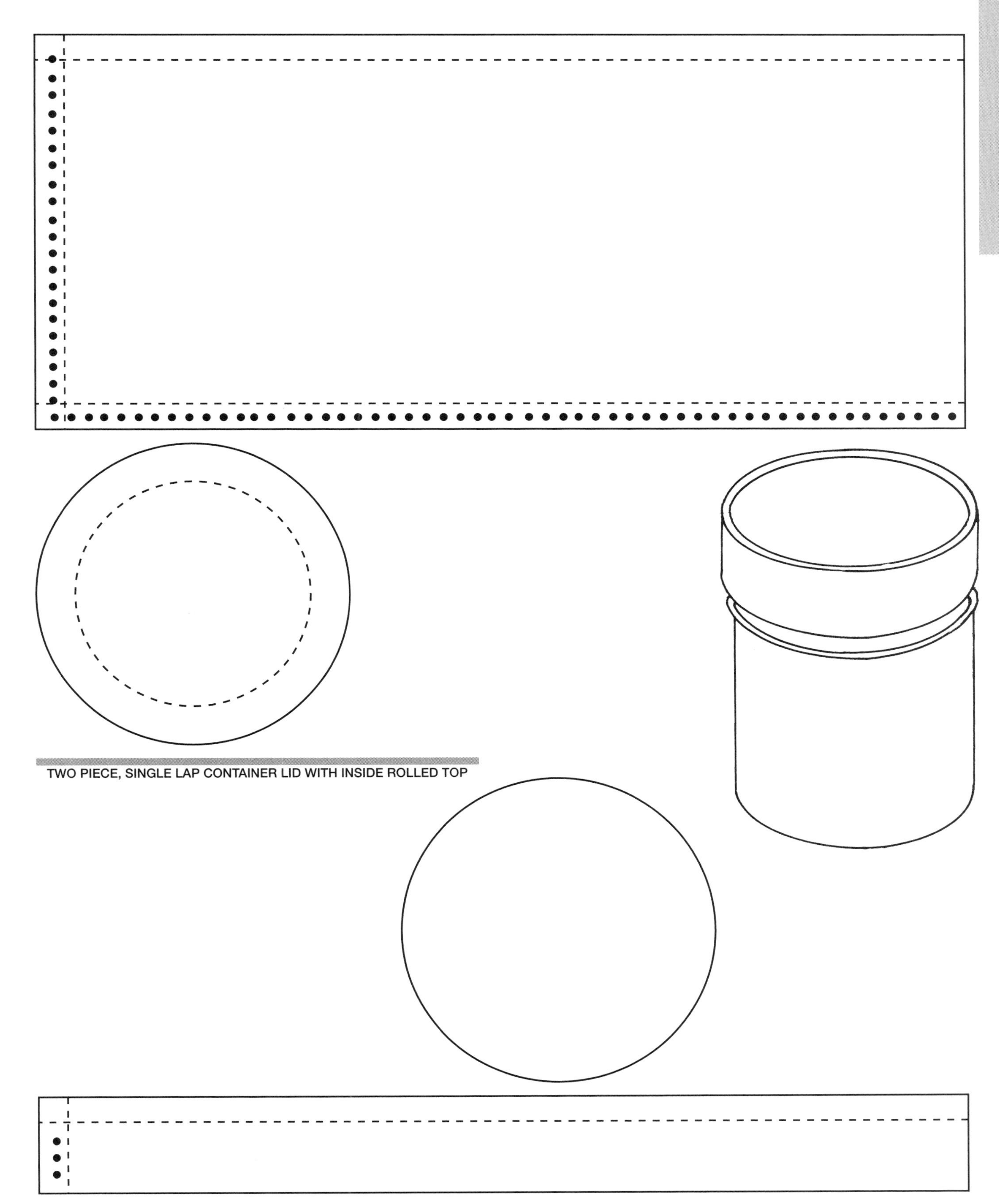

TWO PIECE, SINGLE LAP CONTAINER LID WITH INSIDE ROLLED TOP

TWO PIECE, SINGLE LAP, TAPERED CUP WITH INSIDE FOLDED BOTTOM AND OUTSIDE ROLLED TOP

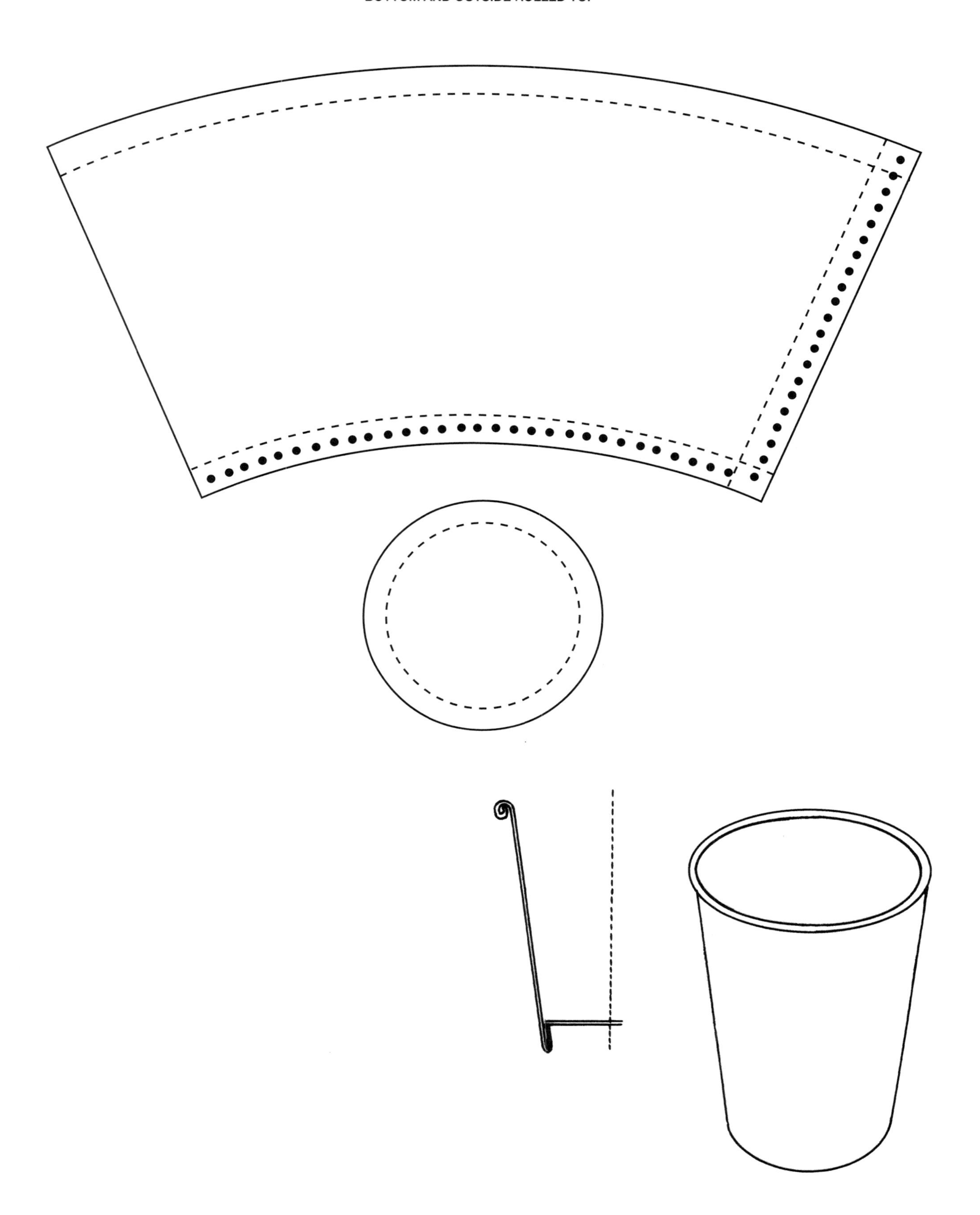

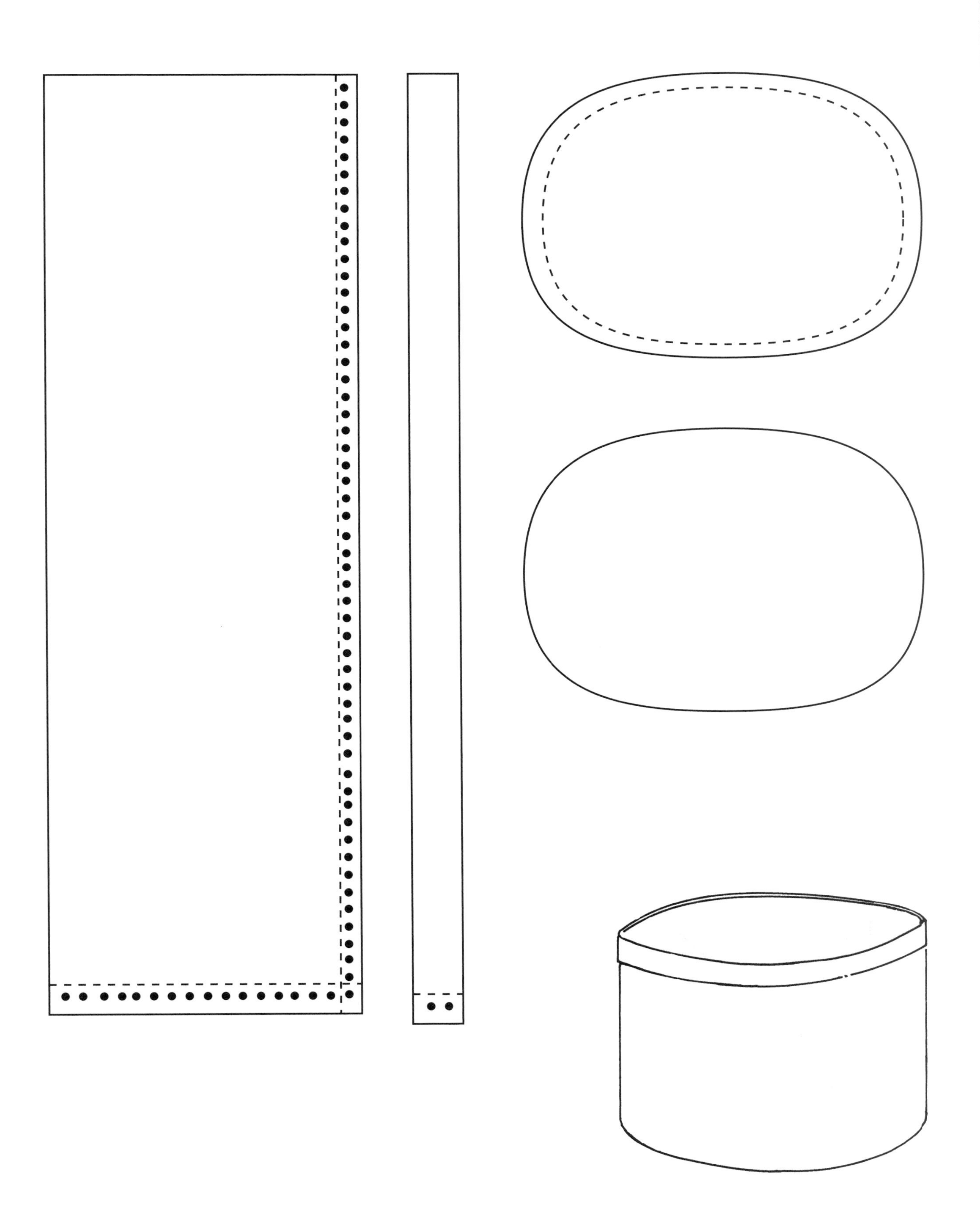

SPIRAL-WOUND TUBE

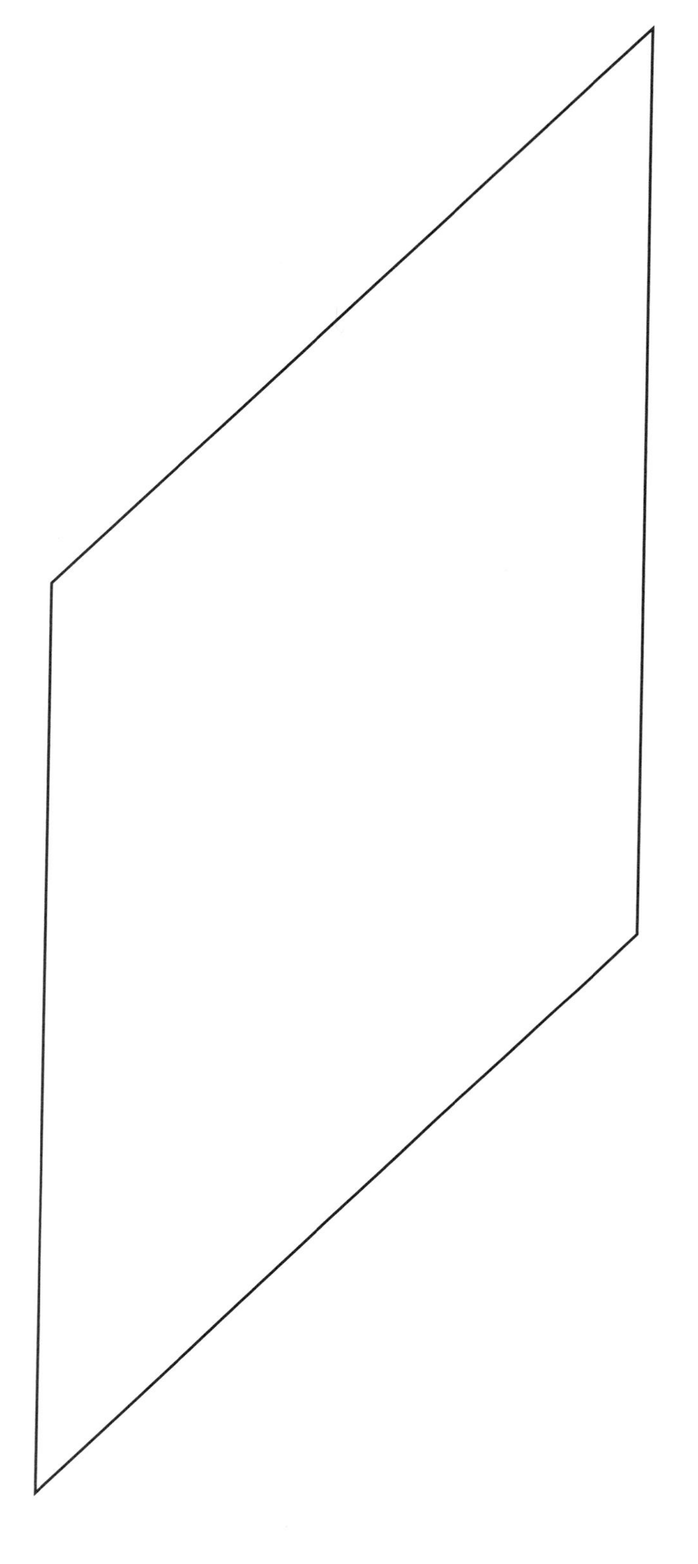

Used predominantly as the core for rolls of paper and mailing tubes.

Spiral-wound tubes may be used as the basis for circular packaging.

Cosmetics and toiletries often make use of this form. The following page shows the use of a tube as the core of a novelty package.

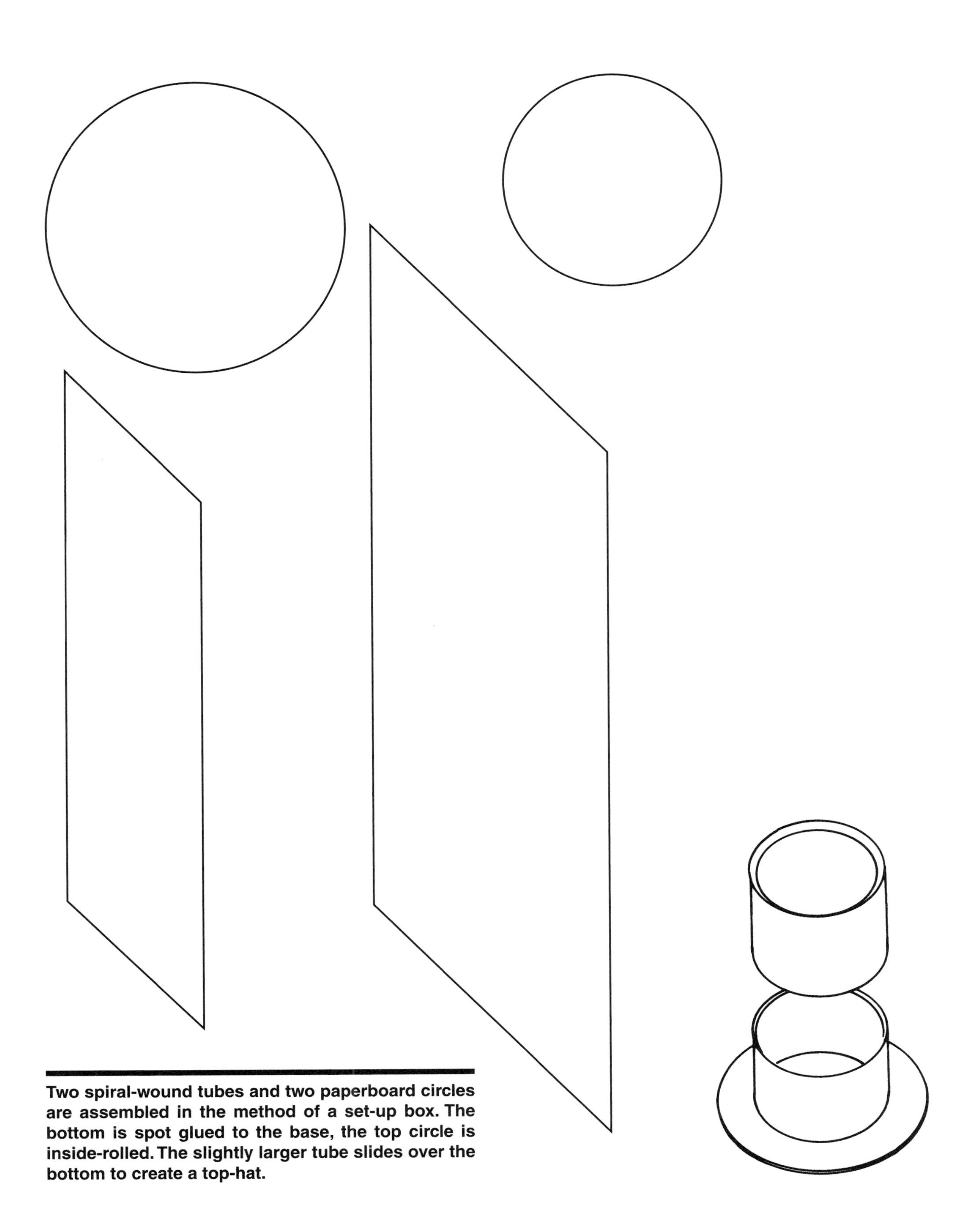

Two spiral-wound tubes and two paperboard circles are assembled in the method of a set-up box. The bottom is spot glued to the base, the top circle is inside-rolled. The slightly larger tube slides over the bottom to create a top-hat.

SIDE LOADED TUBE STYLE SIX-BOTTLE CARRIER

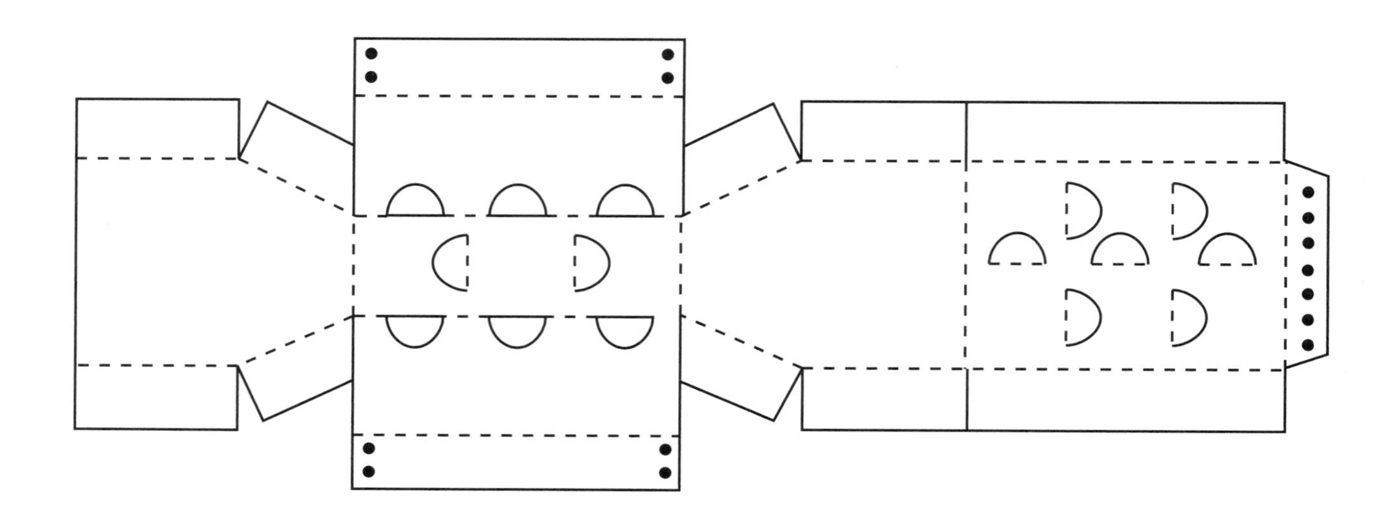

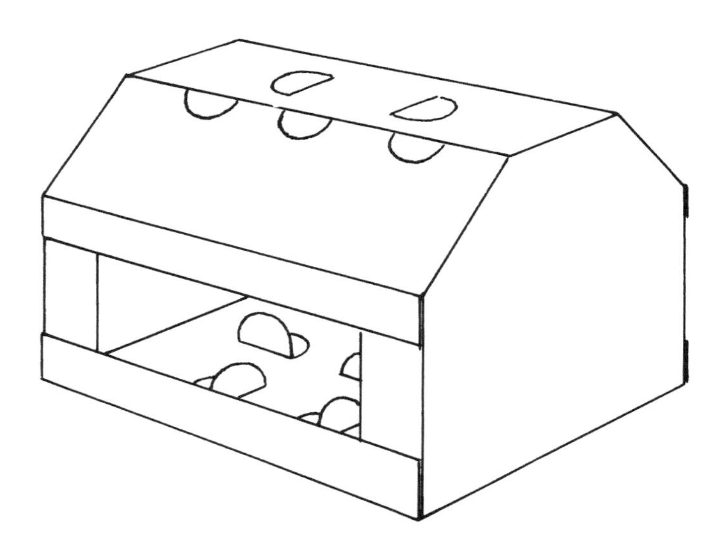

BOTTOM GLUED SLEEVE WITH NECK AND HEEL APERTURE; NECK-THROUGH STYLE

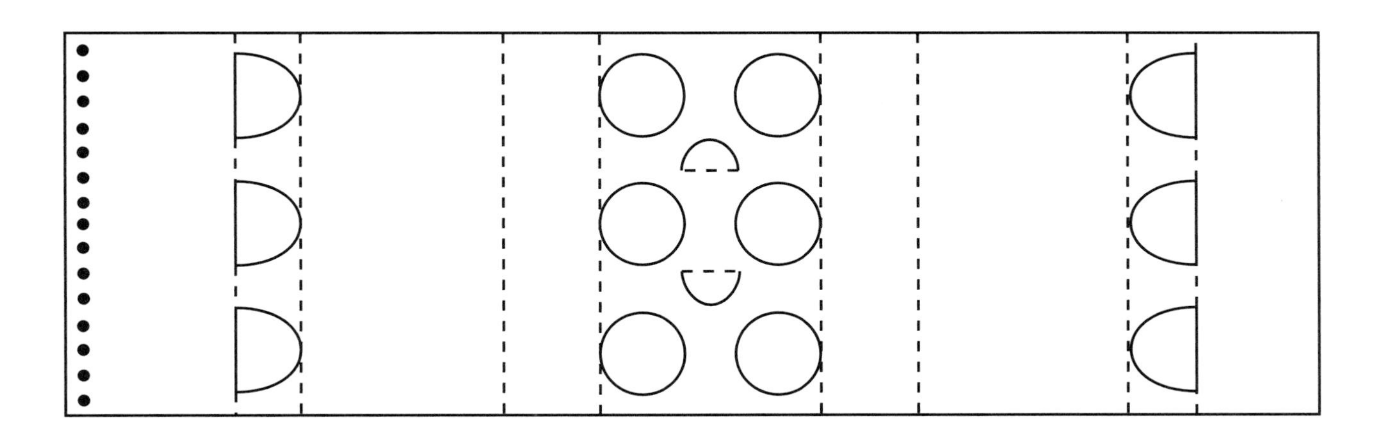

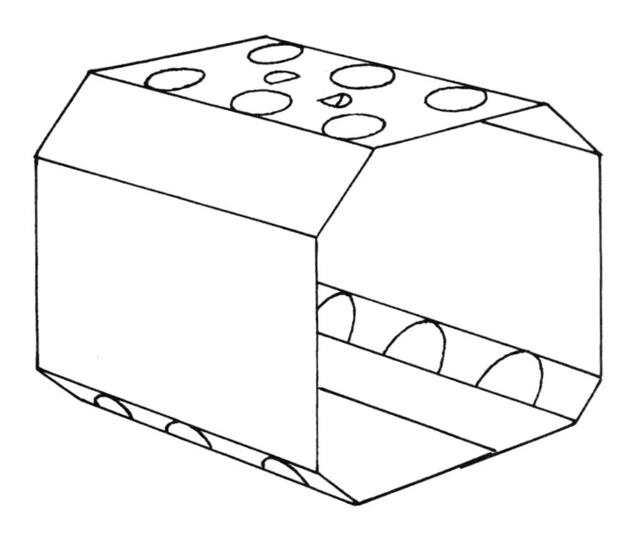

SLEEVE STYLE THREE-COUNT MULTIPACK ACE LOCK BOTTOM

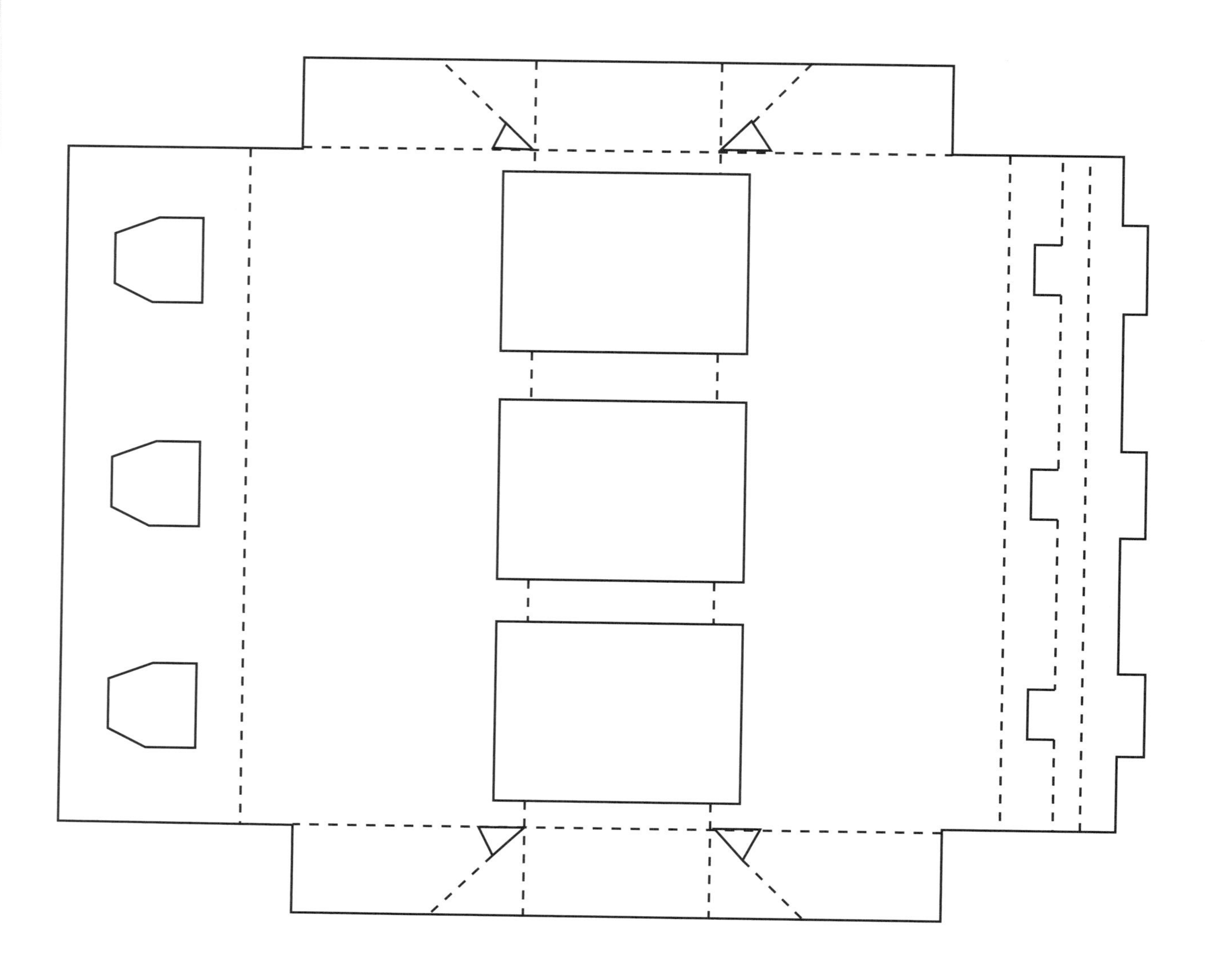

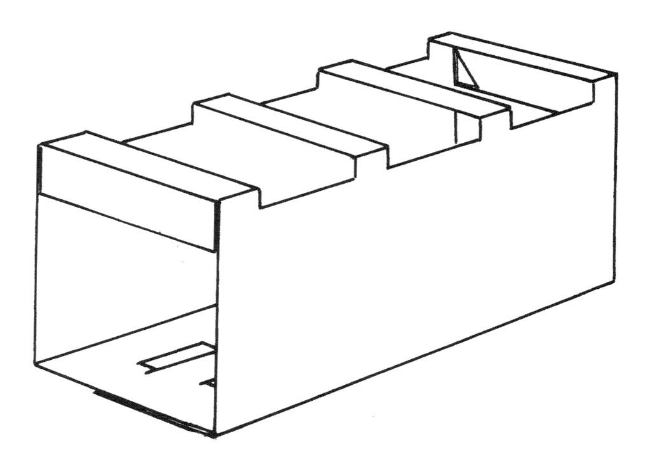

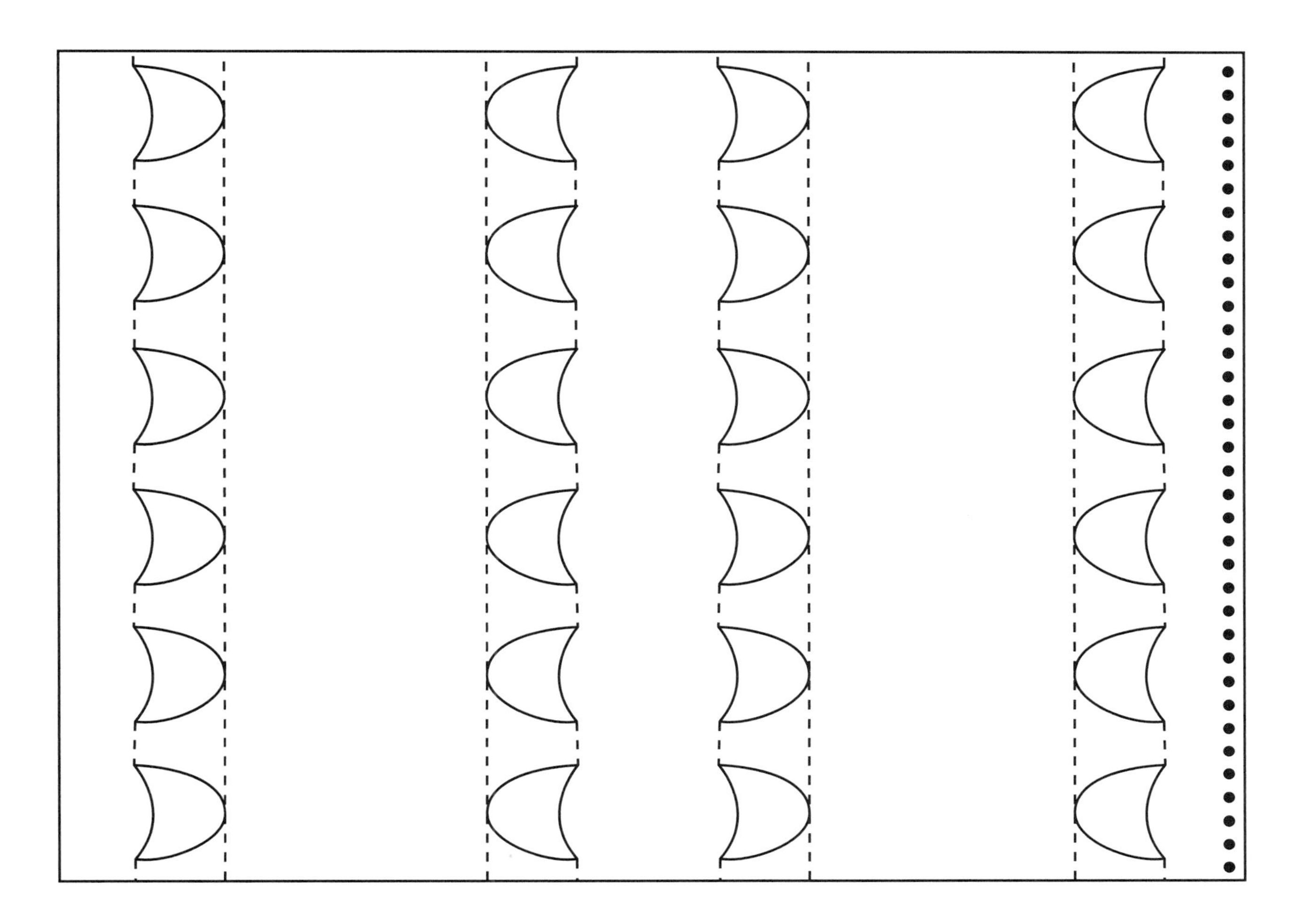

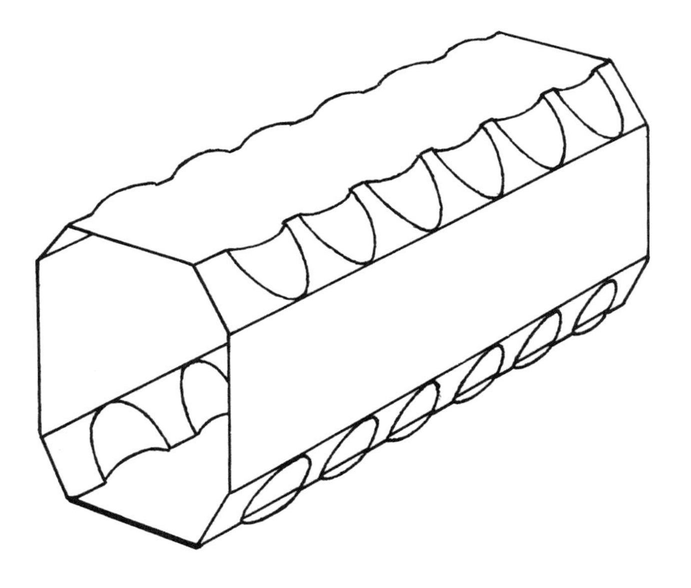

ANDRE-MATIC CAN CARRIER WITH HEMMED EDGE

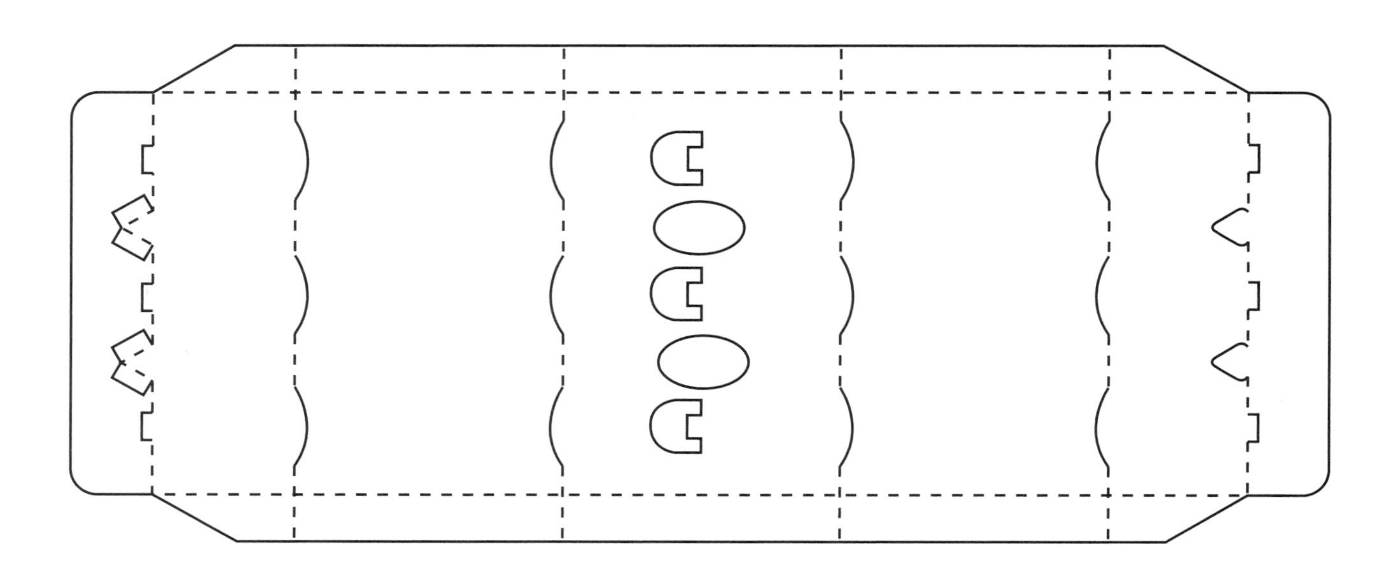

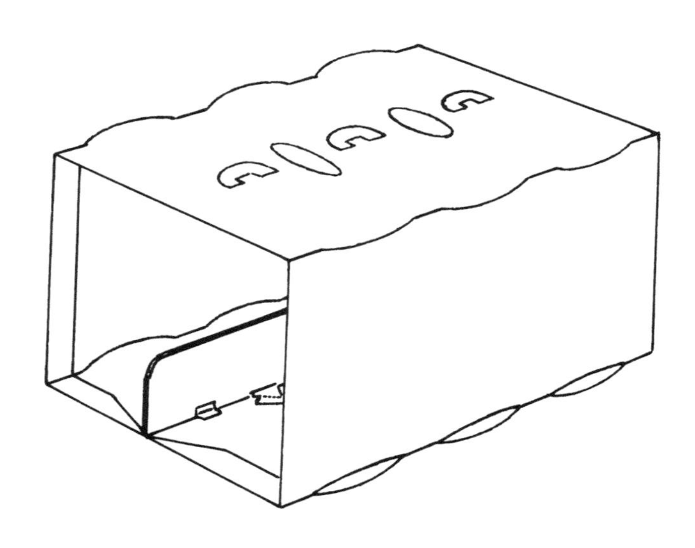

JONES STYLE CARRIER WITH ECONOMY END

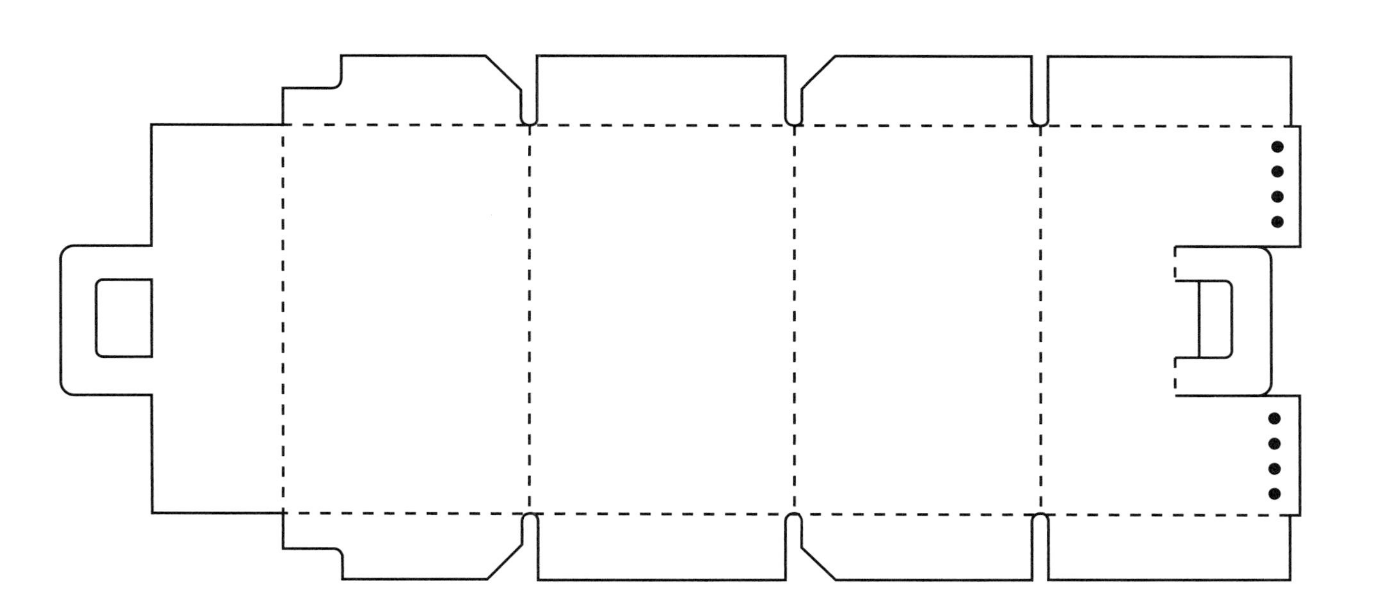

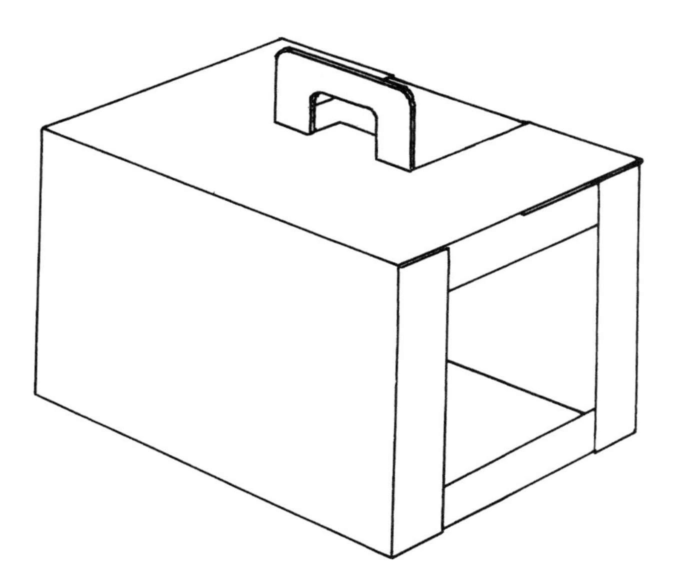

END LOADED TUBE WITH REINFORCED HANDLE

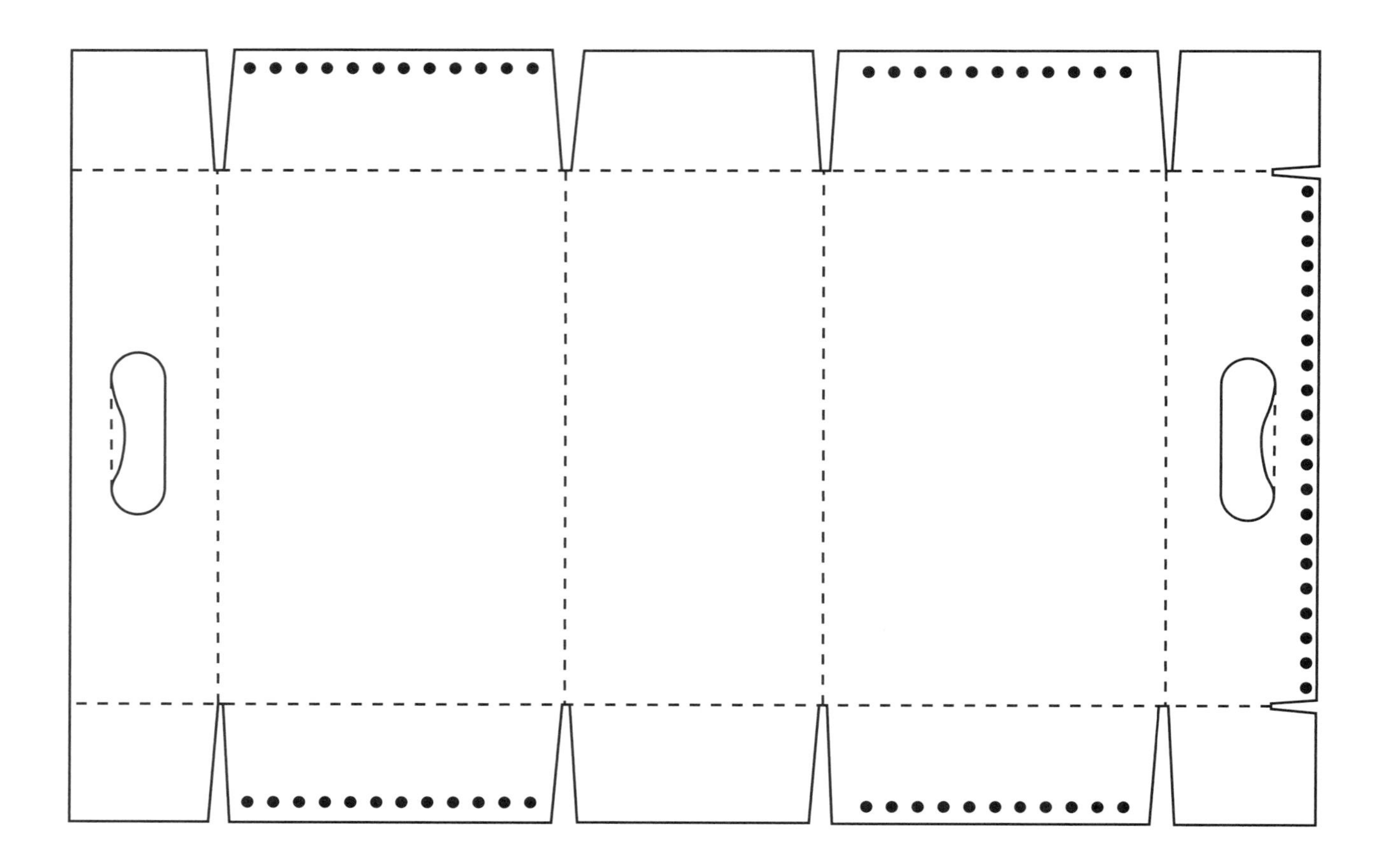

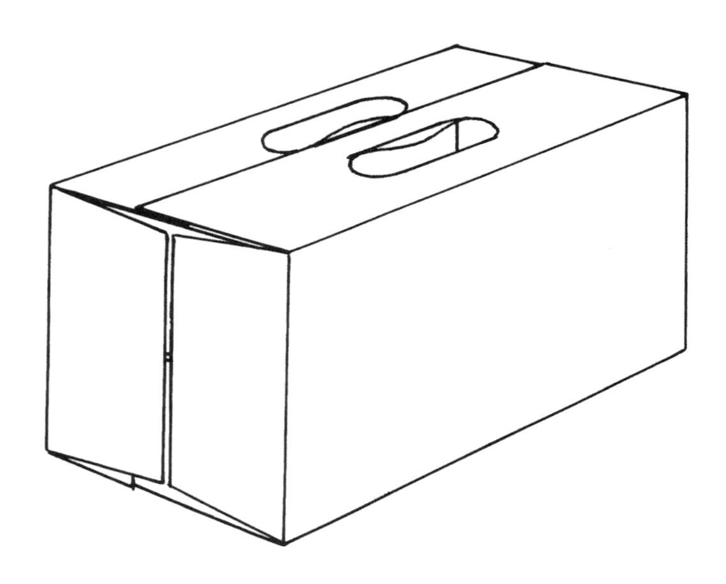

END LOADED BOTTLE CARRIER WITH TAPERED SIDES, REINFORCED HANDLE, AND SEPARATE INSERTS

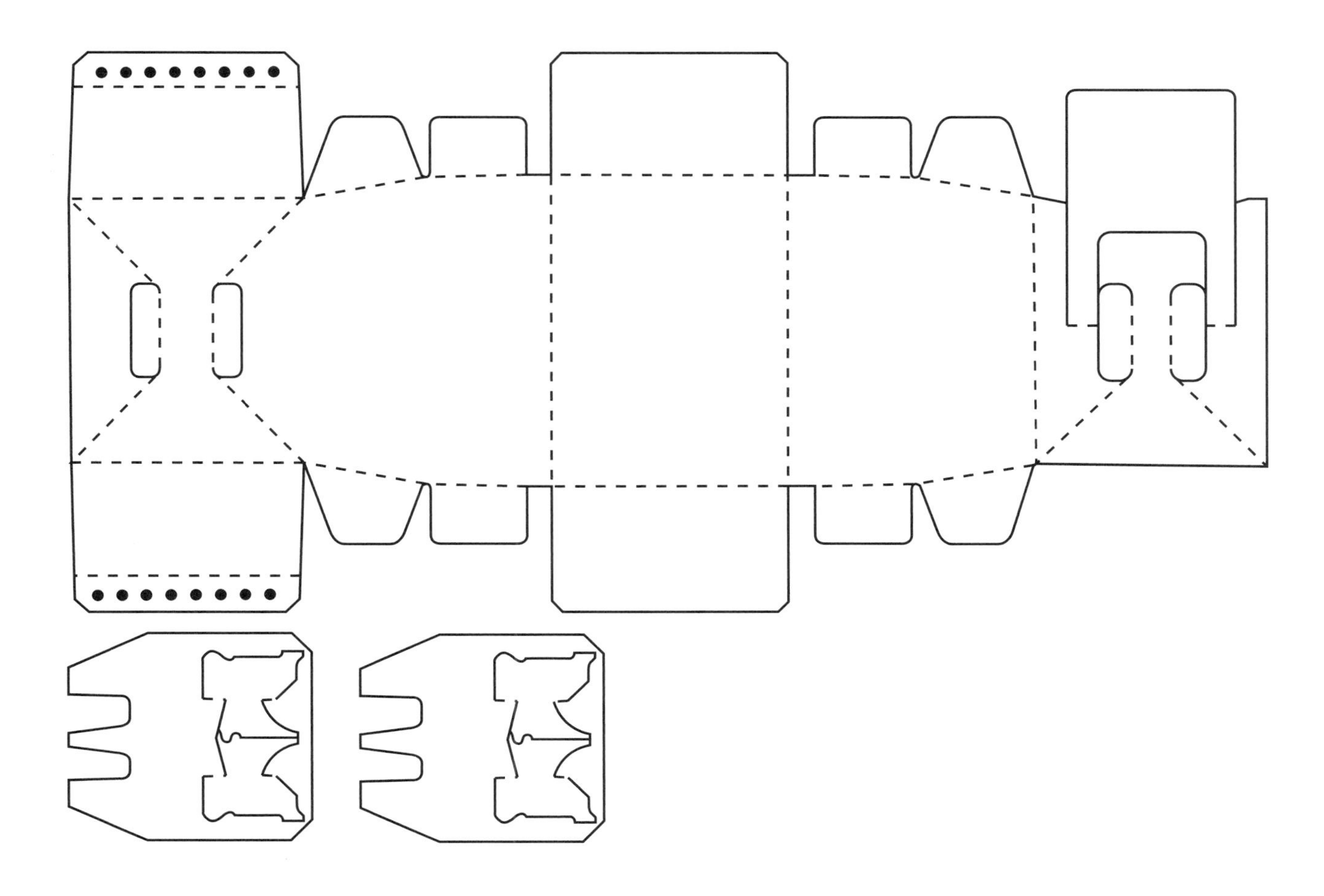

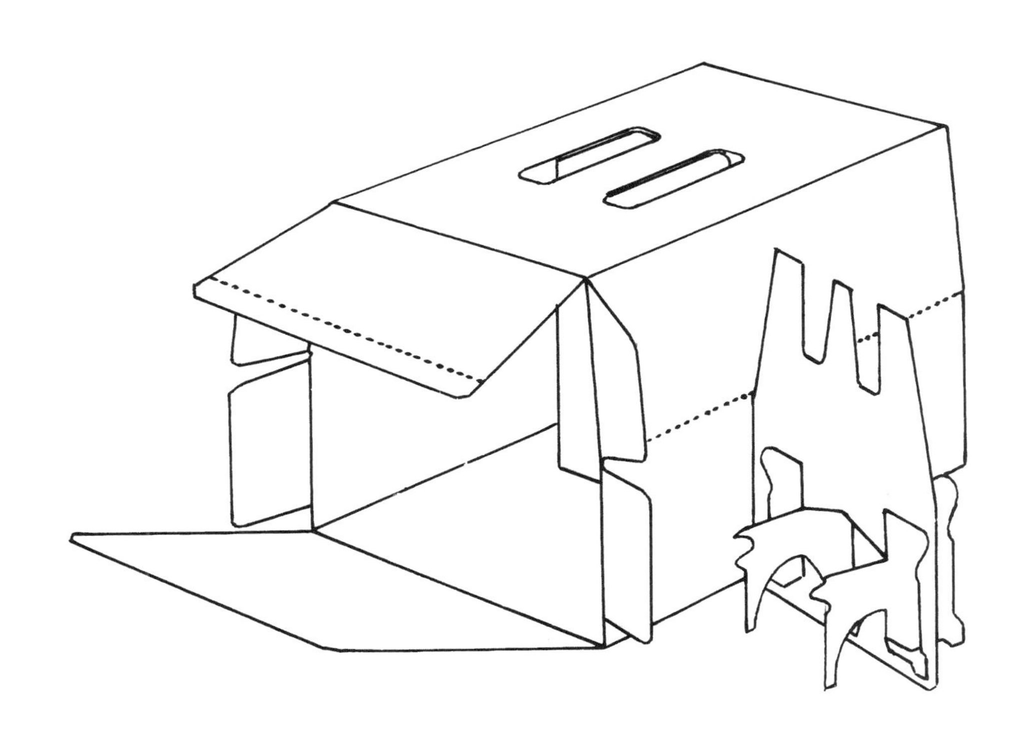

TWELVE-COUNT CAN CARRIER WITH PRE-PERFORATED TEAR-OPEN STRIP
(PAT. #3,955,740)

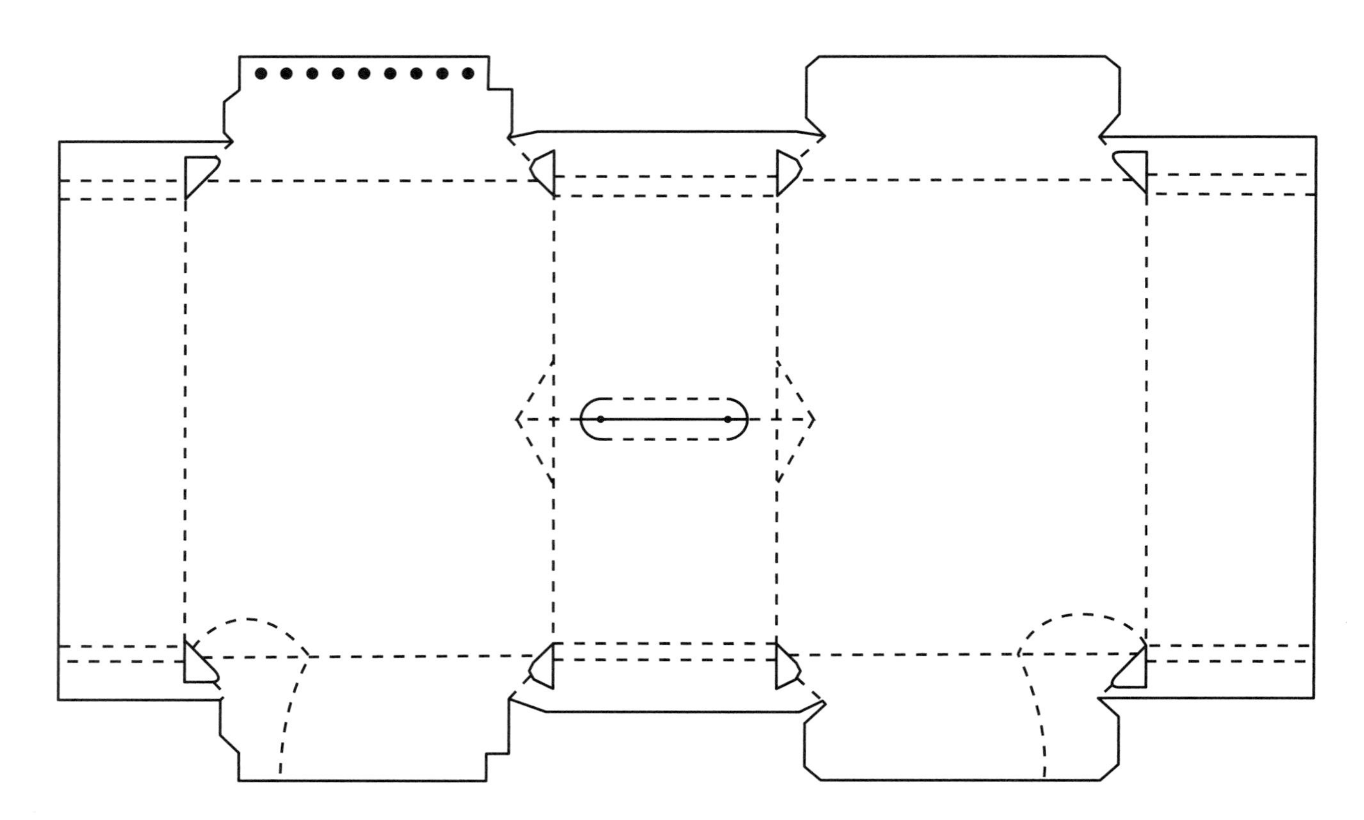

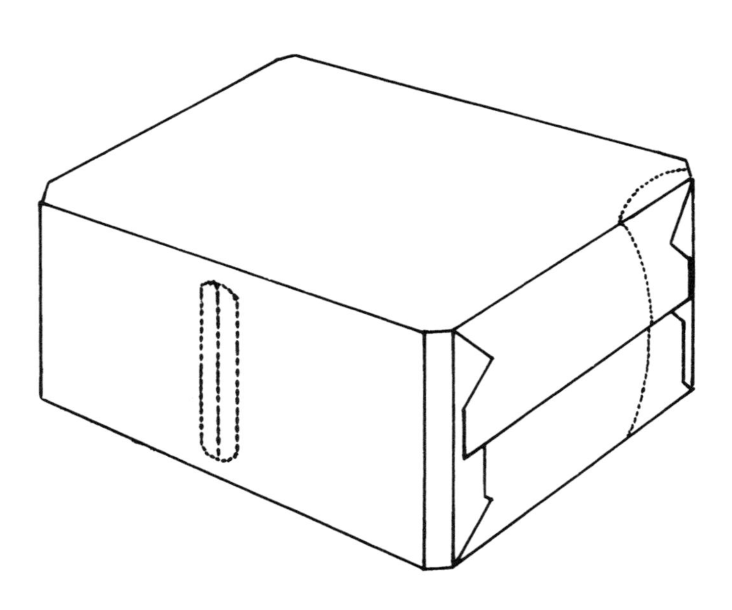

SIX BOTTLE WING LOCKED SLEEVE WITH TOP AND BOTTOM CHIME LOCKS

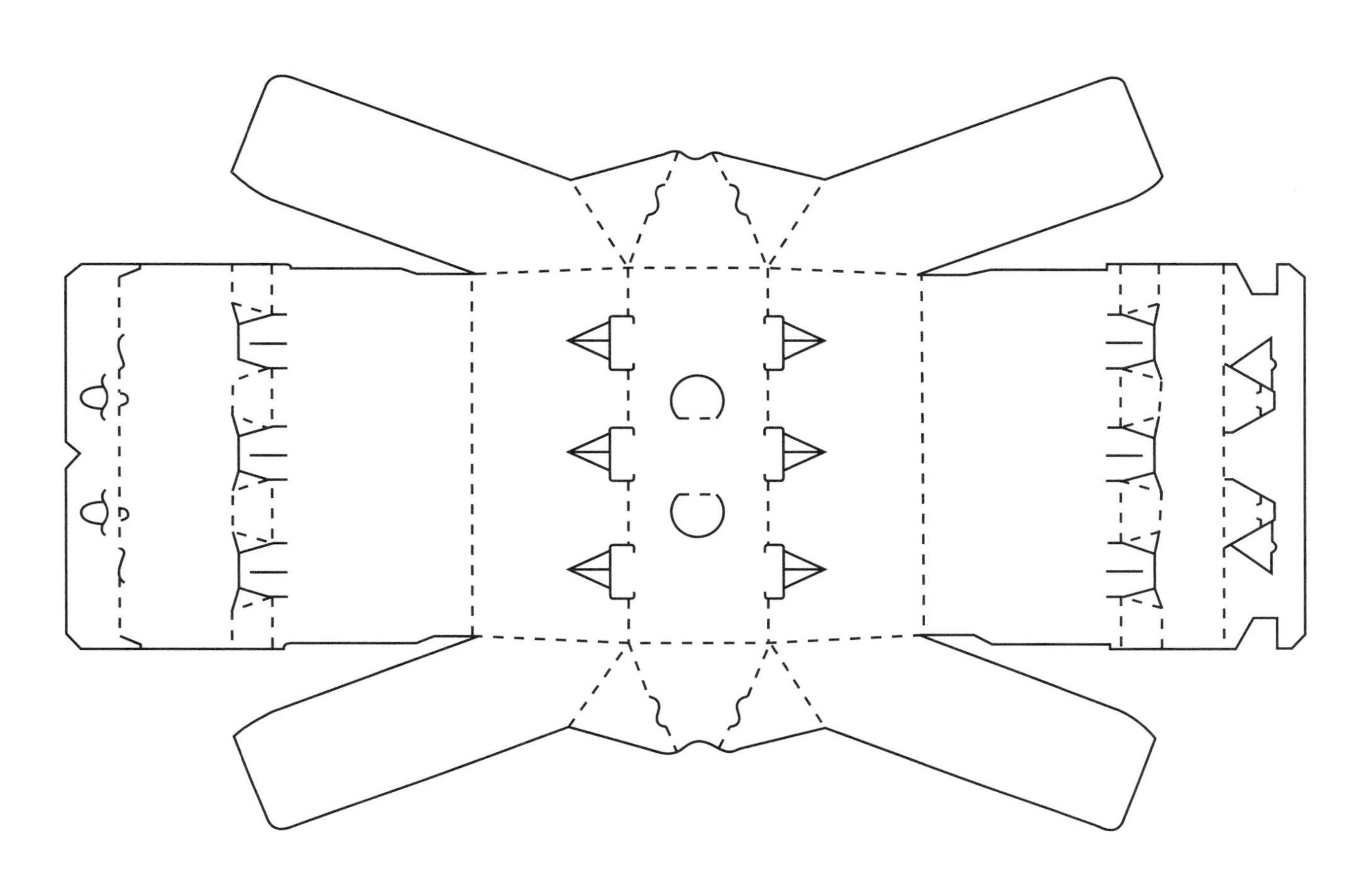

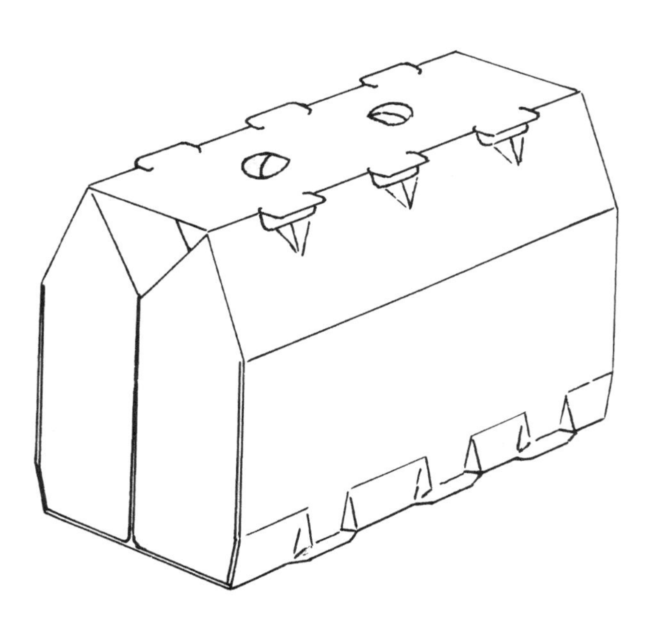

LAYOUT AND CONSTRUCTION OF A SIX-COUNT BOTTLE CARRIER

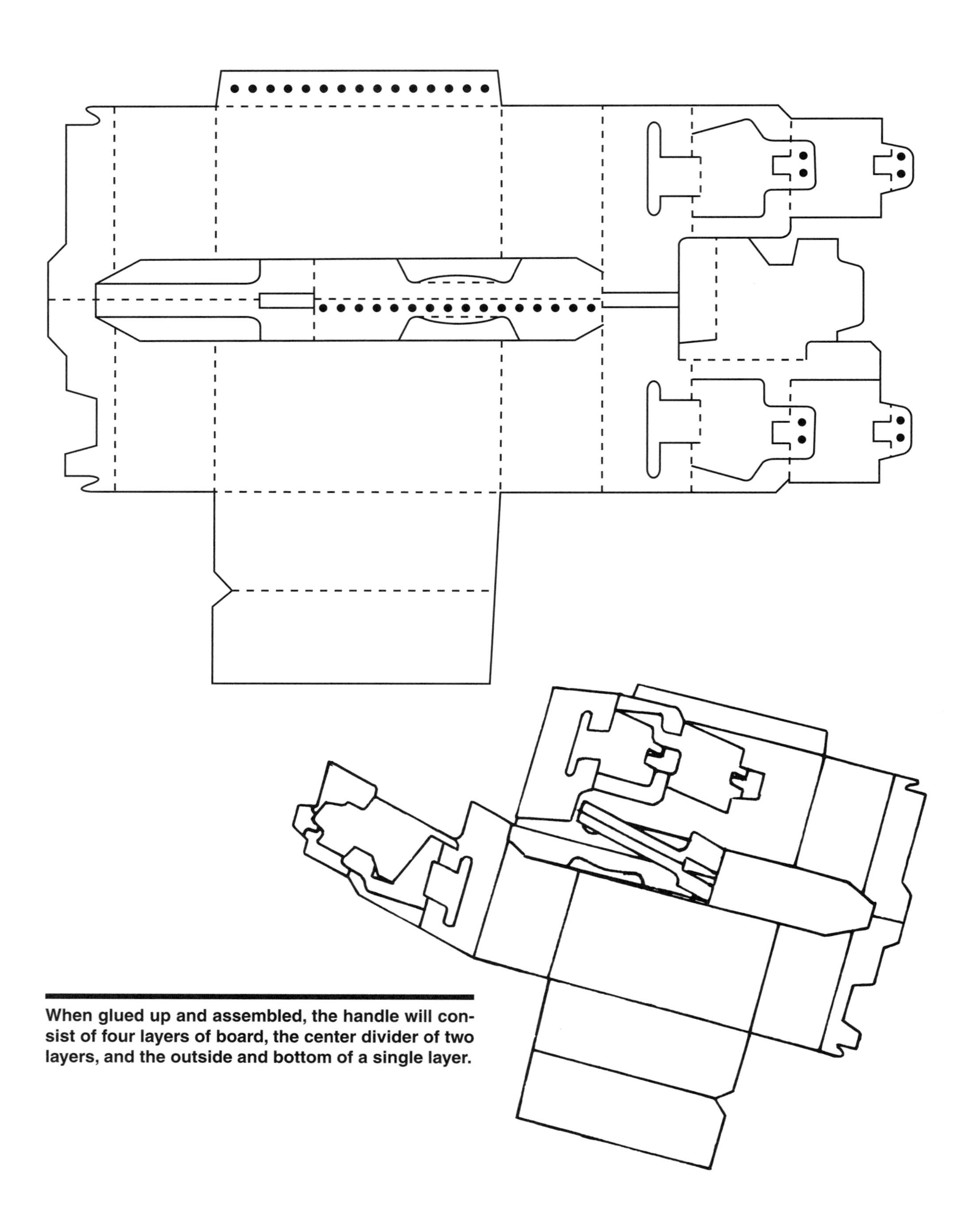

When glued up and assembled, the handle will consist of four layers of board, the center divider of two layers, and the outside and bottom of a single layer.

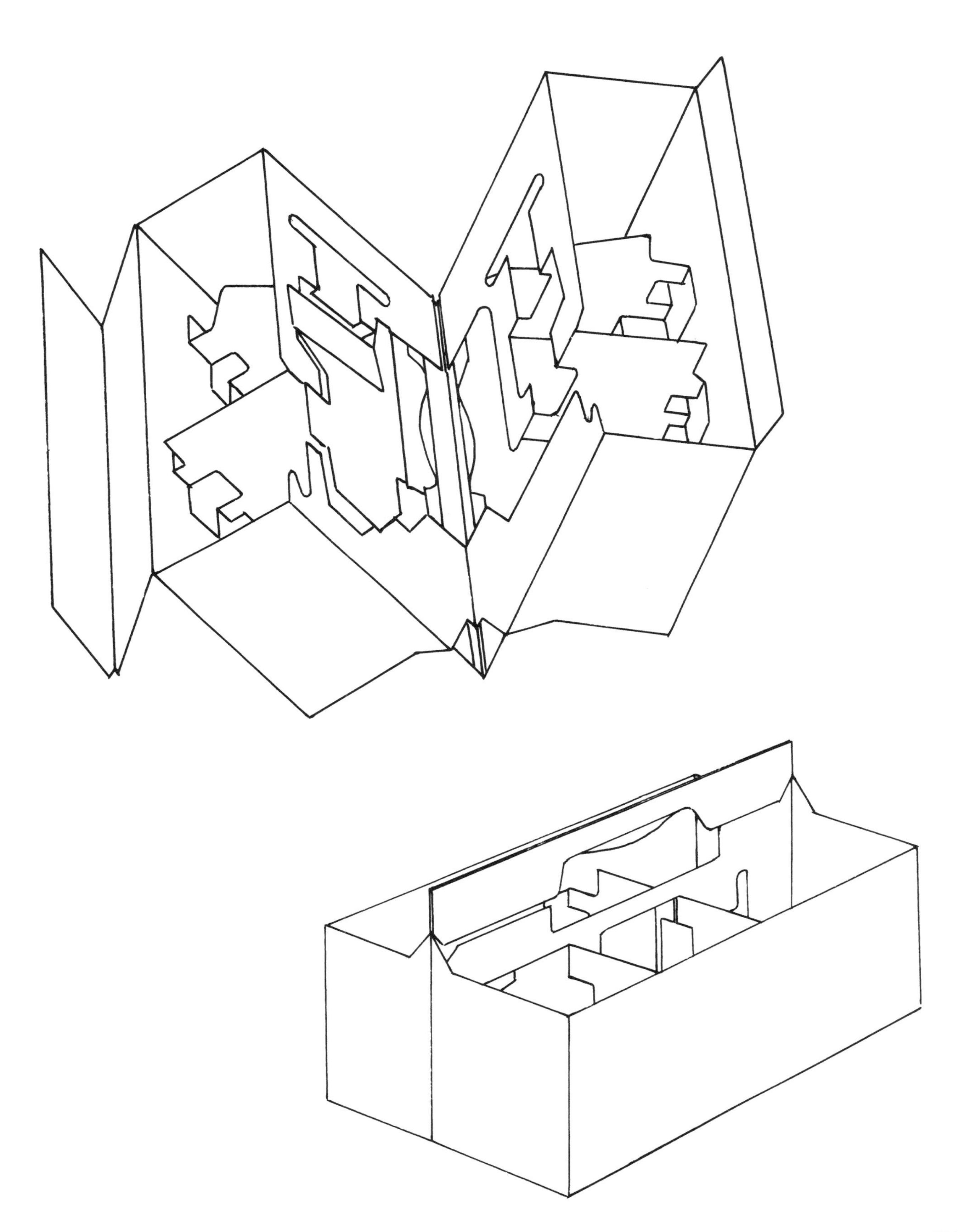

GLUED BOTTOM STRAP BOTTLE CARRIER

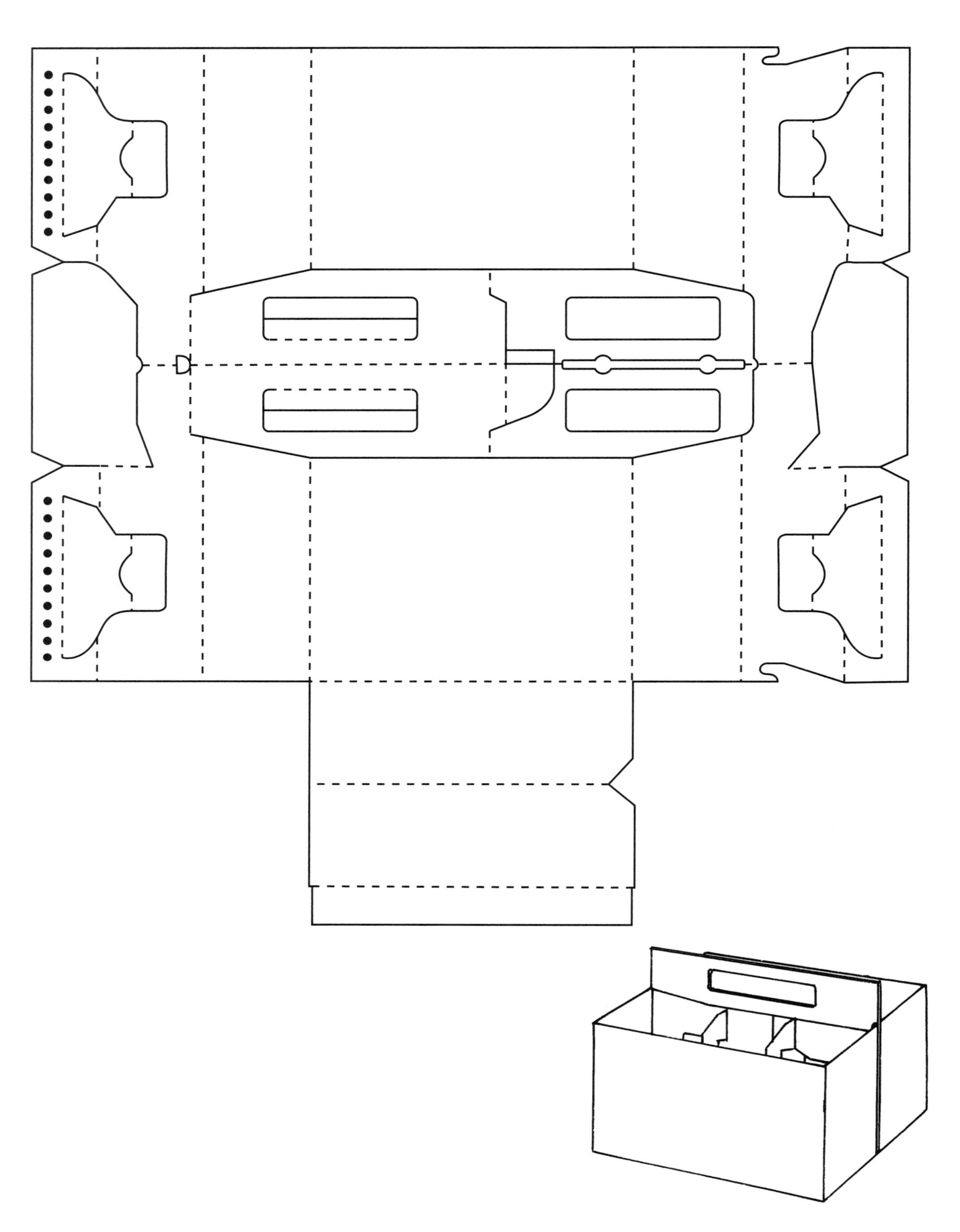

ONE PIECE GLUED BOTTOM STRAP BOTTLE CARRIER

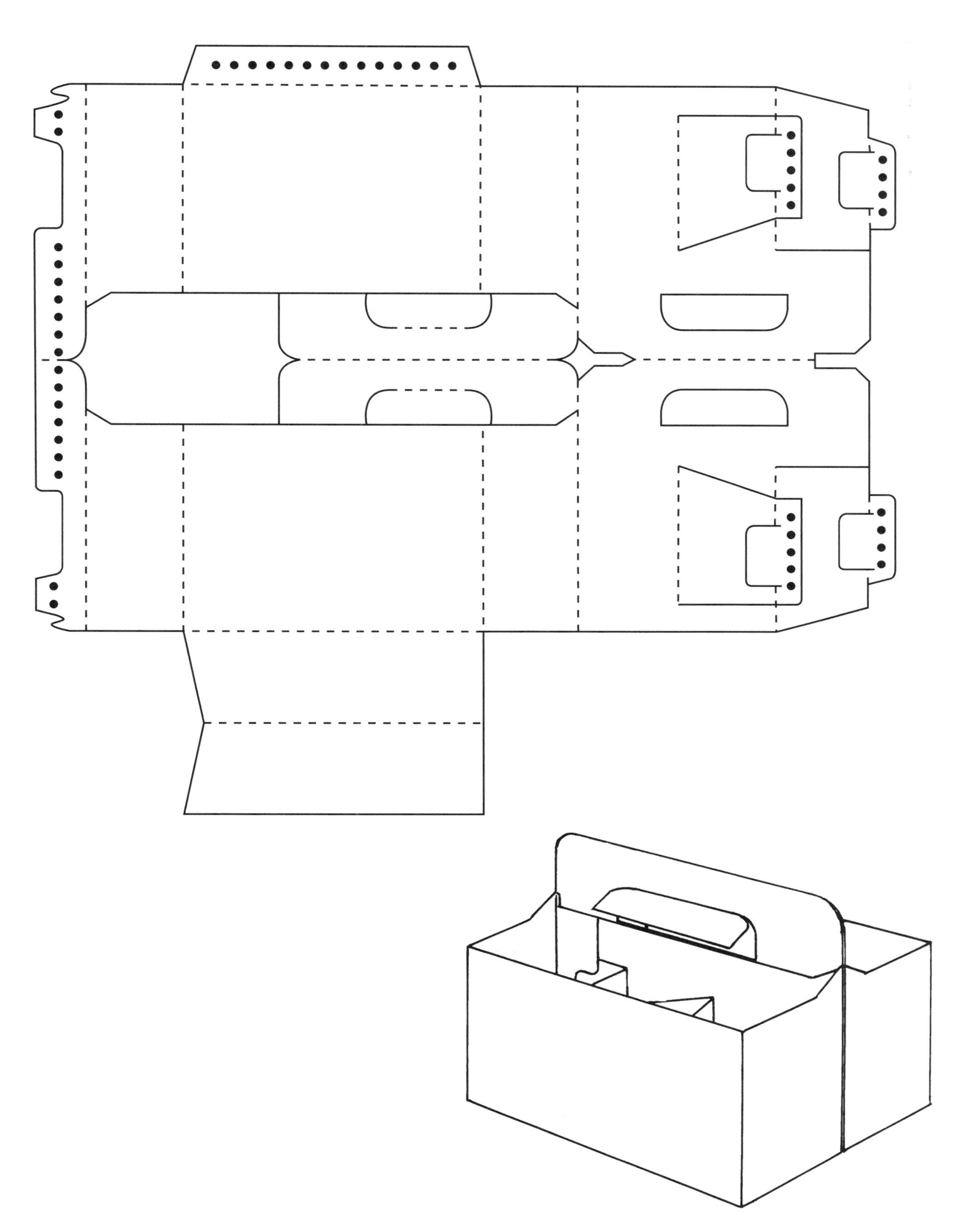

SIX-COUNT BOTTOM GLUED BOTTLE CARRIER

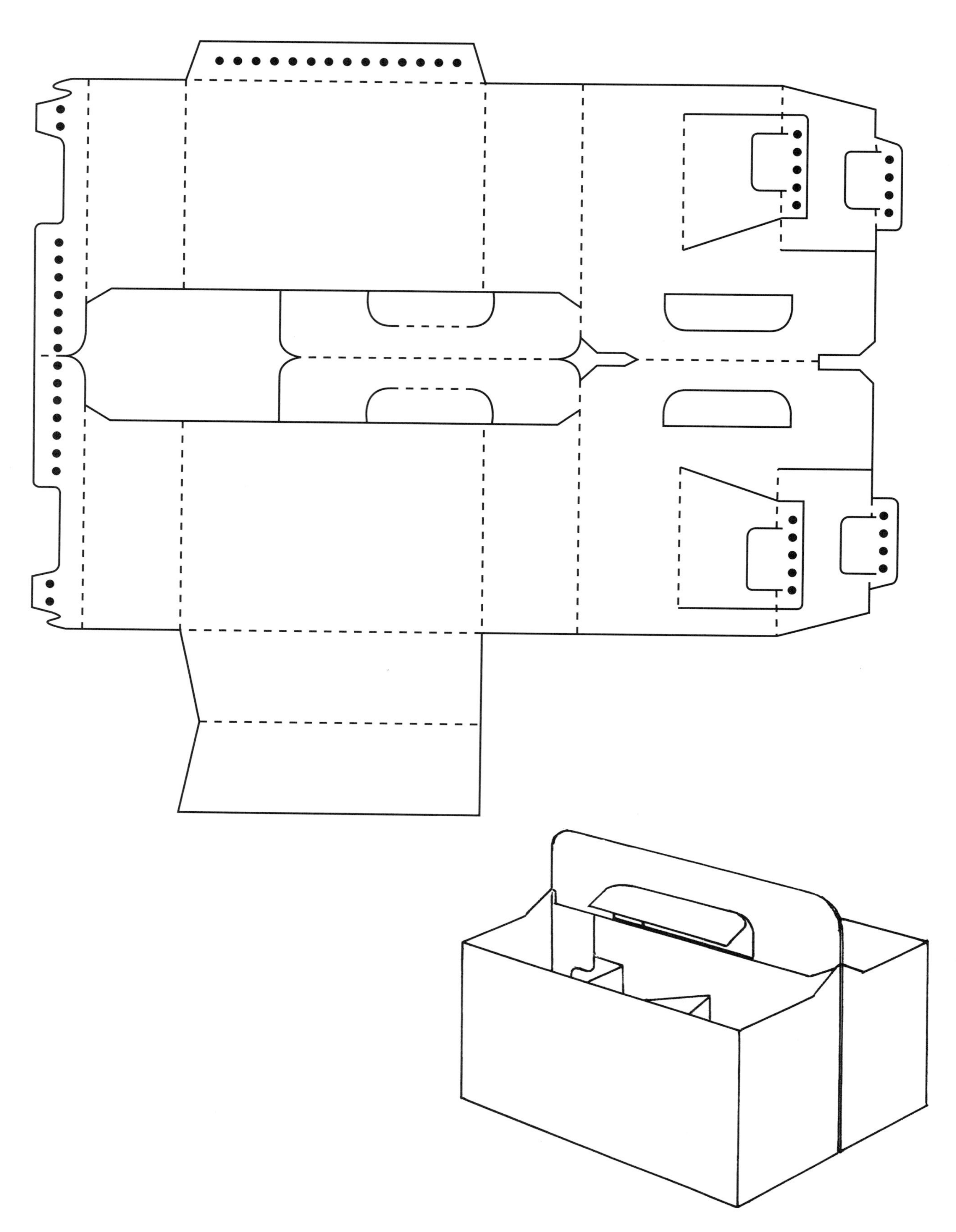

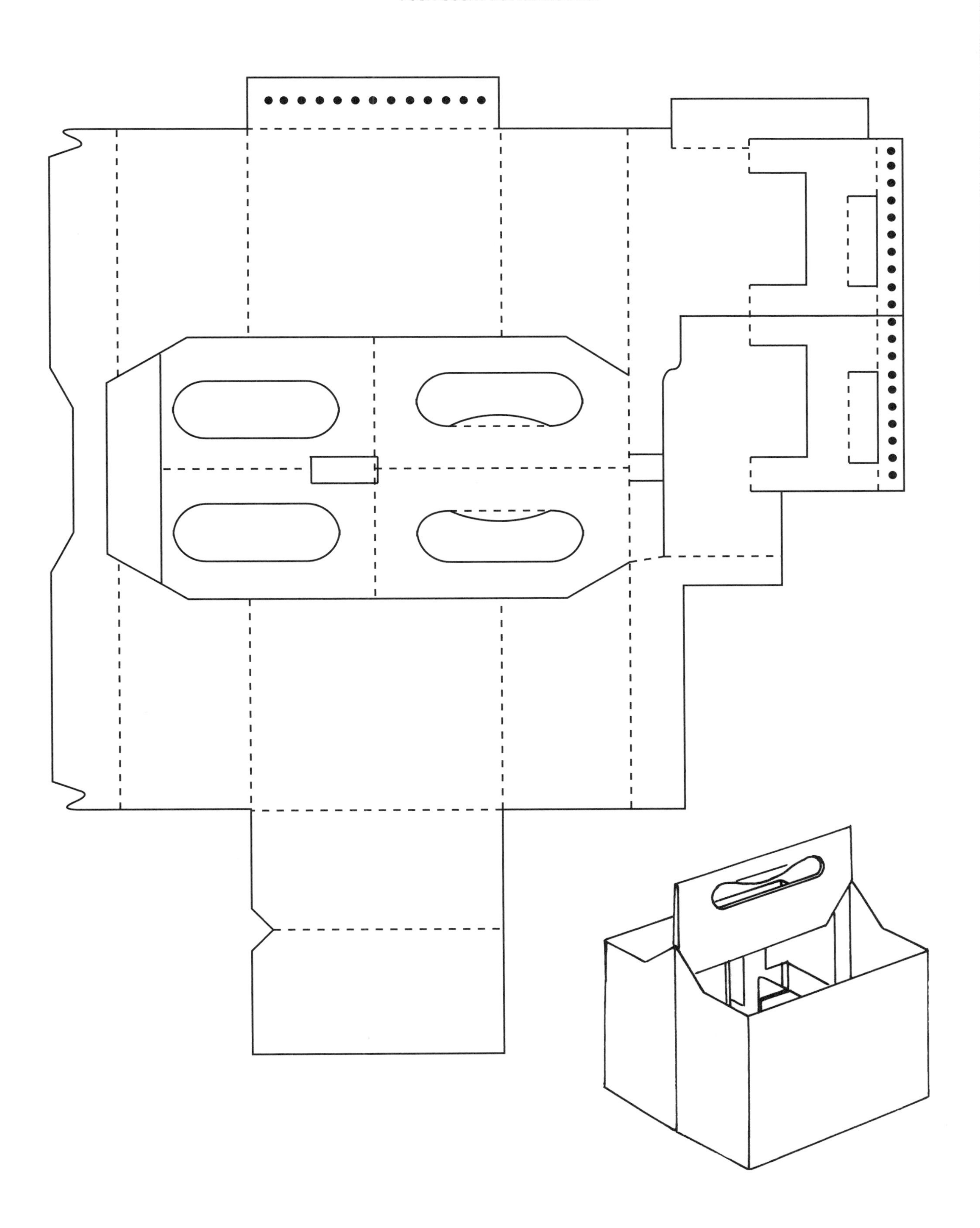

FOUR-COUNT BOTTLE CARRIER (VARIATION)

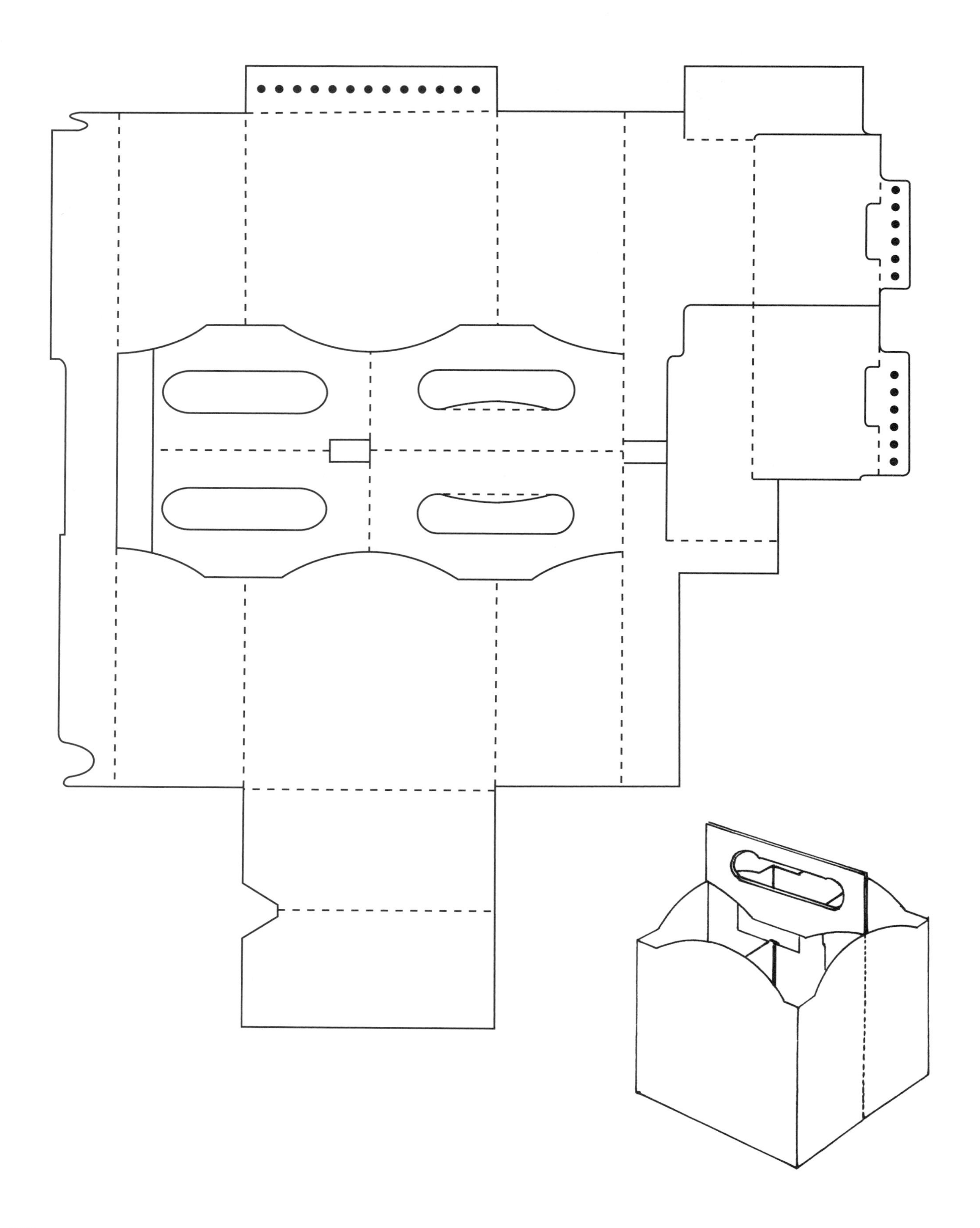

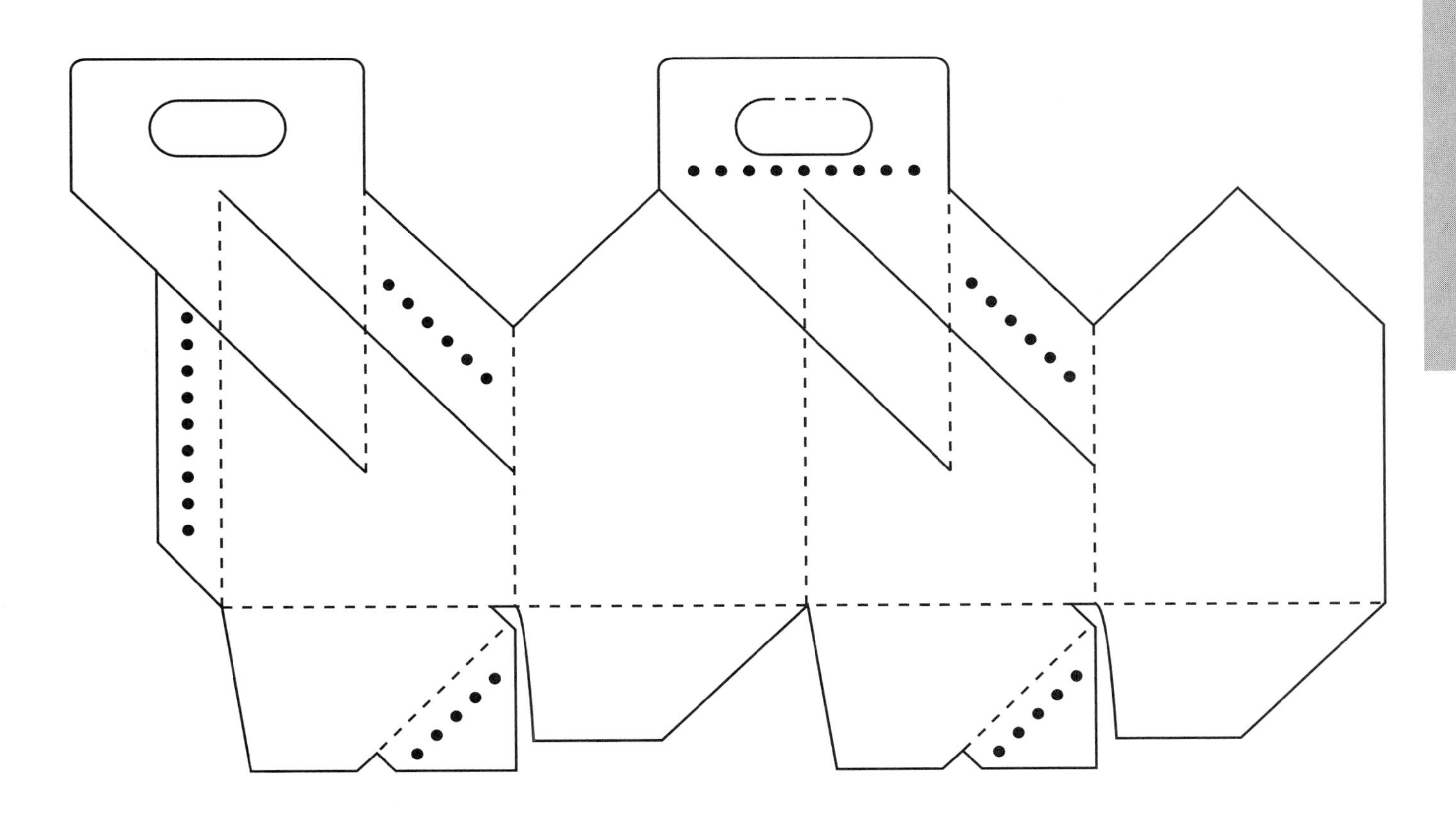

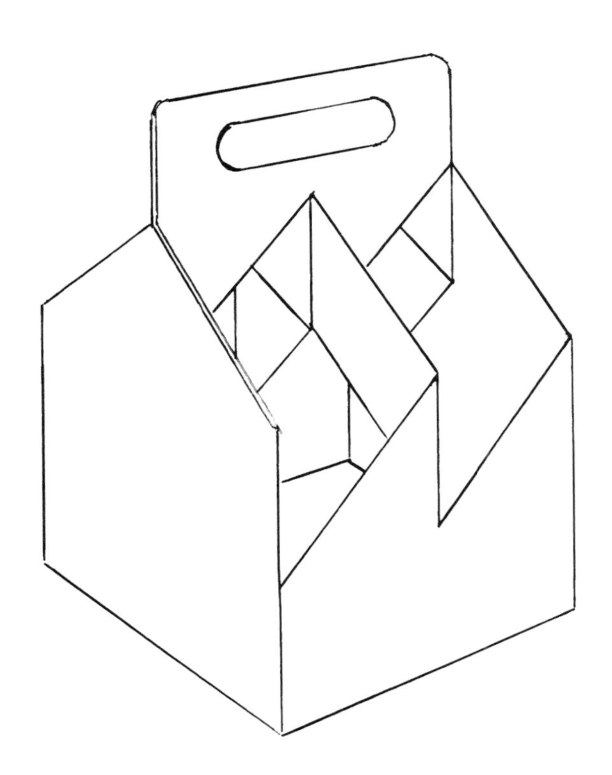

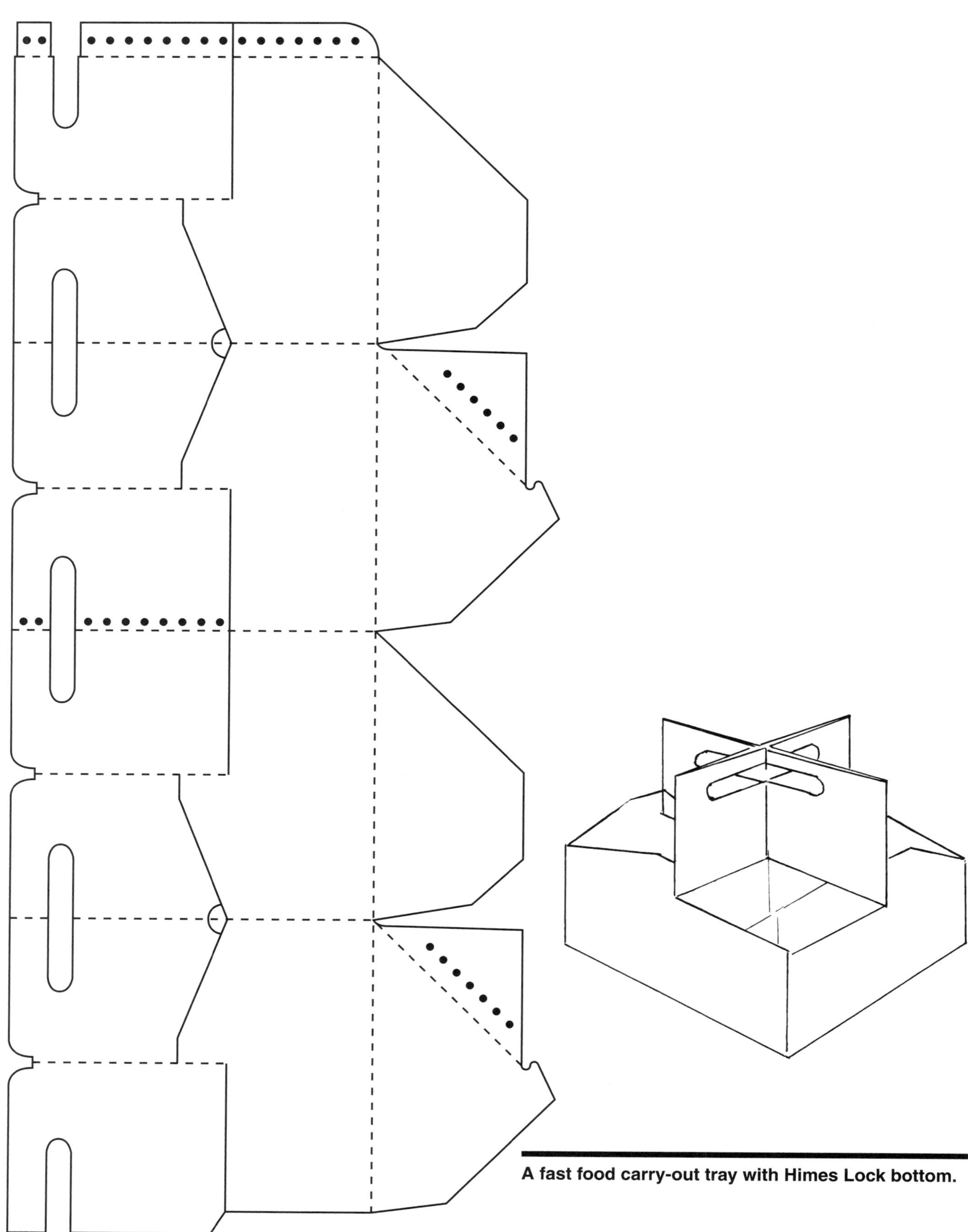

A fast food carry-out tray with Himes Lock bottom.

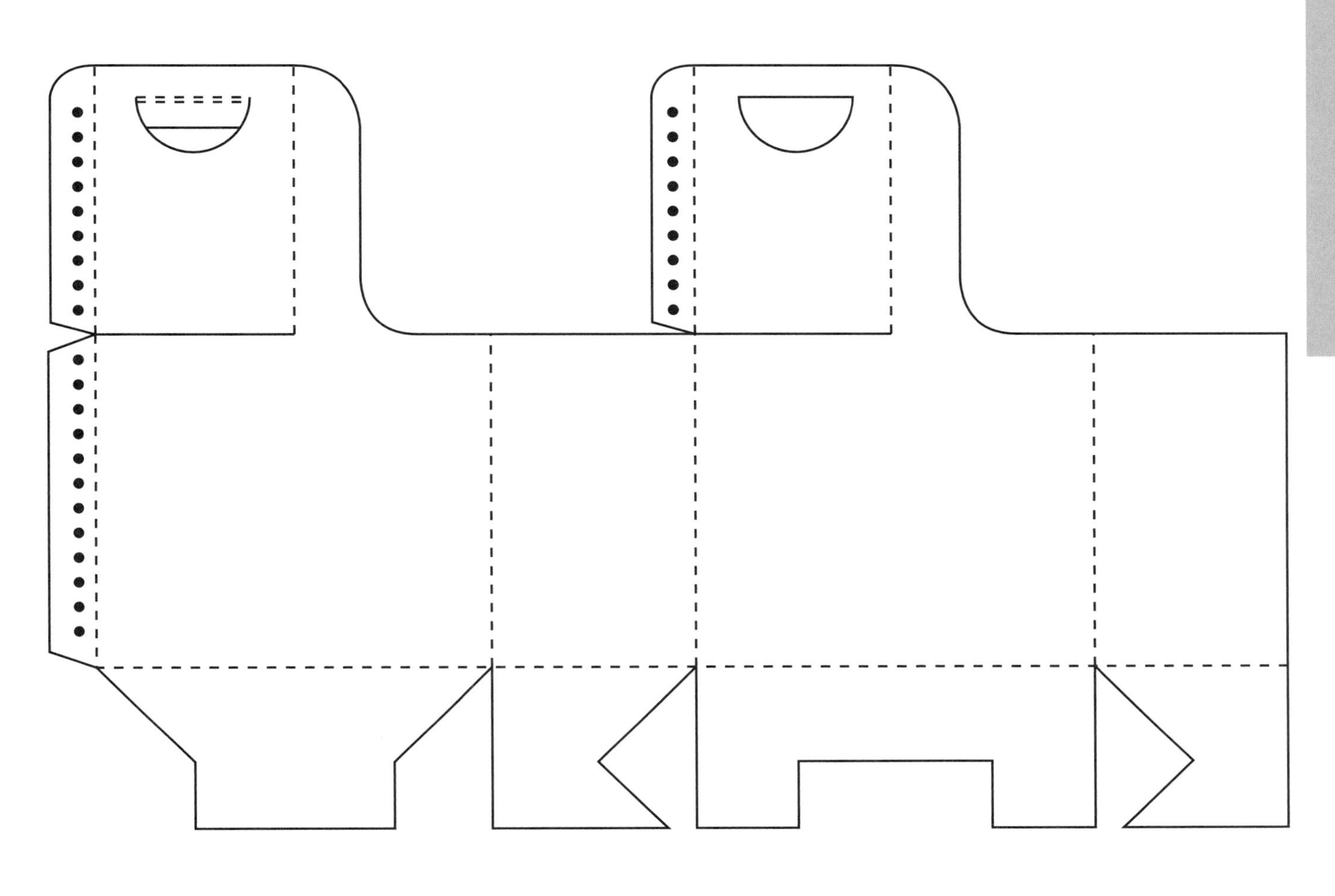

This two compartment carrier with snap lock or Houghland bottom lends itself to many applications.

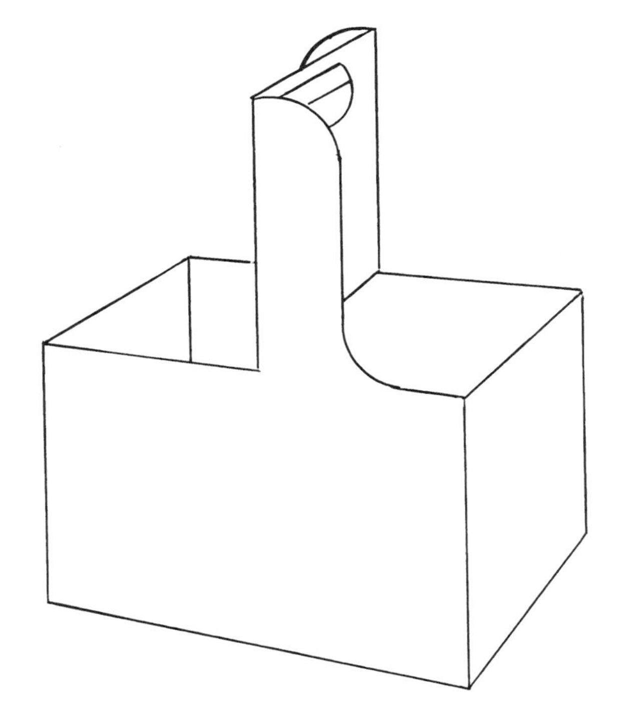

3

The Set-Up or Rigid Paper Box

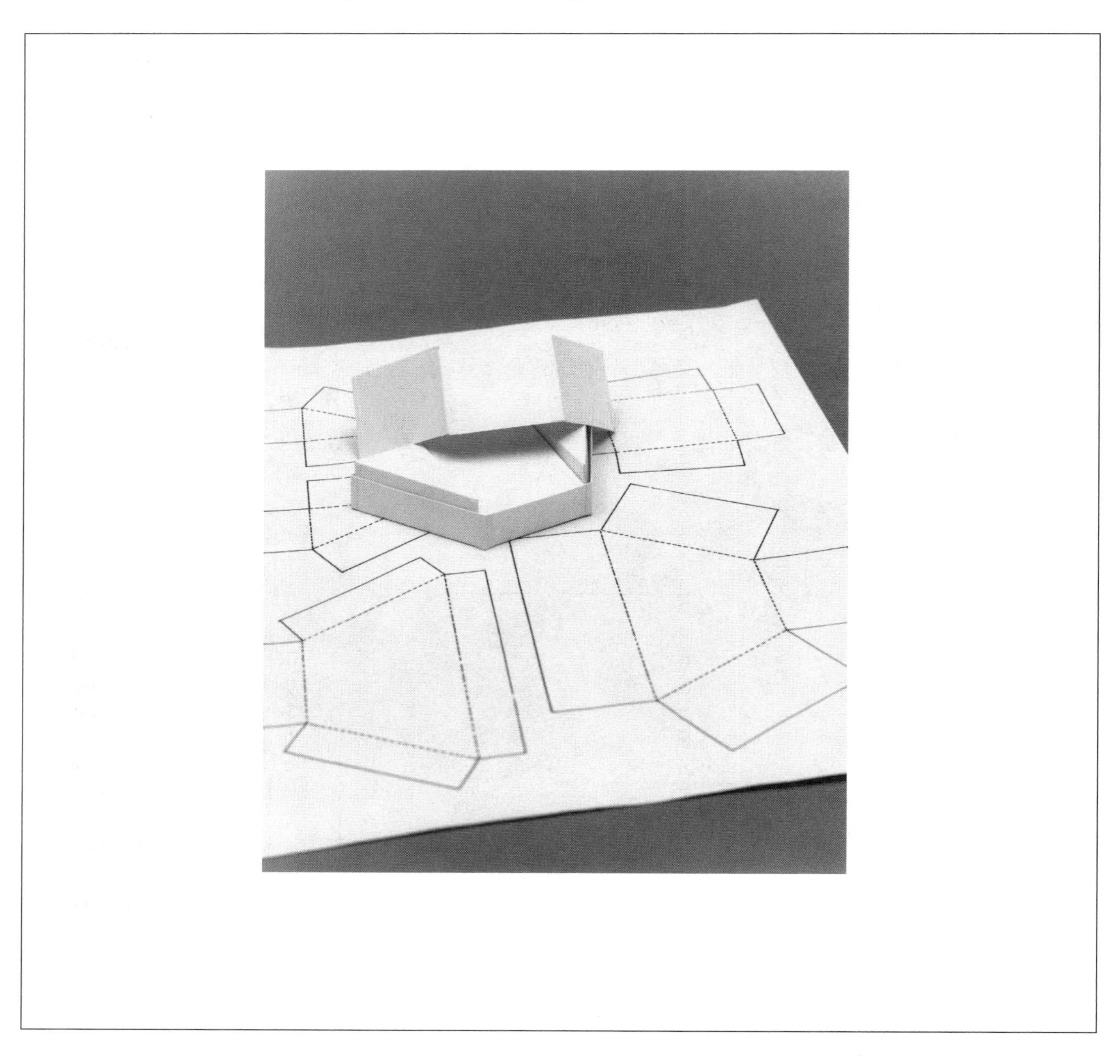

The word *box* is derived from the boxwood baskets made in ancient Rome. By the thirteenth century, the Chinese were making boxes from decorated cardboard made of rice chaff, in which they exported tea. In France, attractive boxes were made for hats, garments, cosmetics, and jewelry in the seventeenth and eighteenth centuries. These boxes were covered with hand-printed decorative papers and fabrics. During the nineteenth century, attractive boxes for toys and games were produced in Germany and England.

The first American boxes using paperboard were made in Boston by Colonel Andrew Dennison in 1844. The name Dennison is still associated with many kinds of paper products.

Store clerks often used crude wooden jigs to form cardboard into "cartons," which were closed and tied with string. From this simple device the set-up box evolved. According to historians, it is possible that an artistic salesperson hand lettered the store's name on the box. This event could be considered the birth of the brand name.

The set-up paper box is produced and delivered as a three-dimensional construction ready to be packed with merchandise. It is manufactured in a series of simple steps that do not require costly dies or machinery. Production runs can be small, medium, or large.

The basic set-up box can be modified to accommodate unusual requirements; transparent plastic domes and platforms, a variety of hinges, embossing, gold stamping, lids, and compartments can be added. The versatility of the set-up box is valuable because it adds to the display and gift appeal of boxed items. It is no accident that set-up boxes are so widely used in the merchandising of cosmetics, jewelry, candy, and gifts. The box not only provides protection but has great promotional value.

Set-up boxes are made from four basic materials: boxboard, corner stays, adhesives, and covering materials. The process is as follows:

1. Flat sheets of boxboard are cut and scored to size. Individual blanks are separated from the sheet and the corners are cut out.
2. The sides of the blank are folded at right angles to the base to form the sides of the tray.
3. The corners of the tray are sealed by adhering paper, fabric, or metal to reinforce the corners.
4. The decorative covering is adhered to the box.

Variations in the basic shape and dimensions of boxes can be made in order to produce an infinite range of designs. Possible shapes include cubes, pyramids, cones, hearts, and stars. Boxes with hinged lids, telescoping boxes, boxes with compartments, boxes with slide-in drawers, and boxes with plastic thermoformed platforms can also be designed. Coverings or wraps may be selected from a wide range of papers or fabrics, which provide both protection and decorative effect.

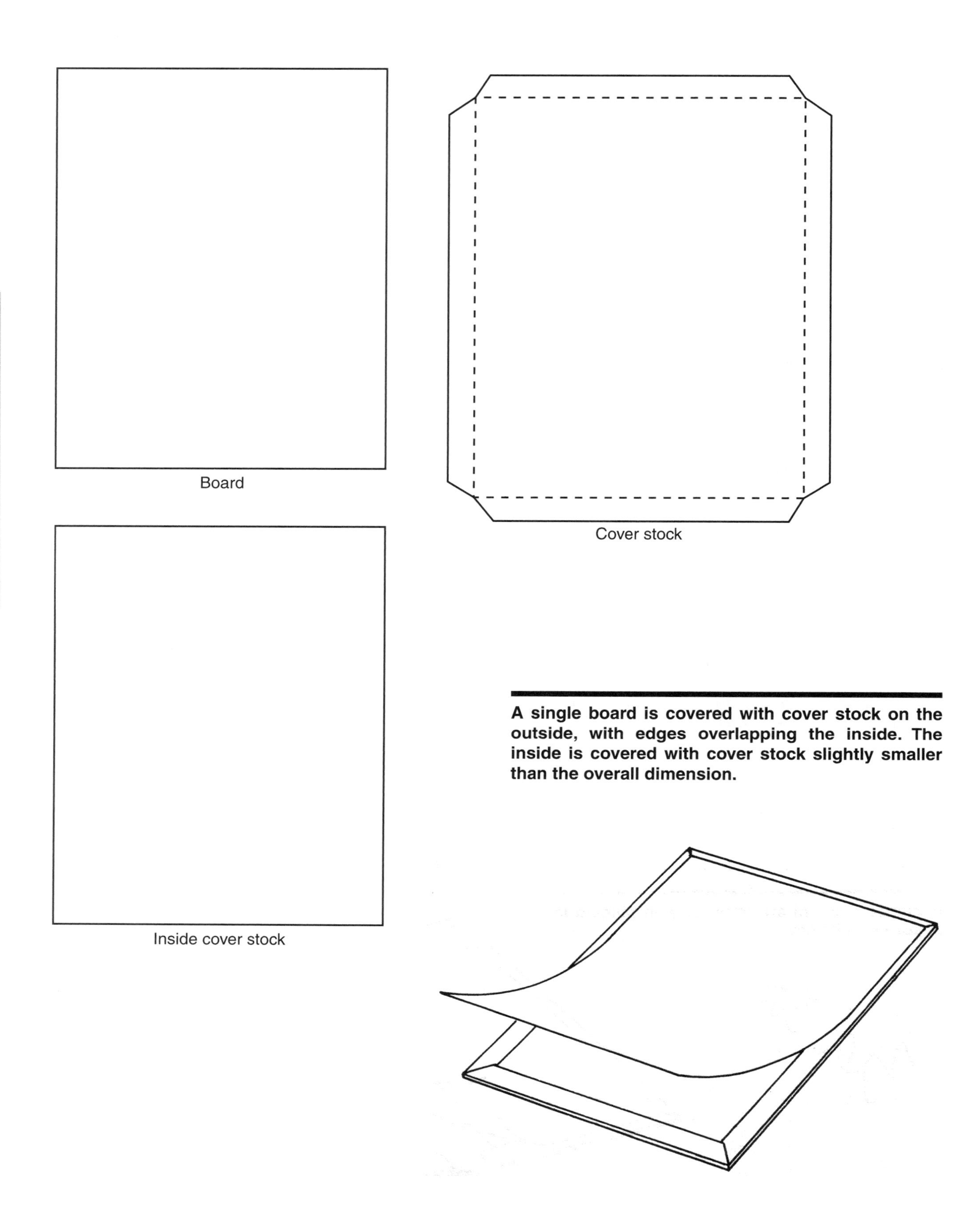

A single board is covered with cover stock on the outside, with edges overlapping the inside. The inside is covered with cover stock slightly smaller than the overall dimension.

Two pieces of board are attached with hinging tape, then covered with cover stock.

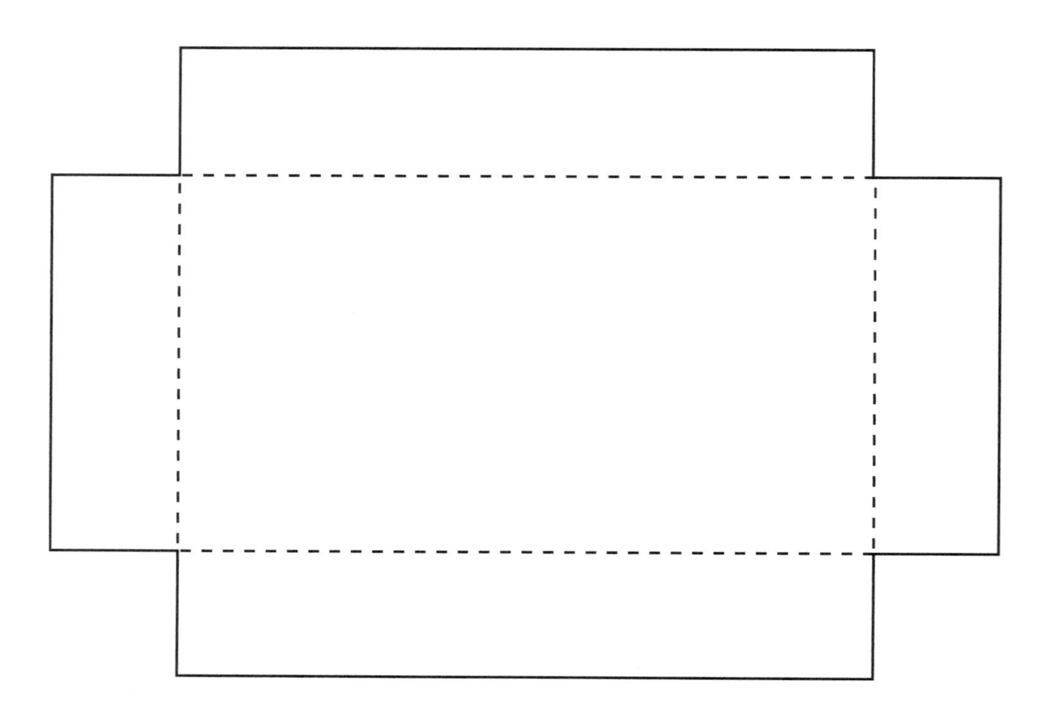

1. The stock is scored with cutting dies that cut approximately halfway through the board. The corners are then cut away so that only the bottom and side panels remain. The side panels are bent up and fastened at the corners with stay tape.

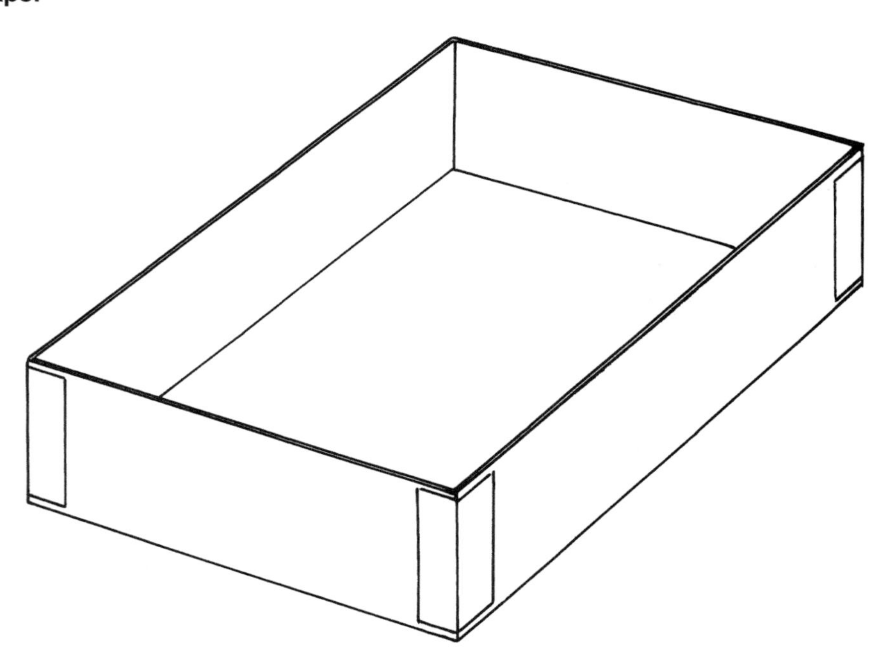

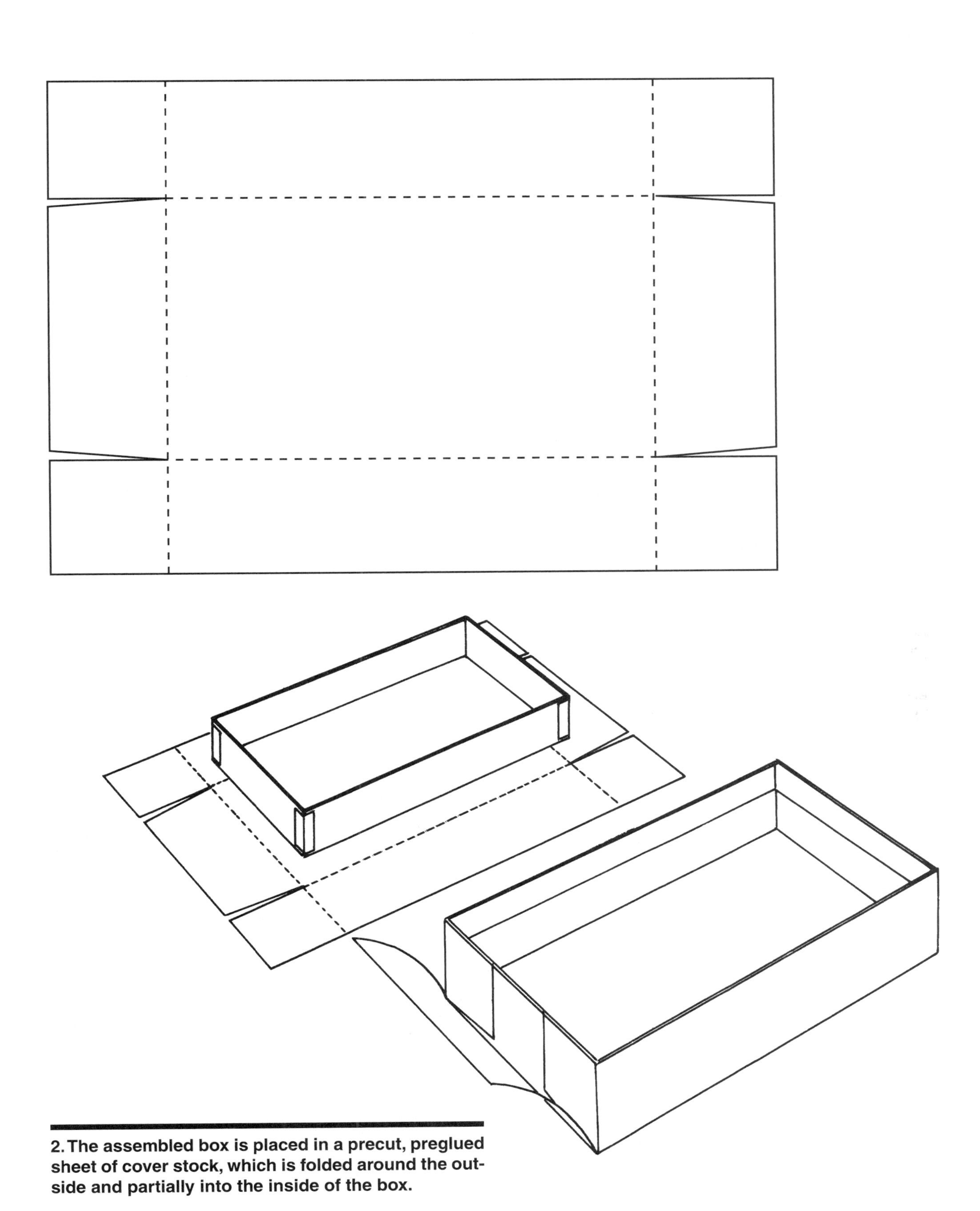

2. The assembled box is placed in a precut, preglued sheet of cover stock, which is folded around the outside and partially into the inside of the box.

1. Since it is somewhat difficult to score the board from which you are preparing the set-up-box model, we often use single-weight illustration board and cut each piece separately. Consider the thickness of the board in the dimensions.

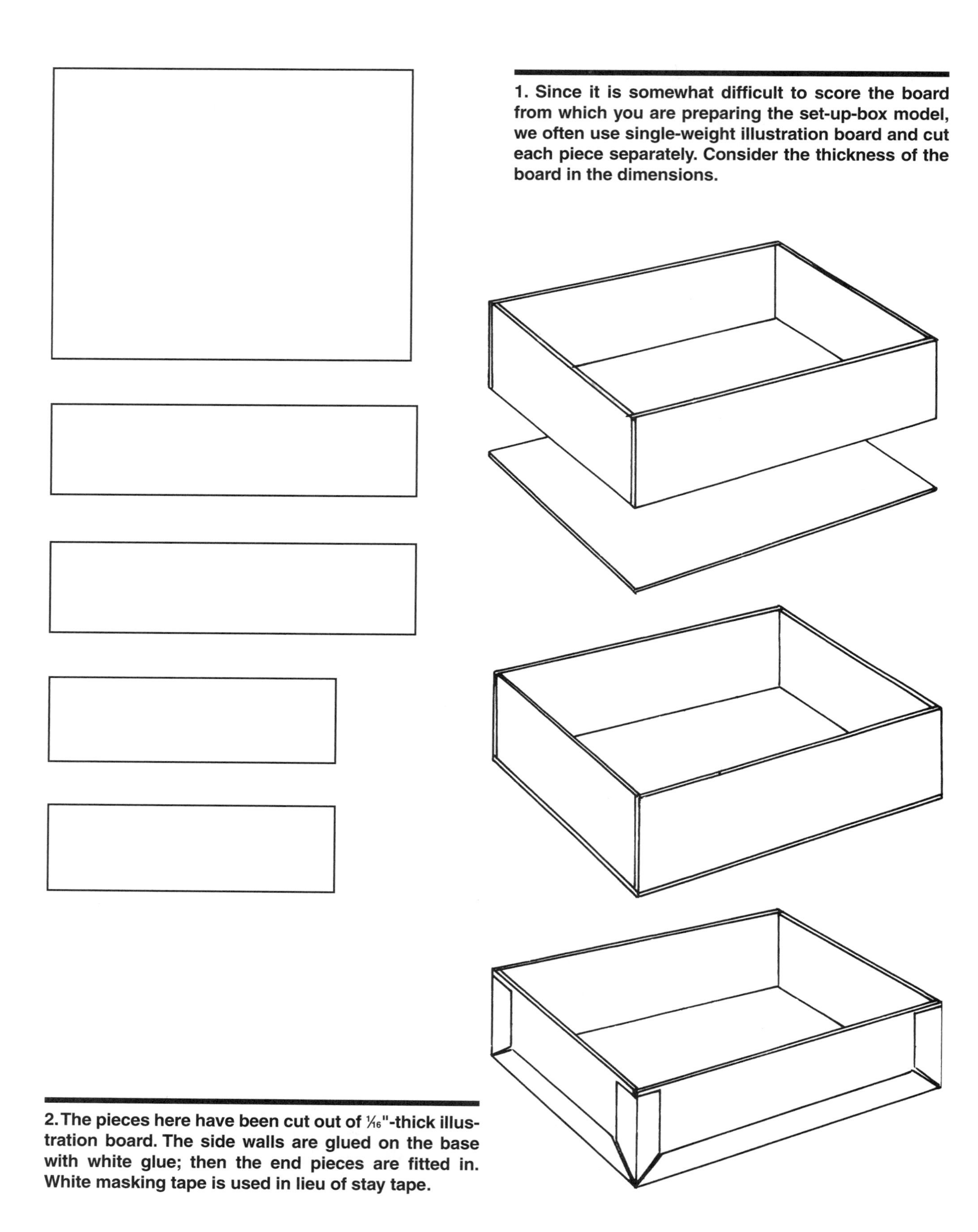

2. The pieces here have been cut out of 1/16"-thick illustration board. The side walls are glued on the base with white glue; then the end pieces are fitted in. White masking tape is used in lieu of stay tape.

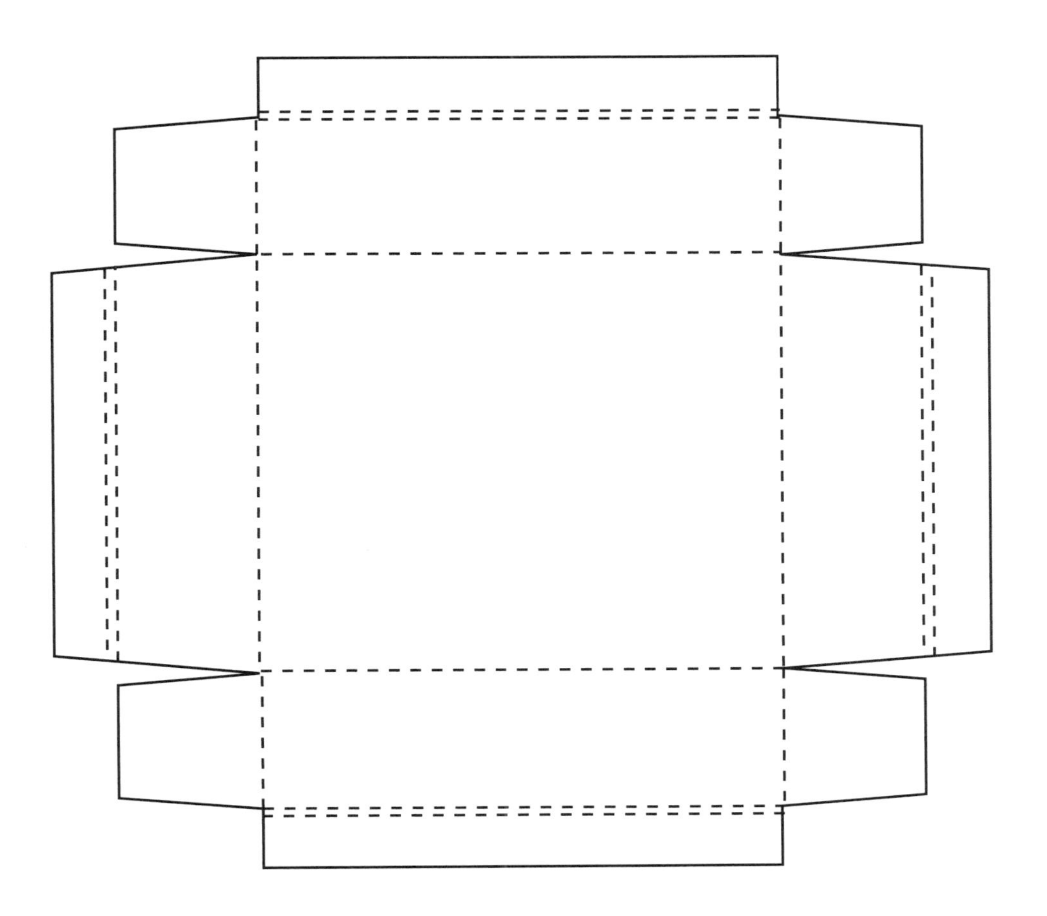

3. The assembly is covered with a nice cover stock. For this you may use rubber cement or, preferably, bookbinder's glue.

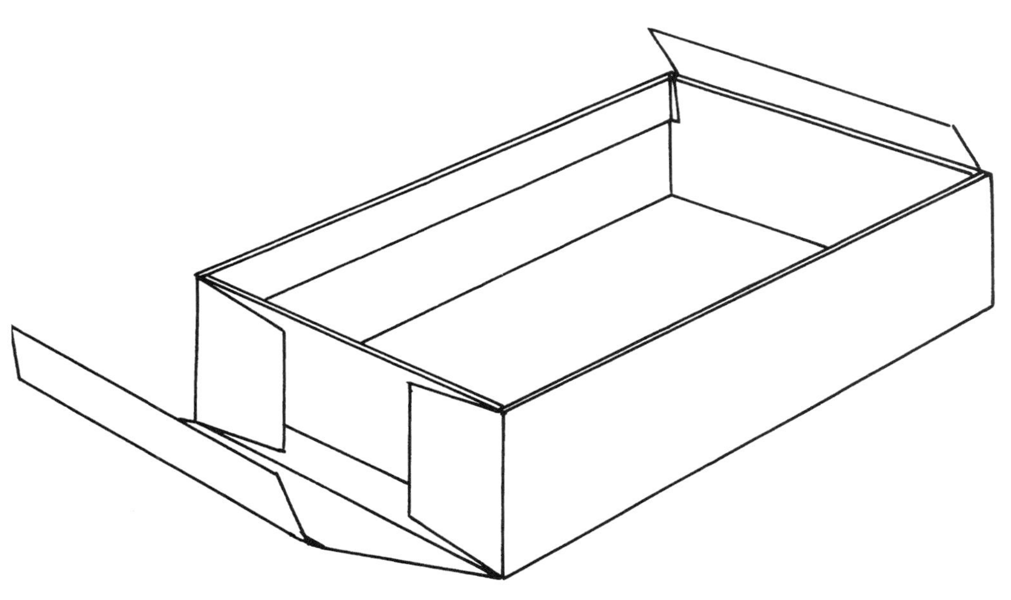

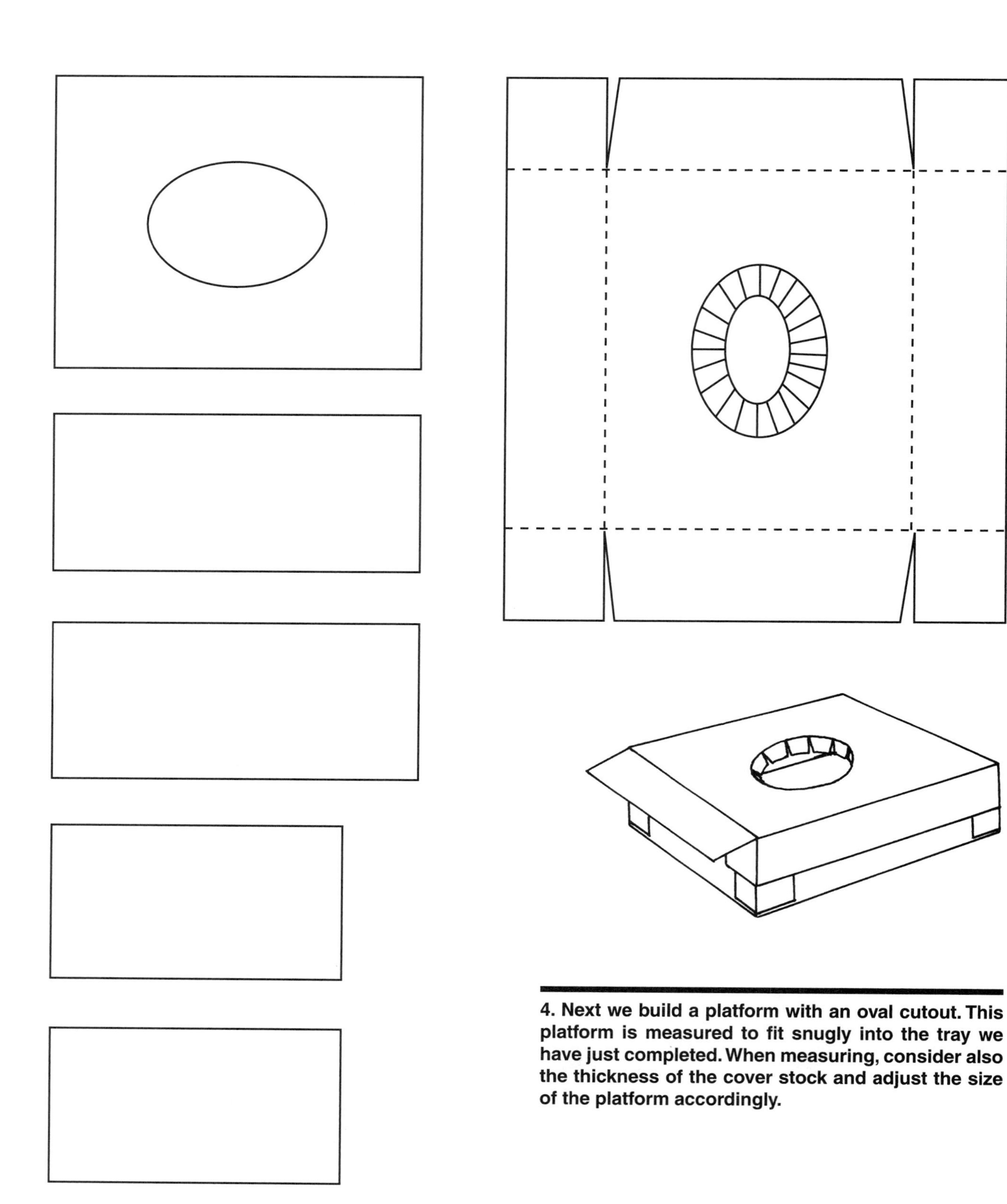

4. Next we build a platform with an oval cutout. This platform is measured to fit snugly into the tray we have just completed. When measuring, consider also the thickness of the cover stock and adjust the size of the platform accordingly.

5. Here we have the top and four sides of the lid, which should have the same width and depth as the base, since it will fit over the platform.

6. This insert is for the lid and need not be covered with cover stock.

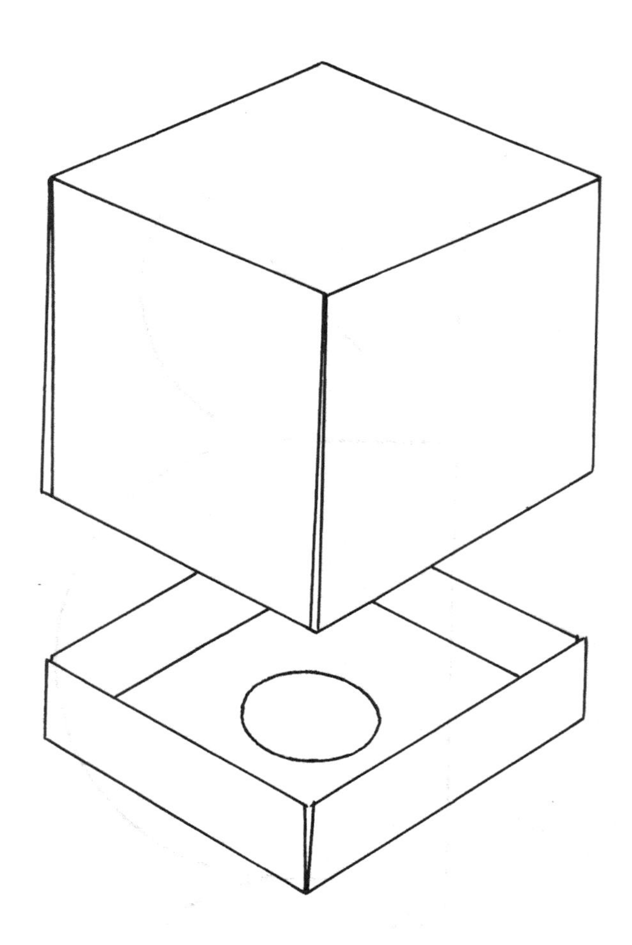

7. Glue the insert into the lid; the circle was cut to fit the cap of the bottle to be packaged.

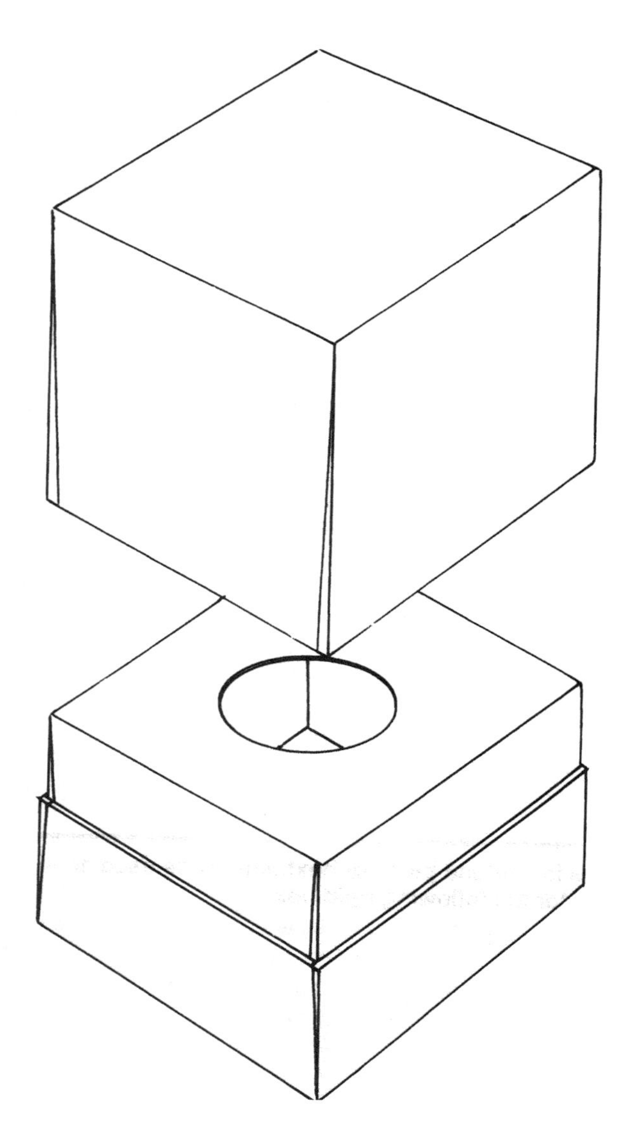

8. The platform may be glued into the base; the result is a handsome rigid or set-up box. If you wish a small part of the platform to show, you need only place an additional liner inside the cover. This will prevent the cover from sliding all the way down to the base.

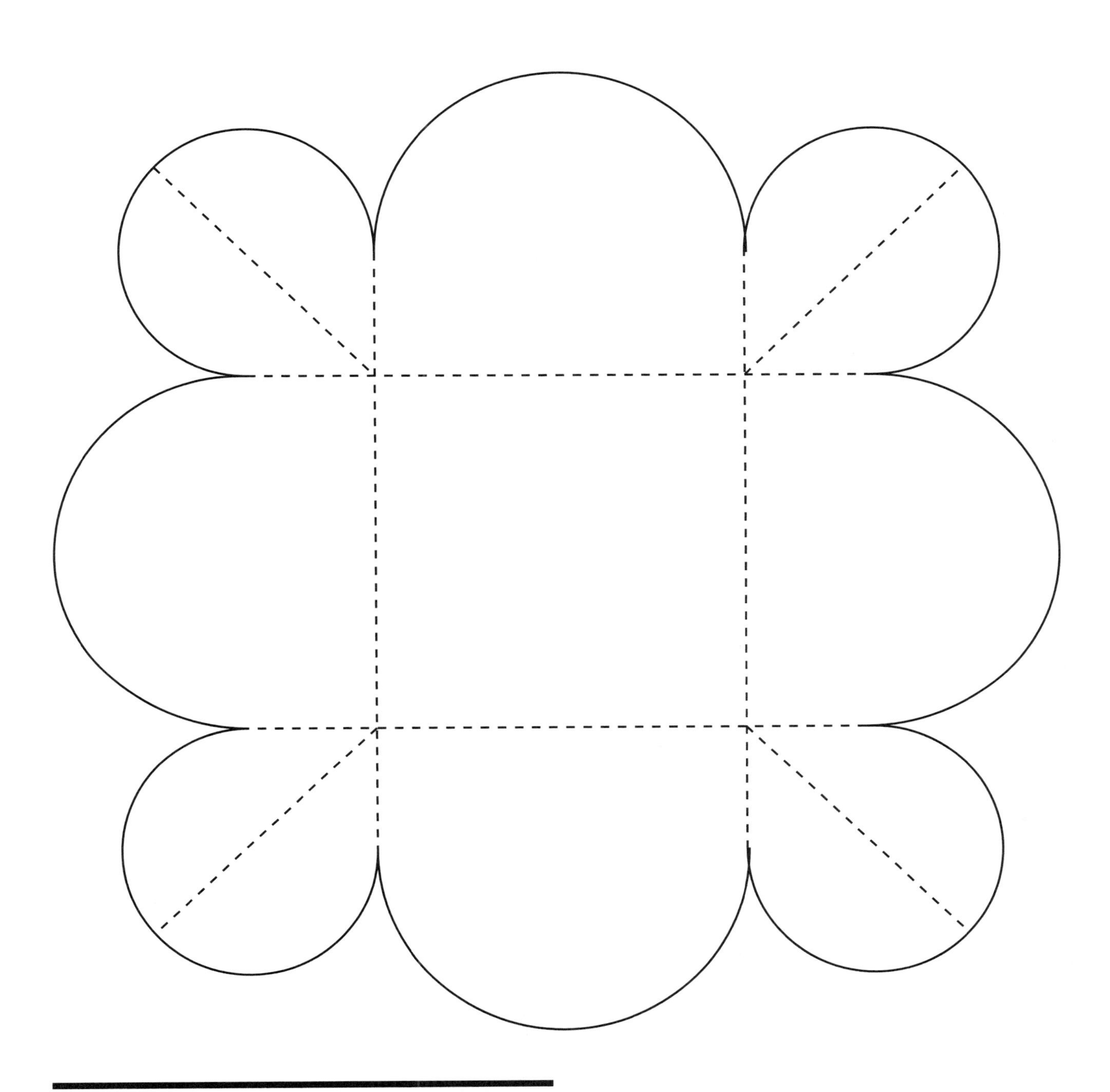

This is a folded piece of boxboard to be used as an insert for the following rigid box.

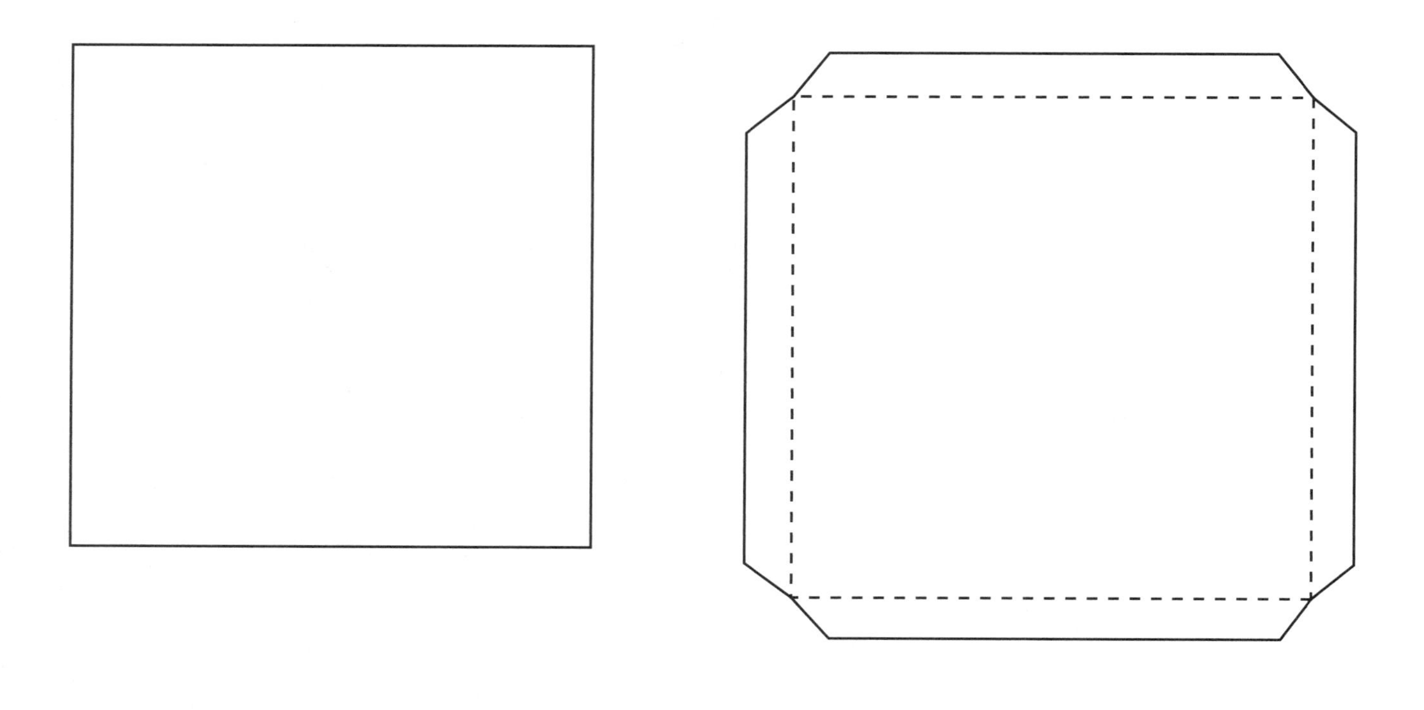

Here the folding insert is glued into the lined base.

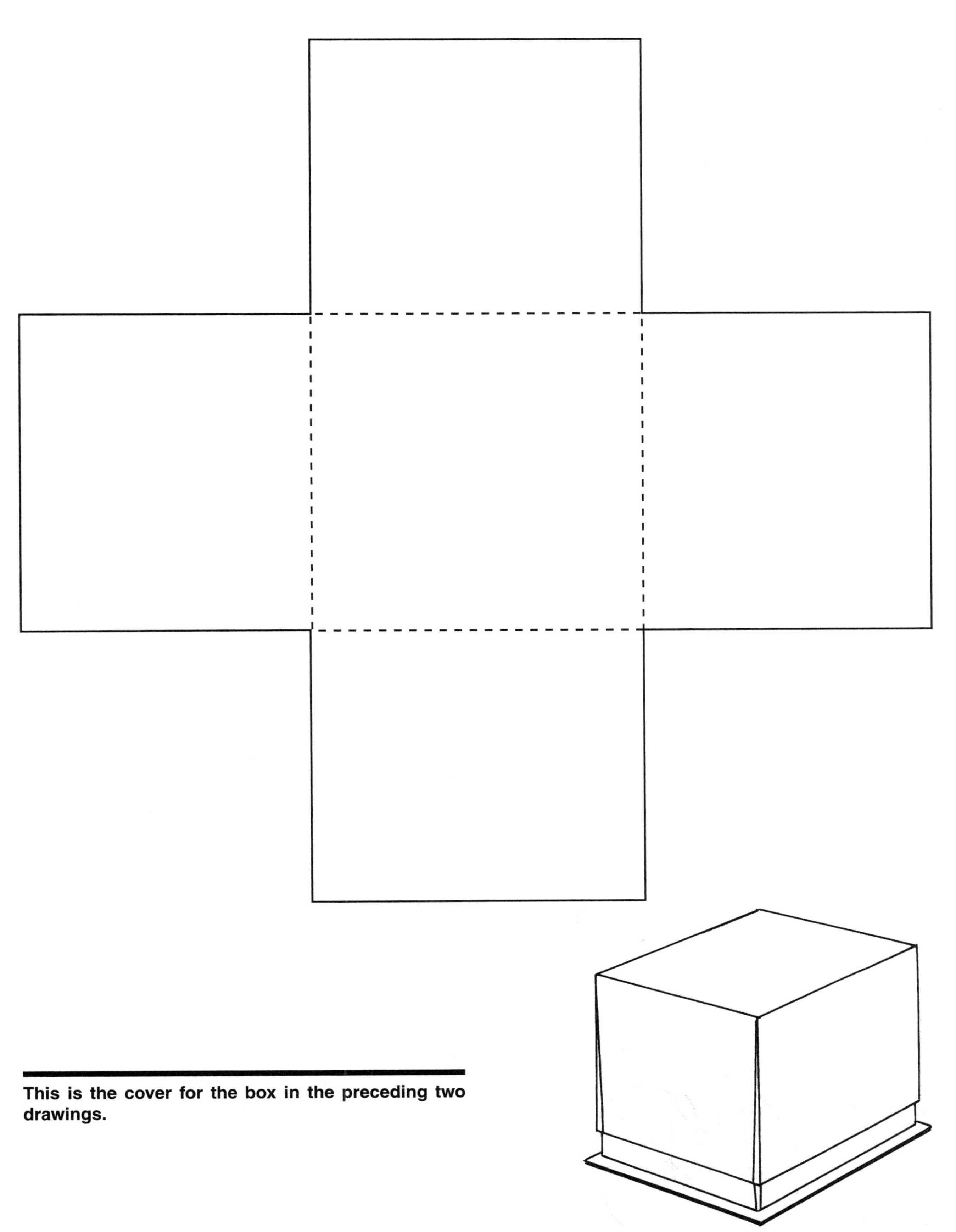

This is the cover for the box in the preceding two drawings.

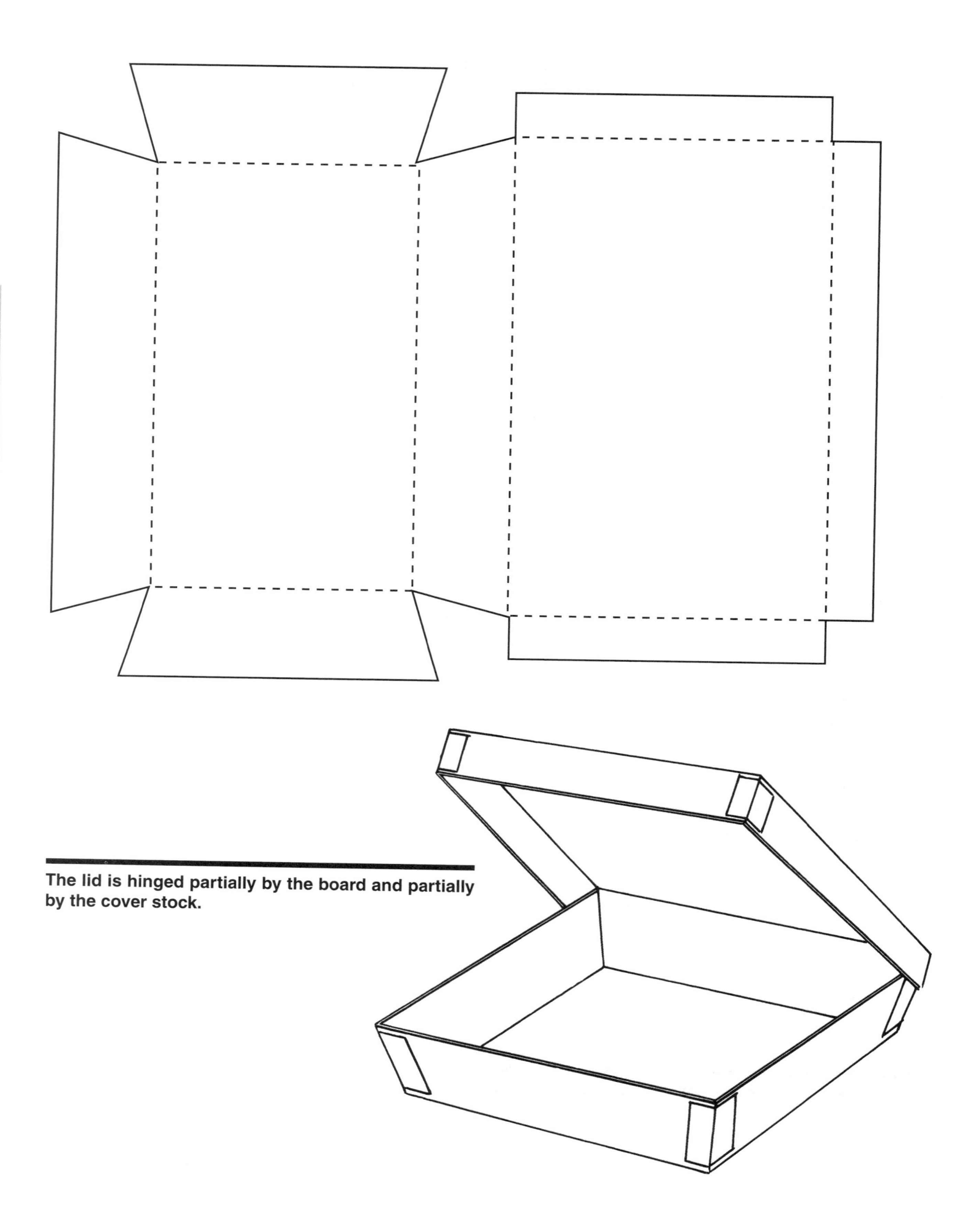

The lid is hinged partially by the board and partially by the cover stock.

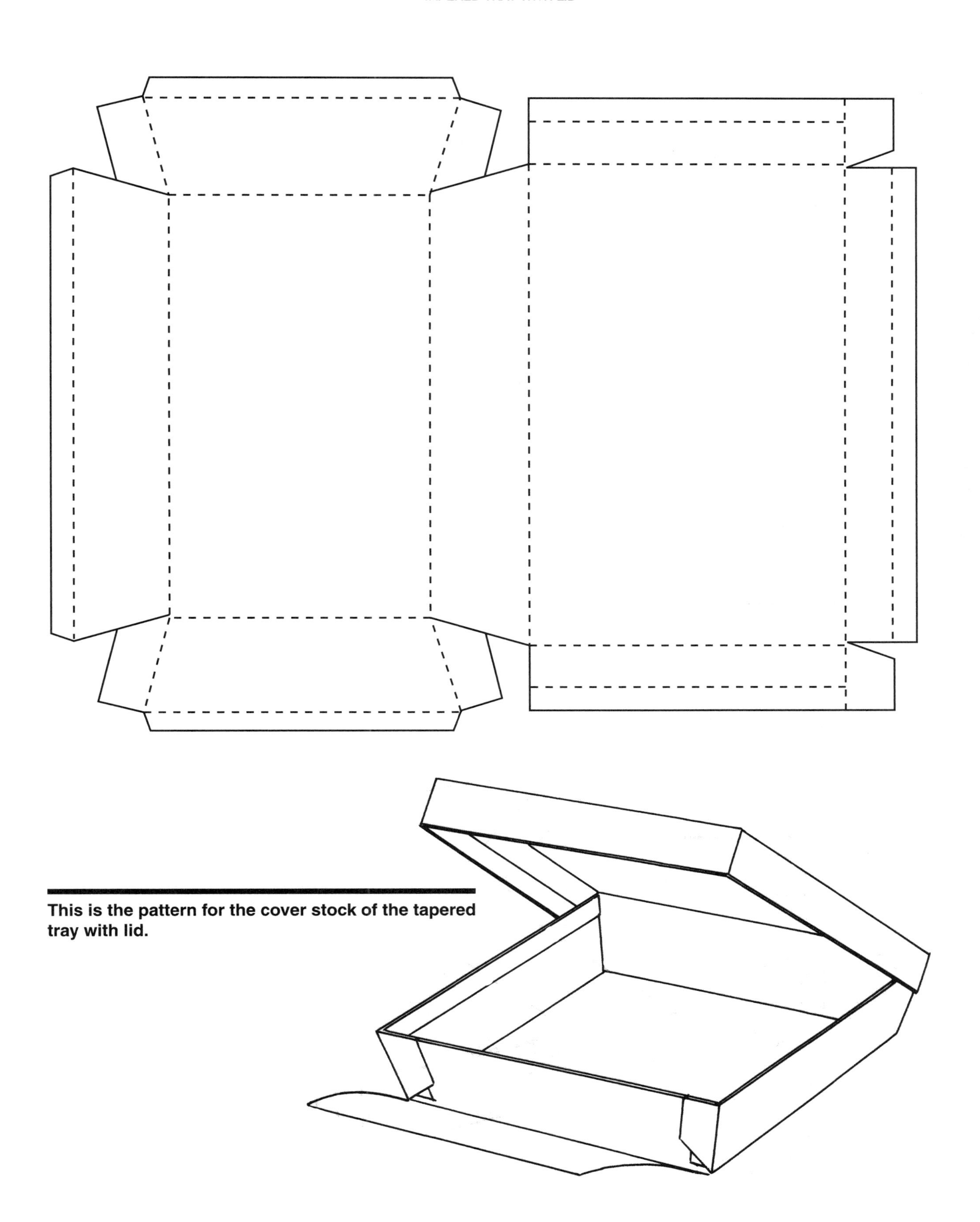

This is the pattern for the cover stock of the tapered tray with lid.

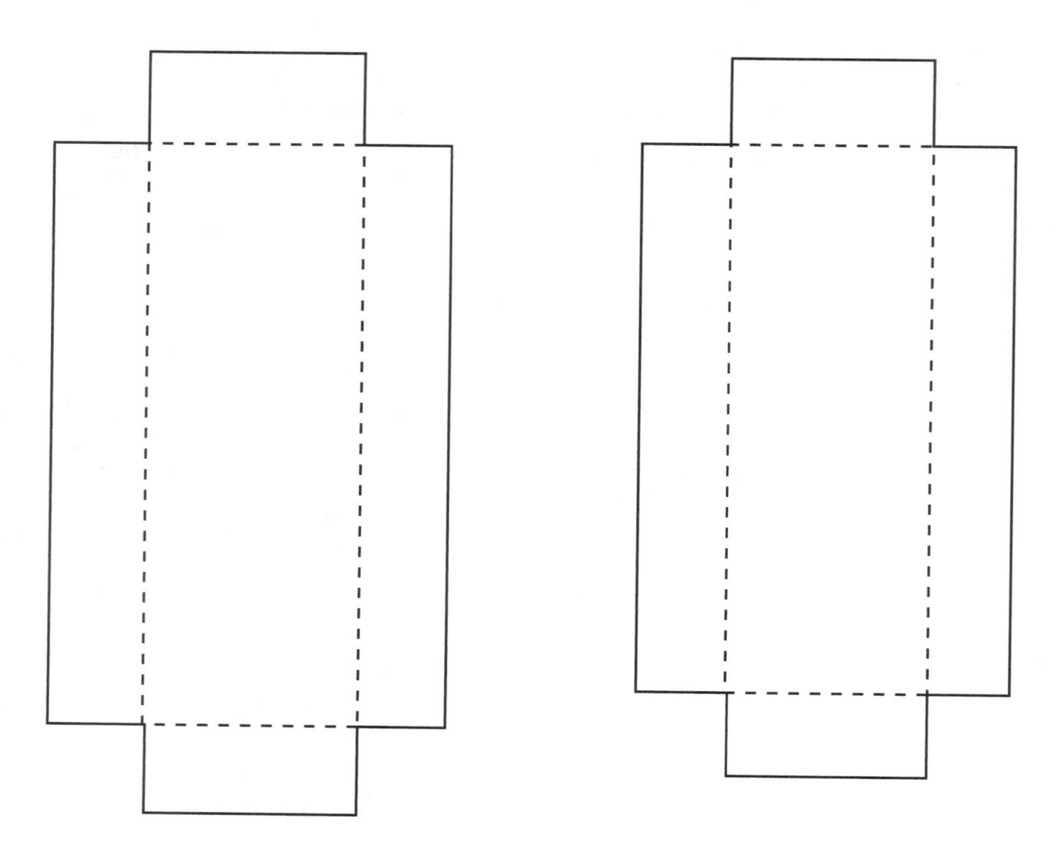

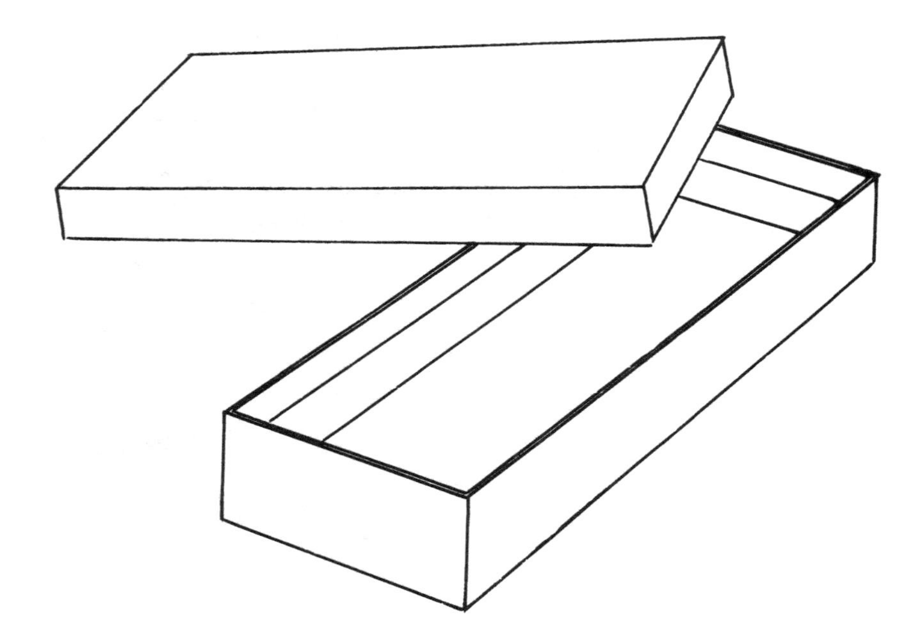

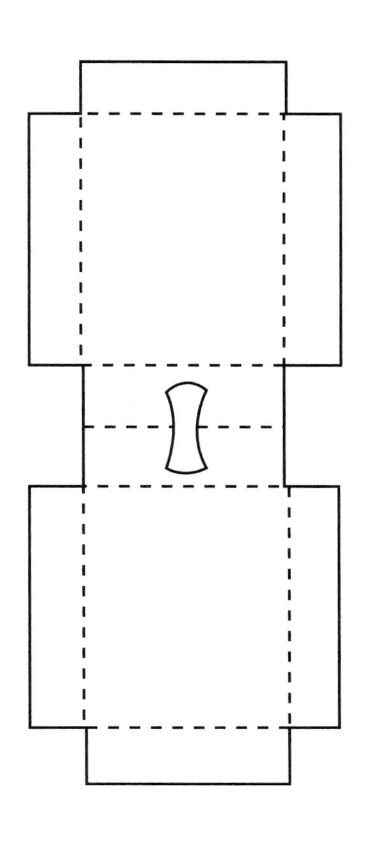

These are inserts for the tray illustrated on the preceding page. The top would be a typical insert for a souvenir spoon. The bottom insert would be for an item such as a pen.

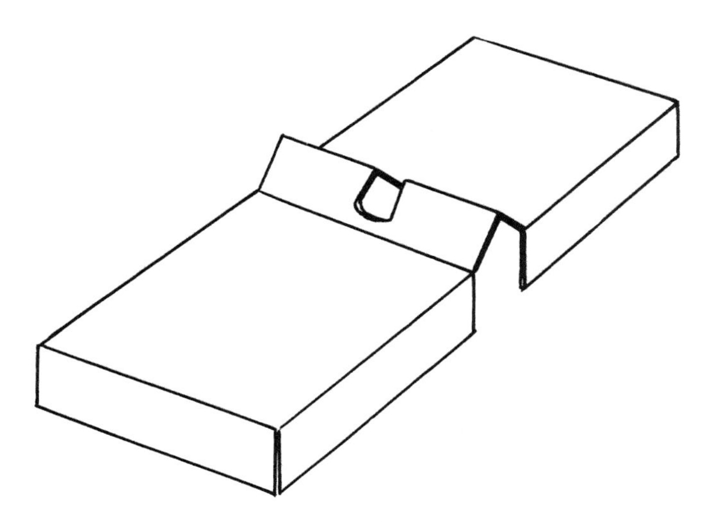

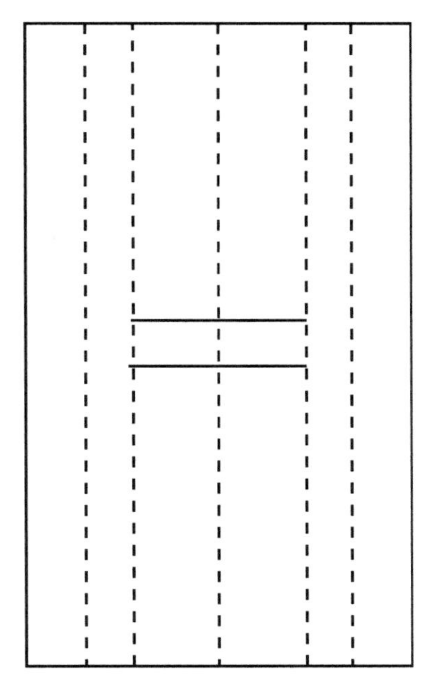

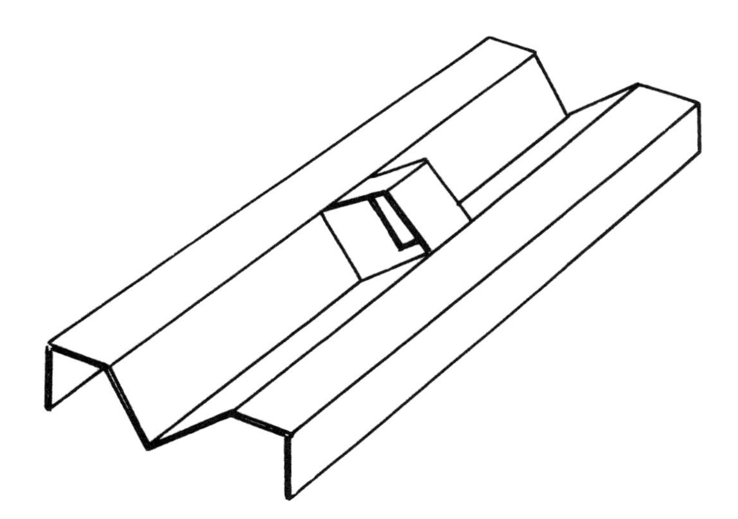

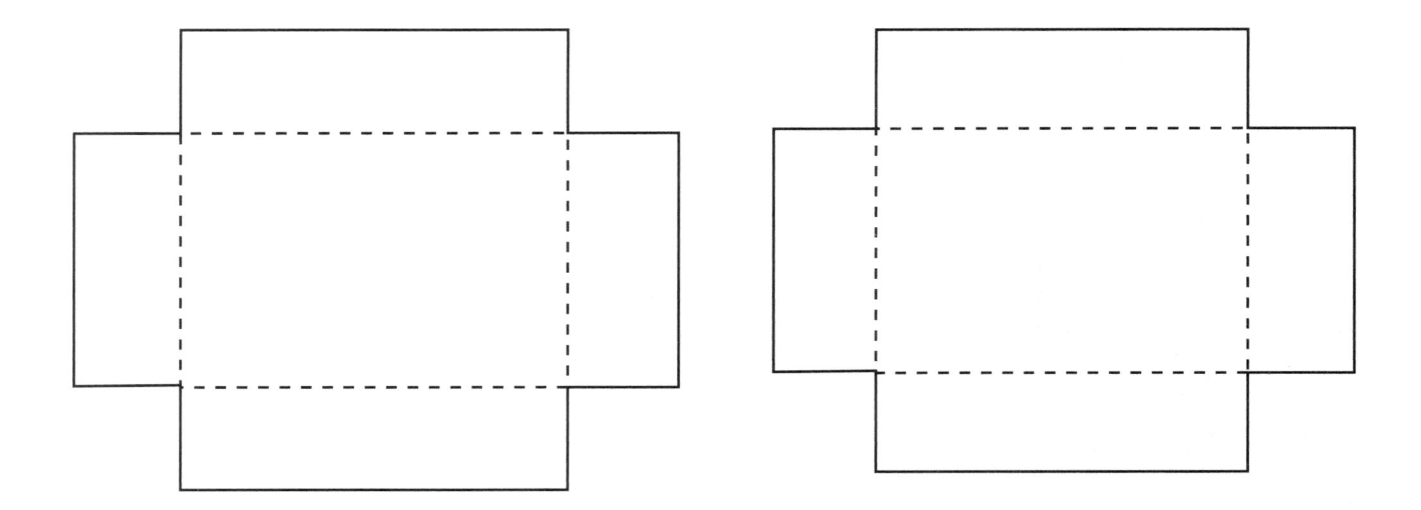

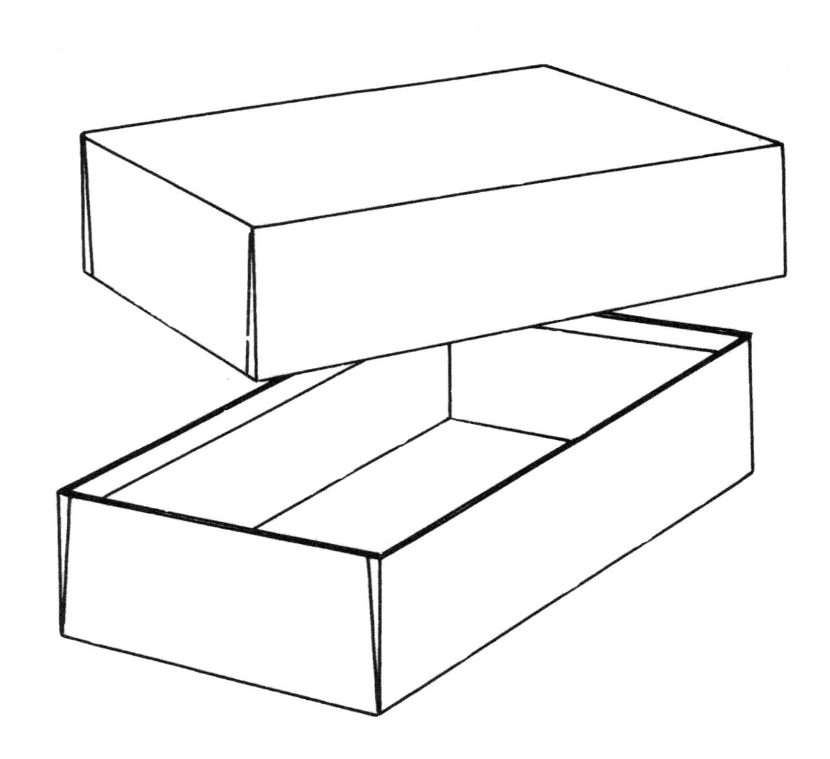

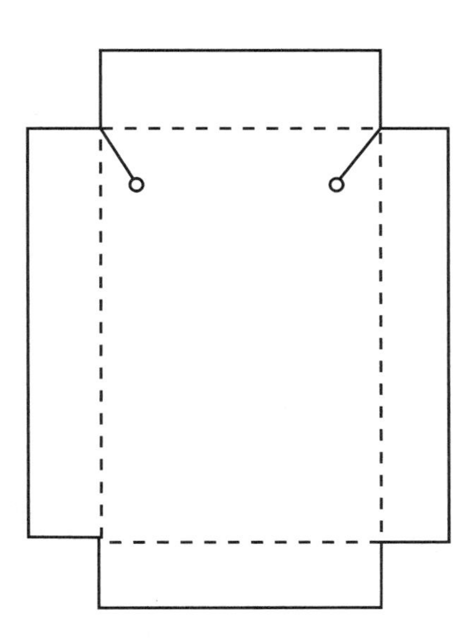

A platform for the box illustrated on the preceding page. This would be ideal for a necklace.

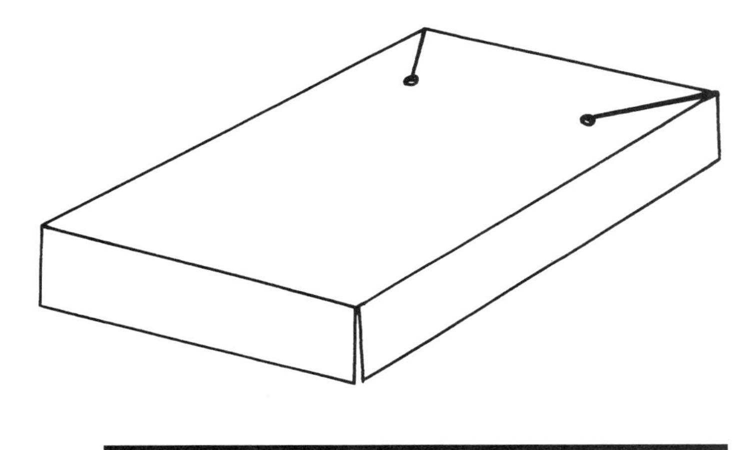

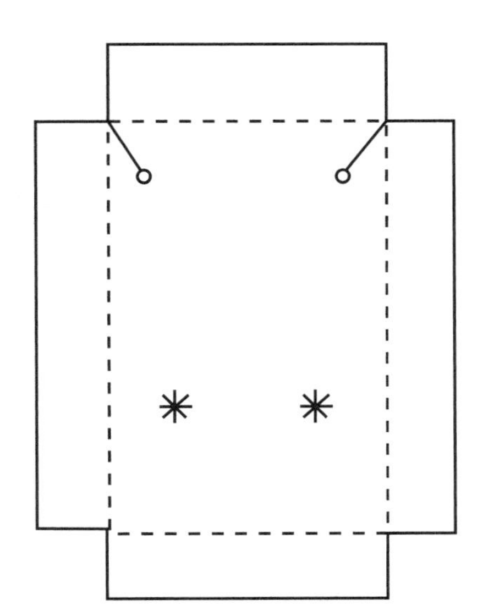

A platform for necklace and earrings.

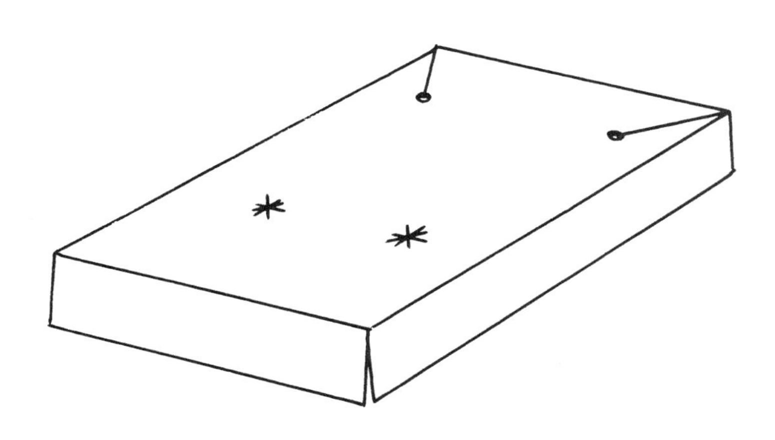

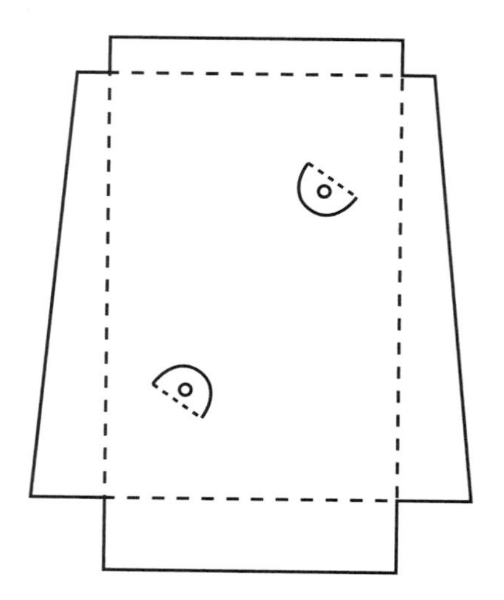

Slanted platforms for jewelry such as a tie pin or stickpin.

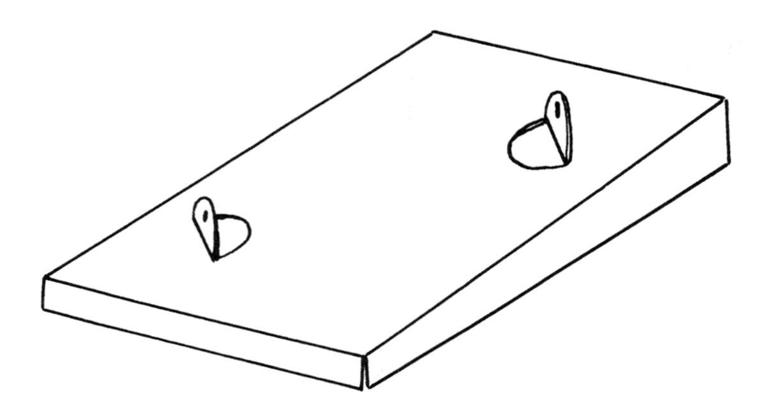

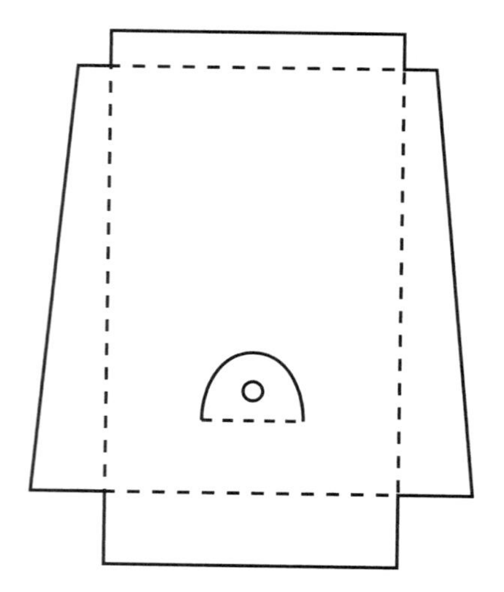

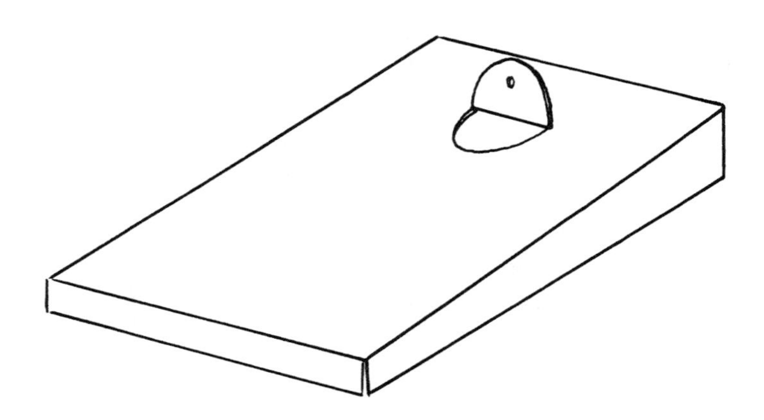

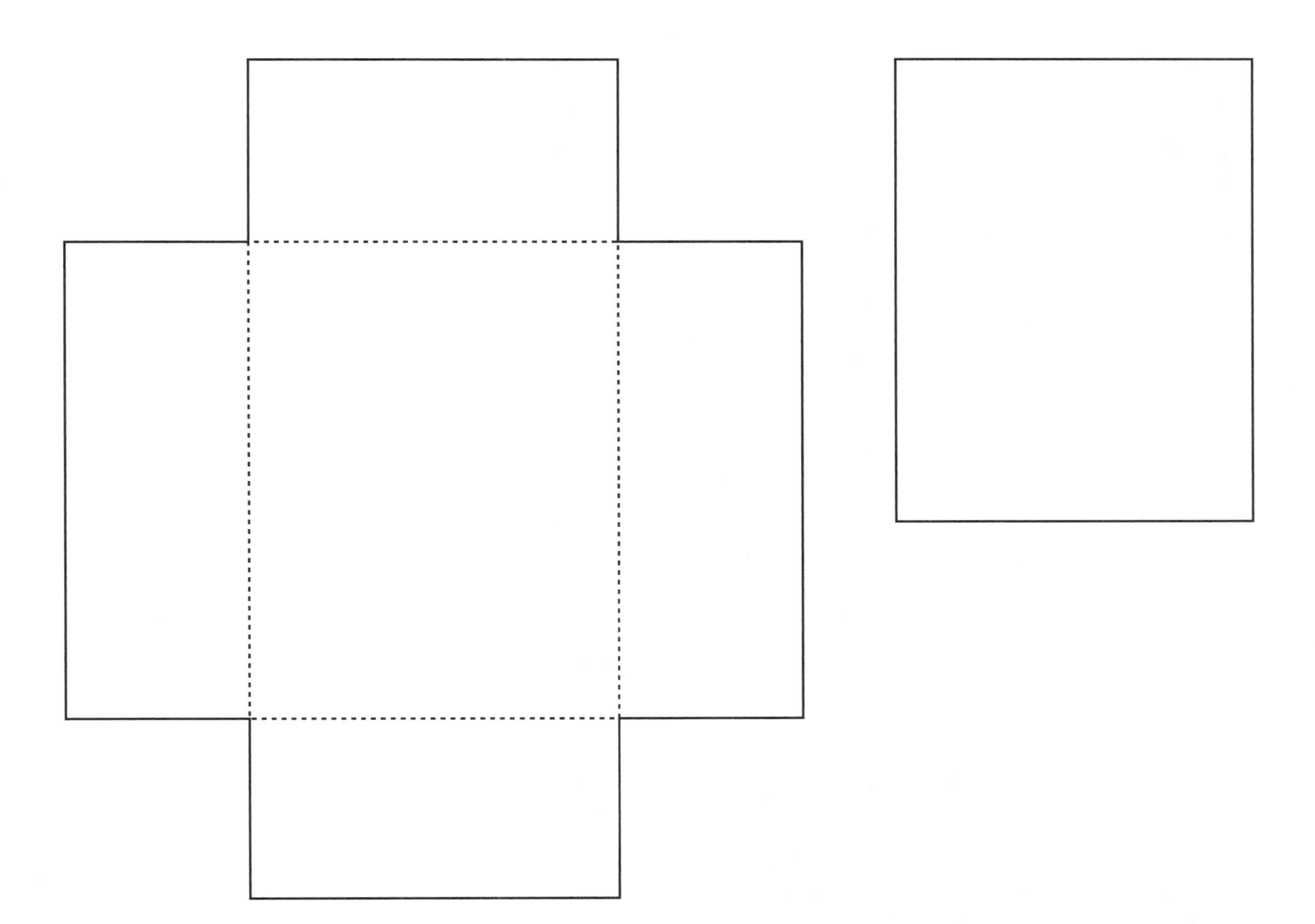

This box has side panels that are slightly higher than the front and back. Once the lid has been attached with hinging tape and/or the cover stock, the front of the lid will rest on the top of the front panel.

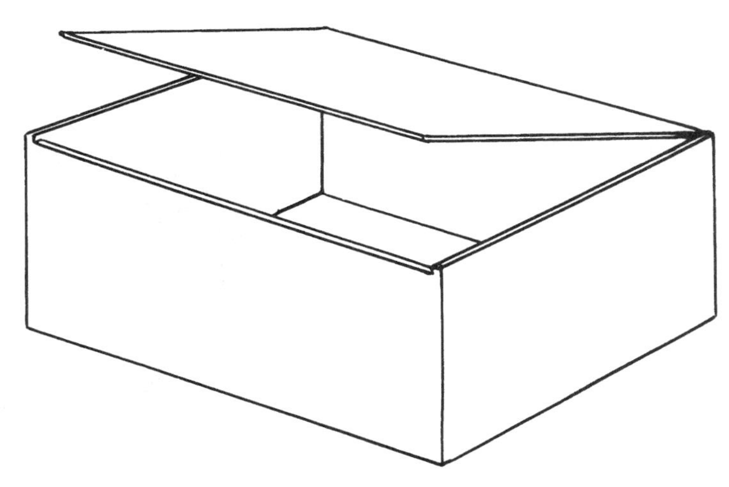

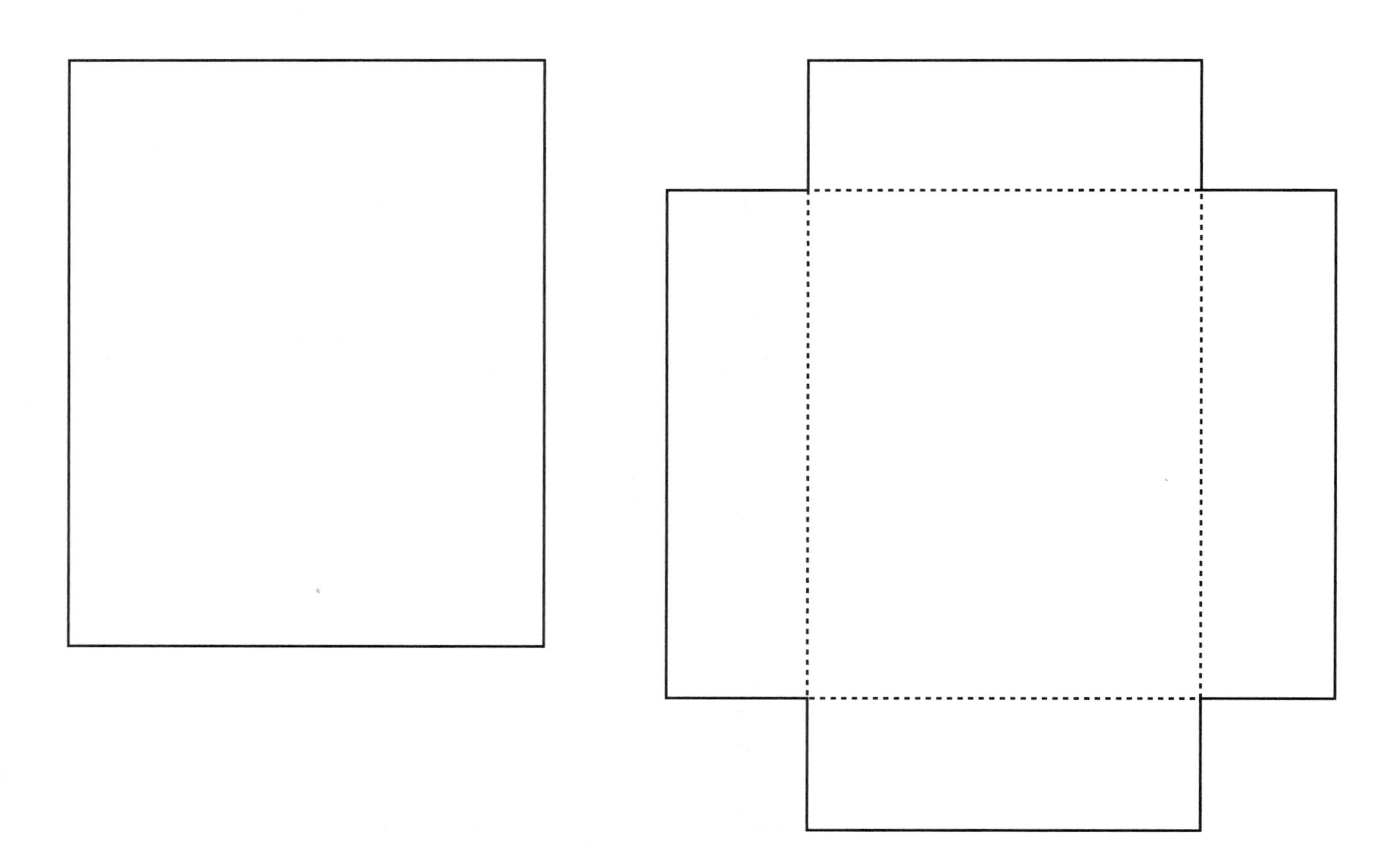

The tray is assembled and glued to the platform. Then the assembly is covered with cover stock.

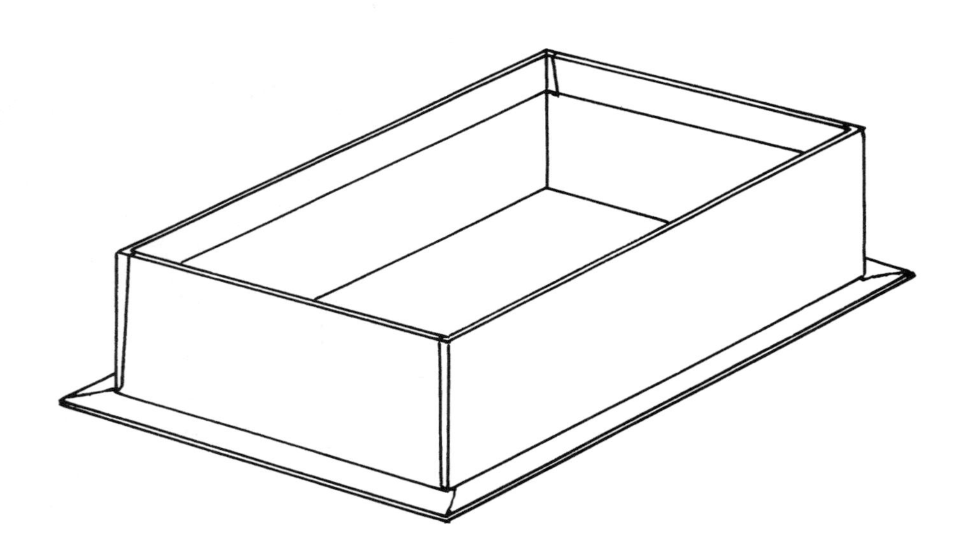

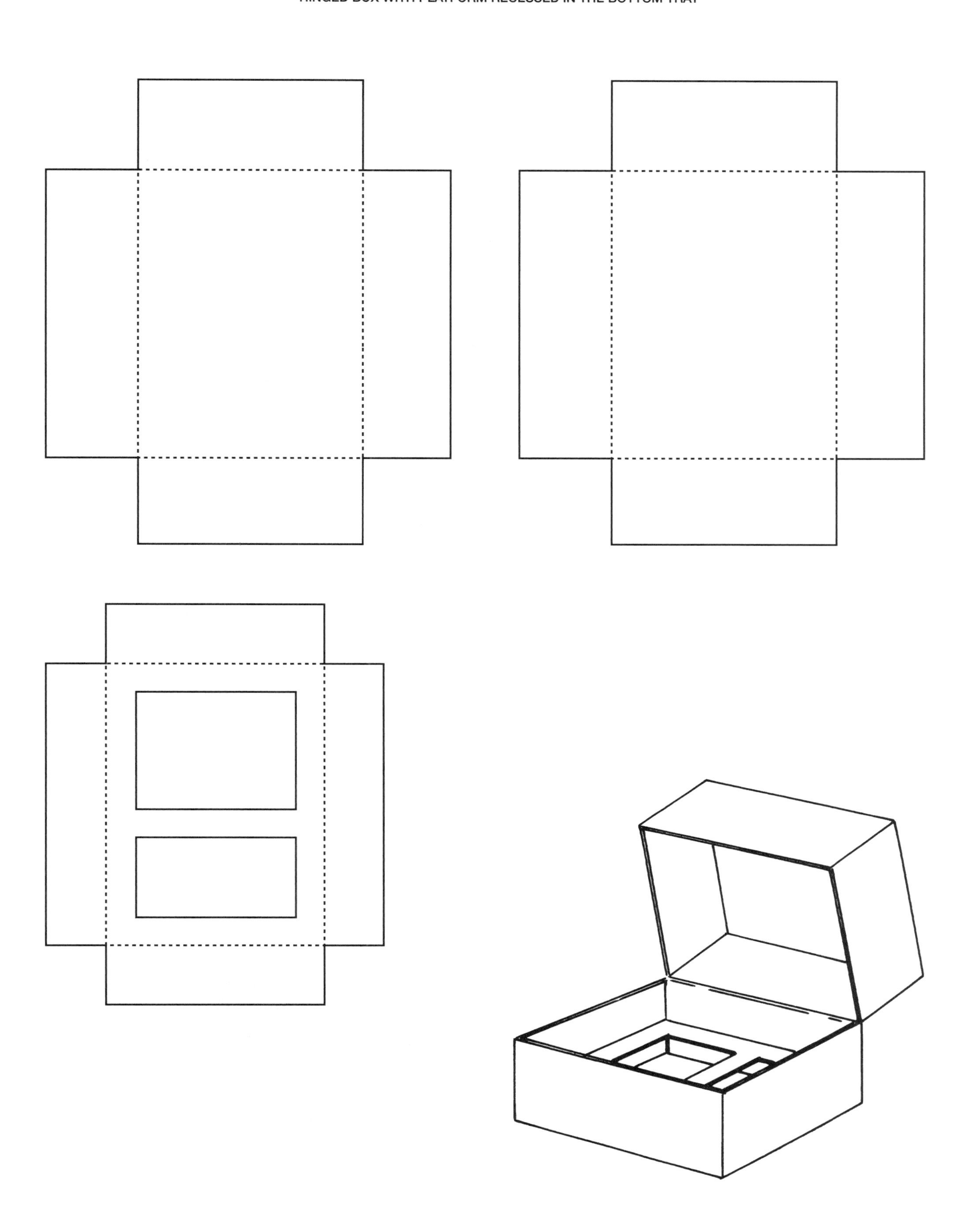

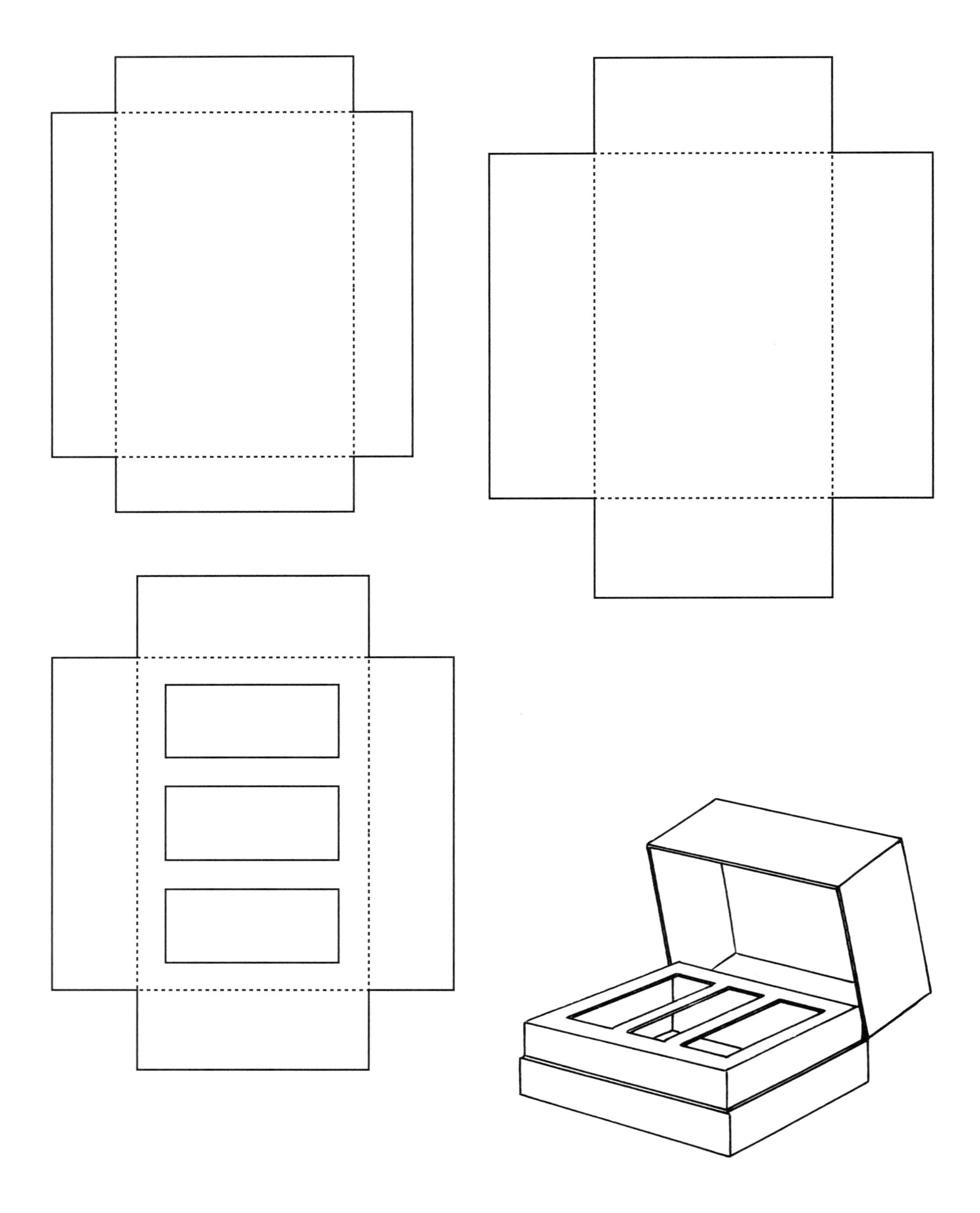

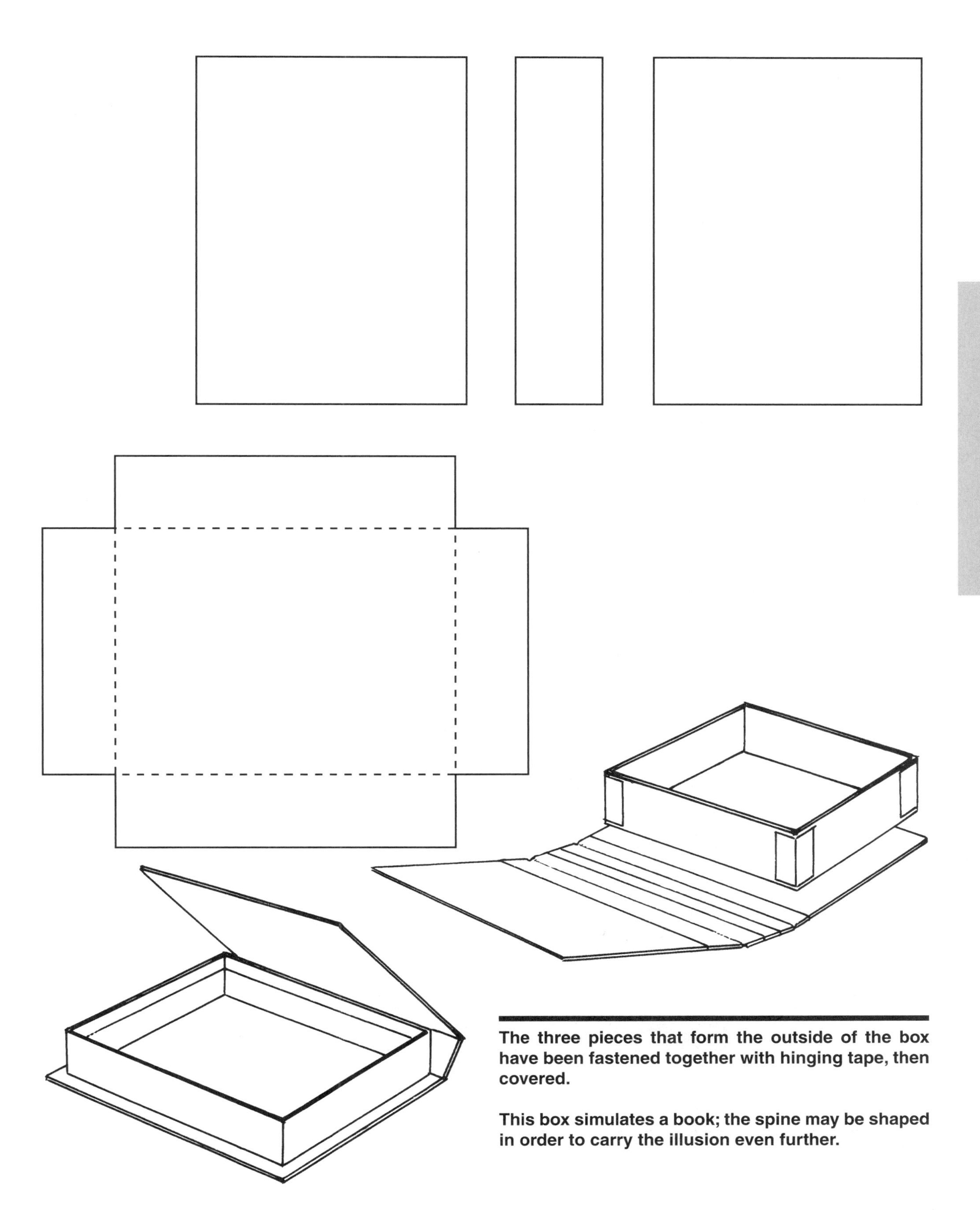

The three pieces that form the outside of the box have been fastened together with hinging tape, then covered.

This box simulates a book; the spine may be shaped in order to carry the illusion even further.

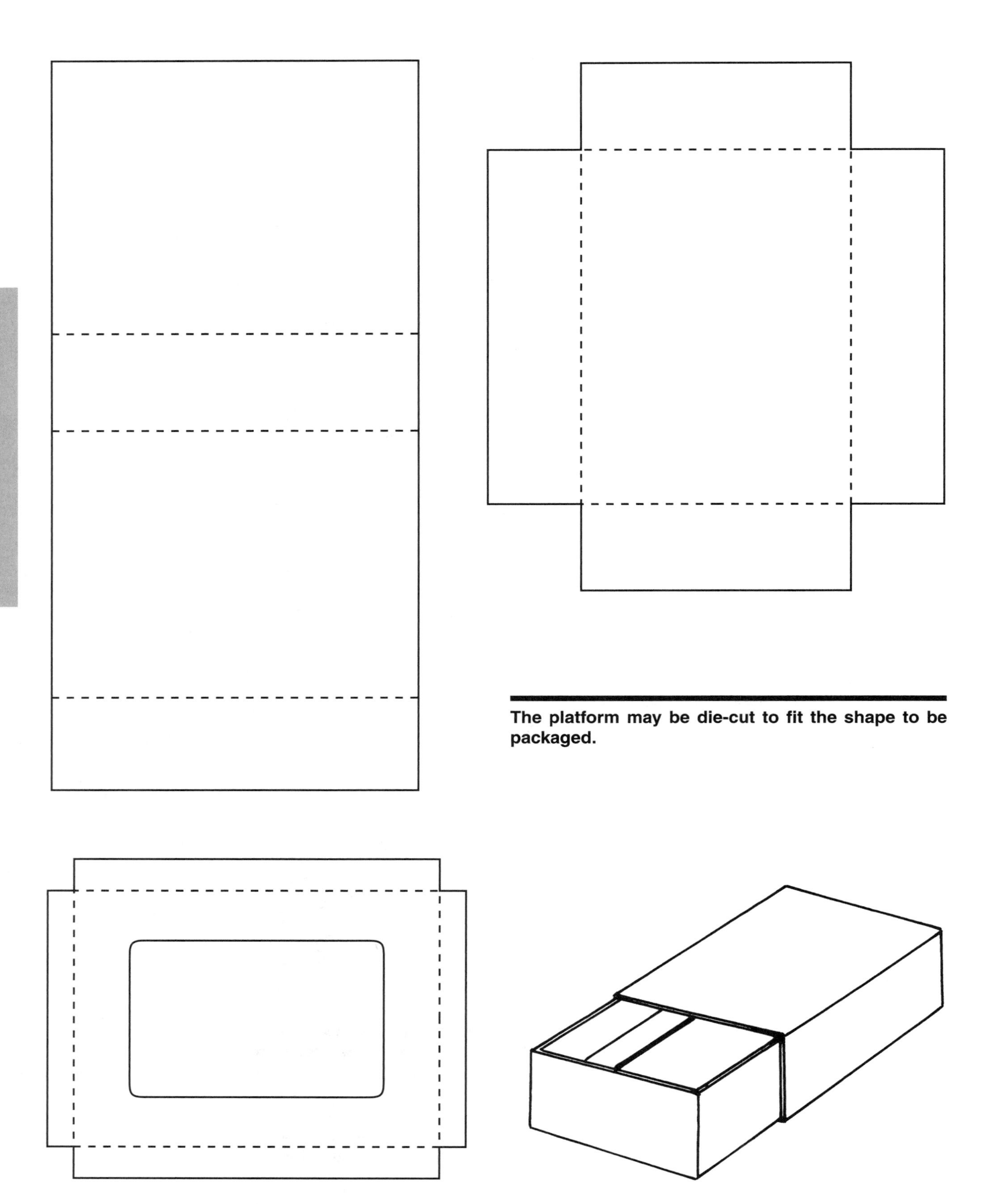

The platform may be die-cut to fit the shape to be packaged.

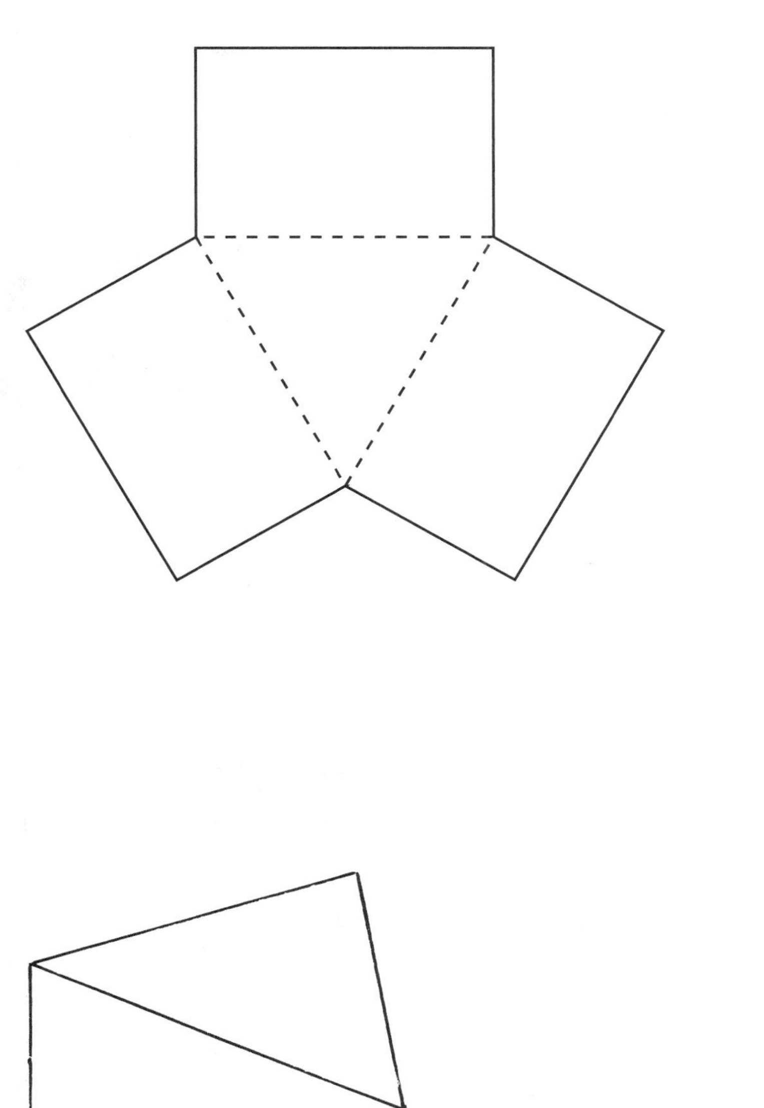

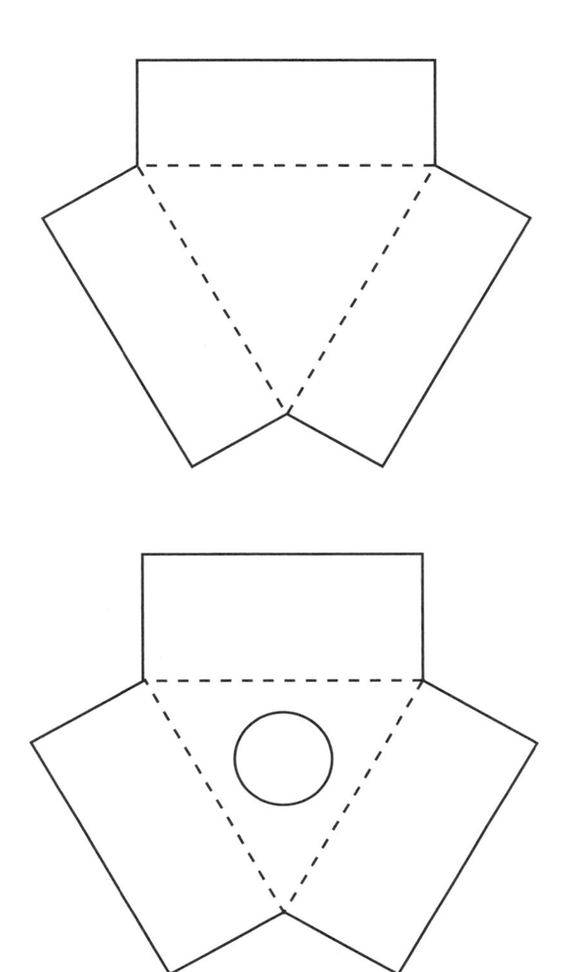

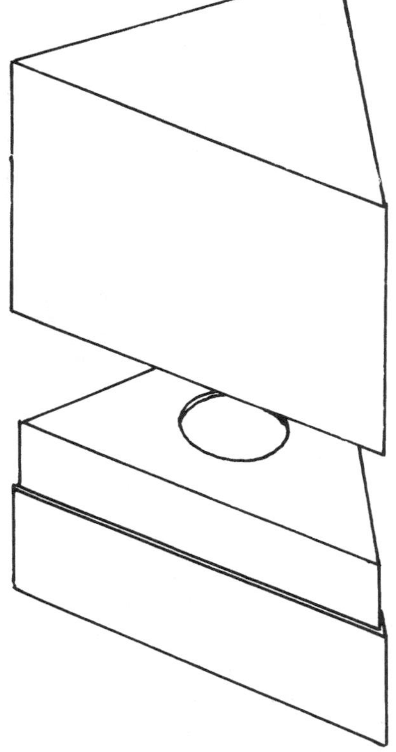

Set-up boxes may be manufactured in many shapes. This triangular box with raised platform is only one of the many geometric possibilities.

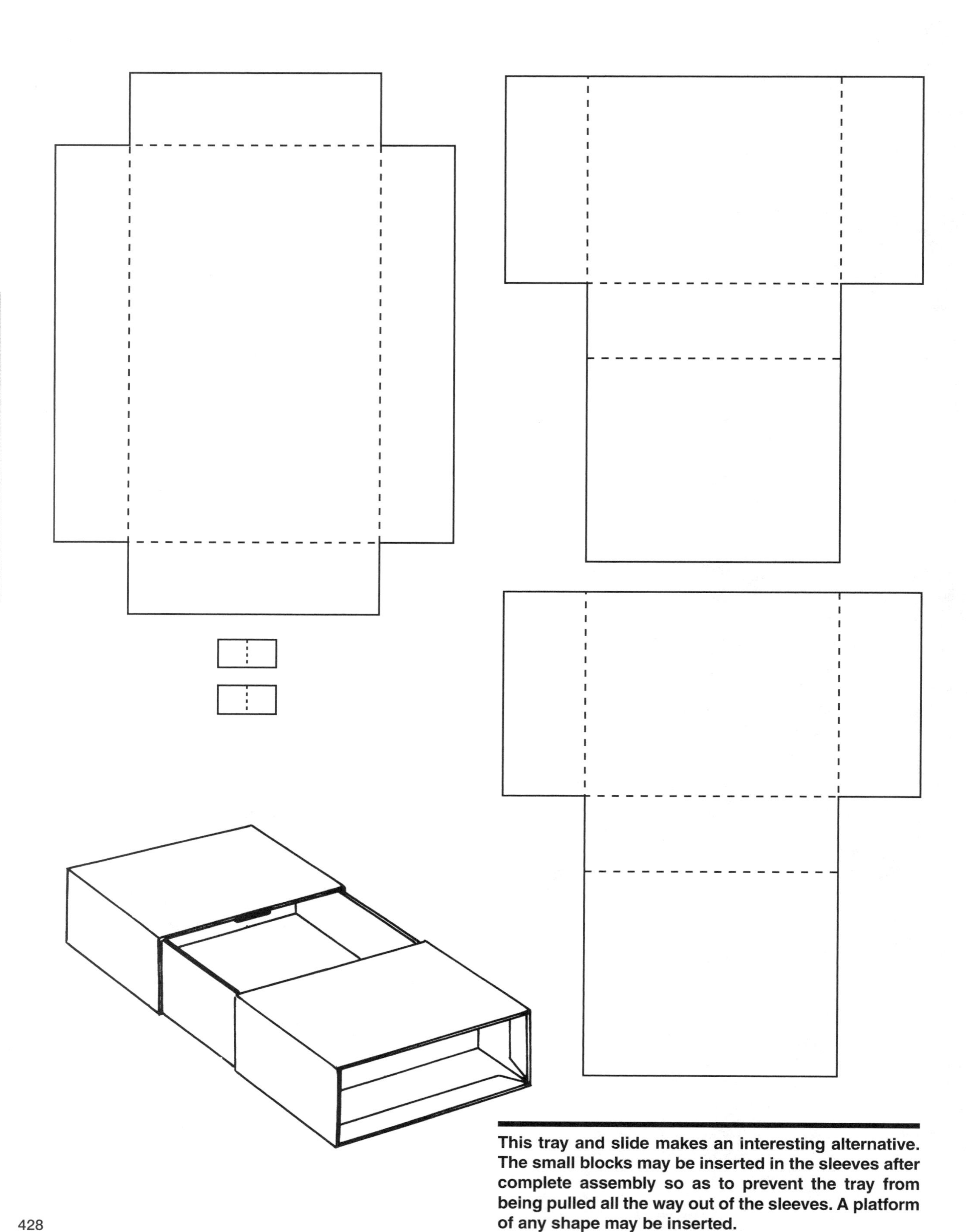

This tray and slide makes an interesting alternative. The small blocks may be inserted in the sleeves after complete assembly so as to prevent the tray from being pulled all the way out of the sleeves. A platform of any shape may be inserted.

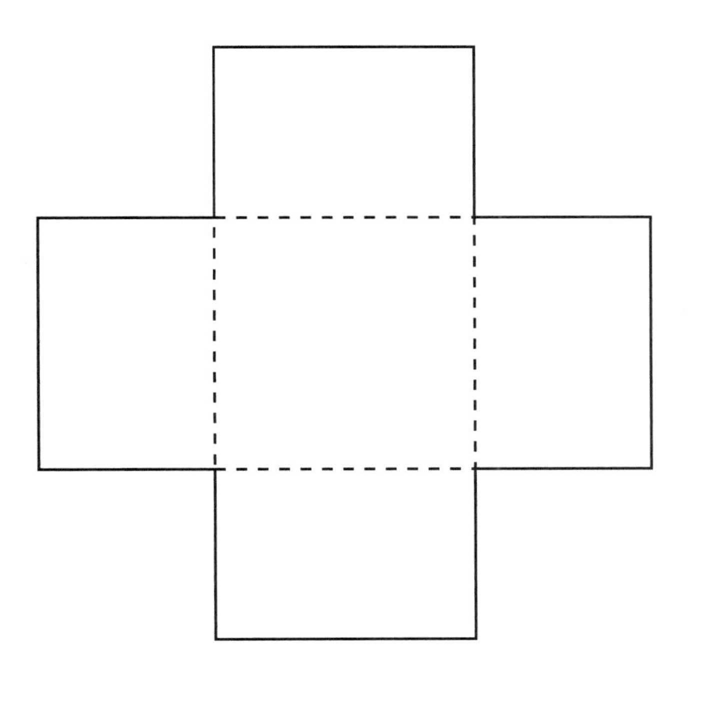

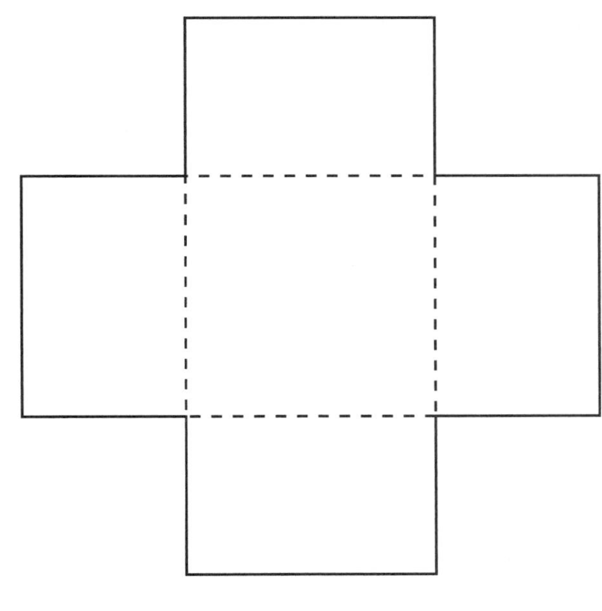

This three piece box has a separate lid that has been attached with hinging tape and rests on the insert when closed. A small piece of ribbon has been attached to the lid under the cover stock as an opening device. Another ribbon could be attached to the lid and side panel to keep the lid from opening too far as well as the keep the lid in an open position.

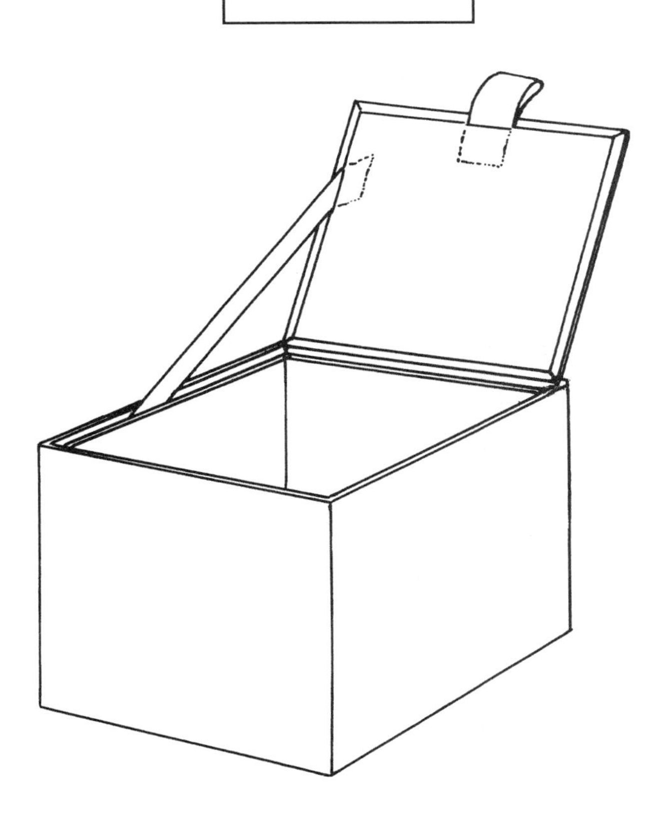

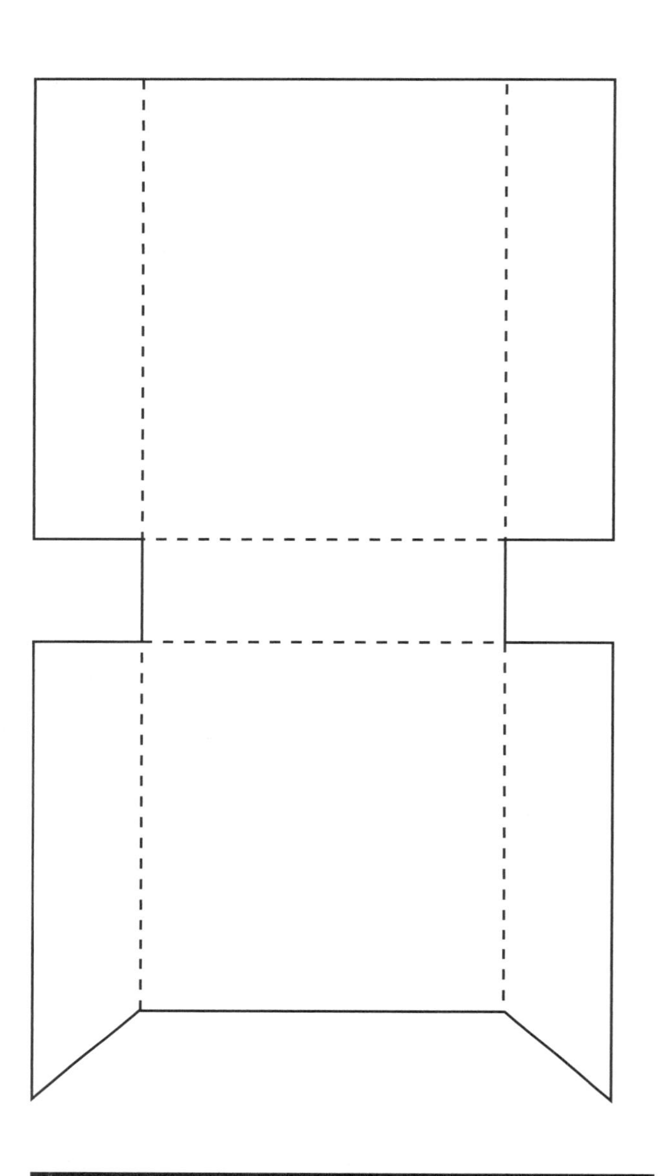

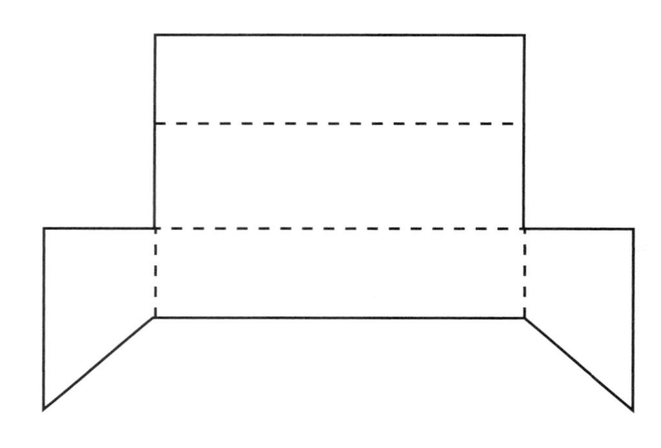

This file box may be used for documents, diskettes, or, with the proper insert platform, cosmetic items.

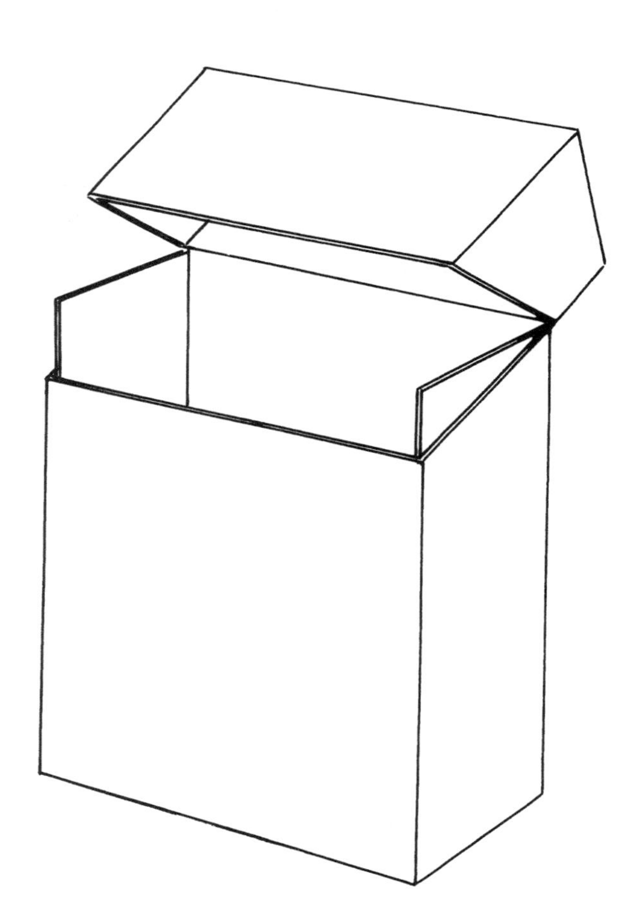

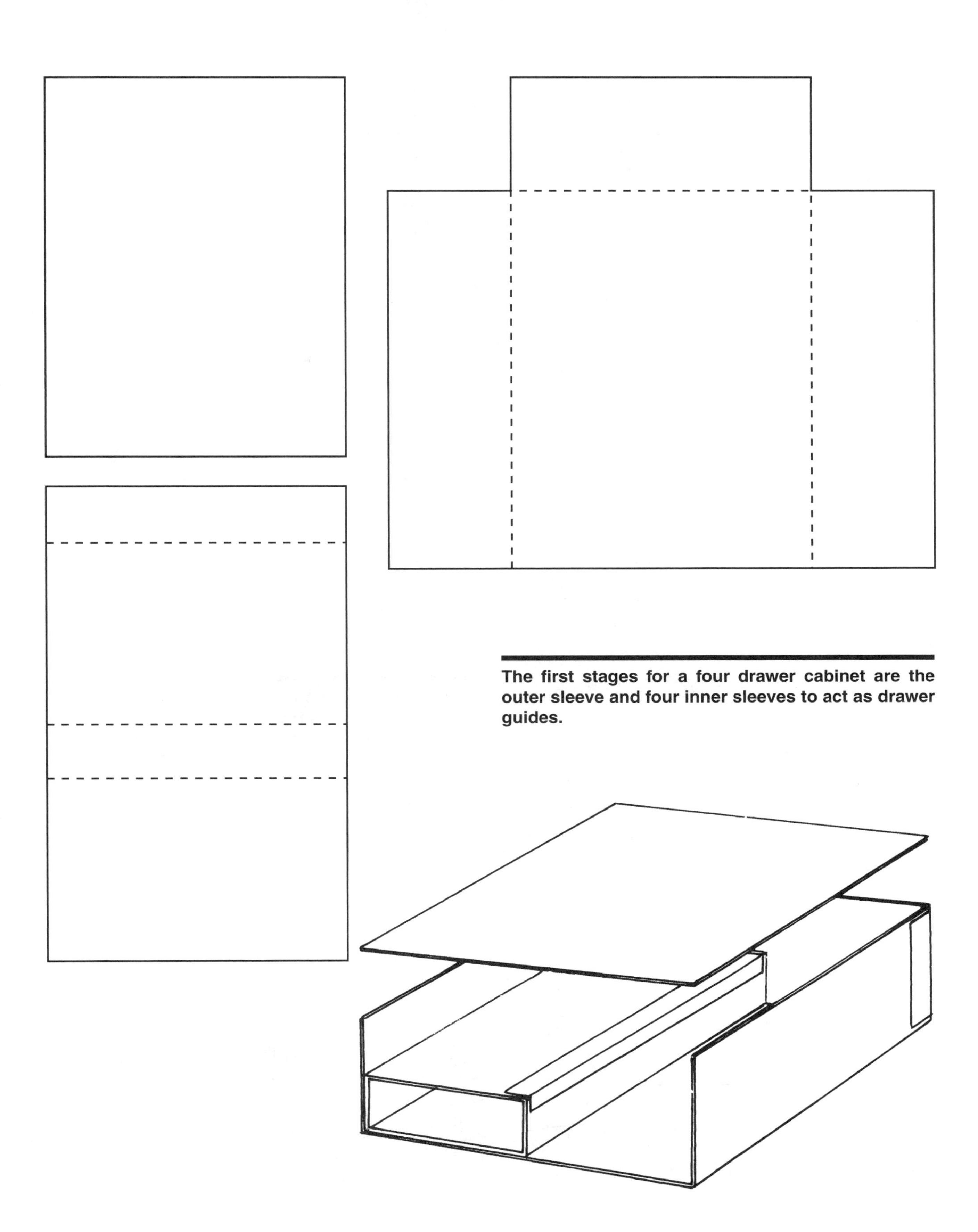

The first stages for a four drawer cabinet are the outer sleeve and four inner sleeves to act as drawer guides.

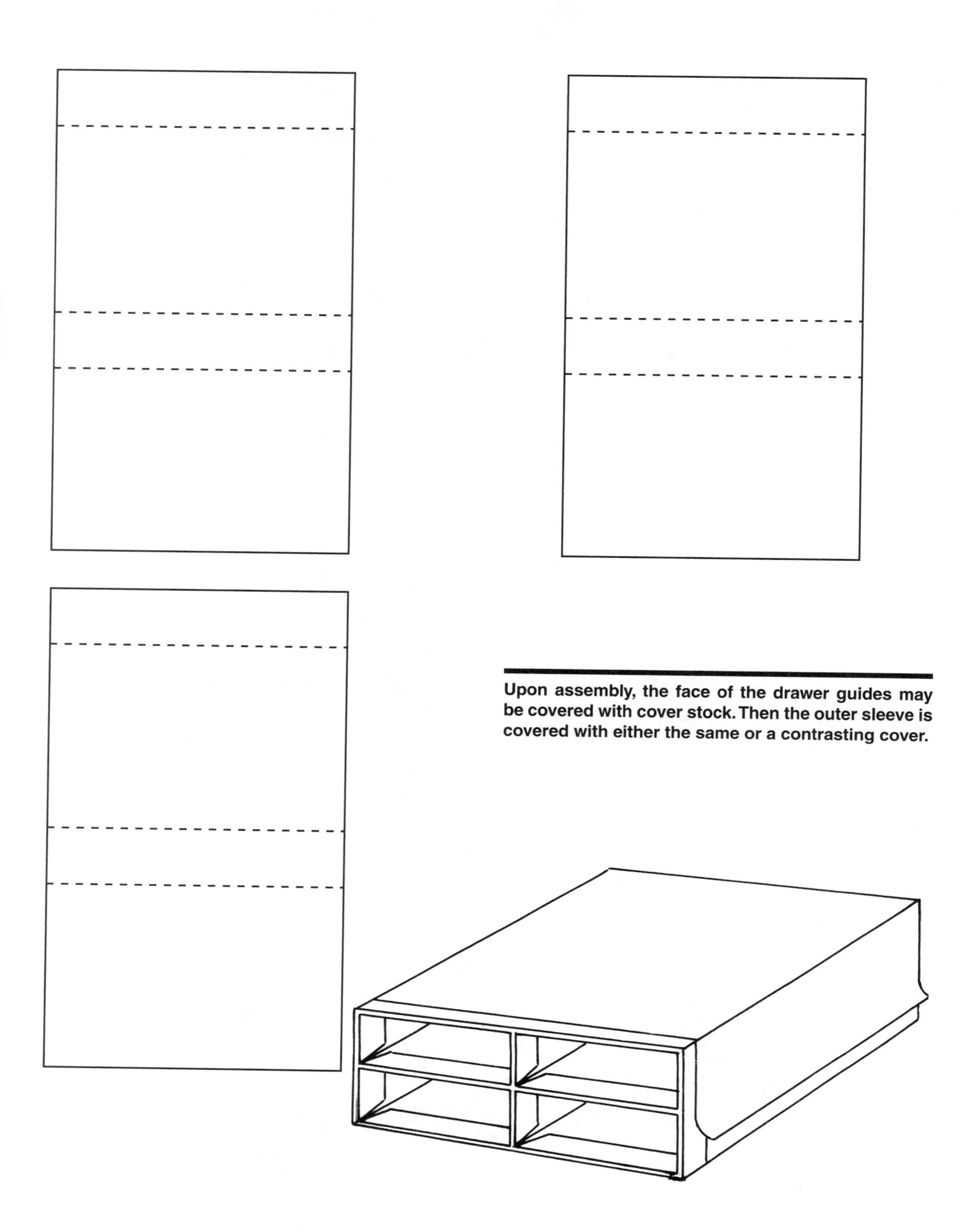

Upon assembly, the face of the drawer guides may be covered with cover stock. Then the outer sleeve is covered with either the same or a contrasting cover.

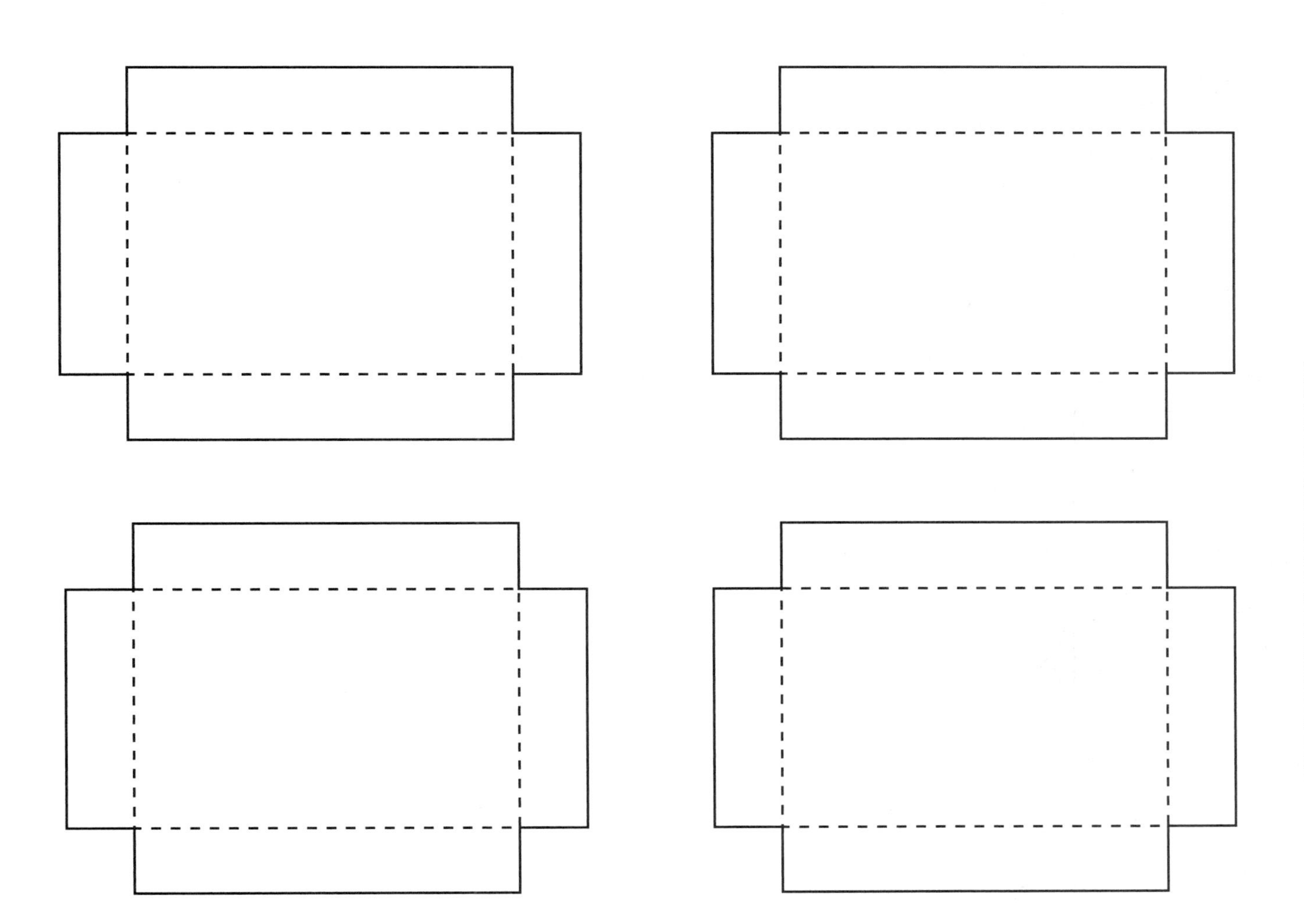

The drawers, when assembled, should be covered and checked for a snug fit. Any available hardware may be used as drawer pulls.

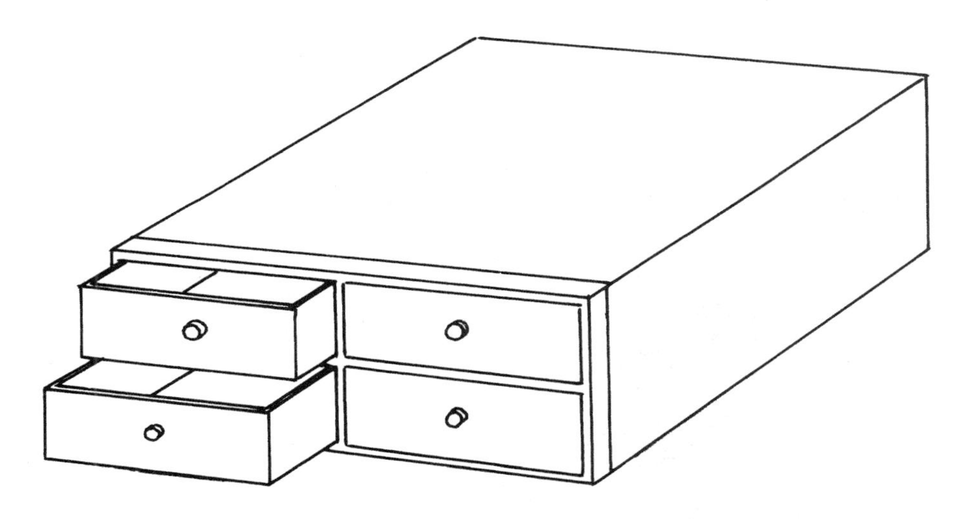

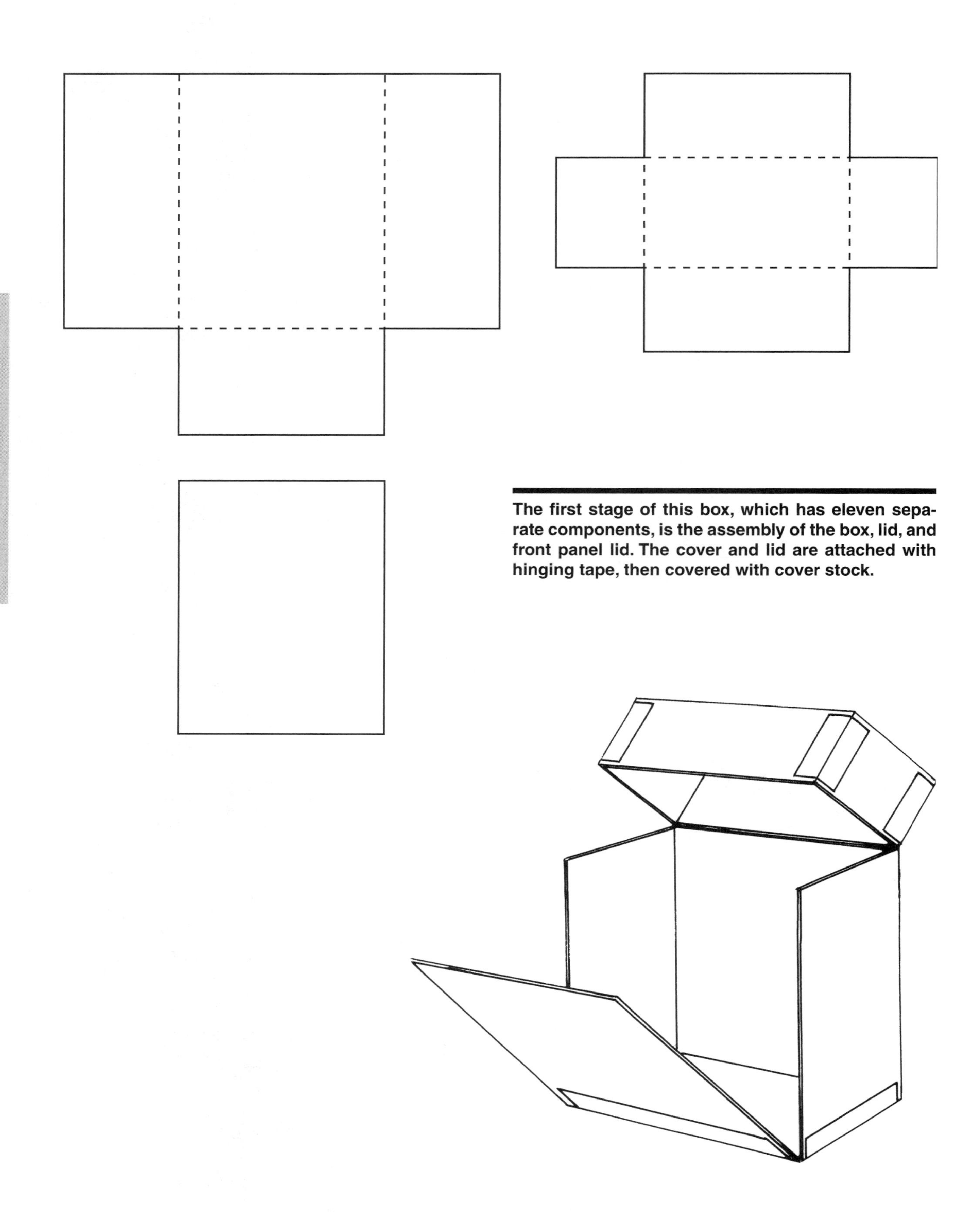

The first stage of this box, which has eleven separate components, is the assembly of the box, lid, and front panel lid. The cover and lid are attached with hinging tape, then covered with cover stock.

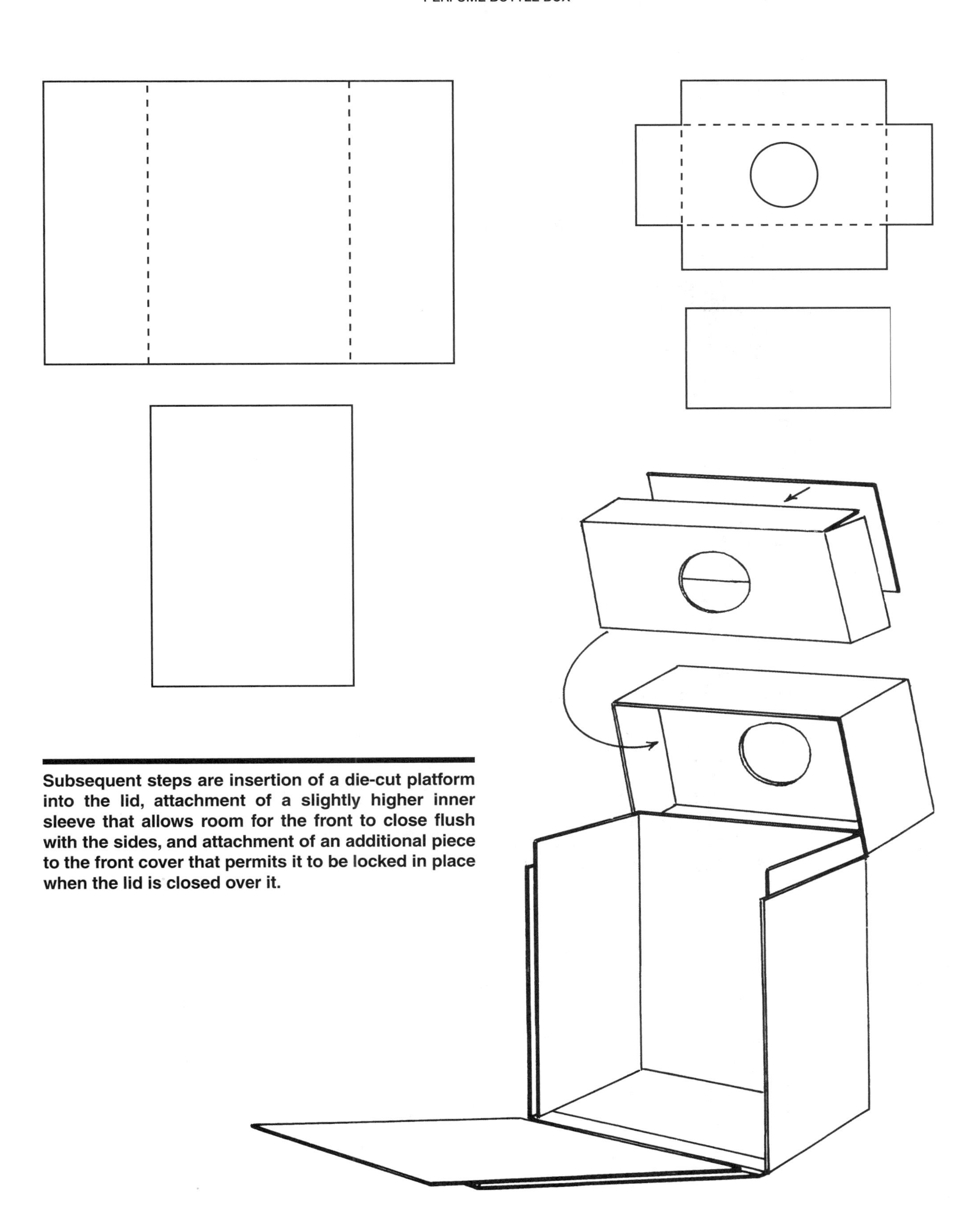

Subsequent steps are insertion of a die-cut platform into the lid, attachment of a slightly higher inner sleeve that allows room for the front to close flush with the sides, and attachment of an additional piece to the front cover that permits it to be locked in place when the lid is closed over it.

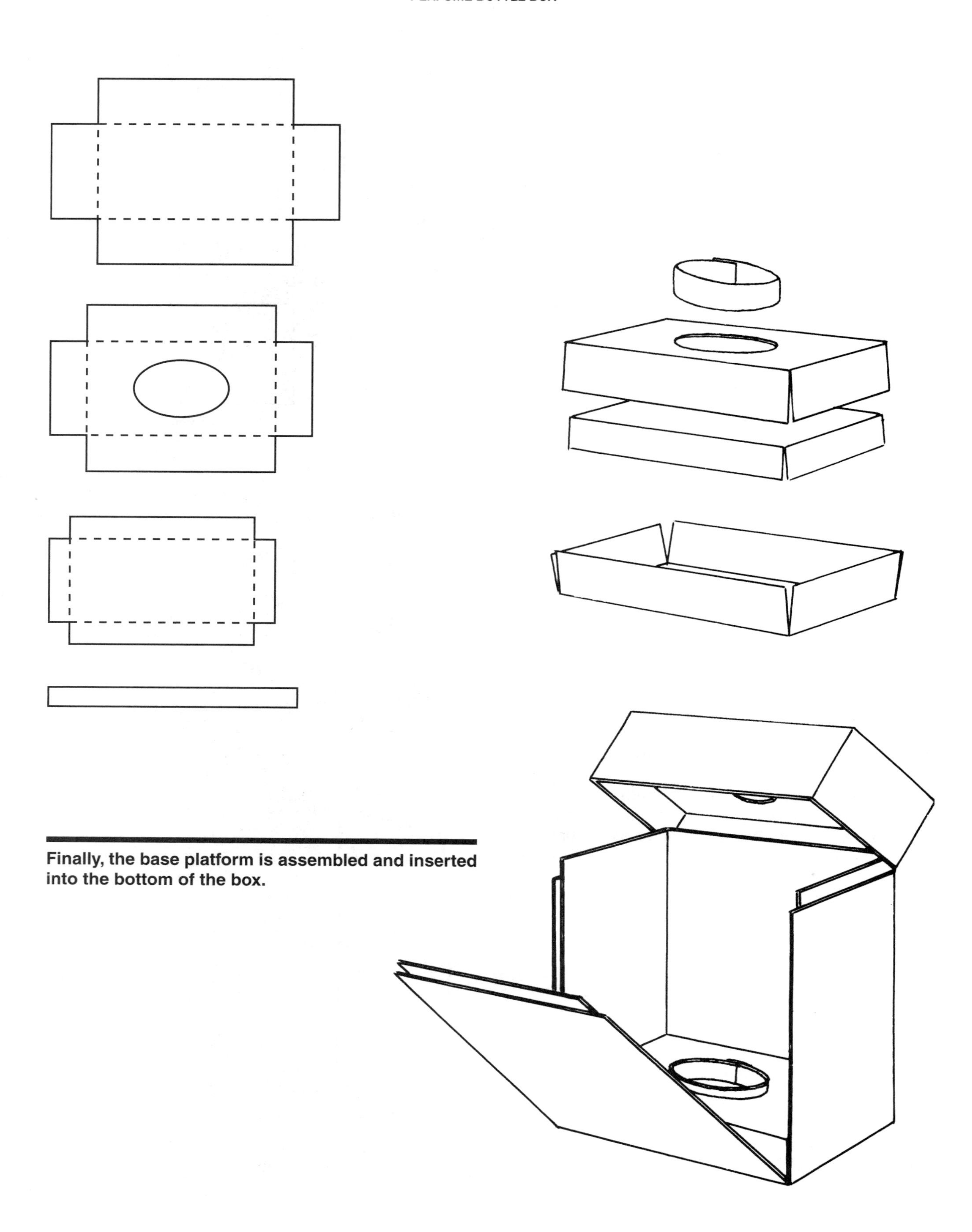

Finally, the base platform is assembled and inserted into the bottom of the box.

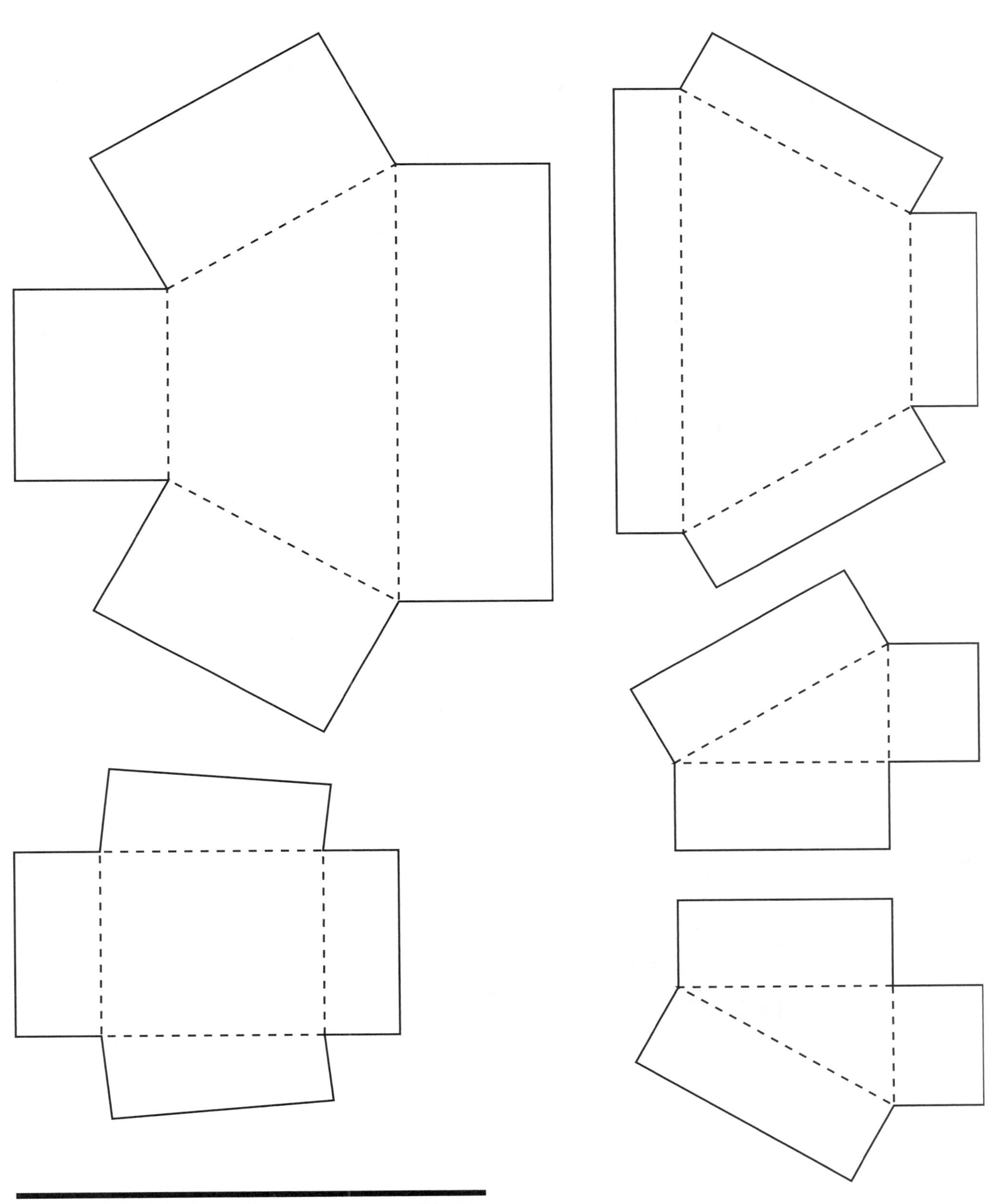

Many varieties of platforms may be assembled in order to create visual excitement.

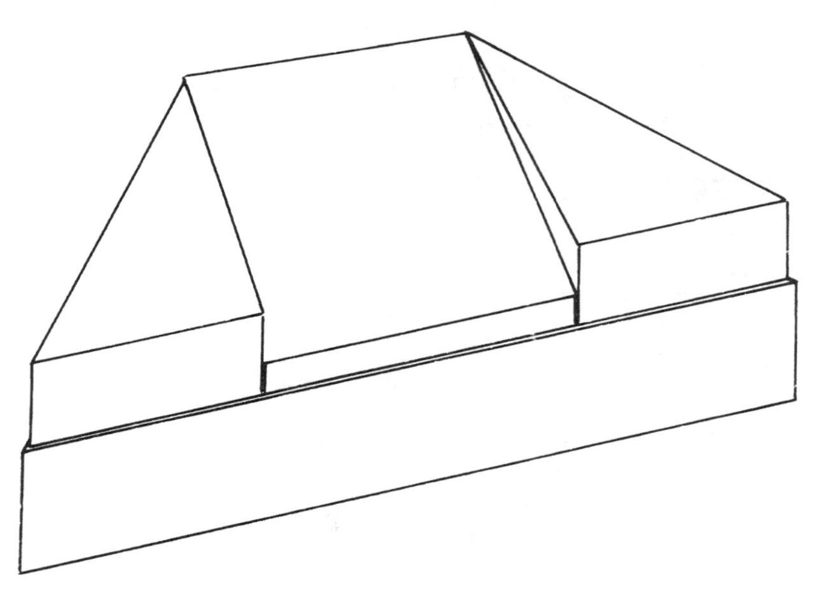

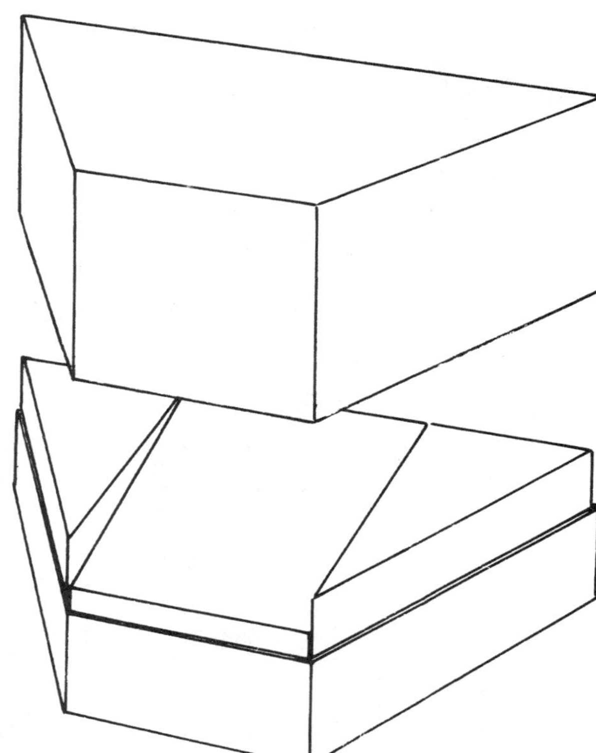

As long as the platforms have been made to the correct size, they may be inserted in various directions. In this example the front of the set-up box may be oriented either way, and the lid may be attached accordingly.

4

Corrugated Containers

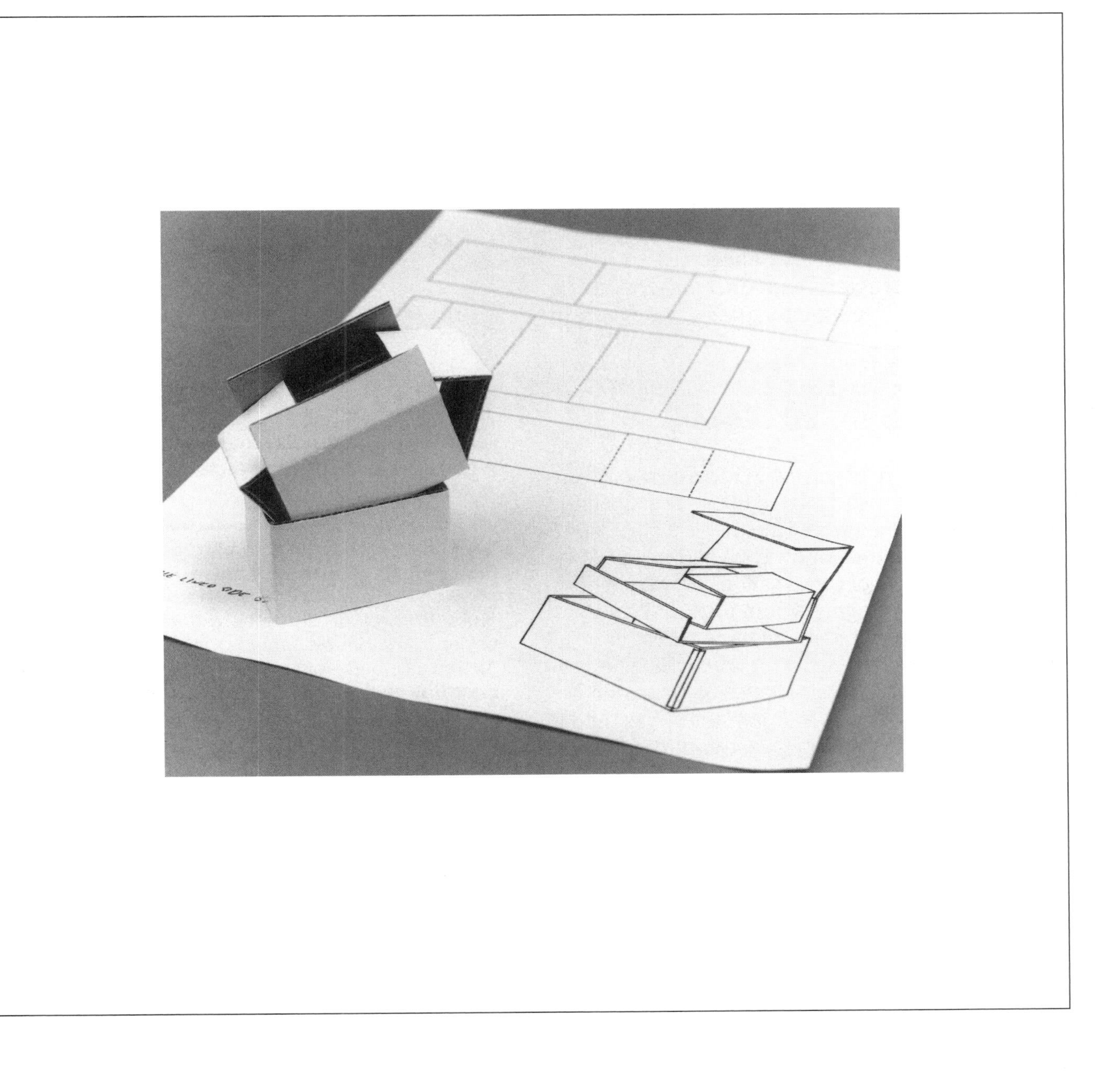

Fluted medium, the heart of the corrugated board, is one of the most unusual of all packaging materials. Originally it was an article of wear. In the mid-nineteenth century, men's hats were fashioned with a sweatband of fluted paper. In 1871 an American, Albert L. Johnes, patented fluted medium for the protection of bottles during storage and shipment. Three years later another American, Oliver Long, patented a process for sandwiching the fluted medium between paperboard sheets. This was the origin of corrugated containers, the workhorse of the packaging industry.

THE STRUCTURE OF CORRUGATED BOARD

The basic structure of corrugated board is simple. It consists of a fluted sheet glued to one or more liners. The most common construction is a sheet of "corrugated medium" sandwiched between two liners. A wide variety of combinations are possible, depending on packaging requirements. Where great strength is required, the three sheets of medium can be combined with appropriate liners.

The structural characteristics of corrugated board are governed by four variables: (1) the strength of the liners, (2) the strength of the corrugated medium, (3) the height and number of flutes per foot, and (4) the number of walls (single, double, or triple).

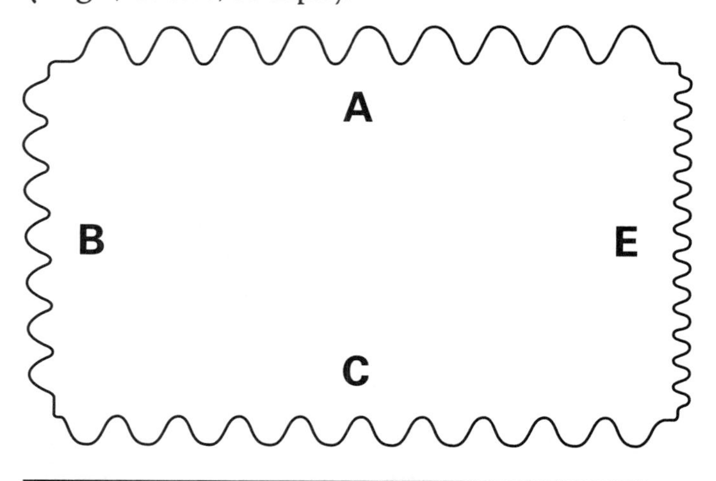

A corrugated flute guage (actual size).

Simplified schematic of a corrugator. The corrugating medium is preconditioned and shaped by the corrugating rolls. The preheated linerboard is then glued to the face of the corrugated paper and conveyed to the double facer for attachment of the liner on the opposite side. The assembly is then dried over hot plates and run through pressure rolls for stability. Once the medium has passed through the cooling section, it is cut to the required size at the delivery section.

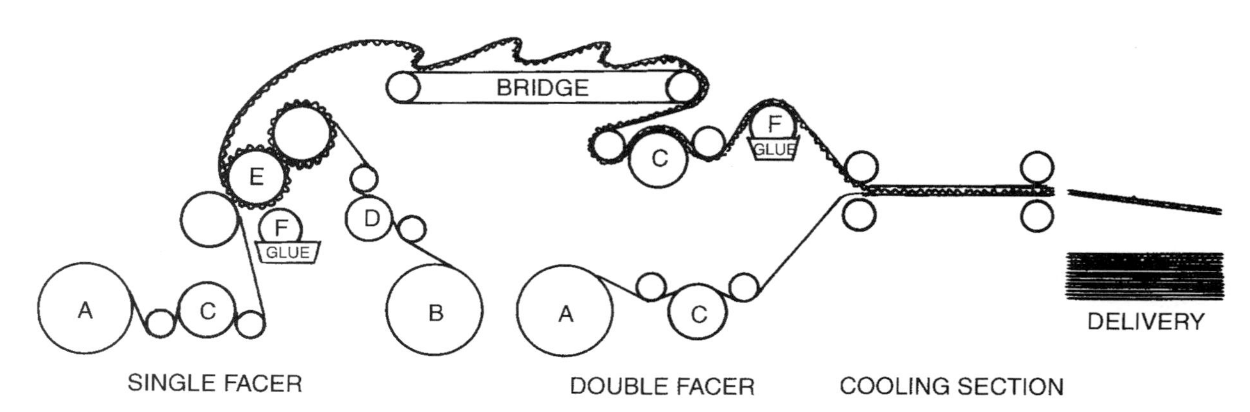

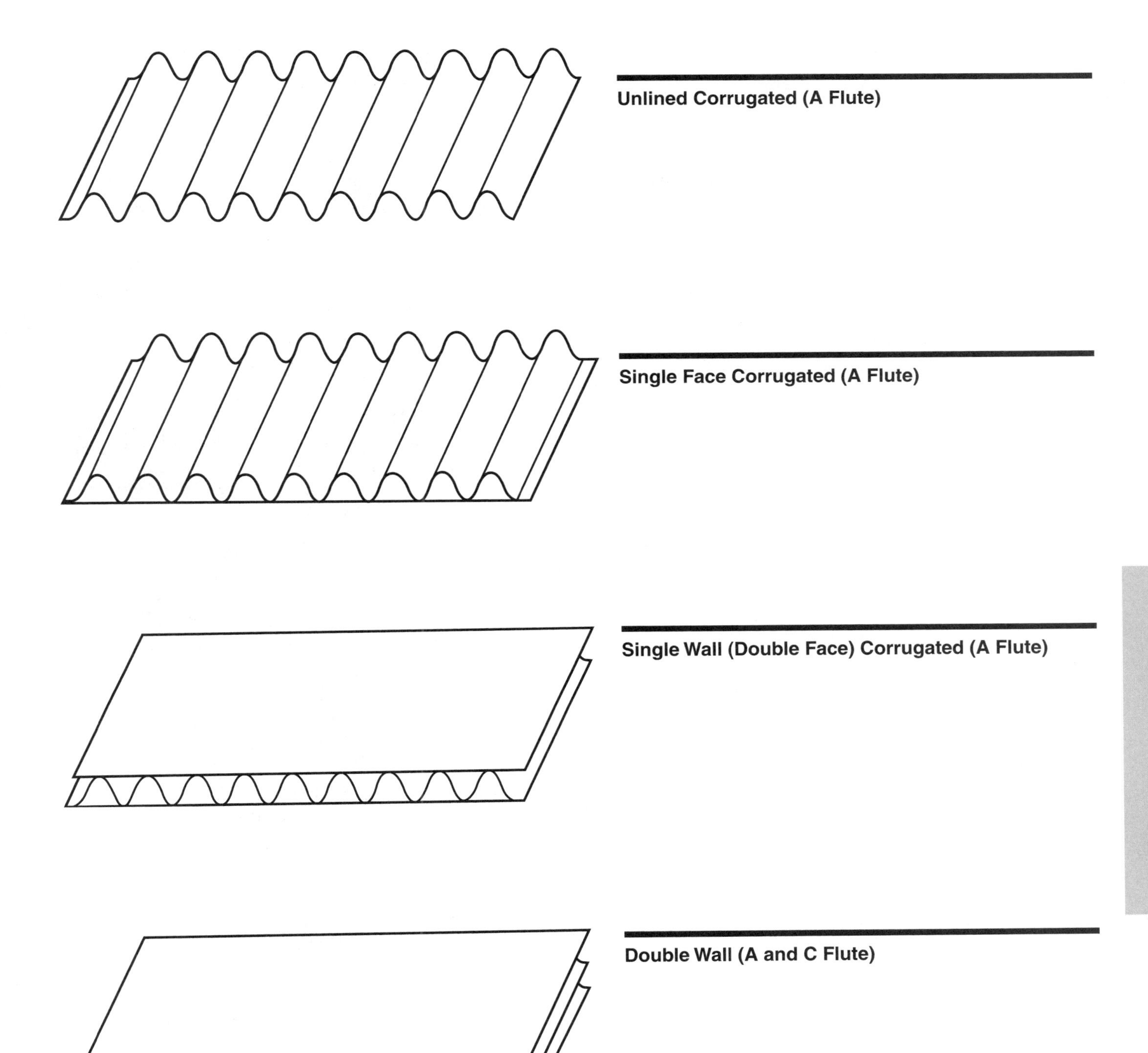
Unlined Corrugated (A Flute)
Single Face Corrugated (A Flute)
Single Wall (Double Face) Corrugated (A Flute)
Double Wall (A and C Flute)

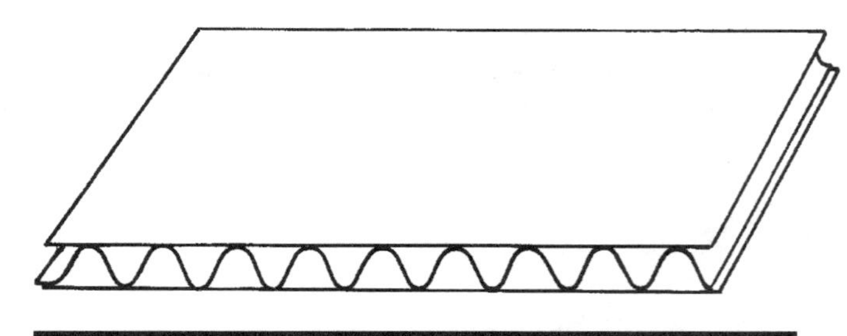

A Flute Corrugated—33 flutes per linear foot. Approximately 3⁄16" (without thickness of facings).

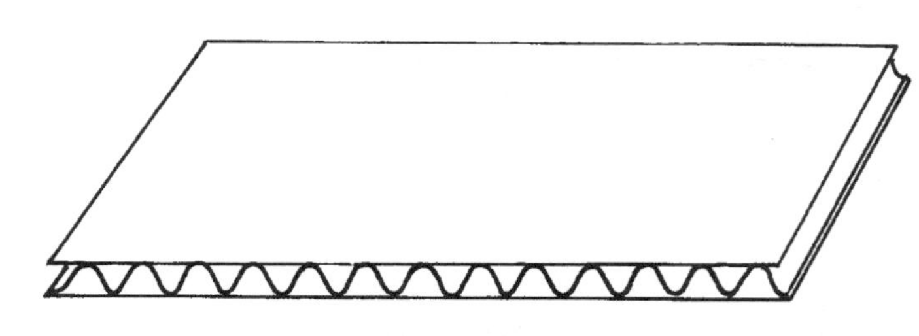

B Flute Corrugated—47 flutes per linear foot. Approximately 3⁄32" (without thickness of facings).

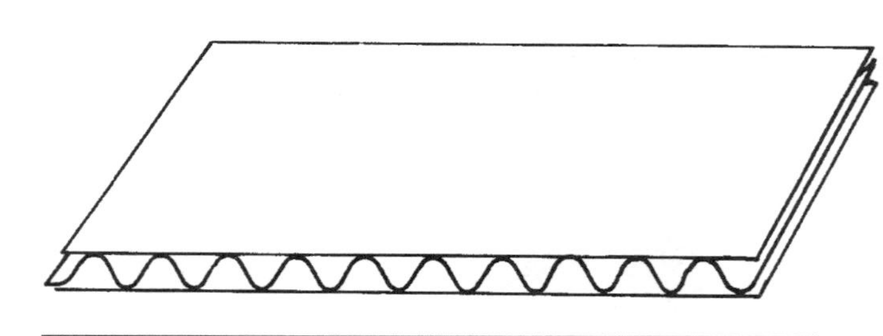

C Flute Corrugated—39 flutes per linear foot. Approximately 9⁄64" (without thickness of facings).

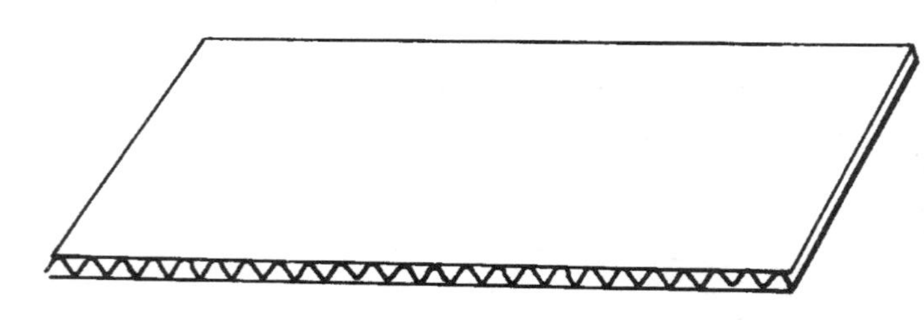

E Flute Corrugated—90 flutes per linear foot. Approximately 3⁄64" (without thickness of facings).

A number of flute structures are available, depending on packaging specifications. *A-flute* has great capacity to absorb shocks owing to the wider spacing of the flutes. *B-flute,* because of the larger number of flutes per foot, provides maximum crush resistance. *C-flute* combines the properties of the first two types, and *E-flute is* used where very thin corrugated board is recommended. In addition, a new grade of corrugated board has recently been developed which has a series of uniquely formed grooves and ridges that are smaller than those in the E-flute and which has 15–20% more ridges per lineal foot than the E-flute.

So universal is the use of corrugated board that its production is viewed as a barometer of the economy as a whole. It is hard to imagine products that cannot be packaged and shipped in corrugated boxes. Today more than 1,160 products—including live fish!—are shipped in corrugated boxes.

Corrugated board is used in multicolored shippers and point-of-purchase displays. It is among the least expensive of all packaging materials.

DESIGNING A CORRUGATED BOX

The design of a corrugated box is a major undertaking. The process of selecting the correct package design for a particular product has grown more complex as new technologies and materials present ever-increasing manufacturing options.

The ways in which corrugated board may be used are practically limitless. Certain basic container styles and designs are suitable for packaging a wide range of products. There are some corrugated interior devices (platforms, padding, or inserts) and plastics (molded polystyrene foam) used to provide reinforcement, bracing, and shock absorption. These are illustrated and described on pages 504-541.

Specialty containers are tailored to the requirements of a particular product. Those requirements may involve everything from the "shipability" of the product itself to how the container is filled, stored, loaded, stacked, braced, chopped, and unpacked. Specialty or custom-made boxes are usually required for special products in large quantities (10,000 or more). Master cartons, which are shipping cartons that hold smaller cartons, are used for food, detergents, housewares, and hardwares.

Government and industry standards and regulations are designed to protect the users of cartons. There are laws pertaining to method of shipment, such as rail, air freight, truck, and parcel post (U.S. Postal Service). In addition, all corrugated materials and cartons must be certified by the manufacturer. Weight, paper content, and puncture and bursting test certificates must be displayed on all corrugated boxes.

A significant trend in corrugated technology is impregnating and coating corrugated board with waxes and plastics. The moisture-resistant coating permits reuse of the carton and shipment of products, such as produce, that were previously shipped in expensive wooded crates and barrels.

Weight, paper contents, and puncture and bursting test certificates must be displayed on all corrugated containers.

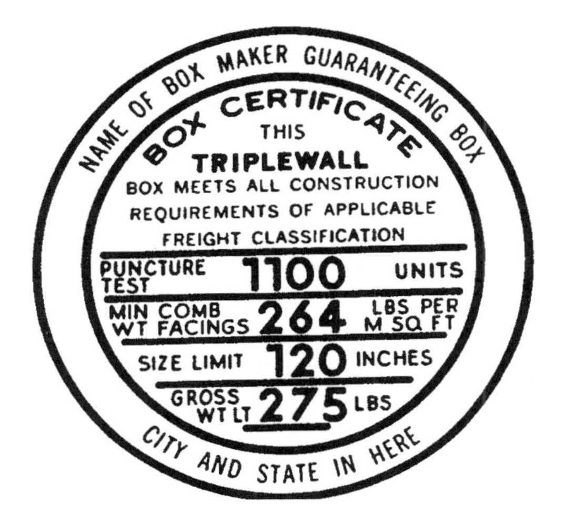

PRINTING ON CORRUGATED BOARD

Direct printing on brown corrugated board, which has a highly absorbent surface, is usually limited to the use of line art. This is the least expensive type of printing usable on corrugated board. Letterpress and, more recently, flexography are the typical printing methods.

Preprint is a term used to refer to the process in which a roll of printed stock is used as the top liner in making corrugated sheet. Flexography and rotogravure are the printing methods used in most preprint processes. The surface may be kraft, white clay-coated kraft, clay-coated bleached liner, or foil. The set-up costs for preprinting are high.

It is difficult to print full color on most corrugated board, with the exception of E-flute. Therefore, labels are often prepared to cover the boxes on one or more of their sides: full-sized labels cover the top and all four sides of a box; partial labels may be used only on the top of a box or on one or two of its sides. There are several variations of label application, including lamination onto the box, depending on size, shape, and cost considerations.

Litho labeling refers to lithographic printing of a sheet of paper that is then adhered (laminated) to corrugated board. Labels can be full or partial. Litho labeling and laminations are used on large boxes for toys, games, housewares, and sporting goods.

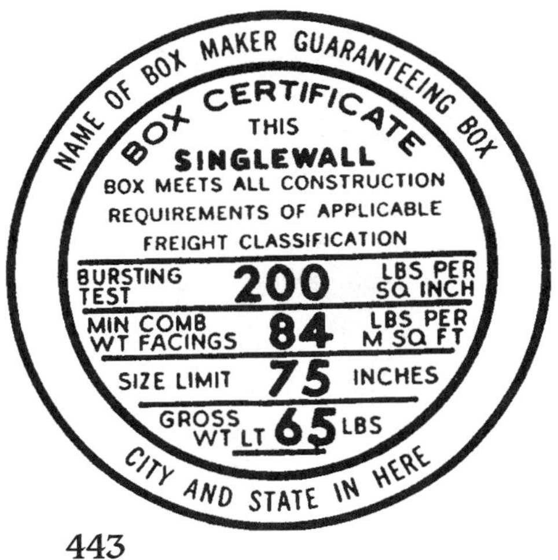

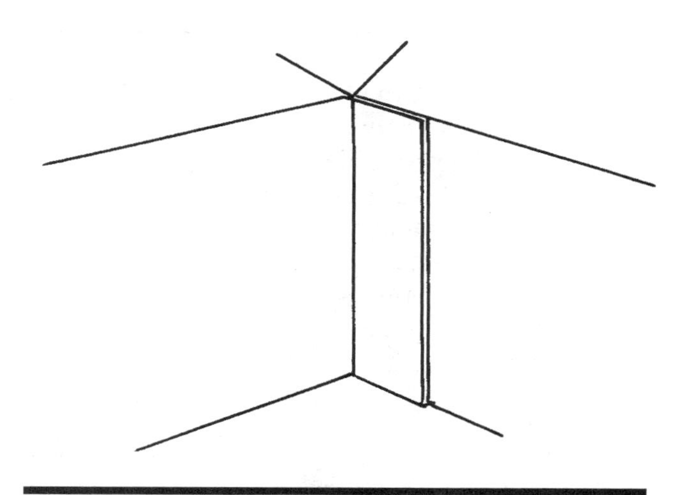

Glued joint: the manufacturer's joint tab is glued to the inside of the box.

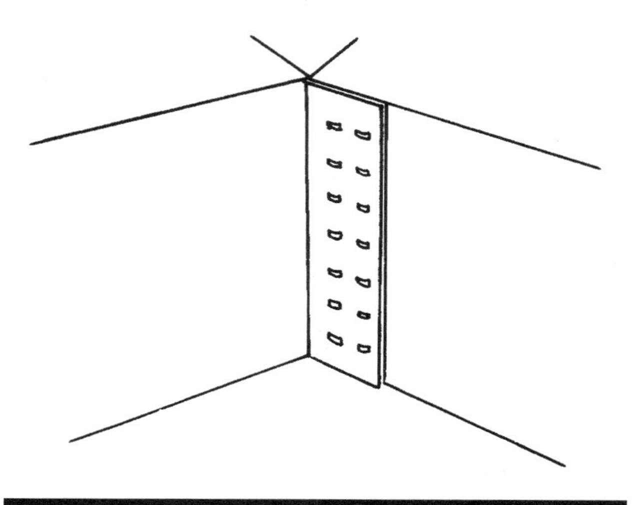

Stitched joint: the lap is stapled to the outside of the box.

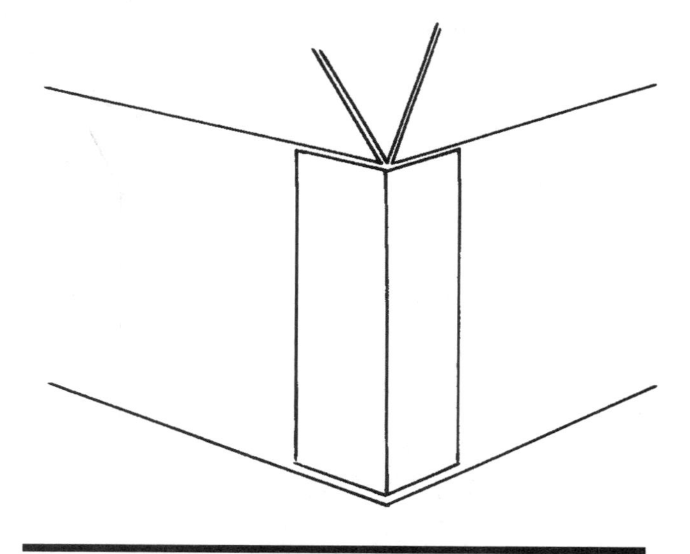

Taped joint: the side and end panels are taped from the outside of the box. No manufacturer's joint tab is needed.

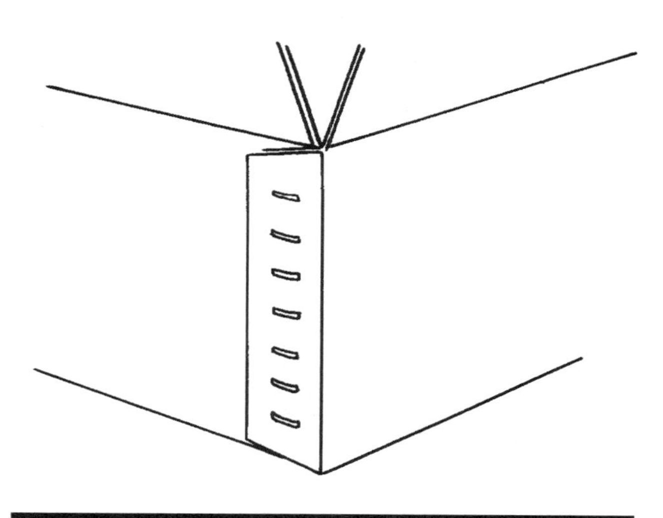

Stitched joint: the lap is stapled to the inside of the box.

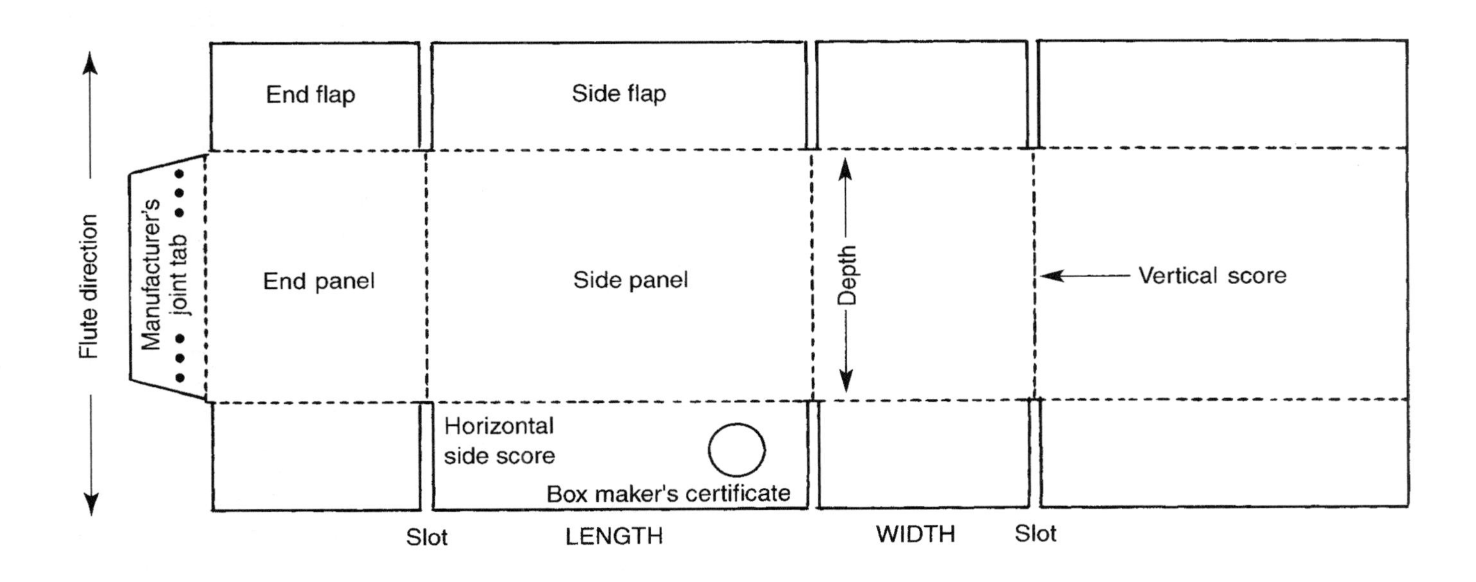

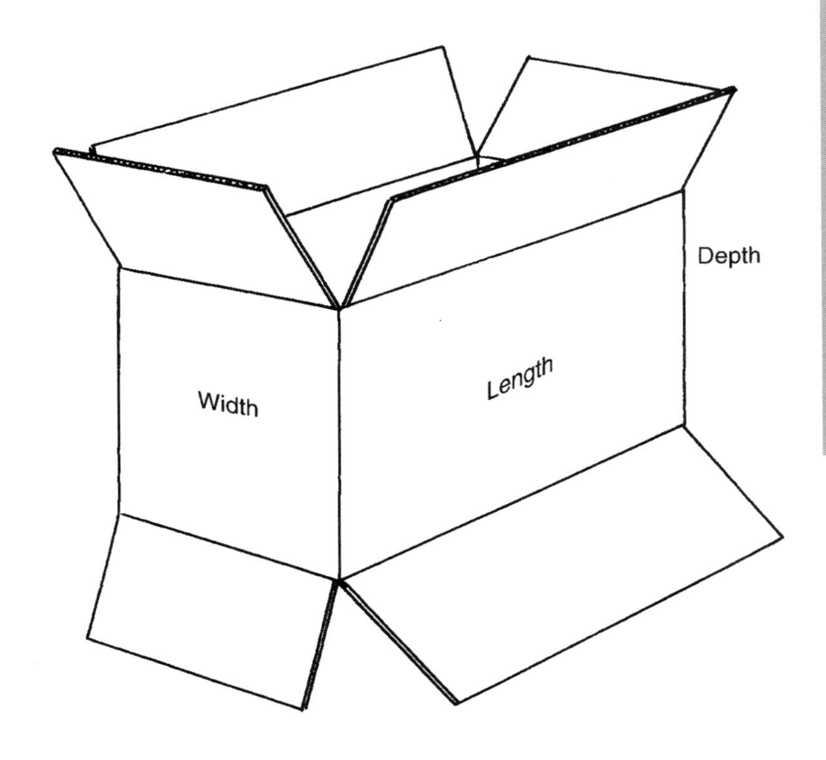

Corrugated Containers

FIVE PANEL FOLDER
(Intl. Box Code 0410)

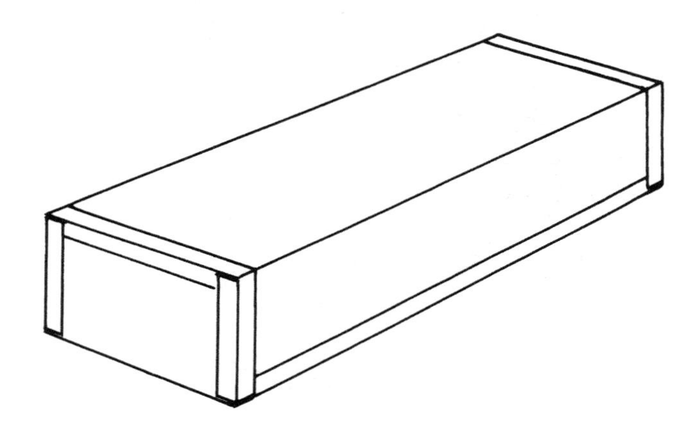

OVERLAP SLOTTED CONTAINER

(Intl. Box Code 0202)

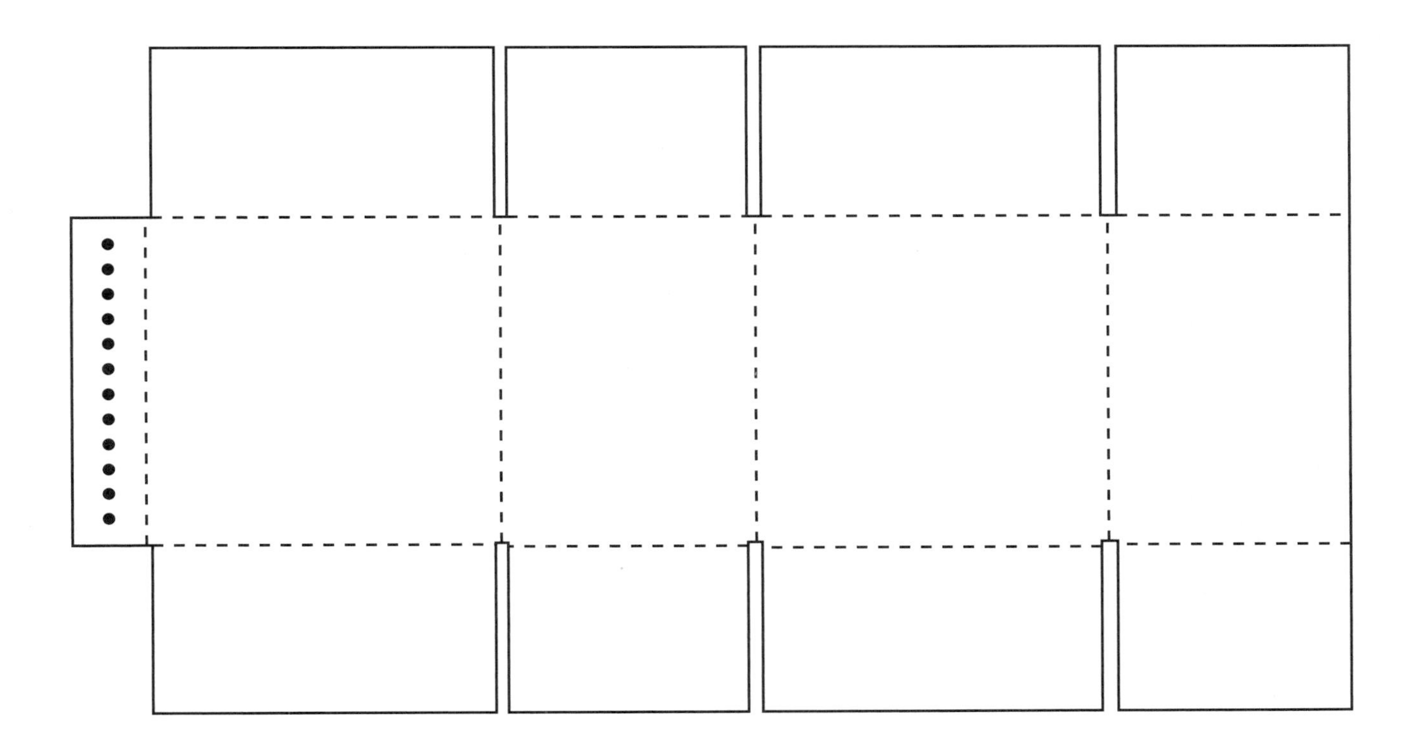

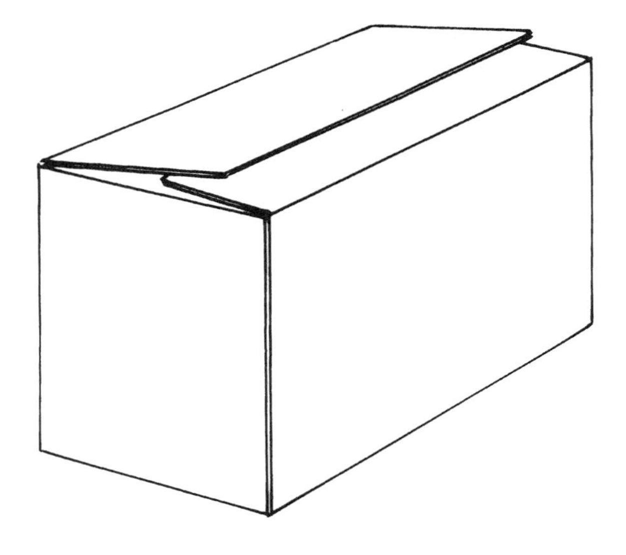

FULL OVERLAP SLOTTED CONTAINER

(Intl. Box Code 0203)

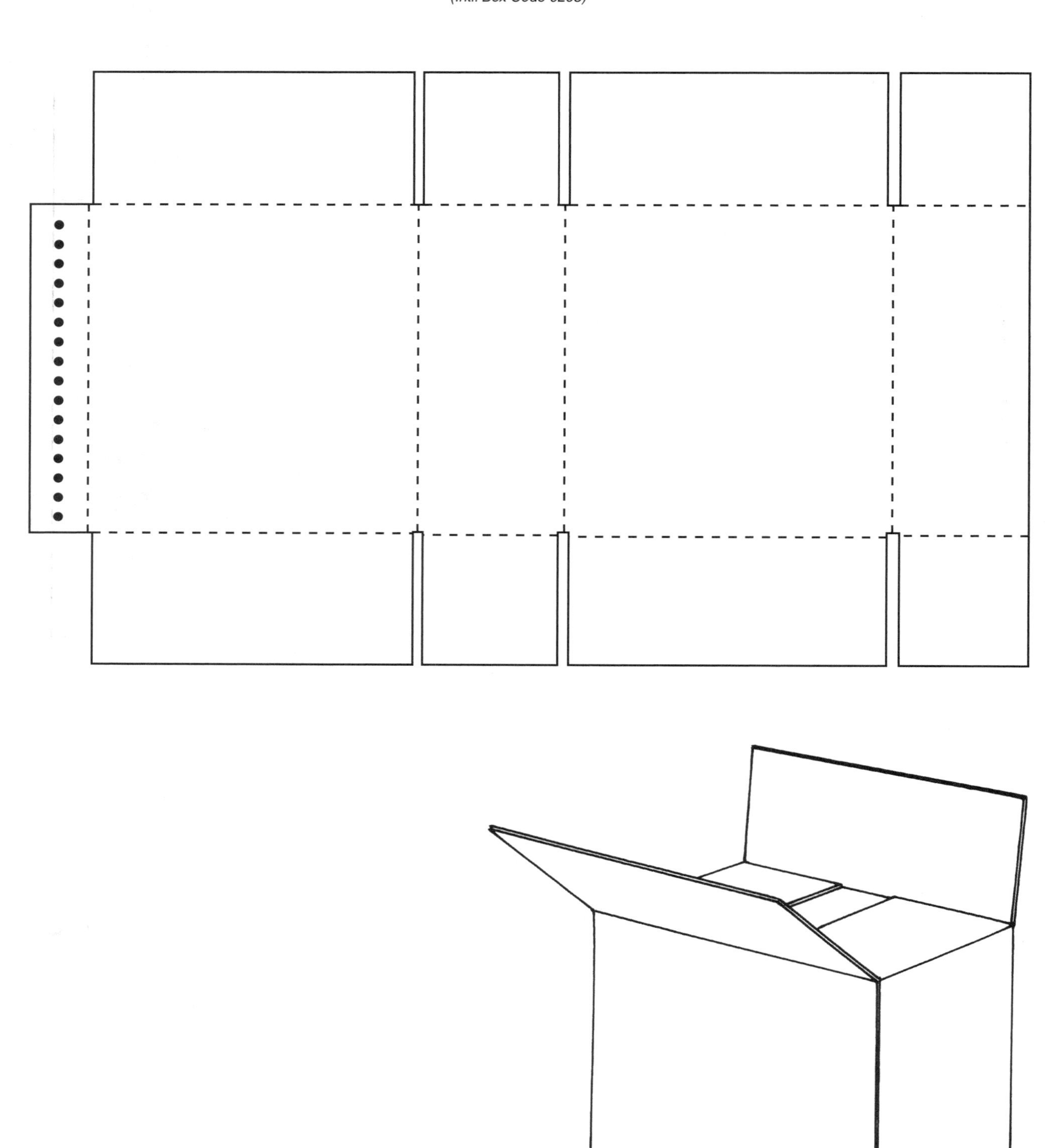

CENTER SPECIAL FULL OVERLAP SLOTTED CONTAINER

(Intl. Box Code 0206)

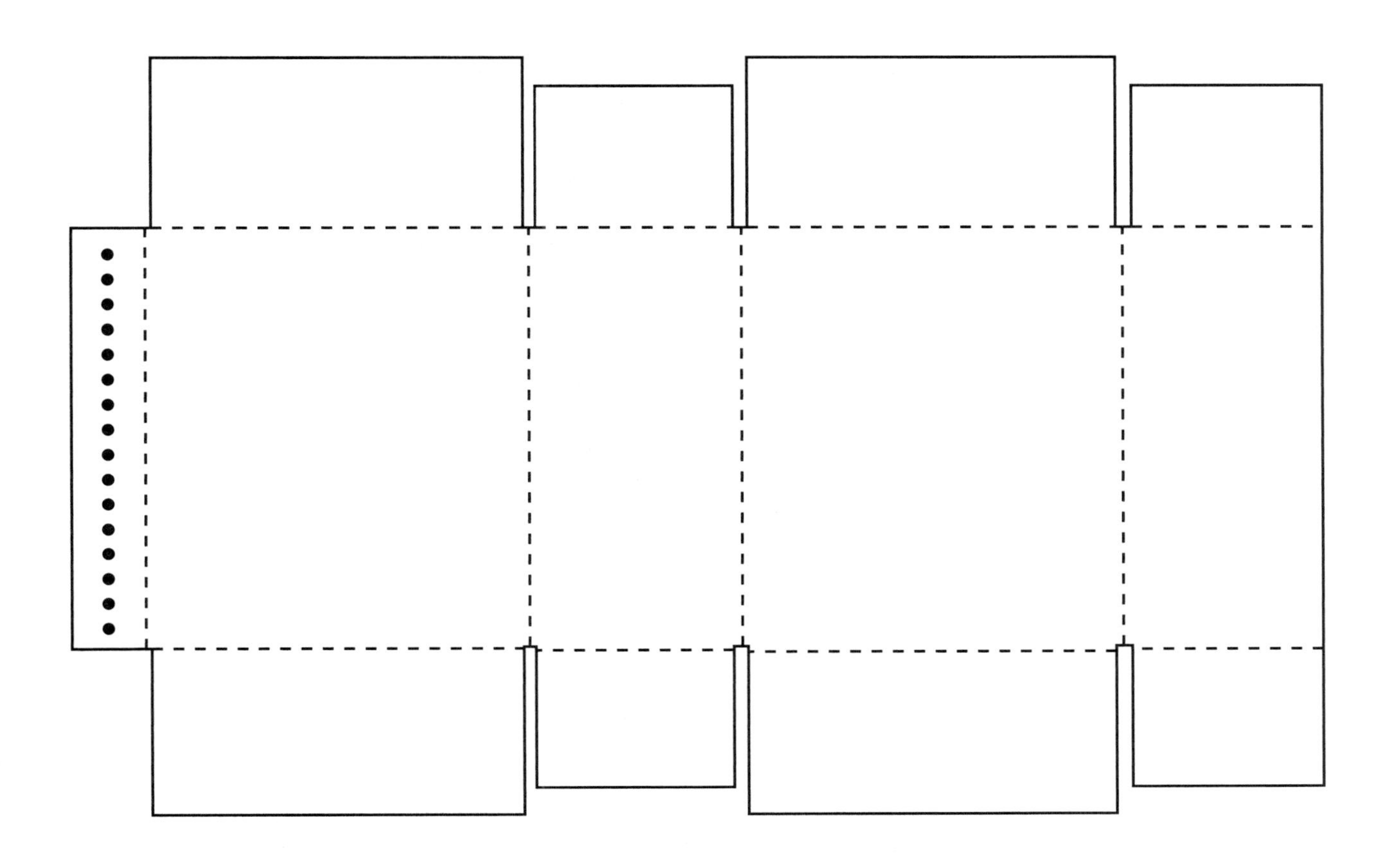

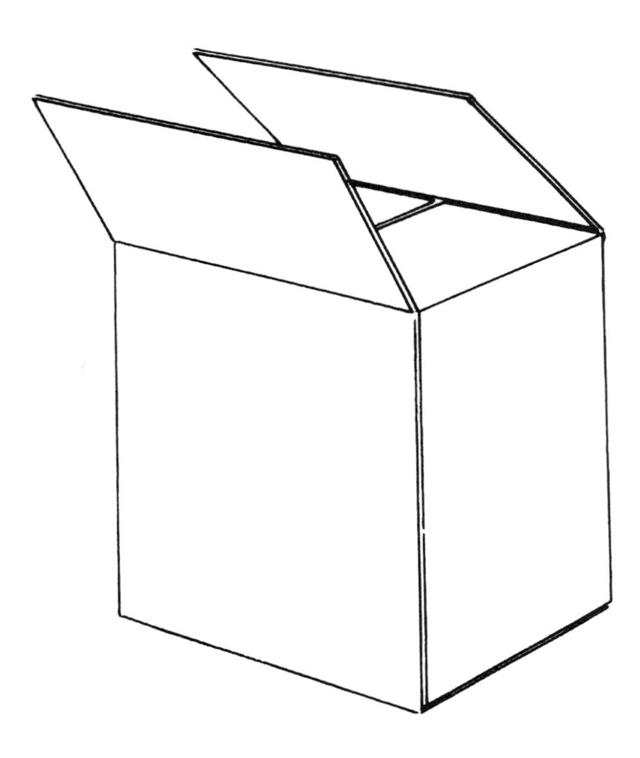

HALF SLOTTED BOX (TAPED) WITH COVER (STAPLED)
(Intl. Box Code 0312)

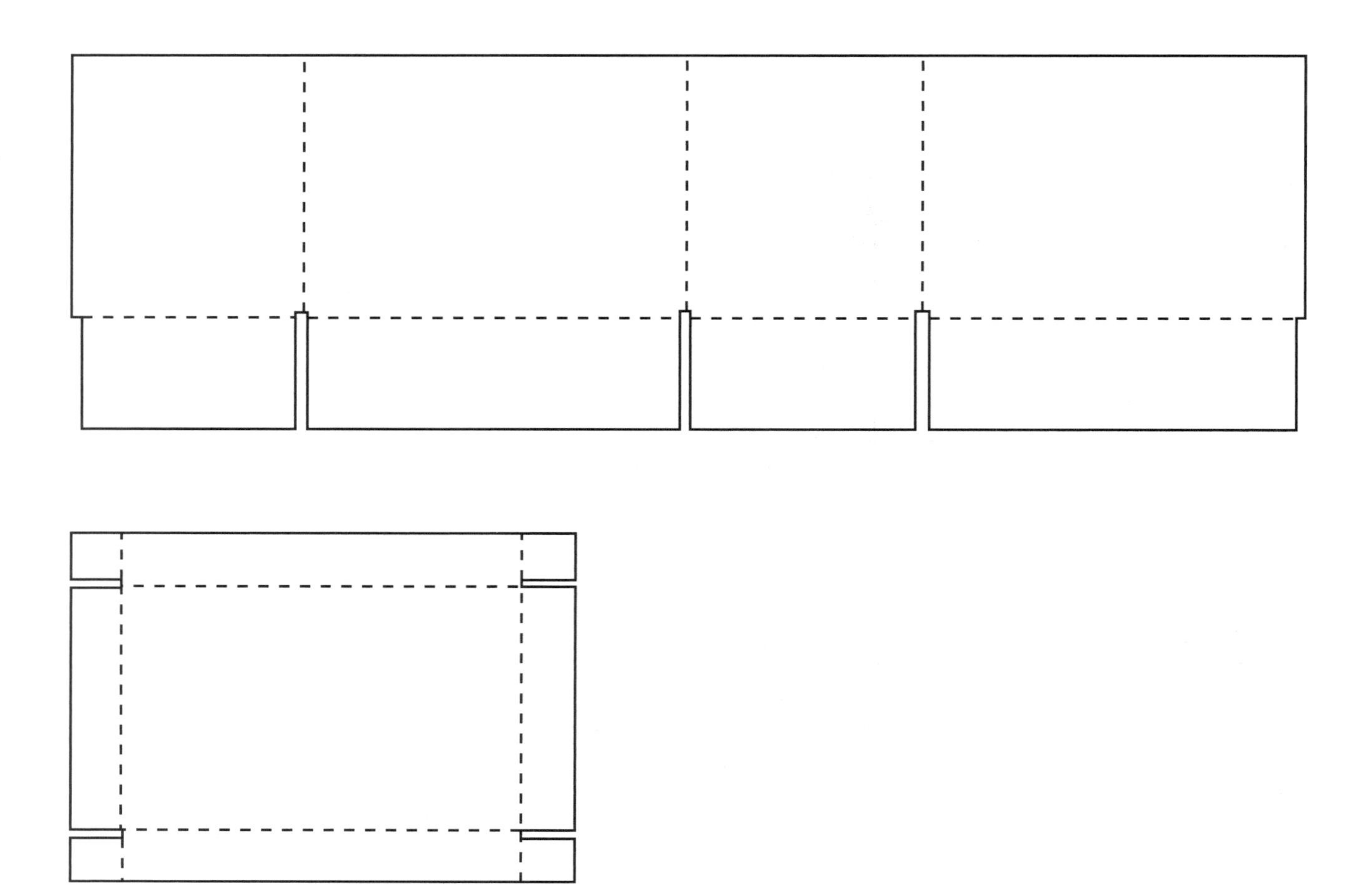

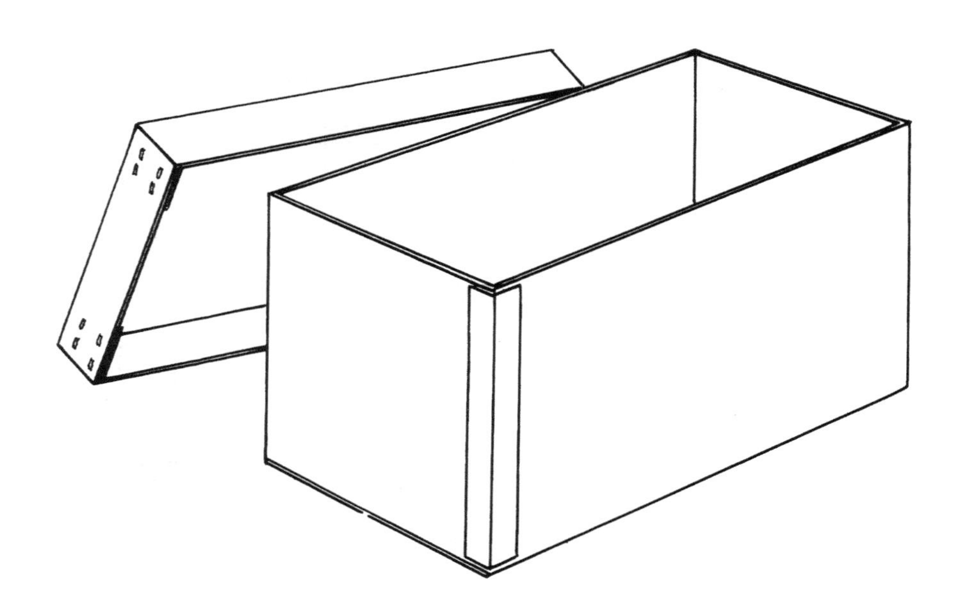

CENTER SPECIAL SLOTTED CONTAINER

(Intl. Box Code 0204)

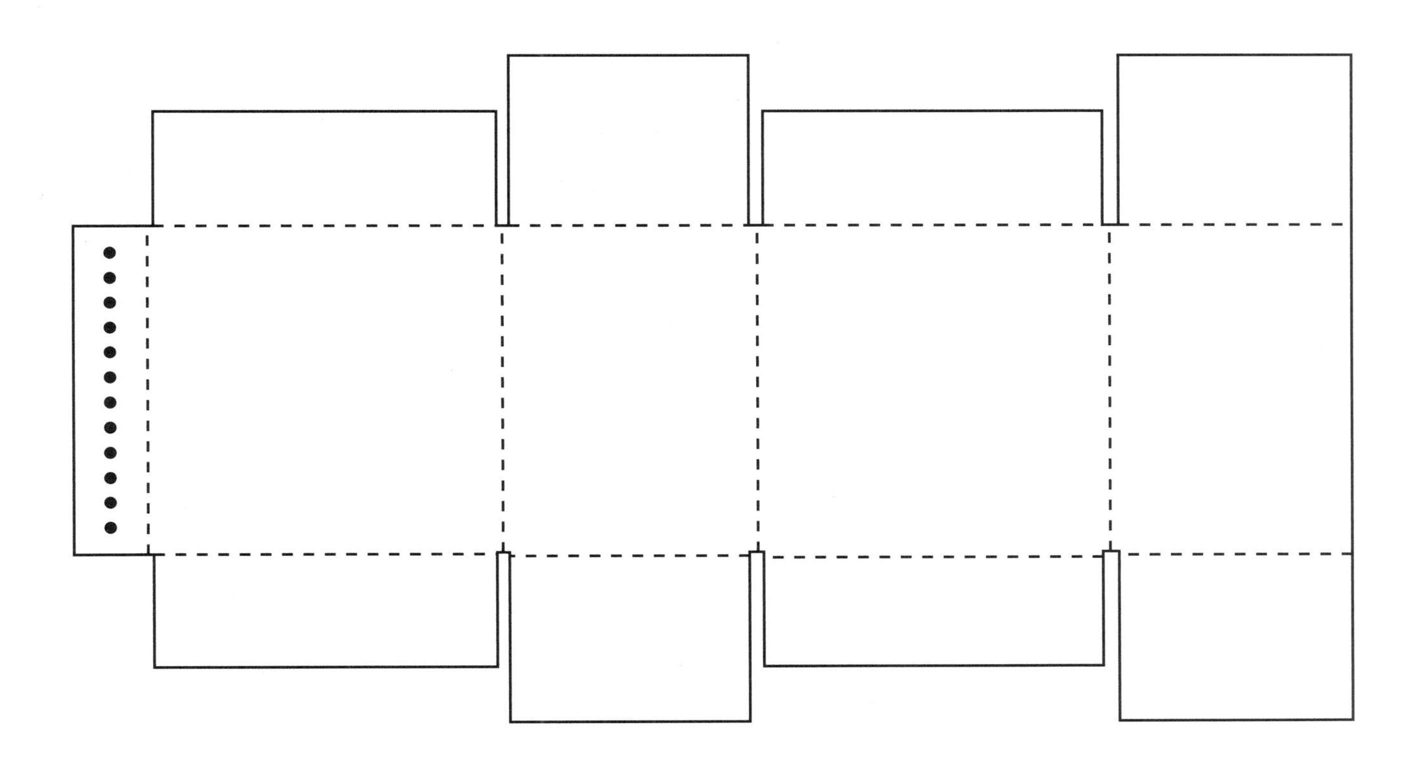

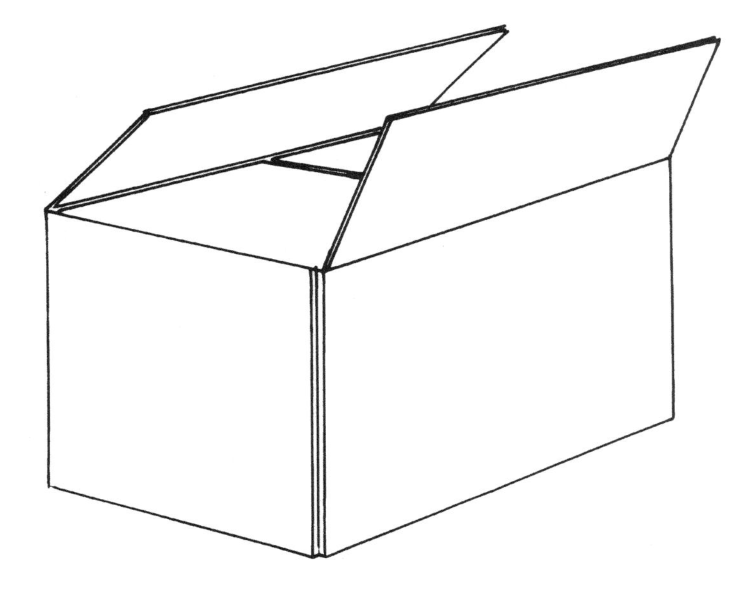

REGULAR SLOTTED CONTAINER

(Intl. Box Code 0201)

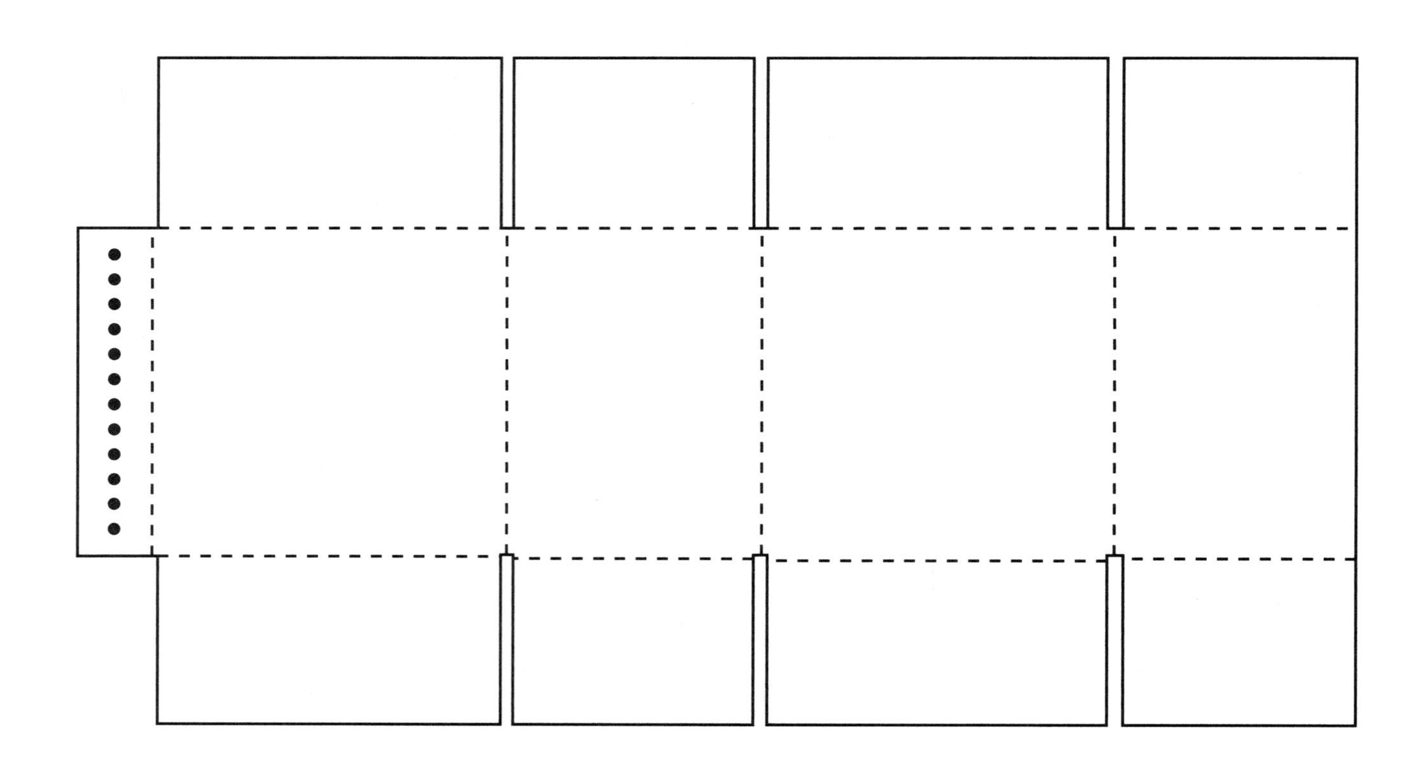

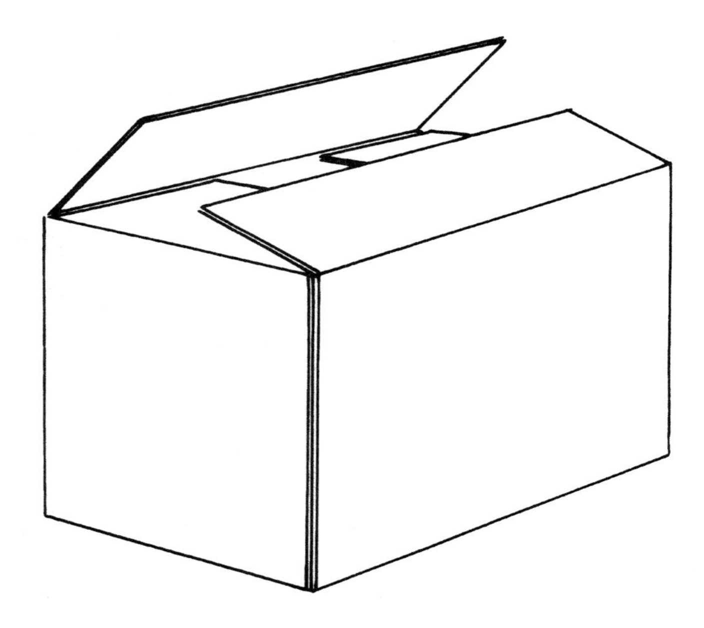

CENTER SPECIAL OVERLAP SLOTTED CONTAINER

(Intl. Box Code 0205)

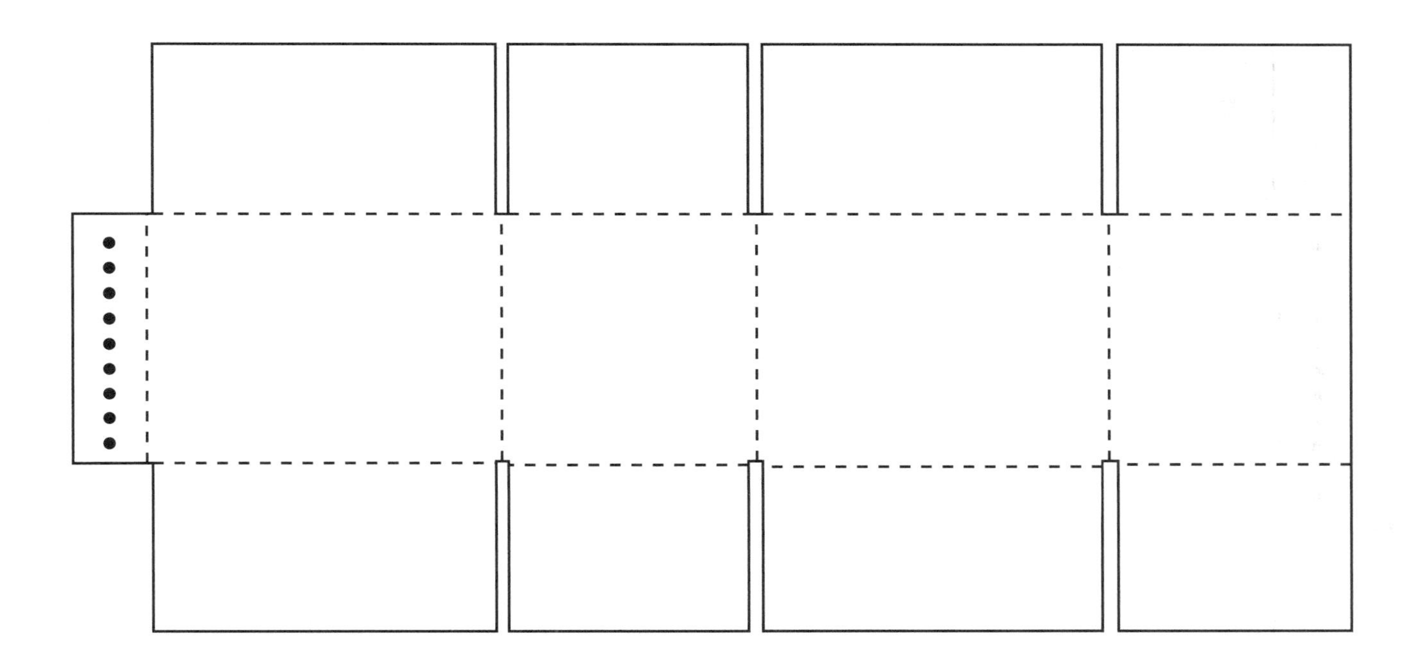

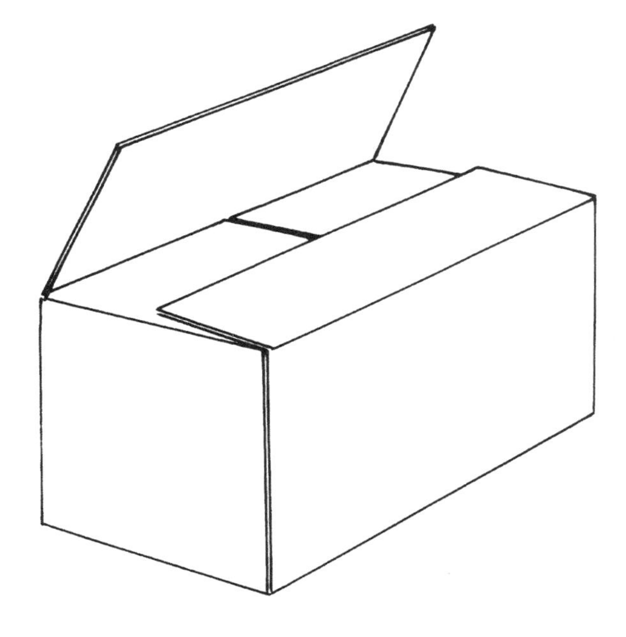

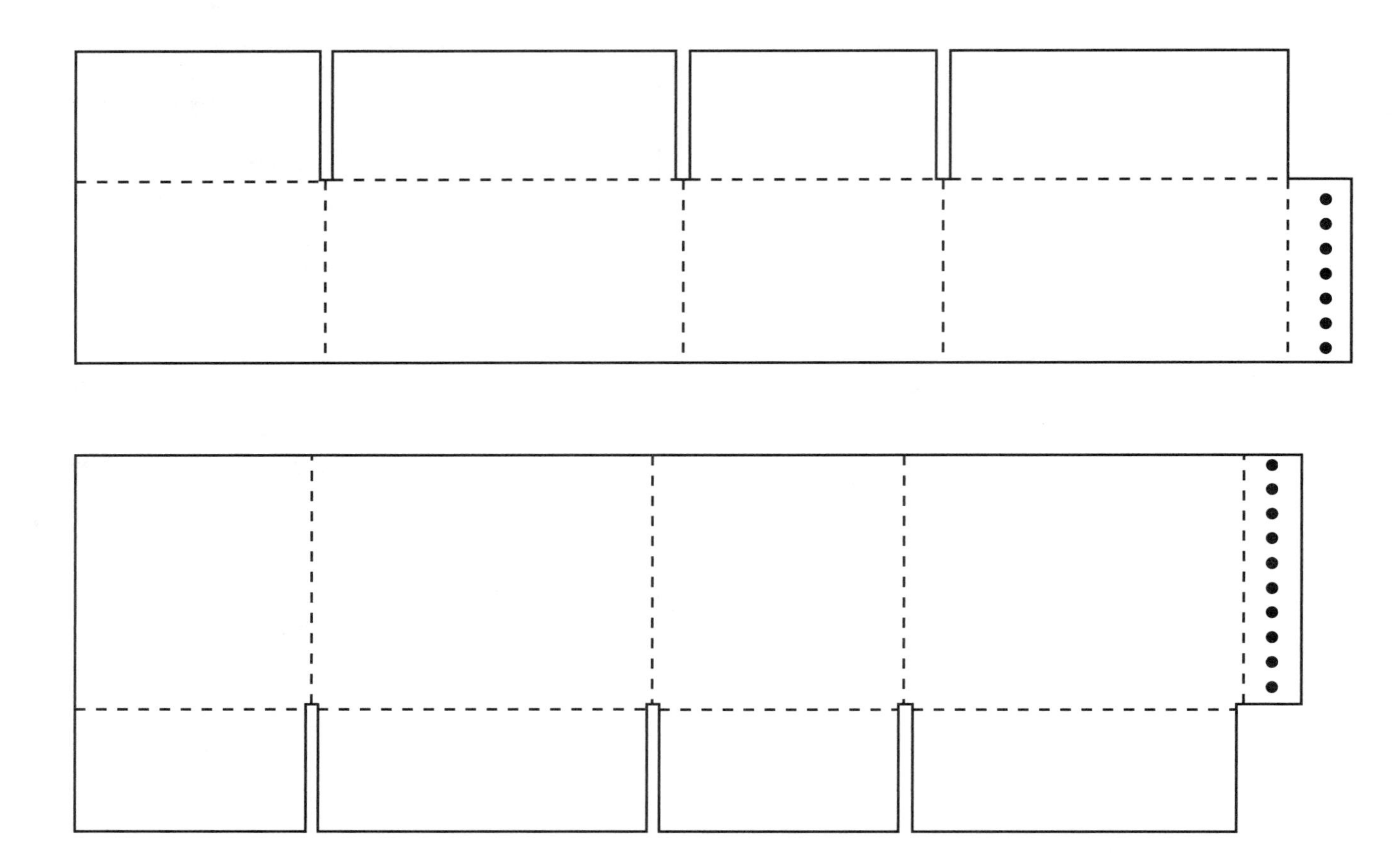

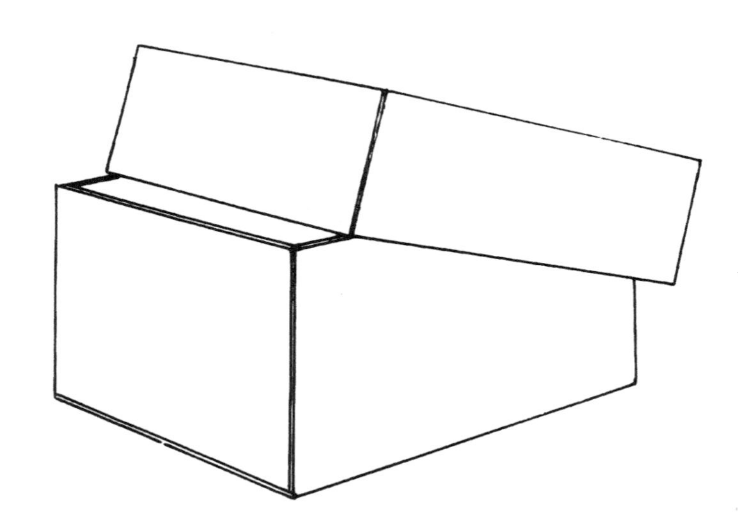

FULL TELESCOPE HALF-SLOTTED CONTAINER

(Intl. Box Code 0320)

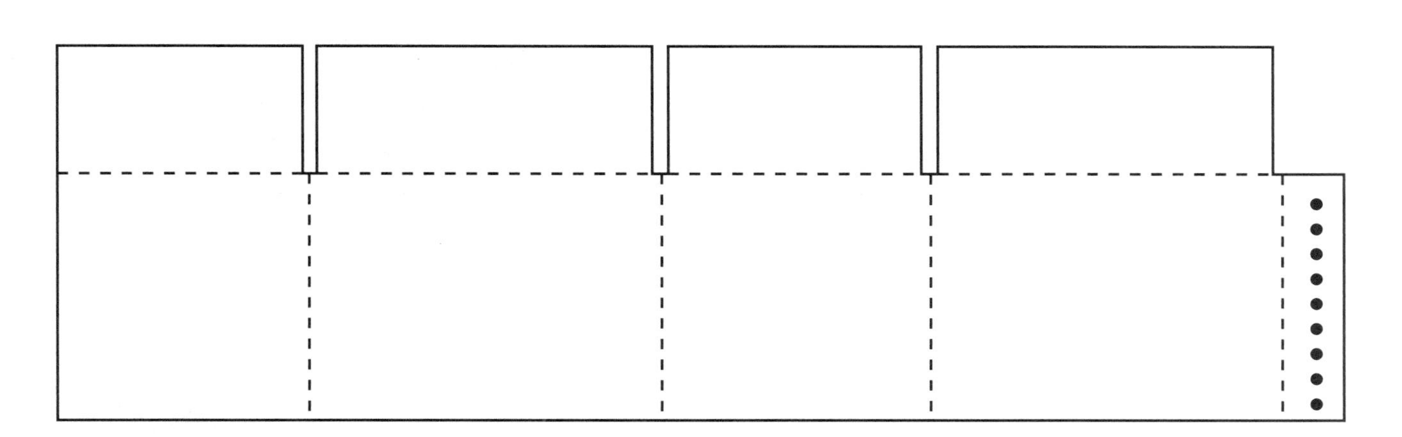

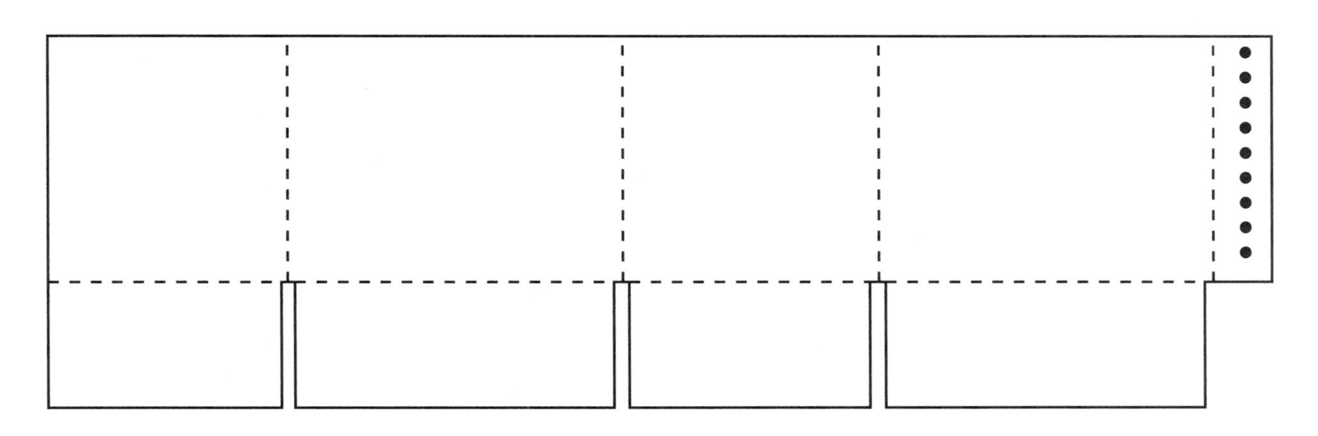

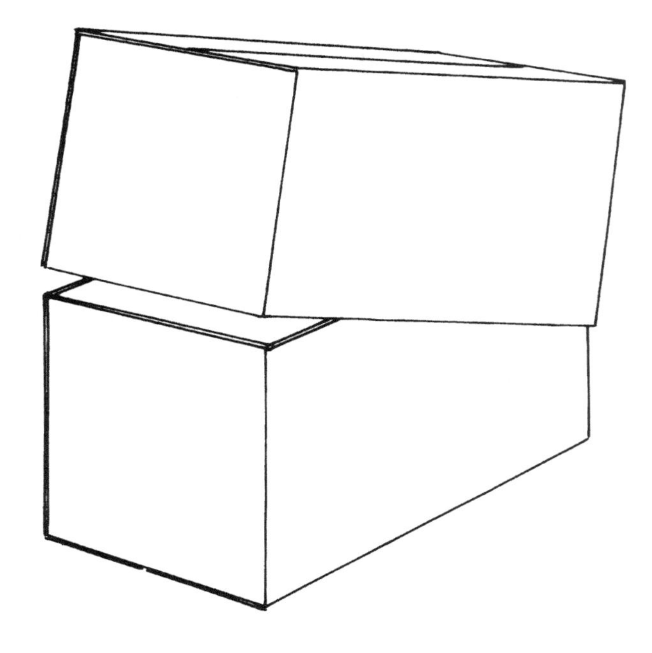

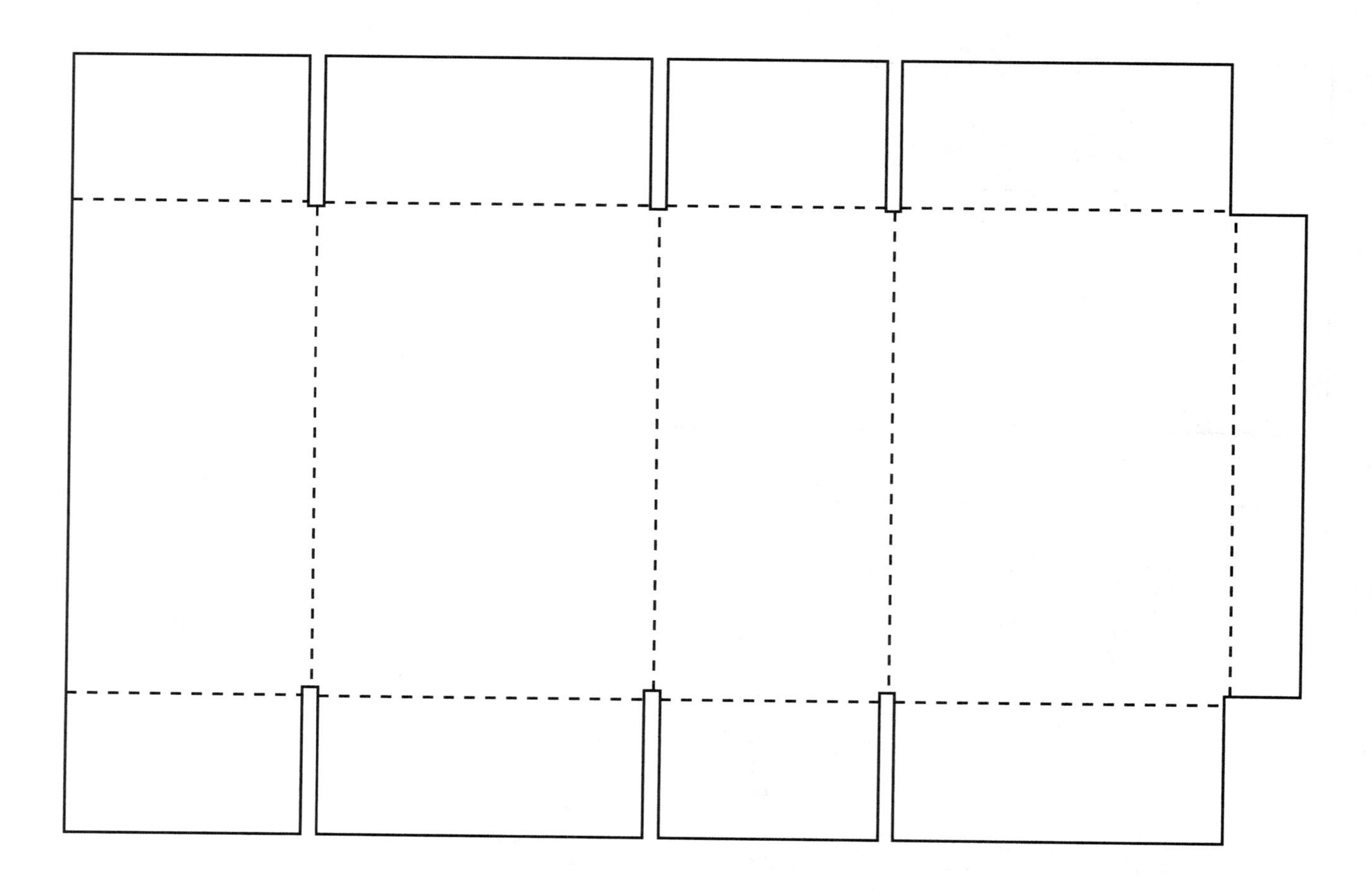

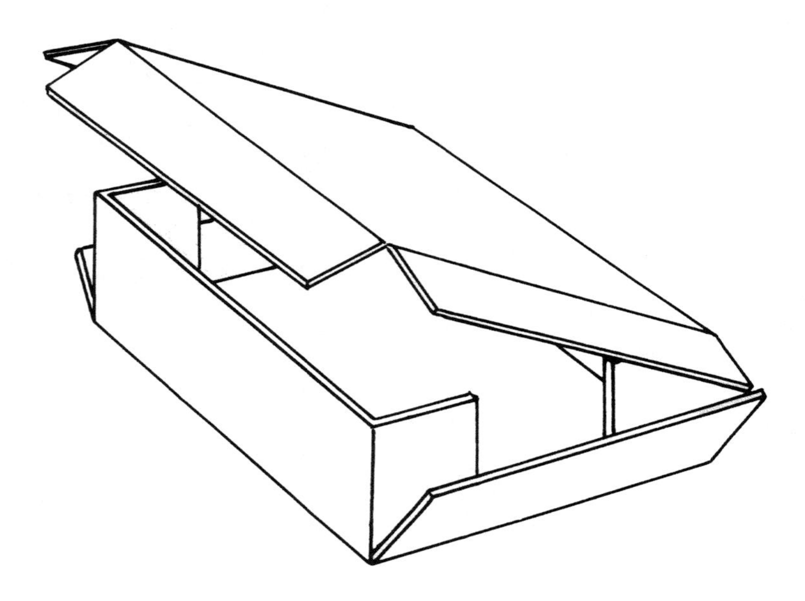

DOUBLE SLIDE BOX

(Intl. Box Code 0510)

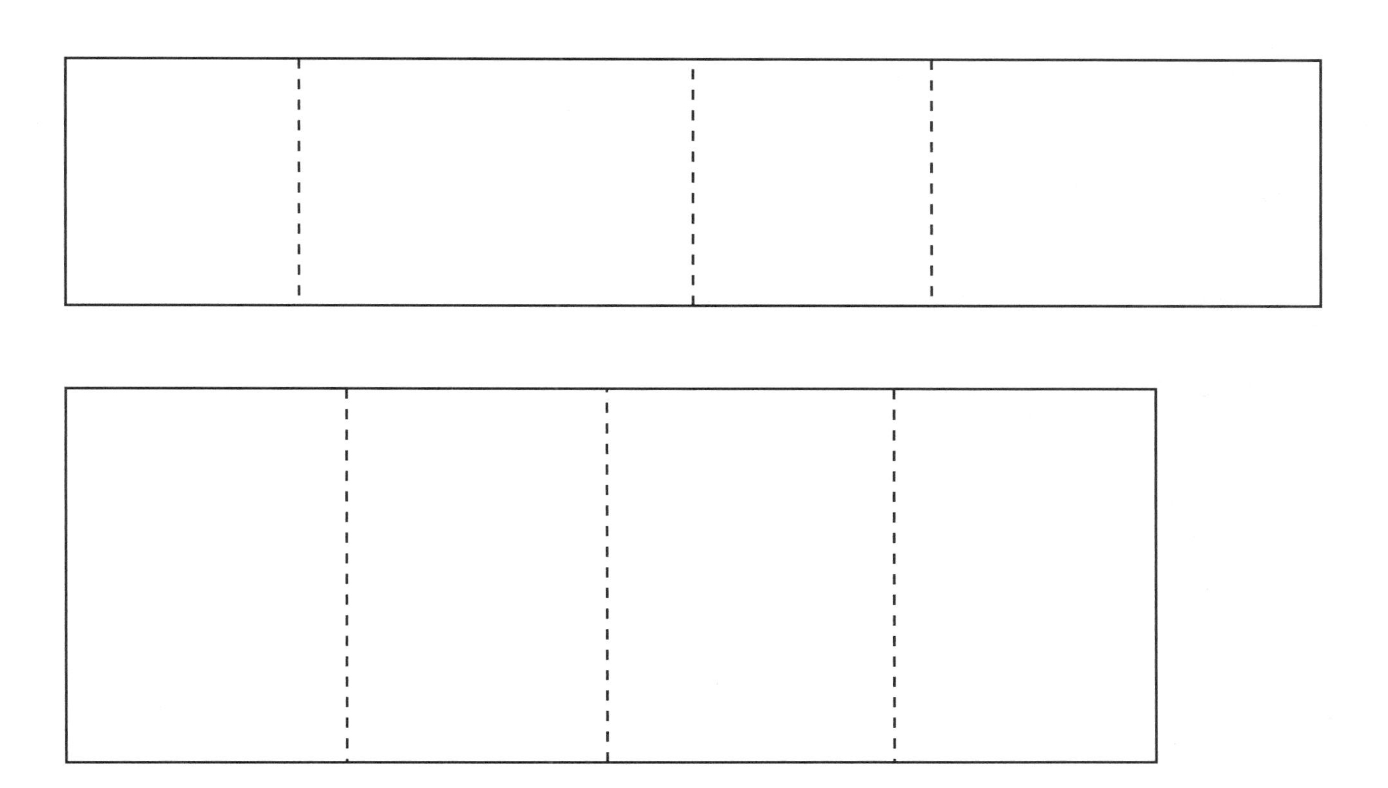

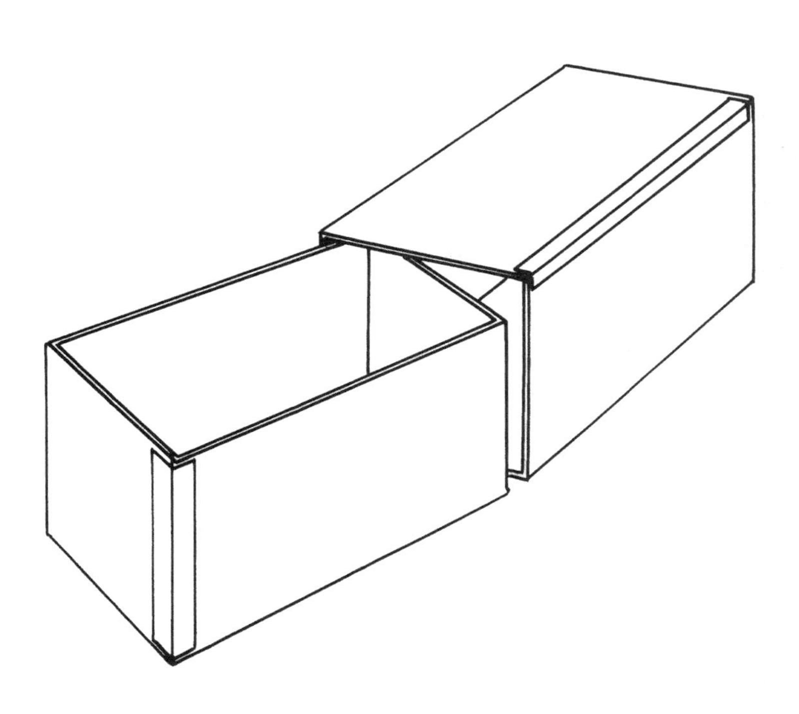

SINGLE LINED SLIDE BOX

(Intl. Box Code 0510)

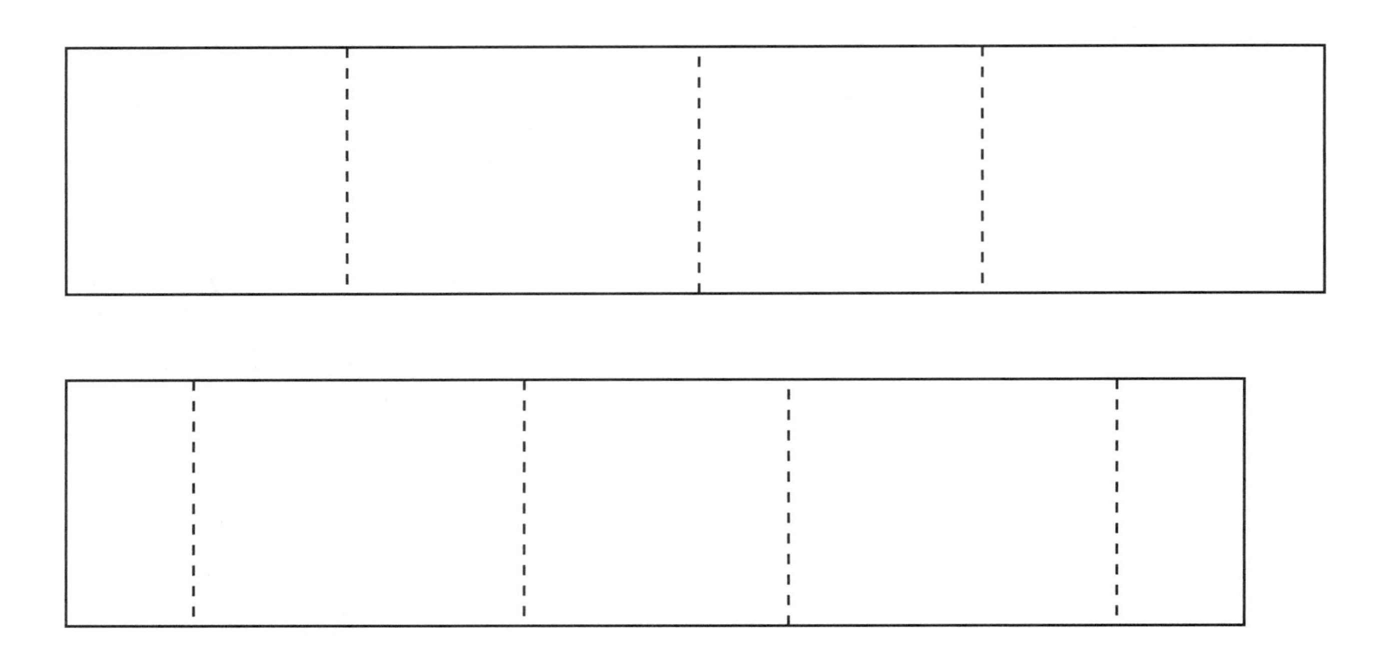

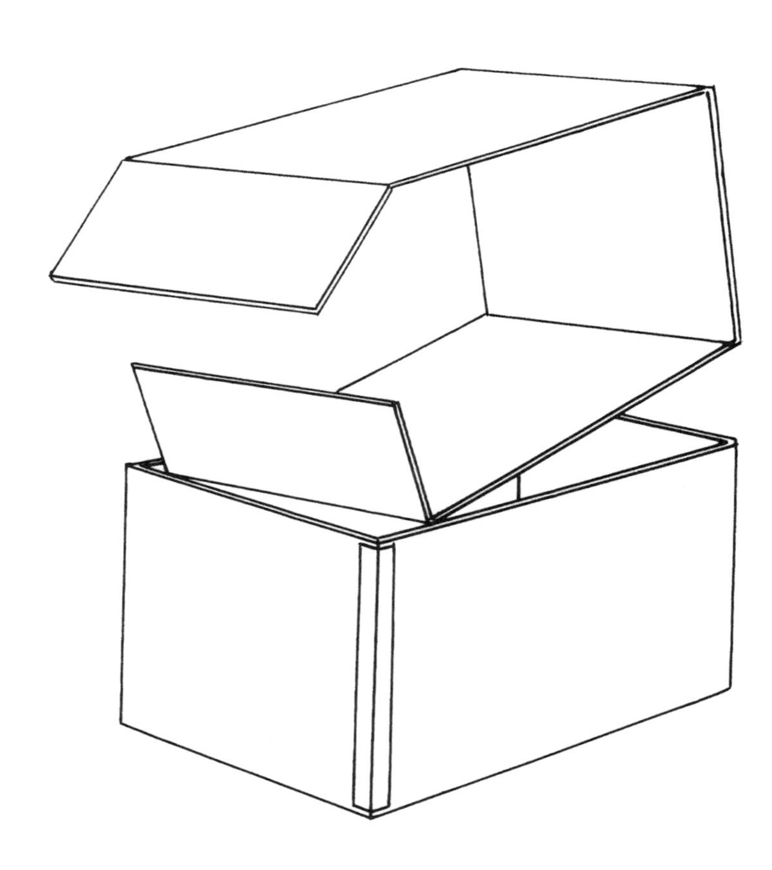

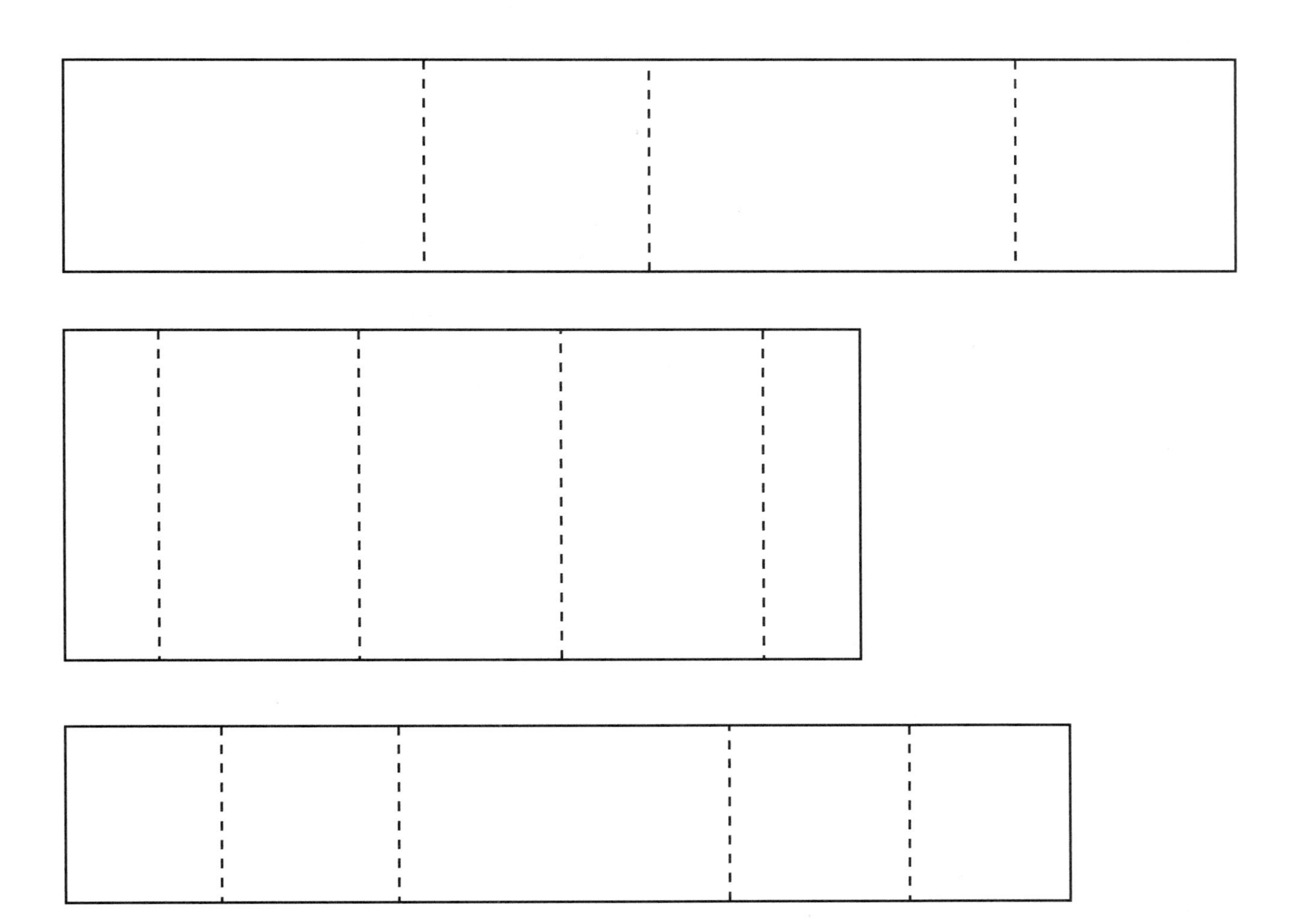

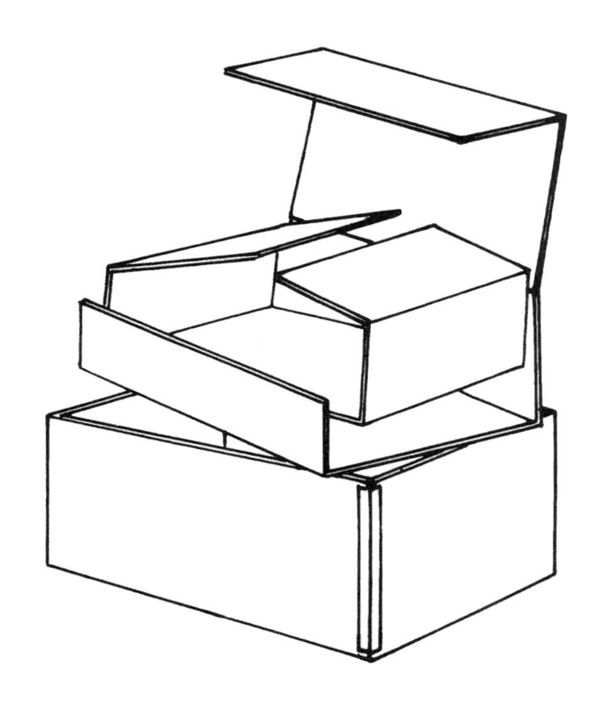

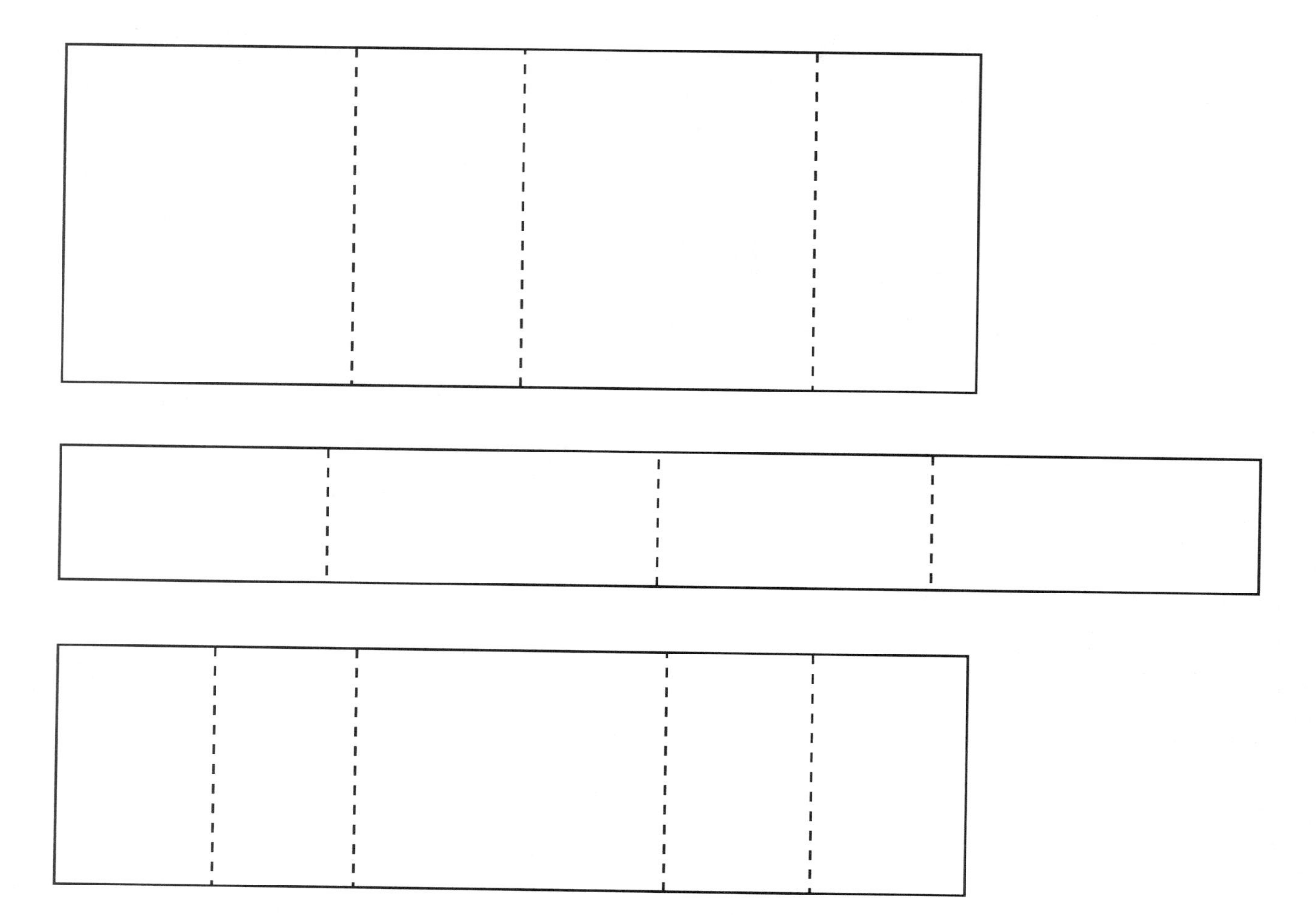

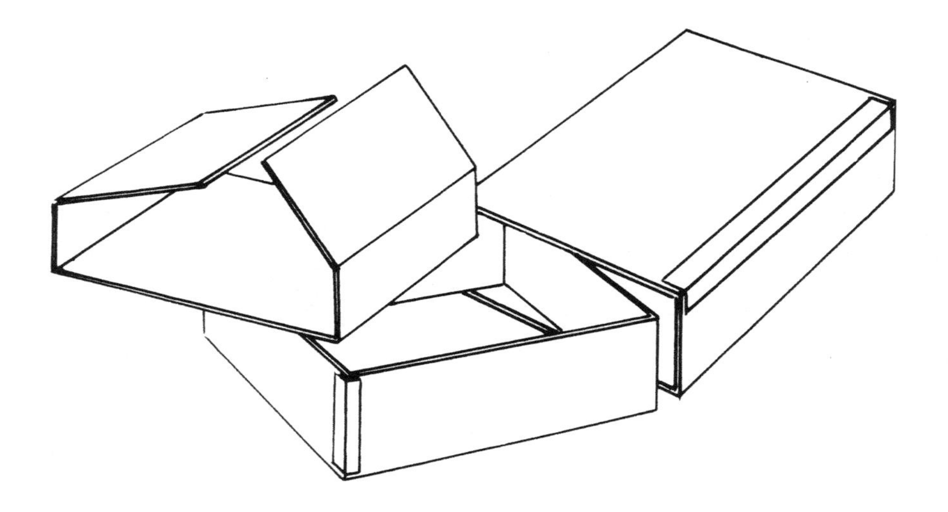

DESIGN STYLE WITH COVER

(Intl. Box Code 0306)

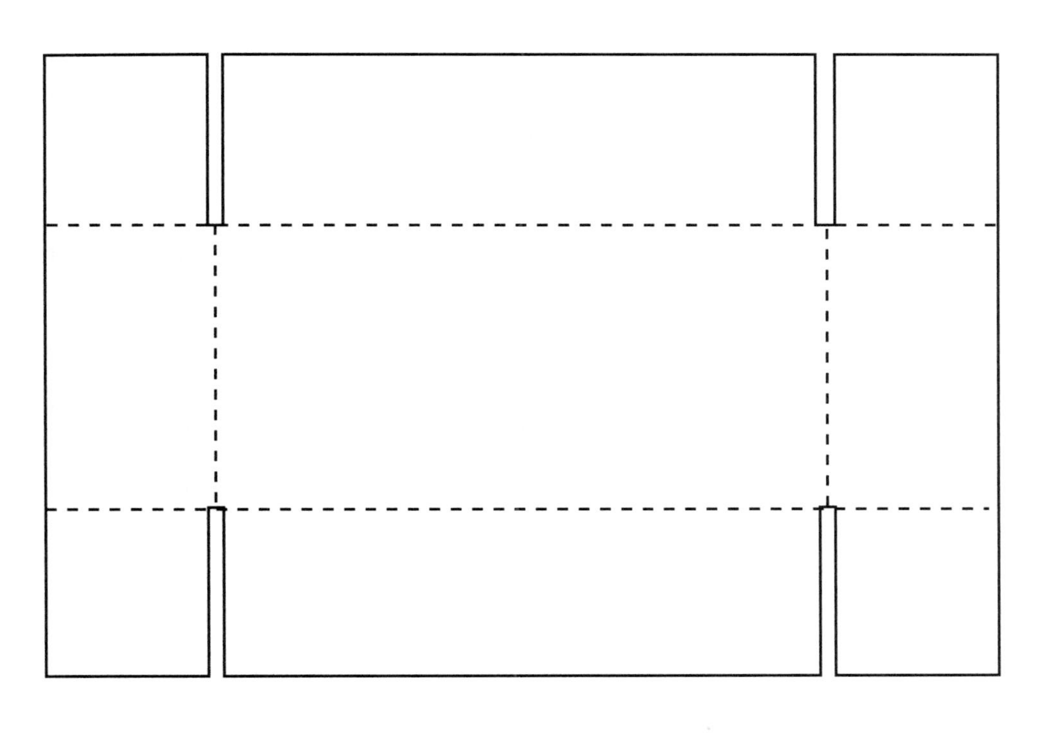

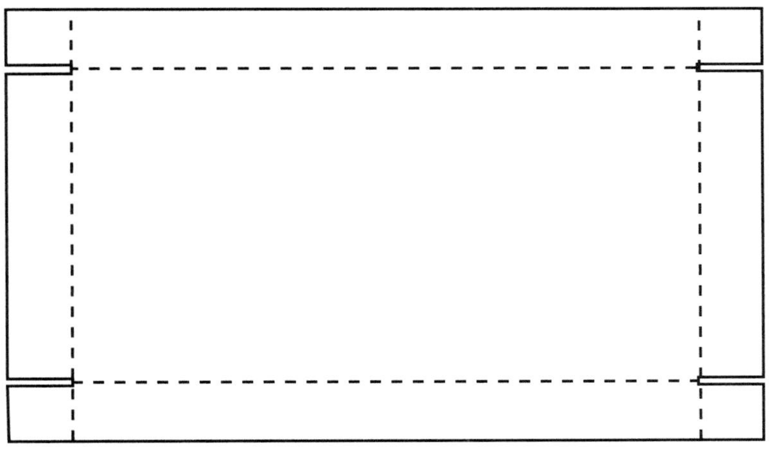

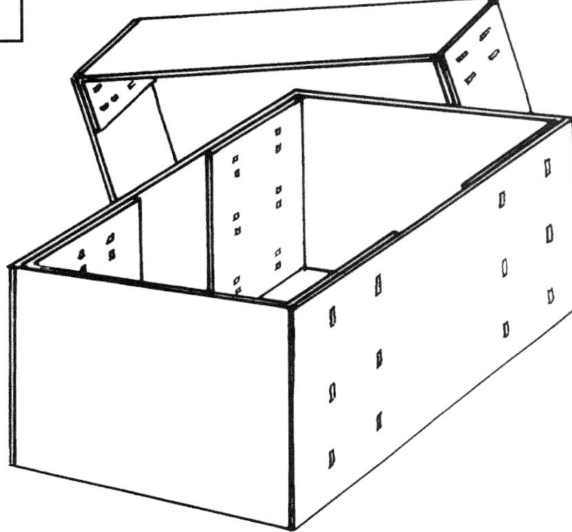

FULL TELESCOPE DESIGN STYLE BOX

(Intl. Box Code 0301)

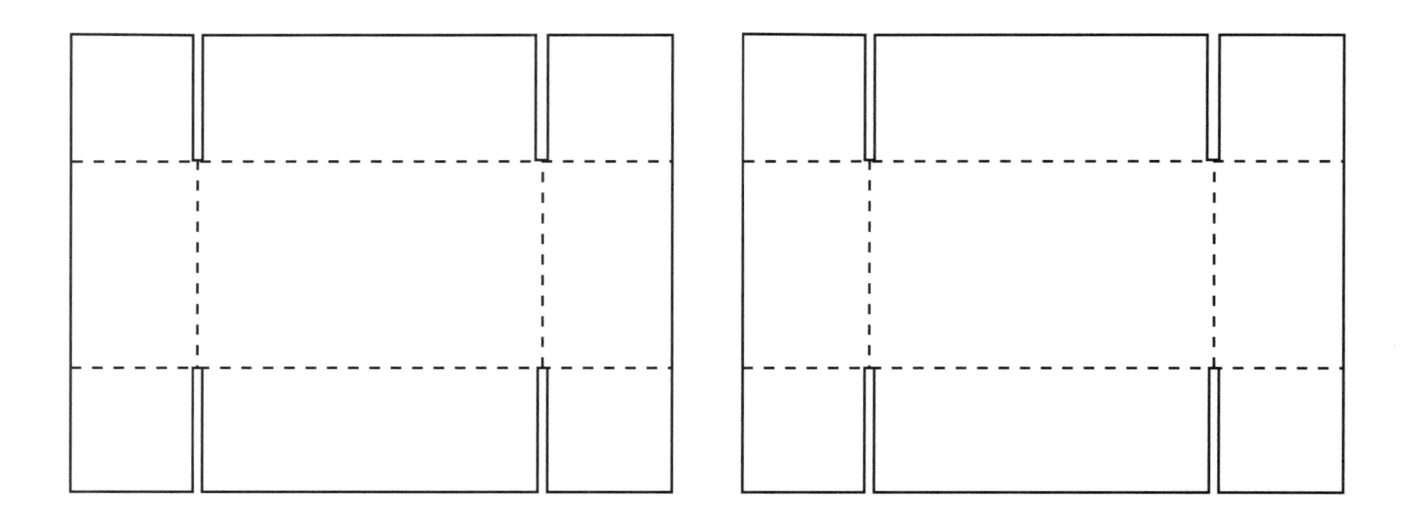

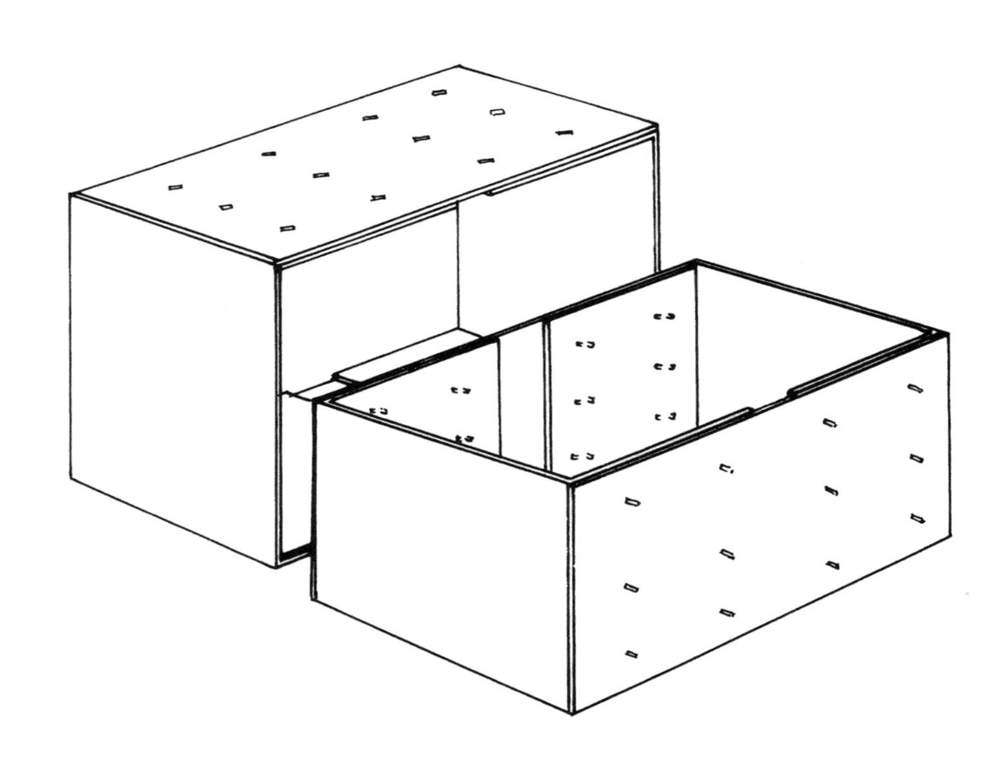

DOUBLE COVER BOX

(Intl. Box Code 0310)

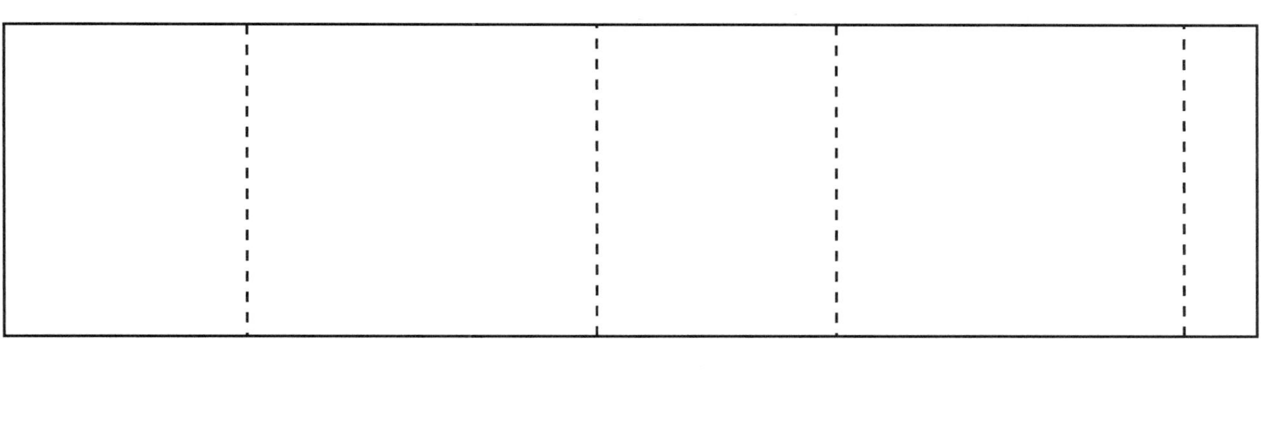

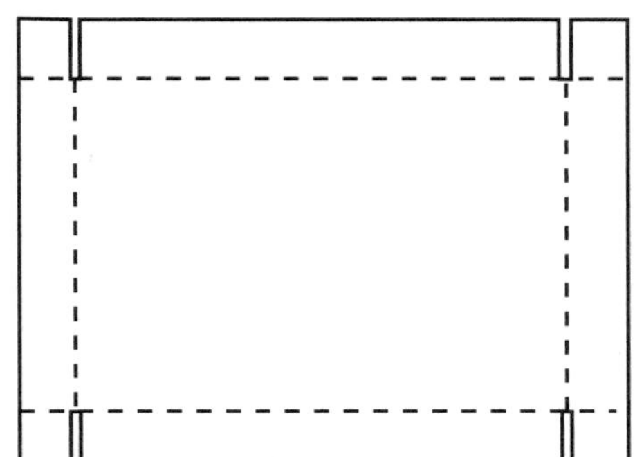

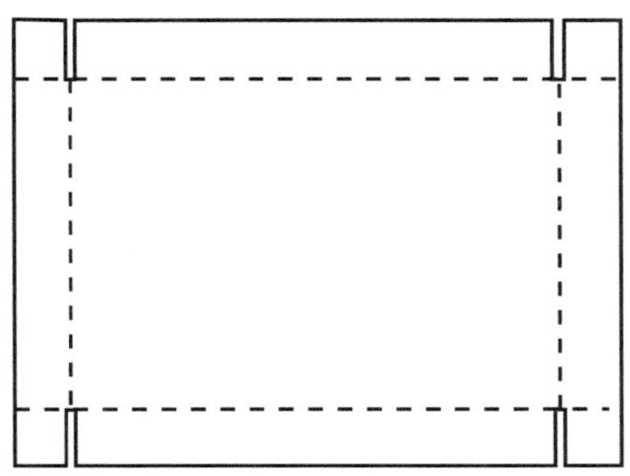

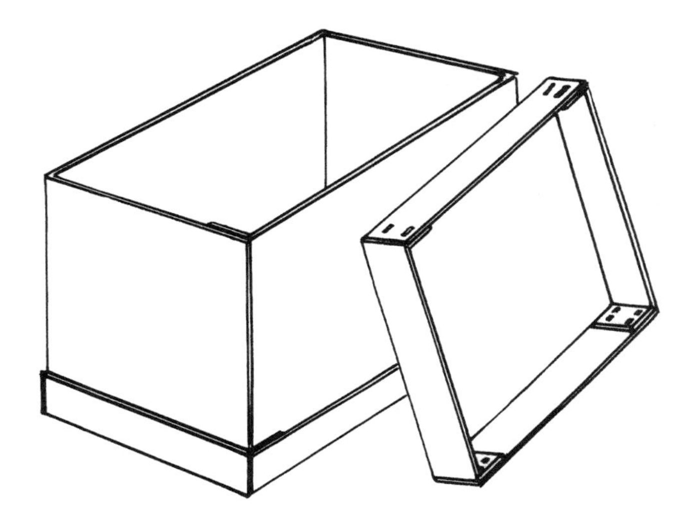

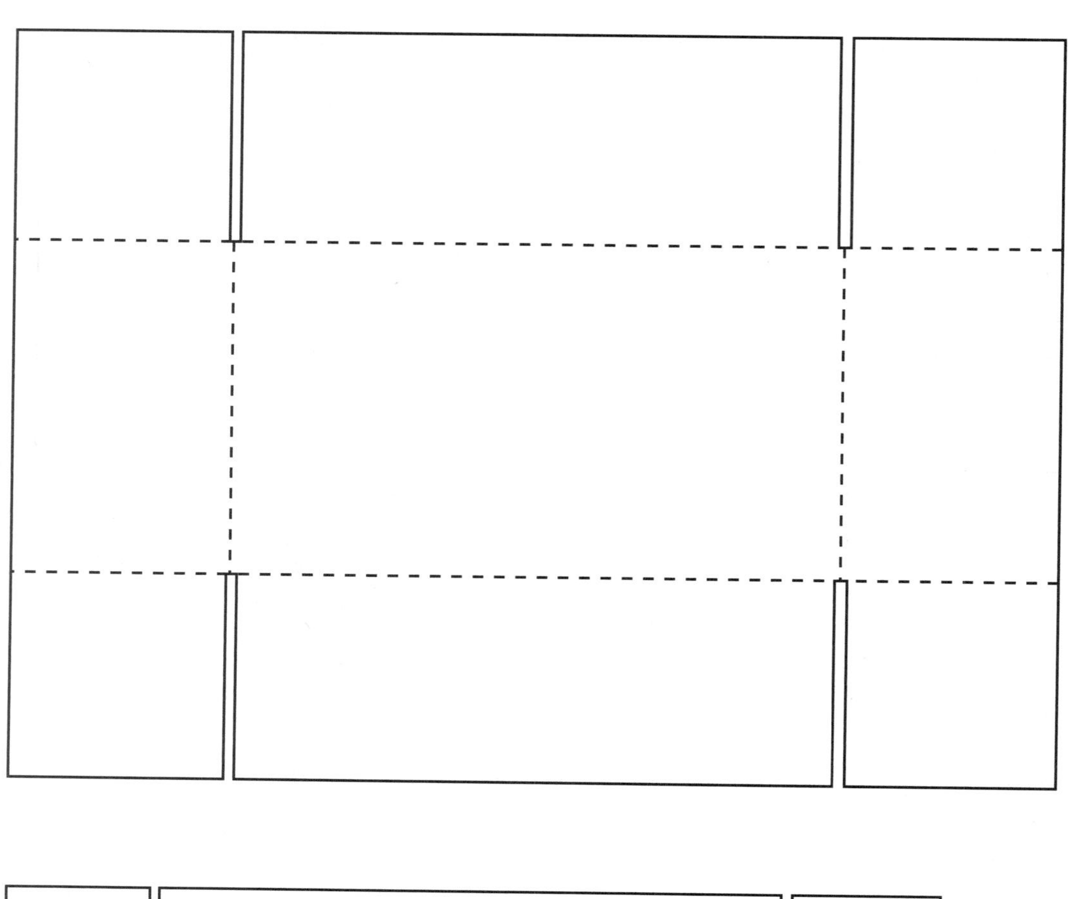

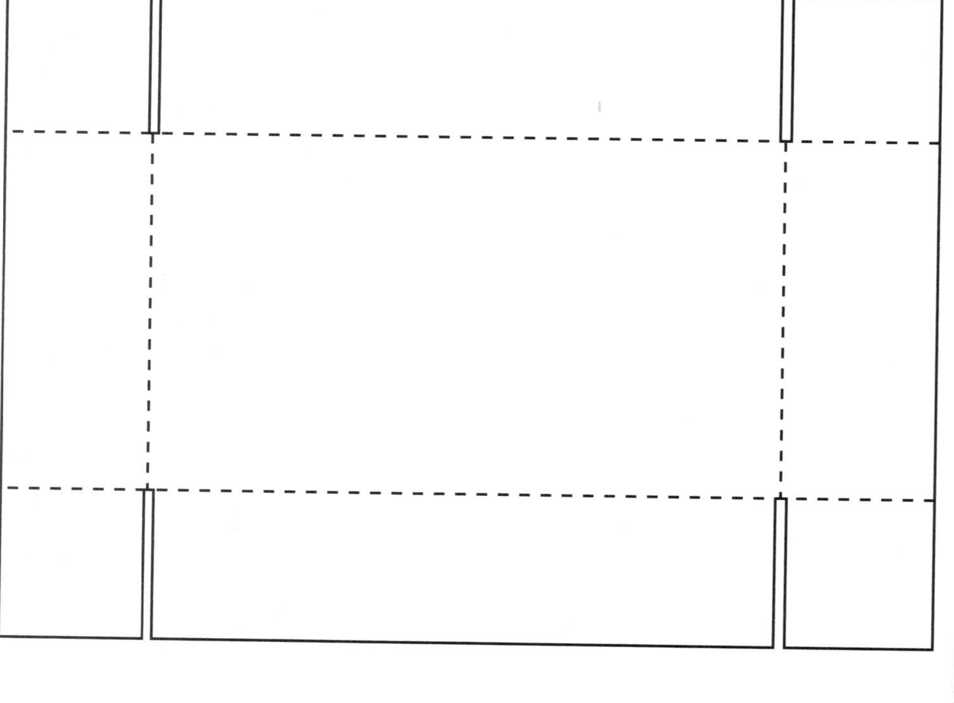

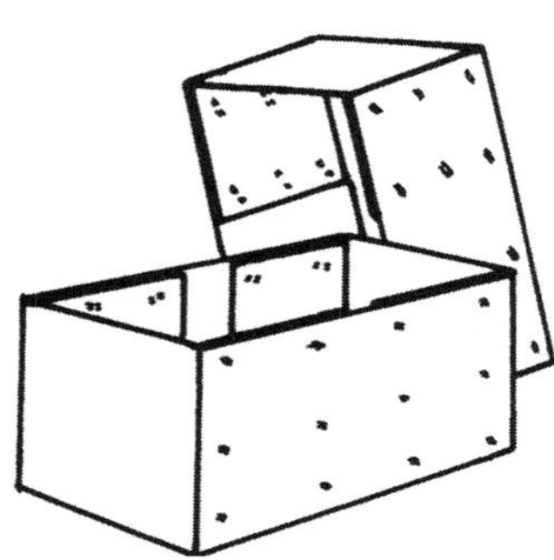

#2 BLISS BOX
(Intl. Box Code 0605)

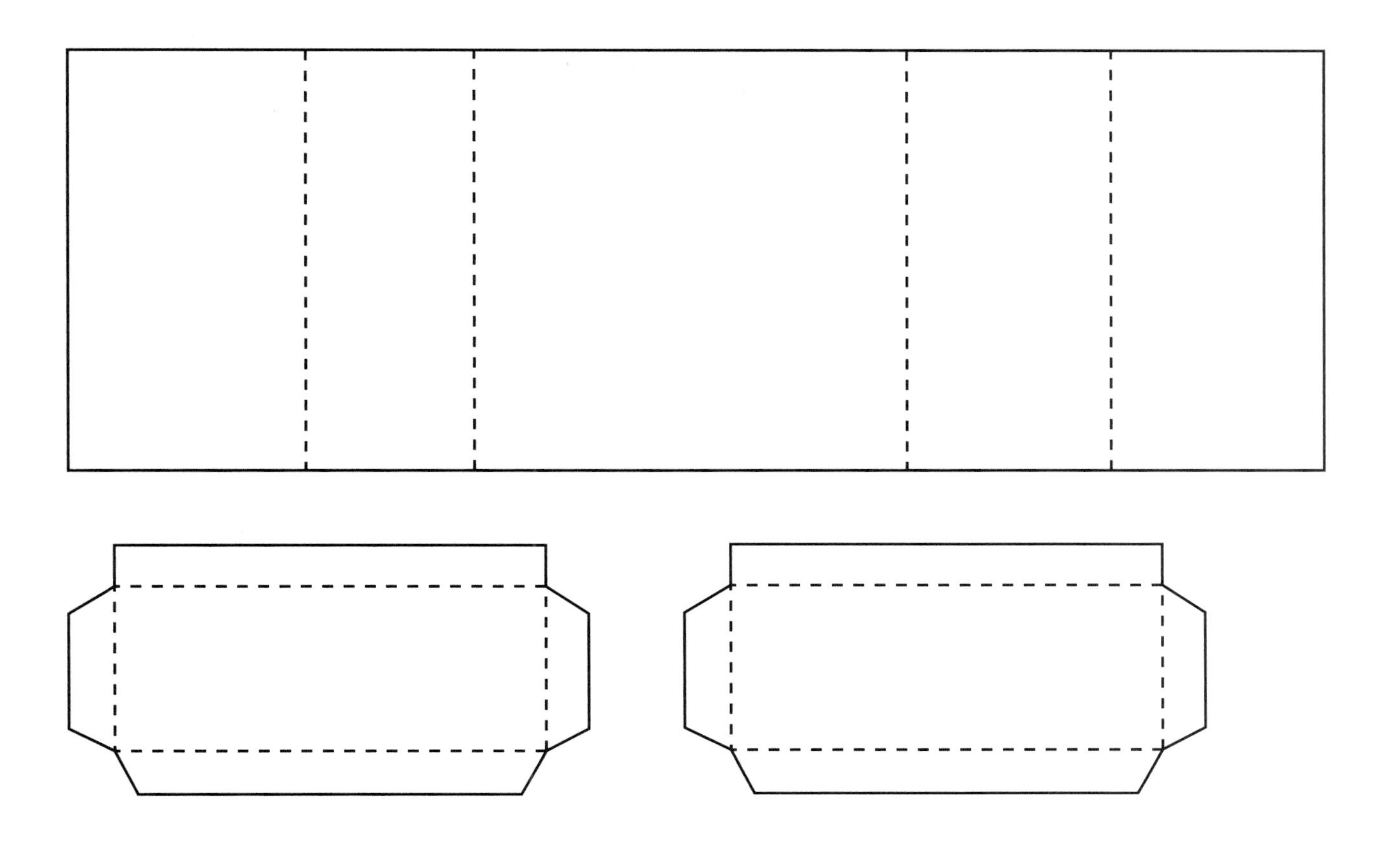

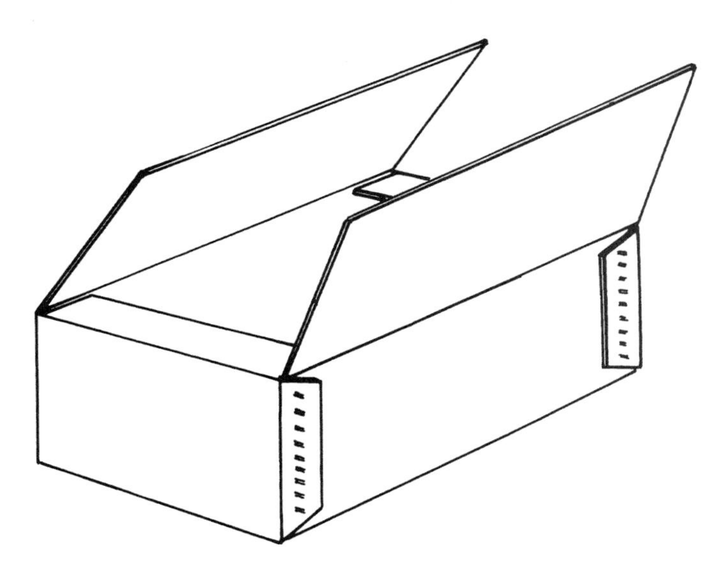

#4 BLISS BOX

(Intl. Box Code 0601)

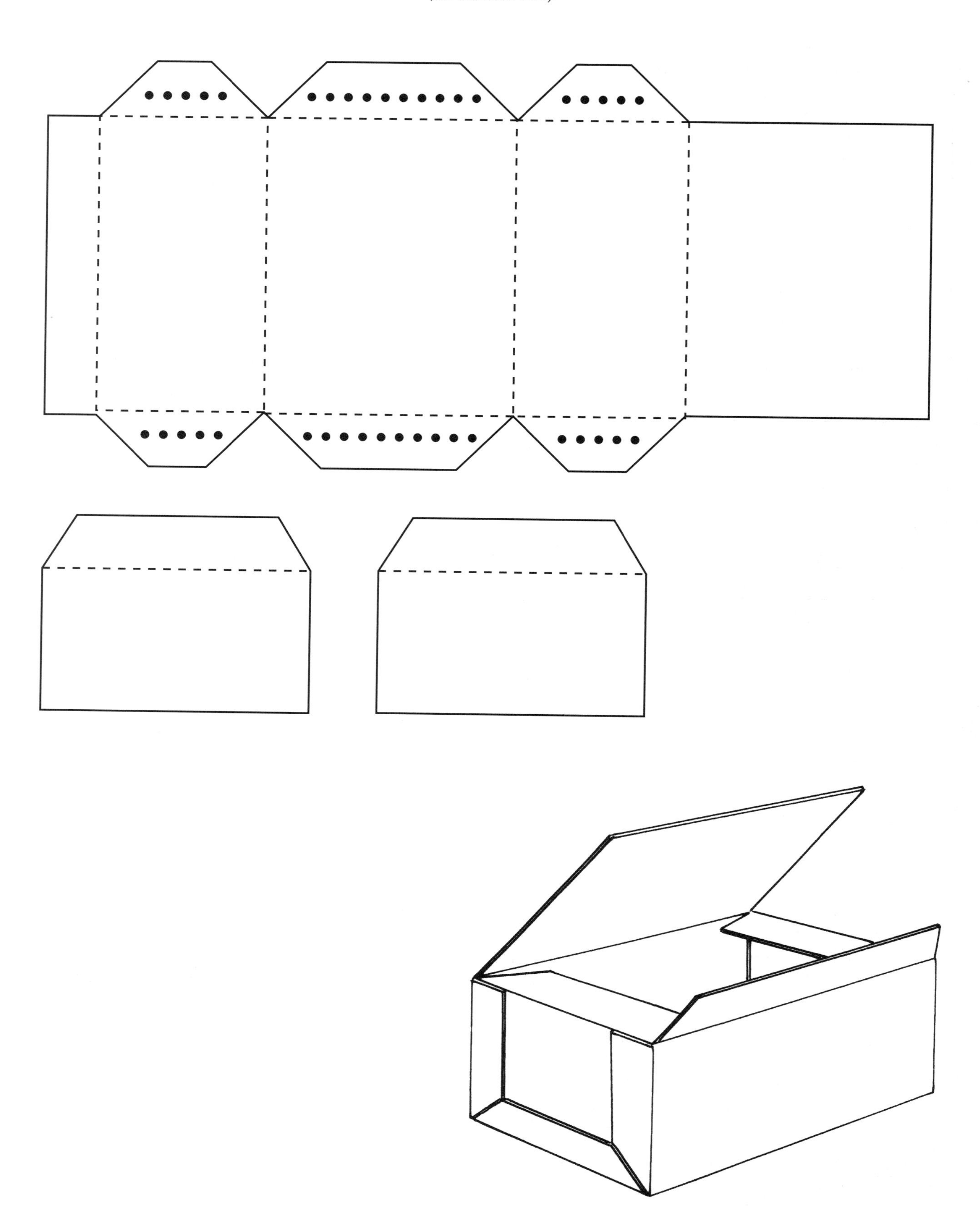

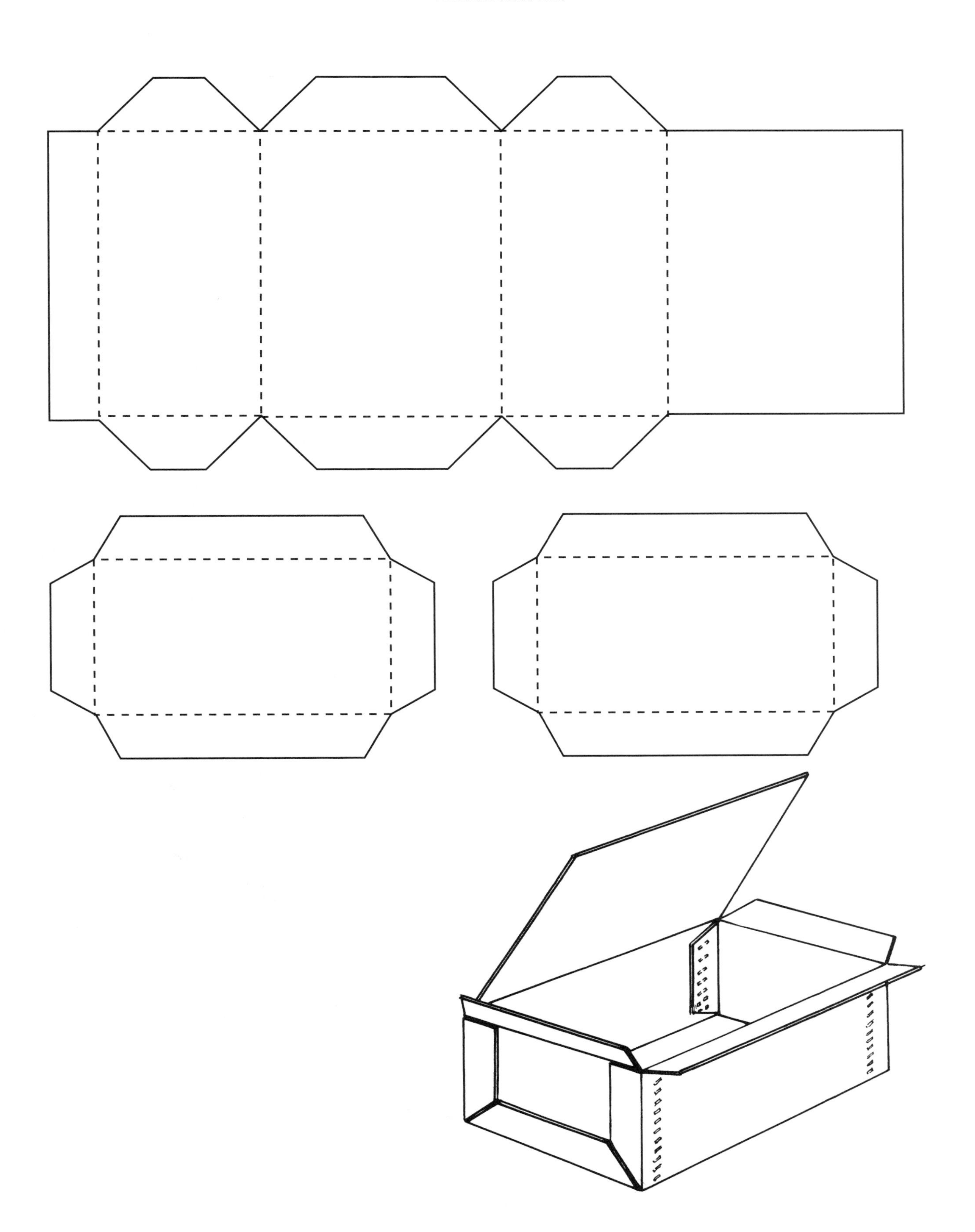

RECESSED END BOX
(Intl. Box Code 0610)

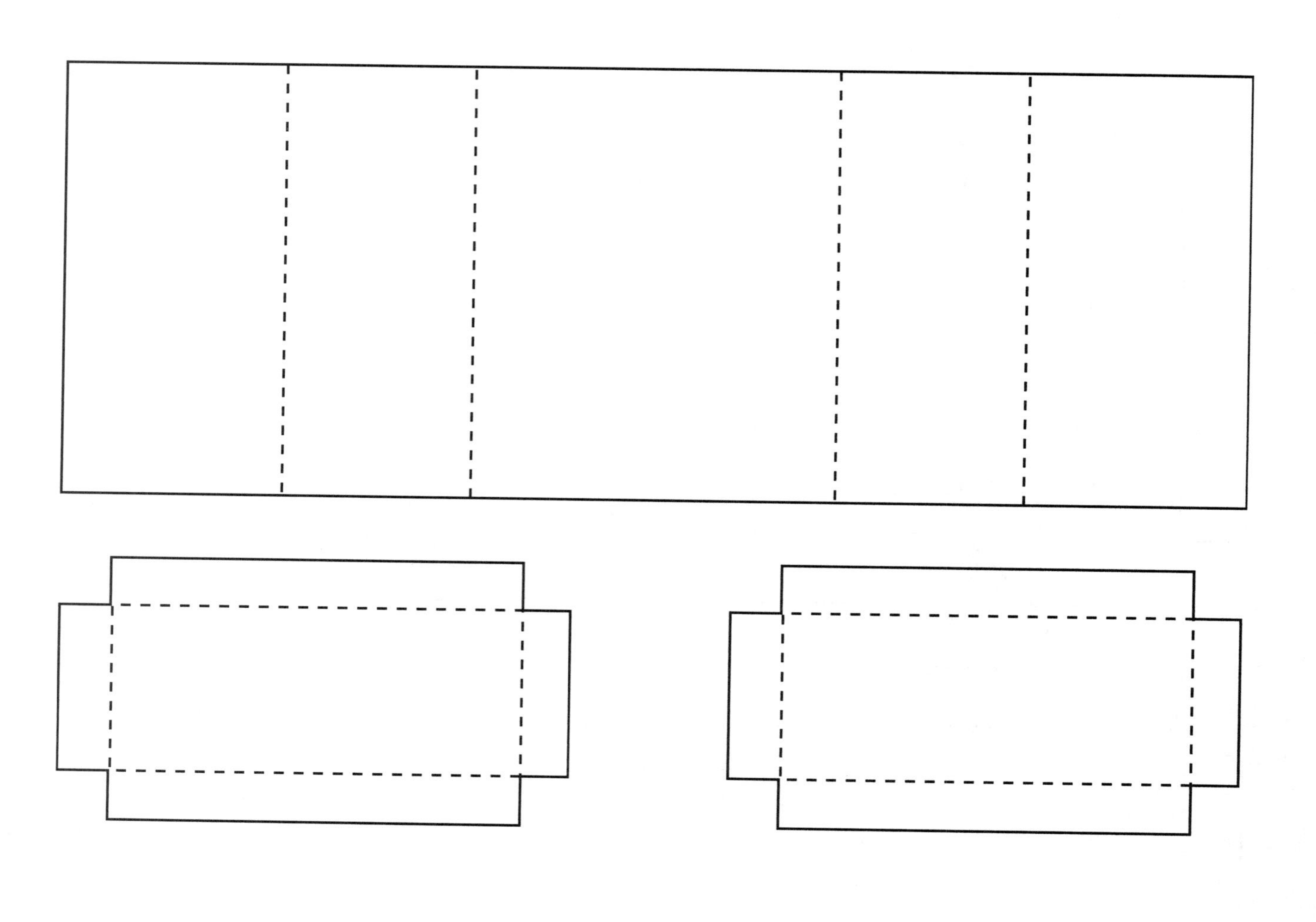

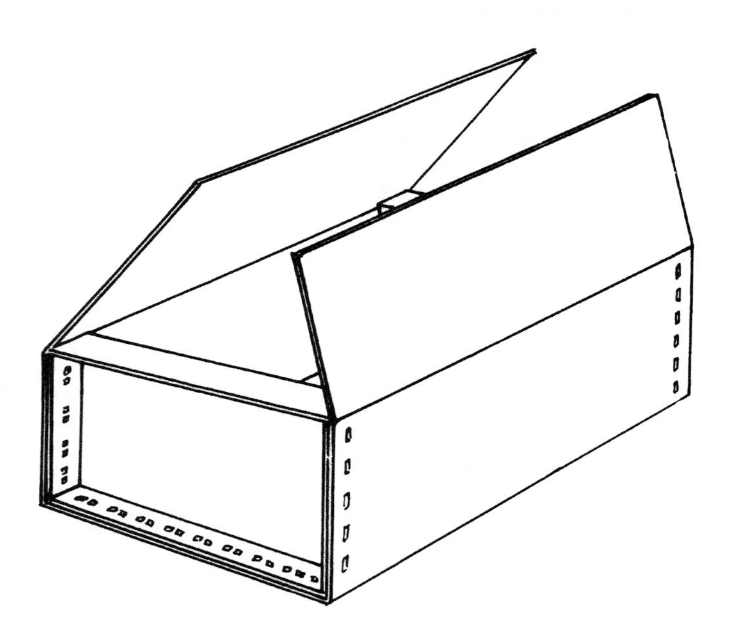

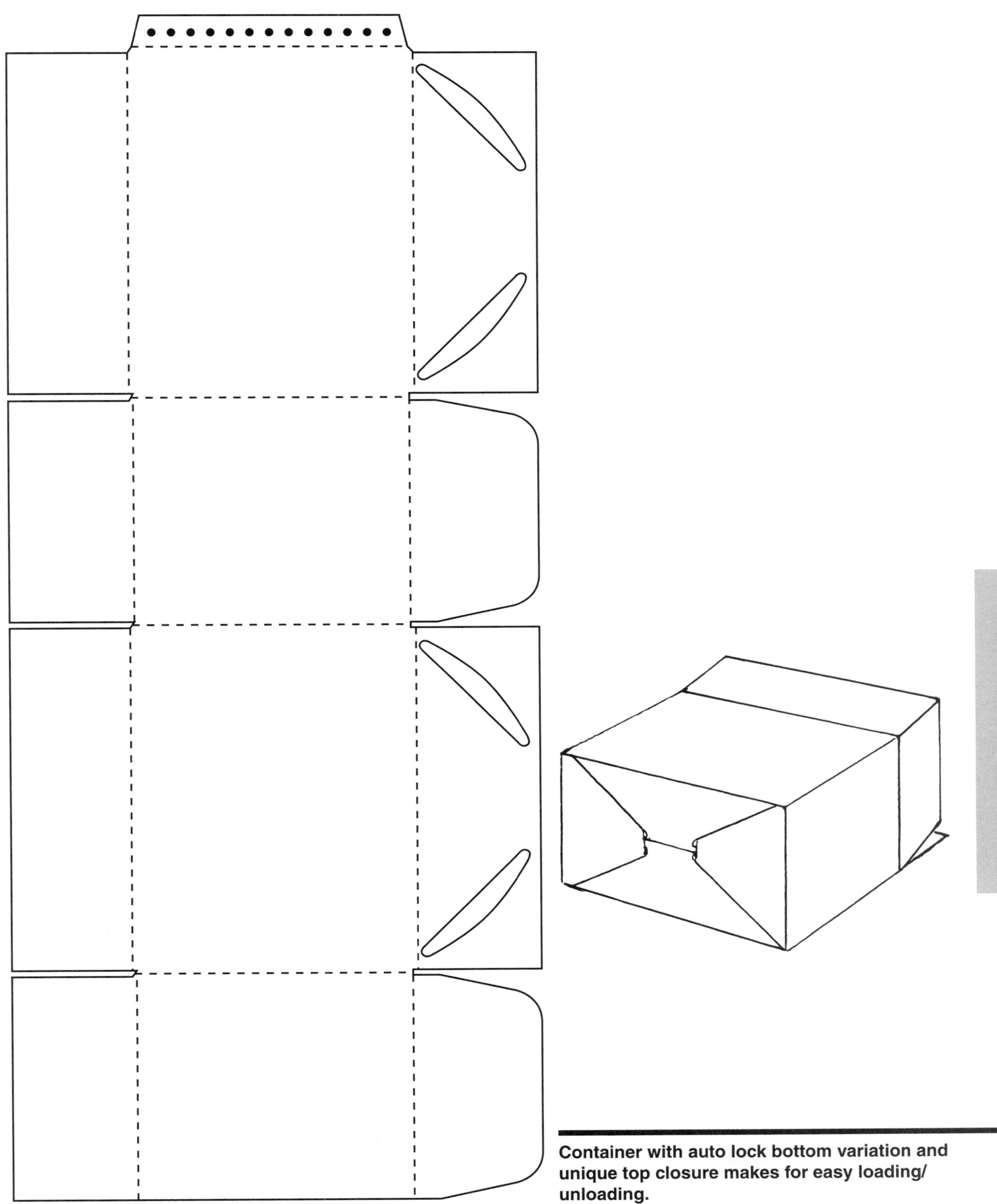

Container with auto lock bottom variation and unique top closure makes for easy loading/unloading.

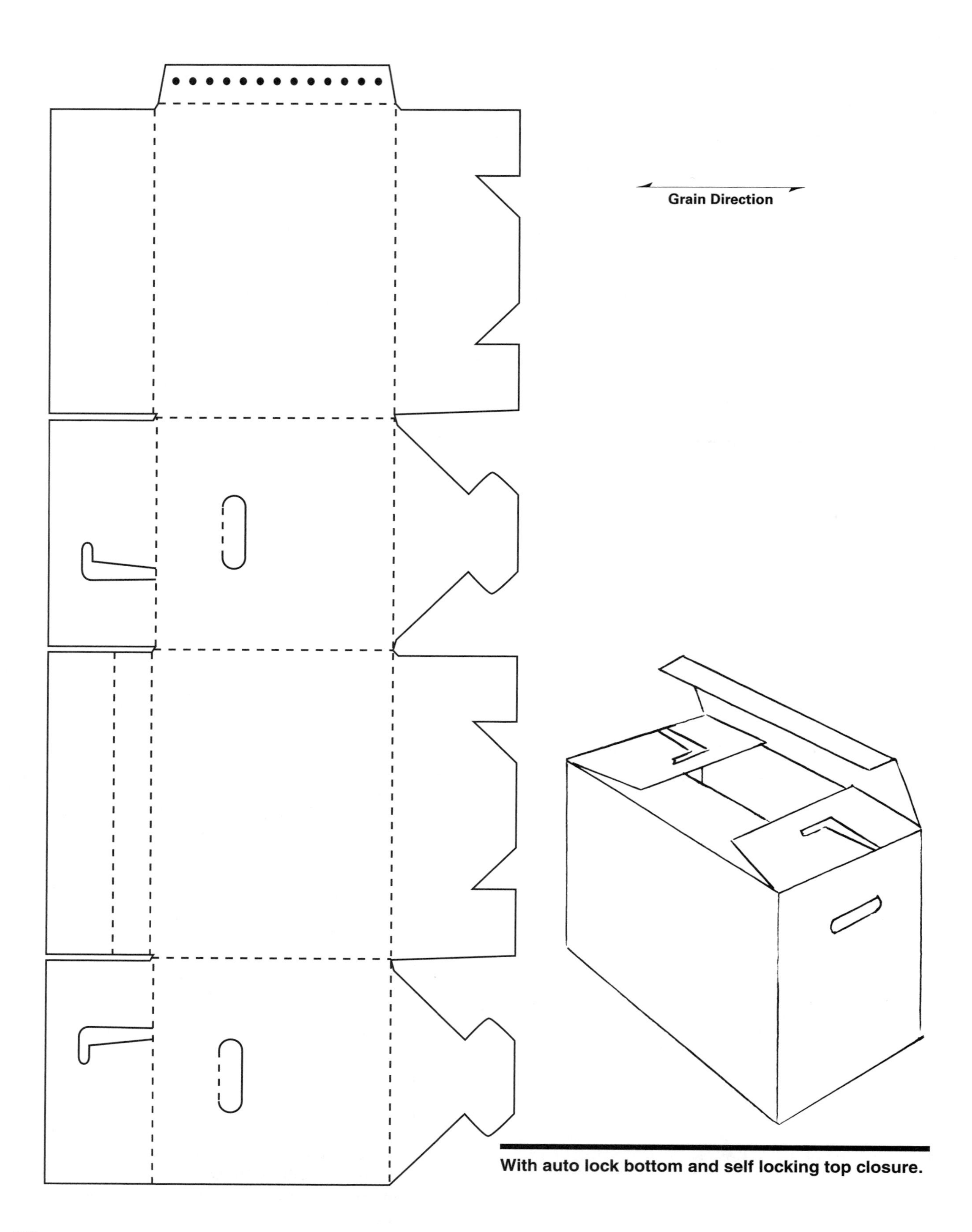

With auto lock bottom and self locking top closure.

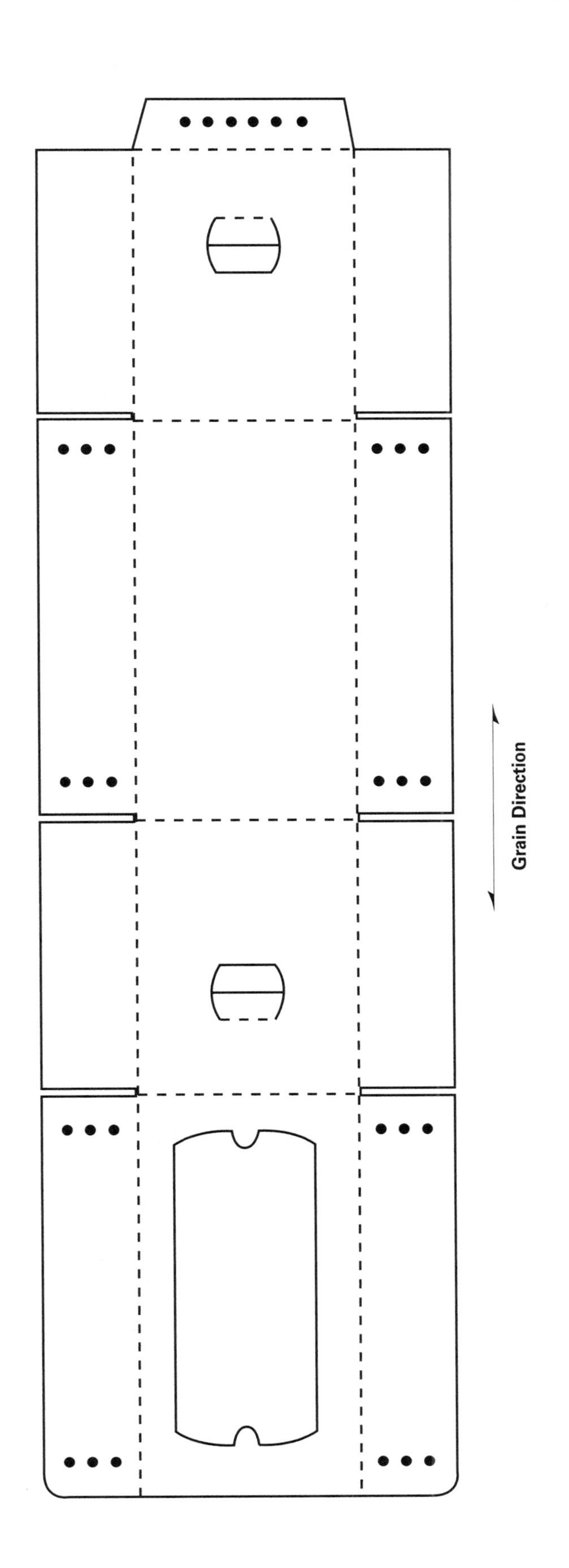

This eight-count two liter soft drink carrier uses a minimum of currugated board and serves as a display unit.

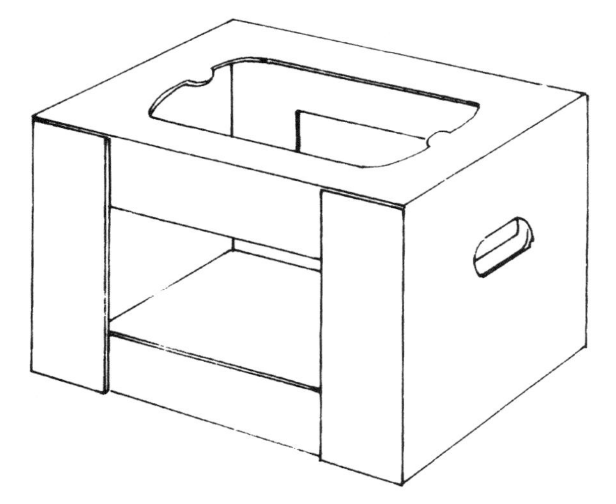

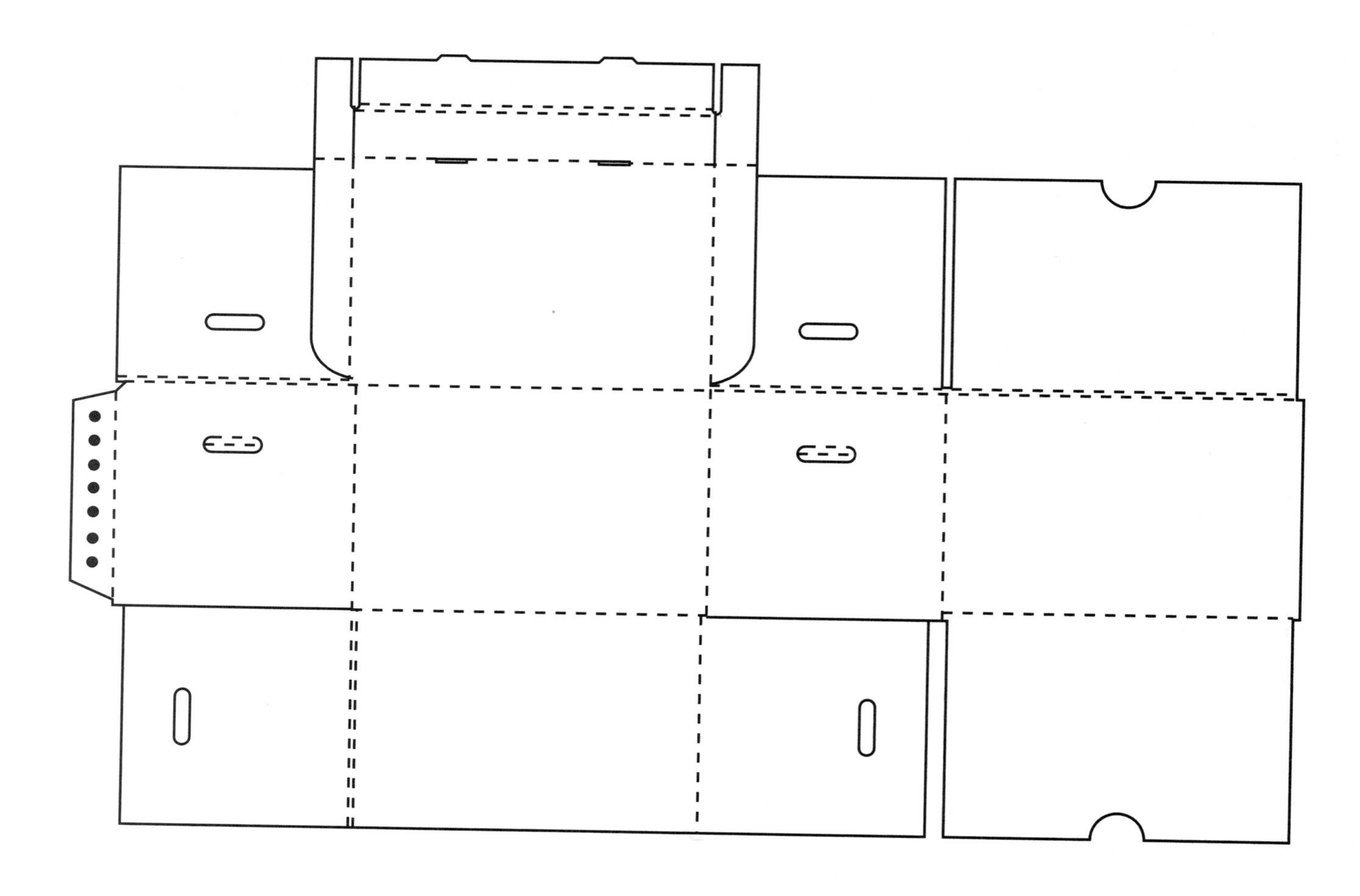

Double-over inner walls, hinged from the top, secure side panels in place. The two layers of the bottom and front wall as well as the triple sidewalls with carrying handles produce an excellent container for files.

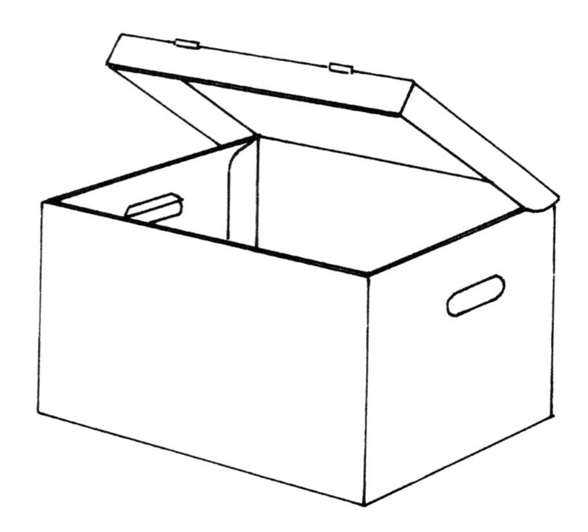

INTERLOCKING DOUBLE COVER BOX

(Intl. Box Code 0325)

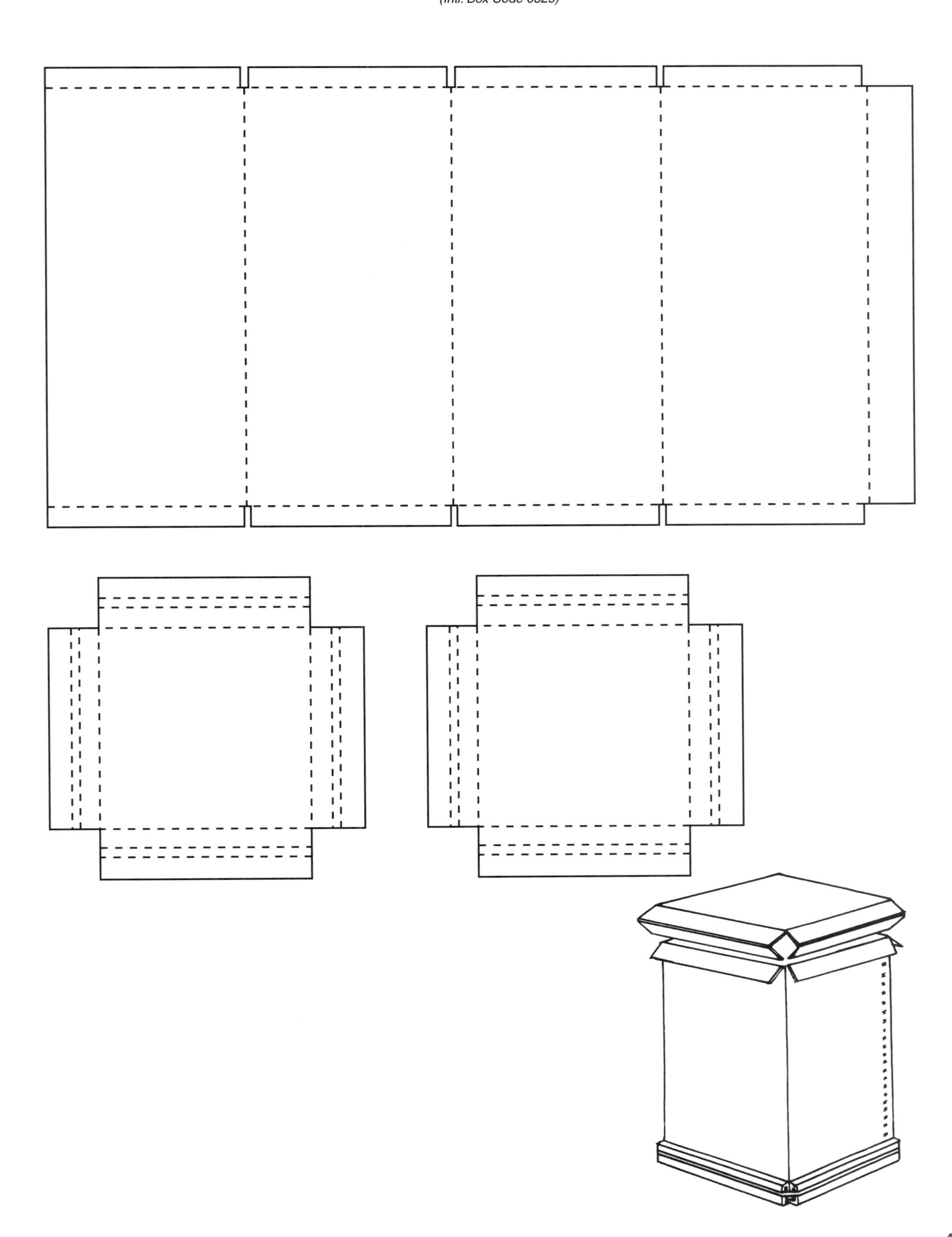

TUCK FOLDER

(Intl. Box Code 0420)

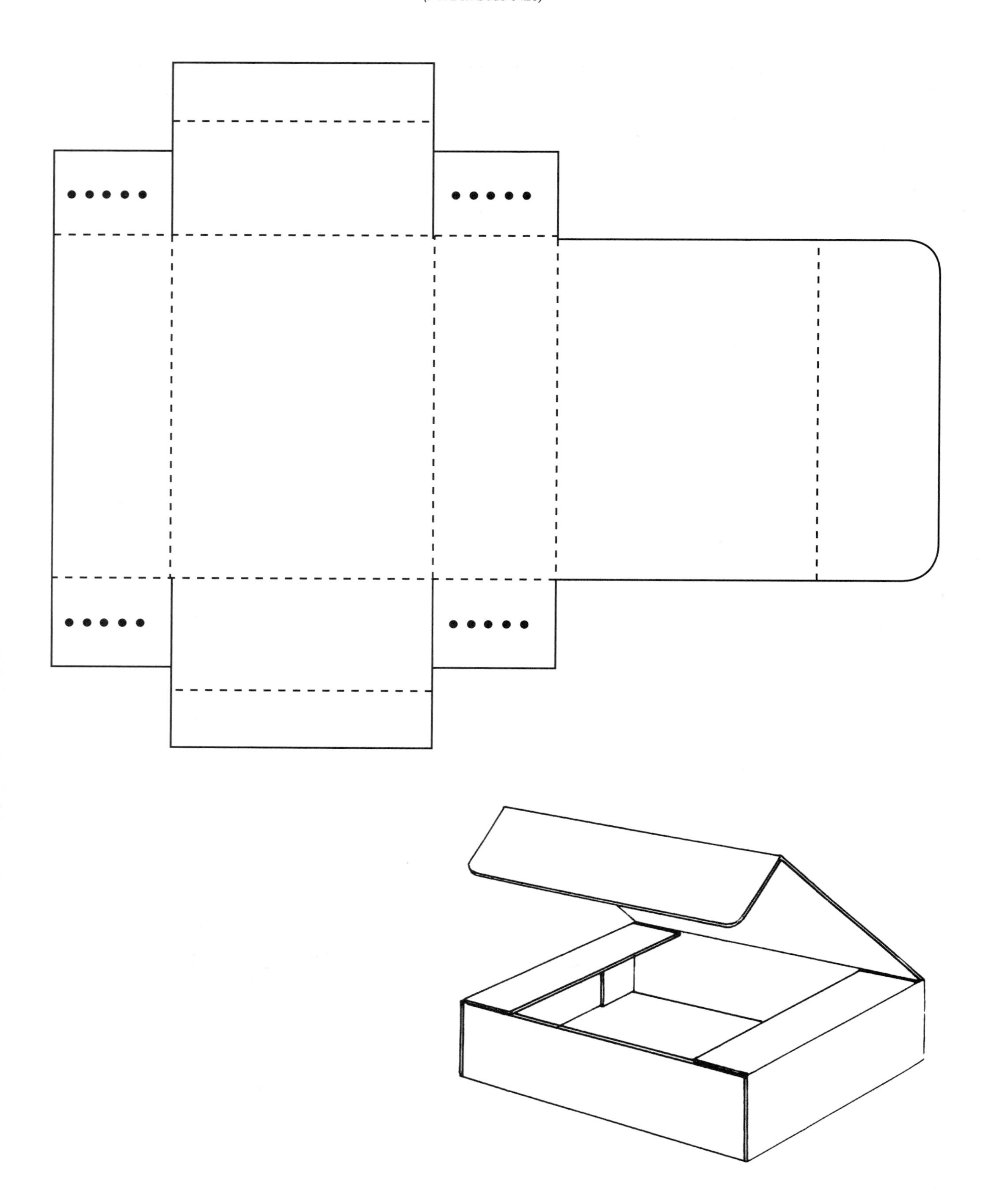

ONE PIECE TELESCOPE BOX

(Intl. Box Code 0409)

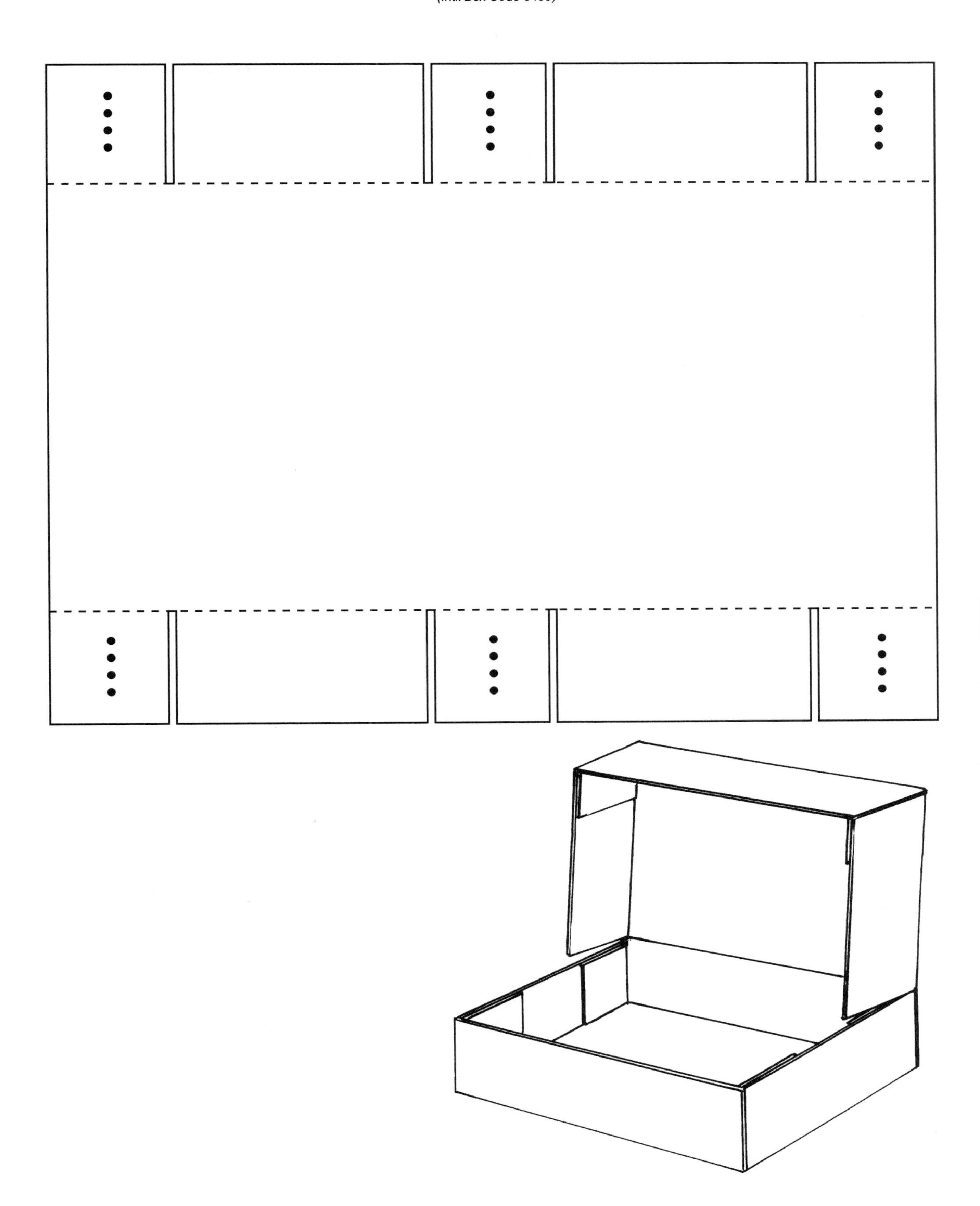

DOUBLE END WALL SELF-LOCKING TRAY

(Intl. Box Code 0422)

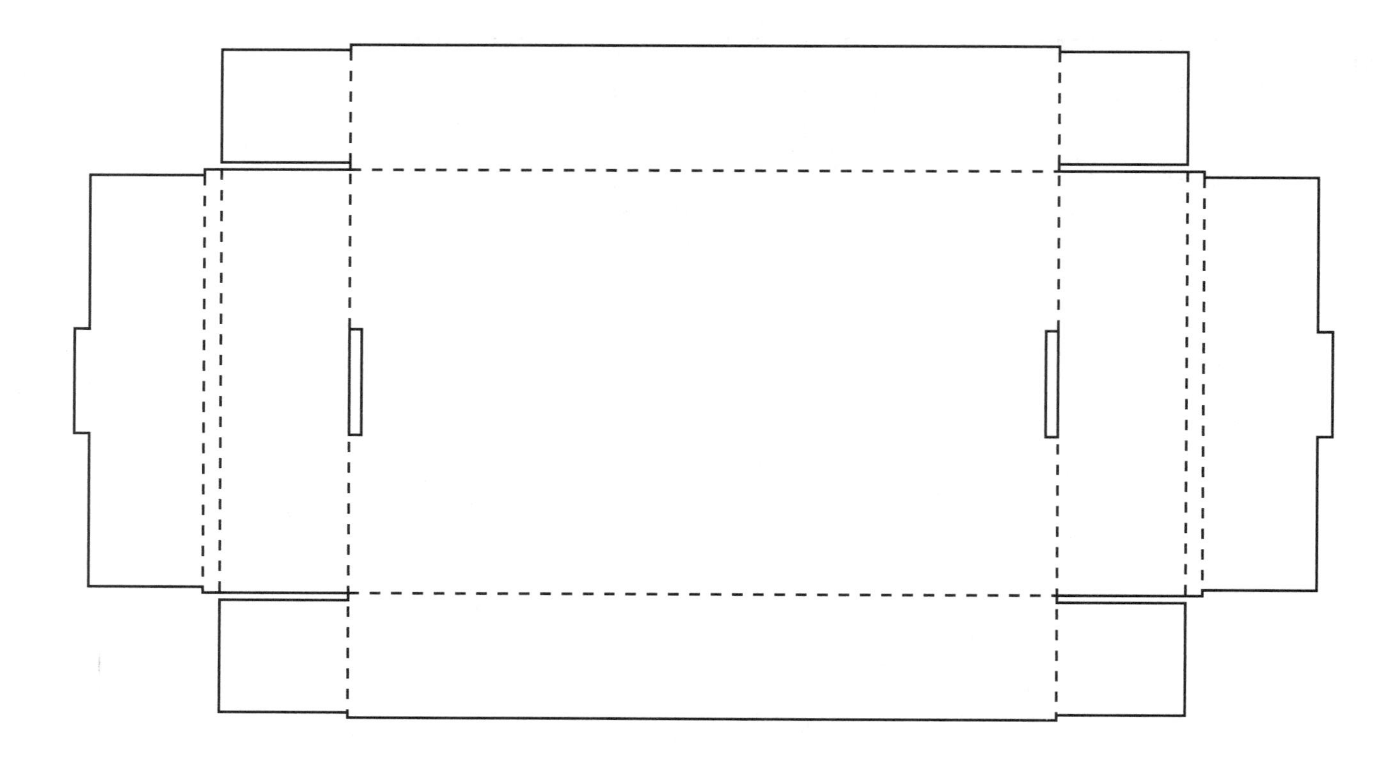

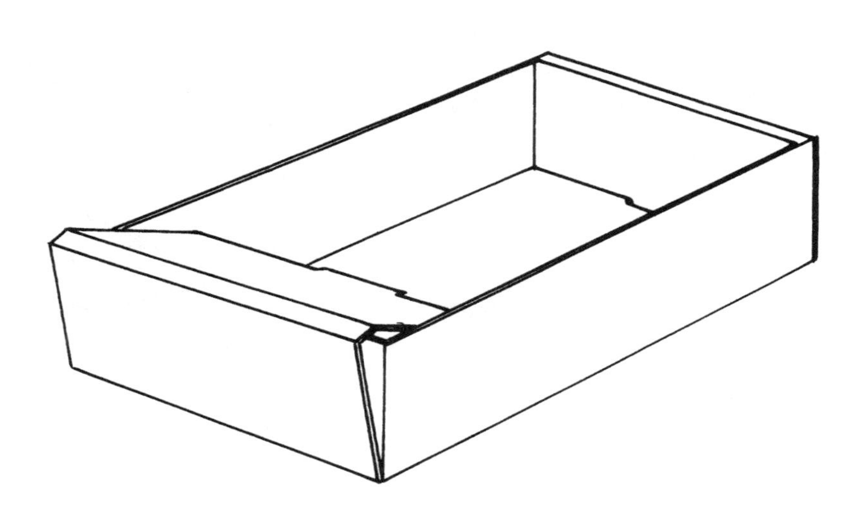

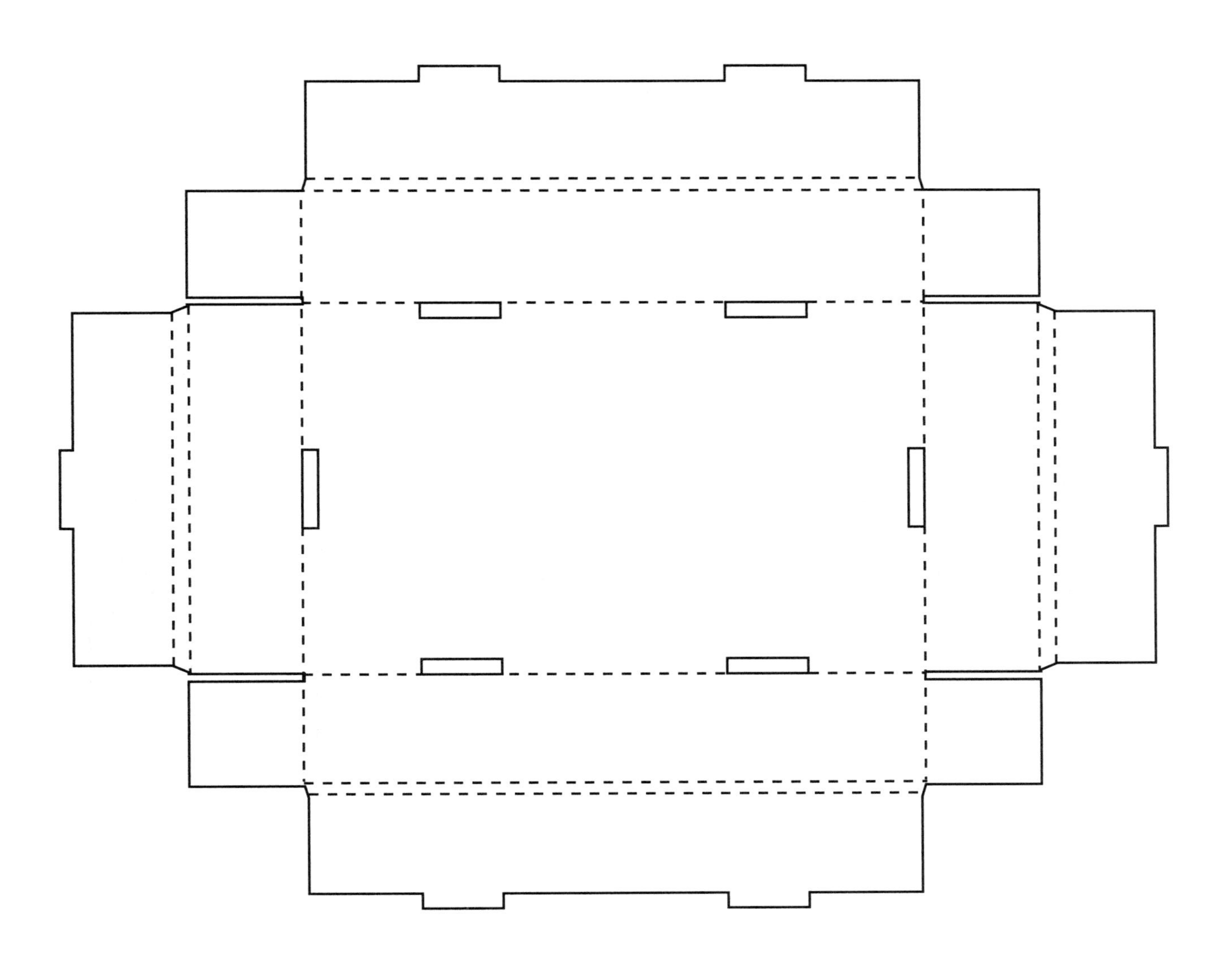

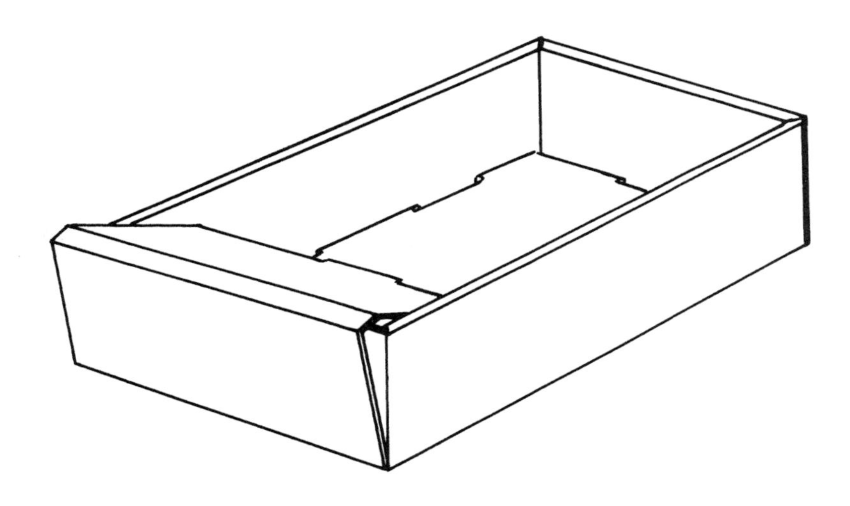

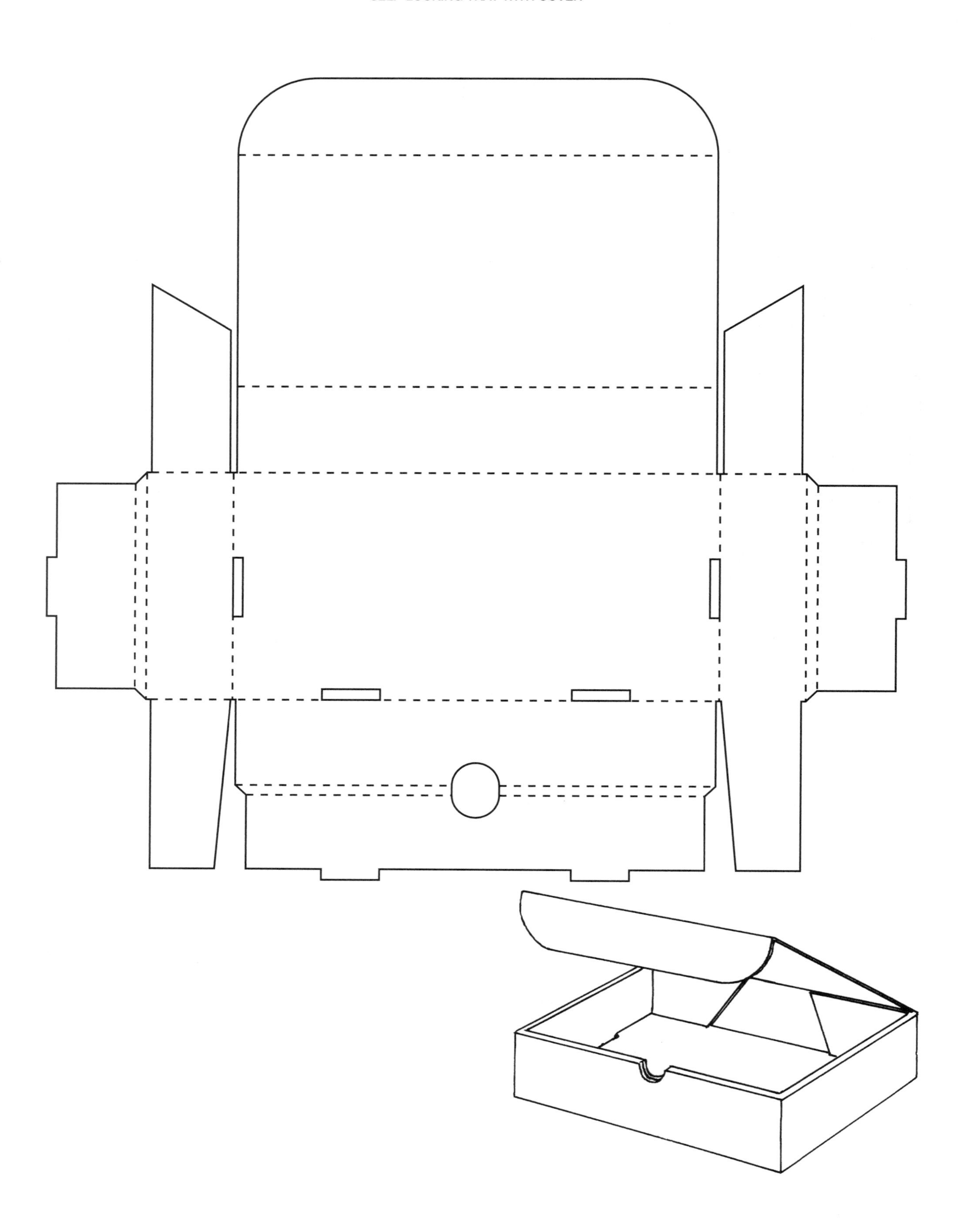

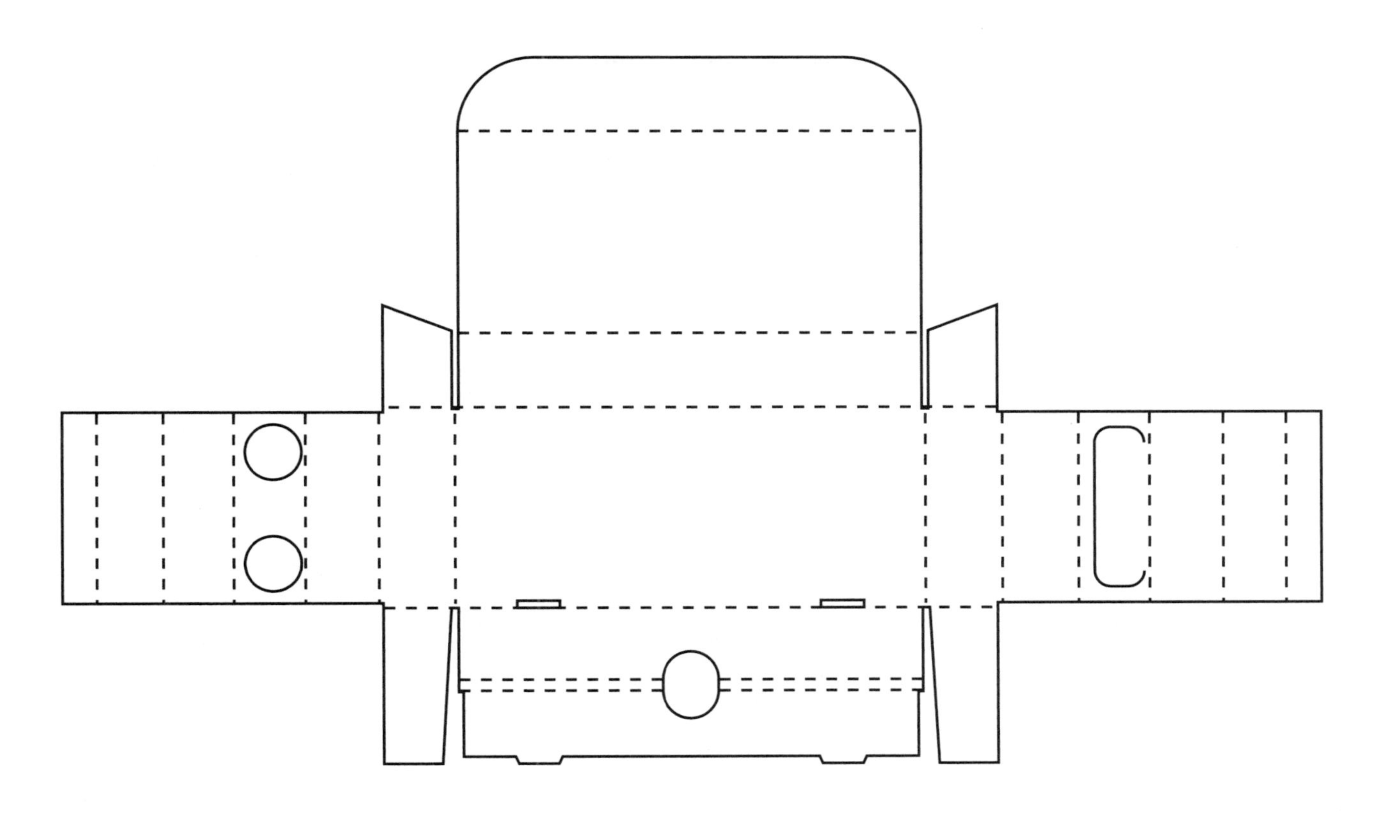

This sample has supports for a U-shaped lightbulb.

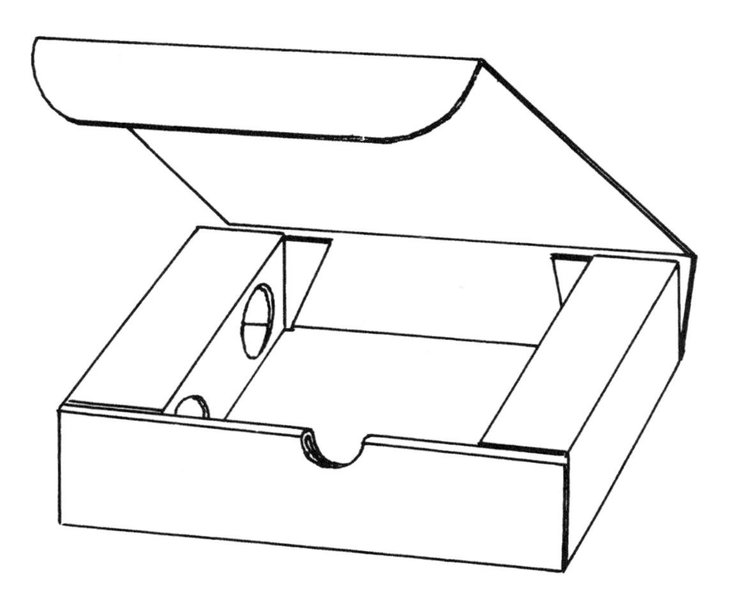

SELF-LOCKING TRAY WITH DOUBLE SIDE WALLS AND TEAR-AWAY FRONT PANEL

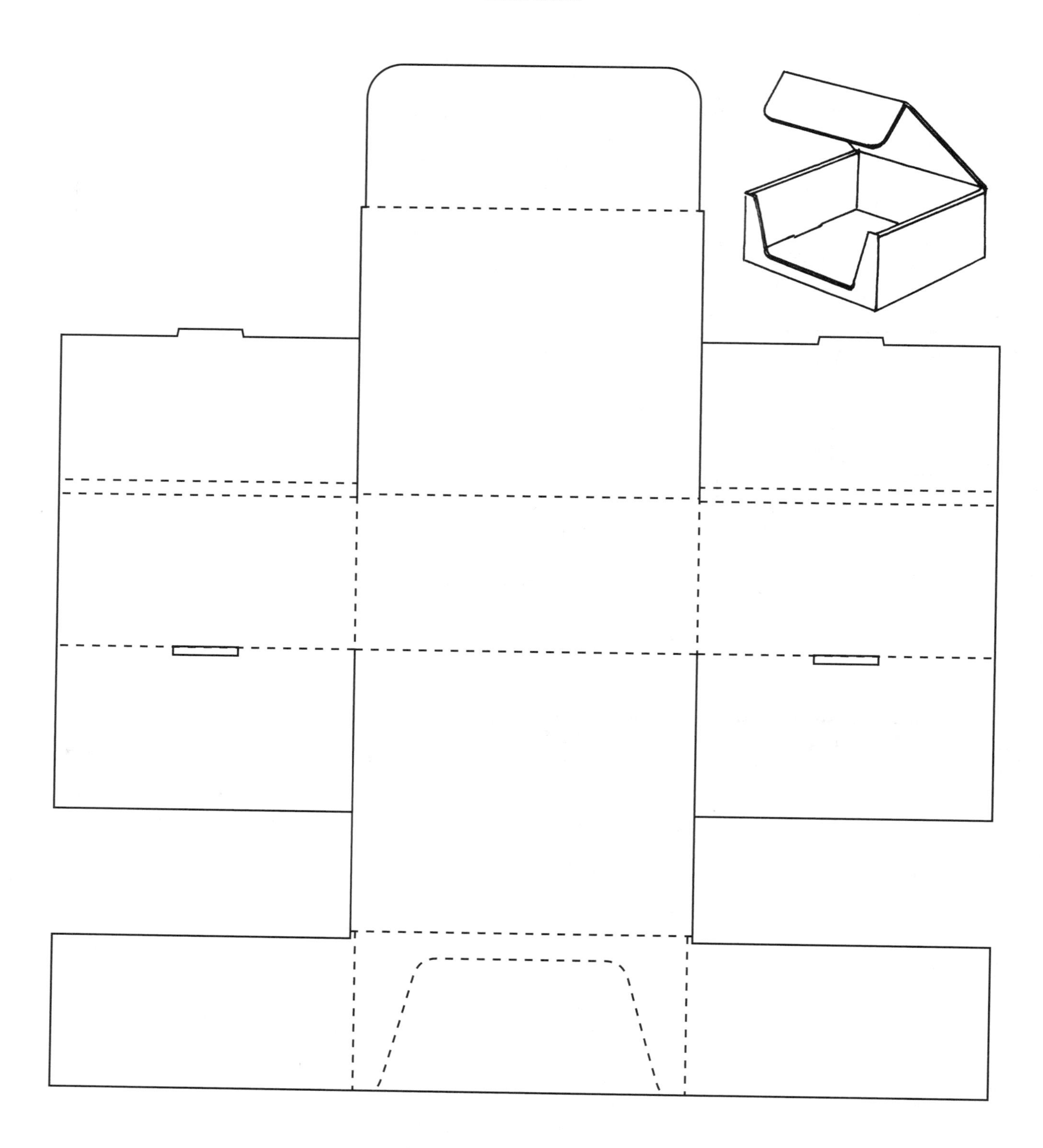

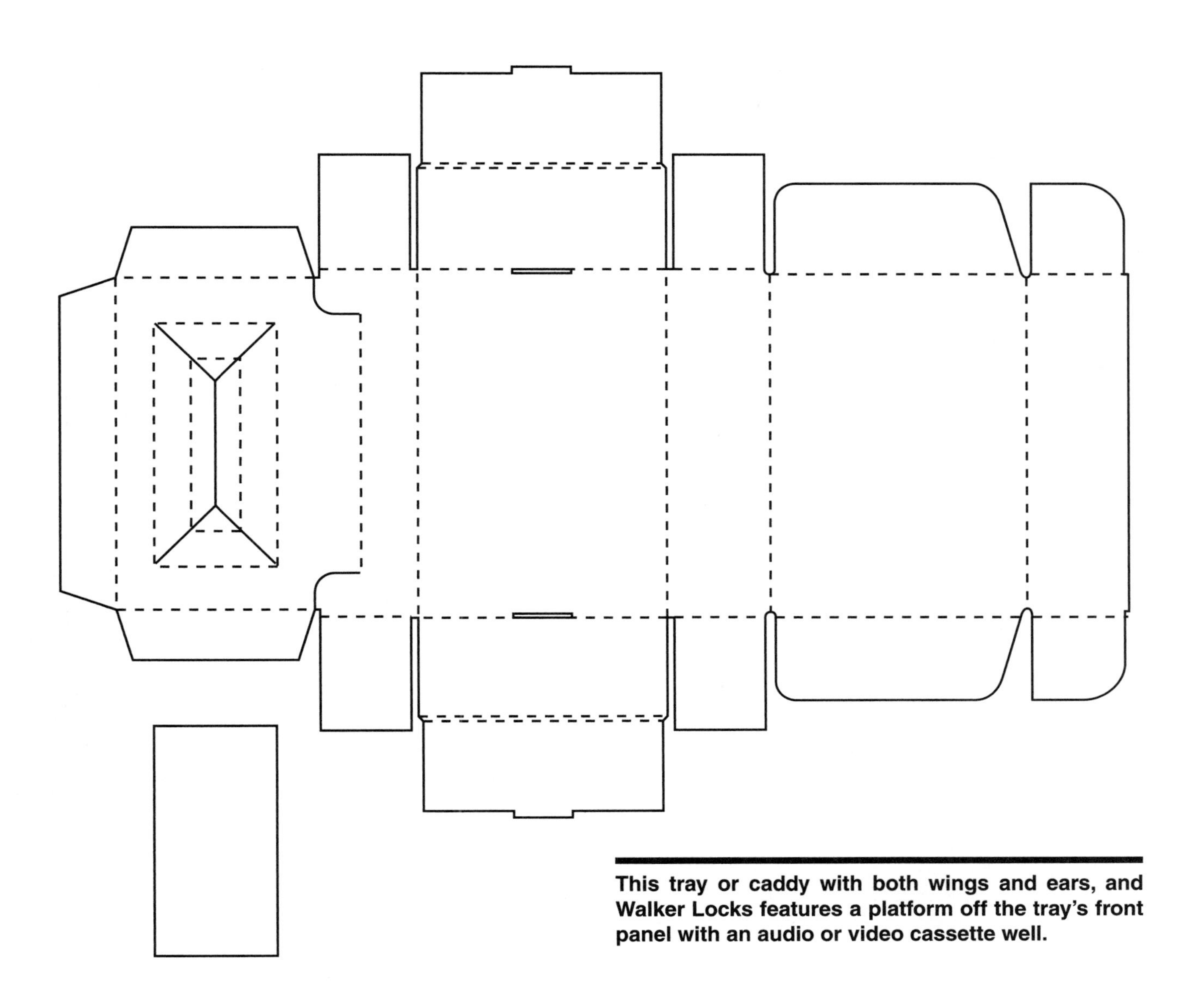

This tray or caddy with both wings and ears, and Walker Locks features a platform off the tray's front panel with an audio or video cassette well.

A pad cut from the waste of the sheet can provide a floor for the cassette well.

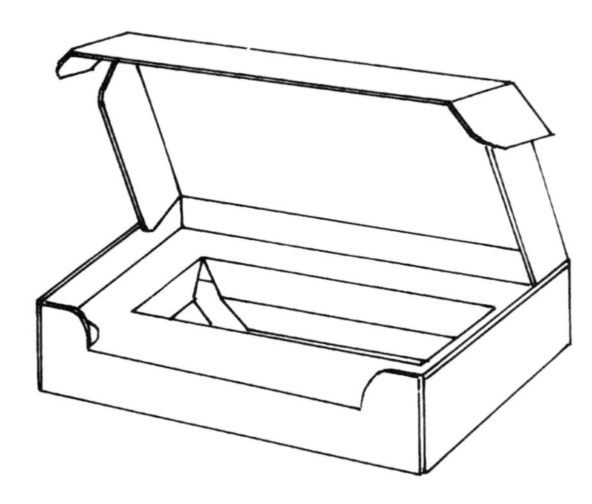

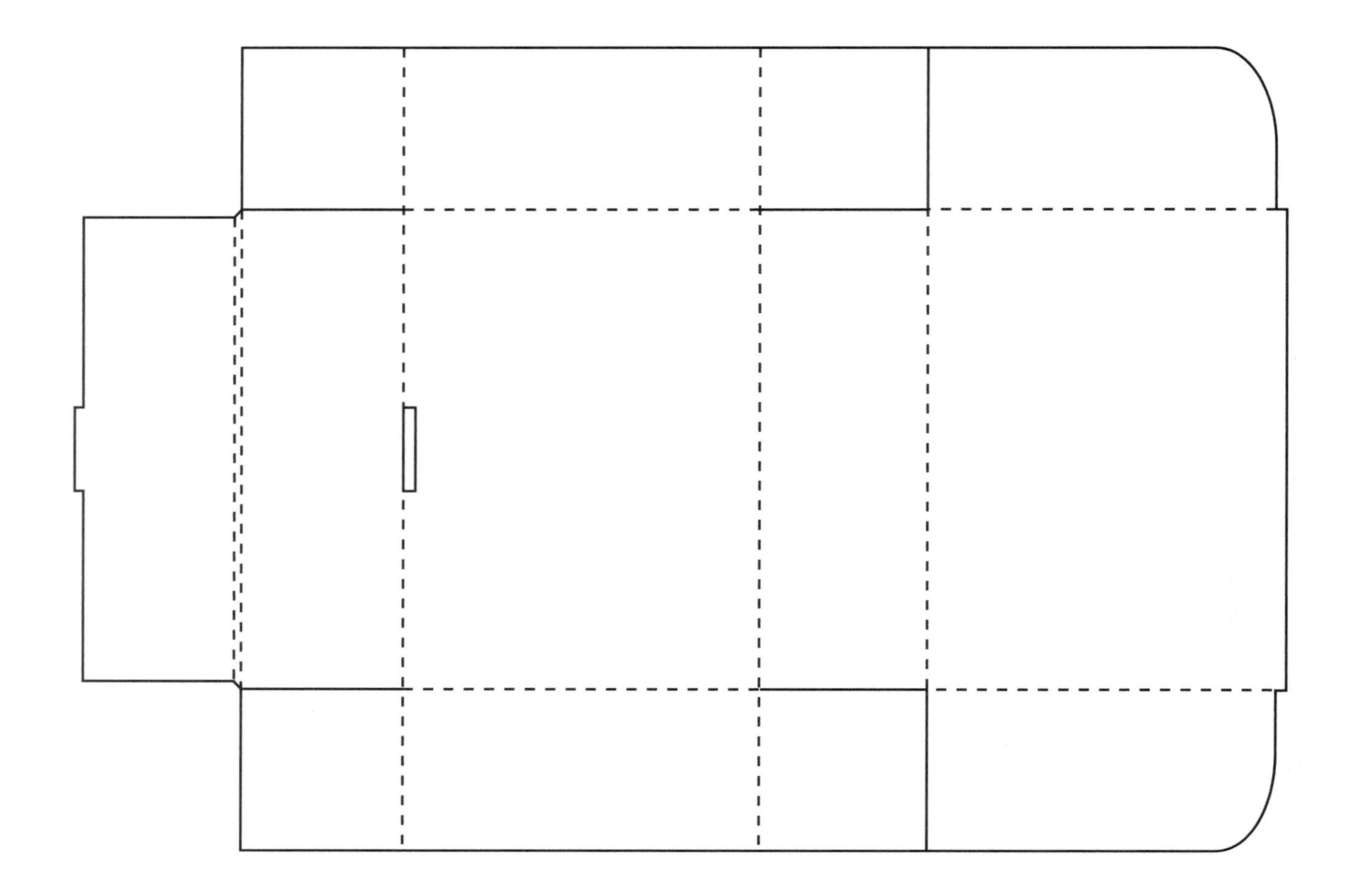

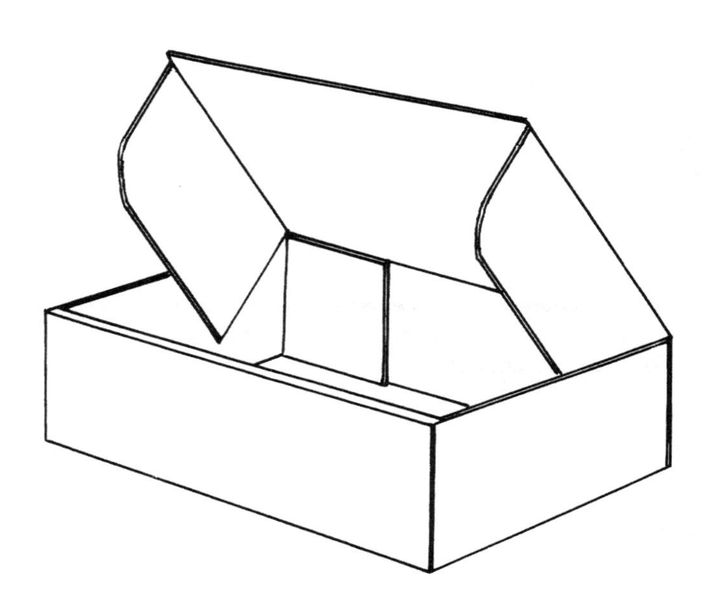

SELF-LOCKING TRAY WITH DOUBLE FRONT AND SIDE WALLS AND LOCKING EARS

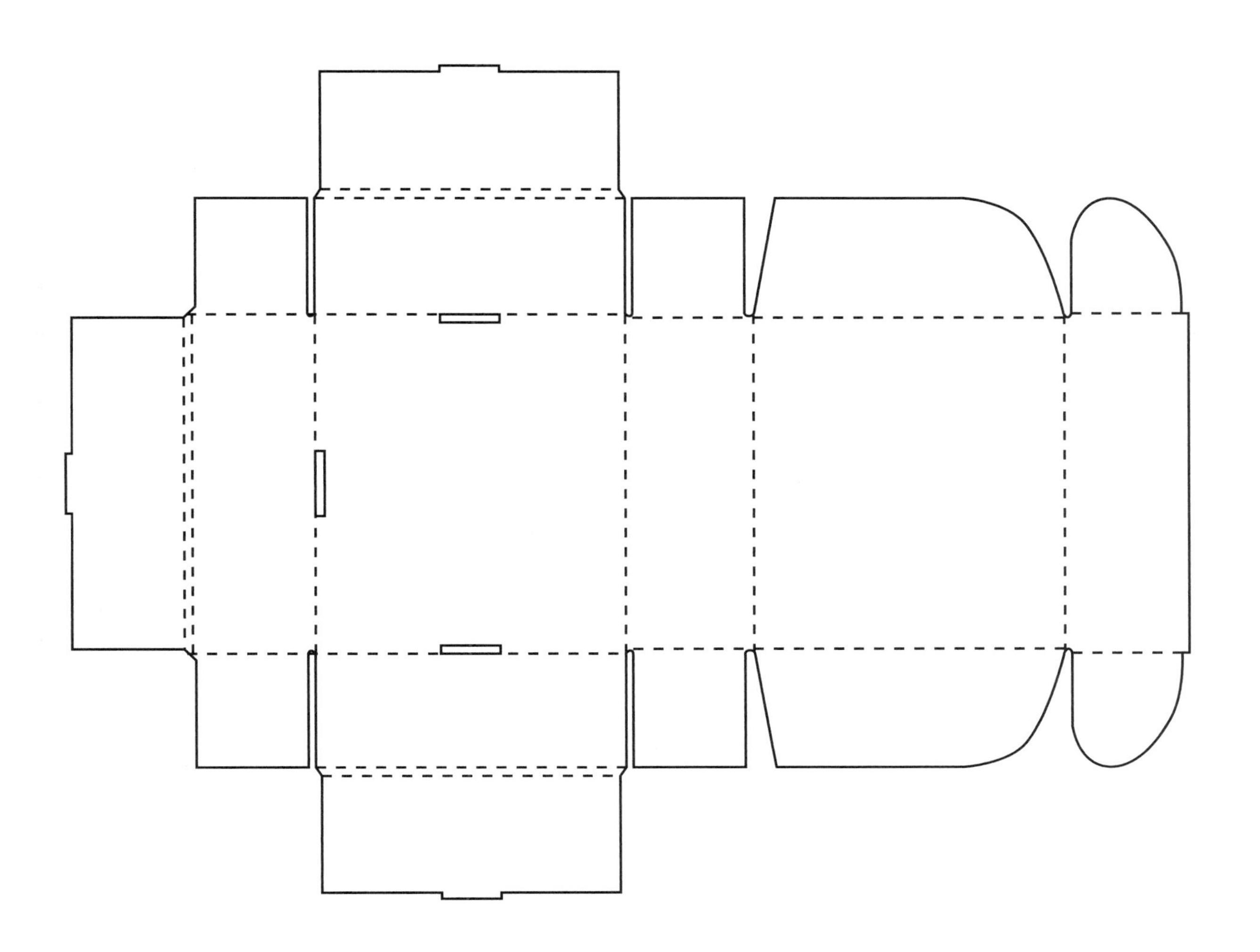

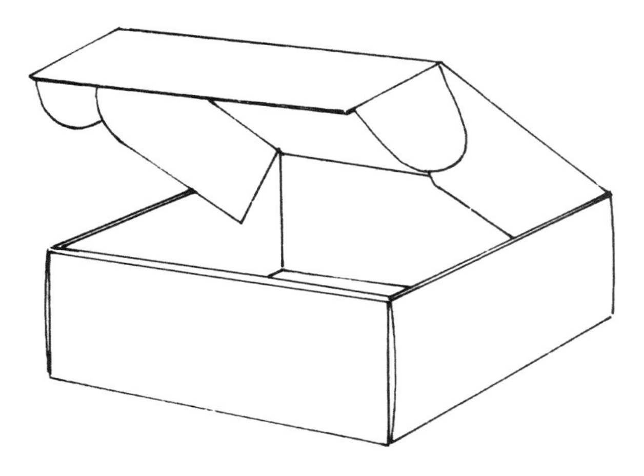

SELF-LOCKING TRAY WITH DOUBLE FRONT WALL AND ZIPPER TEAR STRIP

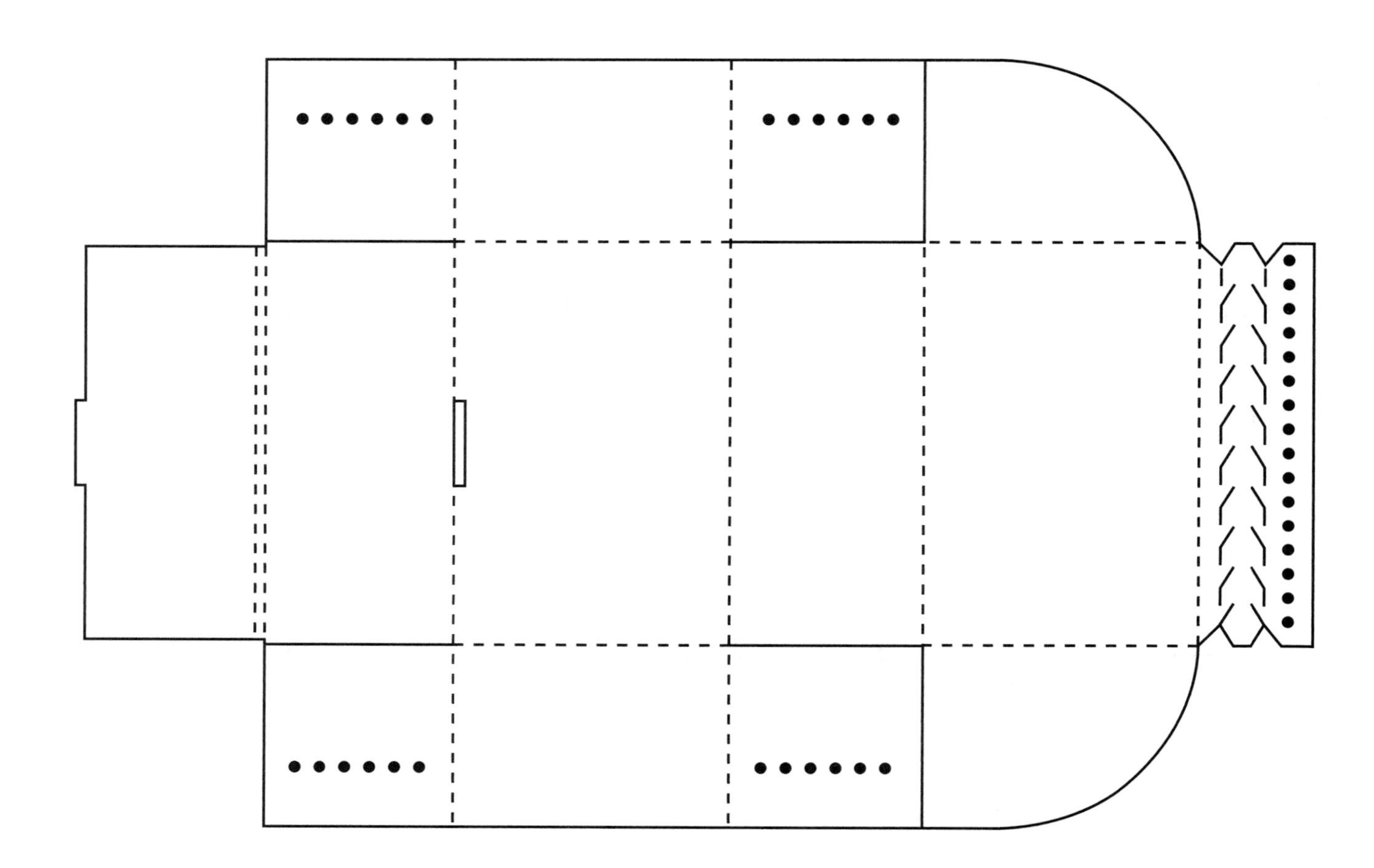

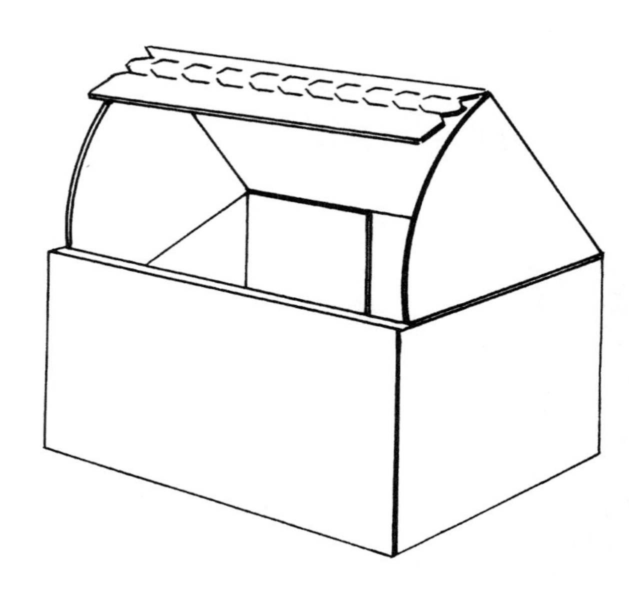

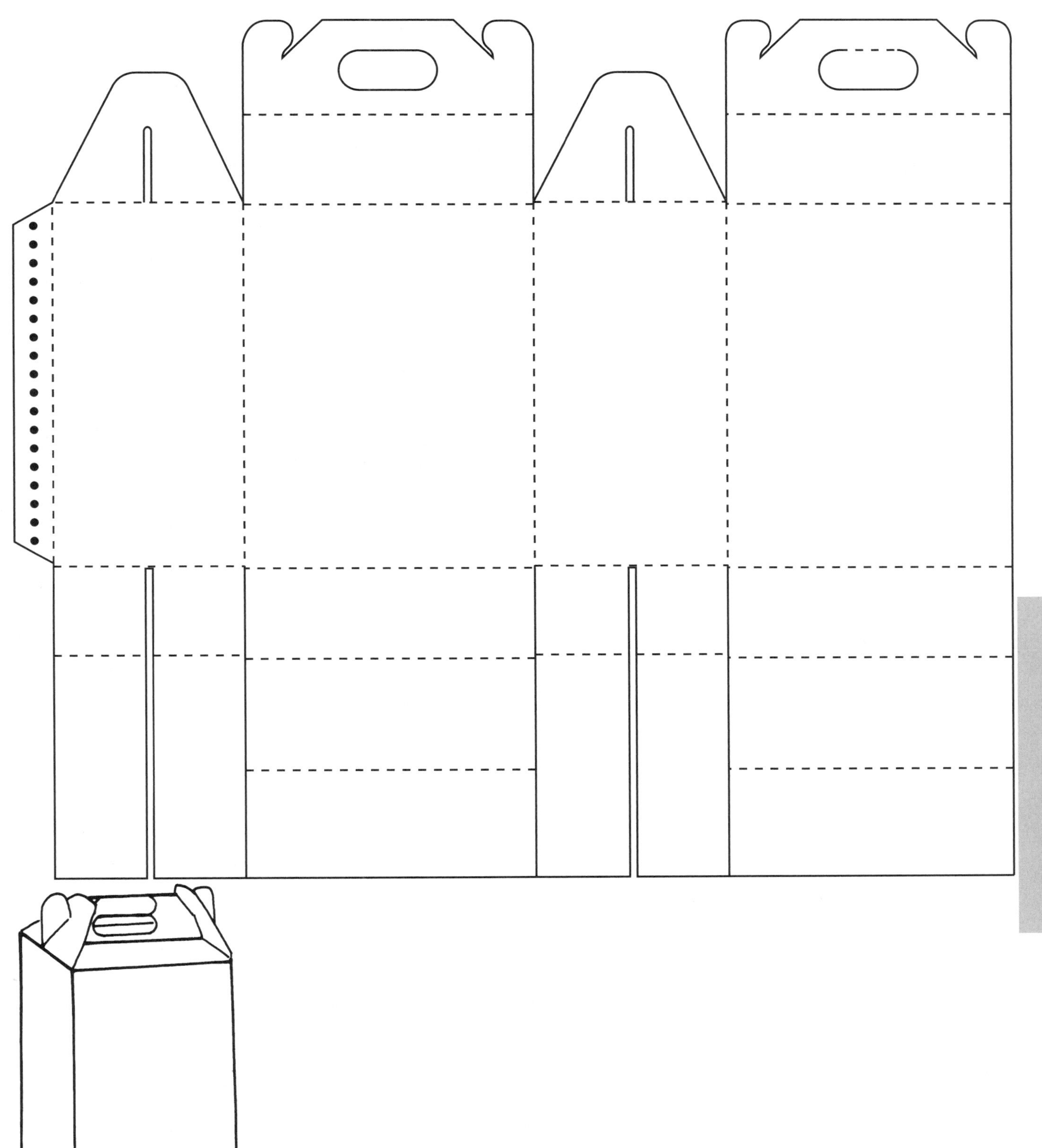

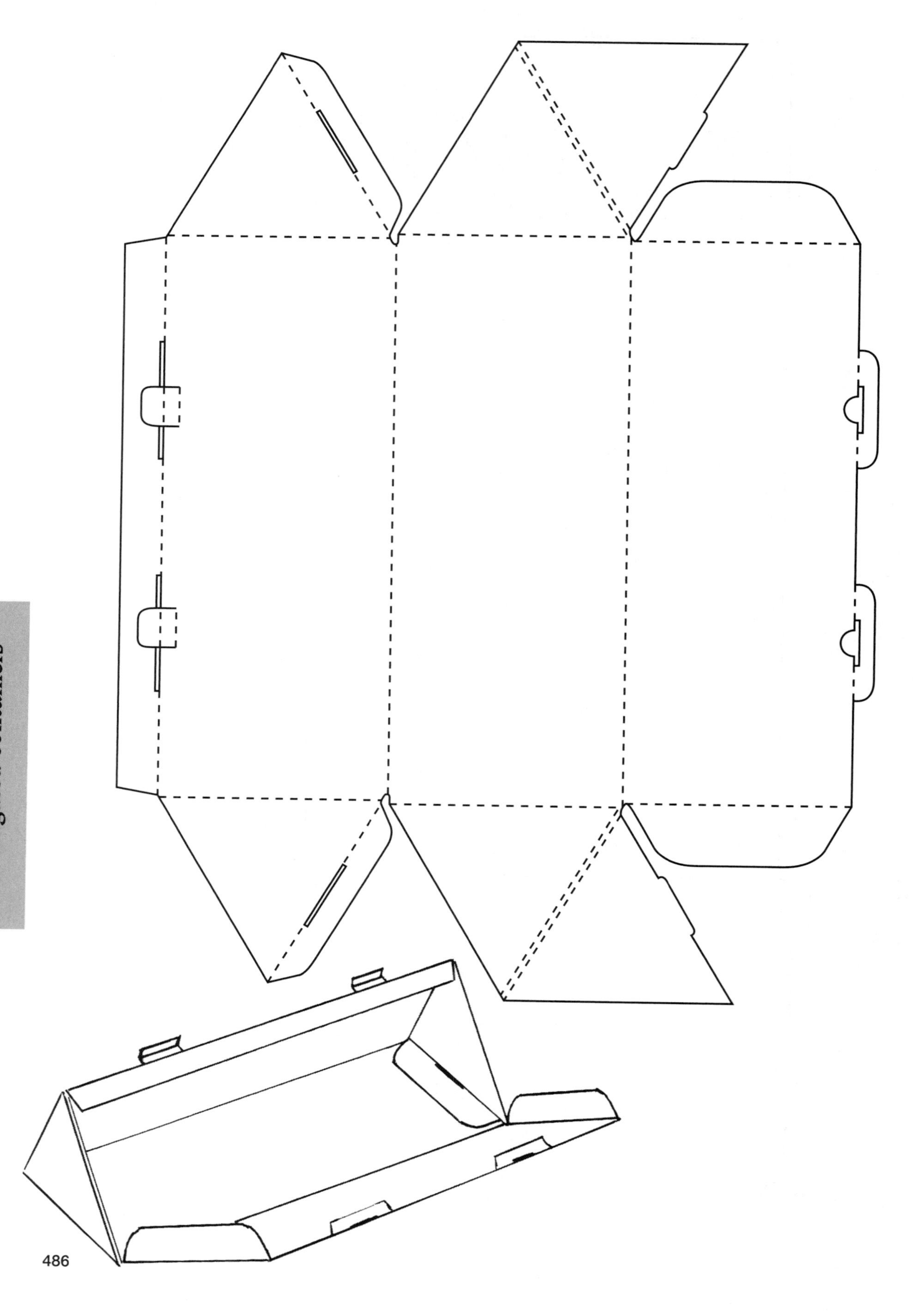

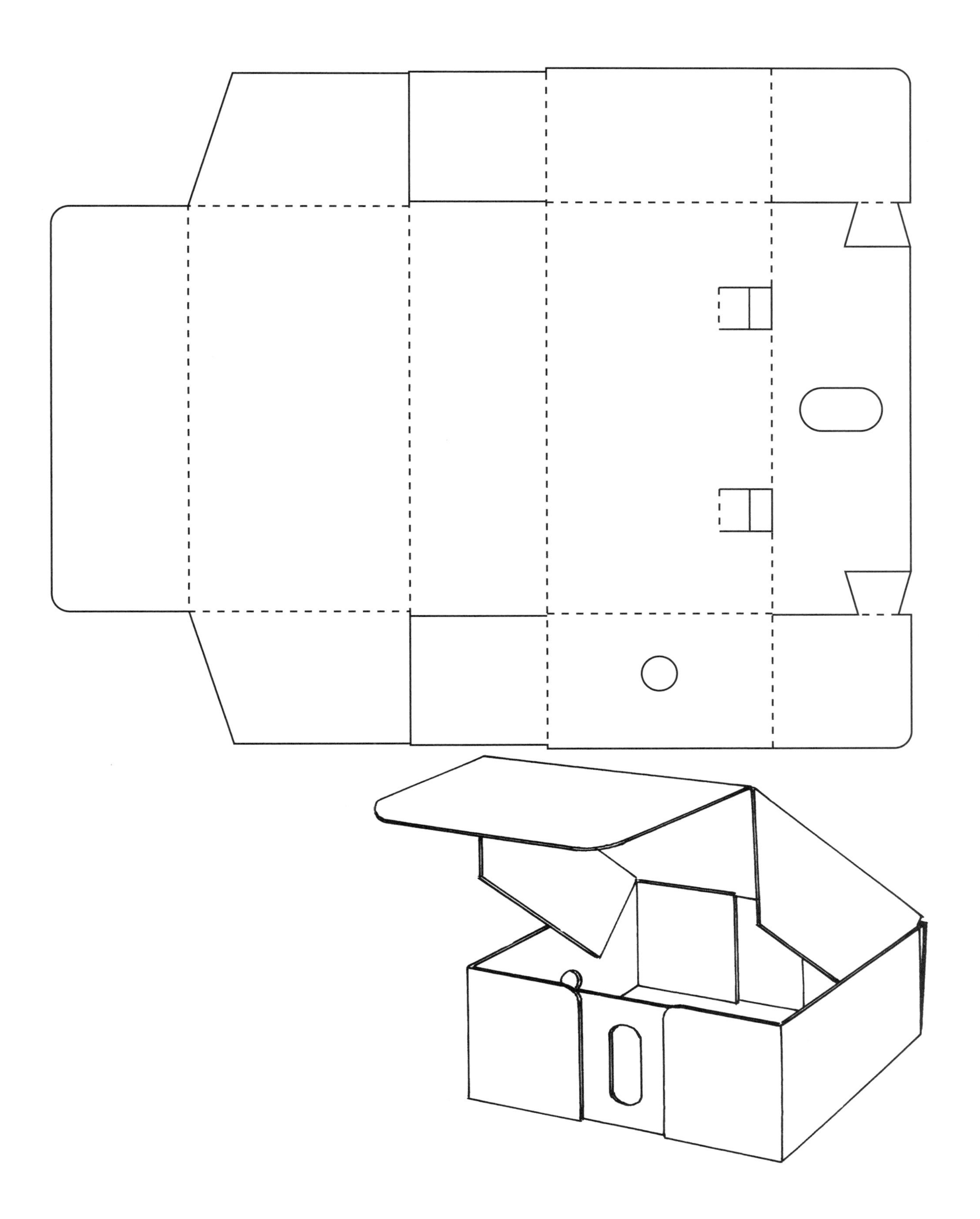

TRIANGULAR SELF-LOCKING TRAY WITH COVER AND PEGBOARD HANGER

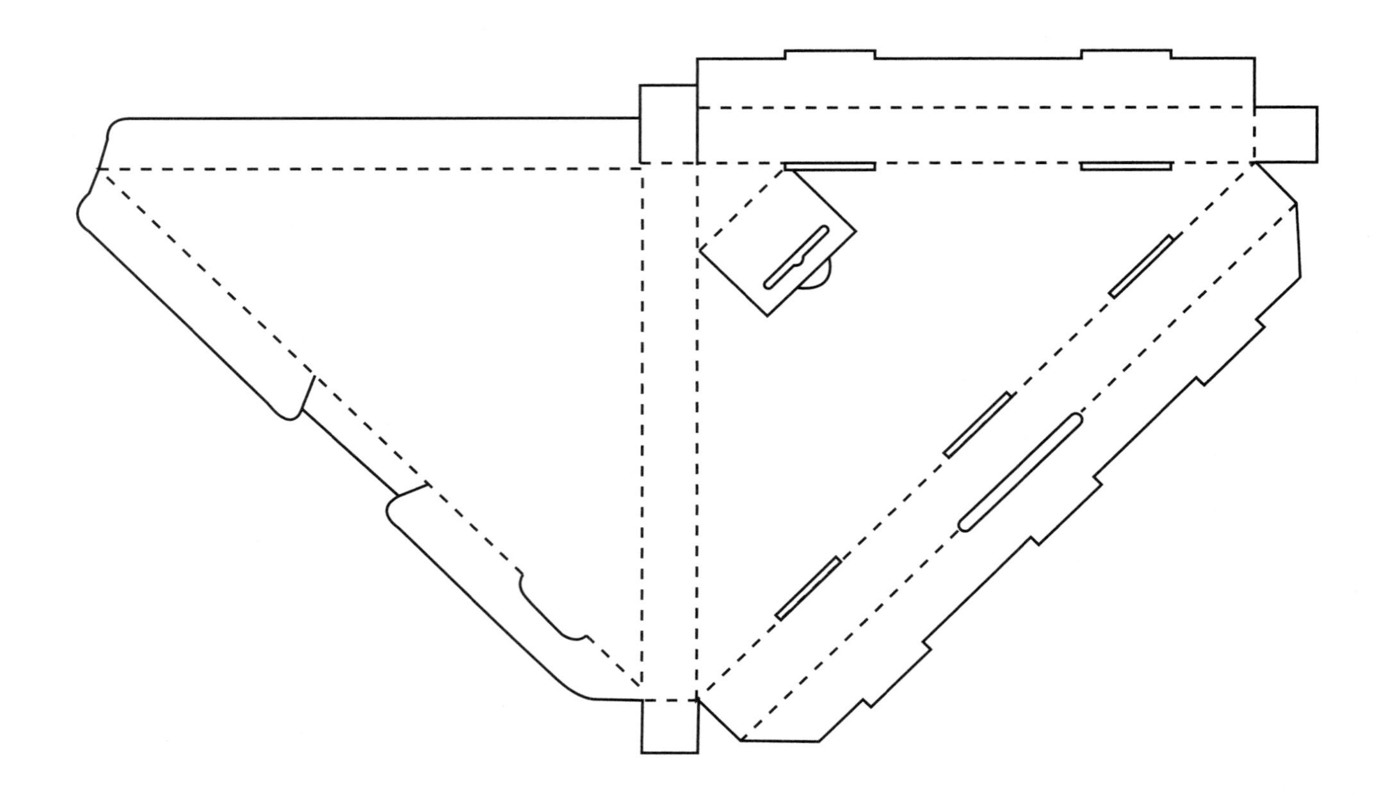

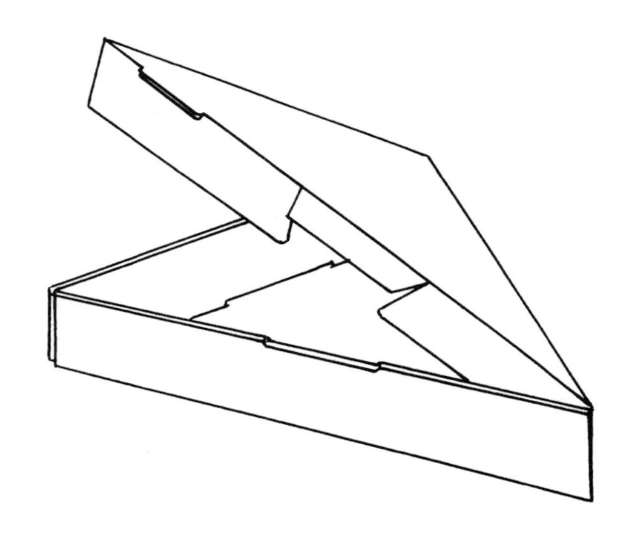

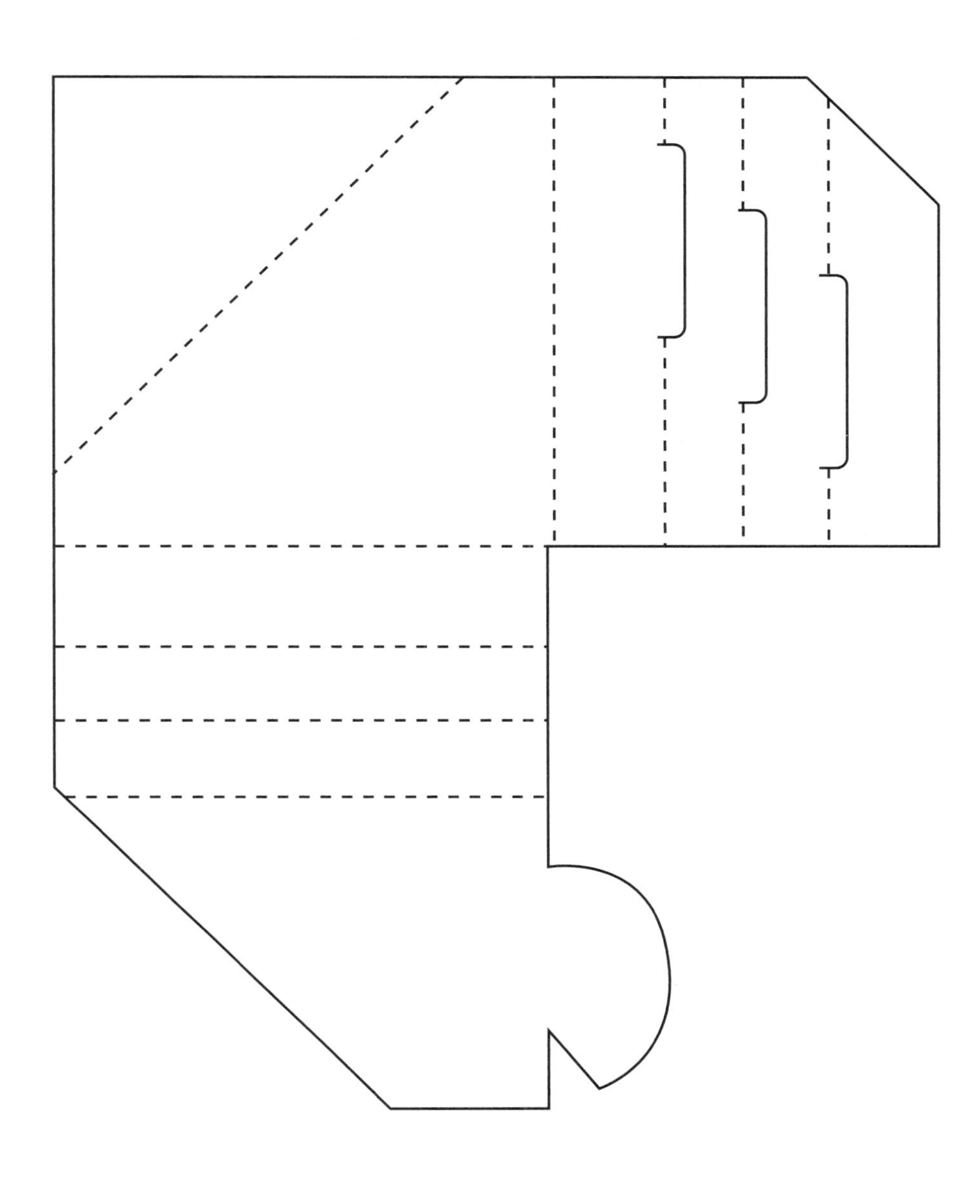

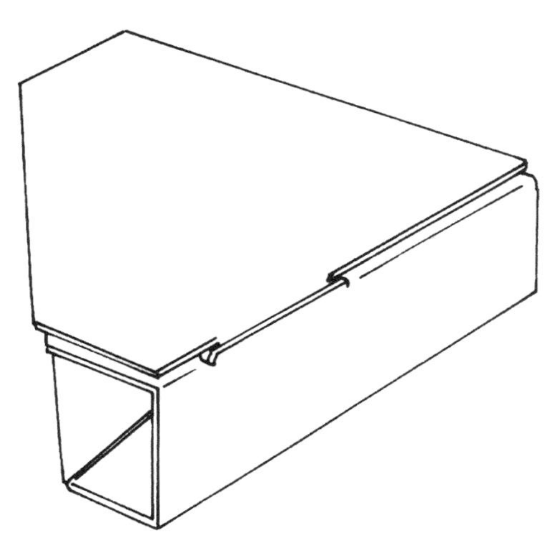

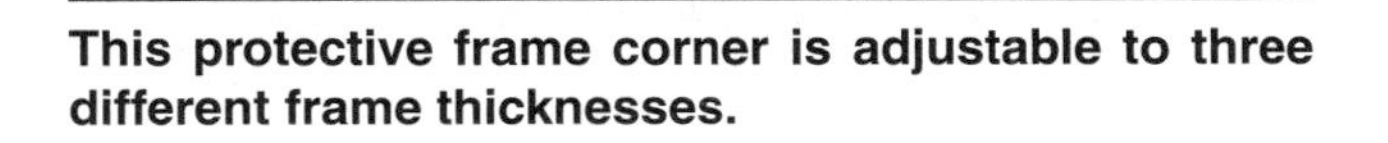

This protective frame corner is adjustable to three different frame thicknesses.

ONE-PIECE FOLDER
(Intl. Box Code 0401)

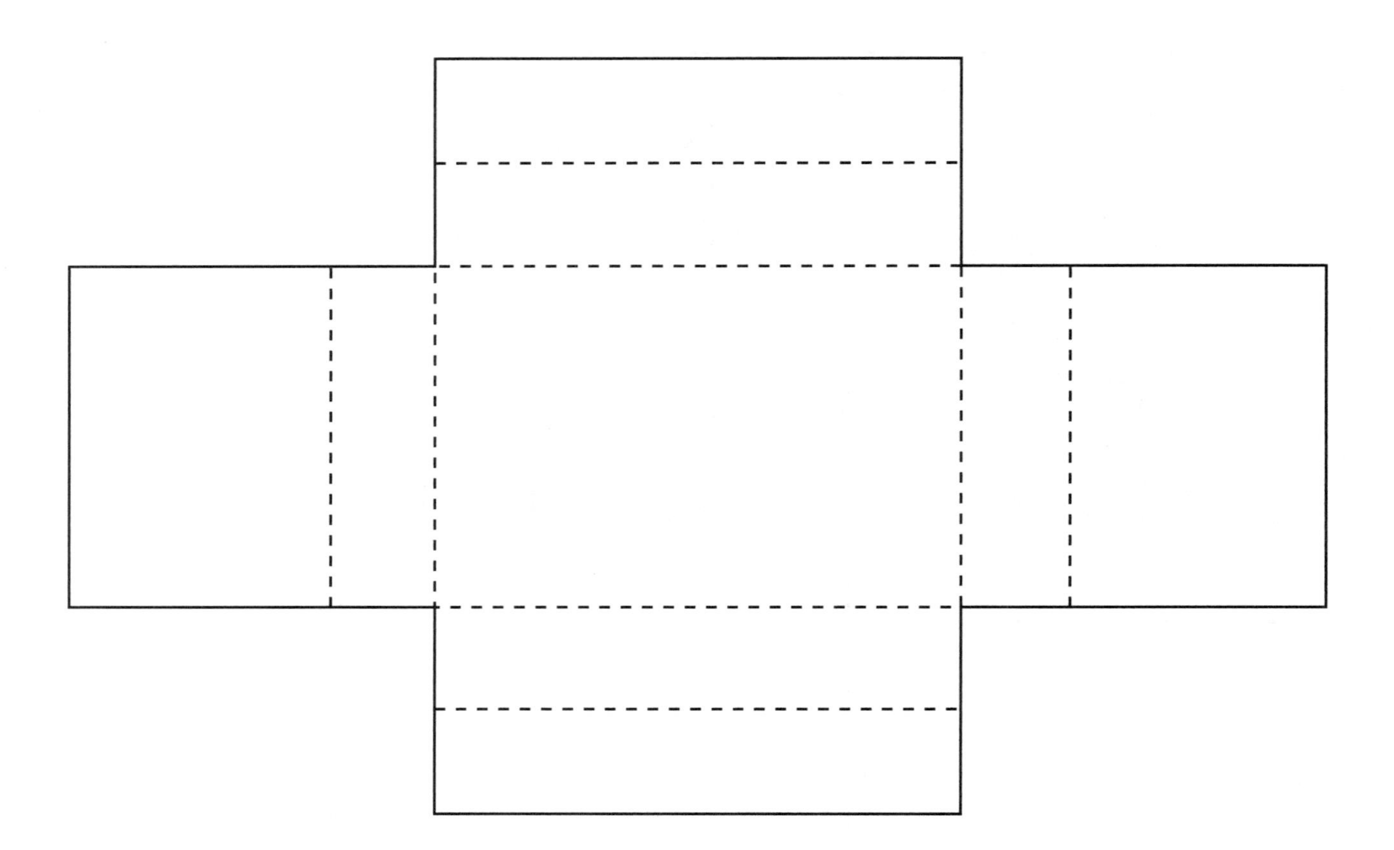

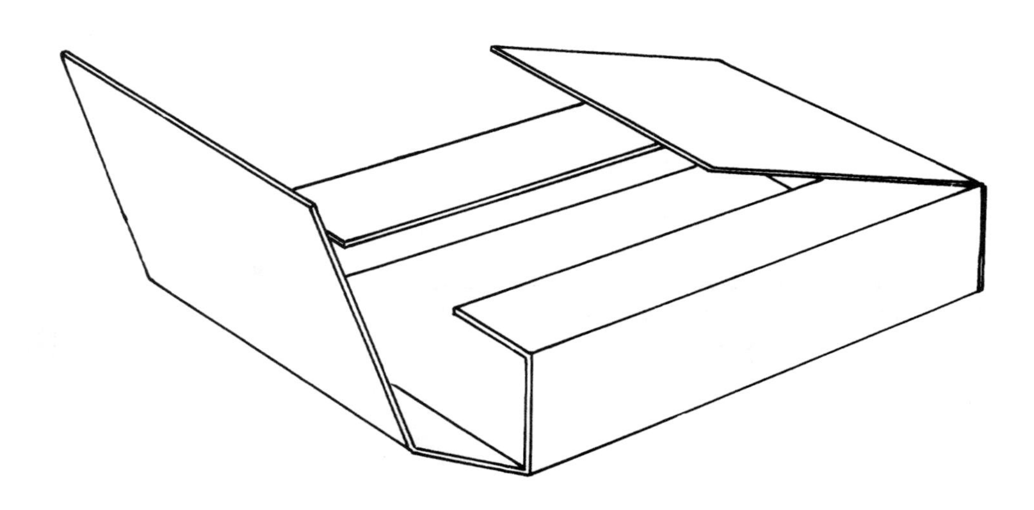

TWO-PIECE FOLDER
(Intl. Box Code 0404)

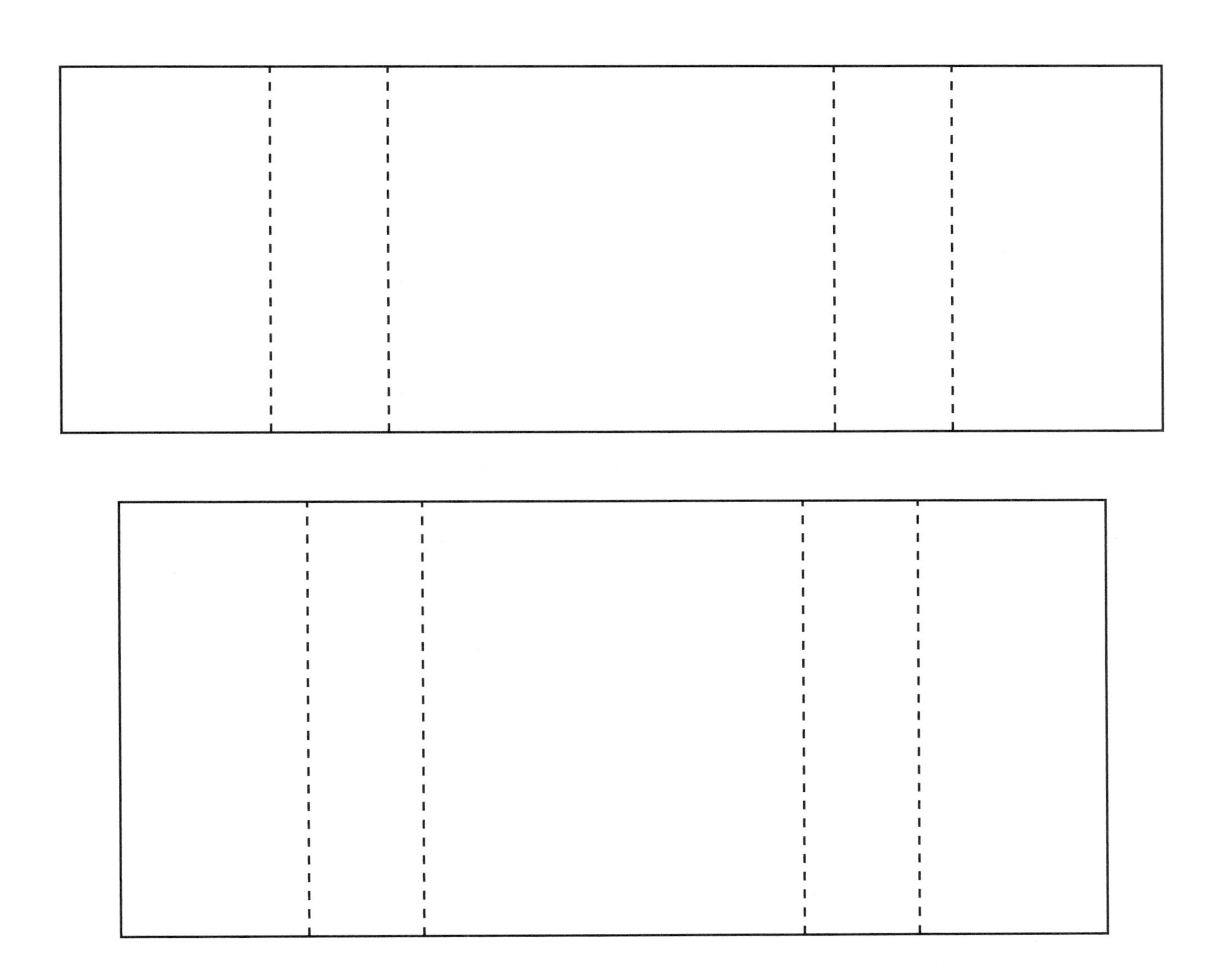

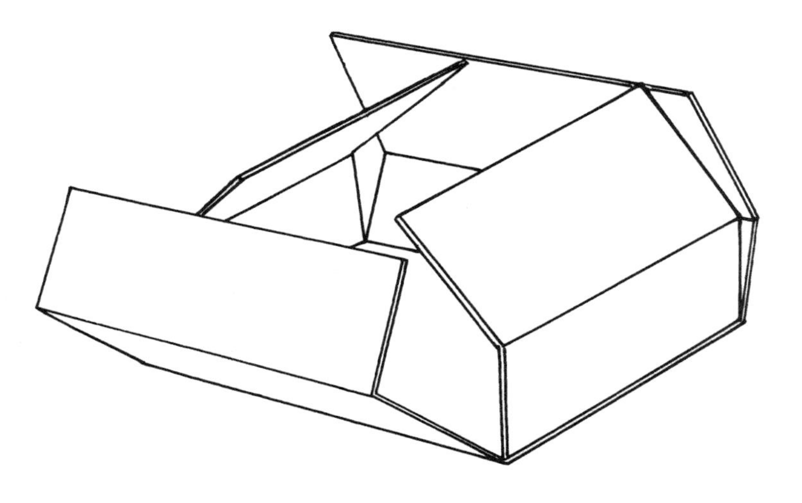

THREE-PIECE FOLDER

(Intl. Box Code 0405)

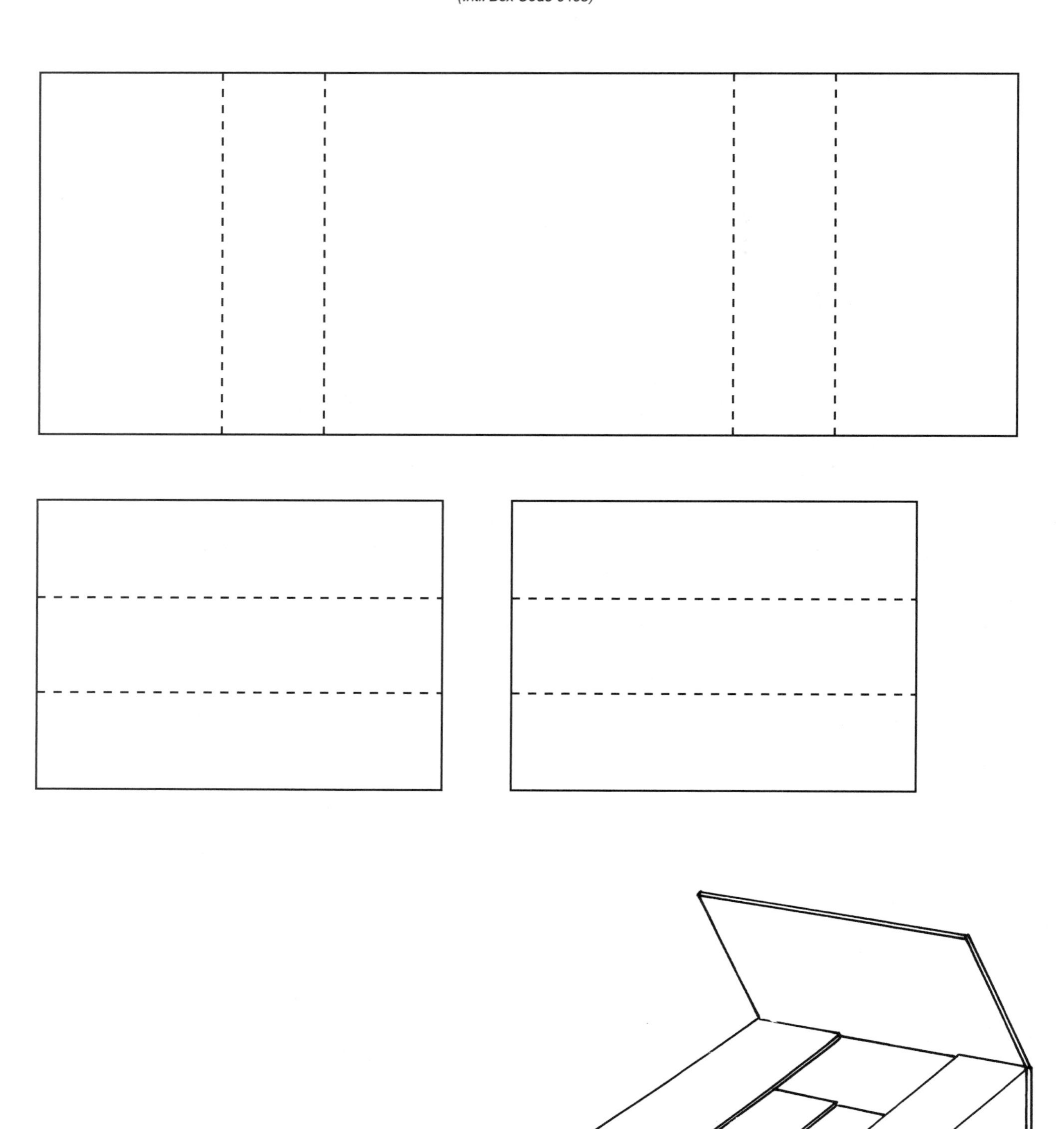

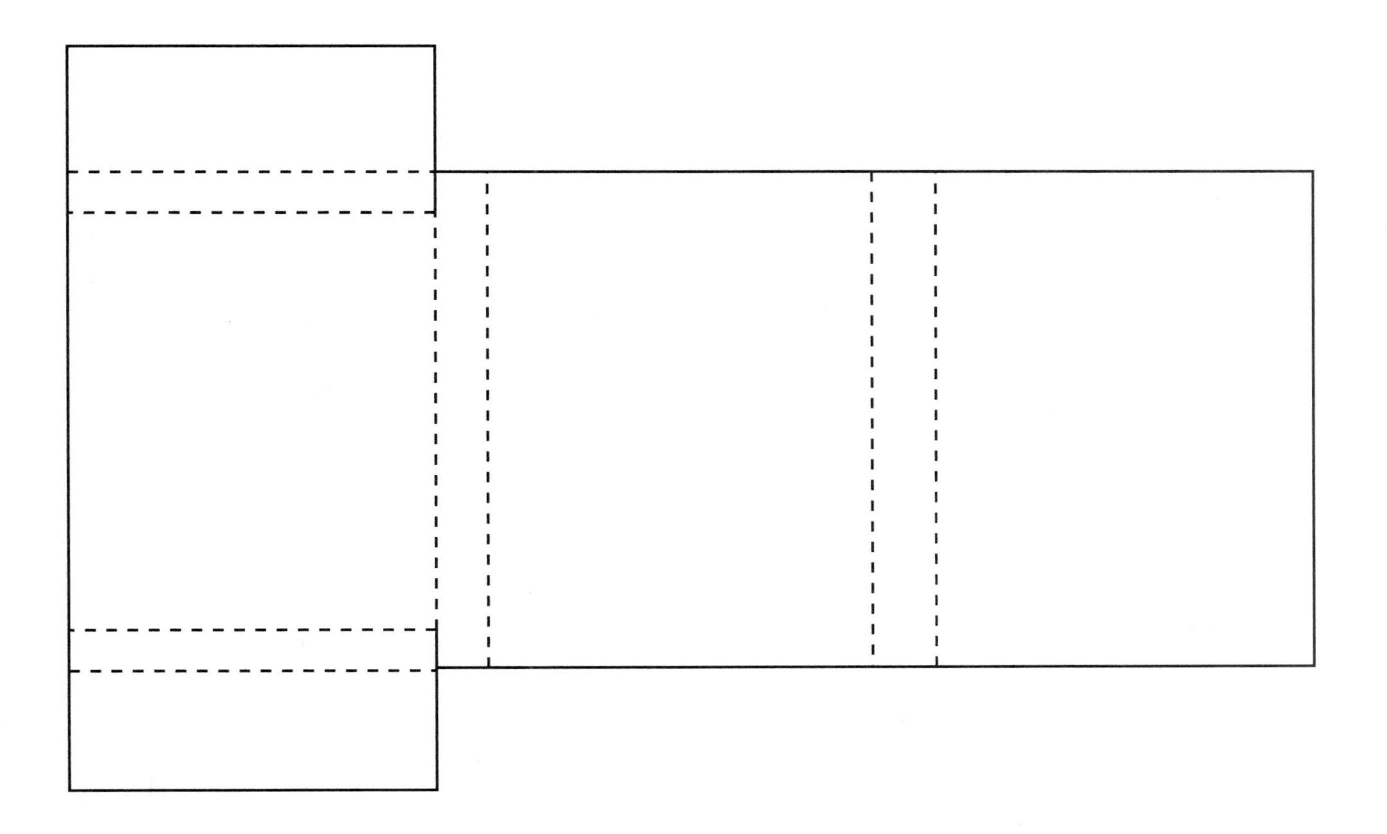

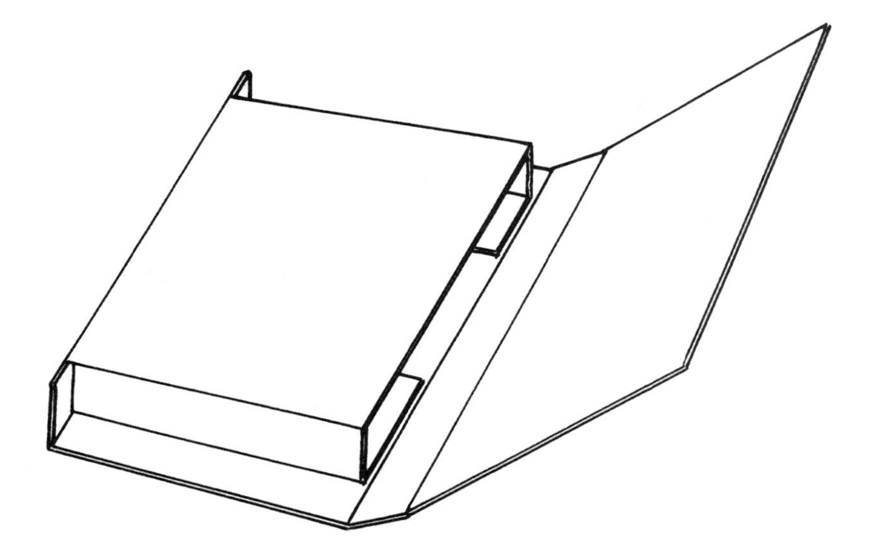

CORRUGATED BOOK SHIPPER WITH RECESSED SELF-LOCKING CLOSURES

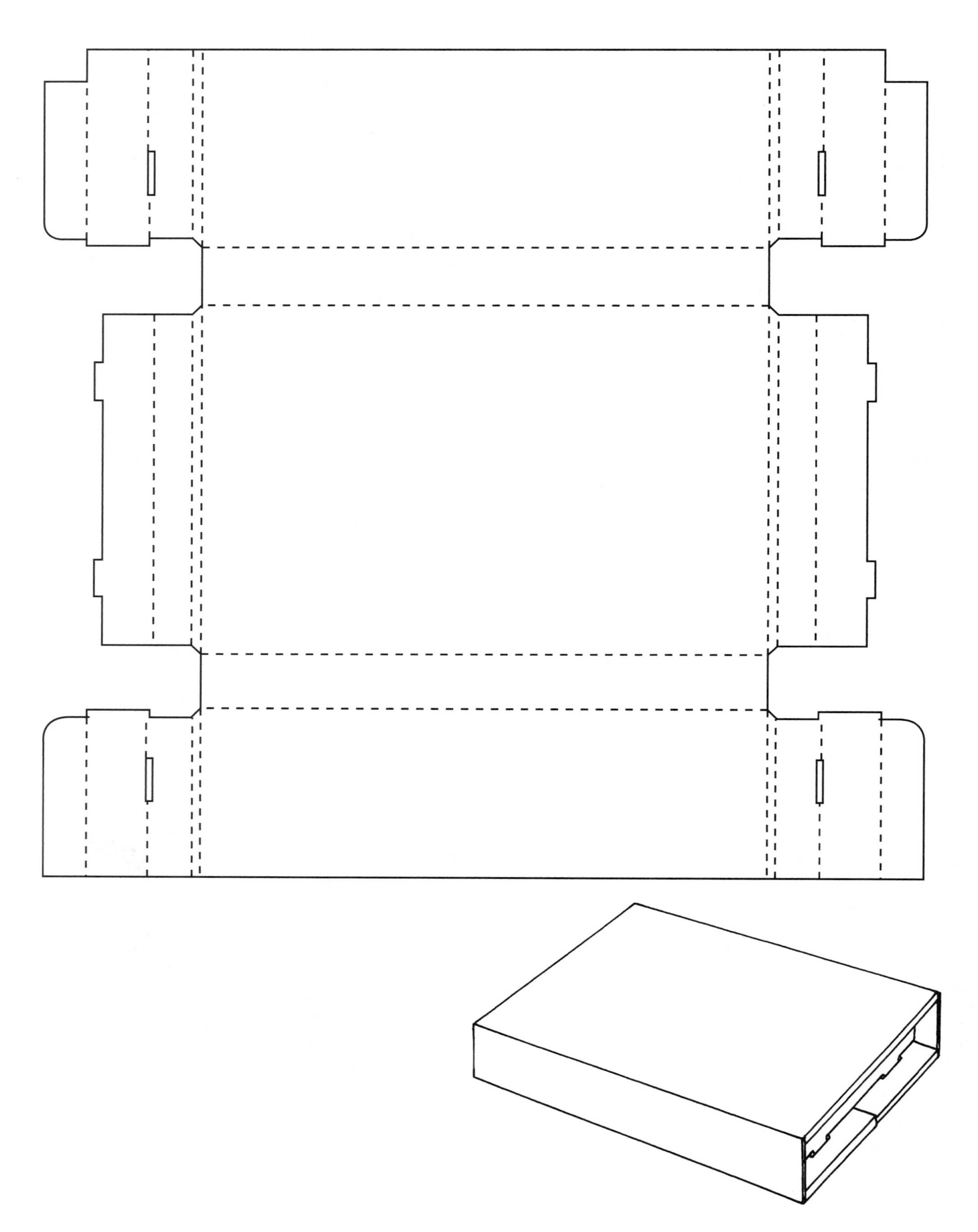

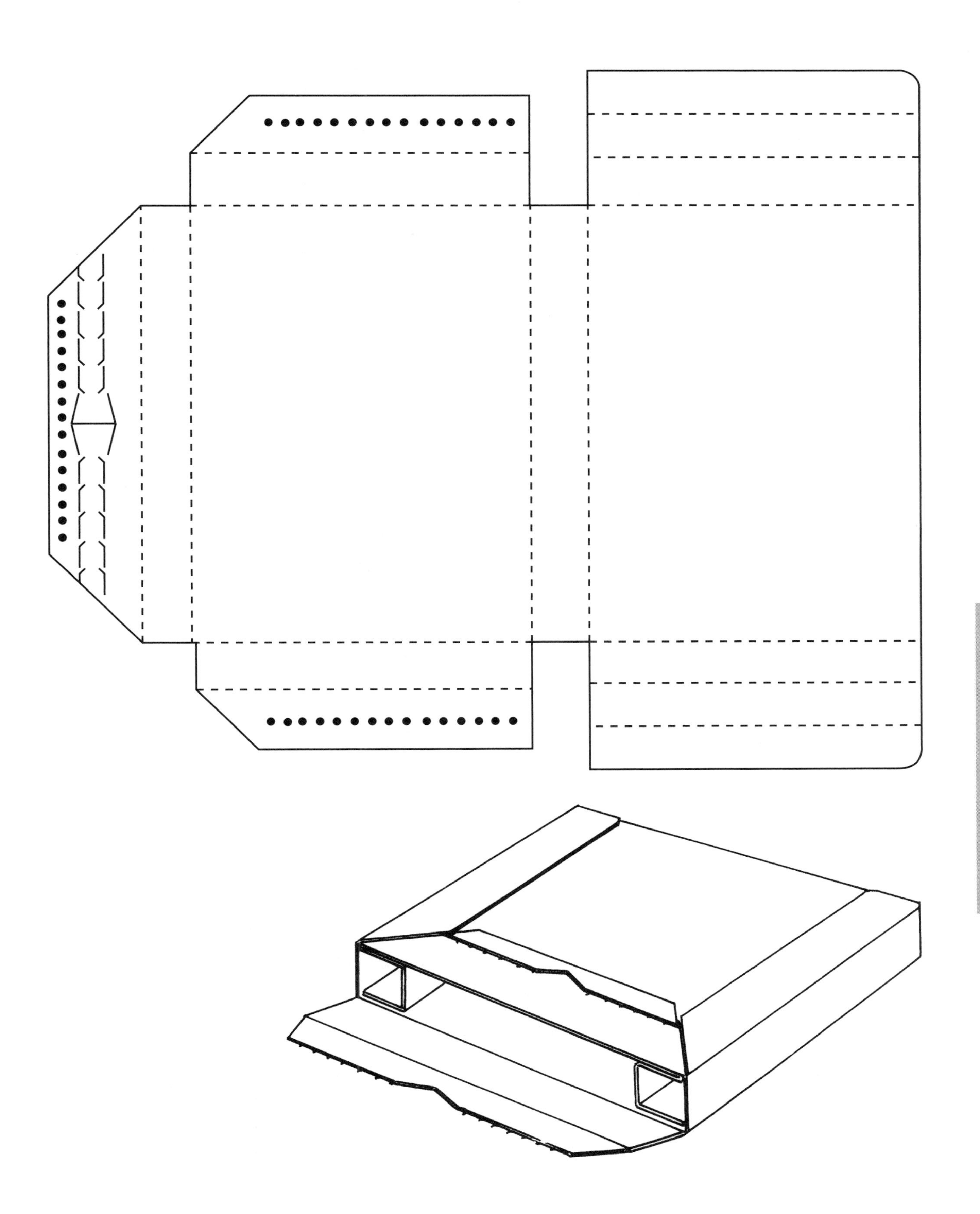

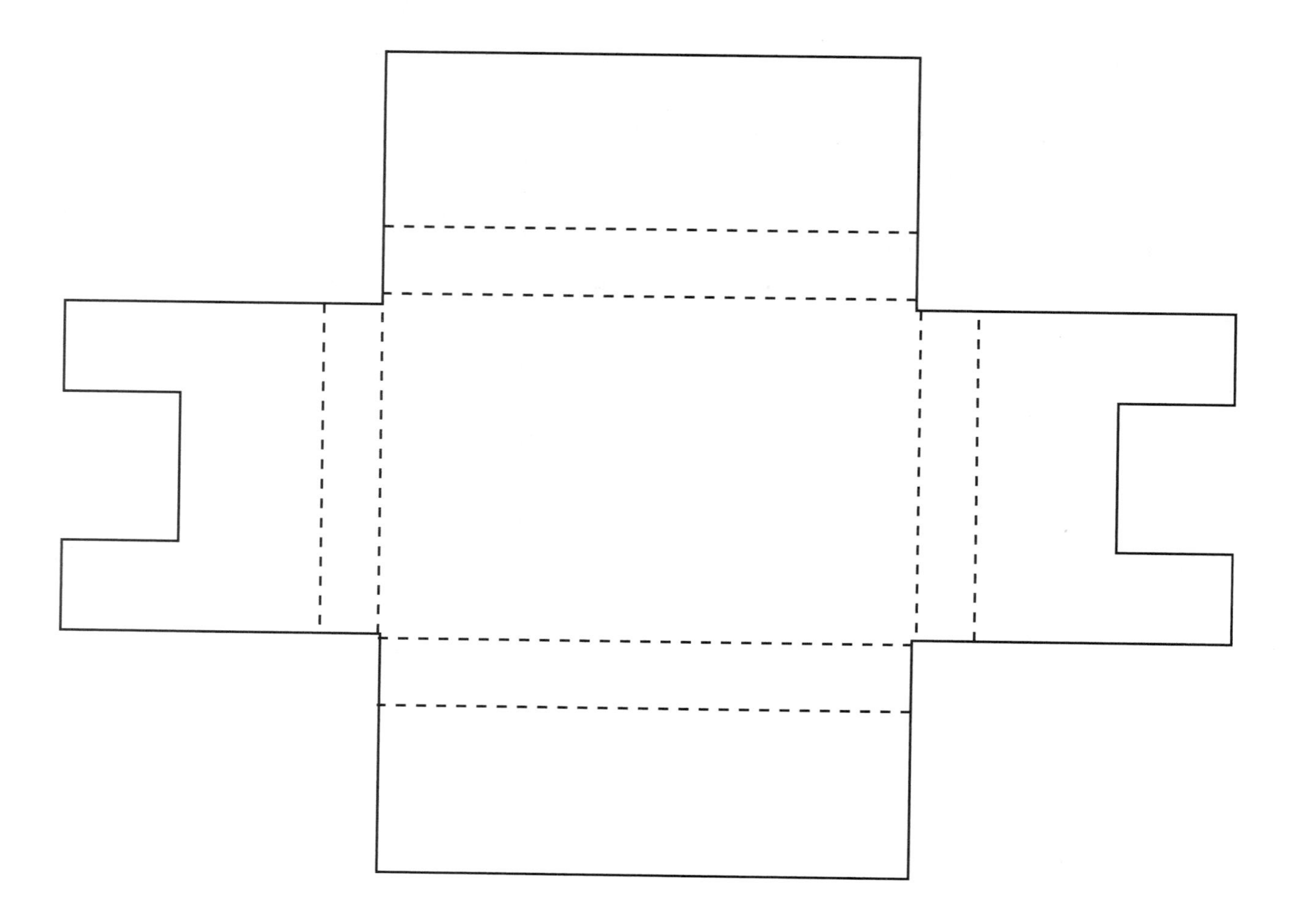

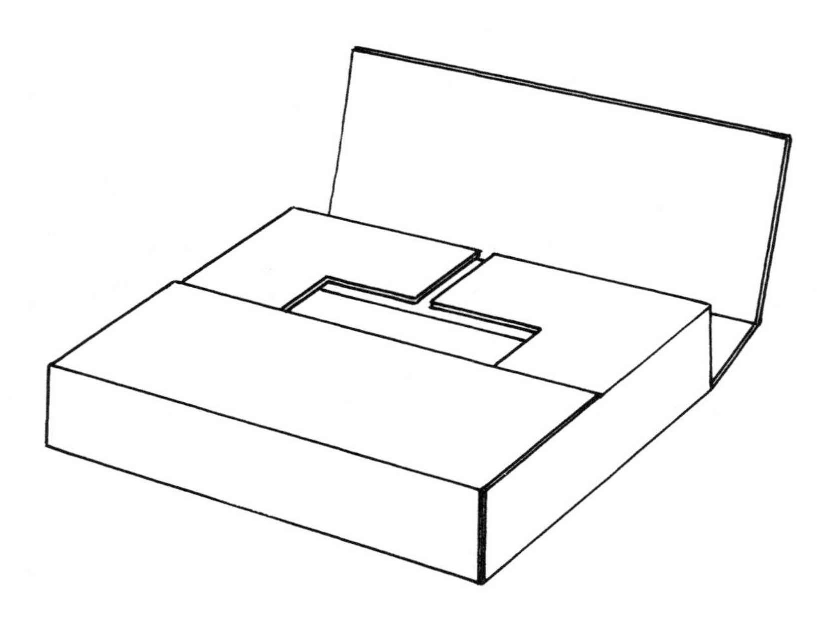

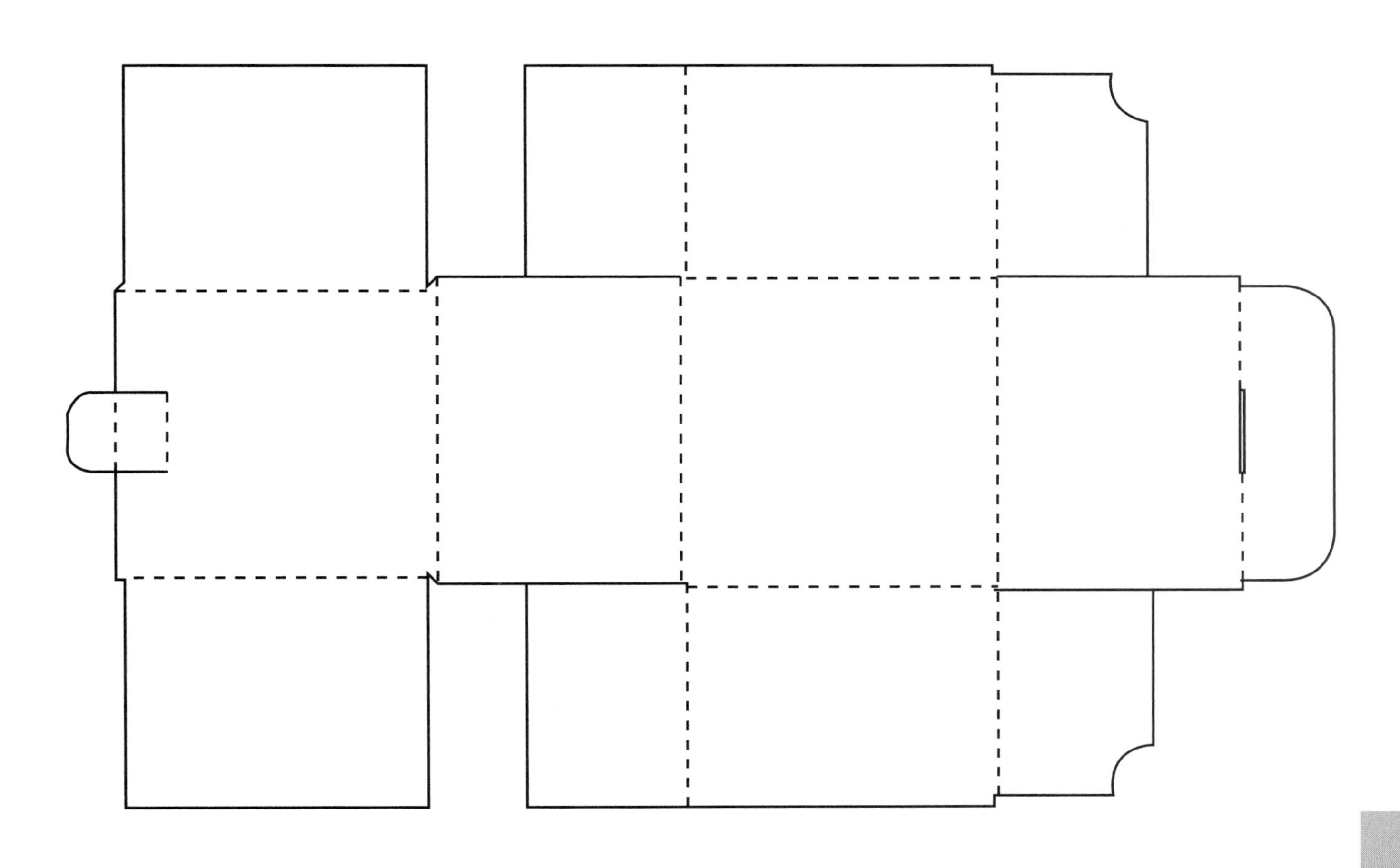

This blank folds into a sturdy container for heavy products, in this case a 5 lb. spool of picture wire.

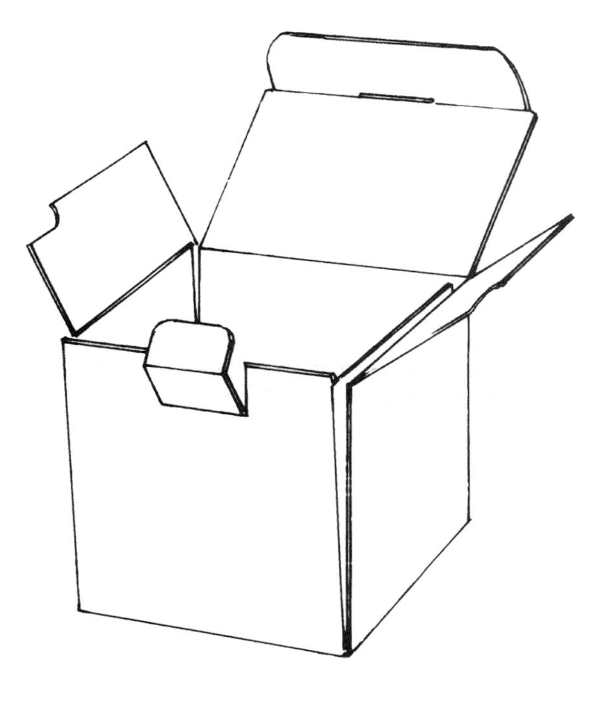

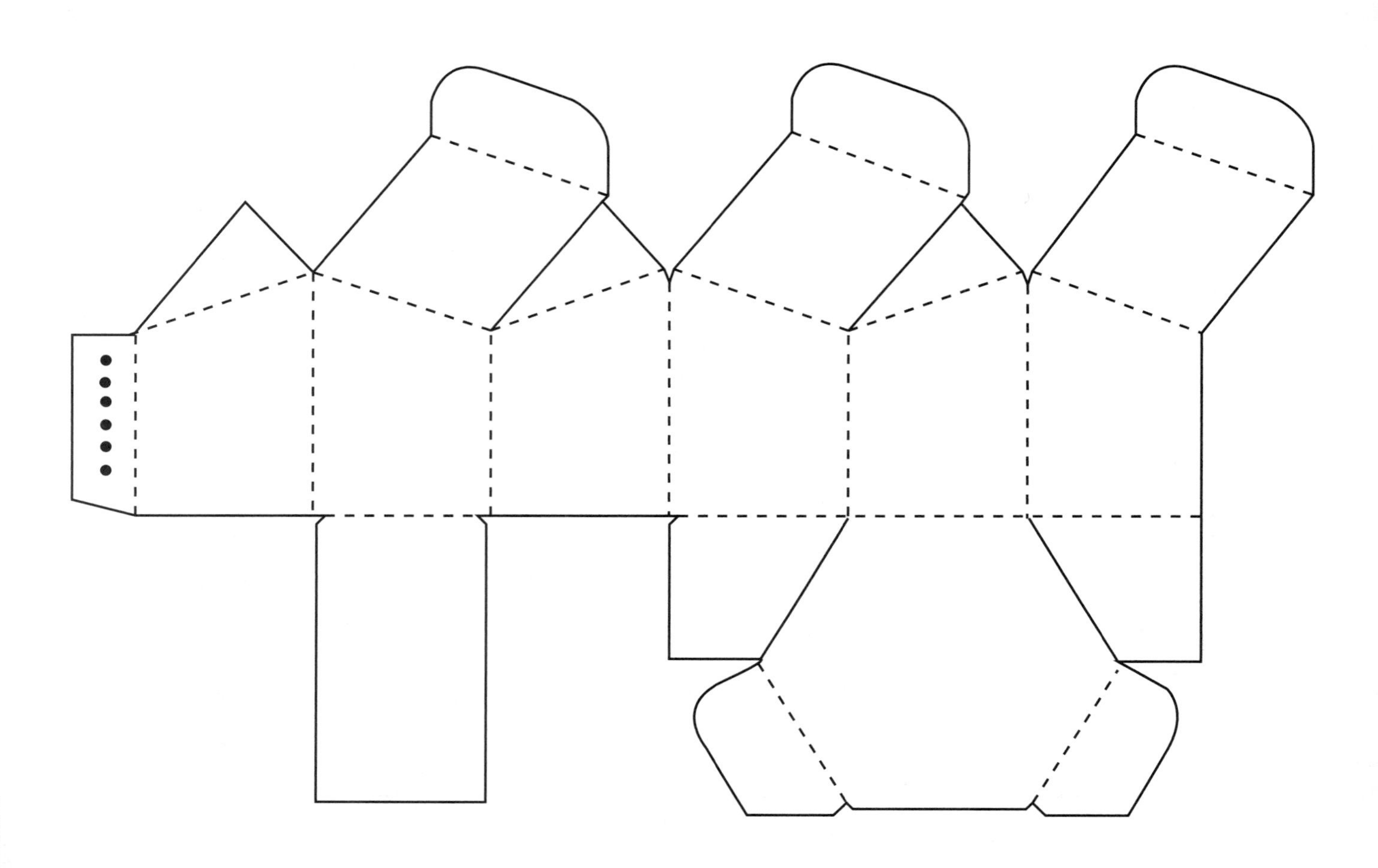

Developed by contemporary designer John Edminster, this structure provides a decorative, reclosable three-faceted top for a hexagonal carton.

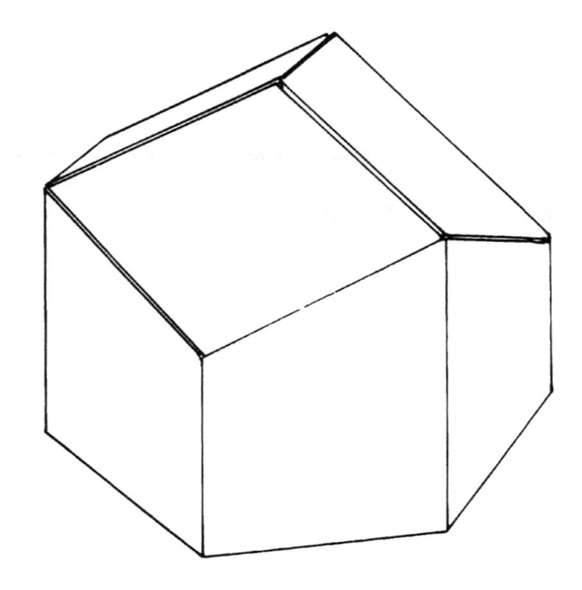

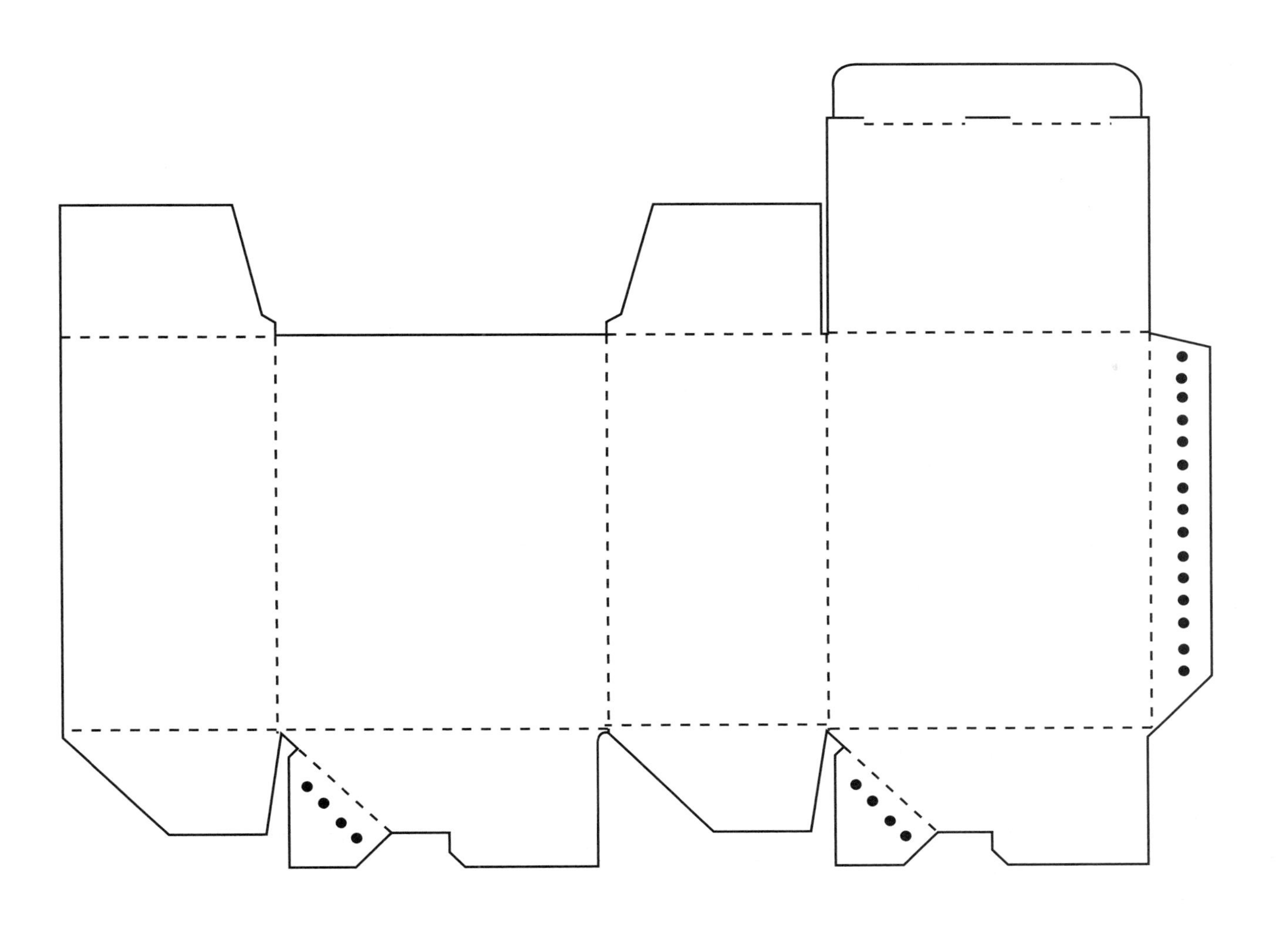

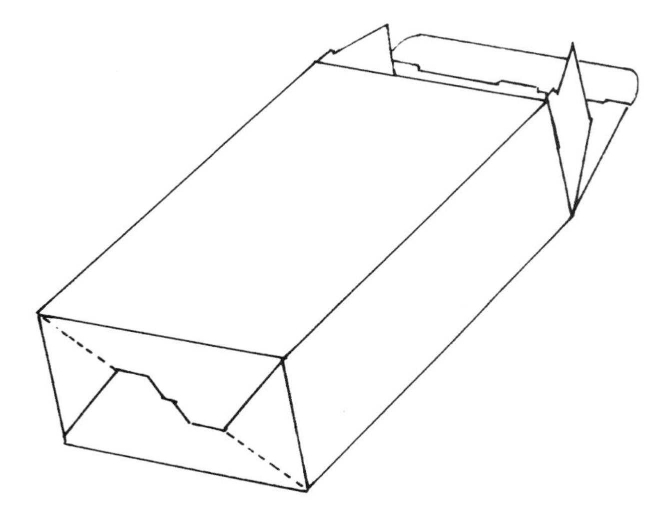

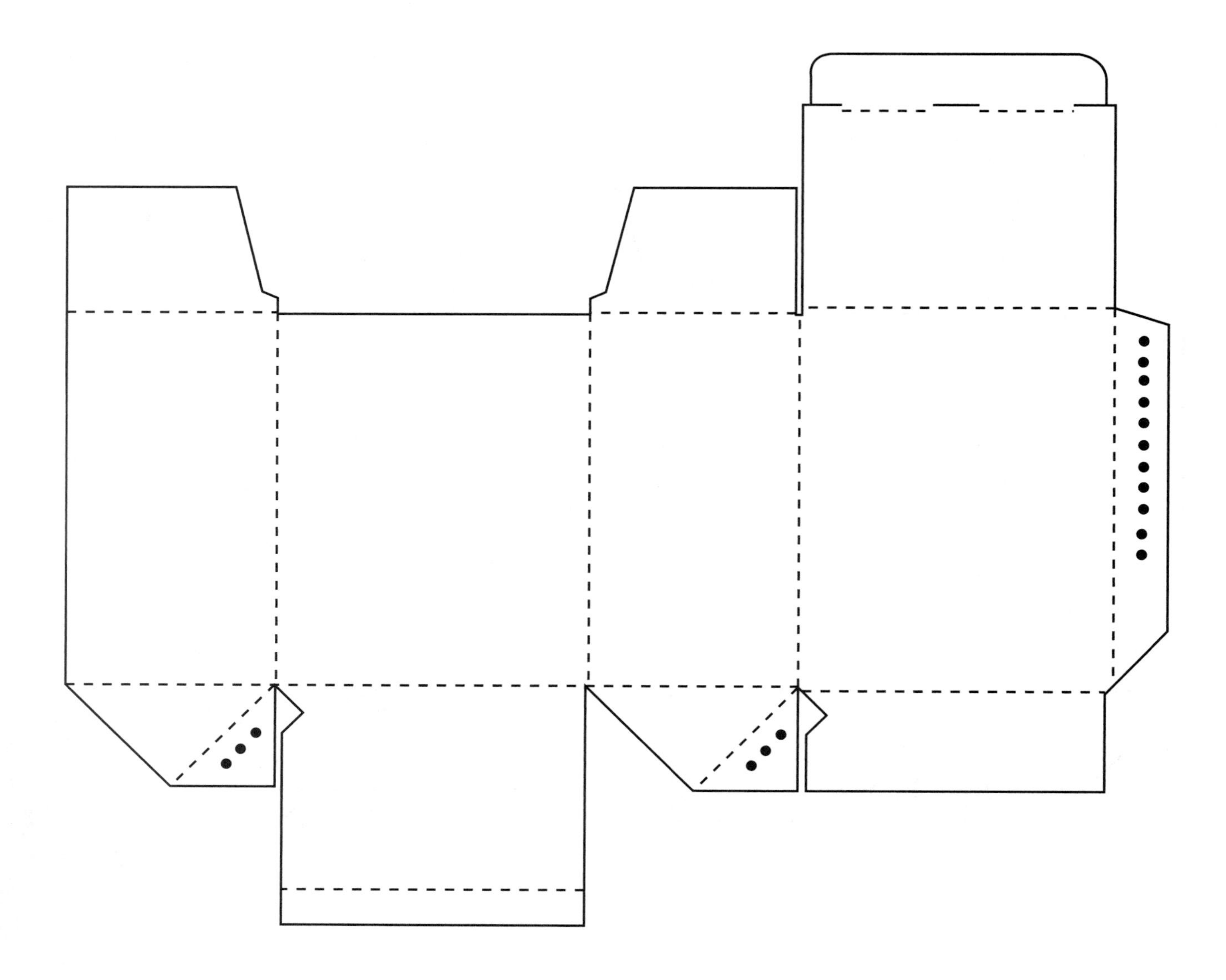

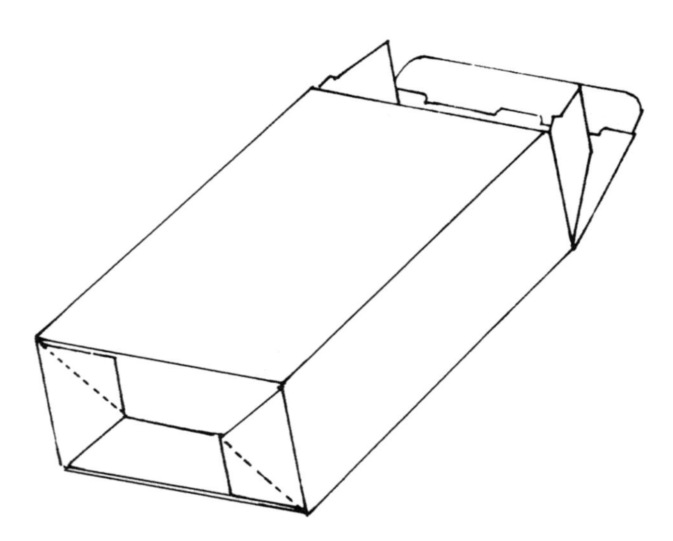

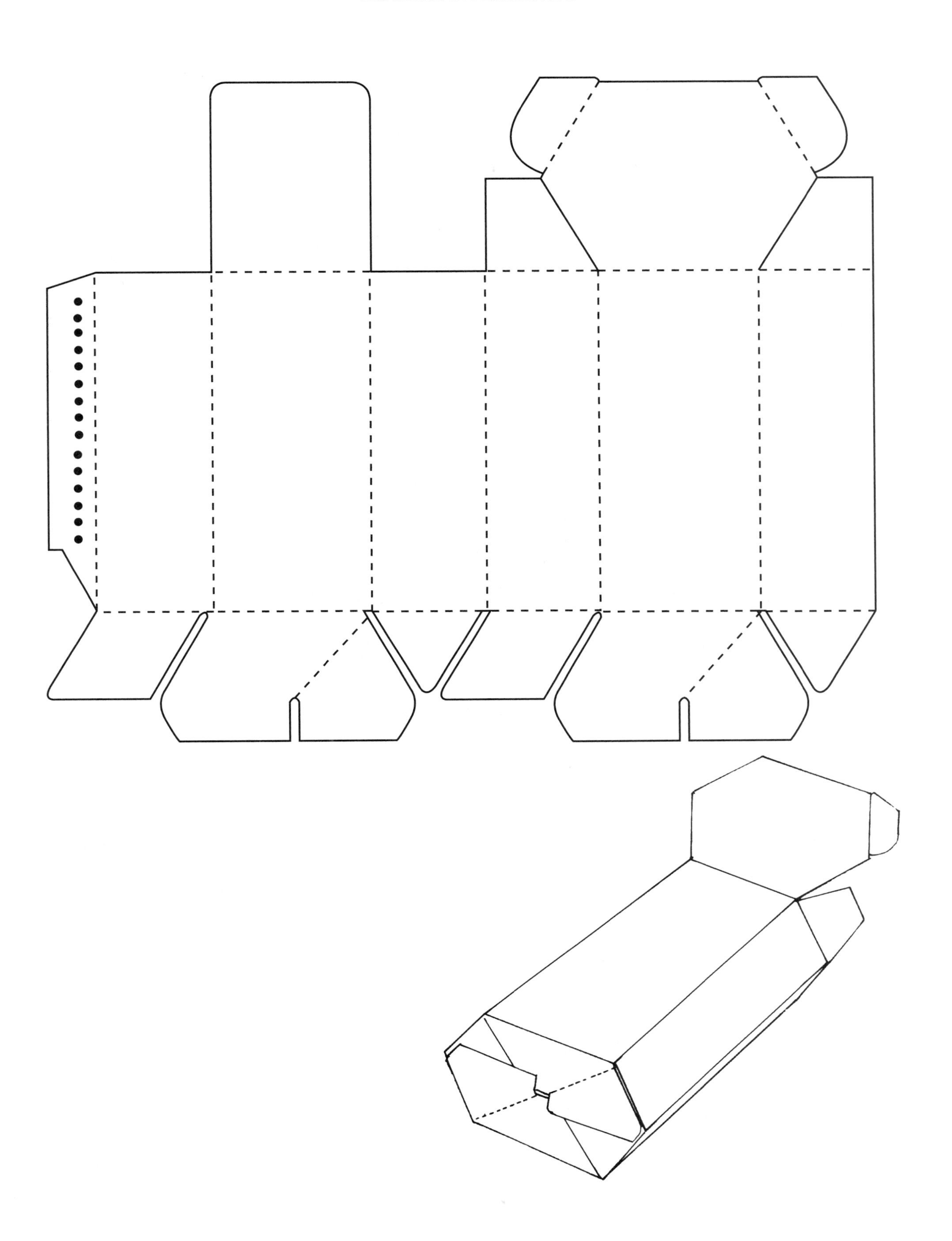

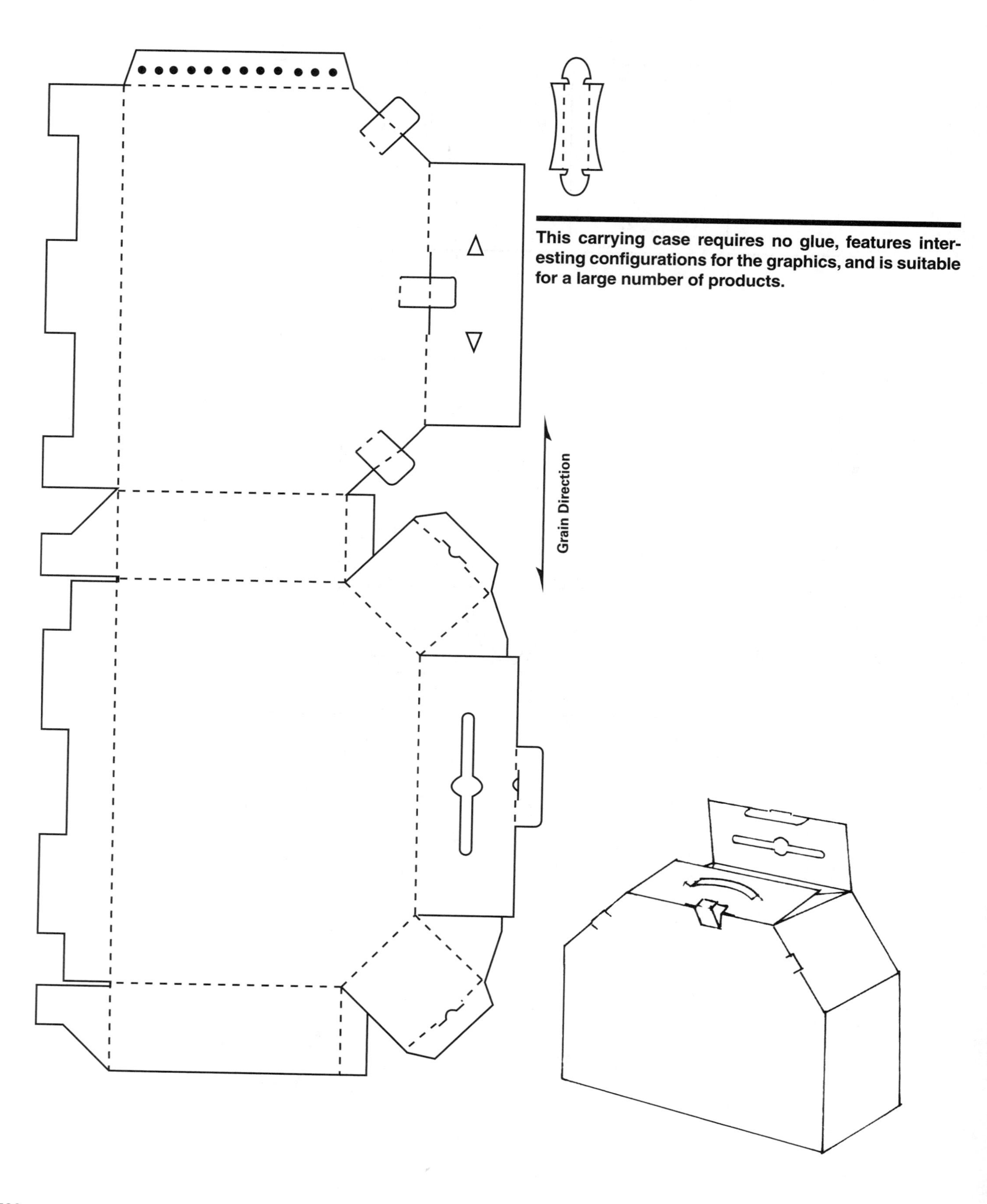

This carrying case requires no glue, features interesting configurations for the graphics, and is suitable for a large number of products.

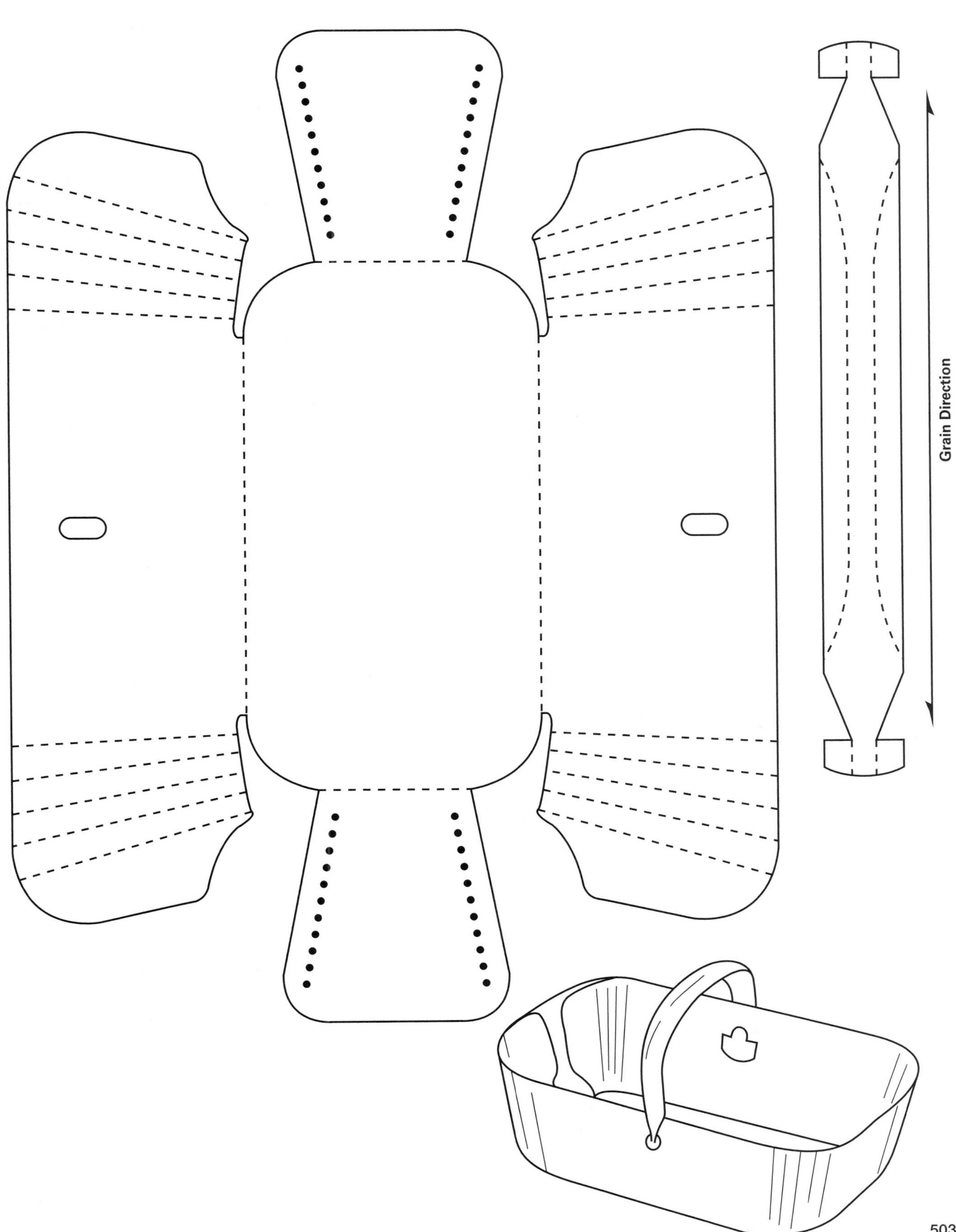
Grain Direction

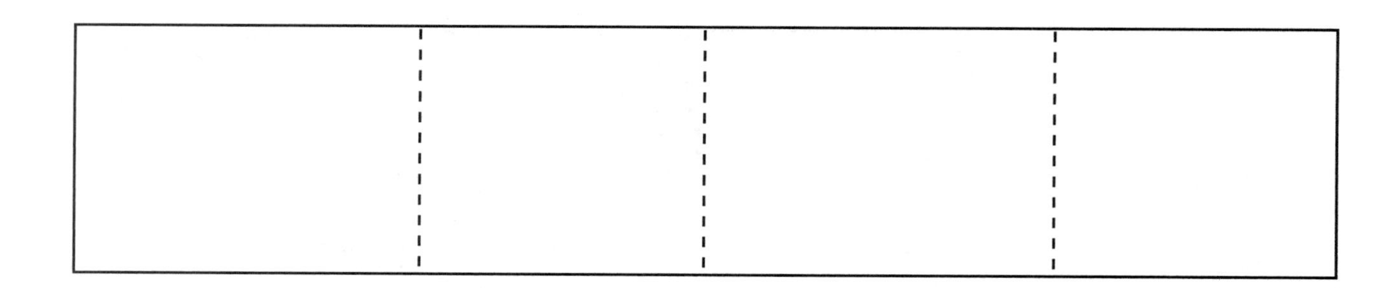

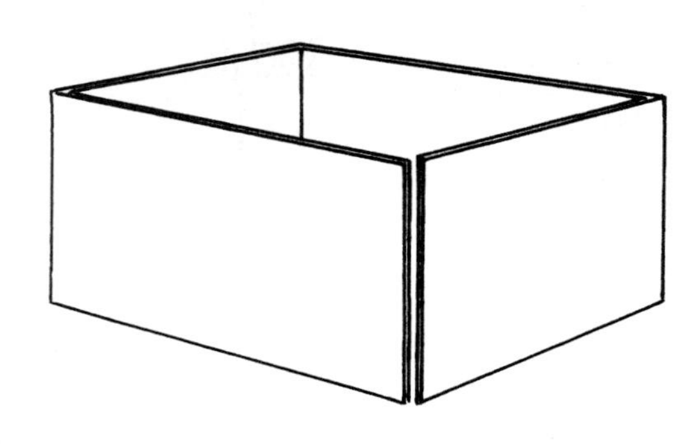

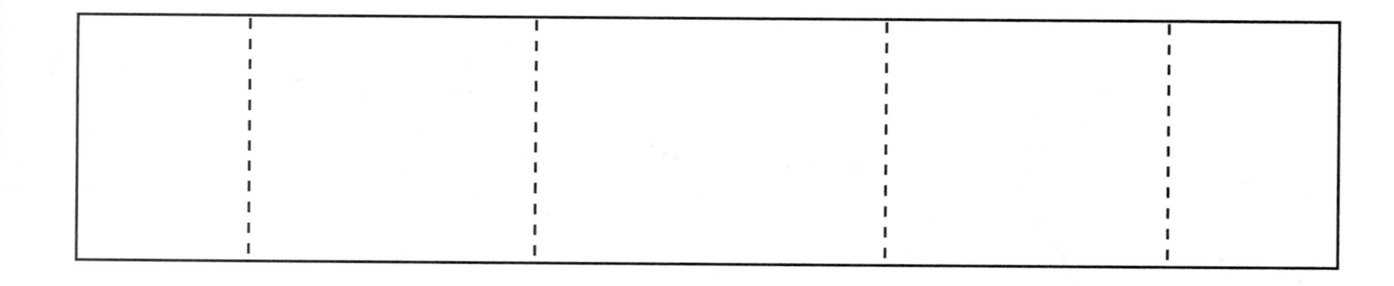

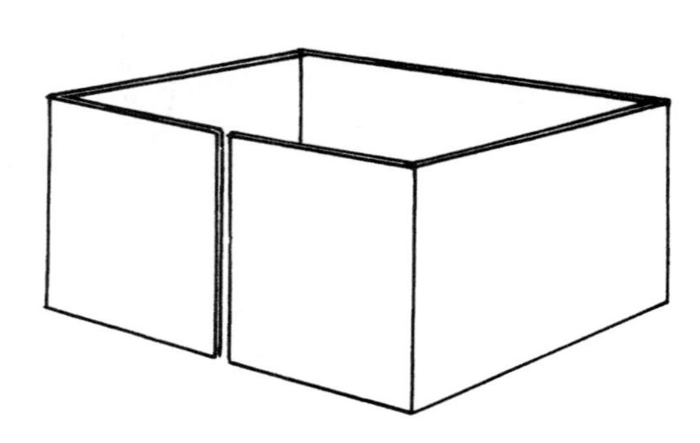

OPEN COMPARTMENT LINERS
(BOTTOM PATTERN SHOWS LINER WITH B BLOCK)

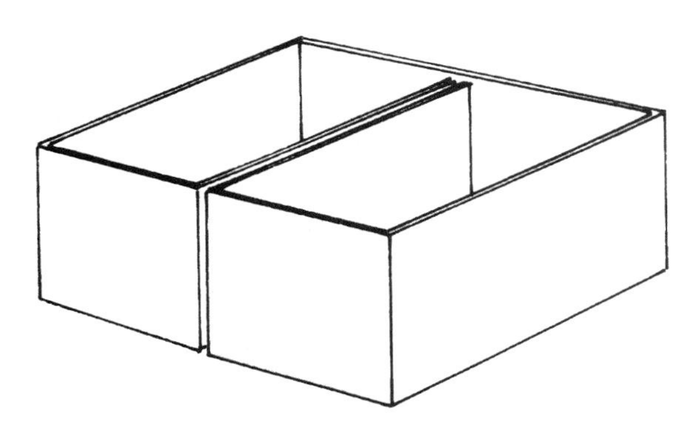

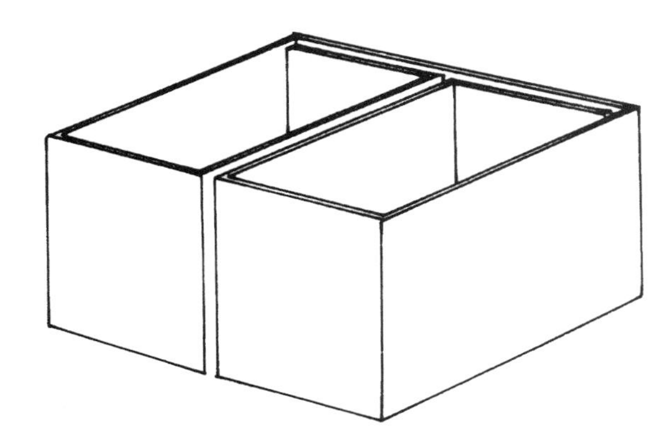

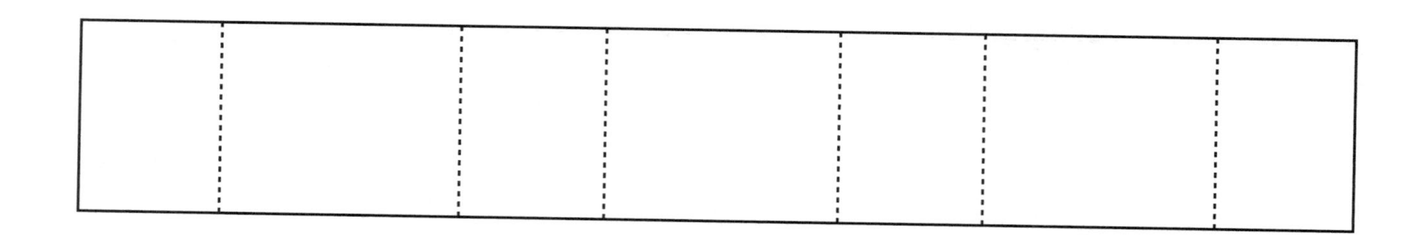

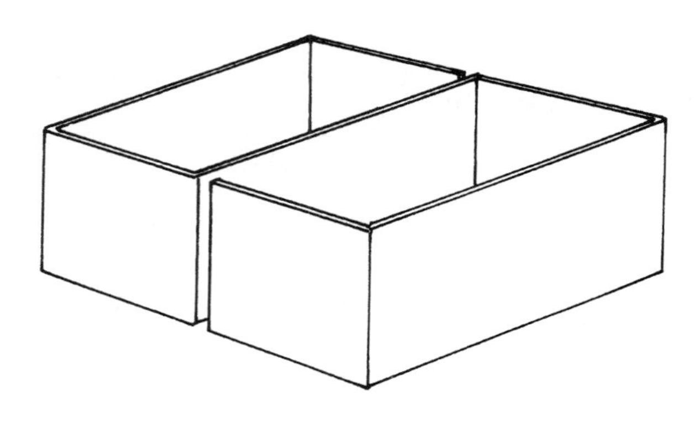

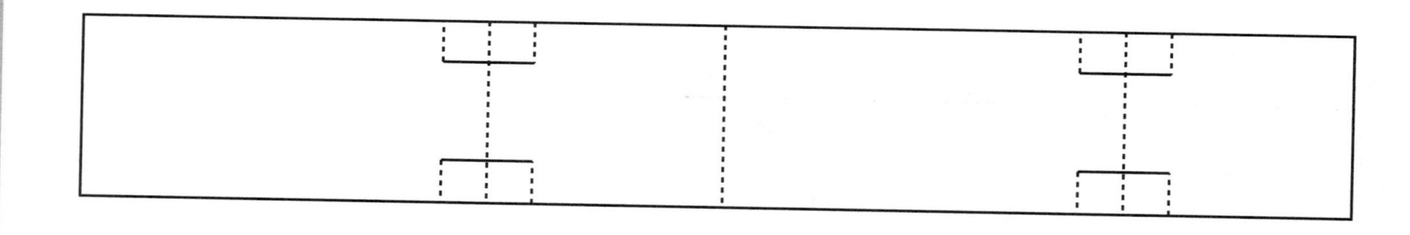

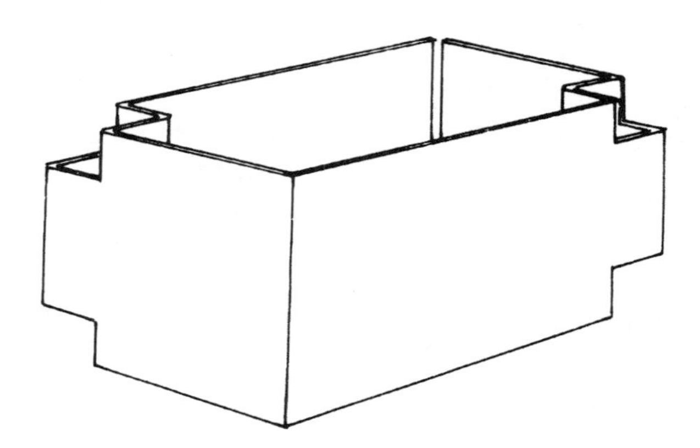

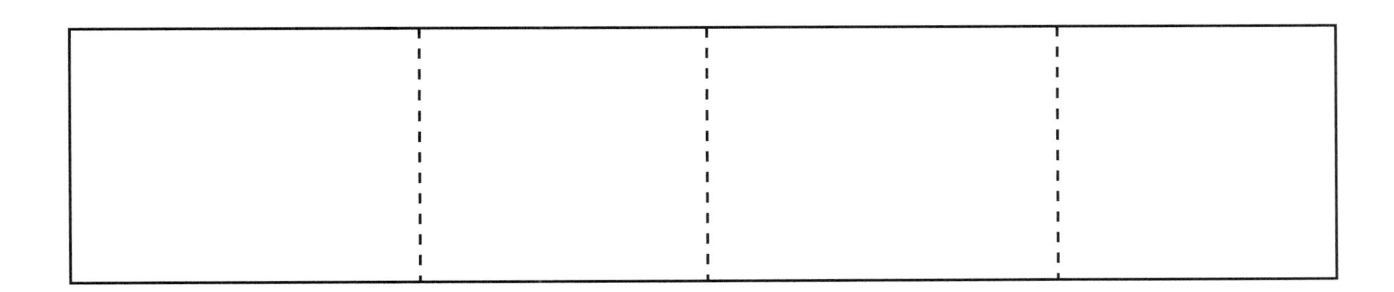

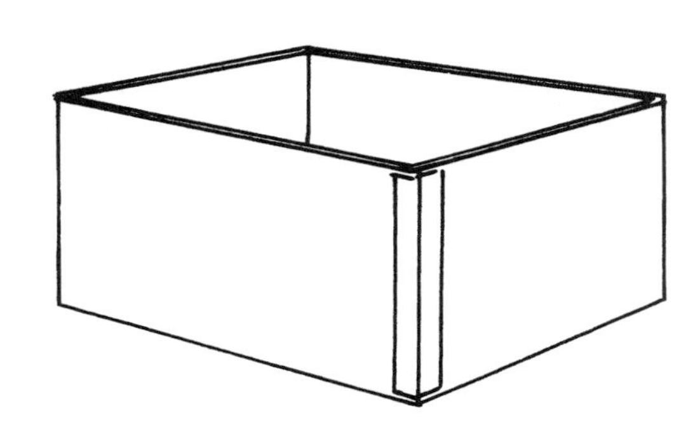

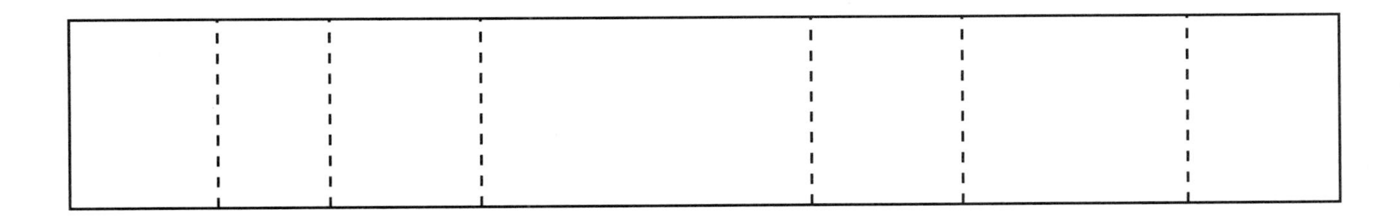

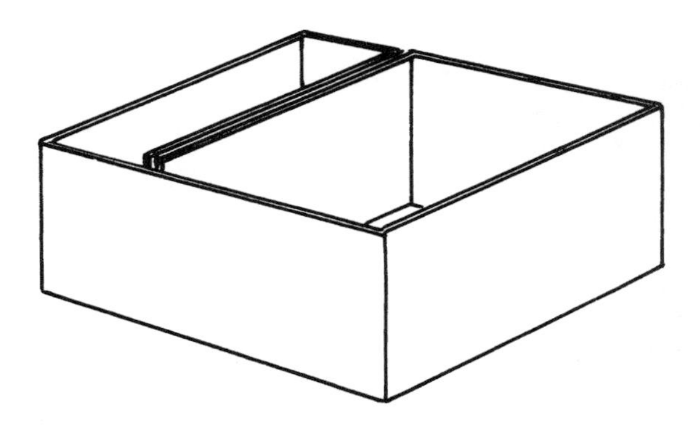

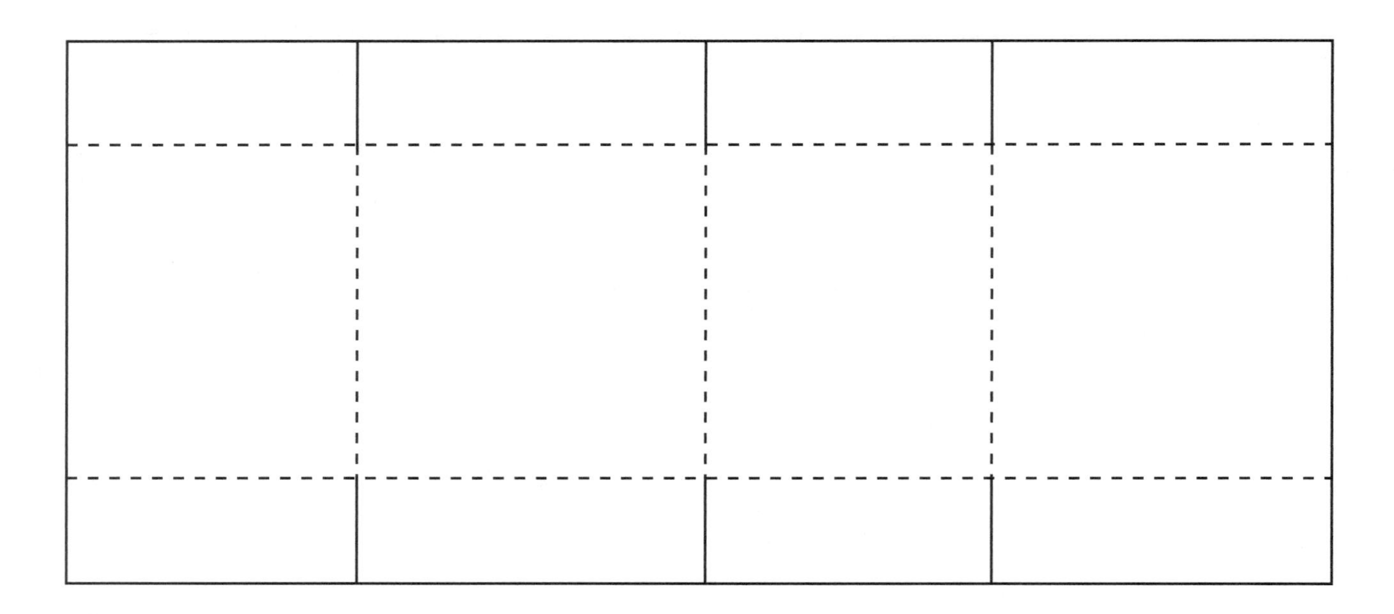

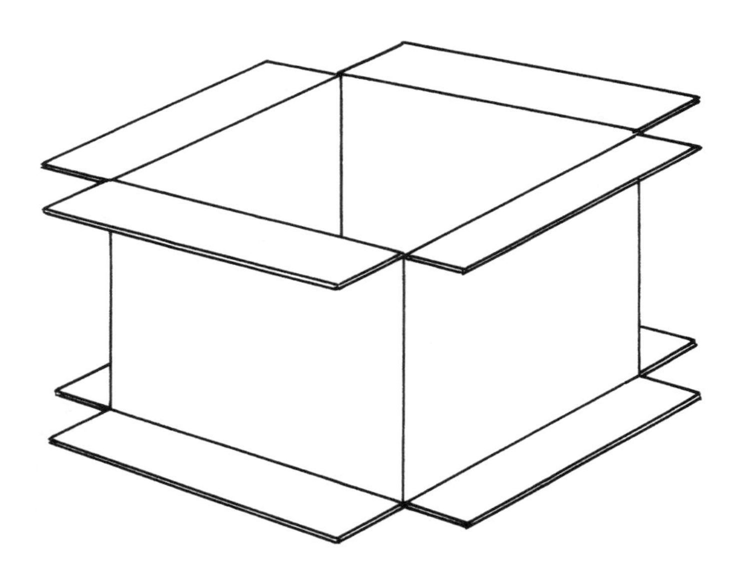

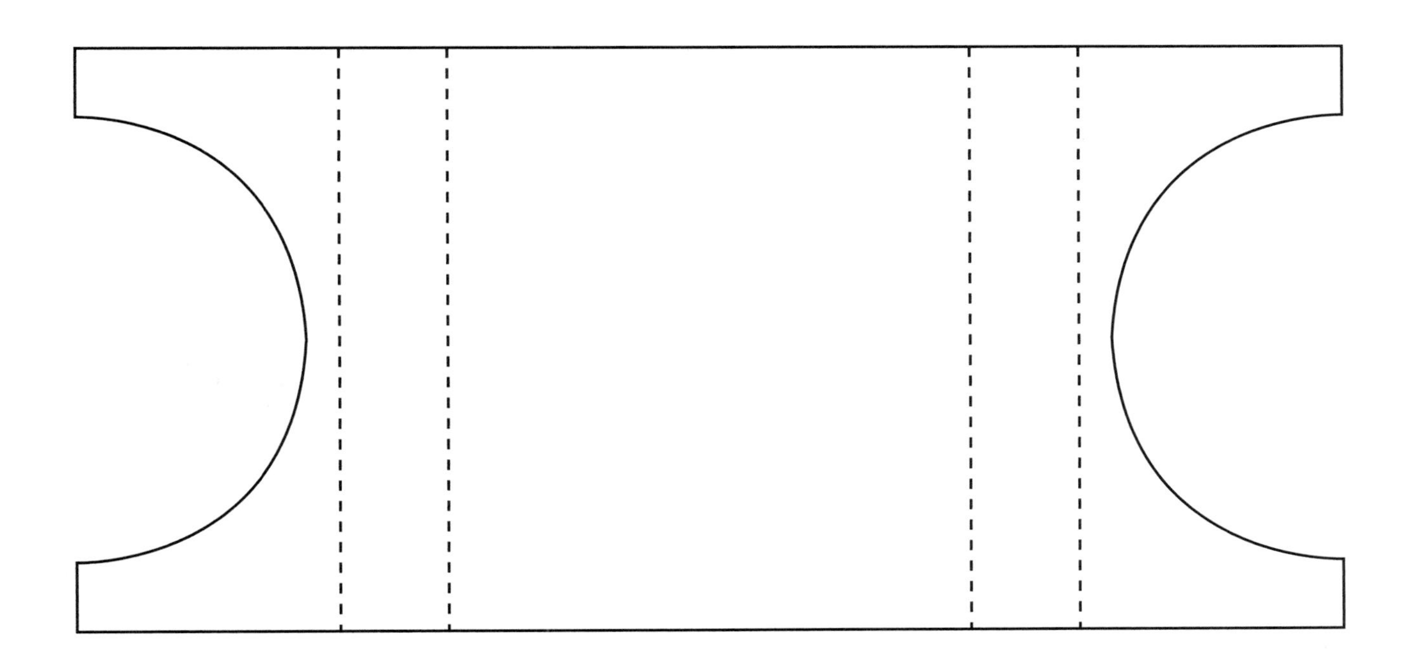

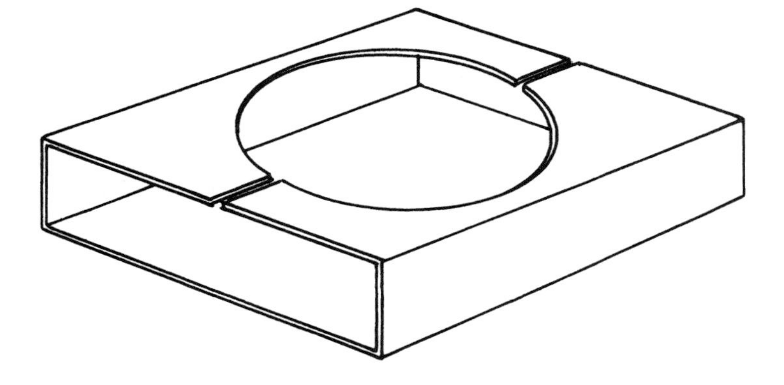

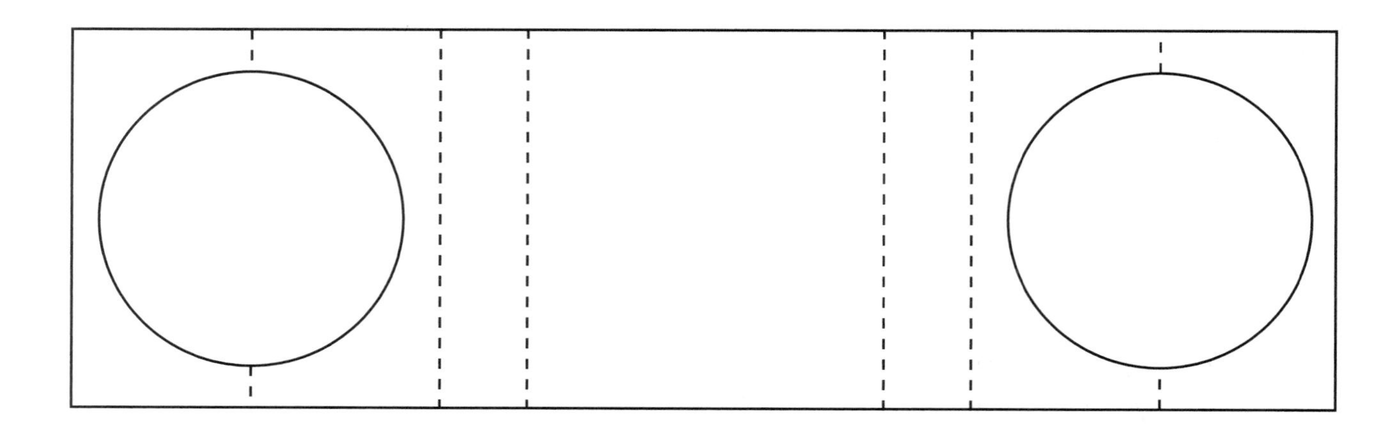

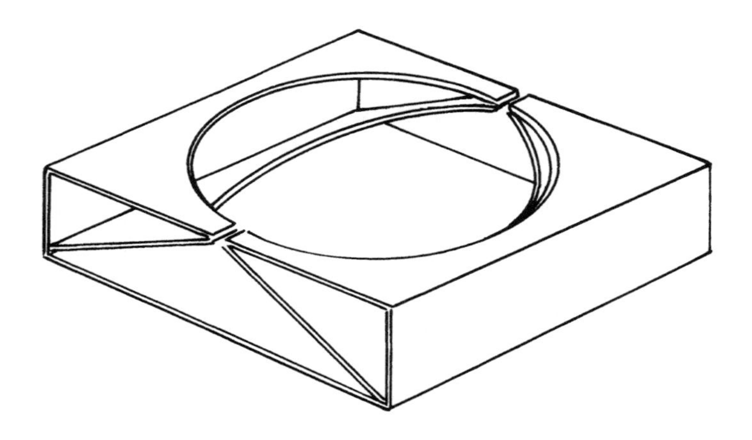

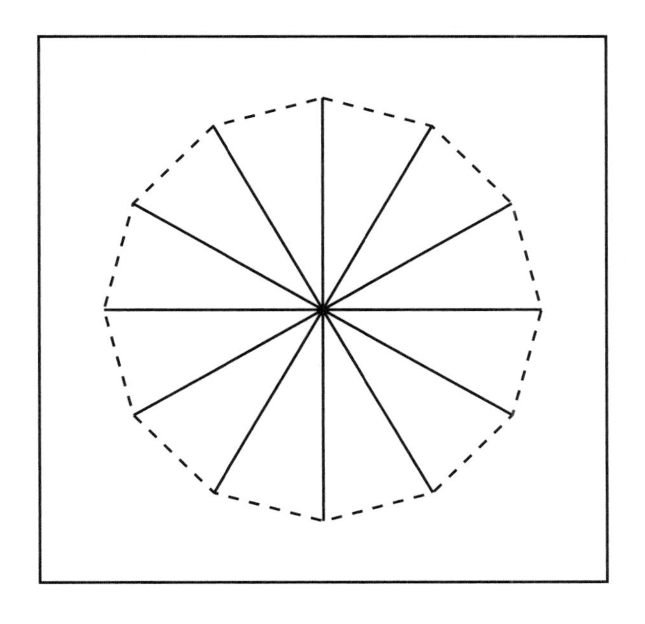

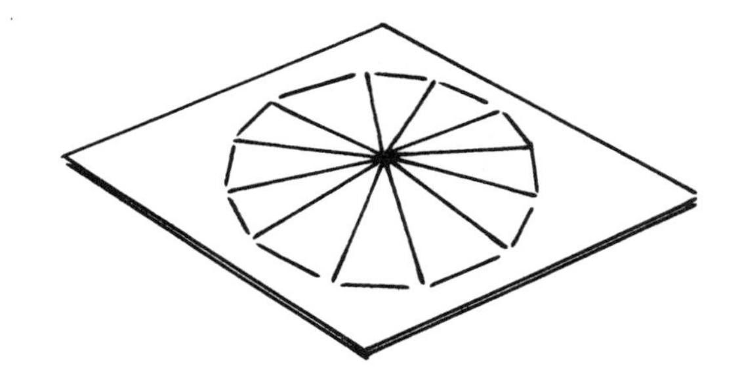

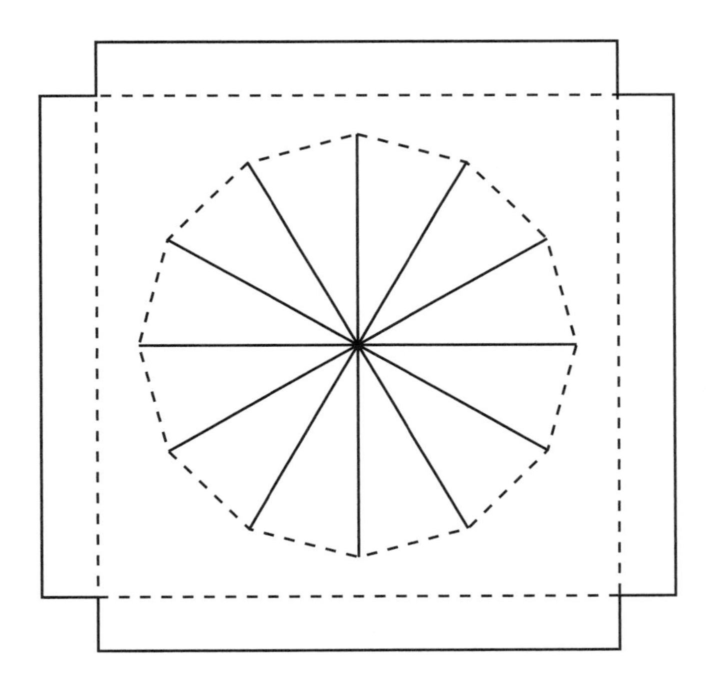

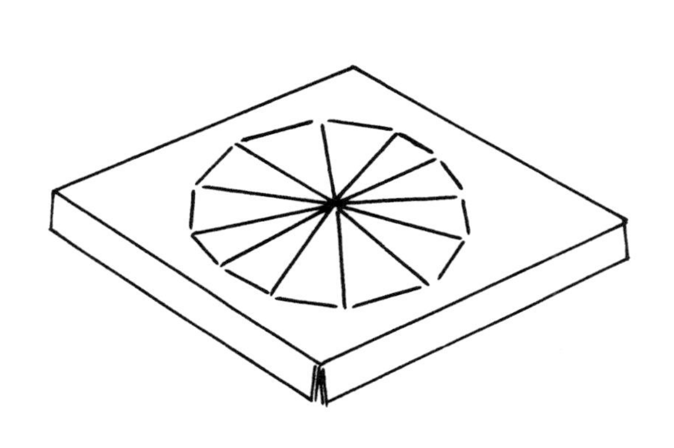

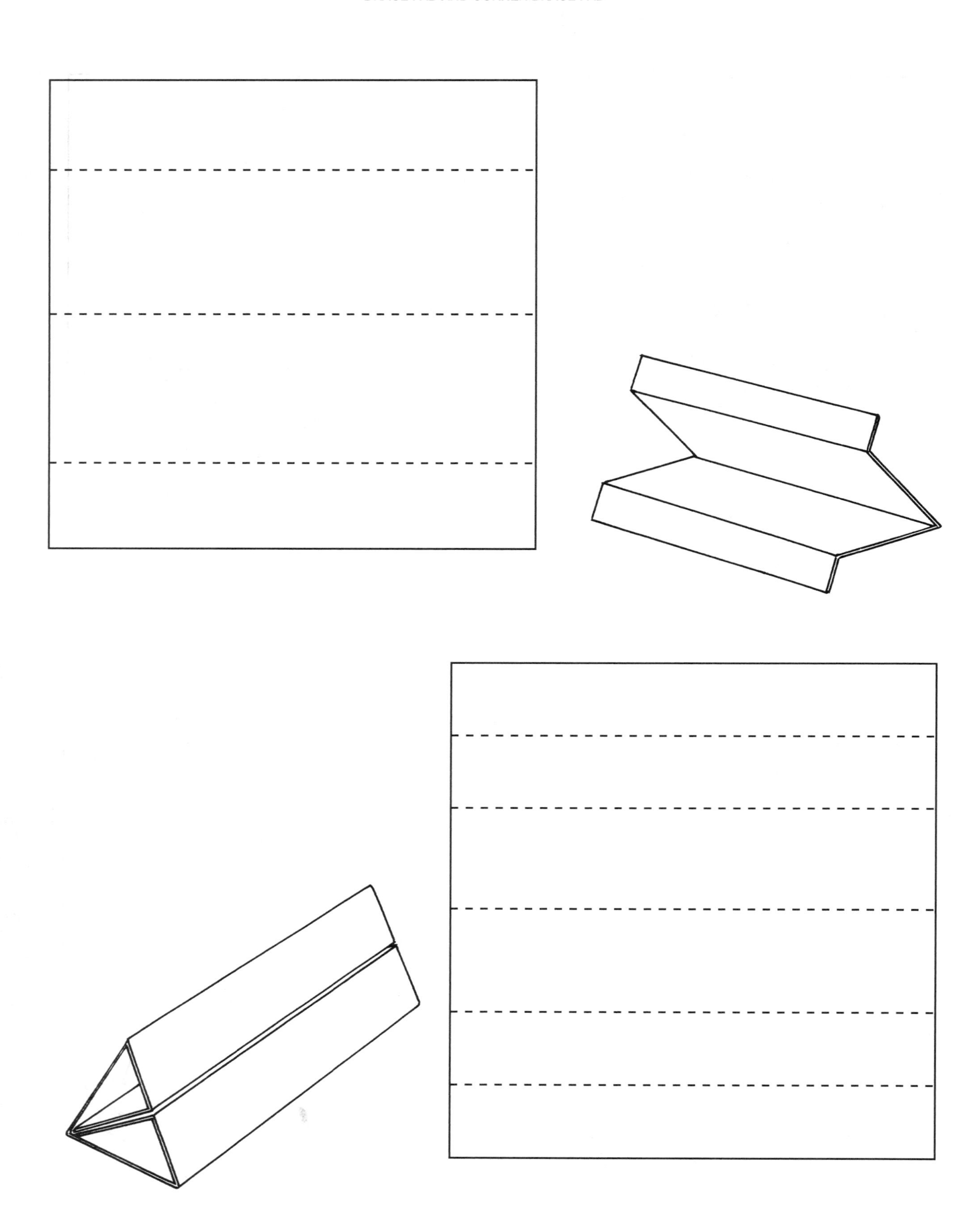

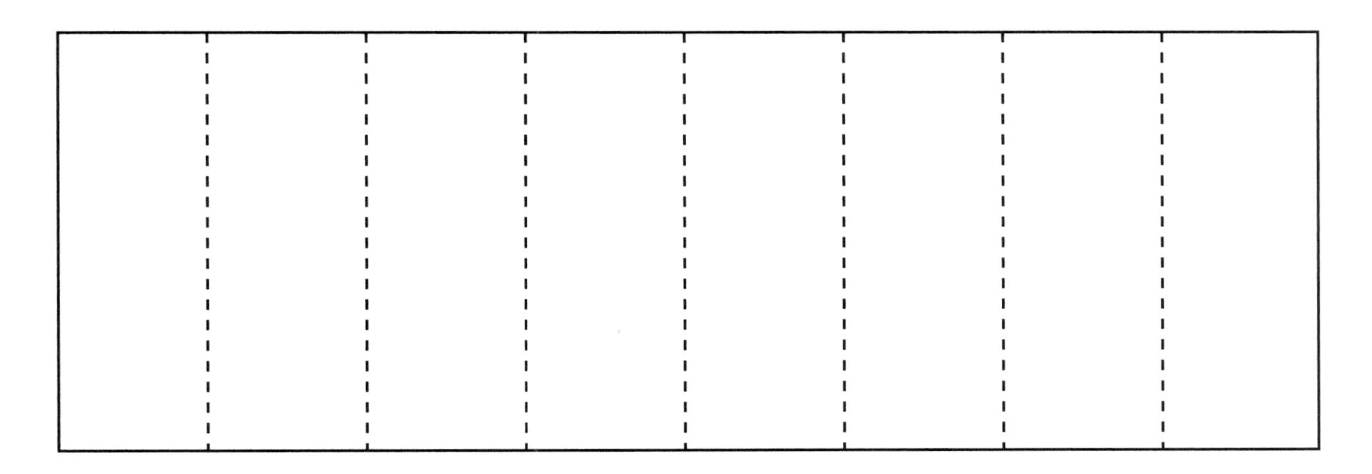

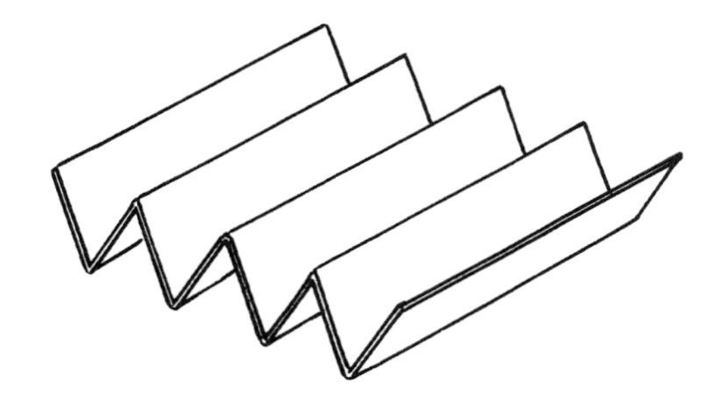

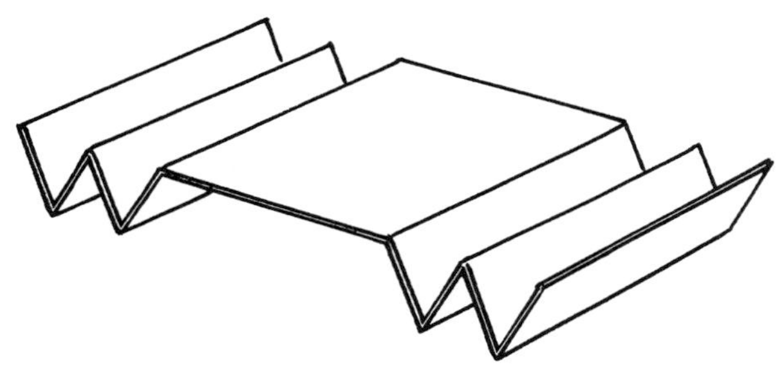

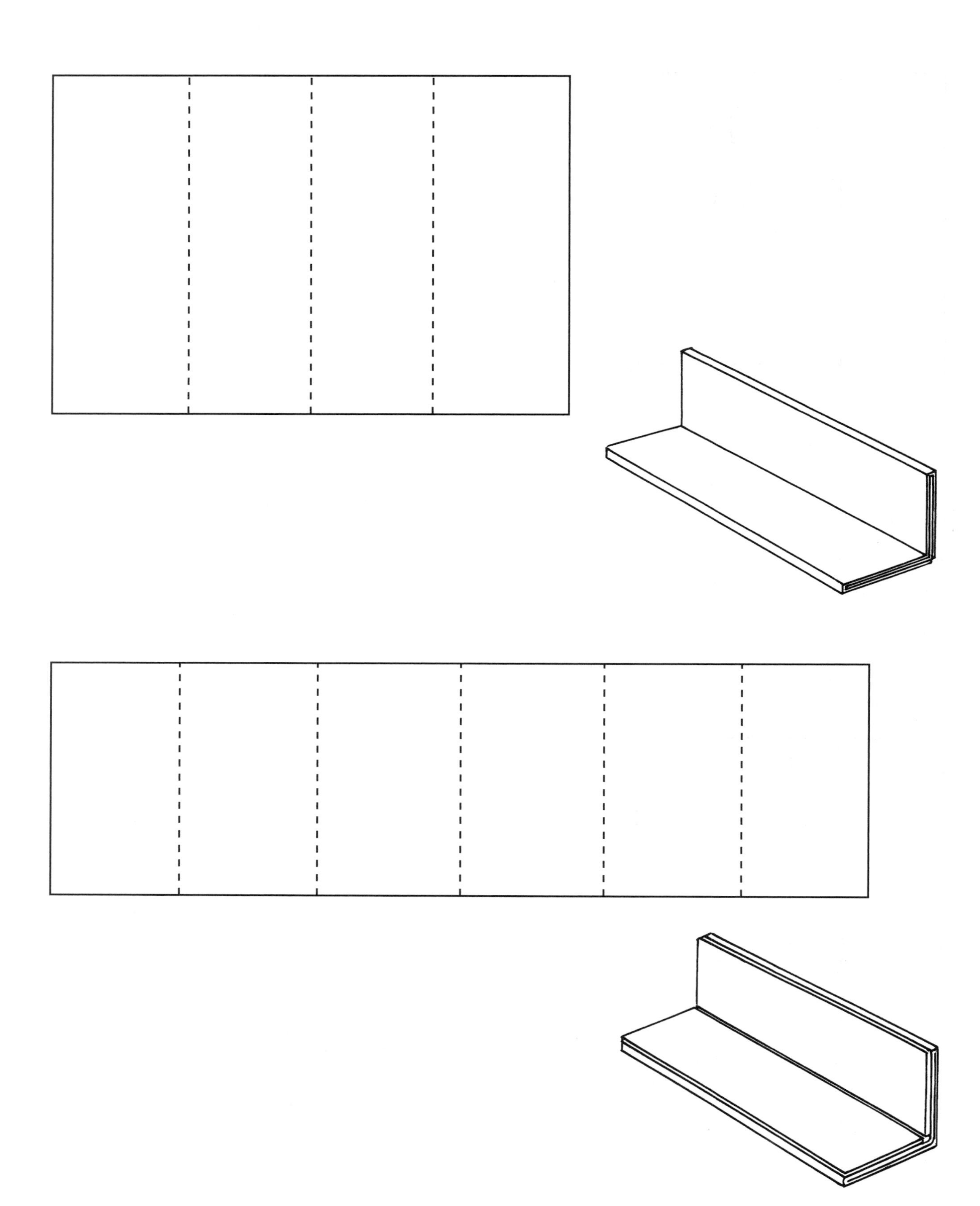

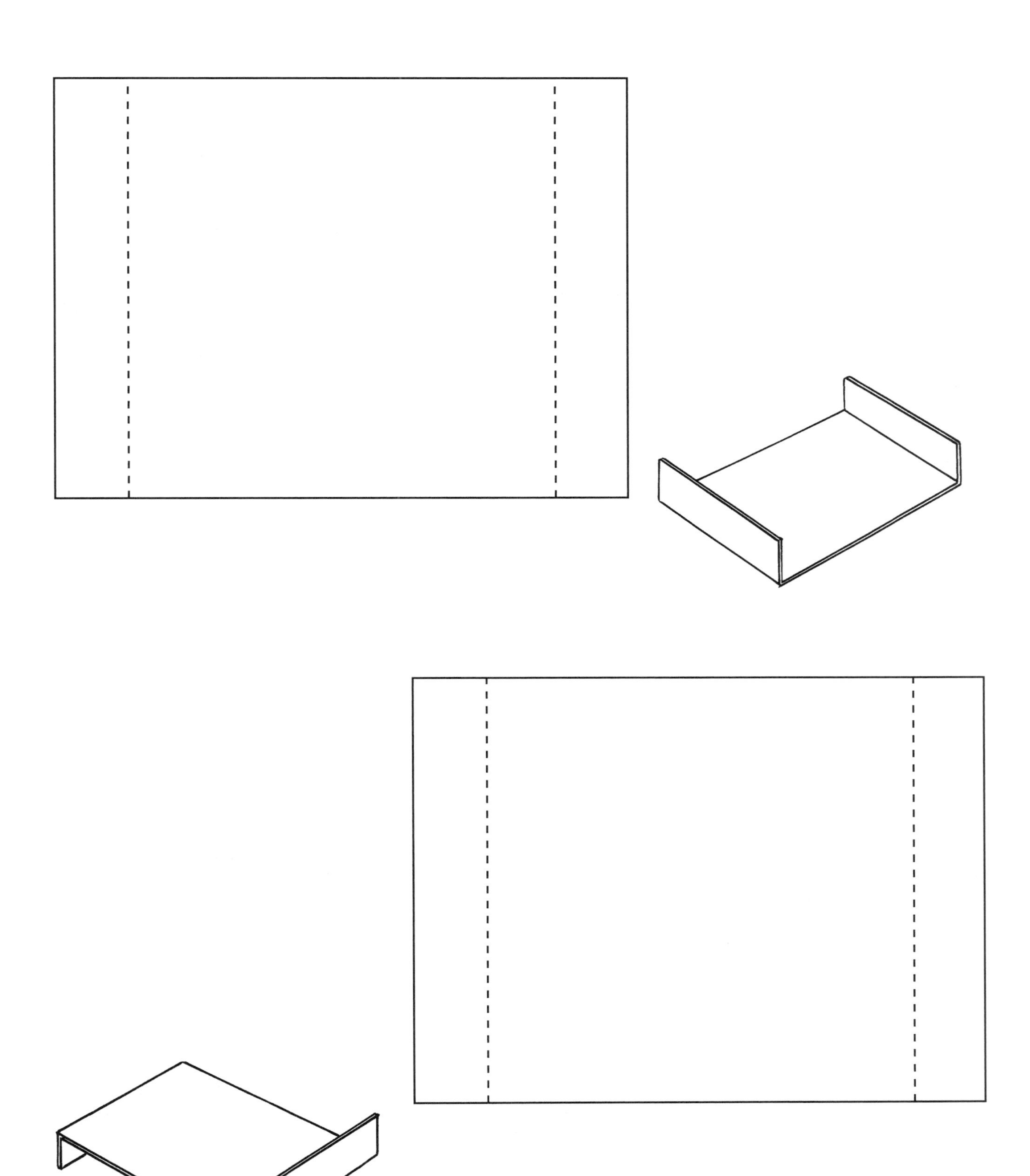

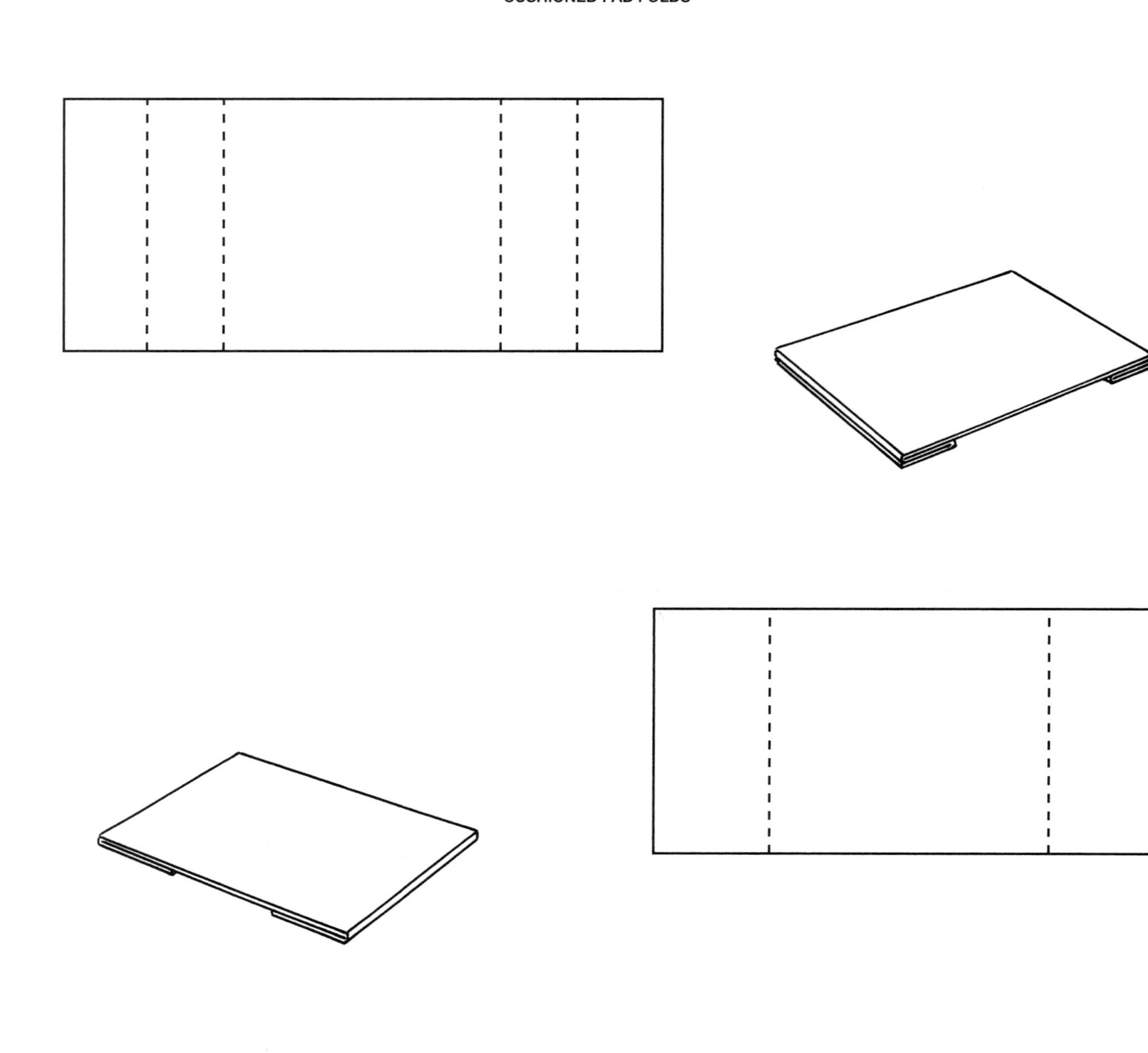

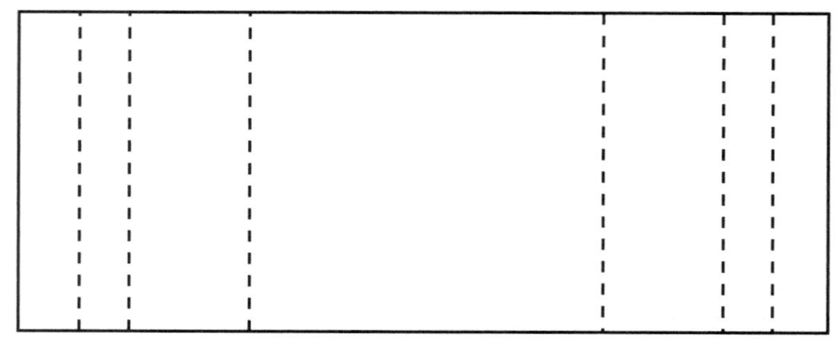

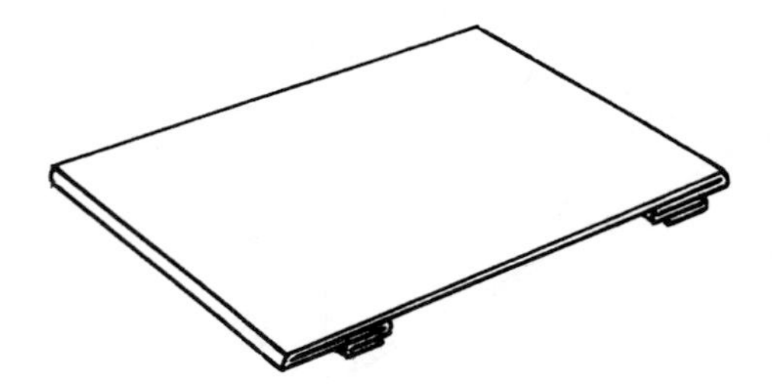

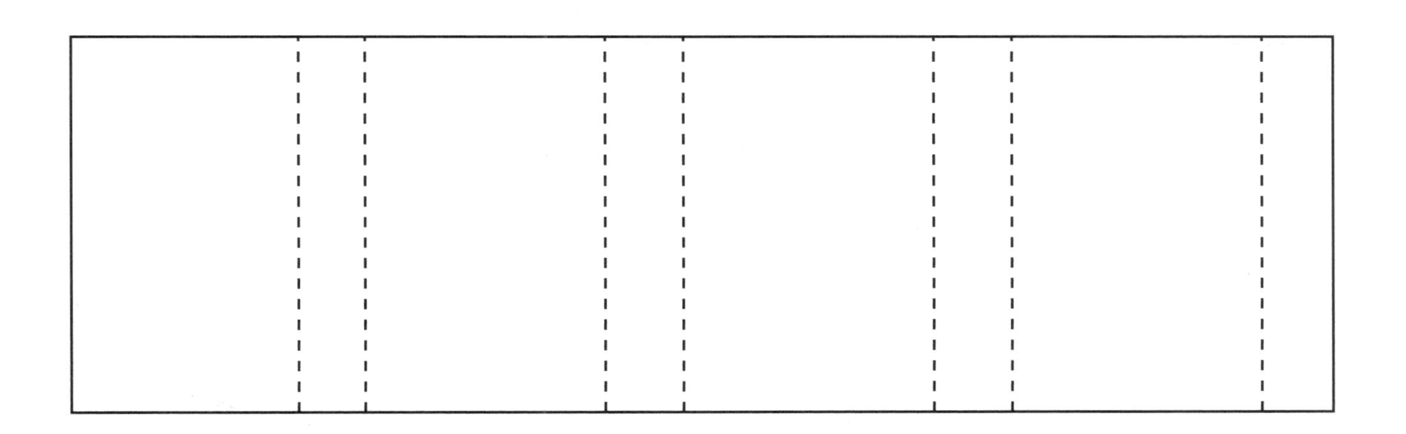

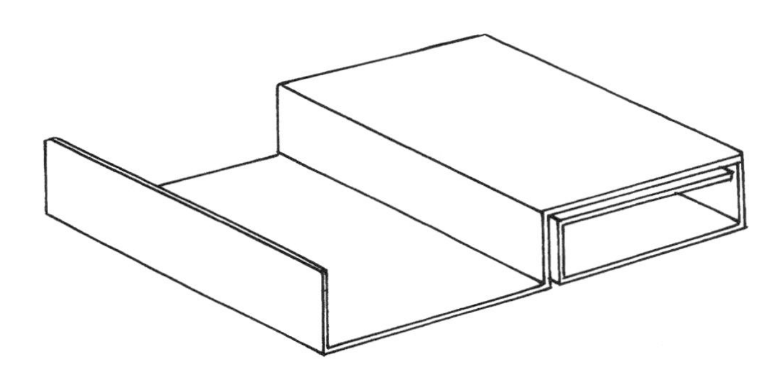

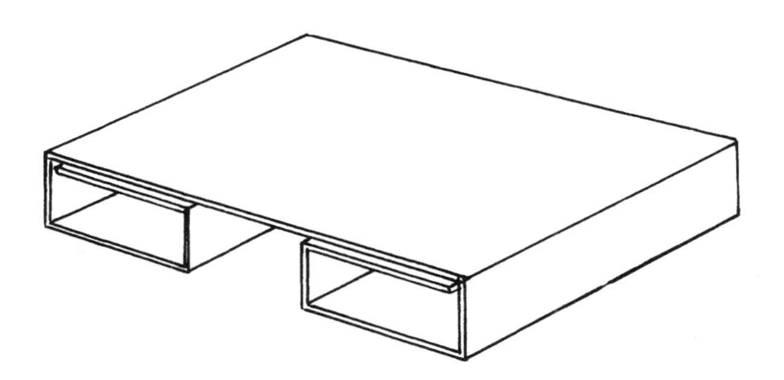

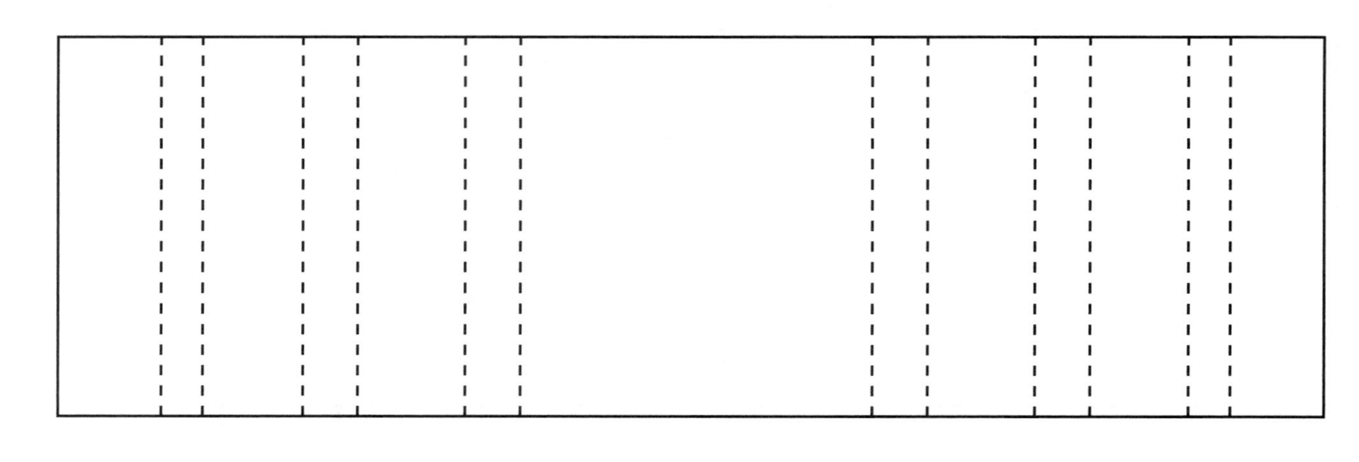

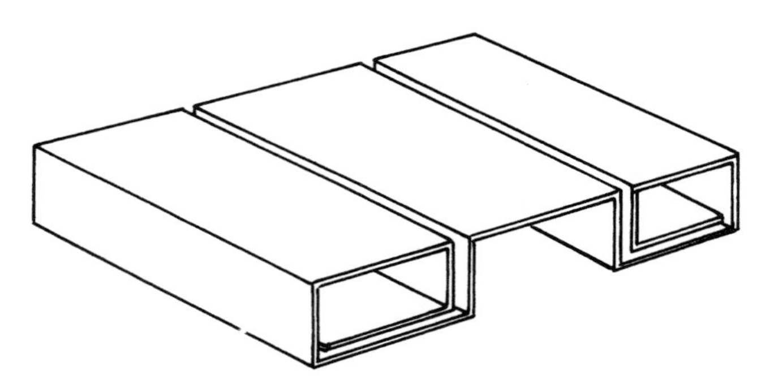

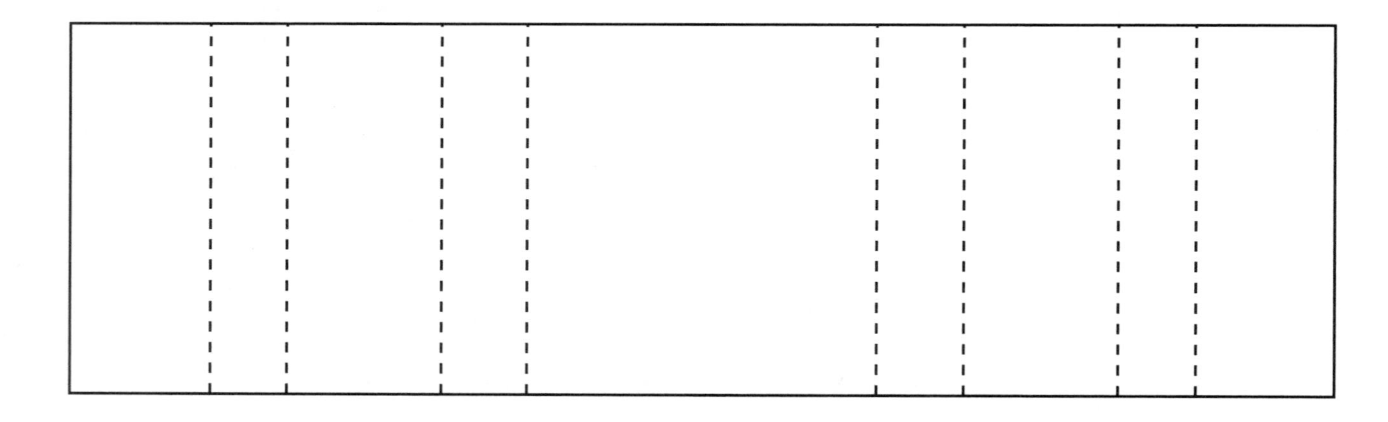

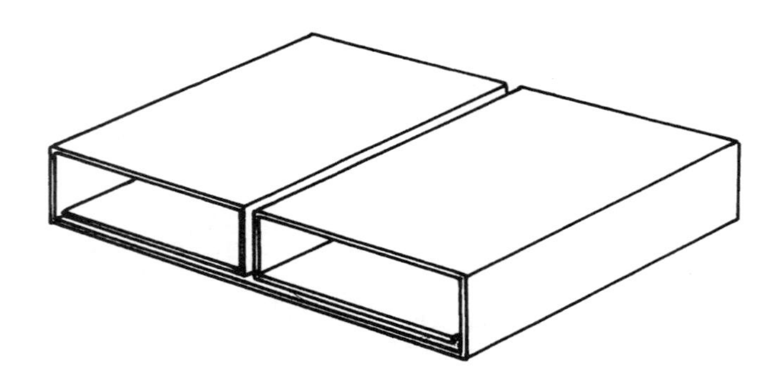

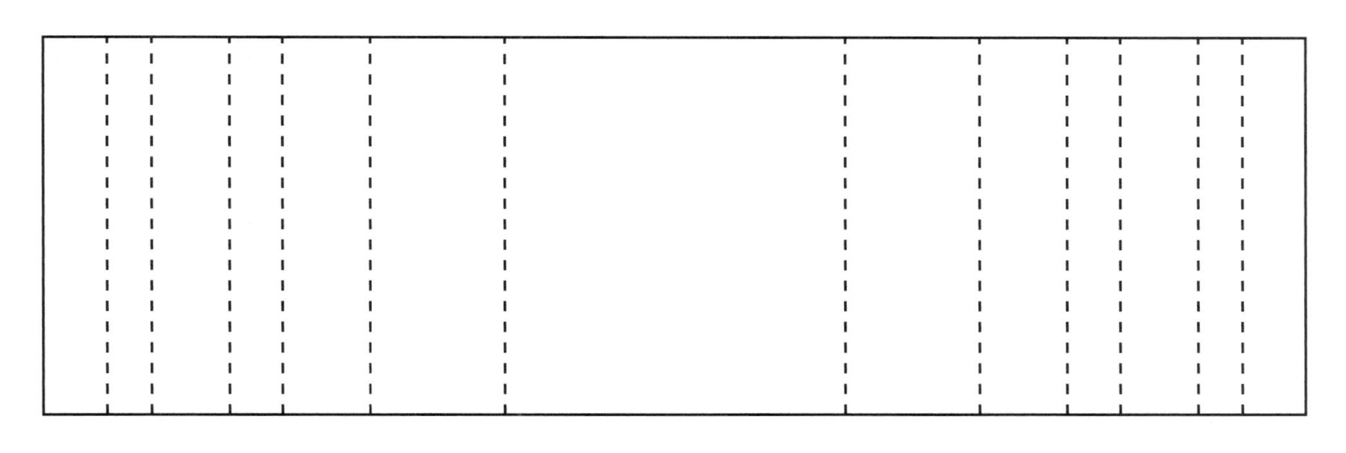

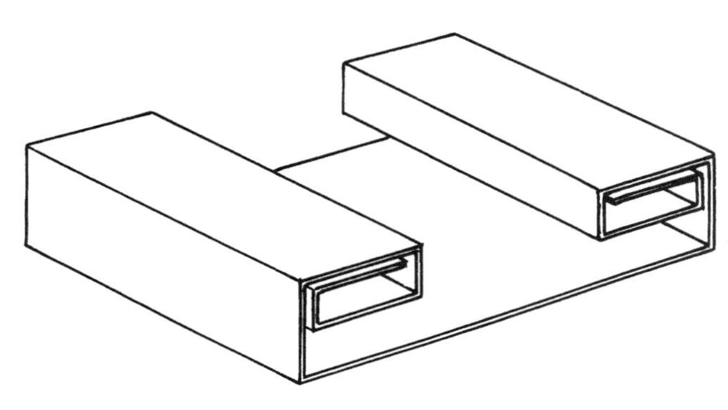

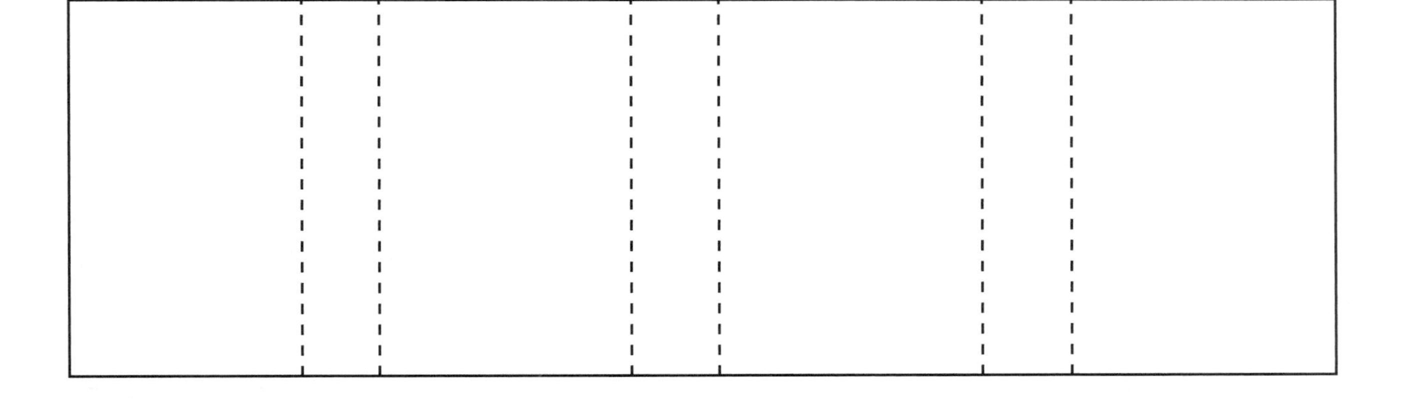

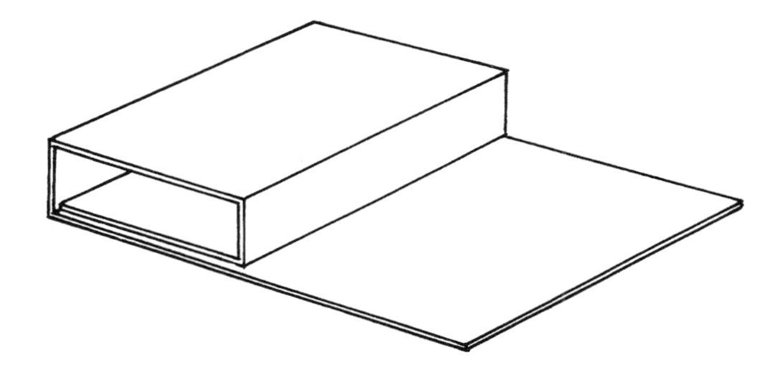

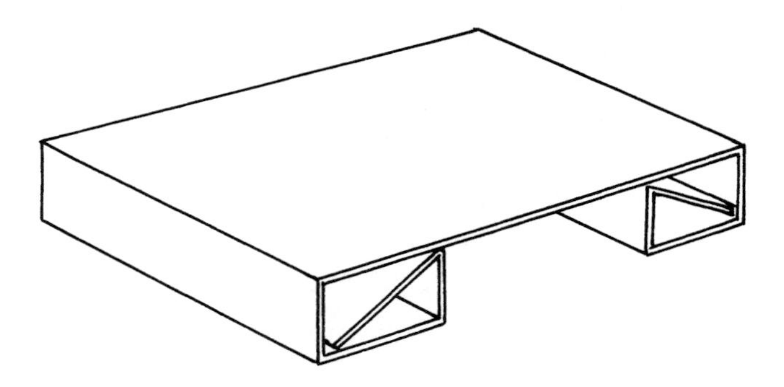

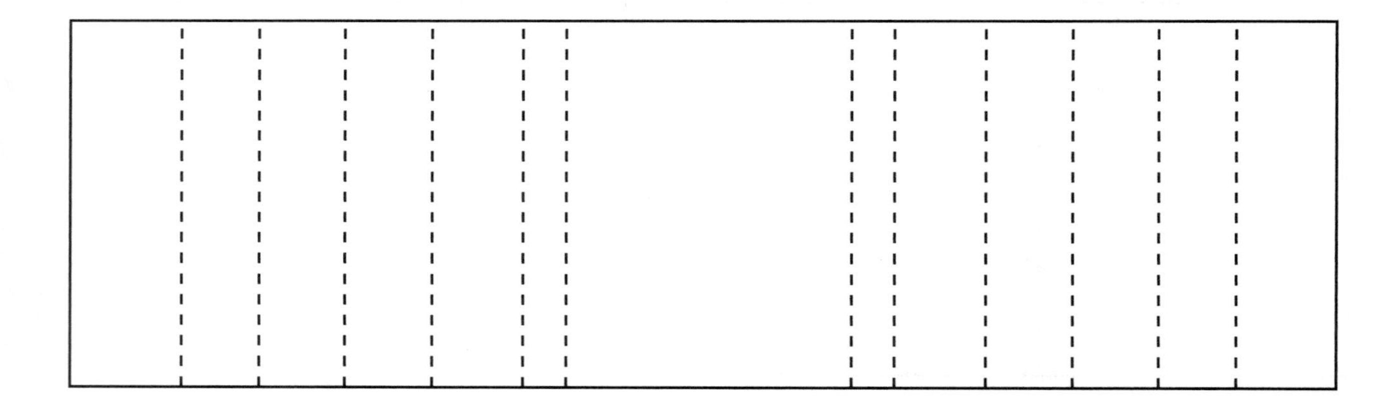

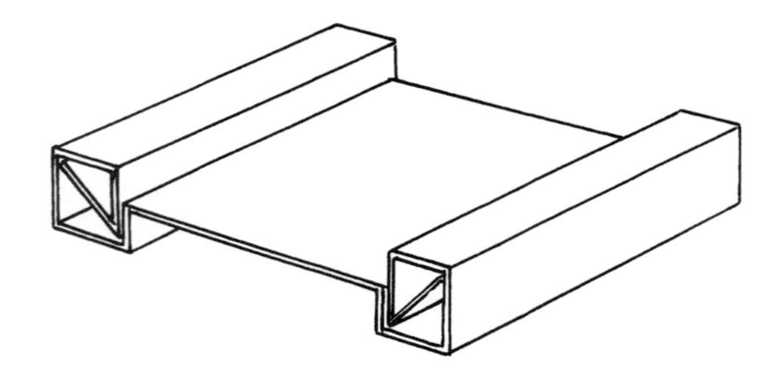

SCORED LINER

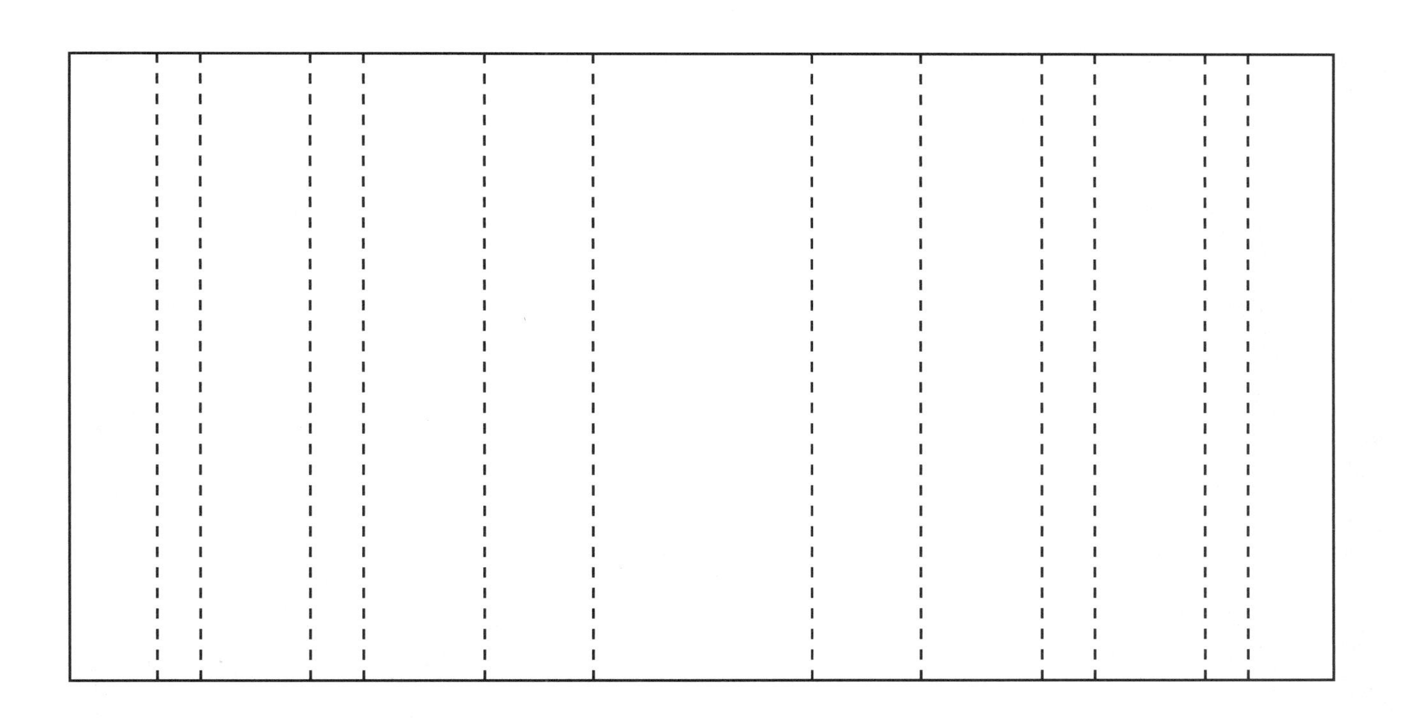

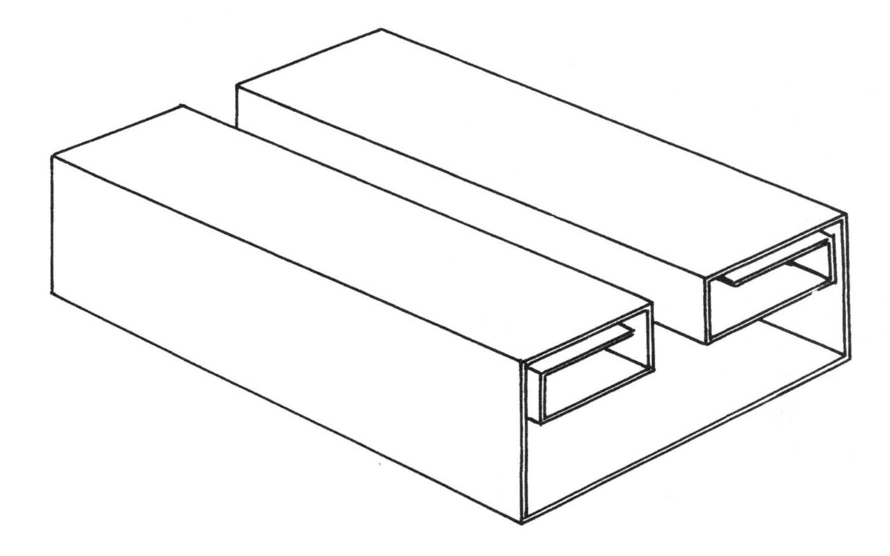

Corrugated Containers

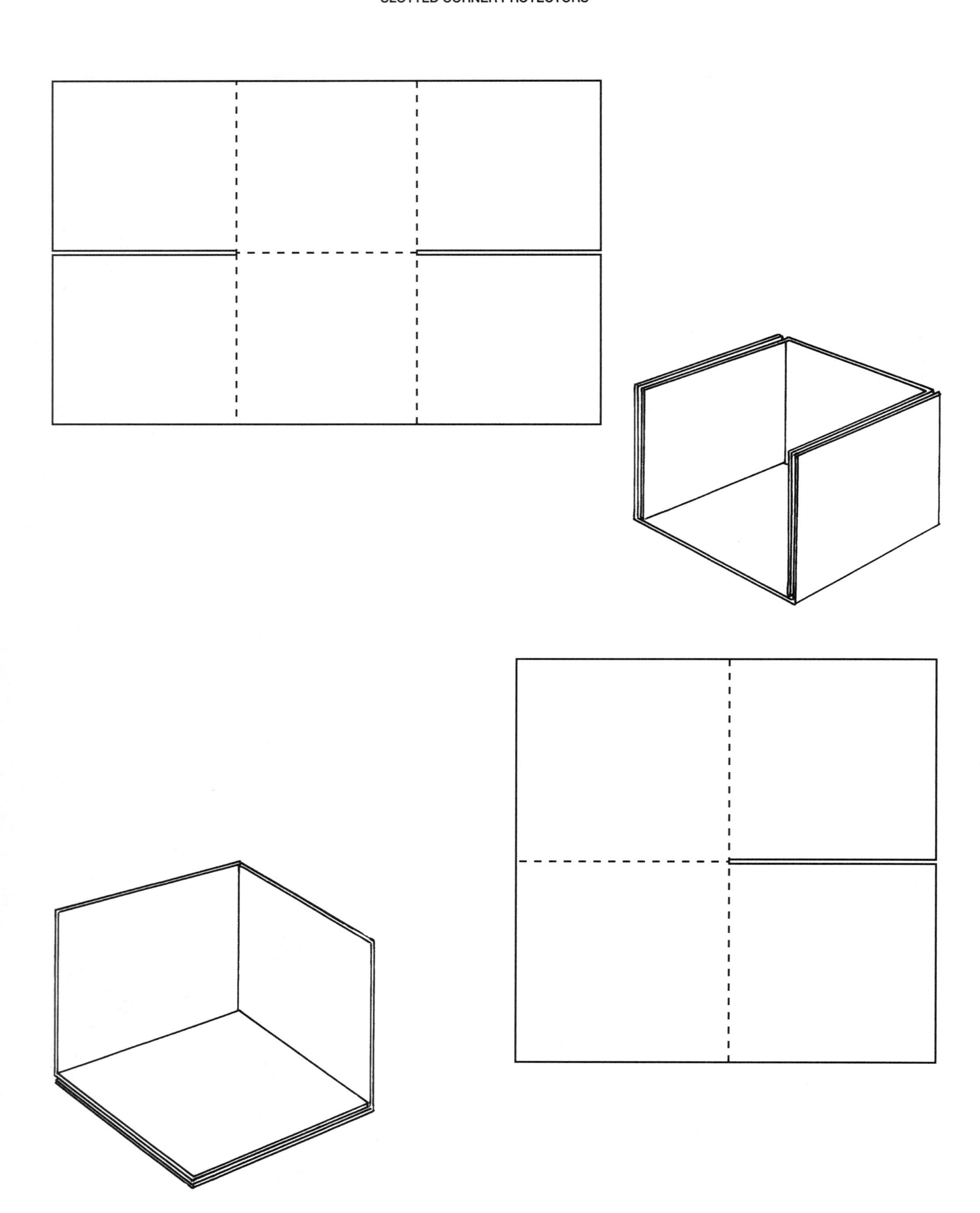

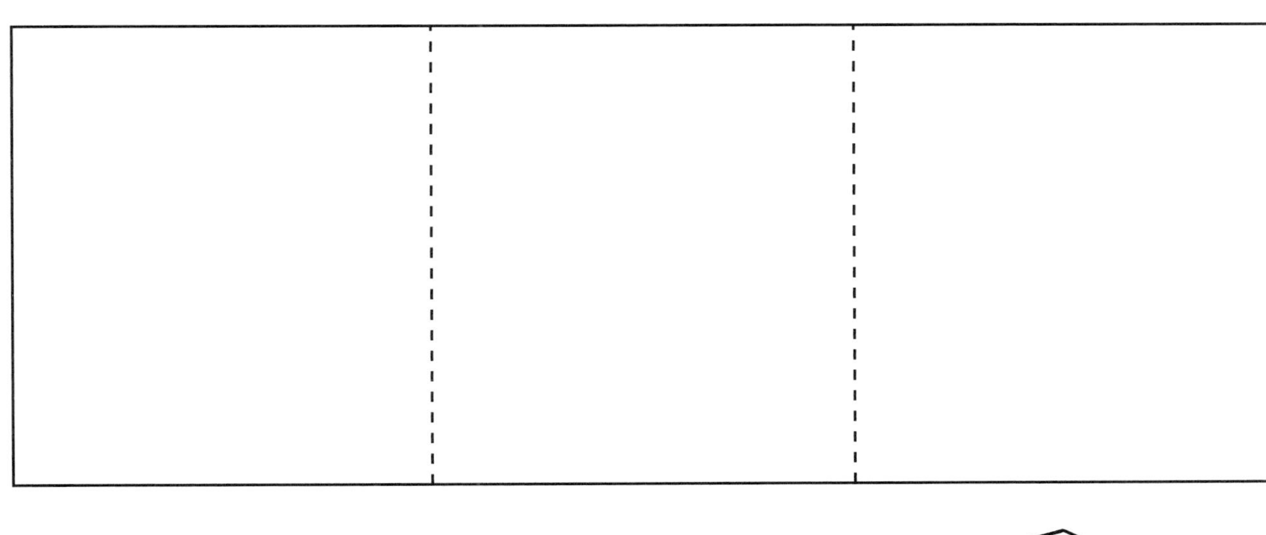

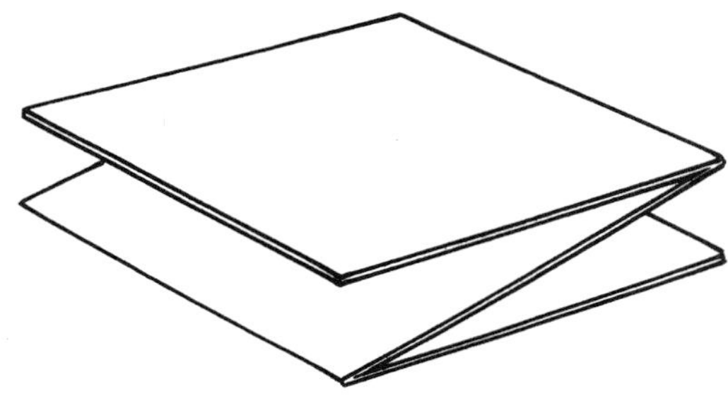

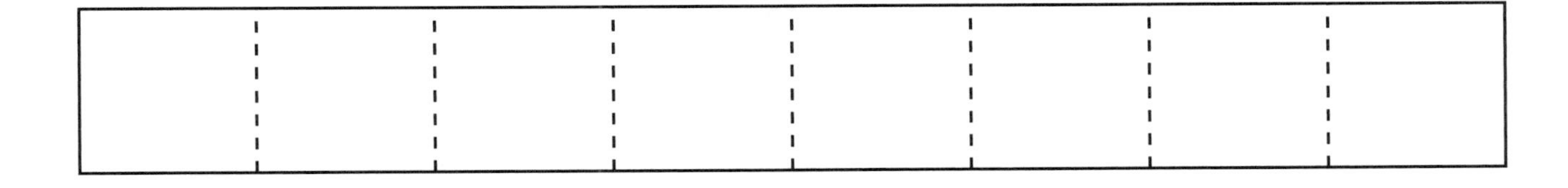

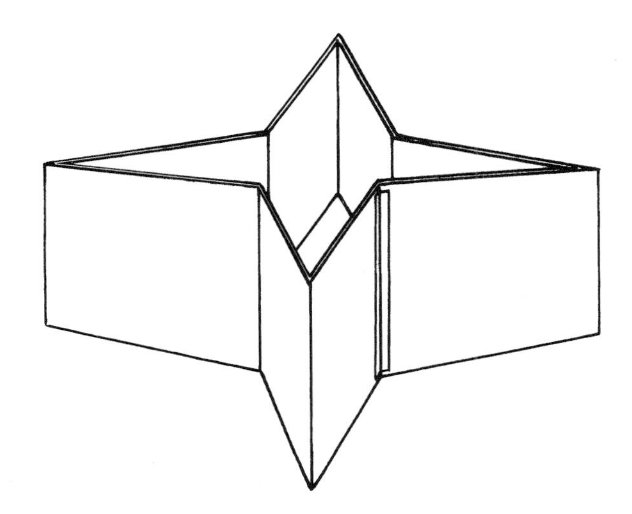

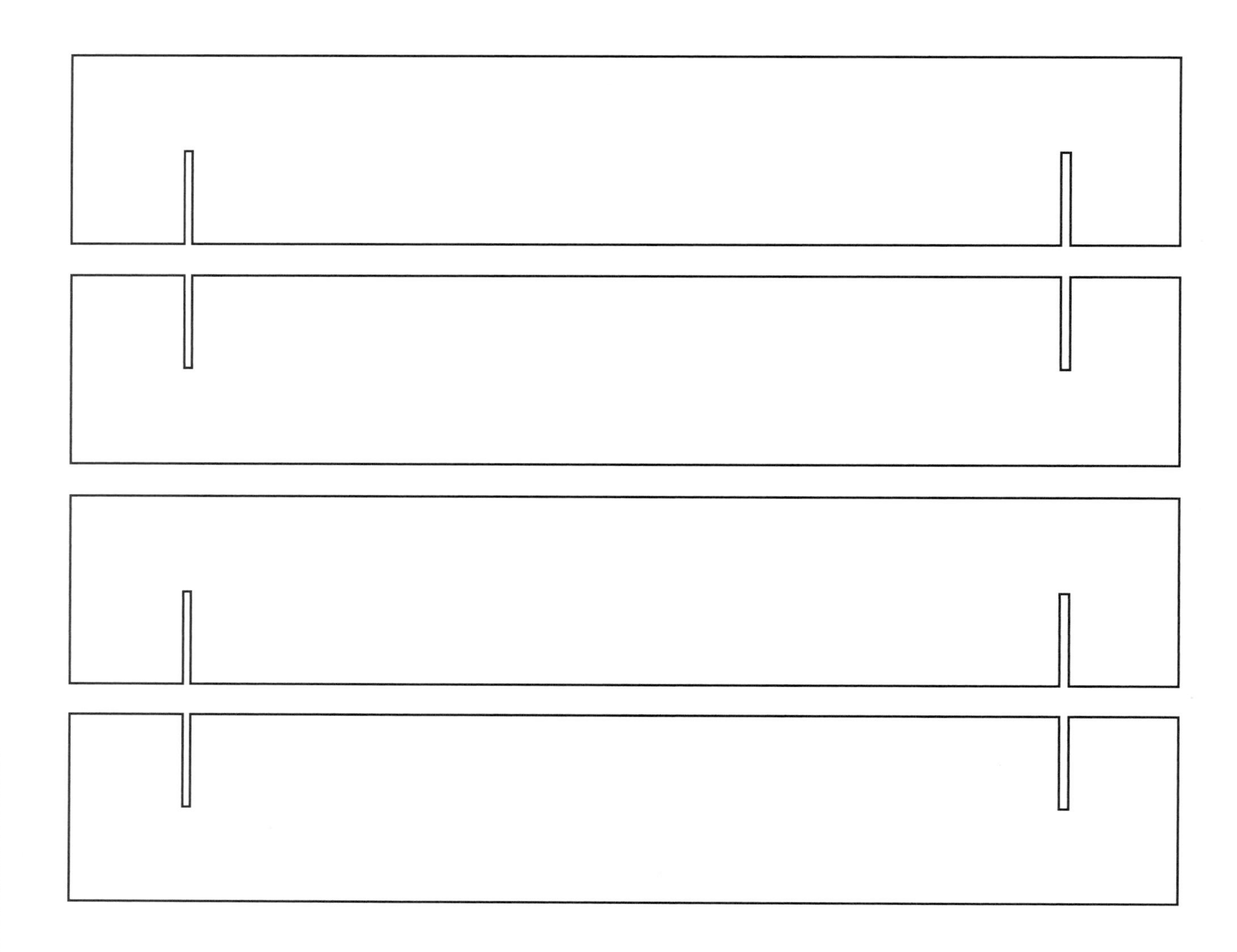

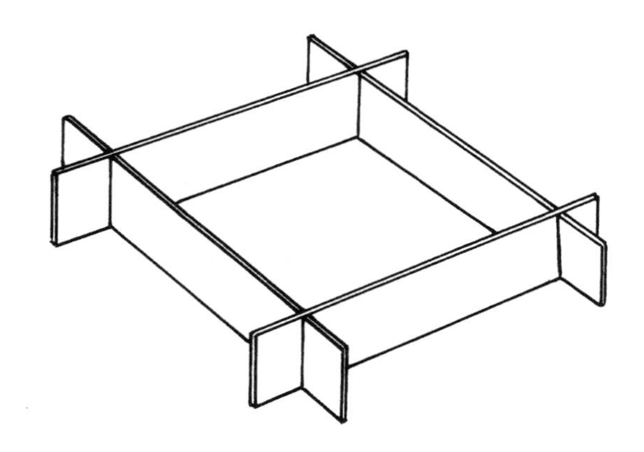

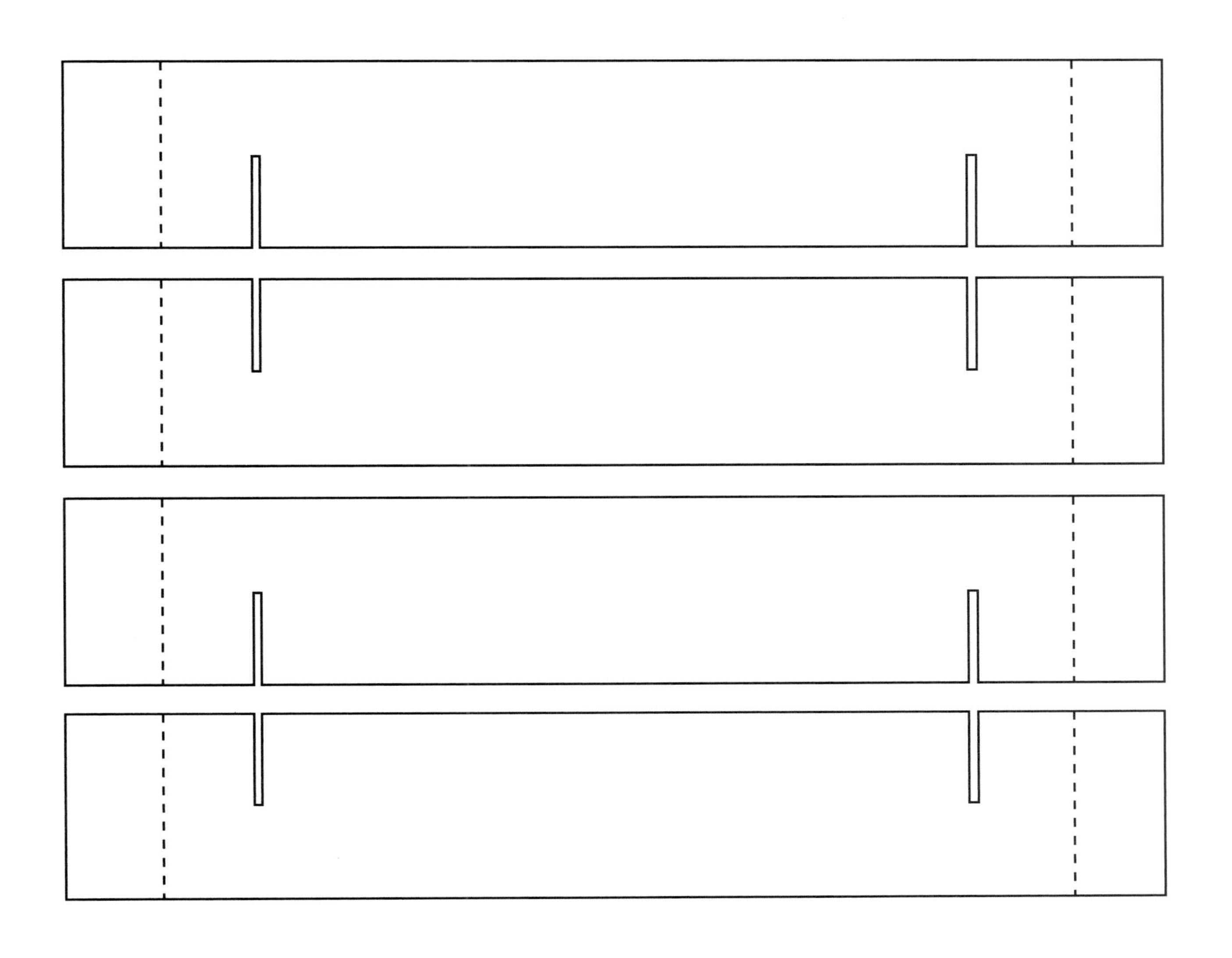

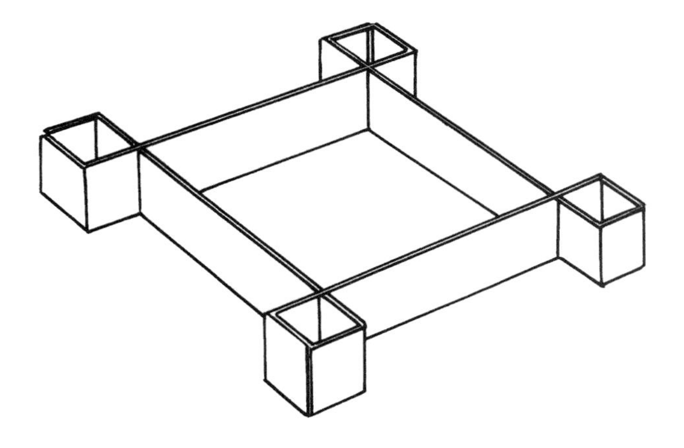

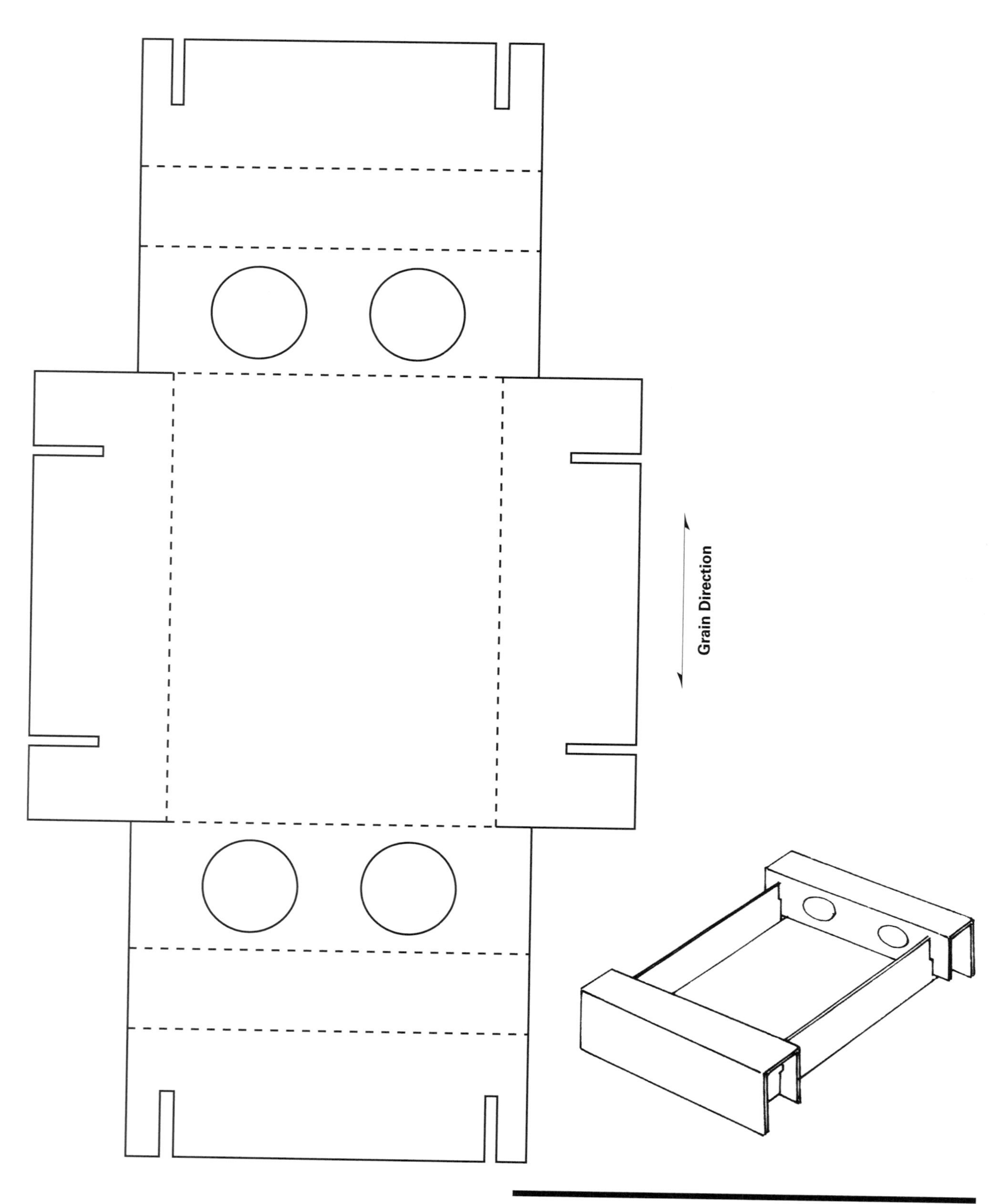

This protective cushioning device may be die-cut to conform to many different fragile objects.

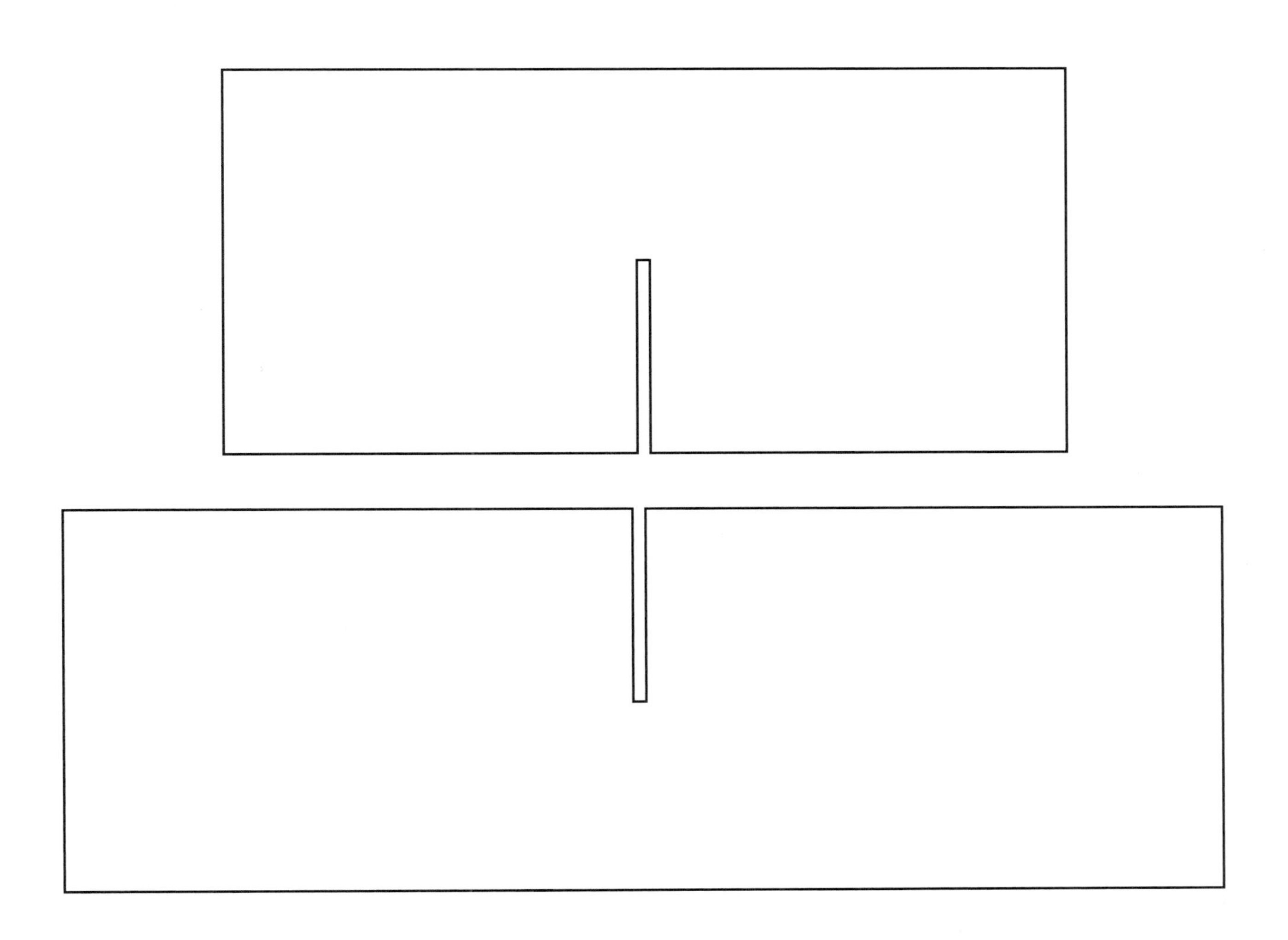

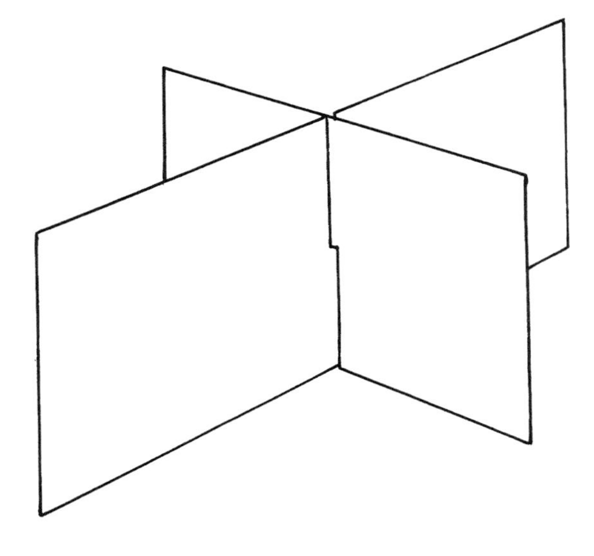

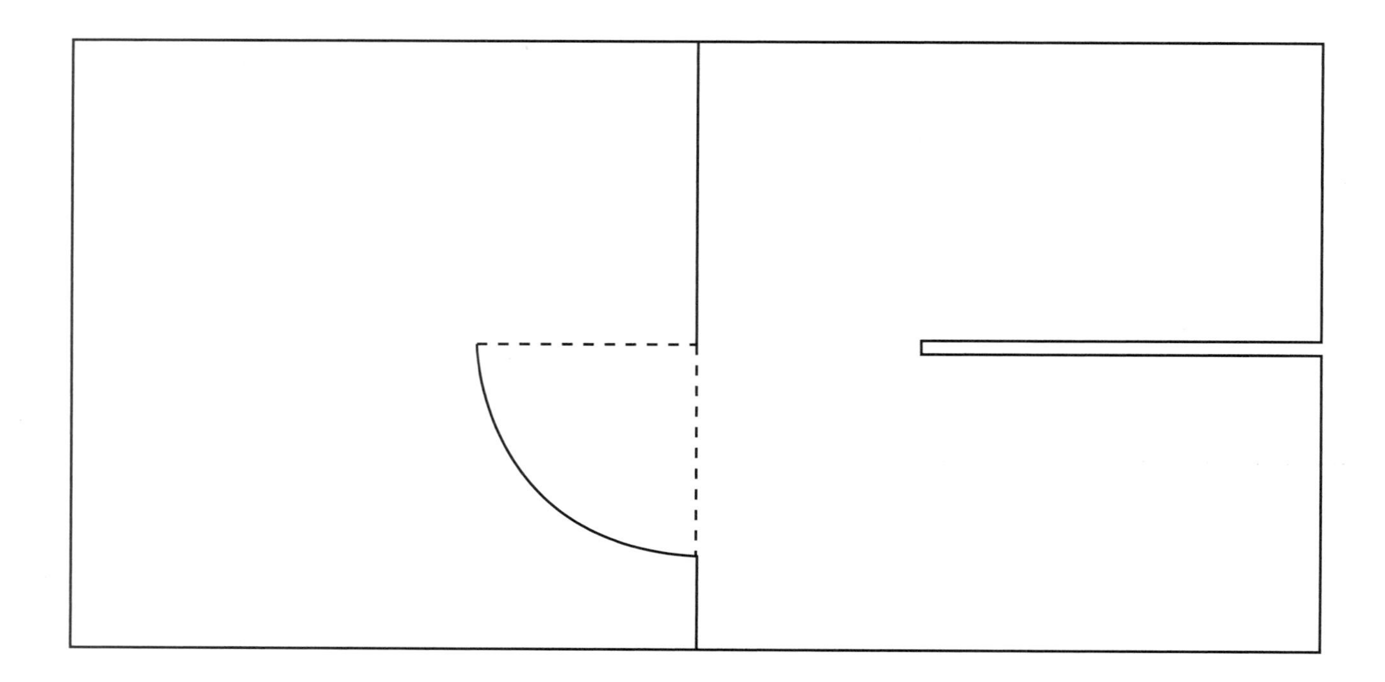

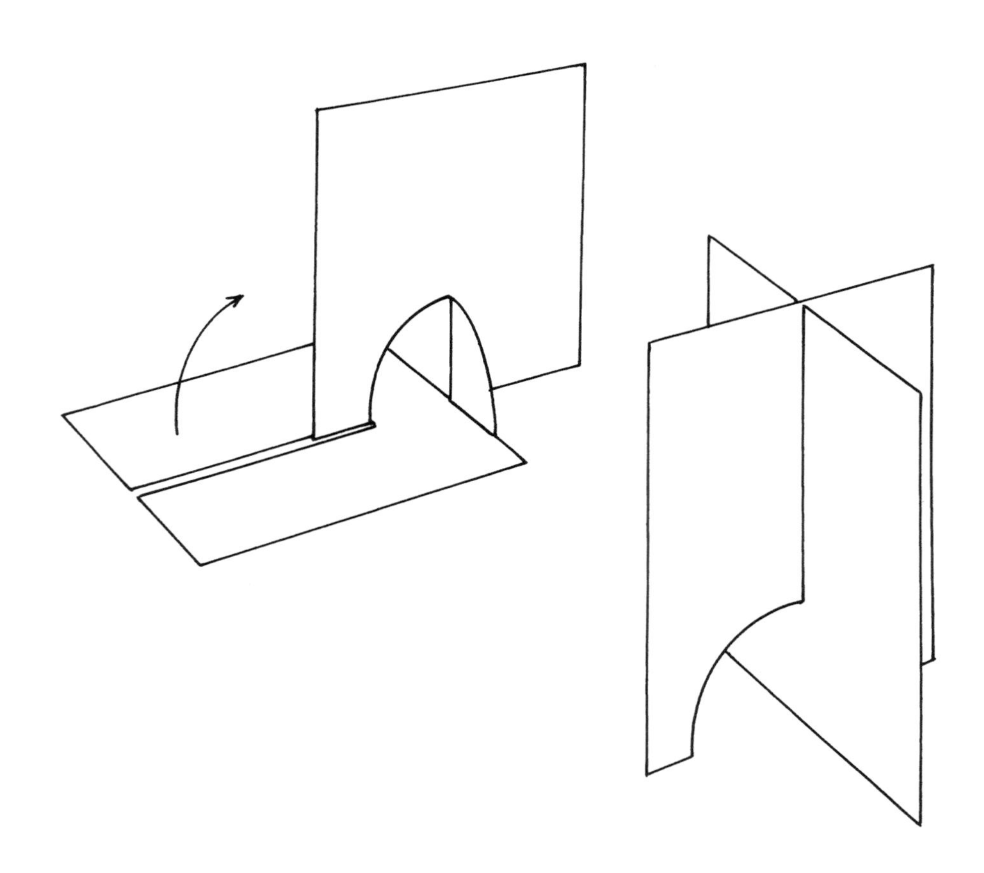

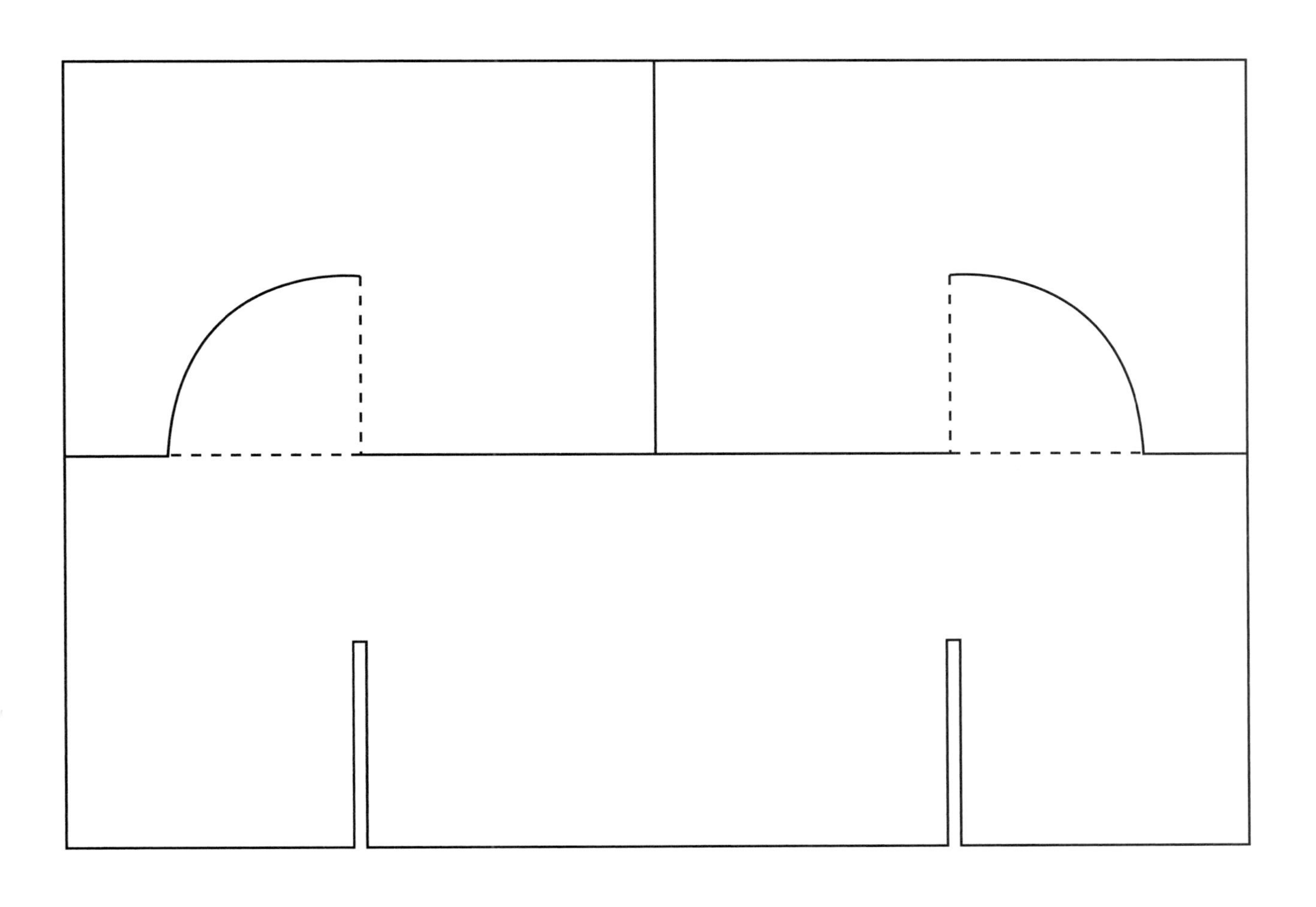

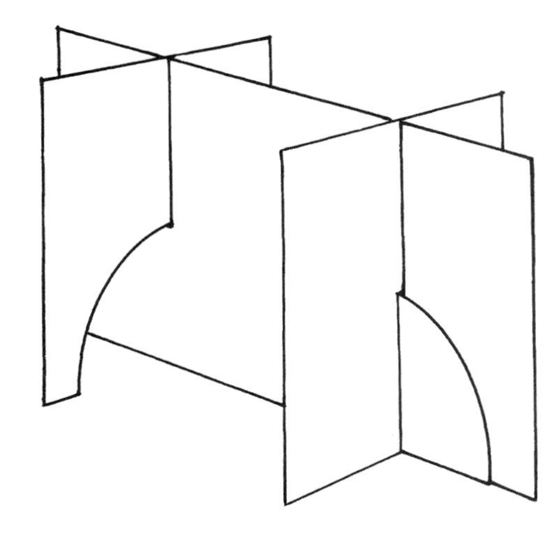

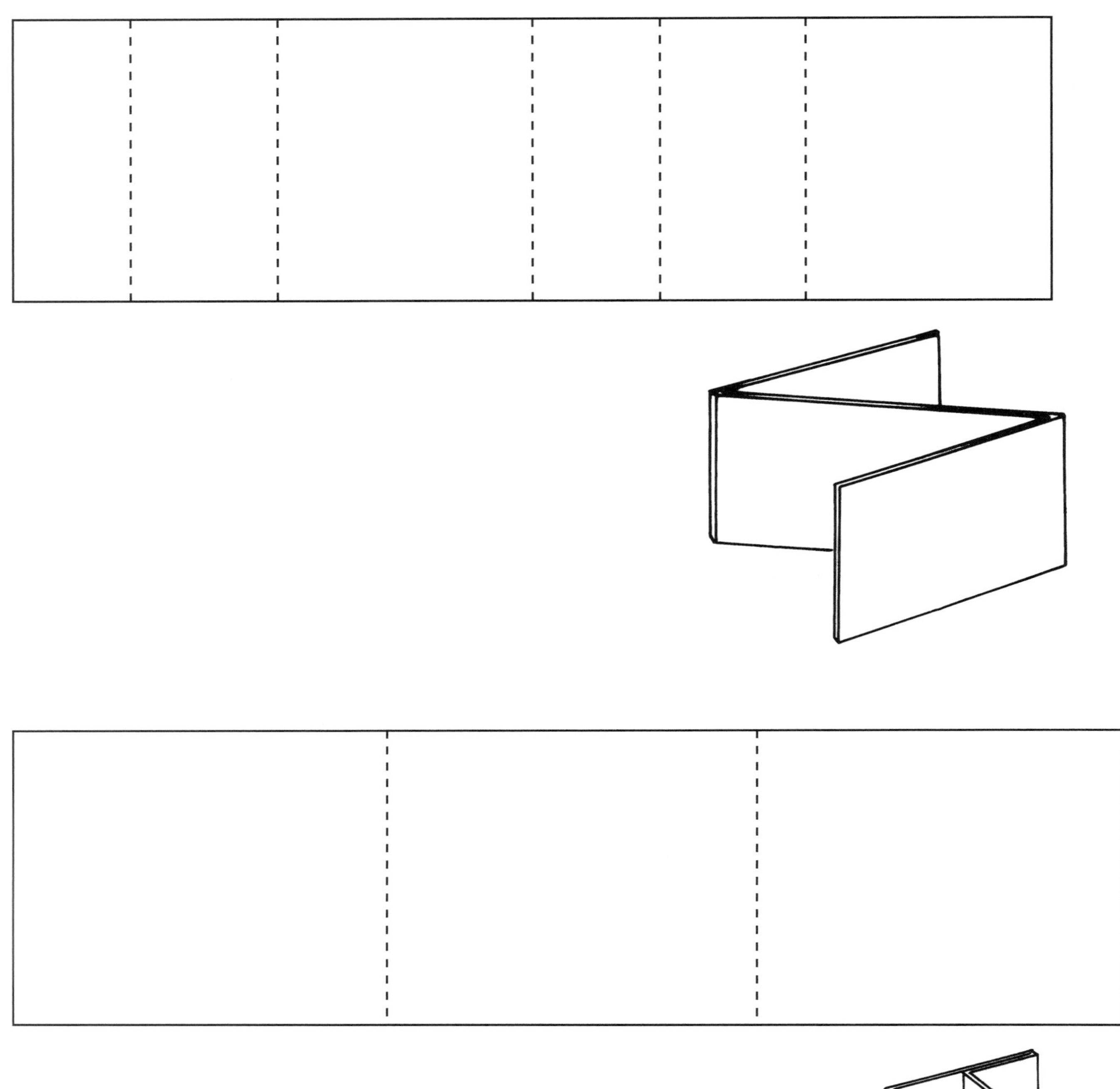

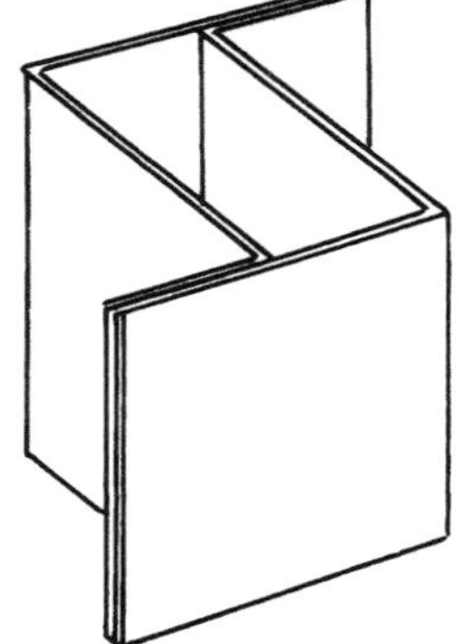

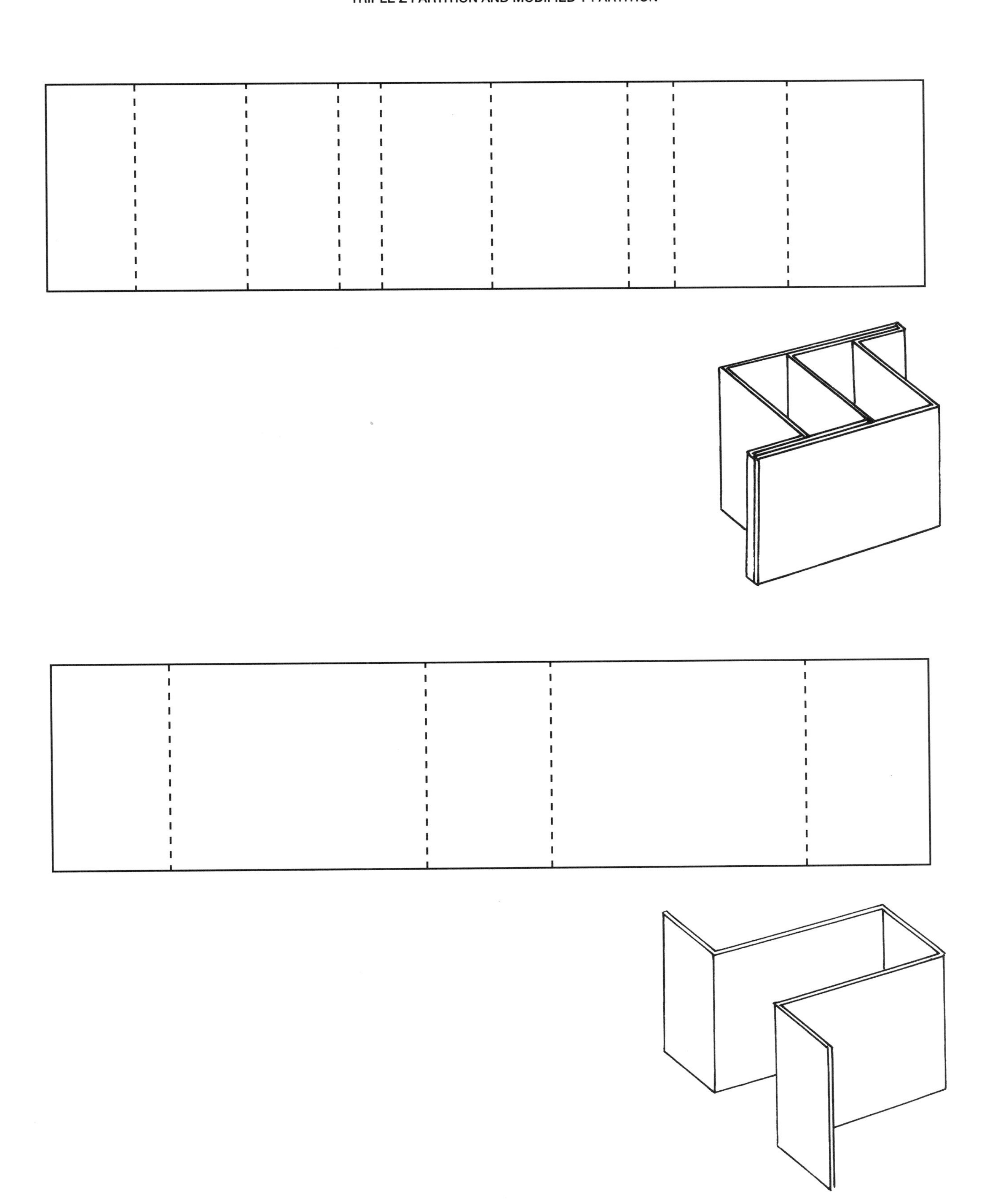

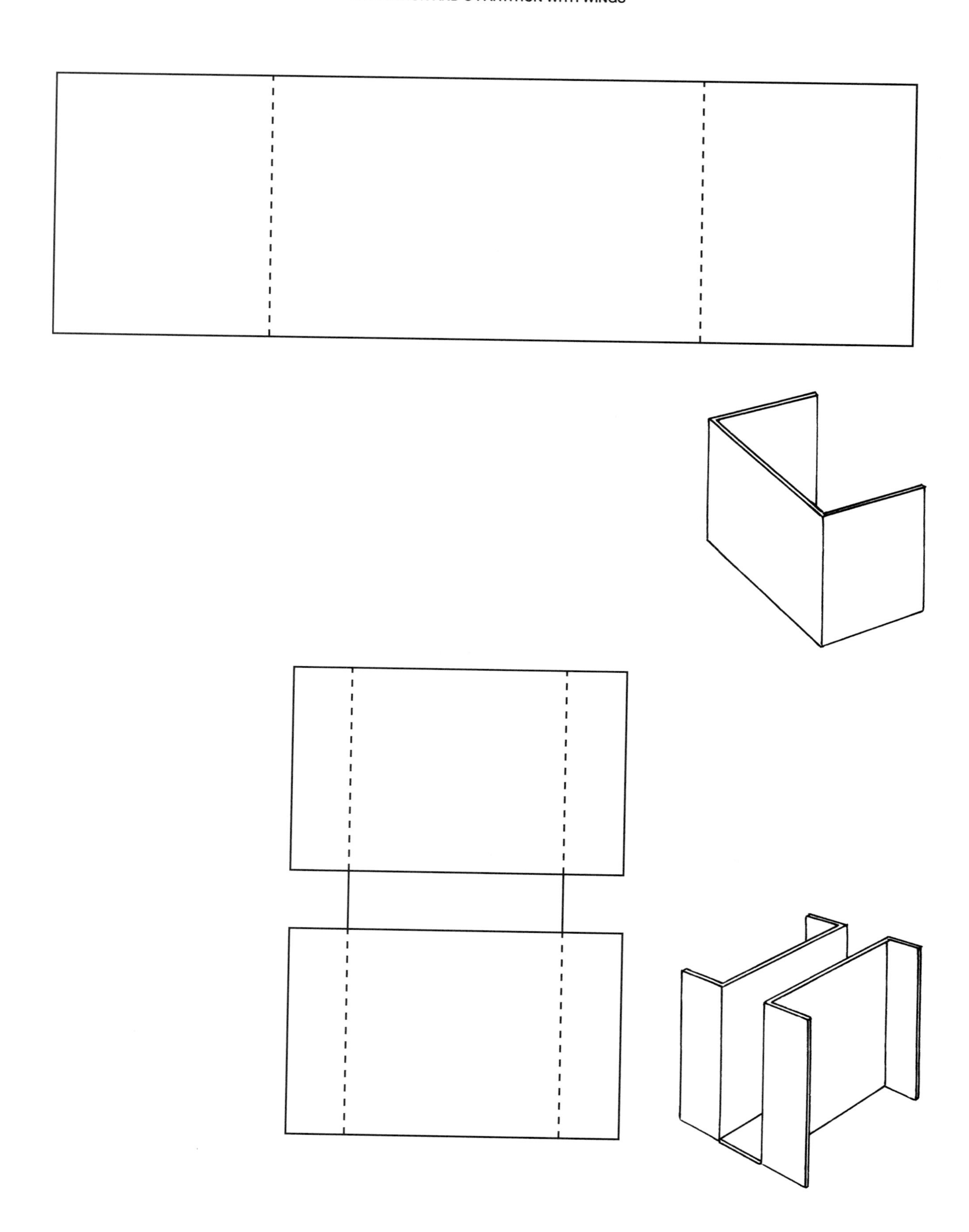

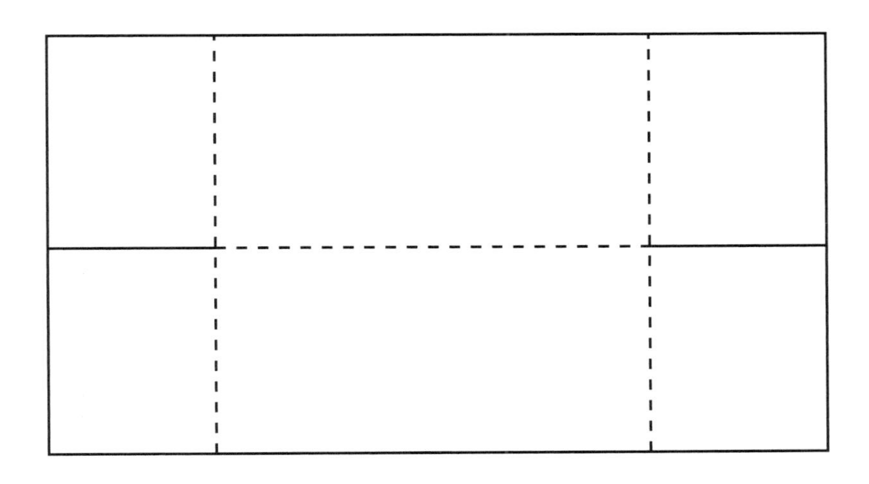

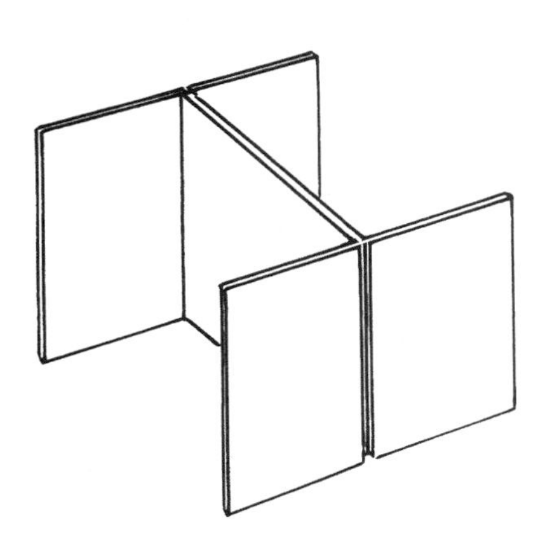

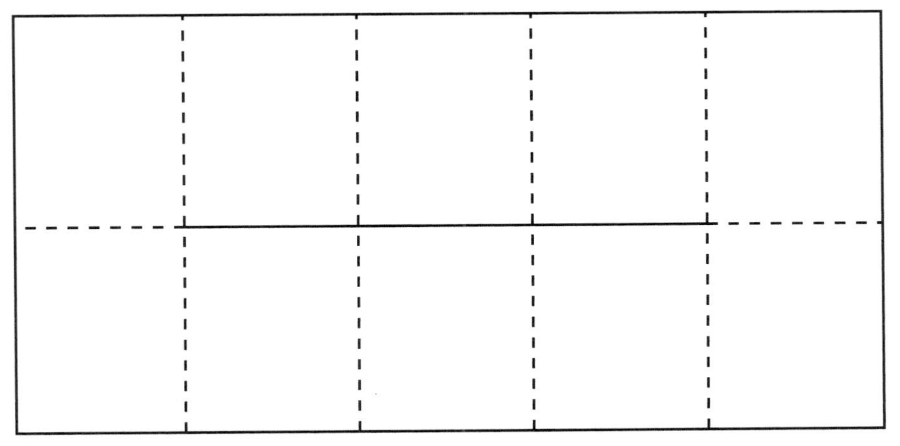

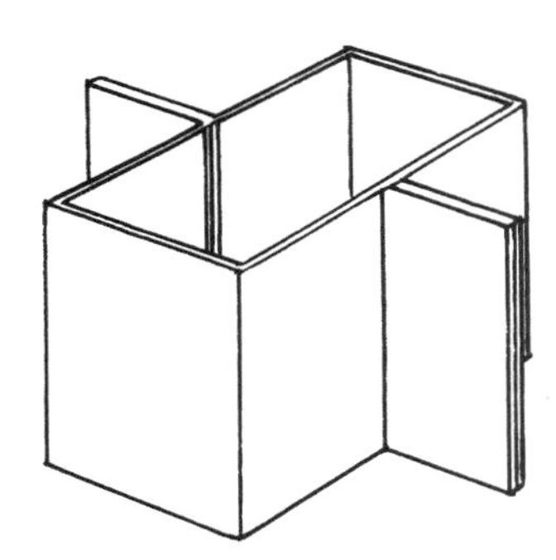

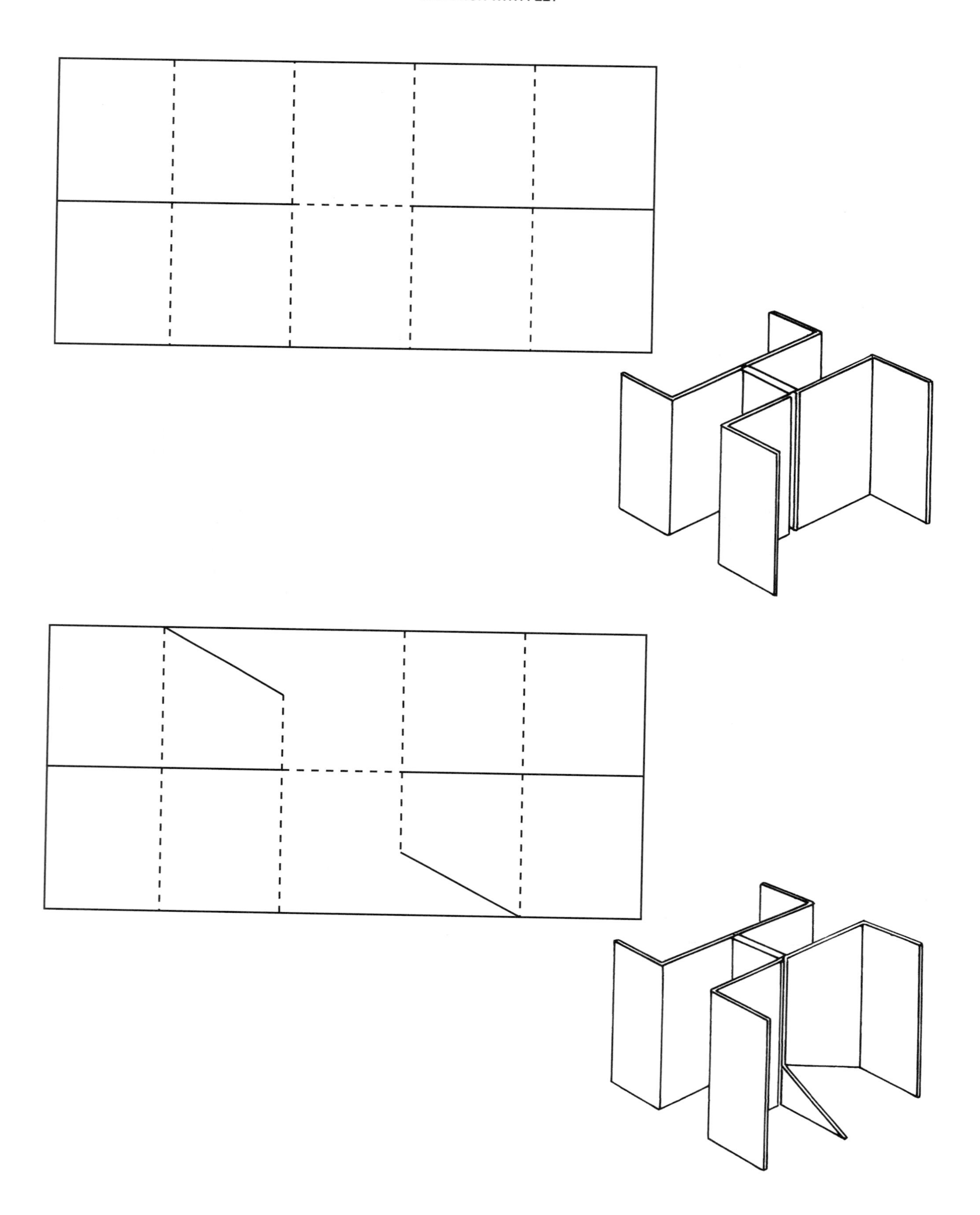

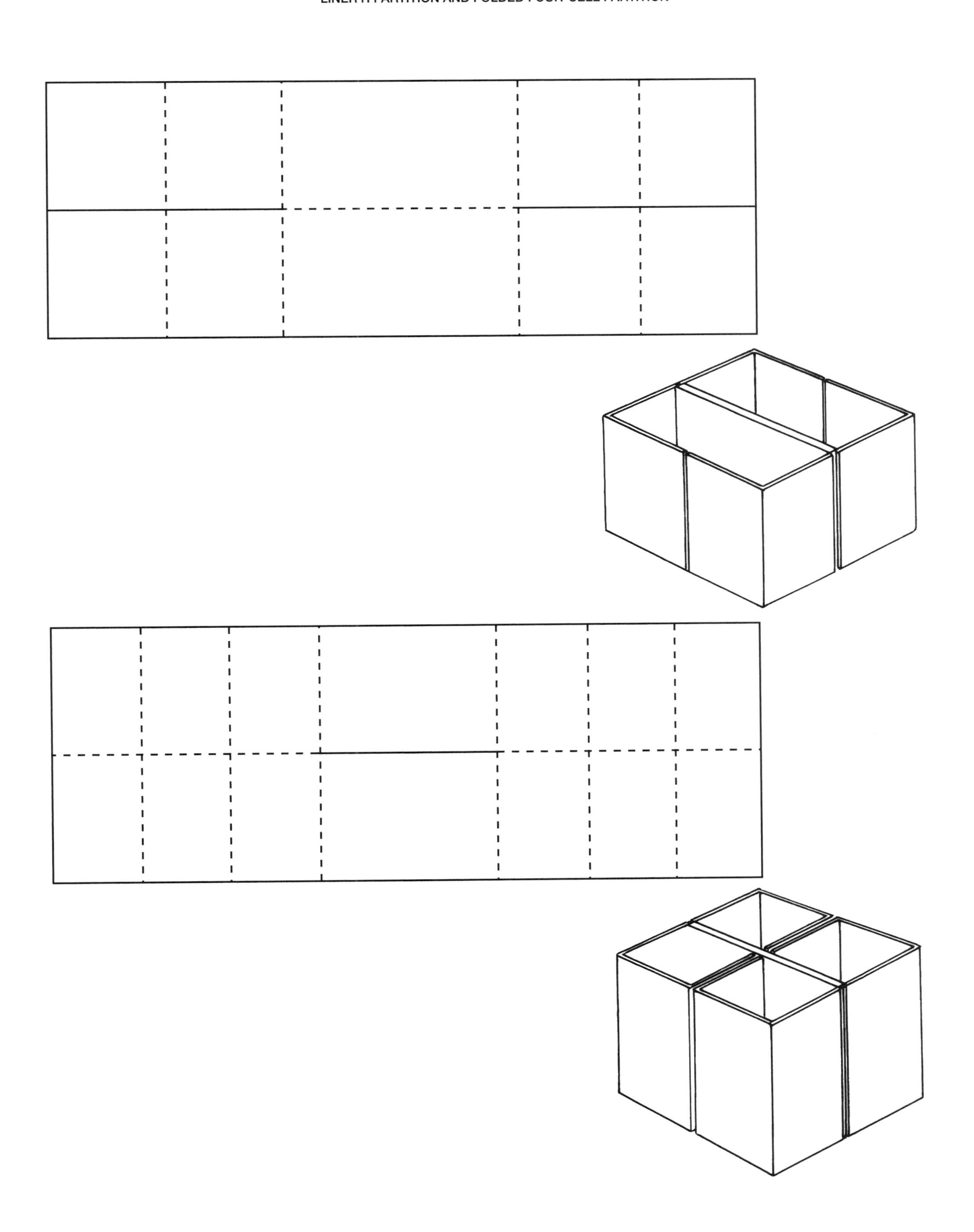

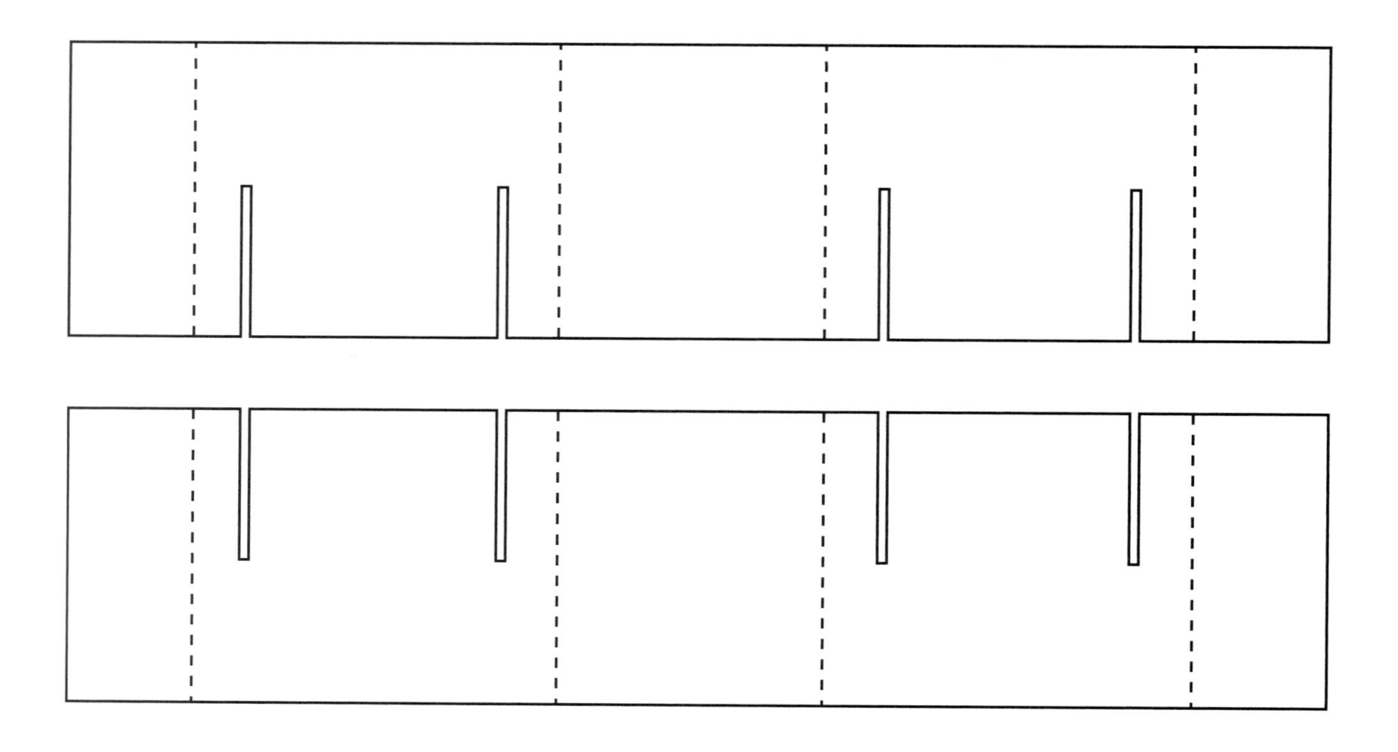

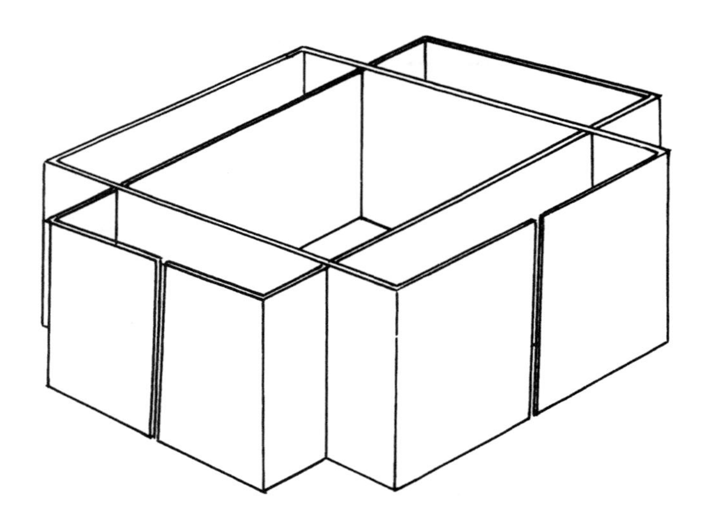

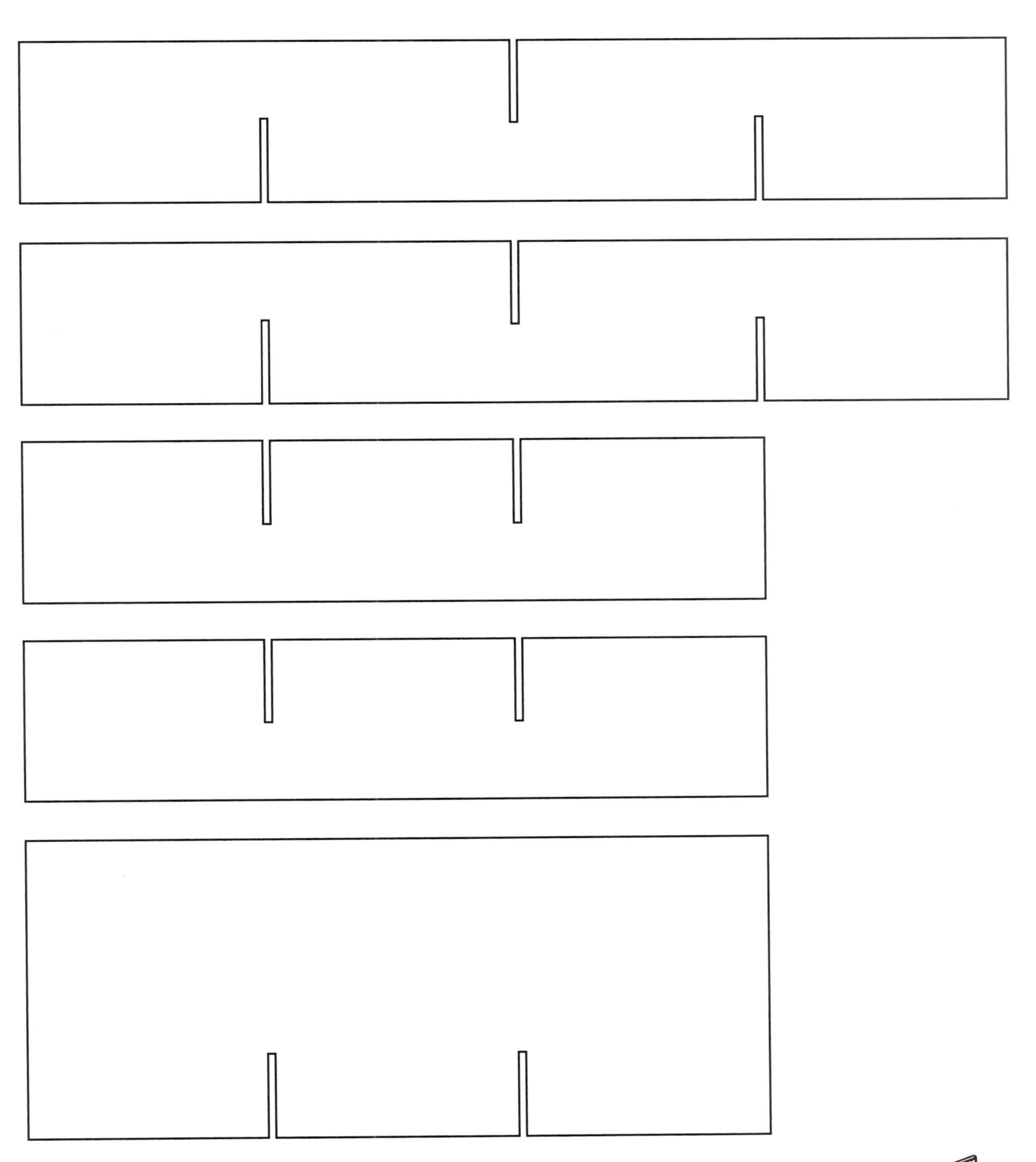

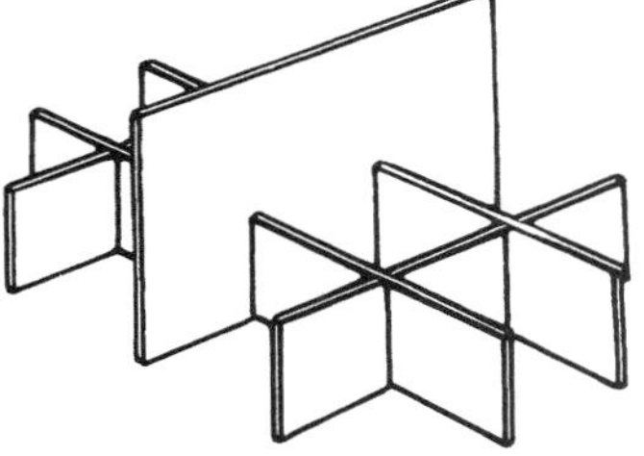

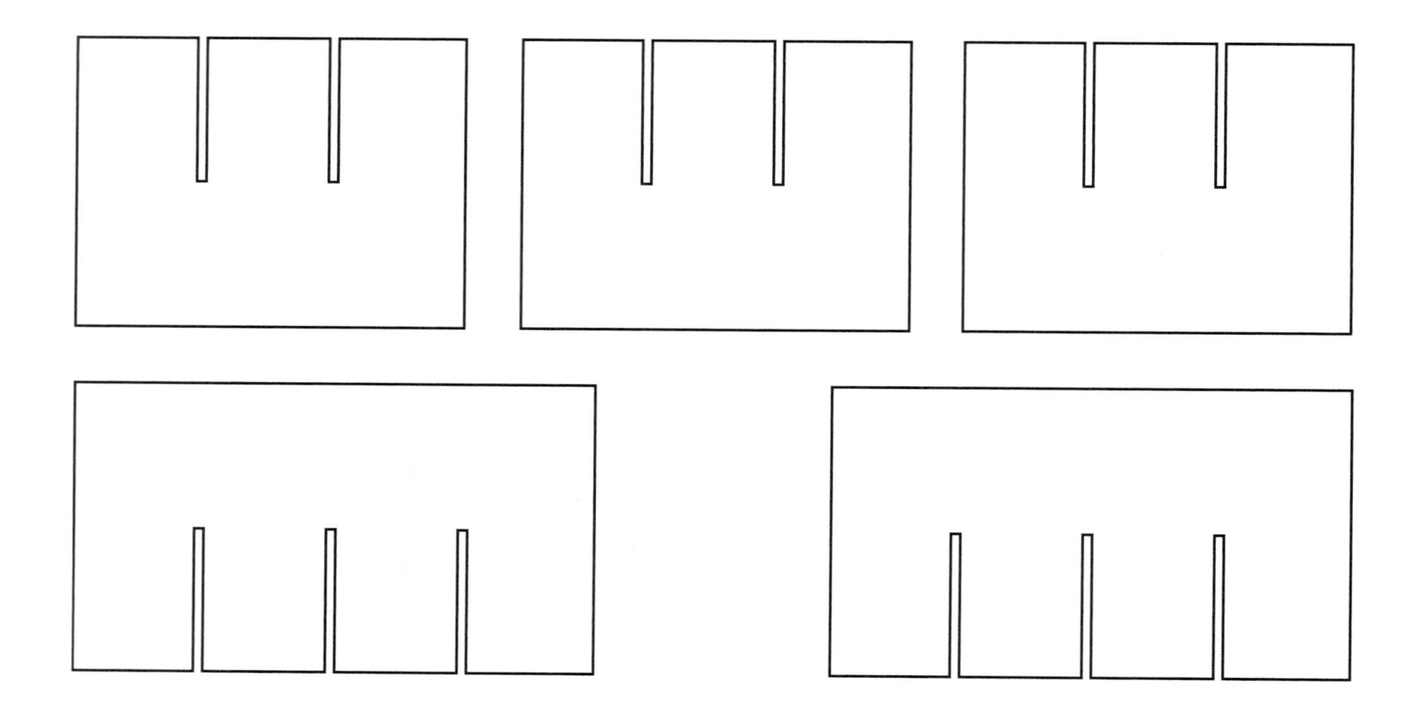

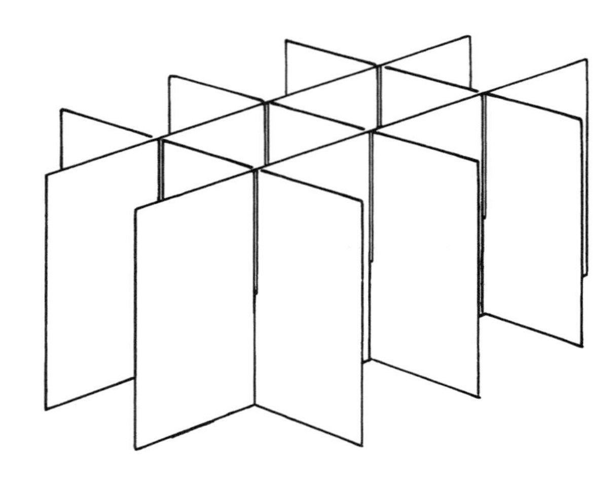

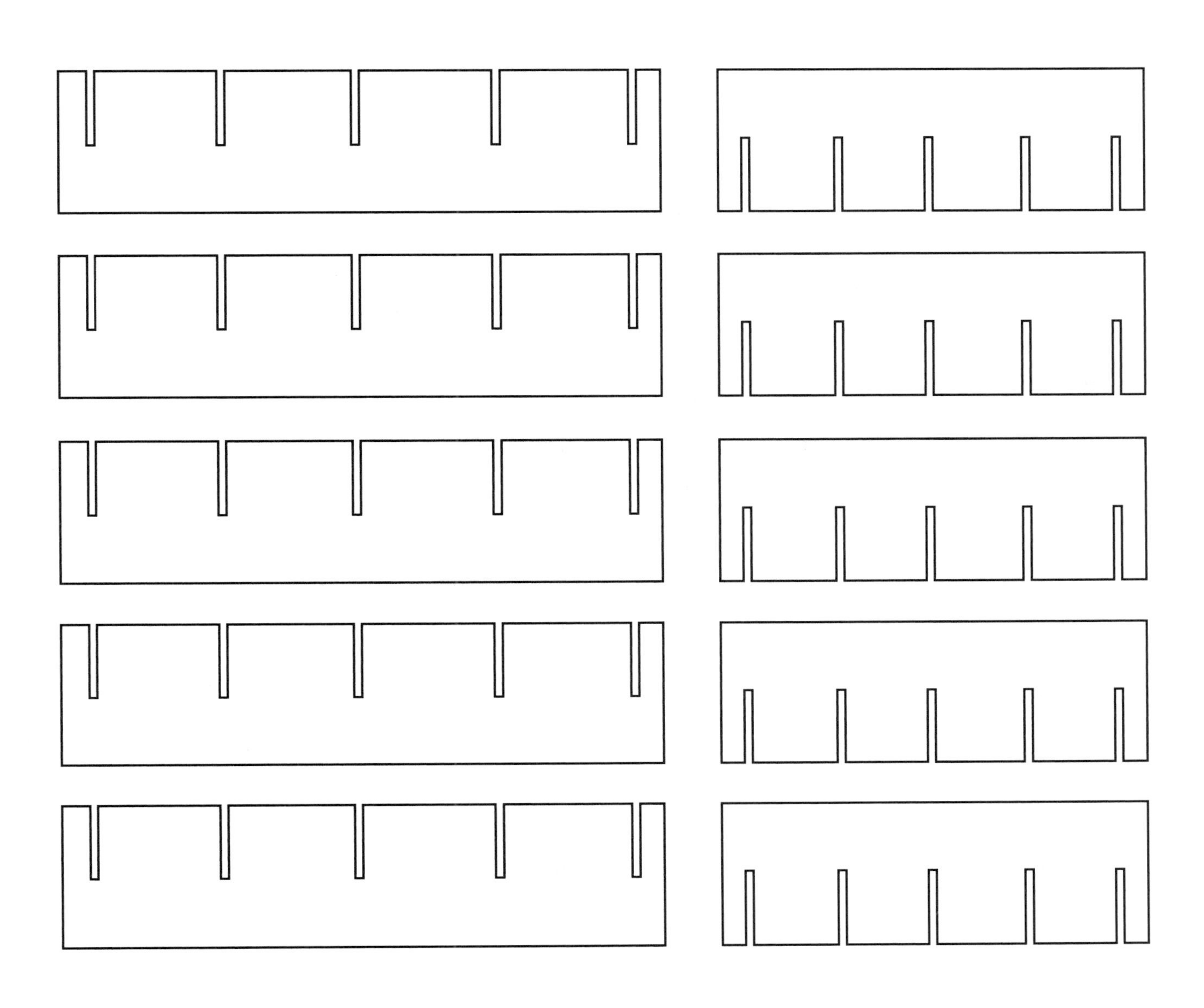

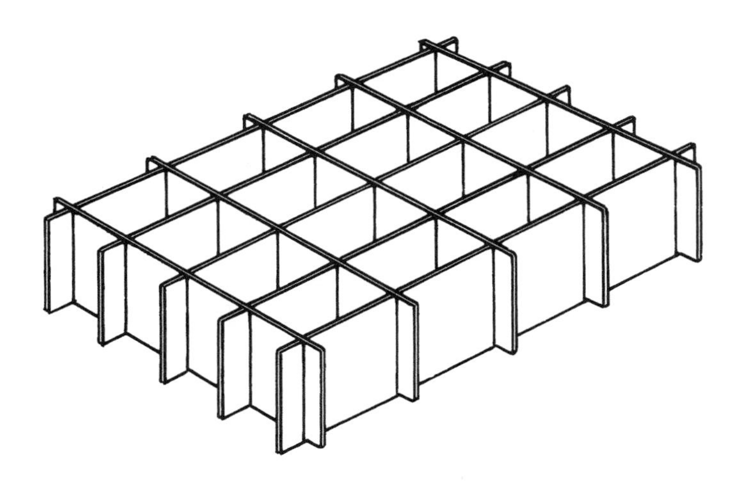

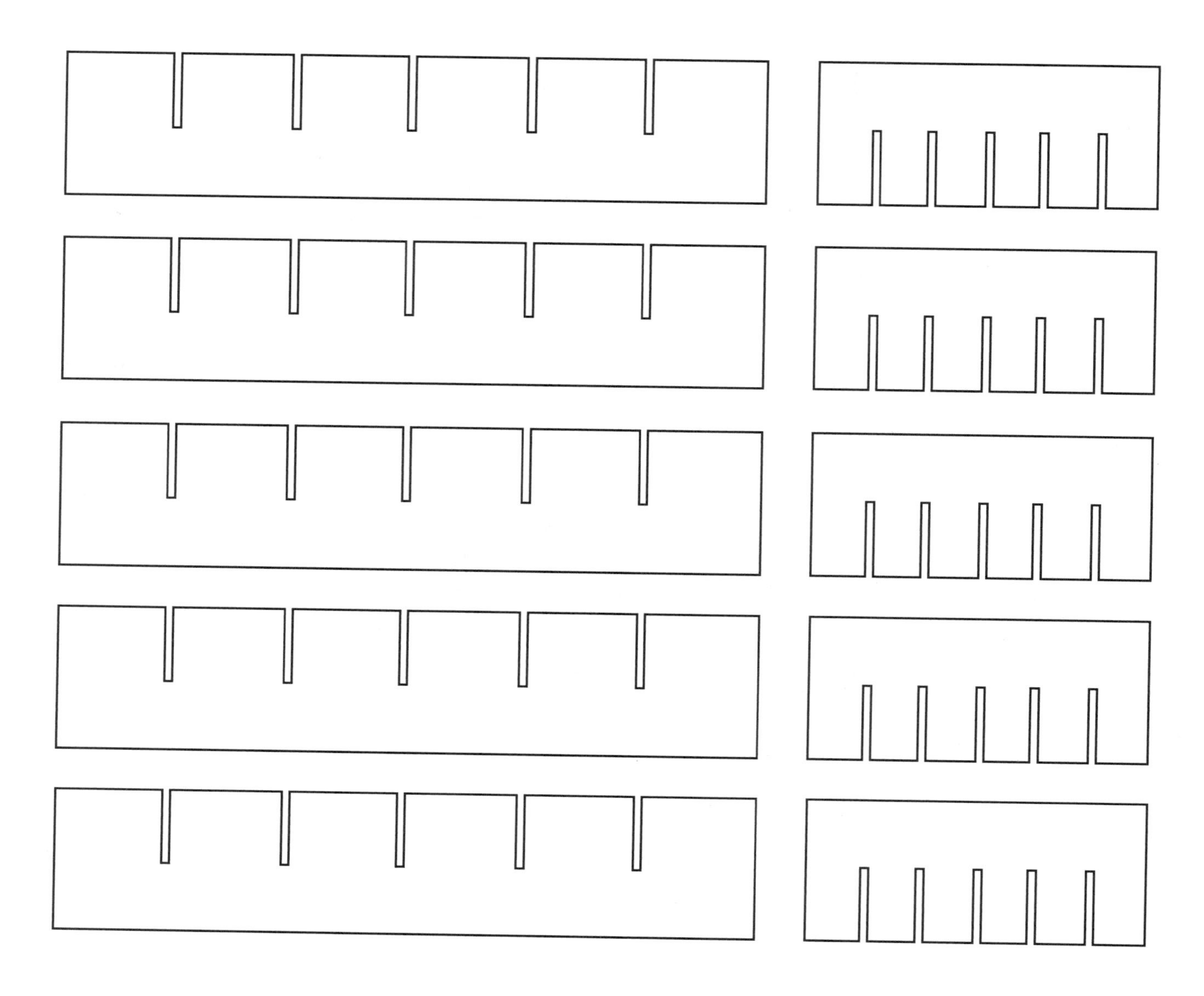

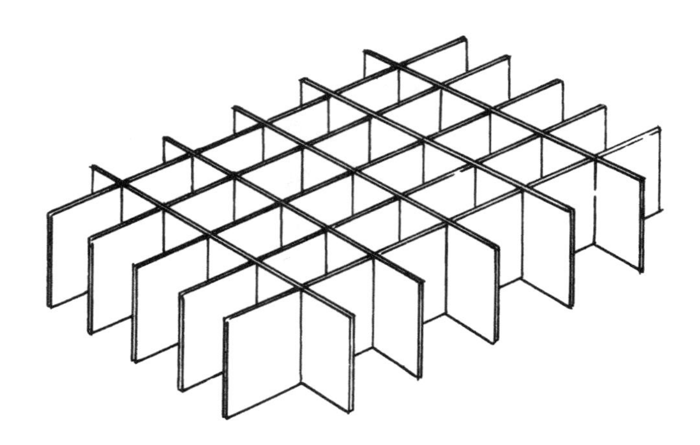

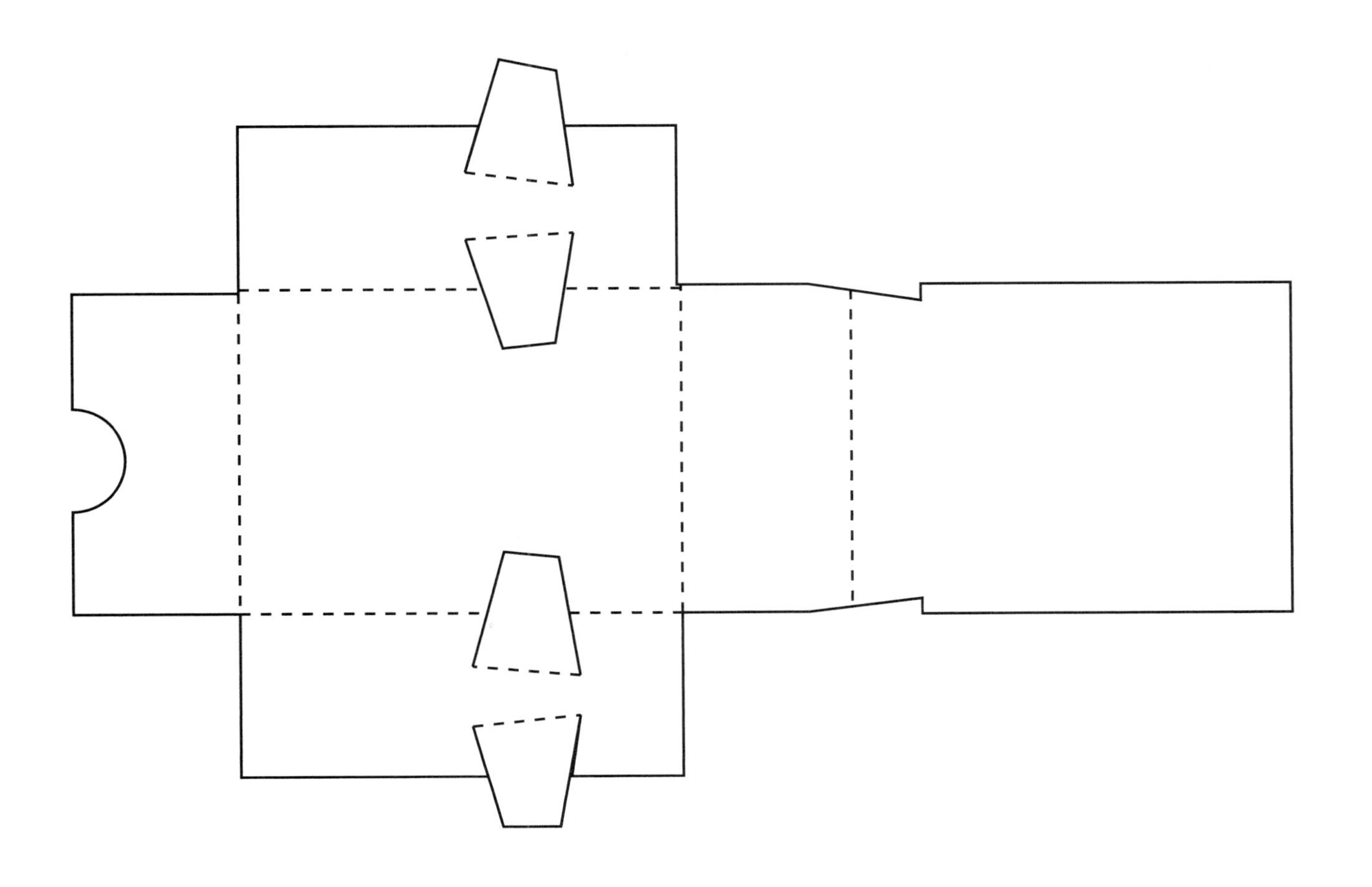

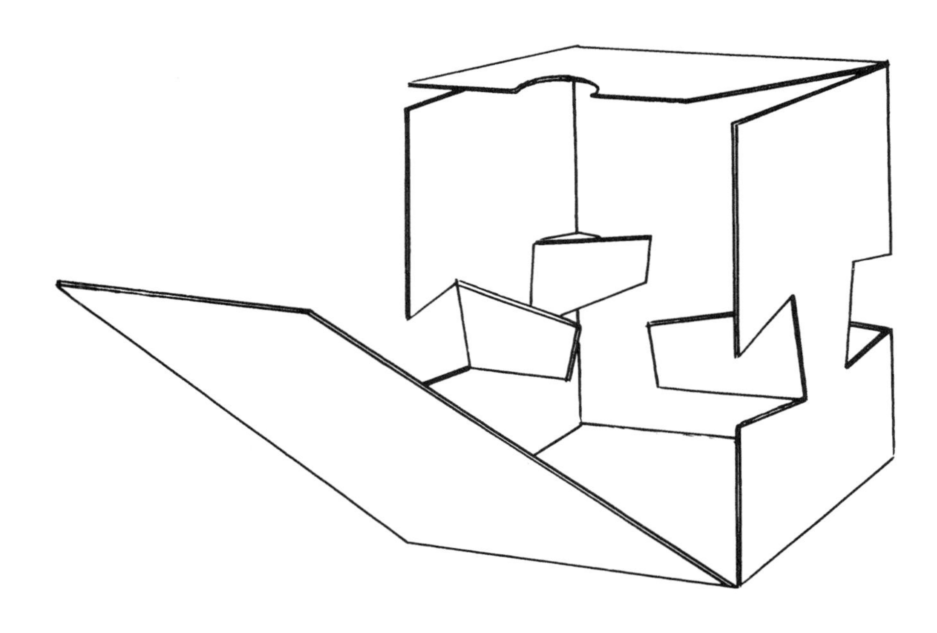

4-CELL AUTO LOCK BOTTOM

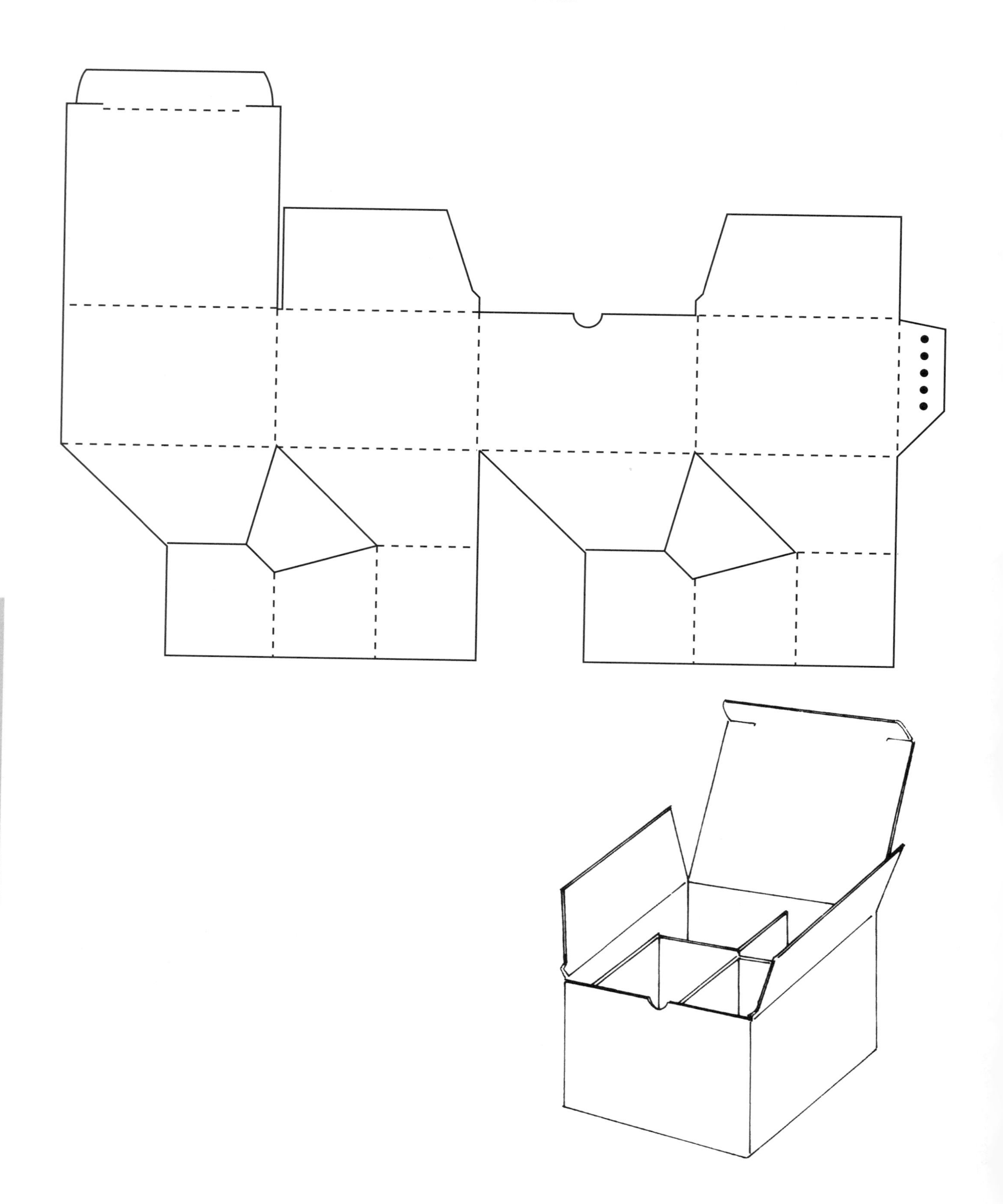

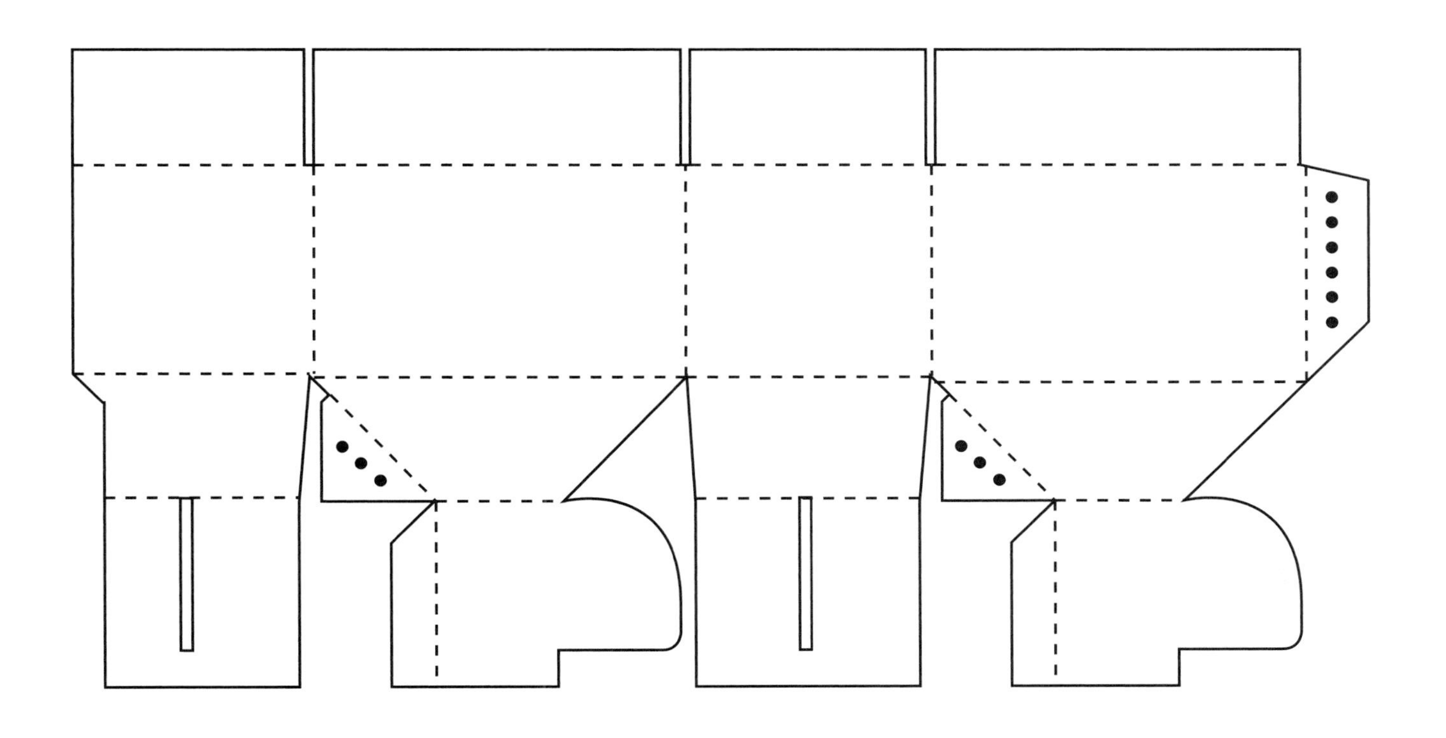

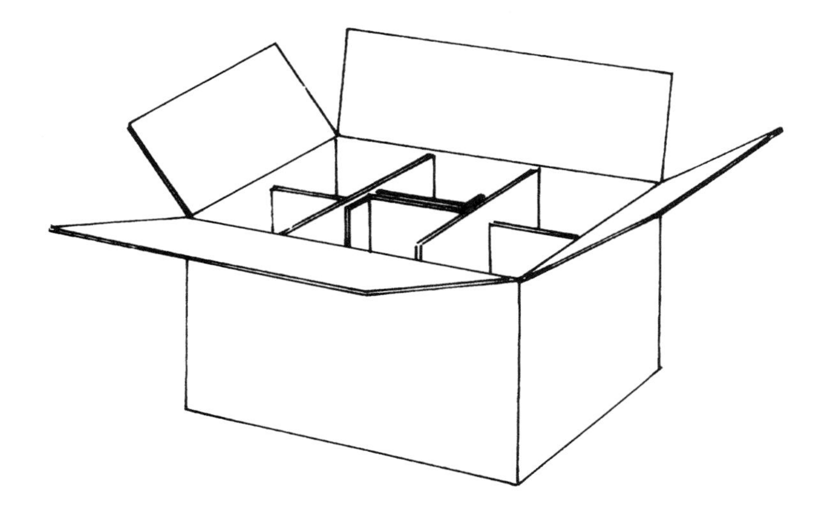

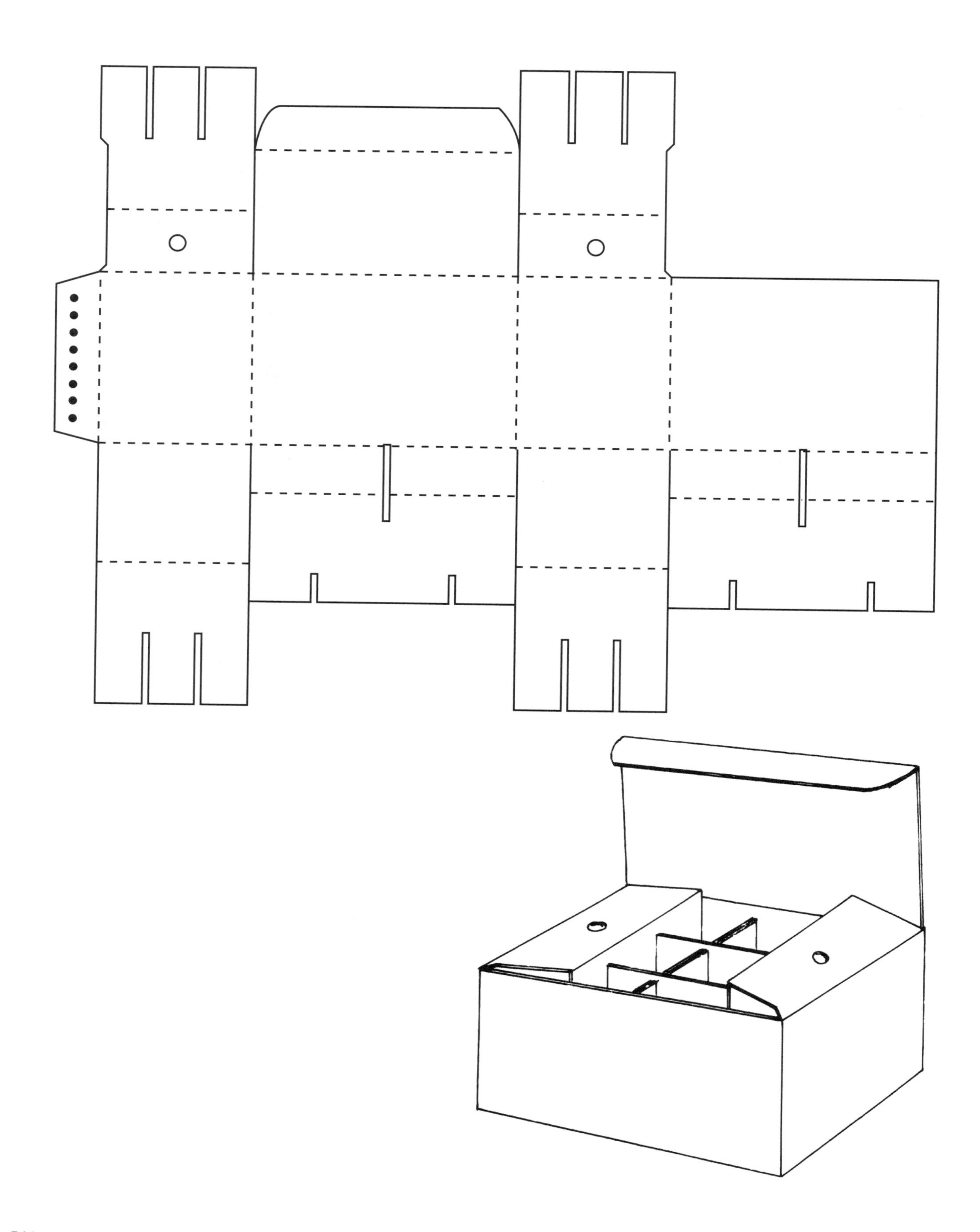

5

Point-of-Purchase Displays

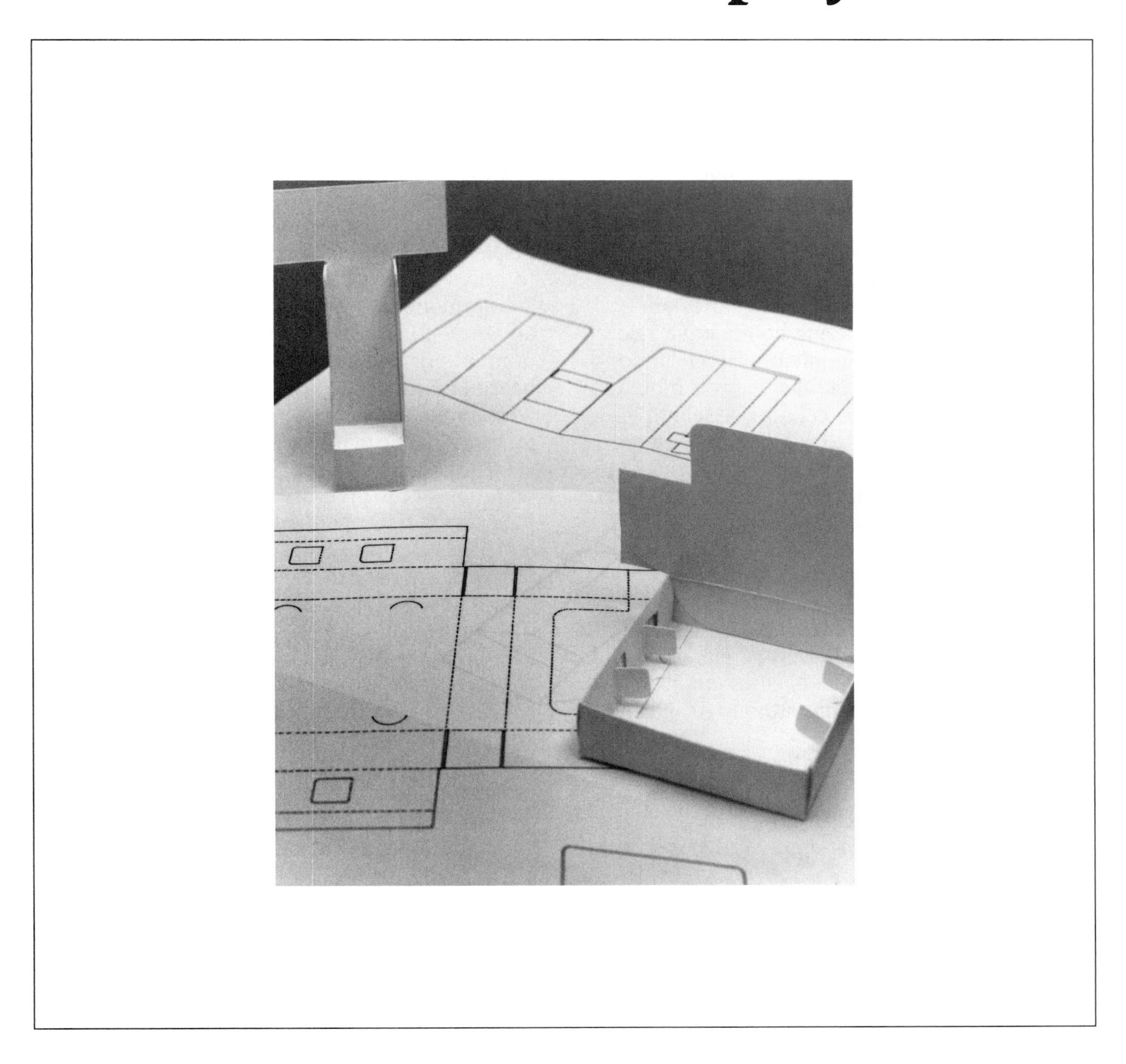

Die-cut three-dimensional displays first appeared in 1910, and photographic full-color displays made their appearance in the late 1920s. With the development of new plastic materials and techniques, point-of-purchase (P.O.P.) displays can now be produced in a variety of styles for practically all consumer products. With the rapid growth of many industries and the increasing variety of products, the point-of-purchase display has become an ever more effective selling aid for retailers.

The expansion of self-service stores and changes in consumer buying habits have also contributed to the development of P.O.P. materials. The modern retail establishment is a busy place in which a large percentage of selling is accomplished through self-service; consumers choose merchandise themselves rather than having it brought to them by salespeople. Often, unplanned buying decisions are made in the store, and an effective, well-designed display may be the deciding factor in a consumer's choice.

Planning and designing a P.O.P. display is a complex job involving a variety of materials and technologies. Paperboards, plastics, and corrugated boards all play an important part in the production of the display.

There are several categories of P.O.P. displays. Among them are display merchandisers, promotional displays, display shippers, counter displays, floor displays, and gravity-fed displays.

(Large supermarket displays, permanent displays, wire displays, dump bins, and motion displays, all of which tend to be larger and more expensive to produce, are not discussed here.)

Display merchandisers play an important part in self-service systems because they are strategically placed in the store, usually near the cash register or checkout counter. The display merchandiser is sometimes called a promotional display because it is designed to be used only for the duration of a particular promotion.

Promotional displays have a short life, usually three to four weeks (i.e., the duration of the promotion). The material used is usually paperboard or E-flute corrugated, often combined with an inexpensive vacuum-formed (thermoformed) plastic platform to hold the product.

The *display shipper* is the most common variation of the promotional display. It is a shipping carton that opens up to form a display with a die-cut *riser* (or reader) panel for art and copy. Shippers are used in the mass merchandising of health and beauty aids, liquor, toys, novelties, pharmaceuticals, and books. The advantage of the shipper is that it combines a shipping carton with a display setup.

Gravity-fed displays are one of the oldest types of display systems. They are used for film, shoe polish, and small packages of various products.

The following pages contain a variety of structural designs and patterns for P.O.P. displays.

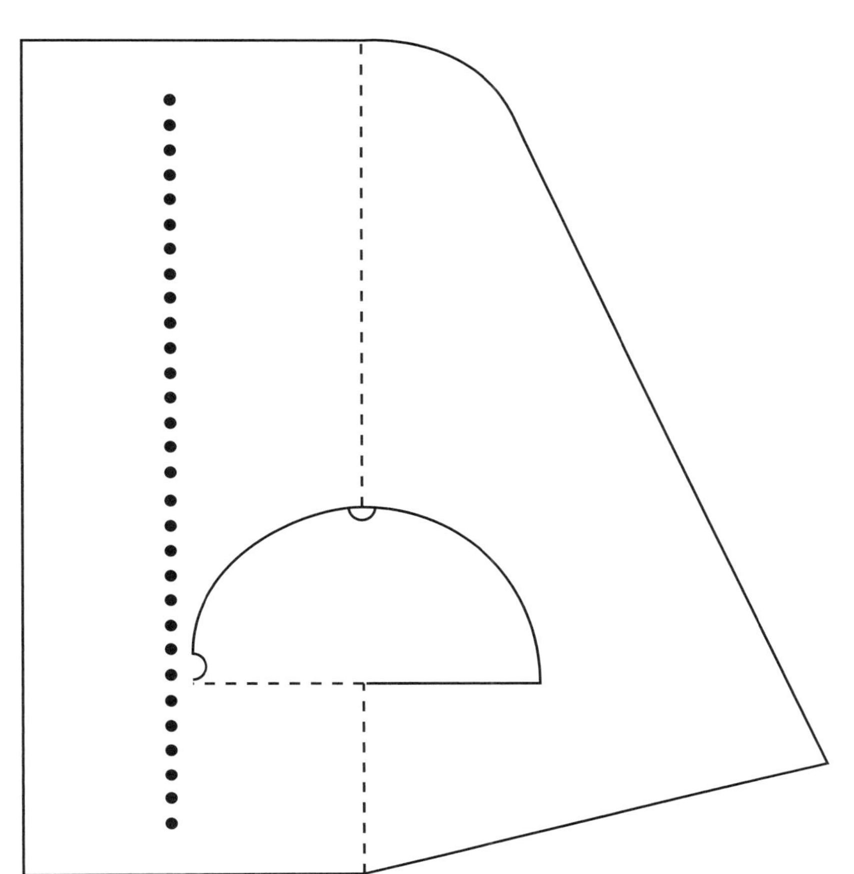

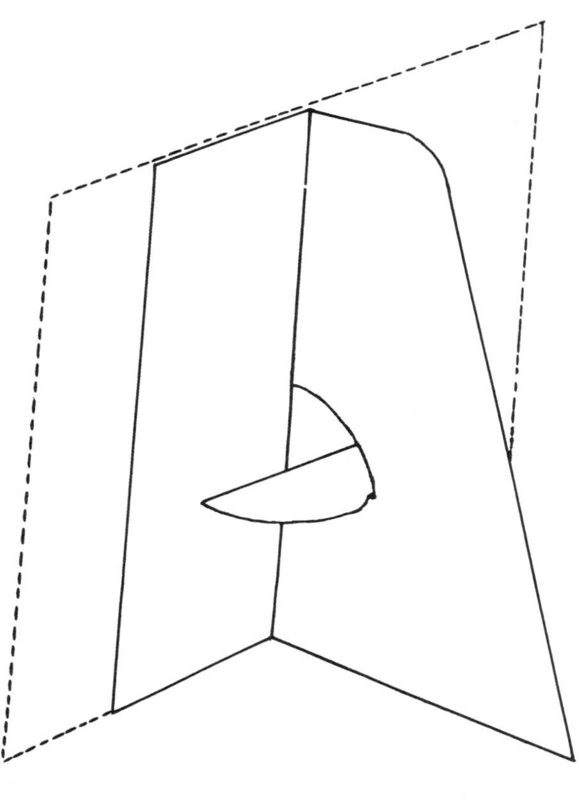

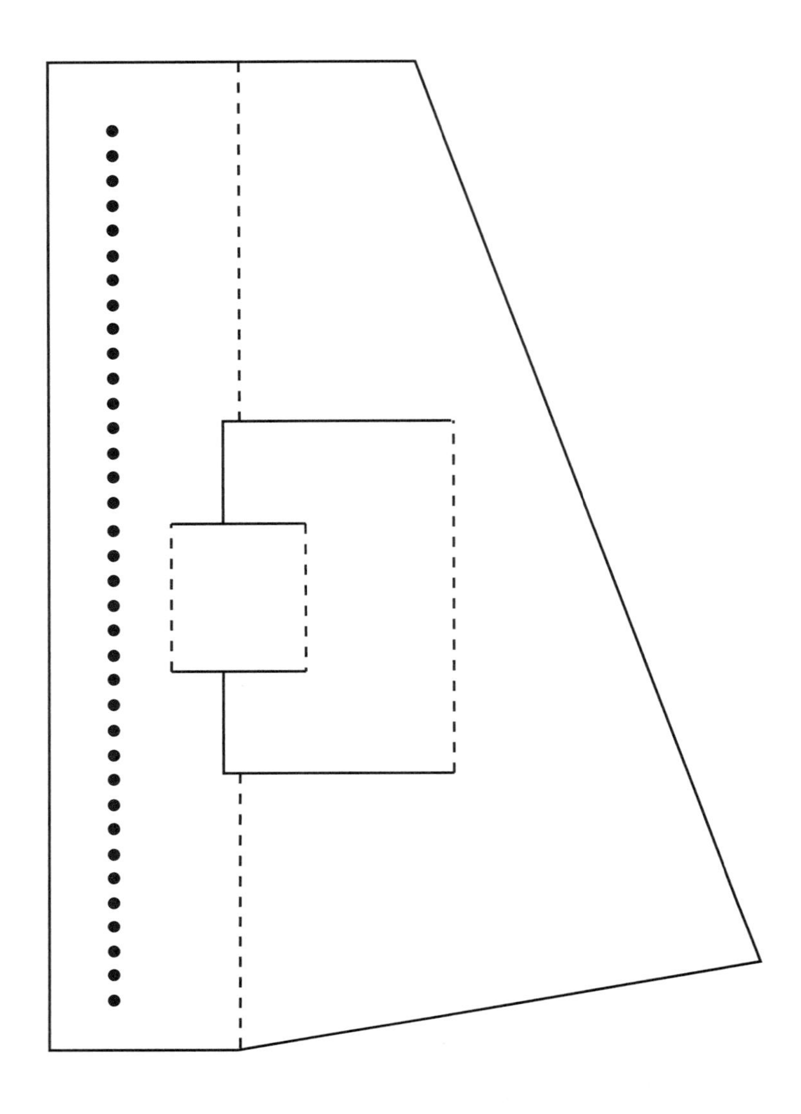

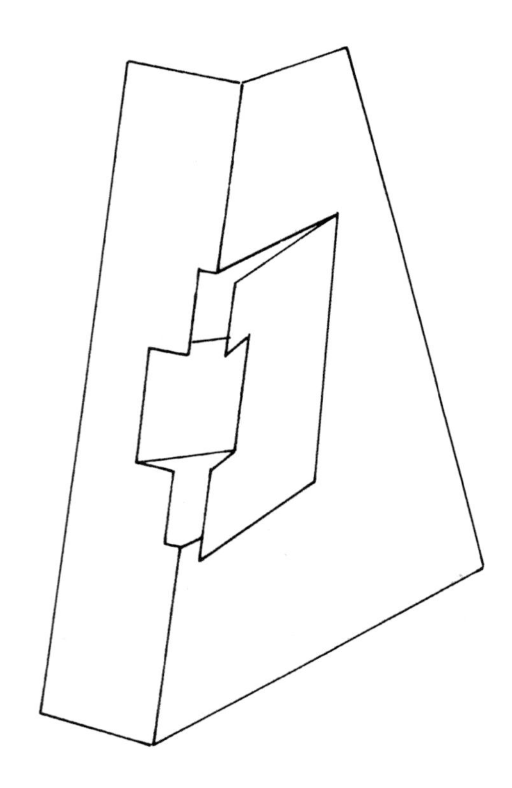

This S.W.E. (single wing easel) invented by Mike Scotti stays erect by spring tension.

ONE PIECE EASEL

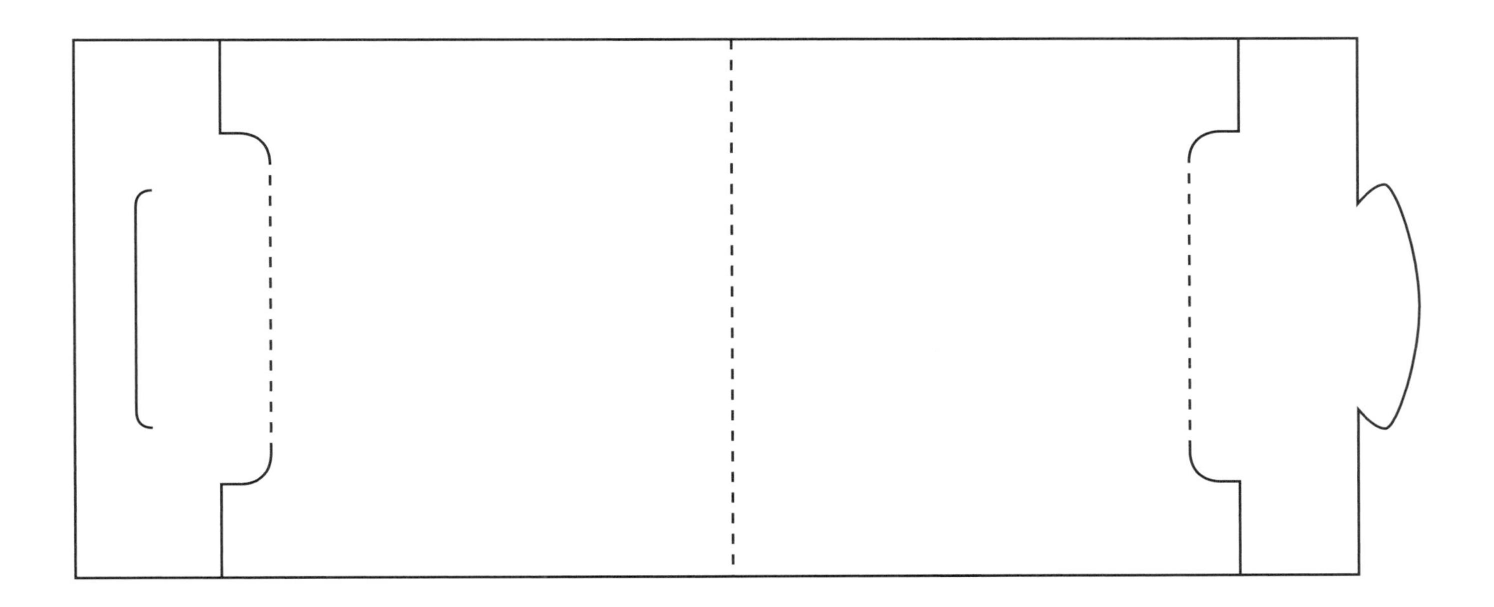

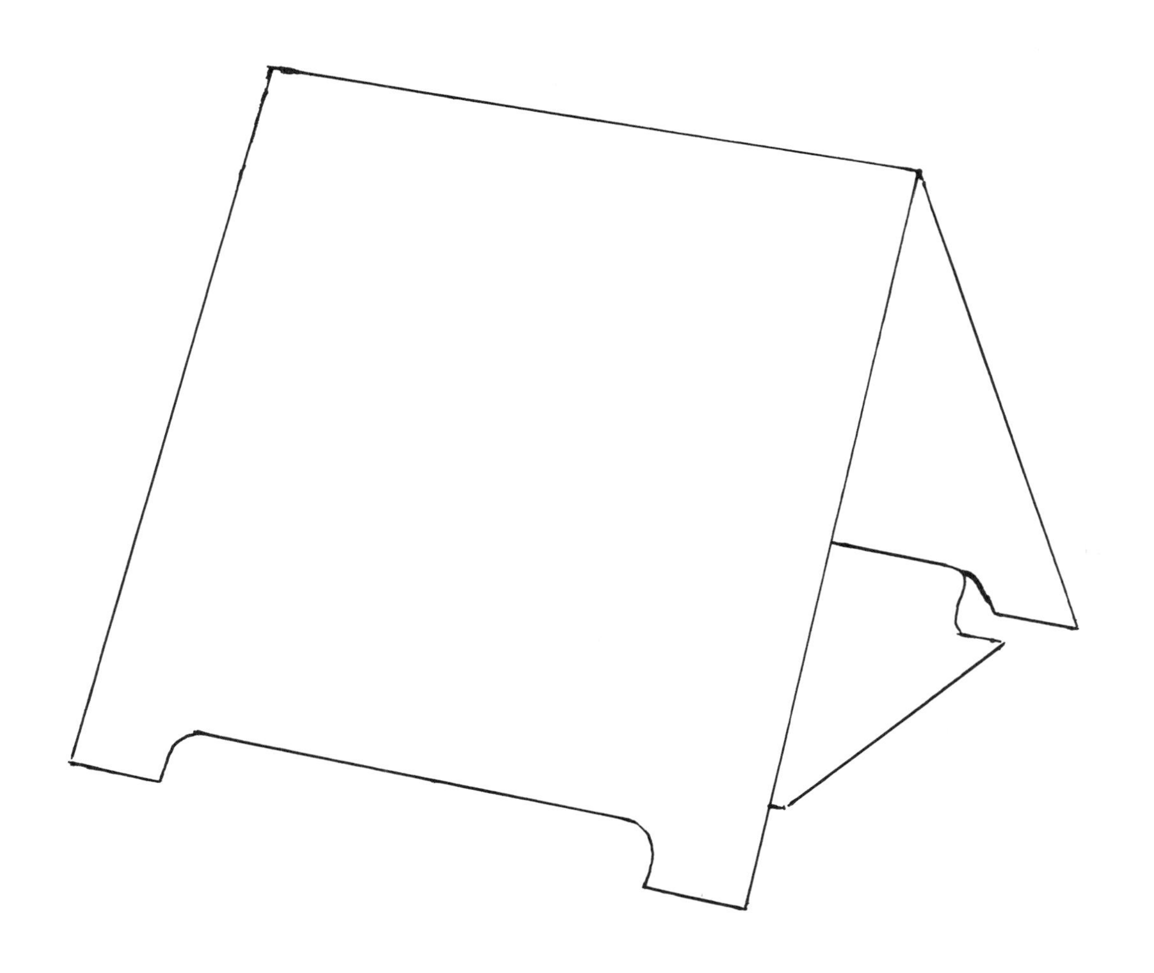

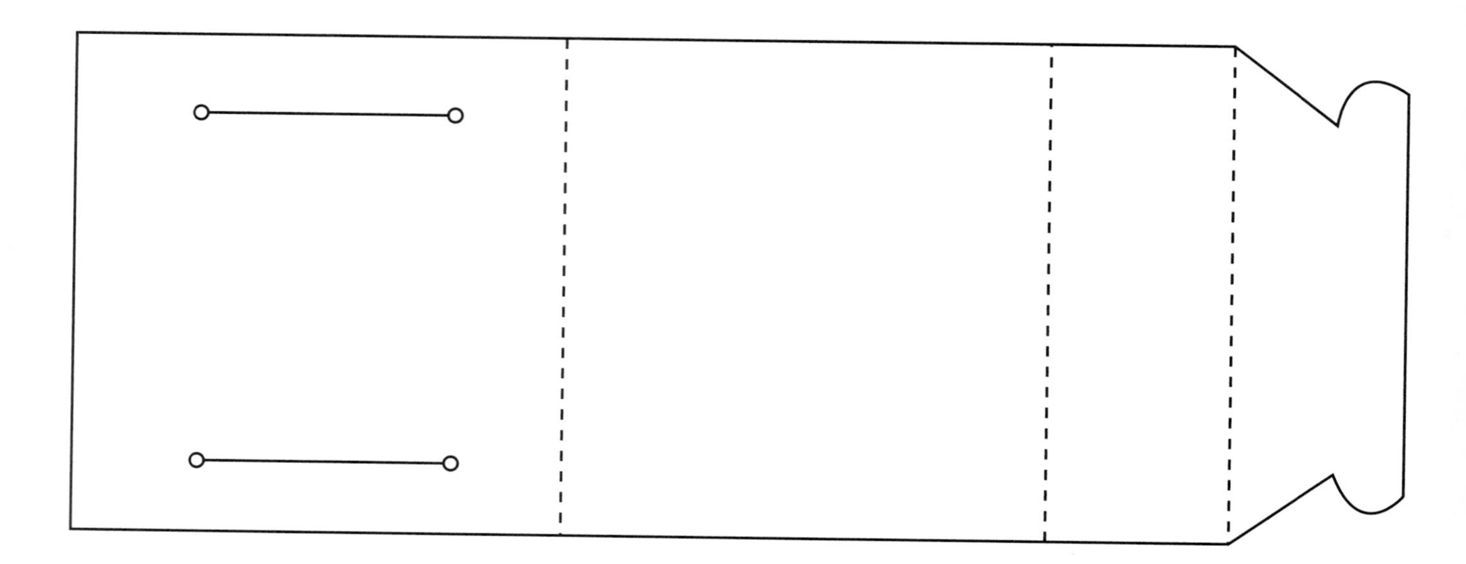

The ears slide in the slots when the easel is set up.

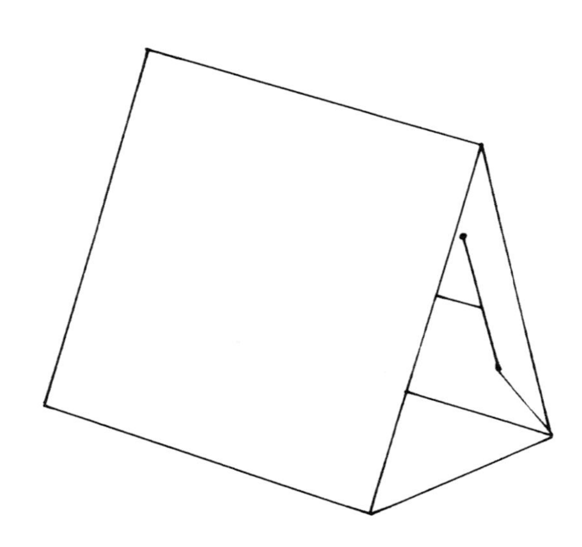

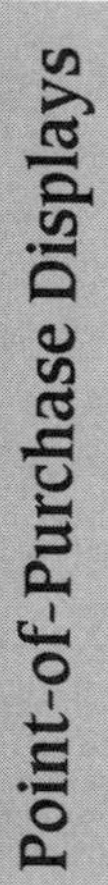

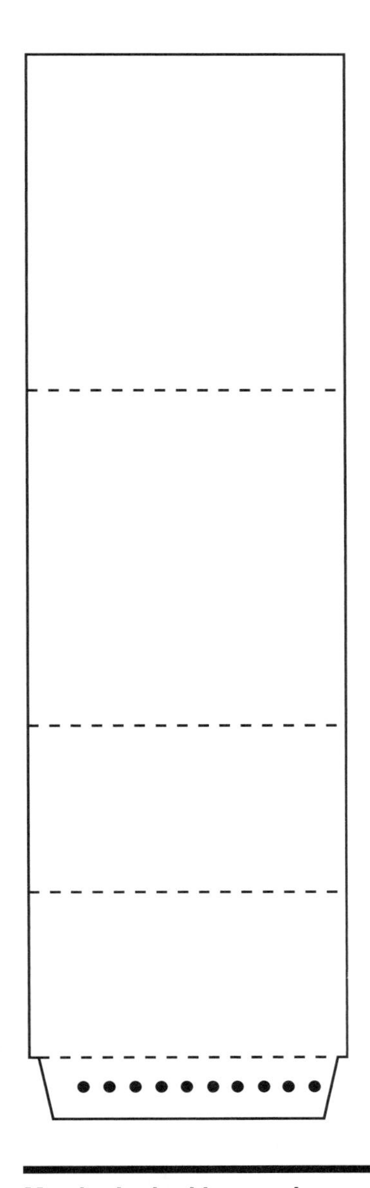

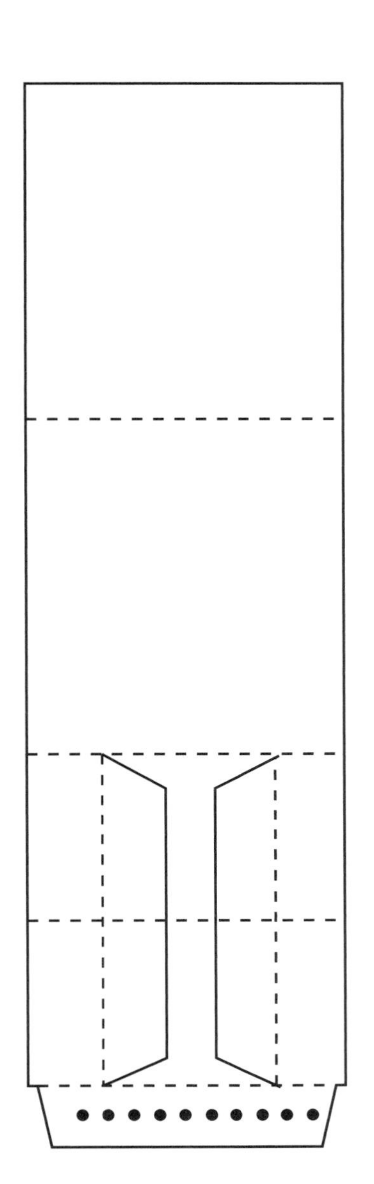

May be locked by pop-ins on the bottom panel.

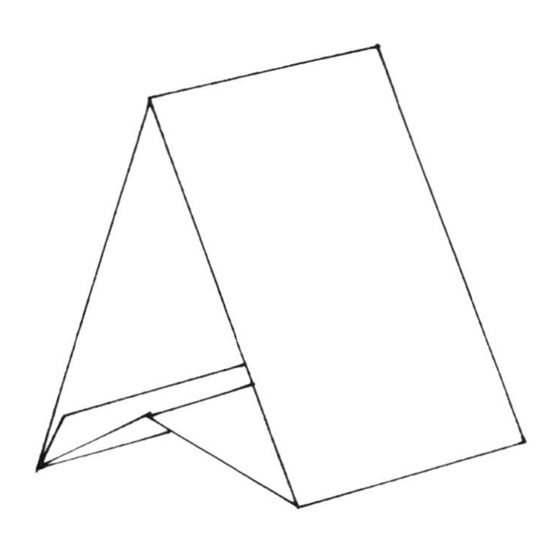

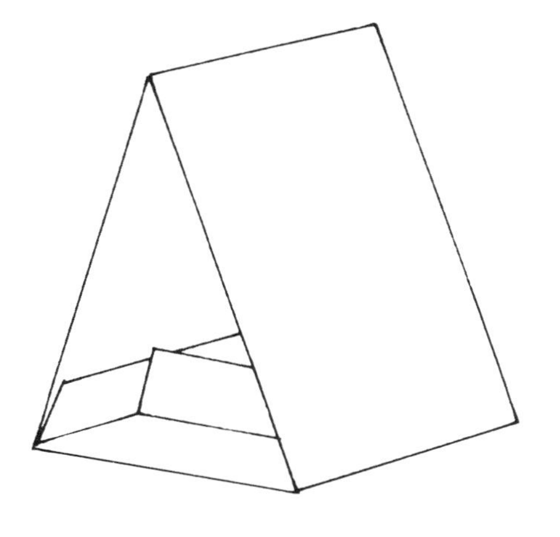

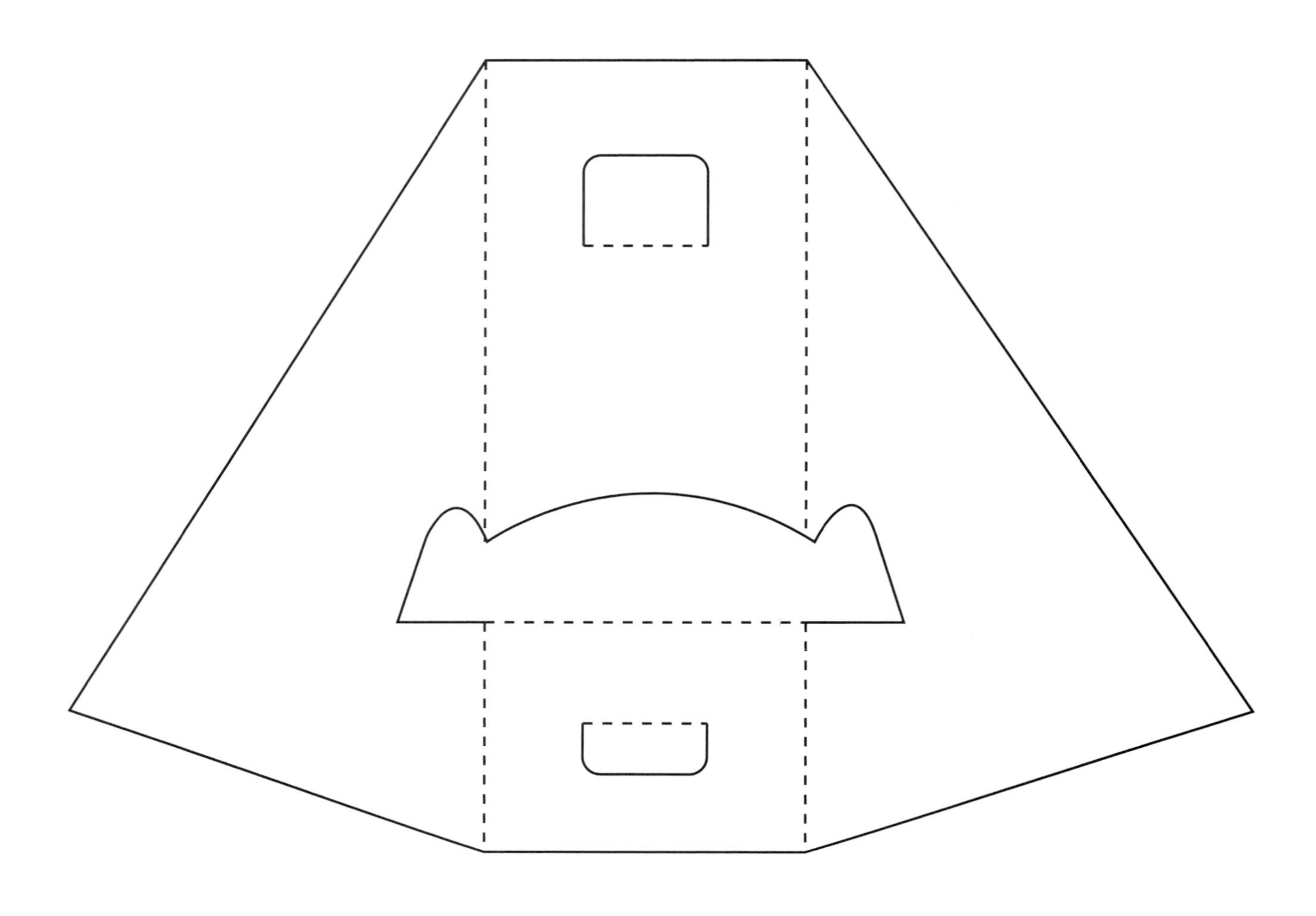

May be glued to back of cards, boxes, etc., or may be keyed with lock tabs to display.

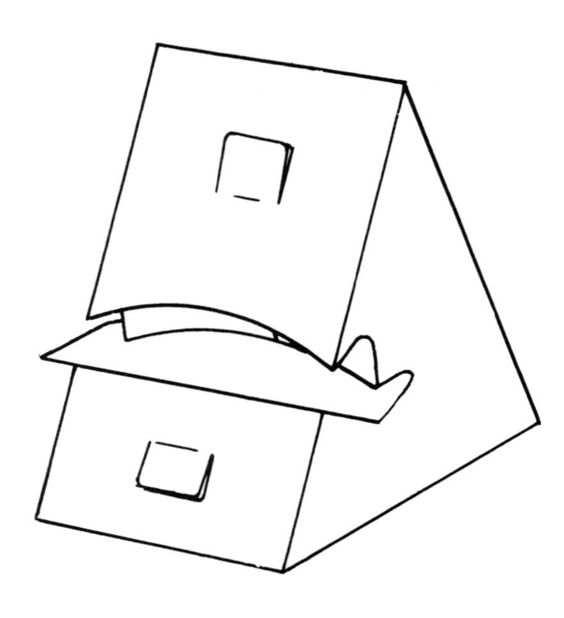

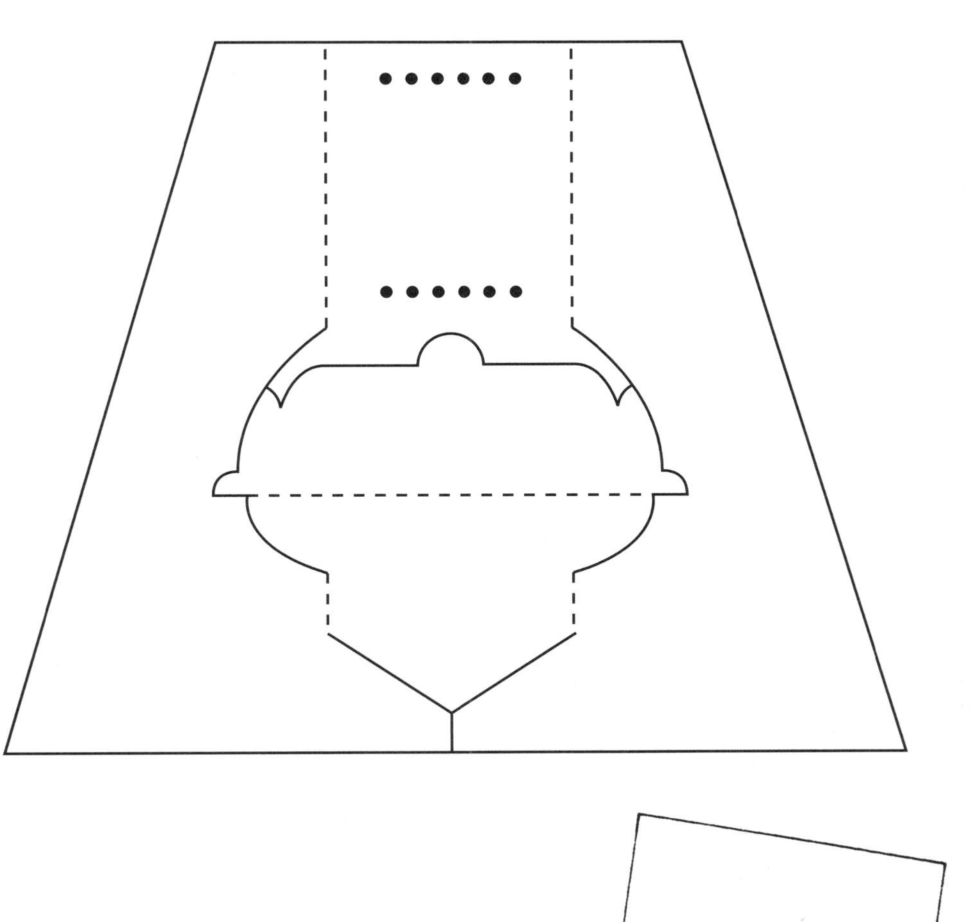

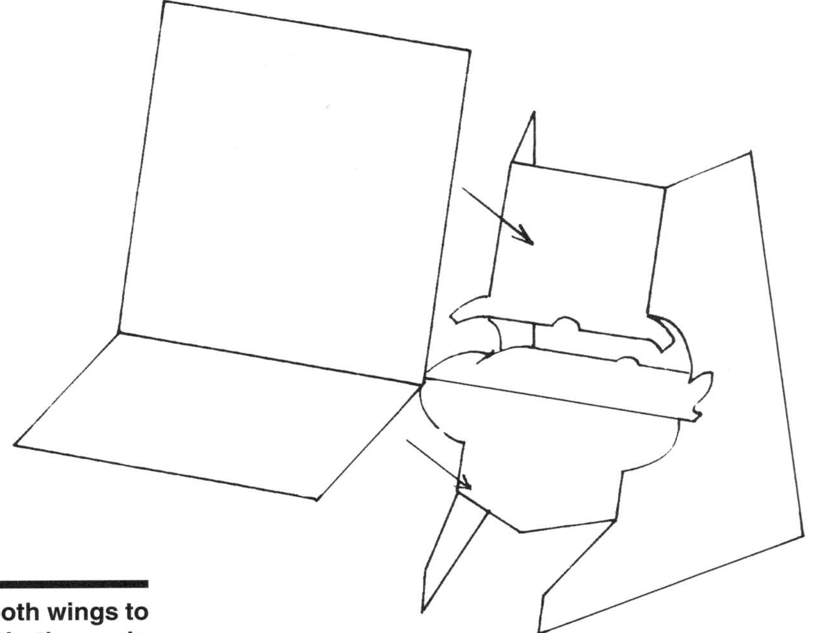

This easel has forward projections off both wings to support a "cowcatcher" panel beneath the main panel of the display.

DISPLAY CARD, BROCHURE, OR SAMPLE HOLDER WITH TWO-WINGED EASEL BUILT IN

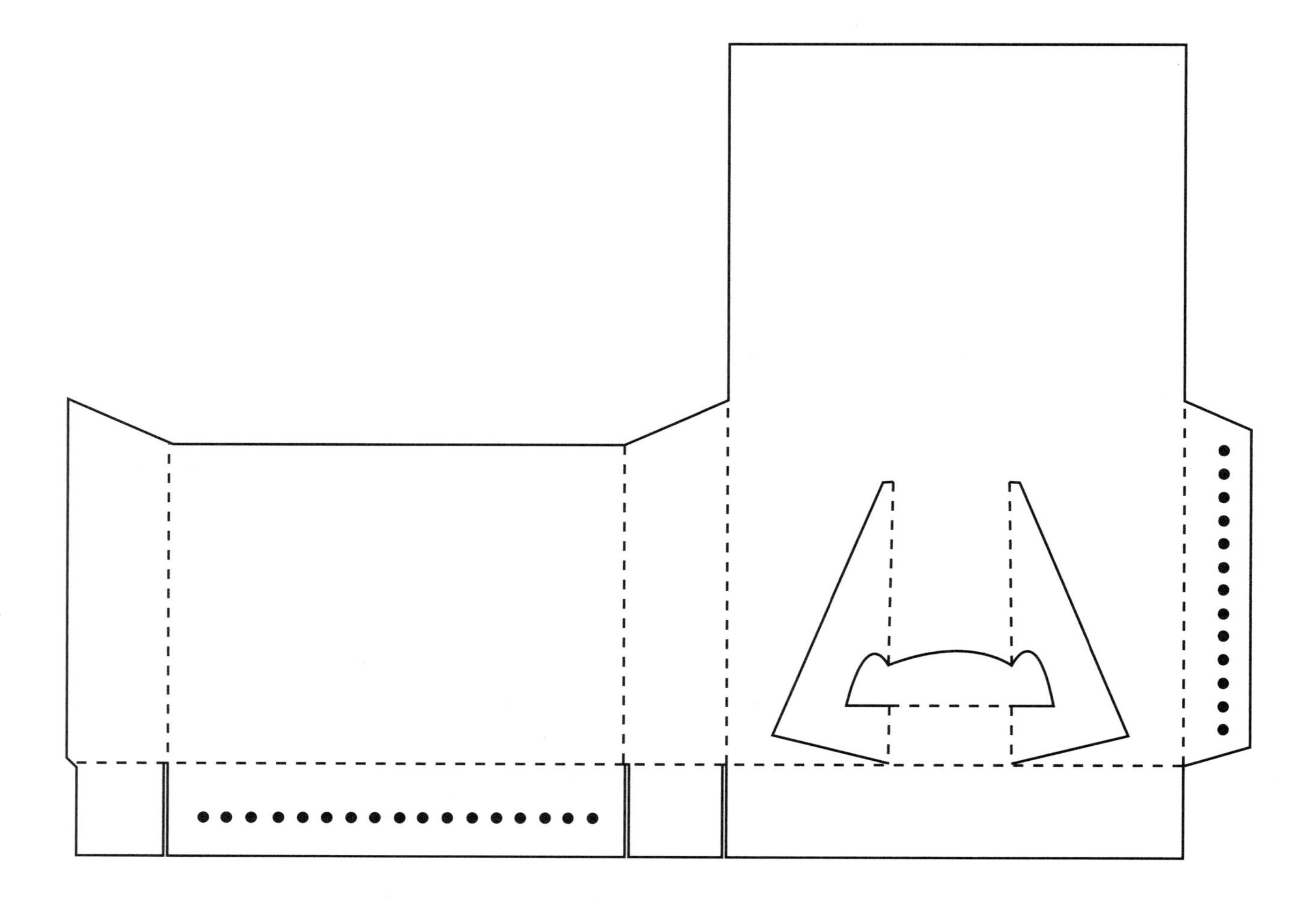

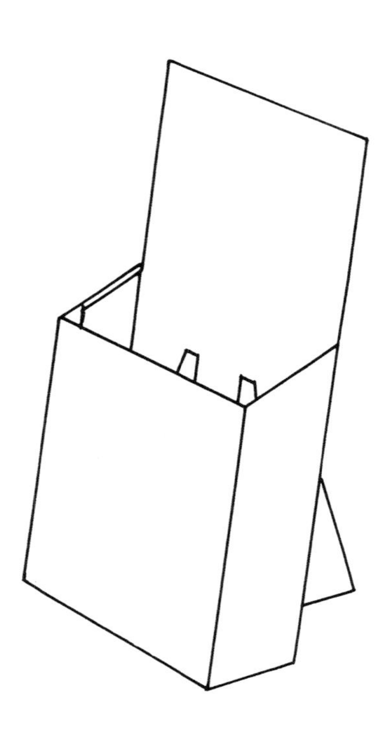

TWO-PIECE DISPLAY EASEL WITH HEADER CARD AND SAMPLE/BROCHURE BIN

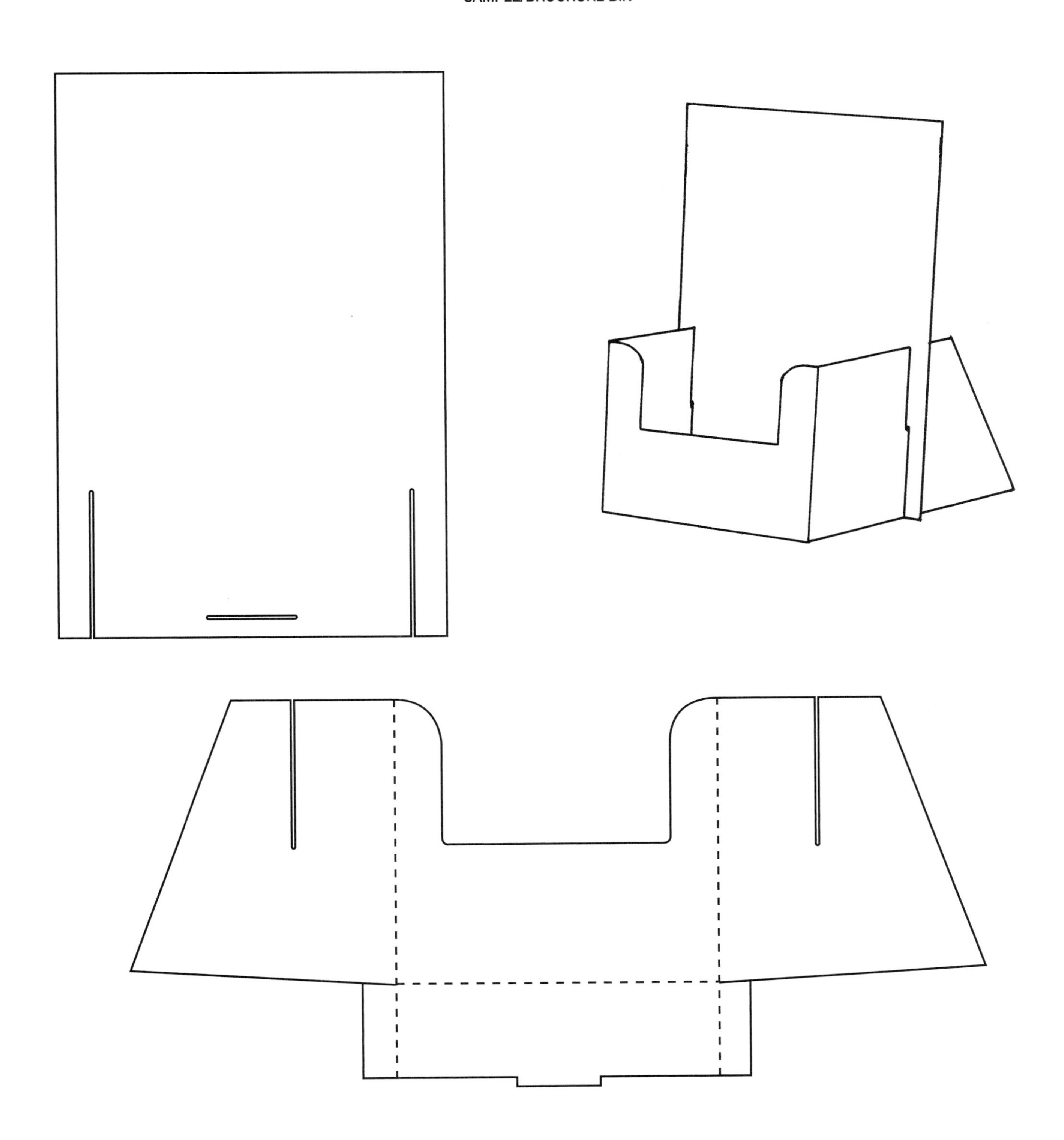

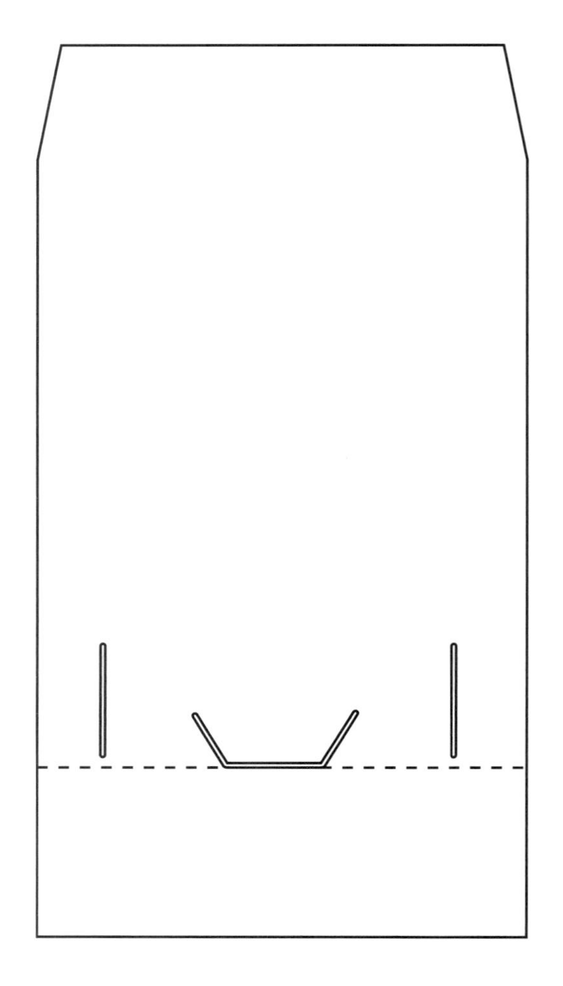

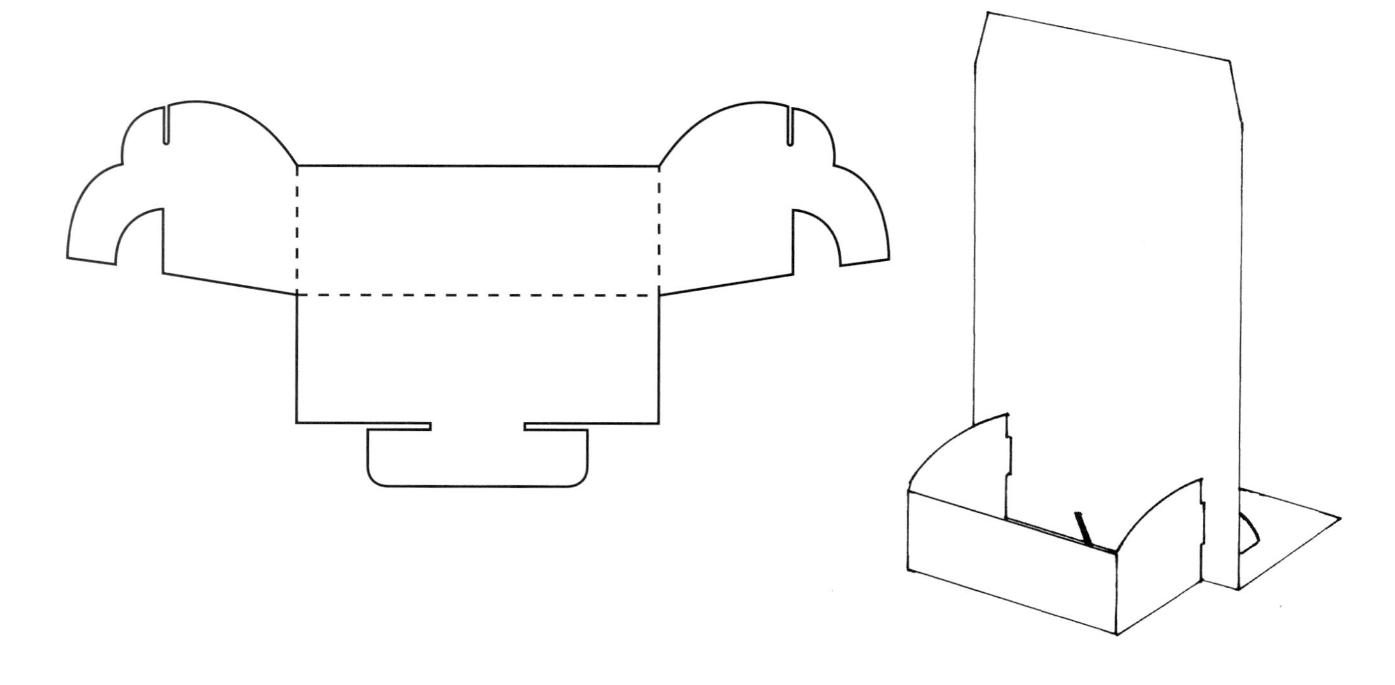

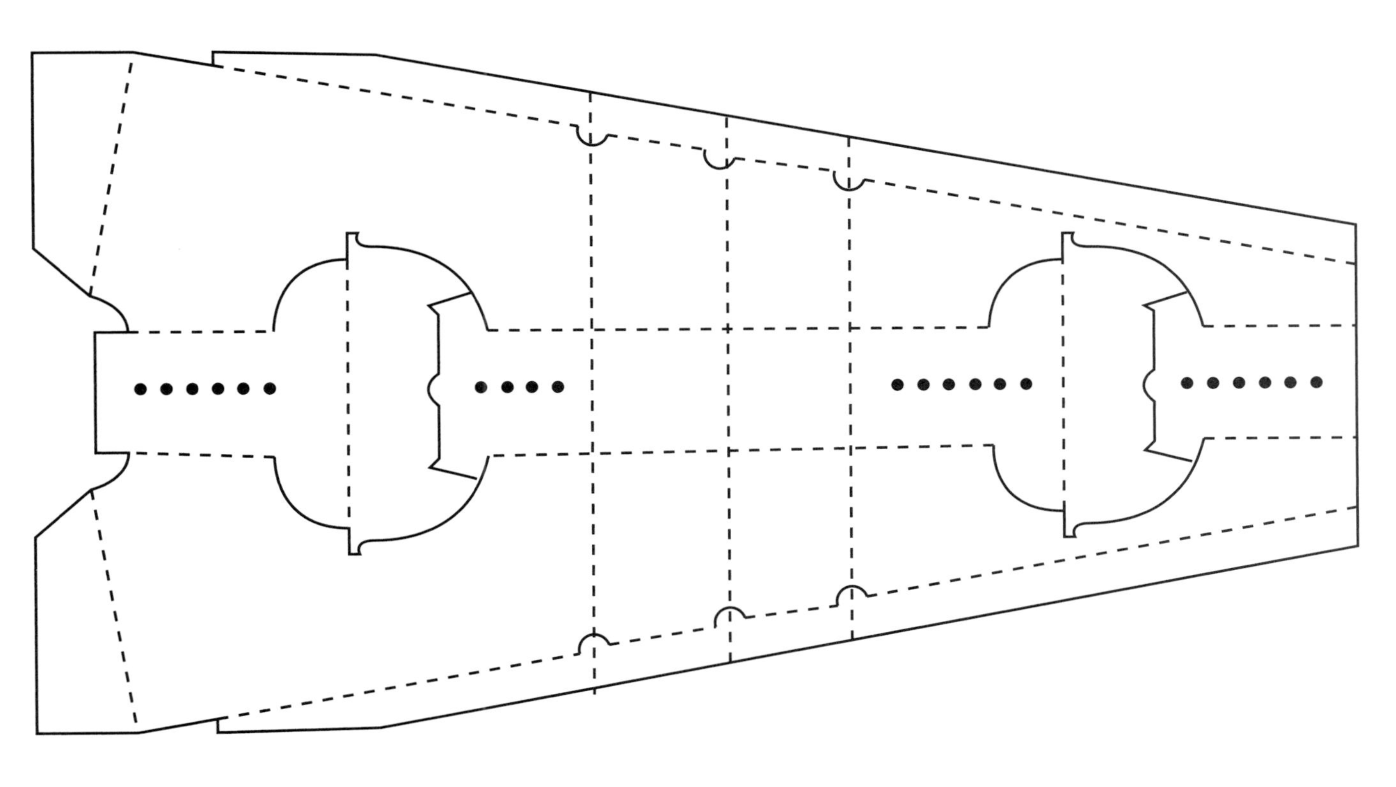

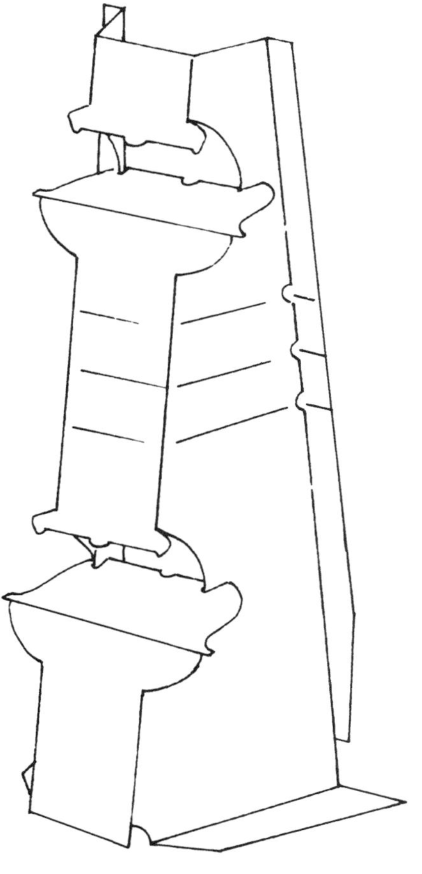

This easel is especially suitable for very tall displays that need to be folded in half for shipping. The middle portion of the easel may be accordion-folded back out of the way of the main lateral crease of the display.

The reinforcing turnbacks lend extra stability.

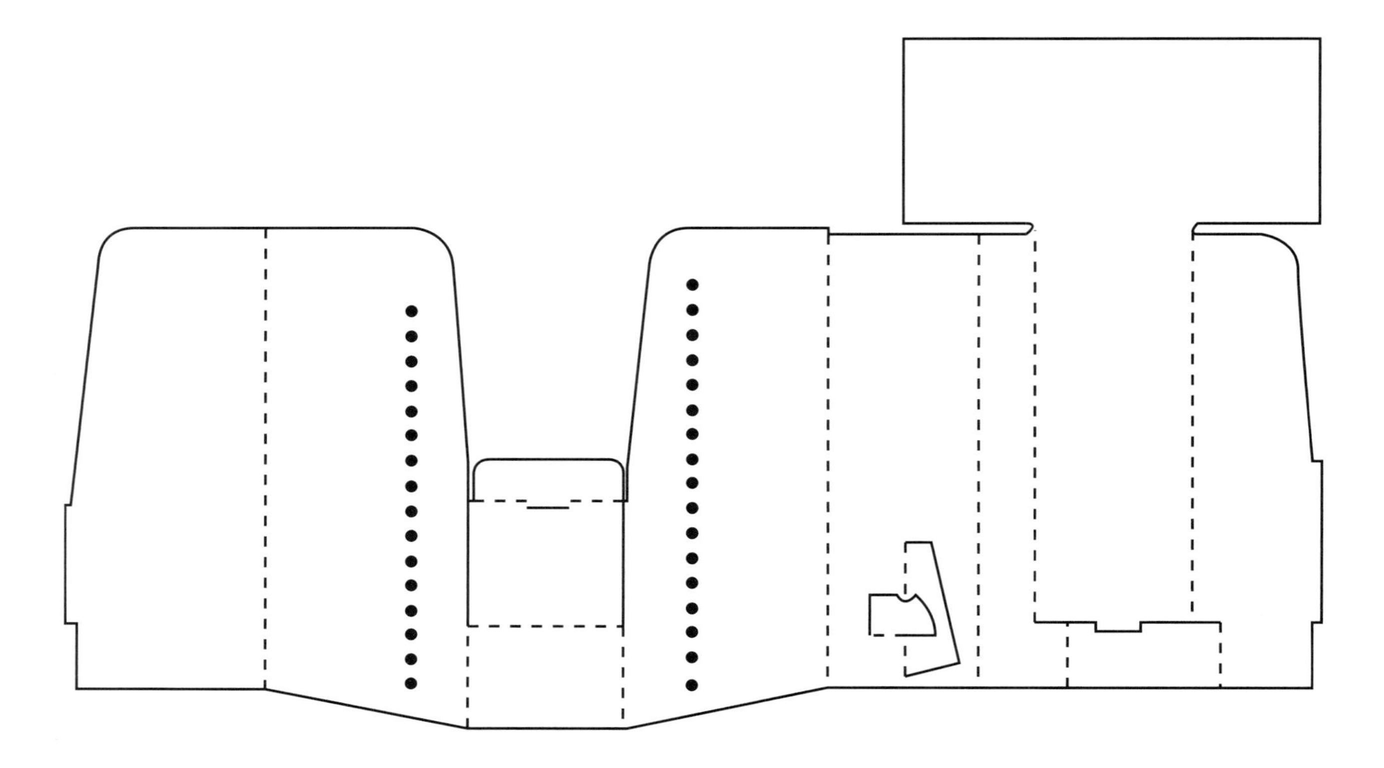

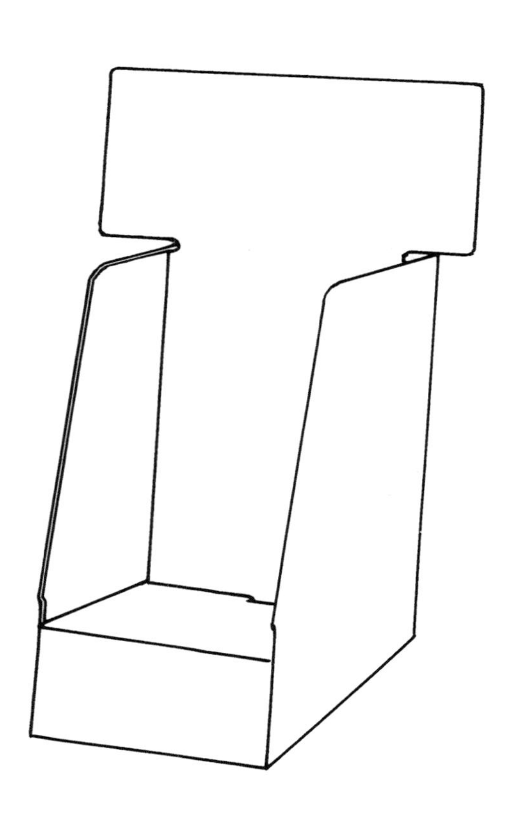

PULL-OUT SLEEVE THAT, WHEN FOLDED BACK, CREATES A SELF-SUPPORTING DISPLAY

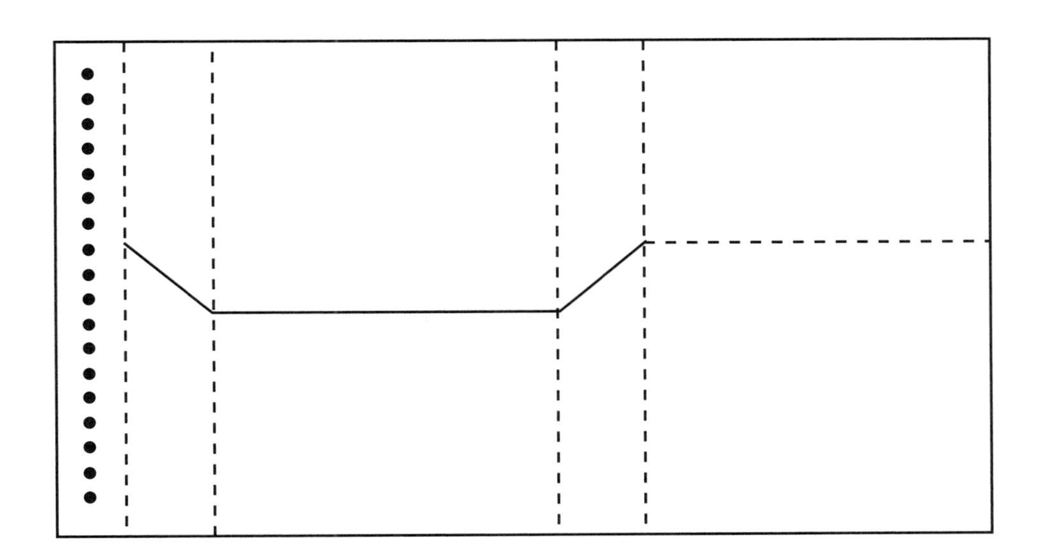

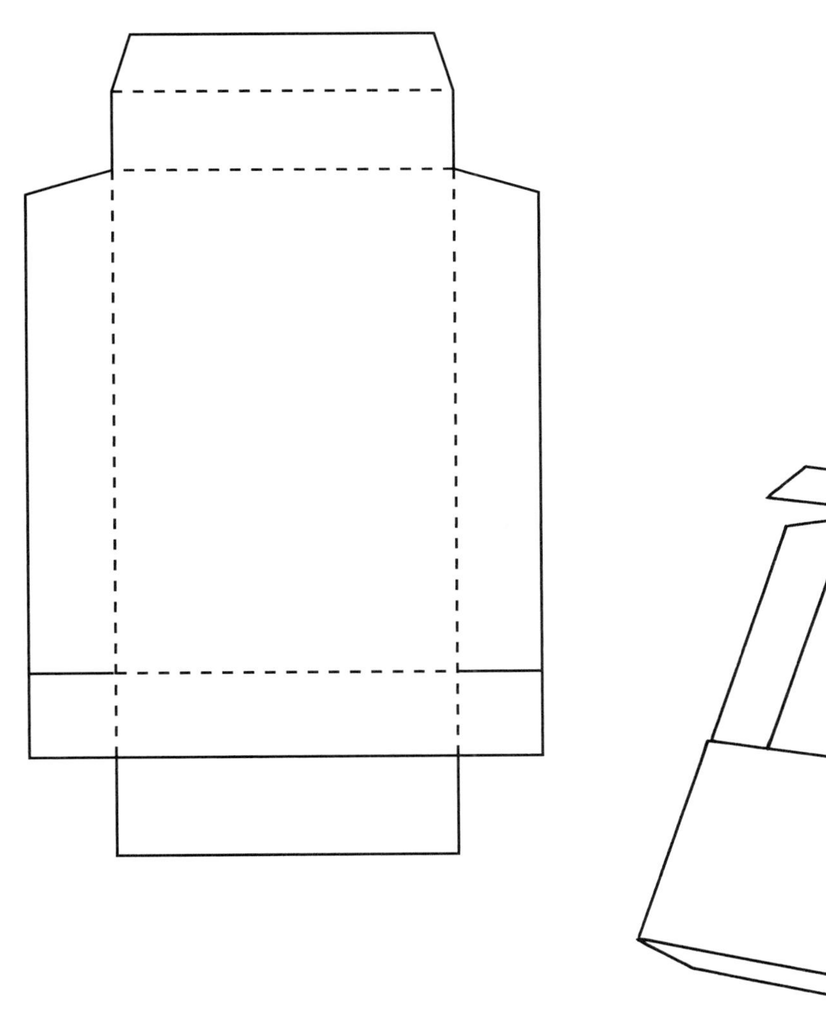

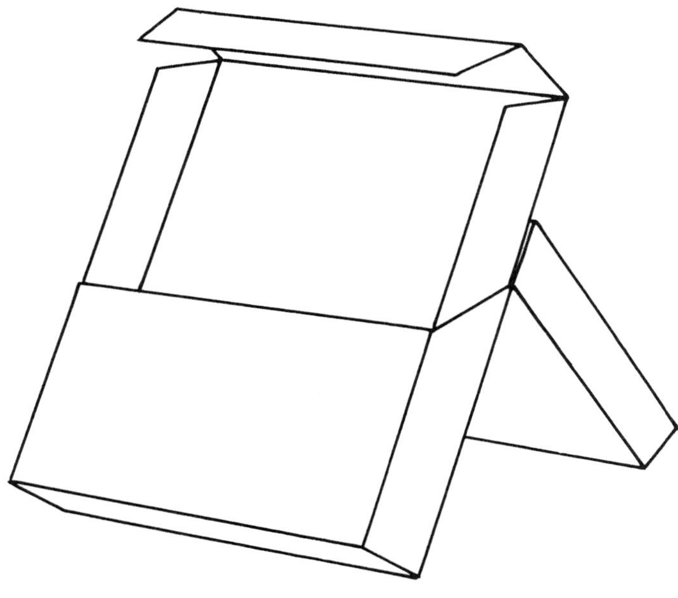

THREE-PIECE RIGID (SET-UP) BOX THAT SERVES AS A DISPLAY WHEN OPENED

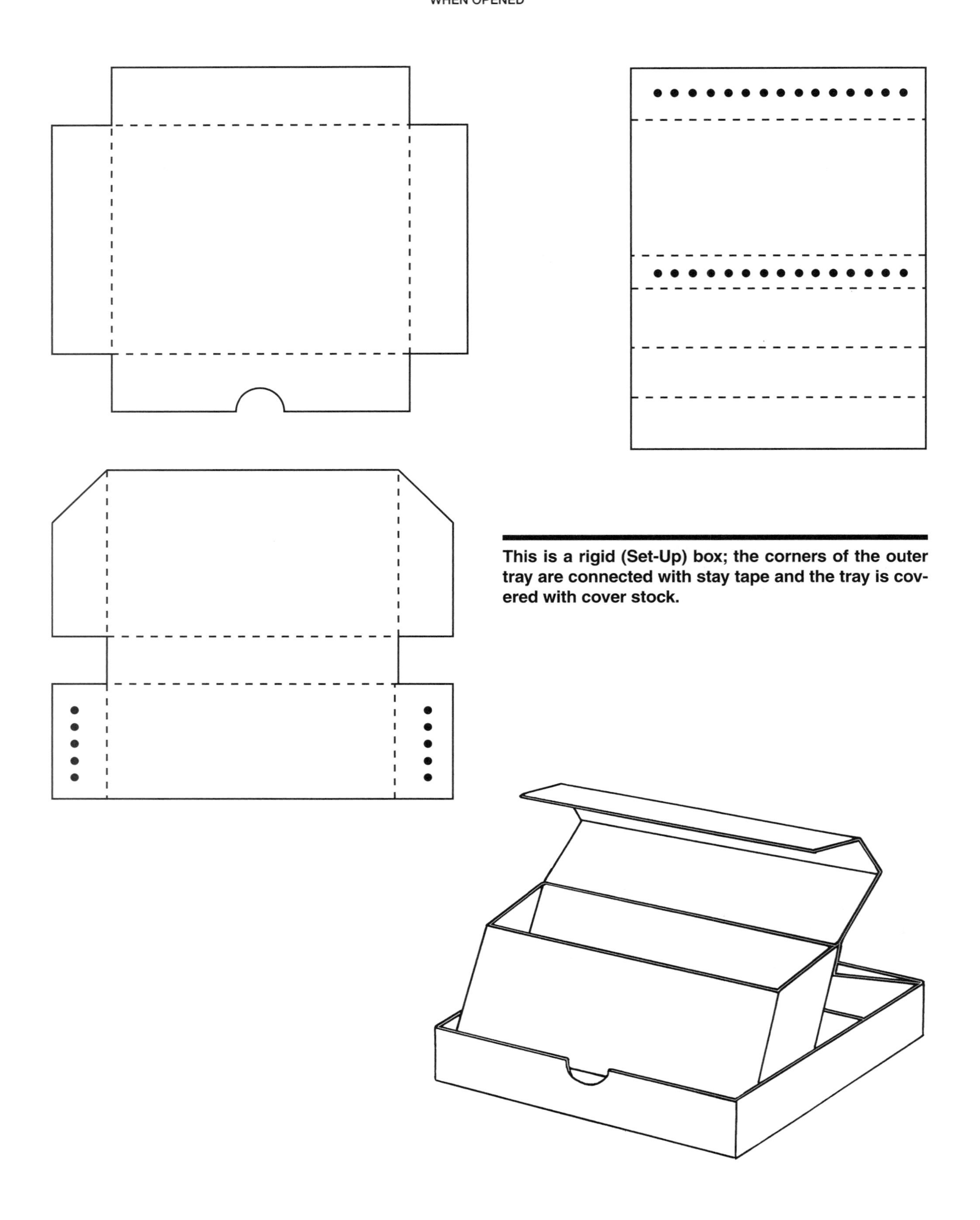

This is a rigid (Set-Up) box; the corners of the outer tray are connected with stay tape and the tray is covered with cover stock.

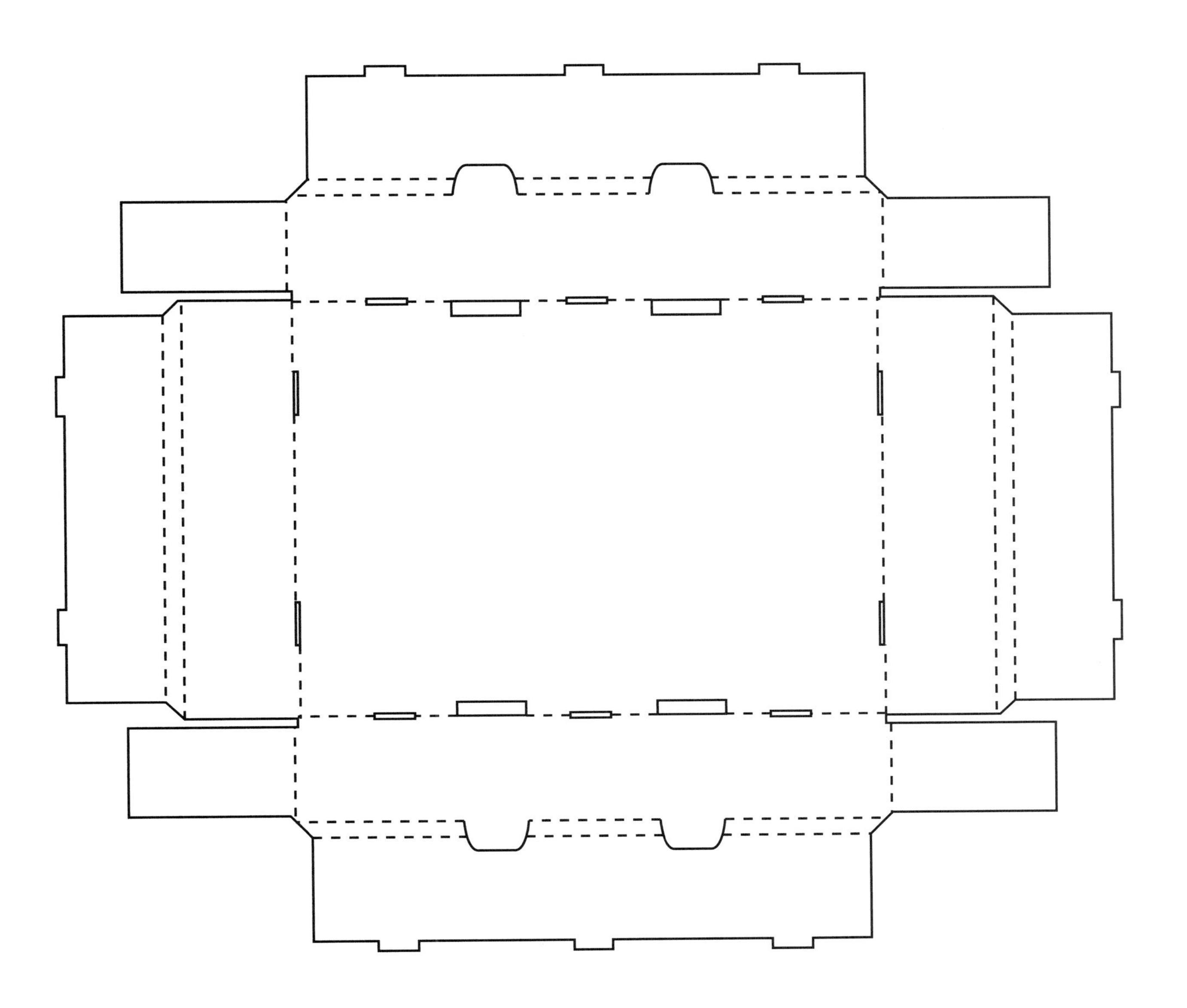

The stacking tabs that protrude above the top surface of the container insert into the slot of the above-stacked container with a compatible footprint.

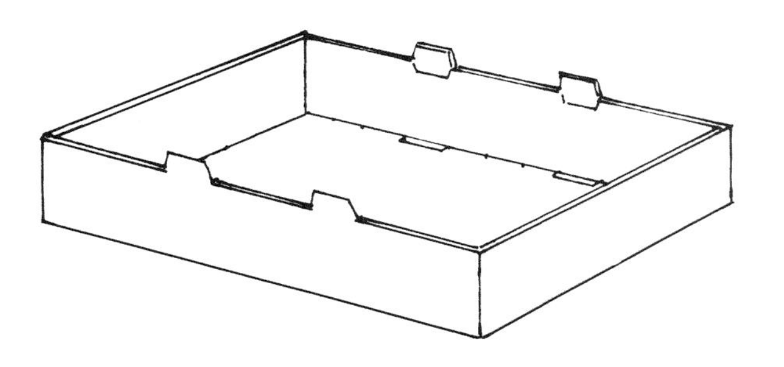

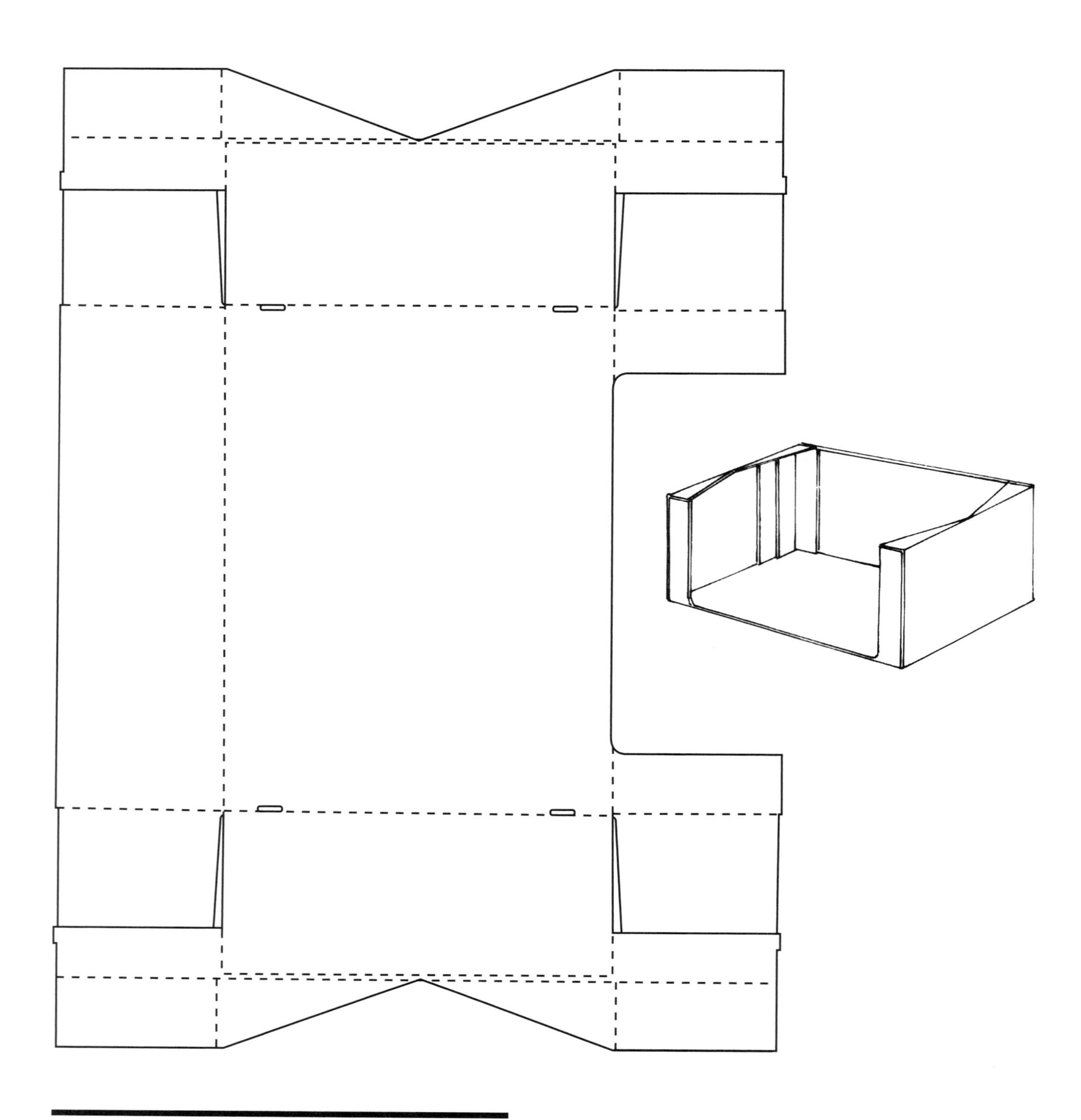

Shown with top and partial front removed, this unit has reinforced corners.

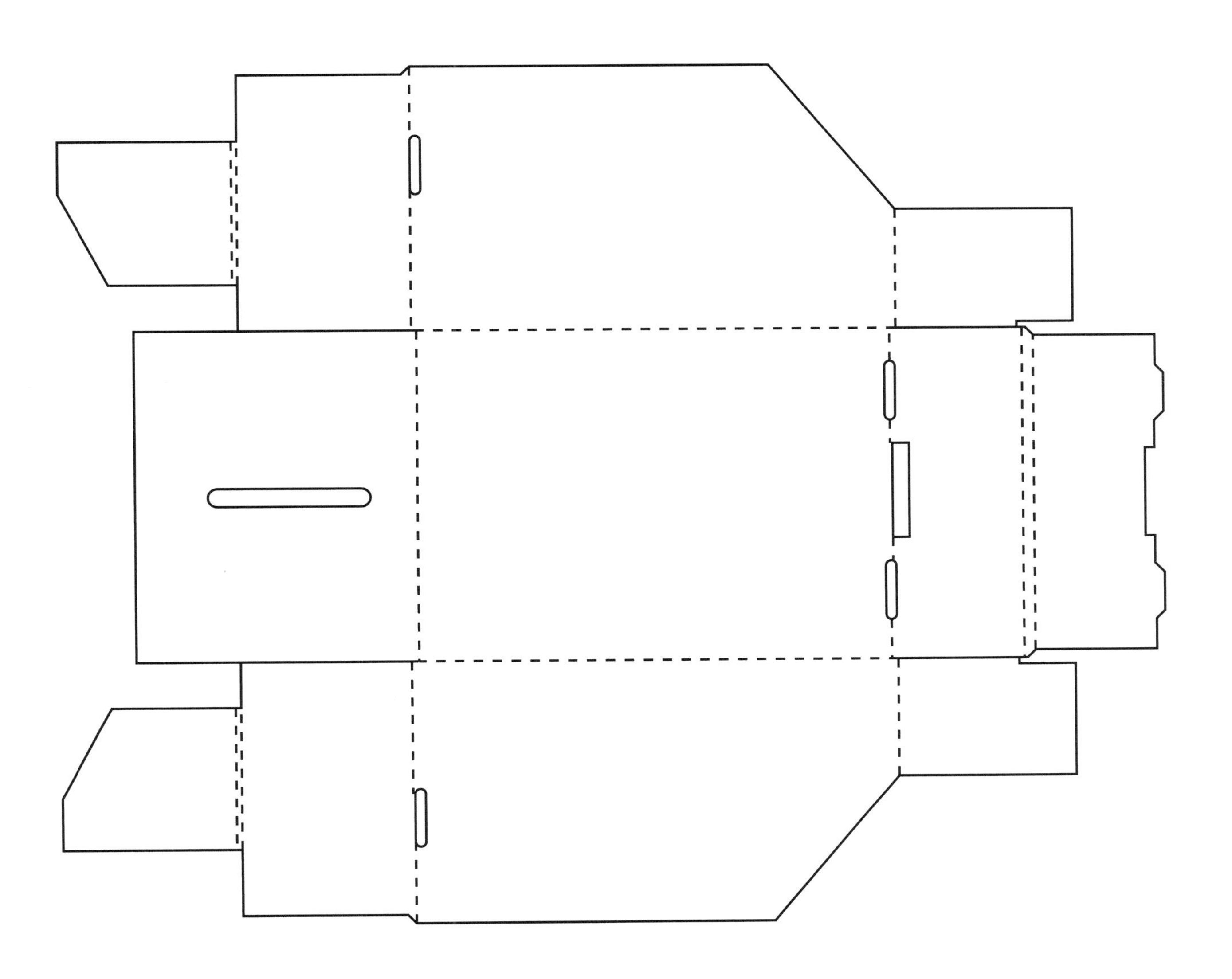

As a shipping container this carton features a lid attached to the back panel (not shown). Once the lid has been torn off, the unit acts as a display bin.

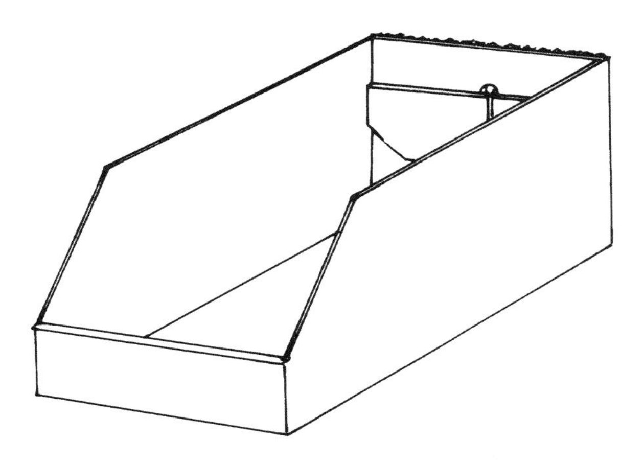

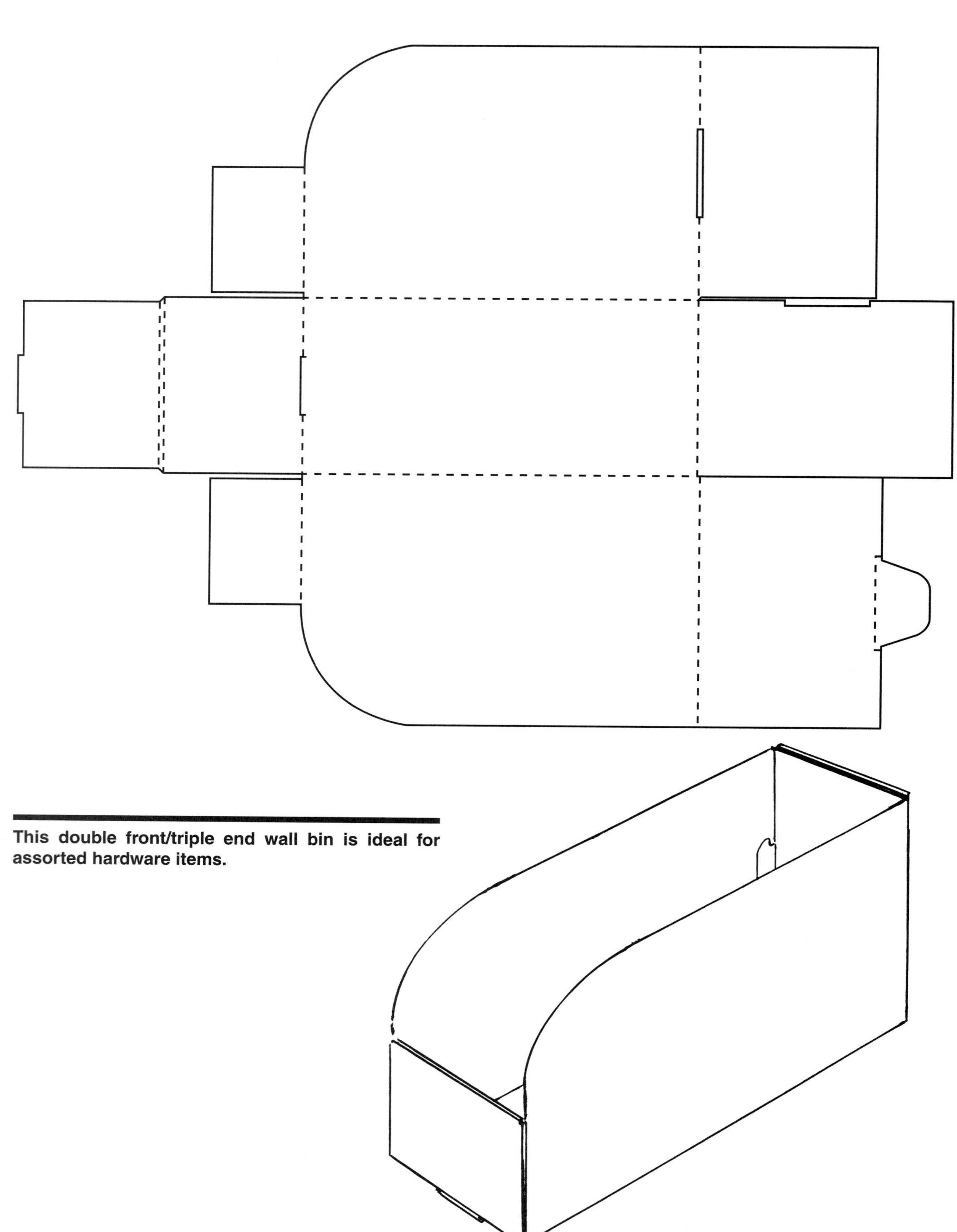

This double front/triple end wall bin is ideal for assorted hardware items.

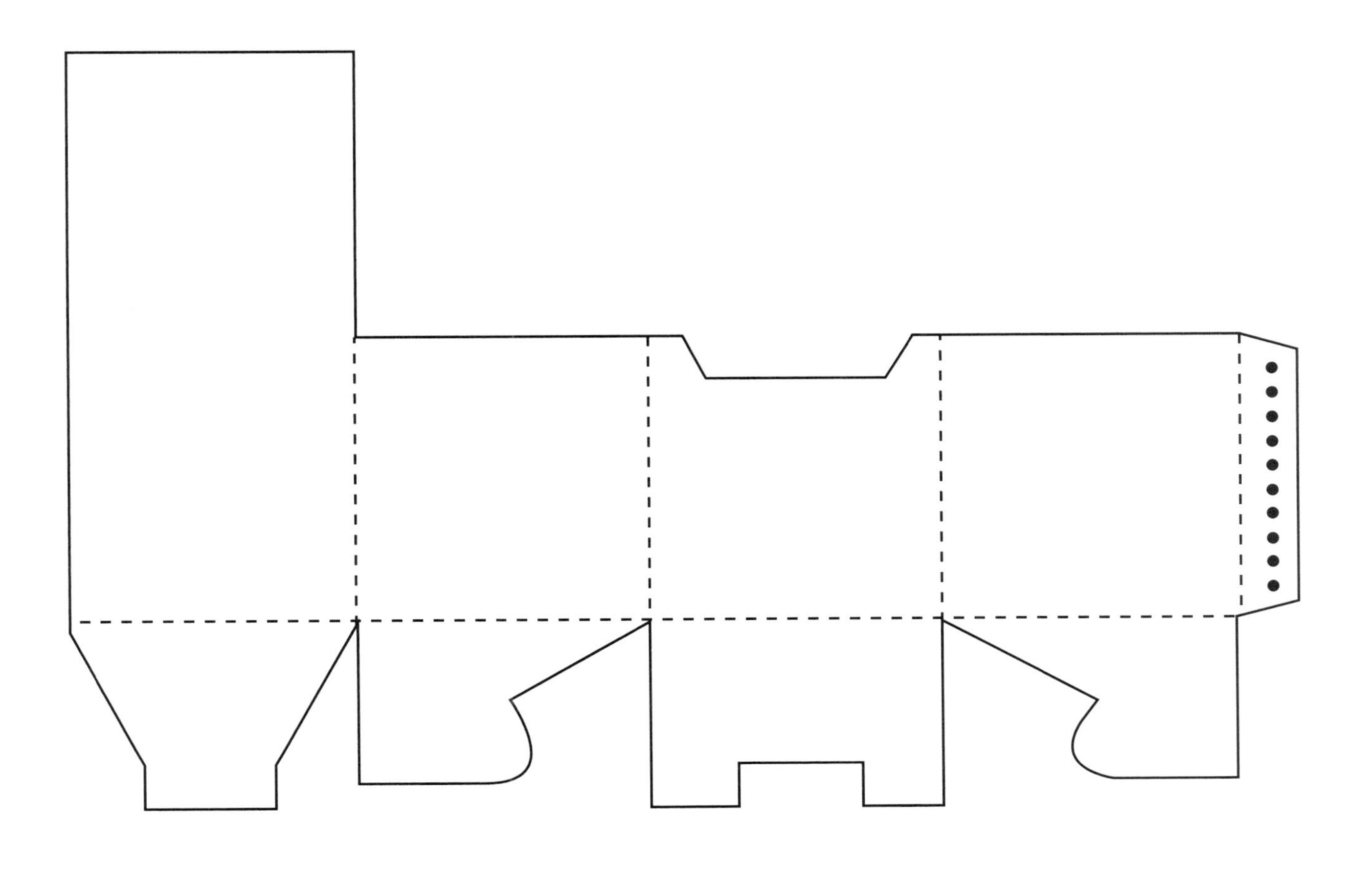

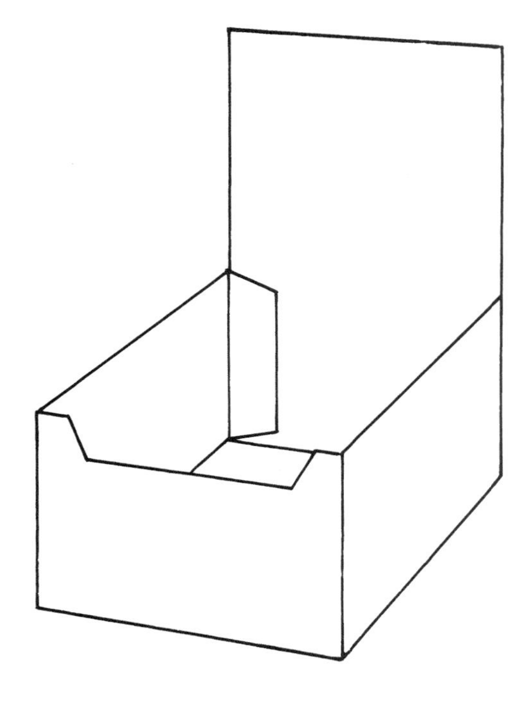

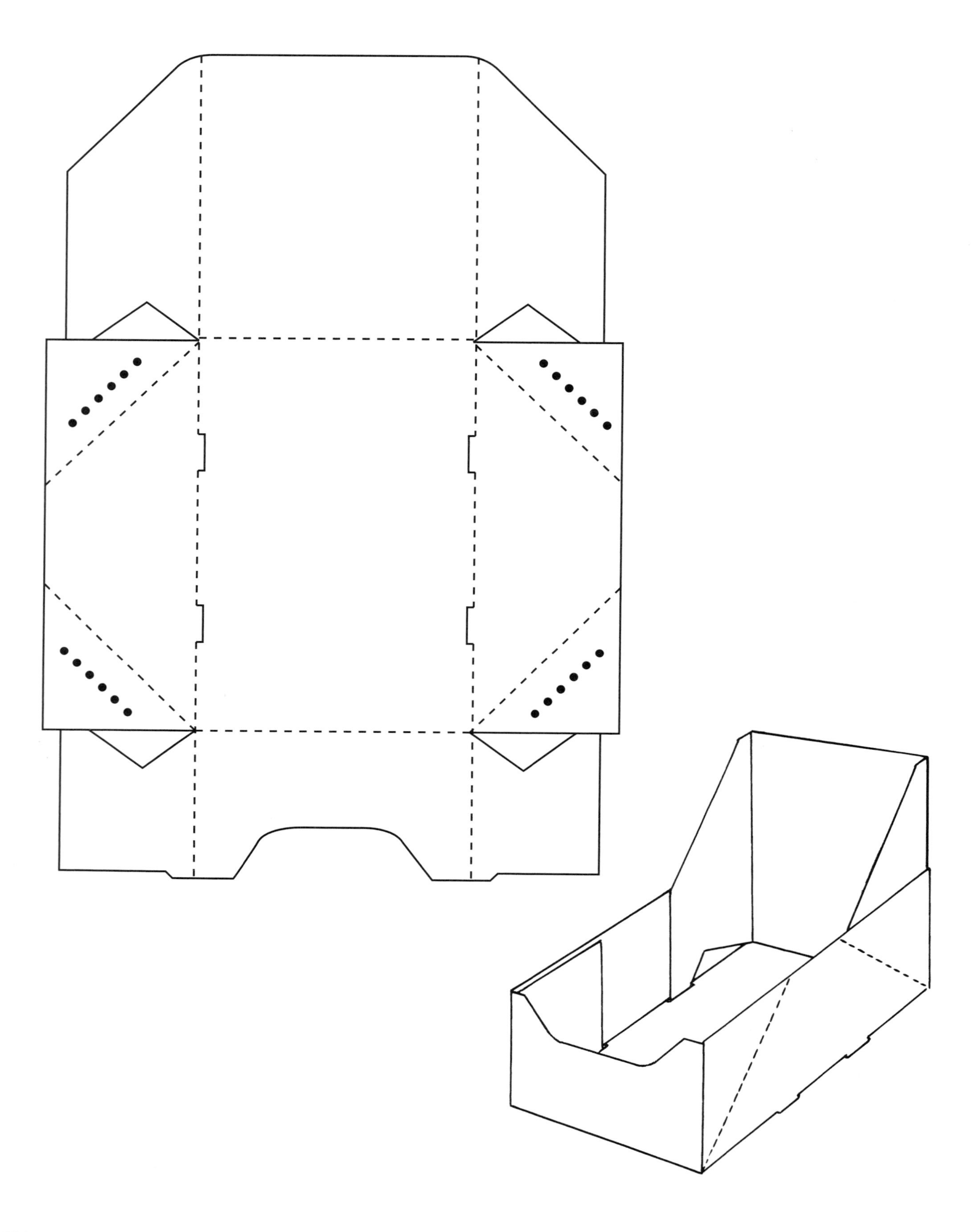

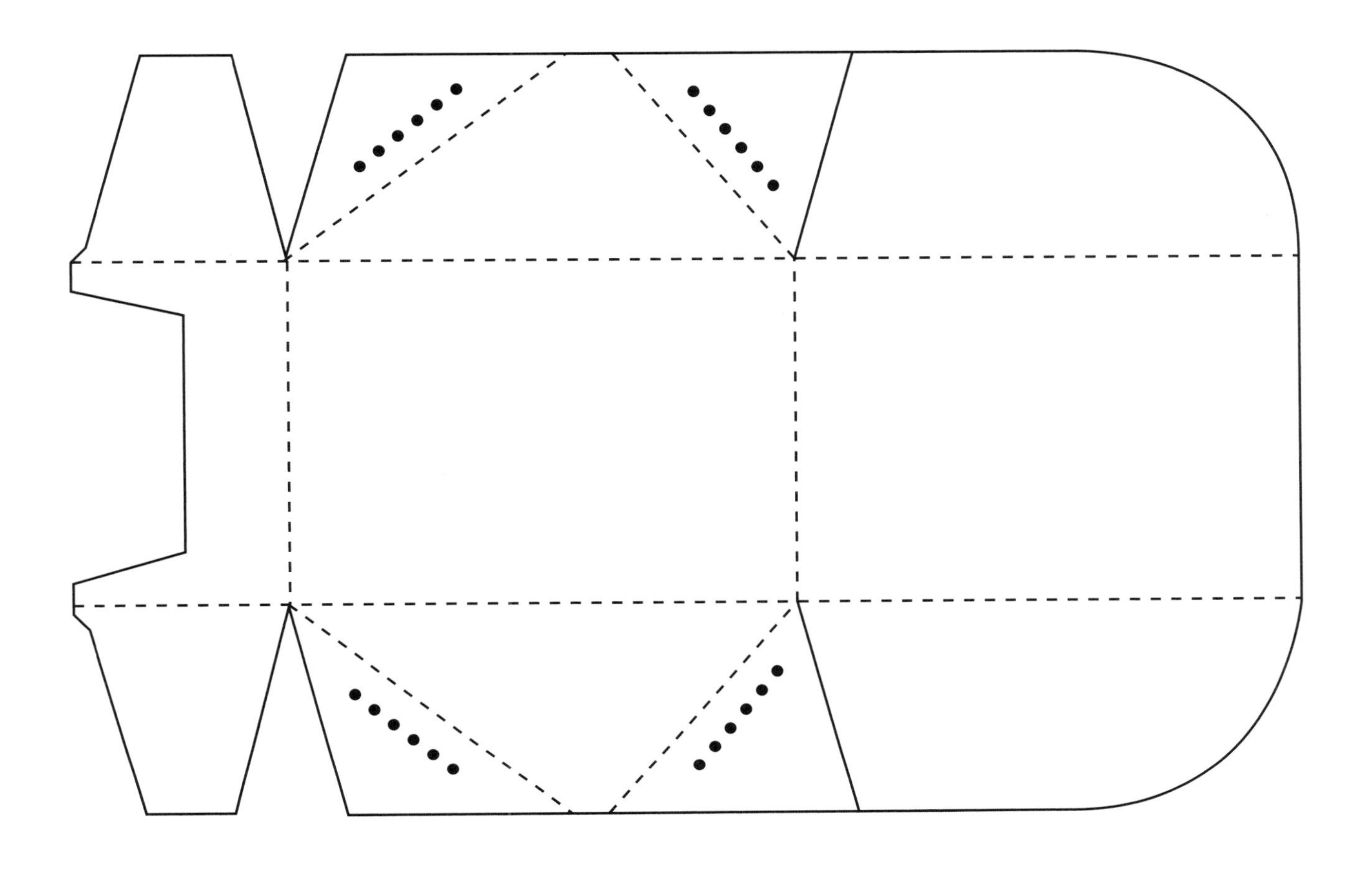

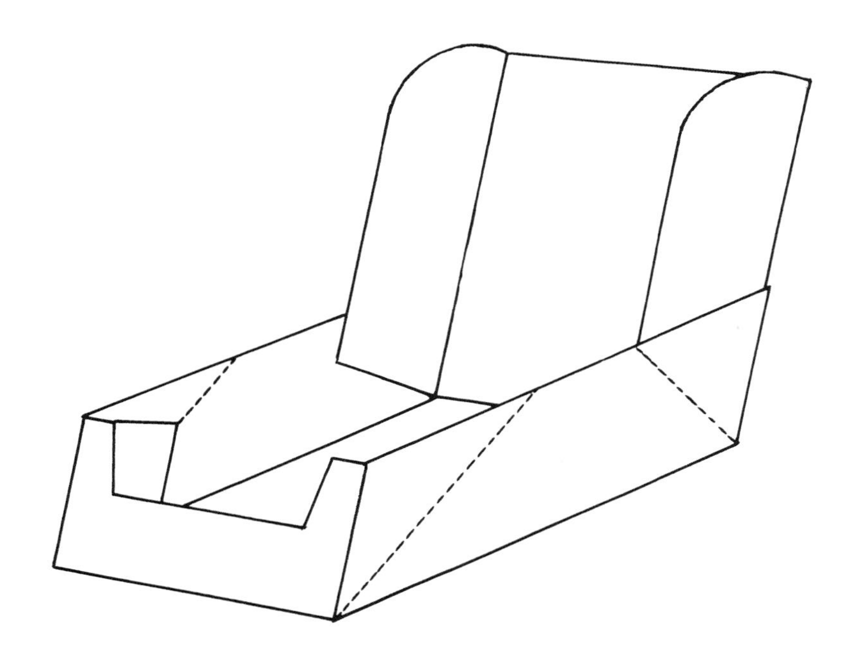

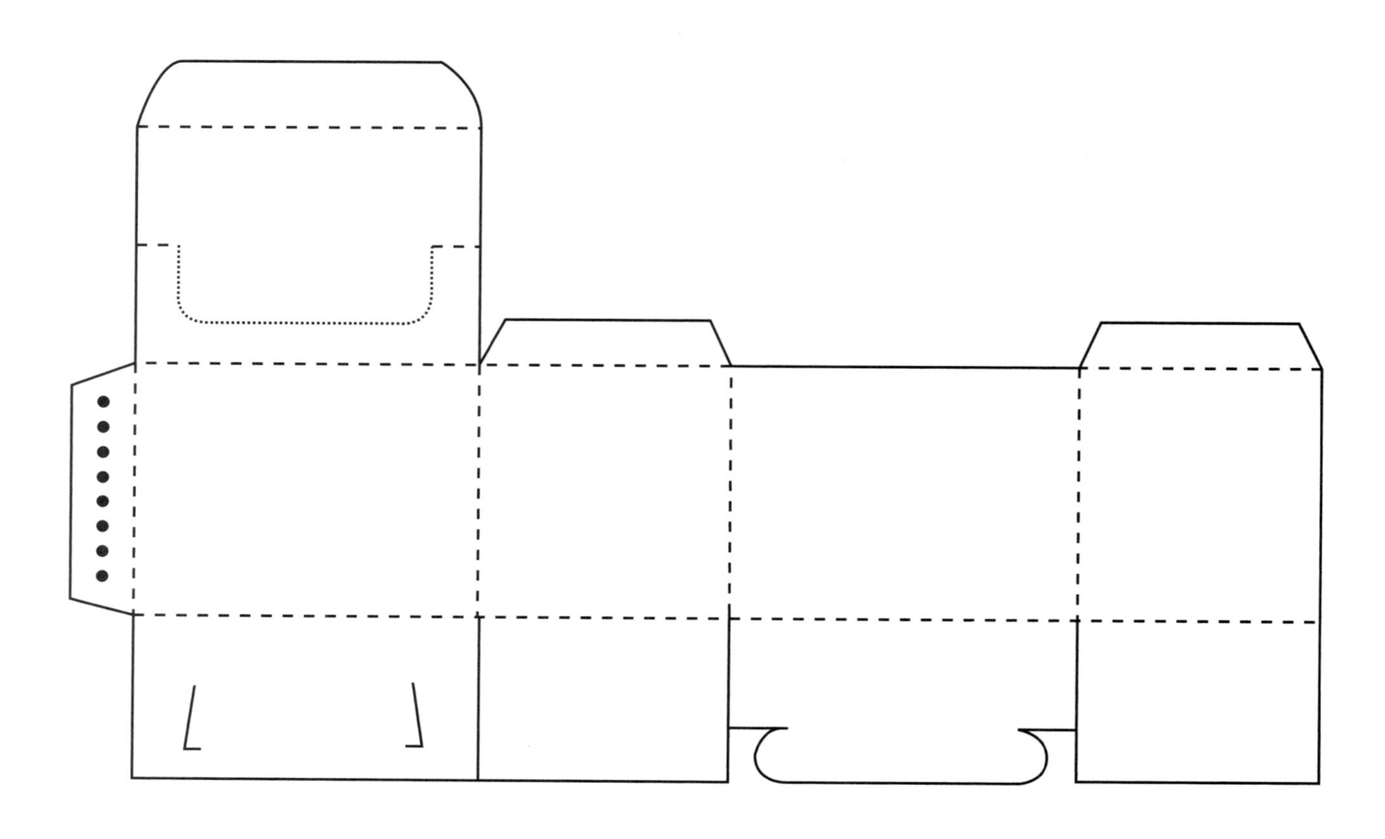

The lid is prescored and perforated to create a display panel when opened.

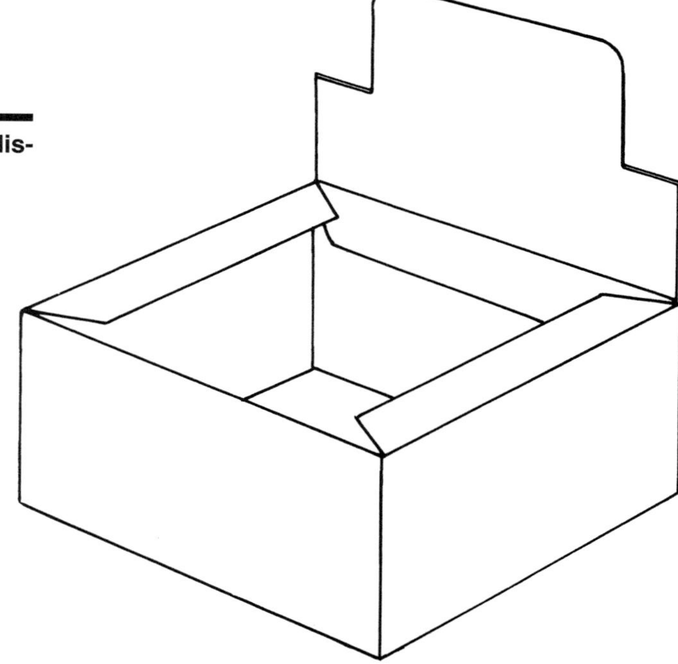

DISPLAY SHIPPER WITH HIMES LOCK BOTTOM AND TONGUE LOCK CLOSURE

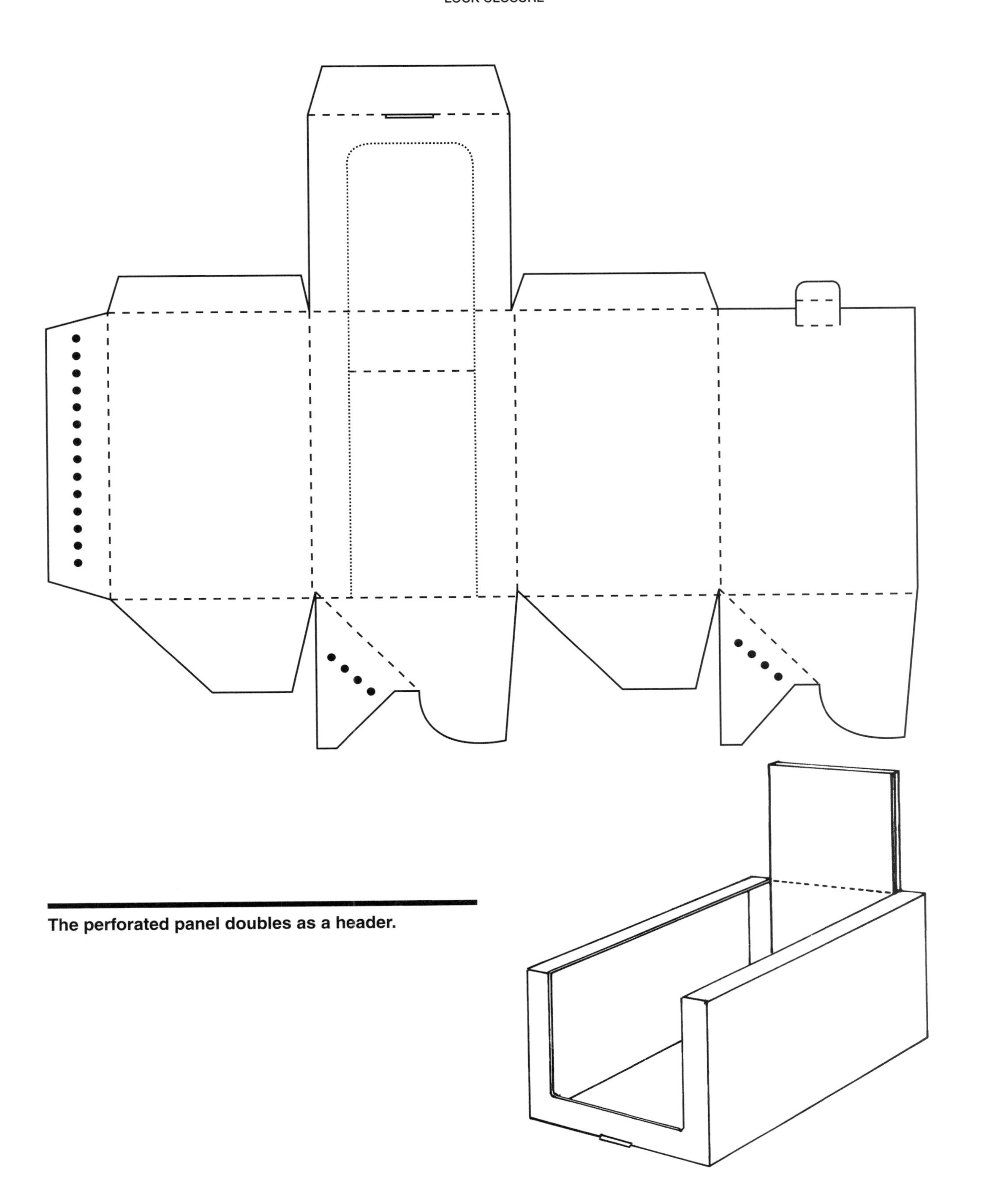

The perforated panel doubles as a header.

RHOMBOID DISPLAY CONTAINER WITH ARTHUR LOCK/LOCK TAB AND SLIT COMBINATION

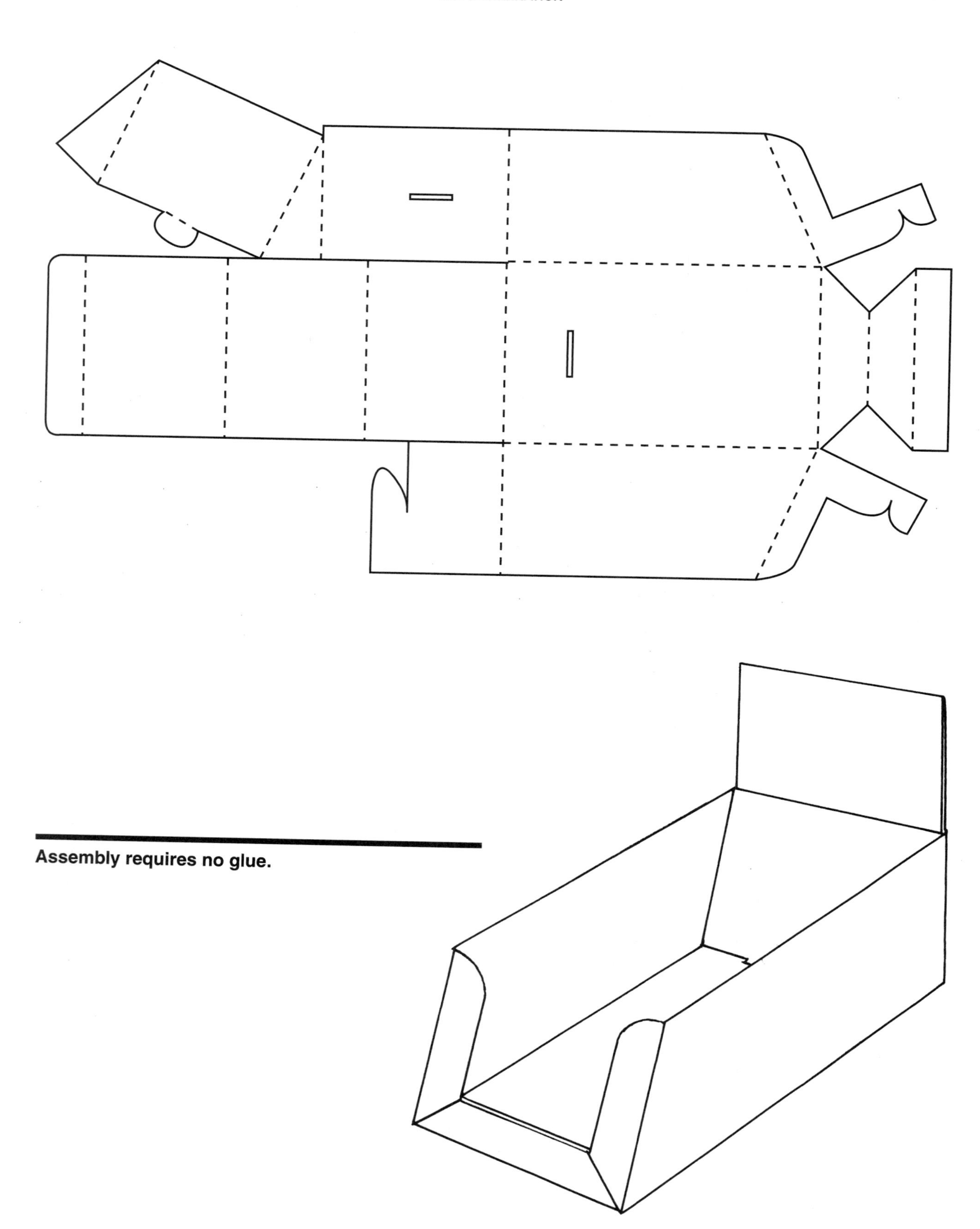

Assembly requires no glue.

ONE-PIECE OVERLAPPING FRONT, LOCK-BACK PANEL, FOLD-OVER DISPLAY

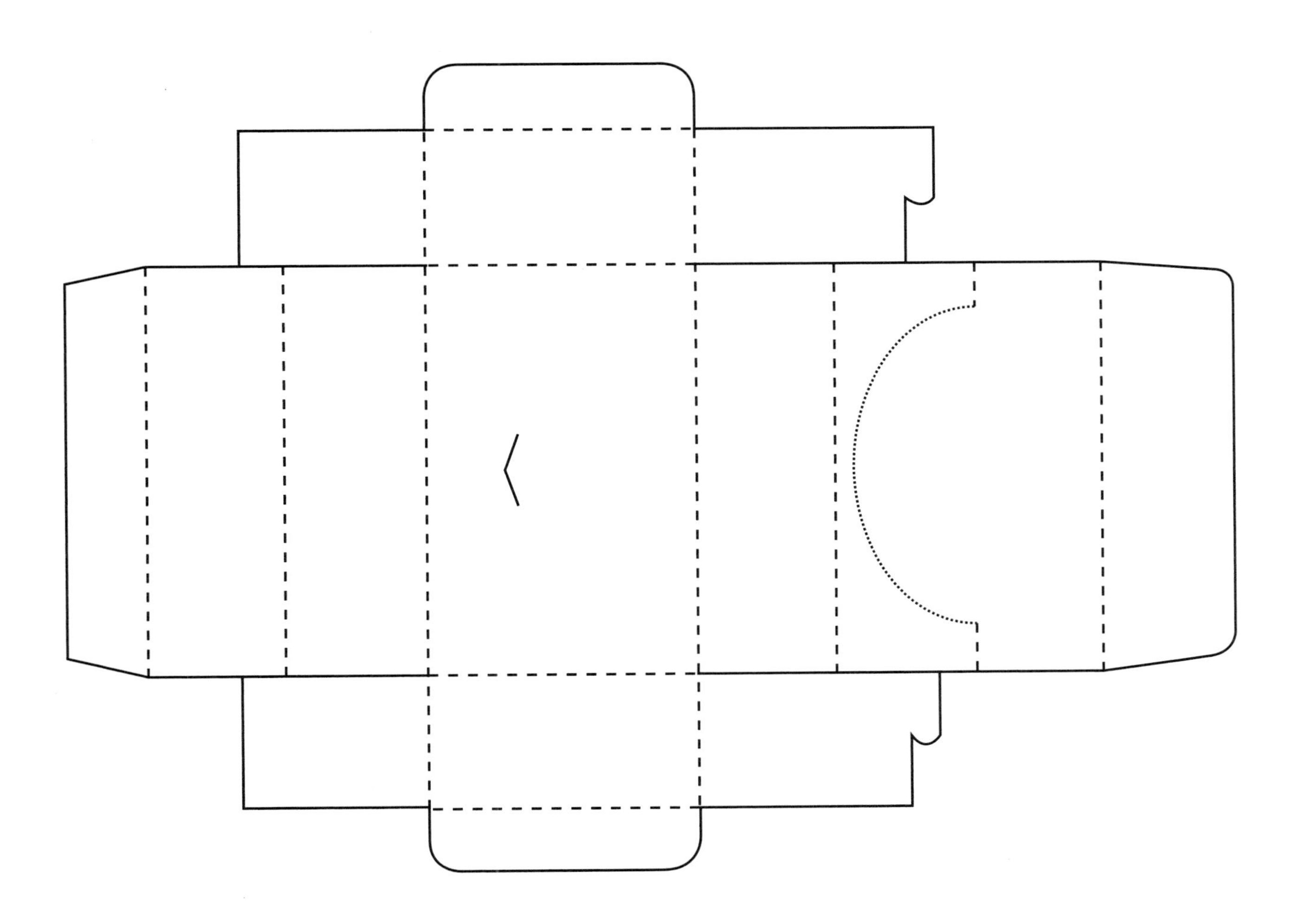

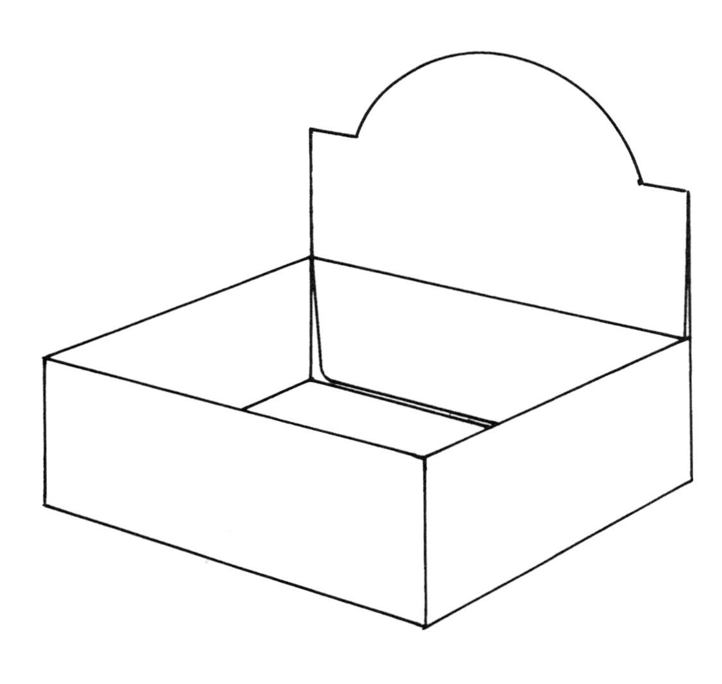

SINGLE FRONT WALL, SINGLE SIDE WALL CONTAINER DISPLAY WITH HIMES LOCK BOTTOM

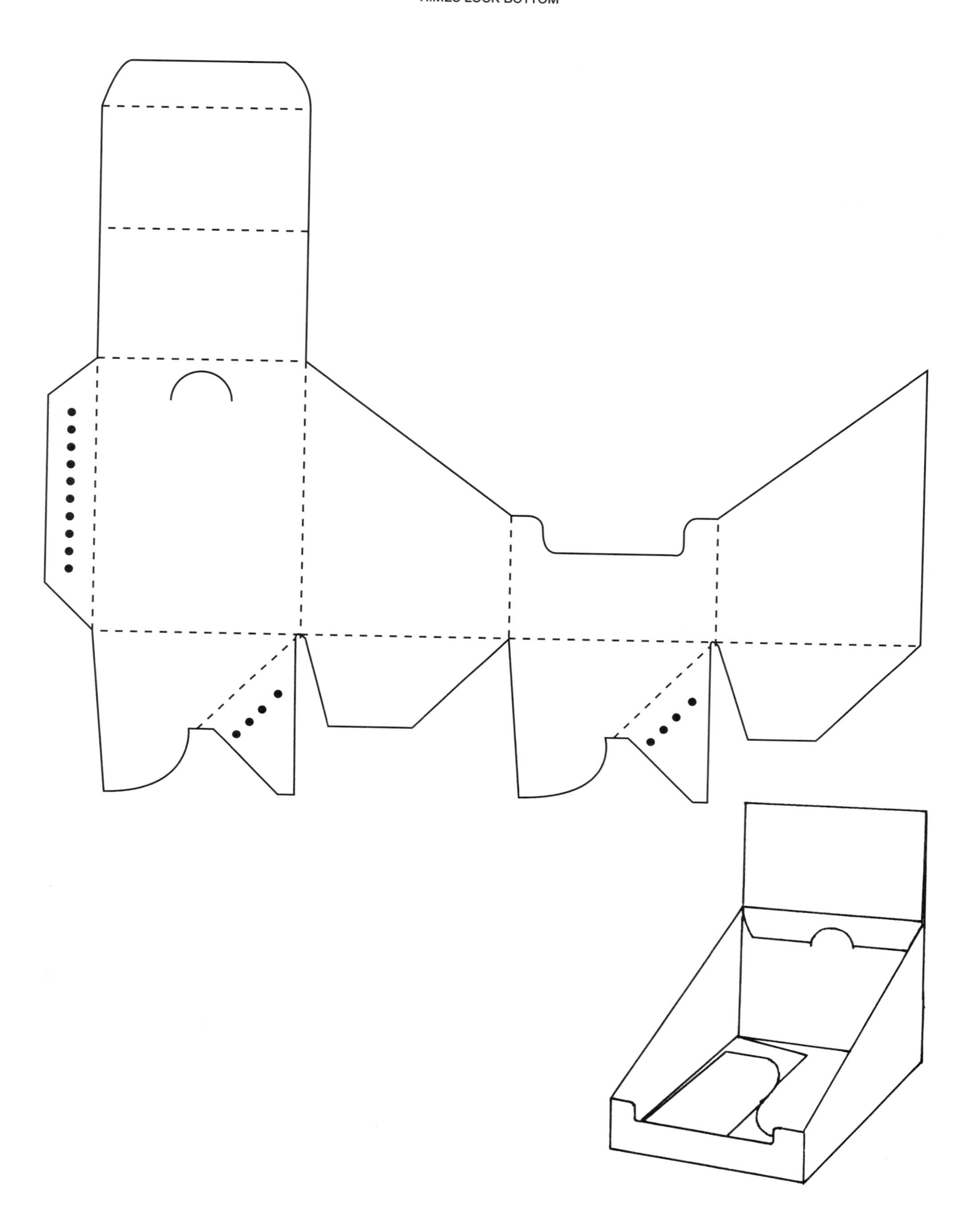

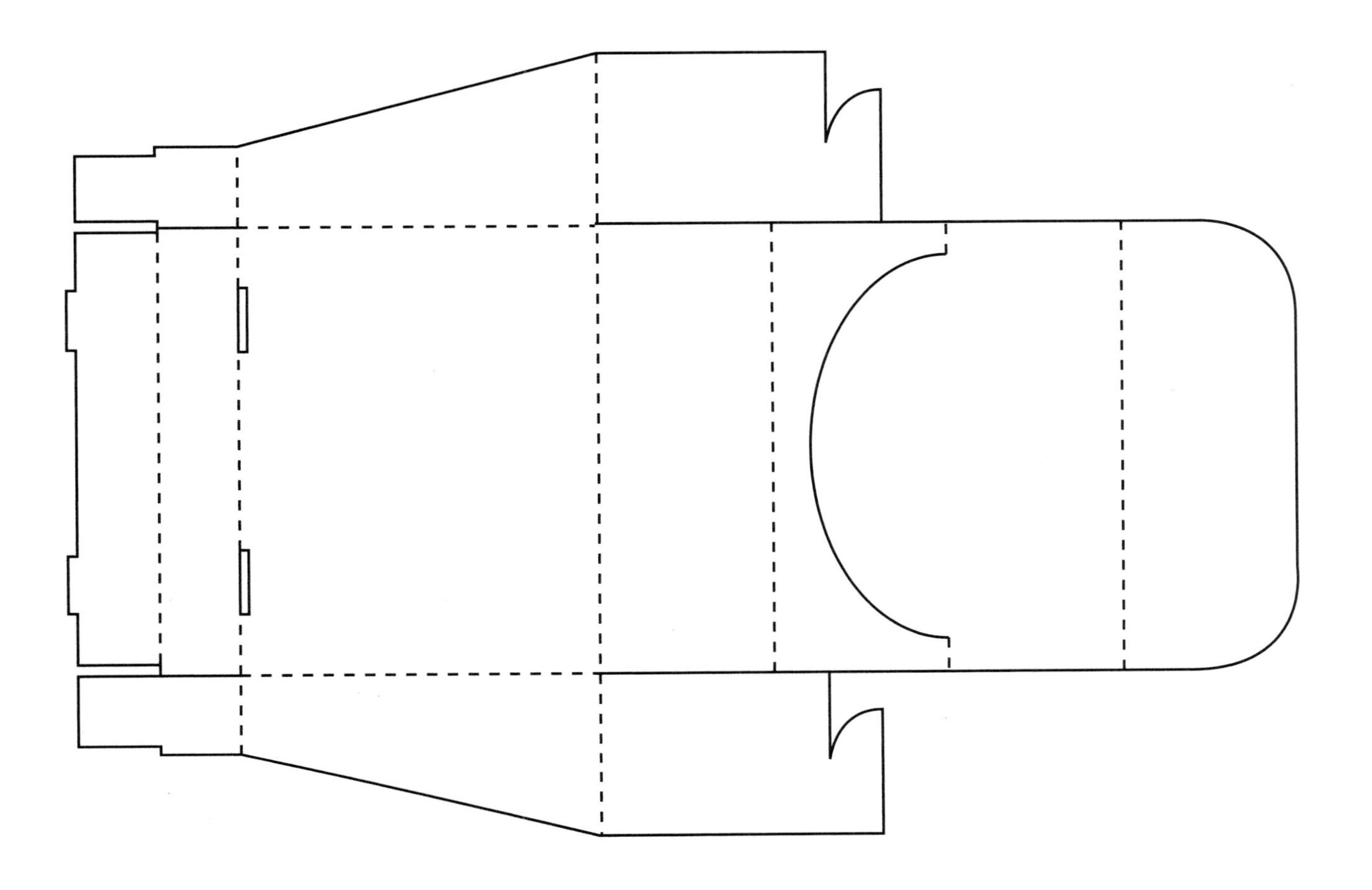

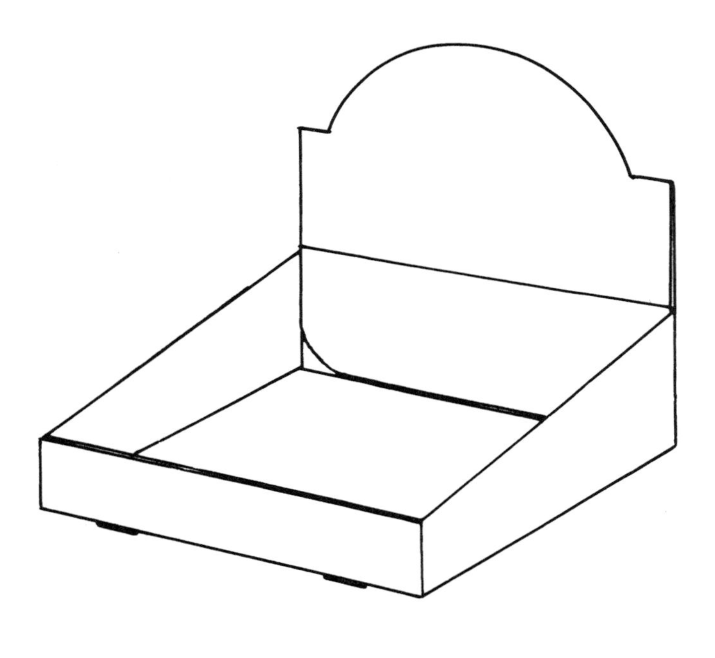

DOUBLE FRONT, DOUBLE SIDE PANEL DISPLAY WITH HOOK LOCK AND WALKER LOCK COMBINATION

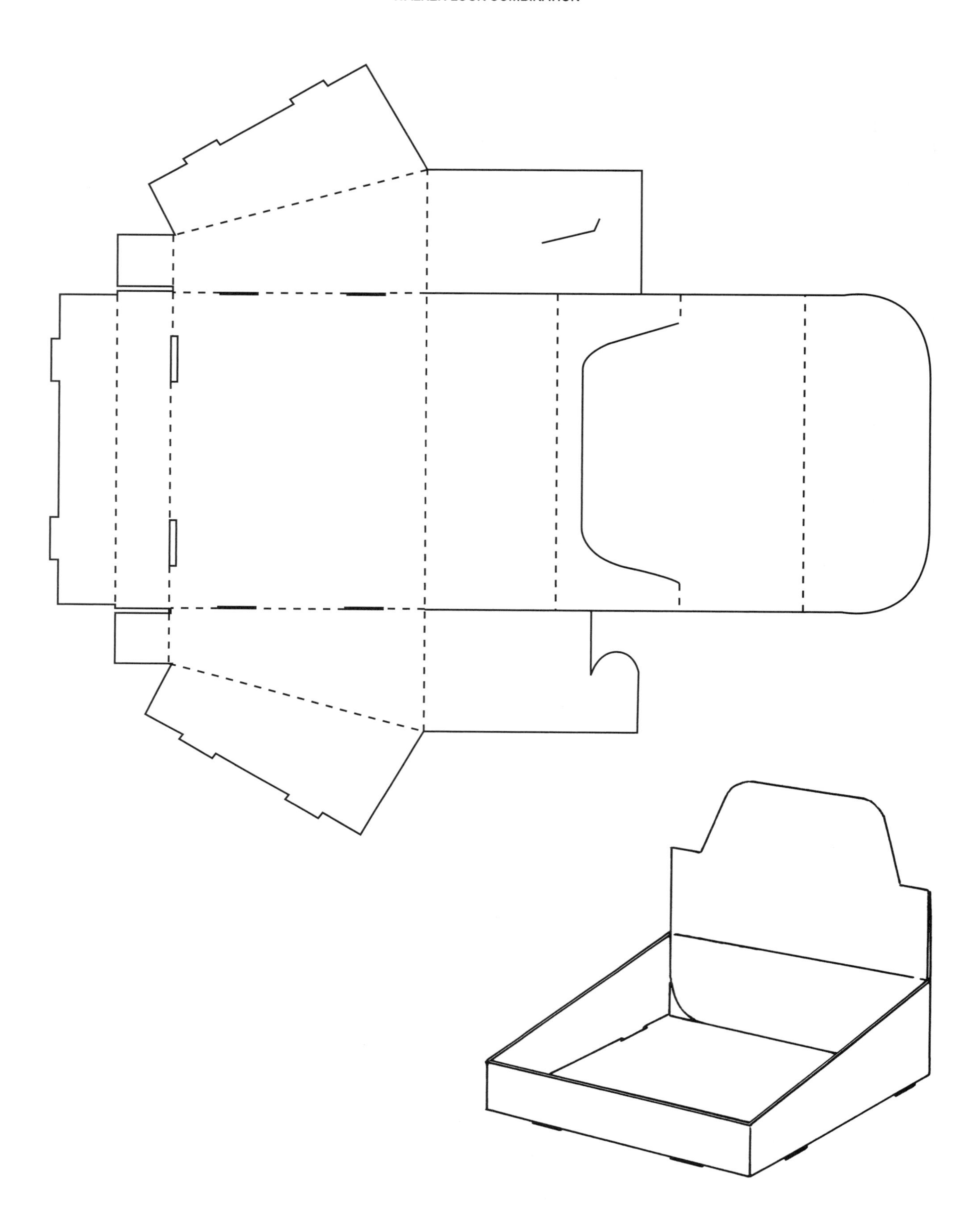

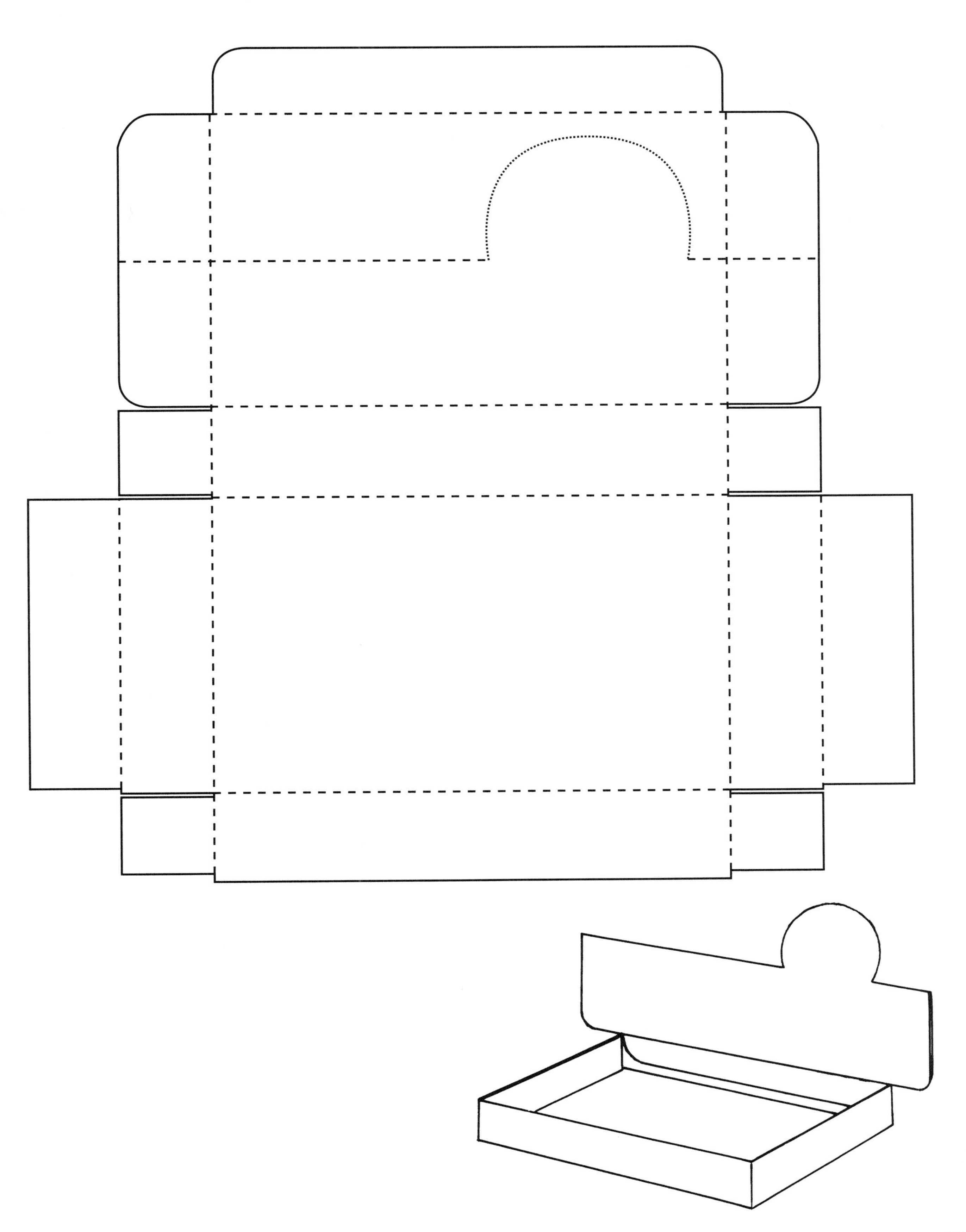

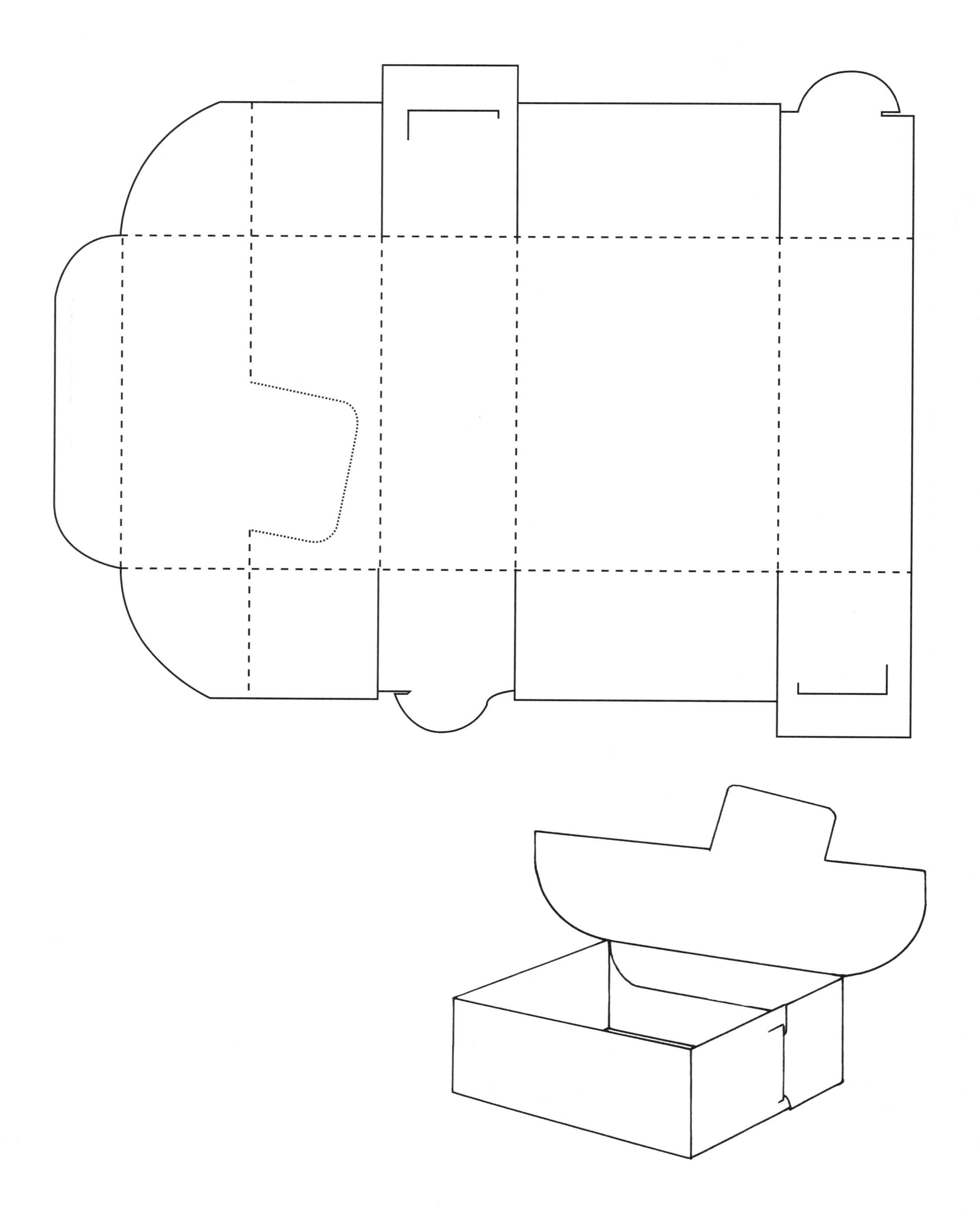

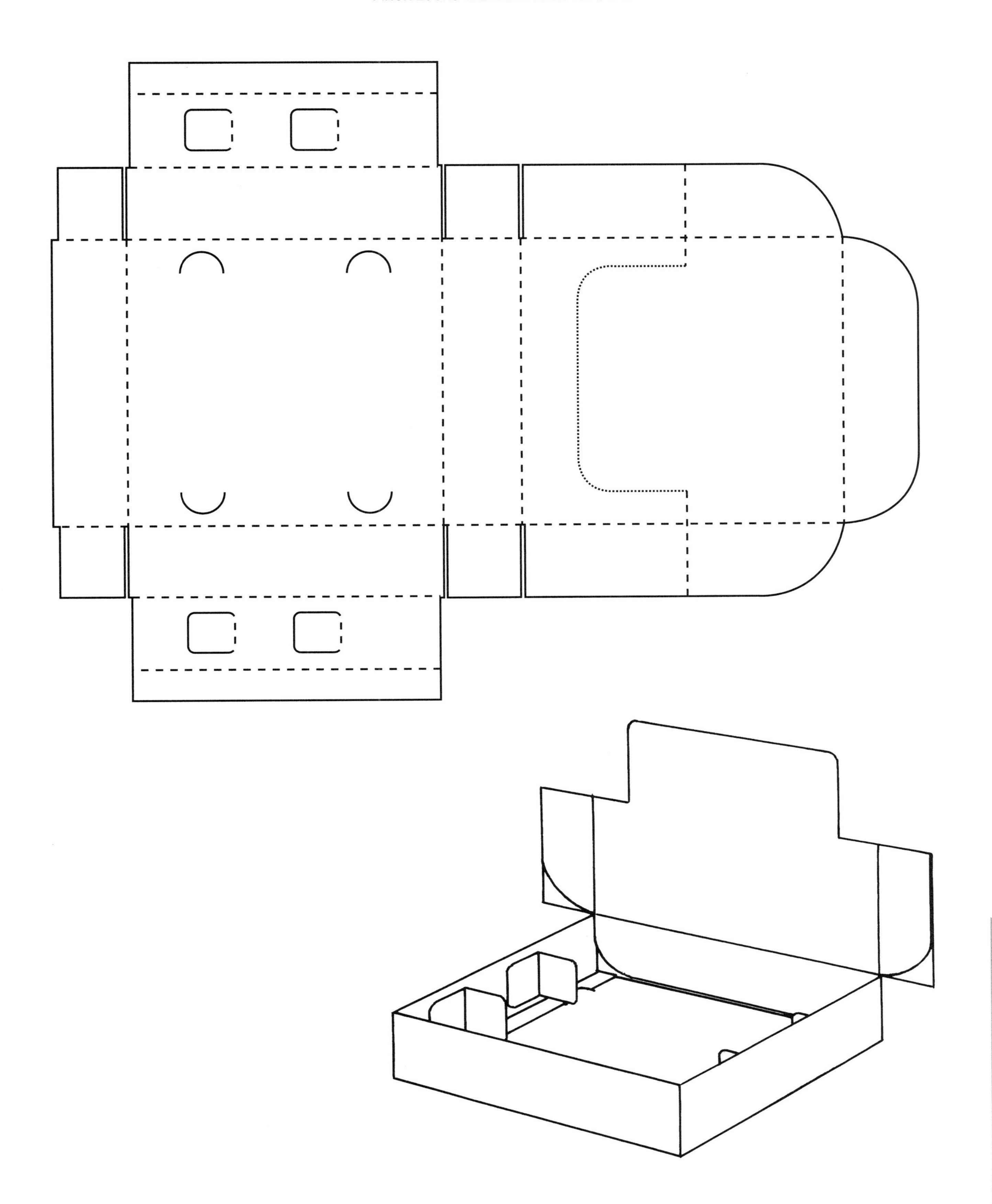

RHOMBOID DISPLAY CONTAINER WITH DOUBLE SIDE WALLS, DOUBLE END WALL, AND ARTHUR LOCKS

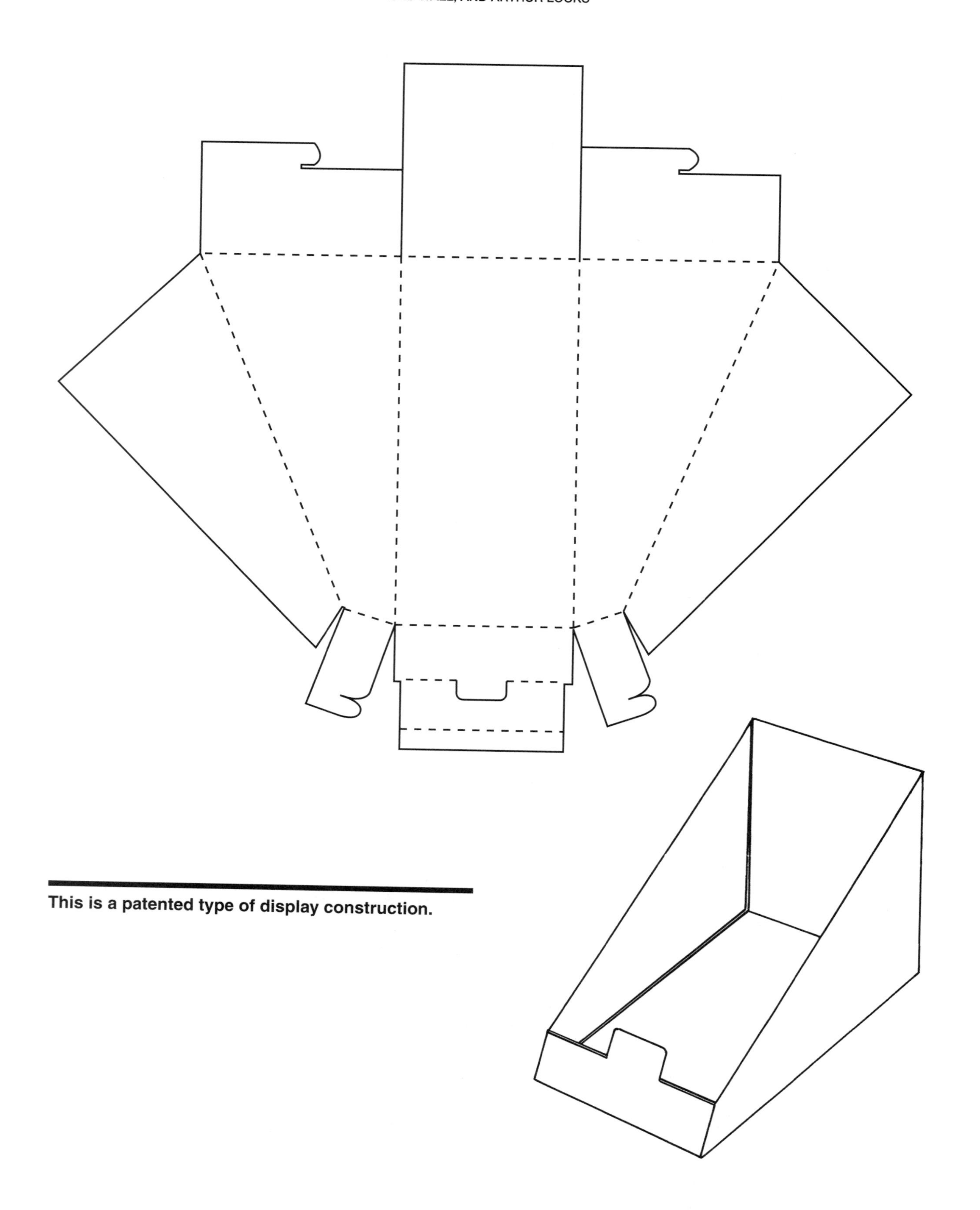

This is a patented type of display construction.

STEP INSERT AND HEADER FOR RHOMBOID DISPLAY ON PREVIOUS PAGE

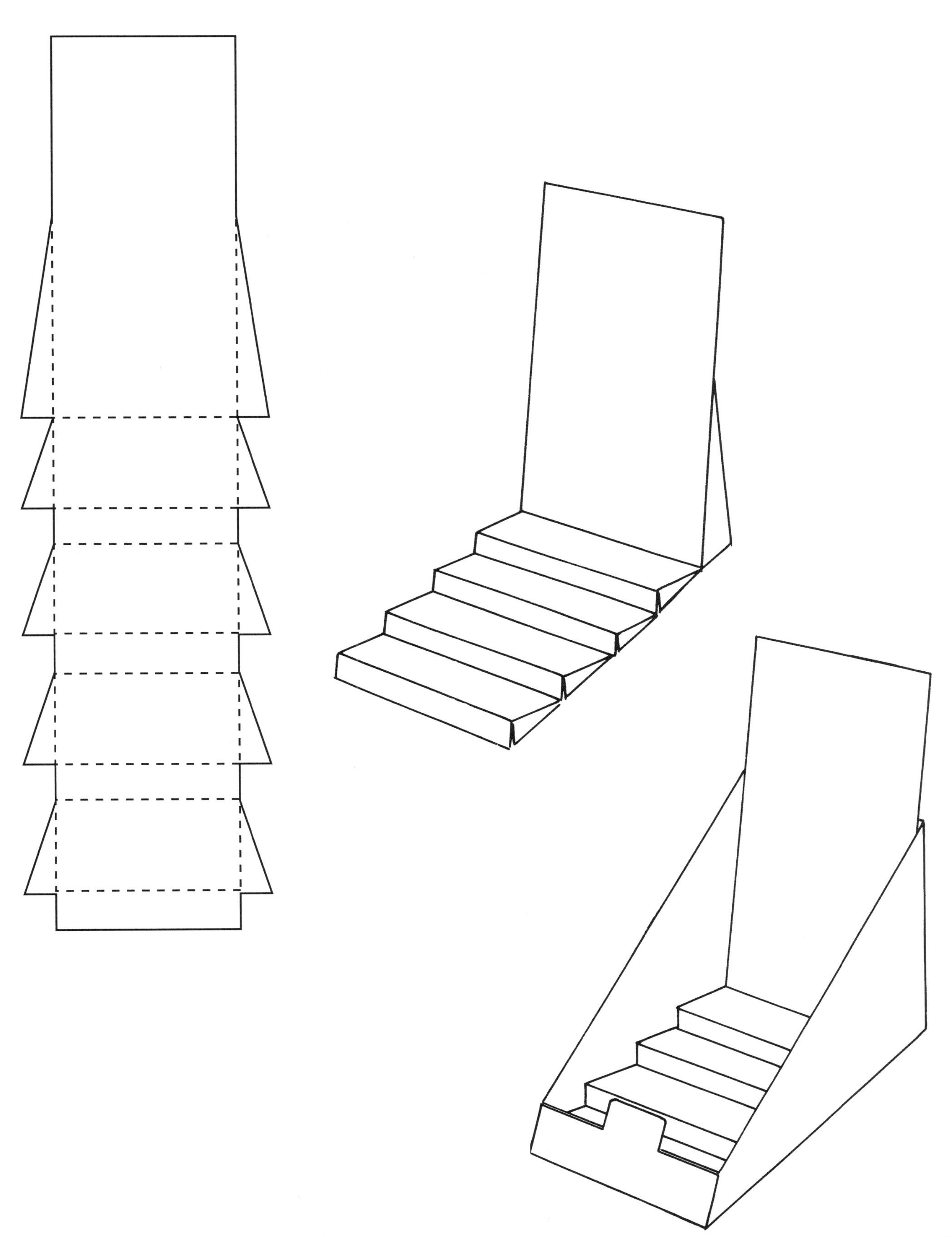

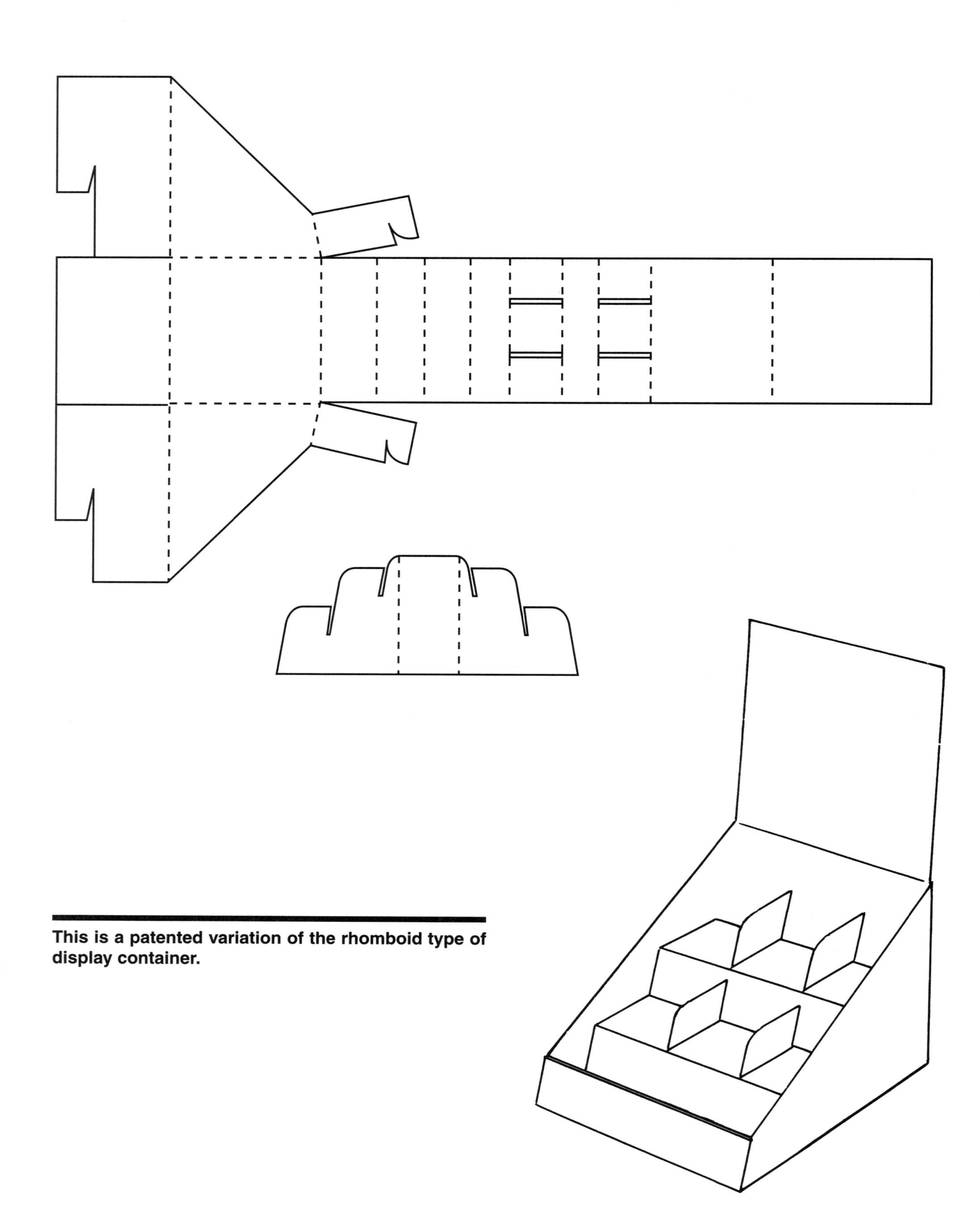

This is a patented variation of the rhomboid type of display container.

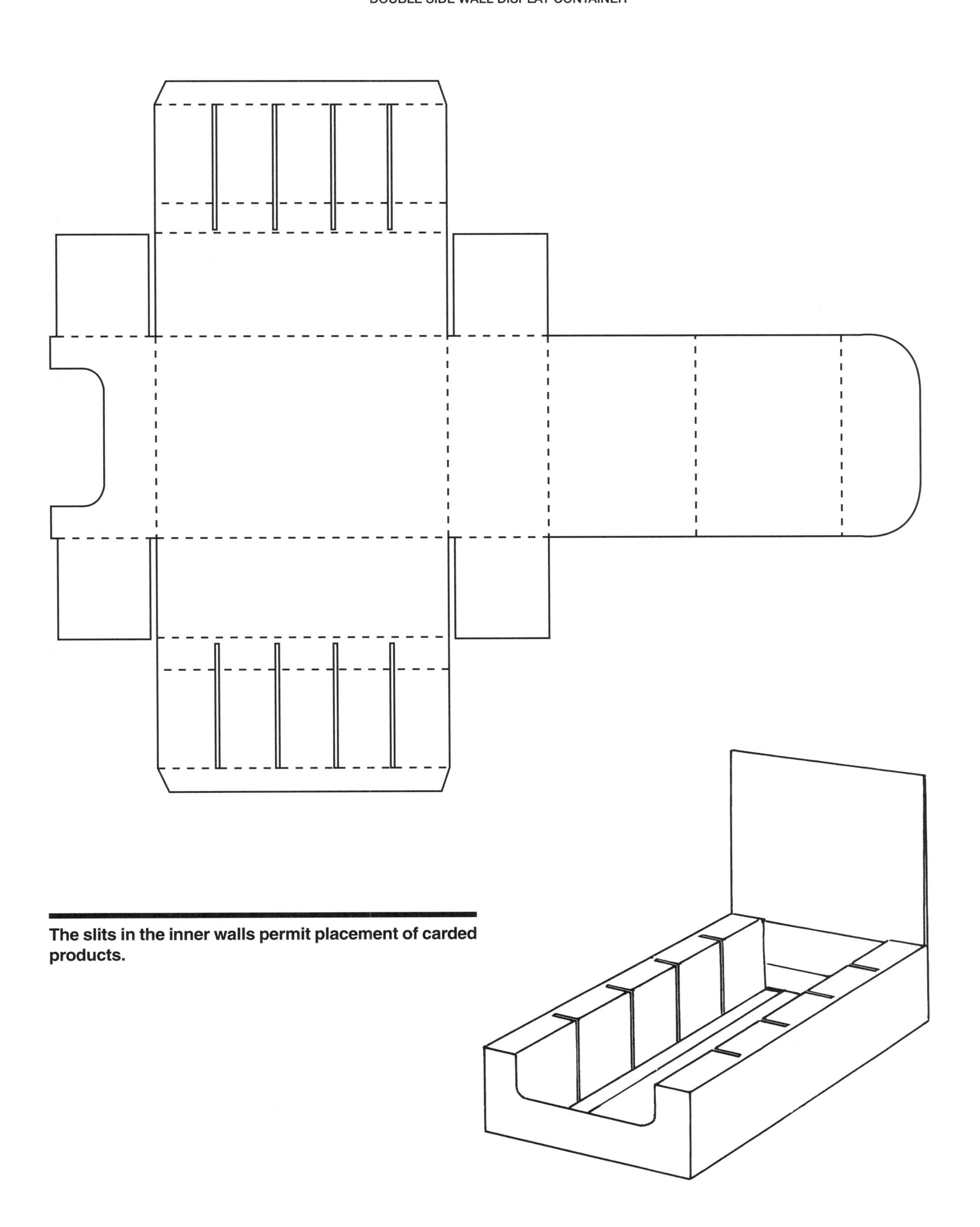

The slits in the inner walls permit placement of carded products.

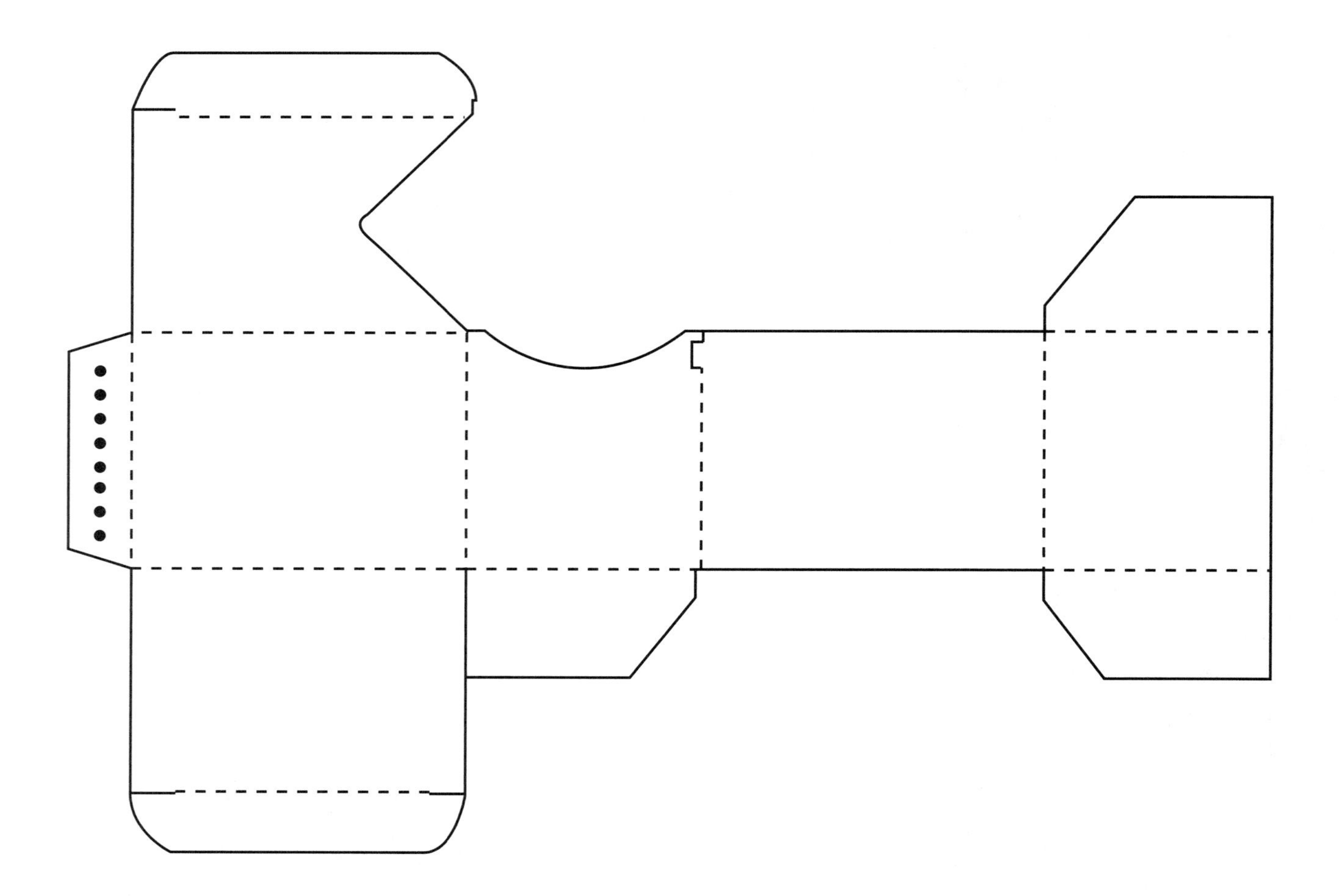

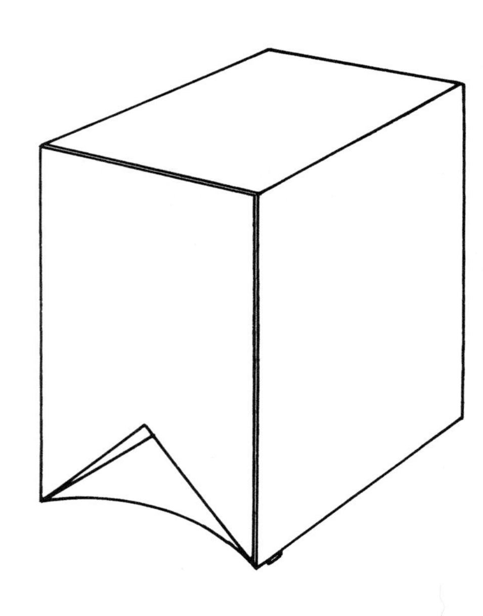

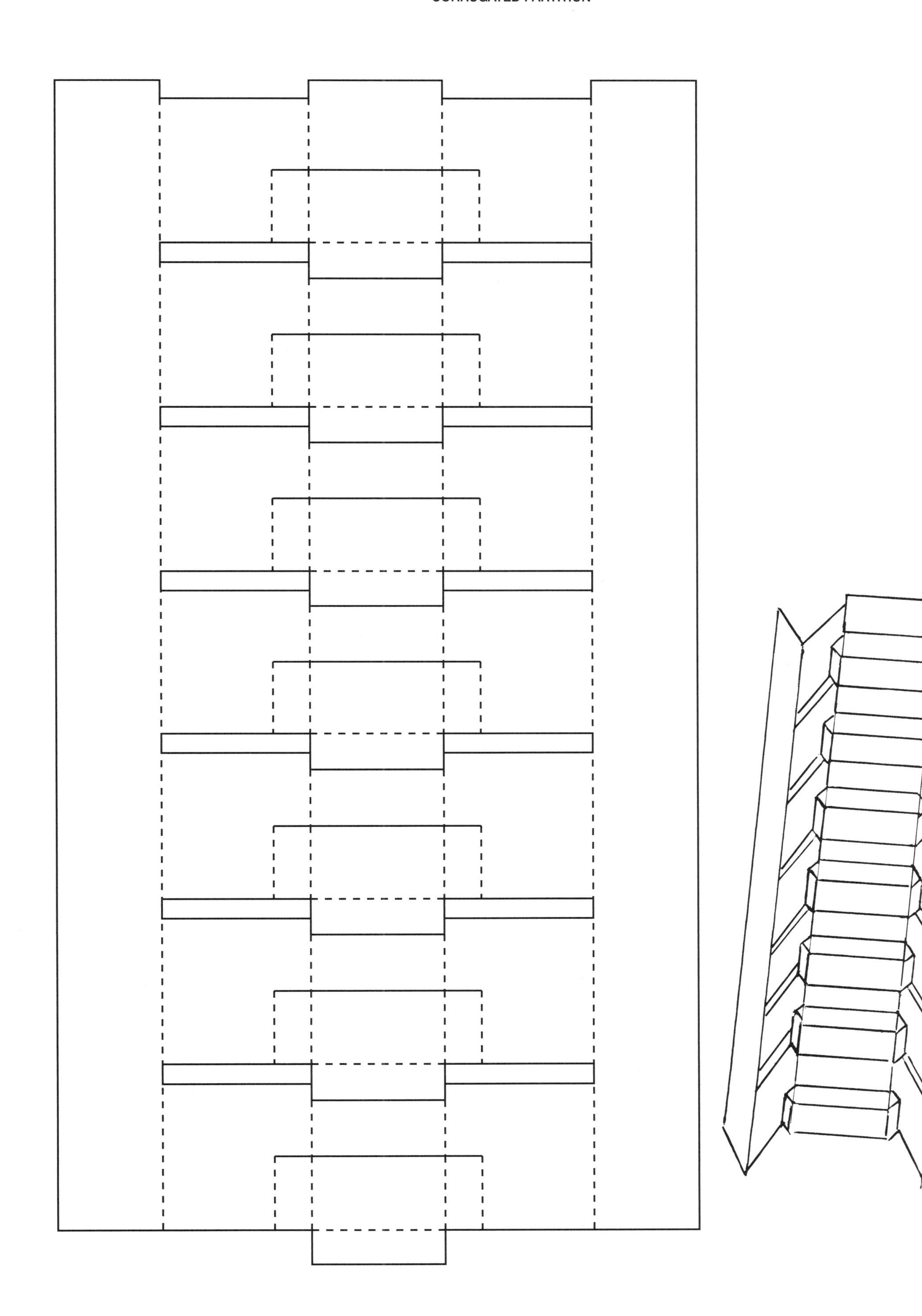

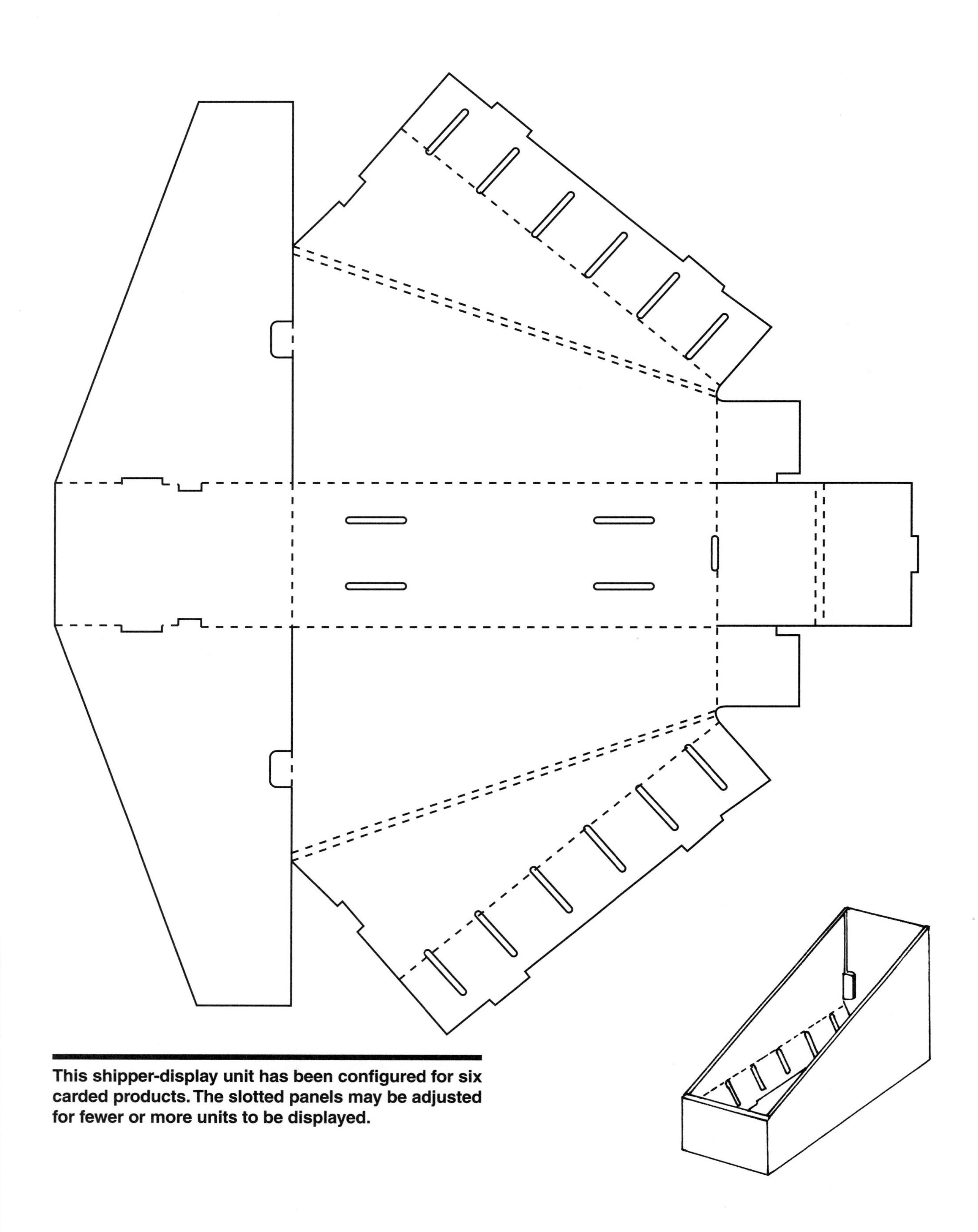

This shipper-display unit has been configured for six carded products. The slotted panels may be adjusted for fewer or more units to be displayed.

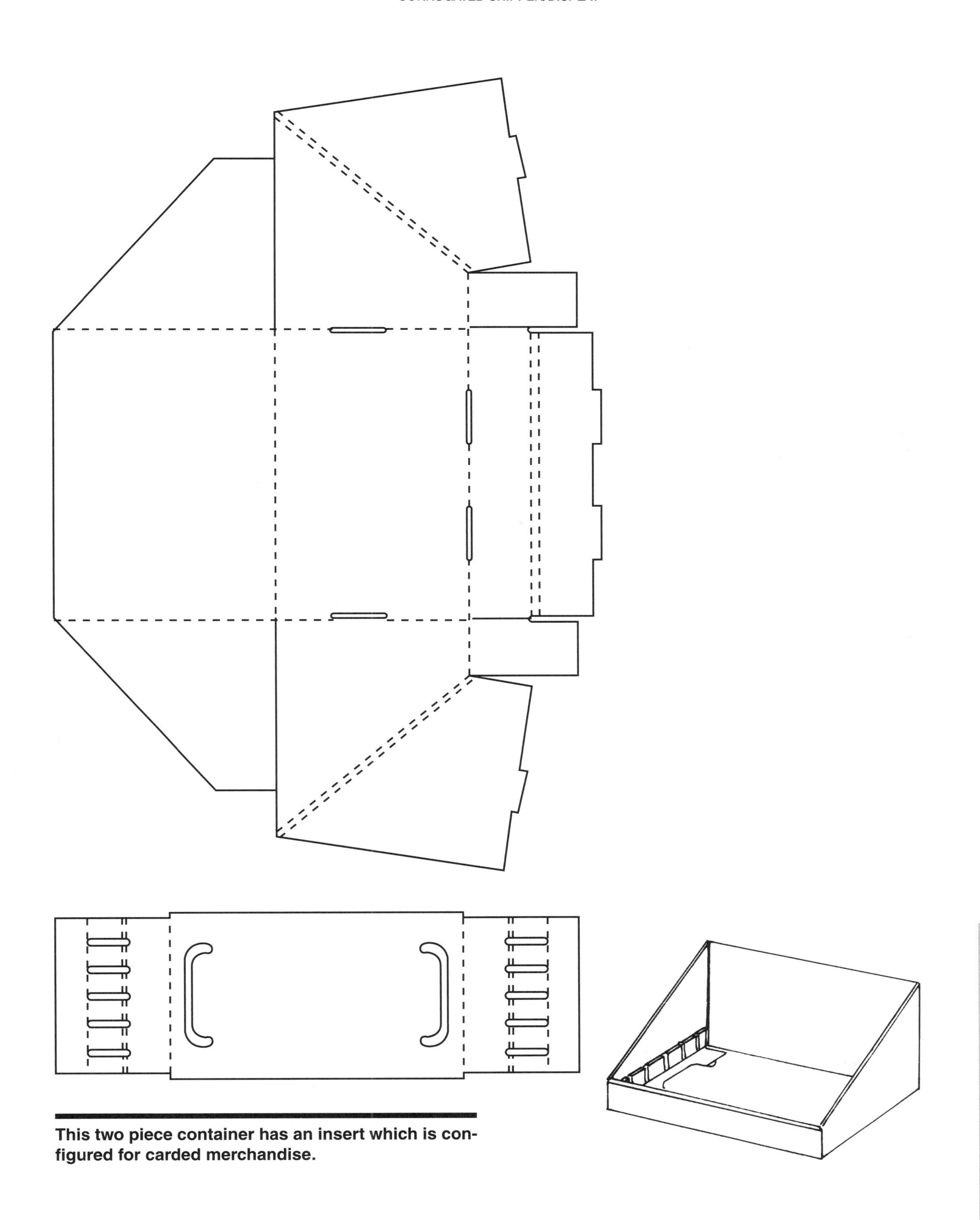

This two piece container has an insert which is configured for carded merchandise.

DOUBLE END WALL, DOUBLE SIDE WALL CORRUGATED SELF-LOCKING DISPLAY

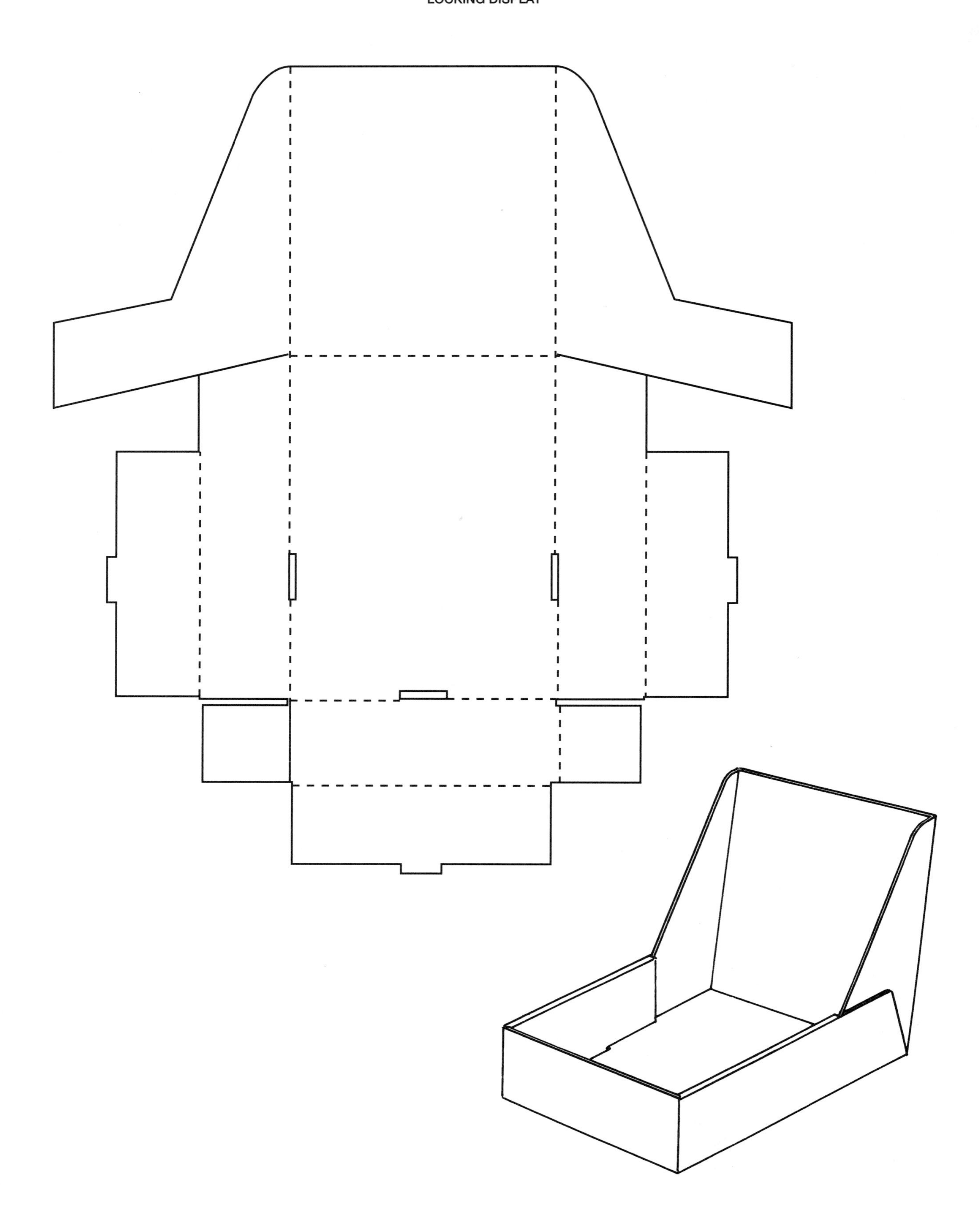

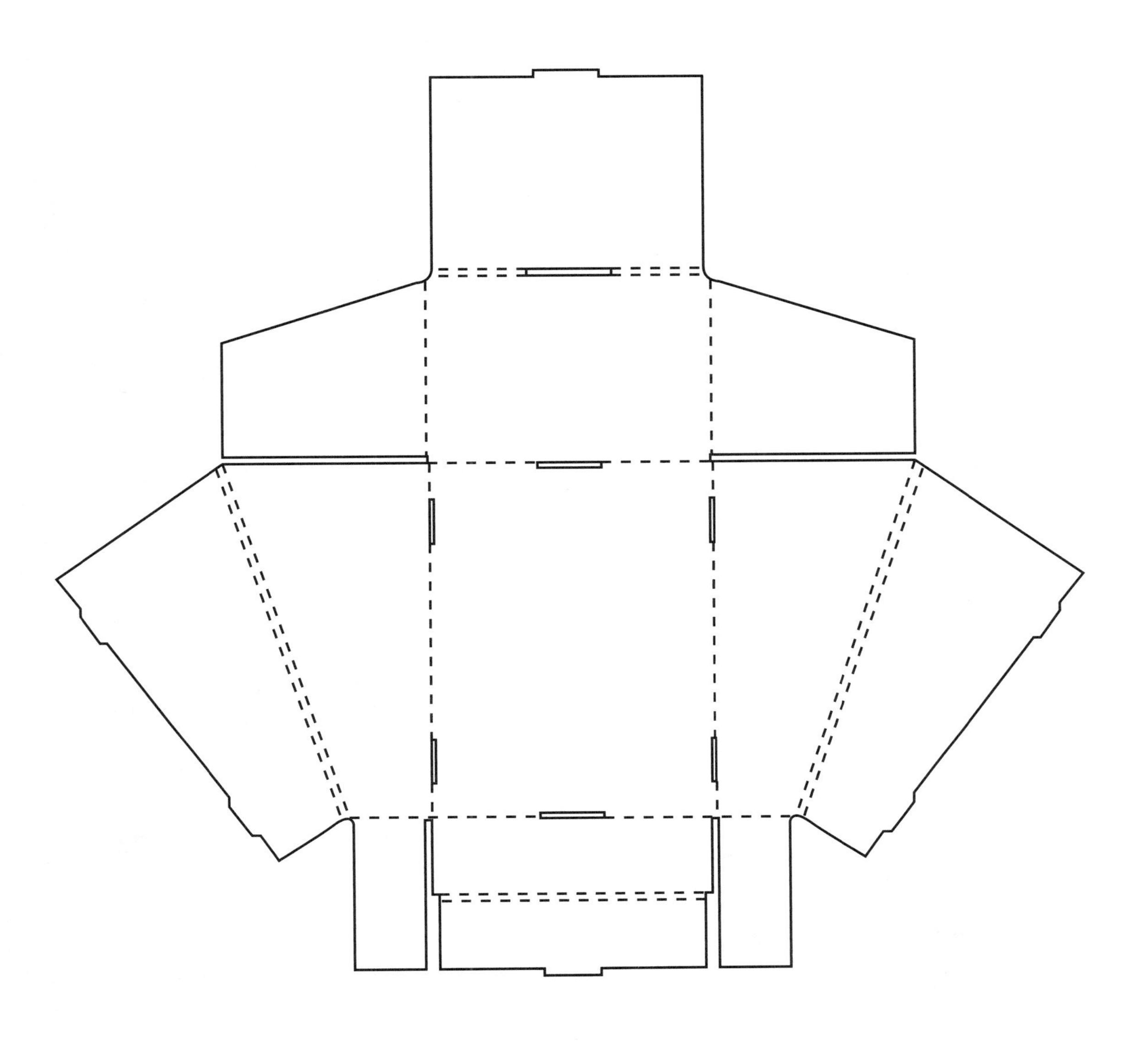

This double front wall, angled double side-wall tray with Walker Locks has a slit in the top of the double back wall to accommodate the separate header or riser illustrated on the next page.

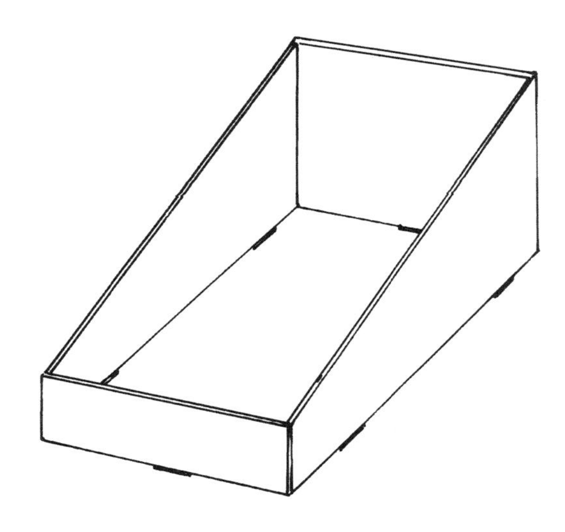

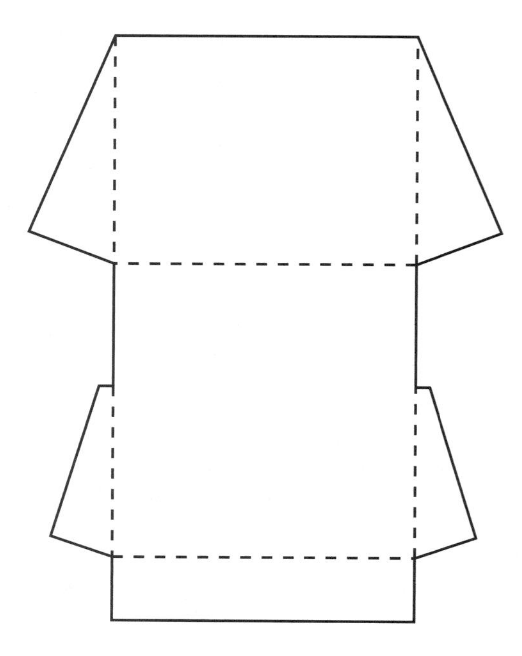

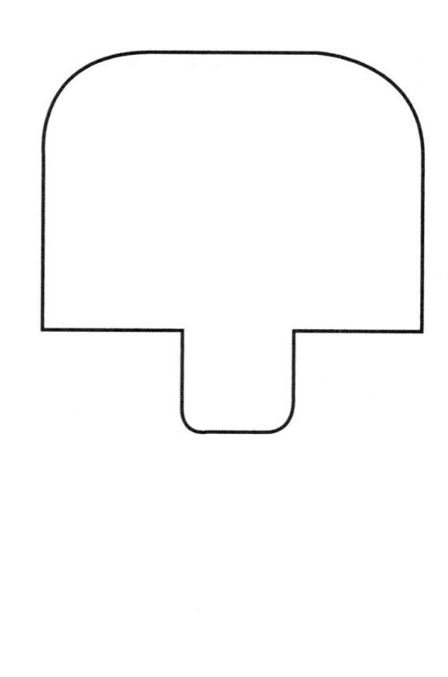

The header may be die-cut into any suitable configuration; the "tongue" has been formatted to fit the slit in the back panel of the corrugated display.

The insert helps the contents of the display lean back.

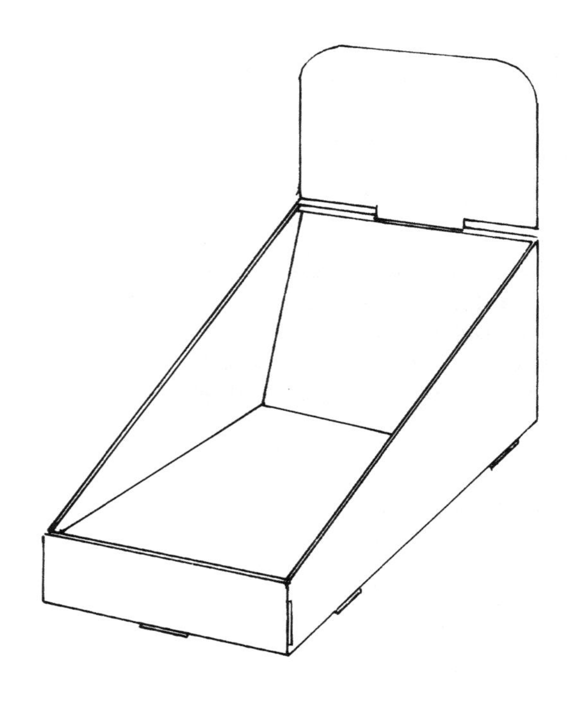

This double front, end, and side wall bin has bottom tabs and corresponding slots in the top to facilitate stacking.

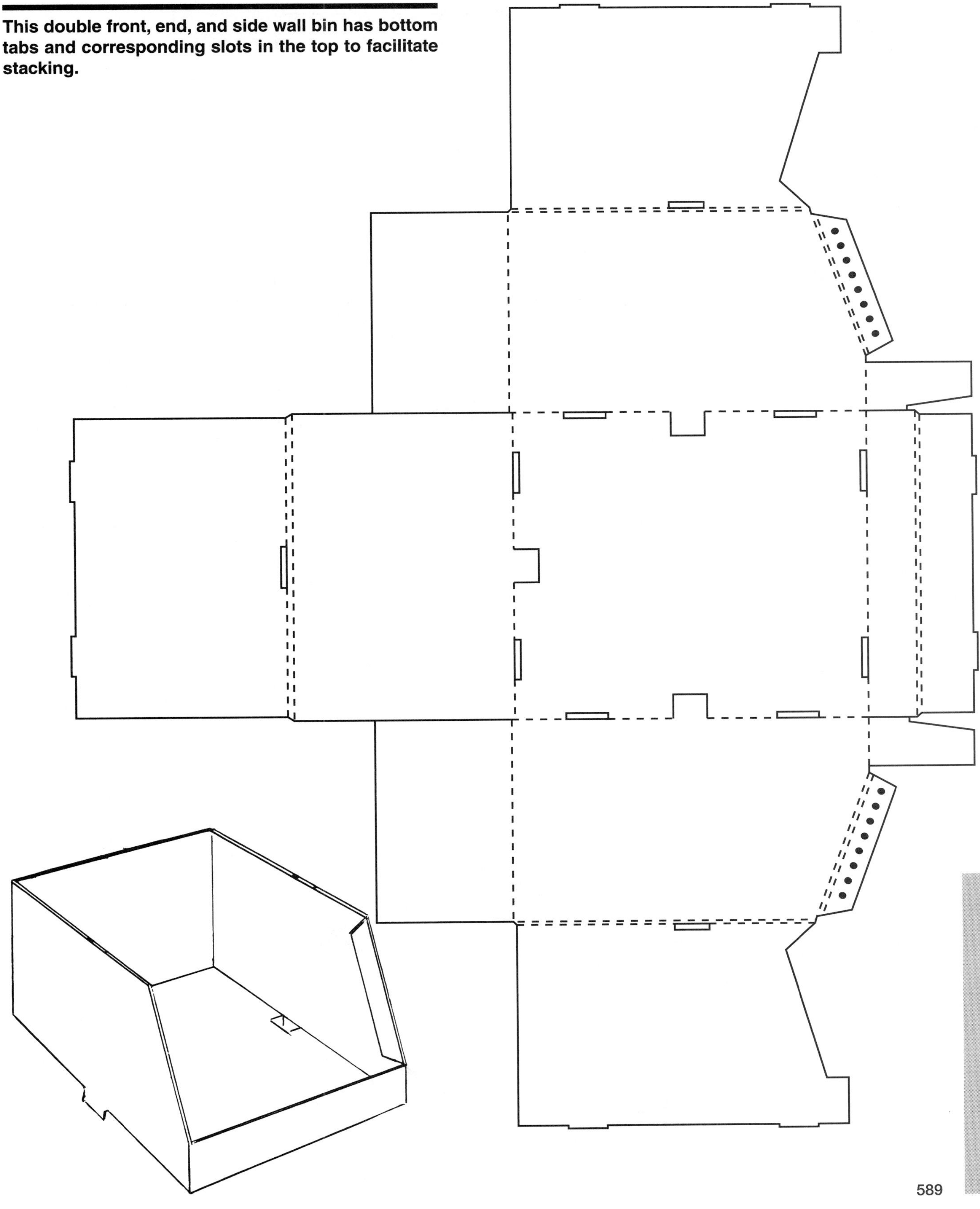

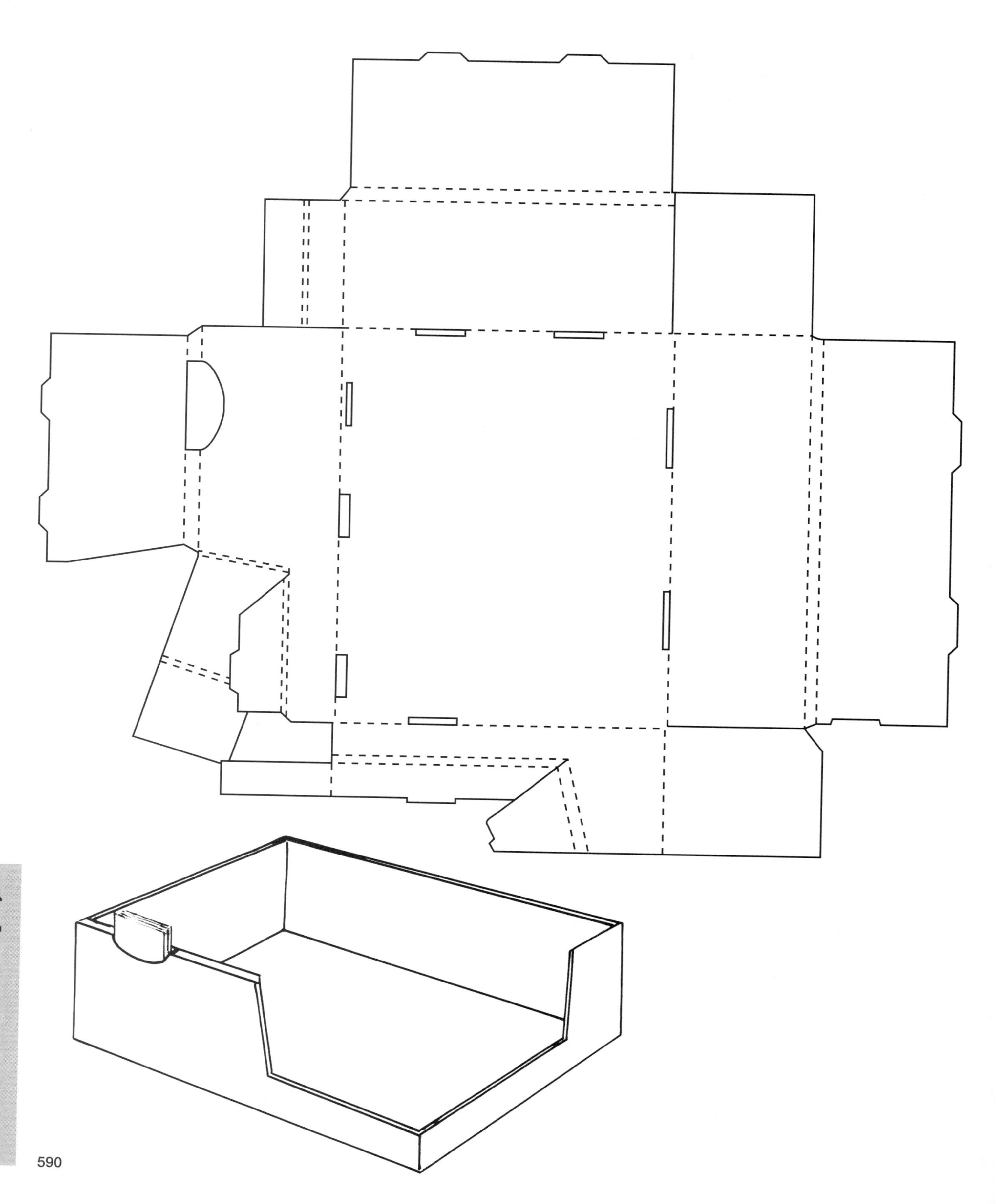

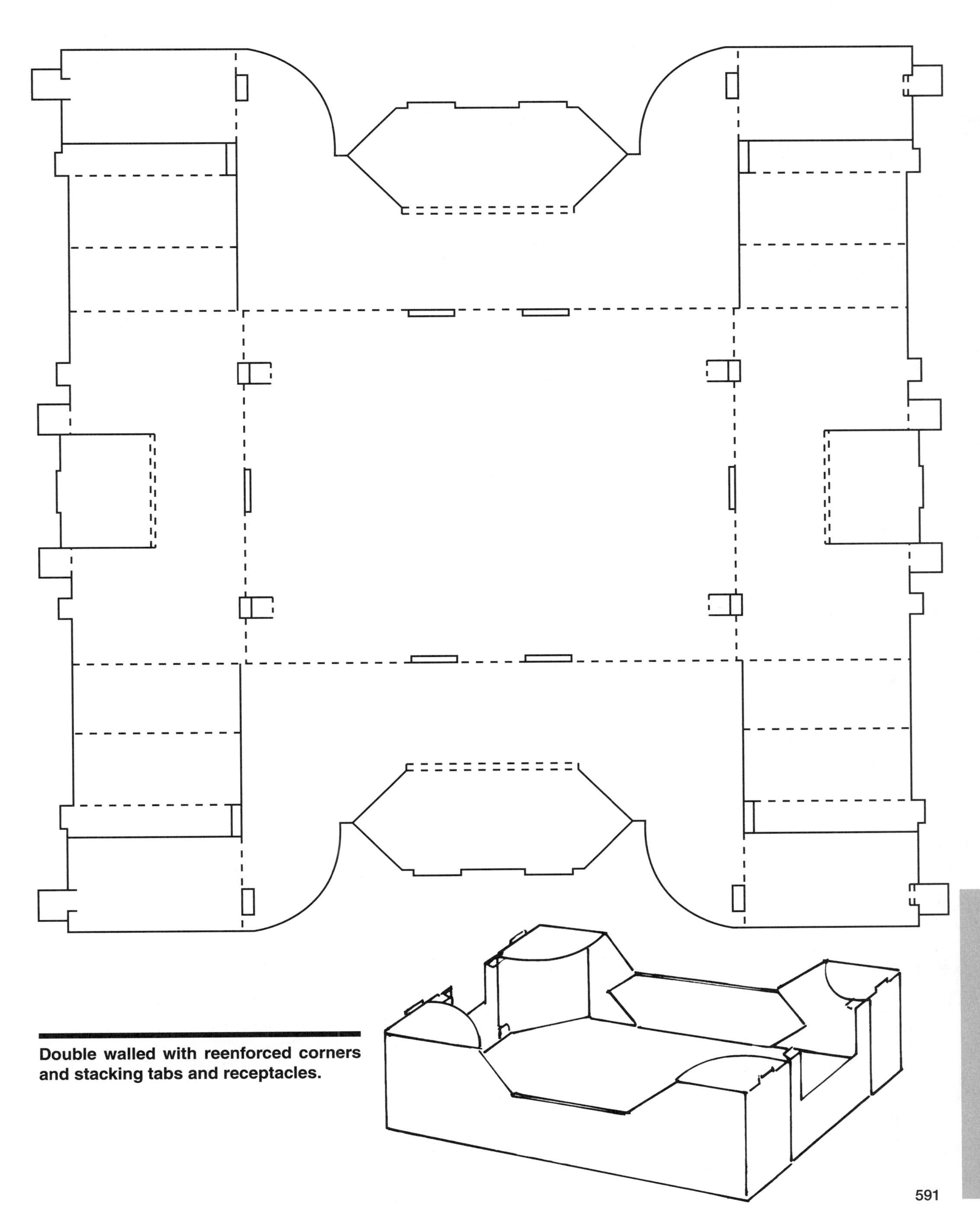

Double walled with reenforced corners and stacking tabs and receptacles.

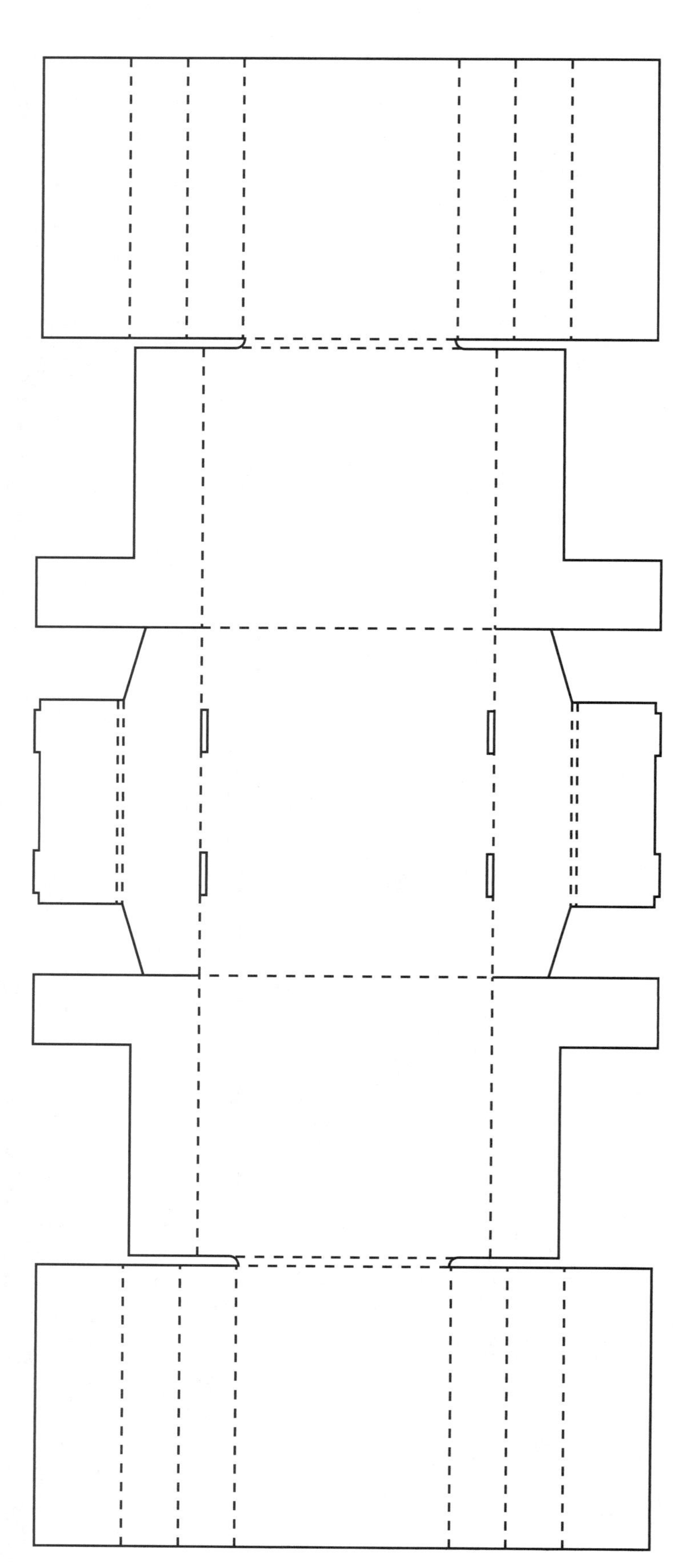

This carton has four reenforced corners to facilitate stacking of heavy products.

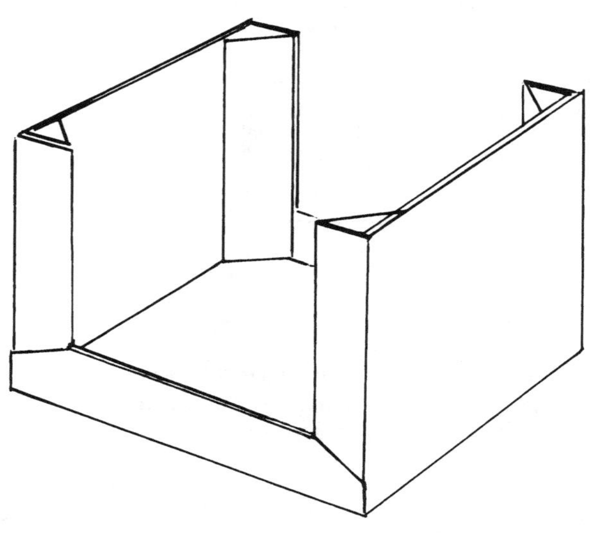

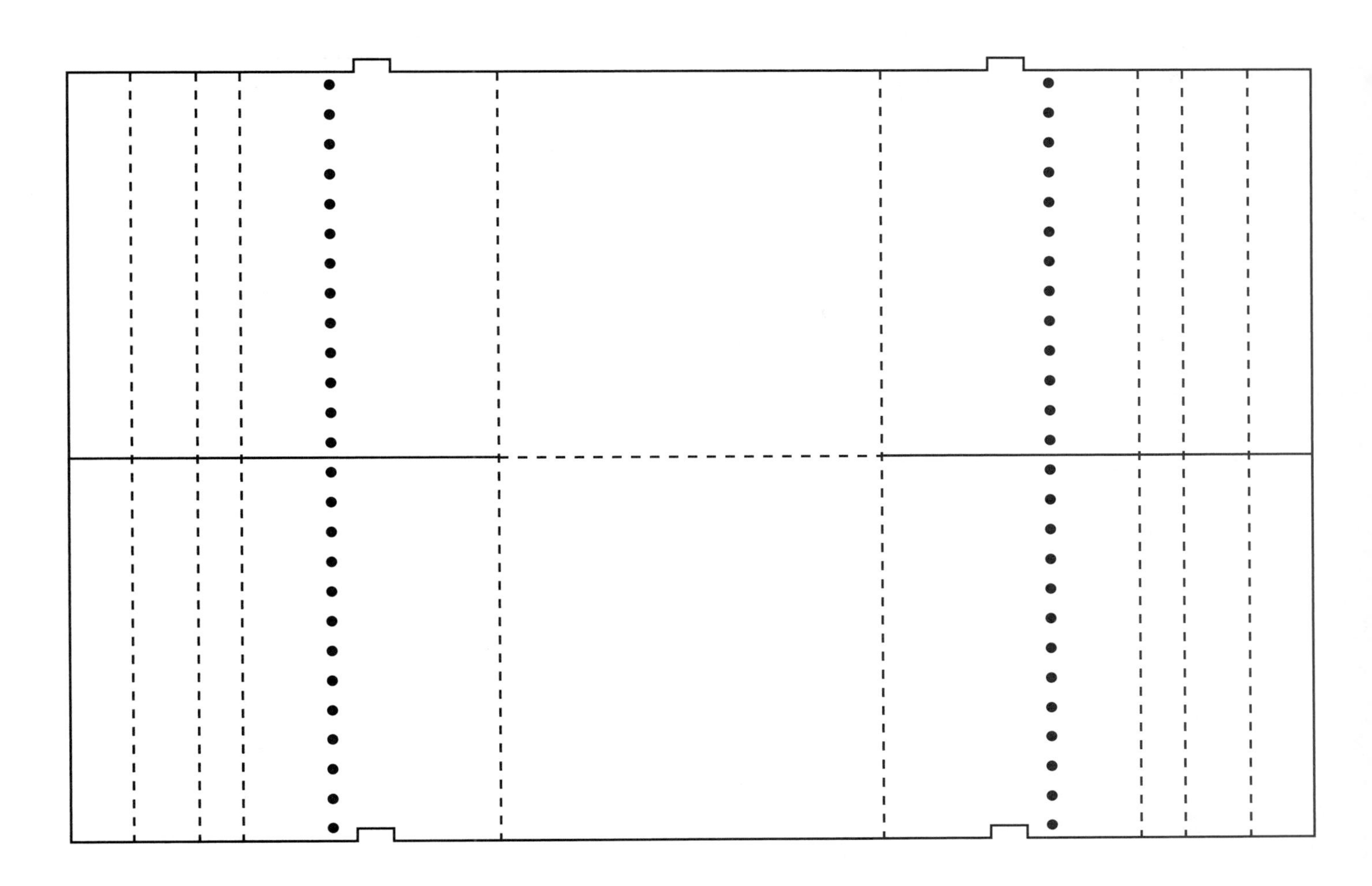

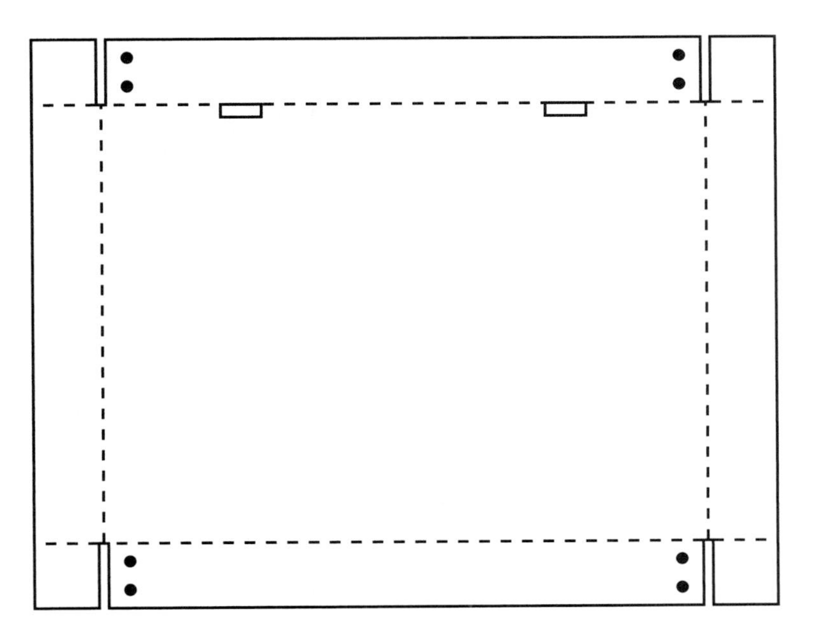

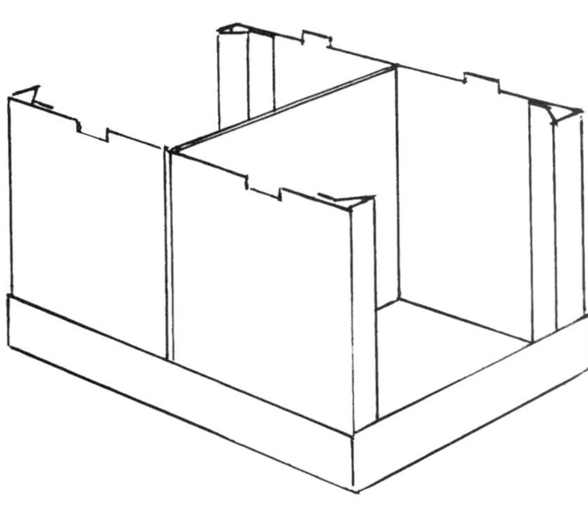

The insert divider facilitates stacking.

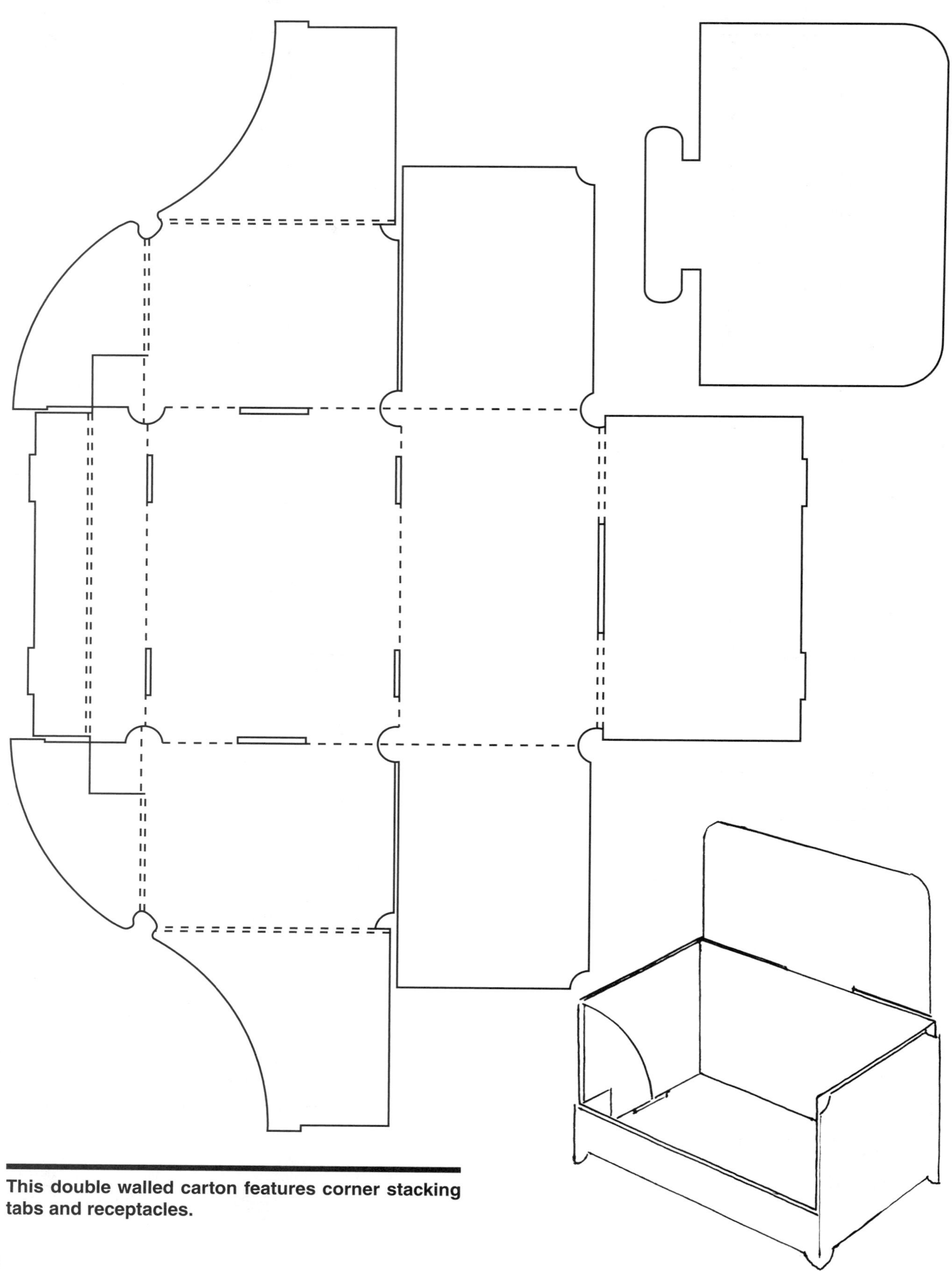

This double walled carton features corner stacking tabs and receptacles.

This insert and product separators fit into the display container on the previous page.

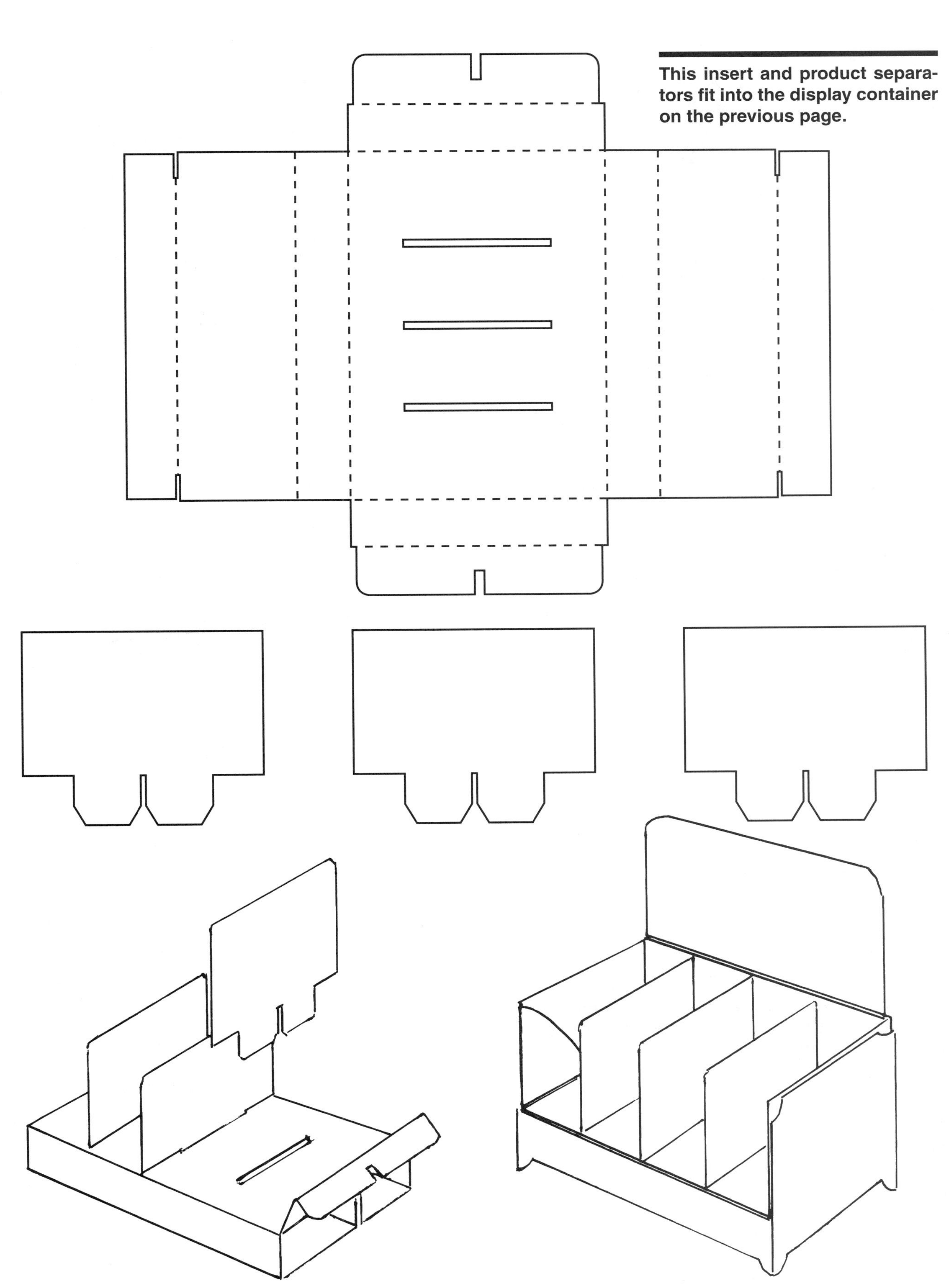

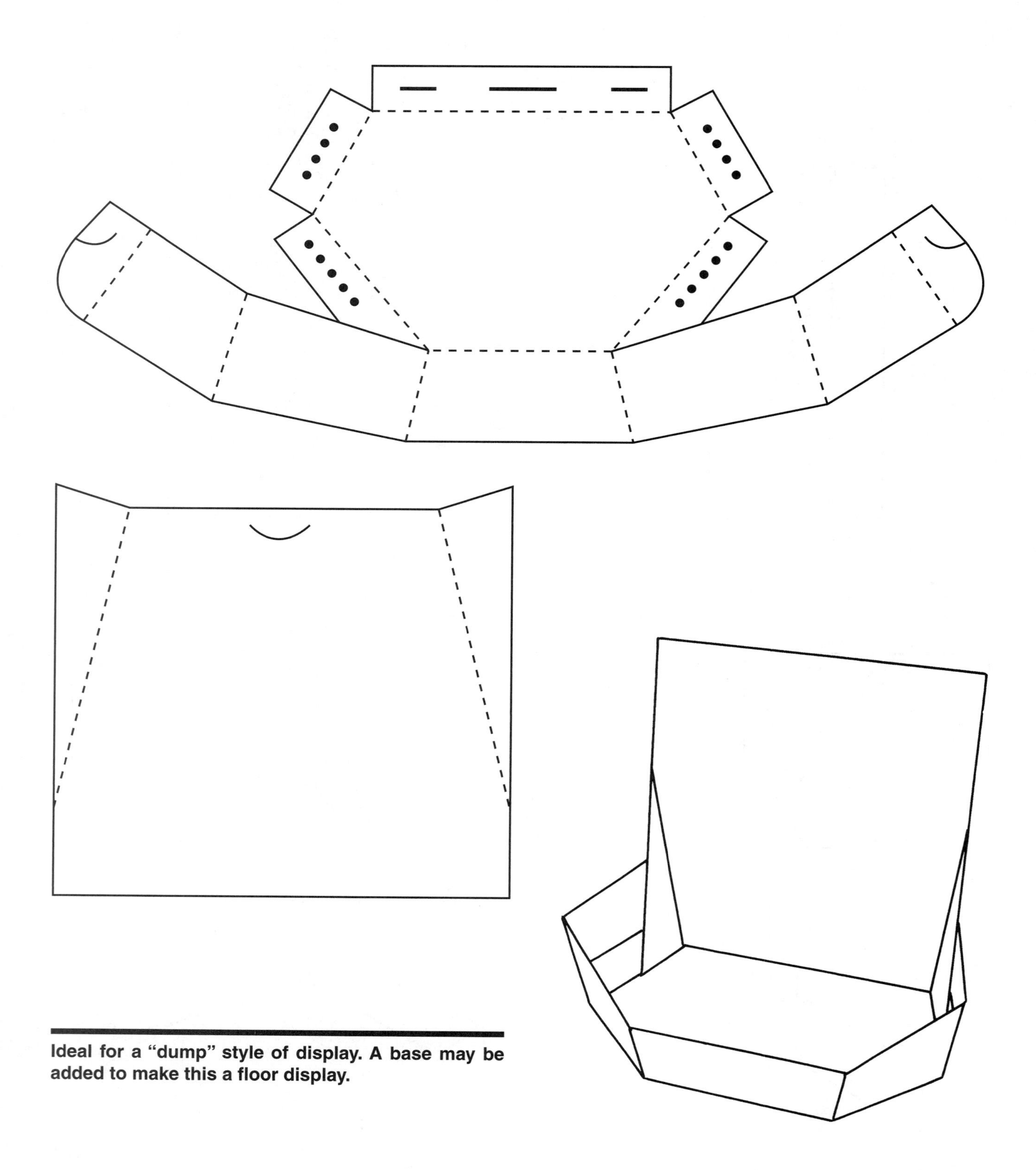

Ideal for a "dump" style of display. A base may be added to make this a floor display.

TWO-PIECE DISPLAY STAND WITH SEPARATE CARD, DIE-CUT TO HOLD A SPECIFIC PRODUCT/PACKAGE

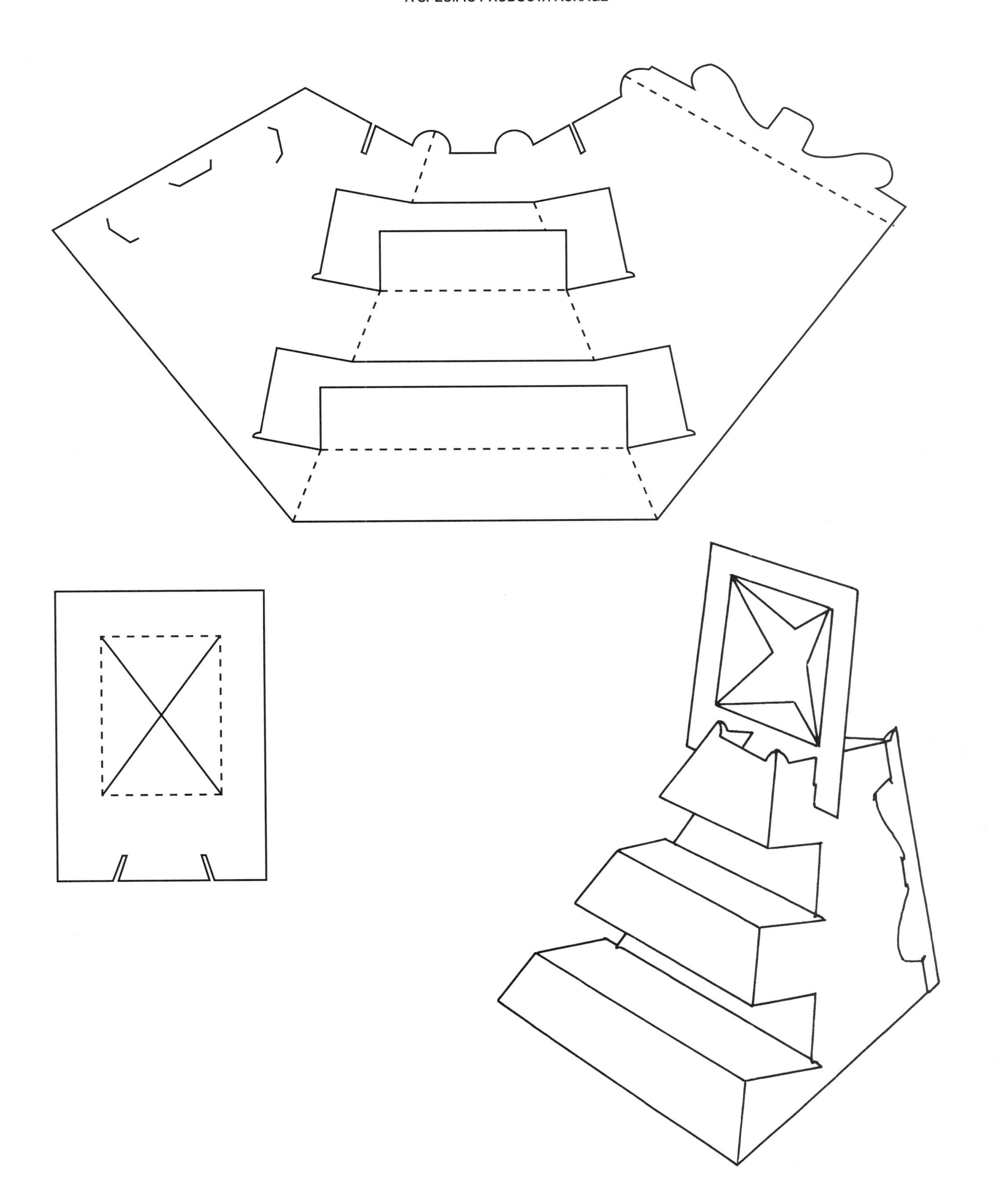

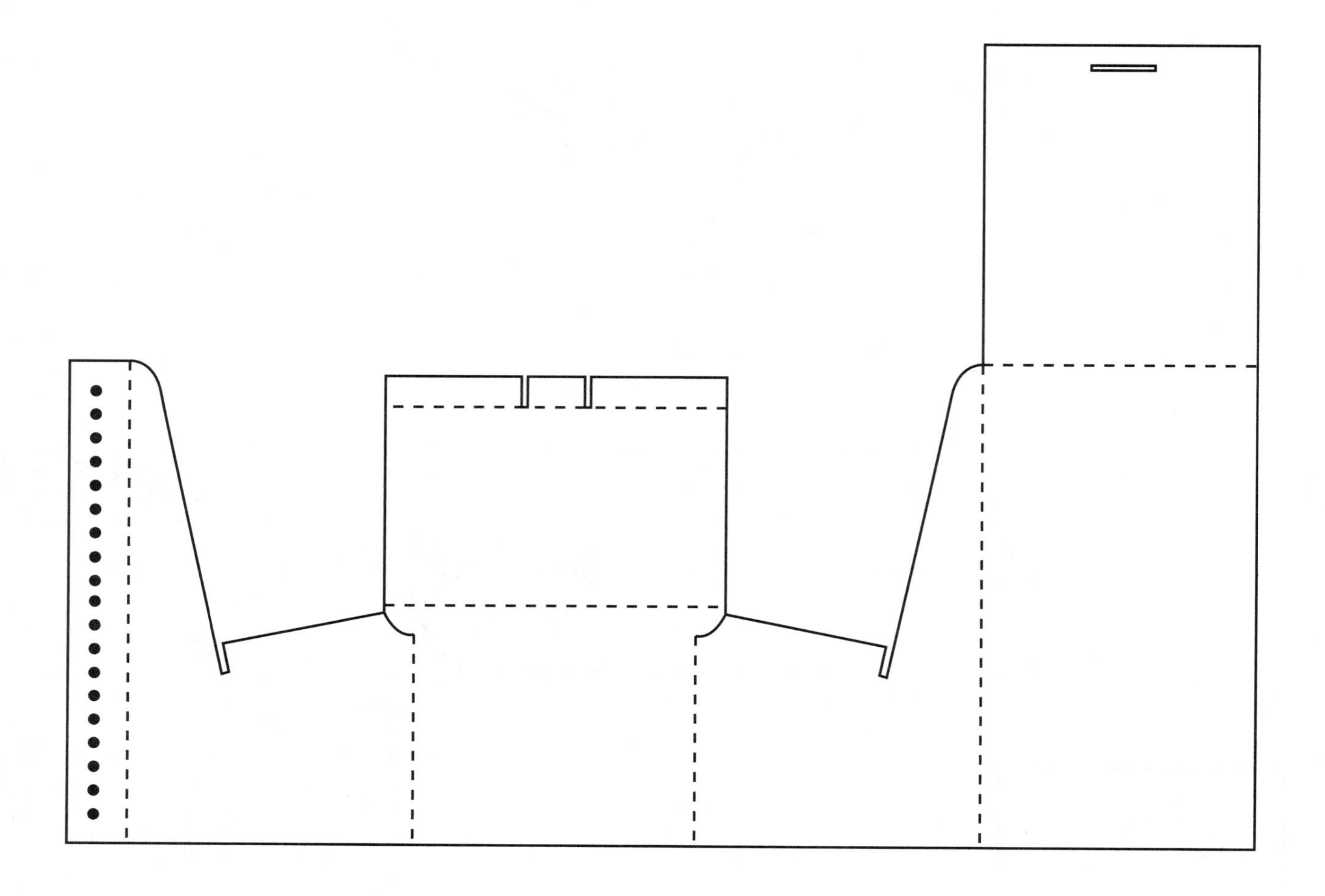

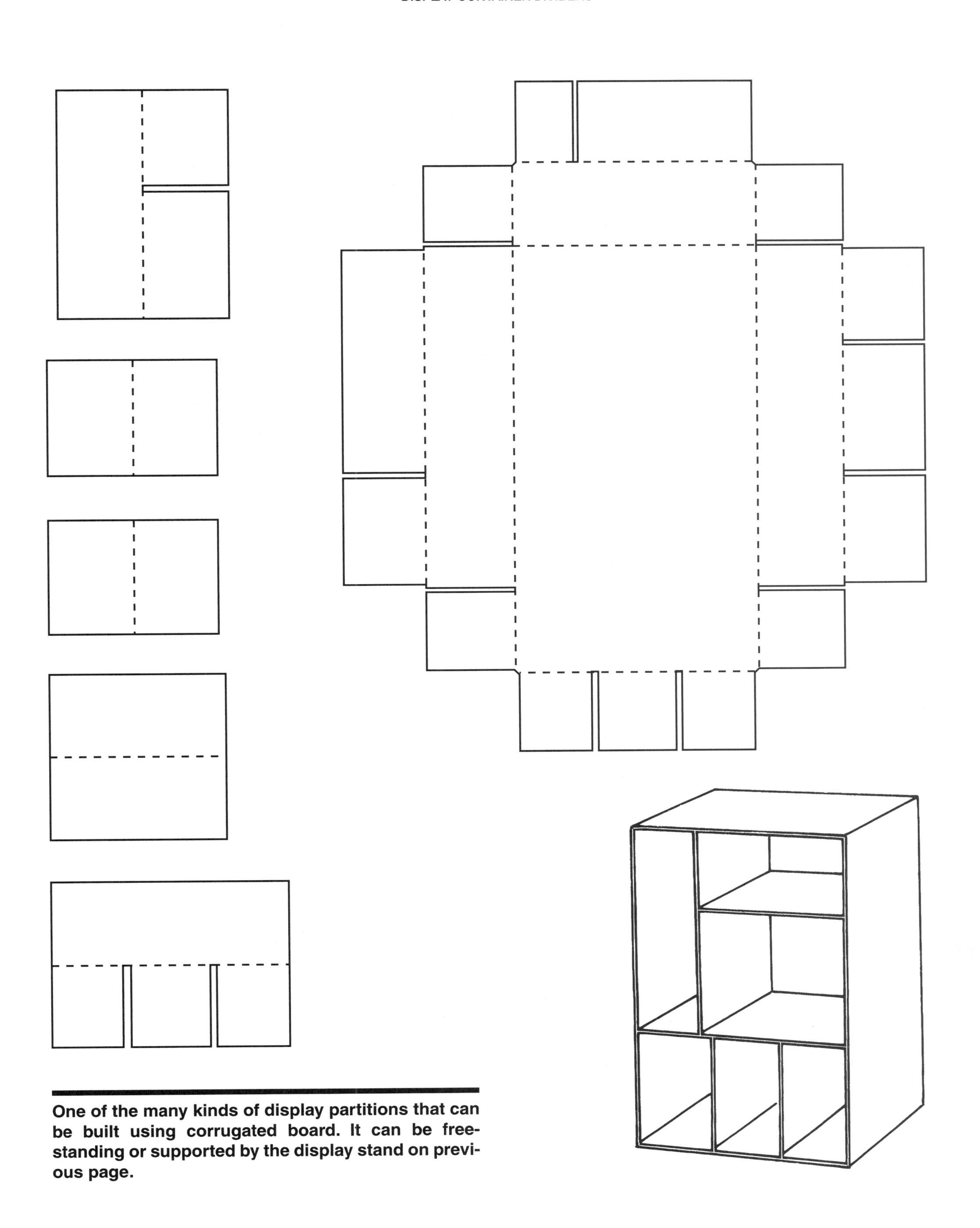

One of the many kinds of display partitions that can be built using corrugated board. It can be free-standing or supported by the display stand on previous page.

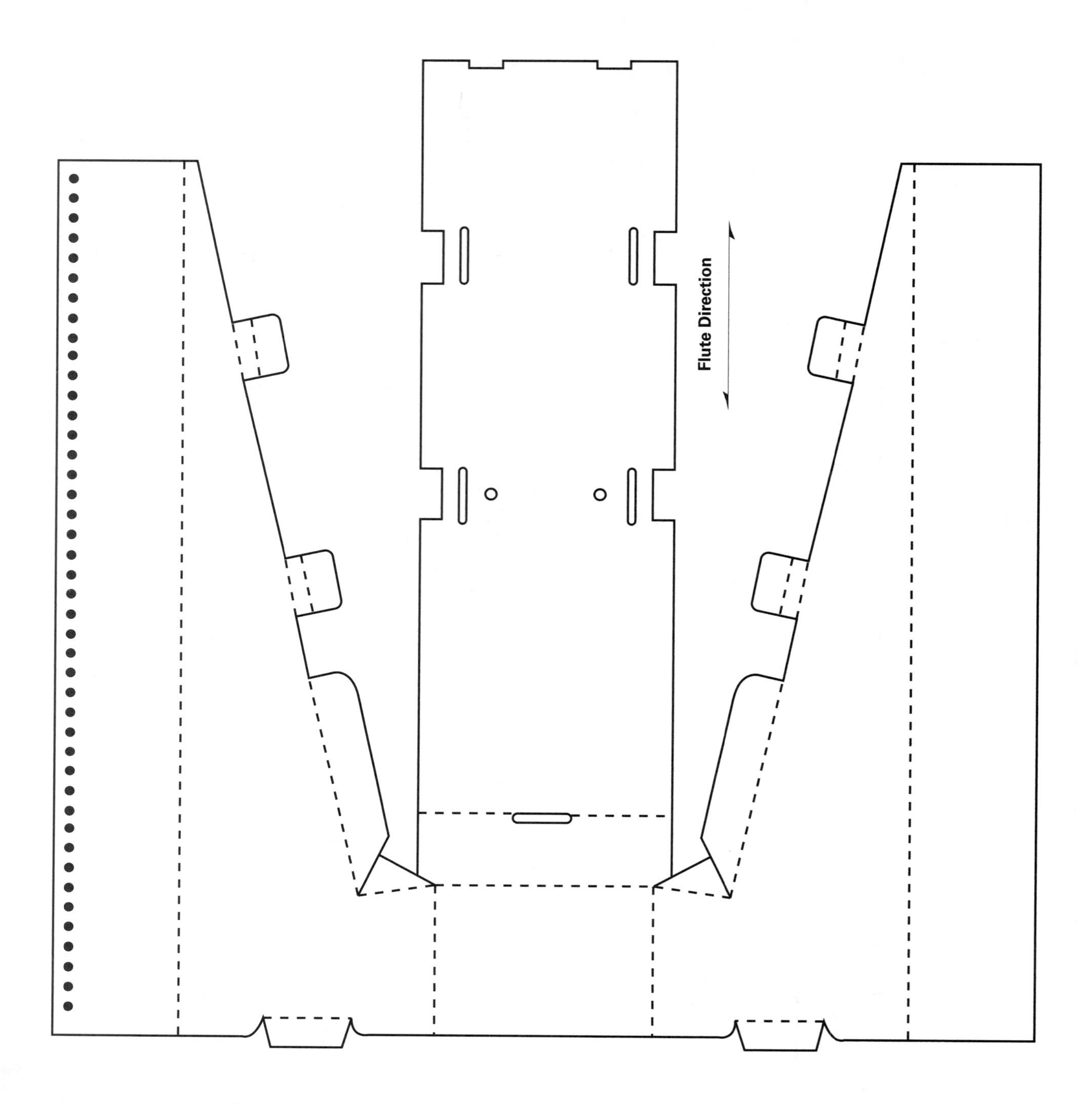

This 50" tall display base may be scored vertically in the center of both side panels and horizontally across all panels for easier shipping and storing. (See page 602.)

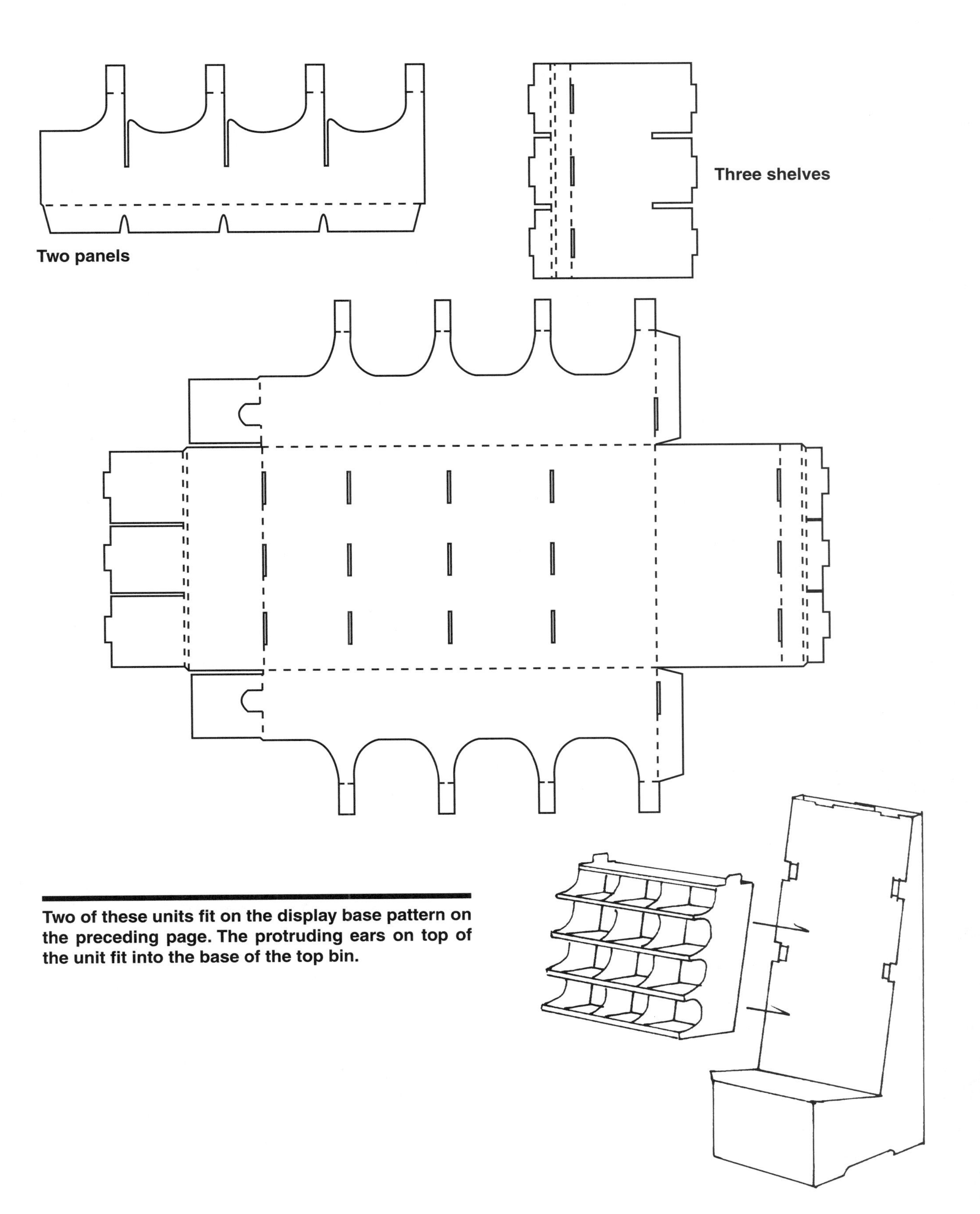

Two of these units fit on the display base pattern on the preceding page. The protruding ears on top of the unit fit into the base of the top bin.

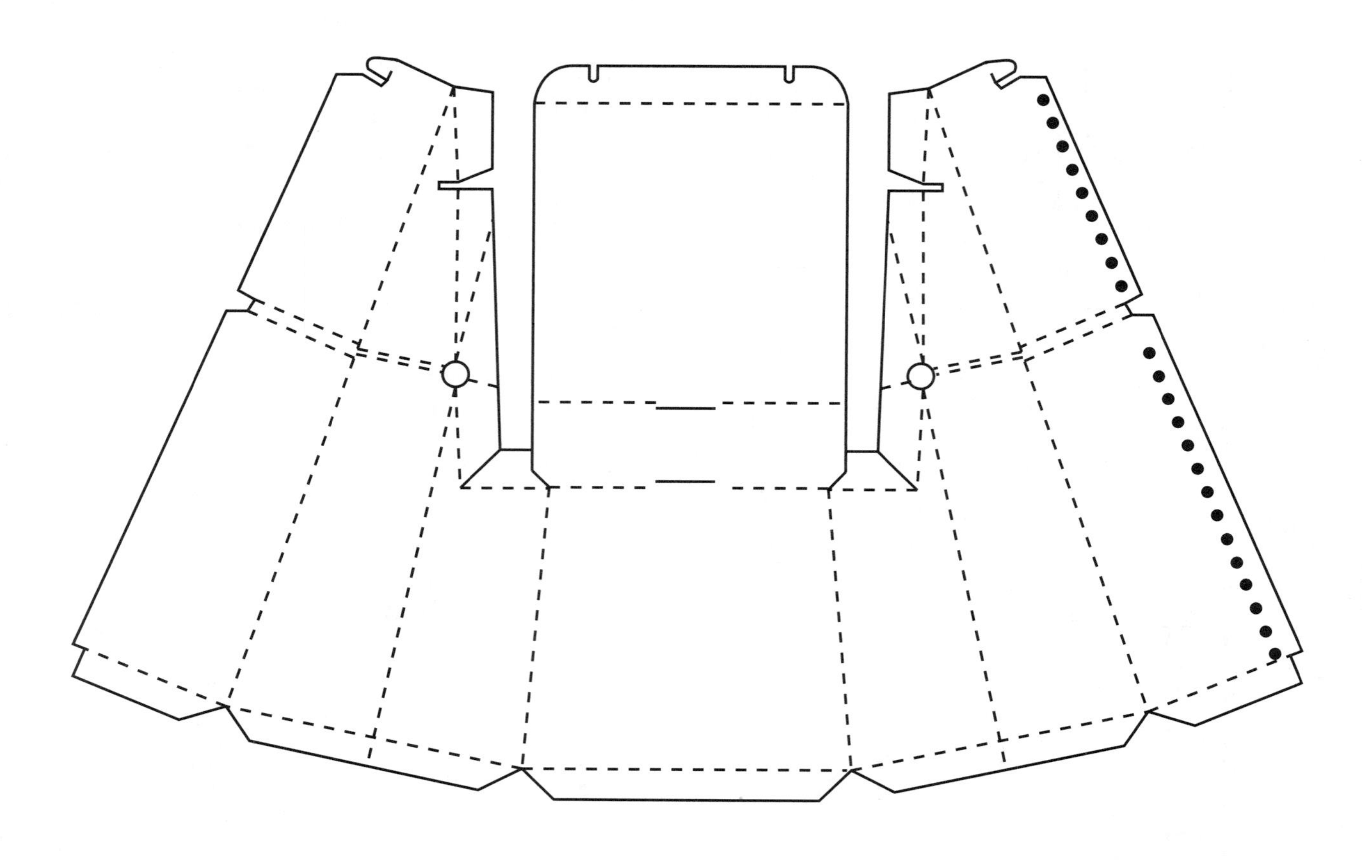

Corrugated trapezoidal display base. The lateral score lines in the side and back panels indicates the *Breakdown Score* which allows the base to be folded in half for shipping.

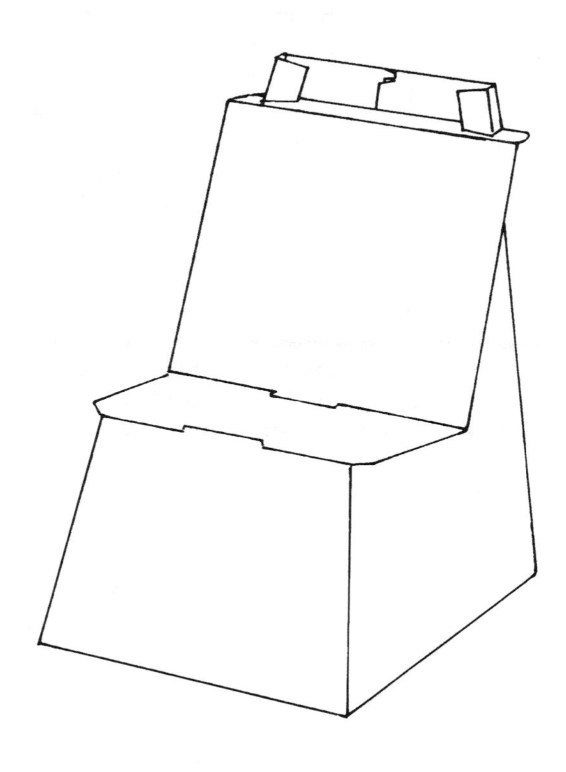

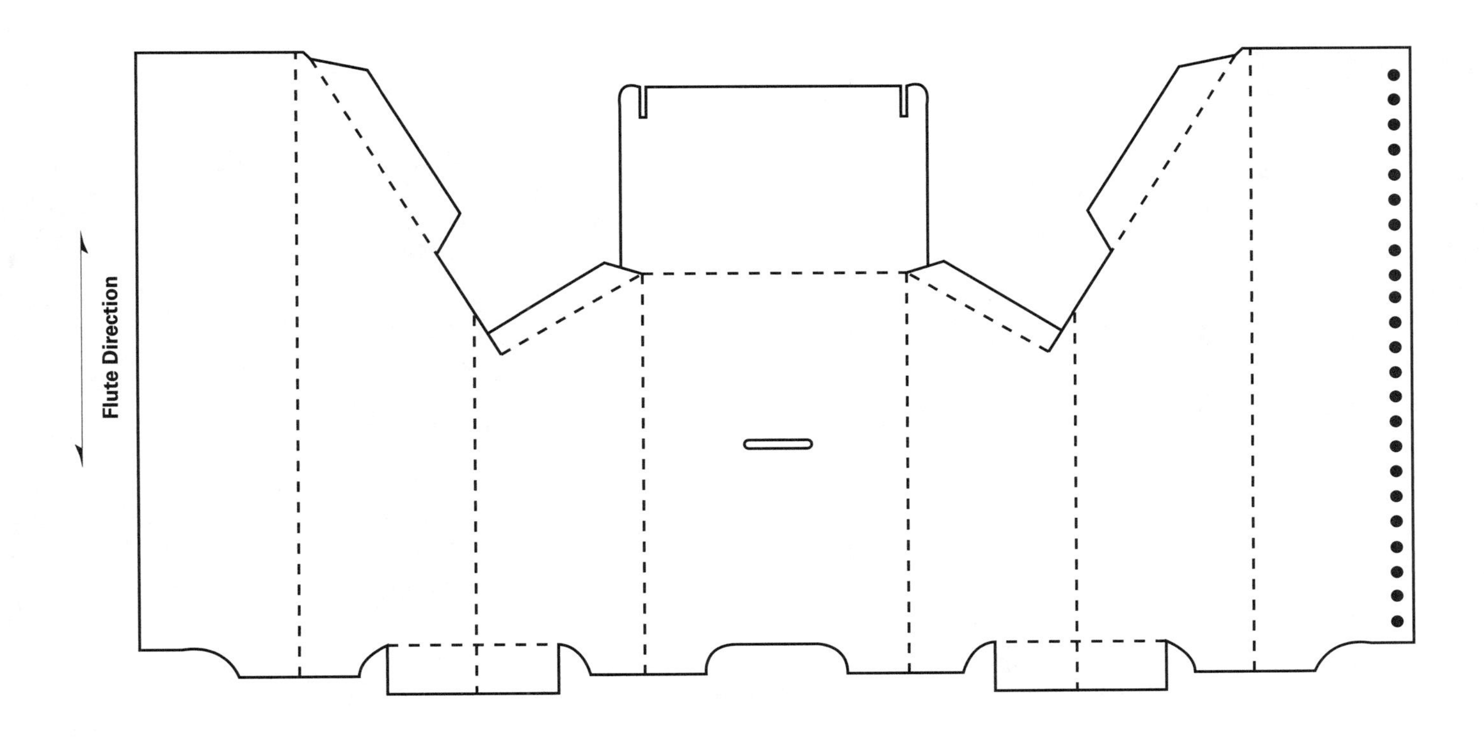

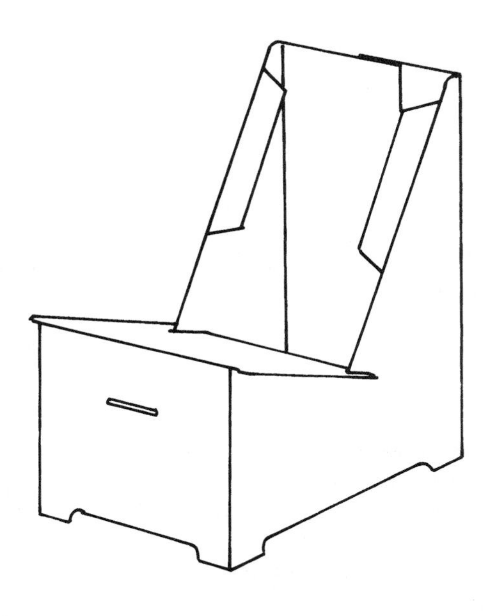

The slot in the bottom front panel mates with the tongue on the inside-out shipper on the following page.

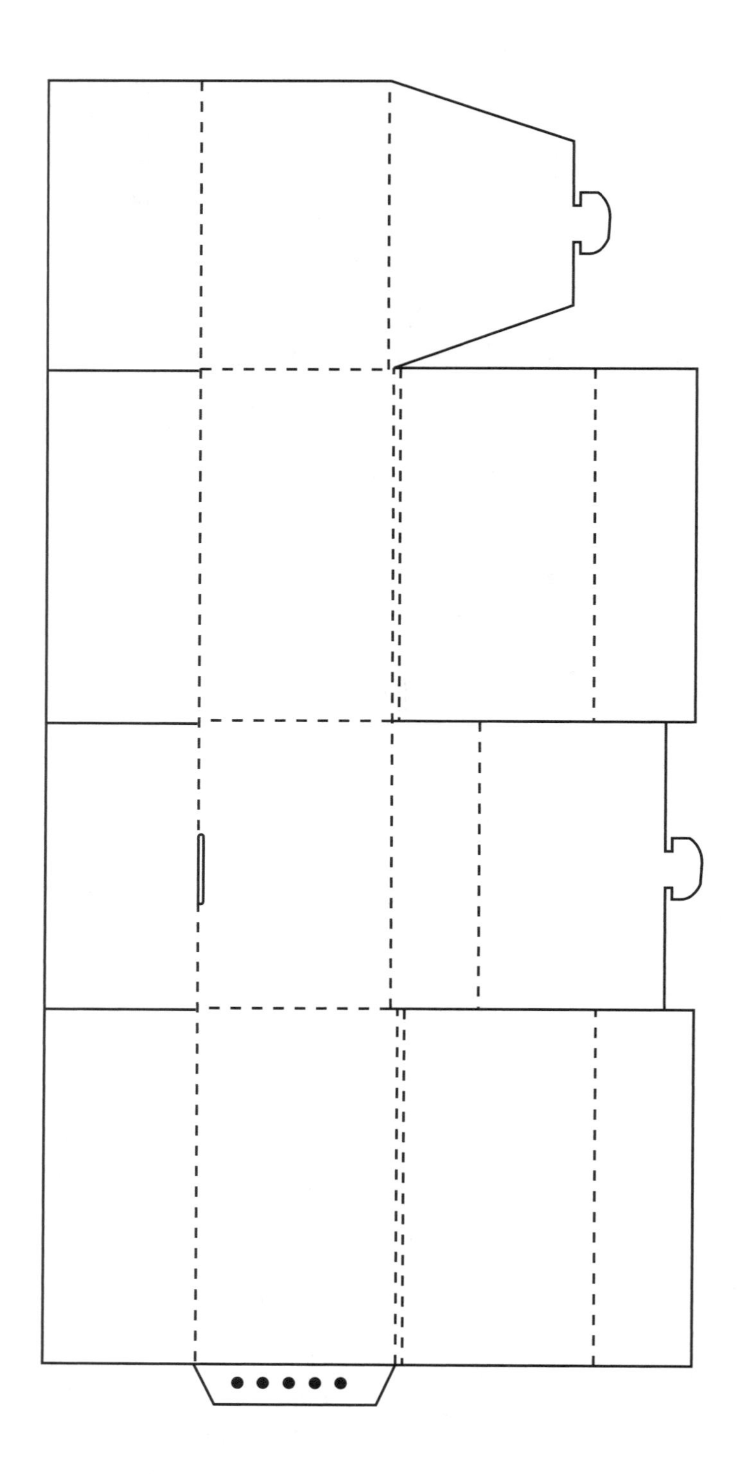

The shipping container has the graphics on the inside panels. The bottom is taped; top panels are folded back and the riser panel is kept in place by the tongue and slot combination.

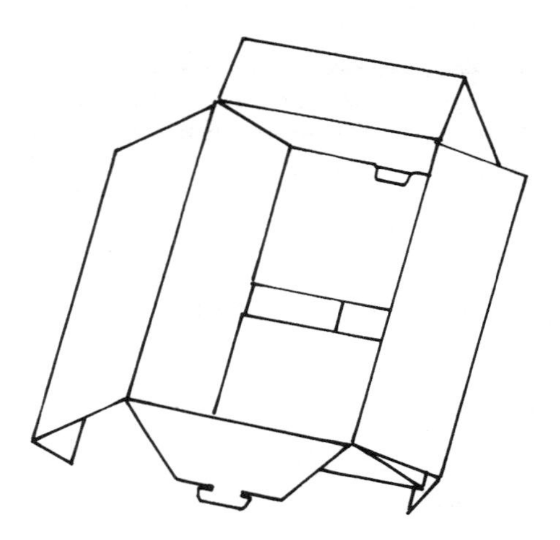

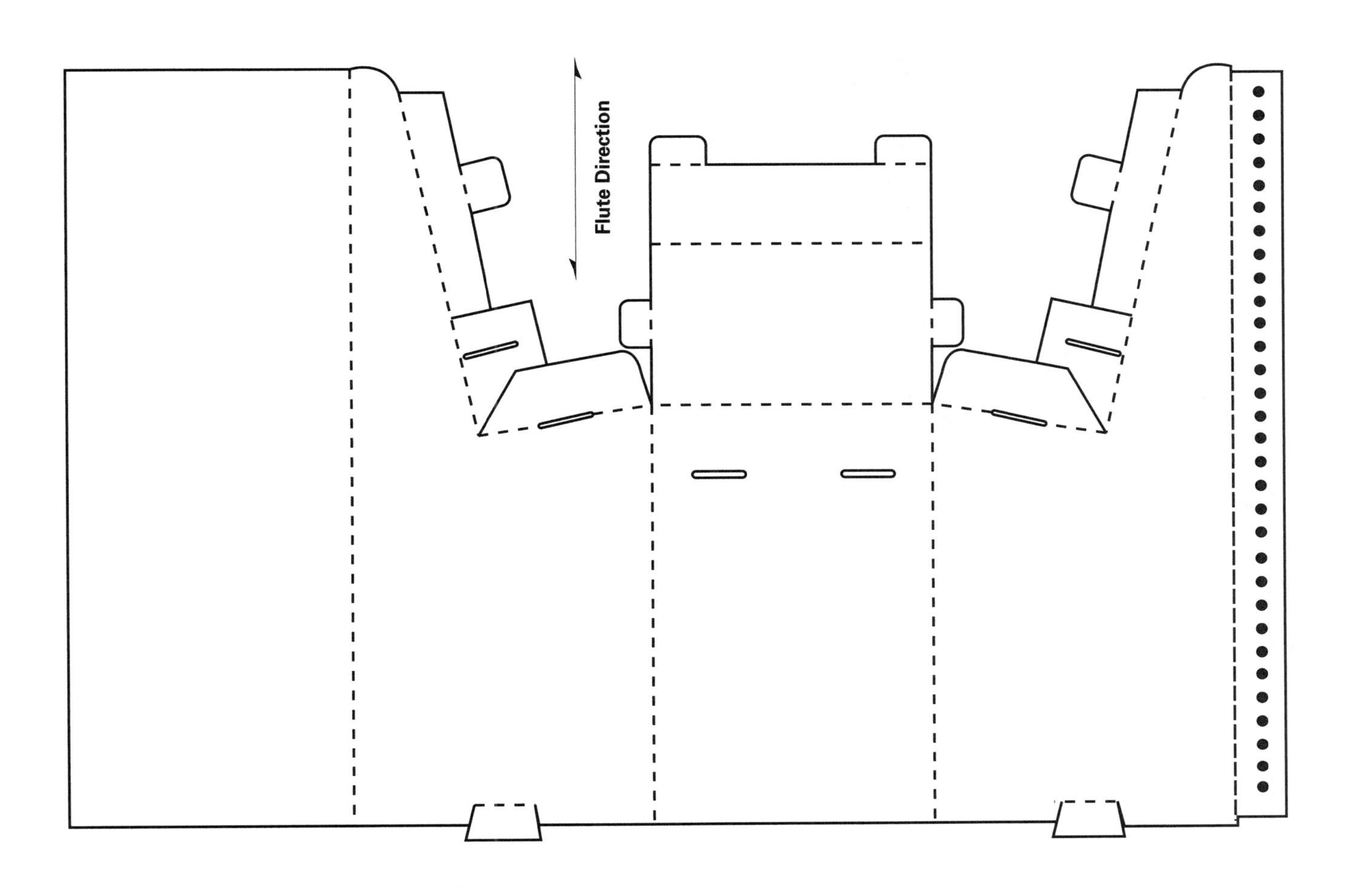

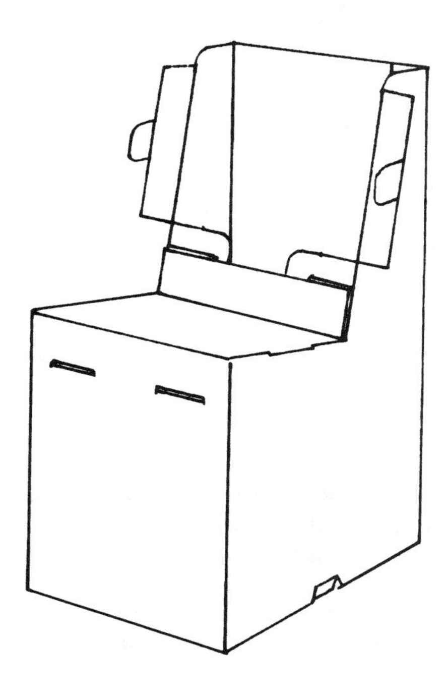

The ears on the front panel lock into the slots on the side panels. The slots in the front panel and ears on the side panels match up with the inside-out shipper/ display container on the following page.

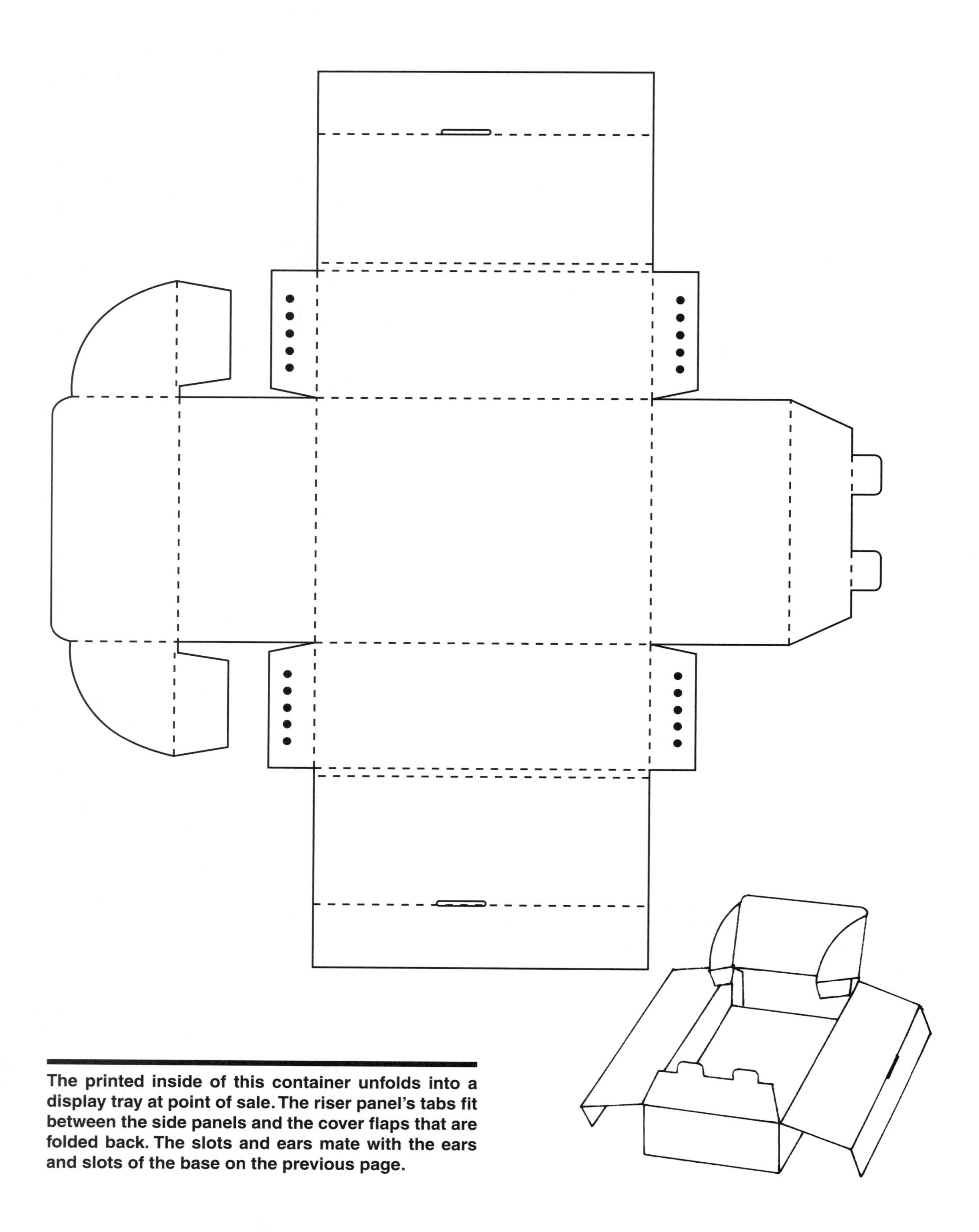

The printed inside of this container unfolds into a display tray at point of sale. The riser panel's tabs fit between the side panels and the cover flaps that are folded back. The slots and ears mate with the ears and slots of the base on the previous page.

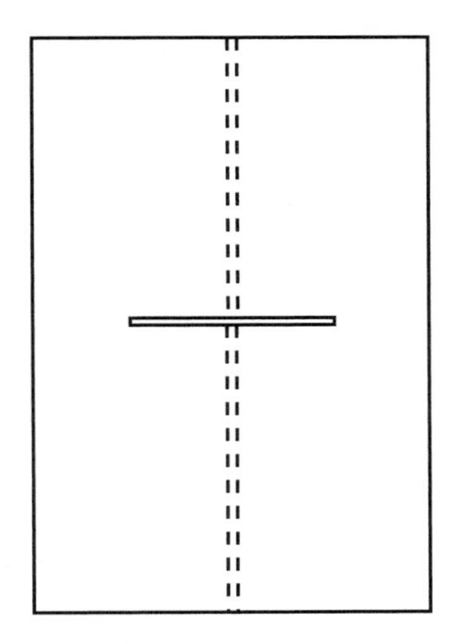

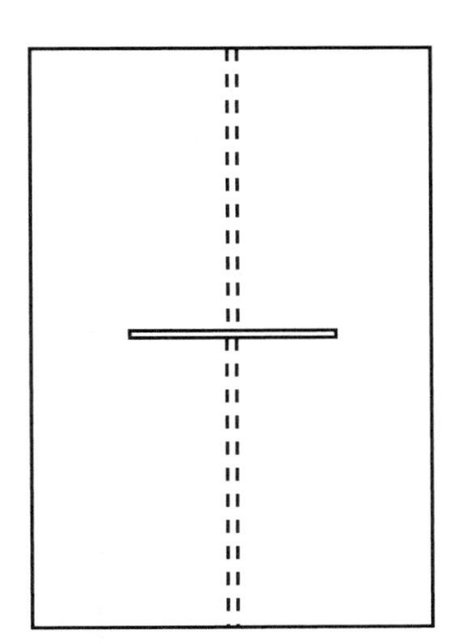

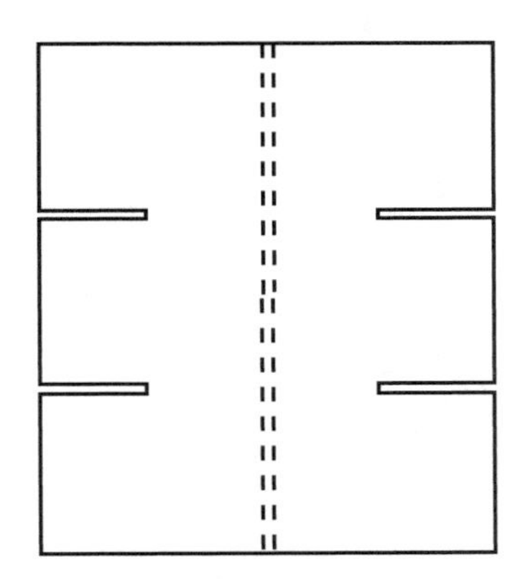

Display containers can be divided into sections fitted to the product(s) to be displayed. The corrugated dividers shown here are slotted, folded, and slide together to form six equal partitions.

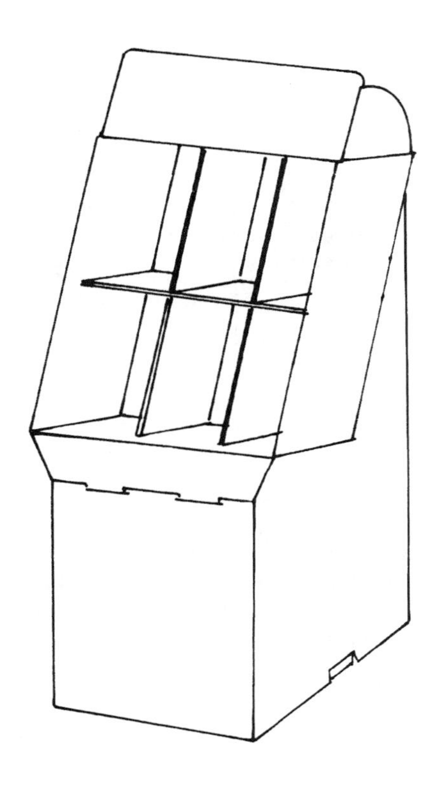

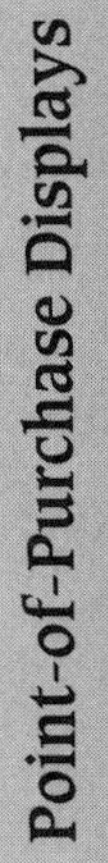

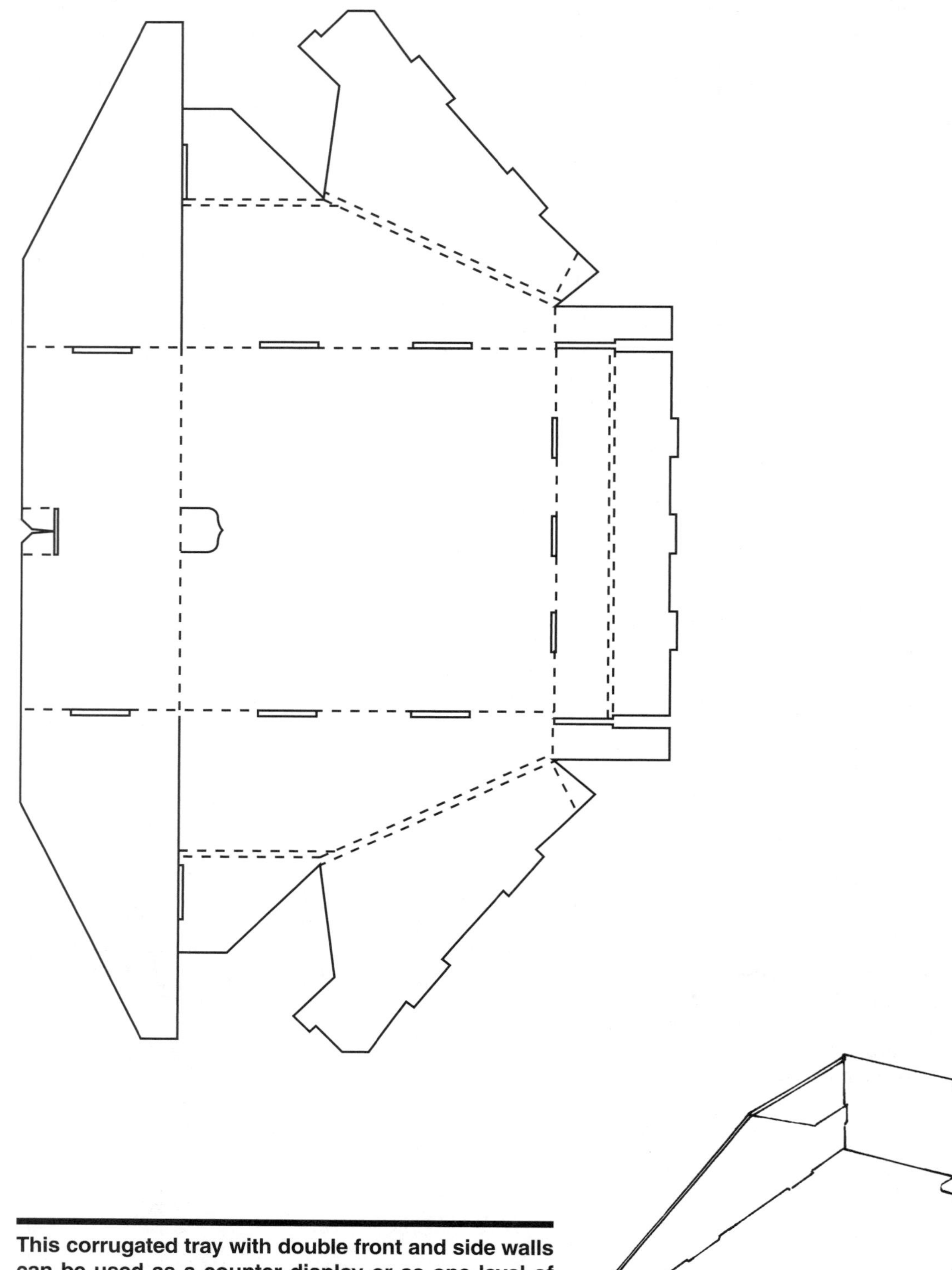

This corrugated tray with double front and side walls can be used as a counter display or as one level of the “waterfall” display shown on the following page.

HOLLOW WALL CORRUGATED SHIPPER/DISPLAY

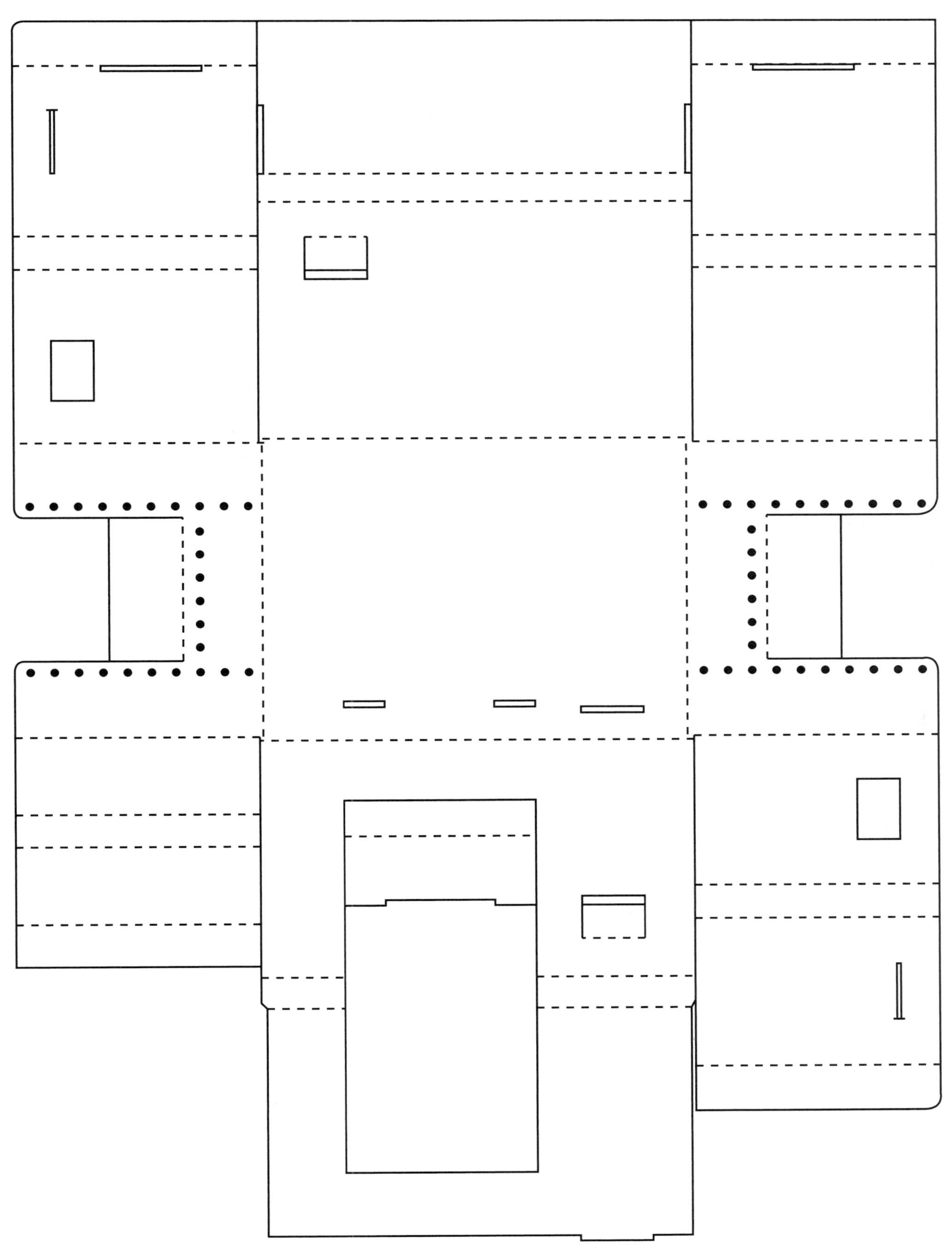

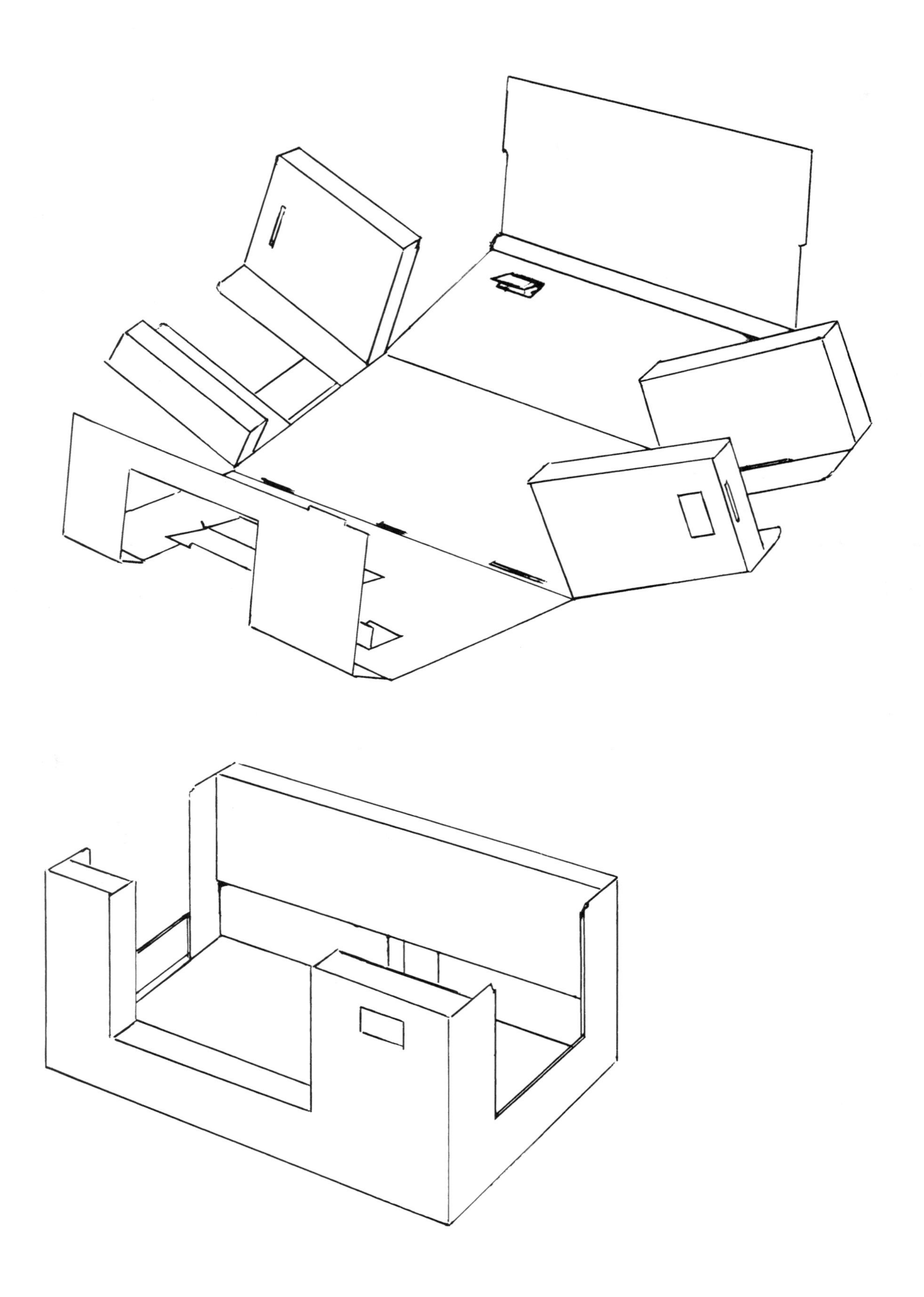

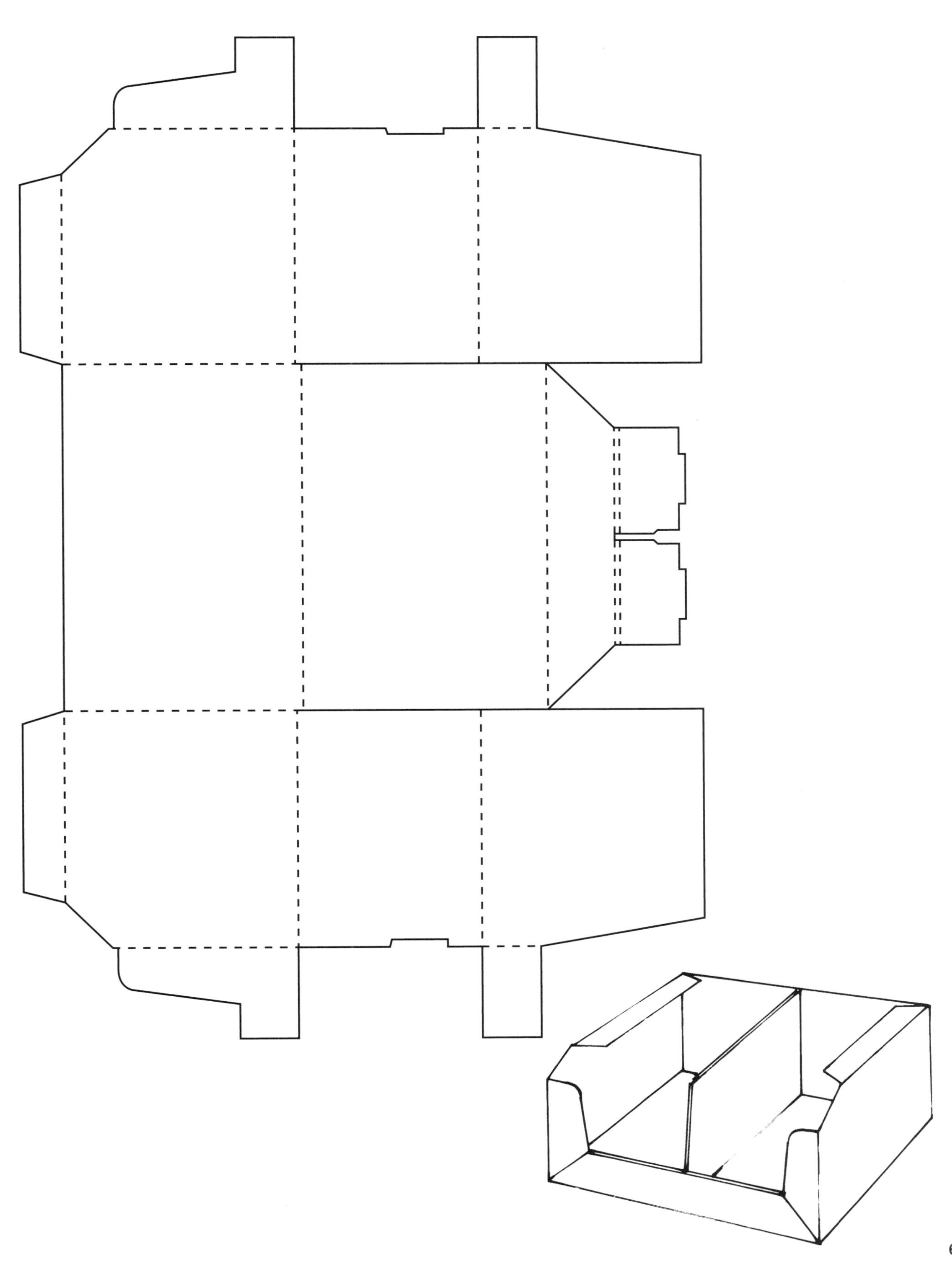

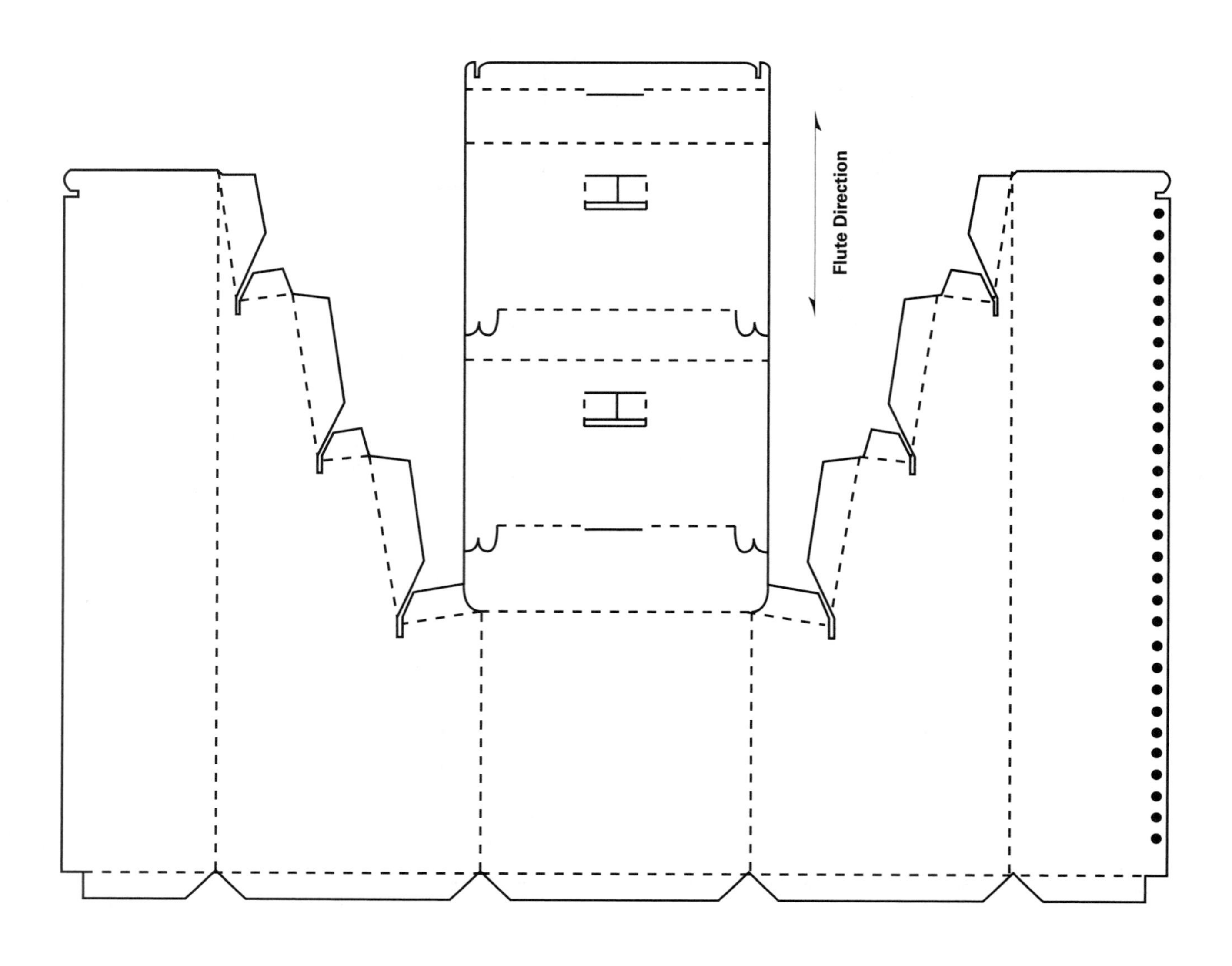

The 3-step waterfall corrugated display stand is ideal for showing one or more products on different levels.

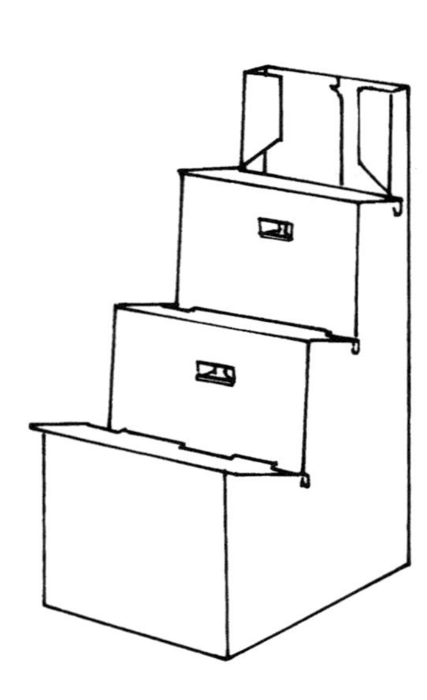

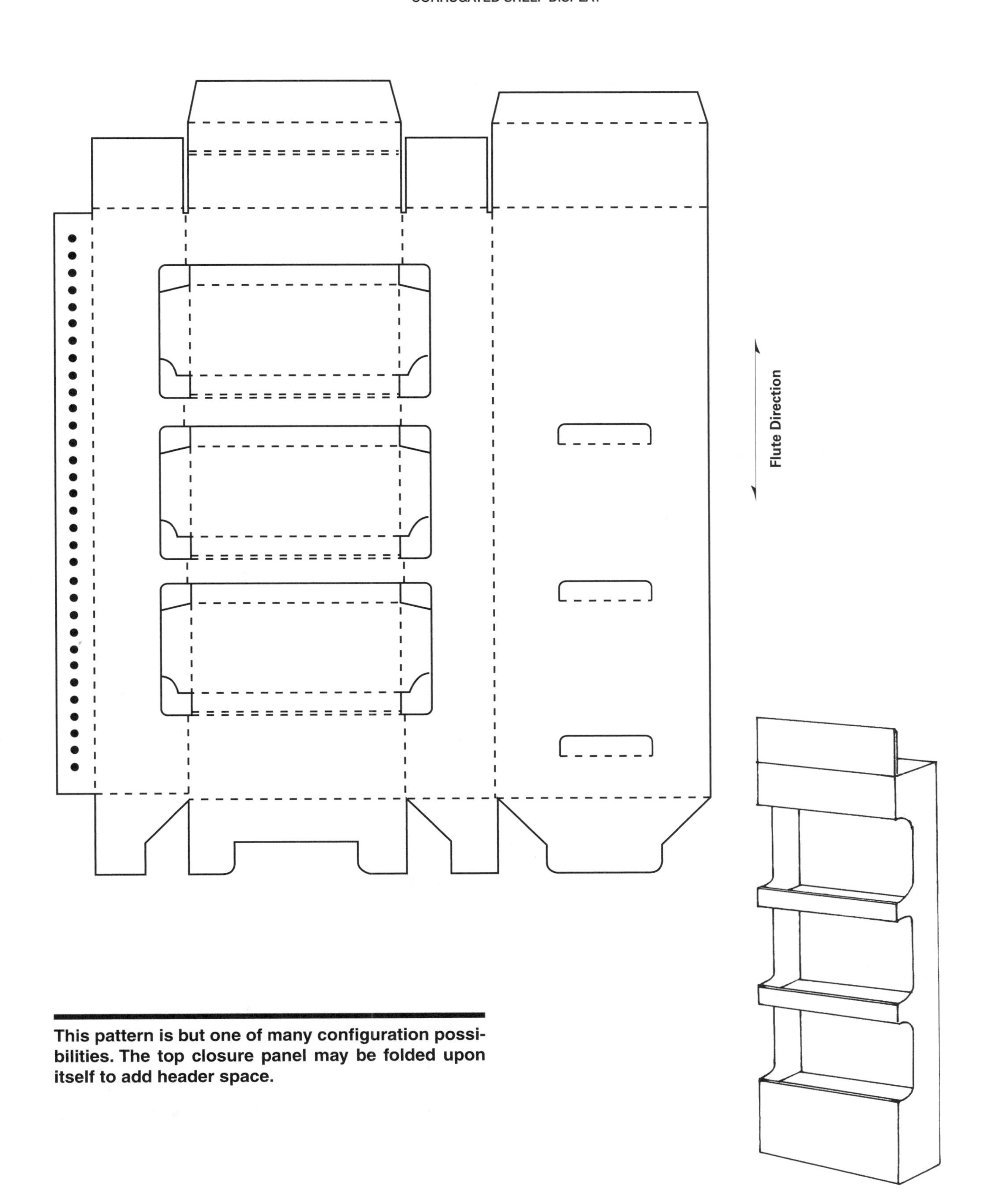

This pattern is but one of many configuration possibilities. The top closure panel may be folded upon itself to add header space.

CORRUGATED DISPLAY BASE

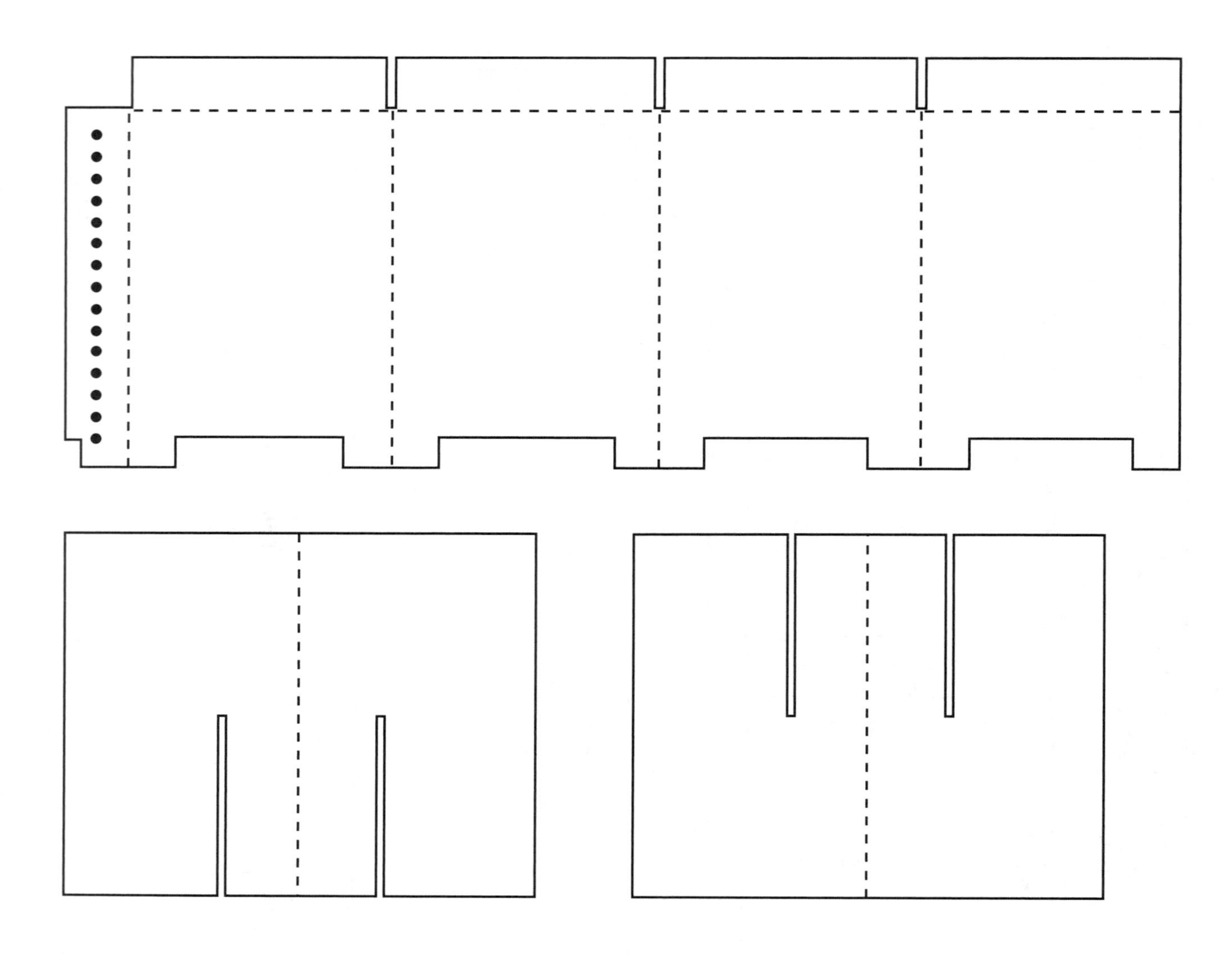

The slotted and scored insert carries the weight.

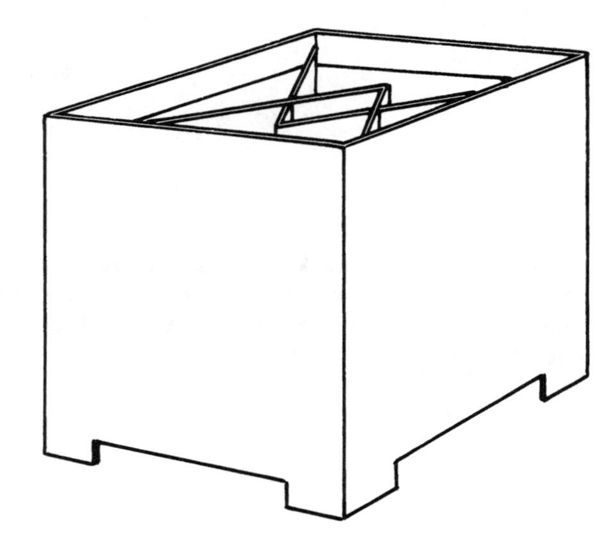

CORRUGATED FLOOR DISPLAY BASE

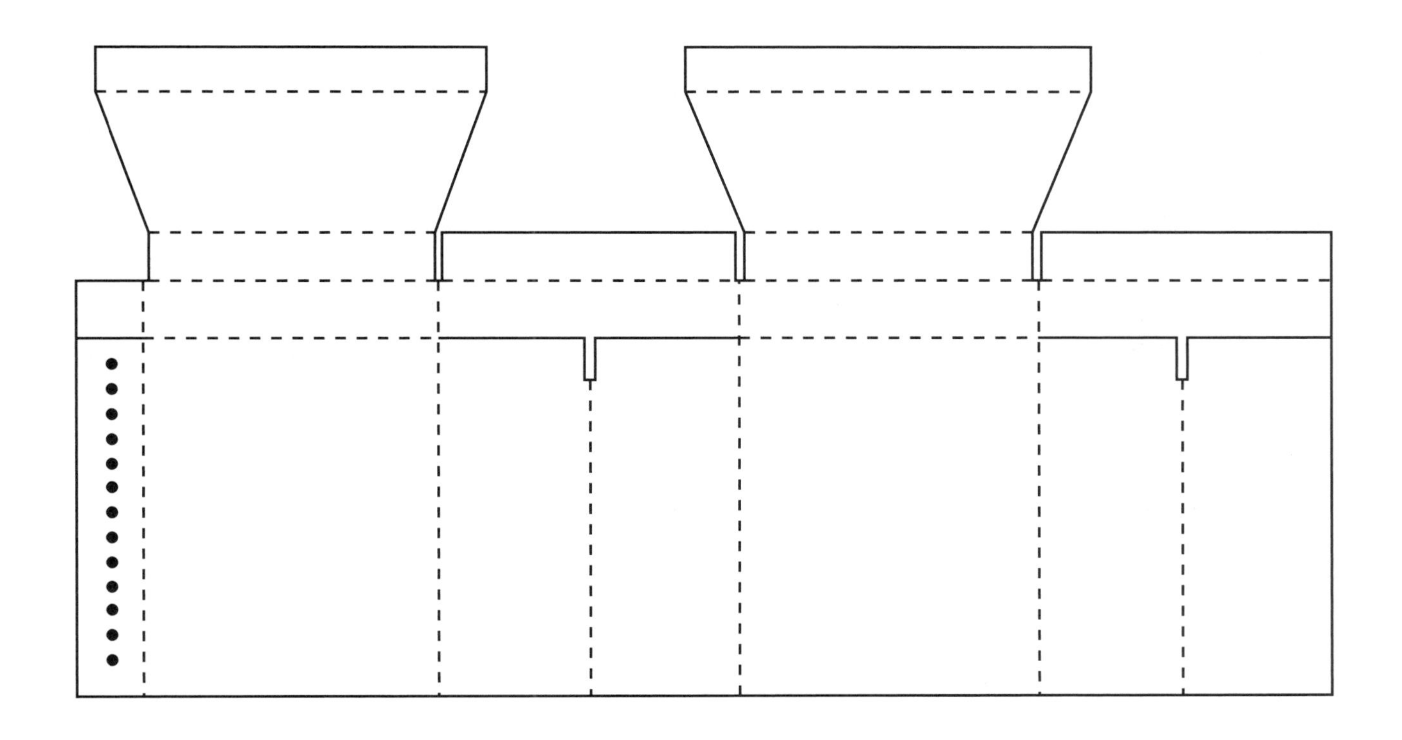

May be used by itself or as a base.

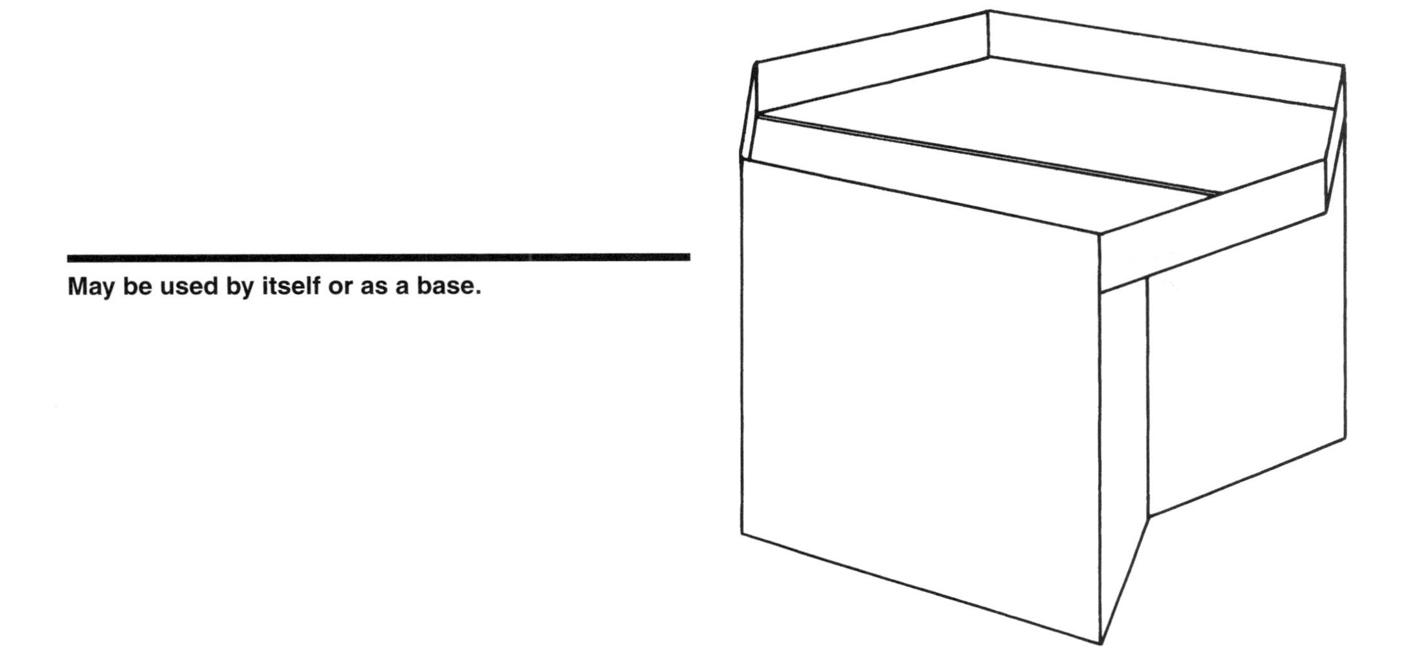

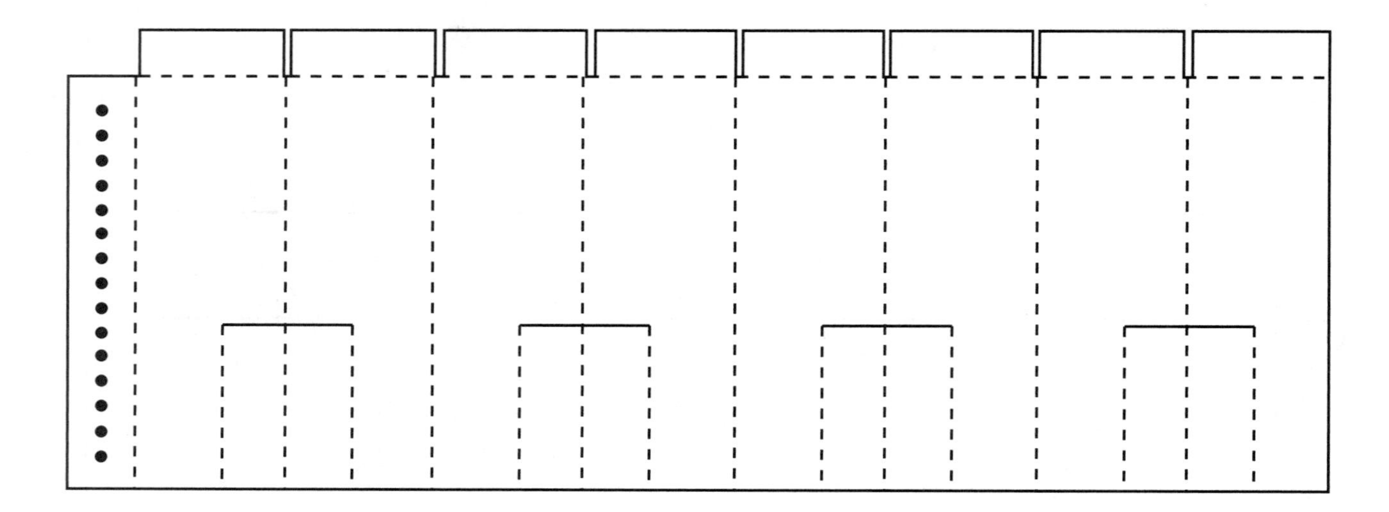

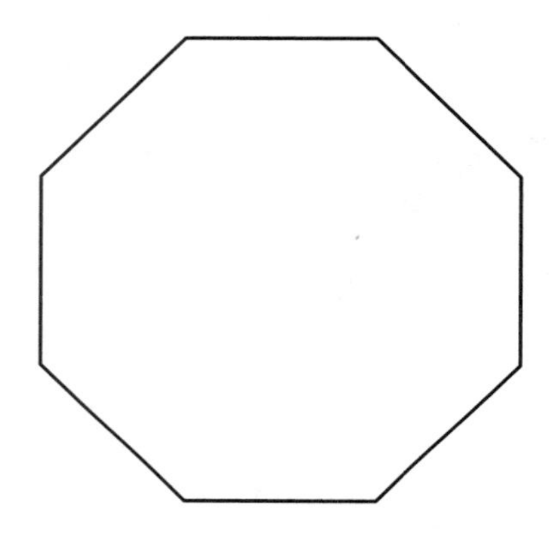

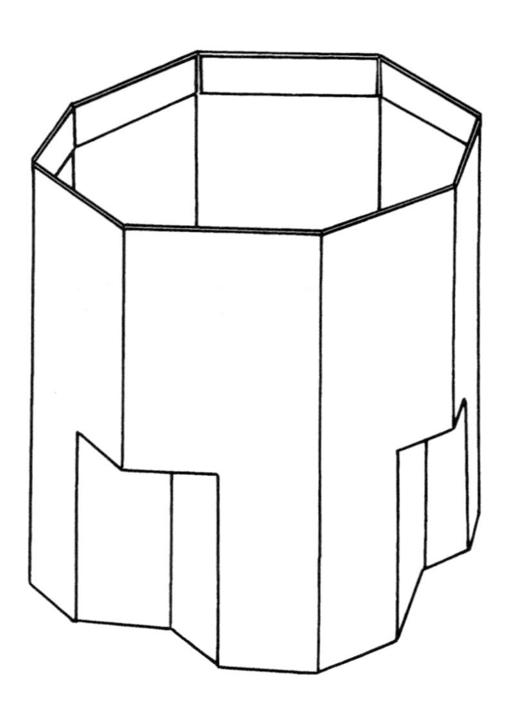

The scored and folded back base supports the platform.

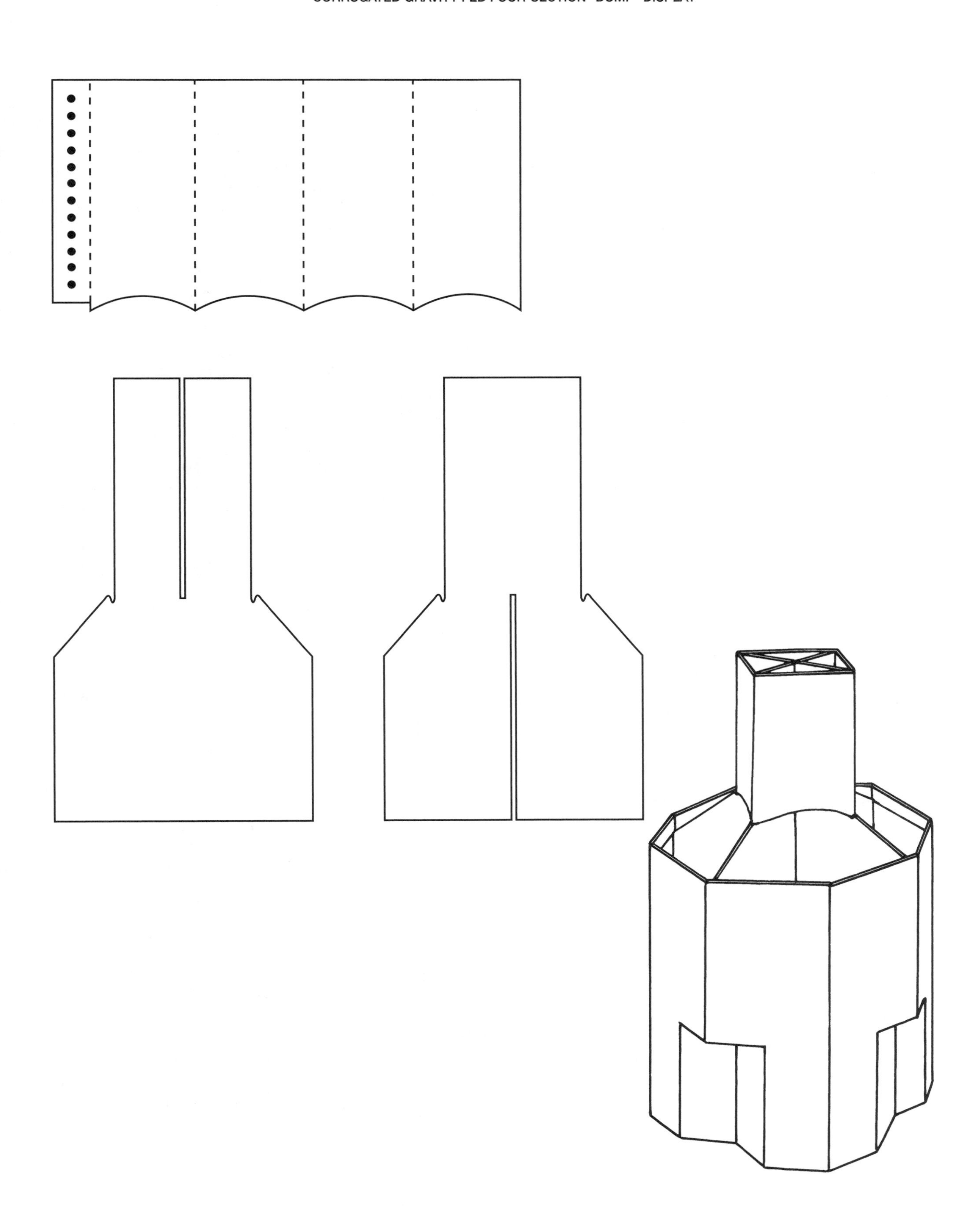

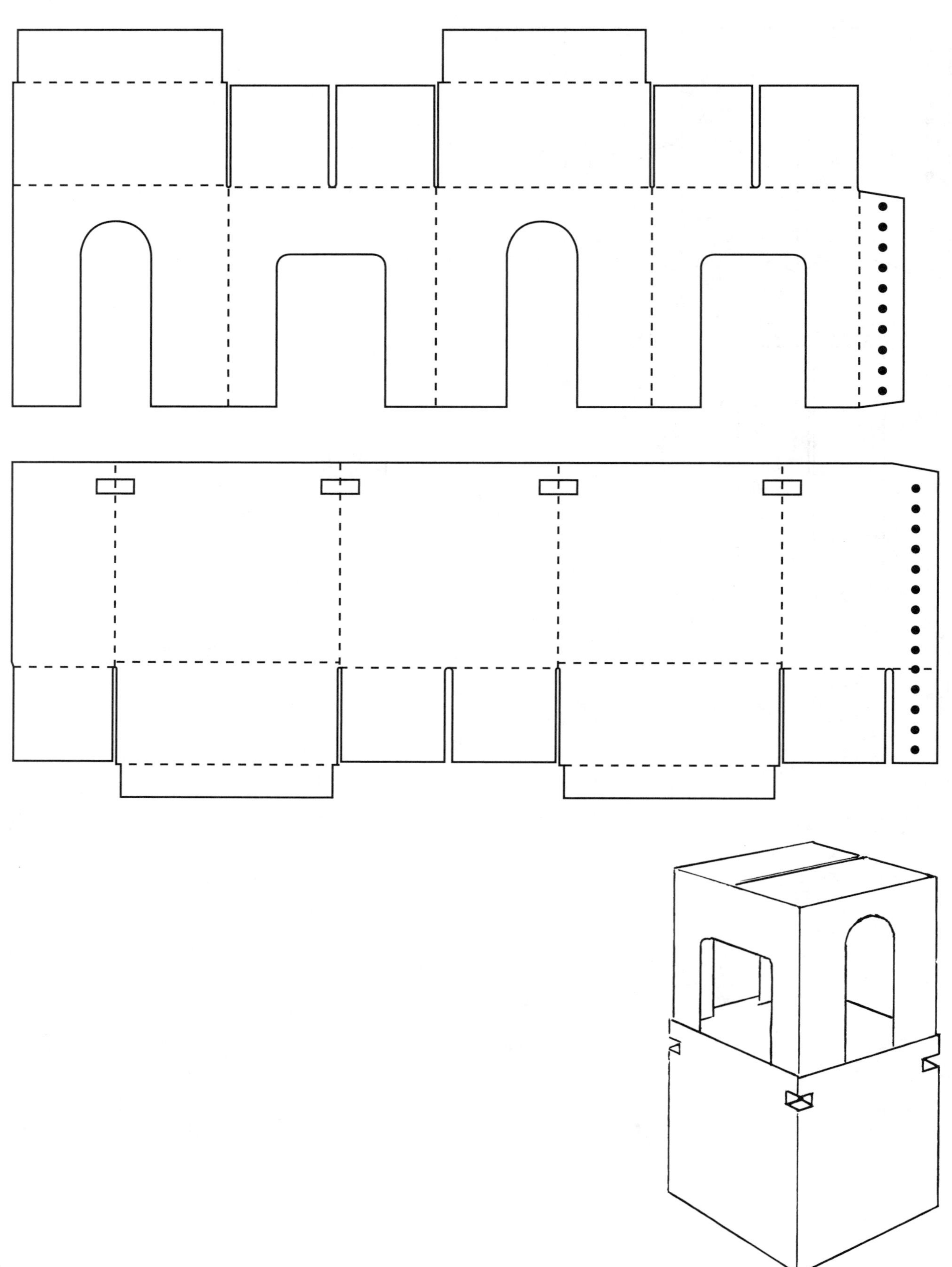

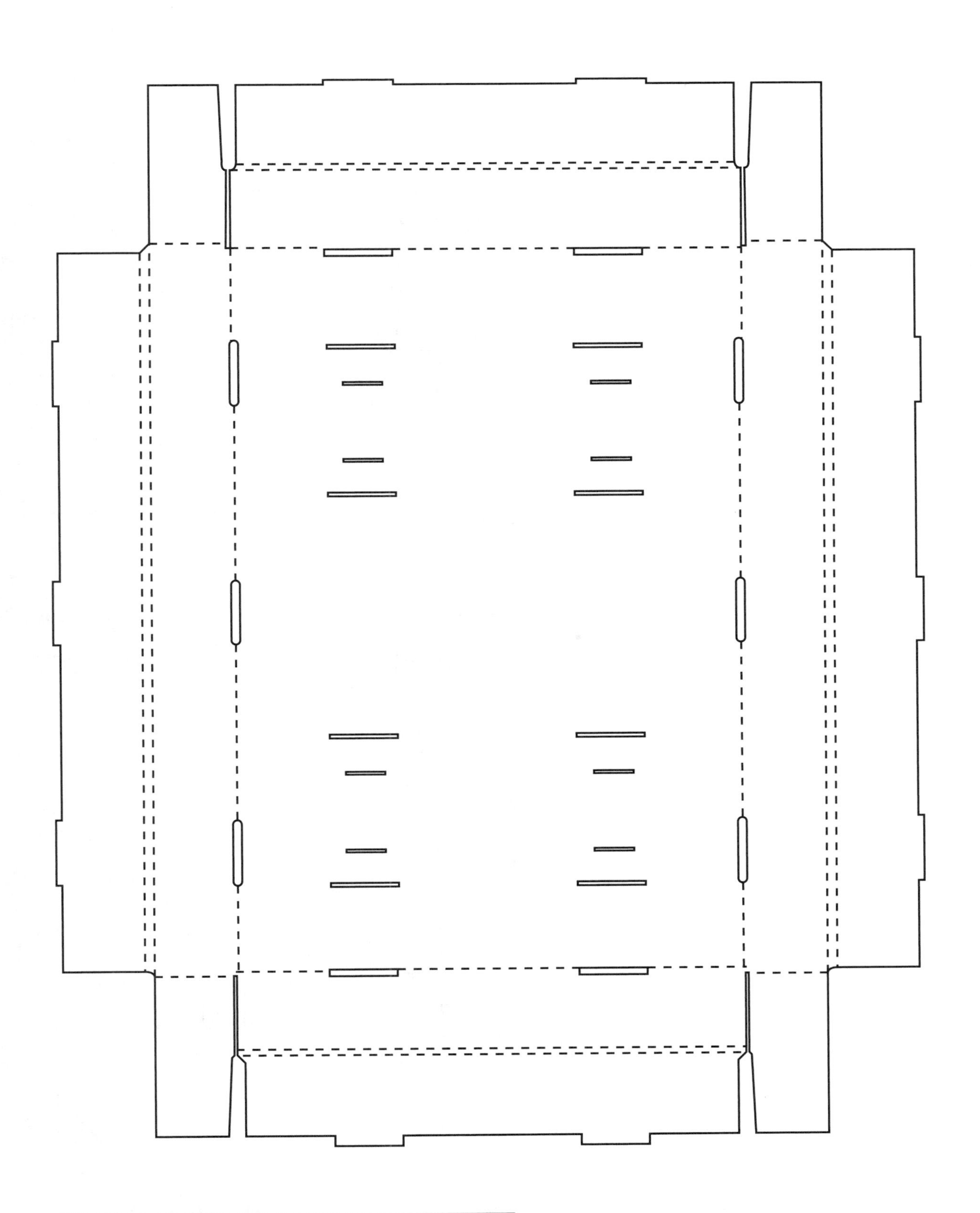

The corrugated speed table top features slots to accommodate the belt locks of the two stands.

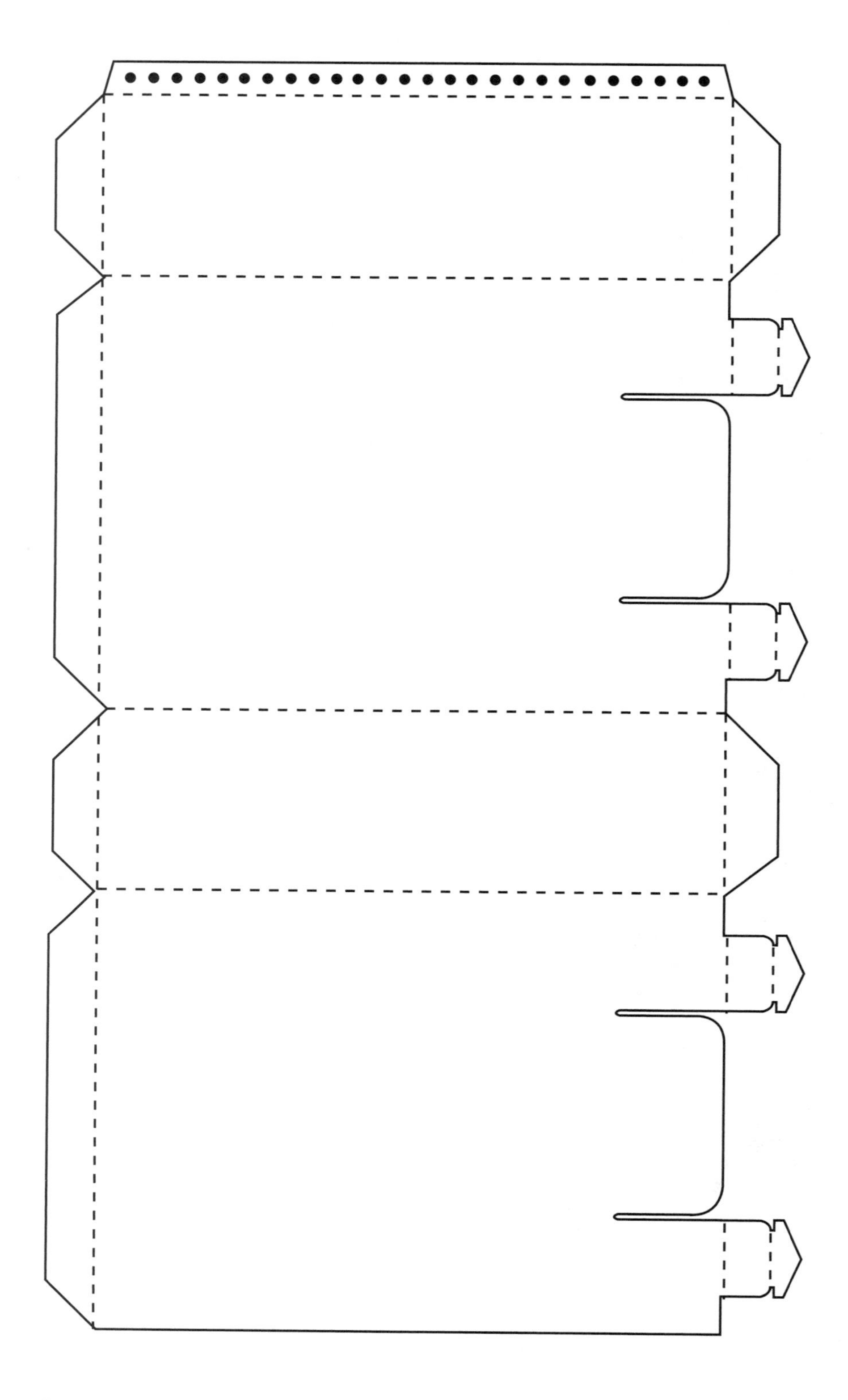

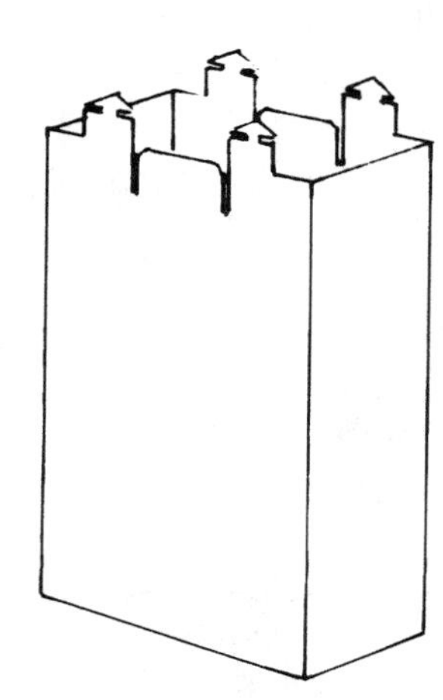

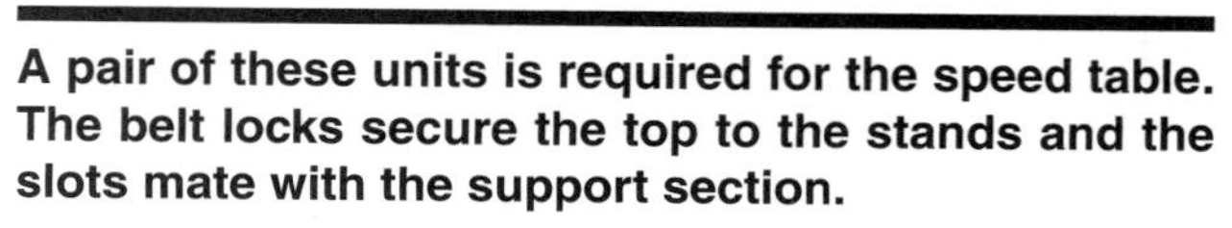

A pair of these units is required for the speed table. The belt locks secure the top to the stands and the slots mate with the support section.

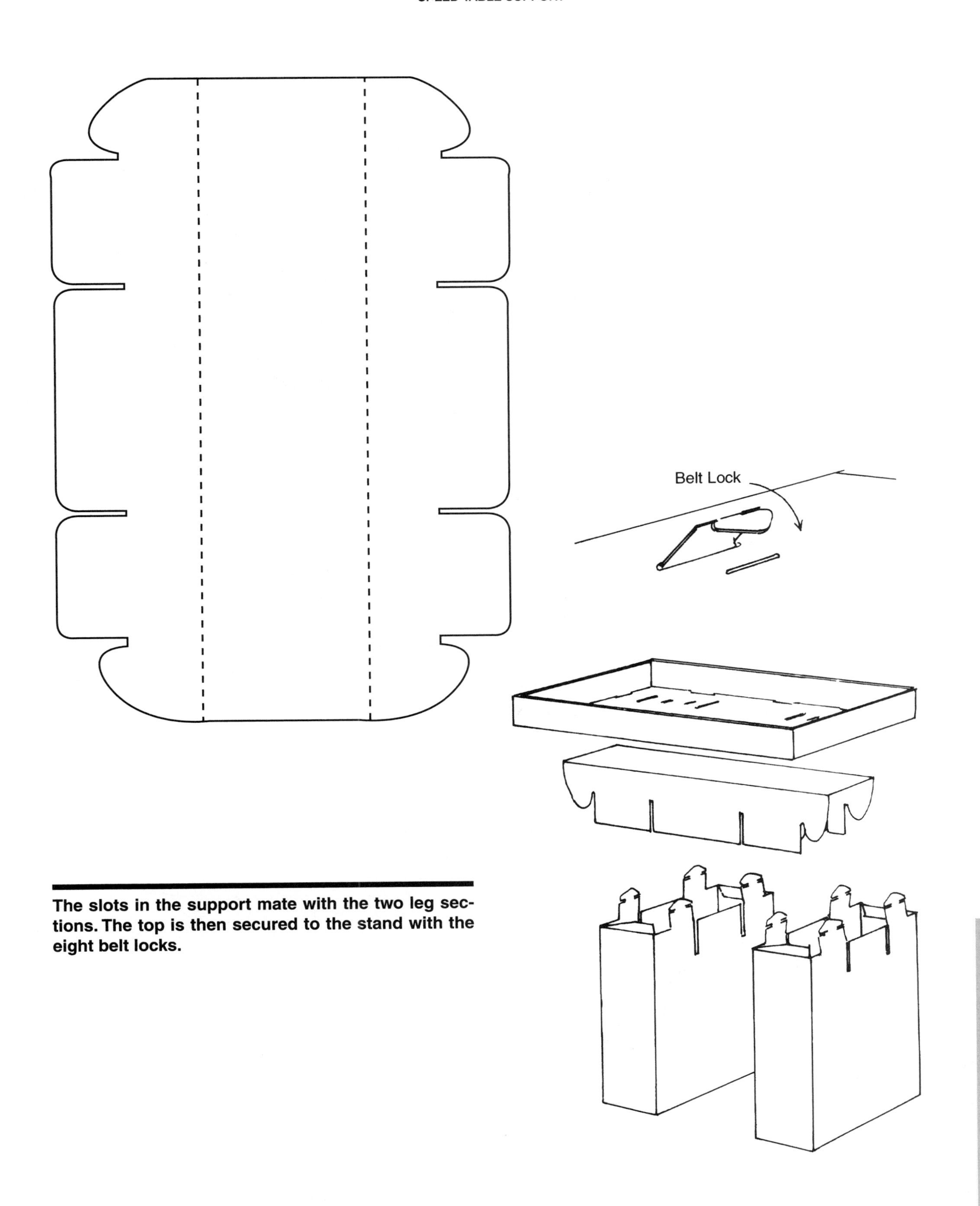

The slots in the support mate with the two leg sections. The top is then secured to the stand with the eight belt locks.

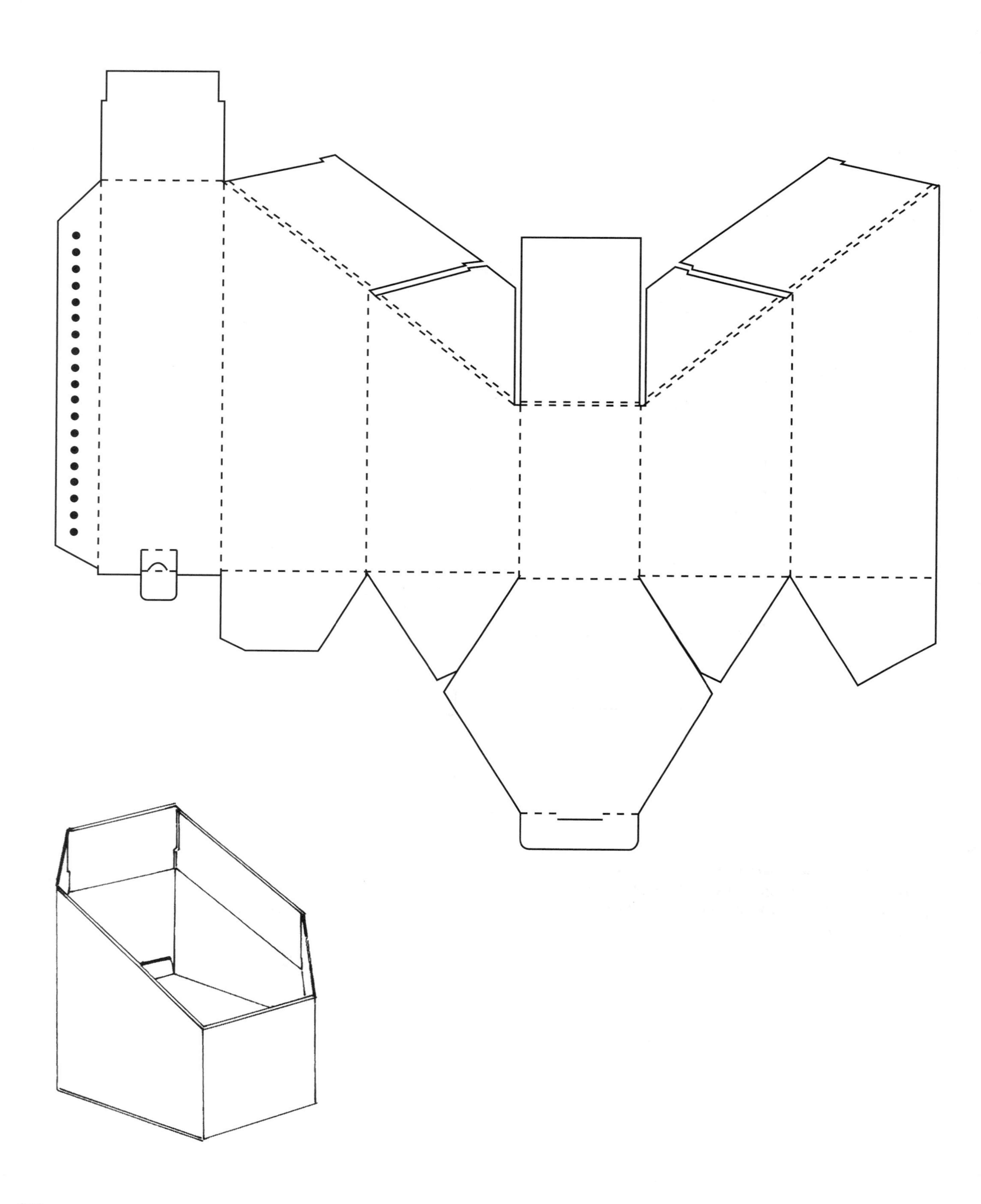

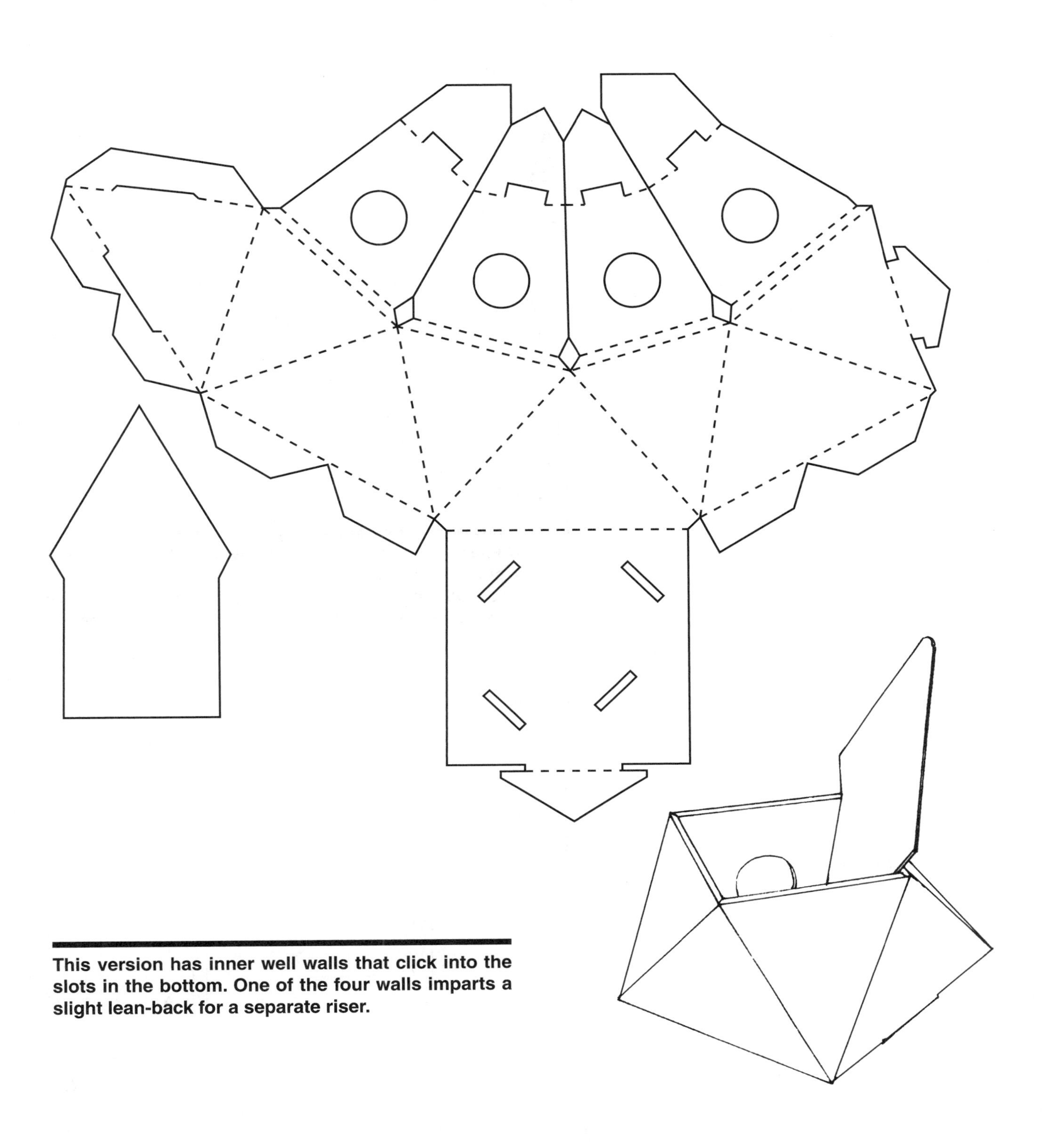

This version has inner well walls that click into the slots in the bottom. One of the four walls imparts a slight lean-back for a separate riser.

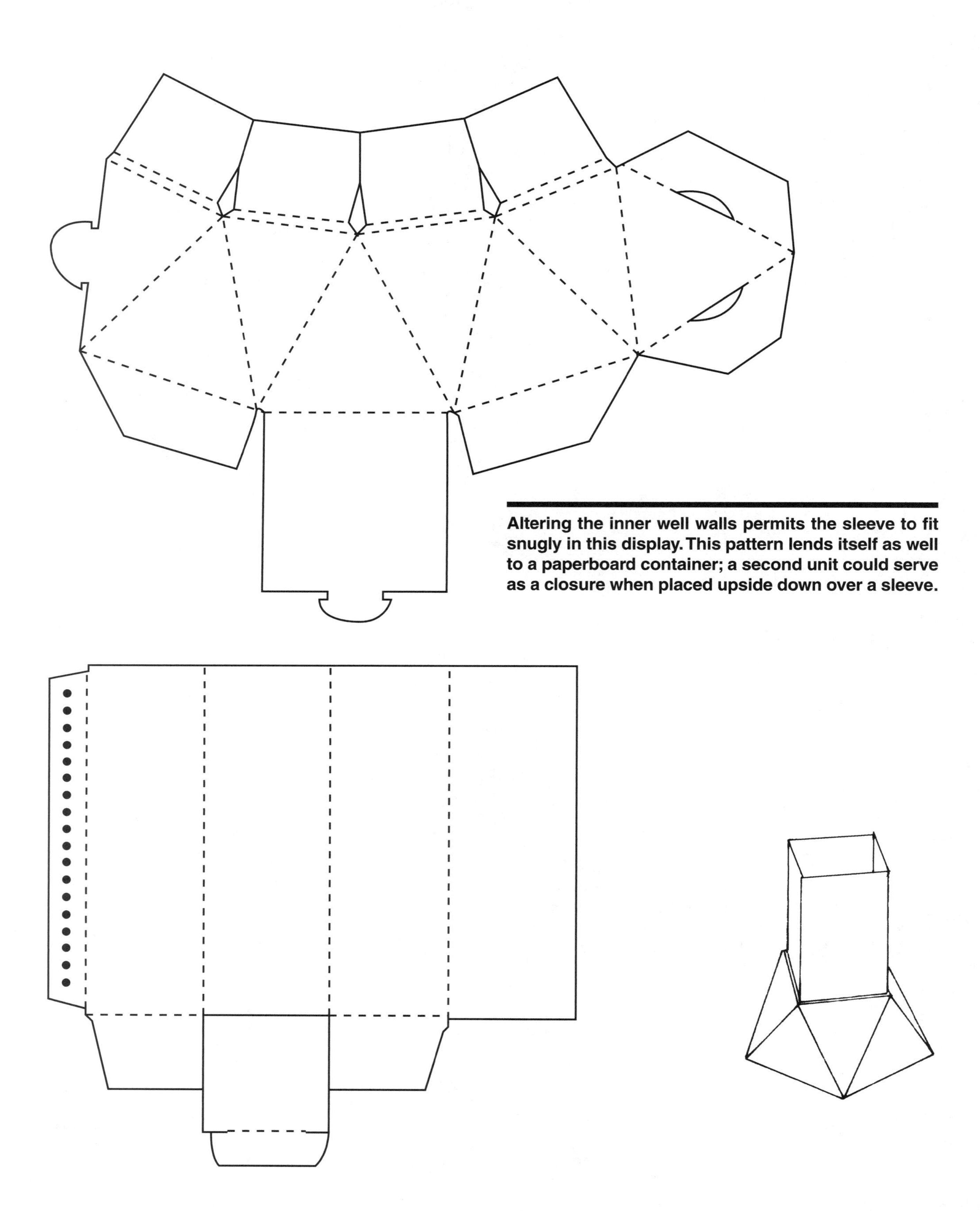

Altering the inner well walls permits the sleeve to fit snugly in this display. This pattern lends itself as well to a paperboard container; a second unit could serve as a closure when placed upside down over a sleeve.

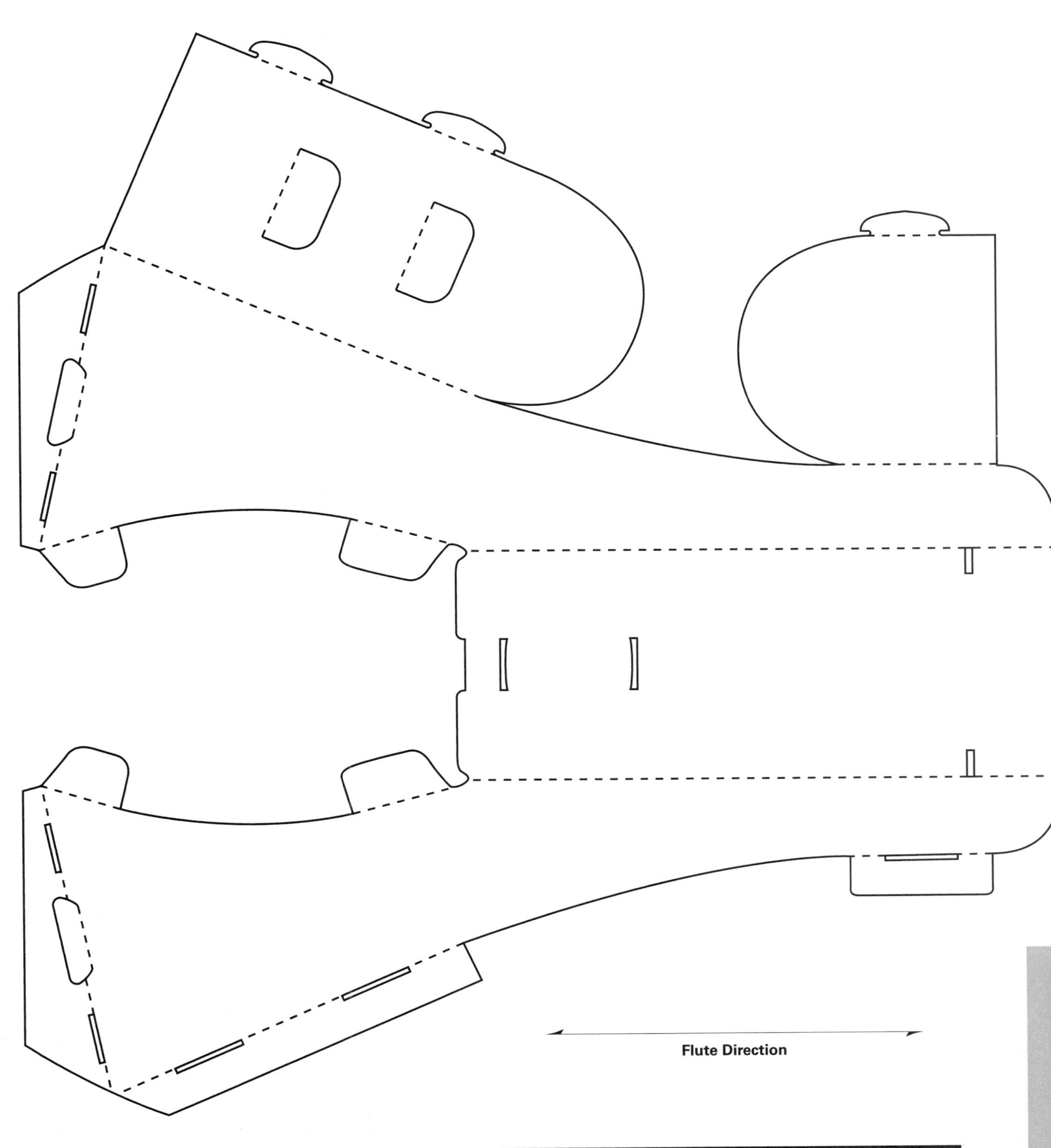

With the following two pages, makes up a four-shelf display.

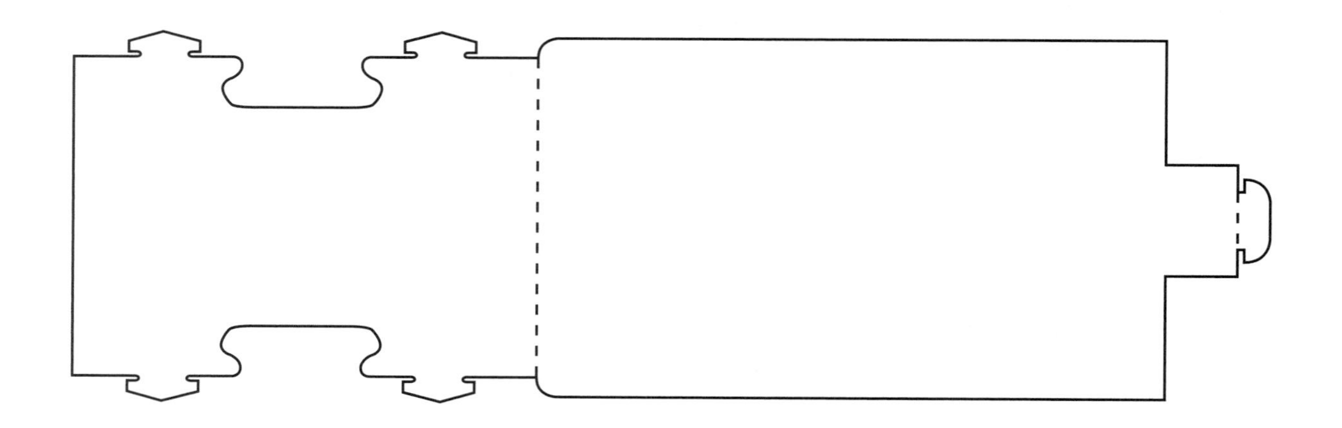

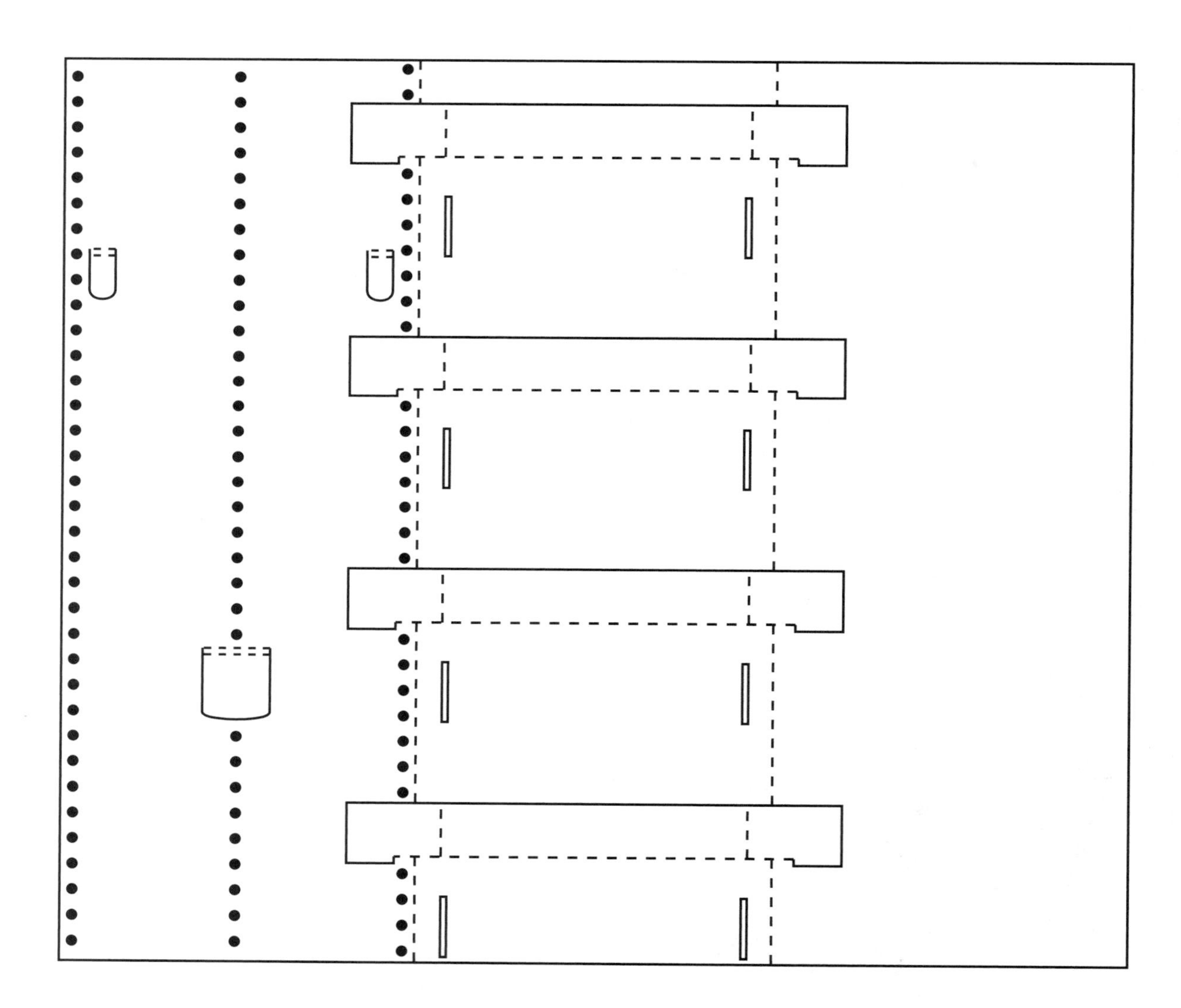

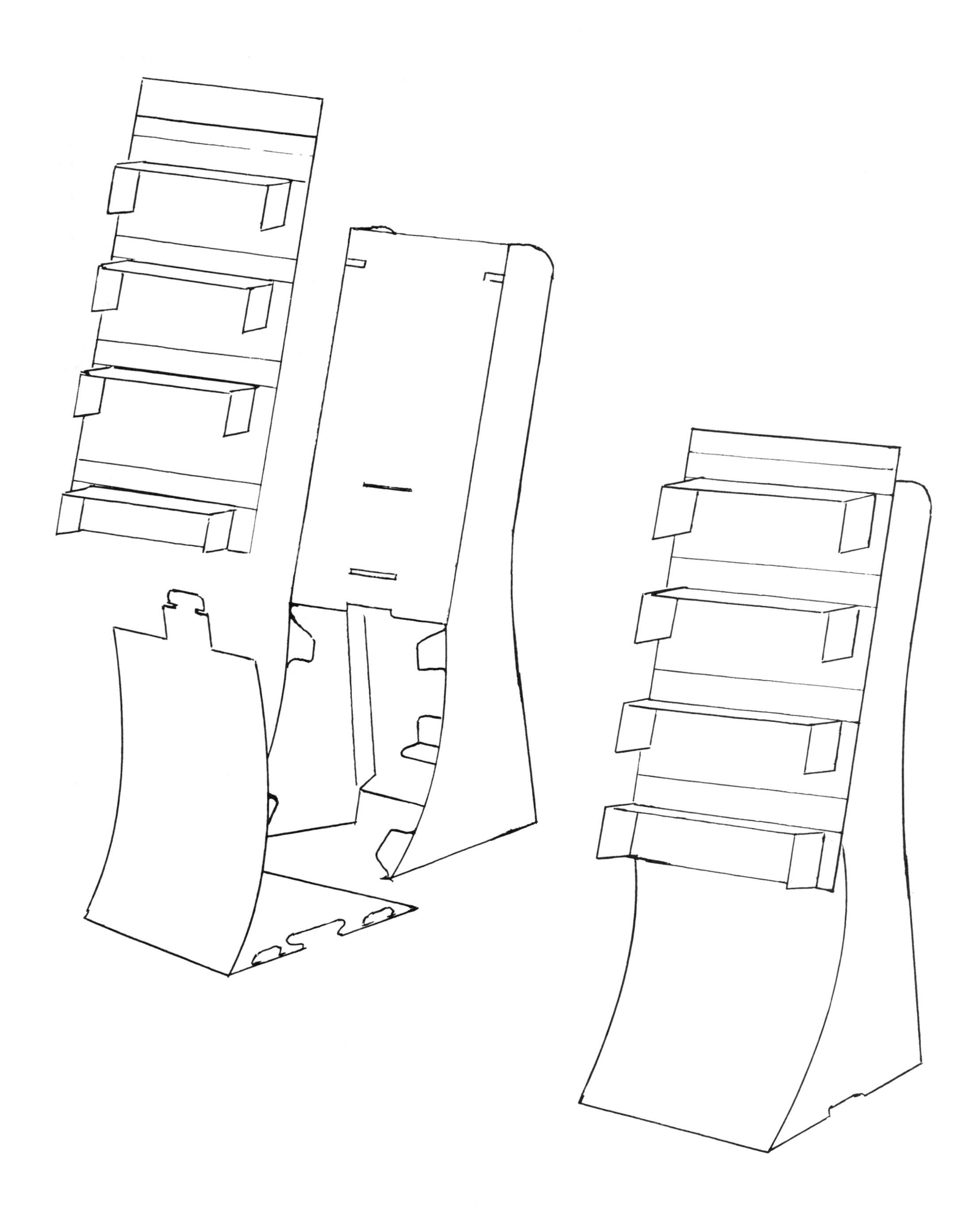

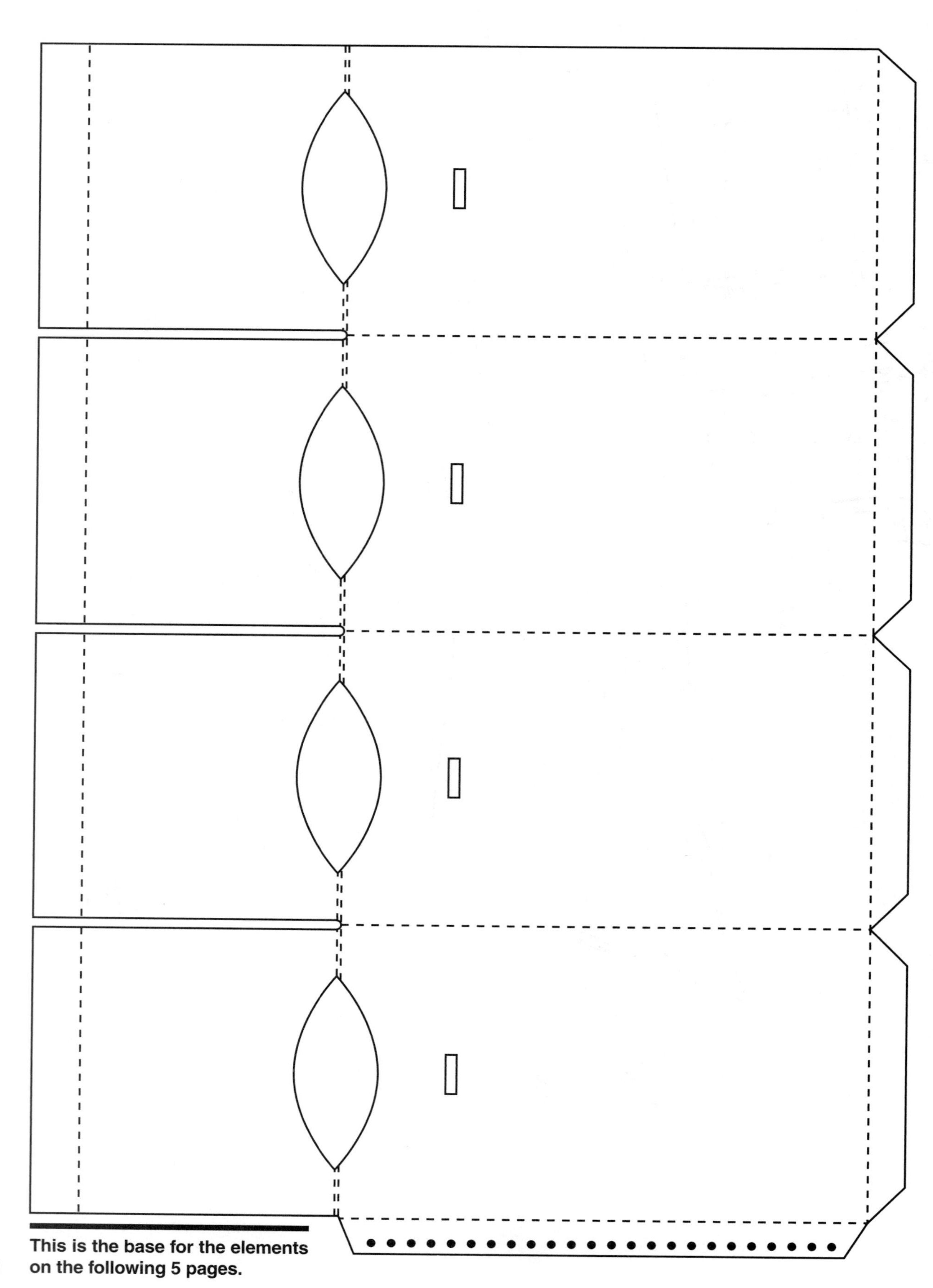

This is the base for the elements on the following 5 pages.

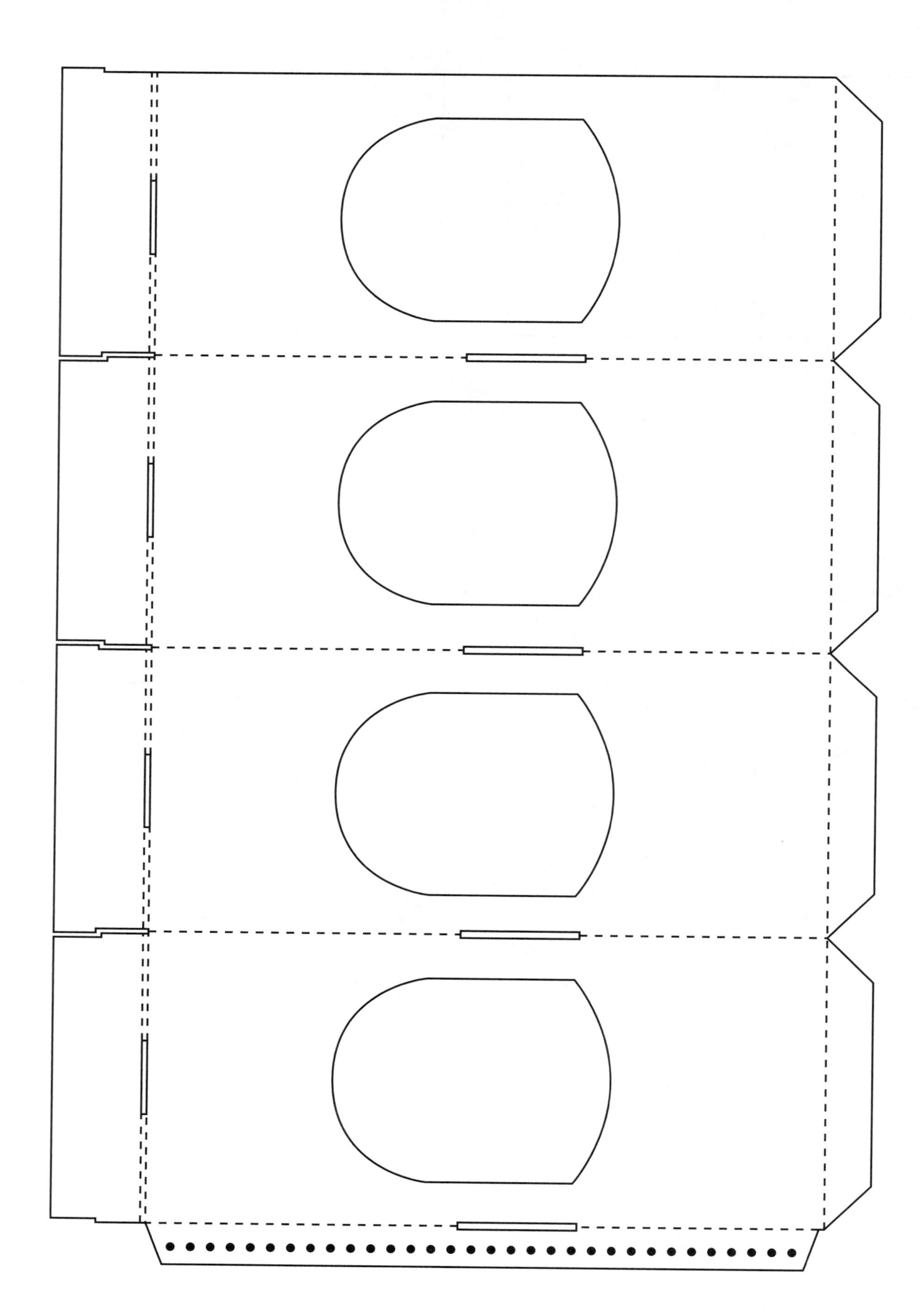

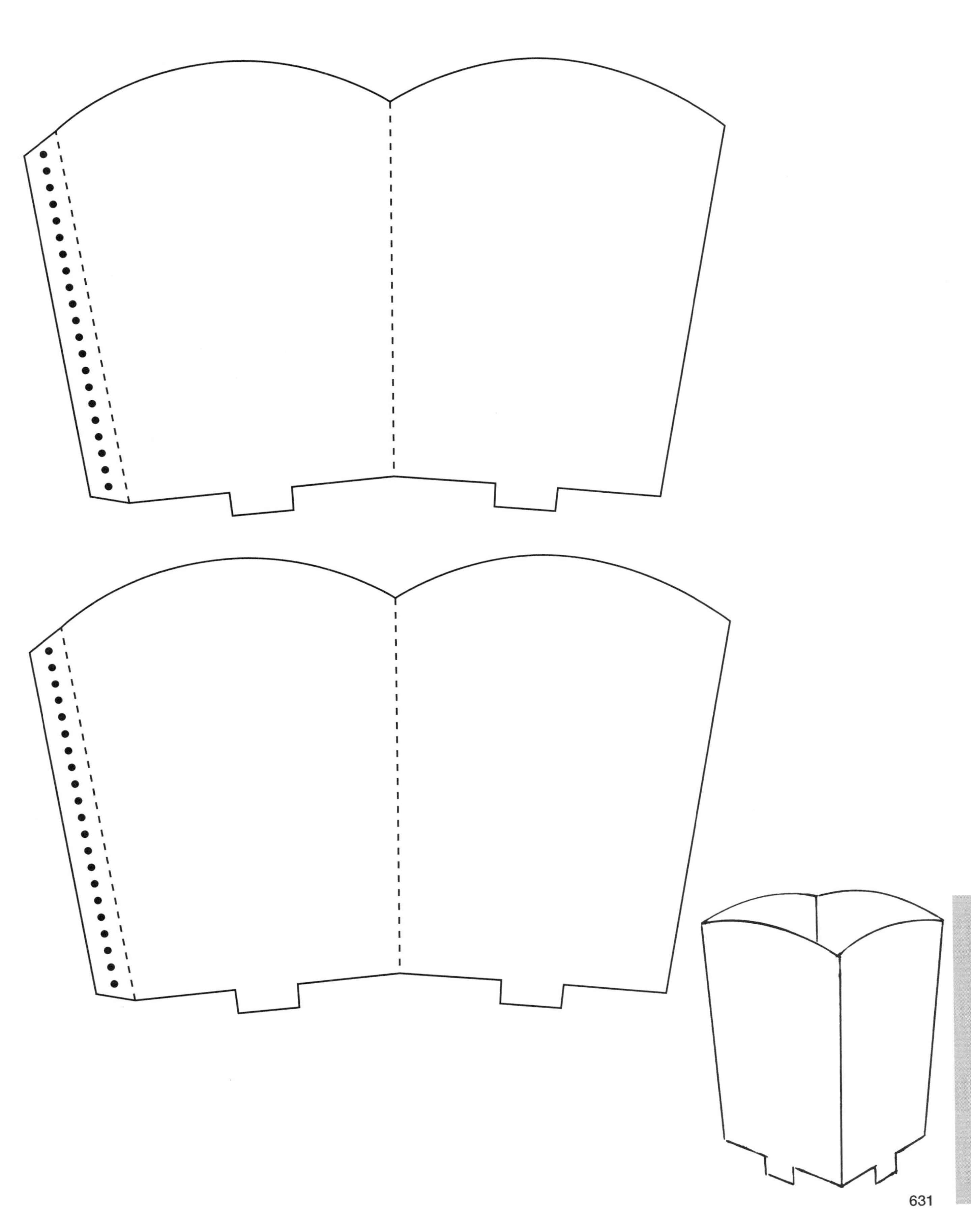

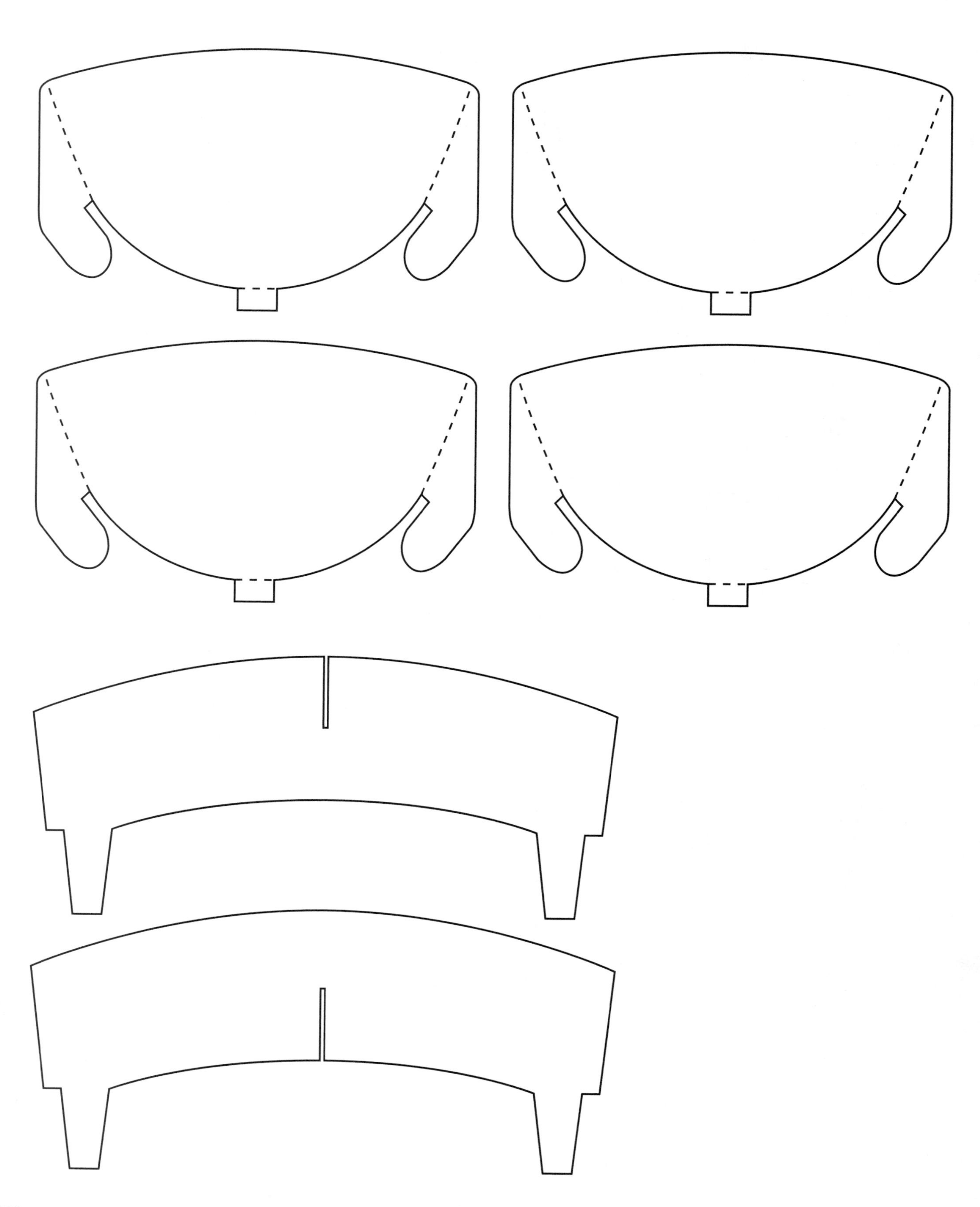

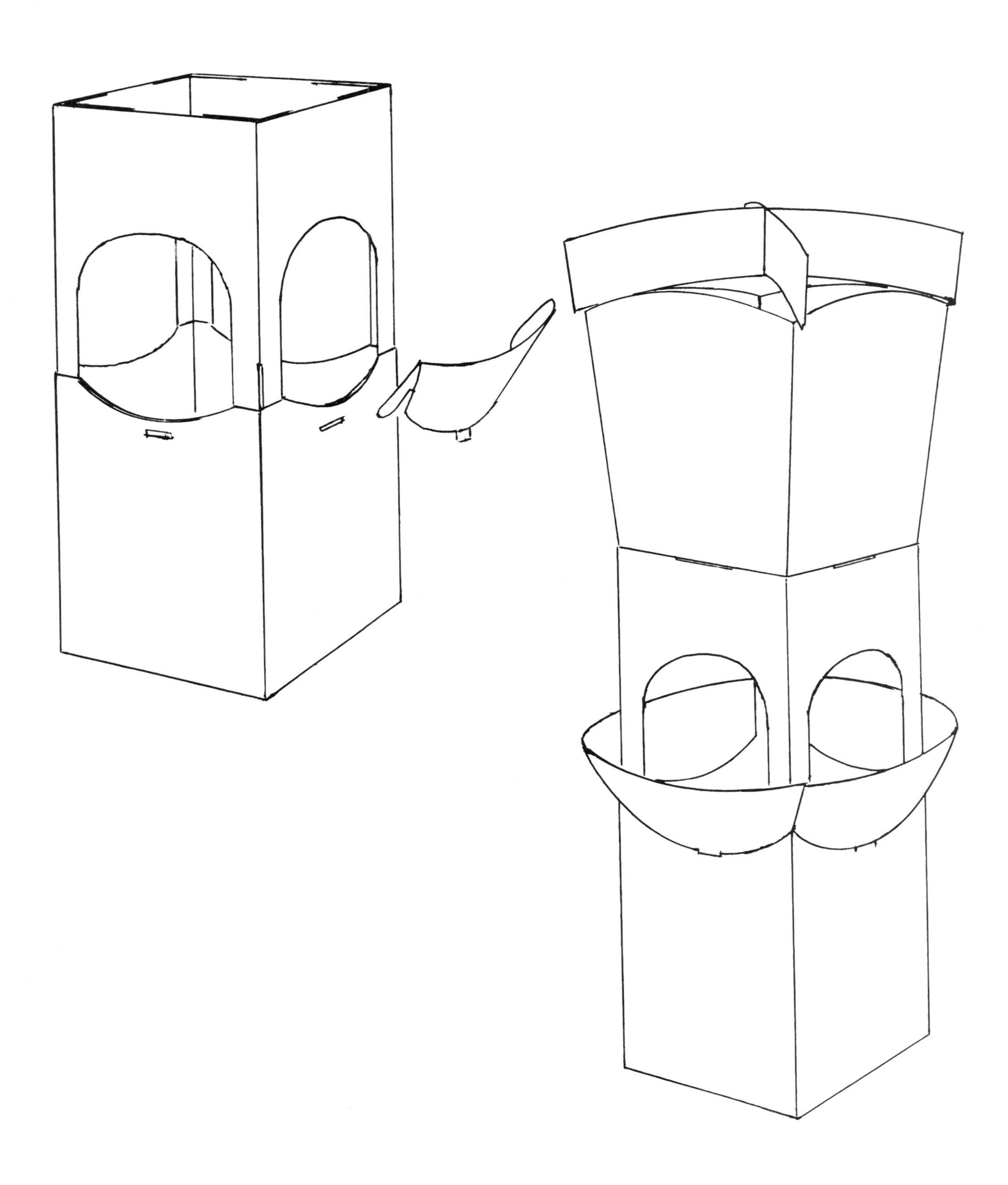

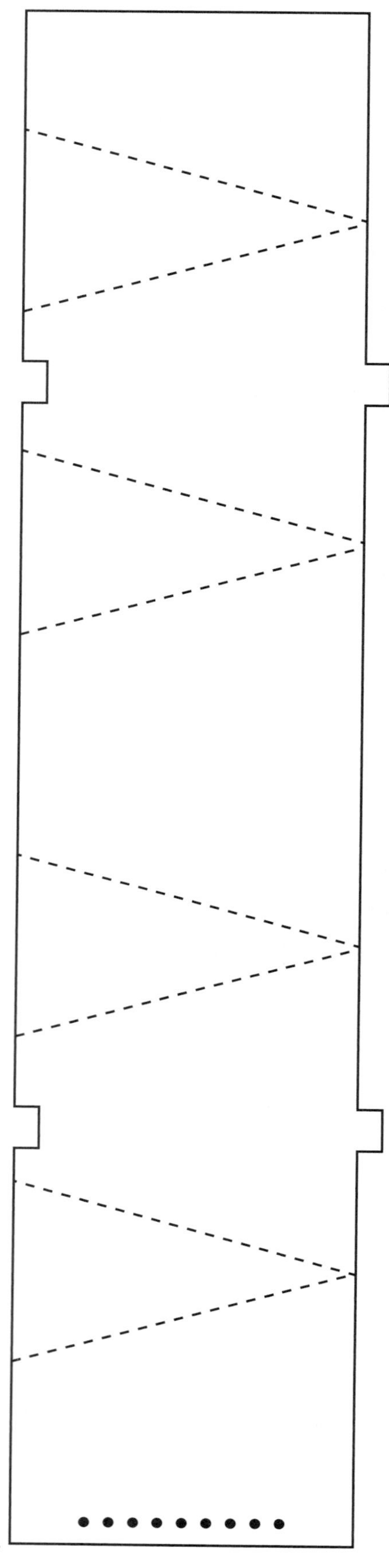

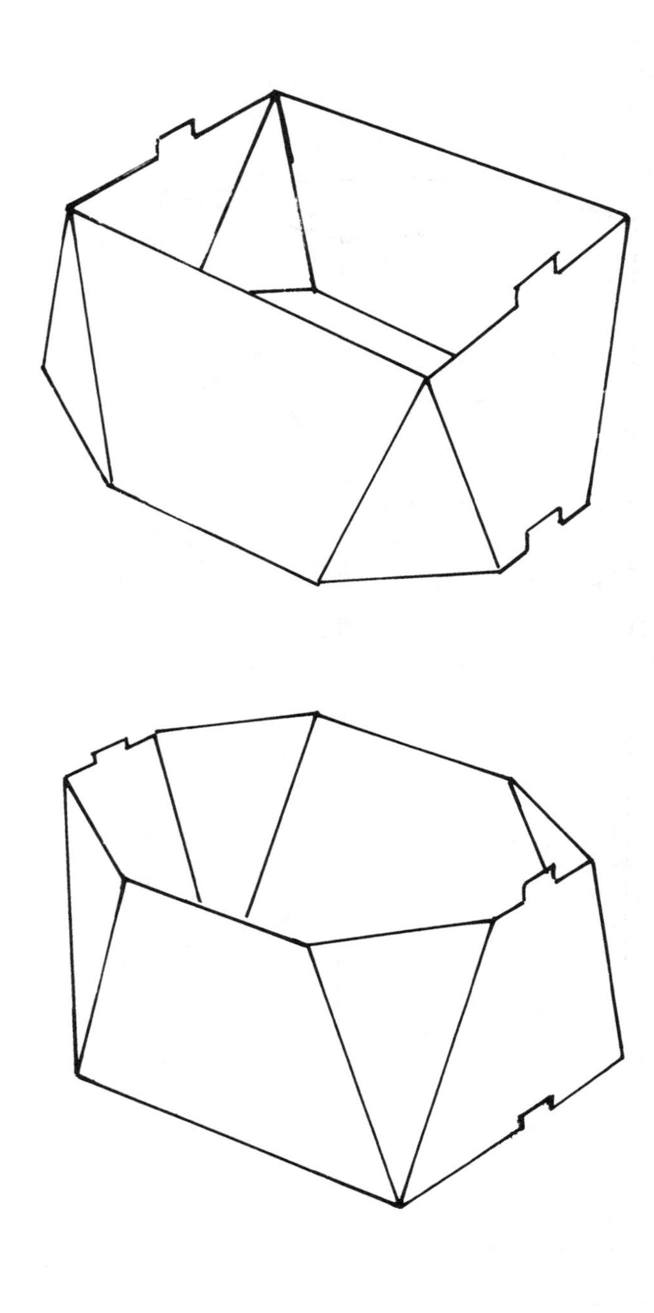

This base may be used either way as illustrated. The stacking tabs would have to be moved to the opposite side. See the next two pages.

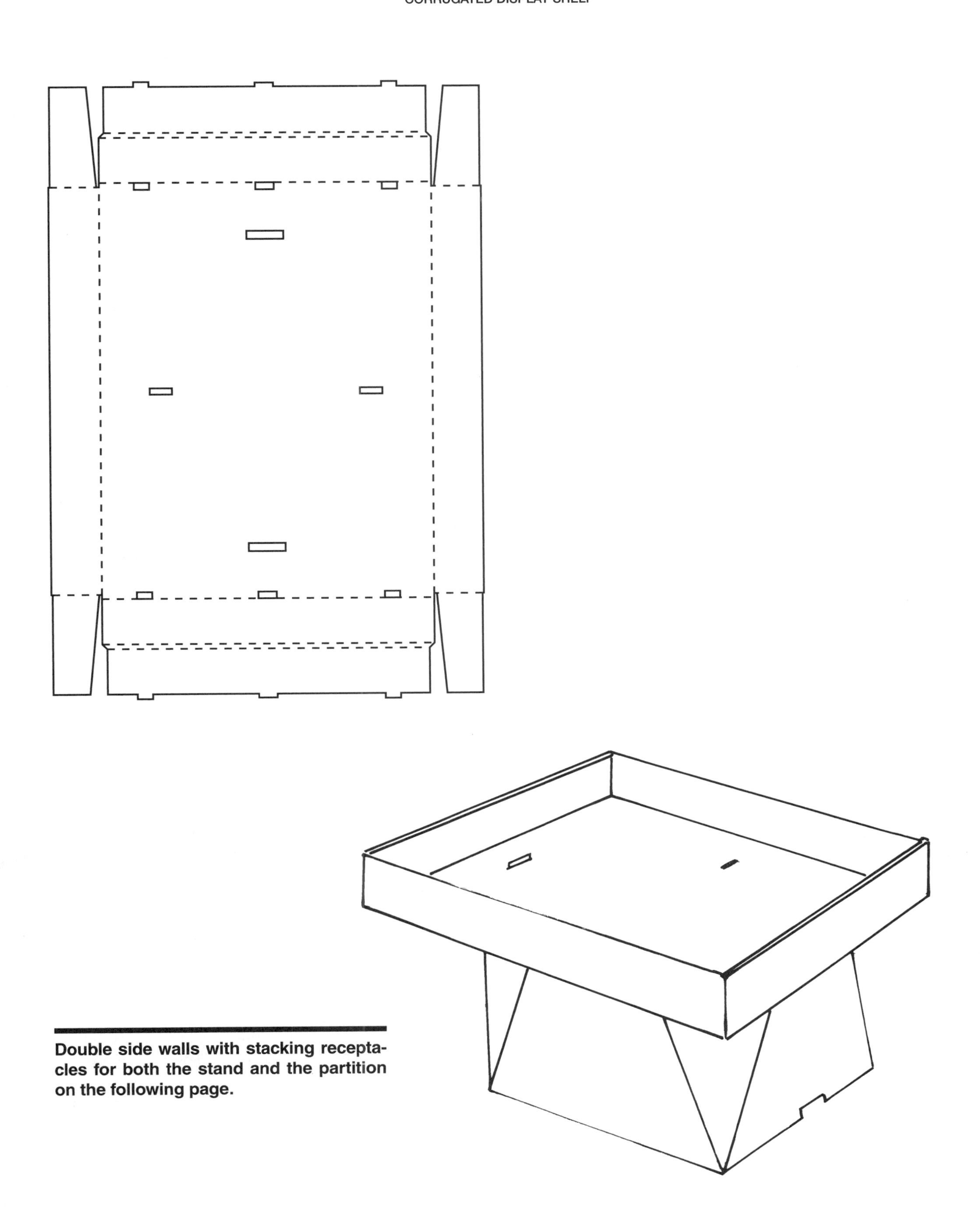

Double side walls with stacking receptacles for both the stand and the partition on the following page.

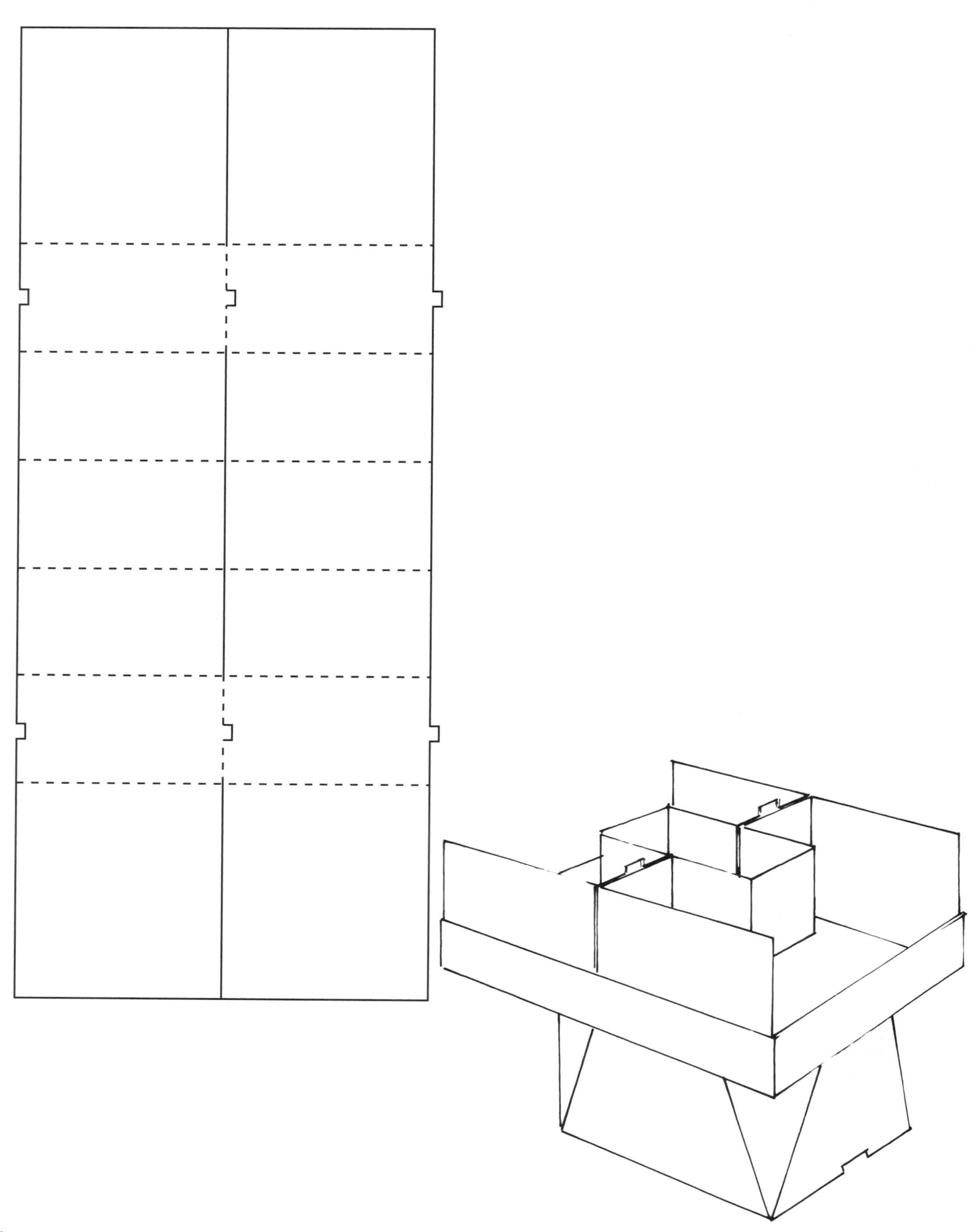

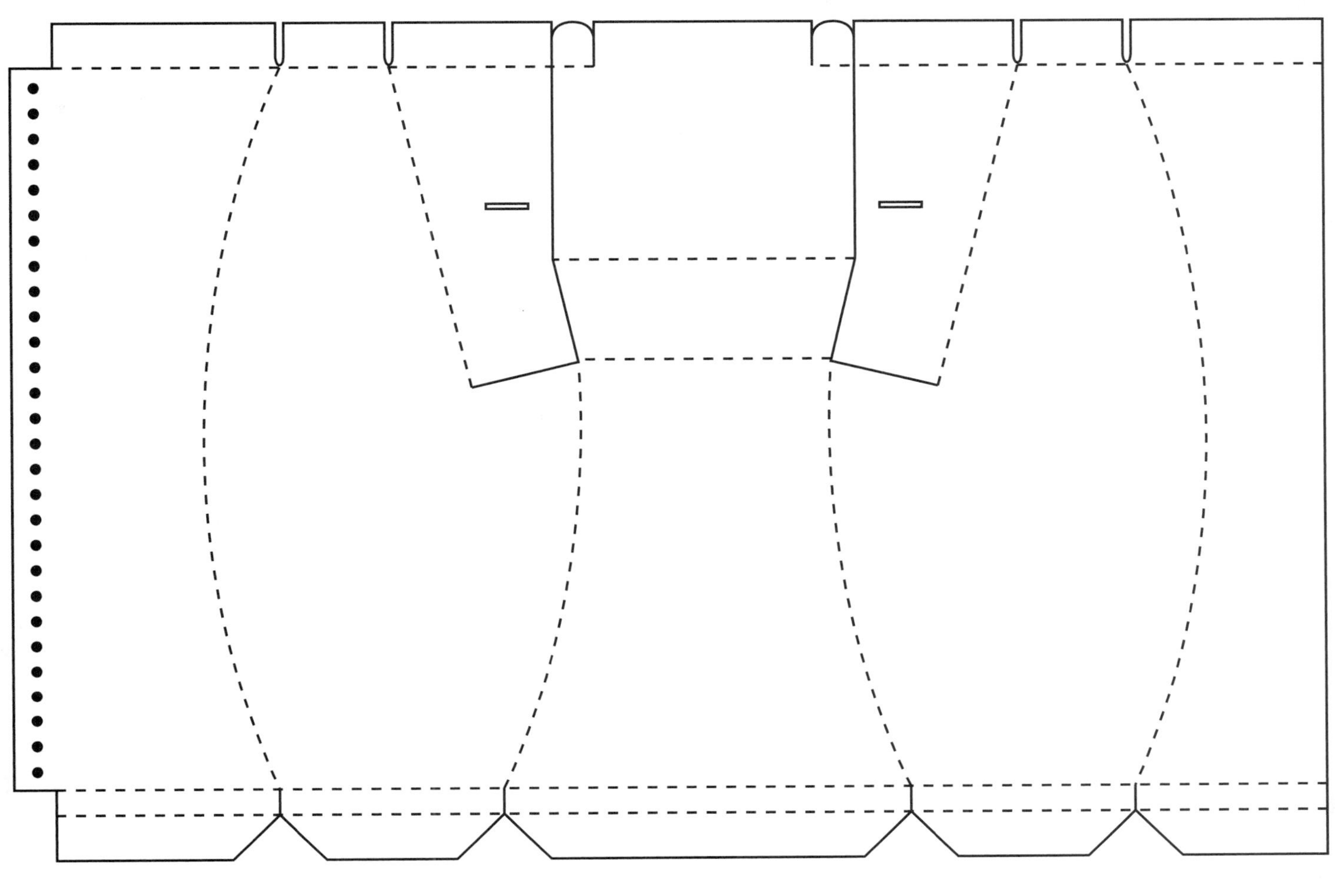

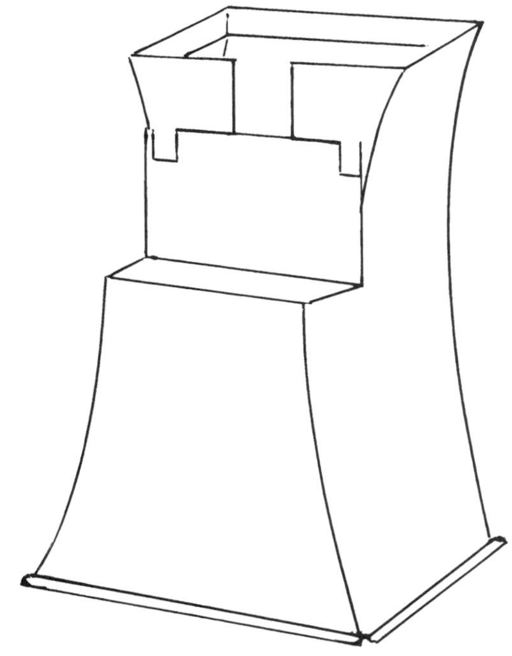

The outfold/infold at the base promote stability of the stand.

Glossary

Adhesion The process of attaching objects (e.g., by pasting, gluing) with adhesive materials.

Aseptic packaging Sterilized containers made of plastic-lined paper (foil and plastic lamination).

Base The lower or receptacle portion of a set-up box.

Blank A folding carton after die-cutting and scoring but before folding and gluing.

Blind embossing See *embossing*.

Box A complete paper box, including base and lid.

Brightwood box A tray-style carton made on the Brightwood machine.

Bundle A unit of boxboard containing 50 pounds of sheets.

Bursting strength test A test that measures the degree of resistance of a material to bursting.

Calendering A process in which paper is pressed between metal rollers to form a continuous sheet under controlled conditions of speed, heat, and pressure.

Caliper points A measure of paper thickness, expressed in units of a thousandth of an inch and usually written in decimals.

Carton A unit container made for bending grades of boxboard.

Chipboard Recycled paperboard.

Compression test A test in which force is applied by two flat surfaces to opposite faces of a box.

Converter A manufacturer that fabricates folding cartons from boxboard and other packaging materials (e.g., sheet plastics).

Corrugated line board and medium test A test that measures the amount of force required to crush a fluted corrugated medium; measures the stiffness of the material.

Corrugated paperboard A material constructed from alternating layers of fluted paperboard. (Flutes are classified as A, B, C, and E.)

Count The actual number of sheets of a given size, weight, and caliper required to make a bundle weighing 50 lbs.

Cutting and creasing die A steel rule form used on a press to cut and score sheets of boxboard into folding carton blanks.

Die-cutting The process of cutting shapes from paperboard using cutting or stamping dies or lasers.

Dividers Loose strips of paper or boxboard used to separate items of merchandise.

Double wall A material consisting of three flat facings and two intermediate corrugated members.

Drop test A mechanical procedure used to test the safety of package contents during shipping; determines the resistance of a filled box to shocks caused by dropping it in certain ways (i.e., on corners, edges, faces, etc.).

Easels Supports attached to a box to hold it upright for display.

Embossing A process in which paper is pressed between metal dies to create an image in relief.

Fiberboard A general term applied to fabricated paperboard utilized in container manufacture; may be either corrugated or solid.

Filled board Boxboard containing center plies consisting of different materials from those used for the top and bottom liners.

Flap lid A lid that has no sides or ends and is hinged to the base of the box.

Flexographic printing Printing from flexible plates made of rubber or plastic.

Flutes The wave shapes in the inner portion of corrugated fiberboard. The flute or corrugation categories (A, B, C, and E) indicate flutes per linear foot.

Folding carton See *carton.*

French lid A lid with an extension edge, sides, and ends; it fits outside the base of the box.

Furnish A mixture of water, pulp, paper scrap, sizing, dyes, and other additives that is fed into the wet end of a paper- or board-making machine.

Grain The direction in which the fibers in paper line up.

Gravure printing A printing process in which specially etched cylinders with cells to accept and store inks rotate the image to the impression cylinder, which transfers the image to paperboard.

Inner packing Materials or parts used in supporting, positioning, or cushioning an item.

Labels Die-cut, often self-adhesive applications used to decorate or identify packages.

Laminating Bonding together two or more layers of material, usually impregnated with thermosetting resin.

Laminator A machine that adheres two or more plies of paper or board.

Laser A device that produces a very narrow beam of extremely intense light; used in industrial processes, medicine, and detailed die-cutting.

Letterpress A printing method in which ink is transferred from a metal plate directly to a sheet of paperboard.

Lid The upper and covering portion of a paper box.

Lifts Strips of ribbon, cloth, tape, or paper attached to an inner tray, allowing it to be easily removed from the base.

Loose wrapping A process in which a lid is covered with paper on a wrapping machine or by hand; the wrapper is a single piece with adhesives only at the edges.

Manufacturer's joint The seam of a carton, where the two edges of the box are joined together by stitching, gluing, or taping.

Master carton A carton that is used to ship smaller cartons.

Neck A shell inserted in the base of a shoulder box, attached by an adhesive and extending above the base into the lid.

Newsback Chipboard that is lined on one side; used for inexpensive printing.

Offset printing (lithography) A printing method in which ink is transferred from a plate to a smooth rubber "blanket" roller that transfers the image to paperboard.

Padded top An extra top, covered with paper or cloth and including one or more layers of wadding or other padding material, that is attached to a top lid.

Paperboard (cardboard) A material made from laminated layers of paper in sheets of 12 points or more.

Paper stock Reclaimed (recycled) material that has been pulped for the use as a finish for paperboard; includes newsprint, reclaimed corrugated containers, and mixed paper and boxboard (kraft) cuttings.

Paster A machine that applies an adhesive to two or more plies of paperboard.

Ply A layer of boxboard formed on a multicylinder machine.

Pulp Basic cellulose fibers resulting from the disintegration of wood, rags, and other organic materials by chemical or mechanical processes; used to make boxboard.

Puncture test A test of the strength of material, expressed in ounces per inch of tear, as measured by the Beach puncture tester.

Recycling The use of previously used materials to make new objects.

Scoring Making an impression or crease in a box blank to facilitate bending, folding, or tearing; feeding boxboard sheets through a scoring unit.

Set-up paperbox A paperbox or rigid construction that has been formed or set up and is ready for use (as opposed to a folding carton or shipping container; see *corrugated box*).

Shell A plain, unprinted carton designed to be overwrapped with plain or printed paper, fabric, or other materials.

Shoulder box A box with a glued neck inserted so that its base, ends, and sides form a shoulder on which the lid rests.

Slide box A box in which the lid is the form of a shell into which the base is inserted at the side or end.

Stay Material used to reinforce the corners of a base, lid, or tray; may be paper, cloth, combination (cloth and paper), wire, metal, or plastic tape.

Stitching, stapling, and taping The application of metal fasteners or tapes to form or close a box.

Telescope box A box in which the sides and ends of the lip slip over the base to either partial or full depth.

Tight wrapping The process of covering a base, lid, or tray with paper on a wrapping machine or by hand; the wrap is a single piece whose entire surface is covered with adhesive.

Tray-style carton A folding carton with a hinged cover extending from the side walls.

Trim Paper or cloth covering to strengthen or decorate base edges and lid or extension edges.

Triple wall A material consisting of four flat facings and three intermediate corrugated members.

Tube-style carton A folding carton that encloses a product with flap, tucks, and locks.

Tuck The end portion of the top of bottom flap of a folding carton, which is inserted into the container to hold the end flaps in place.

Universal Product Code (UPC) A printed code on containers and other forms of packaging that provides information about the product for purposes of inventory control and retail pricing.

Vibration test A method of checking vibration during transportation.

Water absorption test A measure of resistance to water absorption; applied to boxes impregnated with water- and moisture-resistant coatings.

Web A roll of paper or other flexible material as it moves through the machine in the process of being converted.

Bibliography

Ambrose, Gavin and Paul Harris. *This End Up: Original Approaches to Packaging Design.* Gloucester, MA: RotoVision, 2003.

Brody, Aaron L. and Kenneth S. Marsh (eds.), *The Wiley Encyclopedia of Packaging Technology, 2nd Ed.* Hoboken, NJ: John Wiley & Sons, Inc., 1997

Calver, Giles. *What is Packaging Design?* Gloucester, MA: RotoVision SA, 2003.

Camthray, Richard and Edward Denison. *Packaging Prototypes.* Gloucester, MA: RotoVision SA, 1999.

Carter, Davis E. *Power Packaging.* New York, NY: Watson-Guptill Publications, 1999.

Denison, Edward and Guang Yu Ren, *Thinking Green, Packaging Prototypes 3.* Gloucester, MA: RotoVision SA, 2001.

Emblem, Anne and Henry, *Packaging Prototypes 2.* Gloucester, MA: RotoVision SA, 2000

Fibre Box Handbook. Rolling Meadows, IL: Fibre Box Association, 2002.

Fishel, Catharine M., *The Perfect Package.* Gloucester, MA: Rockport Publishers, 2002

Fishel, Catharine M., *Design Secrets: Packaging. 50 Real Life Projects Uncovered.* Gloucester, MA: Rockport Publishers, 2003

Gilleo, Ken, *Area Array Packaging Handbook.* Columbus, OH: McGraw-Hill Professional, 2001.

Hanlon, Joseph, Robert J. Kelsey, and J. F. Hanlon, *Handbook of Packaging Engineering, 3rd ed.* Boca Raton, FL. CRC Press, 1998.

Hine, Thomas, *The Total Package: the Secret History and Hidden Meanings of Boxes, Bottles, Cans, and Other Persuasive Containers.* Boston, MA: Little Brown & Company, 1995.

Ideas and Innovation. a handbook for designers, converters and buyers of paperboard packaging. Alexandria, VA: Paperboard Packaging Council, 2000.

Opie, Robert, *Packaging Sourcebook.* Secaucus, NJ: Chartwell Books, 1998.

Packaging Design in Japan. Cologne, Germany: Taschen Verlag, 1989

Index